McGRAW-HILL
CONCISE
ENCYCLOPEDIA OF
PHYSICS

McGraw-Hill

New York   Chicago   San Francisco   Lisbon   London   Madrid   Mexico City
Milan   New Delhi   San Juan   Seoul   Singapore   Sydney   Toronto

The *McGraw·Hill* Companies

Library of Congress Cataloging in Publication Data

McGraw-Hill concise encyclopedia of physics.
    p.   cm.
  Includes index.
  ISBN 0-07-143955-2
    1. Physics—Encyclopedias. I. Title: Concise encyclopedia of physics.

  QC5.M4245   2004
  530'.03—dc22   2004049911

1 2 3 4 5 6 7 8 9 0   DOC/DOC 0 1 0 9 8 7 6 5
ISBN 0-07-143955-2

This book was printed on acid-free paper.

It was set in Helvetica Black and Souvenir by TechBooks, Fairfax, Virginia.

The book was printed and bound by RR Donnelley, The Lakeside Press.

CONTENTS

EDITORIAL STAFF

EDITING, DESIGN, AND PRODUCTION STAFF

CONSULTING EDITORS

Richard P. Schulz. *American Electric Power Co., Gahanna, Ohio.* ELECTRICAL POWER ENGINEERING.

Prof. Sam Treiman. *Deceased; formerly, Department of Physics, Joseph Henry Laboratories, Princeton University, Princeton, New Jersey.* THEORETICAL PHYSICS.

Prof. Frank M. White. *Department of Mechanical Engineering, University of Rhode Island, Kingston.* FLUID MECHANICS.

Dr. James C. Wyant. *University of Arizona Optical Sciences Center, Tucson.* ELECTROMAGNETIC RADIATION AND OPTICS.

PREFACE

For more than four decades, the *McGraw-Hill Encyclopedia of Science & Technology* has been an indispensable scientific reference work for a broad range of readers, from students to professionals and interested general readers. Found in many thousands of libraries around the world, its 20 volumes authoritatively cover every major field of science. However, the needs of many readers will also be served by a concise work covering a specific scientific or technical discipline in a handy, portable format. For this reason, the editors of the *Encyclopedia* have produced this series of paperback editions, each devoted to a major field of science or engineering.

The articles in this *McGraw-Hill Concise Encyclopedia of Physics* cover all the principal topics of this field. Each one is a condensed version of the parent article that retains its authoritativeness and clarity of presentation, providing the reader with essential knowledge in physics without extensive detail. The authors are international experts, including Nobel Prize winners. The initials of the authors are at the end of the articles; the full names and affiliations are listed in the back of the book.

The reader will find 900 alphabetically arranged entries, many illustrated with images or diagrams. Most include cross references to other articles for background reading or further study. Dual measurement units (U.S. Customary and International System) are used throughout. The Appendix includes useful information complementing the articles. Finally, the Index provides quick access to specific information in the articles.

This concise reference will fill the need for accurate, current scientific and technical information in a convenient, economical format. It can serve as the starting point for research by anyone seriously interested in science, even professionals seeking information outside their own specialty. It should prove to be a much used and much trusted addition to the reader's bookshelf.

MARK D. LICKER
Publisher

ORGANIZATION OF THE ENCYCLOPEDIA

Alphabetization. The approximately 900 article titles are sequenced on a word-by-word basis, not letter by letter. Hyphenated words are treated as separate words. Parenthetical words are disregarded in alphabetizing. In occasional inverted article titles, the comma provides a full stop. The index is alphabetized on the same principles. Readers can turn directly to the pages for much of their research. Examples of sequencing are:

Atom optics	**Conduction (heat)**
Atomic beams	**Conduction band**
Black hole	**Dipole-dipole interaction**
Blackbody	**Dipole moment**

Cross references. Virtually every article has cross references set in CAPITALS AND SMALL CAPITALS. These references offer the user the option of turning to other articles in the volume for related information.

Measurement units. Since some readers prefer the U.S. Customary System while others require the International System of Units (SI), measurements in the Encyclopedia are given in dual units.

Contributors. The authorship of each article is specified at its conclusion, in the form of the contributor's initials for brevity. The contributor's full name and affiliation may be found in the "Contributors" section at the back of the volume.

Appendix. Every user should explore the variety of succinct information supplied by the Appendix, which includes conversion factors, measurement tables, fundamental constants, and a biographical listing of scientists. Users wishing to go beyond the scope of this Encyclopedia will find recommended books and journals listed in the "Bibliographies" section; the titles are grouped by subject area.

Index. The 4800-entry index offers the reader the time-saving convenience of being able to quickly locate specific information in the text, rather than approaching the Encyclopedia via article titles only. This elaborate breakdown of the volume's contents assures both the general reader and the professional of efficient use of the *McGraw-Hill Concise Encyclopedia of Physics*.

A15 phases   A series of intermetallic compounds which have a particular crystal structure and the chemical formula A_3B, where A represents a transition element and B can be either a transition or a nontransition element. Many A15 compounds exhibit the phenomenon of superconductivity at relatively high temperatures in the neighborhood of 20 K ($-424°F$) and in high magnetic fields on the order of several tens of teslas (several hundred kilogauss). High-temperature–high-field superconductivity has a number of important technological applications and is a challenging fundamental research area in condensed-matter physics. *See* SUPERCONDUCTIVITY.

The A15 compounds crystallize in a structure in which the unit cell, the repeating unit of the crystal structure, has the overall shape of a cube. The B atoms are located at the corners and in the center of the cube, while the A atoms are arranged in pairs on the cube faces (*see* illustration). A special characteristic of the A15 crystal structure is that the A atoms form mutually orthogonal linear chains that run throughout the crystal lattice, as shown in the illustration. The extraordinary superconducting properties of the A15 compounds are believed to be primarily associated with these linear chains of transition-element A atoms. *See* CRYSTAL.

Processes have been developed for preparing multifilamentary superconducting wires that consist of numerous filaments of a superconducting A15 compound, such as Nb_3Sn, embedded in a nonsuperconducting copper matrix. Superconducting wires

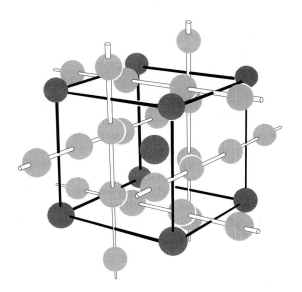

A15 crystal structure of the A_3B intermetallic compounds. The light spheres represent the A atoms; the dark spheres represent the B atoms. The linear chains of A atoms are emphasized.

can be used in electric power transmission lines and to wind electrically lossless coils (solenoids) for superconducting electrical machinery (motors and generators) and magnets. Superconducting magnets are employed to produce intense magnetic fields for laboratory research, confinement of high-temperature plasmas in nuclear fusion research, bending beams of charged particles in accelerators, levitation of high-speed trains, mineral separation, and energy storage. *See* SUPERCONDUCTING DEVICES.

[M.B.Ma.]

Aberration (optics)   A departure of an optical image-forming system from ideal behavior. Ideally, such a system will produce a unique image point corresponding to each object point. In addition, every straight line in the object space will have as its corresponding image a unique straight line. A similar one-to-one correspondence will exist between planes in the two spaces. This type of mapping of object space into image space is called a collinear transformation. When the conditions for a collinear transformation are not met, the departures from that ideal behavior are termed aberrations. They are classified into two general types, monochromatic aberrations and chromatic aberrations. The monochromatic aberrations apply to a single color, or wavelength, of light. The chromatic aberrations are simply the chromatic variation, or variation with wavelength, of the monochromatic aberrations. *See* CHROMATIC ABERRATION; GEOMETRICAL OPTICS; OPTICAL IMAGE.

The monochromatic aberrations can be described in several ways. Wave aberrations are departures of the geometrical wavefront from a reference sphere with its vertex at the center of the exit pupil and its center of curvature located at the ideal image point. The wave aberration is measured along the ray and is a function of the field height and the pupil coordinates of the reference sphere (see illustration).

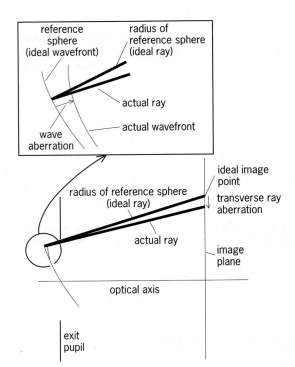

Diagram of the image space of an optical system, showing aberration measures: the wave aberration and the transverse ray aberration.

Transverse ray aberrations are measured by the transverse displacement from the ideal image point to the ray intersection with the ideal image plane. The chief monochromatic aberrations are spherical (aperture) aberrations, coma, astigmatism, curvature of field, and distortion.

Each surface in an optical system introduces aberrations as the light beam passes through the system. The aberrations of the entire system consist of the sum of the surface contributions, some of which may be positive and others negative. The challenge of optical design is to balance these contributions so that the total aberrations of the system are tolerably small. In a well-corrected system the individual surface contributions are many times larger than the tolerance value, so that the balance is rather delicate, and the optical system must be made with a high degree of precision. *See* Lens (OPTICS); OPTICAL SURFACES.   [R.V.S.]

Absolute zero

The temperature at which an ideal gas would exert no pressure. The Kelvin scale of temperatures is defined in terms of the triple point of water, $T_3 = 273.16°$ (where the solid, liquid, and vapor phases coexist), and absolute zero. Temperature is measured most simply via the constant-volume ideal-gas thermometer, in which a small amount of gas is introduced (in order to limit the effect of interactions between molecules) and then sealed off, and the gas pressure P referenced to its value at the triple point $P(T_3)$ is measured. The ideal-gas law applies if the molecules in a gas exert no forces on one another and if they are not attracted to the walls. Absolute zero is the temperature at which the pressure of a truly ideal gas would vanish. *See* TEMPERATURE MEASUREMENT.

According to classical physics, all motion would cease at absolute zero; however, the quantum-mechanical uncertainty principle requires that there be a small amount of residual motion (zero-point motion) even at absolute zero. *See* KINETIC THEORY OF MATTER; UNCERTAINTY PRINCIPLE.

Temperature can also be defined from the Boltzmann distribution. If a collection of spin-1/2 magnetic ions is placed in a magnetic field, the ratio of the occupancy of the lower to the higher energy state is given by the equation below. Here k is

$$\frac{N_L}{N_H} = \exp \frac{|\Delta E|}{kT}$$

Boltzmann's constant, ΔE is the magnitude of the difference in energy between the states, and T is the Kelvin temperature. Thus, at high temperatures the two states have nearly equal occupation probability, while the lower energy state is progressively favored at lower temperatures. At absolute zero, only the lower energy level is occupied. This relation allows for the possibility of negative temperatures when the population of the higher energy state exceeds that of the lower state. *See* BOLTZMANN CONSTANT; BOLTZMANN STATISTICS.

Negative temperatures notwithstanding, the third law of thermodynamics states that the absolute zero of temperature cannot be attained by any finite number of steps. The lowest (and hottest) temperatures that have been achieved are on the order of a picokelvin (10^{-12} K). These are spin temperatures of nuclei which are out of equilibrium with the lattice vibrations and electrons of a solid. The lowest temperatures to which the electrons have been cooled are on the order of 10 microkelvins in metallic systems. *See* LOW-TEMPERATURE PHYSICS; TEMPERATURE.   [J.M.Pa.; D.M.Le.]

Absorption of electromagnetic radiation

The process whereby the intensity of a beam of electromagnetic radiation is attenuated in passing through a material medium by conversion of the energy of the radiation to an equivalent amount of energy appearing within the medium; the radiant energy is converted into heat or some other form of molecular energy. A perfectly transparent medium permits the passage of a beam of radiation without any change in intensity other than that caused by the spread or convergence of the beam, and the total radiant energy emergent from such a medium equals that which entered it, whereas the emergent energy from an absorbing medium is less than that which enters, and, in the case of highly opaque media, is reduced practically to zero. No known medium is opaque to all wavelengths of the electromagnetic spectrum; similarly, no material medium is transparent to the whole electromagnetic spectrum. A medium which absorbs a relatively wide range of wavelengths is said to exhibit general absorption, while a medium which absorbs only restricted wavelength regions of no great range exhibits selective absorption for those particular spectral regions. For example, ordinary window glass is transparent to visible light, but shows general absorption for ultraviolet radiation of wavelengths below about 310 nanometers, while colored glasses show selective absorption for specific regions of the visible spectrum. The color of objects which are not self-luminous and which are seen by light reflected or transmitted by the object is usually the result of selective absorption of portions of the visible spectrum. See COLOR; ELECTROMAGNETIC RADIATION.

The capacity of a medium to absorb radiation depends on a number of factors, mainly the electronic and nuclear constitution of the atoms and molecules of the medium, the wavelength of the radiation, the thickness of the absorbing layer, and the variables which determine the state of the medium, of which the most important are the temperature and the concentration of the absorbing agent. In special cases, absorption may be influenced by electric or magnetic fields. The state of polarization of the radiation influences the absorption of media containing certain oriented structures, such as crystals of other than cubic symmetry. See STARK EFFECT; ZEEMAN EFFECT.

Lambert's law, also called Bouguer's law or the Lambert-Bouguer law, expresses the effect of the thickness of the absorbing medium on the absorption. If I is the intensity to which a monochromatic parallel beam is attenuated after traversing a thickness d of the medium, and I_0 is the intensity of the beam at the surface of incidence (corrected for loss by reflection from this surface), the variation of intensity throughout the medium is expressed by Eq. (1), in which α is a constant for the medium called the absorption

$$I = I_0 e^{-\alpha d} \tag{1}$$

coefficient. This exponential relation can be expressed in an equivalent logarithmic form as in Eq. (2), where $k = \alpha/2.303$ is called the extinction coefficient for radiation

$$\log_{10}(I_0/I) = (\alpha/2.303)d = kd \tag{2}$$

of the wavelength considered. The quantity $\log_{10}(I_0/I)$ is often called the optical density, or the absorbance of the medium.

Beer's law refers to the effect of the concentration of the absorbing medium, that is, the mass of absorbing material per unit of volume, on the absorption. This relation is of prime importance in describing the absorption of solutions of an absorbing solute, since the solute's concentration may be varied over wide limits, or the absorption of gases, the concentration of which depends on the pressure. The effects of thickness d and concentration c on absorption of monochromatic radiation can be combined in a single mathematical expression, given in Eq. (3), in which k' is a constant for a

$$I = I_0 e^{-k'cd} \tag{3}$$

given absorbing substance (at constant wavelength and temperature), independent of the actual concentration of solute in the solution. In logarithms, the relation becomes Eq. (4).

$$\log_{10}(I_0/I) = (k'/2.303)cd = \epsilon cd \tag{4}$$

The values of the constants k' and ϵ in Eqs. (3) and (4) depend on the units of concentration. If the concentration of the solute is expressed in moles per liter, the constant ϵ is called the molar extinction coefficient. Some authors employ the symbol a_M, which is called the molar absorbance index, instead of ϵ.

If Beer's law is adhered to, the molar extinction coefficient does not depend on the concentration of the absorbing solute, but usually changes with the wavelength of the radiation, with the temperature of the solution, and with the solvent.

Absorption of radiation by matter always involves the loss of energy by the radiation and a corresponding gain in energy by the atoms or molecules of the medium. The energy absorbed from radiation appears as increased internal energy, or in increased vibrational and rotational energy of the atoms and molecules of the absorbing medium. As a general rule, translational energy is not directly increased by absorption of radiation, although it may be indirectly increased by degradation of electronic energy or by conversion of rotational or vibrational energy to that of translation by intermolecular collisions.

The energy acquired by matter by absorption of visible or ultraviolet radiation, although primarily used to excite electrons to higher energy states, usually ultimately appears as increased kinetic energy of the molecules, that is, as heat. It may, however, under special circumstances, be reemitted as electromagnetic radiation. Fluorescence is the reemission, as radiant energy, of absorbed radiant energy, normally at wavelengths the same as or longer than those absorbed. The radiant reemission of absorbed radiant energy at wavelengths longer than those absorbed, for a readily observable interval after withdrawal of the exciting radiation, is called phosphorescence. Phosphorescence and fluorescence are special cases of luminescence, which is defined as light emission that cannot be attributed merely to the temperature of the emitting body. *See* FLUORESCENCE; LUMINESCENCE; PHOSPHORESCENCE. [W.W.]

Acceleration The time rate of change of velocity. Since velocity is a directed or vector quantity involving both magnitude and direction, a velocity may change by a change of magnitude (speed) or by a change of direction or both. It follows that acceleration is also a directed, or vector, quantity. If the magnitude of the velocity of a body changes from v_1 ft/s to v_2 ft/s in t seconds, then the average acceleration a has a magnitude given by Eq. (1):

$$a = \frac{\text{velocity change}}{\text{elapsed time}} = \frac{v_2 - v_1}{t_2 - t_1} = \frac{\Delta v}{\Delta t} \tag{1}$$

To designate it fully the direction should be given, as well as the magnitude. *See* VELOCITY.

Instantaneous acceleration is defined as the limit of the ratio of the velocity change to the elapsed time as the time interval approaches zero. When the acceleration is constant, the average acceleration and the instantaneous acceleration are equal.

Whenever a body is acted upon by an unbalanced force, it will undergo acceleration. If it is moving in a constant direction, the acting force will produce a continuous change in speed. If it is moving with a constant speed, the acting force will produce an acceleration consisting of a continuous change of direction. In the general case, the acting force may produce both a change of speed and a change of direction.

Angular acceleration is a vector quantity representing the rate of change of angular velocity of a body experiencing rotational motion. If, for example, at an instant t_1, a rigid body is rotating about an axis with an angular velocity ω_1, and at a later time t_2, it has an angular velocity ω_2, the average angular acceleration α is given by Eq. (2),

$$\overline{\alpha} = \frac{\omega_2 - \omega_1}{t_2 - t_1} = \frac{\Delta\omega}{\Delta t} \tag{2}$$

in radians per second per second. The instantaneous angular acceleration is given by $\alpha = d\omega/dt$.   [R.D.Ru.]

When a body moves in a circular path with constant linear speed at each point in its path, it is also being constantly accelerated toward the center of the circle under the action of the force required to constrain it to move in its circular path. This acceleration toward the center of path is called radial acceleration. The component of linear acceleration tangent to the path of a particle subject to an angular acceleration about the axis of rotation is called tangential acceleration. *See* ROTATIONAL MOTION.
                                                                                    [C.E.H.; R.J.S.]

Acceptor atom   An impurity atom in a semiconductor which can accept or take up one or more electrons from the crystal and become negatively charged. An atom which substitutes for a regular atom of the material but has one less valence electron may be expected to be an acceptor atom. For example, atoms of boron, aluminum, gallium, or indium are acceptors in germanium and silicon (illus. *a*), and atoms of antimony and bismuth are acceptors in tellurium crystals. Acceptor atoms tend to increase the number of holes (positive charge carriers) in the semiconductor (illus. *b*). The energy

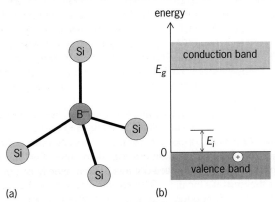

Trivalent acceptor atom, boron (B), in the elemental semiconductor silicon (Si). (*a*) Boron atom in a substitutional position, that is, replacing silicon, a tetravalent host atom, by completing the four tetrahedral covalent bonds with its nearest neighbor silicon atoms. This requires an electron to be accepted from the valence band, thus making boron negatively charged. (*b*) Energy diagram showing that the absence of an electron in the valence band is equivalent to a positive charge carrier, a hole, which is bound to boron via Coulomb attraction with an ionization enzry $E_i \cdot E_g =$ energy gap separating valence band from conduction band.

gained when an electron is taken up by an acceptor atom from the valence band of the crystal is the ionization energy of the atom. *See* DONOR ATOM; SEMICONDUCTOR.

[H.Y.F.; A.K.R.]

Acoustic impedance   At a given surface, the complex ratio of effective sound pressure averaged over the surface to the effective flux (volume velocity or particle velocity multiplied by the surface area) through it. The unit is the $N \cdot s/m^5$ (newton-second/meter^5), or the mks acoustic ohm. In the cgs system the unit is the $dyn \cdot s/cm^5$ (dyne-second/centimeter^5). *See* SOUND PRESSURE.

Specific acoustic impedance is the complex ratio of the effective sound pressure at a point to the effective particle velocity at a point. The unit is the $N \cdot s/m^3$, or the mks rayl. In the cgs system the unit is the $dyn \cdot s/cm^3$, or the rayl. The difference between specific acoustic impedance and acoustic impedance is in the specification of impedance at a point, as compared to the average over a surface.

Characteristic acoustic impedance is the ratio of effective sound pressure at a point to the particle velocity at that point in a free, progressive wave. This ratio is equal to the product of the density of the medium times the speed of sound in the medium. The characteristic impedance of a sound wave is analogous to the characteristic electrical impedance of an infinitely long, dissipationless transmission line. It is common in acoustical analyses to represent specific acoustic impedances in terms of their ratio to the characteristic impedance of air.

Acoustic impedance, being a complex quantity, can have real and imaginary components analogous to those in an electrical impedance. In applying this analogy, the real part of the acoustic impedance is termed acoustic resistance, and the imaginary part is termed acoustic reactance. *See* ELECTRICAL IMPEDANCE.   [W.J.G.]

Acoustic interferometer   An instrument that is sensitive to the interference of two or more acoustic waves. It provides information on acoustic wavelengths that is useful in determining the velocity and absorption of sound in samples of gases, liquids, and materials, and it yields information on the nonlinear properties of solids.

In its simplest form, an acoustic interferometer for use in liquids has a fixed piezo-electric crystal (acting as a transmitter) tuned to the frequency of interest and a parallel reflector at a variable distance from it. Driven by an oscillating electrical voltage, the piezoelectric crystal generates a sound wave, which in turn is reflected by the reflector. The acoustic pressure amplitude on the front face of the crystal depends on the velocity amplitude at the face and the distance to the reflecting surface. The amplitude ratio (radiation impedance) of the acoustic pressure to the velocity and the relative phase shift between the two oscillating quantities depend solely on the distance to the reflecting surface. If the reflector acts as a rigid surface, this amplitude ratio is ideally zero whenever the net round-trip distance between the crystal and the reflector is an odd number of half-wavelengths because the reflected wave is then exactly out of phase with the incident wave at the crystal's location. The crystal then draws the maximum current since the oscillations are unimpeded. *See* ACOUSTIC IMPEDANCE; PIEZOELECTRICITY; WAVE MOTION; WAVELENGTH.

During operation, the current drawn by the crystal is monitored as the reflector is gradually moved away from the crystal. Whenever the reflector position is such that the crystal is at a pressure antinode (place of maximum pressure in a standing wave), there is a strong dip in the current drawn due to the relatively high radiation impedance presented by the standing wave to the crystal face. Consecutive antinodes are a half-wavelength apart. For a given frequency f, a measured distance L between the location of any one antinode and that of its nth successor yields the wavelength $2L/n$

and the speed of sound $c = 2Lf/n$. An acoustic interferometer based on this principle can achieve a precision of 0.01%. Since the current drawn by the crystal is relatively insensitive to the frequency for a given radiation impedance, the sound speed can also be determined by keeping the distance between the crystal and the reflector fixed and gradually sweeping the frequency.

The pressure nodes and antinodes correspond to the local maxima and minima, respectively, in the current drawn. The peak of the current amplitude decreases with the distance traversed by the reflector. If the separation distance is sufficiently large that the exponential decrease associated with absorption dominates any spreading losses, the absorption coefficient for the medium can be derived by measurement of the ratios of current amplitudes at two successive points where the current drawn is a local maximum. See SOUND; SOUND ABSORPTION; ULTRASONICS.   [G.S.K.W.; A.D.P.; S.I.M.]

Acoustic levitation   The use of intense acoustic waves to hold a body that is immersed in a fluid medium against the force of gravity without obvious mechanical support.

Levitation can occur in the presence of fluid flow, including the back-and-forth fluid flow produced by the passage of an acoustic wave. Such acoustically generated forces are extremely small in common experience. But intense acoustic waves are nonlinear in their basic character and, therefore, may exert a net acoustic radiation pressure on an object sufficient to balance the gravitational force and thus levitate it. See ACOUSTIC RADIATION PRESSURE.

The applications of acoustic levitation in air or other gas include an acoustic positioning module that has been carried in the space shuttle and used in fundamental studies of the oscillation and fission of spinning drops. An acoustic levitation furnace has been designed to study the possibility of containerless solidification of molten materials.

Applications of the levitation of objects in liquids have included measurements of the ultimate tensile strengths of liquids, mechanical characterization of superheated and supercooled liquids, the measurement of properties of biological materials (including human red blood cells and lipids from the porpoise dome), the study of shape oscillations and interfacial tension of levitated drops, and the evaporation of charged drop arrays levitated electroacoustically. See SOUND.   [R.E.A.]

Acoustic noise   Unwanted sound. Noise control is the process of obtaining an acceptable noise environment for people in different situations. Understanding noise and its control requires a knowledge of the major sources of noise, sound propagation, human response to noise, and the physics of methods of controlling noise. The continuing increase in noise levels from many different human activities in industrialized societies has led to the term noise pollution.

Noise as an unwanted by-product of an industrialized society affects not only the operators of machines and vehicles, but also other occupants of buildings in which machines are installed, passengers of vehicles, and most importantly the communities in which machines, factories, and vehicles are operated.   [M.J.Cr.]

Acoustic radiation pressure   The net pressure exerted on a surface or interface by an acoustic wave. One might presume that the back-and-forth oscillation of fluid caused by the passage of an acoustic wave will not exert any net force on an object, and this is true for sound waves normally encountered. Intense sound waves, however, can exert net forces in one direction of sufficient magnitude (proportional to the sound intensity) to balance gravitational forces and thus levitate an object in air.

Forces due to acoustic radiation pressure have been used to calibrate acoustic transmitters, to deform and break up liquids, to collect like objects or to separate particles (including biological cells) based on mechanical properties, and to position objects in a sound field, sometimes levitating the sample so that independent studies of the object's properties can be performed. Single bubble sonoluminescence phenomena depend on acoustic radiation forces to maintain a bubble in a zone while its substantial radial oscillations take place. See ACOUSTIC LEVITATION; SOUND; ULTRASONICS.   [R.E.A.]

Acoustic radiometer   A device to measure the acoustic power or intensity of a sound beam by means of the force or torque that the beam exerts on an inserted object or interface. The underlying theory involves the concept of radiation pressure. Such pressure occurs, for example, when a plane sound wave is partially reflected at an interface between two materials, with the nonlinear interaction between the incident and reflected waves giving rise to a steady pressure on the interface. If a narrow beam is incident on the interface and the transmitted wave is fully absorbed by the second material, the magnitude of the radiation force F (area integral of radiation pressure) equals a constant times W/c, where W is the power of the sound beam and c is the sound speed.

A modern acoustic radiometer, used to measure the total power of an ultrasonic sound beam in water and other liquids, employs a vane suspended in the fluid in such a manner that its displacement in a direction normal to its face is proportional to the net force pushing on its front face. The vane is ideally of dimensions somewhat larger than the incident beam's diameter, so that the encountered force is associated with the entire incident beam. To eliminate the possibility of sound being reflected toward the transmitting transducer, the vane is oriented at 45° to the incident sound beam. The vane's horizontal displacement is made to be proportional to the imposed force by fastening the vane at one end of a long pendulum whose rotation from the vertical is opposed by the effect of gravity, such that the apparent spring constant for displacement in a direction at 45° to the face is approximately Mg/L, where M is the apparent mass of the vane (corrected for the presence of water), g is the acceleration of gravity, and L is the length of the pendulum. A nonlinear acoustics theory for such a circumstance yields a proportionality relation between the net horizontal radiation force on the vane and the acoustic power associated with the incident beam. Because the deflection of the vane is proportional to the radiation force, the acoustic power can be determined. See BUOYANCY; PENDULUM.

The concept of the vane device evolved from that of the Rayleigh disk, which was a circular disk that could rotate about its diameter and whose deflection from a nominal 45° orientation was opposed by a torsional spring. The Rayleigh disk was taken to have a radius much smaller than the wavelength, and its use ideally yielded a measurement of the local acoustic intensity that would have existed at the center of the disk were the disk not present. See ACOUSTIC RADIATION PRESSURE; SOUND; SOUND INTENSITY; SOUND PRESSURE.   [A.D.P.; S.I.Ma.]

Acoustic resonator   A device consisting of a combination of elements having mass and compliance whose acoustical reactances cancel at a given frequency. Resonators are often used as a means of eliminating an undesirable frequency component in an acoustical system. In other instances resonators are used to produce an increase in the sound pressure in an acoustic field at a particular frequency.

Resonators are useful most often in the control of low-frequency sound. They are of particular value in reducing the noise from sources having constant frequency excitation.

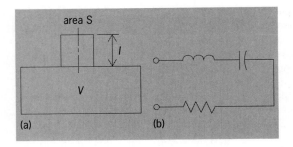

Helmholtz resonator.
(*a*) **Acoustical unit.** (*b*) **Electrical analog.**

Resonators have also found considerable application in architectural acoustics. It is often difficult to obtain adequate control of reverberation time at low frequencies in a large studio or auditorium using conventional acoustical materials. A number of designs for these spaces have included the construction of resonators behind walls or in the ceiling to obtain increased low-frequency absorption and thus provide more satisfactory reverberation characteristics.

The Helmholtz resonator (*see* illustration) is the simplest and most often utilized acoustical resonator. The unit consists of a straight tube of length l and cross-sectional area S, connected to a closed volume V. This combination is directly analogous to the simple series LC electrical circuit. *See* ACOUSTIC IMPEDANCE. [W.J.G.]

Acoustical holography   The recording of sound waves in a two-dimensional pattern (the hologram) and the use of the hologram to reconstruct the entire sound field throughout a three-dimensional region of space. Acoustical holography is an outgrowth of optical holography, invented by Dennis Gabor in 1948. The wave nature of both light and sound make holography possible. Acoustical holography involves reconstruction of the sound field that arises due to radiation of sound at a boundary, such as the vibrating body of a violin, the fuselage of an aircraft, or the surface of a submarine. Both acoustical holography and optical holography rely on the acquisition of an interferogram, a two-dimensional recording at a single frequency of the phase and amplitude of an acoustic or electromagnetic field, usually in a plane. Gabor called this interferogram a hologram. *See* HOLOGRAPHY; INVERSE SCATTERING THEORY.

Two distinct forms of acoustical holography exist. In farfield acoustical holography (FAH), the hologram is recorded far removed from the source. This form of acoustical holography is characterized by the fact that the resolution of the reconstruction is limited to a half-wavelength. This resolution restriction is removed, however, when the hologram is recorded in the acoustic nearfield, an important characteristic of nearfield acoustical holography (NAH), invented by E. G. Williams and J. D. Maynard in 1980.

Nearfield acoustical holography has been used in the automotive industry to study interior noise and tire noise, in musical acoustics to study vibration and radiation of violin-family instruments, and in the aircraft industry to study interior cabin noise and fuselage vibrations. Applications are also found in underwater acoustics, especially in studies of vibration, radiation, and scattering from ships and submarines. *See* ACOUSTIC NOISE; MUSICAL ACOUSTICS; UNDERWATER SOUND.

Typically, temporal acoustic data are acquired by measurement of the acoustic pressure with a single microphone or hydrophone, which scans an imaginary two-dimensional surface. In some cases, an array of microphones is used and the pressure is measured instantaneously by the array.The measured data are processed in a computer to reconstruct the pressure at the surface of the object as well as the vibration of the surface. The measured time data are Fourier-transformed into the frequency domain, creating a set of holograms, one for each frequency bin in the transform. In the inversion process, each hologram is broken up into a set of waves or modes whose propagation characteristics are known from basic principles. Each wave or mode is then back-propagated to the source surface by multiplication by the known inverse propagator, and the field is then recomposed by addition of all these waves or modes.

[E.G.Wi.]

Acoustics The science of sound, which in its most general form endeavors to describe and interpret the phenomena associated with motional disturbances from equilibrium of elastic media. An elastic medium is one such that if any part of it is displaced from its original position with respect to the rest, as for example by an impact, it will return to its original state when the disturbing influence is removed. Acoustics was originally limited to the human experience produced by the stimulation of the human ear by sound incident from the surrounding air. Modern acoustics, however, deals with all sorts of sounds which have no relation to the human ear, for example, seismological disturbances and ultrasonics.

Basic acoustics may be divided into three branches, namely, production, transmission, and detection of sound. Any change of stress or pressure producing a local change in density or a local displacement from equilibrium in an elastic medium can serve as a source of sound. Transmission of sound takes place through an elastic medium by means of wave motion. The most important sound waves are harmonic waves, defined as waves for which the propagated disturbance at any point in its path varies sinusoidally with time with a definite frequency or number of complete cycles per second (the unit being the hertz). Acoustics deals with waves of all frequencies, but not all frequencies are audible by human beings, for whom the average range of audibility extends from 20 to 20,000 Hz. Sound below 20 Hz is referred to as infrasonic, and that above 20,000 Hz is called ultrasonic.

The detection of sound is made possible by the incidence of transmitted sound energy on an appropriate acoustic transducer, such as the ear. For modern applied acoustics, transducers such as the microphone, based on the piezoelectric effect, are widely used. Generally speaking, any transducer used as a source of sound is also available as a detector, though the sensitivity varies considerably with the type.   [R.B.L.]

Acoustooptics The field of science and technology that is concerned with the diffraction of visible or infrared light (usually from a laser) by high-frequency sound in the frequency range of 50–2000 MHz. The term "acousto" is a historical misnomer; sound in this frequency range should properly be called ultrasonic. Such sound cannot be supported by air, but propagates as a mechanical wave disturbance in amorphous or crystalline solids, with a sound velocity ranging from 0.6 to 6 km/s (0.4 to 4 mi/s) and a wavelength from 3 to 100 μm. See LASER; ULTRASONICS.

The sound wave causes a displacement of the solid's molecules either in the direction of propagation (longitudinal wave) or perpendicular to it (shear wave). In either case, it sets up a corresponding wave of refractive-index variation through local dilatation or distortion of the solid medium. It is this wave that diffracts the light by acting as a three-dimensional grating, analogous to x-diffraction in crystals. The fact that the grating is moving is responsible for shifting the frequency of the diffracted light through the Doppler effect. See DIFFRACTION GRATING; DOPPLER EFFECT; REFRACTION OF WAVES; WAVE MOTION; X-RAY DIFFRACTION.

Since the 1960s, acoustooptics has moved from a scientific curiosity to a relevant technology. This evolution was initially driven by the need for fast modulation and deflection of light beams, and later by demands for more general optical processing. It was made possible by the invention of lasers, the development of efficient ultrasonic transducers, and the formulation of realistic models of sound-light interaction.   [A.K.]

Action   Any one of a number of related integral quantities which serve as the basis for general formulations of the dynamics of both classical and quantum-mechanical systems. The term has been associated with four quantities: the fundamental action S, for general paths of a dynamical system; the classical action S_C, for the actual path; the modified action S', for paths restricted to a particular energy; and action variables, for periodic motions.

A dynamical system can be described in terms of some number N of coordinate degrees of freedom that specify its configuration. As the vector q whose components are the degrees of freedom q_1, q_2, ..., q_N varies with time t, it traces a path $q(t)$ in an N-dimensional space. The fundamental action S is the integral of the lagrangian of the system taken along any path $q(t)$, actual or virtual, starting from a specified configuration q_1 at a specified time t_1, and ending similarly at configuration q_2 and time t_2. The value of this action $S[q(t)]$ depends on the particular path $q(t)$. The actual path $q_C(t)$ which is traversed when the system moves according to newtonian classical mechanics gives an extremum value of S, usually a minimum, relative to the other paths. This is Hamilton's least-action principle. The extremum value depends only on the end points and is called the classical action $S_C(q_1, q_2; t_1, t_2)$.

An important variant of Hamilton's principle applies when the virtual paths $q(t)$ are restricted to motions all of the same energy E, but no longer to a specific time interval, $t_1 - t_2$. The modified action $S' = S - E(t_1 - t_2)$ obeys a modified least-action principle, usually called Maupertuis' principle, namely, that the classical path gives again an extremal value of S' relative to all paths of that energy. Maupertuis' principle is closely related to Fermat's principle of least time in classical optics for the path of light rays of a definite frequency through a region of inhomogeneous refractive index. See HAMILTON'S PRINCIPLE; MINIMAL PRINCIPLES.

In quantum mechanics, as originally formulated by E. Schrödinger, the state of particles is described by wave functions which obey the Schrödinger wave equation. States of definite energy in, say, atoms are described by stationary wave functions, which do not move in space. Nonstationary wave functions describe transitory processes such as the scattering of particles, in which the state changes. Both stationary and nonstationary state wave functions are determined, in principle, once the Schrödinger wave propagator (also called the Green function) between any two points q_1 and q_2 is known. In a fundamental restatement of quantum mechanics, R. Feynman showed that all paths from q_1 to q_2, including the virtual paths, contribute to the wave propagator. Each path contributes a complex phase-term $\exp i\,(\phi[q(t)])$, where the phase ϕ is proportional to the action for that path. The resulting sum over paths, appropriately defined, is the path integral (or functional integral) representation of the Schrödinger wave propagator. The path integral has become the general starting point for most formulations of quantum theories of particles and fields. The classical path $q_C(t)$ of least action now plays the role in the wave function as being the path of stationary phase. See PROPAGATOR (FIELD THEORY).   [B.G.]

Adaptive optics   The science of optical systems in which a controllable optical element, usually a deformable mirror, is used to optimize the performance of the system, for example, to maintain a sharply focused image in the presence of wavefront

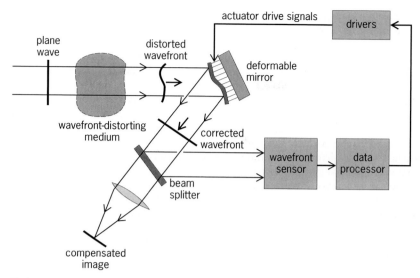

Typical adaptive optics system using discrete components.

aberrations. A distinction is made between active optics, in which optical components are modified or adjusted by external control to compensate slowly changing disturbances, and adaptive optics, which applies to closed-loop feedback systems employing sensors and data processors, operating at much higher frequencies.

In a typical adaptive optics system (see illustration) the distorted light beam to be compensated is reflected from the deformable mirror and is sampled by a beam splitter. The light sample is analyzed in a wavefront sensor that determines the error in each part of the beam. The required corrections are computed and applied to the deformable mirror whose surface forms the shape necessary to flatten the reflected wavefront. The result is to remove the optical error at the sampling point so that the light passing through the beam splitter may be focused to a sharp image. Nonlinear optical devices are also capable of performing some adaptive optics functions; these devices operate at high optical power levels. See ABERRATION (OPTICS); GEOMETRICAL OPTICS; NONLINEAR OPTICS.

The practical development of adaptive optics started in the late 1960s. Its main applications have been to compensate for the effects of atmospheric turbulence in ground-based astronomical telescopes and to improve the beam quality of high-power lasers. Adaptive optics is now used routinely at several astronomical observatories.   [J.W.Ha.]

Adiabatic demagnetization   The removal or diminution of a magnetic field applied to a magnetic substance when the latter has been thermally isolated from its surroundings. The process concerns paramagnetic substances almost exclusively, in which case a drop in temperature of the working substance is produced (magnetic cooling). See PARAMAGNETISM.

Nuclear magnetic moments are one or two thousand times smaller than their ionic (that is, electronic) counterparts, and the characteristic temperature of their mutual interaction lies in the microkelvin rather than millikelvin region. Successful experiments in nuclear adiabatic demagnetization date from the mid-1950s. See ABSOLUTE ZERO; CRYOGENICS; LOW-TEMPERATURE PHYSICS.   [R.P.Hu.]

Adiabatic process   A thermodynamic process in which the system undergoing the change exchanges no heat with its surroundings. An increase in entropy or degree of disorder occurs during an irreversible adiabatic process. However, reversible adiabatic processes are isentropic; that is, they take place with no change in entropy. In an adiabatic process, compression always results in warming, and expansion always results in cooling. *See* ENTROPY; ISENTROPIC PROCESS.

During an adiabatic process, temperature changes are due to internal system fluctuations. For example, the events inside an engine cylinder are nearly adiabatic because the wide fluctuations in temperature take place rapidly, compared to the speed with which the cylinder surfaces can conduct heat. Similarly, fluid flow through a nozzle may be so rapid that negligible exchange of heat between fluid and nozzle takes place. The compressions and rarefactions of a sound wave are rapid enough to be considered adiabatic. *See* SOUND; THERMODYNAMIC PROCESSES.   [P.E.Bl.]

Admittance   The ratio of the current to the voltage in an alternating-current circuit. In terms of complex current I and voltage V, the admittance of a circuit is given by Eq. (1), and is related to the impedance of the circuit Z by Eq. (2). Y is a complex number given by Eq. (3). G, the real part of the admittance, is the conductance of the

$$Y = \frac{I}{V} \tag{1}$$

$$Y = \frac{1}{Z} \tag{2}$$

$$Y = G + jB \tag{3}$$

circuit, and B, the imaginary part of the admittance, is the susceptance of the circuit. The units of admittance are called siemens or mhos (reciprocal ohms). *See* CONDUCTANCE; SUSCEPTANCE.   [J.O.S.]

Aerodynamic force   The force exerted on a body whenever there is a relative velocity between the body and the air. There are only two basic sources of aerodynamic force: the pressure distribution and the frictional shear stress distribution exerted by the airflow on the body surface. The pressure exerted by the air at a point on the surface acts perpendicular to the surface at that point; and the shear stress, which is due to the frictional action of the air rubbing against the surface, acts tangentially to the surface at that point. The distribution of pressure and shear stress represent a distributed load over the surface. The net aerodynamic force on the body is due to the net imbalance between these distributed loads as they are summed (integrated) over the entire surface. *See* BOUNDARY-LAYER FLOW; FLUID FLOW.

For purposes of discussion, it is convenient to consider the aerodynamic force on an airfoil (see illustration). The net resultant aerodynamic force R acting through the center of pressure on the airfoil represents mechanically the same effect as that due to the actual pressure and shear stress loads distributed over the body surface. The velocity of the airflow V_∞ is called the free-stream velocity or the free-stream relative wind. By definition, the component of R perpendicular to the relative wind is the lift, L, and the component of R parallel to the relative wind is the drag D. The orientation of the body with respect to the direction of the free stream is given by the angle of attack, α. The magnitude of the aerodynamic force R is governed by the density ρ_∞ and velocity of the free stream, the size of the body, and the angle of attack. *See* AIRFOIL.

An important measure of aerodynamic efficiency is the ratio of lift to drag, L/D. The higher the value of L/D, the more efficient is the lifting action of the body. The value

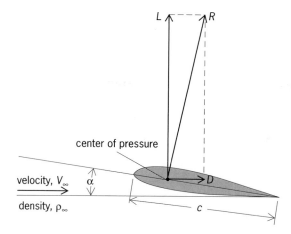

Resultant aerodynamic force (*R*), and its resolution into lift (*L*) and drag (*D*) components.

of L/D reaches a maximum, denoted by $(L/D)_{max}$, at a relatively low angle of attack. Beyond a certain angle the lift decreases with increasing α. In this region, the wing is said to be stalled. In the stall region the flow has separated from the top surface of the wing, creating a type of slowly recirculating dead-air region, which decreases the lift and substantially increases the drag.   [J.D.A.]

Aerodynamic sound   Sound that is generated by the unsteady motion of a gas and its interaction with surrounding surfaces. Aerodynamic sound or noise may be pleasant, such as the sound generated by a flute, or unpleasant, such as the noise of an aircraft on landing or takeoff, or the impulsive noise of a helicopter in descending or forward flight.

Sources of aerodynamic sound may be classified according to their multipole order. Sources associated with unsteady mass addition to the gas are called monopoles. These could be caused by the unsteady mass flux in a jet exhaust or the pulsation of a body. The sound power radiated by monopoles scales with the fourth power of a characteristic source velocity. Sources related to unsteady forces acting on the gas are called dipoles. The singing in telephone wires is related to the nearly periodic lift variations caused by vortex shedding from the wires. Such sources, called dipoles, generate a sound power that scales with the sixth power of the characteristic source velocity. *See* KÁRMÁN VORTEX STREET.

A turbulent fluid undergoes local compression and extension as well as shearing. These events are nearly random on the smallest scales of turbulent motion but may have more organization at the largest scales. The earliest theories of aerodynamic noise, called acoustic analogies, related these unsteady stresses in the fluid to the noise that they would generate in a uniform ambient medium with a constant speed of sound. Such sources are called quadrupoles. The sound power that they radiate scales with the eighth power of the characteristic source velocity. *See* TURBULENT FLOW.

Subsequent extensions of these theories allowed for the motion of these sources relative to a listener. This may result in a Doppler shift in frequency and a convective amplification of the sound if the sources are moving toward the listener. In addition, if the sources are embedded in a sheared flow, such as the exhaust plume of a jet engine, the sound is refracted away from the jet downstream axis. As sound propagates away from the source region, it experiences attenuation due to spherical spreading and real-gas and relaxational effects. The latter effects are usually important only for high-amplitude sound or sound propagation over large distances. *See* ATMOSPHERIC ACOUSTICS; DOPPLER EFFECT; SOUND ABSORPTION.   [P.J.M.]

Aerodynamic wave drag   The force retarding an airplane, especially in supersonic flight, as a consequence of the formation of shock waves. Although the physical laws governing flight at speeds in excess of the speed of sound are the same as those for subsonic flight, the nature of the flow about an airplane and, as a consequence, the various aerodynamic forces and moments acting on the vehicle at these higher speeds differ substantially from those at subsonic speeds. Basically, these variations result from the fact that at supersonic speeds the airplane moves faster than the disturbances of the air produced by the passage of the airplane. These disturbances are propagated at roughly the speed of sound and, as a result, primarily influence only a region behind the vehicle.

The primary effect of the change in the nature of the flow at supersonic speeds is a marked increase in the drag, resulting from the formation of shock waves about the configuration. These strong disturbances, which may extend for many miles from the airplane, cause significant energy losses in the air, the energy being drawn from the airplane. At supersonic flight speeds these waves are swept back obliquely, the angle of obliqueness decreasing with speed. For the major parts of the shock waves from a well-designed airplane, the angle of obliqueness is equal to $\sin^{-1}(1/M)$, where M is the Mach number, the ratio of the flight velocity to the speed of sound. *See* SHOCK WAVE; SUPERSONIC FLOW.

The shock waves are associated with outward diversions of the airflow by the various elements of the airplane. This diversion is caused by the leading and trailing edges of the wing and control surfaces, the nose and aft end of the fuselage, and other parts of the vehicle. Major proportions of these effects also result from the wing incidence required to provide lift.

For a well-designed vehicle, wave drag is usually roughly equal to the sum of the basic skin friction and the induced drag due to lift. *See* AERODYNAMIC FORCE; AIRFOIL.

The wave drag at the zero lift condition is reduced primarily by decreasing the thickness-chord ratios for the wings and control surfaces and by increasing the length-diameter ratios for the fuselage and bodies. Also, the leading edge of the wing and the nose of the fuselage are made relatively sharp. With such changes, the severity of the diversions of the flow by these elements is reduced, with a resulting reduction of the strength of the associated shock waves. Also, the supersonic drag wave can be reduced by shaping the fuselage and arranging the components on the basis of the area rule.

The wave drag can also be reduced by sweeping the wing panels. Some wings intended for supersonic flight have large amounts of leading-edge sweep and little or no trailing-edge sweep. The shape changes required are now determined using very complex fluid-dynamic relationships and supercomputers. *See* COMPUTATIONAL FLUID DYNAMICS.   [R.T.Wh.]

Aerodynamics   The applied science that deals with the dynamics of airflow and the resulting interactions between this airflow and solid boundaries. The solid boundaries may be a body immersed in the airflow, or a duct of some shape through which the air is flowing. Although, strictly speaking, aerodynamics is concerned with the flow of air, in modern times the term has been liberally interpreted as dealing with the flow of gases in general.

Depending on its practical objectives, aerodynamics can be subdivided into external and internal aerodynamics. External aerodynamics is concerned with the forces and moments on, and heat transfer to, bodies moving through a fluid (usually air). Examples are the generation of lift, drag, and moments on airfoils, wings, fuselages, engine nacelles, and whole airplane configurations; wind forces on buildings; the lift and drag on automobiles; and the aerodynamic heating of high-speed aerospace vehicles such as the space shuttle. Internal aerodynamics involves the study of flows moving internally through ducts. Examples are the flow properties inside wind tunnels, jet engines, rocket engines, and pipes. In short, aerodynamics is concerned with the detailed physical properties of a flow field and also with the net effect of these properties in generating an aerodynamic force on a body immersed in the flow, as well as heat transfer to the body. *See* AERODYNAMIC FORCE; AEROTHERMODYNAMICS.

Aerodynamics can also be subdivided into various categories depending on the dominant physical aspects of a given flow. In low-density flow the characteristic size of the flow field, or a body immersed in the flow, is of the order of a molecular mean free path (the average distance that a molecule moves between collisions with neighboring molecules); while in continuum flow the characteristic size is much greater than the molecular mean free path. More than 99% of all practical aerodynamic flow problems fall within the continuum category. *See* RAREFIED GAS FLOW.

Continuum flow can be subdivided into viscous flow, which is dominated by the dissipative effects of viscosity (friction), thermal conduction, and mass diffusion; and inviscid flow, which is, by definition, a flow in which these dissipative effects are negligible. Both viscous and inviscid flows can be subdivided into incompressible flow, in which the density is constant, and compressible flow, in which the density is a variable. In low-speed gas flow, the density variation is small and can be ignored. In contrast, in a high-speed flow the density variation is keyed to temperature and pressure variations, which can be large, so the flow must be treated as compressible. *See* COMPRESSIBLE FLOW; FLUID FLOW; INCOMPRESSIBLE FLOW; VISCOSITY.

In turn, compressible flow is subdivided into four speed regimes: subsonic flow, transonic flow, supersonic flow, and hypersonic flow. These regimes are distinguished by the value of the Mach number, which is the ratio of the local flow velocity to the local speed of sound.

A flow is subsonic if the Mach number is less than 1 at every point. Subsonic flows are characterized by smooth streamlines with no discontinuity in slope. The flow over light, general-aviation airplanes is subsonic.

A transonic flow is a mixed region of locally subsonic and supersonic flow. The flow far upstream of the airfoil can be subsonic, but as the flow moves around the airfoil surface it speeds up, and there can be pockets of locally supersonic flow over both the top and bottom surfaces of the airfoil.

In a supersonic flow, the local Mach number is greater than 1 everywhere in the flow. Supersonic flows are frequently characterized by the presence of shock waves. Across shock waves, the flow properties and the directions of streamlines change discontinuously, in contrast to the smooth, continuous variations in subsonic flow. *See* SUPERSONIC FLOW.

Hypersonic flow is a regime of very high supersonic speeds. A conventional rule is that any flow with a Mach number equal to or greater than 5 is hypersonic. Examples include the space shuttle during ascent and reentry into the atmosphere, and the flight of the X-15 experimental vehicle. The kinetic energy of many hypersonic flows is so high that, in regions where the flow velocity decreases, kinetic energy is traded for internal energy of the gas, creating high temperatures. Aerodynamic heating is a particularly severe problem for bodies immersed in a hypersonic flow. [J.D.A.]

Aerostatics   The science of the equilibrium of gases and of solid bodies immersed in them when under the influence only of natural gravitational forces. Aerostatics is concerned with the balance between the weight of the gases and the weight of any object within them. Archimedes' law that an immersed body experiences a buoyancy force equal to the weight of the fluid displaced is the principal law of aerostatics, if the fluid is air, or of hydrostatics, if the fluid is water. Some phases of meteorology and the flight of balloons and dirigibles are based on aerostatics. In meteorology cloud and fog subsidence and simple pressure and temperature relations with altitude are predicted from aerostatic principles. *See* HYDROSTATICS.   [J.R.Se.]

Aerothermodynamics   Flow of gases in which heat exchanges produce a significant effect on the flow. Traditionally, aerodynamics treats the flow of gases, usually air, in which the thermodynamic state is not far different from standard atmospheric conditions at sea level. In such a case the pressure, temperature, and density are related by the simple equation of state for a perfect gas; and the rest of the gas's properties, such as specific heat, viscosity, and thermal conductivity, are assumed constant. Because fluid properties of a gas depend upon its temperature and composition, analysis of flow systems in which temperatures are high or in which the composition of the gas varies (as it does at high velocities) requires simultaneous examination of thermal and dynamic phenomena. For instance, at hypersonic flight speed the characteristic temperature in the shock layer of a blunted body or in the boundary layer of a slender body is proportional to the square of the Mach number. These are aerothermodynamic phenomena.

Two problems of particular importance require aerothermodynamic considerations: combustion and high-speed flight. Chemical reactions sustained by combustion flow systems produce high temperatures and variable gas composition. Because of oxidation (combustion) and in some cases dissociation and ionization processes, these systems are sometimes described as aerothermochemical. In high-speed flight the kinetic energy used by a vehicle to overcome drag forces is converted into compression work on the surrounding gas and thereby raises the gas temperature. Temperature of the gas may become high enough to cause dissociation (at Mach number ≥ 7) and ionization (at Mach number ≥ 12); thus the gas becomes chemically active and electrically conducting. *See* MACH NUMBER.   [S.Y.C.]

Aharonov-Bohm effect   The predicted effect of an electromagnetic vector or scalar potential in electronic interference phenomena, in the absence of electric or magnetic fields on the electrons.

The fundamental equations of motion for a charged object are usually expressed in terms of the magnetic field $\vec{B}$ and the electric field $\vec{E}$. The force $\vec{F}$ on a charged particle can be conveniently written as in the equations below, where q is the particle's charge,

$$\vec{F} = q\vec{E} \qquad \vec{F} = q\vec{v} \times \vec{B}$$

$\vec{v}$ is its velocity, and the symbol $\times$ represents the vector product. Associated with $\vec{E}$ is a scalar potential V defined at any point as the work W necessary to move a charge from minus infinity to that point, $V = W/q$. Generally, only the difference in potentials between two points matters in classical physics, and this potential difference can be used in computing the electric field. Similarly, associated with $\vec{B}$ is a vector potential $\vec{A}$, a convenient mathematical aid for calculating the magnetic field. *See* ELECTRIC FIELD; POTENTIALS.

In quantum mechanics, however, the basic equations that describe the motion of all objects contain $\vec{A}$ and V directly, and they cannot be simply eliminated. Nonetheless, it was initially believed that these potentials had no independent significance. In 1959, Y. Aharonov and D. Bohm discovered that both the scalar and vector potentials should play a major role in quantum mechanics. They proposed two electron interference experiments in which some of the electron properties would be sensitive to changes of $\vec{A}$ or V, even when there were no electric or magnetic fields present on the charged particles. The absence of $\vec{E}$ and $\vec{B}$ means that classically there are no forces acting on the particles, but quantum-mechanically it is still possible to change the properties of the electron. These counterintuitive predictions are known as the Aharonov-Bohm effect.

Surprisingly, the Aharonov-Bohm effect plays an important role in understanding the properties of electrical circuits whose wires or transistors are smaller than a few micrometers. The electrical resistance in a wire loop oscillates periodically as the magnetic flux threading the loop is increased, with a period of h/e (where h is Planck's constant and e is the charge of the electron), the normal-metal flux quantum. In single wires, the electrical resistance fluctuates randomly as a function of magnetic flux. Both these observations, which were made possible by advances in the technology for fabricating small samples, reflect an Aharonov-Bohm effect. They have opened up a new field of condensed-matter physics because they are a signature that the electrical properties are dominated by quantum-mechanical behavior of the electrons, and that the rules of the classical physics are no longer operative. *See* QUANTUM MECHANICS.　　　[R.A.Web.]

Airfoil　　The cross section of a body that is placed in an airstream in order to produce a useful aerodynamic force in the most efficient manner possible. The cross sections of wings, propeller blades, windmill blades, compressor and turbine blades in a jet engine, and hydrofoils on a high-speed ship are examples of airfoils.

The mean camber line of an airfoil (see illustration) is the locus of points halfway between the upper and lower surfaces as measured perpendicular to the mean camber line itself. The most forward and rearward points of the mean camber line are the leading and trailing edges, respectively. The straight line connecting the leading and trailing edges is the chord line of the airfoil, and the distance from the leading to the trailing edge measured along the chord line is simply designated the chord of the airfoil, represented by c. The thickness of the airfoil is the distance from the upper to the lower surface, measured perpendicular to the chord line, and varies with distance along the chord. The maximum thickness, and where it occurs along the chord, is an important design feature of the airfoil. The camber is the maximum distance between the mean camber line and the chord line, measured perpendicular to the chord line. Both the maximum thickness and the camber are usually expressed in terms of a percentage of

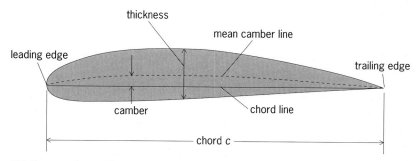

Airfoil nomenclature. The shape shown is an NACA 4415 airfoil.

the chord length; for example, a 12% thick airfoil has a maximum thickness equal to 0.12c.

The airfoil may be imagined as part of a wing which projects into and out of the page, stretching to plus and minus infinity. Such a wing, with an infinite span perpendicular to the page, is called an infinite wing. The aerodynamic force on the airfoil, by definition, is the force exerted on a unit span of the infinite wing. For this reason, airfoil data are frequently identified as infinite wing data.

The flow of air (or any fluid) over the airfoil results in an aerodynamic force (per unit span) on the airfoil, denoted by R. The relative wind is the magnitude and direction of the free-stream velocity far ahead of the airfoil. The angle between the chord line and relative wind is defined as the angle of attack of the airfoil, denoted by α. By definition, the component of R perpendicular to the relative wind is the lift, L; similarly, the component of R parallel to the relative wind is the drag, D.

The airfoil may be visualized as being supported by an axis perpendicular to the airfoil, and taken through any point on the airfoil. The airfoil has a tendency to twist about this axis; that is, there is an aerodynamic moment exerted on the airfoil. By definition, the moment is positive or negative if it tends to increase or decrease respectively the angle of attack (that is, if it tends to pitch the airfoil up or down, respectively).   [J.D.A.]

Alfvén waves   Propagating oscillations in electrically conducting fluids or gases in which a magnetic field is present. Magnetohydrodynamics deals with the effects of magnetic fields on fluids and gases which are efficient conductors of electricity. Molten metals are generally good conductors of electricity, and they exhibit magnetohydrodynamic phenomena. Gases can be efficient conductors of electricity if they become ionized. Ionization can occur at high temperatures or through the ionizing effects of high-energy (usually ultraviolet) photons. A gas which consists of free electrons and ions is called a plasma. Most gases in space are plasmas, and magnetohydrodynamic phenomena are expected to play a fundamental role in the behavior of matter in the cosmos. *See* PLASMA (PHYSICS).

Waves are a particularly important aspect of magnetohydrodynamics. They transport energy and momentum from place to place and may, therefore, play essential roles in the heating and acceleration of cosmical and laboratory plasmas. A wave is a propagating oscillation. If waves are present, a given parcel of the fluid undergoes oscillations about an equilibrium position. The parcel oscillates because there are restoring forces which tend to return it to its equilibrium position. In an ordinary gas, the only restoring force comes from the thermal pressure of the gas. This leads to one wave mode: the sound wave. If a magnetic field is present, there are two additional restoring forces: the tension associated with magnetic field lines, and the pressure associated with the energy density of the magnetic field. These two restoring forces lead to two additional wave modes. Thus there are three magnetohydrodynamic wave modes. However, each restoring force does not necessarily have a unique wave mode associated with it. Put another way, each wave mode can involve more than one restoring force. Thus the usual sound wave, which involves only the thermal pressure, does not appear as a mode in magnetohydrodynamics. The three modes have different propagation speeds, and are named fast mode (F), slow mode (S), and intermediate mode (I). The intermediate mode is sometimes called the Alfvén wave, but some scientists refer to all three magnetohydrodynamic modes as Alfvén waves. The intermediate mode is also called the shear wave. Some scientists give the name magnetosonic mode to the fast mode.

Basic equations. The magnetohydrodynamic wave modes are analyzed by using the magnetohydrodynamic equations for the motion of a conducting fluid in a magnetic field, combined with Maxwell's equations and Ohm's law. *See* MAXWELL'S EQUATIONS.

It is possible to combine Ohm's law with Faraday's law of induction. The resultant equation is called the magnetohydrodynamic induction equation, which is the mathematical statement of the "frozen-in" theorem. This theorem states that magnetic field lines can be thought of as being frozen into the fluid, with the proviso that the fluid is always allowed to slip freely along the field lines. It is the coupling between the fluid and the magnetic field which makes magnetohydrodynamic waves possible. The oscillating magnetic field lines cause oscillations of the fluid parcels, while the fluid provides a mass loading on the magnetic field lines. This mass loading has the effect of slowing down the waves, so that they propagate at speeds much less than the speed of light (which is the propagation speed of waves in a vacuum). *See* ELECTROMAGNETIC RADIATION; LIGHT.

Linearization of equations. Unfortunately, the basic equations are too difficult to be of much use because some of them are nonlinear; that is, they contain products of the quantities for which a solution is sought. Nonlinear magnetohydrodynamics is still only in its infancy, and only a few specialized solutions are known. In order to get solvable equations, scientists accept the limitation of dealing with small-amplitude waves and linearize the equations, so that products of the unknowns are removed. Fortunately, much can still be learned from this procedure; the resulting equations have solutions which are harmonic in time and space. *See* HARMONIC MOTION.

Intermediate mode. The motions in this mode are pure shears. There is no compression of the plasma. The tension in the magnetic field lines is the only restoring force involved in the propagation of the wave. This mode is therefore closely analogous to the propagation of waves on a string.

Because these waves channel energy along magnetic fields, they may be responsible for the observed fact that cosmical plasmas are strongly heated in the presence of magnetic fields.

Fast mode. This mode is difficult to analyze. However, many cosmical and laboratory plasmas satisfy the strong-magnetic-field case where the fast mode is more easily understood.

Fast waves are compressive, and the magnetic field strength fluctuates as well. Thus fast waves are governed by the two restoring forces associated with the tension and pressure in the magnetic field.

The fast mode can propagate energy across the magnetic field.

Slow mode. Like the fast mode, the slow mode is difficult to study in general, and the discussion will again be confined to strong magnetic fields. The slow mode in a strong field is equivalent to sound waves which are guided along the strong magnetic field lines. The strong magnetic field lines can be thought of as a set of rigid pipes which allow free fluid motion along the pipes, but which restrict motion in the other two directions. The motions on the individual pipes are not coupled together, and thus the slow mode is analogous to the sound waves on a set of independent organ pipes. The slow mode channels energy along the magnetic field. Because the sound speed is small, by assumption, the slow mode transmits energy less effectively than the fast or intermediate modes.

Nonlinear effects. Only small-amplitude waves have been considered. Real waves have finite amplitude, and nonlinear effects can sometimes be important. One such effect is the tendency of waves to steepen, ultimately forming magnetohydrodynamic shock waves and magnetohydrodynamic discontinuities. There is an abundance of magnetohydrodynamic discontinuities in the solar wind. *See* SHOCK WAVE.

It is also possible that waves can degenerate into turbulence. There are indications that this too happens in the solar wind. *See* TURBULENT FLOW.

Surface waves. Only waves in a spatially uniform background have been considered. While the analysis of magnetohydrodynamic waves in a nonuniform background is complicated, it is possible to consider an extreme limit, in which the background is uniform except at certain surfaces where it changes discontinuously. Surfaces can support magnetohydrodynamic waves, which are in some respects similar to waves on the surface of a lake. These waves may play important roles in heating cosmical and laboratory plasmas. *See* MAGNETOHYDRODYNAMICS.   [J.V.H.]

Alpha particles   Helium nuclei, which are abundant throughout the universe both as radioactive-decay products and as key participants in stellar fusion reactions. Alpha particles can also be generated in the laboratory, either by ionizing helium or from nuclear reactions. They expend their energy rapidly as they pass through matter, primarily by taking part in ionization processes, and consequently have short penetration ranges. Numerous technological applications of alpha particles can be found in fields as diverse as medicine, space exploration, and geology. Alpha particles are also major factors in the health concerns associated with nuclear waste and other radiation hazards.

The helium nucleus, or alpha particle (α), with mass 4.00150 atomic mass units (u) and charge $+2$, is a strongly bound cluster of two protons (p) and two neutrons (n). Its stability is evident from mass-energy conservation in the hypothetical fusion reaction $2p + 2n \rightarrow \alpha$. The product mass ($= 4.00150$ u) is less than the reactant mass ($= 2 \times 1.00728$ u $+ 2 \times 1.00866$ u) by 0.03038 u. By using Einstein's relation $E = mc^2$ (where c is the speed of light), this decrease in mass m (the alpha-particle binding energy) is equivalent to 28.3 MeV of energy E. The enormous magnitude of this energy is reflected in the fact that the fusion transformation of hydrogen into helium is the main process responsible for the Sun's energy. *See* CONSERVATION OF ENERGY; ENERGY; NUCLEAR BINDING ENERGY.

Alpha radioactivity. Coulombic repulsion between the protons within a nucleus leads to increasingly larger ratios of neutron number N to proton number Z for stable nuclei, as the mass numbers increase. Neutron-deficient nuclei can improve their N/Z ratios by means of alpha decay. The decay occurs because the parent nucleus has a total mass greater than the sum of the masses of the daughter nucleus and the alpha particle. The energy converted from mass energy to kinetic energy, called the Q value, is shared between the daughter nucleus and the alpha particle in accordance with the conservation of momentum. Thus, each radioactive alpha-emitting nuclide emits the alpha with a characteristic kinetic energy, which is one fingerprint in identification of the emitter. *See* NUCLEAR REACTION; RADIOACTIVITY.

There are three major natural series, or chains, through which isotopes of heavy elements decay by successions of alpha decays. Within these series, and with all reaction-produced alpha emitters as well, each isotope decays with a characteristic half-life and emits alpha particles of particular energies and intensities. The presence of these radioactive nuclides in nature depends upon either a continuous production mechanism, for example the interaction of cosmic rays with the atmosphere, or extremely long half-lives of heavy radioactive nuclides produced in past cataclysmic astrophysical events, which accounts for uranium and thorium ores in the Earth. The relative abundances of uranium-238, uranium-235, and their stable final decay products in ores of heavy elements can be used to calculate the age of the ore, and presumably the age of the Earth.

In addition to the study of alpha-particle emitters that appear in nature, alpha decay has provided a useful tool to study artificial nuclei, which do not exist in nature due to their short half-lives. Alpha decay is a very important decay mode for nuclei far from stability with a ratio of protons to neutrons that is too large to be stable, especially for nuclei with atomic mass greater than 150 u. Because of the ease of detecting and interpreting decay alpha particles, their observation has aided tremendously in studying these nuclei far from stability, extending the study of nuclei to the very edge of nuclear existence. Nuclear structure information for more than 400 nuclides has been obtained in this way. In addition, fine structure peaks appear in the alpha-particle spectra for many of these nuclides; each such fine structure peak gives similar information about an excited state in the daughter nucleus.

Interactions with matter. By virtue of their kinetic energy, double positive charge, and large mass, alpha particles follow fairly straight paths in matter, interacting strongly with atomic electrons as they slow down and stop. These electrons may be excited to higher energy states in their host atoms, or they may be ejected, forming ion pairs in which the initial host atom becomes positively charged and the electron leaves. The more energetic ejected electrons, known as delta electrons, cause considerable secondary ionization, which accounts for 60–80% of the total ionization. A cascade of processes occurs along the alpha particle's track, leading to tens of thousands of disruptive events per alpha particle.

The amount of energy expended by an alpha particle to form a single ion pair in passing through a medium is nearly independent of the alpha particle's energy, but it depends strongly on the absorbing medium. While it takes about 35 eV in air and 43 eV in helium to form an ion pair, an energy of only 2.9 eV is required in germanium and 3.6 eV in silicon. The energies expended in gases are roughly correlated to their ionization potentials. For germanium, silicon, and other semiconductors, the lower ion pair energy is, effectively, the amount required to raise an electron to the conduction band. *See* IONIZATION POTENTIAL; SEMICONDUCTOR.

The distance (or range) that an alpha particle travels before it stops depends both on the energy of the particle and on the absorbing medium. The passage of alpha particles through silicon is a particularly important example. The semiconductor industry now produces chips so small that alpha particles from contaminants in the packaging materials can disrupt the memory-array areas of the chips, a serious problem which has been researched in considerable detail.

In biological systems, the ionization and excitation produced by alpha particles can damage or kill cells. By rupturing chemical bonds and forming highly reactive free radicals, alpha particles can be far more destructive than other forms of radiation which interact less strongly with matter. *See* CHARGED PARTICLE BEAMS.

Applications. In the promising medical field of charged-particle radiotherapy, alpha particles are useful in the treatment of inaccessible tumors and vascular disorders. The ionizing power of alpha particles is concentrated near the ends of their paths. Thus they can deliver destructive energy to a tumor while doing little damage to nearby healthy tissue. With proper acceleration, positioning, and dosage, the energy can be delivered so precisely that alpha-particle radiotherapy is uniquely suited for treating highly localized tumors near sensitive normal tissue (for example, the spinal cord).

The element-specific energies of backscattered (Rutherford-scattered) alpha particles are used in remote probes to analyze the mineral composition of geological formations. In particular, alpha particles scattered by light elements transfer more energy than those scattered by heavy elements. In another alpha-particle device, the energy from ^{238}Pu alpha decay is reliably harnessed in batteries based on the Brayton cycle, and used to power scientific equipment left on the Moon. Large power systems of this type are contemplated for use in space stations.   [C.Bin.]

Ammeter   An instrument for the measurement of electric current. The unit of current, the ampere, is the base unit on which rests the International System (SI) definitions of all the electrical units. The operating principle of an ammeter depends on the nature of the current to be measured and the accuracy required. Currents may be broadly classified as direct current (dc), low-frequency alternating current (ac), or radio frequency. At frequencies above about 10 MHz, where the wavelength of the signal becomes comparable with the dimensions of the measuring instrument, current measurements become inaccurate and finally meaningless, since the value obtained depends on the position where the measurement is made. In these circumstances, power measurements are usually used. *See* CURRENT MEASUREMENT.

The measurement of current in terms of the voltage that appears across a resistive shunt through which the current passes has become the most common basis for ammeters, primarily because of the very wide range of current measurement that it makes possible, and more recently through its compatibility with digital techniques. *See* ELECTRICAL UNITS AND STANDARDS; MULTIMETER; VOLTMETER.

The moving-coil, permanent-magnet (d'Arsonval) ammeter remains important for direct-current measurement. Generally they are of modest accuracy, no better than 1%. Digital instruments have taken over all measurements of greater precision because of the greater ease of reading their indications where high resolution is required.

Moving-iron instruments are widely used as ammeters for low-frequency ac applications.

High-frequency currents are measured by the heating effect of the current passing through a physically small resistance element. In modern instruments the temperature of the center of the wire is sensed by a thermocouple, the output of which is used to drive a moving-coil indicator. *See* THERMOCOUPLE.   [R.B.D.K.]

Amorphous solid   A rigid material whose structure lacks crystalline periodicity; that is, the pattern of its constituent atoms or molecules does not repeat periodically in three dimensions. In the present terminology amorphous and noncrystalline are synonymous. A solid is distinguished from its other amorphous counterparts (liquids and gases) by its viscosity: a material is considered solid (rigid) if its shear viscosity exceeds $10^{14.6}$ poise ($10^{13.6}$ Pa · s). *See* CRYSTAL; VISCOSITY.

Oxide glasses, generally the silicates, are the most familiar amorphous solids. However, as a state of matter, amorphous solids are much more widespread than just the oxide glasses. There are both organic (for example, polyethylene and some hard candies) and inorganic (for example, the silicates) amorphous solids. Glasses can be prepared which span a broad range of physical properties. Dielectrics (for example, SiO_2) have very low electrical conductivity and are optically transparent, hard, and brittle. Semiconductors (for example, As_2SeTe_2) have intermediate electrical conductivities and are optically opaque and brittle. Metallic glasses have high electrical and thermal conductivities, have metallic luster, and are ductile and strong.

The obvious uses for amorphous solids are as window glass, container glass, and the glassy polymers (plastics). Less widely recognized but nevertheless established technological uses include the dielectrics and protective coatings used in integrated circuits, and the active element in photocopying by xerography, which depends for its action upon photoconduction in an amorphous semiconductor. In optical communications a highly transparent dielectric glass in the form of a fiber is used as the transmission medium.

It is the changes in short-range order (on the scale of a localized electron), rather than the loss of long-range order alone, that have a profound effect on the properties of amorphous semiconductors. For example, the difference in resistivity between the crystalline and amorphous states for dielectrics and metals is always less than an order of magnitude and is generally less than a factor of 3. For semiconductors, however, resistivity changes of 10 orders of magnitude between the crystalline and amorphous states are not uncommon, and accompanying changes in optical properties can also be large.

One class of amorphous semiconductors is the glassy chalcogenides, which contain one (or more) of the chalcogens sulfur, selenium, or tellurium as major constituents. These materials have application in switching and memory devices. Another group is the tetrahedrally bonded amorphous solids, such as amorphous silicon and germanium. These materials cannot be formed by quenching from the melt (that is, as glasses) but must be prepared by one of the deposition techniques mentioned above.

When amorphous silicon (or germanium) is prepared by evaporation, not all bonding requirements are satisfied, so a large number of dangling bonds are introduced into the material. These dangling bonds create states deep in the gap which limit the transport properties. The number of dangling bonds can be reduced by a thermal anneal below the crystallization temperature, but the number cannot be reduced sufficiently to permit doping. *See* SEMICONDUCTOR.   [B.G.B.]

Ampère's law   A law of electromagnetism which expresses the contribution of a current element of length dl to the magnetic induction (flux density) B at a point near the current. Ampère's law, sometimes called Laplace's law, was derived by A. M. Ampère after a series of experiments during 1820–1825.

Whenever an electric charge is in motion, there is a magnetic field associated with that motion. The flow of charges through a conductor sets up a magnetic field in the surrounding region. Any current may be considered to be broken up into infinitesimal elements of length dl, and each such element contributes to the magnetic induction

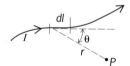

Graphic representation of
Ampère's law.

at every point in the neighborhood. The contribution dB of the element is found to depend upon the current I, the length dl of the element, the distance r of the point P from the current element, and the angle θ between the current element and the line joining the element of the point P (see illustration). Ampère's law expresses the manner of the dependence by Eq. (1). The field near a current may be calculated by finding the

$$dB = k\frac{I\,dl\,\sin\theta}{r^2} \tag{1}$$

vector sum of the contributions of all the various elements that make up the current.

The proportionality factor k depends upon the units used in Eq. (1) and upon the properties of the medium surrounding the current. In the SI system, the factor k is assigned a value of 10^{-7} weber/ampere-meters when the current is in empty space. As in other equations associated with electric and magnetic fields, for example Coulomb's law, it is convenient to replace k by a new factor μ_0 related to k as in Eq. (2). This

$$\mu_0 = 4\pi k \tag{2}$$

substitution removes the factor 4π from many derived equations in which it would otherwise appear. With this substitution Ampère's law becomes Eq. (3). The factor μ_0 is called the permeability of empty space.

$$dB = \frac{\mu_0}{4\pi} \frac{I \; dl \; \sin\theta}{r^2} \qquad (3)$$

The direction of dB at each point may be described in terms of a right-hand rule. If the current element is grasped by the right hand with the thumb pointing in the direction of the current, the fingers encircle the current in the direction of the magnetic induction.

[K.V.M.]

Analog states   States in neighboring nuclear isobars that have the same total angular momentum, parity, and isotopic spin. They also have nearly identical nuclear structure wave functions except for the transformation of one or more neutrons into an equivalent number of protons, which occupy the same single-particle states as the neutrons. Analog states (or isobaric analog states, IAS) have been observed throughout the periodic table, indicating that isotopic spin is a good quantum number. See ANGULAR MOMENTUM; I-SPIN; NUCLEAR STRUCTURE; PARITY (QUANTUM MECHANICS).

Since the nucleon-nucleon interaction has been found to be approximately charge-independent, it is possible to consider protons and neutrons as representing different charge states of a single particle, that is, a nucleon. Thus, a level (commonly referred to as a parent state) in a nucleus with Z protons and N neutrons can be expected to have an analog in the neighboring isobar with $Z + 1$ protons and $N - 1$ neutrons (and the same total number of nucleons, $A = Z + N$), where the protons and neutrons occupy the same orbits as those in the parent state. The energy difference between the parent and analog states predominantly arises from the increased contribution from the electrostatic Coulomb interaction to the total energy arising from the extra proton in the analog state. From this amount must be subtracted the neutron-proton mass difference of 0.782 MeV (energies are given on the atomic mass scale). The agreement between such calculated energies of analog states and their measured values is in general fairly precise but not exact. The reason is that small additional factors influence the level energies, such as electromagnetic effects, a small charge-dependent nuclear interaction, isospin mixing, and nuclear structure effects.

The study of analog states provides important information used to test nuclear theories. For example, the double charge-exchange reactions (π^+, π^-) [where π^+ and π^- represent a pion with positive and negative charge, respectively] have been used to identify double isobaric analog states, that is, analogs in isobars removed by 2 charge units. Such data have been useful for testing various formulas for predicting the relative masses of isobaric multiplets. Single and double charge-exchange reactions utilizing incident pions have also been used to investigate giant resonances built upon analog states. The single-particle structure of parent states can be studied by observing the particle decay of the analog state when the decay resides in the nuclear continuum. Measurements of the widths of analog states provide information pertaining to their fragmentation, for example, their mixing with states of the same spin and parity but with total isospin lower by one unit.

[D.J.Ho.]

Anemometer   An instrument to measure the speed or velocity of gases either in a contained flow, such as airflow in a duct, or in unconfined flows, such as atmospheric wind. To determine the velocity, an anemometer detects change in some physical property of the fluid or the effect of the fluid on a mechanical device inserted into the flow.

An anemometer can measure the total velocity magnitude, the velocity magnitude in a plane, or the velocity component in a particular direction. The cup anemometer, for example, measures the velocity in a plane perpendicular to the axis of its rotation cups. If the cup anemometer is mounted with the shaft perpendicular to the horizontal, it will measure only the component of the wind that is parallel to the ground. Other anemometers, such as the pitot-static tube, are used with the tip aligned with the total velocity vector. Before using an anemometer, it is important to determine how it should be positioned and what component of the total velocity its measurement represents.

An anemometer usually measures gas flows that are turbulent. The cup anemometer, pitot-static tube, and thermal anemometer are mostly used to measure the mean velocity, while the hot-wire, laser Doppler, and sonic anemometers are usually used when turbulence characteristics are being measured. (The term "thermal anemometer" is often used to mean any anemometer that uses a relationship between heat transfer and velocity to determine velocity.)

[D.E.St.]

Angular correlations   An experimental technique that involves measuring the manner in which the likelihood of occurrence (or intensity or cross section) of a particular decay or collision process depends on the directions of two or more radiations associated with the process. Traditionally, these radiations are emissions from the decay or collision process. However, a variant on this technique in which the angular correlations are between an incident and emitted beam of radiation has been widely used; this variant is known as angular distributions.

The fundamental reason for performing such measurements, rather than just scrutinizing a single radiation in a particular direction or measuring the total intensity for a process, is that the angular correlation or angular distribution measurement provides much more information on both the decay or collision process and on the structure and properties of the emitter of the radiation. The technique is used to study a variety of decay and collision processes in atomic and molecular physics, condensed-matter (solid-state) and surface physics, and nuclear and particle physics.

The principal use of this technique in nuclear physics has been to determine the angular momentum, or spin, and parity of excited nuclear states which are radioactive, that is, decay spontaneously, by measuring in coincidence the radiation in specific directions from two successive transitions in the radioactive cascade. The measurements are generally of coincidences between gamma rays, but coincidences between gamma rays and electrons (beta particles) are also used. The form of the angular correlation, the measured intensity as a function of the angle between the two radiations, gives the information about the intermediate excited state in the cascade. *See* NUCLEAR SPECTRA; RADIOACTIVITY.

In atomic and molecular collisions as well as in nuclear and particle collisions, this technique is employed as a means of completely specifying the dynamics of the collision, with the added proviso that the energies of the emitted radiations are also to be measured. Wide use has been made of angular correlations in the impact ionization of atoms by electrons where the directions of both the scattered electron and the ejected electron are measured. *See* ATOMIC STRUCTURE AND SPECTRA; SCATTERING EXPERIMENTS (ATOMS AND MOLECULES).

[S.T.M.]

Angular momentum   In classical physics, the moment of linear momentum about an axis. A point particle with mass m and velocity $\mathbf{v}$ has linear momentum $\mathbf{p} = m\mathbf{v}$. Let $\mathbf{r}$ be an instantaneous position vector that locates the particle from an origin on a specified axis. The angular momentum **L** can be written as the vector cross-product in Eq. (1).

$$\mathbf{L} = \mathbf{r} \times \mathbf{p} \tag{1}$$

See MOMENTUM.

The time rate of change of the angular momentum is equal to the torque **N**. A rigid body satisfies two independent equations of motion (the dynamical equations) given by Eqs. (2) and (3), where d/dt denotes the rate of change, the derivative with respect

$$\frac{d}{dt}\mathbf{p} = \mathbf{F} \tag{2}$$

$$\frac{d}{dt}\mathbf{L} = \mathbf{N} \tag{3}$$

to time t. Only Eq. (2) is required for a point particle. Equation (2) indicates that a rigid body acts as a point particle located at its center of mass. The motion of the center of mass depends upon the net force **F**, which is the vector sum of all applied forces. Equation (3) gives the angular motion about the center of mass. The case of statics occurs when the net force and net torque both vanish. See KINETICS (CLASSICAL MECHANICS); STATICS; TORQUE.

A symmetry is a transformation that leaves a physical system unchanged. A physical quantity is called invariant under a transformation if it remains the same after being transformed. For example, the solutions to Eqs. (2) and (3) are invariant under change of the coordinate origin or orientation of the **i**, **j**, and **k** axes. The freedom to choose any orientation of coordinate axes is called rotational invariance, because one choice of axes can be rotated into another. In physics, the rotational invariance follows from the isotropy and homogeneity of space that has been experimentally established to high accuracy.

The study of symmetry shows that one of the deepest relations in physics is that between dynamics and conservation. A physical quantity is conserved if it is constant in time, although it may vary in space. Noether's theorem states that if a physical system is invariant under a continuous symmetry, a conservation law exists, provided that the observable in question decreases rapidly enough at infinity. Thus, when the force is zero everywhere (the system is invariant under translation in space), the linear momentum is conserved. If the torque is zero everywhere (the system is invariant under rotation), the angular momentum is conserved. If the system is invariant under translations in time, the total energy is conserved. See CONSERVATION LAWS (PHYSICS); CONSERVATION OF ENERGY; CONSERVATION OF MOMENTUM.

Quantum mechanics has a richer and more complicated structure than classical physics. Because of this, the relationship between symmetry and conservation is even more useful.   [B.DeF.]

Anharmonic oscillator   A generalized version of harmonic oscillator in which the relationship between force and displacement is nonlinear. The harmonic oscillator is a highly idealized system that oscillates with a single frequency, irrespective of the amount of pumping or energy injected into the system. Consequently, the harmonic oscillator's fundamental frequency of vibration is independent of the amplitude of the vibrations. Applications of the harmonic oscillator model abound in various fields, but perhaps the most commonly studied system is the Hooke's law mass-spring system. In the Hooke's law system the restoring force exerted on the mass is proportional to the displacement of the mass from its equilibrium position. This linear relationship between force and displacement mandates that the oscillation frequency of the mass will be independent of the amplitude of the displacement. *See* HARMONIC MOTION; HARMONIC OSCILLATOR.

In a mechanical anharmonic oscillator, the relationship between force and displacement is not linear but depends upon the amplitude of the displacement. The nonlinearity arises from the fact that the spring is not capable of exerting a restoring force that is proportional to its displacement because of, for example, stretching in the material comprising the spring. As a result of the nonlinearity, the vibration frequency can change, depending upon the system's displacement. These changes in the vibration frequency result in energy being coupled from the fundamental vibration frequency to other frequencies through a process known as parametric coupling. *See* VIBRATION.

There are many systems throughout the physical world that can be modeled as anharmonic oscillators in addition to the nonlinear mass-spring system. For example, an atom, which consists of a positively charged nucleus surrounded by a negatively charged electronic cloud, experiences a displacement between the center of mass of the nucleus and the electronic cloud when an electric field is present. The amount of that displacement, called the electric dipole moment, is related linearly to the applied field for small fields, but as the magnitude of the field is increased, the field-dipole moment relationship becomes nonlinear, just as in the mechanical system. *See* DIPOLE MOMENT.

Further examples of anharmonic oscillators include the large-angle pendulum, which exhibits chaotic behavior as a result of its anharmonicity; nonequilibrium semiconductors that possess a large hot carrier population, which exhibit nonlinear behaviors of various types related to the effective mass of the carriers; and ionospheric plasmas, which also exhibit nonlinear behavior based on the anharmonicity of the plasma. In fact, virtually all oscillators become anharmonic when their pump amplitude increases beyond some threshold, and as a result it is necessary to use nonlinear equations of motion to describe their behavior. *See* CHAOS; PENDULUM; SEMICONDUCTOR.   [D.R.A.]

Antiferromagnetism   A property possessed by some metals, alloys, and salts of transition elements in which the atomic magnetic moments, at sufficiently low temperatures, form an ordered array which alternates or spirals so as to give no net total moment in zero applied magnetic field. The most direct way of detecting such arrangements is by means of neutron diffraction. *See* NEUTRON DIFFRACTION.

The transition temperature below which the spontaneous antiparallel magnetic ordering takes place is called the Néel temperature. A plot of the magnetic susceptibility of a typical antiferromagnetic powder sample versus temperature is shown in the illustration. Below the Néel point, which is characterized by the sharp kink in the susceptibility, the spontaneous ordering opposes the normal tendency of the magnetic moments to align parallel to the applied field. Above the Néel point, the substance is paramagnetic, and the susceptibility χ obeys the Curie-Weiss law, as in Eq. (1), with a negative para-

$$\chi = C/(T + \theta) \tag{1}$$

magnetic Curie temperature $-\theta$. The Néel temperature is similar to the Curie temperature in ferromagnetism. *See* CURIE-WEISS LAW.

The cooperative transition that characterizes antiferromagnetism is thought to result from an interaction energy U of the form given in Eq. (2), where $\mathbf{S}_i$ and $\mathbf{S}_j$ are the

$$U = -2\Sigma J_{ij}\mathbf{S} \cdot \mathbf{S}_j \tag{2}$$

spin angular momentum vectors associated with the magnetic moments of neighbor atoms i and j, and J_{ij} is an interaction constant. If all J_{ij} are positive, the lowest energy is achieved with all $\mathbf{S}_i$ and $\mathbf{S}_j$ parallel, that is, coupled ferromagnetically. Negative J_{ij} between nearest-neighbor pairs (i,j) may lead to simple antiparallel arrays; if the distant

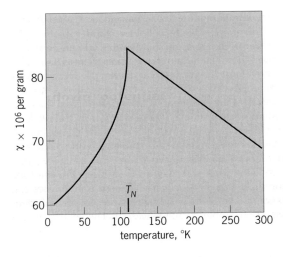

$\chi \times 10^6$ per gram

temperature, °K

T_N

Magnetic susceptibility of powdered manganese oxide. (*After H. Bizette, C. F. Squire, and B. Tsai, 1938*)

neighbors also have sizable negative J_{ij}, a spiral array may have lowest total energy. The interaction constant in Eq. (2) probably arises from superexchange coupling. This is an effective coupling between magnetic spins which is indirectly routed via nonmagnetic atoms in salts and probably via conduction electrons in metals. *See* FERROMAGNETISM; HELIMAGNETISM.

The magnetic moments are known to have preferred direction. Anisotropic effects come from magnetic dipole forces and also from spin-orbit coupling combined with superexchange. Some nearly antiparallel arrays such as Fe_2O_3 show a slight bending (called canting) and exhibit weak ferromagnetism. The anisotropy affects the susceptibility of powder samples and is of extreme importance in antiferromagnetic resonance. *See* MAGNETIC RESONANCE.   [E.A.; F.Ke.]

Antimatter   Matter which is made up of antiparticles. At the most fundamental level every type of elementary particle has its anticounterpart, its antiparticle. The existence of antiparticles was implied by the relativistic wave equation derived in 1928 by P. A. M. Dirac in his successful attempt to reconcile quantum mechanics and special relativity. The antiparticle of the electron (the positron) was first observed in cosmic rays by C. D. Anderson in 1932, while that of the proton (the antiproton) was produced in the laboratory and observed by E. Segré, O. Chamberlain, and their colleagues in 1955. *See* ELECTRON; ELEMENTARY PARTICLE; POSITRON; PROTON; QUANTUM MECHANICS; RELATIVITY.

The mass, intrinsic angular momentum (spin), and lifetime (in the case of unstable particles) of antiparticles and their particles are equal, while their electromagnetic properties, that is, charge and magnetic moment, are equal in magnitude but opposite in sign. Some neutrally charged particles such as the photon and π^0 meson are their own antiparticles. Certain other abstract properties such as baryon number (protons and neutrons are baryons and have baryon number +1) and lepton number (electrons and muons are leptons and have lepton number +1) are reversed in sign between particles and antiparticles. *See* ANGULAR MOMENTUM; BARYON; LEPTON.

The quantum-mechanical operation of turning particles into their corresponding antiparticles is termed charge conjugation (*C*), that of reversing the handedness of particles is parity conjugation (*P*), and that of reversing the direction of time is time reversal (*T*).

A fundamental theorem, the *CPT* theorem, states that correct theories of particle physics must be invariant under the simultaneous operation of *C*, *P*, and *T*. Simply put, the description of physics in a universe of antiparticles with opposite handedness where time runs backward must be the same as the description of the universe. One consequence of the *CPT* theorem is that the above-mentioned properties of antiparticles (mass, intrinsic angular momentum, lifetime, and the magnitudes of charge and magnetic moment) must be identical to those properties of the corresponding particles. This has been experimentally verified to a high precision in many instances. *See* CPT THEOREM; PARITY (QUANTUM MECHANICS).

When a particle and its antiparticle are brought together, they can annihilate into electromagnetic energy or other particles and their antiparticles in such a way that all memory of the nature of the initial particle and antiparticle is lost. Only the total energy and total angular momentum remain. In the reverse process, antiparticles can be produced in particle collisions with matter if the colliding particles possess sufficient energy to create the required mass. For example, a photon with sufficient energy which interacts with a nucleus can produce an electron-positron pair. *See* ELECTRON-POSITRON PAIR PRODUCTION.

Since mesons do not possess baryon or lepton number, only charge, energy, and angular momentum need be conserved in their production. Thus, a process such as a collision of a proton with a proton can produce a single neutral pi meson. Other quantum numbers, such as strangeness and charm, must be conserved if production of mesons possessing these quantum numbers is to proceed through strong or electromagnetic interactions. In these cases a particle with the negative values of the particular quantum number must also be produced. Such a process is termed associated production. *See* CHARM; QUANTUM NUMBERS.

Isolated neutral particles, notably K^0 and B^0 mesons, can spontaneously transform into their antiparticles via the weak interaction. These quantum-mechanical phenomena are termed K–$\overline{K}$ or $\overline{B}$ mixing, respectively. Mixing can lead to particle-antiparticle oscillations wherein a K^0 can become its antiparticle, a $\overline{K}^0$, and later oscillate back to a K^0. It was through this phenomenon that observation of *CP* violation first occurred. That observation, coupled to the *CPT* theorem, implies that physics is not exactly symmetric under time reversal, for example, that the probability of a K^0 becoming a $\overline{K}^0$ is not exactly the same as that in the reverse process.

Experimental observations, both ground- and balloon-based, indicate that the number of cosmic ray antiprotons is less than 1/10,000 that of protons. This number is consistent with the antibaryon production that would be expected from collisions of cosmic protons with the Earth's atmosphere, and is consistent with the lack of appreciable antimatter in the Milky Way Galaxy. Attempts to find antimatter beyond the Milky Way involve searches for gamma radiation resulting from matter-antimatter annihilation in the intergalactic gas that exists between galactic clusters. The null results of these searches suggests that at least the local cluster of galaxies consists mostly of matter. If matter dominates everywhere in the universe, a question arises as to how this came to be. In the standard model of cosmology, the big bang model, the initial condition of the universe was that the baryon number was zero; that is, there was no preference of matter over antimatter. The current theory of how the matter-antimatter asymmetry evolved requires three ingredients: interactions in which baryon number is violated, time reversal (or *CP*) violation, and a lack of thermodynamic equilibrium. The last requirement was satisfied during the first few microseconds after the big bang. Time reversal violation has been observed in the laboratory in K^0 decays, albeit perhaps not of sufficient size to explain the observed baryon-antibaryon asymmetry. But the first ingredient, baryon number violation, has not yet been observed in spite of sensitive searches. Thus, the origin of the dominance of matter over antimatter remains an outstanding mystery of particle and cosmological physics. *See* THERMODYNAMIC PROCESSES.

[M.E.Z.]

Anyons   Particles obeying unconventional forms of quantum statistics. For many years it was believed that only two possible forms of quantum statistics, Bose-Einstein and Fermi-Dirac statistics, were possible, but in fact a continuum of possibilities exists. Elementary excitations (quasiparticles) in the fractional quantum Hall effect are anyons.

In quantum mechanics, in the behavior of identical particles there are important dynamical effects that have no classical analog. Thus, in the case of two indistinguishable particles A and B, the amplitude for the process that leads to A arriving at point x while B arrives at point y must be added to the amplitude for the process that leads to A arriving at y while B arrives at x—the so-called exchange process—because the final states cannot be distinguished. Actually the recipe of adding the amplitude for the exchange process is appropriate only for particles obeying Bose-Einstein statistics (bosons); for particles obeying Fermi-Dirac statistics (fermions), this amplitude must be subtracted. *See* FERMI-DIRAC STATISTICS.

The definition of anyons posits other possible recipes for adding exchange processes, refining the analysis of exchange to take account of the direction in which the exchange takes place. These more general possibilities can be defined only for particles whose motion is restricted to two space dimensions. However, many important materials are effectively two-dimensional, including microelectronic circuitry and the copper oxide layers of high-temperature superconductors. The quantum statistics of the quasiparticles in these systems is under investigation, but the fractional quantized Hall states are known to be anyons. *See* HALL EFFECT; QUANTUM STATISTICS; SUPERCONDUCTIVITY.

[F.Wil.]

Archimedes' principle   The principle that the net fluid force on a body submerged (or floating) in a stationary fluid is an upward force equal to the weight of the fluid displaced by the body. This concept, perhaps the oldest stated principle in fluid mechanics, was first put forth by Archimedes in the third century B.C.

In a static fluid, the weight of the fluid causes an increase in pressure with depth. Thus, at the surface of the fluid, the pressure is atmospheric pressure ($p_0 = 14.7$ lb/in.^{2} = 101 kilonewtons/m^2), while at a depth h the pressure has a larger value of p_1, given by Eq. (1), where γ is the specific weight of the fluid (weight/volume). The differ-

$$p_1 = p_0 + \gamma h \qquad (1)$$

ence in pressure force between the bottom and the top of a water column is therefore given by Eq. (2), where h' and A are the height and area of the column, and

$$(p_b - p_t)A = \gamma h' A \qquad (2)$$

p_b and p_t are the pressures at the bottom and top of the column. This difference is precisely equal to the weight W of the water within the column, given by Eq. (3). If

$$W = \gamma(\text{volume}) = \gamma h' A \qquad (3)$$

the water column were replaced with a solid object, the pressure forces on the object would be the same as on the original water column. That is, the net hydrostatic pressure force on the object, termed the buoyant force, would be equal to the weight of the water displaced (which is the statement of Archimedes' principle). The same concept holds for a body of arbitrary shape, which can be thought of a consisting of many small vertical columns fastened together. Archimedes' principle is valid for submerged or floating bodies in liquids or gases. *See* BUOYANCY; SPECIFIC GRAVITY.   [B.R.M.]

Artificially layered structures   Manufactured, reproducibly layered structures with layer thicknesses approaching interatomic distances. Modern thin-film techniques are at a stage at which it is possible to fabricate these structures, also known as artificial crystals or superlattices, opening up the possibility of engineering new desirable properties into materials. In addition, a variety of solid-state physics problems can be studied which are otherwise inaccessible. The various possibilities include: the application of negative pressure, that is, stretching of the crystalline lattice; the study of dimensional crossover, that is, the transition from a situation in which the layers are isolated and two-dimensional in character to where the layers couple together to form a three-dimensional material; the study of collective behavior, that is, properties which depend on the cooperative behavior of the whole superlattice; and the effect and physics of multiple interfaces and surfaces. For a discussion of semiconductor superlattices *see* CRYSTAL STRUCTURE; SEMICONDUCTOR HETEROSTRUCTURES.

The preparation techniques can be conveniently classified into two groups: evaporation and sputtering. In the evaporation system, two or more particle sources (thermal or electron beam gun) are aimed at a heated substrate where the artificially layered structure is grown. The sputtering method relies on bombarding targets of the proper materials with an inert gas, such as argon, thus producing the beams of the various elements. *See* CRYSTAL GROWTH; MOLECULAR BEAMS.

Once the artificially layered structure is prepared, it is necessary to characterize whether the layer structure is stable at the growth temperature. This is of considerable importance, since the interdiffusion of the constituents in many cases eliminates the layered growth. One of the most successful methods of characterizing layered growth has been x-ray diffraction.

Artificially layered structures are especially useful for the construction of mirrors for soft x-rays since there are no suitable, naturally occurring crystals for this purpose. Superlattices with zero temperature coefficient of resistivity are useful as resistor material, and high-critical-field-magnet tapes using superconducting-insulator superlattices have been proposed. *See* X-RAYS.   [I.K.S.]

Atmospheric acoustics   The science of sound in the atmosphere. The atmosphere has a structure that varies in both space and time, and these variations have significant effects on a propagating sound wave. In addition, when sound propagates close to the ground, the type of ground surface has a strong effect.

Atmospheric sound attenuation. As sound propagates in the atmosphere, several interacting mechanism attenuate and change the spectral or temporal characteristics of the sound received at a distance from the source. The attenuation means that sound propagating through the atmosphere decreases in level with increasing distance between source and receiver. The total attenuation, in decibels, can be approximated as the sum of three nominally independent terms, as given in the equation below,

$$A_{\text{total}} = A_{\text{div}} + A_{\text{air}} + A_{\text{env}}$$

where A_{div} is the attenuation due to geometrical divergence, A_{air} is the attenuation due to air absorption, and A_{env} is the attenuation due to all other effects and includes the effects of the ground, refraction by a nonhomogeneous atmosphere, and scattering effects due to turbulence.

Sound energy spreads out as it propagates away from its source due to geometrical divergence. At distances that are large compared with the effective size of the sound source, the sound level decreases at the rate of 6 dB for every doubling of distance. The phenomenon of geometrical divergence, and the corresponding decrease in sound level with increasing distance from the source, is the same for all acoustic frequencies. In contrast, the attenuation due to the other two terms in the equation depends on frequency and therefore changes the spectral characteristics of the sound.

Air absorption. Dissipation of acoustic energy in the atmosphere is caused by viscosity, thermal conduction, and molecular relaxation. The last arises because fluctuations in apparent molecular vibrational temperatures lag in phase the fluctuations in translational temperatures. The vibrational temperatures of significance are those characterizing the relative populations of oxygen (O_2) and nitrogen (N_2) molecules. Since collisions with water molecules are much more likely to induce vibrational state changes than are collisions with other oxygen and nitrogen molecules, the sound attenuation varies markedly with absolute humidity. See MOLECULAR STRUCTURE AND SPECTRA; VISCOSITY.

The total attenuation due to air absorption increases rapidly with frequency. For this reason, applications in atmospheric acoustics are restricted to sound frequencies below a few thousand hertz it the propagation distance exceeds a few hundred meters. See SOUND ABSORPTION.

Effects of the ground. When the sound source and receiver are above a large flat ground surface in a homogeneous atmosphere, sound reaches the receiver via two paths. There is the direct path from source to receiver and the path reflected from the ground surface. Most naturally occurring ground surfaces are porous to some degree, and their acoustical property can be represented by an acoustic impedance. The acoustic impedance of the ground is in turn associated with a reflection coefficient that is typically less than unity. In simple terms, the sound field reflected from the ground surface suffers a reduction in amplitude and a phase change.

When the source and receiver are both relatively near the ground and are a large distance apart, the direct and reflected fields become nearly equal and cancel each other.

Refraction of sound. Straight ray paths are rarely achieved outdoors. In the atmosphere, both the wind and temperature vary with height above the ground. The velocity of sound relative to the ground is a function of wind velocity and temperature; hence it also varies with height, causing sound waves to propagate along curved paths.

The speed of the wind decreases with decreasing height above the ground because of drag on the moving air at the surface. Therefore, the speed of sound relative to the ground increases with height during downwind propagation, and ray paths curve downward. For propagation upwind, the sound speed decreases with height, and ray paths curve upward (see illustration). In the case of upward refraction, a shadow boundary forms near the ground beyond which no direct sound can penetrate. Some acoustic energy penetrates into a shadow zone via creeping waves that propagate along the ground and that continually shed diffracted rays into the shadow zones. The dominant feature of shadow-zone reception is the marked decrease in a sound's higher-frequency content. The presence of shadow zones explains why sound is generally less audible upwind of a source.

Refraction by temperature profiles is analogous. During the day, solar radiation heats the Earth's surface, resulting in warmer air near the ground. This condition is called a temperature lapse and is most pronounced on sunny days. A temperature lapse is the common daytime condition during most of the year, and also causes ray paths to curve upward. After sunset there is often radiation cooling of the ground, which

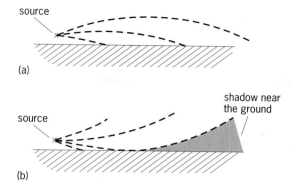

(a)

shadow near
the ground

source

(b)

**Curved ray paths. (*a*) Refraction
downward, during temperature
inversion or downwind
propagation. (*b*) Refraction
upward, during temperature
lapse or upwind propagation.**

produces cooler air near the surface. In summer under clear skies, such temperature inversions begin to form about 2 hours after sunset. Within the temperature inversion, the temperature increases with height, and ray paths curve downward.

The effects of refraction by temperature and wind are additive and produce rather complex sound speed profiles in the atmosphere.

Effects of turbulence. Turbulence in the atmosphere causes the effective sound speed to fluctuate from point to point, so a nominally smooth wave front develops ripples. One result is that the direction of a received ray may fluctuate with time in random manner. Consequently, the amplitude and phase of the sound at a distant point will fluctuate with time. The acoustical fluctuations are clearly audible in the noise from a large aircraft flying overhead. Turbulence in the atmosphere also scatters sound from its original direction. *See* TURBULENT FLOW. [G.A.Da.]

Atom A constituent of matter consisting of z negatively charged electrons bound predominantly by the Coulomb force to a tiny, positively charged nucleus consisting of Z protons and $(A - Z)$ neutrons. Z is the atomic number, and A is the mass or nucleon number. The atomic mass unit is $u = 1.660539 \times 10^{-24}$ g. Electrically neutral atoms ($z = Z$) with the range $Z = 1$ (hydrogen) to $Z = 92$ (uranium) make up the periodic table of the elements naturally occurring on Earth. Isotopes of a given element have different values of A but nearly identical chemical properties, which are fixed by the value of Z. Certain isotopes are not stable; they decay by various processes called radioactivity. Atoms with Z greater than 92 are all radioactive but may be synthesized, either naturally in stellar explosions or in the laboratory using accelerator techniques. *See* ATOMIC MASS UNIT; ELECTRON; ISOTOPE; MASS NUMBER; NUCLEAR STRUCTURE; RADIOACTIVITY.

Atoms with $Z - z$ ranging from 1 to $Z - 1$ are called positive ions. Those having $z - Z = 1$ are called negative ions; none has been found with $z - Z$ greater than 1. *See* NEGATIVE ION. [P.M.K.]

Atom cluster Clusters are aggregates of atoms (or molecules) containing between three and a few thousand atoms that have properties intermediate between those of the isolated monomer (atom or molecule) and the bulk or solid-state material. The study of such species has been an increasingly active research field since about 1980. This activity is due to the fundamental interest in studying a completely new area that can bridge the gap between atomic and solid-state physics and also shows many analogies to nuclear physics. However, the research is also done for its potential technological interest in areas such as catalysis, photography, and epitaxy. A characteristic of clusters which is responsible for many of their interesting properties is the large number of atoms at the surface compared to those in the cluster interior. For many kinds of atomic clusters, all atoms are at the surface for sizes of up to 12 atoms. As the clusters grow further in size, the relative number of atoms at the surface scales as approximately $4N^{-1/3}$, where N is the total number of atoms. Even in a cluster as big as 10^5 atoms, almost 10% of the atoms are at the surface. Clusters can be placed in the following categories:

1. Microclusters have from 3 to 10–13 atoms. Concepts and methods of molecular physics are applicable.

2. Small clusters have from 10–13 to about 100 atoms. Many different geometrical isomers exist for a given cluster size with almost the same energies. Molecular concepts lose their applicability.

3. Large clusters have from 100 to 1000 atoms. A gradual transition is observed to the properties of the solid state.

4. Small particles or nanocrystals have at least 1000 atoms. These bodies display some of the properties of the solid state.

The most favored geometry for rare-gas (neon, argon, and krypton) clusters of up to a few thousand atoms is icosahedral. However, the preferred cluster geometry depends critically on the bonding between the monomers in the clusters. For example, ionic clusters such as those of sodium chloride [(NaCl)$_N$] very rapidly assume the cubic form of the bulk crystal lattice, and for metallic clusters it is the electronic structure rather than the geometric structure which is most important. *See* CRYSTAL STRUCTURE.

There are two main types of sources for producing free cluster beams. In a gas-aggregation source, the atoms or molecules are vaporized into a cold, flowing rare-gas atmosphere. In a jet-expansion source, a gas is expanded under high pressure through a small hole into a vacuum.

In most situations, the valence electrons of the atoms making up the clusters can be regarded as being delocalized, that is, not attached to any particular atom but with a certain probability of being found anywhere within the cluster. The simplest and most widely used model to describe the delocalized electrons in metallic clusters is that of a free-electron gas, known as the jellium model. The positive charge is regarded as being smeared out over the entire volume of the cluster, while the valence electrons are free to move within this homogeneously distributed, positively charged background.   [E.Ca.]

Atom laser   A device that generates an intense coherent beam of atoms through a stimulated process. It does for atoms what an optical laser does for light. The atom laser emits coherent matter waves, whereas the optical laser emits coherent electromagnetic waves. Coherence means, for instance, that atom laser beams can interfere with each other. *See* COHERENCE.

Laser light is created by stimulated emission of photons, a light amplification process. Similarly, an atom laser beam is created by stimulated amplification of matter waves. The conservation of the number of atoms is not in conflict with matter-wave amplification: The atom laser takes atoms out of a reservoir and transforms them into a coherent matter wave similar to the optical laser, which converts energy into coherent electromagnetic radiation (but, in contrast, the number of photons need not be conserved). *See* LASER.

Elements. A laser requires a cavity (resonator), an active medium, and an output coupler (see table).

Cavity. Various analogs of laser cavities for atoms have been realized. The most important ones are magnetic traps (which use the force of an inhomogeneous magnetic

Analogies between an atom laser and the optical laser

Atom laser*	Optical laser
Atoms	Photons
Matter waves	Electromagnetic waves
Atom trap	Laser cavity
Atoms in the Bose condensate	Photons in the lasing mode
Thermal atoms	Gain medium
Evaporative cooling	Excitation of the gain medium
Stimulated scattering of atoms	Stimulated emission of photons
Critical temperature for Bose-Einstein condensation	Laser threshold

* Based on evaporative cooling.

field on the atomic magnetic dipole moment) and optical dipole traps (which use the force exerted on atoms by focused laser beams). *See* PARTICLE TRAP.

Active medium. The active medium is a reservoir of atoms which are transferred to one state of the confining potential, which is the analog of the lasing mode. The reservoir can be atoms confined in other quantum states of the atom cavity or an ultraslow atomic beam. The atoms are transferred to the lasing mode either by collisions or by optical pumping. The transfer of atoms is efficient only for an ultracold sample, which is prepared by laser cooling or evaporative cooling. This cooling ensures that the atoms in the reservoir occupy only a certain range of quantum states which can be efficiently coupled to the lasing mode.

Output coupler. The output coupler extracts atoms from the cavity, thus generating a pulsed or continuous beam of coherent atoms. A simple way to accomplish this step is to switch off the atom trap and release the atoms. This method is analogous to cavity dumping for an optical laser, and extracts all the stored atoms into a single pulse. A more controlled way to extract the atoms requires a coupling mechanism between confined quantum states and propagating modes.

Such a beam splitter for atoms can be realized by applying the Stern-Gerlach effect to atoms in a magnetic trap. Initially, all the atoms have their electron spin parallel to the magnetic field, say spin up, and in this state they are confined in the trap. A short radio-frequency pulse rotates (tilts) the spin of the atoms by a variable angle. Quantum-mechanically, a tilted spin is a superposition of spin up and spin down. Since the spin-down component experiences a repulsive magnetic force, the cloud of atoms is split into a trapped cloud and an out-coupled cloud. By using a series of radio-frequency pulses, a sequence of coherent atom pulses can be formed. These pulses are accelerated downward by gravity and spread out. *See* QUANTUM MECHANICS.

The illustration shows such a sequence of coherent pulses. In this case, sodium atoms are coupled out from a magnetic trap by radio-frequency pulses every 5 ms. The atom pulses are observed by illuminating them with resonant laser light and imaging their shadows, which are caused by absorption of the light. Each pulse contains 10^5–10^6 sodium atoms.

Potential applications. Although a basic atom laser has now been demonstrated, major improvements are necessary before it can be used for applications, especially in terms of increased output power and reduced overall complexity. The atom laser provides ultimate control over the position and motion of atoms at the quantum level, and might find use where such precise control is necessary, for example, for precision measurements of fundamental constants, tests of fundamental symmetries, atom optics

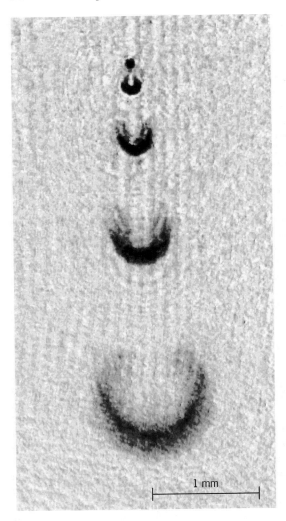

Pulsed atom laser in operation, with pulses of coherent sodium atoms coupled out from a Bose-Einstein condensate that is confined in a magnetic trap.

1 mm

(in particular, atom interferometry and atom holography), and precise deposition of atoms on surfaces. *See* FUNDAMENTAL CONSTANTS; SYMMETRY LAWS (PHYSICS).   [W.Ket.]

Atom optics   The use of laser light and nanofabricated structures to manipulate the motion of atoms in the same manner that rudimentary optical elements control light. The term refers to both an outlook in which atoms in atomic beams are thought of and manipulated like photons in light beams, and a collection of demonstrated techniques for doing such manipulation. Two types of atom optics elements have existed for some time: slits and holes used to collimate molecular beams (the analog of the pinhole camera), and focusing lenses for atoms and molecules (for example, hexapole magnets and quadrupole electrostatic lenses). However, in the 1980s the collection of optical elements for atoms expanded dramatically because of the use of near-resonant laser light and fabricated structures to make several types of mirrors as well as diffraction gratings. The diffraction gratings are particularly interesting because they exploit and demonstrate the (de Broglie) wave nature of atoms in a clear fashion. *See* LASER.

Diffraction gratings. Diffraction gratings for atoms have been made by using either a standing wave of light or a slotted membrane. The standing light wave makes a phase grating (that is, it advances or retards alternate sections of the incident wavefront but does not absorb any of the atom wave), so that the transmitted intensity is high. This approach requires the complexity of a single-mode laser, and introduces the complication that the light acts differently on the various hyperfine states of the atom. The slotted membrane, however, absorbs (or backscatters) atoms which strike the grating bars, but does not significantly alter the phase of the transmitted atoms; it is therefore an amplitude grating. It works for any atom or molecule, regardless of internal quantum state, but with total transmission limited to about 40% by the opacity of the grating bars and requisite support structure. *See* DIFFRACTION GRATING.

Atom interferometers. Atom interferometers have been demonstrated through several different experimental routes, involving both microscopic fabricated structures and laser beams. These interferometers are the first examples of optical systems composed of the elements of atom optics like those discussed above. Atom interferometers, like optical interferometers, are well suited for application to a wide range of fundamental and applied scientific problems. Scientific experiments with atom interferometers divide naturally into three major categories: measurements of atomic and molecular properties, fundamental tests and demonstrations, and inertial effects. *See* FRAME OF REFERENCE; INTERFERENCE OF WAVES; INTERFEROMETRY; OPTICS.   [D.E.P.]

Atomic beams   Unidirectional streams of neutral atoms passing through a vacuum. These atoms are virtually free from the influence of neighboring atoms but may be subjected to electric and magnetic fields so that their properties may be studied. The technique of atomic beams is identical to that of molecular beams. For historical reasons the latter term is most generally used to describe the method as applied to either atoms or molecules.

The method of atomic beams yields extremely accurate spectroscopic data about the energy levels of atoms, and hence detailed information about the interaction of electrons in the atom with each other and with the atomic nucleus, as well as information about the interaction of all components of the atom with external fields. *See* MOLECULAR BEAMS.   [P.Ku.]

Atomic clock   A device that uses an internal resonance frequency of atoms (or molecules) to measure the passage of time. The terms atomic clock and atomic frequency standard are often used interchangeably. A frequency standard generates pulses at regular intervals. It can be made into a clock by the addition of an electronic counter, which records the number of pulses.

Most methods of timekeeping rely on counting some periodic event, such as the rotation of the Earth, the motion of a pendulum in a grandfather clock, or the vibrations of a quartz crystal in a watch. An atomic clock relies on counting periodic events determined by the difference of two different energy states of an atom. A transition between two energy states with energies E_1 and E_2 may be accompanied by the absorption or emission of a photon (particle of electromagnetic radiation). The frequency ν of this radiation is given by the equation

$$h\nu = |E_2 - E_1|$$

where h is Planck's constant. A basic advantage of atomic clocks is that the frequency-determining elements, atoms of a particular isotope, are the same everywhere. Thus, atomic clocks constructed and operated independently will measure the same time interval. *See* ATOMIC STRUCTURE AND SPECTRA; ENERGY LEVEL (QUANTUM MECHANICS); QUANTUM MECHANICS.

An atomic frequency standard can be either active or passive. An active standard uses as a reference the electromagnetic radiation emitted by atoms as they decay from a higher energy state to a lower energy state. A passive standard attempts to match the frequency of an electronic oscillator or laser to the resonant frequency of the atoms by means of a feedback circuit. Either kind of standard requires some kind of frequency synthesis to produce an output near a convenient frequency that is proportional to the atomic resonance frequency. *See* LASER; MASER.

Two different gages of the quality of a clock are accuracy and stability. The accuracy of a frequency standard is defined in terms of the deviation of its frequency from an ideal standard. The stability of frequency standard is defined in terms of the constancy of its average frequency from one interval of time to the next.

The three most commonly used types of atomic clock are the cesium atomic beam, the hydrogen maser, and the rubidium gas cell. The cesium clock has high accuracy and good long-term stability. The hydrogen maser has the best stability for periods of up to a few hours. The rubidium cell is the least expensive and most compact and also has good short-term stability.

The cesium atomic-beam clock uses a 9193-MHz transition between two hyperfine energy states of the cesium-133 atom. Both the atomic nucleus and the outermost electron have magnetic moments; that is, they are like small magnets, with a north and a south pole. The two hyperfine energy states differ in the relative orientations of these magnetic moments. The cesium atoms travel in a collimated beam through a series of evacuated regions, where they are exposed to microwave radiation near their resonance frequency and are deflected into different trajectories by nonuniform magnetic fields. *See* ELECTRON SPIN; HYPERFINE STRUCTURE; MOLECULAR BEAMS; NUCLEAR MOMENTS.

Cesium has become the basis of the international definition of the second; the duration of 9,192,631,770 periods of the radiation corresponding to the transition between the two hyperfine states of the ground state of the cesium-133 atom. The cesium clock is especially well suited for applications such as timekeeping, where absolute accuracy without recalibration is necessary. Measurements from many cesium clocks throughout the world are averaged together to define an international time scale that is uniform to parts in 10^{14}, or about 1 microsecond in a year. *See* PHYSICAL MEASUREMENT.

The hydrogen maser is based on the hyperfine transition of atomic hydrogen, which has a frequency of 1420 MHz. Atoms in the higher hyperfine energy state enter an evacuated storage bulb inside a microwave cavity, and are induced to make a transition to the lower hyperfine state by a process called stimulated emission.

The rubidium gas cell is based on the 6835-MHz hyperfine transition of rubidium-87. The rubidium atoms are contained in a glass cell together with a buffer gas, where they are subjected to optical pumping and microwave radiation at the hyperfine transition frequency; this results in a detectable decrease in the light transmitted through the cell.

Many other kinds of atomic clocks, such as thallium atomic beams and ammonia and rubidium masers, have been demonstrated in the laboratory. The first atomic clock, constructed at the National Bureau of Standards in 1949, was based on a 24-GHz transition in the ammonia molecule. Some laboratories have tried to improve the cesium atomic-beam clock by replacing the magnetic state selection with laser optical pumping and fluorescence detection. One such standard, called NIST-7, is in operation at the U.S. National Institute of Standards and Technology and is the primary frequency standard for the United States. Atomic frequency standards can also be based on optical transitions. One of the best-developed optical frequency standards is the 3.39-micrometer (88-THz) helium-neon laser, stabilized to a transition in the methane molecule. Frequency synthesis chains have been built to link the optical frequency to radio frequencies.

Atomic clocks are used in applications for which less expensive alternatives, such as quartz oscillators, do not provide adequate performance. In addition to maintaining a uniform international time scale, atomic clocks are used to keep time in the Global Positioning System, various digital communications systems, radio astronomy, and navigation of space probes. [W.M.I.]

Atomic mass The mass of an atom or molecule on a scale where the mass of a carbon-12 (^{12}C) atom is exactly 12.0. The mass of any atom is approximately equal to the total number of its protons and neutrons multiplied by the atomic mass unit, $u = 1.660539 \times 10^{-24}$ gram. (Electrons are much lighter, about 0.0005486 u.) No atom differs from this simple formula by more than 1%, and stable atoms heavier than helium all lie within 0.3%. *See* ATOMIC MASS UNIT.

This simplicity of nature led to the confirmation of the atomic hypothesis—the idea that all matter is composed of atoms, which are identical and chemically indivisible for each chemical element. In 1802, G. E. Fischer noticed that the weights of acids needed to neutralize various bases could be described systematically by assigning relative weights to each of the acids and bases. A few years later, John Dalton proposed an atomic theory in which elements were made up of atoms that combine in simple ways to form molecules.

In reality, nature is more complicated, and the great regularity of atomic masses more revealing. Two fundamental ideas about atomic structure come out of this regularity: that the atomic nucleus is composed of charged protons and uncharged neutrons, and that these particles have approximately equal mass. The number of protons in an atom is called its atomic number, and equals the number of electrons in the neutral atom. The electrons, in turn, determine the chemical properties of the atom. Adding a neutron or two does not change the chemistry (or the name) of an atom, but does give it an atomic mass which is 1 u larger for each added neutron. Such atoms are called isotopes of the element, and their existence was first revealed by careful study of radioactive elements. Most naturally occurring elements are mixtures of isotopes, although a single isotope frequently predominates. Since the proportion of the various isotopes is usually about the same everywhere on Earth, an average atomic mass of an element can be defined, and is called the atomic weight. Atomic weights are routinely used in chemistry in order to determine how much of one chemical will react with a given weight of another. *See* ISOTOPE.

In contrast to atomic weights, which can be defined only approximately, atomic masses are exact constants of nature. All atoms of a given isotope are truly identical; they cannot be distinguished by any method. This is known to be true because the quantum mechanics treats identical objects in special ways, and makes predictions that depend on this assumption. One such prediction, the exclusion principle, is the reason that the chemical behavior of atoms with different numbers of electrons is so different. *See* QUANTUM MECHANICS. [F.L.P.; D.E.P.]

Atomic mass unit An arbitrarily defined unit in terms of which the masses of individual atoms are expressed. One atomic mass unit is defined as exactly $^{1}/_{12}$ of the mass of an atom of the nuclide ^{12}C, the predominant isotope of carbon. The unit, also known as the dalton, is often abbreviated amu, and is designated by the symbol u. The relative atomic mass of a chemical element is the average mass of its atoms expressed in atomic mass units.   [J.F.We.]

Atomic nucleus   The central region of an atom. Atoms are composed of negatively charged electrons, positively charged protons, and electrically neutral neutrons. The protons and neutrons (collectively known as nucleons) are located in a small central region known as the nucleus. The electrons move in orbits which are large in comparison with the dimensions of the nucleus itself. Protons and neutrons possess approximately equal masses, each roughly 1840 times that of an electron. The number of nucleons in a nucleus is given by the mass number A and the number of protons by the atomic number Z. Nuclear radii r are given approximately by $r = 1.2 \times 10^{-15}\,\text{m}\,A^{1/3}$.
                                                                    [H.E.D.]

Atomic number   The number of elementary positive charges (protons) contained within the nucleus of an atom. It is denoted by the letter Z. Correspondingly, it is also the number of planetary electrons in the neutral atom.

The concept of atomic number emerged from the work of G. Moseley, done in 1913–1914. He measured the wavelengths of the most energetic rays (K and L lines) produced by using the elements calcium to zinc as targets in an x-ray tube. The square root of the frequency, v, of these x-rays increased by a constant amount in passing from one target to the next. These data, when extended, gave a linear plot of atomic number versus v for all elements studied, using 13 as the atomic number for aluminum and 79 for that of gold.

Moseley's atomic numbers were quickly recognized as providing an accurate sequence of the elements, which the chemical atomic weights had sometimes failed to do. Additionally, the atomic number sequence indicated the positions of elements that had not yet been discovered.

The atomic number not only identifies the chemical properties of an element but facilitates the description of other aspects of atoms and nuclei. Thus, atoms with the same atomic number are isotopes and belong to the same element, while nuclear reactions may alter the atomic number. *See* Isotope; Radioactivity.

When specifically written, the atomic number is placed as a subscript preceding the symbol of the element, while the mass number (A) precedes as a superscript, for example, $^{27}_{13}\text{Al}$, $^{238}_{92}\text{U}$. *See* Mass number.   [H.E.D.]

Atomic physics   The study of the structure of the atom, its dynamical properties, including energy states, and its interactions with particles and fields. These are almost completely determined by the laws of quantum mechanics, with very refined corrections required by quantum electrodynamics. Despite the enormous complexity of most atomic systems, in which each electron interacts with both the nucleus and all the other orbiting electrons, the wavelike nature of particles, combined with the Pauli exclusion principle, results in an amazingly orderly array of atomic properties. These are systematized by the Mendeleev periodic table. In addition to their classification by chemical activity and atomic weight, the various elements of this table are characterized by a wide variety of observable properties. These include electron affinity, polarizability, angular momentum, multiple electric moments, and magnetism. *See* Quantum electrodynamics; Quantum mechanics.

Each atomic element, normally found in its ground state (that is, with its electron configuration corresponding to the lowest state of total energy), can also exist in an infinite number of excited states. These are also ordered in accordance with relatively simple hierarchies determined by the laws of quantum mechanics. The most characteristic signature of these various excited states is the radiation emitted or absorbed when the atom undergoes a transition from one state to another. The systemization and classification of atomic energy levels (spectroscopy) has played a central role in developing an understanding of atomic structure.   [B.B.]

Atomic structure and spectra The idea that matter is subdivided into discrete building blocks called atoms, which are not divisible any further, dates back to the Greek philosopher Democritus. His teachings of the fifth century B.C. are commonly accepted as the earliest authenticated ones concerning what has come to be called atomism by students of Greek philosophy. The weaving of the philosophical thread of atomism into the analytical fabric of physics began in the late eighteenth and the nineteenth centuries. Robert Boyle is generally credited with introducing the concept of chemical elements, the irreducible units of which are now recognized as individual atoms of a given element. In the early nineteenth century John Dalton developed his atomic theory, which postulated that matter consists of indivisible atoms as the irreducible units of Boyle's elements, that each atom of a given element has identical attributes, that differences among elements are due to fundamental differences among their constituent atoms, that chemical reactions proceed by simple rearrangement of indestructible atoms, and that chemical compounds consist of molecules which are reasonably stable aggregates of such indestructible atoms.

Electromagnetic nature of atoms. The work of J. J. Thomson in 1897 clearly demonstrated that atoms are electromagnetically constituted and that from them can be extracted fundamental material units bearing electric charge that are now called electrons. The electrons of an atom account for a negligible fraction of its mass. By virtue of overall electrical neutrality of every atom, the mass must therefore reside in a compensating, positively charged atomic component of equal charge magnitude but vastly greater mass. See ELECTRON.

Thomson's work was followed by the demonstration by Ernest Rutherford in 1911 that nearly all the mass and all of the positive electric charge of an atom are concentrated in a small nuclear core approximately 10,000 times smaller in extent than an atomic diameter. Niels Bohr in 1913 and others carried out some remarkably successful attempts to build solar system models of atoms containing planetary pointlike electrons orbiting around a positive core through mutual electrical attraction (though only certain "quantized" orbits were "permitted"). These models were ultimately superseded by nonparticulate, matter-wave quantum theories of both electrons and atomic nuclei. See QUANTUM MECHANICS.

The modern picture of condensed matter (such as solid crystals) consists of an aggregate of atoms or molecules which respond to each other's proximity through attractive electrical interactions at separation distances of the order of 1 atomic diameter (approximately 10^{-10} m) and repulsive electrical interactions at much smaller distances. These interactions are mediated by the electrons, which are in some sense shared and exchanged by all atoms of a particular sample, and serve as an interatomic glue that binds the mutually repulsive, heavy, positively charged atomic cores together. See SOLID-STATE PHYSICS.

Bohr atom. The hydrogen atom is the simplest atom, and its spectrum (or pattern of light frequencies emitted) is also the simplest. The regularity of its spectrum had defied explanation until Bohr solved it with three postulates, these representing a model which is useful, but quite insufficient, for understanding the atom.

Postulate 1: The force that holds the electron to the nucleus is the Coulomb force between electrically charged bodies.

Postulate 2: Only certain stable, nonradiating orbits for the electron's motion are possible, those for which the angular momentum associated with the motion of an electron in its orbit is an integral multiple of $h/2\pi$ (Bohr's quantum condition on the orbital angular momentum). Each stable orbit represents a discrete energy state.

Postulate 3: Emission or absorption of light occurs when the electron makes a transition from one stable orbit to another, and the frequency ν of the light is such that the difference in the orbital energies equals $h\nu$ (A. Einstein's frequency condition for the photon, the quantum of light).

Here the concept of angular momentum, a continuous measure of rotational motion in classical physics, has been asserted to have a discrete quantum behavior, so that its quantized size is related to Planck's constant h, a universal constant of nature. Velocity v, in rotational motion about a central body, is defined as the product of the component.

Modern quantum mechanics has provided justification of Bohr's quantum condition on the orbital angular momentum. It has also shown that the concept of definite orbits cannot be retained except in the limiting case of very large orbits. In this limit, the frequency, intensity, and polarization can be accurately calculated by applying the classical laws of electrodynamics to the radiation from the orbiting electron. This fact illustrates Bohr's correspondence principle, according to which the quantum results must agree with the classical ones for large dimensions. The deviation from classical theory that occurs when the orbits are smaller than the limiting case is such that one may no longer picture an accurately defined orbit. Bohr's other hypotheses are still valid.

According to Bohr's theory, the energies of the hydrogen atom are quantized (that is, can take on only certain discrete values). These energies can be calculated from the electron orbits permitted by the quantized orbital angular momentum. The orbit may be circular or elliptical, so only the circular orbit is considered here for simplicity. Let the electron, of mass m and electric charge $-e$, describe a circular orbit of radius r around a nucleus of charge $+e$ and of infinite mass. With the electron velocity v, the angular momentum is mvr, and the second postulate becomes Eq. (1). The integer n is called

$$mvr = n\,(h/2\pi) \qquad (n = 1, 2, 3, \ldots) \tag{1}$$

the principal quantum number. The possible energies of the nonradiating states of the atom are given by Eq. (2). Here ε_0 is the permittivity of free space, a constant included

$$E = -\frac{me^4}{8\epsilon_0^2 h^2} \cdot \frac{1}{n^2} \tag{2}$$

in order to give the correct units to the statement of Coulomb's law in SI units.

The same equation for the hydrogen atom's energy levels, except for some small but significant corrections, is obtained from the solution of the Schrödinger equation, as modified by W. Pauli, for the hydrogen atom. *See* QUANTUM NUMBERS.

The frequencies of electromagnetic radiation or light emitted or absorbed in transitions are given by Eq. (3), where E' and E'' are the energies of the initial and final states

$$\nu = \frac{E' - E''}{h} \tag{3}$$

of the atom. Spectroscopists usually express their measurements in wavelength λ or in wave number σ in order to obtain numbers of a convenient size. The wave number of a transition is shown in Eq. (4).

$$\sigma = \frac{\nu}{c} = \frac{E'}{hc} - \frac{E''}{hc} \tag{4}$$

If $T = E/(hc)$, then Eq. (5) results. Here T is called the spectral term.

$$\sigma = T'' - T' \tag{5}$$

The allowed terms for hydrogen, from Eq. (2), are given by Eq. (6). The quantity R

$$T = \frac{me^4}{8\epsilon_0^2 ch^3} \cdot \frac{1}{n^2} = \frac{R}{n^2} \tag{6}$$

is the important Rydberg constant. Its value, which has been measured to a remarkable and rapidly improving accuracy, is related to the values of other well-known atomic constants, as in Eq. (6). *See* RYDBERG CONSTANT.

The effect of finite nuclear mass must be considered, since the nucleus does not actually remain at rest at the center of the atom. Instead, the electron and nucleus revolve about their common center of mass. This effect can be accurately accounted for and requires a small change in the value of the effective mass m in Eq. (6).

In addition to the circular orbits already described, elliptical ones are also consistent with the requirement that the angular momentum be quantized. A. Sommerfeld showed that for each value of n there is a family of n permitted elliptical orbits, all having the same major axis but with different eccentricities. Illustration a shows, for example, the Bohr-Sommerfeld orbits for $n = 3$. The orbits are labeled s, p, and d, indicating values of the azimuthal quantum number $l = 0$, 1, and 2. This number determines the shape of the orbit, since the ratio of the major to the minor axis is found to be $n/(l = 1)$. To a first approximation, the energies of all orbits of the same n are equal. In the case of the highly eccentric orbits, however, there is a slight lowering of the energy due to precession of the orbit (illus. b). According to Einstein's theory of relativity, the

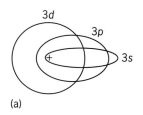

Possible elliptical orbits, according to the Bohr-Sommerfeld theory. (*a*) The three permitted orbits for $n = 3$. (*b*) Precession of the 3*s* orbit caused by the relativistic variation of mass. (*After A. P. Arya, Fundamentals of Atomic Physics, Allyn and Bacon, 1971*)

(a)   (b)

mass increases somewhat in the inner part of the orbit, because of greater velocity. The velocity increase is greater as the eccentricity is greater, so the orbits of higher eccentricity have their energies lowered more. The quantity l is called the orbital angular momentum quantum number or the azimuthal quantum number. *See* RELATIVITY.

Multielectron atoms. In attempting to extend Bohr's model to atoms with more than one electron, it is logical to compare the experimentally observed terms of the alkali atoms, which contain only a single electron outside closed shells, with those of hydrogen. A definite similarity is found but with the striking difference that all terms with $l > 0$ are double. This fact was interpreted by S. A. Goudsmit and G. E. Uhlenbeck as due to the presence of an additional angular momentum of $\frac{1}{2}(h/2\pi)$ attributed to the electron spinning about its axis. The spin quantum number of the electron is $s = \frac{1}{2}$.

The relativistic quantum mechanics developed by P. A. M. Dirac provided the theoretical basis for this experimental observation. *See* ELECTRON SPIN.

Implicit in much of the following discussion is W. Pauli's exclusion principle, first enunciated in 1925, which when applied to atoms may be stated as follows: no more than one electron in a multielectron atom can possess precisely the same quantum numbers. In an independent, hydrogenic electron approximation to multielectron atoms, there are $2n^2$ possible independent choices of the principal (n), orbital (l), and magnetic (m_l, m_s) quantum numbers available for electrons belonging to a given n, and no more. Here m_l and m_s refer to the quantized projections of l and s along some chosen direction. The organization of atomic electrons into shells of increasing radius (the Bohr radius scales as n^2) follows from this principle. *See* EXCLUSION PRINCIPLE.

The energy of interaction of the electron's spin with its orbital angular momentum is known as spin-orbit coupling. A charge in motion through either "pure" electric or "pure" magnetic fields, that is, through fields perceived as "pure" in a static laboratory, actually experiences a combination of electric and magnetic fields, if viewed in the frame of reference of a moving observer with respect to whom the charge is momentarily at rest. For example, moving charges are well known to be deflected by magnetic fields. But in the rest frame of such a charge, there is no motion, and any acceleration of a charge must be due to the presence of a pure electric field from the point of view of an observer analyzing the motion in that reference frame. *See* RELATIVISTIC ELECTRODYNAMICS.

A spinning electron can crudely be pictured as a spinning ball of charge, imitating a circulating electric current. This circulating current gives rise to a magnetic field distribution very similar to that of a small bar magnet, with north and south magnetic poles symmetrically distributed along the spin axis above and below the spin equator. This representative bar magnet can interact with external magnetic fields, one source of which is the magnetic field experienced by an electron in its rest frame, owing to its orbital motion through the electric field established by the central nucleus of an atom. In multielectron atoms, there can be additional, though generally weaker, interactions arising from the magnetic interactions of each electron with its neighbors, as all are moving with respect to each other and all have spin. The strength of the bar magnet equivalent to each electron spin, and its direction in space are characterized by a quantity called the magnetic moment, which also is quantized essentially because the spin itself is quantized. Studies of the effect of an external magnetic field on the states of atoms show that the magnetic moment associated with the electron spin is equal in magnitude to a unit called the Bohr magneton.

The energy of the interaction between the electron's magnetic moment and the magnetic field generated by its orbital motion is usually a small correction to the spectral term, and depends on the angle between the magnetic moment and the magnetic field or, equivalently, between the spin angular momentum vector and the orbital angular momentum vector (a vector perpendicular to the orbital plane whose magnitude is the size of the orbital angular momentum). Since quantum theory requires that the quantum number j of the electron's total angular momentum shall take values differing by integers, while l is always an integer, there are only two possible orientations for s relative to l: s must be either parallel or antiparallel to l.

For the case of a single electron outside the nucleus, the Dirac theory gives Eq. (7) for

$$\Delta T = \frac{R\alpha^2 Z^4}{n^3} \times \frac{j(j+1) - l(l+1) - s(s+1)}{l(2l+1)(l+1)} \tag{7}$$

the spin-orbit correction to the spectral terms. Here $\alpha = e^2/(2\epsilon_0 hc) \cong 1/137$ is called the fine structure constant.

In atoms having more than one electron, this fine structure becomes what is called the multiplet structure. The doublets in the alkali spectra, for example, are due to spin-orbit coupling; Eq. (7), with suitable modifications, can still be applied.

When more than one electron is present in the atom, there are various ways in which the spins and orbital angular momenta can interact. Each spin may couple to its own orbit, as in the one-electron case; other possibilities are orbit-other orbit, spin-spin, and so on. The most common interaction in the light atoms, called *LS* coupling or Russell-Saunders coupling, is described schematically in Eq. (8). This notation indicates that

$$\{(l_1, l_2, l_3, \ldots)(s_1, s_2, s_3, \ldots)\} = \{L, S\} = J \tag{8}$$

the *l* are coupled strongly together to form a resultant *L*, representing the total orbital angular momentum. The s_i are coupled strongly together to form a resultant *S*, the total spin angular momentum. The weakest coupling is that between *L* and *S* to form *J*, the total angular momentum of the electron system of the atom in this state.

Coupling of the *LS* type is generally applicable to the low-energy states of the lighter atoms. The next commonest type is called *jj* coupling, represented in Eq. (9). Each

$$\{(l_1, s_1)(l_2, s_2)(l_3, s_3) \ldots\} = \{j_1, j_2, j_3, \ldots\} = J \tag{9}$$

electron has its spin coupled to its own orbital angular momentum to form a j_i for that electron. The various j_i are then more weakly coupled together to give *J*. This type of coupling is seldom strictly observed. In the heavier atoms it is common to find a condition intermediate between *LS* and *jj* coupling; then either the *LS* or *jj* notation may be used to describe the levels, because the number of levels for a given electron configuration is independent of the coupling scheme.

Nuclear magnetism and hyperfine structure. Most atomic nuclei also possess spin, but rotate about 2000 times slower than electrons because their mass is on the order of 2000 or more times greater than that of electrons. Because of this, very weak nuclear magnetic fields, analogous to the electronic ones that produce fine structure in spectral lines, further split atomic energy levels. Consequently, spectral lines arising from them are split according to the relative orientations, and hence energies of interaction, of the nuclear magnetic moments with the electronic ones. The resulting pattern of energy levels and corresponding spectral-line components is referred to as hyperfine structure. *See* NUCLEAR MOMENTS.

Nuclear properties also affect atomic spectra through the isotope shift. This is the result of the difference in nuclear masses of two isotopes, which results in a slight change in the Rydberg constant. There is also sometimes a distortion of the nucleus, which can be detected by ultrahigh precision spectroscopy. *See* MOLECULAR BEAMS; PARTICLE TRAP.

Doppler spread. In most cases, a common problem called Doppler broadening of the spectral lines arises, which can cause overlapping of spectral lines and make analysis difficult. The broadening arises from motion of the emitted atom with respect to a spectrometer. Several ingenious ways of isolating only those atoms nearly at rest with respect to spectrometric apparatus have been devised. The most powerful employ lasers and either involve saturation spectroscopy, utilizing a saturating beam and probe beam from the same tunable laser, or use two laser photons which jointly drive a single atomic transition and are generated in lasers so arranged that the first-order Doppler shifts of the photons cancel each other. *See* DOPPLER EFFECT.

Radiationless transitions. It would be misleading to think that the most probable fate of excited atomic electrons consists of transitions to lower orbits, accompanied by photon emission. In fact, for at least the first third of the periodic table, the preferred decay mode of most excited atomic systems in most states of excitation and ionization is the electron emission process first observed by P. Auger in 1925 and named after him. For example, a singly charged neon ion lacking a 1s electron is more than 50 times as likely to decay by electron emission as by photon emission. In the process, an outer atomic electron descends to fill an inner vacancy, while another is ejected from the atom to conserve both total energy and momentum in the atom. The ejection usually arises because of the interelectron Coulomb repulsion. *See* AUGER EFFECT.

Cooling and stopping atoms and ions. Despite impressive progress in reducing Doppler shifts and Doppler spreads, these quantities remain factors that limit the highest obtainable spectroscopic resolutions. The 1980s and 1990s saw extremely rapid development of techniques for trapping neutral atoms and singly charged ions in a confined region of space, and then cooling them to much lower temperatures by the application of laser-light cooling techniques. Photons carry not only energy but also momentum; hence they can exert pressure on neutral atoms as well as charged ions. *See* LASER COOLING.

Schemes have been developed to exploit these light forces to confine neutral atoms in the absence of material walls, whereas various types of so-called bottle configurations of electromagnetic fields developed earlier remain the technique of choice for similarly confining ions. Various ingenious methods have been invented to slow down and even nearly stop neutral atoms and singly charged ions, whose energy levels (unlike those of most more highly charged ions) are accessible to tunable dye lasers. These methods often utilize the velocity-dependent light pressure from laser photons of nearly the same frequency as, but slightly less energetic than, the energy separation of two atomic energy levels to induce a transition between these levels.

The magnetooptic trap combines optical forces provided by laser light with a weak magnetic field whose size goes through zero at the geometrical center of the trap and increases with distance from this center. The net result is a restoring force which confines sufficiently laser-cooled atoms near the center. Ingenious improvements have allowed cooling of ions to temperatures as low as 180×10^{-9} K.

For more highly ionized ions, annular storage rings are used in which radial confinement of fast ion beams (with speeds of approximately 10% or more of the speed of light) is provided by magnetic focusing. Two cooling schemes are known to work on stored beams of charged particles, the so-called stochastic cooling method and the electron cooling method. In the former, deviations from mean stored particle energies are electronically detected, and electronic "kicks" that have been adjusted in time and direction are delivered to the stored particles to compensate these deviations. In electron cooling, which proves to be more effective for stored heavy ions of high charge, electron beams prepared with a narrow velocity distribution are merged with the stored ion beams. When the average speeds of the electrons and the ions are matched, the Coulomb interaction between the relatively cold (low-velocity-spread) electrons and the highly charged ions efficiently transfers energy from the warmer ions, thereby reducing the temperature of the stored ions. [I.A.S.]

Atomic theory   The study of the structure and properties of atoms based on quantum mechanics and the Schrödinger equation. These tools make it possible, in principle, to predict most properties of atomic systems. A stationary state of an atom is governed by a time-independent wave function which depends on the position coordinates of all the particles within the atom. To obtain the wave function, the time-independent Schrödinger equation, a second-order differential equation, has to be solved. The potential energy term in this equation contains the Coulomb interaction between all the particles in the atom, and in this way they are all coupled to each other. *See* QUANTUM MECHANICS.

A many-particle system where the behavior of each particle at every instant depends on the positions of all the other particles cannot be solved directly. This is not a problem restricted to quantum mechanics. A classical system where the same problem arises is a solar system with several planets. In classical mechanics as well as in quantum mechanics, such a system has to be treated by approximate methods.

Independent particle model. As a first approximation, it is customary to simplify the interaction between the particles. In the independent particle model the electrons are assumed to move independently of each other in the average field generated by the nucleus and the other electrons. In this case the potential energy operator will be a sum over one-particle operators. The simplest wave function which will satisfy the resulting equation is a product of one-particle orbitals. To fulfill the Pauli exclusion principle, the total wave function must, however, be written in a form such that it will vanish if two particles are occupying the same quantum state. This is achieved with an antisymmetrized wave function, that is, a function which, if two electrons are interchanged, changes sign but in all other respects remains unaltered. The antisymmetrized product wave function is usually called a Slater determinant. *See* EXCLUSION PRINCIPLE.

Hartree-Fock method. In the late 1920s, only a few years after the discovery of the Schrödinger equation, D. Hartree showed that the wave function to a good approximation could be written as a product of orbitals, and also developed a method to calculate the orbitals. Important contributions to the method were also made by V. Fock and J. C. Slater (thus, the Hartree-Fock method). The Hartree-Fock model thus gives the lowest-energy ground state within the assumption that the electrons move independently of each other in an average field from the nucleus and the other electrons.

To simplify the problem even further, it is common to add the requirement that the Hartree-Fock potential should be spherically symmetric. This leads to the central-field model and the so-called restricted Hartree-Fock method.

The Hartree-Fock method gives a qualitative understanding of many atomic properties. Generally it is, for example, able to predict the configurations occupied in the ground states of the elements. Electron binding energies are also given with reasonable accuracy.

Electron correlation. Correlation is commonly defined as the difference between the full many-body problem and the Hartree-Fock model. More specifically, the correlation energy is the difference between the experimental energy and the Hartree-Fock energy. There are several methods developed to account for electron correlation, including the configuration-interaction method, the multiconfiguration Hartree-Fock method, and perturation theory.

Strongly correlated systems. Although the Hartree-Fock model can qualitatively explain many atomic properties, there are systems and properties for which correlation is more important, such as negative ions, doubly-excited states, and some open-shell systems. If the interest is not in calculating the total energy of a state but in understanding some other properties, such as the hyperfine structure, effects beyond the central field model can be more important. *See* HYPERFINE STRUCTURE; NEGATIVE ION.

Relativistic effects. The Schrödinger equation is a nonrelativistic wave equation. In heavy elements the kinetic energy of the electrons becomes very large, and calculations are based on the relativistic counterpart to the Schrödinger equation, the Dirac equation. It is possible to construct a Hartree-Fock model based on the Dirac equation, where the electron-electron interaction is given by the Coulomb interaction, a magnetic contribution, and a term which corrects for the finite speed (retardation) with which the interaction propagates. *See* ANTIMATTER; RELATIVISTIC QUANTUM THEORY.

Radiative corrections. Radiative corrections, which arise when the electromagnetic field is quantized within the theory of quantum electrodynamics, For many-body systems, calculations of radiative effects are usually done within some independent-particle model, and the result is added to a correlated relativistic calculation based on the Dirac equation. *See* ATOMIC STRUCTURE AND SPECTRA; QUANTUM ELECTRODYNAMICS.

[Eva Li.]

Attenuation   The reduction in level of a transmitted quantity as a function of a parameter, usually distance. It is applied mainly to acoustic or electromagnetic waves and is expressed as the ratio of power densities. Various mechanisms can give rise to attenuation. Among the most important are geometrical attenuation, absorption, and scattering.

For unconfined radiation from a point source in free space, the power density (watts per square meter) decreases in proportion to the square of the distance. The power densities, I_1 and I_2, at distances r_1 and r_2 from the source, are related by Eq. (1).

$$I_2 = I_1 \left(\frac{r_1}{r_2} \right)^2 \tag{1}$$

If the signal, in a parallel beam so that there is no geometrical attenuation, passes through a lossy medium, absorption reduces the power level, I, exponentially with distance, x, according to Eq. (2), where a is the attenuation coefficient.

$$I(x) = I(0)e^{-ax} \tag{2}$$

See ABSORPTION OF ELECTROMAGNETIC RADIATION; SOUND ABSORPTION.

Scattering is said to occur if the power is not absorbed in the medium but scattered from inhomogeneities. *See* SCATTERING OF ELECTROMAGNETIC RADIATION.

More complicated situations occur with guided waves, such as acoustic waves in pipes or electromagnetic waves in transmission lines or waveguides, where absorption may take place and irregularities may cause reflection of some power.

In electric circuits, constituent elements are often described as attenuators when they reduce the level of signals passing through them.

Attenuation is usually measured in terms of the logarithm of the power ratio, the units being the neper or the decibel. *See* DECIBEL.   [A.E.Ba.]

Auger effect   One of the two principal processes for the relaxation of an inner-shell electron vacancy in an excited or ionized atom. The Auger effect is a two-electron process in which an electron makes a discrete transition from a less bound shell to the vacant, but more tightly bound, electron shell. The energy gained in this process is transferred, via the electrostatic interaction, to another bound electron which then escapes from the atom. This outgoing electron is referred to as an Auger electron and is labeled by letters corresponding to the atomic shells involved in the process. For example, a KL_IL_{III} Auger electron corresponds to a process in which an L_I electron makes a transition to the K shell and the energy is transferred to an L_{III} electron (illus. a). By the conservation of energy, the Auger electron kinetic energy E is given by $E = E(K) - E(L_I) - E(L_{III})$ where $E(K,L)$ is the binding energy of the various electron shells. Since the energy levels of atoms are discrete and well understood, the Auger energy is a

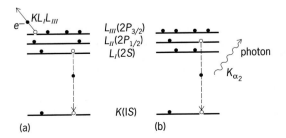

(a)   (b)

Two principal processes for the filling of an inner-shell electron vacancy. (a) Auger emission; a KL_IL_{III} Auger process in which an L_I electron fills the K-shell vacancy with the emission of a KL_IL_{III} Auger electron from the L_{III} shell. (b) Photon emission; a radiative process in which an L_{II} electron fills the K-shell vacancy with the emission of a K_{α_2} photon.

signature of the emitting atom. *See* ELECTRON CONFIGURATION; ENERGY LEVEL (QUANTUM MECHANICS).

The other principal process for the filling of an inner-shell hole is a radiative one in which the transition energy is carried off by a photon (illus. *b*). Inner-shell vacancies in elements with large atomic number correspond to large transition energies and usually decay by such radiative processes; vacancies in elements with low atomic number or outer-shell vacancies with low transition energies decay primarily by Auger processes.

[L.C.F.]

B

Ballistics That branch of applied physics which deals with the motion of projectiles and the conditions governing that motion. Commonly called the science of shooting, it is, for practical purposes, subdivided into exterior and interior ballistics. Exterior ballistics begins at the instant the projectile leaves the muzzle of the gun barrel; interior ballistics, logically, deals with the events preceding this instant, that is, the events inside the gun barrel.   [W.L./D.Wo.]

Band spectrum A spectrum consisting of groups or bands of closely spaced lines. Band spectra are characteristic of molecular gases or chemical compounds. When the light emitted or absorbed by molecules is viewed through a spectroscope with small dispersion, the spectrum appears to consist of very wide asymmetrical lines called bands. These bands usually have a maximum intensity near one edge, called a band head, and a gradually decreasing intensity on the other side. In some band systems the intensity shading is toward shorter waves, in others toward longer waves. Each band system consists of a series of nearly equally spaced bands called progressions; corresponding bands of different progressions form groups called sequences.

When spectroscopes with adequate dispersion and resolving power are used, it is seen that most of the bands obtained from gaseous molecules actually consist of a very large number of lines whose spacing and relative intensities, if unresolved, explain the appearance of bands of continua. For the quantum-mechanical explanations of the details of band spectra *see* MOLECULAR STRUCTURE AND SPECTRA.   [W.F.M./W.W.W.]

Band theory of solids A quantum-mechanical theory of the motion of electrons in solids which predicts certain restricted ranges, or bands, for the electron energies.

If the atoms of a solid are separated from each other to such a distance that they do not interact, the energy levels of the electrons will then be those characteristic of the individual free atoms, and thus many electrons will have the same energy. As the distance between atoms is decreased, the electrons in the outer shells begin to interact, thus altering their energy and broadening the sharp energy level out into a range of possible energy levels called a band. One would expect the process of band formation to be well advanced for the outer, or valence, electrons at the observed interatomic distances in solids. Once the atomic levels have spread into bands, the valence electrons are not confined to individual atoms, but may jump from atom to atom with an ease that increases with the increasing width of the band.

Although energy bands exist in all solids, the term energy band is usually used in reference only to ordered substances, that is, those having well-defined crystal lattices. In such a case, an electron energy state can be classified according to its crystal momentum **p** or its electron wave vector $\mathbf{k} = \mathbf{p}/\hbar$ (where $\hbar$ is Planck's constant h divided by 2π). If the electrons were free, the energy of an electron whose wave vector is **k** would be as shown in the equation below, where E_0 is the energy of the lowest state of a

$$E(\mathbf{k}) = E_0 + \hbar^2 k^2 / 2m_0$$

valence electron and m_0 is the electron mass. In a crystal, however, the electrons are not free because of the effect of the crystal binding and the forces exerted on them by the atoms; consequently, the relation $E(\mathbf{k})$ between energy and wave vector is more complicated. The statement of this relationship constitutes the description of an energy band.

The bands of possible electron energy levels in a solid are called allowed energy bands. There are also bands of energy levels which it is impossible for an electron to have in a given crystal. Such bands are called forbidden bands, or gaps. The allowed energy bands sometimes overlap and sometimes are separated by forbidden bands. The presence of a forbidden band immediately above the occupied allowed states (such as the region A to B in the illustration) is the principal difference in the electronic structures of a semiconductor or insulator and a metal. In the first two substances there is a gap between the valence band or normally occupied states and the conduction band, which is normally unoccupied. In a metal there is no gap between occupied and unoccupied states. The presence of a gap means that the electrons cannot easily be accelerated into higher energy states by an applied electric field. Thus, the substance cannot carry a current unless electrons are excited across the gap by thermal or optical means.

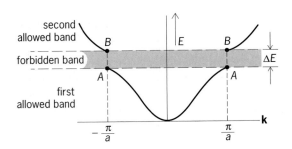

Electron energy E versus wave vector k for a monatomic linear lattice of lattice constant *a*. (*After C. Kittel, Introduction to Solid State Physics, 7th ed., 1995*)

Under external influences, such as irradiation, electrons can make transitions between states in the same band or in different bands. The interaction between the electrons and the vibrations of the crystal lattice can scatter the electrons in a given band with a substantial change in the electron momentum, but only a slight change in energy. This scattering is one of the principal causes of the electrical resistivity of metals. See ELECTRICAL RESISTIVITY.

An external electromagnetic field (for example, visible light) can cause transitions between different bands. Here momentum must be conserved. Because the momentum of a photon $h\nu/c$ (where ν is the frequency of the light and c its velocity) is quite small, the momentum of the electron before and after collision is nearly the same. Such a transition is called vertical in reference to an energy band diagram. Conservation of energy must also hold in the transition, so absorption of light is possible only if there is an unoccupied state of energy $h\upsilon$ available at the same $\mathbf{k}$ as the initial state. These transitions are responsible for much of the absorption of visible and near-infrared light by semiconductors.

The results of energy-band calculations for ordinary metals usually predict Fermi surfaces and other properties that agree rather well with experiment. In addition, cohesive energies and values of the lattice constant in equilibrium can be obtained with reasonable accuracy, and, in the case of ferromagnetic metals, calculated magnetic moments agree with experiment. However, there are significant discrepancies between theoretical calculations and experiments for certain types of systems, including heavy-fermion systems (certain metallic compounds containing rare-earth or actinide elements), superconductors, and Mott-Hubbard insulator (compounds of $3d$ transition elements, for which band calculations predict metallic behavior). Also, band calculations for semiconductors such as silicon, germanium, and gallium arsenide (GaAs) predict values for the energy gap between valence and conduction bands in the range one-half to two-thirds of the measured values. In all these cases, the failures of band theory are attributed to an inadequate treatment of strong electron-electron interactions.

[J.C.]

Barkhausen effect   An effect, due to discontinuities in size or orientation of magnetic domains as a body of ferromagnetic material is magnetized, whereby the magnetization proceeds in a series of minute jumps. *See* FERROMAGNETISM; MAGNETIZATION.

Ferromagnetic materials are characterized by the presence of microscopic domains of some 10^{12} to 10^{15} atoms within which the magnetic moments of the spinning electrons are all parallel. In an unmagnetized specimen, there is random orientation of the various domains. When a magnetic field is applied to the specimen, the domains turn into an orientation parallel to the field, or if parallel to the field, the domains increase in size. During the steep part of the magnetization curve, whole domains suddenly change in size or orientation, giving a discontinuous increase in magnetization. If the specimen being magnetized is within a coil connected to an amplifier and loudspeaker, the sudden changes give rise to a series of clicks or, when there is a rapid change, a hissing sound. This is called the Barkhausen effect; it is an important piece of evidence in support of a domain theory of magnetism.

[K.V.M.]

Barometer   An absolute pressure gage specifically designed to measure atmospheric pressure. This instrument is a type of manometer with one leg at zero pressure absolute. *See* MANOMETER.

The common meteorological barometer (see illustration) is a liquid-column gage filled with mercury. The top of the column is sealed, and the bottom is open and submerged below the surface of a reservoir of mercury. The atmospheric pressure on the reservoir keeps the mercury at a height proportional to that pressure. An adjustable scale, with a vernier scale, allows a reading of column height. Aneroid barometers using metallic diaphragm elements are usually less accurate, though often more sensitive,

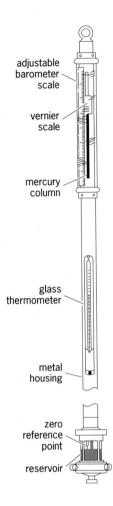

adjustable
barometer
scale

vernier
scale

mercury
column

glass
thermometer

metal
housing

zero
reference
point

reservoir

Mercury barometer.

devices, and not only indicate pressure but may be used to record it. *See* PRESSURE
MEASUREMENT.   [J.H.Z.]

Barretter A bolometer element with a positive temperature coefficient of resistance, used to detect and measure power at radio, microwave, infrared, and optical
frequencies. The temperature of the barretter increases when electromagnetic energy
is absorbed. Barretters are made of metal; therefore, the electrical resistance increases
when the temperature increases. The resulting resistance change of the barretter is
measured by using direct-current or low-frequency instruments. *See* BOLOMETER.

The barretter resistance is selected to absorb most of the power when the barretter
is mounted as a termination in a waveguide or coaxial transmission line. A barretter
can be made to detect power at optical and infrared frequencies by using a very thin
metal ribbon blackened to absorb light.

Barretters with less sensitivity and accuracy for use at radio frequencies can be made
by using low-current fuses made with fine wires.

A meter can be made to measure high-frequency signal amplitudes using a barretter. The temperature and hence the resistance of a barretter can change at audio-frequency rates, but the time constant of a barretter is too great for the resistance to vary at radio-frequency rates. A radio- or microwave-frequency current modulated at a low frequency will cause the barretter resistance to follow the low-frequency signal. If a direct-current voltage is applied to the barretter while the modulated radio-frequency current is also applied, the varying resistance will produce a current which follows the modulation. The low-frequency current can be coupled to the input of an audio amplifier tuned to the modulation frequency by using an audio transformer. The output of the audio amplifier may be rectified to drive a direct-current meter. The meter then indicates the relative amplitude of the radio-frequency or microwave signal. [R.C.Po.]

Baryon The generic name for any hadronic particle with baryon number $B = +1$. By far the most common baryons are the proton and neutron, the two states of the nucleon doublet $N = (p, n)$ [Table 1]. The baryon number of any particular state may be deduced from its production or decay processes, or both, since the total baryon number is conserved (with possible rare exceptions discussed below) and $B = 0$ holds for all mesons and leptons. *See* LEPTON; MESON; NEUTRON; NUCLEON; PROTON.

It is now generally accepted that hadrons are composite, consisting of spin-$1/2$ quarks (q), corresponding antiquarks (q), and some number of gluons, the last being the quanta of the intermediate field which binds the quarks and antiquarks to form hadrons. $B = +\frac{1}{3}$ holds for a quark q, $B = -\frac{1}{3}$ for an antiquark q, while $B = 0$ holds for a gluon. Thus, a baryon consists of three ("valence") quarks, together with some number of quark-antiquark (qq) pairs (called the quark-antiquark sea) and of gluons. The quarks must be assigned fractional charge values, relative to the proton charge (Table 2). *See* GLUONS; HADRON; QUARKS.

Color and quantum chromodynamics. This quark theory of the hadrons has been proposed in a quite specific form, known as quantum chromodynamics (QCD). It is a gauge theory based on a symmetry hypothesized for the hadronic interactions of the quarks, which says that these interactions are invariant with respect to a local (gauge) group of unitary transformations with modulus unity, $SU(3)_C$, acting in an abstract complex three-dimensional space known as color space. Each quark type then has three color states, usually labeled by the suffixes r (red), g (green), and b (blue), corresponding to the three axes of this space. The gauge particle of this symmetry theory is the gluon, a neutral vector particle coupled universally with the currents of color, just as the photon, the gauge particle of quantum electrodynamics (QED), is coupled universally with the electromagnetic current. However, whereas the photon has no charge, the gluon has eight color components, so that it is a color octet. Consequently, there is a gluon contribution to the color currents, and so the gluon field must interact with itself, introducing a nonlinearity into quantum chromodynamics which has no parallel in quantum electrodynamics. This nonlinearity has important implications for quantum chromodynamics, leading to its asymptotic freedom, the property that the coupling of gluon to the color current approaches zero at short distances, which is essential for even qualitative agreement between quantum chromodynamics predictions and the empirical data on high-energy collision processes. *See* COLOR (QUANTUM MECHANICS); QUANTUM CHROMODYNAMICS; QUANTUM ELECTRODYNAMICS; SYMMETRY LAWS (PHYSICS).

An important element in quantum chromodynamics is the confinement dogma, the assertion that only color singlet states have finite energy. This assertion implies that neither a quark nor a gluon can exist in a free state, since the former is a color triplet and the latter a color octet, and indeed no observations of free gluons or quarks have

Table 1. Known stable and semistable baryons and their properties

Baryon	Mass, MeV	Spin parity	Flavors		Lifetime, s	Dominant decay modes	Magnetic moment, n.m.*
			Strangeness (s)	Charm (c)			
p	938.27200 ± 0.00004	$\frac{1}{2}^+$	0	0	$>10^{39}$	—	2.792847
n	939.56533 ± 0.00004	$\frac{1}{2}^+$	0	0	886 ± 1	$p\bar{\nu}_e e^-$	-1.91304
Λ	1115.683 ± 0.006	$\frac{1}{2}^+$	-1	0	$2.63 \pm 0.02 \times 10^{-10}$	$p\pi^-$ (64%) $n\pi^0$ (36%)	-0.613 ± 0.004
Σ^+	1189.37 ± 0.07	$\frac{1}{2}^+$	-1	0	$8.02 \pm 0.03 \times 10^{-11}$	$p\bar{\nu}_e e^-$ $(0.083 \pm 0.002)\%$ $p\pi^0$ (52%) $n\pi^+$ (48%) $p\gamma$ $(0.123 \pm 0.005)\%$	2.46 ± 0.01
Σ^0	1192.64 ± 0.03	$\frac{1}{2}^+$	-1	0	$7.4 \pm 0.7 \times 10^{-20}$	$\Lambda\gamma$	-1.16 ± 0.03
Σ^-	1197.449 ± 0.030	$\frac{1}{2}^+$	-1	0	$1.48 \pm 0.01 \times 10^{-10}$	$n\pi^-$ $n\bar{\nu}_e e^-$ $(0.102 \pm 0.003)\%$	
Ξ^0	1314.8 ± 0.2	$\frac{1}{2}^+$	-2	0	$2.9 \pm 0.1 \times 10^{-10}$	$\Lambda\pi^0$	-1.25 ± 0.02
Ξ^-	1321.31 ± 0.13	$\frac{1}{2}^+$	-2	0	$1.64 \pm 0.02 \times 10^{-10}$	$\Lambda\pi^-$ $\Lambda\bar{\nu}_e e^-$ $(0.056 \pm 0.003)\%$	-0.651 ± 0.003
Ω^-	1672.4 ± 0.3	$(\frac{3}{2}^+?)$	-3	0	$0.82 \pm 0.01 \times 10^{-10}$	ΛK^- (68%) $\Xi^0\pi^-$ (24%) $\Xi^-\pi^0$ (9%)	-2.02 ± 0.05
Λ_c^+	2284.9 ± 0.6	$(\frac{1}{2}^+?)$	0	1	$2.0 \pm 0.1 \times 10^{-13}$	$pK^-\pi^+$ $(5 \pm 1)\%$	—
Σ_c^+	2452 ± 1	$(\frac{1}{2}^+?)$	0	1	hadronic	$\Lambda_c^+\pi^+$ $(+, 0, -)$	—
Ξ_c^+	2466 ± 2	$(\frac{1}{2}^+?)$	-1	1	$4.4 \pm 0.3 \times 10^{-13}$	$\Sigma^+\pi^+\pi^-$, $\Lambda K^-\pi^+\pi^+$	—
Ξ_c^0	2472 ± 2	$(\frac{1}{2}^+?)$	-1	1	$1.0 \pm 0.2 \times 10^{-13}$	$\Xi^-\pi^+$, $\Xi^-\pi^+\pi^+\pi^-$	—
Ω_c^0	2698 ± 3	$(\frac{1}{2}^+?)$	-2	1	$0.6 \pm 0.2 \times 10^{-13}$	$\Xi^- K^-\pi^+\pi^-$	—
Λ_b^0	5624 ± 9	$(\frac{1}{2}^+?)$	0	0 $b=-1$	$1.23 \pm 0.08 \times 10^{-12}$	$\Lambda_c^+\pi^-$	—

*The abbreviation n.m. denotes the unit $e\hbar/2M_p c$ (nuclear magneton).
†Σ_c is included here, although hadronically unstable, because Λ_c, Σ_c, Ξ_c, and Ω_c belong to a common $SU(3)_f$ multiplet; see Fig. 2a.

Table 2. Properties of established quarks and leptons*, arranged in three families

Quark type	d (down)	u (up)	s (strange)	c (charmed)	b (bottom)	t (top)
Charge (Q/e_p)	$-\frac{1}{3}$	$\frac{2}{3}$	$-\frac{1}{3}$	$\frac{2}{3}$	$-\frac{1}{3}$	$\frac{2}{3}$
Mass, GeV†	$\simeq 0.3$	$\simeq 0.3$	$\simeq 0.5$	$\simeq 1.5$	$\simeq 4.7$	$\simeq 174$
Flavor	$I_3 = -\frac{1}{2}$	$I_3 = +\frac{1}{2}$	$s = -1$	$c = +1$	$b = -1$	$t = +1$
Lepton type	ν_e	e^+	μ^+	ν_μ	τ^+	ν_τ

*To each quark and lepton, there exists an antiquark and antilepton with opposite flavor values and with opposite intrinsic parity.
†Quark masses are rough estimates of the "effective mass" of each quark in a hadron.

yet been confirmed. However, no rigorous proof that the dogma follows from quantum chromodynamics has yet been given.

Quantitative predictions of the properties of baryonic states are currently made by using a simplified quark-quark (q-q) potential with the following features: (1) an attractive long-range potential, increasing with separation to ensure confinement, and (2) a spin-dependent potential representing one-gluon exchange, effective at small separation, where the regime of asymptotic freedom holds and perturbation theory is valid. Such predictions have had a great deal of success.

The quark content of the nucleons is given by Eqs. (1).

$$p = (uud) \qquad n = (udd) \tag{1}$$

The replacement of a d quark in the nucleon by an s quark produces a baryon state with spin parity $\frac{1}{2}^+$ and strangeness number $s = -1$, the latter being given by $[n(s) - n(s)]$, where $n(q)$ denotes the number of quarks of type q in the system considered. The states thus reached have the flavor structures of Eqs. (2), the other factors in

$$(\Sigma^+, \Sigma^0, \Sigma^-) = (uus, (ud + du)s/\sqrt{2}, dds) \tag{2a}$$

$$\Lambda = (ud - du)s/\sqrt{2} \tag{2b}$$

their wave functions being identical with those for the nucleons; thus the isotriplet Σ and isosinglet Λ states are obtained. If a u quark and a d quark are each replaced by an s quark in Eqs. (1), the isodoublet Ξ states of Eq. (3) are obtained. The quark has

$$(\Xi^0, \Xi^-) = (uss, dss) \tag{3}$$

$I = 0$, being unaffected by the $SU(2)_\tau$ transformations in the (u, d) space. The flavor wave function (sss) is necessarily symmetric and cannot occur with total spin $S = \frac{1}{2}$. Baryonic states with $s \neq 0$ are collectively termed hyperons. *See* HYPERON; STRANGE PARTICLES.

These eight baryon states (p, n, Σ^+, Σ^0, Σ^-, Λ, Ξ^0, Ξ^-) all have the spin parity $\frac{1}{2}^+$ and the same internal wave functions. It is helpful to use the quantum number $Y = (B + s)$, named hypercharge. Then, if the states are arrayed in the $I_3 - Y$ plane (Fig. 1), the symmetry of their relationship is evident.

The mass difference $\delta m = [m(s) - m(u, d)]$ is quite large, and so $SU(3)_f$ symmetry is much more strongly violated than $SU(2)_\tau$ symmetry. The baryon mass values vary

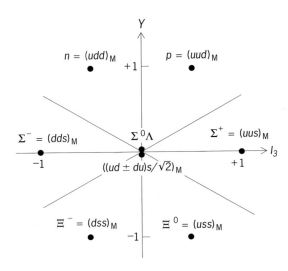

Fig. 1. Baryon octet states, arrayed with respect to I_3 as ordinate and $Y = (B + s)$ as abscissa. The charge number Q is given by $Q = I_3 + Y/2$. There are three axes of symmetry.

widely over the octet; the leading variation is that proportional to the strangeness s, which counts the s-quark content of each baryon. The approximately 75-MeV difference between the mean Σ mass $m(\Sigma)$ and $m(\Lambda)$ has a more subtle origin, but is well accounted for on the basis of the quark-quark (qq) potential from quantum chromodynamics, described above. The small mass differences within each isospin multiplet are believed to be due to the intrinsic (u, d) mass difference and to electromagnetic effects.

The baryon-baryon interactions are of particular interest. That between nucleons gives rise to the existence of atomic nuclei, and has been particularly well studied, both empirically and theoretically. For large separations (greater than 0.8 femtometer), the NN force is due to the exchange of pions and of other known mesons with masses less than about 1 GeV; for small separations (less than 0.4 fm), a strong short-range repulsion is observed, possibly arising from the suppressive effects of the Pauli principle for quarks when the quark structures of the two nucleons overlap. At low energies, the outstanding feature of the NN interaction is its strong noncentral tensor component, which is due to one-pion exchange and is a direct consequence of the pseudoscalar nature of the pion. It also has a strong spin-orbit interaction, observed in NN interactions at higher energy and of much importance for the shell structure of nuclei. See NUCLEAR STRUCTURE.

Many further particle-unstable baryon states, with lifetimes in the range 10^{-22} to 10^{-23} s, have become established, up to mass values of order 2500 MeV, all consistent with the limitations of the three-quark model.

Further baryon states can be formed by replacing one or more of the u, d, and s quarks of the states discussed above by a c quark. If the s and c quarks both had the same mass as the (u, d) quarks, the states formed would correspond to an $SU(4)_f$ symmetry. Extensions of the $1/2^+$ baryon octet (Fig. 2a) and the $3/2^+$ baryon decuplet (Fig. 2b) to arrays in three dimensions are obtained in this way. The lowest plane of each such array consists of the charmless baryon states, in accord with the original octet or decuplet. In reality, the c-quark mass is so large that little quantitative detail of $SU(4)_f$ symmetry can survive in the physical situation, and these arrays have value mainly for general comprehension and for the counting of states.

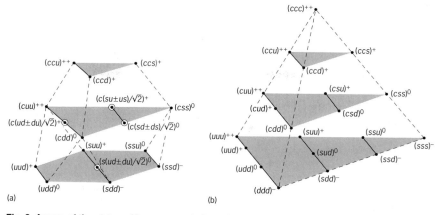

Fig. 2. Arrays of the states of baryons made from three quarks of the types _u_, _d_, _s_, and _c_, with no internal orbital angular momentum. The quark content and charge of each state is specified. The vertical axis specifies the number of _c_ quarks. (_a_) $1/2^+$ baryons. (_b_) $3/2^+$ baryons. The former have flavor symmetry M; the latter have symmetry S.

The fifth quark, named "bottom" (symbol _b_), was discovered in 1978. The baryon $\Lambda_b{}^0$ is well established, with decay mode $\pi^- \Lambda_c{}^+$ and mass 5624 ± 9 MeV. _See_ UPSILON PARTICLES.

The sixth quark, named "top" (symbol _t_), became established in 1994 through top-antitop pair production in proton-antiproton collisions at center-of-mass energy 1800 GeV. Its mass, 174 ± 5 GeV, is remarkably large, far above the bW^+ threshold at ≈ 85 GeV. The decay $t \rightarrow bW^+$ is very rapid, its lifetime being 4×10^{-25} s and its decay width $\Gamma_t \approx 1.5$ GeV. This decay is so much faster than hadronization that there is almost no time for top baryons to form.

No further quark-lepton families are expected. This conclusion comes from counting the number of neutrino species in Z^0 decays, and from astrophysical arguments about light-nucleus formation in the early universe.

For every baryonic state mentioned above, there will exist an antibaryon state with opposite flavor quantum numbers, in particular with $B = -1$. The antiproton $\overline{p}$ was first identified in 1954, and antiproton beams at high-energy proton accelerators are a basic tool in elementary particle research. Most of the expected antibaryon states have been detected and studied in some detail.

In grand unification theories (GUT), which attempt to account for both quarks and leptons, together with their strong, electromagnetic, and weak interactions, quark $\rightarrow$ lepton transitions generally exist, at some level, since such theories assign leptons and quarks to common multiplets. This violation of baryon conservation opens the possibility that the lightest baryon, the proton, might not be absolutely stable but may undergo decay processes such as $p \rightarrow e^+ \pi^0$ at a very low rate. Indeed, cosmology appears to require the existence of nucleon decay processes in order to account for the baryon-antibaryon asymmetry of the universe. Empirically, the partial decay rate $\Gamma(p \rightarrow e^+ \pi^0)$ is less than 0.58×10^{-40} s^{-1} with 90% probability. In the simplest GUT [known as SU(5)], this leads to a proton lifetime greater than 1.7×10^{32} years. Other GUTs may predict smaller decay rates, even zero. Since proton decay offers the possibility of discriminating between various GUTs, the detection of proton decay (and neutron decay involving baryon nonconservation, for example, when bound in a deuteron or alpha particle), or at least the improvement of the present empirical limits on its rate, is an important subject of investigation. *See* ELEMENTARY PARTICLE; FUNDAMENTAL INTERACTIONS; GRAND UNIFICATION THEORIES.   [R.H.D.]

Beam-foil spectroscopy   A technique used in atomic physics to study the structure and dynamics of atomic ions of any element in any state of ionization. For this purpose, a beam of fast ions is sent through a very thin foil. The ion-foil interaction shakes up the electronic shells of the projectile and, after leaving the foil, the ions shed surplus energy by emitting photons, and sometimes electrons. The energies and intensities of these particles yield spectral information on the projectile. *See* PARTICLE ACCELERATOR.

The multitude of collisions inside the foil changes the complement of electrons that travel with the projectile ion; some are ejected and others are captured from the target atoms. The ion beam therefore has a different charge-state composition after passage through the foil. (Higher exit charge states are produced at higher incident beam energies.) The beam-foil interaction efficiently populates atomic levels with a high degree of excitation such as multiply excited and core-excited levels. The richness of the resulting spectra yields a great deal of information on atoms and ions, although it is often difficult to resolve the details of line-rich spectra that reflect the complexity of multiply excited systems.

The ion beam travels in a high vacuum before and after transiting the target foil. This environment minimizes collisional perturbation of the ions. The sudden termination of the ion-foil interaction provides an inherently good time resolution to beam-foil spectroscopy. This property of the source permits lifetime measurements as well as the observation of coherent-excitation phenomena such as quantum beats. Because the ion velocity is constant and measurable, it is sufficient to trace the change in intensity of the fluorescence from the ion beam as a function of distance from the foil in order to determine atomic level lifetimes. *See* FLUORESCENCE.

Beam-foil spectroscopy has developed into many variants which now go under the name of fast-beam spectroscopy. For example, a gas target may be used, a laser, a combination of gas or foil and laser, or a target of free electrons in a heavy-ion storage ring. The ion-foil interaction is capable of producing all ionization stages of all elements from negative ions to U^{91+}. The production of the highest ionization stages, however, requires a beam energy of about 500 MeV/nucleon, which can be reached only at the most energetic accelerators. However, since only the relative motion of electrons and ions is important, the same degree of ionization can be reached by use of 250-keV electrons in an electron-beam ion trap (EBIT). The device offers easier ways to attain high spectroscopic precision because the ions are practically at rest. In beam-foil spectroscopy the ions are rapidly moving, which shifts and broadens the spectral lines. This, in turn, causes problems in wavelength calibration and spectral resolution. However, the inherent time resolution of the foil-excited fast-ion-beam source is unique and remains a great asset in time-resolved spectroscopic measurements. *See* ATOMIC STRUCTURE AND SPECTRA; ION SOURCES; SPECTROSCOPY.   [E.Tr.]

Beat   A variation in the intensity of a composite wave which is formed from two distinct waves with different frequencies. Beats were first observed in sound waves, such as those produced by two tuning forks with different frequencies. Beats also can be produced by other waves. They can occur in the motion of two pendulums of different lengths and have been observed among the different-frequency phonons in a crystal lattice.

One important application of beat phenomena is to use one object with an accurately known frequency to determine the unknown frequency of another such object. The beat-frequency or heterodyne oscillator also operates by producing beats from two frequencies.

[B.DeF.]

Bernoulli's theorem An idealized algebraic relation between pressure, velocity, and elevation for flow of an inviscid fluid. Its most commonly used form is for steady flow of an incompressible fluid, and is given by the equation below, where p is

$$\frac{p}{\rho} + \frac{V^2}{2} + gz = \text{constant}$$

pressure, ρ is fluid density (assumed constant), V is flow velocity, g is the acceleration of gravity, and z is the elevation of the fluid particle. The relation applies along any particular streamline of the flow. The constant may vary across streamlines unless it can be further shown that the fluid has zero local angular velocity.

The above equation may be extended to steady compressible flow (where changes in ρ are important) by adding the internal energy per unit mass, e, to the left-hand side. See COMPRESSIBLE FLOW.

The equation is limited to inviscid flows with no heat transfer, shaft work, or shear work. Although no real fluid truly meets these conditions, the relation is quite accurate in free-flow or "core" regions away from solid boundaries or wavy interfaces, especially for gases and light liquids. Thus Bernoulli's theorem is commonly used to analyze flow outside the boundary layer, flow in supersonic nozzles, flow over airfoils, and many other practical problems. See AERODYNAMICS; BOUNDARY-LAYER FLOW.

[F.M.Wh.]

Beta particles The name first applied in 1897 by Ernest Rutherford to one of the forms of radiation emitted by radioactive nuclei. Beta particles can occur with either negative or positive charge (denoted β^- or β^+) and are now known to be either electrons or positrons, respectively. Electrons and positrons are now referred to as beta particles only if they are known to have originated from nuclear beta decay. Their observed kinetic energies range from zero up to about 5 MeV in the case of naturally occurring radioactive isotopes, but can reach values well over 10 MeV for some artificially produced isotopes. See ALPHA PARTICLES; ELECTRON; GAMMA RAYS; POSITRON; RADIOACTIVITY.

When a nucleus beta-decays, it emits two particles at the same time: One is a beta particle; the other, a neutrino or antineutrino. With this emission, the nucleus itself undergoes a transformation, changing from one element to another. In the case of isotopes that β^+-decay, each decaying nucleus emits a positron and a neutrino, simultaneously reducing its atomic number by one unit; for those isotopes that β^--decay, each nucleus emits an electron and an antineutrino while increasing its atomic number by one. In both classes of decay, the energy released by the nuclear transformation is shared between the two emitted particles. Though the energy released by a particular nuclear transformation is always the same, the fraction of this energy carried away by the beta particle is different for each decaying nucleus. (The neutrino always carries away the remainder, thus conserving energy overall.) When observed collectively, the decaying nuclei of a simple radioactive source emit their beta particles with a continuous distribution of kinetic energies covering the range from zero up to the total nuclear decay energy available.

Radioactive samples often contain several radioactive isotopes. Since each isotope has its own decay energy and beta-particle energy distribution, the energy spectrum of beta particles observed from such a sample would be the sum of a number of distributions, each with a different end-point energy. Indeed, many isotopes, especially those artificially produced with accelerators, can themselves beta-decay by additional paths that also release part of the energy in the form of gamma radiation.

As a beta particle penetrates matter, it loses its energy in collisions with the constituent atoms. Two processes are involved. First, the beta particle can transfer a small fraction of its energy to the struck atom. Second, the beta particle is deflected from its original path by each collision and, since any change in the velocity of a charged particle leads to the emission of electromagnetic radiation, some of its energy is lost in the form of low-energy x-rays (bremsstrahlung). Though the energy lost by a beta particle in a single collision is very small, many collisions occur as the particle traverses matter, causing it to follow a zigzag path as it slows down. *See* BREMSSTRAHLUNG.

The thickness of material that is just sufficient to stop all the beta particles of a particular energy is called the range of those particles. For the continuous energy distribution normally associated with a source of beta particles, the effective range is the one that corresponds to the highest energy in the primary spectrum. That thickness of material stops all of the beta particles from the source. The range depends strongly on the electron energy and the density of the absorbing material.

The slowing-down processes have the same effect on both β^- and β^+ particles. However, as antimatter, the positron (β^+) cannot exist for long in the presence of matter. It soon combines with an atomic electron, with which it annihilates, the masses of both particles being replaced by electromagnetic energy. Usually this annihilation occurs after the positron has come to rest and formed a positronium atom, a bound but short-lived positron-electron system. In that case, the electromagnetic energy that is emitted from the annihilation takes the form of two 511-keV gamma rays that are emitted in opposite directions to conserve momentum. *See* POSITRONIUM.

Beta particles are detected through their interaction with matter. One class of detectors employs gas as the detection medium. Ionization chambers, proportional counters, and Geiger-Müller counters are of this class. In these detectors, after entering through a thin window, the beta particles produce positive ions and free electrons as they collide with atoms of the gas in the process of their slowing down. An electric field applied across the volume of gas causes these ions and electrons to drift along the field lines, causing an ionization current that is then processed in external electronic devices. *See* IONIZATION CHAMBER; PARTICLE DETECTOR.

More precise energy information can be achieved with scintillation detectors. In certain substances, the ion-electron pairs produced by the passage of a charged particle result in the emission of a pulse of visible or near-ultraviolet light. If a clear plastic scintillator is used, it can be mounted on a photomultiplier tube, which converts the transmitted light into a measurable electrical current pulse whose amplitude is proportional to the energy deposited by the incident beta particle. *See* SCINTILLATION COUNTER.

Even better energy information comes from semiconductor detectors, which are effectively solid-state ionization chambers. When a beta particle enters the detector, it causes struck electrons to be raised into the conduction band, leaving holes behind in the valence band. The electrons and holes move under the influence of an imposed electric field, causing a pulse of current to flow. Such detectors are useful mainly for low-energy beta particles. *See* JUNCTION DETECTOR.

Any one of these detectors can be combined with a magnetic spectrometer. Beta particles, like any charged particles, follow curved paths in a perpendicular magnetic field, their radius of curvature being proportional to the square of their energy. Their detected position on exiting the magnetic field can be precisely related to their energy. The best current measurement of the electron antineutrino mass comes from a spectrometer measurement of the tritium beta-decay spectrum. *See* NEUTRINO.   [J.Hard.]

Betatron   A device for accelerating charged particles in an orbit by means of the electric field E from a slowly changing magnetic flux Φ. The electric field is given by $E = -(1/2\pi r_o) \, d\Phi/dt$ (in SI or mks units), where r_o is the orbit radius. The name was chosen because the method was first applied to electrons. In the usual betatron both the accelerating core flux and a guiding magnetic field rise with similar time dependence, with the result that the orbit is circular. However, the orbit can have a changing radius as acceleration progresses. For the long path (usually more than 60 mi or 100 km), variations of axial and radial magnetic field components provide focusing forces, while space charge and space current forces due to the particle beam itself also contribute to the resulting betatron oscillations about the equilibrium orbit. In many other instances of particle beams, the term betatron oscillations is used for the particle oscillations about a beam's path.

Collective effects from self-fields of the beam have been found important and helpful in injecting. Circulating currents of about 3 amperes are contained in the numerous industrial and therapeutic betatrons, although the average currents are below 10^{-7} A. *See* PARTICLE ACCELERATOR.
[D.W.K.]

Biot-Savart law   A law of physics which states that the magnetic flux density (magnetic induction) near a long, straight conductor is directly proportional to the current in the conductor and inversely proportional to the distance from the conductor. The field near a straight conductor can be found by application of Ampère's law. The magnetic flux density near a long, straight conductor is at every point perpendicular to the plane determined by the point and the line of the conductor. Therefore, the lines of induction are circles with their centers at the conductor. Furthermore, each line of induction is a closed line. This observation concerning flux about a straight conductor may be generalized to include lines of induction due to a conductor of any shape by the statement that every line of induction forms a closed path.   ` [K.V.M.]

Birefringence   The splitting which a wavefront experiences when a wave disturbance is propagated in an anisotropic material; also called double refraction. In anisotropic substances the velocity of a wave is a function of displacement direction. Although the term birefringence could apply to transverse elastic waves, it is usually applied only to electromagnetic waves.

In birefringent materials either the separation between neighboring atomic structural units is different in different directions, or the bonds tying such units together have different characteristics in different directions. Many crystalline materials, such as calcite, quartz, and topaz, are birefringent. Diamonds, on the other hand, are isotropic and have no special effect on polarized light of different orientations. Plastics composed of long-chain molecules become anisotropic when stretched or compressed. Solutions of long-chain molecules become birefringent when they flow. This first phenomenon is called photoelasticity; the second, streaming birefringence. *See* CRYSTAL OPTICS; POLARIZED LIGHT; REFRACTION OF WAVES.
[B.H.Bi.]

Blackbody   An ideal energy radiator, which at any specified temperature emits in each part of the electromagnetic spectrum the maximum energy obtainable per unit time from any radiator due to its temperature alone. A blackbody also absorbs all the energy which falls upon it. The radiation properties of real radiators are limited by two extreme cases—a radiator which reflects all incident radiation, and a radiator which absorbs all incident radiation. Neither case is completely realized in nature. Carbon and soot are examples of radiators which, for practical purposes, absorb all radiation.

Both appear black to the eye at room temperature, hence the name blackbody. Often a blackbody is also referred to as a total absorber. *See* HEAT RADIATION. [H.G.S.; P.J.W.]

Bloch theorem A theorem that specifies the form of the wave functions that characterize electron energy levels in a periodic crystal. Electrons that move in a constant potential, that is, a potential independent of the position **r**, have wave functions that are plane waves, having the form $\exp(i\mathbf{k} \cdot \mathbf{r})$. Here, **k** is the wave vector, which can assume any value, and describes an electron having momentum $\hbar\mathbf{k}$. (The quantity $\hbar$ is Planck's constant divided by 2π.) Electrons in a crystal experience a potential that has the periodicity of the crystal lattice. *See* BAND THEORY OF SOLIDS. [A.O.]

Boiling A process in which a liquid phase is converted into a vapor phase. The energy for phase change is generally supplied by the surface on which boiling occurs. Boiling differs from evaporation at predetermined vapor/gas-liquid interfaces because it also involves creation of these interfaces at discrete sites on the heated surface. Boiling is an extremely efficient process for heat removal and is utilized in various energy-conversion and heat-exchange systems and in the cooling of high-energy density components. *See* HEAT TRANSFER.

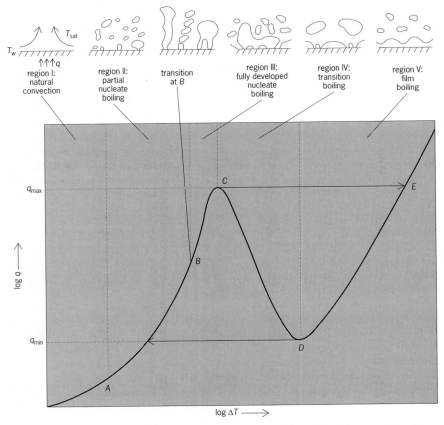

Typical boiling curve, showing qualitatively the dependence of the wall heat flux q on the wall superheat $\triangle T$. Schematic drawings show the boiling process in regions I–V, and transition points A–E.

Boiling is classified into pool and forced-flow. Pool boiling refers to boiling under natural convection conditions, whereas in forced-flow boiling the liquid flow over the heater surface is imposed by external means. Flow boiling is subdivided into external and internal. In external-flow boiling, liquid flow occurs over heated surfaces, whereas internal-flow boiling refers to flow inside tubes. Heat fluxes of 2×10^8 W/m^2, or three times the heat flux at the surface of the Sun, have been obtained in flow boiling. *See* CONVECTION (HEAT).

Pool boiling. The illustration, a qualitative pool boiling curve, shows the dependence of the wall heat flux q on the wall superheat ΔT (the difference between the wall temperature and the liquid's saturation temperature). The plotted curve is for a horizontal surface underlying a pool of liquid at its saturation temperature (the boiling point at a given pressure).

Several heat transfer regimes can be identified on the boiling curve: single-phase natural convection, partial nucleate boiling, fully developed nucleate boiling, transition boiling, and film boiling.

Forced-flow boiling. Forced flow, both external and internal, greatly changes the boiling curve in the illustration. The heat flux is increased by forced convection at temperatures below boiling inception, and after that the nucleate boiling region is extended upward until a flow-enhanced higher maximum flux (corresponding to point C) is achieved. Forced flow boiling in tubes is used in many applications, including steam generators, nuclear reactors, and cooling of electronic components.   [V.K.D.]

Bolometer   A device for detecting and measuring small amounts of thermal radiation. The bolometer is a simple electric circuit, the essential element of which is a slab of material with an electrical property, most often resistance, that changes with temperature. Typical operation involves absorption of radiant energy by the slab, producing a rise in the slab's temperature and thereby a change in its resistance. The electric circuit converts the resistance change to a voltage change, which then can be amplified and observed by various, usually conventional, instruments.

Although bolometers are useful in studying a variety of systems where detection of small amounts of heat is important, their primary application remains as the instrument of choice for measuring weak radiation signals in the infrared and far infrared, that is, at wavelengths from about 1 to 2000 micrometers, from stars and interstellar material. *See* BARRETTER; INFRARED RADIATION; RADIOMETRY; THERMISTOR.   [W.E.K.]

Boltzmann constant   A constant occurring in practically all statistical formulas and having a numerical value of 1.3807×10^{-23} joule/K. It is represented by the letter k. If the temperature T is measured from absolute zero, the quantity kT has the dimensions of an energy and is usually called the thermal energy. At 300 K (room temperature) $kT = 0.0259$ electronvolt.

The value of the Boltzmann constant may be determined from the ideal gas law. For 1 mole of an ideal gas Eq. (1a) holds, where P is the pressure, V the volume, and

$$PV = RT \tag{1a}$$

$$PV = NkT \tag{1b}$$

R the universal gas constant. The value of R, 8.31 J/K mole, may be obtained from equation-of-state data. Statistical mechanics yields for the gas law Eq. (1b). Here N, the number of molecules in 1 mole, is called Avogadro's number and is equal to 6.02×10^{23} molecules/mole. Hence, comparing Eqs. (1a) and (1b), one obtains Eq. (2).

$$k = R/N = 1.3807 \times 10^{-23} \text{J/K} \tag{2}$$

Almost any relation derived on the basis of the partition function or the Bose-Einstein, Fermi-Dirac, or Boltzmann distribution contains the Boltzmann constant. *See* BOLTZMANN STATISTICS; BOSE-EINSTEIN STATISTICS; FERMI-DIRAC STATISTICS; KINETIC THEORY OF MATTER; STATISTICAL MECHANICS.   [M.Dr.]

Boltzmann statistics   To describe a system consisting of a large number of particles in a physically useful manner, recourse must be had to so-called statistical procedures. If the mechanical laws operating in the system are those of classical mechanics, and if the system is sufficiently dilute, the resulting statistical treatment is referred to as Boltzmann or classical statistics. (Dilute in this instance means that the total volume available is much larger than the proper volume of the particles.) A gas is a typical example: The molecules interacting according to the laws of classical mechanics are the constituents of the system, and the pressure, temperature, and other parameters are the overall entites which determine the macroscopic behavior of the gas. In a case of this kind it is neither possible nor desirable to solve the complicated equations of motion of the molecules; one is not interested in the position and velocity of *every* molecule at any time. The purpose of the statistical description is to extract from the mechanical description just those features relevant for the determination of the macroscopic properties and to omit others.

The basic notion in the statistical description is that of a distribution function. Suppose a system of N molecules is contained in a volume V. The molecules are moving around, colliding with the walls and with each other. Construct the following geometrical representation of the mechanical system. Introduce a six-dimensional space (usually called the μ space), three of its coordinate axes being the spatial coordinates of the vessel x, y, z, and the other three indicating cartesian velocity components v_x, v_y, v_z. A molecule at a given time, having a specified position and velocity, may be represented by a point in this six-dimensional space. The state of the gas, a system of N molecules, may be represented by a cloud of N points in this space. In the course of time, this cloud of N points moves through the μ space.

Note that the μ space is actually finite; the coordinates x, y, z of the molecules' position are bounded by the finite size of the container, and the velocities are bounded by the total energy of the system. Imagine now that the space is divided into a large number of small cells, of sizes $w_1, \ldots, w_i, \ldots$. A certain specification of the state of the gas is obtained if, at a given time t, the numbers $n_1(t), \ldots, n_i(t), \ldots$ of molecules in the cells $1, \ldots, i, \ldots$ are given. To apply statistical methods, one must choose the cells such that on the one hand a cell size w is small compared to the macroscopic dimensions of the system, while on the other hand w must be large enough to allow a large number of molecules in one cell. If the cells are thus chosen, the numbers $n_i(t)$, the occupation numbers, will be slowly changing functions of time. The distribution functions $f_i(t)$ are defined by Eq. (1).

$$n_i(t) = f_i(t)w_i \qquad (1)$$

The distribution function f_i describes the state of the gas, and f_i of course varies from cell to cell. Since a cell i is characterized by a given velocity range and position range, and since for appropriately chosen cells f should vary smoothly from cell to cell, f is often considered as a continuous function of the variables x, y, z, vx, vy, vz. The cell size w then can be written as $dxdydzdv_xdv_ydv_z$.

Since a cell i determines both a position and a velocity range, one may associate an energy ϵ_i with a cell. This is the energy a single molecule possesses when it has a representative point in cell i. This assumes that, apart from instantaneous collisions, molecules exert no forces on each other. If this were not the case, the energy of a molecule would be determined by the positions of all other molecules.

Most of the physically interesting quantities follow from a knowledge of the distribution function; the main problem in Boltzmann statistics is to find out what this function is. It is clear that $n_i(t)$ changes in the course of time for three reasons: (1) Molecules located at the position of cell i change their positions and hence move out of cell i; (2) molecules under the influence of outside forces change their velocities and again leave the cell i; and (3) collisions between the molecules will generally cause a (discontinuous) change of the occupation numbers of the cells. Whereas the effect of (1) and (2) on the distribution function follows directly from the mechanics of the system, a separate assumption is needed to obtain the effect of collisions on the distribution function. This assumption, the collision-number assumption, asserts that the number of collisions per unit time, of type $(i, j) \rightarrow (k, l)$ [molecules from cells i and j collide to produce molecules of different velocities which belong to cells k and l], called A_{ij}^{kl}, is given by Eq. (2). Here a_{ij}^{kl} depends on the collision configuration and on the size

$$A_{ij}^{kl} = n_i n_j a_{ij}^{kl} \tag{2}$$

and kind of the molecules but not on the occupation numbers. Gains and losses of the molecules in, say, cell i can now be observed. If the three factors causing gains and losses are combined, the Boltzmann transport equation, written as Eq. (3), is obtained.

$$\frac{\partial f_i}{\partial t} + (\mathbf{v}_i \cdot \Delta_x f_i) + (\mathbf{X}_i \cdot \Delta_v f_i) = \sum_{j,k,l} a_{ij}^{kl} w_j (f_k f_l - f_i f_j) \tag{3}$$

Here $\Delta_x f_i$ is the gradient of f with respect to the positions, $\Delta_v f_i$ refers similarly to the velocities, and $\mathbf{X}_i$ is the outside force per unit mass at cell i. This nonlinear equation determines the temporal evolution of the distribution function. Exact solutions are difficult to obtain. Yet Eq. (3) forms the basis for the kinetic discussion of most transport processes. There is one remarkable general consequence, which follows from Eq. (3). If one defines $H(t)$ as in Eq. (4), one finds by straight manipulation from Eqs. (3) and (4) that Eqs. (5) hold. Hence H is a function which in the course of time always decreases.

$$H(t) = \sum_i n_i \ln f_i \tag{4}$$

$$\frac{dH}{dt} \cdots 0 \quad \frac{dH}{dt} = 0 \quad \text{if } f_i f_i = f_k f_i \tag{5}$$

This result is known as the H theorem. The special distribution which is characterized by Eq. (6) has the property that collisions do not change the distribution in the course

$$f_i f_i = f_k f_l \tag{6}$$

of time; it is an equilibrium or stationary distribution.

The form of the equilibrium distribution may be determined from Eq. (6), with the help of conservation laws. For a gas which as a whole is at rest, it may be shown that the only solution to functional Eq. (6) is given by Eqs. (7a) or (7b). Here A and $\mathcal{B}$ are

$$f_i = Ae^{-\mathcal{B}\epsilon_i} \tag{7a}$$

$$f(\mathbf{x}, \mathbf{v}) = Ae^{(-1/2)\mathcal{B}mv^2 - \mathcal{B}U} \tag{7b}$$

parameters, not determined by Eq. (6), and U is the potential energy at the point x, y, z. Equations (7a) and (7b) are the Maxwell-Boltzmann distribution. Actually A and $\mathcal{B}$ can be determined from the fact that the number of particles and the energy of the system are specified.

The indiscriminate use of the collision-number assumption leads, via the H theorem, to paradoxical results. The basic conflict stems from the irreversible results that appear to emerge as a consequence of a large number of reversible fundamental processes. A careful treatment of the explicit and hidden probability assumptions is the key to the understanding of the apparent conflict. The equilibrium distribution may be thought of as the most probable state of a system. If a system is not in equilibrium, it will most likely (but not certainly) go there; if it is in equilibrium, it will most likely (but not certainly) stay there. By using such probability statements, it may be shown that the paradoxes and conflicts may indeed be removed. A consequence of the probabilistic character of statistics is that the entities computed also possess this characteristic. For example, one cannot really speak definitively of the number of molecules hitting a section of the wall per second, but only about the probability that a given number will hit the wall, or about the average number hitting. In the same vein, the amount of momentum transferred to a unit area of the wall by the molecules per second (this, in fact, is precisely the pressure) is also to be understood as an average. This in particular means that the pressure is a fluctuating entity. The fluctuations in pressure may be demonstrated by observing the motion of a mirror, suspended by a fiber, in a gas. On the average, as many gas molecules will hit the back as the front of the mirror, so that the average displacement will indeed be zero. However, it is easy to imagine a situation where more momentum is transferred in one direction than in another, resulting in a deflection of the mirror. From the knowledge of the distribution function the probabilities for such occurrences may indeed be computed; the calculated and observed behavior agree very well. This clearly demonstrates the essentially statistical character of the pressure. *See* BOLTZMANN TRANSPORT EQUATION; BROWNIAN MOVEMENT; KINETIC THEORY OF MATTER; QUANTUM STATISTICS; STATISTICAL MECHANICS.   [M.Dr.]

Boltzmann transport equation   An equation which is used to study the nonequilibrium behavior of a collection of particles. In a state of equilibrium a gas of particles has uniform composition and constant temperature and density. If the gas is subjected to a temperature difference or disturbed by externally applied electric, magnetic, or mechanical forces, it will be set in motion and the temperature, density, and composition may become functions of position and time; in other words, the gas moves out of equilibrium. The Boltzmann equation applies to a quantity known as the distribution function, which describes this nonequilibrium state mathematically and specifies how quickly and in what manner the state of the gas changes when the disturbing forces are varied. *See* KINETIC THEORY OF MATTER.

Equation (1) is the Boltzmann transport equation shown below, where f is the un-

$$\frac{\partial f}{\partial t} = \left(\frac{\partial f}{\partial t}\right)_{\text{force}} + \left(\frac{\partial t}{\partial t}\right)_{\text{diff}} + \left(\frac{\partial f}{\partial t}\right)_{\text{coll}} \tag{1}$$

known distribution function which, in its most general form, depends on a position vector $\mathbf{r}$, a velocity vector $\mathbf{v}$, and the time t. The quantity $\partial f/\partial t$ on the left side of Eq. (1) is the rate of change of f at fixed values of $\mathbf{r}$ and $\mathbf{v}$. The equation expresses this rate of change as the sum of three contributions: first, $(\partial f/\partial t)_{\text{force}}$ arises when the velocities of the particles change with time as a result of external driving forces; second, $(\partial f/\partial t)_{\text{diff}}$ is the effect of the diffusion of the particles from one region in space to the other; and third, $(\partial f/\partial t)_{\text{coll}}$ is the effect of the collisions of the particles with each other or with other kinds of particles.

The distribution function carries information about the positions and velocities of the particles at any time. The probable number of particles N at the time t within the spatial element $dxdydz$ located at (x, y, z) and with velocities in the element $dv_x dv_y dv_z$

at the point (v_x, v_y, v_z) is given by Eq. (2) or, in vector notation, by Eq. (3). It is assumed

$$N = f(x, y, z, v_x, v_y, v_z, t)dxdydzdv_xdv_ydv_z \qquad (2)$$

$$N = f(\mathbf{r}, \mathbf{v}, t)d^3rd^3v \qquad (3)$$

that the particles are identical; a different distribution function must be used for each species if several kinds of particles are present.

The Boltzmann equation is irreversible in time in the sense that if $f(\mathbf{r}, \mathbf{v}, t)$ is a solution, then $f(\mathbf{r}, -\mathbf{v}, -t)$ is not a solution. Thus if an isolated system is initially not in equilibrium, it approaches equilibrium as time advances; the time-reversed performance, in which the system departs farther from equilibrium, does not occur. This is paradoxical because actual physical systems are reversible in time when looked at on an atomic scale. From a mathematical point of view it is puzzling that one can begin with the exact equations of motion, reversible in time, and by making reasonable approximations arrive at the irreversible Boltzmann equation. The resolution of this paradox lies in the statistical nature of the Boltzmann equation. It does not describe the behavior of a single system, but the average behavior of a large number of systems.

The Boltzmann equation can be used to calculate the electronic transport properties of metals and semiconductors. For example, if an electric field is applied to a solid, one must solve the Boltzmann equation for the distribution function of the electrons. If the electric field is constant, the distribution function is also constant and is displaced in velocity space in such a way that fewer electrons are moving in the direction of the field than in the opposite direction. This corresponds to a current flow in the direction of the field. *See* FREE-ELECTRON THEORY OF METALS.

With the Boltzmann equation one can also calculate the heat current flowing in a solid as the result of a temperature difference, the constant of proportionality between the heat current per unit area and the temperature gradient being the thermal conductivity. In still more generality, both an electric field and a temperature gradient can be applied. The resulting equations describe thermoelectric phenomena, such as the Peltier and Seebeck effects. Finally, if a constant magnetic field is also applied, it is found that the electrical conductivity usually decreases with increasing magnetic field, a behavior known as magnetoresistance. These equations also describe the Hall effect, as well as more complex thermomagnetic phenomena, such as the Ettingshausen and Nernst effects. *See* CONDUCTION (HEAT); GALVANOMAGNETIC EFFECTS; MAGNETORESISTANCE; THERMOELECTRICITY.

Nonequilibrium properties of atomic or molecular gases such as viscosity, thermal conduction, and diffusion have been treated with the Boltzmann equation. Although many useful results, such as the independence of the viscosity of a gas on pressure, can be obtained by simple approximate methods, the Boltzmann equation must be used in order to obtain quantitatively correct results. *See* DIFFUSION; VISCOSITY.

If one proceeds from a neutral gas to a charged gas or plasma, with the electrons partially removed from the atoms, a number of new phenomena appear. As a consequence of the long-range Coulomb forces between the charges, the plasma can exhibit oscillations in which the free electrons move back and forth with respect to the relatively stationary heavy positive ions at the characteristic frequency known as the plasma frequency. A plasma reflects an electromagnetic wave at a frequency lower than the plasma frequency, but transmits the wave at a higher frequency. This fact explains many characteristics of long-distance radio transmission, made possible by reflection of radio waves by the ionosphere, a low-density plasma. *See* PLASMA (PHYSICS).

If a magnetic field is applied to the plasma, its motion can become complex. In an Alfvén wave, which propagates in the direction of the magnetic field, the magnetic field lines oscillate like stretched strings, while waves that propagate in a direction perpendicular to the magnetic field have quite different properties. The outstanding problem in the attainment of a controlled thermonuclear reaction is to design a magnetic field configuration that can contain an extremely hot plasma long enough to allow nuclear reactions to take place. Plasmas in association with magnetic fields also occur in many astronomical phenomena. *See* Magnetohydrodynamics.

Many properties of plasmas can be calculated by studying the motion of individual particles in electric and magnetic fields, or by using hydrodynamic equations or the Vlasov equation, together with Maxwell's equations. However, subtle properties of plasmas, such as diffusion processes and the damping of waves, can best be understood by starting with the Boltzmann equation or the closely related Fokker-Planck equation. *See* Maxwell's equations.   [R.F.]

Bose-Einstein condensation

When a gas of bosonic particles is cooled below a critical temperature, it condenses into a Bose-Einstein condensate. The condensate consists of a macroscopic number of particles, which are all in the ground state of the system. Bose-Einstein condensation is a phase transition, which does not depend on the specific interactions between particles. It is based on the indistinguishability and wave nature of particles, both of which are at the heart of quantum mechanics.

Basic phenomenon in ideal gas. In a simplified picture, particles in a gas may be regarded as quantum-mechanical wavepackets which have a spatial extent on the order of a thermal de Broglie wavelength, given by Eq. (1), where T is the temperature, m the

$$\lambda_{\mathrm{dB}} = \left(\frac{2\pi^2 \hbar^2}{m k_B T}\right)^{1/2} \tag{1}$$

mass of the particle, k_B is the Boltzmann constant, and $\hbar$ is Planck's constant divided by 2π. The wavelength λ_{dB} can be regarded as the position uncertainty associated with the thermal momentum distribution of the particles. At high temperature, λ_{dB} is small, and the probability of finding two particles within this distance of each other is extremely low. Therefore, the indistinguishability of particles is not important, and a classical description applies (namely, Boltzmann statistics). When the gas is cooled to the point where λ_{dB} is comparable to the distance between particles, the individual wavepackets start to overlap and the indistinguishability of particles becomes crucial— an identity crisis can be said to occur. For fermions, the Pauli exclusion principle prevents two particles from occupying the same quantum state; whereas for bosons, quantum statistics (in this case, Bose-Einstein statistics) dramatically increases the probability of finding several particles in the same quantum state. The system undergoes a phase transition and forms a Bose-Einstein condensate, where a macroscopic number of particles occupy the lowest-energy quantum state (Fig. 1). *See* Boltzmann statistics; Exclusion principle; Phase transitions; Quantum mechanics.

Bose-Einstein condensation can be described intuitively in the following way: When the quantum-mechanical wave functions of bosonic particles spatially overlap, the matter waves start to oscillate in concert. A coherent matter wave forms that comprises all particles in the ground state of the system. This transition from disordered to coherent matter waves can be compared to the step from incoherent light to laser light. Indeed, atom lasers based on Bose-Einstein condensation have been realized. *See* Coherence; Laser.

Experimental techniques. The phenomenon of Bose-Einstein condensation is responsible for the superfluidity of helium and for the superconductivity of an electron gas, which involves Bose-condensed electron pairs. However, these phenomena

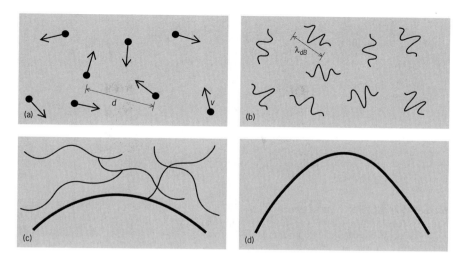

Fig. 1. Criterion for Bose-Einstein condensation in a gas of weakly interacting particles. (*a*) Gas at high temperature, treated as a system of billiard balls, with thermal velocity *v* and density d^{-3}, where *d* is the distance between particles. (*b*) Simplified quantum description of gas at low temperature, in which the particles are regarded as wave packets with a spatial extent of the order of the de Broglie wavelength, λ_{dB}. (*c*) Gas at the transition temperature for Bose-Einstein condensation, when λ_{dB} becomes comparable to *d*. The wave packets overlap and a Bose-Einstein condensate forms (in the case of bosonic particles). (*d*) Pure Bose condensate (giant matter wave), which remains as the temperature approaches absolute zero and the thermal cloud disappears. (*After D. S. Durfee and W. S. Ketterle, Experimental studies of Bose-Einstein condensation, Opt. Express, 2:299–313, Optical Society of America, 1998*)

happen at high density, and their understanding requires a detailed treatment of the interactions. *See* SUPERCONDUCTIVITY; SUPERFLUIDITY.

The quest to realize Bose-Einstein condensation in a dilute weakly interacting gas focused on atomic gases. At ultralow temperatures, all atomic gases liquefy or solidify in thermal equilibrium. Keeping the gas at sufficiently low density can prevent this from occurring. Typical number densities of atoms between 10^{12} and 10^{15} cm^3 imply transition temperatures for Bose-Einstein condensation in the nanokelvin or microkelvin regime.

The realization of Bose-Einstein condensation in atomic gases required techniques to cool gases to such low temperatures, and atom traps to confine the gases at the required

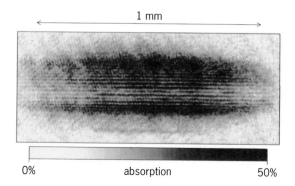

1 mm

0%   absorption   50%

Fig. 2. Interference pattern of two expanding condensates, demonstrating the coherence of Bose-Einstein condensates. This absorption image was observed after a 40-millisecond time of flight. Interference fringes have a spacing of 15 μm. (*After D. S. Durfee and W. S. Ketterle, Experimental studies of Bose-Einstein condensation, Opt. Express, 2:299–313, Optical Society of America, 1998*)

density and keep them away from the much warmer walls of the vacuum chamber. The experiments on alkali vapors (lithium, rubidium, and sodium) use several laser-cooling techniques as precooling, then hold the atoms in a magnetic trap and cool them further by forced evaporative cooling. For atomic hydrogen, the laser-cooling step is replaced by cryogenic cooling.

Macroscopic wave function. In superconductors and liquid helium, the existence of coherence and of a macroscopic wave function is impressively demonstrated through the Josephson effect. In the dilute atomic gases, the coherence has been demonstrated even more directly by interfering two Bose condensates (Fig. 2). The interference fringes typically have a spacing of 15 μm, a huge length for matter waves. (In contrast, the matter wavelength of atoms at room temperature is only 0.05 nm, less than the size of the atoms.)
[W.Ket.]

Bose-Einstein statistics   The statistical description of quantum mechanical systems in which there is no restriction on the way in which particles can be distributed over the individual energy levels. This description applies when the system has a symmetric wave function. This in turn has to be the case when the particles described are of integer spin.

Suppose one describes a system by giving the number of particles n_i in an energy state ϵ_i, where the n_i are called occupation numbers and the index i labels the various states. The energy level ϵ_i, is of finite width, being really a range of energies comprising, say, g_i, individual (nondegenerate) quantum levels. If any arrangement of particles over individual energy levels is allowed, one obtains for the probability of a specific distribution Eq. (1a). In Boltzmann statistics, this same probability would be written as Eq. (1b).

$$W = \prod_i \frac{(n_i + g_i - 1)!}{n_i!(g_i - 1)!} \tag{1a}$$

$$W = \prod_i \frac{g_i^{n_i}}{n_i!} \tag{1b}$$

See BOLTZMANN STATISTICS.

The equilibrium state is defined as the most probable state of the system. To obtain it, one must maximize Eq. (1a) under the conditions given by Eqs. (2a) and (2b), which

$$\sum n_i i = N \tag{2a}$$

$$\sum \epsilon_i n_i = E \tag{2b}$$

express the fact that the total number of particles N and the total energy E are fixed. One finds for the most probable distribution that Eq. (3) holds. Here, A and ß are

$$ni = \frac{gi}{\frac{1}{A}e^{\beta\epsilon_i} + 1} \tag{3}$$

parameters to be determined from Eqs. (2a) and (2b); actually, ß $= 1/kT$, where k is the Boltzmann constant and T is the absolute temperature.

An interesting and important result emerges when Eq. (3) is applied to a gas of photons, that is, a large number of photons in an enclosure. (Since photons have integer spin, this is legitimate.) The formula for the energy density (energy per unit volume) in a given frequency range is found to be the celebrated Planck radiation formula for blackbody radiation. Thus, black-body radiation must be considered as a photon gas, with the photons satisfying Bose-Einstein statistics. *See* HEAT RADIATION.

If an ideal Bose gas, consisting of a fixed number of material particles, is compressed beyond a certain point, some of the particles will condense in a zero state, where they do not contribute to the density or the pressure. If the volume is decreased, this curious condensation phenomenon results, yielding the zero state which has the paradoxical properties of not contributing to the pressure, volume, or density. There is now considerable evidence that many of the superfluid properties exhibited by liquid helium are in fact manifestations of an Einstein condensation. *See* FERMI-DIRAC STATISTICS; QUANTUM STATISTICS; STATISTICAL MECHANICS.

[M.Dr.]

Boundary-layer flow
That portion of a fluid flow, near a solid surface, where shear stresses are significant and the inviscid-flow assumption may not be used. All solid surfaces interact with a viscous fluid flow because of the no-slip condition, a physical requirement that the fluid and solid have equal velocities at their interface. Thus a fluid flow is retarded by a fixed solid surface, and a finite, slow-moving boundary layer is formed. A requirement for the boundary layer to be thin is that the Reynolds number of the body be large, 10^3 or more. Under these conditions the flow outside the boundary layer is essentially inviscid and plays the role of a driving mechanism for the layer. *See* REYNOLDS NUMBER.

A typical low-speed or laminar boundary layer is shown in the illustration. Such a display of the streamwise flow vector variation near the wall is called a velocity profile. The no-slip condition requires that $u(x, 0) = 0$, as shown, where u is the velocity of flow in the boundary layer. The velocity rises monotonically with distance y from the wall, finally merging smoothly with the outer (inviscid) stream velocity $U(x)$. At any point in the boundary layer, the fluid shear stress τ is proportional to the local velocity gradient, assuming a newtonian fluid. The value of the shear stress at the wall is most important, since it relates not only to the drag of the body but often also to its heat transfer. At the edge of the boundary layer, τ approaches zero asymptotically. There is no exact spot where $\tau = 0$; therefore the thickness δ of a boundary layer is usually defined arbitrarily as the point where $u = 0.99U$. *See* LAMINAR FLOW.

When a flow enters a duct or confined region, boundary layers immediately begin to grow on the duct walls. An inviscid core accelerates down the duct center, but soon vanishes as the boundary layers meet and fill the duct with viscous flow. Constrained

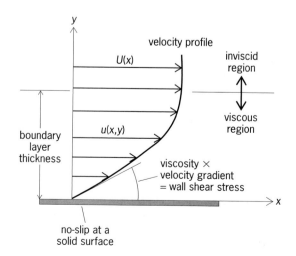

Typical laminar boundary-layer velocity profile.

by the duct walls into a no-growth condition, the velocity profile settles into a fully developed shape which is independent of the streamwise coordinate. The pressure drops linearly downstream, balanced by the mean wall-shear stress. This is a classic and simple case of boundary-layer flow which is well documented by both theory and experiment.

A classic incompressible boundary-layer flow is a uniform stream at velocity U, moving past a sharp flat plate parallel to the stream. In the Reynolds number range 1×10^3 to 5×10^5, the flow is laminar and orderly, with no superimposed fluctuations. The boundary-layer thickness δ grows monotonically with x, and the shape of the velocity profile is independent of x when normalized. The profiles are said to be similar, and they are called Blasius profiles.

The Blasius flat-plate flow results in closed-form algebraic formulas for such parameters as wall-shear stress and boundary-layer thickness as well as for temperature and heat-transfer parameters. These results are useful in estimating viscous effects in flow past thin bodies such as airfoils, turbine blades, and heat-exchanger plates.

The flat plate is very distinctive in that it causes no change in outer-stream velocity U. Most body shapes immersed in a stream flow, such as cylinders, airfoils, or ships, induce a variable outer stream $U(x)$ near the surface. If U increases with x, which means that pressure decreases with x, the boundary layer is said to be in a favorable gradient and remains thin and attached to the surface. If, however, velocity falls and pressure rises with x, the pressure gradient is unfavorable or adverse. The low-velocity fluid near the wall is strongly decelerated by the rising pressure, and the wall-shear stress drops off to zero. Downstream of this zero-shear or separation point, there is backflow and the wall shear is upstream. The boundary layer thickens markedly to conserve mass, and the outer stream separates from the body, leaving a broad, low-pressure wake downstream. Flow separation may be predicted by boundary-layer theory, but the theory is not able to estimate the wake properties accurately.

In most immersed-body flows, the separation and wake occur on the rear or lee side of the body, with higher pressure and no separation on the front. The body thus experiences a large downstream pressure force called pressure drag. This happens to all blunt bodies such as spheres and cylinders and also to airfoils and turbomachinery blades if their angle of attack with respect to the oncoming stream is too large. The airfoil or blade is said to be stalled, and its performance suffers.

All laminar boundary layers, if they grow thick enough and have sufficient velocity, become unstable. Slight disturbances, whether naturally occurring or imposed artificially, tend to grow in amplitude, at least in a certain frequency and wavelength range. The growth begins as a selective group of two-dimensional periodic disturbances, called Tollmien-Schlichting waves, which become three-dimensional and nonlinear downstream and eventually burst into the strong random fluctuations called turbulence. The critical parameter is the Reynolds number. The process of change from laminar to turbulent flow is called transition.

The turbulent flow regime is characterized by random, three-dimensional fluctuations superimposed upon time-mean fluid properties, including velocity, pressure, and temperature. The fluctuations are typically 3–6% of the mean values and range in size over three orders of magnitude, from microscale movements to large eddies of size comparable to the boundary-layer thickness. They are readily measured by modern instruments such as hot wires and laser-Doppler velocimeters. *See* ANEMOMETER.

The effect of superimposing a wide spectrum of eddies on a viscous flow is to greatly increase mixing and transport of mass, momentum, and heat across the flow. Turbulent boundary layers are thicker than laminar layers and have higher heat transfer and friction. The turbulent mean-velocity profile is rather flat, with a steep gradient at the wall. The edge of the boundary layer is a ragged, fluctuating interface which separates the nonturbulent outer flow from large turbulent eddies in the layer. The thickness of such a layer is defined only in the time mean, and a probe placed in the outer half of the layer would show intermittently turbulent and nonturbulent flow.

As the stream velocity U becomes larger, its kinetic energy, $U^2/2$, becomes comparable to stream enthalpy, $c_p T$, where c_p is the specific heat at constant pressure and T is the absolute temperature. Changes in temperature and density begin to be important, and the flow can no longer be considered incompressible. Liquids flow at very small Mach numbers, and compressible flows are primarily gas flows. *See* MACH NUMBER.

In a flow with supersonic stream velocity, the no-slip condition is still valid, and much of the boundary-layer flow near the wall is at low speed or subsonic. The fluid enters the boundary layer and loses much of its kinetic energy, of which a small part is conducted away although most is converted into thermal energy. Thus the near-wall region of a highly compressible boundary layer is very hot, even if the wall is cold and is drawing heat away. The basic difference between low and high speed is the conversion of kinetic energy into higher temperatures across the entire boundary layer.

In a low-speed (incompressible) boundary layer, a cold wall simply means that the wall temperature is less than the free-stream temperature. The heat flow is from high toward lower temperature, that is, into the wall. For a low-speed insulated wall, the boundary-layer temperature is uniform. For a high-speed flow, however, an insulated wall has a high surface temperature because of the viscous dissipation energy exchange in the layer.

Except for the added complexity of having to consider fluid pressure, temperature, and density as coupled variables, compressible boundary layers have similar characteristics to their low-speed counterparts. They undergo transition from laminar to turbulent flow but typically at somewhat higher Reynolds numbers. Compressible layers tend to be somewhat thicker than incompressible boundary layers, with proportionally smaller wall-shear stresses. They tend to resist flow separation slightly better than incompressible flows.

In a supersonic outer stream, shock waves can always occur. Shocks may form in the boundary layer because of obstacles in the layer or downstream, or they may be formed elsewhere and impinge upon a boundary. In either case, the pressure rises sharply behind the shock, an adverse gradient, and this tends to cause early transition to turbulence and early flow separation. Special care must be taken to design aerodynamic surfaces to accommodate or avoid shockwave formation in transonic and supersonic flows. *See* COMPRESSIBLE FLOW.

As boundary layers move downstream, they tend to grow naturally and undergo transition to turbulence. Boundary layers encountering rising pressure undergo flow separation. Both phenomena can be controlled at least partially. Airfoils and hydrofoils can be shaped to delay adverse pressure gradients and thus move separation downstream. Proper shaping can also delay transition. Wall suction removes the low-momentum fluid and delays both transition and separation. Wall blowing into the boundary layer, from downward-facing slots, delays separation but not transition. Changing the wall temperature to hotter for liquids and colder for gases delays transition. Practical systems have been designed for boundary-layer control, but they are often expensive and mechanically complex. *See* AIRFOIL; STREAMLINING; VISCOSITY.   [F.M.Wh.]

Boyle's law   A law of gases which states that at constant temperature the volume of a gas varies inversely with its pressure. This law, formulated by Robert Boyle (1627–1691), can also be stated thus: The product of the volume of a gas times the pressure exerted on it is a constant at a fixed temperature. The relation is approximately true for most gases, but is not followed at high pressure. The phenomenon was discovered independently by Edme Mariotte about 1650 and is known in Europe as Mariotte's law. *See* KINETIC THEORY OF MATTER.   [F.H.R.]

Breakdown potential   The potential difference at which an electrically stressed gas is transformed from an insulator to a conductor. In an electrically stressed gas, as the voltage is increased, the free electrons present in the gas gain energy from the electric field. When the applied voltage is increased to such a level that an appreciable number of these electrons are energetically capable of ionizing the gas, the gas makes the transition from an insulator to a conductor; that is, it breaks down. The potential difference at which this transition occurs is known as the breakdown potential for the particular gaseous medium.

The breakdown potential depends on the nature, number density, and temperature of the gas; on the material, state, and geometry of the electrodes; on the type of voltage applied (steady, alternating, impulsive); and on the degree of preexisting ionization. Areas of surface roughness at the electrodes (especially the cathode) or the presence of conducting particles in the gas greatly reduces the breakdown potential because at such points the electric field is significantly enhanced, increasing the electron energies and thus gas ionization. The breakdown voltage varies considerably from one gaseous medium to another; it is very low for the rare gases, and very high for polyatomic, especially electronegative, gases such as sulfur hexafluoride (SF_6).

The transition of a gas from an insulator to a conductor under an imposed electrical potential occurs in times ranging from milliseconds to nanoseconds, depending on the form of the applied field and the gas density. This transition depends on the behavior of electrons, ions, and photons in the gas, especially the processes which produce or deplete free electrons. Knowledge of these processes often allows prediction of the breakdown voltage of gases and the tailoring of gas mixtures which can withstand high electrical potentials for practical uses. *See* ELECTRICAL BREAKDOWN; ELECTRICAL CONDUCTION IN GASES.

The systematic development of gaseous dielectrics with high dielectric strength (that is, high breakdown potential) is most significant for high-voltage technology, which has a multiplicity of gas insulation needs. Dielectric gases are widely used as insulating media in high-voltage transmission lines, circuit breakers, transformers, substations, high-voltage research apparatus, and other electrical equipment. *See* DIELECTRIC MATERIALS.
[L.G.C.]

Bremsstrahlung   In a narrow sense, the electromagnetic radiation emitted by electrons when they pass through matter. Charged particles radiate when accelerated, and in this case the electric fields of the atomic nuclei provide the force which accelerates the electrons. The continuous spectrum of x-rays from an x-ray tube is that of the bremsstrahlung; in addition, there is a characteristic x-ray spectrum due to excitation of the target atoms by the incident electron beam. The major energy loss of high-energy (relativistic) electrons (energy greater than about 10 MeV, depending somewhat upon material) occurs from the emission of bremsstrahlung, and this is the major source of gamma rays in a high-energy cosmic-ray shower. *See* ELECTROMAGNETIC RADIATION.

In a broader sense, bremsstrahlung is the radiation emitted when any charged particle is accelerated by any force. To a great extent, as a source of photons in the ultraviolet and soft x-ray region for the investigation of atomic structure (particularly in solids), bremsstrahlung from x-ray tubes has been replaced by synchrotron radiation.

Synchrotron radiation is an analog to bremsstrahlung, differing in that the force which accelerates the electron is a macroscopic (large-scale) magnetic field. [C.G.]

Bridge circuit   A circuit composed of a source and four impedances that is used in the measurement of a wide range of physical quantities. The bridge circuit is useful in measuring impedances (resistors, capacitors, and inductors) and in converting signals from transducers to related voltage or current signals. *See* CAPACITOR; INDUCTOR; RESISTOR.

The bridge impedances Z_1, Z_2, Z_3, Z_4, shown in the illustration may be single impedances (resistor, capacitor, or inductor), combinations of impedances, or a transducer with varying impedance. For example, strain gages are resistive transducers whose resistance changes when they are deformed.

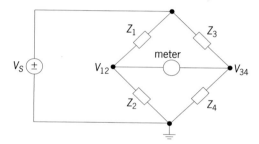

Bridge circuit with source and impedances.

Bridge circuits are often used with transducers to convert physical quantities (temperature, displacement, pressure) to electrical quantities (voltage and current). High-accuracy voltmeters and ammeters are relatively inexpensive, and the voltage form of a signal is usually most convenient for information display, control decisions, and data storage. Another important advantage of the bridge circuit is that it provides greater measurement sensitivity than the transducer.

The bridge circuit is balanced when the output read by the meter is zero. In this condition the voltages on both sides of the meter are identical. The bridge is used in two forms. The null adjustment method requires adjustment of a calibrated impedance to balance it. In this case the meter is usually a highly sensitive current-measuring galvanometer. The null adjustment method is often used to measure impedances, with the output read from a dial attached to the adjustable impedance. The deflection method requires on accurate meter in the bridge to measure the deviation from the balance condition. The deviation is proportional to the quantity being measured.

There are many special forms of the bridge circuit. When all of the impedances are resistive, it is commonly called a Wheatstone bridge. Other common forms use a current source in place of the voltage source, a sinusoidal source in place of a constant (dc) source, or branch impedances which are specific combinations of single passive impedances. The bridge circuit is also used in a variety of electrical applications varying from oscillators to instrumentation amplifier circuits for extremely accurate measurements. *See* WHEATSTONE BRIDGE. [K.D.P.]

Brillouin zone   In the propagation of any type of wave motion through a crystal lattice, the frequency is a periodic function of wave vector **k**. This function may be complicated by being multivalued; that is, it may have more than one branch. Discontinuities may also occur. In order to simplify the treatment of wave motion in a crystal, a zone in **k**-space is defined which forms the fundamental periodic region, such that the frequency or energy for a **k** outside this region may be determined from one of those in it. This region is known as the Brillouin zone (sometimes called the first or the central Brillouin zone). It is usually possible to restrict attention to **k** values inside the zone. Discontinuities occur only on the boundaries. If the zone is repeated indefinitely, all **k**-space will be filled. Sometimes it is also convenient to define larger figures with similar properties which are combinations of the first zone and portions of those formed by replication. These are referred to as higher Brillouin zones.

The central Brillouin zone for a particular solid type is a solid which has the same volume as the primitive unit cell in reciprocal space, that is, the space of the reciprocal lattice vectors, and is of such a shape as to be invariant under as many as possible of the symmetry operations of the crystal. *See* CRYSTALLOGRAPHY.   [J.C.]

Brownian movement   The irregular motion of a body arising from the thermal motion of the molecules of the material in which the body is immersed. Such a body will of course suffer many collisions with the molecules, which will impart energy and momentum to it. Because, however, there will be fluctuations in the magnitude and direction of the average momentum transferred, the motion of the body will appear irregular and erratic.

In principle, this motion exists for any foreign body suspended in gases, liquids, or solids. To observe it, one needs first of all a macroscopically visible body; however, the mass of the body cannot be too large. For a large mass, the velocity becomes small. *See* KINETIC THEORY OF MATTER.   [M.Dr.]

Buoyancy   The resultant vertical force exerted on a body by a static fluid in which it is submerged or floating. The buoyant force F_B acts vertically upward, in opposition to the gravitational force that causes it. Its magnitude is equal to the weight of fluid displaced, and its line of action is through the centroid of the displaced volume, which is known as the center of buoyancy. *See* AEROSTATICS; HYDROSTATICS.

By weighing an object when it is suspended in two different fluids of known specific weight, the volume and weight of the solid may be determined. *See* ARCHIMEDES' PRINCIPLE.

Another form of buoyancy, called horizontal buoyancy, is experienced by models tested in wind or water tunnels. Horizontal buoyancy results from variations in static pressure along the test section, producing a drag in closed test sections and a thrust force in open sections. These extraneous forces must be subtracted from data as a boundary correction. Wind tunnel test sections usually diverge slightly in a downstream direction to provide some correction for horizontal buoyancy.

A body floating on a static fluid has vertical stability. A small upward displacement decreases the volume of fluid displaced, hence decreasing the buoyant force and leaving an unbalanced force tending to return the body to its original position. Similarly, a small downward displacement results in a greater buoyant force, which causes an unbalanced upward force.

A body has rotational stability when a small angular displacement sets up a restoring couple that tends to return the body to its original position. When the center of gravity of the floating body is lower than its center of buoyancy, it will always have rotational stability. Many a floating body, such as a ship, has its center of gravity above its center of buoyancy. Whether such an object is rotationally stable depends upon the shape of the body.   [V.L.S.]

C

Candlepower   Luminous intensity expressed in candelas. The term refers only to the intensity in a particular direction and by itself does not give an indication of the total light emitted. The candlepower in a given direction from a light source is equal to the illumination in footcandles falling on a surface normal to that direction, multiplied by the square of the distance from the light source in feet. The candlepower is also equal to the illumination of metercandles (lux) multiplied by the square of the distance in meters.

The apparent candlepower is the candlepower of a point source which will produce the same illumination at a given distance as produced by a given light source.

The mean horizontal candlepower is the average candlepower of a light source in the horizontal plane passing through the luminous center of the light source.

The mean spherical candlepower is the average candlepower in all directions from a light source as a center. Since there is a total solid angle of 4π (steradians) emanating from a point, the mean spherical candlepower is equal to the total luminous flux (in lumens) of a light source divided by 4π (steradians). See LUMINOUS INTENSITY; PHOTOMETRY.   [R.C.Pu.]

Canonical transformations   Transformations among the coordinates and momenta describing the state of a classical dynamical system which leave the canonical or Hamiltonian form of the equations of motion unchanged. See HAMILTON'S EQUATIONS OF MOTION; HAMILTON'S PRINCIPLE.   [P.M.S.]

Capacitance   The ratio of the charge q on one of the plates of a capacitor (there being an equal and opposite charge on the other plate) to the potential difference v between the plates; that is, capacitance (formerly called capacity) is $C = q/v$.

In general, a capacitor, often called a condenser, consists of two metal plates insulated from each other by a dielectric. The capacitance of a capacitor depends on the geometry of the plates and the kind of dielectric used, since these factors determine the charge which can be put on the plates by a unit potential difference existing between the plates.

In an ideal capacitor, no conduction current flows between the plates. A real capacitor of good quality is the circuit equivalent of an ideal capacitor with a very high resistance in parallel or, in alternating-current (ac) circuits, of an ideal capacitor with a low resistance in series. See CAPACITOR; DIELECTRIC MATERIALS.   [R.P.Wi.]

Capacitance measurement   The measurement of the ratio of the charge induced on a conductor to the change in potential with respect to a neighboring conductor which induces the charge. In a multiconductor system there are capacitances between each pair of conductors. In general, these capacitances are functions of the total geometry, that is, the location of all of the conducting and dielectric bodies. When, as is usually true, only the capacitance between two conductors is of interest, the presence of other conductors is an undesirable complication. It is then customary to distinguish between two-terminal and three-terminal capacitors and capacitance measurements. In a two-terminal capacitor, either one of the conductors of primary interest surrounds the other (in which case the capacitance between them is independent of the location of other bodies except in the vicinity of the terminals); or the somewhat indefinite contributions of the other conductors to the capacitance of interest are accepted.

A three-terminal capacitor consists of two active electrodes surrounded by a third, or shield, conductor. The direct capacitance between the two active electrodes is the capacitance of interest, and, when shielded leads are used, it is independent of the location of all other conductors except the shield.

Every physically realizable capacitor has associated loss in the dielectric and in the metal electrodes. At a single frequency these are indistinguishable, and the capacitor may be represented by either a parallel or series combination of pure capacitance and pure resistance. The measurement of capacitance, then, in general involves the simultaneous measurement of, or allowance for, an associated resistive element. *See* PERMITTIVITY.

Most capacitance measurements involve simply a comparison of the capacitor to be measured with a capacitor of known value. Methods which permit comparison of essentially equal capacitors by simple substitution of one for the other at the same point in a circuit are frequently possible and almost always preferable.

Bridge comparison methods. When capacitors must be compared with high accuracy, bridge methods must be adopted. *See* BRIDGE CIRCUIT; WHEATSTONE BRIDGE.

Resistance-ratio bridges are Wheatstone-bridge configurations in which the potential division of the capacitor being measured and either a parallel combination of a standard loss-free capacitor C_s and a conductance G_s or a series combination of C_s and a resistor R_s is equated, when the detector is nulled, to the ratio of potentials across resistors R_1 and R_2. More commonly now, the reference potential division is that of a variable-ratio autotransformer known as an inductive voltage divider (IVD). *See* INDUCTIVE VOLTAGE DIVIDER.

The Schering bridge yields a measurement of the equivalent series-circuit representation of a capacitor.

The resistance-ratio and Schering bridges are useful for two-terminal capacitance measurements. Their use may be extended to three-terminal measurements and extended in accuracy and range by the introduction of shielding and the addition of the Wagner branch.

Time-constant methods. If a direct voltage is suddenly applied to the series combination of a resistor and an initially discharged capacitor, the charge and the voltage on the capacitor increase exponentially toward their full magnitudes with a time constant equal in seconds to the product of the resistance in ohms and the capacitance in farads. Similarly, when a charged capacitor is discharged through a resistor, the charge and the voltage decay with the same time constant. Various methods are available for the measurement of capacitance by measurement of the time constant of charge or discharge through a known resistor. *See* TIME CONSTANT.

In one such method the time required for the output voltage of an operational amplifier having a capacitor as a feedback component to increase to a value equal to the step-function input voltage applied through a resistor to its input is determined by an electronic voltage-comparison circuit and timer. With the assumption of ideal characteristics for the amplifier, such as infinite gain without feedback, infinite input impedance, and zero output impedance, the measured time interval is equal to the product of the values of the known resistance and the capacitance being measured.   [B.P.K.; F.R.Ko.; G.H.Ra.]

Capacitor   An electrical device capable of storing electrical energy. In general, a capacitor consists of two metal plates insulated from each other by a dielectric. The capacitance of a capacitor depends primarily upon its shape and size and upon the relative permittivity ϵ_r of the medium between the plates. In vacuum, in air, and in most gases, ϵ_r ranges from one to several hundred. *See* CAPACITANCE; PERMITTIVITY.

One classification of capacitors comes from the physical state of their dielectrics, which may be gas (or vacuum), liquid, solid, or a combination of these. Each of these classifications may be subdivided according to the specific dielectric used. Capacitors may be further classified by their ability to be used in alternating-current (ac) or direct-current (dc) circuits with various current levels.

Capacitors are also classified as fixed, adjustable, or variable. The capacitance of fixed capacitors remains unchanged, except for small variations caused by temperature fluctuations. The capacitance of adjustable capacitors may be set at any one of several discrete values. The capacitance of variable capacitors may be adjusted continuously and set at any value between minimum and maximum limits fixed by construction. Trimmer capacitors are relatively small variable capacitors used in parallel with larger variable or fixed capacitors to permit exact adjustment of the capacitance of the parallel combination.

Made in both fixed and variable types, air, gas, and vacuum capacitors are constructed with flat parallel metallic plates (or cylindrical concentric metallic plates) with air, gas, or vacuum as the dielectric between plates. Alternate plates are connected, with one or both sets supported by means of a solid insulating material such as glass, quartz, ceramic, or plastic. Gas capacitors are similarly built but are enclosed in a leakproof case. Vacuum capacitors are of concentric-cylindrical construction and are enclosed in highly evacuated glass envelopes.

The purpose of a high vacuum, or a gas under pressure, is to increase the voltage breakdown value for a given plate spacing. For high-voltage applications, when increasing the spacing between plates is undesirable, the breakdown voltage of air capacitors may be increased by rounding the edges of the plates. Air, gas, and vacuum capacitors are used in high-frequency circuits. Fixed and variable air capacitors incorporating special design are used as standards in electrical measurements. *See* CAPACITANCE MEASUREMENT; ELECTRICAL UNITS AND STANDARDS.

Solid-dielectric capacitors use one of several dieletrics such as a ceramic, mica, glass, or plastic film. Alternate plates of metal, or metallic foil, are stacked with the dielectric, or the dielectric may be metal-plated on both sides.

A large capacitance-to-volume ratio and a low cost per microfarad of capacitance are chief advantages of electrolytic capacitors. These use aluminum or tantalum plates. A paste electrolyte is placed between the plates, and a dc forming voltage is applied. A current flows and by a process of electrolysis builds up a molecule-thin layer of oxide bubbles on the positive plate. This serves as the dieletric. The rest of the electrolyte and the other plate make up the negative electrode. Such a device is said to be polarized and must be connected in a circuit with the proper polarity. Polarized capacitors can be used only in circuits in which the dc component of voltage across the capacitors exceeds the crest value of the ac ripple.

Another type of electrolytic capacitor utilizes compressed tantalum powder and the baking of manganese oxide (MnO_2) as an electrolyte. Nonpolarized electrolytic capacitors can be constructed for use in ac circuits. In effect, they are two polarized capacitors placed in series with their polarities reversed.

Thick-film capacitors are made by means of successive screen-printing and firing processes in the fabrication of certain types of microcircuits used in electronic computers and other electronic systems. They are formed, together with their connecting conductors and associated thick-film resistors, upon a ceramic substrate. Their characteristics and the materials are similar to those of ceramic capacitors.

Thin-film dielectrics are deposited on ceramic and integrated-circuit substrates and then metallized with aluminum to form capacitive components. These are usually single-layer capacitors. The most common dielectrics are silicon nitride and silicon dioxide. [A.Mot.]

Cathode rays   The name given to the electrons originating at the cathodes of gaseous discharge devices. The term has now been extended to include low-pressure devices such as cathode-ray tubes. Furthermore, cathode rays are now used to designate electron beams originating from thermionic cathodes, whereas the term was formerly applied only to cold-cathode devices. [G.H.M.]

Cathodoluminescence   A luminescence resulting from the bombardment of a substance with an electron (cathode-ray) beam. The principal applications of cathodoluminescence are in television, computer, radar, and oscilloscope displays. In these a thin layer of luminescent powder (phosphor) is evenly deposited on the transparent glass faceplate of a cathode-ray tube. After undergoing acceleration, focusing, and deflection by various electrodes in the tube, the electron beam originating in the cathode impinges on the phosphor. The resulting emission of light is observed through the glass faceplate, that is, from the unbombarded side of the phosphor coating.

The luminescence of most phosphors comes from a few sites (activator centers) occupied by selected chemical impurities which have been incorporated into the matrix or host solid. Because of the complex mode of interaction of cathode rays with phosphors, the energy efficiency of light production by cathodoluminescence is lower than the best efficiencies obtainable with photoluminescence. Conversion efficiencies of currently used display phosphors are between 2 and 23%. *See* LUMINESCENCE. [H.N.H.; J.S.H.]

Causality   In physics, the requirement that interactions in any space-time region can influence the evolution of the system only at subsequent times; that is, past events are causes of future events, and future events can never be the causes of events in the past. Causality thus depends on time orientability, the possibility of distinguishing past from future. Not all spacetimes are orientable.

The laws of a deterministic theory (for example, classical mechanics) are such that the state of a closed system (for example, the positions and momenta of particles in the system) at one instant determines the state of that system at any future time. Deterministic causality does not necessarily imply practical predictability. It was long implicitly assumed that slight differences in initial conditions would not lead to rapid divergence of later behavior, so that predictability was a consequence of determinism. Behavior in which two particles starting at slightly different positions and velocities diverge rapidly is called chaotic. Such behavior is ubiquitous in nature, and can lead to the practical impossibility of prediction of future states despite the deterministic character of the physical laws. *See* CHAOS.

Quantum mechanics is deterministic in the sense that, given the state of a system at one instant, it is possible to calculate later states. However, the situation differs from that in classical mechanics in two fundamental respects. First, conjugate variables, for example, position x and momentum p, cannot be simultaneously determined with complete precision. Second, the state variable ψ gives only probabilities that a given eigenstate will be found after the performance of a measurement, and such probabilities are also all that is calculable about a later state ψ' by the deterministic prediction. Despite its probabilistic character, the quantum state still evolves deterministically. However, which eigenvalue (say, of position) will actually be found in a measurement is unpredictable. *See* DETERMINISM; EIGENVALUE (QUANTUM MECHANICS); QUANTUM MECHANICS; QUANTUM THEORY OF MEASUREMENT; UNCERTAINTY PRINCIPLE.

Nonrelativistic mechanics assumes that causal action can be propagated instantaneously, and thus that an absolute simultaneity is definable. This is not true in special relativity. While the state of a system can still be understood in terms of the positions and momenta of its particles, time order, as well as temporal and spatial length, becomes relative to the observer's frame, and there is no possible choice of simultaneous events in the universe that is the same in all reference frames. Only space-time intervals in a fused "spacetime" are invariant with respect to choice of reference frame. The theory of special relativity thus rejects the possibility of instantaneous causal action. Instead, the existence of a maximum velocity of signal transmission determines which events can causally influence others and which cannot. The investigation of a spacetime with regard to which events can causally influence (signal) other regions and which cannot is known as the study of the causal structure of the spacetime. *See* SPACE-TIME.

[D.Sha.]

Cavitation   The formation of vapor- or gas-filled cavities in liquids. If understood in this broad sense, cavitation includes the familiar phenomenon of bubble formation when water is brought to a boil under constant pressure and the effervescence of champagne wines and carbonated soft drinks due to the diffusion of dissolved gases. In engineering terminology, the term cavitation is used in a narrower sense, namely, to describe the formation of vapor-filled cavities in the interior or on the solid boundaries created by a localized pressure reduction produced by the dynamic action of a liquid system without change in ambient temperature. Cavitation in the engineering sense is characterized by an explosive growth and occurs at suitable combinations of low pressure and high speed in pipelines; in hydraulic machines such as turbines, pumps, and propellers; on submerged hydrofoils; behind blunt submerged bodies; and in the cores of vortical structures. This type of cavitation has great practical significance because it restricts the speed at which hydraulic machines may be operated and, when severe, lowers efficiency, produces noise and vibrations, and causes rapid erosion of the boundary surfaces, even though these surfaces consist of concrete, cast iron, bronze, or other hard and normally durable material.

Acoustic cavitation occurs whenever a liquid is subjected to sufficiently intense sound or ultrasound (that is, sound with frequencies of roughly 20 kHz to 10 MHz). When sound passes through a liquid, it consists of expansion (negative-pressure) waves and compression (positive-pressure) waves. If the intensity of the sound field is high enough, it can cause the formation, growth, and rapid recompression of vapor bubbles in the liquid. The implosive bubble collapse generates localized heating, a pressure pulse, and associated high-energy chemistry. *See* SOUND; ULTRASONICS.

Both experiments and calculations show that with ordinary flowing water cavitation commences as the pressure approaches or reaches the vapor pressure, because of impurities in the water. These impurities, called cavitation nuclei, cause weak spots in the liquid and thus prevent it from supporting higher tensions. The exact mechanism of bubble growth is generally described by mathematical relationships which depend upon the cavitation nuclei. Cavitation commences when these nuclei enter a low-pressure region where the equilibrium between the various forces acting on the nuclei surface cannot be established. As a result, bubbles appear at discrete spots in low-pressure regions, grow quickly to relatively large size, and suddenly collapse as they are swept into regions of higher pressure.   [M.L.Bi.]

Cayley-Klein parameters   A set of four complex numbers used to specify the orientation of a body, or equivalently, the rotation R which produces that orientation, starting from some reference orientation. They can be expressed in terms of the Euler angles ψ, θ, and ϕ, as in the equations below.

$$\alpha = \cos\frac{\theta}{2}\, e^{-i(\psi-\phi)/2} \qquad \beta = -i\sin\frac{\theta}{2}\, e^{i(\psi-\phi)/2}$$

$$\gamma = -i\sin\frac{\theta}{2}\, e^{-i(\psi-\phi)/2} \qquad \delta = \cos\frac{\theta}{2}\, e^{i(\psi-\phi)/2}$$

Although these parameters have been used to simplify somewhat the mathematics of spinning top motion, their main use is in quantum mechanics. There they are related to the Pauli spin matrices and represent the change in the spin state of an electron or other particle of half-integer spin under the space rotation R (ψ, θ, ϕ). See SPIN (QUANTUM MECHANICS).   [B.G.]

Center of gravity   A fixed point in a material body through which the resultant force of gravitational attraction acts. The resultant of all forces or attractions produced by the Earth's gravity on a body constitutes its weight. This weight is considered to be concentrated at the center of gravity in mechanical studies of a rigid body. The location of the center of gravity for a body remains fixed in relation to the body regardless of the orientation of the body. If supported at its center of gravity, a body would remain balanced in its initial position. See GRAVITY; RESULTANT OF FORCES.   [N.S.F.]

Center of mass   That point of a material body or system of bodies which moves as though the system's total mass existed at the point and all external forces were applied at the point. The Earth-Moon system moves in the Sun's gravitational field as though both masses were located at a center of mass some 3000 mi (4700 km) from the Earth's geometric center. The function of the center-of-mass concept is to permit analysis of the motion of an entire system as distinguished from that of its individual parts.

Consider a system of mass M composed of n bodies with masses $m_1, m_2, \ldots, m_n$, and radius vectors $r_1, r_2, \ldots, r_n$ measured from some common reference point. Define a point with radius vector R, such that Eq. (1) holds. Then, it is possible to derive Eq. (2), an expression of Newton's second law, which states that the center of mass at

$$MR = \sum_j m_j r_j \qquad (1)$$

$$\frac{d^2 R}{dt^2} = \frac{F}{M} \qquad (2)$$

R moves as though it possessed the total mass of the system and were acted upon by the total external force.

A simplification of the description of collisions can be obtained by using a coordinate system which moves with the velocity of the center of mass before collision. See COLLISION (PHYSICS); RIGID-BODY DYNAMICS.   [J.P.H.]

Center of pressure   A point on a plane surface through which the resultant force due to pressure passes. Such a surface can be supported by a single mounting fixture at its center of pressure if no other forces act. For example, a water gate in a dam can be supported by a single shaft at its center of pressure. *See* RESULTANT OF FORCES.

[N.S.F.]

Central force   A force whose line of action is always directed toward a fixed point. The central force may attract or repel. The point toward or from which the force acts is called the center of force. If the central force attracts a material particle, the path of the particle is a curve concave toward the center of force; if the central force repels the particle, its orbit is convex to the center of force. Undisturbed orbital motion under the influence of a central force satisfies Kepler's law of areas.   [R.L.Du.]

Centrifugal force   A fictitious or pseudo outward force on a particle rotating about an axis which by Newton's third law is equal and opposite to the centripetal force. Like all such action-reaction pairs of forces, they are equal and opposite but do not act on the same body and so do not cancel each other. Consider a mass M tied by a string of length R to a pin at the center of a smooth horizontal table and whirling around the pin with an angular velocity of ω radians per second. The mass rotates in a circular path because of the centripetal force $F_C = M\omega^2R$ which is exerted on the mass by the string. The reaction force exerted by the rotating mass M, the so-called centrifugal force, is $M\omega^2R$ in a direction away from the center of rotation. *See* CENTRIPETAL FORCE.

From another point of view, consider an experimenter in a windowless, circular laboratory that is rotating smoothly about a centrally located vetical axis. No object remains at rest on a smooth surface; all such objects move outward toward the wall of the laboratory as though an outward, centrifugal force were acting. To the experimenter partaking in the rotation, in a rotating frame of reference, the centrifugal force is real. An outside observer would realize that the inward force which the experimenter in the rotating laboratory must exert to keep the object at rest does not keep it at rest, but furnishes the centripetal force required to keep the object moving in a circular path. The concept of an outward, centrifugal force explains the action of a centrifuge.   [C.E.H./R.J.S.]

Centripetal force   The inward force required to keep a particle or an object moving in a circular path. It can be shown that a particle moving in a circular path has an acceleration toward the center of the circle along a radius. *See* ACCELERATION.

This radial acceleration, called the centripetal acceleration, is such that, if a particle has a linear or tangential velocity v when moving in a circular path of radius R, the centripetal acceleration is v^2/R. If the particle undergoing the centripetal acceleration has a mass M, then by Newton's second law of motion the centripetal force F_C is in the direction of the acceleration. This is expressed by the equation below, where ω is the

$$F_C = Mv^2/R = MR\omega^2$$

constant angular velocity and is equal to v/R. From Newton's laws of motion it follows that the natural motion of an object is one with constant speed in a straight line, and that a force is necessary if the object is to depart from this type of motion. Whenever an object moves in a curve, a centripetal force is necessary. In circular motion the tangential speed is constant but is changing direction at the constant rate of ω, so the centripetal force along the radius is the only force involved.   [R.J.S.]

Cerenkov radiation   Light emitted by a high-speed charged particle when the particle passes through a transparent, nonconducting, solid material at a speed greater than the speed of light in the material. The blue glow observed in the water of a nuclear reactor, close to the active fuel elements, is radiation of this kind. The emission of Cerenkov radiation is analogous to the emission of a shock wave by a projectile moving faster than sound, since in both cases the velocity of the object passing through the medium exceeds the velocity of the resulting wave disturbance in the medium.

Particle detectors which utilize Cerenkov radiation are called Cerenkov counters. They are important in the detection of particles with speeds approaching that of light, such as those produced in large accelerators and in cosmic rays, and are used with photomultiplier tubes to amplify the Cerenkov radiation. These counters can emit pulses with widths of about 10^{-10} s, and are therefore useful in time-of-flight measurements when very short times must be measured. They can also give direct information on the velocity of the passing particle. *See* PARTICLE DETECTOR.   [W.B.Fr.]

The properties of Cerenkov radiation have been exploited in the development of a branch of gamma-ray astronomy that covers the energy range of about 10^5–10^8 MeV. A high-energy gamma ray from a source external to the Earth creates in the atmosphere a cascade of secondary electrons and positrons. This cascade is generated by the interplay of two processes: electron-positron pair production from gamma rays, and gamma-ray emission as the electrons and positrons are accelerated by the electric fields of nuclei in the atmosphere (bremsstrahlung). For a primary gamma ray having an energy of 10^{12} eV (1 teraelectronvolt), as many as 1000 or more electrons and positrons will contribute to the cascade. The combined Cerenkov light of the cascade is beamed to the ground over an area a few hundred meters in diameter and marks the arrival direction of the initiating gamma ray to about $1°$. On a clear, dark night this radiation may be detected as a pulse of light lasting a few nanoseconds, by using an optical reflector. *See* BREMSSTRAHLUNG; ELECTRON-POSITRON PAIR PRODUCTION.

This technique offers a means to study regions of the universe where charged particles are accelerated to extreme relativistic energies. Such regions involve highly magnetized, rapidly spinning neutron stars; supernova remnants; and active galactic nuclei. These same motivations drive the satellite observations of the EGRET instrument of the *Compton Gamma-Ray Observatory* at lower gamma-ray energies (up to about 10^4 MeV).   [R.C.La.]

Chain reaction (physics)

Chain reaction (physics)   A succession of generation after generation of acts of division (called fission) of certain heavy nuclei. The fission process releases about 200 MeV (3.2×10^{-4} erg = 3.2×10^{-11} joule) in the form of energetic particles including two or three neutrons. Some of the neutrons from one generation are captured by fissile species (^{233}U, ^{235}U, ^{239}Pu) to cause the fissions of the next generations. The process is employed in nuclear reactors and nuclear explosive devices.   [N.C.R.]

Channel electron multiplier

Channel electron multiplier   A single-particle detector which in its basic form (see illustration) consists of a hollow tube (channel) of either glass or ceramic material with a semiconducting inner surface. The detector responds to one or more primary electron impact events at its entrance (input) by producing, in a cascade multiplication process, a charge pulse of typically 10^4–10^8 electrons at its exit (output). Because particles other than electrons can impact at the entrance of the channel electron multiplier to produce a secondary electron, which is then subsequently multiplied in a cascade, the channel electron multiplier can be used to detect charged particles other than electrons (such as ions or positrons), neutral particles with internal energy (such as metastable excited atoms), and photons as well. As a result, this relatively simple, reliable, and easily applied device is employed in a wide variety of charged-particle and photon spectrometers and related analytical instruments, such as residual gas analyzers, mass spectrometers, and spectrometers used in secondary ion mass spectrometry (SIMS),

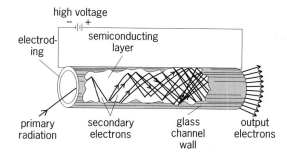

high voltage

electrod-ing

semiconducting layer

primary radiation

secondary electrons

glass channel wall

output electrons

Cutaway view of a straight, single-channel electron multiplier, showing the cascade of secondary electrons resulting from the initial, primary radiation event, which produces an output charge pulse. (*After J. L. Wiza, Microchannel plate detectors, Nucl. Instrum. Meth., 162:587–601, 1979*)

electron spectroscopy for chemical analysis (ESCA), and Auger electron spectroscopy. *See* AUGER EFFECT; PHOTOEMISSION; SPECTROSCOPY.

A related device is the channel electron multiplier array, often called a microchannel plate. The channel electron multiplier array is usually a disk-shaped device with a diameter between 1 and 4 in. (2.5 and 10 cm) and a thickness of a fraction of a millimeter, and consists of millions of miniature channel electron multiplier devices arranged with channel axes perpendicular to the face of the disk. Channel electron multiplier arrays find application as image intensifiers in night vision devices, and are employed to add either large detection area or imaging capabilities, or both, to charged-particle detectors and spectrometers. *See* PARTICLE DETECTOR.   [S.B.E.]

Channeling in solids   The steering of positively charged energetic particles between atomic rows or planes of a crystalline solid. The particles can be positive ions, protons, positrons, or muons. If the angle between the direction of the particle and a particular axis or plane in the crystal is within a small predictable limit (typically a few degrees or less), then the gradually changing electrostatic repulsion between the particle and each successive atomic nucleus of the crystal produces a smooth steering through the crystal lattice. *See* CRYSTAL STRUCTURE.

An obvious consequence of this steered motion is that it prevents violent collisions of the particles with atoms on the lattice sites. Hence, as compared with a randomly directed beam of particles, the channeled beam loses energy more slowly, penetrates more deeply, creates much less damage to the crystal along its track, and is prevented from participating in all close-encounter processes (nuclear reactions, Rutherford scattering, and so forth) with lattice atoms. *See* NUCLEAR REACTION; SCATTERING EXPERIMENTS (NUCLEI).

A related channeling phenomenon is the channeling of energetic electrons or other negative particles. In this case, the particles are attracted to the positively charged atomic nuclei, so that the probability of violent collisions with atoms on lattice sites is enhanced rather than being prevented, and the particles are steered along the rows or planes of nuclei rather than between them.

A closely related phenomenon is called blocking. In this case, the energetic positive particles originate from atomic sites within the crystal lattice by means of fission, alpha-particle decay, or by wide-angle scattering of a nonchanneled external beam in a very close encounter with a lattice atom. Those particles emitted almost parallel to an atomic row or plane will be deflected away from the row by a steering process. Consequently, no particles emerge from the crystal within a certain critical blocking angle of each major crystallographic direction. A piece of film placed some distance from the crystal provides a simple technique for recording blocking patterns. Theoretical considerations show that the same principle is involved in blocking as in channeling; hence, both phenomena exhibit an identical dependence on particle energy, nuclear charge, lattice spacing, and so forth. *See* NUCLEAR FISSION; RADIOACTIVITY.

Applications of channeling include the location of foreign atoms in a crystal, the study of crystal surface structure, and the measurement of nuclear lifetimes. The location of foreign (solute) atoms in a crystal is one of the simplest channeling applications. It is accomplished by measuring the yields of Rutherford backscattered particles, characteristic x-rays, or nuclear reaction products produced by the interaction of channeled particles with the solute atoms. Such yields are enhanced for solute atoms that are displaced into channels of the crystal. This method has been used to determine the lattice positions of solute atoms that have been introduced into crystals, and to determine the amount of lattice damage created by ion implantation. These and similar applications have proved extremely useful in the development of semiconductor devices. *See* LASER-SOLID INTERACTIONS.   [J.A.Da.; M.L.Sw.]

Chaos   System behavior that depends so sensitively on the system's precise initial conditions that it is, in effect, unpredictable and cannot be distinguished from a random process, even though it is deterministic in a mathematical sense.

Throughout history, sequentially using magic, religion, and science, people have sought to perceive order and meaning in a seemingly chaotic and meaningless world. This quest for order reached its ultimate goal in the seventeenth century when newtonian dynamics provided an ordered, deterministic view of the entire universe epitomized in P. S. de Laplace's statement, "We ought then to regard the present state of the universe as the effect of its preceding state and as the cause of its succeeding state."

But if the determinism of Laplace and Newton is totally accepted, it is difficult to explain the unpredictability of a gambling game or, more generally, the unpredictably random behavior observed in many newtonian systems. Commonplace examples of such behavior include smoke that first rises in a smooth, streamlined column from a cigarette, only to abruptly burst into wildly erratic turbulent flow (see illustration); and the unpredictable phenomena of the weather. *See* FLUID FLOW; TURBULENT FLOW.

At a more technical level, flaws in the newtonian view had become apparent by about 1900. The problem is that many newtonian systems exhibit behavior which is so exquisitely sensitive to the precise initial state or to even the slightest outside perturbation that, humanly speaking, determinism becomes a physically meaningless though mathematically valid concept. But even more is true. Many deterministic newtonian-system orbits are so erratic that they cannot be distinguished from a random process even though they are strictly determinate, mathematically speaking. Indeed, in the totality of newtonian-system orbits, erratic unpredictable randomness is overwhelmingly the most common behavior. *See* CLASSICAL MECHANICS; DETERMINISM.

One example of chaos is the evolution of life on Earth. Were this evolution deterministic, the governing laws of evolution would have had built into them anticipation of every natural crisis which has occurred over the centuries plus anticipation of every possible ecological niche throughout all time. Nature, however, economizes and uses the richness of opportunity available through chaos. Random mutations provide choices sufficient to meet almost any crisis, and natural selection chooses the proper one.

Another example concerns the problem that the human body faces in defending against all possible invaders. Again, nature appears to choose chaos as the most economical solution. Loosely speaking, when a hostile bacterium or virus enters the body, defense strategies are generated at random until a feedback loop indicates that the

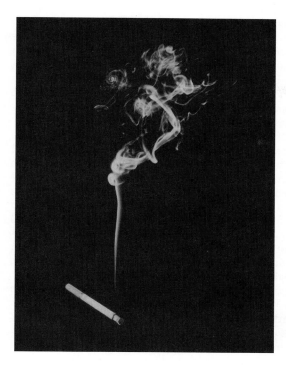

Transition from order to chaos (turbulence) in a rising column of cigarette smoke. The initial smooth streamline flow represents order, while the erratic flow represents chaos.

correct strategy has been found. A great challenge is to mimic nature and to find new and useful ways to harness chaos.

Another matter for consideration is the problem of predicting the weather or the world economy. Both these systems are chaotic and can be predicted more or less precisely only on a very short time scale. Nonetheless, by recognizing the chaotic nature of the weather and the economy, it may eventually be possible to accurately determine the probability distribution of allowed events in the future given the present. At that point it may be asserted with mathematical precision that, for example, there is a 90% chance of rain 2 months from today. Much work in chaos theory seeks to determine the relevant probability distributions for chaotic systems.

Finally, many physical systems exhibit a transition from order to chaos, as exhibited in the illustration, and much work studies the various routes to chaos. Examples include fibrillation of the heart and attacks of epilepsy, manic-depression, and schizophrenia. Physiologists are striving to understand chaos in these systems sufficiently well that these human maladies can be eliminated. *See* PERIOD DOUBLING.

Reduced to basics, chaos and noise are essentially the same thing. Chaos is randomness in an isolated system; noise is randomness entering this previously isolated system from the outside. If the noise source is included to form a composite isolated system, there is again only chaos.   [J.Fo.]

Charge-density wave   A possible ground state of a metal in which the conduction-electron charge density is sinusoidally modulated in space. The periodicity of this extra modulation is unrelated to the lattice periodicity. Instead, it is determined by the dimension of the conduction-electron Fermi surface in momentum space. *See* FERMI SURFACE.

In a quasi-one-dimensional metal, for which conduction electrons are mobile in one direction only, a charge-density wave can be caused by a Peierls instability. This mechanism involves interaction between the electrons and a periodic lattice distortion having a wave vector Q parallel to the conduction axis. The linear-chain metal niobium triselenide ($NbSe_3$) is prototypical.

For isotropic metals, and quasi-two-dimensional metals, Coulomb interactions between electrons are the cause of a charge-density wave instability. The exchange energy, an effect of the Pauli exclusion principle, and the correlation energy, an effect of electron-electron scattering, both act to stabilize a charge-density wave. However, the electrostatic energy attributable to the charge modulation would suppress a charge-density wave were it not for a compensating charge response of the positive-ion lattice. *See* EXCHANGE INTERACTION; EXCLUSION PRINCIPLE.

A wavelike displacement of this lattice will generate a positive-ion charge density that almost cancels the electronic charge modulation of the charge-density wave. A typical value of the displacement amplitude is about 1% of the lattice constant. Ion-ion repulsive interactions must be small in order to permit such a distortion. Consequently, charge-density waves are more likely to occur in metals having small elastic moduli. *See* BAND THEORY OF SOLIDS; CRYSTAL STRUCTURE; SPIN-DENSITY WAVE.   [A.W.O.]

Charged particle beams   Unidirectional streams of charged particles traveling at high velocities. Charged particles can be accelerated to high velocities by electromagnetic fields. They are then able to travel through matter (termed an absorber), interacting with it, losing energy, and causing various effects important in many applications. Examples of charged particles are electrons, positrons, protons, antiprotons, alpha particles, and any ions (atoms with one or several electrons removed or added). In addition, some particles are produced artificially and may be short-lived (pions, muons).

In traveling through matter, charged particles interact with nuclei, producing nuclear reactions and elastic and inelastic collisions with the electrons (electronic collisions) and with entire atoms of the absorber (atomic collisions). Usually, in its travel through matter a charged particle makes few or no nuclear reactions or inelastic nuclear collisions, but many electronic and atomic collisions. The average distance between successive collisions is called the mean free path, λ. In solids, it is of the order of 10 cm (4 in.) for nuclear reactions. It ranges from the diameter of the atoms (about 10^{-10} m) to about 10^{-7} m for electronic collisions. The mean free path, λ, depends on the properties of the particle and, most importantly, on its velocity.

If a charged particle is accelerated, it can emit photons called bremsstrahlung. This process is of great importance for electrons as well as for heavy ions whose kinetic energies are much greater than their rest energies. It is used extensively for the production of x-rays in radiology. *See* BREMSSTRAHLUNG.

In gases, all electrons are bound to individual atoms or molecules in well-defined orbits. These electrons can be moved into other bound orbits (excitation) requiring a well-defined energy. Another possibility is the complete removal of the electron from the atom (ionization), requiring an energy equal to or greater than the ionization energy for the particular electron. In both processes, the charged particle will lose energy and will be deflected very slightly.

There are some major differences between electron beams and beams of heavier particles. In general, the path of an electron will be a zigzag. Angular deflections in the collisions will frequently be large. Electron beams therefore tend to spread out laterally, and the number of primary electrons in the beam at a depth x in the absorber decreases rapidly.

In general, for the same dose (the energy deposited per gram along the beam line) heavy charged particles will produce, because of their higher local ionization, larger biological effects than electrons (which frequently are produced by x-rays).

Electron beams are used in the preservation of food. In medicine, electron beams are used extensively to produce x-rays for both diagnostic and therapeutic (cancer irradiation) purposes. Also, in radiation therapy, deuteron beams incident on Be and ^{3}H targets are used to produce beams of fast neutrons, which in turn produce fast protons, alpha particles, and carbon, nitrogen, and oxygen ions in the irradiated tissue. Energetic pion, proton, alpha, and heavier ion beams can possibly be used for cancer therapy.

Charged particle beams are used in many methods of chemical and solid-state analysis. Nuclear activation analysis can be performed with heavy ions. *See* ELECTRON DIFFRACTION.   [H.Bi.]

Beams of nuclei with lifetimes as short as 10^{-6} s are used for studies in nuclear physics, astrophysics, biology, and materials science. Nuclear beams (or heavy-ion beams) are usually produced by accelerating naturally available stable isotopes. However, radioactive nuclei, most of which do not occur naturally on Earth, must be produced as required in nuclear reactions by using various accelerated beams. Because these radioactive nuclei are produced by the nuclear reactions of primary beams, they are called secondary particles and beams of such nuclei are called radioactive secondary beams. *See* RADIOACTIVITY.

Radioactive secondary beams have made possible the study of the structure of nuclei far from stability. Another important application occurs in the study of nuclear reactions of importance in hot stars and in supernovae, which are crucial for understanding nucleosynthesis in the universe. *See* NUCLEAR STRUCTURE.   [I.T.]

Charged particle optics   The branch of physics concerned with the motion of charged particles under the influence of electric and magnetic fields. A positively charged particle that moves in an electric field experiences a force in the direction of this field. If the particle falls in the field from a potential of U volts to a potential zero, its energy gain, measured in electronvolts, is equal to the product of U and the particle's charge. For example, if a singly and a doubly charged particle are accelerated by a potential drop of 100 V, the two particles will gain energies of 100 eV and 200 eV, respectively. If both particles were initially at rest, they would have final velocities proportional to the square root of K/m, where K is the energy increase and m is the mass of the particle. This relation describes the velocities of energetic particles accurately as long as these velocities are small compared to the velocity of light $c \approx 300{,}000$ km/s (186,000 mi/s), a speed that cannot be exceeded by any particle. *See* ELECTRIC FIELD; ELECTROSTATICS.

If an ensemble of ions of equal energies but of different masses is accelerated simultaneously, the ion masses can be determined from their arrival times after a certain flight distance. Such time-of-flight mass spectrometers have successfully been used, for instance, to investigate the masses of large molecular ions, up to and beyond 350,000 atomic mass units.

If a homogeneous electric field is established between two parallel-plate electrodes at different potentials, a charged particle in the space between the electrodes will experience a force in the direction perpendicular to them. If initially the particle moved parallel to the electrodes, it will be deflected by the electric force and move along a parabolic trajectory. Magnetic fields also deflect charged particles. In contrast to electrostatic fields, however, magnetic fields change only the direction of a particle trajectory and not the magnitude of the particle velocity. Charged particles that enter a magnetic field thus move along circles whose radii increase with the products of their velocities and their mass-to-charge ratios, m/q. If initially all particles start at the same potential U and are accelerated to the potential zero, they will move along radii that are proportional to the square root of $U(m/q)$. Thus, particles of different mass-to-charge ratios can be separated in a magnetic sector field.

A sector-field mass analyzer can be used to determine the masses of atomic or molecular ions in a cloud of such ions. Such systems can also be used to purify a beam of ions that are to be implanted in semiconductors in order to fabricate high-performance transistors and diodes. Finally, such magnetic sector fields are found in large numbers in all types of particle accelerators.

An Einzel lens consists of three cylindrical tubes, the middle one of which is at a higher potential than the outer two. Positively charged particles entering such a device are first decelerated and then accelerated back to their initial energies. Axially symmetric magnetic lenses have also been constructed. Such lenses, also called solenoids, consist mainly of a coil of wire through which an electric current is passed. The charged particles are then constrained to move more or less parallel to the axis of such a coil. Axially symmetric electric and magnetic lenses are used extensively to focus low-energy particle beams. Particularly important applications are in television tubes and in electron microscopes. *See* ELECTROSTATIC LENS; MAGNETIC LENS.

By passing charged particles through electrode or pole-face arrangements, a particle beam can also be focused toward the optic axis. In such quadrupole lenses the electric or the magnetic field strengths, and therefore the forces that drive the charged particles toward or away from the optic axis, increase linearly with the distance from the axis. While quadrupole lenses are found in systems in which low-energy particle beams must be focused, for instance, in mass spectrometers, such lenses have become indispensable for high-energy beams. Consquently, quadrupole lenses, especially magnetic ones, are found in many types of particle accelerators used in research in, for example, nuclear and solid-state physics, as well as in cancer irradiation treatment facilities. *See* CHARGED PARTICLE BEAMS; ELECTRON LENS; PARTICLE ACCELERATOR.   [H.W.]

Charles' law A thermodynamic law, also known as Gay-Lussac's law, which states that at constant pressure the volume of a fixed mass or quantity of gas varies directly with the absolute temperature. Conversely, at constant volume the gas pressure varies directly with the absolute temperature.   [F.H.R.]

Charm A term used to describe a class of elementary particles. Ordinary atoms of matter consist of a nucleus composed of neutrons and protons and surrounded by electrons. Over the years, however, a host of other particles with unexpected properties have been found, associated with both electrons (leptons) and protons (hadrons). The hadrons number in the hundreds, and can be explained as composites of more fundamental constituents, called quarks. The originally simple situation of having an up quark (u) and a down quark (d) has evolved as several more varieties or flavors have had to be added. These are the strange quark (s) with the additional property or quantum number of strangeness to account for the unexpected characteristics of a family of strange particles; the charm quark (c) possessing charm and no strangeness, to explain the discovery of the J/ψ particles, massive statesthree times heavier than the proton; and a fifth quark (b) to explain the existence of the even more massive upsilon (γ) particles. *See* HADRON; QUARKS.

The members of the family of particles associated with charm fall into two classes: those with hidden charm, where the states are a combination of charm and anticharm quarks ($c\bar{c}$), charmonium; and those where the charm property is clearly evident, such as the D^+ ($c\bar{d}$) meson and Λ_c^+ (cud) baryon. Although reasonable progress has been made in the study of charmed states, much work remains to be done. *See* ELEMENTARY PARTICLE.   [N.P.S.]

Chevrel phases
A series of ternary molybdenum chalcogenide compounds. They were reported by R. Chevrel, M. Sergent, and J. Prigent in 1971. The compounds have the general formula $M_xMo_6X_8$, where M represents any one of a large number (nearly 40) of metallic elements throughout the periodic table; x has values between 1 and 4, depending on the M element; and X is a chalcogen (sulfur, selenium or tellurium). The Chevrel phases are of great interest, largely because of their striking superconducting properties.

Most of the ternary molybdenum chalcogenides crystallize in a structure in which the unit cell, that is, the repeating unit of the crystal structure, has the overall shape of a rhombohedron with a rhombohedral angle close to 90°. The building blocks of the Chevrel-phase crystal structure are the M elements and Mo_6X_8 molecular units or clusters. Each Mo_6X_8 unit is a slightly deformed cube with X atoms at the corners, and Mo atoms at the face centers. One of these structures, that of $PbMo_6S_8$, is shown in the illustration. *See* CRYSTAL; CRYSTALLOGRAPHY.

Several of the Chevrel-phase compounds have relatively high values of the super-conducting transition temperature, T_c, the maximum being about 15 K ($-433°$F) for $PbMo_6S_8$. The Chevrel-phase $PbMo_6S_8$ has a value of the upper critical magnetic field near absolute zero $H_{c2}(O)$ of about 60 teslas (600 kilogauss), which was the largest value observed prior to the discovery of high-temperature ceramic superconductors in 1986. A number of Chevrel-phase compounds of the form RMo_6X_8, where R is a

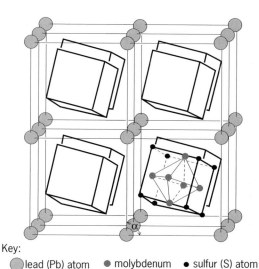

Key:

⬤ lead (Pb) atom   ● molybdenum (Mo) atom   • sulfur (S) atom

Crystal structure of $PbMo_6S_8$. Each lead atom Is surrounded by eight Mo_6S_8 units, the structure of which is shown In the lower right-hand part of the figure. The rhombohedral angle α Is Indicated.

rare-earth element with a partially filled 4f electron shell and X is S or Se, display magnetic order at low temperatures in addition to superconductivity. *See* SUPERCON-DUCTIVITY.   [M.B.Ma.]

Child-Langmuir law   A law governing space-charge-limited flow of electron current between two plane parallel electrodes in vacuum when the emission velocities of the electrons can be neglected. It is often called the three-halves power law, and is expressed by the equation below.

$$j\,(\mathrm{A/cm^2}) = 2.33 \times 10^{-6}\,\frac{V\,(\mathrm{volts})^{3/2}}{d\,(\mathrm{cm})^2}$$

Here V is the potential difference between the two electrodes, d their separation, and j the current density at the collector electrode, or anode. The potential difference V is the applied voltage reduced by the difference in work function of the collector and emitter. The Child-Langmuir law applies, to a close approximation, to other electrode geometries as well. Thus for coaxial cylinders with the inner cylinder the cathode, it leads to a deviation from the true value of the current density of 13% at most. *See* SPACE CHARGE.   [E.G.R.]

Choked flow   Fluid flow through a restricted area whose rate reaches a maximum when the fluid velocity reaches the sonic velocity at some point along the flow path. The phenomenon of choking exists only in compressible flow and can occur in several flow situations. *See* COMPRESSIBLE FLOW.

   Through varying-area duct. Choked flow can occur through a convergent flow area or nozzle attached to a huge reservoir. Flow exits the reservoir through the nozzle if the back pressure is less than the reservoir pressure. When the back pressure is decreased slightly below the reservoir pressure, a signal from beyond the nozzle exit is transmitted at sonic speed to the reservoir. The reservoir responds by sending fluid through the nozzle. Further, the maximum velocity of the fluid exists at the nozzle throat where the area is smallest.

   When the back pressure is further decreased, fluid exits the reservoir more rapidly. Eventually, however, the velocity at the throat reaches the sonic velocity. Then the fluid velocity at the throat is sonic, and the velocity of the signal is also sonic. Therefore, further decreases in back pressure are not sensed by the reservoir, and correspondingly will not induce any greater flow to exit the reservoir. The nozzle is thus said to be choked, and the mass flow of fluid is a maximum. *See* MACH NUMBER; SOUND.

   With friction. Choked flow can also occur through a long constant-area duct attached to a reservoir. As fluid flows through the duct, friction between the fluid and the duct wall reduces the pressure acting on the fluid. As pressure is reduced, other fluid properties are affected, such as sonic velocity, density, and temperature. The maximum Mach number occurs at the nozzle exit, and choked flow results when this Mach number reaches 1.

   With heat addition. A reservoir with a constant-area duct attached may also be considered in the case that the flow through the duct is assumed to be frictionless but heat is added to the system along the duct wall. *See* FLUID FLOW; GAS DYNAMICS.   [W.S.J.]

Chromatic aberration   The type of error in an optical system in which the formation of a series of colored images occurs, even though only white light enters the system. Chromatic aberrations are caused by the fact that the refraction law determining the path of light through an optical system contains the refractive index, which is a function of wavelength. Thus the image position and the magnification of an optical system are not necessarily the same for all wavelengths, nor are the aberrations the same for all wavelengths. *See* ABERRATION (OPTICS); REFRACTION OF WAVES.   [M.J.H.]

Classical field theory   The mathematical discipline that studies the behavior of distributions of matter and energy when their discrete nature can be ignored; also known as continuum physics or continuum mechanics. The discrete nature of matter refers to its molecular nature, and that of energy to the quantum nature of force fields and of the mechanical vibrations that exist in any sample of matter. The theory is normally valid when the sample is of laboratory size or larger, and when the number of quanta present is also very large. *See* PHONON; PHOTON; QUANTUM MECHANICS.

Classical field theories can be formulated by the molecular approach, which seeks to derive the macroscopic (bulk) properties by taking local averages of microscopic quantities, or by the phenomenological approach, which ignores the microscopic nature of the sample and uses properties directly measurable with laboratory equipment. Although the microscopic treatment sometimes yields profounder insights, the phenomenological approach can use partial differential equations since neglecting the microscopic structure allows quantities such as density and pressure to be expressed by continuously varying numbers.

Examples of classical field theories include the deformation of solids, flow of fluids, heat transfer, electromagnetism, and gravitation. Solving the equations has produced a vast body of mathematics. Computers have aided in special calculations, but many mathematicians are working on the analytical theory of partial differential equations, and new results continue to be produced.   [D.P.]

Classical mechanics   The science dealing with the description of the positions of objects in space under the action of forces as a function of time. Some of the laws of mechanics were recognized at least as early as the time of Archimedes (287?–212 B.C.). In 1638, Galileo stated some of the fundamental concepts of mechanics, and in 1687, Isaac Newton published his *Principia*, which presents the basic laws of motion, the law of gravitation, the theory of tides, and the theory of the solar system. This monumental work and the writings of J. D'Alembert, J. L. Lagrange, P. S. Laplace, and others in the eighteenth century are recognized as classic works in the field of mechanics. Jointly they serve as the base of the broad field of study known as classical mechanics, or Newtonian mechanics. This field does not encompass the more recent developments in mechanics, such as statistical, relativistic, or quantum mechanics.

In the broad sense, classical mechanics includes the study of motions of gases, liquids, and solids, but more commonly it is taken to refer only to solids. In the restricted reference to solids, classical mechanics is subdivided into statics, kinematics, and dynamics. Statics considers the action of forces that produce equilibrium or rest; kinematics deals with the description of motion without concern for the causes of motion; and dynamics involves the study of the motions of bodies under the actions of forces upon them. For some of the more important areas of classical mechanics *see* BALLISTICS; COLLISION (PHYSICS); DYNAMICS; ENERGY; FORCE; GRAVITATION; KINEMATICS; LAGRANGE'S EQUATIONS; MASS; MOTION; RIGID-BODY DYNAMICS; STATICS; WORK.   [N.S.G.]

Clock paradox   The phenomenon occurring in the special theory of relativity wherein two observers who start together with identical clocks and then undergo different motions can have different total elapsed time on their clocks when they rejoin later. This effect is a well-defined, mathematically consistent prediction of special relativity which has been verified by experiment but, historically, it has been referred to as a paradox because of erroneous reasoning in the manner in which the effect is commonly analyzed. The clock-paradox phenomenon arises because there is no notion of absolute simultaneity in the theory of special relativity.

The clock-paradox effect is illustrated by the following hypothetical example. (In the context of the example the effect usually is referred to as the twin paradox.) Two identical twins are separated as young adults. One of the twins remains on Earth and, for the purposes of the discussion, is assumed to undergo inertial motion. (In fact, in the context of general relativity, the twin on Earth would be viewed as accelerating—only freely falling observers would be viewed as inertial—but in this example the corrections made to the effect by performing a proper, general-relativistic analysis would be negligible.) The other twin is placed in a rocket ship which accelerates very rapidly away from Earth until it is receding from Earth at nearly the speed of light. After coasting for a while at this speed, the rocket ship turns around and accelerates very rapidly back toward Earth, so that it soon is traveling at nearly the speed of light. The rocket ship then lands on Earth and the twins rejoin. The twin on Earth is old (and everything else on Earth has aged considerably), but the twin in the rocket ship (and everything else therein) has barely aged at all.

The clock-paradox phenomenon has been observed directly in an experiment performed in 1971 by J. C. Hafele and R. E. Keating, who observed differences in elapsed times of atomically stabilized clocks flown in airplanes as compared with ones on the ground. In this experiment the special relativistic effect is so small, since the velocities achieved by airplanes are much smaller than c, that the tiny corrections due to general relativity cannot be neglected, so the experiment actually must be viewed as verifying the analogous clock effect in general relativity rather than purely the clock effect of special relativity. *See* RELATIVITY; SPACE-TIME.   [R.M.Wal.]

Coherence   The attribute of two or more waves, or parts of a wave, whose relative phase is constant during the resolving time of the observer. The concept has been developed most extensively in optics, but is applicable to all wave phenomena.

Consider two waves, with the same mean angular frequency ω, given by Eqs. (1) and (2). It is convenient, and no restriction, to choose both A and B real. These ex-

$$\Psi_A(x,t) = A \exp\{i[k(\omega)x - \omega t - \delta_A(t)]\} \tag{1}$$

$$\Psi_B(x,t) = B \exp\{i[k(\omega)x - \omega t - \delta_B(t)]\} \tag{2}$$

pressions as they stand could describe de Broglie waves in quantum mechanics. For real waves, such as components of the electric field in light or radio beams, or the pressure oscillations in sound, it is necessary to retain only the real parts of these and subsequent expressions. The frequency spectrum is assumed to be narrow, in the sense that a Fourier analysis of expressions (1) and (2) gives appreciable contributions only for angular frequencies close to ω. This assumption means that, on the average, $\delta_A(t)$ and $\delta_B(t)$ do not change much per period. *See* ELECTROMAGNETIC RADIATION; QUANTUM MECHANICS; SOUND.

Suppose that the waves are detected by an apparatus with resolving time T, that is, T is the shortest interval between two events for which the events do not seem to be simultaneous. For the human eye and ear, T is about 0.1 s, while a fast electronic device might have a T of 10^{-10} s. If the relative phase $\delta(t)$, given by Eq. (3), does not, on the

$$\delta(t) = \delta_B(t) - \delta_A(t) \tag{3}$$

average, change noticeably during T, then the waves are coherent. If during T there are sufficient random fluctuations for all values of $\delta(t)$, modulus 2π, to be equally probable, then the waves are incoherent. If during T there are noticeable random fluctuations in

$\delta(t)$, but not enough to make the waves completely incoherent, then the waves are partially coherent. These distinctions are not useful unless T is specified. On the one hand, only waves that have existed forever and that fill all of space can have absolutely fixed frequency and phase. On the other hand, two independent sound waves in the phases change appreciably in 0.01 s would seem incoherent to the human ear, but would seem highly coherent to a fast electronic device.

The degree of coherence is related to the interference patterns that can be observed when the two beams are combined. *See* INTERFERENCE OF WAVES.

Coherence is also used to describe relations between phases within the same beam. Suppose that a wave represented by Eq. (1) is passing a fixed observer characterized by a resolving time T. The phase δ_A may fluctuate, perhaps because the source of the wave contains many independent radiators. The coherence time Δt_W of the wave is defined to be the average time required for $\delta_A(t)$ to fluctuate appreciably at the position of the observer. If Δt_W is much greater than T, the wave is coherent; if Δt_W is of the order of T, the wave is partially coherent; and if Δt_W is much less than T, the wave is incoherent. These concepts are very close to those developed above.

Extended sources give partial coherence and produce interference fringes with visibility V less than unity. A. A. Michelson exploited this fact with his stellar interferometer, a modified double-slit arrangement with movable mirrors that permit adjustment of the effective separation D' of the slits. It can be shown that if the source is a uniform disk of angular diameter θ, then the smallest value of D' that gives zero V is $1.22\lambda/\theta$. The same approach has also been applied in radio astronomy. A different technique, developed by R. Hanbury Brown and R. Q. Twiss, measures the correlation between the intensifies received by separated detectors with fast electronics. *See* INTERFEROMETRY.

Because they are highly coherent sources, lasers and masers provide very large intensities per unit frequency. *See* LASER.   [R.G.Wi.]

Photon statistics is concerned with the probability distribution describing the number of photons incident on a detector, or present in a cavity. By extension, it deals with the correlation properties of beams of light.

According to the quantum theory of electromagnetism, quantum electrodynamics, light is made up of particles called photons, each of which possesses an energy E of $\hbar\omega$, where $\hbar$ is Planck's constant divided by 2π and ω is the angular frequency of the light (the frequency multiplied by 2π). In general, however, the photon number is an intrinsically uncertain quantity. It is impossible to precisely specify both the phase $\phi = \omega t$ of a wave and the number of photons $n \approx E/(\hbar\omega)$ that it contains; the uncertainties of these two conjugate variables must satisfy $\Delta n \Delta\phi \geq \frac{1}{2}$. For a beam to be coherent in the sense of having a well-defined phase, it must not be describable in terms of a fixed number of particles. (Lacking a fixed phase, a single photon may interfere only with itself, not with other photons.) *See* PHOTON; QUANTUM ELECTRODYNAMICS.

The most familiar example of this uncertainty is shot noise, the randomness of the arrival times of individual photons. There is no correlation between photons in the coherent state emitted by a classical source such as an ideal laser or radio transmitter, so the number of photons detected obeys Poisson statistics, displaying an uncertainty equal to the square root of the mean. The shot noise constitutes the dominant source of noise at low light levels, and may become an important factor in optical communications as well as in high-precision optical devices (notably those that search for gravitational radiation).   [A.M.S.]

Cohesion (physics)

The tendency of atoms or molecules to coalesce into extended condensed states. This tendency is practically universal. In all but exceptional cases, condensation occurs if the temperature is sufficiently low; at higher temperatures, the thermal motions of the constituents increase, and eventually the solid assumes gaseous form. The cohesive energy is the work required to separate the condensed phase into its constituents or, equivalently, the amount by which the energy of the condensed state is lower than that of the isolated constituents. The science of cohesion is the study of the physical origins and manifestations of the forces causing cohesion, as well as those opposing it. It is thus closely related to the science of chemical bonding in molecules, which treats small collections of atoms rather than extended systems. See INTERMOLECULAR FORCES.

The origin and magnitude of the attractive forces depend on the chemical nature of the constituent atoms or molecules. Strong attractive interactions are usually associated with constituents having valence electron shells which are partly filled or open; if the valence electron shells are completely filled or closed, the interactions are weaker.

For open-shell constituents, as the atoms approach, the electron energy levels on different atoms begin to interact, forming a complex of energy levels in the solid. Some of these are below the atomic energy levels and some above. Since the atomic shells are partly filled, the lower energy levels in the solid are filled, but at least some of the higher levels are empty. Thus the average energy of the occupied levels in the solid is lower than that in the isolated atoms, resulting in an attractive force. Bonding in open-shell systems can be approximately divided into three categories, although most cases involve a combination. See BAND THEORY OF SOLIDS; FERMI-DIRAC STATISTICS; SOLID-STATE PHYSICS; VALENCE BAND.

1. *Covalent bonding.* This type of bonding is most similar to the molecular bond. The electron energy levels in the solid are split into a lower and a higher portion, with the states in the lower one filled and the higher one empty. Covalent bonds are strongly directional, with electron charge accumulating around the bond centers. Materials bonded in this fashion typically form structures with low coordination numbers, prototypical materials elements in group 14 of the periodic table, the insulator carbon, and the semiconductors silicon and germanium. See SEMICONDUCTOR.

2. *Metallic bonding.* In this case, there is no split between the lower and higher states of the electrons in the solid; rather, they occupy levels from the bottom up to a cutoff point known as the Fermi level. For example, in transition metals, the electron states in the solid derived from the atomic d orbitals form a complex which is gradually filled with increasing atomic number. The bulk of the cohesive energy is due to this complex. The metallic bond is less directional than the covalent bond, with a more uniform distribution of electronic charge. Metals usually form closely packed structures. See FERMI SURFACE; FREE-ELECTRON THEORY OF METALS.

3. *Ionic bonding.* This occurs in compounds having at least two distinct types of atoms. One or more of the species of atoms (the cations) have only a small number of electrons in their valence shells, whereas at least one species (the anions) has a nearly filled valence shell. As the atoms approach each other, electrons drop from the cation valence states into holes in the anion valence shell, forming a closed-shell configuration in the solid. The different types of atoms in the solid have net charges; a strong attractive force results from the interaction between unlike charges. For example, in sodium chloride (NaCl), the sodium atoms acquire positive charges, and the chlorine atoms acquire negative charges. The closest interatomic separations in the solid are between sodium and chlorine, so that the attractive electrostatic interactions outweigh the repulsive ones. See IONIC CRYSTALS.

In closed-shell constituents, the above effects are greatly reduced because the atomic or molecular shells are basically inert. The constituents retain their separate identities in the solid environment. If the constituents are atomic, as in rare-gas solids, the cohesion is due to the van der Waals forces. The positions of the electrons in an atom fluctuate over time, and at any given time their distribution is far from spherical. This gives rise to fluctuating long-ranged electric fields, which average zero over time, but can still have appreciable effects on neighboring atoms. The electrons on these atoms move in the direction of the force exerted by the electric field. The net result is that the interactions between unlike charges (electrons and nuclei) are increased in the solid, whereas the interactions between like charges are reduced. Thus the solid has a lower energy than the isolated atoms.

In solids made up of molecules, there are additional electrostatic interactions due to the nonspherical components of the molecular charge density. These interactions are strongest if the molecules are polar. This means that the center of the positive charge on the molecule is at a different point in space from that of the negative charge. Polar molecules, such as water (H_2O), form structures in which the positive charge on a molecule is close to the negative charges of its neighbors. For nonpolar molecules, the electrostatic interactions are usually weaker than the van der Waals forces. The nonspherical interactions in such cases are often so weak that the molecules can rotate freely at elevated temperatures, while the solid is still held together by the van der Waals forces.

The repulsive forces in the condensed phase are a dramatic illustration of the combined action of two quantum-mechanical principles, the exclusion principle and the uncertainty principle.

The exclusion principle states that the quantum-mechanical wave function for the electrons in the solid must be antisymmetric under the interchange of the coordinates of any two electrons. Consequently, two electrons of the same spin are forbidden from being very close to each other. See EXCLUSION PRINCIPLE.

The uncertainty principle states that if the motion of an electron is confined, its kinetic energy must rise, resulting in a repulsive force opposing the confinement. The kinetic energy due to the confinement is roughly inversely proportional to the square of the radius of the region of confinement. According to the exclusion principle, the motion of an electron in a solid is partially confined because it is forbidden from closely approaching other electrons of the same spin. Thus the uncertainty principle in turn implies a repulsive force. See UNCERTAINTY PRINCIPLE.   [A.E.Ca.]

Coil   One or more turns of wire used to introduce inductance into an electric circuit. At power line and audio frequencies a coil has a large number of turns of insulated wire wound close together on a form made of insulating material, with a closed iron core passing through the center of the coil. This is commonly called a choke and is used to pass direct current while offering high opposition to alternating current.

At higher frequencies a coil may have a powdered iron core or no core at all. The electrical size of a coil is called inductance and is expressed in henries or millihenries. In addition to the resistance of the wire, a coil offers an opposition to alternating current, called reactance, expressed in ohms. The reactance of a coil increases with frequency. See INDUCTOR.   [J.Mar.]

Collision (physics)   Any interaction between particles, aggregates of particles, or rigid bodies in which they come near enough to exert a mutual influence, generally with exchange of energy. The term collision, as used in physics, does not necessarily imply actual contact.

In classical mechanics, collision problems are concerned with the relation of the magnitudes and directions of the velocities of colliding bodies after collision to the velocity vectors of the bodies before collision. When the only forces on the colliding bodies are those exerted by the bodies themselves, the principle of conservation of momentum states that the total momentum of the system is unchanged in the collision process. This result is particularly useful when the forces between the colliding bodies act only during the instant of collision. The velocities can then change only during the collision process, which takes place in a short time interval. Under these conditions the forces can be treated as impulsive forces, the effects of which can be expressed in terms of an experimental parameter known as the coefficient of restitution. *See* CONSERVATION OF MOMENTUM; IMPACT.

The study of collisions of molecules, atoms, and nuclear particles is an important field of physics. Here the object is usually to obtain information about the forces acting between the particles. The velocities of the particles are measured before and after collision. Although quantum mechanics instead of classical mechanics should be used to describe the motion of the particles, many of the conclusions of classical collision theory are valid. *See* SCATTERING EXPERIMENTS (ATOMS AND MOLECULES); SCATTERING EXPERIMENTS (NUCLEI).

Collisions can be classed as elastic and inelastic. In an elastic collision, mechanical energy is conserved; that is, the total kinetic energy of the system of particles after collision equals the total kinetic energy before collision. For inelastic collisions, however, the total kinetic energy after collision is different from the initial total kinetic energy.

In classical mechanics the total mechanical energy after an inelastic collision is ordinarily less than the initial total mechanical energy, and the mechanical energy which is lost is converted into heat. However, an inelastic collision in which the total energy after collision is greater than the initial total energy sometimes can occur in classical mechanics. For example, a collision can cause an explosion which converts chemical energy into mechanical energy. In molecular, atomic, and nuclear systems, which are governed by quantum mechanics, the energy levels of the particles can be changed during collisions. Thus these inelastic collisions can involve either a gain or a loss in mechanical energy.   [P.W.S.]

Colloidal crystals   Periodic arrays of suspended colloidal particles. Common colloidal suspensions (colloids) such as milk, blood, or latex are polydisperse; that is, the suspended particles have a distribution of sizes and shapes. However, suspensions of particles of identical size, shape, and interaction, the so-called monodisperse colloids, do occur. In such suspensions, a new phenomenon that is not found in polydisperse systems, colloidal crystallization, appears: under appropriate conditions, the particles can spontaneously arrange themselves into spatially periodic structures. This ordering is analogous to that of identical atoms or molecules into periodic arrays to form atomic or molecular crystals. However, colloidal crystals are distinguished from molecular crystals, such as those formed by very large protein molecules, in that the individual particles do not have precisely identical internal atomic or molecular arrangements. On the other hand, they are distinguished from periodic stackings of macroscopic objects like cannonballs in that the periodic ordering is spontaneously adopted by the system through the thermal agitation (brownian motion) of the particles. These conditions limit the sizes of particles which can form colloidal crystals to the range from about 0.01 to about 5 micrometers. *See* BROWNIAN MOVEMENT; KINETIC THEORY OF MATTER.

The most spectacular evidence for colloidal crystallization is the existence of naturally occurring opals. The ideal opal structure is a periodic close-packed three-dimensional array of silica microspheres with hydrated silica filling the spaces not occupied by particles. Opals are the fossilized remains of an earlier colloidal crystal suspension. Another important class of naturally occurring colloidal crystals are found in concentrated suspensions of nearly spherical virus particles, such as *Tipula* iridescent virus and tomato bushy stunt virus. Colloidal crystals can also be made from the synthetic monodisperse colloids, suspensions of plastic (organic polymer) microspheres. Such suspensions have become important systems for the study of colloidal crystals, by virtue of the controllability of the particle size and interaction. *See* CRYSTAL.   [N.A.C.]

Color   That aspect of visual sensation enabling a human observer to distinguish differences between two structure-free fields of light having the same size, shape, and duration. Although luminance differences alone permit such discriminations to be made, the term color is usually restricted to a class of differences still perceived at equal luminance. These depend upon physical differences in the spectral compositions of the two fields, usually revealed to the observer as differences of hue or saturation.

Color discriminations are possible because the human eye contains three classes of cone photoreceptors that differ in the photopigments they contain and in their neural connections. Two of these, the R and G cones, are sensitive to all wavelengths of the visible spectrum from 380 to 700 nanometers. (Even longer or shorter wavelengths may be effective if sufficient energy is available.) R cones are maximally sensitive at about 570 nm, G cones at about 540 nm. The ratio R/G of cone sensitivities is minimal at 465 nm and increases monotonically for wavelengths both shorter and longer than this. This ratio is independent of intensity, and the red-green dimension of color variation is encoded in terms of it. The B cones, whose sensitivity peaks at about 440 nm, are not appreciably excited by wavelengths longer than 540 nm. The perception of blueness and yellowness depends upon the level of excitation of B cones in relation to that of R and G cones. No two wavelengths of light can produce equal excitations in all three kinds of cones. It follows that, provided they are sufficiently different to be discriminable, no two wavelengths can give rise to identical sensations.

Different complex spectral distributions usually, but not always, look different. Suitable amounts of short-, middle-, and long-wavelength lights, if additively mixed, can for example excite the R, G, and B cones exactly as does a light containing equal energy at all wavelengths. As a result, both stimuli look the same. This is an extreme example of the subjective identity of physically different stimuli known as chromatic metamerism. Additive mixture is achievable by optical superposition, rapid alternation at frequencies too high for the visual system to follow, or (as in color television) by the juxtaposition of very small elements which make up a field structure so fine as to exceed the limits of visual acuity. *See* LIGHT.

Although colors are often defined by appeal to standard samples, the trivariant nature of color vision permits their specification in terms of three values. Ideally these might be the relative excitations of the R, G, and B cones. Because too little was known about cone action spectra in 1931, the International Commission on Illumination (CIE) adopted at that time a different but related system for the prediction of metamers (the CIE system of colorimetry). This widely used system permits the specification of tristimulus values X, Y, and Z, which make almost the same predictions about color matches as do calculations based upon cone action spectra. If, for fields 1 and 2, $X_1 = X_2$, $Y_1 = Y_2$, and $Z_1 = Z_2$, then the two stimuli are said to match (and therefore have the same color) whether they are physically the same (isometric) or different (metameric).

Colors are often specified in a two-dimensional chart known as the CIE chromaticity diagram, which shows the relations among tristimulus values independently of luminance. In this plane, y is by convention plotted as a function of x, where $y = Y/(X + Y + Z)$ and $x = x/(x + Y + Z)$. [The value $z = Z/(X + Y + Z)$ also equals $1 - (x + y)$ and therefore carries no additional information.] Such a diagram is shown in the

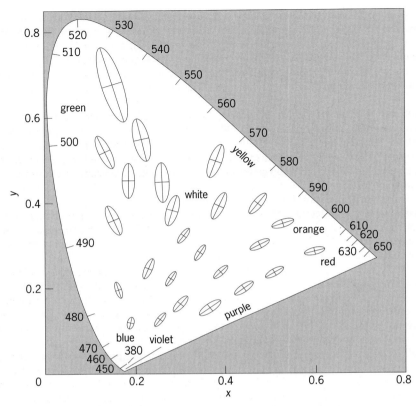

The 1931 CIE chromaticity diagram showing discrimination ellipses enlarged 10 times.

illustration, in which the continuous locus of spectrum colors is represented by the outermost contour. All nonspectral colors are contained within an area defined by this boundary and a straight line running from red to violet. The diagram also shows discrimination data for 25 regions, which plot as ellipses represented at 10 times their actual size. A discrimination unit is one-tenth the distance from the ellipse's center to its perimeter. Predictive schemes for interpolation to other regions of the CIE diagram have been worked out.

A chromaticity diagram has some very convenient properties. Chief among them is the fact that additive mixtures of colors plot along straight lines connecting the chromaticities of the colors being mixed. Although it is sometimes convenient to visualize colors in terms of the chromaticity chart, it is important to realize that this is not a psychological color diagram. Rather, the chromaticity diagram makes a statement about the results of metameric color matches, in the sense that a given point on the diagram represents the locus of all possible metamers plotting at chromaticity coordinates x, y. However, this does not specify the appearance of the color, which can be dramatically altered by preexposing the eye to colored lights (chromatic adaptation) or, in the complex scenes of real life, by other colors present in nearby or remote areas (color contrast and assimilation). Nevertheless, within limits, metamers whose color appearance is thereby changed continue to match.

For simple, directly fixated, and unstructured fields presented in an otherwise dark environment, there are consistent relations between the chromaticity coordinates of a color and the color sensations that are elicited. Therefore, regions of the chromaticity diagram are often denoted by color names, as shown in the illustration.

Although the CIE system works rather well in practice, there are important limitations. Normal human observers do not agree exactly about their color matches, chiefly because of the differential absorption of light by inert pigments in front of the photoreceptors. Much larger individual differences exist for differential colorimetry, and the system is overall inappropriate for the 4% of the population (mostly males) whose color vision is abnormal. The system works only for an intermediate range of luminances, below which rods (the receptors of night vision) intrude, and above which the bleaching of visual photopigments significantly alters the absorption spectra of the cones.   [R.M.Bo.]

Color (quantum mechanics)
A term used to refer to a hypothetical quantum number carried by the quarks which are thought to make up the strongly interacting elementary particles. It has nothing to do with the ordinary, visual use of the word color.

The quarks which are thought to make up the strongly interacting particles have a spin angular momentum of one-half unit of h (Planck's constant). According to a fundamental theorem of relativity combined with quantum mechanics, they must therefore obey Fermi-Dirac statistics and be subject to the Pauli exclusion principle. No two quarks within a particular system can have exactly the same quantum numbers. *See* EXCLUSION PRINCIPLE; FERMI-DIRAC STATISTICS.

However, in making up a baryon, it often seemed necessary to violate this principle. The Ω^- particle, for example, is made of three strange quarks, and all three had to be in exactly the same state. O. W. Greenberg is responsible for the essential idea for the solution to this paradox. In 1964 he suggested that each quark type (u, d, and s) comes in three varieties identical in all measurable qualities but different in an additional property, which has come to be known as color. The exclusion principle could then be satisfied and quarks could remain fermions, because the quarks in the baryon would not all have the same quantum numbers. They would differ in color even if they were the same in all other respects. *See* BARYON; ELEMENTARY PARTICLE; MESON; QUARKS.   [T.Ap.]

Color centers
Atomic and electronic defects of various types which produce optical absorption bands in otherwise transparent crystals such as the alkali halides, alkaline earth fluorides, or metal oxides. They are general phenomena found in a wide range of materials. Color centers are produced by gamma radiation or x-radiation, by addition of impurities or excess constituents, and sometimes through electrolysis. A well-known example is that of the F-center in alkali halides such as sodium chloride, NaCl. The designation F-center comes from the German word *Farbe*, which means color. F-centers in NaCl produce a band of optical absorption toward the blue end of the visible spectrum; thus the colored crystal appears yellow under transmitted light. On the other hand, KCl with F-centers appears magenta, and KBr appears blue. *See* CRYSTAL.

Color centers have been under investigation for many years. Theoretical studies guided by detailed experimental work have yielded a deep understanding of specific centers. The crystals in which color centers appear tend to be transparent to light and to microwaves. Consequently, experiments which can be carried out include optical spectroscopy, luminescence and Raman scattering, magnetic circular dichroism, magnetic

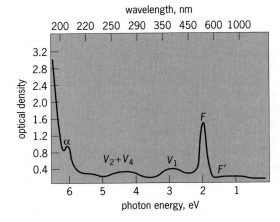

Absorption bands produced in a KBr crystal by exposure to x-rays at 81 K. Bands are designated by letters. Optical density is equal to $\log_{10} (I_0/I)$, where I_0 is the intensity of incident light and I the intensity of transmitted light. Steep rise in optical density at far left is due to intrinsic absorption in crystals (modified slightly by existence of F-centers).

resonance, and electromodulation. Color centers find practical application in radiation dosimeters; schemes have been proposed to use color centers in high-density memory devices; and tunable lasers have been made from crystals containing color centers.

The illustration shows the absorption bands due to color centers produced in potassium bromide by exposure of the crystal at the temperature of liquid nitrogen (81 K) to intense penetrating x-rays. Several prominent bands appear as a result of the irradiation. The F-band appears at 600 nanometers and the so-called V-bands appear in the ultraviolet.

Color bands such as the F-band and the V-band arise because of light absorption at defects dispersed throughout the lattice. This absorption is caused by electronic transitions at the centers. On the other hand, colloidal particles, each consisting of many atoms, dispersed through an optical medium also produce color bands. In this case, if the particles are large enough, the extinction of light is due to both light scattering and light absorption. Colloidal gold is responsible for the color of some types of ruby glass. Colloids may also form in alkali halide crystals—for example, during heat treatment of an additively colored crystal with an excess of alkali metal.

Atomically dispersed centers such as F-centers are part of the general phenomena of trapped electrons and holes in solids. The accepted model of the F-center is an electron trapped at a negative ion vacancy. Many other combinations of electrons, holes, and clusters of lattice vacancies have been used to explain the various absorption bands in ionic crystals.

Impurities can play an important role in color-center phenomena. Certain impurities in ionic crystals produce color bands characteristic of the foreign ion. For example, hydrogen can be incorporated into the alkali halides with resultant appearance of an absorption band (the U-band) in the ultraviolet. In this case, the U-centers interact with other defects. The rate at which F-centers are produced by x-irradiation is greatly increased by the incorporation of hydrogen, the U-centers being converted into F-centers with high efficiency.   [F.C.Br.]

Color filter   An optical element that partially absorbs incident radiation, often called an absorption filter. The absorption is selective with respect to wavelength, or color, limiting the colors that are transmitted by limiting those that are absorbed. Color filters absorb all the colors not transmitted. They are used in photography, optical instruments, and illuminating devices to control the amount and spectral composition of the light.

Color filters are made of glass for maximum permanence, of liquid solutions in cells with transparent faces for flexibility of control, and of dyed gelatin or plastic (usually cellulose acetate) for economy, convenience, and flexibility. The plastic filters are often of satisfactory permanence, but they are sometimes cemented between glass plates for greater toughness and scratch resistance. They do not have as good quality as gelatin filters.

Color filters are sometimes classified according to their type of spectral absorption: short-wavelength pass, long-wavelength pass or band-pass; diffuse or sharp-cutting; monochromatic or conversion. The short-wavelength pass transmits all wavelengths up to the specified one and then absorbs. The long-wavelength pass is the opposite. Every filter is a band-pass filter when considered generally. Even an ordinary piece of glass does not transmit in the ultraviolet or infrared parts of the spectrum. Color filters, however, are usually discussed in terms of the portion of the visible part of the spectrum. Sharp and diffuse denote the sharpness of the edges of the filter band pass. Monochromatic filters are very narrow band-pass filters. Conversion filters alter the spectral response or distribution of one selective detector or source to that of another, for example, from that of a light bulb to that of the Sun. *See* ABSORPTION OF ELECTROMAGNETIC RADIATION; COLOR.

[W.L.Wo.]

Compressible flow   Flow in which density changes are significant. Pressure changes normally occur throughout a fluid flow, and these pressure changes, in general, induce a change in the fluid density. In a compressible flow, the density changes that result from these pressure changes have a significant influence on the flow. The changes in the flow that result from the density changes are often termed compressibility effects. All fluids are compressible. However, compressibility effects are more frequently encountered in gas flows than in liquid flows.

An important dimensionless parameter in compressible flows is the Mach number, M. This is defined by Eq. (1), where a is the speed of sound and V is the velocity of

$$M = \frac{V}{a} \tag{1}$$

the flow. For a gas, the speed of sound is given by Eq. (2), where R is the gas constant,

$$a = \sqrt{kRT} \tag{2}$$

$k = c_p/c_v$, c_p and c_v being the specific heats at constant pressure and constant volume respectively, and T is the temperature. If $M < 0.3$ in a flow, the density changes in the flow will usually be negligible; that is, the flow can be treated as incompressible. Compressible flows are, therefore, as a rough guide, associated with Mach numbers greater than 0.3.

When $M < 1$, the flow is said to be subsonic; when $M = 1$, the flow is said to be sonic; when M varies from slightly below 1 to slightly above 1, the flow is said to be transonic; and if $M > 1$, the flow is said to be supersonic. When the Mach number is very high, this usually being taken to mean $M > 5$, the flow is said to be hypersonic.

Compressible flows can have features that do not occur in low-speed flows. For example, shock waves and expansion waves can occur in supersonic flows. Another important phenomenon that can occur due to compressibility is choking, where the mass flow rate through a duct system may be limited as a result of the Mach number being equal to 1 at some point in the flow. *See* CHOKED FLOW; SHOCK WAVE; SONIC BOOM.

Another effect of compressibility is associated with the acceleration of a gas flow through a duct. In incompressible flow, an increase in velocity is associated with a decrease in the cross-sectional area of the duct, this in fact being true as long as $M < 1$. However, when $M > 1$, that is, when the flow is supersonic, the opposite is true; that is, an increase in the velocity is associated with an increase in the cross-sectional area. Therefore, in order to accelerate a gas flow from subsonic to supersonic velocities in a duct, it is necessary first to decrease the area and then, once the Mach number has reached 1, to increase the area, that is, to use a so-called convergent-divergent nozzle. An example is the nozzle fitted to a rocket engine. *See* FLUID FLOW; MACH NUMBER; SUPERSONIC FLOW.   [P.H.Oo.]

Compton effect   The increase in wavelength of electromagnetic radiation, observed mainly in the x-ray and gamma-ray region, on being scattered by material objects. This increase in wavelength is caused by the interaction of the radiation with the weakly bound electrons in the matter in which the scattering takes place. The Compton effect illustrates one of the most fundamental interactions between radiation and matter and displays in a very graphic way the true quantum nature of electromagnetic radiation. Together with the laws of atomic spectra, the photoelectric effect, and pair production, the Compton effect has provided the experimental basis for the quantum theory of electromagnetic radiation. *See* ANGULAR MOMENTUM; ATOMIC STRUCTURE AND SPECTRA; ELECTRON-POSITRON PAIR PRODUCTION; LIGHT; PHOTOEMISSION; QUANTUM MECHANICS; UNCERTAINTY PRINCIPLE.

Perhaps the greatest significance of the Compton effect is that it demonstrates directly and clearly that in addition to its wave nature with transverse oscillations, electromagnetic radiation has a particle nature and that these particles, the photons, behave quite like material particles in collisions with electrons. This discovery by A. H. Compton and P. Debye led to the formulation of quantum mechanics by W. Heisenberg and E. Schrödinger and provided the basis for the beginning of the theory of quantum electrodynamics, the theory of the interactions of electrons with the electromagnetic field.

The Compton effect has played a significant role in several diverse scientific areas. Compton scattering (often referred to as incoherent scattering, in contrast to Thomson scattering or also Rayleigh scattering, which are called coherent scattering) is important in nuclear engineering (radiation shielding), experimental and theoretical nuclear physics, atomic physics, plasma physics, x-ray crystalloghaphy, elementary particle physics, and astrophysics, to mention some of these areas. In addition the Compton effect provides an important research tool in some branches of medicine, in molecular chemistry and solid-state physics, and in the use of high-energy electron accelerators and charged-particle storage rings.   [E.N.H.]

The development of high-resolution silicon and germanium semiconductor radiation detectors opened new areas for applications of Compton scattering. Semiconductor detectors make it possible to measure the separate probabilities for Rayleigh and Compton scattering. An effective atomic number has been assigned to compounds that appears to successfully correlate theory with Rayleigh-Compton ratios.

Average density can be measured by moving to higher energies where Compton scattering does not have to compete with Rayleigh scattering. At these energies, Compton scattering intensity has been successfully correlated with mass density. An appropriate application is the measurement of lung density in living organisms.

The ability to put large detectors in orbit above the Earth' atmosphere has created the field of gamma-ray astronomy. This field is now based largely on the data from the *Compton Gamma-Ray Observatory*, all of whose detectors made use of

the Compton effect (although not exclusively). *See* GAMMA-RAY DETECTORS; GAMMA RAYS; X-RAYS.
[I.K.M.]

Computational fluid dynamics

The numerical approximation to the solution of mathematical models of fluid flow and heat transfer. Computational fluid dynamics is one of the tools (in addition to experimental and theoretical methods) available to solve fluid-dynamic problems. With the advent of modern computers, computational fluid dynamics evolved from potential-flow and boundary-layer methods and is now used in many diverse fields, including engineering, physics, chemistry, meteorology, and geology. The crucial elements of computational fluid dynamics are discretization, grid generation and coordinate transformation, solution of the coupled algebraic equations, turbulence modeling, and visualization.

Numerical solution of partial differential equations requires representing the continuous nature of the equations in a discrete form. Discretization of the equations consists of a process where the domain is subdivided into cells or elements (that is, grid generation) and the equations are expressed in discrete form at each point in the grid by using finite difference, finite volume, or finite element methods. The finite difference method requires a structured grid arrangement (that is, an organized set of points formed by the intersections of the lines of a boundary-conforming curvilinear coordinate system), while the finite element and finite volume methods are more flexible and can be formulated to use both structured and unstructured grids (that is, a collection of triangular elements or a random distribution of points).

There are a variety of approaches for resolving the phenomena of fluid turbulence. The Reynolds-averaged Navier-Stokes (RANS) equations are derived by decomposing the velocity into mean and fluctuating components. An alternative is large-eddy simulation, which solves the Navier-Stokes equations in conjunction with a subgrid turbulence model. The most direct approach to solving turbulent flows is direct numerical simulation, which solves the Navier-Stokes equations on a mesh that is fine enough to resolve all length scales in the turbulent flow. Unfortunately, direct numerical simulation is limited to simple geometries and low-Reynolds-number flows because of the limited capacity of even the most sophisticated supercomputers. *See* TURBULENT FLOW.

The final step is to visualize the results of the simulation. Powerful graphics workstations and visualization software permit generation of velocity vectors, pressure and velocity contours, streamline generation, calculation of secondary quantities (such as vorticity), and animation of unsteady calculations. Despite the sophisticated hardware, visualization of three-dimensional and unsteady flows is still particularly difficult. Moreover, many advanced visualization techniques tend to be qualitative, and the most valuable visualization often consists of simple x-y plots comparing the numerical solution to theory or experimental data.

Computational fluid dynamics has wide applicability in such areas as aerodynamics, hydraulics, environmental fluid dynamics, and atmospheric and oceanic dynamics, with length and time scales of the physical processes ranging from millimeters and seconds to kilometers and years. Vehicle aerodynamics and hydrodynamics, which have provided much of the impetus in the development of computational fluid dynamics, are primarily concerned with the flow around aircraft, automobiles, and ships. *See* AERODYNAMIC FORCE; AERODYNAMICS; FLUID FLOW; HYDRODYNAMICS.
[E.Pa.; F.Ste.]

Condensation

A phase-change process in which vapor converts into liquid when the temperature of the vapor is reduced below the saturation temperature corresponding to the pressure in the vapor. For a pure vapor this pressure is the total pressure,

Steam at atmospheric pressure condensing on a vertical copper surface. Film condensation is visible on the right side, and dropwise condensation in the presence of a promoter is visible on the left side. The horizontal tube is a thermocouple. (*J. F. Welch and J. W. Westwater, Department of Chemical Engineering, University of Illinois, Urbana*)

whereas in a mixture of a vapor and a noncondensable gas it is the partial pressure of the vapor. Sustaining the process of condensation on a cold surface in a steady state requires cooling of the surface by external means. Condensation is an efficient heat transfer process and is utilized in various industrial applications. Condensation of vapor on a cold surface can be classified as filmwise or dropwise. Direct-contact condensation refers to condensation of vapor (bubbles or a vapor stream) in a liquid or condensation on liquid droplets entrained in the vapor. If vapor temperature falls below its saturation temperature, condensation can occur in the bulk vapor. This phenomenon is called homogeneous condensation (formation of fog) and is facilitated by foreign particles such as dust. *See* HEAT TRANSFER.

In film condensation, a thin film of liquid forms upon condensation of vapor on a cold surface that is well wetted by the condensate. The liquid film flows downward as a result of gravity.

In dropwise condensation, on surfaces that are not well wetted, vapor may condense in the form of droplets (see illustration). The droplets form on imperfections such as cavities, dents, and cracks on the surface. The droplets of 10–100 μm diameter contribute most to the heat transfer rate. As a droplet grows to a size that can roll down the surface because of gravity, it wipes the surface of the droplets in its path. In the wake behind the large droplet, numerous smaller droplets form and the process repeats. The heat transfer coefficients with dropwise condensation can be one to two orders of magnitude greater then that for film condensation.

Direct-condensation involves condensation of vapor bubbles in a host liquid and condensation on droplets entrained in vapor. Both are also very efficient heat transfer processes, especially when the vapor-liquid interface oscillates.   [V.K.D.]

Conductance   The real part of the admittance of an alternating-current circuit. The admittance Y of an alternating-current circuit is a complex number given by Eq. (1).

$$Y = G + jB \qquad (1)$$

The real part G is the conductance. The units of conductance, like those of admittance, are called siemens or mhos. Conductance is a positive quantity. The conductance of a resistor R is given by Eq. (2).

$$G = \frac{1}{R} \qquad (2)$$

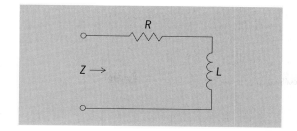

Circuit with a resistor and inductor in series.

In general the conductance of a circuit may depend on the capacitors and inductors in the circuit as well as on the resistors. For example, the circuit in the illustration has impedance at frequency ω given by Eq. (3) and admittance given by Eq. (4), so that the conductance, given by Eq. (5), depends on the inductance L as

$$Z = R + jL\omega \tag{3}$$

$$Y = \frac{1}{R + jL\omega} \tag{4}$$

$$G = \frac{R}{R^2 + L^2\omega^2} \tag{5}$$

well as the resistance R. See ADMITTANCE; ELECTRICAL IMPEDANCE. [J.O.S.]

Conduction (electricity) The passage of electric charges due to a force exerted on them by an electric field. Conductivity is the measure of the ability of a conductor to carry electric current; it is defined as the ratio of the amount of charge passing through unit area of the conductor (perpendicular to the current direction) per second divided by the electric field intensity (the force on a unit charge). Conductivity is the reciprocal of resistivity and is therefore commonly expressed in units of siemens per meter, abbreviated S/m. See ELECTRICAL RESISTIVITY.

In metals and semiconductors (such as silicon, of which transistors are made) the charges that are responsible for current are free electrons and holes (which, as missing electrons, act like positive charges). These are electrons or holes not bound to any particular atom and therefore able to move freely in the field. Conductivity due to electrons is known as n-type conductivity; that due to holes is known as p-type. See HOLE STATES IN SOLIDS; SEMICONDUCTOR.

The conductivity of metals is much higher than that of semiconductors because they have many more free electrons or holes. The free electrons or holes come from the metal atoms. Semiconductors differ from metals in two important respects. First, the semiconductor atoms do not contribute free electrons or holes unless thermally excited, and second, free electrons or holes can also arise from impurities or defects.

An exception to some of the rules stated above has been found in conjugated polymers. Polyacetylene, for example, although a semiconductor with extremely high resistance when undoped, can be doped so heavily with certain nonmetallic impurities (iodine, for example) that it attains a conductivity comparable to that of copper.

In metals, although the number of free carriers does not vary with temperature, an increase in temperature decreases conductivity. The reason is that increasing temperature causes the lattice atoms to vibrate more strongly, impeding the motion of the free carriers in the field. This effect also occurs in semiconductors, but the increase in number of free carriers with temperature is usually a stronger effect. At low temperatures the thermal vibrations are weak, and the impediment to the motion of free carriers in the field comes from imperfections and impurities, which in metals usually does not vary with temperature. At the lowest temperatures, close to absolute zero, certain metals become superconductors, possessing infinite conductivity. *See* SUPER-CONDUCTIVITY.

Electrolytes conduct electricity by means of the positive and negative ions in solution. In ionic crystals, conduction may also take place by the motion of ions. This motion is much affected by the presence of lattice defects such as interstitial ions, vacancies, and foreign ions. *See* IONIC CRYSTALS.

Electric current can flow through an evacuated region if electrons or ions are supplied. In a vacuum tube the current carriers are electrons emitted by a heated filament. The conductivity is low because only a small number of electrons can be "boiled off" at the normal temperatures of electron-emitting filaments. *See* ELECTRON EMISSION.   [E.M.Co.]

Conduction (heat)   The flow of thermal energy through a substance from a higher- to a lower-temperature region. Heat conduction occurs by atomic or molecular interactions. Conduction is one of the three basic methods of heat transfer, the other two being convection and radiation. *See* CONVECTION (HEAT); HEAT RADIATION; HEAT TRANSFER.

Steady-state conduction is said to exist when the temperature at all locations in a substance is constant with time, as in the case of heat flow through a uniform wall. Examples of essentially pure transient or periodic heat conduction and simple or complex combinations of the two are encountered in the heat-treating of metals, air conditioning, food processing, and the pouring and curing of large concrete structures. Also, the daily and yearly temperature variations near the surface of the Earth can be predicted reasonably well by assuming a simple sinusoidal temperature variation at the surface and treating the Earth as a semi-infinite solid. The widespread importance of transient heat flow in particular has stimulated the development of a large variety of analytical solutions to many problems. The use of many of these has been facilitated by presentation in graphical form.

For an example of the conduction process, consider a gas such as nitrogen which normally consists of diatomic molecules. The temperature at any location can be interpreted as a quantitative specification of the mean kinetic and potential energy stored in the molecules or atoms at this location. This stored energy will be partly kinetic because of the random translational and rotational velocities of the molecules, partly potential because of internal vibrations, and partly ionic if the temperature (energy) level is high enough to cause dissociation. The flow of energy results from the random travel of high-temperature molecules into low-temperature regions and vice versa. In colliding with molecules in the low-temperature region, the high temperature molecules give up some of their energy. The reverse occurs in the high-temperature region. These processes take place almost instantaneously in infinitesimal distances, the result being a quasi-equilibrium state with energy transfer. The mechanism for energy flow in liquids and solids is similar to that in gases in principle, but different in detail.   [W.H.Gi.]

Conduction band   The electronic energy band of a crystalline solid which is partially occupied by electrons. The electrons in this energy band can increase their energies by going to higher energy levels within the band when an electric field is applied to accelerate them or when the temperature of the crystal is raised. These electrons are called conduction electrons, as distinct from the electrons in filled energy bands, which, as a whole, do not contribute to electrical and thermal conduction. In metallic conductors the conduction electrons correspond to the valence electrons (or a portion of the valence electrons) of the constituent atoms. In semiconductors and insulators at sufficiently low temperatures, the conduction band is empty of electrons. Conduction electrons come from thermal excitation of electrons from a lower energy band or from impurity atoms in the crystal. *See* BAND THEORY OF SOLIDS; ELECTRIC INSULATOR; SEMICONDUCTOR; VALENCE BAND.   [H.Y.F.]

Conformal optics   Conformal optical systems have outer surfaces whose shape is chosen to optimize the interaction with the environment in which the optical system is being used. The imaging through such conformal optical windows is likely to suffer from

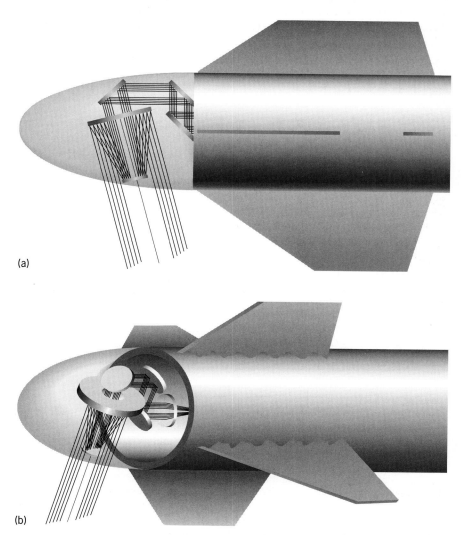

(a)

(b)

Geometric optics of an optical tracker behind a conformal window that is shaped to provide an improved drag profile for a missile. Paths of rays through the optical system are shown. (*a*) Side view. (*b*) Oblique view. (*Optical Research Associates*)

extreme aberration, requiring special techniques for correction. Computer-intensive methods of design, fabrication, and testing of optics have reached a level where the development of cost-effective methods for insertion of these conformal optics concepts into operational systems appears to be practical. *See* ABERRATION (OPTICS).

Important applications of conformal optics are found in missile and aircraft systems. Missiles and aircraft carry optical sensors for imaging, detection, and ranging that must look at the world through the outer skin of the vehicle. Traditionally, the windows for viewing through the skin of missiles and aircraft have had simple optical forms, such as flats or spheres, that enable the optical tracking systems to operate by using well-known technology. But these optically advantageous windows degrade the performance of the vehicle through increased drag, aerodynamic heating, or other undesirable effects. One example is the use of an optical tracker or seeker on the front end of a missile. The use of a conformal window, whose shape conforms more closely to the optimal, pointed ogival shape, reduces the drag of the missile and provides significant gains in the missile's performance. Such ogival shapes produce considerable optical aberration, however (see illustration).

The approach to design with conformal optics does not call for complete abandonment of current understanding of the image-formation process. Optical design methods are based on the description of the wavefront passing through surfaces by use of numerical ray tracing. The understanding of the aberrations arising at surfaces is obtained from an analytic method for describing the surface and the wavefront to stated levels of accuracy. The aberrations produced by general aspheric surfaces defy simple analytic descriptions but can be obtained by fitting of the numerical ray-tracing results. *See* GEOMETRICAL OPTICS.   [R.R.S.]

Conservation laws (physics)

Conservation laws (physics)   Principles which state that the total values of specified quantities remain constant in time for an isolated system. Conservation laws occupy enormously important positions both at the foundations of physics and in its applications.

Realization in classical mechanics. There are three great conservation laws of mechanics: the conservation of linear momentum, often referred to simply as the conservation of momentum; the conservation of angular momentum; and the conservation of energy.

The linear momentum, or simply momentum, of a particle is equal to the product of its mass and velocity. It is a vector quantity. The total momentum of a system of particles is simply the sum of the momenta of each particle considered separately. The law of conservation of momentum states that this total momentum does not change in time. *See* CONSERVATION OF MOMENTUM; MOMENTUM.

The angular momentum of a particle is more complicated. It is defined by the vector product of the position and momentum vectors. The law of conservation of angular momentum states that the total angular momentum of an isolated system is constant in time. *See* ANGULAR MOMENTUM.

The conservation of energy is perhaps the most important law of all. Energy is a scalar quantity, and takes two forms: kinetic and potential. The kinetic energy of a particle is defined to be one-half the product of its mass and the square of its velocity. The potential energy is loosely defined as the ability to do work. The total energy is the sum of the kinetic and potential energies, and according to the conservation law it remains constant in time for an isolated system.

The essential difficulty in applying the conservation of energy law can be appreciated by considering the problem of two colliding bodies. In general, the bodies emerge from the collision moving more slowly than when they entered. This phenomenon seems to violate the conservation of energy, until it is recognized that the bodies involved may consist of smaller particles. Their random small-scale motions will require kinetic energy, which robs kinetic energy from the overall coherent large-scale motion of the bodies that are observed directly. One of the greatest achievements of nineteenth-century physics was the recognition that small-scale motion within macroscopic bodies could be identified with the perceived property of heat. *See* CONSERVATION OF ENERGY; ENERGY; KINETIC THEORY OF MATTER.

Position in modern physics. As physics has evolved, the great conservation laws have likewise evolved in both form and content, but have never ceased to be important guiding principles.

In order to account for the phenomena of electromagnetism, it was necessary to go beyond the notion of point particles, to postulate the existence of continuous electric and magnetic fields filling all space. To obtain valid conservation laws, energy, momentum, and angular momentum must be ascribed to the electromagnetic fields. *See* ELECTROMAGNETIC RADIATION; MAXWELL'S EQUATIONS; POYNTING'S VECTOR.

In the special theory of relativity, energy and momentum are not independent concepts. Einstein discovered perhaps the most important consequence of special relativity, that is, the equivalence of mass and energy, as a consequence of the conservation laws. The "law" of conservation of mass is understood as an approximate consequence of the conservation of energy. *See* CONSERVATION OF MASS; RELATIVITY.

A remarkable, beautiful, and very fruitful connection has been established between symmetries and conservation laws. Thus the law of conservation of linear momentum is understood as a consequence of the homogeneity of space, the conservation of angular momentum as a consequence of the isotropy of space, and the conservation of energy as a consequence of the homogeneity of time. *See* SYMMETRY LAWS (PHYSICS).

The development of general relativity, the modern theory of gravitation, necessitates attention to a fundamental question for the conservation laws: The laws refer to an "isolated system," but it is not clear that any system is truly isolated. This is a particularly acute problem for gravitational forces, which are long range and add up over cosmological distances. It turns out that the symmetry of physical laws is actually a more fundamental property than the conservation laws themselves, for the symmetries remain valid while the conservation laws, strictly speaking, fail.

In quantum theory, the great conservation laws remain valid in a very strong sense. Generally, the formalism of quantum mechanics does not allow prediction of the outcome of individual experiments, but only the relative probability of different possible outcomes. One might therefore entertain the possibility that the conservation laws were valid only on the average. However, momentum, angular momentum, and energy are conserved in every experiment. *See* QUANTUM MECHANICS; QUANTUM THEORY OF MEASUREMENT.

Conservation laws of particle type. There is another important class of conservation laws, associated not with the motion of particles but with their type. Perhaps the most practically important of these laws is the conservation of chemical elements. From a modern viewpoint, this principle results from the fact that the small amount of energy involved in chemical transformations is inadequate to disrupt the nuclei deep within atoms. It is not an absolute law, because some nuclei decay spontaneously, and at sufficiently high energies it is grossly violated. *See* RADIOACTIVITY.

Several conservation laws in particle physics are of the same character: They are useful even though they are not exact because, while known processes violate them, such processes are either unusually slow or require extremely high energy. *See* ELEMENTARY PARTICLE.   [F.Wil.]

Conservation of energy   The principle of conservation of energy states that energy cannot be created or destroyed, although it can be changed from one form to another. Thus in any isolated or closed system, the sum of all forms of energy remains constant. The energy of the system may be interconverted among many different forms—mechanical, electrical, magnetic, thermal, chemical, nuclear, and so on—and as time progresses, it tends to become less and less available; but within the limits of small experimental uncertainty, no change in total amount of energy has been observed in any situation in which it has been possible to ensure that energy has not entered or left the system in the form of work or heat. For a system that is both gaining and losing energy in the form of work and heat, as is true of any machine in operation, the energy principle asserts that the net gain of energy is equal to the total change of the system's internal energy. *See* THERMODYNAMIC PRINCIPLES.

There are many ways in which the principle of conservation of energy may be stated, depending on the intended application. Of particular interest is the special form of the principle known as the principle of conservation of mechanical energy which states that the mechanical energy of any system of bodies connected together in any way is conserved, provided that the system is free of all frictional forces, including internal friction that could arise during collisions of the bodies of the system.

J. P. Joule and others demonstrated the equivalence of heat and work by showing experimentally that for every definite amount of work done against friction there always appears a definite quantity of heat. The experiments usually were so arranged that the heat generated was absorbed by a given quantity of water, and it was observed that a given expenditure of mechanical energy always produced the same rise of temperature in the water. The resulting numerical relation between quantities of mechanical energy and heat is called the Joule equivalent, or is also known as mechanical equivalent of heat.

In view of the principle of equivalence of mass and energy in the restricted theory of relativity, the classical principle of conservation of energy must be regarded as a special case of the principle of conservation of mass-energy. However, this more general principle need be invoked only when dealing with certain nuclear phenomena or when speeds comparable with the speed of light (1.86×10^5 mi/s or 3×10^8 m/s) are involved.
                                                                            [D.E.R./L.N.]

Conservation of mass   The notion that mass, or matter, can be neither created nor destroyed. According to conservation of mass, reactions and interactions which change the properties of substances leave unchanged their total mass; for instance, when charcoal burns, the mass of all of the products of combustion, such as ashes, soot, and gases, equals the original mass of charcoal and the oxygen with which it reacted.

The special theory of relativity of Albert Einstein, which has been verified by experiment, has shown, however, that the mass of a body changes as the energy possessed by the body changes. Such changes in mass are too small to be detected except in subatomic phenomena. Furthermore, matter may be created, for instance, by the materialization of a photon (quantum of electromagnetic energy) into an electron-positron pair; or it may be destroyed, by the annihilation of this pair of elementary

particles to produce a pair of photons. *See* ELECTRON-POSITRON PAIR PRODUCTION; RELATIVITY.
[L.N.]

Conservation of momentum
The principle that, when a system of masses is subject only to forces that masses of the system exert on one another, the total vector momentum of the system is constant. Since vector momentum is conserved, in problems involving more than one dimension the component of momentum in any direction will remain constant. The principle of conservation of momentum holds generally and is applicable in all fields of physics. In particular, momentum is conserved even if the particles of a system exert forces on one another or if the total mechanical energy is not conserved. Use of the principle of conservation of momentum is fundamental in the solution of collision problems. *See* COLLISION (PHYSICS); MOMENTUM.
[P.W.S.]

Constraint
A restriction on the natural degrees of freedom of a system. If n and m are the numbers of the natural and actual degrees of freedom, the difference $n - m$ is the number of constraints. In principle $n = 3N$, where N is the number of particles, for example, atoms. In practice n is determined by the number of effectively rigid components.

A holonomic system is one in which the n original coordinates can be expressed in terms of m independent coordinates and possibly also the time. It is characterized by frictionless contacts and inextensible linkages. The new coordinates are called generalized coordinates. *See* LAGRANGE'S EQUATIONS.

Nonholonomic systems cannot be reduced to independent coordinates because the constraints are not on the n coordinate values themselves but on their possible changes. For example, an ice skate may point in all directions but at each position it must point along its path. *See* DEGREE OF FREEDOM (MECHANICS).
[B.G.]

Contact potential difference
An electrostatic potential that exists between samples of two dissimilar electrically conductive materials (metals or semiconductors with different electron work functions) that have been brought into thermal equilibrium with each other, usually through a physical contact. Although normally measured between two surfaces which are not in contact, this potential is called the contact potential difference. Initially it is expected that mobile charge carriers (electrons or holes) will migrate from one sample to the other. If there is a net flow of electrons from material A to material B (see illustration), material B will become negatively charged and material A will become positively charged, assuming that they were originally neutral. This process is self limiting because a potential difference between the two samples will develop due to the charge separation and will grow to a value sufficient to stop further motion of the electrons from A to B.

In a metal or a semiconductor, the electrons are distributed in energy such that virtually all of them exist at or below a level called the Fermi level. When any combination of metals and semiconductors are put into equilibrium with one another, the Fermi levels in all will coincide. The contact potential difference between materials is that value necessary to raise or lower the potential energies of the electrons to produce a common Fermi level. Since they are then at the same energy, electrons in either material will have no net force on them, that is, no reason to travel to the other material. Because it causes no net force on the equilibrium distribution of electrons, contact potential difference cannot be directly measured with an ordinary voltmeter.

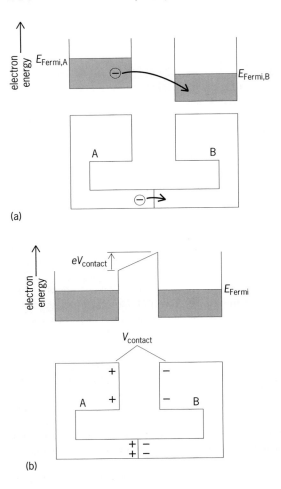

(a)

(b)

Development of a contact potential as two conductive materials are brought into thermal equilibrium. (*a*) Initial charge transfer. (*b*) Thermal equilibrium. Diagrams show corresponding electron energy distributions. E_{Fermi} = Fermi level; $V_{contact}$ = contact potential difference; e = electron charge.

Nevertheless, it profoundly affects the behavior of a number of electronic devices. *See* FREE-ELECTRON THEORY OF METALS; SEMICONDUCTOR; WORK FUNCTION (ELECTRONICS).

[J.E.No.]

Convection (heat)   The transfer of thermal energy by actual physical movement from one location to another of a substance in which thermal energy is stored. A familiar example is the free or forced movement of warm air throughout a room to provide heating. Technically, convection denotes the nonradiant heat exchange between a surface and a fluid flowing over it. Although heat flow by conduction also occurs in this process, the controlling feature is the energy transfer by flow of the fluid—hence the name convection. Convection is one of the three basic methods of heat transfer, the other two being conduction and radiation. *See* CONDUCTION (HEAT); HEAT RADIATION; HEAT TRANSFER.

Natural convection is exemplified by the cooling of a vertical surface in a large quiescent body of air of temperature t_∞. The lower-density air next to a hot vertical surface moves upward because of the buoyant force of the higher-density cool air farther away from the surface. At any arbitrary vertical location x, the actual variation of

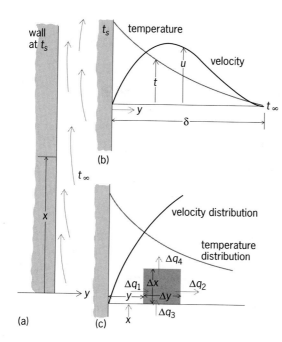

(b)

(a)   (c)

Temperature and velocity distributions in air near a heated vertical surface at arbitary vertical location. The distance δ is that distance at which the velocity and the temperature reach ambient surrounding conditions.

velocity u with distance y from the surface will be similar to that in illus. b, increasing from zero at the surface to a maximum, and then decreasing to zero as ambient surrounding conditions are reached. In contrast, the temperature t of the air decreases from the heated wall value t's to the surrounding air temperature. These temperature and velocity distributions are clearly interrelated, and the distances from the wall through which they exist are coincident because, when the temperature approaches that of the surrounding air, the density difference causing the upward flow approaches zero.

The region in which these velocity and temperature changes occur is called the boundary layer. Because velocity and temperature gradients both approach zero at the outer edge, there will be no heat flow out of the boundary layer by conduction or convection. See BOUNDARY-LAYER FLOW.

When air is blown across a heated surface, forced convection results. Although the natural convection forces are still present in this latter case, they are clearly negligible compared with the imposed forces. The process of energy transfer from the heated surface to the air is not, however, different from that described for natural convection. The major distinguishing feature is that the maximum fluid velocity is at the outer edge of the boundary layer. This difference in velocity profile and the higher velocities provide more fluid near the surface to carry along the heat conducted normal to the surface. Consequently, boundary layers are very thin.

Heat convection in turbulent flow is interpreted similarly to that in laminar flow. Rates of heat transfer are higher for comparable velocities, however, because the fluctuating velocity components of the fluid in a turbulent flow stream provide a macroscopic exchange mechanism which greatly increases the transport of energy normal to the main flow direction. Because of the complexity of this type of flow, most of the information regarding heat transfer has been obtained experimentally. See LAMINAR FLOW; TURBULENT FLOW.

Convection heat transfer which occurs during high-speed flight or high-velocity flow over a surface is known as aerodynamic heating. This heating effect results from the conversion of the kinetic energy of the fluid as it approaches a body to internal energy as it is slowed down next to the surface. In the case of a gas, its temperature increases, first, because of compression as it passes through a shock and approaches the stagnation region, and second, because of frictional dissipation of kinetic energy in the boundary layer along the surface.

The phenomena of condensation and boiling are important phase-change processes involving heat release or absorption. Because vapor and liquid movement are present, the energy transfer is basically by convection. Local and average heat-transfer coefficients are determined and used in the Newton cooling-law equation for calculating heat rates which include the effects of the latent heat of vaporization. [W.H.Gi.]

Correspondence principle   A fundamental hypothesis according to which classical mechanics can be understood as a limiting case of quantum mechanics; or conversely, many characteristic features in quantum mechanics can be approximated on the basis of classical mechanics, provided classical mechanics is properly reinterpreted. This idea was first proposed by N. Bohr in the early 1920s as a set of rules for understanding the spectra of simple atoms and molecules.

The classical motions in simple dynamical systems can be understood as composed of independent partial motions, each with its own degree of freedom. Each degree of freedom accumulates its own classical action-integral. The frequency of the classical motion for any particular degree of freedom is given by the partial derivative of the energy function with respect to the corresponding action. Bohr noticed that this classical result yields the correct quantum-theoretical result for the light frequency in a transition from one energy level to another, provided the derivative is replaced by the difference in the energies. Moreover, precise information about the possibility of such transitions and their intensities is obtained by analyzing the related classical motion. This information becomes better as the quantum numbers involved be come larger. The apparent inconsistencies in Bohr's quantum theory are thereby overcome by a set of rules that came to be called the correspondence principle. *See* ACTION.

After 1925, the success of the new quantum mechanics, particularly wave mechanics, reduced the correspondence principle to a somewhat vague article of faith among physicists. However, the appeal to classical mechanics is still convenient for some rather crude estimates such as the total number of levels below a given energy. Such estimates help in finding the approximate shape of large atoms and large nuclei in the Thomas-Fermi model.

The correspondence principle, however, has assumed a more profound significance. Experimental techniques in atomic, molecular, mesoscopic, and nuclear physics have improved dramatically. High-precision data for many thousands of energy levels are available where the traditional methods of quantum mechanics are not very useful or informative. However, the basic idea behind the correspondence principle must still be valid: Quantum mechanics must be understandable in terms of classical mechanics for the highly excited states, even in difficult cases like the three-body problem, where the overall behavior seems unpredictable and chaotic. The wider application of Bohr's correspondence principle allows many basic but difficult problems to be seen in a new light. *See* ATOMIC STRUCTURE AND SPECTRA; CHAOS; MESOSCOPIC PHYSICS; MOLECULAR STRUCTURE AND SPECTRA; QUANTUM MECHANICS. [M.C.G.]

Coulomb excitation   Nuclear excitation caused by the time-dependent long-ranged electric field acting between colliding nuclei. Theoretically, the Coulomb force between the positively charged colliding nuclei is well understood, and the interaction is exactly calculable. Coulomb excitation usually is the dominant reaction in nuclear scattering, and even occurs at low bombarding energies where the separation of the nuclei is sufficiently large that the short-ranged nuclear force does not act. See COULOMB'S LAW.

Coulomb excitation plays a vital role in probing the response of both shape and volume collective modes of motion as well as the interplay of single-particle degrees of freedom of the nuclear many-body system. The goal of this work is to develop better models of nuclear structure and to elucidate the underlying nuclear force. See NUCLEAR STRUCTURE.

[D.Cl.]

Coulomb explosion   A process in which a molecule moving with high velocity strikes a solid and the electrons that bond the molecule are torn off rapidly in violent collisions with the electrons of the solid; as a result, the molecule is suddenly transformed into a cluster of charged atomic constituents that then separate under the influence of their mutual Coulomb repulsion. See COULOMB'S LAW.

Coulomb explosions are most commonly studied using a particle accelerator, normally employed in nuclear physics research (Van de Graaff generator, cyclotron, and so forth), to produce a beam of fast molecular ions that are directed onto a solid-foil target. The Coulomb explosion of the molecular projectiles begins within the first few tenths of a nanometer of penetration into the foil, continues during passage of the projectiles through the foil, and runs to completion after emergence of the projectiles into the vacuum downstream from the foil. Detectors located downstream make precise measurements of the energies and charges of the molecular fragments together with their angles of emission relative to the beam direction. The Coulomb explosion causes the fragment velocities to be shifted in both magnitude and direction from the beam velocity. See PARTICLE ACCELERATOR.

Coulomb explosion experiments serve two main purposes. First, they yield valuable information on the interactions of fast ions with solids. For example, it is known that a fast ion generates a polarization wake that trails behind it as it traverses a solid. This wake can be studied in detail by using diatomic molecular-ion beams, since the motion of a trailing fragment is influenced not only by the Coulomb explosion but also by the wake of its partner. Second, Coulomb-explosion techniques can be used to determine the stereochemical structures of molecular-ion projectiles. See ELECTRON WAKE; MOLECULAR STRUCTURE AND SPECTRA.

[D.S.Ge.]

Coulomb's law   For electrostatics, Coulomb's law states that the direct force F of point charge q_1 on point charge q_2, when the charges are separated by a distance r, is given by $F = k_0 q_1 q_2/r^2$, where k_0 is a constant of proportionality whose value depends on the units used for measuring F, q, and r. It is the basic quantitative law of electrostatics. In the rationalized meter-kilogram-second (mks) system of units, $k_0 = 1/(4\pi\epsilon_0)$, where ϵ_0 is called the permittivity of empty space and has the value 8.85×10^{-12} farad/m. Thus, Coulomb's law in the rationalized mks system is as in the equation below, where q_1 and q_2 are expressed in coulombs, r is expressed in meters, and F is

$$F = \frac{1}{4\pi\epsilon_0}\frac{q_1 q_2}{r^2}$$

given in newtons. See ELECTRICAL UNITS AND STANDARDS.

The direction of F is along the line of centers of the point charges q_1 and q_2, and is one of attraction if the charges are opposite in sign and one of repulsion if the charges have the same sign. For a statement of Coulomb's law as applied to point magnet poles.

Experiments have shown that the exponent of r in the equation is very accurately the number 2. Lord Rutherford's experiments, in which he scattered alpha particles by atomic nuclei, showed that the equation is valid for charged particles of nuclear dimensions down to separations of about 10^{-12} cm. Nuclear experiments have shown that the forces between charged particles do not obey the equation for separations smaller than this. *See* ELECTROSTATICS.   [R.P.Wi.]

Couple   A system of two parallel forces of equal magnitude and opposite sense. Under a couple's action a rigid body tends only to rotate about a line normal to the couple's plane. This tendency reflects the vector properties of a couple.

The total force of a couple is zero. The total moment **C** of a couple is identical about any point. Accordingly, **C** is the moment of either force about a point on the other and is perpendicular to the couple's plane. *See* RESULTANT OF FORCES; STATICS.

The moment of a couple about a directed line is the component of its total moment in the line's direction. Couples are equivalent whose total moments are equal.   [N.S.F.]

CPT theorem   A fundamental ingredient in quantum field theories, which dictates that all interactions in nature, all the force laws, are unchanged (invariant) on being subjected to the combined operations of particle-antiparticle interchange (so-called charge conjugation, C), reflection of the coordinate system through the origin (parity, P), and reversal of time, T. In other words, the *CPT* operator commutes with the hamiltonian. The operations may be performed in any order; *TCP*, *TPC*, and so forth, are entirely equivalent. If an interaction is not invariant under any one of the operations, its effect must be compensated by the other two, either singly or combined, in order to satisfy the requirements of the theorem. *See* QUANTUM FIELD THEORY.

The *CPT* theorem appears implicitly in work by J. Schwinger in 1951 to prove the connection between spin and statistics. Subsequently, G. Lüders and W. Pauli derived more explicit proofs, and it is sometimes known as the Lüders-Pauli theorem. The proof is based on little more than the validity of special relativity and local interactions of the fields. The theorem is intrinsic in the structure of all the successful field theories. *See* QUANTUM STATISTICS; RELATIVITY; SPIN (QUANTUM MECHANICS).

CPT assumed paramount importance in 1957, with the discovery that the weak interactions were not invariant under the parity operation. Almost immediately afterward, it was found that the failure of P was attended by a compensating failure of C invariance. Initially, it appeared that CP invariance was preserved and, with the application of the *CPT* theorem, invariance under time reversal. Then, in 1964 an unmistakable violation of CP was discovered in the system of neutral K mesons. *See* PARITY (QUANTUM MECHANICS).

One question immediately posed by the failure of parity and charge conjugation invariance is why, as one example, the π^+ and π^- mesons, which decay through the weak interactions, have the same lifetime and the same mass. It turns out that the equality of particle-antiparticle masses and lifetimes is a consequence of *CPT* invariance and not C invariance alone. *See* ELEMENTARY PARTICLE; MESON; SYMMETRY LAWS (PHYSICS).   [V.F.]

Creeping flow   Fluid at very low Reynolds number. In the flow of fluids, a Reynolds number (density · length · velocity/viscosity) describes the relative importance of inertia effects to viscous effects. In creeping flow the Reynolds number is very small (less than 1) such that the inertia effects can be ignored in comparison to the viscous resistance. Creeping flow at zero Reynolds number is called Stokes flow.

Mathematically, viscous fluid flow is governed by the Navier-Stokes equation. In creeping flow the nonlinear momentum terms are unimportant, and the Navier-Stokes equation can be linearized. *See* FLUID FLOW; FLUID MECHANICS; NAVIER-STOKES EQUATION; REYNOLDS NUMBER; VISCOSITY.

Examples of creeping flow include very small objects moving in a fluid, such as the settling of dust particles and the swimming of microorganisms. Other examples include the flow of fluid (ground water or oil) through small channels or cracks, such as in hydrodynamic lubrication or the seepage in sand or rock formations. The flow of high-viscosity fluids may also be described by creeping flow, such as the extrusion of melts or the transport of paints, heavy oils, or food-processing materials.   [C.Y.W.]

Critical phenomena The unusual physical properties displayed by substances near their critical points. The study of critical phenomena of different substances is directed toward a common theory.

Ideally, if a certain amount of water (H_2O) is sealed inside a transparent cell and heated to a high temperature T, for instance, $T > 647$ K ($374°C$ or $705°F$), the enclosed water exists as a transparent homogeneous substance. When the cell is allowed to cool down gradually and reaches a particular temperature, namely the boiling point, the enclosed water will go through a phase transition and separate into liquid and vapor phases. The liquid phase, being more dense, will settle into the bottom half of the cell. This sequence of events takes place for water at most moderate densities. However, if the enclosed water is at a density close to 322.2 kg $\cdot$ m^{-3}, rather extraordinary phenomena will be observed. As the cell is cooled toward 647 K ($374°C$ or $705°F$), the originally transparent water will become increasingly turbid and milky, indicating that visible light is being strongly scattered. Upon slight additional cooling, the turbidity disappears and two clear phases, water and vapor, are found. This phenomenon is called the critical opalescence, and the water sample is said to have gone through the critical phase transition. The density, temperature, and pressure at which this transition

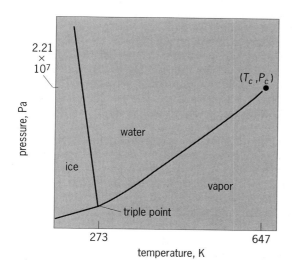

Phase diagram of water (H_2O) on pressure-temperature (P-T) plane.

happens determine the critical point and are called respectively the critical density ρ_c, the critical temperature T_c, and the critical pressure P_c. For water $\rho_c = 322.2$ kg $\cdot$ m^{-3}, $T_c = 647$ K (374°C or 705°F), and $P_c = 2.21 \times 10^7$ pascals.

Different fluids, as expected, have different critical points. Although the critical point is the end point of the vapor pressure curve on the pressure-temperature (P-T) plane (see illustration), the critical phase transition is qualitatively different from that of the ordinary boiling phenomenon that happens along the vapor pressure curve. In addition to the critical opalescence, there are other highly unusual phenomena that are manifested near the critical point; for example, both the isothermal compressibility and heat capacity diverge to infinity as the fluid approaches T_c. See THERMODYNAMIC PROCESSES.

Many other systems, for example, ferromagnetic materials such as iron and nickel, also have critical points. The ferromagnetic critical point is also known as the Curie point. As in the case of fluids, a number of unusual phenomena take place near the critical point of ferromagnets, including singular heat capacity and divergent magnetic susceptibility. The study of critical phenomena is directed toward describing the various anomalous and interesting types of behavior near the critical points of these diverse and different systems with a single common theory. See CURIE TEMPERATURE; FERRO-MAGNETISM.   [M.H.W.C.]

Cryogenics   The science and technology of phenomena and processes at low temperatures, defined arbitrarily as below 150 K (−190°F). Phenomena that occur at cryogenic temperatures include liquefaction and solidification of ambient gases; loss of ductility and embrittlement of some structural materials such as carbon steel; increase in the thermal conductivity to a maximum value, followed by a decrease as the temperature is lowered further, of relatively pure metals, ionic compounds, and crystalline dielectrics (diamond, sapphire, solidified gases, and so forth); decrease in the thermal conductivity of metal alloys and plastics; decrease in the electrical resistance of relatively pure metals; decrease in the heat capacity of solids; decrease in thermal noise and disorder of matter; and appearance of quantum effects such as superconductivity and superfluidity. See ELECTRICAL RESISTIVITY; SPECIFIC HEAT; SUPERCONDUCTIVITY; SUPERFLUIDITY; THERMAL CONDUCTION IN SOLIDS.

Low-temperature environments are maintained with cryogens (liquefied gases) or with cryogenic refrigerators. The temperature afforded by a cryogen ranges from its triple point to slightly below its critical point. Commonly used cryogens are liquid helium-4 (down to 1 K), liquid hydrogen, and liquid nitrogen. Less commonly used because of their expense are liquid helium-3 (down to 0.3 K) and neon. The pressure maintained over a particular cryogen controls its temperature. Heat input—both the thermal load and the heat leak due to imperfect insulation—boils away the cryogen, which must be replenished. See LIQUID HELIUM; THERMODYNAMIC PROCESSES.

A variety of techniques are available for prolonged refrigeration. Down to about 1.5 K, refrigeration cycles involve compression and expansion of appropriately chosen gases. At lower temperatures, liquid and solids serve as refrigerants. Adiabatic demagnetization of paramagnetic ions in solid salts is used in magnetic refrigerators to provide temperatures from around 4 K down to 0.003 K. Nuclear spin demagnetization of copper can achieve 5×10^{-8} K. Helium-3/helium-4 dilution refrigerators are frequently used for cooling at temperatures between 0.3 and 0.002 K, and adiabatic compression of helium-3 (Pomeranchuk cooling) can create temperatures down to 0.001 K. See ADIABATIC DEMAGNETIZATION.

Both the latent heat of vaporization and the sensible heat of the gas (heat content of the gas) must be removed to liquefy a gas. Of the total heat that must be removed to liquefy the gas, the latent heat is only 1.3% for helium and 46% for nitrogen. Consequently, an efficient liquefier must supply refrigeration over the entire temperature range between ambient and the liquefaction point, not just at the liquefaction temperature. The Collins-Claude refrigeration cycle forms the basis (with a multitude of variations) of most modern cryogenic liquefiers. Gas is compressed isothermally and cooled in a counterflow heat exchanger by the colder return stream of low-pressure gas. During this cooling, a fraction of the high-pressure stream (equal to the rate of liquefaction) is split off and cooled by the removal of work (energy) in expansion engines or turbines. This arrangement provides the cooling for the removal of the sensible heat. At the end of the counterflow cooling, the remaining high-pressure stream is expanded in either a Joule-Thomson valve or a wet expander to give the liquid product and the return stream of saturated vapor. See LIQUEFACTION OF GASES.

The work input required to produce refrigeration is commonly given in terms of watts of input power per watt of cooling, that is, W/W. Cooling with a refrigerator is more efficient (that is, requires a lower W/W) than cooling with evaporating liquid supplied from a Dewar because the refrigerator does not discard the cooling available in the boil-off gas.

[D.E.D.]

Crystal   A solid in which the atoms or molecules are arranged periodically. Within a crystal, many identical parallelepiped unit cells, each containing a group of atoms, are packed together to fill all space (see illustration). In scientific nomenclature, the term crystal is usually short for single crystal, a single periodic arrangement of atoms. Most gems are single crystals. However, many materials are polycrystalline, consisting of many small grains, each of which is a single crystal. For example, most metals are polycrystalline. See SINGLE CRYSTAL.

In electronics, the term crystal is restricted to mean piezoelectric crystal. Piezoelectric crystals contract or expand under application of electric voltages, and conversely they generate voltages when compressed. They are used for oscillators, pressure sensors, and position actuators. See PIEZOELECTRICITY.

The anisotropic microscopic structure of a crystal is often reflected in its external form, consisting of flat faces and sharp edges. Crystal structure is generally determined via

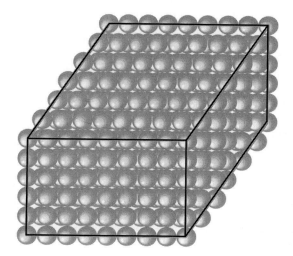

Structure of a simple crystal. Spheres, representing atoms, are packed together into a cubic lattice. This crystal has 4-fold symmetry axes passing through the front face; after a 90° rotation the structure appears unchanged.

diffraction of x-rays, neutrons, or electrons. Unlike disordered materials such as glasses or liquids, the diffraction pattern of a periodic array of atoms consists of individual sharp spots. The symmetry and structure of the crystal can be inferred from the symmetry of the diffraction pattern and the intensities of the diffracted beams. *See* ELECTRON DIFFRACTION; NEUTRON DIFFRACTION; X-RAY DIFFRACTION.

A crystal can be characterized by the symmetry operations that leave its structure invariant. These can include rotation about an axis through a specific angle, reflection through a plane, inversion through a point, translations by a unit cell dimension, and combinations of these. For a periodic structure, the only allowable rotational symmetries are 2-fold, 3-fold, 4-fold, and 6-fold. A quasicrystal is a solid which yields a sharp diffraction pattern but has rotational symmetries (such as 5-fold or 10-fold) which are inconsistent with a periodic arrangement of atoms. *See* QUASICRYSTAL.

A plastic crystal is generally composed of organic molecules which are rotationally disordered. The centers of the molecules lie at well-defined, periodically spaced positions, but the orientations of the molecules are random. Plastic crystals are often very soft and may flow under their own weight.

A liquid crystal is a material which is intermediate in structure between a liquid and a solid. Liquid crystals usually flow like liquids but have some degree of internal order. They are generally composed of rodlike organic molecules, although in some cases they are composed of disklike molecules. In a nematic liquid crystal, the rods all have the same general orientation, but the positions of the rods are disordered. In a smectic liquid crystal, rodlike molecules are ordered into sheets, within which there is only liquidlike order. A smectic can thus be thought of as being crystalline in one dimension and liquid in the other two. In a discotic liquid crystal, disklike molecules are ordered into columnar arrays; there is short-range liquidlike order within the columns, but the columns form a two-dimensional crystal. *See* CRYSTAL DEFECTS; CRYSTAL GROWTH; CRYSTAL STRUCTURE; CRYSTALLOGRAPHY.   [P.A.He.]

Crystal absorption spectra   The wavelength or energy dependence of the attenuation of electromagnetic radiation as it passes through a crystal, due to its conversion to other forms of energy in the crystal. When atoms are grouped into an ordered array to form a crystal, their interaction with electromagnetic radiation is greatly modified. Free atoms absorb electromagnetic radiation by transitions between a few electronic states of well-defined energies, leading to absorption spectra consisting of sharp lines. In a crystal, these states are broadened into bands, and the cores of the atoms are constrained to vibrate about equilibrium positions. The ability of electromagnetic radiation to transfer energy to bands and ionic vibrations leads to broad absorption spectra that bear little resemblance to those of the free parent atoms. *See* ABSORPTION OF ELECTROMAGNETIC RADIATION; ATOMIC STRUCTURE AND SPECTRA; BAND THEORY OF SOLIDS; LATTICE VIBRATIONS.   [D.E.As.]

Crystal counter   A device, more correctly described as a crystal detector, that detects ionizing radiation of all types and is adaptable to measuring neutrons. The sensitive element is a single crystal with a dc resistance normally higher than 10^{12} ohms. The crystals are small and are cut or grown to volumes ranging from less than 1 mm^3 to approximately 200 mm^3.

Crystal detectors fall into two categories: Certain crystals act as thermoluminescent detectors, of which lithium fluoride (LiF), lithium borate ($Li_2B_4O_7$), and calcium sulfate ($CaSO_4$) are among the best known. Other crystals, for example, cadmium telluride (CdTe) and mercury iodide (HgI_2), act as conduction detectors, delivering either pulses or a dc signal, depending upon the associated electronic circuitry. *See* IONIZATION CHAMBER; THERMOLUMINESCENCE.

Diamond is a unique crystal that functions as a thermoluminescent detector or, if suitable contacts are made, as a conduction detector. The efficiency of the diamond detector in the thermoluminescent or conduction mode is strongly dependent on the impurity atoms included within the crystal lattice, with nitrogen and boron playing dominant roles. Not all diamonds are good detectors; only the rare and expensive natural types IB or IIA are appropriate. Besides being stable and nontoxic, diamond has an additional attractive feature as a detector. As an allotrope of carbon, it has the atomic number $Z = 6$. Human soft tissue has an effective $Z = 7.4$, so that diamond is a close tissue-equivalent material, an essential characteristic for biological dosimetry, for example, in measurements in living organisms.

Good crystal detectors are insulators and therefore have significant band-gap energies. A large band gap impedes the spontaneous excitation of charge carriers between the valence and conduction bands, thus lowering leakage currents and movement of charge carriers to trapping centers. Room temperature devices are consequently possible. *See* BAND THEORY OF SOLIDS; ELECTRIC INSULATOR; TRAPS IN SOLIDS.

In thermoluminescent detectors, the crystal is heated at a controlled rate on a metal tray by means of an electric current. The photon emission from the crystal is monitored by a photomultiplier the output of which is directed toward an appropriate recording device. The result is a "glow curve," the area of which correlates with the number of traps depopulated, which are in turn directly related to the radiation-field intensity. The integrated light output therefore becomes a direct measure of the total radiation dose.

In a conduction detector, a charged particle entering the crystal transfers its kinetic energy to the bulk of the crystal by creating charge carriers (electron-hole pairs). A photon of sufficient energy interacts with the crystal atoms, losing all or part of its energy through the photoelectric effect, the Compton effect, or pair production. In each of these processes, electrons are either liberated or created, and they in turn have their energy dissipated in the bulk of the crystal by creating charge carriers. When the carrier pair is created, the individual carriers move under the influence of the electric field toward the oppositely charged contacts. On arriving at the contacts, the charges can be measured at the output point either as a dc or as a pulse signal, depending upon the circuitry. It is, however, necessary for full efficiency of the counting system that both types of carriers are collected equally. *See* COMPTON EFFECT; ELECTRON-POSITRON PAIR PRODUCTION; GAMMA-RAY DETECTORS; PHOTOEMISSION.

In the thermoluminescent mode the crystals measure the total dose of the applied radiation, whereas in the conduction mode they measure the instantaneous dose rate; in both cases it is ultimately the crystal itself that limits the sensitivity and resolution of the system. Present methods of synthesis for crystals permit the detection of radiation fields down to nearly background values (0.1 microgray/h or 10^{-5} rad/h) even with crystals as small as 1 mm^3. Small crystals make detectors possible that are capable of very high spatial resolution. This feature is important in electron radiation therapy. *See* PARTICLE DETECTOR.

[R.J.K.]

Crystal defects Departures of a crystalline solid from a regular array of atoms or ions. A "perfect" crystal of NaCl, for example, would consist of alternating Na$^+$ and Cl$^-$ ions on an infinite three-dimensional simple cubic lattice, and a simple defect (a vacancy) would be a missing Na$^+$ or Cl$^-$ ion. There are many other kinds of possible defects, ranging from simple and microscopic, such as the vacancy and other structures

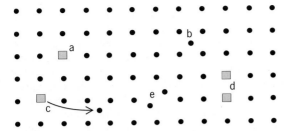

Key:
a = vacancy (Schottky defect)
b = interstitial
c = vacancy-interstitial pair (Frenkel defect)
d = divacancy
e = split interstitial
▨ = vacant site

Some simple defects in a lattice.

shown in the illustration, to complex and macroscopic, such as the inclusion of another material, or a surface.

Natural crystals always contain defects, due to the uncontrolled conditions under which they were formed. The presence of defects which affect the color can make these crystals valuable as gems, as in ruby (Cr replacing a small fraction of the Al in Al_2O_3). Crystals prepared in the laboratory will also always contain defects, although considerable control may be exercised over their type, concentration, and distribution.

The importance of defects depends upon the material, type of defect, and properties which are being considered. Some properties, such as density and elastic constants, are proportional to the concentration of defects, and so a small defect concentration will have a very small effect on these. Other properties, such as the conductivity of a semiconductor crystal, may be much more sensitive to the presence of small numbers of defects. Indeed, while the term defect carries with it the connotation of undesirable qualities, defects are responsible for many of the important properties of materials, and much of solid-state physics and materials science involves the study and engineering of defects so that solids will have desired properties. A defect-free silicon crystal would be of little use in modern electronics; the use of silicon in devices is dependent upon small concentrations of chemical impurities such as phosphorus and arsenic which give it desired electronic properties.

An important class of crystal defect is the chemical impurity. The simplest case is the substitutional impurity, for example, a zinc atom in place of a copper atom in metallic copper. Impurities may also be interstitial; that is, they may be located where atoms or ions normally do not exist. In metals, impurities usually lead to an increase in the electrical resistivity. Impurities in semiconductors are responsible for the important electrical properties which lead to their widespread use. The energy levels associated with impurities and other defects in nonmetals may also lead to optical absorption in interesting regions of the spectrum.

Even in a chemically pure crystal, structural defects will occur. These may be simple or extended. One type of simple defect is the vacancy, but other types exist (see illustration). The atom which left a normal site to create a vacancy may end up in an interstitial position, a location not normally occupied. Or it may form a bond with a normal atom in such a way that neither atom is on the normal site, but the two are symmetrically displaced from it. This is called a split interstitial. The name Frenkel defect is given to a vacancy-interstitial pair, whereas an isolated vacancy is a Schottky defect.

The simplest extended structural defect is the dislocation. An edge dislocation is a line defect which may be thought of as the result of adding or subtracting a half-plane of atoms. A screw dislocation is a line defect which can be thought of as the result of cutting partway through the crystal and displacing it parallel to the edge of the cut. Dislocations are of great importance in determining the mechanical properties of crystals. A dislocation-free crystal is resistant to shear, because atoms must be displaced over high-potential-energy barriers from one equilibrium position to another. It takes relatively little energy to move a dislocation (and thereby shear the crystal), because the atoms at the dislocation are barely in stable equilibrium. Such plastic deformation is known as slip.

For both scientific and practical reasons, much of the research on crystal defects is directed toward the dynamic properties of defects under particular conditions, or defect chemistry. Much of the motivation for this arises from the often undesirable effects of external influences on material properties, and a desire to minimize these effects. Examples of defect chemistry abound, including one as familiar as the photographic process, in which incident photons cause defect modifications in silver halides or other materials. Properties of materials in nuclear reactors is another important case.

[W.B.F.]

Crystal growth   The growth of crystals, which can occur by natural or artificial processes. Crystal growth generally comes about by means of the following processes occurring in series: (1) diffusion of the atoms (or molecules, in the case of molecular crystals such as hydrocarbons or polymers) of the crystallizing substance through the surrounding environment (or solution) to the surface of the crystal, (2) diffusion of these atoms over the surface of the crystal to special sites on the surface, (3) incorporation of atoms into the crystal at these sites, and (4) diffusion of the heat of crystallization away from the crystal surface. The rate of crystal growth may be limited by any of these four steps. The initial formation of the centers from which crystal growth proceeds is known as nucleation.

During its growth into a fluid phase, a crystal often develops and maintains a definite polyhedral form which may reflect the characteristic symmetry of the microscopic pattern of atomic arrangement in the crystal. The bounding faces of this form are those which are perpendicular to the directions of slowest growth. How this comes about is illustrated in Fig. 1, in which it is seen that the faces *b*, normal to the faster-growing direction, disappear, and the faces *a*, normal to the slower-growing directions, become predominant. Growth forms, like that shown, are not necessarily equilibrium forms, but they are likely to be most regular when the departures from equilibrium are not large. *See* CRYSTAL STRUCTURE.

The atomic binding sites on the surface of a crystal can be of several kinds. Thus an atom must be more weakly bound on a perfectly developed plane of atoms at the crystal surface (site A) than at a ledge formed by an incomplete plane one atom thick (site B). Atom A binds with only three neighboring atoms, whereas atom B binds to five neighbors. (An atom in the middle of the island monolayer has bonds with nine neighbors.) Therefore, the binding of atoms in an island monolayer on the crystal surface will be less per atom than it would be within a completed surface layer.

The potential energy of a crystal is most likely to be minimum in forms containing the fewest possible ledge sites. This means that, in a regime of regular crystal growth, dilute fluid, and moderate departure from equilibrium, the crystal faces of the growth form are likely to be densely packed and atomically smooth. There will be a critical size of monolayer, which will be a decreasing function of supersaturation, such that all

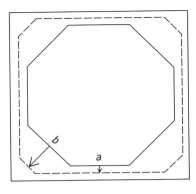

a
b

Fig. 1. Schematic representation of cross section of crystal at three stages of regular growth.

monolayers smaller than the critical size tend to shrink out of existence, and those which are larger will grow to a complete layer. The critical monolayers form by a fluctuation process. Kinetic analyses indicate that the probability of critical fluctuations is so small that in finite systems perfect crystals will not grow, except at substantial departures from equilibrium. That, in ordinary experience, finite crystals do grow in a regular regime only at infinitesimal departures from equilibrium is explained by the screw dislocation theory. According to this theory, growth is sustained by indestructible surface ledges which result from the emergence of screw dislocations in the crystal face. See CRYSTAL DEFECTS.

When the departures from equilibrium (supersaturation or undercooling) are sufficiently large, the more regular growth shapes become unstable and cellular (grasslike) or dendritic (treelike) morphologies develop. Essentially, the development of protuberances on an initially regular crystal permits more efficient removal of latent heat or of impurities, but at the cost of higher interfacial area and the associated excess surface energy. When the supersaturation becomes so great that the energy associated with the increase in interfacial area is unimportant, protuberances proliferate and the crystal grows in a multibranched form that is even more complicated; its shape is characterized by fractal geometry. See FRACTALS.   [M.J.A.; D.T.]

The advent of semiconductor-based technology generated a demand for large, high-quality single crystals, not only of semiconductors but also of associated electronic materials. With increasing sophistication of semiconductor devices, an added degree of freedom in materials properties was obtained by varying the composition of major components of the semiconductor crystal over very short distances. Thin, multilayered single-crystal structures, and even structures that vary in composition both normal and lateral to the growth direction, are often required.

Bulk single crystals are usually grown from a liquid phase. The liquid may have approximately the same composition as the solid; it may be a solution consisting primarily of one component of the crystal; or it may be a solution whose solvent constitutes at most a minor fraction of the crystal's composition. The most important bulk crystal growth technique is the crystal-pulling or Czochralski method, in which a rotating seed crystal is dipped into the melt (Fig. 2). Rotation reduces radial temperature gradients, and slow withdrawal of the rotating seed results in growth of a cylinder of single-crystal material. Crystal diameter and length depend upon the details of the temperature and pulling rate, and the dimensions of the melt container. Crystal quality depends very critically upon minimization of temperature gradients that enhance the formation of dislocations. Pulled silicon crystals 6 in. (15 cm) in diameter are important for the semiconductor industry. Ruby, sapphire, and group 13–15 compound semiconductor

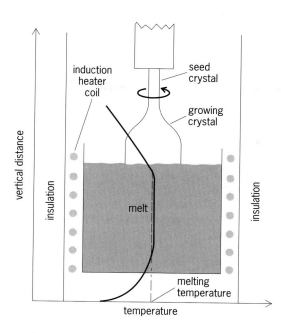

Fig. 2. Czochralski crystal growth and temperature distribution.

crystals are among the many crystals that are routinely grown by the Czochralski technique.

The evolution of methods for the growth of very thin but very high-quality epitaxial layers has resulted largely from the need for such layers of semiconductors and magnetic garnets. The technique most closely related to the methods used for bulk crystal growth is liquid-phase epitaxy. For a typical binary semiconductor, growth is done onto a substrate single-crystal wafer from a solution rich in the component with the lowest partial pressure. For a binary compound, the grown layers may differ only in impurity concentrations to modify their electrical characteristics. More often, multilayered structures with layers differing in major component composition but having the same crystal structure and lattice parameter are required. The simplest example of liquid-phase epitaxy with major composition changes in layers is the growth of layers of aluminum gallium arsenide ($Al_xGa_{1-x}As$; $1 > x > 0$) on a gallium arsenide (GaAs) substrate.

Growth by liquid-phase epitaxy is done in an apparatus in which the substrate wafer is sequentially brought into contact with solutions that are at the desired compositions and may be supersaturated or cooled to achieve growth. For crystalline solid solutions other than $Al_xGa_{1-x}As$, very precise control over solution compositions is required to achieve a lattice match. Typically, structures grown by liquid-phase epitaxy have four to six layers ranging widely in composition and having thickness from 10^{-7} to 10^{-6} m.

The desirability of highly reproducible growth and even thinner epitaxial layers of 13–15 compounds on large wafer areas has led to the development of molecular-beam epitaxy and several forms of chemical-vapor deposition. Molecular-beam epitaxy is an ultrahigh-vacuum technique in which beams of atoms or molecules of the constituent elements of the crystal provided by heated effusion ovens, impinge upon a heated substrate crystal. It has been used for epitaxial layers as thin as 0.5 nanometer. Molecular-beam epitaxy has also been used for group 12–16 compounds and for silicon. *See*

Artificially layered structures; Molecular beams; Semiconductor; Semiconductor heterostructures.
[M.B.P.]

Crystal optics   The study of the propagation of light, and associated phenomena, in crystalline solids. For a simple cubic crystal the atomic arrangement is such that in each direction through the crystal the crystal presents the same optical appearance. The atoms in anisotropic crystals are closer together in some planes through the material than in others. In anisotropic crystals the optical characteristics are different in different directions. In classical physics the progress of an electromagnetic wave through a material involves the periodic displacement of electrons. In anisotropic substances the forces resisting these displacements depend on the displacement direction. Thus the velocity of a light wave is different in different directions and for different states of polarization. The absorption of the wave may also be different in different directions. *See* Dichroism; Trichroism.

   In an isotropic medium the light from a point source spreads out in a spherical shell. The light from a point source embedded in an anisotropic crystal spreads out in two wave surfaces, one of which travels at a faster rate than the other. The polarization of the light varies from point to point over each wave surface, and in any particular direction from the source the polarization of the two surfaces is opposite. The characteristics of these surfaces can be determined experimentally by making measurements on a given crystal.

   In the most general case of a transparent anisotropic medium, the dielectric constant is different along each of three orthogonal axes. This means that when the light vector is oriented along each direction, the velocity of light is different. One method for calculating the behavior of a transparent anisotropic material is through the use of the index ellipsoid, also called the reciprocal ellipsoid, optical indicatrix, or ellipsoid of wave normals. This is the surface obtained by plotting the value of the refractive index in each principal direction for a linearly polarized light vector lying in that direction (see illustration). The different indices of refraction, or wave velocities associated with a given propagation direction, are then given by sections through the origin of the coordinates in which the index ellipsoid is drawn. These sections are ellipses, and the major and minor axes of the ellipse represent the fast and slow axes for light proceeding along the normal to the plane of the ellipse. The length of the axes represents the refractive indices for the fast and slow wave, respectively. The most asymmetric type of ellipsoid has three unequal axes. It is a general rule in crystallography that no property of a crystal will have less symmetry than the class in which the crystal belongs.

   Accordingly, there are many crystals which, for example, have four- or sixfold rotation symmetry about an axis, and for these the index ellipsoid cannot have three unequal

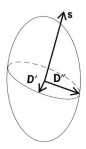

Index ellipsoid, showing construction of directions of vibrations of *D* vectors belonging to a wave normal *s*. (*After M. Born and E. Wolf, Principles of Optics, 7th ed., Cambridge University Press, 1999*)

axes but is an ellipsoid of revolution. In such a crystal, light will be propagated along this axis as though the crystal were isotropic, and the velocity of propagation will be independent of the state of polarization. The section of the index ellipsoid at right angles to this direction is a circle. Such crystals are called uniaxial and the mathematics of their optical behavior is relatively straightforward.

In crystals of low symmetry the index ellipsoid has three unequal axes. These crystals are termed biaxial and have two directions along which the wave velocity is independent of the polarization direction. These correspond to the two sections of the ellipsoid which are circular. See CRYSTALLOGRAPHY.

The normal to a plane wavefront moves with the phase velocity. The Huygens wavelet, which is the light moving out from a point disturbance, will propagate with a ray velocity. Just as the index ellipsoid can be used to compute the phase or wave velocity, so can a ray ellipsoid be used to calculate the ray velocity. The length of the axes of this ellipsoid is given by the velocity of the linearly polarized ray whose electric vector lies in the axis direction. See PHASE VELOCITY.

The refraction of a light ray on passing through the surface of an anisotropic uniaxial crystal can be calculated with Huygens wavelets in the same manner as in an isotropic material. For the ellipsoidal wavelet this results in an optical behavior which is completely different from that normally associated with refraction. The ray associated with this behavior is termed the extraordinary ray. At a crystal surface where the optic axis is inclined at an angle, a ray of unpolarized light incident normally on the surface is split into two beams: the ordinary ray, which proceeds through the surface without deviation; and the extraordinary ray, which is deviated by an angle determined by a line drawn from the center of one of the Huygens ellipsoidal wavelets to the point at which the ellipsoid is tangent to a line parallel to the surface. The two beams are oppositely linearly polarized.

[B.H.Bi]

Crystal structure   The arrangement of atoms, ions, or molecules in a crystal. Crystals are solids having, in all three dimensions of space, a regular repeating internal unit of structure.

Crystals have been studied using x-rays, which excite signals from the atoms. The signals are of different strengths and depend on the electron density distribution about atomic cores. Light atoms give weaker signals and hydrogen is invisible to x-rays. However, the mutual atomic arrangements that are called crystal structures can be derived once the chemical formulas and physical densities of solids are known, based on the knowledge that atomic positions are not arbitrary but are dictated by crystal symmetry, and that the diffraction signals received are the result of systematic constructive interference between the scatterers within the regularly repeating internal unit of pattern. See CRYSTALLOGRAPHY; POLYMORPHISM (CRYSTALLOGRAPHY); X-RAY CRYSTALLOGRAPHY; X-RAY DIFFRACTION.

Crystals are defined in terms of space, population, and mutual arrangement. Crystal space is represented as an indefinitely extended lattice of periodically repeating points. The periodicity of the lattice is defined by the lengths and mutual orientations of three lattice vectors that enclose the pattern. Population is defined as the total number and kind of fundamental units of structure that form the pattern. The order and periodicity of crystals must extend to about 100 nanometers in all three dimensions of space to give the sharply defined diffraction signals required for mapping structural details by x-rays. Intermediate states of order are seen in liquid crystals, which have long molecules as fundamental units of structure. These are arranged with their lengths parallel to each other, but without periodicity, in the nematic state. In the smectic state there is

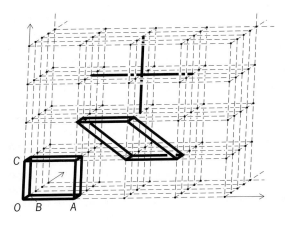

A space lattice, two possible unit cells, and the environment of a point.

orientation in equally spaced planes but no sideways periodicity, like traffic moving freely on a multilane highway.

In reality crystal space is not, in general, perfect. The growth process is characterized by constraints and turbulences, and by the dynamic interaction between the crystal and its environment. The process is reflected within the structures formed as an assemblage of atoms is collected and made relatively immobile by releasing the energy known as the heat of crystallization. The resulting crystal lattices resemble a mosaic of slightly misaligned adjacent regions. This is fortunate for research in x-ray crystallography. Perfect alignment would result in subtraction of energy by interference with the primary beam, due to a 180° phase reversal of the reflected beam (primary extinction). Internally diffracted beams would also be attenuated by internal reflection from regions above them (secondary extinction). *See* CRYSTAL GROWTH.

Each of the spatially misaligned mosaic blocks of a single crystal is assumed to maintain lattice periodicity within it. This assumption is confirmed by the sharp diffraction patterns observed. There are some "wrong" atoms, vacant lattice sites, trapped gas atoms, and so forth, and the atomic occupants jiggle about while also vibrating cooperatively and synchronously in complex internal modes of motion. Intricate patterns of electron exchange are enacted, and systematic changes in spin orientations can occur for an atom with a magnetic moment. Details like these are important for understanding the relationships between structure determination on the atomic and molecular levels and the cooperative behavior that determines bulk properties and functions. *See* CRYSTAL DEFECTS; LATTICE VIBRATIONS.

A rectangular space lattice with two possible cells is outlined in the illustration. These have the same cell volumes but different symmetries. Since crystallographic unit cells are completely defined by three lattice vectors, the crystal symmetry referenced to this lattice can be no higher than orthorhombic: $a \neq b \neq c$ ($OA \neq OB \neq OC$), and all angles equal to 90°. This and a possible monoclinic cell, with the same vectors a and b (OA and OB) and one angle not equal to 90°, are outlined. If the OAB plane is rotated and the vector a (OA) is extended to terminate at the next lattice point, then all angles differ from 90° and the crystal symmetry represented becomes triclinic. The mutual arrangement and atom coordinates of the cell population must be such that the environment, seen from every point of the space lattice, remains the same.

Screw axes combine the rotation of an ordinary symmetry axis with a translation parallel to it and equal to a fraction of the unit distance in this direction. If screw axes are present in crystals, it is clear that the displacements involved are of the order of a few tenths of nanometer and that they cannot be distinguished macroscopically from ordinary symmetry axes. The same is true for glide mirror planes, which combine the mirror image with a translation parallel to the mirror plane over a distance that is half the unit distance in the glide direction. The handedness of screw axes is a very important feature of many biological and mineral structures.

Space groups are indefinitely extended arrays of symmetry elements disposed on a space lattice. A space group acts as a three-dimensional kaleidoscope: An object submitted to its symmetry operations is multiplied and periodically repeated in such a way that it generates a number of interpenetrating identical space lattices. Space groups are denoted by the Hermann-Mauguin notation preceded by a letter indicating the Bravais lattice on which it is based. For example, P $2_1 2_1 2_1$ is an orthorhombic space group; the cell is primitive and three mutually perpendicular screw axes are the symmetry elements. J. D. H. Donnay and D. Harker have shown that it is possible to deduce the space group from a detailed study of the external morphology of crystals.

In general, metallic structures are relatively simple, characterized by a close packing and a high degree of symmetry. Manganese, gallium, mercury, and one form of tungsten are exceptions. A characteristic of metallic structures is the frequent occurrence of allotropic forms; that is, the same metal can have two or more different structures which are most frequently stable in a different temperature range. The forces which link the atoms together in metallic crystals are nondirectional. This means that each atom tends to surround itself by as many others as possible. This results in a close packing, similar to that of spheres of equal radius, and yields three distinct systems: close-packed (face-centered) cubic, hexagonal close-packed, and body-centered cubic.

Simple crystal structures are usually named after the compounds in which they were first discovered (diamond or zinc sulfide, cesium chloride, sodium chloride, and calcium fluoride). Many compounds of the types $A^+ X^-$ and $A^{2+} X_2^-$ have such structures. They are highly symmetrical, the unit cell is cubic, and the atoms or ions are disposed at the corners of the unit cell and at points having coordinates that are combinations of 0, 1, $^1/_2$, or $^1/_4$.

The sodium chloride structure is an arrangement in which each positive ion is surrounded by six negative ions, and vice versa. The centers of the positive and the negative ions each form a face-centered cubic lattice. Systematic study of the dimensions of the unit cells of compounds having this structure has revealed that:

1. Each ion can be assigned a definite radius. A positive ion is smaller than the corresponding atom and a negative ion is larger.

2. Each ion tends to surround itself by as many others as possible of the opposite sign because the binding forces are nondirectional.

In the cesium chloride structure each of the centers of the positive and negative ions forms a primitive cubic lattice; the centers are mutually shifted. Contact of the ions of opposite sign here is along the cube diagonal. In the diamond structure, each atom is in the center of a tetrahedron formed by its nearest neighbors. The 4-coordination follows from the well-known bonds of the carbon atoms.

The calcium fluoride structure is divided into eight equal cubelets, calcium ions are situated at corners and centers of the faces of the cell. The fluorine ions are at the centers of the eight cubelets.

[D.Ev.]

Crystal whiskers Single crystals that have grown in filamentary form. Some grow from their base: either these are extruded to relieve pressure in the base material or they grow as a result of a chemical reaction at the base. In both cases, the growth occurs at a singularity in the base material. Other crystal whiskers grow at their tip, into a supersaturated medium, and the filamentary growth results from a strong anisotropy in the growth rate. *See* Single crystal.

Great interest in whiskers developed after it was discovered that the strength exhibited by some whiskers approached that expected theoretically for perfect crystals. This great strength results from the internal and surface perfection of the whiskers, whereas the strength of most materials is limited by defects. The interest in the high strength of the whiskers centered on the possibility of using them in composites to increase the strength of more ductile matrix materials. Fibers of silica, boron, and carbon, which are much easier to fabricate in large quantity than whiskers, exhibit similarly high strengths, and are now used in composites. *See* Crystal defects.   [K.A.J.; R.S.Wa.]

Crystallography   The branch of science that deals with the geometric forms of crystals. How to describe, classify, and measure such forms are the first questions of crystallography. Revealing the forces that made them and the activities within them are the modern directions of the field. Crystallography is essential to progress in the applied sciences and technology and developments in all materials areas, including metals and alloys, ceramics, glasses, and polymers, as well as drug design. It is equally vital to progress in fundamental physics and chemistry, mineralogy and geology, and computer science, and to understanding of the dynamics and processes of living systems. *See* Crystal structure; Polymorphism (crystallography).

The external morphology of crystals reflects their growth rates in different directions. These directions remain constant during the course of the growth process, and are represented mathematically as the normals to sets of parallel planes that are imagined as being added on as growth proceeds. The faces that meet and define an edge belong to a zone, a zone being a set of planes that share one common direction, the zone axis. The invariance of interfacial angles, measured by rotation about an axis that is defined by the zone direction, was discovered in the seventeenth century. *See* Crystal growth.

Interfacial angles are calculated from spherical geometry. Figure 1 illustrates the procedure for a crystal having well-developed faces of which three are mutually

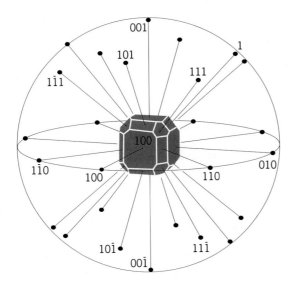

Fig. 1. Spherical projection of normals to crystal faces.

perpendicular. The normals to these faces are the natural directions for constructing an orthogonal frame of reference for measurement. The crystal is imagined to be shrunk and placed at the center of a sphere with coordinates (0,0,0). The face normals, labeled [100], [010], and [001], define the directions of an orthogonal reference system. Normals to the same set of planes, but oppositely directed, are labeled [$\bar{1}$00], [0$\bar{1}$0], [00$\bar{1}$]. The reversal of sign indicates that the crystal must be rotated 180° to obtain the same view. Rotation about the [001] direction interchanges the positions of [100] and [$\bar{1}$00] faces and their bounding edges. Rotation about [010] turns these faces upside down. Correct designations for group movements and symmetry operations are clearly essential for establishing and maintaining orientation in crystal space. The directions of face normals determine points at which the imagined sphere is pierced. The solid angles between an array of such points, all lying on the same great circle of the sphere, belong to a zone.

Optical measurements and stereographic projections established the constancy of interfacial angles, independent of how well developed the faces are. Such properties as the cleavage of large rhombohedral crystals of calcite ($CaCO_3$) into little rhombs suggested that the large crystal could be represented by geometrically identical smaller units stacked together, by translation, to fill space. The 14 lattices of Bravais (Fig. 2) enlarged the seven crystal systems of optical mineralogy by adding centering points to

Bravais lattice cells	Axes and interaxial angles	Examples
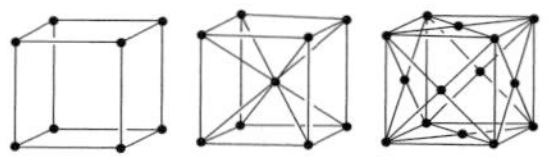 Cubic P Cubic I Cubic F	Three axes at right angles; all equal: $a = b = c$; $\alpha = \beta = \gamma = 90°$	Copper (Cu), silver (Ag), sodium chloride (NaCl)
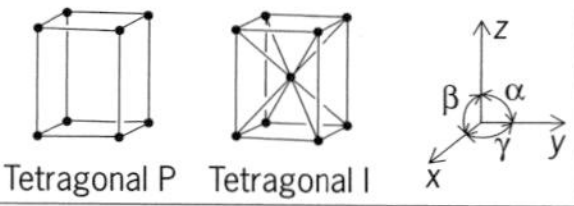 Tetragonal P Tetragonal I	Three axes at right angles; two equal: $a = b \neq c$; $\alpha = \beta = \gamma = 90°$	White tin (Sn), rutile (TiO_2), β-spodumene ($LiAlSi_2O_6$)
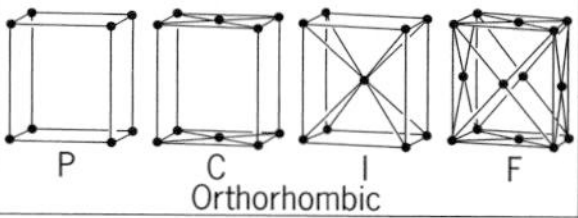 P C I F Orthorhombic	Three axes at right angles; all unequal: $a \neq b \neq c$; $\alpha = \beta = \gamma = 90°$	Gallium (Ga), perovskite ($CaTiO_3$)
 Monoclinic P Monoclinic C	Three axes, one pair not at right angles, of any lengths: $a \neq b \neq c$; $\alpha = \gamma = 90° \neq \beta$	Gypsum ($CaSO_4 \cdot 2H_2O$)
 Triclinic P	Three axes not at right angles, of any lengths: $a \neq b \neq c$; $\alpha \neq \beta \neq \gamma \neq 90°$	Potassium chromate (K_2CrO_7)
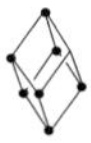 Trigonal R (rhombohedral)	Rhombohedral: three axes equally inclined, not at right angles; all equal: $a = b = c$; $\alpha = \beta = \gamma \neq 90°$	Calcite ($CaCO_3$), arsenic (As), bismuth (Bi)
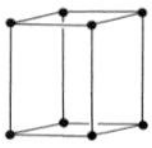 Trigonal and hexagonal C (or P)	Hexagonal: three equal axes coplanar at 120°, fourth axis at right angles to these: $a_1 = a_2 = a_3 \neq c$; $\alpha = \beta = 90°$, $\gamma = 120°$	Zinc (Zn), cadmium (Cd), quartz (SiO_2) [P]

Fig. 2. The 14 Bravais lattices, derived by centering of the seven crystal classes (P and R) defined by symmetry operators.

them: body (I), face (F), and base (C) centers. The 14 lattices define three-dimensional distributions of mathematical points such that the environments of all points of the lattice are identical. They also define the symmetries of frameworks for constructing mathematical models to represent the observed and measured realities—models made from cells of the smallest volume, but also highest symmetry, that stack together by translation to fill space.

Stacking of model cells does not imply that a crystal grows by stacking identical bricks; a lattice of identically surrounded mathematical points does not imply that any real objects, atoms or molecules, are located at the points; and filling space by translation of identical cells does not imply that the space defined by the cells is filled. Rather, the Bravais lattices are a formalism for representing observed geometries and symmetries of real crystals by three-dimensional lattices of identically surrounded points.

The lattices also provide the means to identify imaginary planes within the cell; these are called Miller indices (h, k, l). They consist of small whole numbers. For example, each of the six faces of a simple cube, with the origin of a coordinate frame of reference at the cube body center, is normal to one of the reference axes and parallel to the plane defined by the two others. The six faces are indexed as their normals in Fig. 1—(100), ($\bar{1}$00), (010), (0$\bar{1}$0), (001), and (00$\bar{1}$)—to represent a face that intercept the x, x axis but not the y and z; the y, y axis but not the x and z; and so forth. Hypothetical parallel planes with 1/2 the interplanar spacing are represented as (200), ($\bar{2}$00), (020), and so forth.

A complete mathematical formalism exists for modeling an external morphological form and the symmetry relations between imagined units of structure within it. The symmetry operators include rotation axes, glide and mirror planes, and left- and right-handed screw axes which will simultaneously rotate and translate a three-dimensional object to create its clone in a different spatial position and orientation. The operators minimize the detail required to specify the spatial arrangements of patterns and objects that fill two-dimensional and three-dimensional space. The so-called color space groups of crystallography greatly increase the number of distinguishably different symmetries beyond the classical 230 by adding a fourth coordinate to the three space coordinates. This is done to encode a real difference that will be manifested in some property. The different directions of the magnetic moments of chemically identical atoms of an element such as iron provide an example of the need for representing a difference on the atomic level between cells that are otherwise identical.

The sharp x-ray line spectra characteristic of the bombarded element are the primary probes for determining interior structural detail of crystals. Cameras with cylindrical film and enclosed powdered samples record all diffraction lines as arcs of concentric circles. This fundamental powder method has endured since 1917 and is now employed with improved beam purity and optics, improved diffractometers which couple sample and detector rotation, electronic detection, rapid sequential recording, and computer indexing programs that provide patterns of compounds, mixtures of phases, and dynamic changes that occur when crystals are subjected and respond to external stress. The method is applied to single crystals, polycrystalline aggregates, and multiple-phase mixtures, randomly disposed either in space or in geometrically designed composite materials. *See* X-RAY CRYSTALLOGRAPHY; X-RAY POWDER METHODS.

The dynamics of living systems, the difficulties in distinguishing light elements, and the inherent ambiguities of measuring, decoding, and mapping crystal structures are continuing challenges. Major achievements of crystallography include the determination of the structures of deoxyribonucleic acid (DNA), proteins, other biological compounds, and boranes; the development of direct methods of phase determination; and the determination of the structure and mechanism of a photosynthetic center.   [D.Ev.]

Curie temperature   The critical or ordering temperature for a ferromagnetic or a ferrimagnetic material. The Curie temperature Tc is the temperature below which there is a spontaneous magnetization M in the absence of an externally applied magnetic field, and above which the material is paramagnetic. In the disordered state above the Curie temperature, thermal energy overrides any interactions between the local magnetic moments of ions. Below the Curie temperature, these interactions are predominant and cause the local moments to order or align so that there is a net spontaneous magnetization.

In the ferromagnetic case, as temperature T increases from absolute zero, the spontaneous magnetization decreases from M_0, its value at $T = 0$. At first this occurs gradually, then with increasing rapidity until the magnetization disappears at the Curie temperature. In ferrimagnetic materials the course of magnetization with temperature may be more complicated, but the spontaneous magnetization disappears at the Curie temperature.

In antiferromagnetic materials the corresponding ordering temperature is termed the Néel temperature. Below the Néel temperature the magnetic sublattices have a spontaneous magnetization, though the net magnetization of the material is zero. Above the Néel temperature the material is paramagnetic. *See* ANTIFERROMAGNETISM.

The ordering temperatures for magnetic materials vary widely. The ordering temperature for ferroelectrics is also termed the Curie temperature, below which the material shows a spontaneous electric moment. *See* FERROMAGNETISM; PYROELECTRICITY.   [J.F.Di.]

Curie-Weiss law   A relation between magnetic or electric susceptibilities and the absolute temperature which is followed by ferromagnets, antiferromagnets, nonpolar ferroelectrics and antiferroelectrics, and some paramagnets. The Curie-Weiss law is usually written as the equation below, where χ is the susceptibility, C is a constant for

$$\chi = C/(T - \theta)$$

each material, T is the absolute temperature, and θ is called the Curie temperature. Antiferromagnets and antiferroelectrics have a negative Curie temperature. The Curie-Weiss law refers to magnetic and electric behavior above the transition temperature of the material in question. It is not always precisely followed, and it breaks down in the region very close to the transition temperature. Often the susceptibility will behave according to a Curie-Weiss law in different temperature ranges with different values of C and θ. *See* CURIE TEMPERATURE; ELECTRIC SUSCEPTIBILITY; MAGNETIC SUSCEPTIBILITY.

[E.A.; F.Ke.]

Current comparator   An instrument for determining the ratio of two currents, based on Ampère's laws. The application of the current comparator, unlike the voltage ratio transformer, is not limited to alternating currents but can be used with direct currents as well.

The current comparator is based on Ampère's circuital law, which states that the integral of the magnetizing force around a closed path is equal to the sum of the currents which are linked with that path. Thus, if two currents are passed through a toroid by two windings of known numbers of turns, and the integral of the magnetizing force around the toroid is equal to zero, the current ratio is exactly equal to the inverse of the turns ratio.

For alternating-current comparators, the ampere-turn unbalance is given by the voltage at the terminals of a uniformly distributed detection winding on a magnetic core. Direct-current comparators use two magnetic cores which are modulated by alternating current in such a way that the dc ampere-turn unbalance is indicated by the presence of even harmonics in the voltage. Neither method is an exact measure of the integral of the magnetizing force, and various design features are added to overcome this deficiency. The most important of these are the magnetic shields, which are configured as hollow toroids. They protect the magnetic core and detection winding from the leakage fluxes of the current-carrying windings and ambient magnetic fields, and they are responsible for an improvement in accuracy of about three orders of magnitude. The copper shields supplement this action at higher frequencies and also provide mechanical protection for the magnetic steels.

Applications of the ac comparator include current and voltage transformer calibration, measurement of losses in large capacitors, inductive reactors and power transformers, resistance measurement, and the calibration of active and reactive power and energy meters. Further, the current comparator, like the current transformer, has the advantage of being applicable to measurements involving high-voltage, high-power circuits as well as being the basis for very accurate high-current-transconductance amplifiers.

In dc applications, the current comparator provides the means to resolve the first three or four most significant digits of a balance by turns-count, thus eliminating problems associated with switch contact resistance and thermal electromotive forces. Its uses include eight-decade resistance and thermometry bridges, a seven-decade potentiometer, and high current ratio standards. It also provides the means for generating highly accurate direct-current voltages and currents. *See* CURRENT MEASUREMENT; ELECTRICAL MEASUREMENTS.   [W.J.M.M.; N.L.K.; E.So.]

Current density   A vector quantity equal in magnitude to the instantaneous rate of flow of electric charge, perpendicular to a chosen surface, per unit area per unit time. If a wire of cross-sectional area A carries a current I, the current density J is I/A. The units of J in the rationalized meter-kilogram-second system are amperes per square meter.   [J.W.St.]

Current measurement   The measurement of the rate of passage of electric charges in a circuit. The unit of measurement, the ampere (A), is one of the base units of the International System of Units (SI). It is defined as that constant current which, if maintained in two straight parallel conductors of infinite length, of negligible circular cross section, and placed 1 m apart in vacuum, would produce between these conductors a force equal to 2×10^{-7} newton per meter of length.

In order to establish an electrical unit in accordance with the SI definition, it is necessary to carry out an experimental determination. The ampere cannot be realized exactly as defined. Electromagnetic theory has to be used to relate a practical experiment to the definition.

Since January 1, 1990, working standards of voltage and resistance have provided the foundations of practical electrical measurements. The standard of voltage is based on the alternating-current (ac) Josephson effect, in which voltage is related to frequency. By international agreement the value 483 597.9 GHz/V for the Josephson constant is now used throughout the world. The working unit of resistance is maintained through the quantum Hall effect, with an agreed value of 25 812.807 ohms for the voltage-to-current ratio obtained under certain defined experimental conditions. These values have been chosen to provide the best known approximations to the SI units and have the advantage of reproducibility at the level of 1 part in 10^8. The working standard of current is derived from measurements of voltage across a known resistor. *See* ELECTRICAL UNITS AND STANDARDS; HALL EFFECT; JOSEPHSON EFFECT.

The moving-coil (d'Arsonval) meter measures direct currents (dc) from 10 microamperes to several amperes. The accuracy is likely to be a few percent of the full-scale indication, although precision instruments can achieve 0.1% or even better. Above 1 milliampere a shunt usually carries the major part of the current; only a small fraction is used to deflect the meter. Since the direction of deflection depends on the direction of the current, the d'Arsonval movement is suitable for use only with unidirectional currents. Rectifiers are used to obtain dc and drive the meter from an ac signal. The resulting combination is sensitive to the rectified mean value of the ac waveform.

In the moving-iron meter, two pieces of soft magnetic material, one fixed and one movable, are situated inside a single coil. When current flows, both pieces become magnetized in the same direction and accordingly repel each other. The moving piece is deflected against a spring or gravity restoring force, the displacement being indicated by a pointer. As the repulsive force is independent of current direction, the instrument responds to low-frequency ac as well as dc. The natural response of such a movement is to the root-mean-square (rms) value of the current. The accuracy of moving-iron meters is less than that of moving-coil types. *See* AMMETER.

For radio-frequency applications it is essential that the sensing element be small and simple to minimize inductive and capacitive effects. In a thermocouple meter the temperature rise of a short, straight heater wire is measured by a thermocouple and the corresponding current is indicated by a d'Arsonval movement. In a hot-wire ammeter the thermal expansion of a wire heated by the current is mechanically enhanced and used to deflect a pointer. Both instruments, based on heating effects, respond to the rms value of the current. Above 100 MHz, current measurements are not made directly, as the value of current is likely to change with position owing to reflections and standing waves. *See* MICROWAVE MEASUREMENTS; THERMOCOUPLE.

Above 50 A the design of shunts becomes difficult. For ac, current transformers can be used to reduce the current to a level convenient for measurement. At the highest accuracy, current comparators may be used in which flux balance is detected when the magnetizing ampere-turns from two signals are equal and opposite. Direct-current comparators are available in which dc flux balance is maintained and any unbalance is used to servo a second, or slave, current signal. For the highest accuracy, second-harmonic modulators are used, and for lower precision, Hall effect sensors. Electronically balanced ac and dc current comparators make clip-around ammeters possible, in which an openable magnetic core can be closed around a current-carrying conductor. This allows the meter to be connected into and removed from the circuit without breaking it or interrupting the current. *See* CURRENT COMPARATOR.

The obvious method for measuring a very small current is to determine the voltage drop across a large resistor. A sensitive voltage detector having very low offset current is required, for example, an electrometer. Electrometers based on MOSFET (metal-oxide-semiconductor field-effect transistor) devices have overtaken other designs where the very highest resolution is required, as they can have offset current drifts less than 10^{-16} A. In order to provide a low impedance to the measured current, it is preferable to use the electrometer device in an operational-amplifier configuration. The input becomes a virtual ground, and so stray capacitance across the input connection does not degrade the rate of response of the circuit as seriously as in the simple connection. *See* TRANSISTOR.

[R.B.D.K.]

Cyclotron resonance experiments

The measurement of charge-to-mass ratios of electrically charged particles from the frequency of their helical motion in a magnetic field. Such experiments are particularly useful in the case of conducting crystals, such as semiconductors and metals, in which the motions of electrons and holes are strongly influenced by the periodic potential of the lattice through which they move. Under such circumstances the electrical carriers often have "effective masses" which differ greatly from the mass in free space; the effective mass is often different for motion in different directions in the crystal. Cyclotron resonance is also observed in gaseous plasma discharges and is the basis for a class of particle accelerators. *See* BAND THEORY OF SOLIDS; PARTICLE ACCELERATOR.

Cyclotron resonance is most easily understood as the response of an individual charged particle; but, in practice, the phenomenon involves excitation of large numbers of such particles. Their net response to the electromagnetic radiation may significantly affect the overall dielectric behavior of the material in which they move. Thus, a variety of new wave propagation mechanisms may be observed which are associated with the cyclotron motion, in which electromagnetic energy is carried through the solid by the spiraling carriers. These collective excitations are generally referred to as plasma waves. In general, for a fixed input frequency, the plasma waves are observed to travel through the conducting solid at magnetic fields higher than those required for cyclotron resonance. The most easily observed of these excitations is a circularly polarized wave, known as a helicon, which travels along the magnetic field lines. It has an analog in the ionospheric plasma, known as the whistler mode and frequently detected as radio interference. There is, in fact, a fairly complete correspondence between the resonances and waves observed in conducting solids and in gaseous plasmas. Cyclotron resonance is more easily observed in such low-density systems since collisions are much less frequent there than in solids. In such systems the resonance process offers a means of transferring large amounts of energy to the mobile ions, a necessary condition if nuclear fusion reactions are to occur. *See* NUCLEAR FUSION; PLASMA (PHYSICS).   [W.M.W.]

D'Alembert's principle   The principle that the resultant of the external forces **F** and the kinetic reaction acting on a body equals zero. The kinetic reaction is defined as the negative of the product of the mass m and the acceleration **a**. The principle is therefore stated as $\mathbf{F} - m\mathbf{a} = 0$. While D'Alembert's principle is merely another way of writing Newton's second law, it has the advantage of changing a problem in kinetics into a problem in statics. The techniques used in solving statics problems may then provide relatively simple solutions to some problems in dynamics; D'Alembert's principle is especially useful in problems involving constraints. *See* CONSTRAINT.   [P.W.S.]

Dalitz plot   Pictorial representation in high-energy nuclear physics for data on the distribution of certain three-particle configurations. Many elementary-particle decay processes and high-energy nuclear reactions lead to final states consisting of three particles (which may be denoted by a, b, c, with mass values m_a, m_b, m_c). Well-known examples are provided by the K-meson decay processes, Eqs. (1) and (2), and by the K- and $\overline{K}$-meson reactions with hydrogen, Eqs. (3) and (4). For definite total energy E

$$K^+ \rightarrow \pi^+ + \pi^+ + \pi^- \tag{1}$$

$$K^+ \rightarrow \pi^0 + \mu^+ + \nu \tag{2}$$

$$K^+ + p \rightarrow K^0 + \pi^+ + p \tag{3}$$

$$K^- + p \rightarrow \Lambda + \pi^+ + \pi^- \tag{4}$$

(measured in the barycentric frame), these final states have a continuous distribution of configurations, each specified by the way this energy E is shared among the three particles. (The barycentric frame is the reference frame in which the observer finds zero for the vector sum of the momenta of all the particles of the system considered.) *See* ELEMENTARY PARTICLE.

   If the three particles have kinetic energies T_a, T_b, and T_c (in the barycentric frame), Eq. (5) is obtained. As shown in the illustration, this energy sharing may be represented

$$T_a + T_b + T_c = E - m_a c^2 - m_b c^2 - m_c c^2 = Q \tag{5}$$

uniquely by a point F within an equilateral triangle LMN of side $2Q/\sqrt{3}$, such that the perpendiculars FA, FB, and FC to its sides are equal in magnitude to the kinetic energies T_a, T_b, and T_c. The most important property of this representation is that the area occupied within this triangle by any set of configurations is directly proportional to their volume in phase space.

   Not all points F within the triangle LMN correspond to configurations realizable physically, since the a, b, c energies must be consistent with zero total momentum for the three-particle system. With nonrelativistic kinematics and with equal masses m

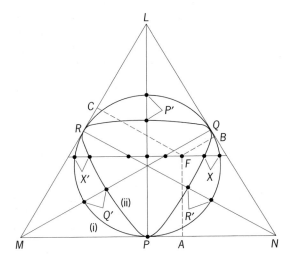

Configuration of a three-particle system (*abc*) in its barycentric frame is specified by a point *F* such that the three perpendiculars *FA*, *FB*, and *FC* to the sides of an equilateral triangle *LMN* (of height *Q*) are equal in magnitude to the kinetic energies T_a, T_b, T_c, where *Q* denotes their sum.

for *a*, *b*, *c*, the only allowed configurations are those corresponding to points *F* lying within the circle inscribed within the triangle, shown as (i) in the illustration. More generally, with relativistic kinematics, the limiting boundary is distorted as illustrated by the boundary curve (ii), drawn for the $\omega \rightarrow 3\pi$ decay process, where the final masses are equal.

[R.H.D.]

Dalton's law   The total pressure of a mixture of gases is the sum of the partial pressures of each gas in the mixture. The law was established by John Dalton (1766–1844). In his original formulation, the partial pressure of a gas is the pressure of the gas if it alone occupied the container at the same temperature. Dalton's law may be expressed as $P = P_A + P_B + \cdots$, where P_J is the partial pressure of the gas J, and P is the total pressure of the mixture; this formulation is strictly valid only for mixtures of ideal gases. For real gases, the total pressure is not the sum of the partial pressures (except in the limit of zero pressure) because of interactions between the molecules.

In modern physical chemistry the partial pressure is defined as $P_J = x_J P$, where x_J is the mole fraction of the gas J, the ratio of its amount in moles to the total number of moles of gas molecules present in the mixture. With this definition, the total pressure of a mixture of any kind of gases is the sum of their partial pressures. However, only for an ideal gas is the partial pressure (as defined here) the pressure that the gas would exert if it alone occupied the container. *See* KINETIC THEORY OF MATTER; THERMODYNAMIC PRINCIPLES.

[P.W.A.]

Damping   A term broadly used to denote either the dissipation of energy in, and the consequent decay of, oscillations of all types or the extent of the dissipation and decay. The energy losses arise from frictional (or analogous) forces which are unavoidable in any system or from the radiation of energy to space or to other systems. For sufficiently small oscillations, the analogous forces are proportional to the velocity of the vibrating member and oppositely directed thereto; the ratio of force to velocity is $-R$, the mechanical resistance. For the role of damping in the case of forced oscillations, where it is decisive for the frequency response, *see* FORCED OSCILLATION; RESONANCE (ACOUSTICS AND MECHANICS). *See also* HARMONIC MOTION; OSCILLATION; VIBRATION.

An undamped system of mass m and stiffness s oscillates at an angular frequency $\omega_0 = (s/m)^{1/2}$. The effect of a mechanical resistance R is twofold: It produces a change in the frequency of oscillation, and it causes the oscillations to decay with time. If u is one of the oscillating quantities (displacement, velocity, acceleration) of amplitude A, then Eq. (1) holds in the damped case, whereas in the undamped case Eq. (2) holds.

$$u = Ae^{-\alpha t} \cos \omega_d t \tag{1}$$

$$u = A \cos \omega_0 t \tag{2}$$

The reciprocal time $1/\alpha$ in Eq. (1) may be called the damping constant.

The damped angular frequency ω_d in Eq. (1) is always less than ω_0. According to Eq. (1), the amplitude of the oscillation decays exponentially; the time required for the amplitude to decrease to the fraction $1/e$ of its initial value is equal to $1/\alpha$.

A common measure of the damping is the logarithmic decrement δ, defined as the natural logarithm of the ratio of two successive maxima of the decaying sinusoid. If T is the period of the oscillation, then Eq. (3) holds. Then $1/\delta$ is the number of cycles

$$\delta = \alpha T \tag{3}$$

required for the amplitude to decrease by the factor $1/e$ in the same way that $1/\alpha$ is the time required.

The Q of a system is a measure of damping usually defined from energy considerations. The Q is π times the ratio of peak energy stored to energy dissipated per cycle and is equal to π/δ.

If α in Eq. (1) exceeds ω_0, then the system is not oscillatory and is said to be overdamped. If the mass is displaced, it returns to its equilibrium position without overshoot, and the return is slower as the ratio α/ω_0 increases. If $\alpha = \omega_0$ (that is, $Q = 1/2$), the oscillator is critically damped. In this case, the motion is again nonoscillatory, but the return to equilibrium is faster than for any overdamped case. [M.Gr.]

De Broglie wavelength The wavelength $\gamma = h/p$ associated with a beam of particles (or with a single particle) of momentum p; $h = 6.626 \times 10^{34}$ joule-second is Planck's constant. The same formula gives the momentum of an individual photon associated with a light wave of wavelength γ. This formula, along with the profound proposition that all matter has wavelike properties, was first put forth by Louis de Broglie in 1924, and is fundamental to the modern theory of matter and its interaction with electromagnetic radiation. *See* QUANTUM MECHANICS. [E.G.]

Decibel A logarithmic unit used to express the magnitude of a change in level of power, voltage, current, or sound intensity. A decibel (dB) is 1/10 bel.

In acoustics a step of 1 bel is too large for most uses. It is therefore the practice to express sound intensity in decibels. The level of a sound of intensity I in decibels relative to a reference intensity I_R is given by notation (1). Because sound intensity is

$$10 \log_{10} \frac{I}{I_R} \tag{1}$$

proportional to the square of sound pressure P, the level in decibels is given by Eq. (2). The reference pressure is usually taken as 0.0002 dyne/cm^2 or 0.0002 microbar.

$$10 \log_{10} \frac{P^2}{P_R{}^2} = 20 \log_{10} \frac{P}{P_R} \tag{2}$$

(The pressure of the Earth's atmosphere at sea level is approximately 1 bar.) *See* SOUND PRESSURE.

The neper is similar to the decibel but is based upon natural (napierian) logarithms. One neper is equal to 8.686 dB.   [K.D.K.]

Deep inelastic collisions   Either highly energetic collisions of elementary particles, namely, leptons and nucleons, which probe the nucleons' internal structure; or collisions between two heavy ions in which the two nuclei interact strongly while their nuclear surfaces overlap.

Elementary-particle collisions. Deep inelastic collisions of elementary particles are very energetic collisions between leptons such as electrons or neutrinos and nucleons (that is, protons or neutrons, typically in a nucleus) in which the target nucleon breaks up into many particles and the lepton is scattered through a large angle in the center-of-mass frame. These collisions are akin to the Rutherford scattering experiments in which most alpha particles went through a thin gold foil undeflected but some were deflected through large angles. In both cases, the explanation for large deflections is that the incident particle encounters not a uniform sphere of material but a few hard or pointlike objects inside the target. The alpha particles encounter gold nuclei, while leptons strike quarks inside the nucleons. *See* ALPHA PARTICLES; LEPTON; NUCLEON; QUARKS; SCATTERING EXPERIMENTS (NUCLEI).

Deep inelastic scattering experiments are conducted to study the structure of protons and neutrons. In each collision the fraction x of the nucleon's momentum carried by the struck quark is measured, and thus the x distributions of quarks inside a proton are directly measured. These are known as structure functions. Studies of these have shown, among other things, that the momentum of a proton is not carried entirely by quarks. In fact, only about half the momentum can be ascribed to quarks. The other half is believed to be carried by gluons, which are carriers of the strong force which binds the quarks within nucleons and other hadrons. *See* GLUONS.

Modern-day experiments in deep inelastic scattering often use neutrinos or muons as probes. Neutrinos have the advantage that they are not affected by the electric charge of the target nucleus and hence scatter directly off the quarks. Muons are easy to detect and identify. However, the highest-energy deep inelastic collisions are carried out by using an electron-proton collider. These experiments aim to study structure functions at very low values of x where some models predict new behavior. The highest-energy deep inelastic collisions use electrons as probes to search for quark substructure.   [M.V.P.]

Heavy-ion collisions. Deep inelastic collisions of heavy ions are characterized by features that are intermediate between those of comparatively simple quasielastic, few-nucleon transfer reactions and those of highly complex compound-nucleus reactions. These deep inelastic or damped collisions often occur for heavy-ion reactions at center-of-mass energies less than 5 MeV per nucleon above the Coulomb barrier. During the brief encounter of the two nuclei, large amounts of kinetic energy of radial and orbital motion can be dissipated. The lifetime of the dinuclear complex (analogous to a chemical molecule) corresponds to the time required for the intermediate system to make a partial rotation (10^{-22} s to 5×10^{-21} s). On separation, the final total kinetic energies of the two reaction fragments can be well below those corresponding to the Coulomb repulsion of spheres, indicating that the fragments are highly deformed in the exit channel, as is known to be the case for fission fragments. *See* NUCLEAR FISSION.   [J.R.Hu.]

Degeneracy (quantum mechanics)   A term referring to the fact that two or more stationary states of the same quantum-mechanical system may have the same energy even though their wave functions are not the same. In this case the common energy level of the stationary states is degenerate. The statistical weight of the level is proportional to the order of degeneracy, that is, to the number of states with the same energy; this number is predicted from Schrödinger's equation. In quantum mechanics and in other branches of mathematical physics, the term degeneracy is employed also to characterize the eigenvalues of operators other than the energy operator. *See* EIGENVALUE (QUANTUM MECHANICS).

[E.G.]

Degree of freedom (mechanics)   Any one of the number of independent ways in which the space configuration of a mechanical system may change. A material particle confined to a line in space can be displaced only along the line, and therefore has one degree of freedom. A particle confined to a surface can be displaced in two perpendicular directions and accordingly has two degrees of freedom. A particle free in physical space has three degrees of freedom corresponding to three possible perpendicular displacements. A system composed of two free particles has six degrees of freedom, and one composed of N free particles has $3N$ degrees. If a system of two particles is subject to a requirement that the particles remain a constant distance apart, the number of degrees of freedom becomes five. Any requirement which diminishes by one the degrees of freedom of a system is called a holonomic constraint. *See* CONSTRAINT.

[R.A.Fi.]

De Haas–van Alphen effect   An oscillatory behavior of the magnetic moment of a pure metal crystal with changes in the applied magnetic field B, at very low temperatures. It is named after its discoverers, W. J. de Haas and P. M. van Alphen. This effect has its origin in the Bohr-Sommerfeld quantization of the orbits of conduction electrons under the influence of the magnetic field.

A measurement of the temperature dependence of the oscillation amplitude permits a determination of the cyclotron frequency, or equivalently the electron mass. In a metal, as in a semiconductor, electrons behave with an effective mass m^* rather than the free electron mass, and the de Haas–van Alphen effect thus allows a measurement of m^* in metals. *See* KINETIC THEORY OF MATTER.

Application of the Bohr-Sommerfeld quantization rules shows that the "period" of the de Haas–van Alphen oscillations (it is a period in magnetic field, not time) is given by the equation below. Here a_p is the area of the orbit in momentum space of an

$$\Delta\left(\frac{1}{B}\right) = \frac{2\pi \hbar e}{a_p}$$

electron at the Fermi level, that is, the cross section of the Fermi surface; e is the charge of the electron; and $\hbar$ is Planck's constant divided by 2π. By studying the dependence of a_p on the direction B of the magnetic field, sufficient information can be obtained to construct the detailed shape of the Fermi surfaces of a metal.

Virtually all metals available in sufficient purity have been studied by the de Haas–van Alphen effect. The effect is also the most powerful probe of Fermi surface properties in alloys and intermetallic compounds. *See* BAND THEORY OF SOLIDS; FERMI SURFACE.

[J.B.K.]

Delta electrons   Energetic electrons ejected from atoms in matter by the passage of ionizing particles. In every primary ionizing collision between a charged particle and an atom, one or more electrons are ejected. Delta electrons are, by definition, that small fraction of these emitted electrons having energies which are large compared to the ionization potential. The name is a traditional one—comparable to alpha particles, for energetic helium nuclei, and beta particles, for energetic electrons emitted in radioactive decays. *See* ALPHA PARTICLES; BETA PARTICLES; IONIZATION.

Delta electrons are responsible for the "hairy" appearance of charged particle tracks when they are observed in cloud chambers or in photographic emulsions. In studies of super-high-energy particles in cosmic radiation and from the highest-energy accelerators, observation of the number of delta electrons per centimeter of path length has been shown to lead to a reliable determination of the charge of the energetic particle.   [D.A.B.]

Delta resonance   A member of a class of subatomic particles called baryons, which exists in four electric charge states and has a total spin of $J = {}^{3}/_{2}$. In the underlying quark model, the delta resonance (Δ) consists of three quarks whose intrinsic spins of $^{1}/_{2}$ are lined up in the same direction. The Δ is closely related to the more familiar nucleon constituents of atomic nuclei, the neutrons (n) and protons (p). *See* NUCLEON; QUARKS.

The Δ was first observed as a resonant interaction of a beam of pi mesons (π) with a proton target. The probability of a scattering interaction between the π and the proton is strongly dependent on energy, attaining a maximum at the Δ mass of 1236 MeV/c^2 (where c is the speed of light). The formation of the very short-lived Δ (with a lifetime on the order of 10^{-23} s) followed immediately by its decay back into pion and nucleon. *See* MESON; SCATTERING EXPERIMENTS (NUCLEI).

The Δ plays an important role in a wide variety of nuclear phenomena, even under conditions of low energy and momentum transfer. The study of these phenomena reveals much about the presence of pions in nuclei, in addition to neutrons and protons. *See* BARYON; ELEMENTARY PARTICLE; NUCLEAR STRUCTURE.   [C.D.]

Demagnetization   The reduction or elimination of the magnetic moment in an object; that is, the reverse of magnetization. It is commonly encountered as a procedure for eliminating the inadvertent magnetization of iron (or other ferromagnetic) parts of a sensitive mechanical device that would otherwise result in a malfunction. A suitably intense magnetic field applied in a direction opposite to that of the existing magnetization will serve to reduce or destroy that magnetization. (Alternatively, the material could, if practical, be heated to a temperature above its Curie point, then returned to room temperature, in the absence of any external magnetic field.) The adiabatic (isentropic) demagnetization of paramagnetic materials is a technique used to produce temperatures very near absolute zero. It has been used to cool and study a magnetic substance itself or, through thermal contact, a secondary substance (refrigeration). *See* ADIABATIC DEMAGNETIZATION; FERROMAGNETISM; MAGNETIZATION.   [R.P.Hu.]

Density   The mass per unit volume of a material. The term is applicable to mixtures and pure substances and to matter in the solid, liquid, gaseous, or plasma state. Density of all matter depends on temperature; the density of a mixture may depend on its composition, and the density of a gas on its pressure. Common units of density are grams per cubic centimeter, and slugs or pounds per cubic foot. The specific gravity of a material is defined as the ratio of its density to the density of some standard material, such as water at a specified temperature, for example, 60°F (15.6°C), or, for gases the basis may be air at standard temperature and pressure. Another related concept is weight density, which is defined as the weight of a unit volume of the material. *See* MASS; WEIGHT.   [L.N.]

Density matrix A matrix which is constructed as the most general statistical description of the states of a many-particle quantum-mechanical system. The state of a quantum system is described by a normalized wave function $\psi(x, t)$ [where x stands for all coordinates of the system, and t for the time], which satisfies the Schrödinger equation (1), where H is the hamiltonian of the system, and $\hbar$ is Planck's constant divided

$$H\psi(x, t) = \frac{h}{i} \frac{\partial \psi(x, t)}{\partial t} \tag{1}$$

by 2π. Furthermore, $\psi(x, t)$ may be expanded in terms of a complete orthonormal set $\{\varphi(x)\}$, as in Eq. (2). Then, the density matrix is defined by Eq. (3), and this density

$$\psi(x, t) = \sum_n a_n(t)\varphi_n(x) \tag{2}$$

$$\rho_{mn}(t) = a_n^*(t)a_m(t) \tag{3}$$

matrix describes a pure state. Examples of pure states are a beam of polarized electrons and the photons in a coherent beam emitted from a laser. *See* LASER; QUANTUM MECHANICS.

In quantum statistics, one deals with an ensemble of N systems which have the same hamiltonian. If the αth member of the ensemble is in the state ψ^α in Eq. (4), the density matrix is defined as the ensemble average, Eq. (5). In general, this density

$$\psi^\alpha(x, t) = \sum_n a_n{}^\alpha(t)\varphi_n(x) \tag{4}$$

$$\rho_{mn}(t) = \frac{1}{N} \sum_\alpha [a_n{}^\alpha(t)]^* \, a_m{}^\alpha(t) \tag{5}$$

matrix describes a mixed state, for example, a beam of unpolarized electrons or the photons emitted from an incoherent source such as an incandescent lamp. The pure state is a special case of the mixed state when all members of the ensemble are in the same state. *See* STATISTICAL MECHANICS. [S.H.L.]

Desorption A process in which atomic and molecular species residing on the surface of a solid leave the surface and enter the surrounding gas or vacuum. In stimulated desorption studies, species residing on a surface are made to desorb by incident electrons or photons. Measurements of these species provide insight into the ways that radiation affects matter, and are useful analytical probes of surface physics and chemistry. In thermal desorption studies, adsorbed surface species are caused to desorb as the sample is heated under controlled conditions. These measurements can provide information on surface-bond energies, the species present on the surface and their coverage, the order of the desorption process, and the number of bonding states or sites.

Stimulated desorption. Stimulated desorption from surfaces is initiated by electronic excitation of the surface bond by incident electrons or photons. The classical model of desorption is an adaptation of the theory of gas-phase dissociation, in which desorption results from excitation from a bonding state to an antibonding state.

Another model which is more applicable to the phenomenon of ion desorption was first observed in studies of the desorption of positively ionized oxygen (O^+) from the surface of titanium(IV) oxide (TiO_2). Here it is found that O^+ is desorbed not by valence level excitation, but by ionization of the titanium and oxygen core levels. These levels, of course, have little to do with bonding. Furthermore, the fact that the oxygen is desorbed as an O^+ ion (whereas it is nominally at O^{2-} on the surface) implies a large (three-electron) charge-transfer preceding desorption. This mechanism for desorption can also be effective for covalently bonded surface species.

Stimulated desorption studies are finding wide use. First, they can show the ways in which radiation affects the structure of solids. This will have important applications in the areas of radiation-induced damage and chemistry. Second, as an analytical tool, they offer a unique new way to study the physics and chemistry of atoms on surfaces which, when combined with the many other surface techniques based largely on electron spectroscopy, can provide new insight. Finally, models of the surface bond are put to a much sterner test in attempting to explain desorption phenomena.

An additional important discovery is that ion angular distributions from stimulated desorption are not isotropic, but show that ions are emitted in relatively narrow cones which project along the nominal ground-state bond directions. Thus this technique provides a direct display of the surface-bonding geometry.

Thermal desorption. Thermal desorption mass spectroscopy is possibly the oldest technique for the study of adsorbates on surfaces. Three primary forms of the thermal desorption experiment involve measurement of (1) the rate of desorption from a surface during controlled heating (temperature-programmed thermal desorption), (2) the rate of desorption at constant temperature (isothermal desorption), and (3) surface life-times and diffusion under exposure to a pulsed beam of adsorbates (molecular-beam experiments). Of the three, temperature-programmed thermal desorption is by far the most widely applied. The most straightforward information provided is the nature of the desorbed species from mass analysis, and a determination of the absolute coverage by the adsorbate, which is very difficult to obtain with other techniques. The technique can also provide important kinetic parameters of the desorption process.

While the thermal desorption techniques are among the simplest of surface probes, they remain indispensable because of their directness and the variety of information they convey. Thus while surface science moves to detailed methods involving extremely sophisticated apparatus, the simple thermal desorption methods remain an important part of the overall picture. *See* SURFACE PHYSICS.   [M.L.K.]

Determinism   The principle that nature follows exact laws, so that what will happen in the future is a necessary consequence of the state of the world at any given moment in the past. This view, if fully adopted, implies that events which seem to occur by chance would be fully understood if more was known about them, and that apparently free thoughts and choices are explainable and in principle predictable in terms of neuroscience. In a looser sense, determinism refers to claims that mental freedom is much more restricted than people are inclined to suppose.

The question of determinism in physical science cannot be considered apart from the philosophical problem; this gives it added importance and forces its consideration in a very critical spirit.

The idea that the world is composed of atoms moving under the influence of certain forces according to certain laws can be traced back to the Greek philosopher Leucippus. Deterministic philosophy was prominent in the work of the seventeenth-century thinker René Descartes, and became widely known through his influence. Isaac Newton carried out a large part of the cartesian program. His theory explained so many natural processes that it began to appear that the universe since the time of Creation might actually have run its course in a deterministic fashion like a machine, without divine intervention. A century after Newton, Pierre Simon de Laplace argued that an Omniscient Calculator, provided with exact knowledge of the state of the universe at

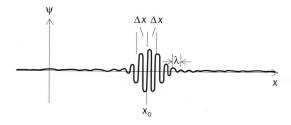

A wave packet—of wavelength approximately λ, and produced perhaps by opening and closing a shutter—traveling through space.

present, would be able to predict the entire future. *See* CLASSICAL MECHANICS; NEWTON'S LAWS OF MOTION.

The quantum mechanics of the 1920s introduced the paradox of particles which are, at the same time, waves. A wave of length λ was supposed to accompany, or describe, a particle of momentum $p = h/\lambda$, where h is Planck's constant. The probability of finding a free particle is expressed by a (complex) wave packet (see illustration), and the particle has appreciable probability of being found only where the wave function is large; that is, within a region of size roughly Δx on each side of x_0 in the illustration. A theorem of Fourier analysis shows that Δp, the range of momenta present in the wave packet, is related to Δx as shown below. This means that the more

$$\Delta x \, \Delta p \geq h/(2\pi)$$

precisely the specification of a particle's position x is attempted by means of a localized wave function, the less precisely can its momentum be specified. W. Heisenberg's principle of uncertainty, or indeterminacy, states that it is impossible to measure, and therefore to know, x and p any more accurately than is allowed by this relation, and there are also other pairs of dynamical variables similarly related.

There remains the question as to whether Heisenberg's principle is merely an unfortunate limitation on an experimenter's ability to know or whether it goes deeper. The general opinion of physicists is that of N. Bohr: the principle expresses a limitation of the precision with which concepts such as position and momentum can be applied at all. Therefore Laplace's Omniscient Calculator cannot predict the future. *See* QUANTUM MECHANICS; UNCERTAINTY PRINCIPLE.

Perceptive mathematicians have warned that determinism is not as obvious a consequence of newtonian physics as it might appear. A series of mathematical results have been proven whose general effect is that for the vast majority of dynamical systems any error in the initial conditions, however small, will be amplified, in general exponentially, and so quickly that the predicted result will soon bear no relation to reality. Thus unless it is assumed that initial conditions are known with perfect accuracy, and perfectly accurate computation takes place thereafter, the Omniscient Calculator will wind up getting everything wrong. *See* CHAOS.

Very few people now think that all events in the natural world are exactly determined. Experiments suggest that some human and animal behavior can reliably be predicted and controlled, but nobody knows the limits within which this can be done.   [D.P.]

Deuteron   The nucleus of the atom of heavy hydrogen, ^{2}H (deuterium). The deuteron d is composed of a proton and a neutron; it is the simplest multinucleon nucleus. Its binding energy is 2.227 MeV; that is, this is the amount of energy which must be added to a deuteron for it to dissociate into a proton and a neutron. Deuterons are much used as projectiles in nuclear bombardment experiments. *See* NUCLEAR REACTION.
                                                        [H.E.D.]

Diamagnetism   That branch of magnetism which treats of diamagnetic phenomena and of the properties of diamagnetic bodies. Diamagnetism is a property exhibited by substances with a negative magnetic susceptibility, that is, by substances which magnetize in a direction opposite to that of an applied magnetic field. A diamagnetic substance has a magnetic permeability less than 1, and is repelled when placed near a magnet. The magnetization of diamagnetic substances is associated with the currents induced on application of a magnetic field. *See* MAGNETIC SUSCEPTIBILITY.

Although all matter exhibits diamagnetism, only those substances in which paramagnetism is absent are referred to as diamagnetic. This is because paramagnetism, if present, usually predominates, and the gross magnetic response of the material is paramagnetic. Important exceptions are the alkali and alkaline earth metals. The condition for pure diamagnetism is that all electronic spins be paired and all orbital moments either be zero or effectively cancel one another. *See* PARAMAGNETISM.

As stated previously, the diamagnetic response of a substance is small; only a very small fraction of the applied magnetic field is shielded from the interior of the substance by the induced diamagnetic currents. There is one case, however, in which the inducing field is completely shielded (except for small surface effects). This is the perfect diamagnetism exhibited by superconductors, and is known as the Meissner effect. *See* MEISSNER EFFECT; SUPERCONDUCTIVITY.   [E.A.; F.Ke.]

Dichroism   In certain anisotropic materials, the property of having different absorption coefficients for light polarized in different directions. There are few natural materials which exhibit strong dichroism. One of the first to be discovered was tourmaline. Light transmitted by thin plates of dark forms of tourmaline is almost completely polarized. *See* POLARIZED LIGHT.

If the absorption in a dichroic material is different for different linear states of polarization, the material is termed linear dichroic. If it is different for right and left circularly polarized light, it is termed circular dichroic. Similarly, there can be elliptically dichroic crystals.   [B.H.Bi.]

The study of dichroism allows conclusions as to the submicroscopic fine structure of cells. In visible light only a few cellular components, such as chloroplasts, show absorption. An absorption can, however, be produced by staining. The dichroic staining of plant fibers is especially simple. The elongate stain particles of benzidine dyes, for example, congo red, are deposited in an oriented manner in the spaces between the microfibrils and produce an intrinsic dichroism of the fiber: colored for a vibration plane parallel, colorless for a plane perpendicular to the stain particles and fibrils. Therefore, the direction of strongest absorption indicates the course (parallel or helical) of the microfibrils in the fiber.

Ultraviolet dichroism gives direct information as to the orientation of the absorbing molecules or molecular groups in cell structures. The method has been especially helpful for studies of orientation of deoxyribonucleic acid in nuclei and chromosomes. Lignifed plant cell walls show ultraviolet dichroism. It is pure form dichroism, a fact which eliminates the possibility that lignin is in an anisotropic state in the wall.

By irradiation with ultraviolet light, various compounds of the cell are caused to fluoresce. The fluorescent light is polarized if the object is anisotropic. This phenomenon, called difluorescence, is observable in lignifed cell walls, and leads to the same conclusions as to lignin deposition as emerge from dichroism studies.   [F.Ru.]

Dielectric materials   Materials which are electrical insulators or in which an electric field can be sustained with a minimal dissipation of power. Dielectrics are employed as insulation for wires, cables, and electrical equipment, as polarizable media for capacitors, in apparatus used for the propagation or reflection of electromagnetic waves, and for a variety of artifacts, such as rectifiers and semiconductor devices, piezoelectric transducers, dielectric amplifiers, and memory elements. The term dielectric, though it may be used for all phases of matter, is usually applied to solids and liquids.

The ideal dielectric material does not exhibit electrical conductivity when an electric field is applied. In practice, all dielectrics do have some conductivity, which generally increases with increase in temperature and applied field. If the applied field is increased to some critical magnitude, the material abruptly becomes conducting, a large current flows (often accompanied by a visible spark), and local destruction occurs to an extent dependent upon the amount of energy which the source supplies to the low-conductivity path. This critical field depends on the geometry of the specimen, the shape and material of the electrodes, the nature of the medium surrounding the dielectric, the time variation of the applied field, and other factors. Temperature instability can occur because of the heat generated through conductivity or dielectric losses, causing thermal breakdown. Breakdown can be brought about by a variety of different causes, sometimes by a number of them acting simultaneously. Nevertheless, under carefully specified and controlled experimental conditions, it is possible to measure a critical field which is dependent only on the inherent insulating properties of the material itself in those conditions. This field is called the intrinsic electric strength of the dielectric. *See* ELECTRICAL BREAKDOWN.

Many of the traditional industrial dielectric materials are still in common use, and they compete well in some applications with newer materials regarding their electrical and mechanical properties, reliability, and cost. For example, oil-impregnated paper is still used for high-voltage cables. Various types of pressboard and mica, often as components of composite materials, are also in use. Elastomers and press-molded resins are also of considerable industrial significance. However, synthetic polymers such as polyethylene, polypropylene, polystyrene, polytetrafluoroethylene, polyvinyl chloride, polymethyl methacrylate, polyamide, and polyimide have become important, as has polycarbonate because it can be fabricated into very thin films. Generally, polymers have crystalline and amorphous regions, increasing crystallinity causing increased density, hardness, and resistance to chemical attack, but often producing brittleness. Many commercial plastics are amorphous copolymers, and often additives are incorporated in polymers to achieve certain characteristics or to improve their workability.   [J.H.Ca.]

Dielectric measurements   Measurements of the dielectric properties of a material, which are characterized by its complex relative permittivity ϵ_r. For all materials except ferroelectrics, this quantity does not depend on applied field: the general behavior is linear, and so voltage of any convenient magnitude can be used for measurement. *See* FERROELECTRICS; PERMITTIVITY.

Bridge methods. The most commonly used apparatus for measuring ϵ_r is the alternating-current (ac) bridge. These bridges are readily available in the operating range 10–10^6 Hz, and sometimes outside it; ultralow-frequency bridges can go as low as 10^{-3} Hz. Most specimen holders for solids are essentially parallel-plate capacitors with the specimen filling all the space between the plates; for liquids, a test cell with cylindrical electrodes is usually employed. The bridges most commonly used are of the Wheatstone type, the most versatile for dielectric measurements being the Schering bridge.

Resonance methods. Resonance methods, useful for frequencies greater than 1 MHz, involve the injection of voltage or current by one of several methods into an *LC* (inductance-capacitance) resonant circuit. Measurements over a range of frequencies may be made by using coils with different inductance values, but ultimately the inductance required becomes impracticably small, and in the range 10^8-10^9 Hz reentrant cavities are often used. These are hybrid devices in which the plates holding the specimen still form a lumped capacitor, but the inductance and capacitances are distributed along a coaxial line. At higher frequencies, the wavelength is comparable to the dimensions of the apparatus, and transmission methods in coaxial lines and waveguides must be used.

Transmission methods. Coaxial lines are used in the frequency range 300 MHz–3 GHz, and waveguides in the range 3–30 GHz. The transmission characteristics are determined by the complex permittivity of the material filling the line or guide. Many different measurement techniques have been devised, but all derive values of the complex relative permittivity ϵ_r from its relationship to the complex propagation factor γ.

In practice, traveling waves are rarely used as the basis of measurement, except for high-loss materials. Usually, reflections from terminations set up standing waves, the amplitude of which in the case of a liquid-filled line can be measured by a suitable probe. The ratios of the field magnitudes at adjacent maxima, and the distance between them, give the information required. *See* WAVELENGTH.

Submillimeter measurements. Dielectric measurements are difficult to carry out in the frequency range 30–300 GHz, for which λ_0 is in the range 1 cm–1 mm, but for λ_0 less than 1 mm, methods related to infrared spectroscopy are used. Broadband continuous spectra result from Fourier transform spectroscopy, which in its simplest form is equivalent to normal infrared spectroscopy, with the specimen in one of the two passive arms of the interferometer, between the beam divider and either the source or the detector. In the more sophisticated dispersive Fourier transform spectroscopy, the specimen is in one of the active arms, that is, between the beam divider and either mirror. Discrete-point spectra also may be obtained by the use of a Mach-Zehnder interferometer and a laser source. By using interferometric techniques, the frequency range can be extended up to about 5 THz. *See* INFRARED SPECTROSCOPY; INTERFEROMETRY; SPECTROSCOPY.

Time-domain methods. If a constant direct-current (dc) voltage is suddenly applied to a dielectric specimen, in principle the charging current is related through the Fourier integral transformation to the steady-state ac current which would flow if the applied voltage were sinusoidal at any particular frequency. If the dc voltage is suddenly removed, a similar relationship holds between the discharge current and ac current. Thus the variation of complex permittivity with frequency can in principle be derived from a transient signal in the time domain. Because of various limitations, the method is not capable of giving results of an accuracy at all frequencies comparable to those obtainable from a single frequency measurement. Nevertheless, with the aid of computer analysis, the response over a large frequency range can be obtained much more quickly than would be possible by using point-by-point measurement methods.   [J.H.Ca.]

Diffraction   The bending of light, or other waves, into the region of the geometrical shadow of an obstacle. More exactly, diffraction refers to any redistribution in space of the intensity of waves that results from the presence of an object that causes variations of either the amplitude or phase of the waves. Most diffraction gratings cause a periodic modulation of the phase across the wavefront rather than a modulation of the amplitude. Although diffraction is an effect exhibited by all types of wave motion, this article will deal only with electromagnetic waves, especially those of visible light. For discussion of the phenomenon as encountered in other types of waves *see* ELECTRON DIFFRACTION; NEUTRON DIFFRACTION; SOUND.

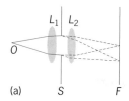 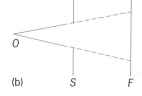

(a)   S   F   (b)   S   F

Fig. 1. Observation of the two principal types of diffraction, in the case of a circular aperture. (*a*) Fraunhofer and (*b*) Fresnel diffraction.

Diffraction is a phenomenon of all electromagnetic radiation, including radio waves; microwaves; infrared, visible, and ultraviolet light; and x-rays. The effects for light are important in connection with the resolving power of optical instruments. *See* X-RAY DIFFRACTION.

There are two main classes of diffraction, which are known as Fraunhofer diffraction and Fresnel diffraction. The former concerns beams of parallel light, and is distinguished by the simplicity of the mathematical treatment required and also by its practical importance. The latter class includes the effects in divergent light, and is the simplest to observe experimentally.

To illustrate the difference between the methods of observation of the two types of diffraction, Fig. 1 shows the experimental arrangements required to observe them for a circular hole in a screen S. The light originates at a very small source O, which can conveniently be a pinhole illuminated by sunlight. In Fraunhofer diffraction, the source lies at the principal focus of a lens L_1 which renders the light parallel as it falls on the aperture. A second lens L_2 focuses parallel diffracted beams on the observing screen F, situated in the principal focal plane of L_2. In Fresnel diffraction, no lenses intervene. The diffraction effects occur chiefly near the borders of the geometrical shadow, indicated by the broken lines. An alternative way of distinguishing the two classes, therefore, is to say that Fraunhofer diffraction concerns the effects near the focal point of a lens or mirror, while Fresnel diffraction concerns those effects near the edges of shadows. Photographs of diffraction patterns are shown in Fig. 2.

Fraunhofer diffraction. This class of diffraction is characterized by a linear variation of the phases of the Huygens secondary waves with distance across the wavefront, as they arrive at a given point on the observing screen. At the instant that the incident plane wave occupies the plane of the diffracting screen, it may be regarded as sending out, from each element of its surface, a multitude of secondary waves, the joint effect of which is to be evaluated in the focal plane of the lens L_2. The analysis

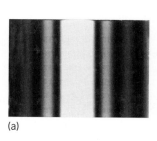 

(a)   (b)

Fig. 2. Diffraction patterns, photographed with visible light. (*a*) Fraunhofer pattern, for a slit; (*b*) Fresnel pattern, circular aperture.

of these secondary waves involves taking account of both their amplitudes and their phases. The simplest way to do this is to use a graphical method, the method of the so-called vibration curve, which can readily be extended to cases of Fresnel diffraction. *See* HUYGENS' PRINCIPLE.

The vibration curve results from the addition of a large (really infinite) number of infinitesimal vectors, each representing the contribution of the Huygens secondary waves from an element of surface of the wavefront. If these elements are assumed to be of equal area, the magnitudes of the amplitudes to be added will all be equal. They will, however, generally differ in phase, so that if the elements were small but finite each would be drawn at a small angle with the preceding one, as shown in Fig. 3a. The resultant of all elements would be the vector A. When the individual vectors represent the contributions from infinitesimal surface elements (as they must for the Huygens wavelets), the diagram becomes a smooth curve, the vibration curve, shown in Fig. 3b. The intensity on the screen is then proportional to the square of this resultant amplitude. In this way, the distribution of the intensity of light in any Fraunhofer diffraction pattern may be determined.

The intensity distribution for Fraunhofer diffraction by a slit as a function of the angle θ from the center may be simply calculated by the method of the vibration curve. The intensity at any angle is given by Eq. (1), where I_0 is the intensity at the center of the pattern, and ß is given by Eq. (2), where b is the width of the slit and λ is the wavelength.

$$I = I_0 \frac{\sin^2 ß}{ß^2} \tag{1}$$

$$ß = (\pi b \sin \theta)\lambda \tag{2}$$

The central maximum is twice as wide as the subsidiary ones, and is about 21 times as intense as the strongest of these. A photograph of this pattern is shown in Fig. 2a.

Fraunhofer diffraction by a circular aperture determines the resolving power of instruments such as telescopes, cameras, and microscopes, in which the width of the light beam is usually limited by the rim of one of the lenses. The method of the vibration curve may be extended to find the angular width of the central diffraction maximum for this case. An exact construction of the curve or, better, a mathematical calculation shows that the extreme phase differences required are $\pm 1.220\pi$, yielding Eq. (3) for

$$\sin \theta \approx \theta = \pm \frac{1.220\lambda}{d} \tag{3}$$

the angle θ at the first zero of intensity. Here d is the diameter of the circular aperture. This pattern has circular symmetry and consists of a diffuse central disk, called the Airy disk, surrounded by faint rings. The angular radius of the disk, given by Eq. (3), may be extremely small for an actual optical instrument, but it sets the ultimate limit to the sharpness of the image, that is, to the resolving power. *See* RESOLVING POWER (OPTICS).

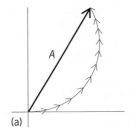
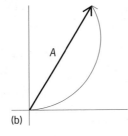

Fig. 3. Vibration curves. *(a)* Addition of many equal amplitudes differing in phase by equal amounts. *(b)* Equivalent curve, when amplitudes and phase differences become infinitesimal.

Fresnel diffraction. The diffraction effects obtained when the source of light or the observing screen are at a finite distance from the diffracting aperture or obstacle come under the classification of Fresnel diffraction. This type of diffraction requires for its observation only a point source, a diffracting screen of some sort, and an observing screen. The latter is often advantageously replaced by a magnifier or a low-power microscope. The observed diffraction patterns generally differ according to the radius of curvature of the wave and the distance of the point of observation behind the screen. If the diffracting screen has circular symmetry, such as that of an opaque disk or a round hole, a point source of light must be used. If it has straight, parallel edges, it is desirable from the standpoint of brightness to use an illuminated slit parallel to these edges. In the latter case, it is possible to regard the wave emanating from the slit as a cylindrical one. For the purpose of deriving the vibration curve, the appropriate way of dividing the wavefront into infinitesimal elements is to use annular rings in the first case, and strips parallel to the axis of the cylinder in the second case.

The zone plate is a special screen designed to block off the light from every other half-period zone, and represents an interesting application of Fresnel diffraction. The Fresnel half-period zones are drawn, with radii proportional to the square roots of whole numbers, and alternate ones are blackened. The drawing is then photographed on a reduced scale. When light from a point source is sent through the negative, an intense point image is produced, much like that formed by a lens.   [F.A.J./W.W.W.]

Diffraction grating   An optical device consisting of an assembly of narrow slits or grooves, which by diffracting light produces a large number of beams which can interfere in such a way as to produce spectra. Since the angles at which constructive interference patterns are produced by a grating depend on the lengths of the waves being diffracted, the waves of various lengths in a beam of light striking the grating will be separated into a number of spectra, produced in various orders of interference on either side of an undeviated central image. By controlling the shape and size of the diffracting grooves when producing a grating and by illuminating the grating at suitable angles, a beam of light can be thrown into a single spectrum whose purity and brightness may exceed that produced by a prism. Gratings can now be made with much larger apertures than prisms, and in such form that they waste less light and give higher intrinsic dispersion and resolving power. *See* DIFFRACTION.

Transmission gratings consist of a large number of narrow transparent and opaque slits alternating side by side in regular order and with uniform separation, through which a beam of light will appear as a series of spectra in various orders of interference. Reflection gratings, either plane or concave, are used in most spectrographs. Such a grating may consist of an original ruling or of a metal-coated replica from an original. Large grating replicas can now be made which are practically indistinguishable in performance or permanence from an original.

Gratings are engraved by highly precise ruling engines, which use a diamond tool to press into a highly polished mirror surface a series of many thousands of fine shallow burnished grooves. If a grating is to give resolution approaching the theoretical limit, its grooves must be ruled straight, parallel, and equally spaced to within a few tenths of the shortest incident wavelength. Scattered light and false images may arise from local spacing error and groove shape variations of only a few hundredths of the diffracted wavelength.

A grating spectroscope usually consists of a slit, a lens or mirror to collimate the light sent through the slit into a parallel beam, a transmission or reflection grating to disperse the light, a lens or mirror to focus the light into spectrum lines (which are monochromatic images of the slit in the light of each wavelength passing through it), and an eyepiece for viewing the spectrum. If a camera is substituted for the telescope, the instrument becomes a grating spectrograph. If a photoelectric cell, a thermocouple, or other radiation-detecting device is used instead of a camera or telescope, the device becomes a grating spectrometer. *See* INFRARED SPECTROSCOPY.   [G.R.H.]

Diffusion   The transport of matter from one point to another by random molecular motions. It occurs in gases, liquids, and solids.

Diffusion plays a key role in processes as diverse as permeation through membranes, evaporation of liquids, dyeing textile fibers, drying timber, doping silicon wafers to make semiconductors, and transporting of thermal neutrons in nuclear power reactors. Rates of important chemical reactions are limited by how fast diffusion can bring reactants together or deliver them to reaction sites on enzymes or catalysts. The forces between molecules and molecular sizes and shapes can be studied by making diffusion measurements. *See* SEMICONDUCTOR.

Molecules in fluids (gases and liquids) are constantly moving. Even in still air, for example, nitrogen and oxygen molecules ricochet off each other at bullet speeds. Molecular diffusion is easily demonstrated by pouring a layer of water over a layer of ink in a narrow glass tube. The boundary between the ink and water is sharp at first, but it slowly blurs as the ink diffuses upward into the clear water. Eventually, the ink spreads evenly along the tube without any help from stirring.

Gases. A number of techniques are used to measure diffusion in gases. In a two-bulb experiment, two vessels of gas are connected by a narrow tube through which diffusion occurs. Diffusion is followed by measuring the subsequent changes in the composition of gas in each vessel. Excellent results are also obtained by placing a lighter gas mixture on top of a denser gas mixture in a vertical tube and then measuring the composition along the tube after a timed interval.

Rates of diffusion in gases increase with the temperature (T) approximately as $T^{3/2}$ and are inversely proportional to the pressure. The interdiffusion coefficients of gas mixtures are almost independent of the composition.

Kinetic theory shows that the self-diffusion coefficient of a pure gas is inversely proportional to both the square root of the molecular weight and the square of the molecular diameter. Interdiffusion coefficients for pairs of gases can be estimated by taking averages of the molecular weights and collision diameters. Kinetic-theory predictions are accurate to about 5% at pressures up to 10 atm (1 megapascal). Theories which take into account the forces between molecules are more accurate, especially for dense gases. *See* KINETIC THEORY OF MATTER.

Liquids. The most accurate diffusion measurements on liquids are made by layering a solution over a denser solution and then using optical methods to follow the changes in refractive index along the column of solution. Excellent results are also obtained with cells in which diffusion occurs between two solution compartments through a porous diaphragm. Many other reliable experimental techniques have been devised.

Room-temperature liquids usually have diffusion coefficients in the range 0.5–5 $\times$ 10^{-5} cm^2 s^{-1}. Diffusion in liquids, unlike diffusion in gases, is sensitive to changes in composition but relatively insensitive to changes in pressure. Diffusion of high-viscosity, syrupy liquids and macromolecules is slower. The diffusion coefficient of aqueous serum albumin, a protein of molecular weight 60,000 atomic mass units, is only 0.06 $\times$ 10^{-5} cm^2 s^{-1} at 25°C (77°F).

When solute molecules diffuse through a solution, solvent molecules must be pushed out of the way. For this reason, liquid-phase interdiffusion coefficients are inversely proportional to both the viscosity of the solvent and the effective radius of the solute molecules. Accurate theories of diffusion in liquids are still under development. *See* VISCOSITY.

[D.G.L.]

Solids. Diffusion in solids is an important topic of physical metallurgy and materials science since diffusion processes are ubiquitous in solid matter at elevated temperatures. They play a key role in the kinetics of many microstructural changes that occur during the processing of metals, alloys, ceramics, semiconductors, glasses, and polymers. Typical examples of such changes include nucleation of new phases, diffusive phase transformations, precipitation and dissolution of a second phase, recrystallization, high-temperature creep, and thermal oxidation. Direct technological applications concern diffusion doping during the fabrication of microelectronic devices, solid electrolytes for battery and fuel cells, surface hardening of steels through carburization or nitridation, diffusion bonding, and sintering. *See* PHASE TRANSITIONS.

The atomic mechanisms of diffusion are closely connected with defects in solids. Point defects such as vacancies and interstitials are the simplest defects and often mediate diffusion in an otherwise perfect crystal. Dislocations, grain boundaries, phase boundaries, and free surfaces are other types of defects in a crystalline solid. They can act as diffusion short circuits because the mobility of atoms along such defects is usually much higher than in the lattice. *See* CRYSTAL DEFECTS.

[H.Me.]

Dimensional analysis   A technique that involves the study of dimensions of physical quantities. Dimensional analysis is used primarily as a tool for obtaining information about physical systems too complicated for full mathematical solutions to be feasible. It enables one to predict the behavior of large systems from a study of small-scale models. It affords a convenient means of checking mathematical equations. Finally, dimensional formulas provide a useful cataloging system for physical quantities.

All the commonly used systems of units in physical science have the property that the number representing the magnitude of any quantity (other than purely numerical ratios) varies inversely with the size of the unit chosen. Thus, if the length of a given piece of land is 300 ft, its length in yards is 100. The ratio of the magnitude of 1 yd to the magnitude of 1 ft is the same as that of any length in feet to the same length in yards, that is, 3. The ratio of two different lengths measured in yards is the same as the ratio of the same two lengths measured in feet, inches, miles, or any other length units. This universal property of unit systems, often known as the absolute significance of relative magnitude, determines the structure of all dimensional formulas. *See* UNITS OF MEASUREMENT.

In defining a system of units for a branch of science such as mechanics or electricity, certain quantities are chosen as fundamental and others as secondary, or derived. The choice of the fundamental units is always arbitrary and is usually made on the basis of convenience in maintaining standards. In mechanics the fundamental units most often chosen are mass, length, and time.

It can be proved that every secondary quantity which satisfies the condition of the absolute significance of relative magnitude is expressible as a product of powers of the primary quantities. Such an expression is known as the dimensional formula of the secondary quantity. There is no requirement that the exponents be integral.

The technique of dimensional analysis has several important applications. It is intuitively obvious that only terms whose dimensions are the same can be equated. The equation 10 kg = 10 m/s, for example, makes no sense. A necessary condition for the correctness of any equation is that the two sides have the same dimensions. This is often a help in the verification of complicated analytic expressions. Of course, an equation can be correct dimensionally and still be wrong by a purely numerical factor.

The application of dimensional analysis to the derivation of unknown relations depends upon the concept of completeness of equations. An expression which remains formally true no matter how the sizes of the fundamental units are changed is said to be complete. Assume a group of n physical quantities $x_1, x_2, \ldots, x_n$, for which there exists one and only one complete mathematical expression connecting them, namely, $\phi(x_1, x_2, \ldots, x_n) = 0$. Some of the quantities $x_1, x_2, \ldots, x_n$ may be dimensional constants. Assume further that the dimensional formulas of the n quantities are expressed in terms of m fundamental quantities $\alpha, \beta, \gamma, \ldots$. Then it will always be found that this single relation ϕ can be expressed in terms of some arbitrary function F of $n - m$ independent dimensionless products $\pi_1, \pi_2, \ldots, \pi_{n-m}$, made up from among the n variables, as in the equation below.

$$F(\pi_1, \pi_2, \ldots, \pi_{n-m}) = 0$$

This is known as the π theorem. It was first rigorously proved by E. Buckingham. The main usefulness of the π theorem is in the deduction of the form of unknown relations. The procedure is particularly useful in hydraulics and aeronautical engineering, where detailed solutions are often extremely complicated.

A further application of dimensional analysis is in model design. Often the behavior of large complex systems can be deduced from studies of small-scale models at a great saving in cost. In the model each parameter is reduced in the same proportion relative to its value in the original system.

Dimensional formulas also provide a convenient shorthand notation for representing the definitions of secondary quantities and are helpful in changing units from one system to another.   [J.W.St.]

Dimensionless groups   A dimensionless group is any combination of dimensional or dimensionless quantities possessing zero overall dimensions. Dimensionless groups are frequently encountered in engineering studies of complicated processes or as similarity criteria in model studies. A typical dimensionless group is the Reynolds number (a dynamic similarity criterion), $N_{Re} = VD\rho/\mu$. Since the dimensions of the quantities involved are velocity V: $[L/\theta]$; characteristic dimension D: $[L]$; density ρ: $[M/L^3]$; and viscosity μ: $[M/L\theta]$ (with M, L, and θ as the fundamental units of mass, length, and time), the Reynolds number reduces to a dimensionless group and can be represented by a pure number in any coherent system of units. See DYNAMIC SIMILARITY.

Many important problems in applied science and engineering are too complicated to permit completely theoretical solutions to be found. However, the number of interrelated variables involved can be reduced by carrying out a dimensional analysis to group the variables as dimensionless groups. See DIMENSIONAL ANALYSIS.

The advantages of using dimensionless groups in studying complicated phenomena include:

1. A significant reduction in the number of "variables" to be investigated; that is, each dimensionless group, containing several physical variables, may be treated as a single compound "variable," thereby reducing the number of experiments needed as well as the time required to correlate and interpret the experimental data.

2. Predicting the effect of changing one of the individual variables in a process (which it may be impossible to vary much in available equipment) by determining the effect of varying the dimensionless group containing this parameter (this must be done with some caution, however).

3. Making the results independent of the scale of the system and of the system of units being used.

4. Simplifying the scaling-up or scaling-down of results obtained with models of systems by generalizing the conditions which must exist for similarity between a system and its model.

5. Deducing variation in importance of mechanisms in a process from the numerical values of the dimensionless groups involved; for instance, an increase in the Reynolds number in a flow process indicates that molecular (viscous) transfer mechanisms will be less important relative to transfer by bulk flow ("inertia" effects), since the Reynolds number is known to represent a measure of the ratio of inertia forces to viscous forces. See FROUDE NUMBER; KNUDSEN NUMBER; MACH NUMBER; REYNOLDS NUMBER. [J.Ca.; G.D.F.]

Dimensions (mechanics) Length, mass, time, or combinations of these quantities serving as an indication of the nature of a physical quantity. Quantities with the same dimensions can be expressed in the same units. For example, although speed can be expressed in various units such as miles/hour, feet/second, and meters/second, all these speed units involve the ratio of a length unit to a time unit; hence, the dimensions of speed are the ratio of length L to time T, usually stated as LT^{-1}. The dimensions of all mechanical quantities can be expressed in terms of L, T, and mass M. The validity of algebraic equations involving physical quantities can be tested by a process called dimensional analysis; the terms on the two sides of any valid equation must have the same dimensions. See DIMENSIONAL ANALYSIS; UNITS OF MEASUREMENT. [D.Wi.]

Diopter A measure of the power of a lens or a prism. The diopter (also called dioptrie) is usually abbreviated D. Its dimension is a reciprocal length, and its unit is the reciprocal of 1 m (3.28 ft). See FOCAL LENGTH; LENS (OPTICS).

The dioptric power of a prism is defined as the measure of the deviation of a ray going through a prism measured at the distance of 1 m. A prism that deviates a ray by 1 cm in a distance of 1 m is said to have a power of one prism diopter. See OPTICAL PRISM.

Spectacle lenses in general consist of thin lenses, which are either spherical, to correct the focus of the eye for near and far distances, or cylindrical or toric, to correct the astigmatism of the eye. An added prism corrects a deviation of the visual axis. The diopter thus gives a simple method for prescribing the necessary spectacle for the human eye. [M.J.H.]

Dipole Any object or system that is oppositely charged at two points or poles, such as a magnet, a polar molecule, or an antenna element. The properties of a dipole are determined by its dipole moment, that is, the product of one of the charges by their separation directed along an axis through the centers of charge. See DIPOLE MOMENT.

An electric dipole consists of two electric charges of equal magnitude but opposite polarity, separated by a short distance (see illustration); or more generally, a localized distribution of positive and negative electricity without net charge whose mean positions of positive and negative charge do not coincide.

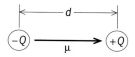

Electric dipole with moment $\mu = Qd$.

Molecular dipoles which exist in the absence of an applied field are called permanent dipoles, while those produced by the action of a field are called induced dipoles. *See* POLAR MOLECULE.

The term magnetic dipole originally referred to the fact that a magnet has two poles and, because of these two poles, experiences a torque in a magnetic field if its axis is not along a magnetic flux line of the field. It is now generalized to include electric circuits which, because of the current, also experience torques in magnetic fields. *See* MAGNET.

An electric dipole whose moment oscillates sinusoidally radiates electromagnetic waves and is known as a hertzian dipole; it is of interest in developing the theory of electromagnetic radiation. For practical purposes, a half-wave dipole, consisting of two collinear conducting rods, fed at the center, whose combined length equals half the wavelength of the radiation to be transmitted or received, is often used as an antenna element, either by itself or in an array, or as a feed for a reflector. *See* ELECTROMAGNETIC RADIATION.   [A.E.Ba.]

Dipole-dipole interaction   The interaction of two atoms, molecules, or nuclei by means of their electric or magnetic dipole moments. This is the first term of the multipole-multipole series of invariants. More precisely, the interaction occurs when one dipole is placed in the field of another dipole. The interaction energy depends on the strength and relative orientation of the two dipoles, as well as on the distance between the centers and the orientation of the radius vector connecting the centers with respect to the dipole vectors. The electric dipole-dipole interaction and magnetic dipole-dipole interaction must be distinguished.

The center of the negative charge distribution of a molecule may fail to coincide with its center of gravity, thus creating a dipole moment. An example is the water molecule. If such molecules are close together, there will be a (electric) dipole-dipole interaction between them. Atoms do not have permanent dipole moments, but a dipole moment may be induced by the presence of another atom nearby; this is called the induced dipole-dipole interaction. Induced dipole-dipole forces between atoms and molecules are known by many different names: van der Waals forces, London forces, or dispersion forces. These induced dipole-dipole forces are responsible for cohesion and surface tension in liquids. They also act between unlike molecules, resulting in the adsorption of atoms on macroscopic objects. *See* COHESION (PHYSICS); INTERMOLECULAR FORCES; VAN DER WAALS EQUATION.

The magnetic dipole-dipole interaction is found both on a macroscopic and on a microscopic scale. Two compass needles within reasonable proximity of each other illustrate clearly the influence of the dipole-dipole interaction. In quantum mechanics, the magnetic moment is partially due to a current arising from the motion of the electrons in their orbits, and partially due to the intrinsic moment of the spin. The same interaction exists between nuclear spins. Magnetic dipole-dipole forces are particularly important in low-temperature solid-state physics, the interaction between the spins of the ions in paramagnetic salts being a crucial element in the use of such salts as thermometers and as cooling substances. *See* ADIABATIC DEMAGNETIZATION; DIPOLE; ELECTRON; LOW-TEMPERATURE THERMOMETRY; MAGNET; MAGNETIC THERMOMETER; NUCLEAR MOMENTS.   [P.H.E.M.]

Dipole moment   A mathematical quantity characteristic of a dipole unit equal to the product of one of its charges times the vector distance separating the charges.

The dipole moment μ associated with a distribution of electric charges q_i is given by

$$\mu = \sum_i q_i \mathbf{r}_i$$

where $\mathbf{r}_i$ is the vector to the charge q_i. For systems with a net charge (for example, positive), the origin is taken at the mean position of the positive charges (and vice versa). Dipole moments have the dimensions coulomb-meters. Molecular dipole moments were previously expressed in debye units, where 1 debye $= 3.336 \times 10^{-30}$ C · m. *See* DIPOLE.

[R.D.W.]

Dispersion (radiation)   The separation, by refraction, interference, scattering, or diffraction, of acoustic and electromagnetic radiation or energy into its constituent wavelengths or frequencies. For a refracting, transparent substance, such as a prism of glass, the dispersion is characterized by the variation of refractive index with change in wavelength of the radiation. Refractive index (n) is defined as the ratio of the velocity of the radiation in free space (air at standard temperature and pressure for sound, and a vacuum for electromagnetic radiation) to the velocity in the substance in question. I. Newton used a small hole in a window shade and a glass prism to disperse sunlight into a visible spectrum, from violet through red. Using a second prism, he showed that no further decomposition of any of the spectral colors could be achieved. *See* OPTICAL PRISM; REFRACTION OF WAVES.

The condition where the refractive index decreases as wavelength increases is termed normal dispersion. The opposite condition is termed anomalous dispersion, and almost always occurs in regions outside the range of visible wavelengths.   [R.A.Buc.]

Dispersion relations   Relations between the real and imaginary parts of a response function. A response function relates a cause and its effect through an integral equation. The term dispersion refers to the fact that the index of refraction of a medium is a function of frequency. In 1926 H. A. Kramers and R. Kronig showed that the imaginary part of an index of refraction (that is, the absorptivity) determines the real part (that is, the refractivity); this is called the Kramers-Kronig relation. The term dispersion relation is now used for the analogous relation between the real and imaginary parts of the values of any response function.   [C.J.G.]

Displacement current   The name given by J. C. Maxwell to the term $\partial \mathbf{D}/\partial t$ which must be added to the current density $\mathbf{i}$ to extend to time-varying fields the magnetostatic result of A. M. Ampère that $\mathbf{i}$ equals the curl of the magnetic intensity $\mathbf{H}$. In integral form this result is given by the equation below, where the unit vector $\mathbf{n}$ is

$$\oint \mathbf{H} \cdot d\mathbf{s} = \int_S \left(\mathbf{i} + \frac{\partial \mathbf{D}}{\partial t} \right) \cdot \mathbf{n} \, dS$$

perpendicular to the surface dS. The concept of displacement current has important consequences for insulators and for free space where $\mathbf{i}$ vanishes. For conductors, however, the difference between the above equation and Ampère's result is negligible. *See* MAXWELL'S EQUATIONS.

If one defines current as a transport of charge, the term displacement current is certainly a misnomer when applied to a vacuum where no charges exist. If, however, current is defined in terms of the magnetic fields it produces, the expression is legitimate.

[W.R.Sm.]

Domain (electricity and magnetism)

A region in a solid within which elementary atomic or molecular magnetic or electric moments are uniformly aligned.

Ferromagnetic domains are regions of parallel-aligned magnetic moments. Each domain may be thought of as a tiny magnet pointing in a certain direction. The relatively thin boundary region between two domains is called a domain wall. Within a wall the magnetic moments rotate from the direction of one of the domains to the direction in the adjacent domain.

A ferromagnet generally consists of a large number of domains. For example, a sample of pure iron at room temperature contains many domains whose directions are distributed randomly, making the sample appear to be unmagnetized as a whole. Iron is called magnetically soft since the domain walls move easily if a magnetic field is applied. In a magnetically hard or permanent magnet material a net macroscopic magnetization is introduced by exposure to a large external magnetic field, but thereafter domain walls are difficult to either form or move, and the material retains its overall magnetization.

Antiferromagnetic domains are regions of antiparallel-aligned magnetic moments. They are associated with the presence of grain boundaries, twinning, and other crystal inhomogeneities.

Ferroelectric domains are electrical analogs of ferromagnetic domains. *See* ANTIFERROMAGNETISM; FERROELECTRICS; FERROMAGNETISM; MAGNETIC MATERIALS; MAGNETIZATION; TWINNING (CRYSTALLOGRAPHY).   [J.F.He.]

Donor atom

An impurity atom in a semiconductor which can contribute or donate one or more conduction electrons to the crystal by becoming ionized and positively charged. For example, an atom of column 5 of the periodic table substituting for a regular atom of a germanium or silicon crystal is a donor because it has one or more valence electrons which can be detached and added to the conduction band of the crystal. Donor atoms thus tend to increase the number of conduction electrons in the semiconductor. The ionization energy of a donor atom is the energy required to dissociate the electron from the atom and put it in the conduction band of the crystal. *See* ACCEPTOR ATOM; SEMICONDUCTOR.   [H.Y.F.]

Doppler effect

The change in the frequency of a wave observed at a receiver whenever the source or the receiver of the wave is moving relative to the other or to the carrier of the wave (the medium). The effect was predicted in 1842 by C. Doppler, and first verified for sound waves in 1845 from experiments conducted on a moving train.

The Doppler effect for sound waves is now a commonplace experience: If one is passed by a fast car or a plane, the pitch of its noise is considerably higher in approaching than in parting. The same phenomenon is observed if the source is at rest and the receiver is passing it. The linear optical Doppler effect was first observed in 1905 from a shift of spectral lines emitted by a beam of fast ions (canal rays) emerging from a hole in the cathode of a gas discharge tube run at high voltage. Still, their velocity was several orders of magnitude below that of light in vacuum. The precise interferometric experiments of A. A. Michelson and E. W. Morley (1887) showed clearly that the velocity of light is not bound to any ether, but is measured to be the same in any moving system. This result was a crucial check for A. Einstein's theory of special relativity (1905), which also makes a clear prediction for the optical Doppler effect.

The Doppler effect has important applications in remote-sensing, high-energy physics, astrophysics, and spectroscopy.

Let a wave from a sound source or radar source, or from a laser, be reflected from a moving object back to the source, which may itself move as well. Then a frequency shift is observed by a receiver connected to the source. The measurement provides an excellent means for the remote sensing of velocities of any kind of object, including cars, ships, planes, satellites, flows of fluids, or winds.

The light from distant stars and galaxies shows a strong Doppler shift to the red, indicating that the universe is rapidly expanding. However, this effect can be mixed up with the gravitational redshift that results from the energy loss which a light quantum suffers when it emerges from a strong gravitational field.

The Doppler width and Doppler shift of spectral lines in sunlight (Fraunhofer lines) are important diagnostic tools for the dynamics of the Sun's atmosphere, indicating its temperature and turbulence. [E.W.O.]

Duality (physics)

Duality (physics) The state of having two natures, which is often applied in physics. The classic example is wave-particle duality. The elementary constituents of nature—electrons, quarks, photons, gravitons, and so on—behave in some respects like particles and in others like waves.

Duality is often used in a more precise sense. It indicates that two seemingly different, theoretical descriptions of a physical system are actually mathematically equivalent. Such an occurrence is very useful. Various properties and phenomena are clearer in one or the other of the descriptions, and calculations that are difficult or impossible in one description may be simple in the other. In the case of wave-particle duality, the wave description corresponds to a theory of quantized fields, where the field variables are governed by an uncertainty principle. The particle description corresponds to a Feynman integral over all particle paths in spacetime. The quantized-field and path integral theories sound very different but are mathematically equivalent, making identical predictions. *See* FEYNMAN INTEGRAL; QUANTUM FIELD THEORY; QUANTUM MECHANICS; UNCERTAINTY PRINCIPLE; WAVE MECHANICS.

Weak-strong duality. In some systems, there is weak-strong duality, meaning that when the coupling constant g of the original description is large that of the dual description, g', is small; for example $g' = 1/g$. When g is large, so the interactions in the original description are strong and the perturbation theory in this description is highly inaccurate, then perturbation theory in the dual description gives a very accurate description.

Duality in superstring theory. It is believed that a complete theory of all particles and interactions must be based on quantization of one-dimensional objects (loops) rather than points: this is superstring theory. In superstring theory there is again the problem that perturbation theory is the main tool, giving an incomplete description of the physics. The situation has greatly improved with the discovery that weak-strong duality is a general property of string theory. In fact, there are five known string theories, and all are dual to one another. A notable feature in string theory is that in addition to strings and solitons, duality requires certain other objects as well: D-branes, which are local disturbances to which strings become fixed. Remarkably, the same methods have also been used to solve some long-standing problems regarding the quantum mechanics of black holes. *See* QUANTUM GRAVITATION; SUPERSTRING THEORY. [J.Pol.]

Dynamic nuclear polarization

Dynamic nuclear polarization The creation of assemblies of nuclei whose spin axes are not oriented at random, and which are in a steady state that is not a state of thermal equilibrium. Under commonly occurring conditions, the spin axes of nuclei (with nonzero spin) are oriented at random; where this is not so, the nuclei are said to be polarized. Assemblies of polarized nuclei are not in a state of thermal equilibrium except under rather extreme conditions (for example, temperatures below 10 millikelvins and magnetic fields greater than several teslas), and therefore schemes have been devised to produce polarized assemblies, in a steady state which is not a state of thermal equilibrium, under less extreme conditions of temperature and so forth. Such schemes constitute dynamic nuclear polarization.

Among the many applications of polarized nuclei are the following. Nuclear forces are spin-dependent, and although the spin-dependent part can be found by using unpolarized assemblies, the experiments are simpler and their interpretation is clearer if polarized nuclei are used. Assemblies of polarized nuclei have a lower geometrical symmetry than assemblies of randomly oriented nuclei, and so these have been used to investigate the fundamental symmetries of nature. Polarized nuclei have been used to enhance the signal in free precession magnetometers and similar instruments, and the use of an assembly of polarized nuclei as a gyroscope has also been suggested. *See* NUCLEAR ORIENTATION; PARITY (QUANTUM MECHANICS); SPIN (QUANTUM MECHANICS).
[J.M.D.]

Dynamic similarity   A relationship existing between two fluid flows when they have identical types of forces that are parallel at all corresponding points, with magnitudes related by a constant scale factor. Dynamic similarity makes it possible to scale results from model tests to predict corresponding results for the full-scale prototype.

Dynamic similarity requires faithful reproduction of detail on the model (geometric similarity); the same flow pattern, including boundary shapes (kinematic similarity); and test conditions that match relevant dimensionless ratios between model and prototype. Dynamically similar flows are said to be homologous. It may not be possible in a practical test to match all dimensionless parameters. It is most important to match parameters that represent the dominant physical effects. Thus, correct simulation of viscous effects requires that Reynolds number be matched; Mach number may be ignored if compressibility effects are not important. In ship model tests, Froude number must be matched to duplicate wave patterns; the effect of Reynolds number on viscous drag may be predicted analytically. *See* DIMENSIONAL ANALYSIS; DIMENSIONLESS GROUPS; FLUID MECHANICS; FROUDE NUMBER; MACH NUMBER; REYNOLDS NUMBER.   [A.T.McD.]

Dynamics   That branch of mechanics which deals with the motion of a system of material particles under the influence of forces, especially those which originate outside of the system under consideration. From Newton's third law of motion, namely, to every action there is an equal and opposite reaction, the internal forces cancel in pairs and do not contribute to the motion of the system as a whole, although they determine the relative motion, if any, of the several parts.

Particle dynamics refers to the motion of a single particle under the influence of external forces, particularly electromagnetic and gravitational forces. The dynamics of a rigid body is the study of the motion, under given forces, of a system of particles, the distances between which are postulated to be constant throughout the motion.

In classical dynamics the basic relation that enables the motion to be determined once the force is known is Newton's second law of motion, which states that the resultant force on a particle is equal to the product of the mass of the particle times its acceleration. For a many-particle system it becomes impracticable to write and solve this equation for each individual particle and, in general, the motion may be computed only on a statistical basis (that is, by the methods of statistical mechanics) unless, as for a few particles or a rigid body, the number of degrees of freedom is sufficiently small. *See* DEGREE OF FREEDOM (MECHANICS); KINEMATICS; KINETICS (CLASSICAL MECHANICS); NEWTON'S LAWS OF MOTION; RIGID-BODY DYNAMICS; STATISTICAL MECHANICS.   [H.C.Co./B.G.]

E

Echo   A sound wave which has been reflected or otherwise returned with sufficient magnitude and time delay to be perceived in some manner as a sound wave distinct from that directly transmitted. Multiple echo describes a succession of separately indistinguishable echos arising from a single source. When the reflected waves occur in rapid succession, the phenomenon is often termed a flutter echo. Echoes and flutter echoes are generally detrimental to the quality of the acoustics of rooms. They may be minimized through the proper selection of room dimensions, room shape, and distribution of sound-absorbing materials.

Echoes have been put to a variety of uses in measurement problems. For example, the distance between two points can be measured by timing the duration required for a direct sound originating at one location to strike an object at the other point and to return an echo to the location of the initial source. Ultrasonic echo techniques have achieved considerable success in nondestructive testing of materials. *See* REFLECTION OF SOUND; SOUND; ULTRASONICS.

[W.J.G.]

Eddy current   An electric current induced within the body of a conductor when that conductor either moves through a nonuniform magnetic field or is in a region where there is a change in magnetic flux. It is sometimes called Foucault current. Although eddy currents can be induced in any electrical conductor, the effect is most pronounced in solid metallic conductors. Eddy currents are utilized in induction heating and to damp out oscillations in various devices.

It is possible to reduce the eddy currents by laminating the conductor, that is, by building the conductor of many thin sheets that are insulated from each other rather than making it of a single solid piece. The laminations do not reduce the induced emfs, but if they are properly oriented to cut across the paths of the eddy currents, they confine the currents largely to single laminae, where the paths are long, making higher resistance.

[K.V.M.]

Eigenfunction   One of the solutions of an eigenvalue equation. A parameter-dependent equation that possesses nonvanishing solutions only for particular values (eigenvalues) of the parameter is an eigenvalue equation, the associated solutions being the eigenfunctions (sometimes eigenvectors). In older usage the terms characteristic equation and characteristic values (functions) are common. Eigenvalue equations appear in many contexts, including the solution of systems of linear algebraic equations (matrix equations), differential or partial differential equations, and integral equations. The importance of eigenfunctions and eigenvalues in applied mathematics results from the widespread applicability of linear equations as exact or approximate descriptions of physical systems. However, the most fundamental application of these concepts is in quantum mechanics where they enter into the definition and physical interpretation of the theory. Only linear eigenvalue equations will be discussed. *See* EIGENVALUE (QUANTUM MECHANICS); ENERGY LEVEL (QUANTUM MECHANICS); QUANTUM MECHANICS.   [J.M.R.]

Eigenvalue (quantum mechanics)

If an equation containing a variable parameter possesses nontrivial solutions only for certain special values of the parameter, these solutions are called eigenfunctions and the special values are called eigenvalues.

The eigenfunction-eigenvalue relation is of particular importance in quantum mechanics because of its prominence in the equations which relate the mathematical formalism of the theory with physical results. *See* QUANTUM MECHANICS.   [D.P.]

Electric charge

A basic property of elementary particles of matter. One does not define charge but takes it as a basic experimental quantity and defines other quantities in terms of it.

According to modern atomic theory, the nucleus of an atom has a positive charge because of its protons, and in the normal atom there are enough extranuclear electrons to balance the nuclear charge so that the normal atom as a whole is neutral. Generally, when the word charge is used in electricity, it means the unbalanced charge (excess or deficiency of electrons), so that physically there are enough "nonnormal" atoms to account for the positive charge on a "positively charged body" or enough unneutralized electrons to account for the negative charge on a "negatively charged body."

In line with this usage, the total charge q on a body is the total unbalanced charge possessed by the body. For example, if a sphere has a negative charge of 1×10^{-10} coulomb, it has 6.24×10^{8} electrons more than are needed to neutralize its atoms. The coulomb is the unit of charge in the meter-kilogram-second (mks) system of units. *See* COULOMB'S LAW; ELECTRICAL UNITS AND STANDARDS; ELECTROSTATICS.   [R.P.Wi.]

Electric current

The net transfer of electric charge per unit time. It is usually measured in amperes. The passage of electric current involves a transfer of energy. Except in the case of superconductivity, a current always heats the medium through which it passes.

On the other hand, a stream of electrons or ions in a vacuum, which also may be regarded as an electric current, produces no local heating. Measurable currents range in magnitude from the nearly instantaneous 10^{5} or so amperes in lightning strokes to values of the order of 10^{-16} ampere, which occur in research applications.

All matter may be classified as conducting, semiconducting, or insulating, depending upon the ease with which electric current is transmitted through it. Most metals, electrolytic solutions, and highly ionized gases are conductors. Transition elements, such as silicon and germanium, are semiconductors, while most other substances are insulators. *See* CONDUCTION (ELECTRICITY); DISPLACEMENT CURRENT; ELECTRIC INSULATOR; SEMICONDUCTOR; SUPERCONDUCTIVITY.   [J.W.St.]

Electric field

A condition in space in the vicinity of an electrically charged body such that the forces due to the charge are detectable. An electric field (or electrostatic field) exists in a region if an electric charge at rest in the region experiences a force of electrical origin. Since an electric charge experiences a force if it is in the vicinity of a charged body, there is an electric field surrounding any charged body.

The electric field intensity (or field strength) **E** at a point in an electric field has a magnitude given by the quotient obtained when the force acting on a test charge q' placed at that point is divided by the magnitude of the test charge q'. Thus, it is force per unit charge. A test charge q' is one whose magnitude is small enough so it does not alter the field in which it is placed. The direction of **E** at the point is the direction of the force **F** on a positive test charge placed at the point. Thus, **E** is a vector point function, since it has a definite magnitude and direction at every point in the field, and its defining equation is Eq. (1).

$$\mathbf{E} = \mathbf{F}/q' \tag{1}$$

Electric flux density or electric displacement **D** in a dielectric (insulating) material is related to **E** by either of the equivalent equations shown as Eqs. (2), where **P** is the

$$\mathbf{D} = \epsilon_0\mathbf{E} + \mathbf{P} \quad \mathbf{D} = \epsilon\mathbf{E} \tag{2}$$

polarization of the medium, and ϵ is the permittivity of the dielectric which is related to ϵ_0, by the equation $\epsilon = k\epsilon_0$, k being the relative dielectric constant of the dielectric. In empty space, $\mathbf{D} = \epsilon_0\mathbf{E}$.

In addition to electrostatic fields produced by separations of electric charges, an electric field is also produced by a changing magnetic field. *See* ELECTRIC CHARGE; ELECTROMAGNETIC INDUCTION; POTENTIALS.   [R.P.Wi.]

Electric insulator   A material that blocks the flow of electric current across it. Insulators are to be distinguished from electrolytes, which are electronic insulators but ionic conductors. Electric insulators are used to insulate electric conductors from one another and to confine electric currents to specified pathways, as in the insulation of wires, electric switchgear, and electronic components. They provide an electrical, mechanical, and heat-dissipation function. The electrical function of an insulator is characterized by its resistivity, its dielectric strength (breakdown voltage), and its dielectric constant (relative permittivity). Insulators can be solid, liquid, or gaseous.

The resistivity of a material is a measure of the electric current density that flows across it in response to an applied electric field. Solid and liquid insulators have direct-current resistivities of 10^{10} ohm-meters at room temperature as compared to 10^{-8} Ω-m for a good metal or 10^{-3} Ω-m for a fast ionic conductor. *See* BAND THEORY OF SOLIDS; ELECTRICAL RESISTANCE; ELECTRICAL RESISTIVITY; SEMICONDUCTOR.

A material used in the electrical industry for insulation, capacitors, or encapsulation may be exposed to large voltage gradients that it must withstand for the operating life of the system. Failure occurs if an electric short-circuit develops across the material. Such a failure is called dielectric breakdown. The breakdown voltage gradient, expressed in kilovolts per millimeter, is a measure of the dielectric strength. Dielectric breakdown of a solid is destructive; liquids and gases are self-healing.

An insulator is also known as a dielectric. The dielectric constant k is defined as the ratio of the capacitance of a flat-plate condenser, or capacitor, with a dielectric between the plates to that with a vacuum between the plates; this ratio is also the relative permittivity of the dielectric. The dielectric constant is a measure of the ability of the insulator to increase the stored charge on the plates of the condenser as a result of the displacement of charged species within the insulator. *See* CAPACITANCE; CAPACITOR; DIELECTRIC MATERIALS; PERMITTIVITY.

Successful application of solid insulating materials also depends on their mechanical properties. Insulation assemblies commonly must withstand thermal-expansion mismatch, tension, compression, flexing, or abrasion as well as a hostile chemical-thermal environment. The introduction of cracks promotes the penetration of moisture and other contaminants that promote failure, and the presence of pores may cause damaging corona discharge on the surface of a high-voltage conductor. As a result, composite materials and engineered shapes must be tailored to meet the challenges of a particular operational environment.

For example, glasses and varnishes are used as sealants, and oil is used to impregnate high-voltage, paper-insulated cables to eliminate air pockets. Porcelain is a commonly used material for the suspension of high-voltage overhead lines, but it is brittle. Therefore, a hybrid insulator was developed that consists of a cylindrical porcelain interior covered by a mastic sealant and a silicone elastomer sheath heat-shrunk onto the porcelain core. The circular fins of the outer sheath serve to shed water. However, twisted-pair cables insulated with poly(tetrafluoroethylene) are used for high-speed data transmission where a small dielectric constant of the insulator material is needed to reduce signal attenuation.

Liquid or gas insulation provides no mechanical strength, but it may provide a cheap, flexible insulation not subject to mechanical failure. Biphenyls are used as insulating liquids in capacitors; alkyl benzenes in oil-filled cables; and polybutenes for high-pressure cables operating at alternating-current voltages as high as 525 kV. Sulfur hexafluoride (SF_6) is a nonflammable, nontoxic electron-attracting gas with a breakdown voltage at atmospheric pressure more than twice that of air. Fluorocarbons such as C_2F_6 and C_4F_8 as well as the Freons are also used, and breakdown voltages have been increased significantly in gas mixtures through a synergistic effect. Used as an insulating medium in high-voltage equipment at pressures up to 10 atm (1 megapascal), sulfur hexafluoride can reduce the size of electrical substations by a factor of 10 over that of air-insulated equipment. Enclosure of metal cable in a metallic conduit filled with sulfur hexafluoride gas has been used to shield nearby workers from exposure to high electric fields.

Finally, the ability to transfer heat may be an overriding consideration for the choice of an electric insulator. Electrical machines generate heat that must be dissipated. In electronic devices, the thermal conductivity of the solid substrate is a primary consideration. Where mechanical considerations permit, circulating liquid or gaseous insulation is commonly used to carry away heat. Liquids are particularly good transporters of heat, but they are subject to oxidation. In transformers, for example, the insulators are generally mineral or synthetic oils that are circulated, in some places with gaseous nitrogen to inhibit oxidation, to carry away the heat generated by the windings and magnetic core.   [J.B.G.]

Electric spark   A transient form of gaseous conduction. This type of discharge is difficult to define, and no universally accepted definition exists. It can perhaps best be thought of as the transition between two more or less stable forms of gaseous conduction. For example, the transitional breakdown which occurs in the transition from a glow to an arc discharge may be thought of as a spark. *See* ELECTRICAL CONDUCTION IN GASES.

Electric sparks play an important part in many physical effects. Usually these are harmful and undesirable effects, ranging from the gradual destruction of contacts in a conventional electrical switch to the large-scale havoc resulting from lightning discharges. Sometimes, however, the spark may be very useful. Examples are its function in the ignition system of an automobile, its use as an intense short-duration illumination source in high-speed photography, and its use as a source of excitation in spectroscopy. In the second case the spark may actually perform the function of the camera shutter, because its extinction renders the camera insensitive. *See* SPECTROSCOPY.   [G.H.M.]

Electric susceptibility A dimensionless parameter measuring the ease of polarization of a dielectric. In a vacuum, the electric flux density **D** (measured in coulombs/m^2) and electric field strength **E** (volts/m) are related by Eq. (1) where ϵ_0

$$\mathbf{D} = \epsilon_0 \mathbf{E} \tag{1}$$

is the permittivity of free space, having the value 8.854×10^{-12} farad/m. In a dielectric material, polarization occurs, and the flux density has the value given by Eq. (2), where

$$\mathbf{D} = \epsilon_0 \epsilon_r \mathbf{E} = \epsilon_0 \mathbf{E} + \mathbf{P} \tag{2}$$

ϵ_r is the relative permittivity of the material and **P** is the polarization flux density. This can be written as Eq. (3), where $\chi_e = \epsilon_r - 1$ is known as the electric susceptibility of

$$\mathbf{P} = \epsilon_0(\epsilon_r - 1)\mathbf{E} = \epsilon_0 \chi_e \mathbf{E} \tag{3}$$

the dielectric material. It is a measure of that part of the relative permittivity which is due to the material itself.

The electric susceptibility can be related to the polarizability α by expressing the polarization in terms of molecular parameters. Thus Eqs. (4) hold, where N is the

$$\mathbf{P} = N\langle\mu\rangle_{\text{avg}} = N\alpha\mathbf{E}_L$$
$$\chi_e = \frac{N\alpha\mathbf{E}_L}{\epsilon_0 \mathbf{E}} \tag{4}$$

number of molecules per unit volume, $\langle\mu_{\text{avg}}\rangle$ is their average dipole moment, and $\mathbf{E}_L$ is the local electric field strength at a molecular site. At low concentrations, $\mathbf{E}_L$ approaches **E**, and the susceptibility is proportional to the concentration N. For a discussion of the properties and measurement of electric susceptibility *see* PERMITTIVITY; POLARIZATION OF DIELECTRICS.

[R.D.W.; A.E.Ba.]

Electrical breakdown A large, usually abrupt rise in electric current in the presence of a small increase in electric voltage. Breakdown may be intentional and controlled or it may be accidental. Lightning is the most familiar example of breakdown.

In a gas, such as the atmosphere, the potential gradient may become high enough to accelerate the naturally present ions to velocities that cause further ionization upon collision with atoms. If the region of ionization does not extend between oppositely charged electrodes, the process is corona discharge. If the region of ionization bridges the gap between electrodes, thereby breaking down the insulation provided by the gas, the process is ionization discharge. When controlled by the ballast of a fluorescent lamp, for example, the process converts electric power to light. In a gas tube the process provides controlled rectification.

In a solid, such as an insulator, when the electric field gradient exceeds 10^6 volts/cm, valence bonds between atoms are ruptured and current flows. Such a disruptive current heats the solid abruptly. In a semiconductor if the applied backward or reverse potential across a junction reaches a critical level, current increases rapidly with further rise in voltage. This avalanche characteristic is used for voltage regulation in the Zener diode. In a transistor the breakdown sets limits to the maximum instantaneous voltage that can safely be applied between collector and emitter. *See* BREAKDOWN POTENTIAL; TRANSISTOR.

[F.H.R.]

Electrical conduction in gases The process by means of which a net charge is transported through a gaseous medium. It encompasses a variety of effects and modes of conduction, ranging from the Townsend discharge at one extreme to the arc discharge at the other. The current in these two cases ranges from a fraction of 1 microampere in the first to thousands of amperes in the second. It covers a pressure range from less than 10^{-4} atm (10 pascals) to greater than 1 atm (100 kilopascals).

In general, the feature which distinguishes gaseous conduction from conduction in a solid or liquid is the active part which the medium plays in the process. Not only does the gas permit the drift of free charges from one electrode to the other, but the gas itself may be ionized to produce other charges which can interact with the electrodes to liberate additional charges. Quite apparently, the current voltage characteristic may be nonlinear and multivalued. *See* SEMICONDUCTOR.

The applications of the effects encountered in this area are of significant commercial and scientific value. A few commercial applications are thyratrons, gaseous rectifiers, ignitrons, glow tubes, and gas-filled phototubes. These tubes are used in power supplies, control circuits, pulse production, voltage regulators, and heavy-duty applications such as welders. In addition, there are gaseous conduction devices widely used in research problems. Some of these are ion sources for mass spectrometers and nuclear accelerators, ionization vacuum gages, radiation detection and measurement instruments, and thermonuclear devices for the production of power.   [G.H.M.]

Electrical impedance   The measure of the opposition that an electrical circuit presents to the passage of a current when a voltage is applied. In quantitative terms, it is the complex ratio of the voltage to the current in an alternating current (ac) circuit.

A generalized ac circuit may be composed of the interconnection of various types of circuit elements. The impedance of the circuit is given by $Z = V/I$, where Z is a complex number given by $Z = R + jX$. R, the real part of the impedance, is the resistance of the circuit, and X, the imaginary part of the impedance, is the reactance of the circuit. The units of impedance are ohms. *See* ELECTRICAL RESISTANCE; REACTANCE.   [J.O.S.]

Electrical measurements   Measurements of the many quantities by which the behavior of electricity is characterized. Measurements of electrical quantities extend over a wide dynamic range and frequencies ranging from 0 to 10^{12} Hz. The International System of Units (SI) is in universal use for all electrical measurements. Electrical measurements are ultimately based on comparisons with realizations, that is, reference standards, of the various SI units. These reference standards are maintained by the National Institute of Standards and Technology in the United States, and by the national standards laboratories of many other countries. *See* ELECTRICAL UNITS AND STANDARDS.

Direct-current (dc) measurements include measurements of resistance, voltage, and current in circuits in which a steady current is maintained. Resistance is defined as the ratio of voltage to current. For many conductors this ratio is nearly constant, but depends to a varying extent on temperature, voltage, and other environmental conditions. The best standard resistors are made from wires of special alloys chosen for low dependence on temperature and for stability.

The SI unit of resistance, the ohm, is realized by means of a quantized Hall resistance standard. This is based upon the value of the ratio of fundamental constants h/e^2, where h is Planck's constant and e is the charge of the electron, and does not vary with time. *See* HALL EFFECT.

The principal instruments for accurate resistance measurement are bridges derived from the basic four-arm Wheatstone bridge, and resistance boxes. Many multirange digital electronic instruments measure resistance potentiometrically, that is, by measuring the voltage drop across the terminals to which the resistor is connected when a known current is passed through them. The current is then defined by the voltage drop across an internal reference resistor. For high values of resistance, above a megohm, an alternative technique is to measure the integrated current into a capacitor (over a suitably defined time interval) by measuring the final capacitor voltage. Both methods are capable of considerable refinement and extension. *See* ELECTRICAL RESISTANCE; HALL EFFECT; OHMMETER; RESISTANCE MEASUREMENT; WHEATSTONE BRIDGE.

The SI unit of voltage, the volt, is realized by using arrays of Josephson junctions. This standard is based on frequency and the ratio of fundamental constants e/h, so the accuracy is limited by the measurement of frequency. Josephson arrays can produce voltages between 200 μV and 10 V. At the highest levels of accuracy, higher voltages are measured potentiometrically, by using a null detector to compare the measured voltage against the voltage drop across a tapping of a resistive divider, which is standardized (in principle) against a standard cell. *See* JOSEPHSON EFFECT.

The Zener diode reference standard is the basis for most commercial voltage measuring instruments, voltage standards, and voltage calibrators. The relative insensitivity to vibration and other environmental and transportation effects makes the diodes particularly useful as transfer standards. Under favorable conditions these devices are stable to a few parts per million per year.

Most dc digital voltmeters, which are the instruments in widest use for voltage measurement, are essentially analog-to-digital converters which are standardized by reference to their built-in reference diodes. The basic range in most digital voltmeters is between 1 and 10 V, near the reference voltage. Other ranges are provided by means of resistive dividers, or amplifiers in which gain is stabilized by feedback resistance ratios. In this way these instruments provide measurements over the approximate range from 10 nanovolts to 10 kV. *See* VOLTAGE MEASUREMENT; VOLTMETER.

The most accurate measurements of direct currents less than about 1 A are made by measuring the voltage across the potential terminals of a resistor when the current is passed through it. Higher currents, up to about 50 kA, are best measured by means of a dc current comparator, which accurately provides the ratio of the high current to a much lower one which is measured as above. At lower accuracies, resistive shunts may be used up to about 5000 A, but the effective calibration of such shunts is a difficult process. *See* CURRENT COMPARATOR; CURRENT MEASUREMENT.

Alternating-current (ac) voltages are established with reference to the dc voltage standards by the use of thermal converters. These are small devices, usually in an evacuated glass envelope, in which the temperature rise of a small heater is compared by means of a thermocouple when the heater is operated sequentially by an alternating voltage and by a reference (dc) voltage. Resistors, which have been independently established to be free from variation with frequency, permit direct measurement of power frequency voltages up to about 1 kV. Greater accuracy is provided by multijunction (thermocouple) thermal converters, although these are much more difficult and expensive to make. Improvements in digital electronics have led to alternative approaches to ac measurement. For example, a line frequency waveform may be analyzed by using fast sample-and-hold circuits and, in principle, be calibrated relative to a dc reference standard. Also, electronic root-mean-square detectors may now be used instead of thermal converters as the basis of measuring instruments. *See* THERMAL CONVERTERS.

Voltages above a few hundred volts are usually measured by means of a voltage transformer, which is an accurately wound transformer operating under lightly loaded conditions.

The principal instrument for the comparison and generation of variable alternating voltages below about 1 kV is the inductive voltage divider, a very accurate and stable device. They are widely used as the variable elements in bridges or measurement systems. *See* INDUCTIVE VOLTAGE DIVIDER.

Alternating currents of less than a few amperes are measured by the voltage drop across a resistor, whose phase angle has been established as adequately small by bridge methods. Higher currents are usually measured through the use of current transformers, which are carefully constructed (often toroidal) transformers operating under near-short-circuited conditions. The performance of a current transformer is established by calibration against an ac current comparator, which establishes precise current ratios by the injection of compensating currents to give an exact flux balance.

Commercial instruments for measurement of ac quantities are usually dc measuring instruments, giving a reading of the voltage obtained from some form of ac-dc transducer. This may be a thermal converter, or a series of diodes arranged to have a square-law response, in which case the indication is substantially the root-mean-square value. Some lower-grade instruments measure the value of the rectified signal, which is usually more nearly related to the peak value.

There has been a noticeable trend toward the use of automated measurement systems for electrical measurements, facilitated by the readiness with which modern digital electronic instruments may be interfaced with computers. Many of these instruments have built-in microprocessors, which improve their convenience in use, accuracy, and reliability. For power measurements. For measurements at frequencies above about 300 MHz, *see* MICROWAVE MEASUREMENTS   [R.G.Jon.; O.C.J.]

Electrical resistance   Opposition of a circuit to the flow of electric current. Ohm's law states that the current I flowing in a circuit is proportional to the applied potential difference V. The constant of proportionality is defined as the resistance R. Hence, Eq. (1) holds. If V and I are measured in volts and amperes, respectively, R

$$V = I R \qquad (1)$$

is measured in ohms. Microscopically, resistance is associated with the impedance to flow of charge carriers offered by the material. For example, in a metallic conductor the charge carriers are electrons moving in a polycrystalline material in which their journey is impeded by collisions with imperfections in the local crystal lattice, such as impurity atoms, vacancies, and dislocations. In these collisions the carriers lose energy to the crystal lattice, and thus Joule heat is liberated in the conductor, which rises in temperature. The Joule heat P is given by Eq. (2).

$$P = I^2 R = I V = \frac{V^2}{R} \qquad (2)$$

See CRYSTAL DEFECTS; ELECTRICAL RESISTIVITY; JOULE'S LAW; OHM'S LAW.   [P.A.S.]

Electrical resistivity   The electrical resistance offered by a homogeneous unit cube of material to the flow of a direct current of uniform density between opposite faces of the cube. Also called specific resistance, it is an intrinsic, bulk (not thin-film) property of a material. Resistivity is usually determined by calculation from the measurement of electrical resistance of samples having a known length and uniform cross section according to the following equation, where ρ is the resistivity, R the measured resistance,

A the cross-sectional area, and l the length. In the mks system (SI), the unit of resistivity is the ohm-meter. Therefore, in the equation below, resistance is expressed in ohms, and the sample dimensions in meters.

$$\rho = RA/l$$

The room-temperature resistivity of pure metals extends from approximately 1.5×10^{-8} ohm-meter for silver, the best conductor, to 135×10^{-8} ohm-meter for manganese, the poorest pure metallic conductor. Most metallic alloys also fall within the same range. Insulators have resistivities within the approximate range of 10^8 to 10^{16} ohm-meters. The resistivity of semiconductor materials, such as silicon and germanium, depends not only on the basic material but to a considerable extent on the type and amount of impurities in the base material. Large variations result from small changes in composition, particularly at very low concentrations of impurities. Values typically range from 10^{-4} to 10^5 ohm-meters. See ELECTRICAL RESISTANCE; SEMICONDUCTOR.

The temperature coefficients (changes with temperature) of resistivity of pure metallic conductors are positive. Resistivity increases by about 0.4%/K at room temperature and is nearly proportional to the absolute temperature over wide temperature ranges. As the temperature is decreased toward absolute zero, resistivity decreases to a very low residual value for some metals. The resistivity of other metals abruptly changes to zero at some temperature above absolute zero, and they become superconductors.

Metals, and some semiconductors in particular, exhibit a change in resistivity when placed in a magnetic field. Theoretical relations to explain the observed phenomena have not been well developed.   [C.E.A.]

Electrical units and standards

The process of measurement consists in finding out how many times the quantity to be measured contains a fixed quantity of the same kind, called a unit. The definitions of the units often involve complex physical theory and do not lend themselves readily to practical realization. The concrete representations of units are known as measurement standards. In practice, measurements are made by using an instrument calibrated against a local reference standard, which itself has been calibrated either directly or by several links in a traceability chain against the national standard held by the national standards laboratory.

Electrical and magnetic units. A proposal by W. E. Weber in 1851 led to the absolute cgs system in which all units of quantities to be measured could be derived from the base units of length, mass, and time—the centimeter, gram, and second. This system was widely adopted although it had three weaknesses: the size of the units was inconvenient for practical use; it was difficult to realize the units from their definitions; and there were separate sets of units for electrostatic and electromagnetic quantities, based respectively on the inverse-square laws of force between electric charges and between magnetic poles.

International units. The first weakness was resolved by international agreement in 1881 to fix the practical units—the volt, the ohm, and the ampere—at 10^8, 10^9, and 0.1 times the respective cgs electromagnetic units. The other weaknesses were avoided by the decisions of the 1908 International Congress in London, where realizations of these units in terms of easily reproducible standards were defined.

The mksa units. A more fundamental change resulted from a proposal by G. Giorgi in 1902. This led to the adoption of the mksa system of units, in which there are four base units: the meter, the kilogram, the second, and the ampere. Use of the meter and the kilogram instead of the centimeter and the gram gave units of a size more convenient for practical use, and use of the ampere as a base unit resolved the conflict between electrostatic and electromagnetic units while maintaining the magnitudes of the widely used practical units. This was a truly coherent system, in the sense that other units were derived from the base units without the need for factors of proportionality other than unity.

SI units. From the mksa system the present-day SI (Système Internationale), formally adopted in 1954, has developed, by the addition of further base units to include other fields of measurement. The seven base units of SI are the kilogram (kg; mass); second (s; time); meter (m; length); ampere (A; electric current); kelvin (K; thermodynamic temperature); candela (cd; luminous intensity); and mole (m; amount of substance). The units of other physical quantities (derived units) are derived from the base units by simple numerical relations.

The SI base unit for electrical measurements is the ampere (A), the unit of electric current. It is defined in terms of a hypothetical experiment as that constant current which, if maintained in two straight parallel conductors of infinite length, of negligible circular cross section, and placed 1 meter apart in vacuum, would produce between these conductors a force equal to 2×10^{-7} newton per meter of length.

The volt (V) is the unit of potential difference and of electromotive force. It is defined as the potential difference between two points of a conducting wire carrying a constant current of 1 ampere when the power dissipated between these points is equal to 1 watt. From the ampere and the volt, the ohm is derived by Ohm's law, and the other derived quantities follow in a similar manner by the application of known physical laws. *See* OHM'S LAW.

The remaining units of electrical and magnetic quantities are:

Coulomb (C): The unit of electric charge, equal to 1 ampere-second. The coulomb is the quantity of electricity carried in 1 second by a current of 1 ampere.

Farad (F): The unit of capacitance, equal to 1 coulomb per volt. The farad is the capacitance of a capacitor between the plates of which there appears a potential difference of 1 volt when it is charged by a quantity of electricity of 1 coulomb.

Henry (H): The unit of inductance, equal to 1 weber per ampere. The henry is the inductance of a closed circuit in which an electromotive force of 1 volt is produced when the electric current in the circuit varies uniformly at the rate of 1 ampere per second.

Ohm (Ω): The unit of electrical resistance, equal to 1 volt per ampere. The ohm is defined as the resistance between two points of a conductor when a constant potential difference of 1 volt, applied to these points, produces in the conductor a current of 1 ampere, the conductor not being the seat of any electromotive force.

Siemens (S): The unit of electrical conductance (the reciprocal of resistance), equal to 1 ampere per volt. It was formerly known as the mho.

Tesla (T): The unit of magnetic flux density, equal to 1 weber per square meter.

Weber (Wb): The unit of magnetic flux, equal to 1 volt-second. The weber is the magnetic flux which, linking a circuit of one turn, would produce in it an electromotive force of 1 volt if it were reduced to zero at a uniform rate in 1 second.

The mechanical units of frequency (hertz), energy or work (joule), and power (watt) are frequently involved in expressing electrical and magnetic quantities. The cgs units, such as the gauss, gilbert, maxwell, and oersted, formerly used, are not part of the SI and are now obsolete. *See* UNITS OF MEASUREMENT.

Electrical standards. Realization of the values of the electrical and other units from their SI definitions involves great experimental difficulties. For this reason, it is customary for national standards laboratories to maintain stable primary standards of the units against which other reference standards can be compared. From time to time, absolute determinations of the values of these primary standards are made in terms of their definitions. By the late 1980s, the Josephson effect and the quantum Hall effect had made possible the standardization of the volt and the ohm by relation to fundamental physical constants. The recommendation by the CCE of the values to be adopted for these constants in 1990 led to a complete change in primary electrical standards and the method of handling them; and all the major national laboratories, and the BIPM, now use this method. *See* FUNDAMENTAL CONSTANTS; HALL EFFECT; JOSEPHSON EFFECT.

For many years the primary standards maintained by most laboratories were the volt, in terms of the mean electromotive force of a group of Weston cells, and the ohm, using a group of standard resistors. A range of reference standards of other quantities are derived from these, including direct-current (dc) voltage and resistance at a variety of levels; alternating-current (ac) voltage, resistance, and power; capacitance and inductance; radio-frequency (rf) and microwave quantities; magnetic quantities and properties of materials; dielectric properties; and other quantities. These secondary standards are used for day-to-day measurements and for the calibration of local reference standards of other users in the national measurement system. *See* CAPACITANCE MEASUREMENT; INDUCTANCE MEASUREMENT; MAGNETIC MATERIALS; MICROWAVE MEASUREMENTS; PERMITTIVITY; RESISTANCE MEASUREMENT; VOLTAGE MEASUREMENT.   [A.E.Ba.]

Electricity   Physical phenomena involving electric charges, their motions, and their effects. The motion of a charge is affected by its interaction with the electric field and, for a moving charge, the magnetic field. The electric field acting on a charge arises from the presence of other charges and from a time-varying magnetic field. The magnetic field acting on a moving charge arises from the motion of other charges and from a time-varying electric field. Thus electricity and magnetism are ultimately inextricably linked. In many cases, however, one aspect may dominate, and the separation is meaningful. *See* ELECTRIC CHARGE; ELECTRIC FIELD; MAGNETISM.

The quantitative development of electricity began late in the eighteenth century. J. B. Priestley in 1767 and C. A. Coulomb in 1785 discovered independently the inverse-square law for stationary charges. This law serves as a foundation for electrostatics. *See* COULOMB'S LAW; ELECTROSTATICS.

In 1800 A. Volta constructed and experimented with the voltaic pile, the predecessor of modern batteries. It provided the first continuous source of electricity. In 1820 H. C. Oersted demonstrated magnetic effects arising from electric currents. The production of induced electric currents by changing magnetic fields was demonstrated by M. Faraday in 1831. In 1851 he also proposed giving physical reality to the concept of lines of force. This was the first step in the direction of shifting the emphasis away from the charges and onto the associated fields. *See* ELECTROMAGNETIC INDUCTION; ELECTROMAGNETISM; LINES OF FORCE.

In 1865 J. C. Maxwell presented his mathematical theory of the electromagnetic field. This theory, which proposed a continuous electric fluid, not only synthesized a unified theory of electricity and magnetism, but also showed optics to be a branch of electromagnetism. *See* ELECTROMAGNETIC RADIATION; MAXWELL'S EQUATIONS.

The developments of theories about electricity subsequent to Maxwell have all been concerned with the microscopic realm. Faraday's experiments on electrolysis in 1833 had indicated a natural unit of electric charge, thus pointing toward a discrete rather than continuous charge. The existence of electrons, negatively charged particles, was postulated by A. Lorenz in 1895 and demonstrated by J. J. Thomson in 1897. The existence of positively charged particles (protons) was shown shortly afterward (1898) by W. Wien. Since that time, many particles have been found having charges numerically equal to that of the electron. The question of the fundamental nature of these particles remains unsolved, but the concept of a single elementary charge unit is apparently still valid. *See* BARYON; ELECTRON; ELEMENTARY PARTICLE; HYPERON; MESON; PROTON; QUARKS.

The sources of electricity in modern technology depend strongly on the application for which they are intended.

The principal use of static electricity today is in the production of high electric fields. Such fields are used in industry for testing the ability of components such as insulators and condensers to withstand high voltages, and as accelerating fields for charged-particle accelerators. The principal source of such fields today is the Van de Graaff generator. *See* PARTICLE ACCELERATOR.

The major use of electricity arises in devices using direct current and low-frequency alternating current. The use of alternating current, introduced by S. Z. de Ferranti in 1885–1890, allows power transmission over long distances at very high voltages with a resulting low-percentage power loss followed by highly efficient conversion to lower voltages for the consumer through the use of transformers. *See* ELECTRIC CURRENT.

Large amounts of direct current are used in the electrodeposition of metals, both in plating and in metal production, for example, in the reduction of aluminum ore.

The principal sources of low-frequency electricity are generators based on the motion of a conducting medium through a magnetic field. The moving charges interact with the magnetic field to give a charge motion that is normal to both the direction of motion and the magnetic field. In the most common form, conducting wire coils rotate in an applied magnetic field. The rotational power is derived from a water-driven turbine in the case of hydroelectric generation, or from a gas-driven turbine or reciprocating engine in other cases.

Many high-frequency devices, such as communications equipment, television, and radar, involve the consumption of only moderate amounts of power, generally derived from low-frequency sources. If the power requirements are moderate and portability is needed, the use of ordinary chemical batteries is possible. Ion-permeable membrane batteries are a later development in this line. Fuel cells, particularly hydrogen-oxygen systems, are being developed. They have already found extensive application in earth satellite and other space systems. The successful use of thermoelectric generators based on the Seebeck effect in semiconductors has been reported. *See* THERMOELECTRICITY.

The solar battery, also a semiconductor device, has been used to provide charging current for storage batteries in telephone service and in communications equipment in artificial satellites.

Direct conversion of mechanical energy into electrical energy is possible by utilizing the phenomena of piezoelectricity and magnetostriction. These have some application in acoustics and stress measurements. Pyroelectricity is a thermodynamic corollary of piezoelectricity. *See* MAGNETOSTRICTION; PIEZOELECTRICITY; PYROELECTRICITY.   [W.Ar.]

Electrodynamics   The study of the relations between electrical, magnetic, and mechanical phenomena. This includes considerations of the magnetic fields produced by currents, the electromotive forces induced by changing magnetic fields, the forces on currents in magnetic fields, the propagation of electromagnetic waves, and the behavior of charged particles in electric and magnetic fields. Classical electrodynamics deals with fields and charged particles in the manner first systematically described by J. C. Maxwell, whereas quantum electrodynamics applies the principles of quantum mechanics to electrical and magnetic phenomena. Relativistic electrodynamics is concerned with the behavior of charged particles and fields when the velocities of the particles approach that of light. Cosmic electrodynamics is concerned with electromagnetic phenomena occurring on celestial bodies and in space. *See* ELECTROMAGNETISM; QUANTUM ELECTRODYNAMICS; RELATIVISTIC ELECTRODYNAMICS.

[J.W.St.]

Electroluminescence   A general term for the luminescence excited by the application of an electric field to a system, usually in the solid state. Solid-state electroluminescent systems can be made quite thin, leading to applications in thin-panel area light sources and flat screens to replace cathode-ray tubes for electronic display and image formation. *See* LUMINESCENCE.

Modern interest in electroluminescence dates from the discovery by G. Destriau in France in 1936 that when a zinc sulfide (ZnS) phosphor powder is suspended in an insulator (oil, plastic, or glass ceramic) and an intense alternating electric field is applied with capacitorlike electrodes, visible light is emitted. The phosphor, prepared from zinc sulfide by addition of a small amount of copper impurity, was later shown to contain particles of a copper sulfide (Cu_2S) phase in addition to copper in its normal role as a luminescence activator in the zinc sulfide lattice. The intensification of the applied electric field by the sharp conductive or semiconductive copper sulfide inhomogeneities is believed to underlie the mechanism of Destriau-type electroluminescence. Minority carriers are ejected from these high-field spots into the low- or moderate-field regions of the phosphor, where they recombine to excite the activator centers. The structure of a Destriau-type electroluminescent cell is shown in the illustration; the light is observed through the transparent indium–tin oxide electrode.

The application of electroluminescence to display and image formation received great impetus from work in the late 1960s and mid-1970s on thin-film electroluminescence (TFEL), giving rise to devices that are different in structure and mechanism from the Destriau conditions. The phosphor in these devices is not a powder but a thin (about 500 nanometers) continuous film prepared by sputtering or vacuum evaporation. The luminescence activators are manganese or rare-earth ions, atomic species with internal electronic transitions that lead to characteristic luminescence. The phosphor film does not contain copper sulfide or any other separate phase, and is sandwiched between two thin (about 200 nm) transparent insulating films also prepared by evaporative means.

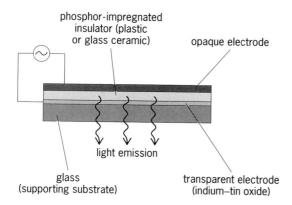

phosphor-impregnated insulator (plastic or glass ceramic)

opaque electrode

light emission

glass (supporting substrate)

transparent electrode (indium–tin oxide)

Structure of powdered-phosphor (Destriau) electroluminescent cell, edge view (not to scale).

Conducting electrodes are applied to the outside of each insulating film; one of the electrodes is again a transparent coating of indium–tin oxide on glass, which serves as supporting substrate. If an imaging matrix is desired, both electrodes consist of grids of parallel lines, with the direction of the grid on one insulator (row) orthogonal to the other grid (column). By approximate circuitry the entire matrix can be scanned, applying voltage where desired to a phosphor element that is located between the intersection of a row and column electrode. A thin-film electroluminescent device acts like a pure capacitor at low applied voltage; no light is emitted until the voltage reaches a threshold value determined by the dielectric properties of the insulator and phosphor films. Above this threshold a dissipative current flows, and light emission occurs. The brightness increases very steeply with the applied voltage but is finally saturated. The light output, or average brightness, is roughly proportional to the frequency up to at least 5 kHz, and also depends on the waveform of the applied voltage.

The best thin-film electroluminescent phosphor is manganese-activated zinc sulfide, which emits yellow light peaking at 585 nm. Activation of zinc sulfide and certain alkaline earth sulfides with different rare earths has yielded many other promising electroluminescent phosphors emitting blue, green, red, and white, and making full-color matrix-addressed thin-film electroluminescent displays possible. The light output of thin-film electroluminescent displays has been very reliable, with typically only 10% loss after tens of thousands of hours of operation.

Injection electroluminescence results when a semiconductor *pn* junction or a point contact is biased in the forward direction. This type of emission, first observed from silicon carbide (SiC) in 1907, is the result of radiative recombination of injected minority carriers, with majority carriers being a material. Such emission has been observed in a large number of semiconductors. The wavelength of the emission corresponds to an energy equal, at most, to the forbidden band gap of the material, and hence in most of these materials the wavelength is in the infrared region of the spectrum. If a *pn* junction is biased in the reverse direction, so as to produce high internal electric fields, other types of emission can occur, but with very low efficiency. *See* SEMICONDUCTOR; SEMICONDUCTOR DIODE.

Light emission may also occur when electrodes of certain metals, such as Al or Ta, are immersed in suitable electrolytes and current is passed between them. In many cases this galvanoluminescence is electroluminescence generated in a thin oxide layer formed on the electrode by electrolytic action. In addition to electroluminescence proper, other interesting effects (usually termed electrophotoluminescence) occur when electric fields are applied to a phosphor which is concurrently, or has been previously, excited by other means. These effects include a decrease or increase in steady-state photoluminescence brightness when the field is applied, or a burst of afterglow emission if the field is applied after the primary photoexcitation is removed. *See* PHOTOLUMINESCENCE.   [J.H.S.; C.C.K.]

Electromagnetic field A changing magnetic field always produces an electric field, and conversely, a changing electric field always produces a magnetic field. This interaction of electric and magnetic forces gives rise to a condition in space known as an electromagnetic field. The characteristics of an electromagnetic field are expressed mathematically by Maxwell's equation. *See* ELECTRIC FIELD; ELECTROMAGNETIC RADIATION; ELECTROMAGNETIC WAVE.   [J.Be.]

Electromagnetic induction The production of an electromotive force either by motion of a conductor through a magnetic field in such a manner as to cut across the magnetic flux or by a change in the magnetic flux that threads a conductor. *See* ELECTROMOTIVE FORCE (EMF).

If the flux threading a coil is produced by a current in the coil, any change in that current will cause a change in flux, and thus there will be an induced emf while the current is changing. This process is called self-induction. The emf of self-induction is proportional to the rate of change of current.

The process by which an emf is induced in one circuit by a change of current in a neighboring circuit is called mutual induction. Flux produced by a current in a circuit A threads or links circuit B. When there is a change of current in circuit A, there is a change in the flux linking coil B, and an emf is induced in circuit B while the change is taking place. Transformers operate on the principle of mutual induction.

The phenomenon of electromagnetic induction has a great many important applications in modern technology.

[K.V.M.]

Electromagnetic radiation   Energy transmitted through space or through a material medium in the form of electromagnetic waves. The term can also refer to the emission and propagation of such energy. Whenever an electric charge oscillates or is accelerated, a disturbance characterized by the existence of electric and magnetic fields propagates outward from it. This disturbance is called an electromagnetic wave. The frequency range of such waves is tremendous, as is shown by the electromagnetic spectrum in the table. The sources given are typical, but not mutually exclusive.

In theory, any electromagnetic radiation can be detected by its heating effect. This method has actually been used over the range from x-rays to radio. Ionization

Electromagnetic spectrum

Frequency, Hz	Wavelength, m	Nomenclature	Typical source
10^{23}	3×10^{-15}	Cosmic photons	Astronomical
10^{22}	3×10^{-14}	γ-rays	Radioactive nuclei
10^{21}	3×10^{-13}	γ-rays, x-rays	
10^{20}	3×10^{-12}	x-rays	Atomic inner shell
		Positron-electron annihilation	
10^{19}	3×10^{-11}	Soft x-rays	Electron impact on a solid
10^{18}	3×10^{-10}	Ultraviolet, x-rays	Atoms in sparks
10^{17}	3×10^{-9}	Ultraviolet	Atoms in sparks and arcs
10^{16}	3×10^{-8}	Ultraviolet	Atoms in sparks and arcs
10^{15}	3×10^{-7}	Visible spectrum	Atoms, hot bodies, molecules
10^{14}	3×10^{-6}	Infrared	Hot bodies, molecules
10^{13}	3×10^{-5}	Infrared	Hot bodies, molecules
10^{12}	3×10^{-4}	Far-infrared	Hot bodies, molecules
10^{11}	3×10^{-3}	Microwaves	Electronic devices
10^{10}	3×10^{-2}	Microwaves, radar	Electronic devices
10^{9}	3×10^{-1}	Radar	Electronic devices
		Interstellar hydrogen	
10^{8}	3	Television, FM radio	Electronic devices
10^{7}	30	Short-wave radio	Electronic devices
10^{6}	300	AM radio	Electronic devices
10^{5}	3000	Long-wave radio	Electronic devices
10^{4}	3×10^{4}	Induction heating	Electronic devices
10^{3}	3×10^{5}		Electronic devices
100	3×10^{6}	Power	Rotating machinery
10	3×10^{7}	Power	Rotating machinery
1	3×10^{8}		Commutated direct current
0	Infinity	Direct current	Batteries

effects measured by cloud chambers, photographic emulsions, ionization chambers, and Geiger counters have been used in the γ- and x-ray regions. Direct photography can be used from the γ-ray to the infrared region.

Fluorescence is effective in the x-ray and ultraviolet ranges. Bolometers, thermocouples, and other heat-measuring devices are used chiefly in the infrared and microwave regions. Crystal detectors, vacuum tubes, and transistors cover the microwave and radio frequency ranges. *See* DIFFRACTION; ELECTROMAGNETIC WAVE; GAMMA RAYS; HEAT RADIATION; INFRARED RADIATION; INTERFERENCE OF WAVES; LIGHT; MAXWELL'S EQUATIONS; POLARIZATION OF WAVES; RADIATION; REFLECTION OF ELECTROMAGNETIC RADIATION; REFRACTION OF WAVES; SCATTERING OF ELECTROMAGNETIC RADIATION; ULTRAVIOLET RADIATION; WAVE MOTION; X-RAYS. [W.R.Sm.]

Electromagnetic wave   A disturbance, produced by the acceleration or oscillation of an electric charge, which has the characteristic time and spatial relations associated with progressive wave motion. A system of electric and magnetic fields moves outward from a region where electric charges are accelerated, such as an oscillating circuit or the target of an x-ray tube. The wide wavelength range over which such waves are observed is shown by the electromagnetic spectrum. The term electric wave, or hertzian wave, is often applied to electromagnetic waves in the radar and radio range. Electromagnetic waves may be confined in tubes, such as wave guides, or guided by transmission lines. They were predicted by J. C. Maxwell in 1864 and verified experimentally by H. Hertz in 1887. *See* ELECTROMAGNETIC RADIATION.   [W.R.Sm.]

Electromagnetism   The branch of science dealing with the observations and laws relating electricity to magnetism. Electromagnetism is based upon the fundamental observations that a moving electric charge produces a magnetic field and that a charge moving in a magnetic field will experience a force. The magnetic field produced by a current is related to the current, the shape of the conductor, and the magnetic properties of the medium around it by Ampère's law. The magnetic field at any point is described in terms of the force that it exerts upon a moving charge at that point. The electrical and magnetic units are defined in terms of the ampere, which in turn is defined from the force of one current upon another. The association of electricity and magnetism is also shown by electromagnetic induction, in which a changing magnetic field sets up an electric field within a conductor and causes the charges to move in the conductor. *See* EDDY CURRENT; ELECTRICITY; ELECTROMAGNETIC INDUCTION; FARADAY'S LAW OF INDUCTION; HALL EFFECT; INDUCTANCE; LENZ'S LAW; MAGNETISM; RELUCTANCE.   [K.V.M.]

Electrometer   A highly sensitive instrument which measures all or some of the following variables: current, charge, voltage, and resistance. There are two classes of electrometers, mechanical and electronic. The mechanical instruments have been largely replaced by electronic types. *See* CURRENT MEASUREMENT; ELECTRICAL MEASUREMENTS; VOLTAGE MEASUREMENT.

Mechanical electrometers rely for their operation on the mechanical forces associated with electrostatic fields. Attracted-disk instruments, in which the attractive force between two plates, with a potential difference between them, is measured in terms of the fundamental units of mass and length, are sometimes termed absolute electrometers. They are widely used as electrostatic voltmeters for measuring potentials greater than 1 kilovolt. *See* ELECTRICAL UNITS AND STANDARDS.

The quadrant electrometer consists of a cylindrical metal box divided into quadrants which stand on insulating pillars. Opposite quadrants are connected electrically, and a light, thin metal vane of large area is suspended by a conducting torsion fiber inside the quadrants. An unknown potential is applied across the two quadrant pairs, and electrostatic forces on the vane cause a deflection proportional to the potential. Potentials as low as 10 millivolts can be measured. Small charges and currents can also be measured if the capacitance between vane and quadrants is known. See ELECTROSCOPE; ELECTROSTATICS; VOLTMETER.

Electronic electrometers utilize some form of electronic amplifier, typically an operational amplifier with a field-effect-transistor input stage to minimize the input current. In the most sensitive applications, problems arise due to drift in the amplifier characteristics and electrical noise present in the circuit components. To obviate these effects, electrometers employing a vibrating capacitor or varactor diodes are used. The signal to be measured is converted to an alternating-current (ac) signal and subsequently amplified by an ac amplifier which is less susceptible to drift and noise. The amplified signal is finally reconverted to direct current (dc). See TRANSISTOR.   [R.W.J.B.]

Electromotive force (emf)   A measure of the strength of a source of electrical energy. The term is often shortened to emf. It is not a force in the usual mechanical sense (and for this reason has sometimes been called electromotance), but it is a conveniently descriptive term for the agency which drives current through an electric circuit. In the simple case of a direct current I (measured in amperes) flowing through a resistor R (in ohms), Ohm's law states that there will be a voltage drop (or potential difference) of $V = IR$ (in volts) across the resistor. To cause this current to flow requires a source with emf (also measured in volts) $E = V$. More generally, Kirchhoff's voltage law states that the sum of the source emf's taken around any closed path in an electric circuit is equal to the sum of the voltage drops. This is equivalent to the statement that the total emf in a closed circuit is equal to the line integral of the electric field strength around the circuit. See ELECTRIC CURRENT; ELECTRIC FIELD; ELECTRICAL RESISTANCE; OHM'S LAW.

An emf may be steady (direct), as for a battery, or time-varying, as for a charged capacitor discharging through a resistor. Emf's may be generated by a variety of physical, chemical, and biological processes. Some of the more important are:

1. Electrochemical reactions, as used in direct-current (dc) batteries, in which the emf results from the reactions between electrolyte and electrodes.

2. Electromagnetic induction, in which the emf results from a change in the magnetic flux linking the circuit. This finds application in alternating-current rotary generators and transformers, providing the basis for the electricity supply industry. See ELECTROMAGNETIC INDUCTION; FARADAY'S LAW OF INDUCTION.

3. Thermoelectric effects, in which a temperature difference between different parts of a circuit produces an emf. The main use is for the measurement of temperature by means of thermocouples; there are some applications to electric power generation. See THERMOCOUPLE; THERMOELECTRICITY.

4. The photovoltaic effect, in which the absorption of light (or, more generally, electromagnetic radiation) in a semiconductor produces an emf. This is widely used for scientific purposes in radiation detectors and also, increasingly, for the generation of electric power from the Sun's radiation. See PHOTOVOLTAIC EFFECT; RADIOMETRY.

5. The piezoelectric effect, in which the application of mechanical stress to certain types of crystal generates an emf. There are applications in sound recording, in ultrasonics, and in various types of measurement transducer. *See* KIRCHHOFF'S LAWS OF ELECTRIC CIRCUITS; PIEZOELECTRICITY; ULTRASONICS.   [A.E.Ba.]

Electron   An elementary particle which is the negatively charged constituent of ordinary matter. The electron is the lightest known particle which possesses an electric charge. Its rest mass is $m_e \cong 9.1 \times 10^{-28}$ g, about $1/1836$ of the mass of the proton or neutron, which are, respectively, the positively charged and neutral constituents of ordinary matter. Discovered in 1895 by J. J. Thomson in the form of cathode rays, the electron was the first elementary particle to be identified. *See* ELECTRIC CHARGE; ELEMENTARY PARTICLE; NUCLEAR STRUCTURE.

The charge of the electron is $-e \cong -4.8 \times 10^{-10}$ esu $= -1.6 \times 10^{-19}$ coulomb. The sign of the electron's charge is negative by convention, and that of the equally charged proton is positive. This is a somewhat unfortunate convention, because the flow of electrons in a conductor is thus opposite to the conventional direction of the current.

Electrons are emitted in radioactivity (as beta rays) and in many other decay processes; for instance, the ultimate decay products of all mesons are electrons, neutrinos, and photons, the meson's charge being carried away by the electrons. The electron itself is completely stable. Electrons contribute the bulk to ordinary matter; the volume of an atom is nearly all occupied by the cloud of electrons surrounding the nucleus, which occupies only about 10^{-13} of the atom's volume. The chemical properties of ordinary matter are determined by the electron cloud. *See* MESON; RADIOACTIVITY.

The electron obeys the Fermi-Dirac statistics, and for this reason is often called a fermion. One of the primary attributes of matter, impenetrability, results from the fact that the electron, being a fermion, obeys the Pauli exclusion principle; the world would be completely different if the lightest charged particle were a boson, that is, a particle that obeys Bose-Einstein statistics. *See* BOSE-EINSTEIN STATISTICS; EXCLUSION PRINCIPLE; FERMI-DIRAC STATISTICS; POSITRON.   [C.J.G.]

Magnetic moment. The electron has magnetic properties by virtue of (1) its orbital motion about the nucleus of its parent atom and (2) its rotation about its own axis. The magnetic properties are best described through the magnetic dipole moment associated with 1 and 2. The classical analog of the orbital magnetic dipole moment is the dipole moment of a small current-carrying circuit. The electron spin magnetic dipole moment may be thought of as arising from the circulation of charge, that is, a current, about the electron axis; but a classical analog to this moment has much less meaning than that to the orbital magnetic dipole moment. The magnetic moments of the electrons in the atoms that make up a solid give rise to the bulk magnetism of the solid.

Spin. That property of an electron which gives rise to its angular momentum about an axis within the electron. Spin is one of the permanent and basic properties of the electron. Both the spin and the associated magnetic dipole moment of the electron were postulated by G. E. Uhlenbeck and S. Goudsmit in 1925 as necessary to allow the interpretation of many observed effects, among them the so-called anomalous Zeeman effect, the existence of doublets (pairs of closely spaced lines) in the spectra of the alkali atoms, and certain features of x-ray spectra. *See* SPIN (QUANTUM MECHANICS).

The spin quantum number is s, where s is always $1/2$. This means that the component of spin angular momentum along a preferred direction, such as the direction of a magnetic field, is $\pm 1/2\hbar$, where $\hbar$ is Planck's constant h divided by 2π. The spin angular momentum of the electron is not to be confused with the orbital angular momentum of the electron associated with its motion about the nucleus. In the latter case the maximum component of angular momentum along a preferred direction is $l\hbar$, where l is the angular momentum quantum number and may be any positive integer or zero. *See* QUANTUM NUMBERS.

The electron has a magnetic dipole moment by virtue of its spin. The approximate value of the dipole moment is the Bohr magneton μ_0 which is equal to $eh/4\pi mc = 9.27 \times 10^{-21}$ erg/oersted, where e is the electron charge measured in electrostatic units, m is the mass of the electron, and c is the velocity of light. (In SI units, $\mu_0 = 9.27 \times 10^{-24}$ joule/tesla.) The orbital motion of the electron also gives rise to a magnetic dipole moment μ_l, that is equal to μ_0 when $l = 1$.   [Ar.R.]

Electron capture   The process in which an atom or ion passing through a material medium either loses or gains one or more orbital electrons. In the passage of charged particles (defined here as nuclei having more or less than Z atomic electrons, where Z is the atomic number) through matter, the capture (and loss) of electrons is an important process in the slowing down of the particles and therefore has a strong influence on their range. Thus a neutral hydrogen atom loses only about half as much energy per centimeter as the positively charged proton in passing through matter consisting of light elements.

For the ordinary charged particles (alpha particles and protons) the capture process is important only at low energies, when the particle velocity is of the order of electron velocities in the stopping material, and thus is important at the end of the range. For fission fragments, however, which initially have a large excess of positive charges, electron capture occurs immediately and continues throughout the slowing-down process. This fact causes the energy-loss mechanisms at the latter part of the range to be different for fission fragments and protons or alpha particles. *See* NUCLEAR FISSION.

The nuclear capture of electrons (K capture) occurs by a process quite different from atomic capture and is in fact a consequence of the general beta interaction. This general interaction includes β^- decay (the oldest known beta transformation and hence the name), β^+ decay (or positron decay), and K capture, the latter so called because the electron captured by the nucleus is taken from the K shell (the shell nearest the nucleus) of atomic electrons. A second-order process, called L capture, can also occur, in which (to speak pictorially and thus somewhat imprecisely) an s electron (from the K shell) is captured with the simultaneous transition of a p electron (from the L shell) to the K shell with the emission of gamma radiation. *See* RADIOACTIVITY.   [McA.H.H.]

Electron configuration   The orbital arrangement of an atom's electrons. Negatively charged electrons are attracted to a positively charged nucleus to form an atom or ion. Although such bound electrons exhibit a high degree of quantum-mechanical wavelike behavior, there still remain particle aspects to their motion. Bound electrons occupy orbitals that are somewhat concentrated in spatial shells lying at different distances from the nucleus. As the set of electron energies allowed by quantum mechanics is discrete, so is the set of mean shell radii. Both these quantized physical quantities are primarily specified by integral values of the principal, or total, quantum number n. The full electron configuration of an atom is correlated with a set of values for all the quantum numbers of each and every electron. In addition to n, another important quantum number is l, an integer representing the orbital angular momentum of an electron in units of $h/2\pi$, where h is Planck's constant. The values 1, 2, 3, 4, 5, 6, 7 for n and 0, 1, 2, 3 for l together suffice to describe the electron configurations of all

Distribution of electrons in the atoms

Element and atomic number		K 1,0 1s	L 2,0 2s	L 2,1 2p	M 3,0 3s	M 3,1 3p	M 3,2 3d	N 4,0 4s	N 4,1 4p	N 4,2 4d	N 4,3 4f	O 5,0 5s	O 5,1 5p	O 5,2 5d	O 5,3 5f	Ground term	Ionization potential, eV
H	1	1	—	—	—	—	—	—	—	—	—	—	—	—	—	$^2S_{1/2}$	13.5981
He	2	2	—	—	—	—	—	—	—	—	—	—	—	—	—	$^1S_0$	24.5868
Li	3	2	1	—	—	—	—	—	—	—	—	—	—	—	—	$^2S_{1/2}$	5.3916
Be	4	2	2	—	—	—	—	—	—	—	—	—	—	—	—	$^1S_0$	9.322
B	5	2	2	1	—	—	—	—	—	—	—	—	—	—	—	$^2P^\circ_{1/2}$	8.298
C	6	2	2	2	—	—	—	—	—	—	—	—	—	—	—	$^3P_0$	11.260
N	7	2	2	3	—	—	—	—	—	—	—	—	—	—	—	$^4S^\circ_{3/2}$	14.534
O	8	2	2	4	—	—	—	—	—	—	—	—	—	—	—	$^3P_2$	13.618
F	9	2	2	5	—	—	—	—	—	—	—	—	—	—	—	$^2P^\circ_{3/2}$	17.422
Ne	10	2	2	6	—	—	—	—	—	—	—	—	—	—	—	$^1S_0$	21.564
Na	11	Neon configuration			1	—	—	—	—	—	—	—	—	—	—	$^2S_{1/2}$	5.139
Mg	12				2	—	—	—	—	—	—	—	—	—	—	$^1S_0$	7.646
Al	13				2	1	—	—	—	—	—	—	—	—	—	$^2P^\circ_{1/2}$	5.986
Si	14				2	2	—	—	—	—	—	—	—	—	—	$^3P_0$	8.151
P	15				2	3	—	—	—	—	—	—	—	—	—	$^4S^\circ_{3/2}$	10.486
S	16				2	4	—	—	—	—	—	—	—	—	—	$^3P_2$	10.360
Cl	17				2	5	—	—	—	—	—	—	—	—	—	$^2P^\circ_{3/2}$	12.967
Ar	18				2	6	—	—	—	—	—	—	—	—	—	$^1S_0$	15.759
K	19	Argon configuration					—	1	—	—	—	—	—	—	—	$^2S_{1/2}$	4.341
Ca	20						—	2	—	—	—	—	—	—	—	$^1S_0$	6.113
Sc	21						1	2	—	—	—	—	—	—	—	$^2D_{3/2}$	6.54
Ti	22						2	2	—	—	—	—	—	—	—	$^3F_2$	6.82
V	23						3	2	—	—	—	—	—	—	—	$^4F_{3/2}$	6.74
Cr	24						5	1	—	—	—	—	—	—	—	$^7S_3$	6.765
Mn	25						5	2	—	—	—	—	—	—	—	$^6S_{5/2}$	7.432
Fe	26						6	2	—	—	—	—	—	—	—	$^5D_4$	7.870
Co	27						7	2	—	—	—	—	—	—	—	$^4F_{9/2}$	7.86
Ni	28						8	2	—	—	—	—	—	—	—	$^3F_4$	7.635
Cu	29						10	1	—	—	—	—	—	—	—	$^2S_{1/2}$	7.726
Zn	30						10	2	—	—	—	—	—	—	—	$^1S_0$	9.394
Ga	31						10	2	1	—	—	—	—	—	—	$^2P^\circ_{1/2}$	5.999
Ge	32						10	2	2	—	—	—	—	—	—	$^3P_0$	7.899
As	33						10	2	3	—	—	—	—	—	—	$^4S^\circ_{3/2}$	9.81
Se	34						10	2	4	—	—	—	—	—	—	$^3P_2$	9.752
Br	35						10	2	5	—	—	—	—	—	—	$^2P^\circ_{3/2}$	11.814
Kr	36						10	2	6	—	—	—	—	—	—	$^1S_0$	13.999
Rb	37	Krypton configuration							—	—	1	—	—	—	—	$^2S_{1/2}$	4.177
Sr	38								—	—	2	—	—	—	—	$^1S_0$	5.693
Y	39								1	—	2	—	—	—	—	$^2D_{3/2}$	6.38
Zr	40								2	—	2	—	—	—	—	$^3F_2$	6.84
Nb	41								4	—	1	—	—	—	—	$^6D_{1/2}$	6.88
Mo	42								5	—	1	—	—	—	—	$^7S_3$	7.10
Tc	43								5	—	2	—	—	—	—	$^6S_{5/2}$	7.28
Ru	44								7	—	1	—	—	—	—	$^5F_5$	7.366
Rh	45								8	—	1	—	—	—	—	$^4F_{9/2}$	7.46
Pd	46								10	—	—	—	—	—	—	$^1S_0$	8.33

(continued)

known normal atoms and ions, that is, those that have their lowest possible values of total electronic energy. The first seven shells are also given the letter designations K, L, M, N, O, P, and Q respectively. Electrons with l equal to 0, 1, 2, and 3 are designated s, p, d, and f, respectively. *See* QUANTUM MECHANICS; QUANTUM NUMBERS.

In any configuration the number of equivalent electrons (same n and l) is indicated by an integral exponent (not a quantum number) attached to the letters s, p, d, and f.

Distribution of electrons in the atoms (cont.)

Element and atomic number	Configuration of inner shells	N	O				P			Q	Ground term	Ionization potential, eV
		4,3 4f	5,0 5s	5,1 5p	5,2 5d	5,3 5f	6,0 6s	6,1 6p	6,2 6d	7,0 7s		
Ag 47	Palladium configuration	—	1	—	—	—	—	—	—	—	$^2S_{1/2}$	7.576
Cd 48		—	2	—	—	—	—	—	—	—	$^1S_0$	8.993
In 49		—	2	1	—	—	—	—	—	—	$^2P^\circ_{1/2}$	5.786
Sn 50		—	2	2	—	—	—	—	—	—	$^3P_0$	7.344
Sb 51		—	2	3	—	—	—	—	—	—	$^4S^\circ_{3/2}$	8.641
Te 52		—	2	4	—	—	—	—	—	—	$^3P_2$	9.01
I 53		—	2	5	—	—	—	—	—	—	$^2P^\circ_{3/2}$	10.457
Xe 54		—	2	6	—	—	—	—	—	—	$^1S_0$	12.130
Cs 55	The shells 1s to 4d contain 46 electrons	—	The shells 5s to 5p contain 8 electrons		—	—	1	—	—	—	$^2S_{1/2}$	3.894
Ba 56		—			—	—	2	—	—	—	$^1S_0$	5.211
La 57		—			1	—	2	—	—	—	$^2D_{3/2}$	5.5770
Ce 58		1			1	—	2	—	—	—	$^1G_4$	5.466
Pr 59		3			—	—	2	—	—	—	$^4I^\circ_{9/2}$	5.422
Nd 60		4			—	—	2	—	—	—	$^5I_4$	5.489
Pm 61		5			—	—	2	—	—	—	$^6H_{5/2}$	5.554
Sm 62		6			—	—	2	—	—	—	$^7F_0$	5.631
Eu 63		7			—	—	2	—	—	—	$^8S^\circ_{7/2}$	5.666
Gd 64		7			1	—	2	—	—	—	$^9D^\circ_2$	6.141
Tb 65		9			—	—	2	—	—	—	$^6H^\circ_{15/2}$	5.852
Dy 66		10			—	—	2	—	—	—	$^5I_8$	5.927
Ho 67		11			—	—	2	—	—	—	$^4I^\circ_{15/2}$	6.018
Er 68		12			—	—	2	—	—	—	$^3H_6$	6.101
Tm 69		13			—	—	2	—	—	—	$^2F^\circ_{7/2}$	6.184
Yb 70		14			—	—	2	—	—	—	$^1S_0$	6.254
Lu 71		14			1	—	2	—	—	—	$^2D_{3/2}$	5.426
Hf 72	The shells 1s to 5p contain 68 electrons				2	—	2	—	—	—	$^3F_2$	6.865
Ta 73					3	—	2	—	—	—	$^4F_{3/2}$	7.88
W 74					4	—	2	—	—	—	$^5D_0$	7.98
Re 75					5	—	2	—	—	—	$^6S_{5/2}$	7.87
Os 76					6	—	2	—	—	—	$^5D_4$	8.5
Ir 77					7	—	2	—	—	—	$^4F_{9/2}$	9.1
Pt 78					9	—	1	—	—	—	$^3D_3$	9.0
Au 79	The shells 1s to 5d contain 78 electrons					—	1	—	—	—	$^2S_{1/2}$	9.22
Hg 80						—	2	—	—	—	$^1S_0$	10.43
Tl 81						—	2	1	—	—	$^2P^\circ_{1/2}$	6.108
Pb 82						—	2	2	—	—	$^3P_0$	7.417
Bi 83						—	2	3	—	—	$^4S^\circ_{3/2}$	7.289
Po 84						—	2	4	—	—	$^3P_2$	8.43
At 85						—	2	5	—	—	$^2P^\circ_{3/2}$	
Rn 86						—	2	6	—	—	$^1S_0$	10.749
Fr 87						—	2	6	—	1	$^2S_{1/2}$	
Ra 88						—	2	6	—	2	$^1S_0$	5.278
Ac 89						—	2	6	1	2	$^2D_{3/2}$	5.17
Th 90						—	2	6	2	2	$^3F_2$	6.08
Pa 91						2	2	6	1	2	$^4K_{11/2}$	5.89
U 92						3	2	6	1	2	$^5L_6$	6.05
Np 93						4	2	6	1	2	$^6L_{11/2}$	6.19
Pu 94						6	2	6	—	2	$^7F_0$	6.06
Am 95						7	2	6	—	2	$^8S^\circ_{7/2}$	5.993
Cm 96						7	2	6	1	2	$^9D^\circ_2$	6.02
Bk 97						9	2	6	0	2	$^6H^\circ_{5/2}$	6.23
Cf 98						10	2	6	0	2	$^5I_8$	6.30
Es 99						11	2	6	0	2	$^4I^\circ_{15/2}$	6.42
Fm 100						12	2	6	0	2	$^3H_6$	6.50
Md 101						13	2	6	0	2	$^2F^\circ_{7/2}$	6.58
No 102						14	2	6	0	2	$^1S_0$	6.65
Lw 103						(14)	2	6	(1)	(2)		

According to the Pauli exclusion principle, the maximum is s^2, p^6, d^{10}, and f^{14}. *See* EXCLUSION PRINCIPLE.

An electron configuration is categorized as having even or odd parity, according to whether the sum of p and f electrons is even or odd. Strong spectral lines result only from transitions between configurations of unlike parity. *See* PARITY (QUANTUM MECHANICS).

Insofar as they are known from spectroscopic investigations, the electron configurations characteristic of the normal or ground states of the first 103 chemical elements are shown in the table.

In the next-to-last column of the table, the spectral term of the energy level with lowest total electronic energy is shown. The main part of the term symbol is a capital letter, S, P, D, F, and so on, that represents the total electronic orbital angular momentum. Attached to this is a superior prefix, 1, 2, 3, 4, and so on, that indicates the multiplicity, and an anterior suffix, 0, $^1/_2$, 1, $^3/_2$, 2, $^5/_2$, and so on, that shows the total angular momentum, or J value, of the atom in the given state. A sign $^\circ$ above the J value signifies that the spectral term and electron configuration have odd parity.

The last column of the table presents the first ionization potential of the atom, the energy required to remove from an atom its least firmly bound electron and transform a neutral atom into a singly charged ion. *See* ATOMIC STRUCTURE AND SPECTRA; IONIZATION POTENTIAL.   [J.E.B]

Electron diffraction   The phenomenon associated with interference processes that occur when electrons are scattered by atoms to form diffraction patterns. The wave character of electrons is shown most strikingly, and doubtless most conclusively, by the phenomena of interference. For this reason, the diffraction of electrons presents the most obvious confirmation of quantum mechanics. Because of the dependence of the diffraction pattern on the distances between the atoms, electron diffraction is also an important tool for the study of the structure of crystals and of free molecules, analogous to the use of x-rays for these purposes. *See* X-RAY CRYSTALLOGRAPHY; X-RAY DIFFRACTION.

According to energy $E = eV$ (where e is electron charge and V is potential difference), two major techniques of structure analysis with electron beams are distinguished: low-energy electron diffraction (LEED) [$E \simeq$ 5–500 eV] and high-energy electron diffraction (HEED) [$E \simeq$ 5–500 keV]. In addition, electrons generated in condensed matter by incident electrons or x-ray photons are diffracted (in Auger electron diffraction and photoelectron diffraction). Unlike neutrons and x-rays, electrons penetrate matter only for a very short distance before they lose energy (by inelastic scattering) or are scattered elastically (diffracted). *See* COHERENCE; DIFFRACTION; INTERFERENCE OF WAVES; MEAN FREE PATH; QUANTUM MECHANICS.

Low-energy electron diffraction. LEED is used mainly for the study of the structure of single-crystal surfaces and of processes on such surfaces that are associated with changes in the lateral periodicity of the surface. A monochromatic, nearly parallel electron beam, of 10^{-4} to 10^{-3} m (4×10^{-3} to 4×10^{-2} in.) in diameter, strikes the surface, usually at normal incidence. The elastically backscattered electrons are separated from all other electrons by a retarding field and detected with a suitable movable collector or, more frequently on a hemispherical fluorescent screen with the crystal in its center. The intensity of the diffraction spots can be measured as a function of the energy of the incident electrons to obtain so-called $I(V)$ curves.

The most important contribution of LEED is to the understanding of chemisorption, which precedes corrosion and, in many cases, epitaxy. Here, not only the structure of many adsorption systems, mainly of gases on metals, or metals on other metals and semiconductors, has been studied, but also the kinetics of the adsorption and desorption process as well as changes in the adsorption layer upon heating. The combination of LEED with Auger electron spectroscopy (AES) and with work-function measurements has proven particularly powerful in these studies, because such methods give the coverage and information on the location of the adsorbed atoms normal to the surface. Combining LEED with other complementary techniques such as ion scattering spectroscopy, electron energy loss spectroscopy, or photoelectron spectroscopy has become increasingly popular and can enable the elimination of ambiguities in the interpretation of many LEED results. *See* AUGER EFFECT; SURFACE PHYSICS.

High-energy electron diffraction. HEED is used mainly for the study of the structure of thin foils, films, and small particles (thickness or diameter of 10^{-9} to 10^{-6} m or 4×10^{-8} to 4×10^{-5} in.), of molecules, and also of the surfaces of crystalline materials. A monochromatic, usually nearly parallel, electron beam with a diameter of 10^{-3} to 10^{-8} m (4×10^{-2} to 4×10^{-7} in.) is incident on the target. The forward-scattered electrons (backscattering is negligible) are detected by means of a fluorescent screen, a photoplate, or some other current-sensitive detector, usually without the inelastically scattered electrons being eliminated.

Similar to LEED, reflection HEED (RHEED) can be used for the determination of the lateral arrangement of the atoms in the topmost layers of the surface, including the structure of adsorbed layers. Although it is more convenient to deduce the periodicity of the atomic arrangement parallel to the surface from LEED patterns than from RHEED patterns, LEED frequently becomes inapplicable when the surface is rough. This usually occurs in the later stages of corrosion or in precipitation. In such investigations RHEED is far superior to LEED because the fast electrons can penetrate the asperities and produce a transmission HEED (THEED) pattern. RHEED has become particularly important for thin-film growth monitoring via the specular beam intensity oscillations caused by monolayer-by-monolayer growth. *See* CRYSTAL GROWTH.

In scanning HEED (SHEED) the diffracted electrons are not recorded on photographic film but are directly measured electronically with sensitive detectors. By moving the detector across the diffraction pattern or by deflecting the diffracted electrons across a stationary detector (scanning), the intensity distribution in the diffraction pattern can be displayed quantitatively on an XY recorder. The main application of SHEED is in the study of processes which are accompanied by changes of the intensity distribution, such as the growth of thin films and annealing and corrosion processes.

The technological importance of thin film and interface devices has led to an upsurge of thin film growth studies by conventional transmission HEED (THEED), usually combined with transmission electron microscopy. Information obtained this way has been mainly on the orientation of the crystallites composing the film.   [E.B.]

Diffraction in gases and liquids. Electron diffraction in gases and liquids is similar in principle to that in solids; the differences arise from the lack in gases and liquids of any highly regular arrangement of the component atoms. In gases the low density makes it possible to study diffraction by individual atoms and molecules. The results obtained from monatomic gases represent the density of electronic charge in the atom as a function of the distance from the nucleus. The results from gaseous polyatomic molecules represent the equilibrium distances between the atomic nuclei and the average amplitudes of vibration associated with these distances. Liquids have been studied much less thoroughly than have gases. *See* NEUTRON DIFFRACTION; SCATTERING EXPERIMENTS (ATOMS AND MOLECULES); SCATTERING EXPERIMENTS (NUCLEI).   [L.O.B.]

Electron emission   The liberation of electrons from a substance into vacuum.

Since all substances are built up of atoms and since all atoms contain electrons, any substance may emit electrons; usually, however, the term refers to emission of electrons from the surface of a solid.

The process of electron emission is analogous to that of ionization of a free atom, in which the latter parts with one or more electrons. The energy of the electrons in an atom is lower than that of an electron at rest in vacuum; consequently, in order to ionize an atom, energy must be supplied to the electrons in some way or other. By the same token, a substance does not emit electrons spontaneously, but only if some of the electrons have energies equal to, or larger than, that of an electron at rest in vacuum. This may be achieved by various means, such as by heating, irradiation with light (photoemission), bombardment with charged particles (secondary emission), or use of a strong electric field (field, or cold, emission). *See* FIELD EMISSION; PHOTOEMISSION; SECONDARY EMISSION; THERMIONIC EMISSION.   [A.J.D.]

Electron-hole recombination   The process in which an electron, which has been excited from the valence band to the conduction band of a semiconductor, falls back into an empty state in the valence band, which is known as a hole. *See* BAND THEORY OF SOLIDS.

Light with photon energies greater than the band gap can be absorbed by the crystal, exciting electrons from the filled valence band to the empty conduction band (illus. *a*). The state in which an electron is removed from the filled valence band is known as a hole. It is analogous to a bubble in a liquid. The hole can be thought of as being mobile and having positive charge. The excited electrons and holes rapidly lose energy (in about 10^{-12} s) by the excitation of lattice phonons (vibrational quanta). The excited electrons fall to near the bottom of the conduction band, and the holes rise to near the top of the valence band, and then on a much longer time scale (of 10^{-9} to 10^{-6} s) the electron drops across the energy gap into the empty state represented by the hole. This is known as electron-hole recombination. An energy approximately equal to the band gap is released in the process. Electron-hole recombination is radiative if the released energy is light and nonradiative if it is heat. *See* PHONON.

Electron-hole recombination requires an excited semiconductor in which both electrons and holes occupy the same volume of the crystal. This state can be produced by purely electrical means by forward-biasing a *pn* junction. The current passing through a *pn* diode in electrons per second equals the rate of electron-hole recombination (illus. *b*). A major application of this phenomenon is the light-emitting diode. *See* LIGHT-EMITTING DIODE; LUMINESCENCE; SEMICONDUCTOR DIODE.

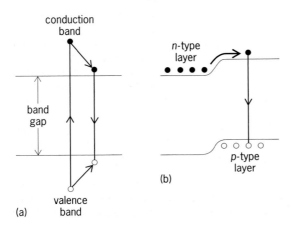

Recombination of electrons and holes generated by (*a*) optical absorption and (*b*) a forward-biased *pn* junction.

Efficient radiative recombination between free electrons and holes takes place only in direct-bandgap semiconductors. During an optical transition, momentum is conserved, and since the photon carries away negligible momentum, transitions take place only between conduction-band and valence-band states having the same momentum. This is easily satisfied in direct-bandgap semiconductors, because electrons and holes collect at the conduction band at minimum and the valence band at maximum, and both extrema have the same momentum. However, for indirect-bandgap semiconductors, the conduction-band minimum and valence-band maximum have very different momenta, and consequently optical transitions between free electrons and holes are forbidden. Radiative electron-hole recombination is possible in indirect-band-gap semiconductors when the transition is assisted by lattice phonons and impurities. *See* CRYSTAL.

Apart from its application in light-emitting diodes and laser operation, radiative recombination, especially at low temperatures (approximately 2 K or −456°F), has been a very important tool for studying the interaction of electrons and holes in semiconductor crystals. *See* EXCITON.

Competing with radiative recombination are the nonradiative recombination processes of multiphonon emission and Auger recombination. It is suspected that nonradiative recombination by multiphonon emission drives the movement of atoms at room temperature that are responsible for device degradation phenomena such as the climb of dislocations found in GaAs light-emitting diodes and lasers. Auger recombination has been shown to limit the performance of long-wavelength (1.3–1.6 micrometer) lasers and light-emitting diodes used in optical communication systems. *See* AUGER EFFECT; LASER; SEMICONDUCTOR. [C.H.He.]

Electron lens An electric or magnetic field, or a combination thereof, which acts upon an electron beam in a manner analogous to that in which an optical lens acts upon a light beam. Electron lenses find application for the formation of sharply focused electron beams, as in cathode-ray tubes, and for the formation of electron images, as in infrared converter tubes, various types of television camera tubes, and electron microscopes.

Any electric or magnetic field which is symmetrical about an axis is capable of forming either a real or a virtual electron image of an object on the axis which either emits electrons or transmits electrons from another electron source. Hence, an axially symmetric electric or magnetic field is analogous to a spherical optical lens.

The lens action of an electric and magnetic field of appropriate symmetry can be derived from the fact that it is possible to define an index of refraction for electron paths in such fields. This index depends on the field distribution and the velocity and direction of the electrons.

Electron lenses differ from optical lenses both in that the index of refraction is continuously variable within them and in that it covers an enormous range. Furthermore, in the presence of a magnetic field, the index of refraction depends both on the position of the electron in space and on its direction of motion. It is not possible to shape electron lenses arbitrarily. *See* ELECTROSTATIC LENS; MAGNETIC LENS. [E.G.R.]

Electron-positron pair production A process in which an electron and a positron are simultaneously created in the vicinity of a nucleus or subatomic particle. Electron-positron pair production is an example of the materialization of energy predicted by special relativity and is accurately described by quantum electrodynamics. Pair production usually refers to external pair production, in which the positron (positively charged antielectron) and electron are created from a high-energy gamma ray as it passes through matter. Electron-positron pairs are also produced from internal pair conversions in nuclei, decays of unstable subatomic particles, and collisions between charged particles. *See* QUANTUM ELECTRODYNAMICS.

In external conversion, the energy of an incoming gamma ray (a high-energy electromagnetic photon) is directly converted into the mass of the electron-positron pair. The photon energy $h\nu$ (where h is Planck's constant and ν is the photon frequency) must therefore exceed twice the rest mass of the electron $2m_0c^2$, equal to 1.022 MeV (m_0 is the electron mass, c the velocity of light). In order to conserve both energy and momentum in this process, the pair must be created near a nucleus, which recoils to balance the momentum of the incoming photon with the momenta of the created electron and positron. Because the nucleus is so much heavier than the electron, it carries away almost no energy from the pair, and the energy of the photon in excess of $2m_0c^2$ is shared unequally as kinetic energy by the positron and electron. Individually, the electron and positron each exhibit a distribution of kinetic energies ranging from zero to the maximum available energy, $E_{max} = h\nu - 2m_0c^2$, correlated with one another so that their sum is equal to E_{max}. Similarly, the positron and electron are emitted over a broad range of angles, although they exhibit a tendency to move in the same direction, which reflects the momentum of the incoming photon. For incident photon energies above 5 MeV, external pair production is the dominant mechanism by which gamma rays are absorbed in matter. *See* GAMMA RAYS; PHOTON.

Internal pair creation differs from external conversion in that the positron and electron are created directly from energy liberated by the deexcitation of an excited nucleus (produced, for example, in radioactive decay or nuclear collisions) to a state of lower energy, if the transition energy exceeds the pair mass threshold of $2m_0c^2$. Internal pair creation usually occurs only 10^{-3} times as often as deexcitation by gamma-ray emission, although the exact pair creation probability, as well as the angular correlation between the emitted positron and electron, depends on the nuclear charge and upon the energy and multipolarity of the nuclear transition. *See* RADIOACTIVITY.

Many unstable subatomic particles, such as the neutral Z boson and J/psi meson, decay into a positron-electron pair alone or with other particles. Since the decaying parent particle is massive, momentum is conserved without the presence of an additional nucleus as is required in external conversion. Decay into a single pair alone creates a positron and electron with equal and opposite momenta.   [T.E.C.]

Electron spin   That property of an electron which gives rise to its angular momentum about an axis within the electron. Spin is one of the permanent and basic properties of the electron. Both the spin and the associated magnetic dipole moment of the electron were postulated by G. E. Uhlenbeck and S. Goudsmit in 1925 as necessary to allow the interpretation of many observed effects, among them the so-called anomalous Zeeman effect, the existence of doublets (pairs of closely spaced lines) in the spectra of the alkali atoms, and certain features of x-ray spectra. *See* SPIN (QUANTUM MECHANICS).

The spin quantum number is s, which is always $^1/_2$. This means that the component of spin angular momentum along a preferred direction, such as the direction of a magnetic field, is $\pm^1/_2\hbar$, where $\hbar = h/2\pi$ and h is Planck's constant. The spin angular momentum of the electron is not to be confused with the orbital angular momentum of the electron associated with its motion about the nucleus. In the latter case the maximum component of angular momentum along a preferred direction is $l\hbar$, where l is the angular momentum quantum number and may be any positive integer or zero. *See* ANGULAR MOMENTUM; QUANTUM NUMBERS.

Electron magnetic moment. The electron has a magnetic dipole moment by virtue of its spin. The approximate value of the dipole moment is the Bohr magneton μ_0 which is equal, in SI units, to $eh/4\pi m = 9.27 \times 10^{-24}$ joule/tesla, where e is the electron charge measured in coulombs, and m is the mass of the electron. The orbital motion of the electron also gives rise to a magnetic dipole moment μ_l that is equal to μ_0 when $l = 1$. *See* MAGNETON.

The orbital magnetic moment of an electron can readily be deduced with the use of the classical statements of electromagnetic theory in quantum-mechanical theory; the simple classical analog of a current flowing in a loop of wire describes the magnetic effects of an electron moving in an orbit. The spin of an electron and the magnetic properties associated with it are, however, not possible to understand from a classical point of view.

In the Landé g factor, g is defined as the negative ratio of the magnetic moment, in units of μ_0, to the angular momentum, in units of $\hbar$. For the orbital motion of an electron, $g_l = 1$. For the spin of the electron the appropriate g value is $g_s \simeq 2$; that is, unit spin angular momentum produces twice the magnetic moment that unit orbital angular momentum produces. The total electronic magnetic moment of an atom depends on the state of coupling between the orbital and spin angular momenta of the electron.

Atomic beam measurements. With the development of spectroscopy by the atomic beam method, a new order of precision in the measurement of the frequencies of spectral lines became possible. By using the atomic-beam techniques, it became possible to measure g_s/g_l directly, with the result $g_s/g_l = 2(1.001168 \pm 0.000005)$. The magnetic moment of the electron therefore is not μ_0 but $1.001168\mu_0$, or equivalently the g factor of the electron departs from 2 by the so-called g factor anomaly defined as $a = (g_2 - 2)/2$ so that $\mu = (1 + a)_0$. Thus the first molecular beam work gave $a = 0.001168$. *See* MOLECULAR BEAMS.

Calculation of g-factor anomaly. It is not possible to give a qualitative description of the effects which give rise to the g-factor anomaly of the electron. The detailed theoretical calculation of the quantity is in the domain of quantum electrodynamics, and involves the interaction of the zero-point oscillation of the electromagnetic field with the electron. Comparison of theoretical determination of a with its experimental measurement constitutes the most accurate and direct existing test of the theory of quantum electrodynamics. *See* ATOMIC STRUCTURE AND SPECTRA; GYROMAGNETIC RATIO; QUANTUM ELECTRODYNAMICS; QUANTUM MECHANICS. [A.Ri.; T.Ki.]

Electron tube A device in which electrons can travel through a sealed chamber containing at least two electrodes and gas at a very low pressure. The gas pressure usually ranges from about 10^{-6} to 10^{-9} atm (10^{-1} to 10^{-4} pascal). At the low extreme of this pressure range, electron tubes are sometimes referred to as vacuum tubes, and at the high extreme as gas tubes.

At least one of the electrodes must emit electrons, and at least one must collect electrons. The emitting electrode, the cathode, may emit electrons through one or more of four mechanisms: thermionic or primary emission, secondary emission photoelectric emission, or field emission. Electrons must acquire more energy than they have in the conduction band of a metal in order to escape from the surface of a metal. They acquire this energy, respectively, in the four mechanisms listed above, from heat, electron or ion impact, a photon impact, or an external electric field. Photoelectric emission is used in light-sensing devices, often in combination with secondary electron multiplication to amplify the current. Secondary emission, sometimes in combination with thermionic emission, plays an important role in magnetrons and in crossed-field amplifiers. Field emission is used in some experimental amplifiers, flat-panel display devices, and x-ray tubes, but by far the most common type of emitting electrode used in electron tubes is the thermionic cathode. *See* FIELD EMISSION; PHOTOEMISSION; SECONDARY EMISSION; THERMIONIC EMISSION; X-RAY TUBE.

A diode is a two-electrode tube, with a cathode and a collecting electrode. A. Fleming (1904) developed the first thermionic diode using an oxide cathode. Because the collecting electrode is usually operated at a positive potential with respect to the cathode in order to collect much of the available electron current from the cathode, it is called an anode. Even so, because of the thermal energy of thermionic electrons, the anode can collect some electrons when it has a slightly negative potential.

L. DeForest (1906) added a third electrode to a diode in order to control the current flow from cathode to anode. This third electrode, the grid, took the form of a fairly open array or mesh made of wires with a diameter small compared to their spacing. In this geometry, much of the electric field from the anode terminates on the grid, and the field from the grid that terminates on the cathode exerts a primary influence on the space-charge current that flows to or through the grid. When the grid is at a negative potential with respect to the cathode, current flows due to the anode field that leaks through the grid, but the grid can collect no current. When the grid and anode are both positive, much more current flows and divides between the grid and anode.

Unfortunately, in triode amplifiers at high frequencies, the capacitance between the anode and grid electrodes, in combination with typical grid circuit reactances, can cause positive feedback, regeneration, or oscillation unless circuits that provide compensating negative feedback are used. For this reason W. Schottky (1919) invented the tetrode, which has a second or screen grid between the first or control grid and the anode. This grid was operated at a constant positive potential and effectively shielded the control grid from the anode. At large signal levels, it also created a problem by collecting secondary electrons emitted from the anode as a result of primary electron impacts when the instantaneous voltage on the anode was less than the screen grid voltage. This problem was dealt with in two ways. G. Jobst and D. H. Tellegen (1926) introduced the pentode, which has a third very open suppressor grid between the screen grid and the anode. It was connected to the cathode. This created an electric field which returned secondary electrons to the anode. A more elegant solution to the secondary electron problem was provided in the beam-power tetrode. In these tetrodes the anode was placed far enough from the screen grid that the charge of the electrons traveling between the screen grid and anode actually depressed the potential in the space between the screen and anode enough to return secondary electrons to the anode.

The tubes discussed so far act as valves that control the flow of a current to a load. The potential energy of the current is derived from a direct-current power source. There is another class of electron tubes, most of which are referred to as microwave tubes, in which electrons are accelerated to a velocity at which they have a kinetic energy that is equivalent to the full voltage of the power supply that was used to accelerate them. If these electrons are bunched periodically in time, they can be made to give up their energy to the electric field in a gap or gaps in a very high frequency or microwave circuit. Microwave tubes include the inductive output tube, the klystron, traveling-wave tubes, crossed-field devices, and cyclotron-resonance devices.

In the inductive output tube, invented by A. V. Haeff (1939), an electron beam is amplitude-modulated with a grid and then accelerated through a hole in the first accelerating electrode to form the high-velocity beam of electrons that passes through a gap in the center conductor of the coaxial external cavity resonator and into the collector. Inductive output tubes are used in many television transmitters operating between 470 and 900 MHz.

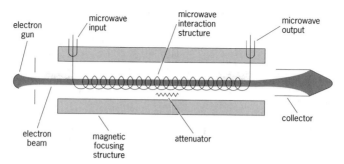

**Basic elements of a typical traveling-wave tube. (*After D. Christiansen,
ed., Electronic Engineers' Handbook, 4th ed., McGraw-Hill, 1996*)**

The klystron, invented by R. Varian and S. Varian (1939), has a similar output cavity and collector, but has a beam which is first accelerated in a diode electron gun and then velocity-modulated in another reentrant cavity gap. Fast electrons overtake slowed electrons and yield an intensity-modulated beam by the time the electrons reach the output cavity. Additional cavities may be interposed between the input and output cavities to provide very high gain (often as high as 60 dB).

In traveling-wave tubes (see illustration) invented by R. Kompfner (1946), a high-velocity electron beam is velocity-modulated by, and gives up its energy to, periodically loaded or helical waveguides which slow the electromagnetic wave to a velocity nearly equal to that of the electron beam. Again very high gain is possible.

Input-output tubes, klystrons, and traveling-wave tubes are used in television broadcasting, satellite communications systems, radar, scientific accelerators, medical accelerators used for cancer therapy, and military countermeasures equipment.

In magnetrons and crossed-field amplifiers, electrons circulate about a cylindrical cathode in a radial direct-current electric field and an axial magnetic field. Concentric with, and outside, the cathode is a periodically loaded transmission line that propagates a wave having components that travel in synchronism with the rotating electron cloud. The electrons follow orbits that allow them to take energy from the radial direct-current electric field and transfer it to the circumferential radio-frequency electric field of the wave on the circuit. Magnetrons are used in huge quantities in household microwave ovens. They and crossed-field amplifiers are also used in ground-based, shipboard, and airborne radars.

Cyclotron-resonance devices including gyrotrons, gyroklystrons, and gyro-traveling-wave tubes again employ electrons that have been accelerated to the full energy provided by the electrical power supply. The beam is formed in a magnetic field so that it has a great deal of momentum perpendicular to the magnetic field, and the electrons follow helical paths. A radio-frequency electric field perpendicular to the axis of the electron trajectories will modulate the energy of the electrons and hence the relativistic mass and the cyclotron frequency. This azimuthal velocity modulation causes the electrons to draw into rodlike bunches that can give up their energy to a circuit supporting either the same alternating electric field that bunched them (in a gyrotron), or to an alternating electric field in another circuit (in a gyroklystron). Cyclotron-resonance devices can be built using very long circuits producing very weak electric fields, and as a result, having very low losses at very high frequencies. Efficient gyrotrons have been built at frequencies as high as several hundred gigahertz and have produced continuous power of hundreds of kilowatts.   [R.S.Sy.]

Electron wake   The pattern of electron density fluctuation and electromagnetic disturbance set up by the passage of a swift ion through condensed matter. In dense media that can sustain well-defined resonance oscillations at a frequency ω_0, wakes of periodic character will form behind swift charged particles having speed v. The periodicity in space, λ, the distance between troughs of the wake, is given by

$$\lambda = \frac{2\pi v}{\omega_0} \qquad (1)$$

The oscillations trail behind the ion, move with the ion velocity, and have the frequency ω_0. In addition, close collisions between the ion and electrons of the medium cause electrons to recoil to form the analog of a bow wave ahead of the ion. See RESONANCE (QUANTUM MECHANICS).

The wake at the position of the guiding ion is of special significance. The electric field there times the ion charge represents the reaction of the medium to the ion and yields the stopping power of the medium for the ion, that is, the energy loss per unit path length.

When molecular ions are injected into a solid with speeds greater than $v_0 = 2.2 \times 10^6$ m/s, the so called Bohr speed (the speed of an electron in the ground state of hydrogen according to the Bohr model), the valence electrons are stripped, leaving atomic ions to propagate as clusters of correlated charged particles through the medium. A dicluster is composed of two atomic ions travelling close together at nearly the same velocity. A wake is formed given by a (generally nonlinear) superposition of wakes due to the individual ions of the cluster. The dynamically modified Coulomb repulsion between its constituents causes the cluster, in effect, to explode. A pair of ions traveling with the same initial velocity, and created exactly abreast of one another, will recede rapidly from one another because of the Coulomb force acting on them. In typical experiments, the cluster-particle interaction probes the slope of the cluster wake potential near the origin, because the foils used in most experiments are so thin that the separation of the ions after traveling through the foil does not greatly exceed their initial separation. An important effect of the wake interaction is to cause the cluster to lose energy at a faster rate than would its isolated constituents traveling at the same speed, because the wake field of a given ion in a cluster acts, in most experiments, to retard the other ions of the cluster.

Measurements of the angular deflections and energy losses of protons resulting from diclusters formed from swift $(HeH)^+$ or $(OH)^+$ ions bombarding thin foils yield angular distributions that have a circular character due to the action of a Coulomb explosion. Such distributions generally have a large peaked region on the perimeter due to the trailing protons that are focused by the wake of the other ion in the Coulomb explosion. There is also a much smaller peak due to protons that lead the ion. Such experiments have been important in establishing the structure of molecular ions that were formerly not well known.

In other work with diclusters, the oscillatory character of the wake is vividly displayed in a two-foil experiment. A cluster enters the first foil and, after passing through a vacuum separating the carbon foils, enters the second one. The trailing ion experiences the wake force of the leading one. The dependence of the yield of secondary electrons from a final target on the distance between carbon foils shows the characteristic oscillatory behavior. See CHARGED PARTICLE BEAMS; COULOMB EXPLOSION.   [R.H.Ri.]

Electronvolt   A unit of energy used for convenience in atomic systems. Specifically, it is the change in energy of an electron, or of any particle having a charge numerically equal to that of an electron, when it is moved through a difference of potential of 1 mks volt. Its value (in mks units) is obtained from the equation $W = qV$, where W is energy in joules, q the charge in coulombs, and V the potential difference in volts. For a potential difference of 1 volt and the electronic charge of 1.602×10^{-19} coulomb, the electronvolt is 1.602×10^{-19} joule. *See* ELECTRON; IONIZATION POTENTIAL.

[G.H.M.]

Electrooptics   The branch of physics which deals with the influence of an electric field on the optical properties of matter, especially in its crystalline form. These properties include transmission, emission, and absorption of light.

An electric field applied to a transparent crystal can change its refractive indexes and, therefore, alter the state of polarization of light propagating through it. When the refractive-index changes are directly proportional to the applied field, the phenomenon is termed the Pockels effect. When they are proportional to the square of the applied field, it is called the Kerr effect. *See* KERR EFFECT; POLARIZED LIGHT; REFRACTION OF WAVES.

The Pockels effect is used in a light modulator called the Pockels cell. This device (see illustration) consists of a crystal C (usually potassium dihydrogen phosphate, or KDP) placed between two polarizers P_1 and P_2 whose axes are crossed. Ring electrodes bonded to two crystal faces allow an electronic driver V to apply an electric field parallel to the axis OZ along which a light beam (for example, a laser beam) is made to propagate. Pockels cells can be switched on and off in well under 1 nanosecond. *See* LASER; OPTICAL MODULATORS.

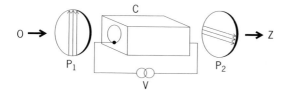

Pockels cell light modulator. The bold arrows represent a light beam.

The linearity and high-speed response of the Pockels effect within an electrooptic crystal make possible a unique optical technique for measuring the amplitude of repetitive high-frequency (greater than 1 GHz) electric signals that cannot be measured by conventional means. The technique, known as electrooptic sampling, employs a special traveling-wave Pockels cell between crossed polarizers. It is used to analyze ultrafast electric signals such as those generated by high-speed transistors and optical detectors.

[M.A.D.; J.A.V.]

A self-electrooptic-effect device (SEED) is a combination of a quantum-well electrooptic modulator with a photodetector which, when light shines on it, changes the voltage on the modulator. Although the device relies internally on an electrooptic effect, the output from the modulator is controlled by the light shining on the photodetector, giving an optically controlled device with an optical output. Most of these devices rely on the quantum-confined Stark effect in semiconductor quantum-well heterostructures as the electrooptic mechanism and utilize the changes in optical absorption resulting from this mechanism. *See* SEMICONDUCTOR HETEROSTRUCTURES; STARK EFFECT.   [D.A.B.M.]

Electroscope   An instrument for detecting the presence and sign of an electric charge. It is the simplest type of ionization chamber. *See* IONIZATION CHAMBER.

The illustration shows a common type of simple gold-leaf electroscope. Gold leaf (L) is used because it is an extremely thin conducting foil which has low mass per unit area and is very flexible. Hence, it responds quickly and vigorously to small electrostatic

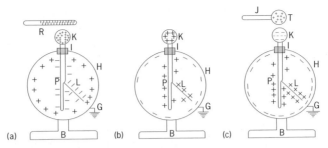

Electroscope. (P is a metal support terminating in knob K; B is the base; I is an insulator; and H is a cylindrical metal housing with flat ends and windows.) (*a*) Being charged by induction by negative charge on hard-rubber rod R. (*b*) Positive charge left on its leaf after induction process is complete. (*c*) Testing the sign of an unknown charge on test ball T.

forces. In the illustration H serves as a grounded electrostatic shield, as well as a shield against air currents. The hard-rubber rod R (illus. *a*) with its negative charge has set up the charge distribution by the process of electrostatic induction. The response shown is a test for the fact that R has a charge. *See* ELECTROSTATICS.

To leave the electroscope with a net charge, a grounded conductor is touched to K so that the surplus electrons on P and L go off to ground, leaving the bound positive charge on K. The ground connection is then broken and R is removed. At this stage (illus. *b*) the electroscope is said to have a positive charge because there is a positive charge on its leaf system.

If an electroscope has a charge of known sign, as in illus. *b*, it can be used to test the sign of an unknown charge, as in illus. *c*, where the metal test ball (T), with its insulating handle (J), has the unknown charge. In the situation pictured, L moves farther away from P as T is brought slowly up toward K, showing that T has a positive charge. If T had a negative charge, L would move toward P, as T slowly approaches K. The converse situation, if the leaf system in illustration *c* had a negative charge initially, can be readily visualized.

Although electroscopes have been built with a wide variety of geometries, the principle of operation is essentially the same for all. If an electroscope has a scale, permitting quantitative measurements, it is called an electrometer or electrostatic voltmeter. For information on electrometers *see* VOLTMETER   [R.P.Wi.]

Electrostatic lens   An electrostatic field with axial or plane symmetry which acts upon beams of charged particles of uniform velocity as glass lenses act on light beams. The action of electrostatic fields with axial symmetry is analogous to that of spherical glass lenses, whereas the action of electrostatic fields with plane symmetry is analogous to that of cylindrical glass lenses. Plane symmetry as used here signifies that the electrostatic potential is constant along any normal to a family of parallel planes.

The action of an electrostatic lens on the paths of charged particles passing through it is most readily visualized with the aid of an equipotential plot of the fields in a plane of symmetry of the lens. The equipotential lines in the plot indicate the intersection with the plane of the drawing of surfaces on which the electrostatic potential is a constant. The paths of charged particles in the electrostatic field are bent toward the normals of the equipotentials as the particles are accelerated, and away from the normals as the particles are decelerated.

Axially symmetric lenses are generally formed at or between circular apertures and cylinders maintained at suitable potentials. For any of these it is possible to define focal points, principal planes, and focal lengths in the same manner as for light lenses and to determine with their aid image magnification for any object position.

Lenses of plane symmetry, analogous to cylindrical glass lenses, are formed between parallel planes and at slits, replacing the circular cylinders and apertures of lenses with axial symmetry.

[E.G.R.]

Electrostatics   The class of phenomena recognized by the presence of electrical charges, either stationary or moving, and the interaction of these charges, this interaction being solely by reason of the charges and their positions and not by reason of their motion. *See* ELECTRIC CHARGE.

At least 90% of the topics that are normally classified as electrostatics are concerned with the manipulation of charged particles by electric fields. When a particle becomes charged by rubbing or other means, it has either a surplus or a deficit of electrons. A body with a surplus of electrons is said to be negatively charged; a body with a deficiency, positively charged. The amount or quantity of charge on a body is expressed in coulombs (positive or negative). A coulomb is an enormous amount of charge, and in most electrostatic situations charge levels of a small fraction of a coulomb give rise to significant effects. Electrostatic forces always exist between charged bodies. Bodies with like charge experience repulsive forces, while oppositely charged bodies experience attraction.

Principles. If two bodies are charged to Q_1 and Q_2 coulombs and are separated in vacuum by a distance of r meters, the force F in newtons between them is given by Coulomb's law, Eq. (1).

$$F = \frac{Q_1 Q_2}{4\pi \epsilon_0 r^2} \tag{1}$$

In electrical science, ϵ_0 is an important constant known as the permittivity or dielectric constant of free space, and is also sometimes called the primary electric constant. It has the value $\epsilon_0 = 8.85416 \times 10^{-12}$ farad per meter. *See* COULOMB'S LAW; PERMITTIVITY.

Coulomb's law shows that a body charged to Q_1 experiences a force due to the presence of another body charged to Q_2. Q_2 may be considered to influence the whole of space surrounding it, because if Q_1 were to be positioned anywhere it would experience a force due to the presence of Q_2. The property of a charge to influence the whole of space can be modeled by a three-dimensional force field permeating the whole of the space surrounding the charge Q_2. This field is called the electric field. When there are many charged bodies present in an environment, the force that would be exerted on a charged particle at any location can be found by calculating the field at the location due to the presence of each charged body separately, and the net field is obtained by adding up the individual components. *See* ELECTRIC FIELD.

A system of charged particles or bodies is unstable unless the particles are prevented from moving, since the like-charged particles will repel each other until they are infinitely far apart, and oppositely charged bodies will attract one another and come together. The system has potential energy. The potential energy of two charged particles separated by a distance r can be shown to be given by Eq. (2).

$$\text{Potential energy} = \frac{Q_1 Q_2}{4\pi \epsilon_0 r} \text{ joules} \tag{2}$$

See ENERGY; POTENTIALS.

Charging methods. The three principal methods of applying electric charge to objects are corona charging, induction charging, and tribocharging. *See* ENERGY.

The corona-charging method relies upon the impact of charged atoms or molecules (ions) on charged bodies. Copious quantities of ions may be generated by a corona discharge, which is a region in which an intense electric field acts upon air molecules and ionizes them so that free ions are produced. A sharply pointed electrode maintained at a high positive or negative potential induces a stream of positive or negative ions which may be used for charging surfaces. The stream of ions from a corona point is usually so intense that neutral air molecules become entrained in the flow to produce a corona wind which can deflect a candle flame. Ions from a corona discharge may be used to charge isolated bodies, insulating surfaces or particles by simply directing a corona wind onto the surface to be charged. In the case of particles, it is normally sufficient for them to pass through a corona discharge region to receive a significant charge from ion-particle collisions.

Surfaces may be charged by exposure to an electrostatic field. If the surface is a liquid and it is disrupted into droplets, they will be electrically charged. Induction charging of equipment and personnel may occur when they are exposed to an electric field. Personnel charged in this way may generate electrostatic discharges when approaching grounded surfaces. Sensitive microelectronic devices can be damaged and computer data can be corrupted by such discharges.

Applications. Electrostatics is put to good use in a wide variety of applications. For examples, the electrostatic precipitator enables smoke emissions from power-station chimneys, smelting plants, and other industrial plants to be reduced to relatively low, acceptable levels. On a smaller scale, efficient filters exist for removing dust from the air in offices, public places, and the home. In some filters, dust particles undergo corona charging as they are sucked by a fan through a duct, and are then collected on grounded electrodes; in others, permanently electrified filter material is used, made from thin plastic sheets which have been treated by surface bombardment from a corona ion source.

In several applications which utilize electrostatics, solid or liquid particles are charged and sprayed onto grounded objects. Dry powder coating is used in many industries in preference to the wet-paint-spraying process. Crop spraying is another important application in which electrostatic forces help to efficiently apply herbicides or insecticides. Research into electrospraying, sometimes called electrohydrodynamic atomization, is leading to new applications for the deposition of ceramic, glass, and polymer films and for powder particle production of special materials. The electrospraying of materials is also used for analysis by means of mass spectrometry, as the electrospray process is gentle and does not disrupt delicate complex molecules.

In electrophotography an optical system is used to project the image to be copied onto a light-sensitive semiconducting surface precharged by a corona source. Exposure of the surface to light reduces the electrical conductivity of the material and allows surface charge to leak away to a back plate in proportion to the intensity of the light, so that bright parts of the image are regions that have lost most of the original charge while dark zones remain fully charged. A mixture of very fine black toner particles and coarser carrier particles is then brought into contact with the charged surface. Transfer of only charged toner particles onto the latent charged surface occurs. A sheet of paper is then laid over the toner-covered surface, and transfer of toner to paper occurs so that an image remains on the paper when it is peeled off the surface. Some ink-jet printers also utilize electrostatic principles; by ensuring that ink drops are formed in the presence of an electrostatic field, they become charged and may be deflected electrostatically to a printing surface.

Another development being commercially exploited is the production of metallic ion or droplet beams using electrostatic forces acting upon a liquid-metal surface. Considerable success has been achieved with many molten metals including gold and silver. Either ion or charged droplet beams may be formed depending on the operating conditions of the source. The beams so formed may be very well defined and directed with great accuracy onto targets where they can be used for ion implantation or for the formation of conducting tracks in the fabrication of microelectronic circuits.

Electrostatic treators using electric fields have been used to separate water droplets from crude oil as well as move and deposit inorganic particles of sand, mud, and clay and organic particles.

Ion engines which produce thrust by electrostatically accelerating mercury or cesium ions have been successfully operated in space. Colloid thrusters, operating on exactly the same principles as electrostatic paint or crop sprayers, have also been developed. In these a propellant such as glycerol is atomized and accelerated from a nozzle by an electrostatic field.

[A.G.B.]

Electrostriction   A form of elastic deformation of a dielectric induced by an electric field; specifically, the term applies to those components of strain which are independent of reversal of the field direction. Electrostriction is a property of all dielectrics and is thus distinguished from the converse piezoelectric effect, a field-induced strain which changes sign upon field reversal and which occurs only in piezoelectric materials. See DIELECTRIC MATERIALS; PIEZOELECTRICITY.

The electrostrictive effect in certain ceramics is employed for commercial purposes in electromechanical transducers for sonic and ultrasonic applications.

[R.D.W.]

Electroweak interaction   One of the three basic forces of nature, along with the strong nuclear interaction and the gravitational interaction. The terms "force" and "interaction between particles" are used interchangeably in this context. All of the known forces, such as atomic, nuclear, chemical, or mechanical forces, are manifestations of one of the three basic interactions.

Until the early 1970s, it was believed that there were four fundamental forces: strong nuclear, electromagnetic, weak nuclear, and gravity. It was by the work of S. Glashow, S. Weinberg, and A. Salam that the electromagnetic and the weak nuclear forces were unified and understood as a single interaction, called the electroweak interaction. This unification was a major step in understanding nature, similar to the achievement of J. C. Maxwell and others a century earlier in unifying the electric forces and magnetic forces into the electromagnetic interactions. A goal of theoretical physics is to achieve a further simplification in understanding nature and describe the presently known three basic interactions in a unified way, usually referred to as the grand unified theory (GUT). Whether this is possible remains to be seen. See ELECTROMAGNETISM; FUNDAMENTAL INTERACTIONS; GRAVITATION; MAXWELL'S EQUATIONS; STRONG NUCLEAR INTERACTIONS; WEAK NUCLEAR INTERACTIONS.

Some of the properties of the basic interactions are summarized in the table. The strong nuclear forces are the strongest, electroweak is intermediate, and gravity the most feeble by a huge factor. The ranges, that is, the distances over which the forces act, also differ greatly. The strong nuclear and the weak interactions have a very short range, while electromagnetism and gravity act over very large distances. Thus, at very short subatomic distances the strong nuclear force, which holds the atomic nucleus together and governs many interactions of the subnuclear particles, dominates. At

Basic forces in nature				
Interaction	Relative strength	Property acted on	Force carrier	Range
Strong nuclear	1	Color charge (r, g, b)	Gluon (g)	10^{-13} cm
Electroweak { Electromagnetic	10^{-2}	Electric charge (q)	Photon (γ)	∞
Electroweak { Weak nuclear	10^{-6}	Weak charges (t_3, y)	Bosons $(W^{\pm}, Z^0)$	10^{-16} cm
Gravity	10^{-40}	Mass (m)	Graviton (G)	∞

larger distances the electromagnetic forces dominate, and hold the atom together and govern chemical and most mechanical forces in everyday life. At even larger scales, objects such as planets, stars, and galaxies are electrically neutral (have an exact balance of positive and negative electric charges) so that the electromagnetic forces between them are negligible, and thus the gravitational forces dominate in astronomical and cosmological situations.

Each of the basic forces acts on, or depends on, different properties of matter. Gravity acts on mass, and electromagnetic forces act on electric charges that come in two kinds, positive and negative. The strong nuclear forces act on a much less well-known property, called color charge, which come in three kinds, r, b, and g (often called red, blue, and green). The weak nuclear forces act on equally esoteric properties called weak isospin t and hypercharge y. While the mass and the electric charge are properties that are recognized in everyday situations, the color charge and the weak isospin and hypercharge have no correspondence in the large-scale everyday world. *See* COLOR (QUANTUM MECHANICS); ELECTRIC CHARGE; HYPERCHARGE; I-SPIN; MASS.

All known forms of matter are made of molecules and atoms, which are made up of the nucleus (protons and neutrons) and orbital electrons. These in turn can be understood to be made up of the fundamental constituents, the quarks and the leptons. Each of these comes in six kinds. All of the quarks and leptons have gravitational and weak interactions since they have nonzero values of mass and weak isospin and hypercharge. The particles with zero electric charge have no electromagnetic interactions, and the leptons have no strong nuclear interactions since they carry no color charge. *See* LEPTON; QUARKS.

The present understanding is that the basic forces are not contact forces but act over distances larger than the sizes of the particles (action at a distance). In this picture, based on field theory, the forces are carried or mediated by intermediate particles that are called gauge bosons. For example, the electromagnetic force between an electron and a proton is carried by the quantum of the electromagnetic field called the photon (γ). The strong nuclear force is carried by the gluon (g), and the gravitational force is carried by the graviton (G). The weak nuclear force comes in two categories: the charge-changing (charged-current, for short) mediated by the $W^{\pm}$ bosons, and the neutral-current weak interactions mediated by the Z^0 boson. *See* GLUONS; GRAVITON; INTERMEDIATE VECTOR BOSON; PHOTON.

All of the fundamental constituents, the quarks and the leptons, carry one-half unit of angular momentum (spin = 1/2) as if they were spinning around their own axis. (Such particles are called fermions.) By the rules of quantum mechanics, the direction of this spin is quantized to be either parallel or antiparallel to the direction of motion of the particle. Particles with spin direction parallel to their direction of motion have helicity $+1$ and are called right-handed, and particles with antiparallel spin have helicity

−1 and are called left-handed. One of the symmetries of nature is called parity, which is a symmetry between right-handed and left-handed coordinate systems. If parity symmetry holds, left-handed and right-handed particles must have the same interactions. In 1956 T. D. Lee and C. N. Yang proposed that parity symmetry is violated in the weak interactions, and this proposal was soon verified experimentally. It was found that the left-handed and the right-handed particles have different weak interactions. In particular, the right-handed particles have no weak isospin, and thus only the left-handed particles participate in the charged-current weak interactions. See ANGULAR MOMENTUM; HELICITY (QUANTUM MECHANICS); PARITY (QUANTUM MECHANICS); SPIN (QUANTUM MECHANICS).

Until the early 1970s, the electromagnetic and the weak interactions were believed to be separate basic interactions. At that time the Weinberg-Salam-Glashow model was proposed to understand these two interactions in a unified way. In its original form, this model, based on an unbroken gauge symmetry, led to some physically unacceptable features such as zero masses for all the constituent particles and predictions of infinities for some measurable quantities. Through the pioneering work of G. 'tHooft, M. Veltman, and others, it was shown that the theory can be made renormalizable, removing the infinities and providing masses to the particles, by spontaneous breaking of the gauge symmetry and the introduction of one new particle, the Higgs boson. See HIGGS BOSON; RENORMALIZATION; SYMMETRY BREAKING.

The neutral gauge bosons, the W^0 and B^0, form a quantum-mechanical mixture, which produces the two physically observable gauge bosons, the γ and the Z^0, as given by Eqs. (1). The γ is the well-known photon that mediates the electromagnetic

$$\gamma = \sin\theta \; W^0 + \cos\theta \; B^0 \qquad Z^0 = \cos\theta \; W^0 - \sin\theta \; B^0 \qquad (1)$$

interactions. The Z^0 mediates the neutral-current weak interactions, and the $W^\pm$ mediate the charged-current weak interactions. In this way, all of these interactions are described by a common unified theory. The mixing angle θ in Eqs. (1), forming the γ and the Z^0, is called the weak mixing angle and is the fundamental parameter of the theory. The strength and nature of the interactions of the particles are determined by the vector and axial vector coupling constants g_v and g_A. In the electroweak model all of these couplings are given in terms of the single parameter of the theory, the weak mixing angle, and the properties of the leptons and quarks. The model also gives a relationship between the electric charge q and the weak charges t_3 and y, Eq. (2).

$$q = t_3 + 1/2y \qquad (2)$$

See NONRELATIVISTIC QUANTUM THEORY; QUANTUM MECHANICS.

The coupling constants that govern the electroweak interactions of all of the particles can be summarized as:

1. Electromagnetic interactions: $g_v = q$, $g_A = 0$
2. Charged current weak interactions: $g_v = -g_A = t$
3. Neutral current weak interactions: $g_v = t_3 - 2q \sin^2\theta$, $g_A = -t_3$

In the above expressions, t stands for the magnitude of the weak isospin, and t_3 is its projection along an axis of quantization.

The electroweak theory has great predictive power. Its first and most striking prediction was the existence of neutral-current weak interactions mediated by the Z^0 boson. Until the time of this prediction, the weak interactions were believed to be of charged-current nature only, with no neutral-current component.

A second major triumph for the electroweak theory was the discovery of the W and Z bosons in 1983 at the proton-antiproton collider at the CERN Laboratory in Geneva, with masses very close to the values predicted by the theory. At this time the validity of the theory was considered firmly established. *See* PARTICLE ACCELERATOR.

The successful electroweak theory, combined with quantum chromodynamics (QCD), the theory describing the strong nuclear interactions, forms the so-called standard model of particle physics. This standard model has been brilliantly successful in accurately predicting and describing all experimental results over a huge energy range, varying from the electronvolt energies of atomic physics to the 100-GeV energy range of the largest existing particle colliders. It represents a landmark achievement of both experimental and theoretical physics. *See* QUANTUM CHROMODYNAMICS; STANDARD MODEL.

In spite of these great successes, two major problems remain to be solved in this field. The first one is that the standard model in its present form cannot explain the masses of the fundamental constituents, the quarks and leptons. These masses vary over a large range, from a few electronvolts to 174 GeV. The basic gauge symmetry on which the standard model is based would indicate that these masses should all be the same. There must therefore be an additional piece of the puzzle, usually referred to as the source of the electroweak symmetry breaking, that remains to be found. Hypothetical ideas about this missing element of the model range from the prediction of a single additional particle, the Higgs boson, to complicated models such as supersymmetry that predict dozens of new elementary particles. *See* HIGGS BOSON; SUPERSYMMETRY; SYMMETRY BREAKING.

The second outstanding problem is the search for a theory that not only describes the strong nuclear and the electroweak interactions but includes gravity as well. The standard model is based on the principles of quantum mechanics, while the current understanding of the gravitational forces is based on Einstein's theory of general relativity. No one so far has been able to combine these two theories; that is, a quantum theory of gravity does not, as yet, exist. The search for such a grand unified theory is a major focus of activity in theoretical physics. *See* ELEMENTARY PARTICLE; QUANTUM GRAVITATION; RELATIVITY.   [C.B.]

Elementary particle   A particle that is not a compound of other particles. At one time the elementary particles of matter were the atoms of the chemical elements, but the atoms are now known to be compounds of the electron, proton, and neutron. In turn, the proton and neutron, and likewise all the other hadrons (strongly interacting particles), are now known to be compounds of quarks. It is convenient, however, to continue to call hadrons elementary particles to distinguish them from their compounds (atomic nuclei, for instance); this usage is also justified by the fact that quarks are not strictly particles, because, as far as is known, they cannot be isolated. The term fundamental particle can be used to denote particles that are truly fundamental constituents of matter and are not compounds in any sense. *See* ELECTRON; HADRON; NEUTRON; PROTON; QUARKS.

The known fundamental particles (see table) fall into two categories: the gauge bosons, comprising the photon, gluon, and weak bosons; and the fermions, comprising the quarks and leptons. The graviton, the quantum of the gravitational field, has been omitted from table since it plays no role in high-energy particle physics: it is firmly predicted by theory, but the prospect of direct observation is exceedingly remote. Of the gauge bosons, the photon has been known since the beginning of quantum mechanics. The heavy gauge bosons $W^{\pm}$ and Z^0 were observed in 1983; their properties had been deduced from the weak interactions, for which they are responsible. The lightest (and stable) lepton, the electron (e), is the first known fundamental particle. The next found was the muon (μ, originally called the mu meson). The fundamental fermions are grouped into three families. Gluons and quarks are never seen as free particles; this phenomenon is known as confinement. Particles that are composed of

The fundamental particles[a]

Gauge bosons	$J^P_C = 1^-_-$	Self-conjugate except $\overline{W^+} = W^-$.		

Name	Symbol	Charge[b]	Mass and width, GeV	Couplings
Photon	γ	0	0	$A \Rightarrow \gamma A$
Gluon[c]	g	0	0	$A \Rightarrow gA'$
Weak bosons				
Charged[d]	$W^{\pm}$	± 1	80.4, 2.1	$U \Rightarrow W^+D$
Neutral[e]	Z^0	0	91.2, 2.49	$A \Rightarrow Z^0 A$

Fermions[f]	$J = 1/2$	All have distinct antiparticles, except perhaps the neutrinos.				

Name	Charge[b]	Symbol and mass, GeV		Symbol and mass, GeV		Symbol and mass, GeV
Leptons						
Neutrinos	0	ν_e	10^{-8}	ν_μ	<.0003	ν_τ <.02
Charged						
leptons[g]	-1	e	.00051	μ	.106[h]	τ 1.78[h]
Quarks[c]						
Up type	$2/3$	u	.005	c	1.4	t 175[i]
Down type	$-1/3$	d	.01	s	.15	b 4.8

[a] The graviton, with $J^P_C = 2^+_+$, has been omitted, since it plays no role in high-energy particle physics.
[b] In units of the proton charge.
[c] The gluon is a color SU_3 octet (8); each quark is a color triplet (3). These colored particles are confined constituents of hadrons; they do not appear as free particles.
[d] The branching ratios (%) of the decay modes of the W^+ are:

ud, cs	34 each
$\nu_e e^+$, $\nu_\mu \mu^+$, $\nu_\tau \tau^+$	11 each

[e] The branching ratios (%) of the decay modes of the Z^0 are:

$d\bar{d}$, $s\bar{s}$, $b\bar{b}$	16.8 each
$u\bar{u}$, $c\bar{c}$	9.7 each
$\nu_e\bar{\nu}_e$, $\nu_\mu\bar{\nu}_\mu$, $\nu_\tau\bar{\nu}_\tau$	6.7 each
e^+e^-, $\mu^+\mu^-$, $\tau^+\tau^-$	3.4 each

[f] The three known families (generations) of fermions are displayed in three columns.
[g] Any further charged leptons have mass greater than 40 GeV.
[h] The μ and τ leptons are unstable, with the following mean life and principal decay modes (branching ratios in %):

μ	$\tau_\mu = 2.2 \times 10^{-6}$ s	$e\bar{\nu}_e\nu_\mu$ 100
τ	$\tau_\tau = 3 \times 10^{-13}$ s	$\mu\bar{\nu}_\mu\nu_\tau$ 18, $e\bar{\nu}_e\nu_\tau$ 18, (hadrons)$\bar{\nu}_\tau$ 64

[i] The t quark has a width ≈ 2 GeV, with dominant decay to Wb.

quarks and gluons are called hadrons; essentially, mesons are composed of a quark-antiquark pair $q\bar{q}$, and baryons are three quarks qqq, bound together by the exchange of gluons. *See* BARYON; GLUONS; GRAVITON; INTERMEDIATE VECTOR BOSON; LEPTON; MESON; PHOTON.

Particles with the properties of the quarks of the quark model (charges $\pm\frac{2}{3}e$ or $\pm\frac{1}{3}e$ and masses less than 300 MeV) have never been observed. Direct evidence both for quarks and for their confinement is given by the phenomenon of hadronic jets. For example, in high-energy deep-inelastic electron-proton scattering, in which the electron loses a sizable fraction of its energy, the observed cross section shows that the charge of the proton is carried by pointlike (radius less than 10^{-1} femtometer) particles of small mass. However, no such particles are seen in the final state of this process, or indeed of any other high-energy collision. What is seen is a narrow shower of hadrons. The interpretation is that the electron scatters off one of the quarks in the proton and gives it a large energy and momentum, the quark responding as though it were a free particle of mass much less than 100 MeV, consistent with the masses of the u and d

quarks (see table). Later, through the production of quark-antiquark pairs, the energy and momentum of the struck quark is divided among a number of hadrons, mostly pions, a process called hadronization or fragmentation of the quark, which is to be distinguished from the decay of a free particle. The resulting shower of hadrons, whose total momentum vector is roughly that of the original quark, is called a hadronic jet (like a jet of water which breaks up into a spray of droplets). Such jets are also seen in other high-energy reactions, such as e^+e^- annihilation into hadrons, and also in pp collisions; they are the closest available phenomenon to the actual observation of a quark as a free particle.

To each kind of particle there corresponds an antiparticle, or conjugate particle, which has the same mass and spin, belongs to the conjugate representation (multiplet) of internal symmetry, and has opposite values of charge, I_3, strangeness, and so forth (quantum numbers which are conserved additively). The product of the space parities of a particle and its antiparticle is $+1$ if the particle is a boson, -1 if a fermion. For instance, the electron e and its antiparticle, the positron e^-, have the same masses and spins, and opposite charges and lepton number, and an S-wave state of e and e^- has parity -1. Particles for which the antiparticle is the same as the particle are called self-conjugate; examples are the photon γ and the neutral pion π^0. The equality of masses implies the equality of lifetimes of particle and antiparticle. Thus the positron is stable; however, in the presence of ordinary matter it soon annihilates with an electron, and thus is not a component of ordinary matter. *See* ANTIMATTER; POSITRON.

The interactions of particles are responsible for their scattering and transformations (decays and reactions). Because of interactions, an isolated particle may decay into other particles. Two particles passing near each other may transform, perhaps into the same particles but with changed momenta (elastic scattering) or into other particles (inelastic scattering). The rates or cross sections of these transformations, and so also the interactions responsible for them, fall into three groups: strong (typical decay rates of 10^{21}–10^{23} s^{-1}), electromagnetic (10^{16}–10^{19} s^{-1}), and weak ($<10^{15}$ s^{-1}). Strong interactions occur only between hadrons. Electromagnetic interactions result from the coupling of charge to the electromagnetic field. Weak interactions are usually unobservable in competition with strong or electromagnetic interactions. They are observable only when they do something which those much stronger interactions cannot do (forbidden by the selection rules); for instance, by changing flavors they can make a particle decay which would otherwise be stable, and by making parity-violating transition amplitudes they can produce an otherwise absent asymmetry in the angular distribution of a reaction. *See* SELECTION RULES (PHYSICS).

Most particles are unstable and decay into smaller-mass particles. The only particles which appear to be stable are the massless particles (graviton, photon), the neutrinos (possibly massless), the electron, the proton, and the ground states of stable nuclei, atoms, and molecules. It is speculated that some or all of the neutrinos may be massive and unstable and that the proton (and therefore all nuclei) may be unstable. The present view is that the only massive particles which are strictly stable are the electron and the lightest neutrino(s). The electron is the lightest charged particle; its decay would be into neutral particles and could not conserve charge. Likewise, the lightest neutrino is the lightest fermion; its decay would be into bosons and could not conserve angular momentum. *See* NEUTRINO.

The unstable elementary particles must be studied within a short time of their creation, which occurs in the collision of a fast (high-energy) particle with another particle. Such fast particles exist in nature, namely the cosmic rays, but their flux is small; thus

most elementary particle research is based on high-energy particle accelerators. *See*
Nuclear reaction; Particle accelerator; Particle detector.

Hadrons can be divided into the quasistable (or hadronically stable) and the unstable.
The quasistable hadrons are simply those that are too light to decay into other hadrons
by way of the strong interactions, such decays being restricted by the requirement that
isobaric spin I and flavors be conserved.

The unstable hadrons are also called particle resonances. Their lifetimes, of the order
of 10^{-23} s, are much too short to be observed directly. Instead they appear, through
the uncertainty principle, as spreads in the masses of the particles—that is, in their
widths—just as in the case of nuclear resonances. *See* Uncertainty principle.

A characteristic of the hadrons is that they are grouped into i-spin multiplets (for
example, n, p; $\pi^{-}, \pi^{0}, \pi^{+}$); the masses of the particles in each multiplet differ by only a
few megaelectronvolts (MeV). The i-spin multiplets of hadrons themselves form groups
(called supermultiplets) which were recognized in 1961 as multiplets (representations)
of the group SU_3 (now referred to as $SU_3{}^{\text{flavor}}$ to distinguish this physical symmetry
from $SU_3{}^{\text{color}}$). For instance, the lightest mesons ($\eta, K, \bar{K}, \pi$) and baryons ($\Lambda, N, \Xi,
\Sigma$) are each a set of eight particles having i-spins $I = (0, \frac{1}{2}, \frac{1}{2}, 1)$ and hypercharges
$Y = (0, 1, -1, 0)$ respectively; this pattern is that of the octet, $\{8\}$, representation of
the group SU_3. Again, the lowest-mass $J^P = {}^3/_2{}^+$ baryons ($\Delta, \Sigma^*, \Xi^*, \Omega$), ten particles
with $I = ({}^3/_2, 1, \frac{1}{2}, 0)$ and $Y = (1, 0, -1, -2)$, form a decuplet, $\{10\}$, representation of
SU_3. The spread of the masses in these groups is about a hundred times greater than
in the i-spin multiplets, a few hundred MeV compared to a few MeV. According to the
quark model, this SU_3 symmetry and the pattern of charges in the SU_3 multiplets result
simply from the existence of a third kind (flavor) of quark, the s (strange) quark, with
charge the same as the d quark, namely $\frac{1}{3}$, together with the flavor independence of
the glue force; that is, all three quarks u, d, and s have the same interaction with the
glue field. The resulting flavor SU_3 symmetry is broken by the relatively large mass of
the s, approximately 150 MeV. The three quarks make up the fundamental triplet, $\{3\}$,
representation of SU_3.

Hadrons are known which contain yet more massive quarks, the c and the b (see
the table). The resulting symmetry is badly broken, and the supermultiplets hardly
recognizable.

It appears that the "glue" field which binds quarks together to make hadrons is a
Yang-Mills (that is, a non-abelian) gauge field of an SU_3 symmetry group, $SU_3{}^{\text{color}}$.
This is an exact symmetry of nature. The quanta of the field are called gluons, and
its quantum theory is called quantum chromodynamics (QCD). The gluon field resembles the electromagnetic field, but has an internal symmetry index (octet index)
which runs over eight values; that is, there are really eight fields, corresponding to
the eight parameters needed to specify an SU_3 transformation. Just as the electromagnetic field is coupled to (that is, photons are emitted and absorbed by) the density and current of a conserved quantity, charge, the gluon field is coupled to color.
The coupling of the gluon to a particle is fixed by the color of the particle (that is,
what member of what color multiplet) and just one universal coupling constant g,
analogous to the electronic unit of charge e. (The analogy breaks down in quantum theory, as discussed below; the quantity g is no longer constant but it is still
universal.)

Since the long-range forces observed between hadrons are no different than those
between other particles, hadrons must be colorless, that is, color singlet combinations
of quarks, their colored constituents. The two simplest combinations of quarks which
can be colorless are $\bar{q}_1 q_2$ and $q_1 q_2 q_3$; these are found in nature as the basic structure
of mesons and baryons, respectively. The exchange of gluons between any of the

quarks in these colorless combinations gives rise to an attractive force, which binds them together.

Gluons are not colorless, and therefore they are coupled to themselves. This situation is very different from electromagnetism, where the photon does not carry charge. The consequence of this self-coupling of massless particles is a severe infrared (small momentum transfer or large distance) divergence of perturbation theory. In particular, the interaction between two colored particles through the gluon field, which in lowest order is an inverse-square Coulomb force, proportional to g^2/r^2 (where r is the distance between the particles), becomes stronger than this inverse-square force at larger r. A way of describing this is to say that the coupling constant g is effectively larger at larger r; this defines the so-called running coupling constant $g(r)$. According to the first-order radiative correction, $g(r)$ becomes infinite at a certain distance, the so-called scale parameter r_c.

A specific form for the gluonic force between two colored particles, at large r, namely that it falls to a nonzero constant value λ, of the order of $\hbar c r_c^{-2}$ (where $\hbar$ is Planck's constant divided by 2π, and c is the speed of light), is suggested by a model, the superconductor analogy. This force is confining.

The conjecture is that the vacuum is like a superconductor with respect to color, with the interchange, however, of electric and magnetic quantities. That is, the vacuum acts like a color magnetic superconductor which confines color flux into bundles which have a diameter of order r_c and an energy per unit length equal to λ of order $\hbar c r_c^{-2}$. The color flux bundles run between colored particles; they can also form closed loops. These flux bundles are often idealized as having vanishing diameter and are then called strings. This idealization is obviously good only if the flux bundles are long compared to r_c, and if their local radius of curvature is always much larger than r_c.

According to the so-called naive quark model, hadrons are bound states of nonrelativistic (slowly moving) quarks, analogous to nuclei as bound states of nucleons. The interactions between the quarks are taken qualitatively from QCD, namely a confining central potential and (exactly analogous to electrodynamic interactions) spin-spin (hyperfine) and spin-orbit potentials; quantitatively, these potentials are adjusted to make the energy levels of the model system fit the observed hadron masses. This model should be valid for hadrons composed of heavy quarks but not for hadrons containing light quarks (u, d, s), but in fact it succeeds in giving a good description of many properties of all hadrons. One reason is that many of these properties follow from so-called angular physics, that is, symmetry-based physical principles that transcend the specific model. A meson is a bound state of a quark and an antiquark, $q_1 q_2$. A baryon is a bound state of three quarks, $q_1 q_2 q_3$.

The known heavy quarks are the c (charm), b (bottom), and t (top) quarks, whose masses are larger than the natural energy scale of QCD, ≈ 1 GeV. But because the width of the t is also larger than 1 GeV, the t quark decays before the QCD force acts on it, and thus before any well-defined hadron forms. So in the present context "heavy quarks" mean only c and b. A hadron which contains a single heavy quark resembles an atom; the heavy quark sits nearly at rest at the center, and is a static source of the color field, just as the atomic nucleus is a static source of the electric field. Just as an atom is changed very little (except in mass) if its nucleus is replaced by another of the same charge (an isotope), a heavy-quark hadron is changed very little (except in mass) if its heavy quark is replaced by another of the same color. This is called heavy-quark symmetry. So, for example, the D, D^*, B, and B^* mesons are similar, except in mass. This plays an important role in the quantitative analysis of their weak decays.

If a hadron contains two heavy quarks, then in a not too highly excited state the heavy quarks move slowly, compared to the speed of light c, and so the effect of the exchange of gluons between the quarks can be approximated (up to radiative corrections) by a potential energy which depends only on the positions of the quarks (local static potential); further, the wave function of the system satisfies the ordinary nonrelativistic Schrödinger equation. Consequently, the properties of hadrons composed of heavy quarks are rather easily calculated.

Mesons with the composition $c\bar{c}$ and $b\bar{b}$ are called charmonium and bottomonium, respectively. These names are based on the model of positronium, ee^-; the generic name for flavorless mesons, $q\bar{q}$, is quarkonium. Since both heavy quarkonium and positronium are systems of a fermion bound to its antifermion by a central force, they are qualitatively very similar.

The electroweak theory, starting from the observation that both the electromagnetic and weak interactions result from the exchange of vector (spin-1) bosons, has unified these interactions into a spontaneously broken gauge theory. Similarly, the observation that the strong (hadronic) interactions are also due to the exchange of vector bosons (gluons) suggests that all these vector bosons (the photon, the three weak bosons, and the eight gluons) are quanta of the components of the gauge field of a large symmetry group, SU_5 or larger. Such theories are called grand unification theories (GUTs). The large symmetry group of the grand unification theory must be spontaneously broken, making all the gauge bosons massive except the gluon octet and the photon, leaving $SU_3 \times U_1$ (color $\times$ electromagnetism) as the apparent gauge symmetry of the world. *See* GRAND UNIFICATION THEORIES.

In these theories, the leptons and quarks occur together in multiplets of the large symmetry group. These multiplets are called families (or generations). The known fundamental fermions do seem to fall into three families (see table). Each family consists of a weak i-spin doublet of leptons (neutrino [charge 0] and charged lepton [charge $+e$]), and a color triplet of weak i-spin doublets of quarks (up-type [charge $\frac{2}{3}e$] and down-type [charge $-\frac{1}{3}e$]).

[C.J.G.]

Ellipsometry   A technique for determining the properties of a material from the characteristics of light reflected from its surface. The materials studied include thin films, semiconductors, metals, and liquids.

When an electromagnetic wave passes through a medium, it causes the electrons associated with the atoms of the medium to oscillate at the frequency of the wave. As a result, the wave is slowed so that its velocity in the medium is less than its velocity in empty space. Another result may be a transfer of energy from the wave to the electrons, thereby causing the amplitude of the wave to decrease as it penetrates into the material. These two processes are described phenomenologically by the complex refractive index. *See* ABSORPTION OF ELECTROMAGNETIC RADIATION; REFRACTION OF WAVES.

When an electromagnetic wave is incident on a medium, only part of it is transmitted into the medium. The fraction reflected depends on the complex refractive index, the angle of incidence, and the polarization state of the wave. For multilayers with different complex refractive indices, the fraction also depends on the layer thicknesses. The two basic types of polarization are parallel, designated p, and perpendicular, designated s. These terms refer to the orientation of the electric vector with respect to the plane of incidence, which is defined by the directions of the incident and reflected waves. The (intensity-independent) ratios of the amplitudes and phases of the reflected and incident p- and s-polarized electric fields are described by the complex reflectances r_p and r_s. *See* POLARIZATION OF WAVES; POLARIZED LIGHT; REFLECTION OF ELECTROMAGNETIC RADIATION.

The (also intensity-independent) ratio of the p- to the s-polarized component of such a wave is termed the polarization state. Simple examples include linear polarization, where the p- and s-polarized components are in phase, and circular polarization, where the p- and s-polarized amplitudes are equal but the phases differ by 90°. The geometric terms refer to the locus of the p and s (or y and x) components of the electric field when plotted in the complex plane. The general polarization state is elliptical.

Since the complex reflectances r_p and r_s depend on the properties of the medium, the medium can be investigated by determining its reflectance for either p- or s-polarized light, that is, by determining the ratio of reflected and incident intensities $I_{refl}/I_{inc} = R = |r|^2$. This is the objective of reflectometry. Alternatively, because r_p and r_s are different, the complex reflectance ratio $\rho = r_p/r_s$, which is equal to the ratio of reflected and incident polarization states, can also be determined. This is the objective of ellipsometry. The ratio ρ is traditionally expressed in terms of angles ψ and Δ as in the equation below.

$$\rho = \frac{r_p}{r_s} = \tan \psi \, e^{j\Delta}$$

Because it deals with complex, intensity-independent quantities, an ellipsometric measurement is analogous to an impedance measurement. This gives ellipsometry certain advantages relative to reflectometry, such as higher accuracy and higher information content in a single measurement. A standard experimental approach, now used almost exclusively in spectroscopic applications, is to determine ρ by establishing a known state of polarization for the incident beam, for example, by passing it through a fixed linear polarizer, then determining the polarization state of the beam after reflection by passing it through a rotating linear polarizer, called an analyzer. The rotating analyzer essentially unrolls the polarization ellipse, allowing the azimuth angle of its major axis and its minor-major axis ratio to be determined by the phase and amplitude of the alternating-current component of the detected intensity. *See* ELECTRICAL IMPEDANCE.

The primary application of ellipsometry is materials analysis, particularly the nondestructive analysis of thin films in semiconductor technology. Developing applications include the real-time monitoring and control of dynamic processes such as material deposition and etching. Spectroellipsometry, where the complex refractive index is measured and analyzed as a function of wavelength, has almost exclusively replaced reflectometry in materials analysis. Deposition and etching involve kinetic ellipsometry, where single-wavelength data are monitored as a function of time. *See* SEMICONDUCTOR HETEROSTRUCTURES.   [D.E.As.]

Emissivity   The ratio of the radiation intensity of a nonblack body to the radiation intensity of a blackbody. This ratio, which is usually designated by the Greek letter ϵ, is always less than or just equal to one. The emissivity characterizes the radiation or absorption quality of nonblack bodies. Published values are readily available for most substances. Emissivities vary with temperature and also vary throughout the spectrum. For an extended discussion of blackbody radiation and related information *see* HEAT RADIATION.

A spectral emissivity of zero means that the heat radiator emits no radiation at this wavelength. Strongly selective radiators, such as insulators or ceramics, have spectral emissivities close to 1 in some parts of the spectrum, and close to zero in other parts. Carbon has a high spectral emissivity throughout the visible and infrared spectrum, exceeding 0.90 in certain portions; thus carbon is a good blackbody radiator. Tantalum is the only metal with a spectral emissivity greater than 0.5 in the visible spectrum. All other metals have a lower spectral emissivity. Tungsten is a relatively good emitter, with a spectral emissivity of 0.43–0.47 within the visible region of the spectrum. *See* BLACKBODY.

[H.G.S.; P.J.W.]

Energy   The ability of one system to do work on another system. There are many kinds of energy: chemical energy from fossil fuels, electrical energy distributed by a utility company, radiant energy from the Sun, and nuclear energy from a reactor. The units of energy include ergs, joules, foot-pounds, and foot-poundals. Work and heat have the same units as energy, but are entirely different physical concepts. *See* HEAT; WORK.

Any particle or system of particles subject to conservative forces has two kinds of energy, potential energy and kinetic energy. Potential energy is the energy due to position or configuration, and kinetic energy is the energy due to motion.

Energy is conserved for all isolated mechanical systems. This is because if a system A is isolated, there is no other system B that it can give any energy to, and its total energy must remain constant. This system A can convert kinetic energy to potential energy, and it can convert one form of potential energy to another, but the total energy must remain the same. The meaning of conserved total energy is that the system has the same value of total energy at all times. *See* CONSERVATION OF ENERGY.

In 1905 A. Einstein showed that at high velocities near the speed of light important modifications must be made in physical concepts. One particularly radical idea which he advanced was that space and time are not independent, but rather are two aspects of the same object, a space-time manifold. This necessitated a reexamination of the concept of energy and led to the conclusion that the inertia, or mass m, depends upon its energy through the mass-energy relation shown below, where c is the speed of

$$E = mc^2$$

light in vacuum. Furthermore, energy and momentum conservation become joined in a single four-momentum conservation law in special relativity. *See* INTERNAL ENERGY; RELATIVITY.

[B.DeF.]

Energy level (quantum mechanics)   One of the allowed values of the internal energy of an isolated physical system. This energy is not free to vary continuously above its minimum value, as predicted by classical mechanics, but is constrained to lie among a set or spectrum of particular values. This spectrum may consist of both an isolated discrete portion and a continuous component of restricted range. The term energy level usually refers to one of the allowed values in the discrete set.

The primary indication for the existence of discrete energy levels came from the study of the spectrum of emissions of energetically excited atomic systems. Historically, the most important such spectrum is that of the simplest atom, hydrogen, a system of one proton and one electron bound together by their electromagnetic attraction. Within the framework of classical physics, the structure of the hydrogen atom poses fundamental problems. The first is the existence of a stable ground state: An electron in orbit around a proton is in constant acceleration, and therefore, according to Maxwell's classical electromagnetic theory, should continuously radiate away energy. Furthermore, the radiation emitted as the atom decays to a lower energy state should form a continuous spectrum of frequencies. However, the hydrogen atom both possesses a stable ground state and emits radiation at only a discrete set of frequencies.

In 1913 N. Bohr made a fundamental advance by postulating that the angular momentum of the electron-proton system could take on only a discrete set of values.

The angular momentum is said to be quantized. A consequence is that the excitation energies of the hydrogen atom also have a discrete spectrum. Bohr made the further postulate that the atom decays from an excited level, E_k, only by making a transition to a lower energy level, E_j, emitting a single light quantum (photon) in the process. The energy, E_γ, of this photon is given by the conservation of energy, $E_\gamma = E_k - E_j$. Although Bohr's postulates are in many ways without real foundation, they were later justified and extended by the development of quantum mechanics. *See* ATOMIC STRUCTURE AND SPECTRA.

The quantization of the allowed energy values that occurs in quantum mechanics has analogs for other physical quantities as well, such as angular momentum. The basic reason why such quantization occurs for bound systems of particles in quantum mechanics but not in classical mechanics is that in quantum mechanics particles have associated wavelike attributes, specifically a wave function which encodes the dynamical state of the particle. (This is the content of wave-particle duality.) The wave function of a bound state satisfies an equation similar in many ways to the equation describing waves on a guitar string or drumhead of finite extent. Such musical instruments produce only certain specific notes, or frequencies, for a given length of string or size of drumhead. In other words, the frequencies are quantized. Similarly, the modes of oscillation of the wave function for a quantum system of finite extent are also quantized, leading to discrete energy levels, and so forth. An unbound quantum system, however, is analogous to a string of infinite length, which can play a continuous range of notes.

Energy levels are of great importance for many systems other than simple atoms such as hydrogen. For instance, they determine the interactions and binding of molecules in chemistry and biochemistry, the stability or decay of nuclei, and the macroscopic properties of solids, such as the optical properties of dyes or semiconductors. The observed spectroscopy of the energy levels of a system can also elucidate the properties of a new force, just as the study of hydrogen led to the development of quantum mechanics and quantum field theory.   [J.M.R.]

Enthalpy   For any system, that is, the volume of substance under discussion, enthalpy is the sum of the internal energy of the system plus the system's volume multiplied by the pressure exerted by the system on its surroundings. The sum is given the special symbol H primarily as a matter of convenience because this sum appears repeatedly in thermodynamic discussion. Previously, enthalpy was referred to as total heat or heat content, but these terms are misleading and should be avoided. Enthalpy is, from the viewpoint of mathematics, a point function, as contrasted with heat and work, which are path functions. Point functions depend only on the initial and final states of the system undergoing a change; they are independent of the paths or character of the change. For change in enthalpy with pressure or temperature *see* THERMODYNAMIC PRINCIPLES; ENTROPY; THERMODYNAMIC PROCESSES.   [H.C.W.; W.A.S.]

Entropy   A function first introduced in classical thermodynamics to provide a quantitative basis for the common observation that naturally occurring processes have a particular direction. Subsequently, in statistical thermodynamics, entropy was shown to be a measure of the number of microstates a system could assume. Finally, in communication theory, entropy is a measure of information. Each of these aspects will be considered in turn. Before the entropy function is introduced, it is necessary to discuss reversible processes.

Reversible processes. Any system under constant external conditions is observed to change in such a way as to approach a particularly simple final state called an equilibrium state. For example, two bodies initially at different temperatures are connected by a metal wire. Heat flows from the hot to the cold body until the temperatures of both bodies are the same. It is common experience that the reverse processes never occur if the systems are left to themselves; that is, heat is never observed to flow from the cold to the hot body. Max Planck classified all elementary processes into three categories: natural, unnatural, and reversible. Natural processes do occur, and proceed in a direction toward equilibrium. Unnatural processes move away from equilibrium and never occur. A reversible process is an idealized natural process that passes through a continuous sequence of equilibrium states.

Entropy function. The state function entropy S puts the foregoing discussion on a quantitative basis. Entropy is related to q, the heat flowing into the system from its surroundings, and to T, the absolute temperature of the system. The important properties for this discussion are:

1. $dS > q/T$ for a natural change.
   $dS = q/T$ for a reversible change.
2. The entropy of the system S is made up of the sum of all the parts of the system so that $S = S_1 + S_2 + S_3 \cdots$. *See* HEAT; TEMPERATURE.

Nonconservation. In his study of the first law of thermodynamics, J. P. Joule caused work to be expended by rubbing metal blocks together in a large mass of water. By this and similar experiments, he established numerical relationships between heat and work. When the experiment was completed, the apparatus remained unchanged except for a slight increase in the water temperature. Work (W) had been converted into heat (Q) with 100% efficiency. Provided the process was carried out slowly, the temperature difference between the blocks and the water would be small, and heat transfer could be considered a reversible process. The entropy increase of the water at its temperature T is $\Delta S = Q/T = W/T$. Since everything but the water is unchanged, this equation also represents the total entropy increase. The entropy has been created from the work input, and this process could be continued indefinitely, creating more and more entropy. Unlike energy, entropy is not conserved. *See* CONSERVATION OF ENERGY; THERMODYNAMIC PROCESSES.

Degradation of energy. Energy is never destroyed. But in the Joule friction experiment and in heat transfer between bodies, as in any natural process, something is lost. In the Joule experiment, the energy expended in work now resides in the water bath. But if this energy is reused, less useful work is obtained than was originally put in. The original energy input has been degraded to a less useful form. The energy transferred from a high-temperature body to a lower-temperature body is also in a less useful form. If another system is used to restore this degraded energy to its original form, it is found that the restoring system has degraded the energy even more than the original system had. Thus, every process occurring in the world results in an overall increase in entropy and a corresponding degradation in energy. [W.F.J.]

Measure of information. The probability characteristic of entropy leads to its use in communication theory as a measure of information. The absence of information about a situation is equivalent to an uncertainty associated with the nature of the situation. This uncertainty is the entropy of the information about the particular situation. [F.H.R.]

Environmental fluid mechanics The study of the flows of air and water, of the species carried by them, and of their interactions with geological, biological, social, and engineering systems in the vicinity of a planet's surface. The environment on the Earth is intimately tied to the fluid motion of air (atmosphere), water (oceans), and species concentrations (air quality). In fact, the very existence of the human race depends upon its abilities to cope within the Earth's environmental fluid systems.

Meteorologists, oceanologists, geologists, and engineers study environmental fluid motion. Weather and ocean-current forecasts are of major concern, and fluid motion within the environment is the main carrier of pollutants. Biologists and engineers examine the effects of pollutants on humans and the environment, and the means for environmental restoration. Air quality in cities is directly related to the airborne spread of dust particles and of exhaust gases from automobiles. The impact of pollutants on drinking-water quality is especially important in the study of ground-water flow. Likewise, flows in porous media are important in oil recovery and cleanup. Lake levels are significantly influenced by climatic change, a relationship that has become of some concern in view of the global climatic changes that may result from the greenhouse effect (whereby the Earth's average temperature increases because of increasing concentrations of carbon dioxide in the atmosphere).

Scales of motion. Environmental fluid mechanics deals with the study of the atmosphere, the oceans, lakes, streams, surface and subsurface water flows (hydrology), building exterior and interior airflows, and pollution transport within all these categories. Such motions occur over a wide range of scales, from eddies on the order of centimeters to large recirculation zones the size of continents. This range accounts in large part for the difficulties associated with understanding fluid motion within the environment. In order to impart motion (or inertia) to the atmosphere and oceans, internal and external forces must develop. Global external forces consist of gravity, Coriolis, and centrifugal forces, and electric and magnetic fields (to a lesser extent). The internal forces of pressure and friction are created at the local level, that is, on a much smaller spatial scale; likewise, these influences have different time scales. The winds and currents arise as a result of the sum of all these external and internal forces.

Governing equations. The foundations of environmental fluid mechanics lie in the same conservation principles as those for fluid mechanics, that is, the conservation of mass, momentum (velocity), energy (heat), and species concentration (for example, water, humidity, other gases, and aerosols). The differences lie principally in the formulations of the source and sink terms within the governing equations, and the scales of motion. These conservation principles form a coupled set of relations, or governing equations, which must be satisfied simultaneously. The governing equations consist of nonlinear, independent partial differential equations that describe the advection and diffusion of velocity, temperature, and species concentration, plus one scalar equation for the conservation of mass. In general, environmental fluids are approximately newtonian, and the momentum equation takes the form of the Navier-Stokes equation. An important added term, neglected in small-scale flow analysis, is the Coriolis acceleration, $2\Omega \times V$, where Ω is the angular velocity of the Earth and V is the flow velocity. *See* CONSERVATION LAWS (PHYSICS); CONSERVATION OF ENERGY; CONSERVATION OF MASS; CONSERVATION OF MOMENTUM; DIFFUSION; NAVIER-STOKES EQUATION; NEWTONIAN FLUID.

Fortunately, not every term in the Navier-Stokes equation is important in all layers of the environment. The key to being able to obtain solutions to the Navier-Stokes equation lies in determining which terms can be neglected in specific applications. For convenience, problems can be classified on the basis of the order of importance of the terms in the equations utilizing nondimensional numbers based on various ratios of values. *See* DIMENSIONLESS GROUPS.

Measurements. Because of the scales of motion and time associated with the environment, and the somewhat random nature of the fluid motion, it is difficult to conduct full-scale, extensive experimentation. Likewise, some quantities (such as vorticity or vertical velocity) resist direct observations. It is necessary to rely on the availability of past measurements and reports (as sparse as they may be) to establish patterns, especially for climate studies. However, some properties can be measured with confidence.

Modeling. There are two types of modeling strategies: physical and mathematical. Physical models are small-scale (laboratory) mockups that can be measured under variable conditions with precise instrumentation. Such modeling techniques are effective in examining wind effects on buildings and species concentrations within city canyons (flow over buildings). Generally, a large wind tunnel is needed to produce correct atmospheric parameters (such as Reynolds number) and velocity profiles. Mathematical models (algebra- and calculus-based) can be broken down further into either analytical models, in which an exact solution exists, or numerical models, whereby approximate numerical solutions are obtained using computers. [D.W.P.]

Equilibrium of forces   In a mechanical system the condition under which no acceleration takes place. Newtonian mechanics today is based upon two definitions which modify, but are essentially equivalent to, Newton's three fundamental laws. These definitions postulate the action of forces on particles. A particle is defined as a conceptual volume element that has mass and is sufficiently small to have point location. A body is defined as a system of particles. To develop the mechanics of a body, these definitions are applied to each of its particles and their influences summed. *See* ACCELERATION.

The law of motion is that, in a newtonian frame of reference (with few exceptions, a frame of reference fixed with respect to Earth is considered to be newtonian), a particle of mass m acted on by resultant force $\mathbf{F}$ has acceleration $\mathbf{a}$ in accordance with the equation $\mathbf{F} = km\mathbf{a}$. Therein, k is a positive constant whose value depends upon the units in which $\mathbf{F}$, m, and $\mathbf{a}$ are measured. The action-reaction law states that when one particle exerts force on another, the other particle exerts on the one a collinear force equal in magnitude but oppositely directed.

A body acted upon by force is in equilibrium when its constituent particles are in equilibrium. The forces exerted on its particles (and therefore on the body) are either internal or external to the body. An internal force is one exerted by one particle on another in the same body. An external force is one exerted on a particle or the body by a particle not of the body. *See* DYNAMICS; KINETICS (CLASSICAL MECHANICS); STATICS. [N.S.F.]

Euler angles   Three angular parameters that specify the orientation of a body with respect to reference axes. They are used for describing rotating systems such as gyroscopes, tops, molecules, and nonspherical nuclei. They are not symmetrical in the three angles but are simpler to use than other rotational parameters.

Unfortunately, different definitions of Euler's angles are used, and therefore it is confusing to compare equations in different references. The definition given here is the majority convention according to H. Margenau and G. Murphy.

Let $OXYZ$ be a right-handed cartesian (right-angled) set of fixed coordinate axes and $Oxyz$ a set attached to the rotating body (see illustration).

The orientation of $Oxyz$ can be produced by three successive rotations about the fixed axes starting with $Oxyz$ parallel to $OXYZ$. Rotate through (1) the angle ψ counterclockwise about OZ, (2) the angle θ counterclockwise about OX, and (3) the angle ϕ

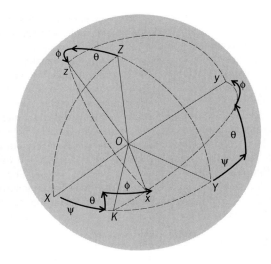

Euler angles. The successive movements of the axes on a unit sphere described in the text are shown by arrows. The complete rotation may also be obtained by a different sequence of rotations, namely, first through ϕ about *OZ*, then through θ about the displaced x axis (which is *OK*), then through ψ about *OZ*.

counterclockwise about *OZ* again. The line of intersection *OK* of the xy and XY planes is called the line of nodes.

Denote a rotation about *OZ*, for example, by Z (angle). Then the complete rotation is, symbolically, given by the equation below where the rightmost operation is done first.

$$R(\psi, \theta, \phi) = Z(\phi)X(\theta)Z(\psi)$$

[B.G.]

Euler's equations of motion
A set of three differential equations expressing relations between the force moments, angular velocities, and angular accelerations of a rotating rigid body. They are equations of motion in the usual dynamical sense, of forms (1)–(3).

$$I_1(d\omega_1/dt) + (I_3 - I_2)\omega_2\omega_3 = M_1 \tag{1}$$

$$I_2(d\omega_2/dt) + (I_1 - I_3)\omega_3\omega_1 = M_2 \tag{2}$$

$$I_3(d\omega_3/dt) + (I_2 - I_1)\omega_1\omega_2 = M_3 \tag{3}$$

The formulation employs as coordinate axes the three principal axes of rotational inertia of the body that can rotate about a body-fixed point, which is the center of mass if constraints are absent. These reference axes, which form a righthand set, are indicated by subscripts 1, 2, and 3 in the equations, where I_1, I_2, and I_3 represent the principal moments of inertia; ω_1, ω_2, and ω_3 the angular velocities about the axes; M_1, M_2, and M_3 the corresponding force moments; and t the time. *See* RIGID-BODY DYNAMICS.

[R.A.Fi.]

Exchange interaction
A quantum-mechanical phenomenon that gives the energy of two elementary particles. Exchange effects arise for all kinds of elementary particles, but these effects were first introduced into physics in consideration of atomic structure and the energy of the electrons in an atom. In this context, they arise as a consequence of two facts: electrons are indistinguishable, and they obey the Pauli exclusion principle. *See* ELEMENTARY PARTICLE.

Indistinguishability demands that the wave function describing two electrons either is unchanged or changes sign when the labels of the two electrons are exchanged (that is, it is either symmetric or antisymmetric to this exchange). In the case of electrons, the wave function is antisymmetric; this is a more precise statement of the Pauli exclusion principle. *See* ELECTRON; EXCLUSION PRINCIPLE.

When the spins of the two electrons are parallel the spatial part of the wave function is antisymmetric under exchange, and when they are opposed it is symmetric. The distribution in space of the two electrons is different in these two states, and so their mutual electrostatic energy is different. This difference in the electrostatic energy, called the exchange energy, appears as an interaction between the two electrons which depends on their relative orientation (although this dependence is incidental).

Exchange is the mechanism by which the electron spins in many magnetic materials are lined up parallel or opposed (for example, in α-iron and the ferrites). It is also the mechanism whereby the electron spins are parallel in the first excited state of the helium atom, and the spins of the two electrons are opposed in the carbon-carbon covalent bond. *See* ATOMIC STRUCTURE AND SPECTRA; ELECTRON CONFIGURATION; FERROMAGNETISM; QUANTUM MECHANICS; SPIN (QUANTUM MECHANICS). [J.M.D.]

Excitation potential   The difference in potential between an excited atomic or molecular state and the ground state. The term is most generally used in connection with electron excitation, but it can be applied to excited molecular vibrational and rotational states.

A closely related term is excitation energy. If the unit of potential is taken as the volt and the unit of energy as the electron volt, then the two are numerically equal. According to the Bohr theory, there is a relationship between the wavelength of the photon associated with the transition and the excitation energies of the two states. Thus the basic equation for the emission or absorption of energy is as shown below, where h

$$\frac{hc}{\lambda} = E_i - E_f$$

is Planck's constant, c the velocity of light, λ the wavelength of the photon, and E_f and E_i the energies of the final and initial states, respectively. *See* EXCITED STATE; GROUND STATE; IONIZATION POTENTIAL. [G.H.M.]

Excited state   In quantum mechanics, a stationary state of higher energy than the lowest stationary state or ground state of a particle or a system of particles. Customarily, only bound stationary states, which generally are at most denumerably infinite in number, are spoken of as excited, although the formal quantum theory often treats the noncountable unbound stationary states on an equal footing with the bound states. *See* GROUND STATE; METASTABLE STATE. [E.G.]

Exciton   A fundamental quantum of electronic excitation in condensed matter, consisting of a negatively charged electron and a positively charged hole bound to each other by electrostatic attraction. Excitons exist in all kinds of condensed matter, whenever it is possible for an electron to be excited from a filled energy level to an empty one, leaving behind a hole. Unlike an excitation in a single atom or molecule, an exciton can in general move through the solid like a particle. Excitons transport energy, not charge or mass. Typically, an exciton is created when a photon is absorbed in a solid; the exciton then moves through the crystal; and finally the electron and hole recombine, resulting in the emission of another photon, often at a wavelength different from that of the original photon. Excitons can also be created by injection of free electrons into excited states via an electric current. *See* ELECTRON-HOLE RECOMBINATION; HOLE STATES IN SOLIDS; LUMINESCENCE.

Excitons fall into two broad classes, Wannier (or Wannier-Mott) excitons and Frenkel excitons, based on their size relative to the interatomic or intermolecular distances in the material. In Wannier excitons, typically observed in covalent semiconductors and insulators, the electron and hole are separated by a distance much larger than the atomic spacing, so that the effect of the crystal lattice on the exciton can be taken into account primarily via an average permittivity. In Frenkel excitons, typically seen in molecular or rare-gas crystals, the electron and hole are separated by a distance comparable to the atomic spacing, so that the exciton is localized to a single site at any given time. Wannier excitons move essentially like free particles, while motion of Frenkel excitons is envisioned as hopping from one site to another. *See* ELECTRIC INSULATOR; SEMICONDUCTOR.   [D.Sn.]

Exclusion principle   No two electrons may simultaneously occupy the same quantum state. This principle, often called the Pauli principle, was first formulated by Wolfgang Pauli in 1925 and, for time-independent quantum states, it means that no two electrons may be described by state functions which are characterized by exactly the same quantum numbers. In addition to electrons, all known particles having half-integer intrinsic angular momentum, or spin, obey the exclusion principle. It plays a central role in the understanding of many diverse phenomena, including the periodic table of the elements and their chemical activities, the electron contribution to the specific heat of metals, the shell structure in the atomic nucleus analogous to that of electrons in atoms, and certain symmetries in the scattering of identical particles. *See* ANGULAR MOMENTUM; QUANTUM NUMBERS; SPIN (QUANTUM MECHANICS).

Using the fact that a system will try to occupy the state of lowest possible energy, the electron configuration of atoms may be understood by simply filling the single-particle energy levels according to the Pauli principle. This is the basis of Niels Bohr's explanation of the periodic table. *See* ATOMIC STRUCTURE AND SPECTRA; ELECTRON CONFIGURATION.
                                                                                [S.A.Wi.]

Exotic nuclei   Nuclei with ratios of neutron number N to proton number Z much larger or much smaller than those of nuclei found in nature. Studies of nuclear matter under extreme conditions, in which the nuclei are quite different in some way from those found in nature, are at the forefront of nuclear research. Such extreme conditions include nuclei at high temperature and at high density (several times normal nuclear density), as well as those with larger or smaller N/Z ratios. The N/Z ratio depends on the nature of the attractive nuclear force that binds the protons and neutrons in the nucleus and its competition and complex interplay with the disruptive Coulomb or electrical force that pushes the positively charged protons apart.

A chart of the nuclides is shown in the illustration. The squares are the stable nuclei ($Z \leq 83$) and the very long-lived nuclei (with half-lives of the order of 10^9 years or more) found in nature. The first jagged lines to either side are the limits of the presently observed nuclei; very little is known about those at the edges. The light, stable nuclei (such as ^{4}He) have $Z = N$, reflecting the preference of the nuclear forces for $N = Z$ symmetry, but as Z increases, the strength of the Coulomb force demands more neutrons than protons to make a particular element stable, and $N \approx 1.6\,Z$ for the heaviest long-lived nuclei (such as ^{238}U). For $Z > 92$, the Coulomb force causes most nuclei to spontaneously fission. *See* NUCLEAR FISSION.

Stable nuclei lie in the so-called valley of beta stability. As the N/Z ratio decreases (proton-rich nuclei) or increases (neutron-rich nuclei) compared to that of the stable isotopes, there is, respectively, energy for a proton or neutron in the nucleus to undergo beta (β^+, β^-) decay to move the nucleus back toward stability. Most knowledge of

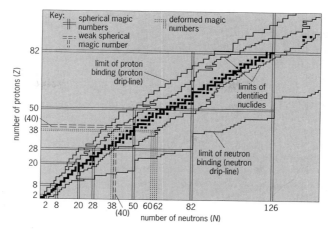

Chart of the nuclides. The squares are the stable and very long-lived nuclei found in nature.

nuclear structure and decay has been gained from nuclei in or near the valley of beta stability. *See* RADIOACTIVITY.

The spherical shell model was developed to explain the so-called spherical magic numbers for protons and neutrons, which give nuclei with these numbers a very stable structure and spherical shape. Spherical magic Z and N of 2, 8, 20, 28, 50, 82, and 126, and the weaker magic Z and N at 40, are shown in the illustration as horizontal and vertical lines. The nuclear shell model resembles the atomic shell model, where the noble gases (helium, neon, argon, and so forth) have filled shells and then there is a gap in the binding energy to the next electron shell (or orbit). A major question concerns what happens to these spherical magic proton and neutron gaps (orbits) in exotic nuclei. *See* NUCLEAR STRUCTURE.

Another major question concerns the decay modes whereby a nucleus rids itself of excess energy and returns to the stable forms of nuclear matter. As N/Z decreases or increases from the stable values, a point is reached for a given Z where, if one more neutron is pulled out, one proton becomes unbound (an isotope of that element cannot exist with that number of neutrons), or, if one more neutron is added, that neutron is not bound to that nucleus. These limits define, respectively, the proton and neutron drip-lines (the jagged lines furthest from the valley of stability in the illustration).

Answers to the above questions give significant insights into the structure and decay modes of nuclear matter that make it possible to test and extend theoretical models of the nucleus and the understanding of the nature of both the strong and weak nuclear forces. These insights could not be gained by studies of nuclei in and near the valley of stability.

These insights include the discovery of nuclear shape coexistence. Competing bands of levels occur in one nucleus, which overlap in energy but are quite separate in their decays because they are built on quite different coexisting nuclear shapes. Shape coexistence is now known to be important in many nuclei throughout the periodic table.

A significant advance was made in the nuclear shell model with the discovery of new magic numbers associated with shell gaps in the energies of the proton and neutron orbitals. These new numbers may be called deformed magic numbers because they stabilize a nucleus in a deformed shape, just as the spherical magic numbers give stability to a spherical shape. The deformed magic numbers (shell gaps) identified so far include N and Z of 38 and N of 60 and 62.

Exotic nuclei exhibit decay modes not seen near stability, such as proton radioactivity and beta-delayed particle emission. (After beta decay, the highly excited nucleus can emit one or more particles, such as one or two protons, an alpha particle, or one or two neutrons.)   [J.H.H.]

Extended x-ray absorption fine structure (EXAFS)   The structured absorption on the high-energy side of an x-ray absorption edge. The absorption edges for an element are abrupt increases in x-ray absorption that occur when the energy of the incident x-ray matches the binding energy of a core electron (typically a $1s$ or a $2p$ electron). For x-ray energies above the edge energy, a core electron is ejected from the atom. The ejected core electron can be thought of as a spherical wave propagating outward from the absorbing atom. The photoelectron wavelength is determined by its kinetic energy, which is in turn determined by the difference between the incident x-ray energy and the core-electron binding energy. As the x-ray energy increases, the kinetic energy of the photoelectron increases, and thus its wavelength decreases. *See* ABSORPTION OF ELECTROMAGNETIC RADIATION; LIGHT; PHOTOEMISSION; QUANTUM MECHANICS.

The x-ray-excited photoelectron will be scattered by the neighboring atoms surrounding the absorbing atom. The portion of the photoelectron wave that is scattered back in the direction of the absorbing atom is responsible for the EXAFS oscillations. If the outgoing and backscattered photoelectron waves are out of phase and thus interfere destructively, there is a local minimum in the x-ray absorption cross section. At a higher x-ray energy (shorter photoelectron wavelength), constructive interference leads to a local maximum in x-ray absorption (see illustration). EXAFS thus arises from photoelectron scattering, making it a spectroscopically detected scattering method. *See* INTERFERENCE OF WAVES; SCATTERING OF ELECTROMAGNETIC RADIATION.

EXAFS typically refers to structured absorption from approximately 50 to 1000 eV or more above the absorption edge. X-ray absorption near edge structure (XANES) is often used to refer to the structure in the near (around 50 eV) region of the edge. X-ray absorption fine structure (XAFS) has gained some currency as a reference to the entire structured absorption region (XANES+EXAFS).

EXAFS spectra contain structural information comparable to that obtained from single-crystal x-ray diffraction. The principal advantage of EXAFS in comparison with

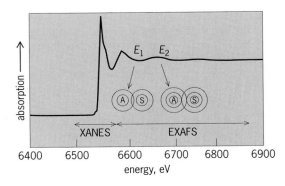

X-ray absorption spectrum for manganese, showing XANES and EXAFS regions. As the x-ray energy increases from E_1 to E_2, the interference of the outgoing and backscattered photoelectron wave (shown schematically by concentric circles around the absorbing, *A*, and scattering, *S*, atoms) changes from destructive to constructive.

crystallography is that EXAFS is a local structure probe and does not require the presence of long-range order. This means that EXAFS can be used to determine the local structure in noncrystalline samples. *See* X-RAY CRYSTALLOGRAPHY.   [J.P.Ha.]

Eyepiece   A lens or optical system which offers to the eye the image originating from another system (the objective), at a suitable viewing distance. The image can be virtual. *See* OPTICAL IMAGE.

In modern instruments, most eyepieces (also called oculars) are not independently corrected for all errors. They are designed to balance out certain residual aberrations of the objective or (as in the microscope) of a group of objectives, for instance, chromatic difference of magnification. *See* ABERRATION (OPTICS).

The Ramsden eyepiece consists of two planoconvex lenses, the field lens and the eye lens, with their plane sides out. Both of these lenses have the same power and focal length; their separation is equal to their common focal length. *See* GEOMETRICAL OPTICS; LENS (OPTICS).

The Huygens eyepiece also consists of two planoconvex lenses, but the plane sides of both lenses face the eye, The focal length of the field lens is in general three times that of the eye lens, and the separation is twice the focal length of the eye lens.   [M.J.H.]

F

Faraday effect   Rotation of the plane of polarization of a beam of linearly polarized light when the light passes through matter in the direction of the lines of force of an applied magnetic field. Discovered by M. Faraday in 1846, the effect is often called magnetic rotation. *See* MAGNETOOPTICS.

The Faraday effect is particularly simple in substances having sharp absorption lines, that is, in gases and in certain crystals, particularly at low temperatures. Here the effect can be fully explained from the fundamental properties of the atoms and molecules involved. In other substances the situation may be more complex, but the same principles furnish the explanation.

Rotation of the plane of polarization occurs when there is a difference between the indices of refraction n^+ for right-handed polarized light and n^- for left-handed polarized light. Most substances do not show such a difference without a magnetic field, except optically active substances such as crystalline quartz or a sugar solution. It should be noted that the index of refraction in the vicinity of an absorption line changes with the frequency. *See* POLARIZED LIGHT.   [G.H.Di.; W.W.W.]

Faraday's law of induction   A statement relating an induced electromotive force (emf) to the change in magnetic flux that produces it. For any flux change that takes place in a circuit, Faraday's law states that the magnitude of the emf ξ induced in the circuit is proportional to the rate of change of flux as in the expression below.

$$\xi \propto \frac{d\Phi}{dt}$$

The time rate of change of flux in this expression may refer to any kind of flux change that takes place. If the change is motion of a conductor through a field, $d\Phi/dt$ refers to the rate of cutting flux. If the change is an increase or decrease in flux linking a coil, $d\Phi/dt$ refers to the rate of such change. It may refer to a motion or to a change that involves no motion. *See* ELECTROMAGNETIC INDUCTION.   [K.V.M.]

Fermi-Dirac statistics   The statistical description of particles or systems of particles that satisfy the Pauli exclusion principle. This description was first given by E. Fermi, who applied the Pauli exclusion principle to the translational energy levels of a system of electrons. It was later shown by P. A. M. Dirac that this form of statistics is also obtained when the total wave function of the system is antisymmetrical. *See* EXCLUSION PRINCIPLE.

Such a system is described by a set of occupation numbers $\{n_i\}$ which specify the number of particles in energy levels ϵ_i. It is important to keep in mind that ϵ_i represents a finite range of energies, which in general contains a number, say g_i, of nondegenerate quantum states. In the Fermi statistics, at most one particle is allowed in a nondegenerate state. (If spin is taken into account, two particles may be contained in such a state.) This is simply a restatement of the Pauli exclusion principle, and means that $n_i \leqq g_i$. The probability of having a set $\{n_i\}$ distributed over the levels ϵ_i, which contain g_i nondegenerate levels, is described by Eq. (1), which gives just the number of

$$W = \prod_i \frac{g_i!}{(g_i - n_i)!\, n_i!} \tag{1}$$

ways that n_i can be picked out of g_i, which is intuitively what one expects for such a probability. The equilibrium state which actually exists is the set of n's that makes W a maximum, under the auxiliary conditions given in Eqs. (2a) and (2b). These conditions ex-

$$\sum_i n_i = N \tag{2a}$$

$$\sum_i n_i \epsilon_i = E \tag{2b}$$

press the fact that the total energy E and the total number of particles N are given. Equation (3) holds for this most probable distribution. Here A and β are parameters, to be

$$n_i = \frac{g_i}{\frac{1}{A} \epsilon^{\beta \epsilon^i} + 1} \tag{3}$$

determined from Eq. (3); in fact, $\beta = 1/kT$, where k is Boltzmann's constant and T is the absolute temperature. When the 1 in the denominator may be neglected, Eq. (3) goes over into the Boltzmann distribution.

Classical conditions pertain when the volume per particle is much larger than the volume associated with the de Broglie wavelength λ of a particle. For electrons in a metal at 300 K, the ratio of the volume per particle to λ^3 has the value 10^{-4}, showing that classical statistics fail altogether. When the classical distribution fails, a degenerate Fermi distribution results. A somewhat lengthy calculation yields the result that in this case the contribution of the electrons to the specific heat is negligible. This resolves an old paradox, for, according to the classical equipartition law, the electronic specific heat C should be $(3/2)Nk$, whereas in reality it is very small. *See* BOSE-EINSTEIN STATISTICS; KINETIC THEORY OF MATTER; QUANTUM STATISTICS; STATISTICAL MECHANICS.   [M.Dr.]

Fermi surface   The surface in the electronic wavenumber space of a metal that separates occupied from empty states. Every possible state of an electron in a metal can be specified by the three components of its momentum, or wavenumber. The name derives from the fact that half-integral spin particles, such as electrons, obey Fermi-Dirac statistics and at zero temperature fill all levels up to a maximum energy called the Fermi energy, with the remaining states empty. *See* FERMI-DIRAC STATISTICS.

The fact that such a surface exists for any metal, and the first direct experimental determination of a Fermi surface (for copper) in 1957, were central to the development of the theory of metals. A surprise arising from the earliest determined Fermi surfaces was that many of the shapes were close to what would be expected if the electrons interacted only weakly with the crystalline lattice. The long-standing free-electron theory of metals was based upon this assumption, but most physicists regarded it as a serious oversimplification. *See* FREE-ELECTRON THEORY OF METALS.

The momentum p of a free electron is related to the wavelength λ of the electronic wave by the equation below, where $\hbar$ is Planck's constant divided by 2π. The ratio

$$p = \frac{2\pi \hbar}{\lambda}$$

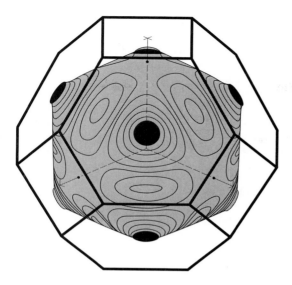

Fermi surface of copper, as
determined in 1957; two shapes
were found to be consistent
with the original data, and the
other, slightly more deformed
version turned out to be correct.

$2\pi/\lambda$, taken as a vector in the direction of the momentum, is called the wavenumber k. If the electron did not interact with the metallic lattice, the energy would not depend upon the direction of k, and all constant-energy surfaces, including the Fermi surface, would be spherical.

The Fermi surface of copper was found to be distorted (see illustration) but was still a recognizable deformation of a sphere. The polyhedron surrounding the Fermi surface in the illustration is called the Brillouin zone. It consists of Bragg-reflection planes, the planes made up of the wavenumbers for which an electron can be diffracted by the periodic crystalline lattice. The square faces, for example, correspond to components of the wavenumber along one coordinate axis equal to $2\pi/a$, where a is the cube edge for the copper lattice. For copper the electrons interact with the lattice so strongly that when the electron has a wavenumber near to the diffraction condition, its motion and energy are affected and the Fermi surface is correspondingly distorted. The Fermi surfaces of sodium and potassium, which also have one conducting electron per atom, are very close to a sphere. These alkali metals are therefore more nearly free-electron-like. *See* BRILLOUIN ZONE.

In transition metals there are electrons arising from atomic d levels, in addition to the free electrons, and the corresponding Fermi surfaces are more complex than those of the nearly free-electron metals. However, the Fermi surfaces exist and have been determined experimentally for essentially all elemental metals.

The motion of the electrons in a magnetic field provides the key to experimentally determining the Fermi surface shapes. The simplest method conceptually derives from ultrasonic attenuation. Sound waves of known wavelength pass through the metal and a magnetic field is adjusted, yielding fluctuations in the attenuation as the orbit sizes match the sound wavelength. This measures the diameter of the orbit and Fermi surface. The most precise method uses the de Haas-van Alphen effect, based upon the quantization of the electronic orbits in a magnetic field. Fluctuations in the magnetic susceptibility give a direct measure of the cross-sectional areas of the Fermi surface. *See* DE HAAS-VAN ALPHEN EFFECT; SKIN EFFECT (ELECTRICITY); ULTRASONICS.

[W.A.H.]

Ferrimagnetism   A specific type of ordering in a system of magnetic moments or the magnetic behavior resulting from such order. In some magnetic materials the magnetic ions in a crystal unit cell may differ in their magnetic properties. This is clearly so when some of the ions are of different species. It is also true for similar ions occupying crystallographically inequivalent sites. Such ions differ in their interactions with other ions, because the dominant exchange interaction is mediated by the neighboring nonmagnetic ions. They also experience different crystal electric fields, and these affect the magnetic anisotropy of the ion. A collection of all the magnetic sites in a crystal with identical behavior is referred to as a magnetic sublattice. A material is said to exhibit ferrimagnetic order when, first, all moments on a given sublattice point in a single direction and, second, the resultant moments of the sublattices lie parallel or antiparallel to one another. The notion of such an order is due to L. Néel, who showed in 1948 that its existence would explain many of the properties of the magnetic ferrites. *See* FERROMAGNETISM.

In general, there is a net moment, the algebraic sum of the sublattice moments, just as for a normal ferromagnet. However, its variation with temperature rarely exhibits the very simple behavior of the normal ferromagnet. For example, in some materials, as the temperature is raised over a certain range, the magnetization may first decrease to zero and then increase again. Ferrimagnets can be expected, in their bulk properties, measured statically or at low frequencies, to resemble ferromagnets with unusual temperature characteristics. *See* CURIE TEMPERATURE.   [L.R.W.]

Ferroelectrics   Crystalline substances which have a permanent spontaneous electric polarization (electric dipole moment per cubic centimeter) that can be reversed by an electric field. In a sense, ferroelectrics are the electrical analog of the ferromagnets, hence the name. The spontaneous polarization is the so-called order parameter of the ferroelectric state. The names Seignette-electrics or Rochelle-electrics, which are also widely used, are derived from the name of the first substance found to have this property, Seignette salt or Rochelle salt. *See* FERROMAGNETISM.

From a practical standpoint ferroelectrics can be divided into two classes. In ferroelectrics of the first class, spontaneous polarization can occur only along one crystal axis; that is, the ferroelectric axis is already a unique axis when the material is in the paraelectric phase. Typical representatives of this class are Rochelle salt, monobasic potassium phosphate, ammonium sulfate, guanidine aluminum sulfate hexahydrate, glycine sulfate, colemanite, and thiourea.

In ferroelectrics of the second class, spontaneous polarization can occur along several axes that are equivalent in the paraelectric phase. The following substances belong to this class: barium(IV) titanate-type (or perovskite-type) ferroelectrics; cadmium niobate; lead niobate; certain alums, such as methyl ammonium alum; and ammonium cadmium sulfate.

From a scientific standpoint, one can distinguish proper ferroelectrics and improper ferroelectrics. In proper ferroelectrics, the structure change at the Curie temperature can be considered a consequence of the spontaneous polarization. In improper ferroelectrics, the spontaneous polarization can be considered a by-product of another structural phase transition. Examples of such systems are gadolinium molybdate and boracites.

The spontaneous polarization can occur in at least two equivalent crystal directions; thus, a ferroelectric crystal consists in general of regions of homogeneous polarization that differ only in the direction of polarization. These regions are called ferroelectric domains. Ferroelectrics of the first class consist of domains with parallel and antiparallel polarization, whereas ferroelectrics of the second class can assume much more complicated domain configurations. The region between two adjacent domains is called a domain wall. Within this wall, the spontaneous polarization changes its direction.

As a rule, the dielectric constant ϵ measured along a ferroelectric axis increases in the paraelectric phase when the Curie temperature is approached. In many ferroelectrics, this increase can be approximated by the Curie-Weiss law. *See* CURIE-WEISS LAW.

Ferroelectrics can be divided into two groups according to their piezoelectric behavior. The ferroelectrics in the first group are already piezoelectric in the unpolarized phase. Those piezoelectric moduli which relate stresses to polarization along the ferroelectric axis have essentially the same temperature dependence as the dielectric constant along this axis, and hence become very large near the Curie point. The spontaneous polarization gives rise to a large spontaneous piezoelectric strain which is proportional to the spontaneous polarization.

The ferroelectrics in the second group are not piezoelectric when they are in the paraelectric phase. However, the spontaneous polarization lowers the symmetry so that they become piezoelectric in the polarized phase. This piezoelectric activity is often hidden because the piezoelectric effects of the various domains can cancel. However, strong piezoelectric activity of a macroscopic crystal or even of a polycrystalline sample occurs when the domains have been aligned by an electric field. The spontaneous strain is proportional to the square of the spontaneous polarization. *See* PIEZOELECTRICITY.

Antiferroelectric crystals are characterized by a phase transition from a state of lower symmetry (generally low-temperature phase) to a state of higher symmetry (generally high-temperature phase). The low-symmetry state can be regarded as a slightly distorted high-symmetry state. It has no permanent electric polarization, in contrast to ferroelectric crystals. The crystal lattice can be regarded as consisting of two interpenetrating sublattices with equal but opposite electric polarization. This state is referred to as the antipolarized state. In a certain sense, an antiferroelectric crystal is the electrical analog of an antiferromagnetic crystal.

The piezoelectric effect of ferroelectrics (and certain antiferroelectrics) finds numerous applications in electromechanical transducers. The large electrooptical effect (birefringence induced by an electric field) is used in light modulators. In certain ferroelectrics, light can induce changes of the refractive indices. These substances can be used for optical information storage and in real-time optical processors. The temperature dependence of the spontaneous polarization corresponds to a strong pyroelectric effect which can be exploited in thermal and infrared sensors. [W.K.]

Ferromagnetism　A property exhibited by certain metals, alloys, and compounds of the transition (iron group), rare-earth, and actinide elements in which, below a certain temperature called the Curie temperature, the atomic magnetic moments tend to line up in a common direction. Ferromagnetism is characterized by the strong attraction of one magnetized body for another.

Atomic magnetic moments arise when the electrons of an atom possess a net magnetic moment as a result of their angular momentum. The combined effect of the atomic magnetic moments can give rise to a relatively large magnetization, or magnetic moment per unit volume, for a given applied field. Above the Curie temperature, a ferromagnetic substance behaves as if it were paramagnetic: Its susceptibility approaches the Curie-Weiss law. The Curie temperature marks a transition between order and

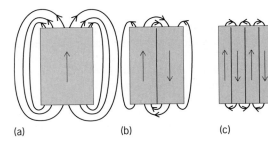

Lowering of magnetic field energy by domains. (*a*) Lines of force for a single domain. (*b*) Shortening of lines of force by division into two domains. (*c*) Reduction of field energy by further subdivision.

disorder of the alignment of the atomic magnetic moments. Some materials having atoms with unequal moments exhibit a special form of ferromagnetism below the Curie temperature called ferrimagnetism. *See* CURIE TEMPERATURE; CURIE-WEISS LAW; ELECTRON SPIN; FERRIMAGNETISM; MAGNETIC SUSCEPTIBILITY; PARAMAGNETISM.

The characteristic property of a ferromagnet is that, below the Curie temperature, it can possess a spontaneous magnetization in the absence of an applied magnetic field. Upon application of a weak magnetic field, the magnetization increases rapidly to a high value called the saturation magnetization, which is in general a function of temperature. For typical ferromagnetic materials, their saturation magnetizations, and Curie temperatures, *see* MAGNETIZATION.

Small regions of spontaneous magnetization, formed at temperatures below the Curie point, are known as domains. As shown in the illustration, domains originate in order to lower the magnetic energy. In illus. *b* it is shown that two domains will reduce the extent of the external magnetic field, since the magnetic lines of force are shortened. On further subdivision, as in, this field is still further reduced.

Another way to describe the energy reduction is to note that the interior demagnetizing fields, coming from surface poles, are much smaller in the long, thin domains of illus. *c* than in the "fat" domain of illus. *a*.

The question arises as to how long this subdivision process continues. With each subdivision there is a decrease in field energy, but there is also an increase in Heisenberg exchange energy, since more and more magnetic moments are aligning antiparallel. Finally a state is reached in which further subdivision would cause a greater increase in exchange energy than decrease in field energy, and the ferromagnet will assume this state of minimum total energy.

Materials easily magnetized and demagnetized are called soft; these are used in alternating-current machinery. The problem of making cheap soft materials is complicated by the fact that readily fabricated metals usually have many crystalline boundaries and crystal grains oriented in many directions. The ideal cheap soft material would be an iron alloy fabricated by some inexpensive technique which results in all crystal grains being oriented in the same or nearly the same direction. Various complicated rolling and annealing methods have been discovered in the continued search for better grain-oriented or "cube-textured" steels.

Materials which neither magnetize nor demagnetize easily are called hard; these are used in permanent magnets. A number of permanent-magnet materials have enjoyed technological importance. The magnet steels contain carbon, chromium, tungsten, or cobalt additives, serving to impede domain wall motion and thus to generate coercivity. Alnicos are aluminum-nickel-iron alloys containing finely dispersed, oriented, elongated particles precipitated by thermal treatment in a field. Hard ferrite magnets are based on the oxides $BaFe_{12}O_{19}$ and $SrFe_{12}O_{19}$. Hard ferrite magnets are relatively inexpensive and are used in a great variety of commercial applications. Rare earth–transition metal materials whose rare-earth component provides huge magne-tocrystalline anisotropy can be translated into large coercivity in a practical magnet, while the magnetization arises chiefly from the transition-metal component. Examples include samarium-cobalt magnets based on the $SmCo_5$ or Sm_2Co_{17} intermetallic compounds.

[E.A.; F.Ke.; J.F.He.]

Feynman diagram   A pictorial representation of elementary particles and their interactions. Feynman diagrams show paths of particles in space and time as lines, and interactions between particles as points where the lines meet.

The illustration shows Feynman diagrams for electron-electron scattering. In each diagram, the straight lines represent space-time trajectories of noninteracting electrons, and the wavy lines represent photons, particles that transmit the electromagnetic interaction. External lines at the bottom of each diagram represent incoming particles (before the interactions), and lines at the top, outgoing particles (after the interactions).

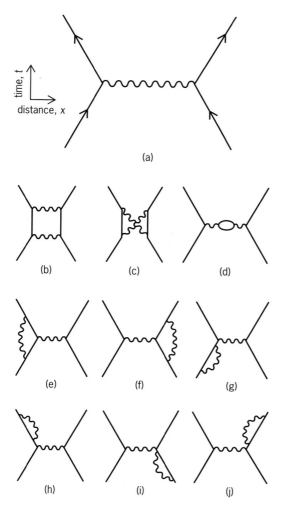

Feynman diagrams for electron-electron (Møller) scattering: (*a*) second-order diagram (two-vertices); (*b–j*) fourth-order diagrams.

Interactions between photons and electrons occur at the vertices where photon lines meet electron lines. *See* ELECTRON; PHOTON.

Each Feynman diagram corresponds to the probability amplitude for the process depicted in the diagram. The set of all distinct Feynman diagrams with the same incoming and outgoing lines corresponds to the perturbation expansion of a matrix element of the scattering matrix in field theory. This correspondence can be used to formulate the rules for writing the amplitude associated with a particular diagram. The perturbation expansion and the associated Feynman diagrams are useful to the extent that the strength of the interaction is small, so that the lowest-order terms, or diagrams with the fewest vertices, give the main contribution to the matrix element. *See* PERTURBATION (QUANTUM MECHANICS); SCATTERING MATRIX.

Since their introduction in quantum electrodynamics, Feynman diagrams have been widely applied in other field theories. They are employed in studies of electroweak interactions, certain situations in quantum chromodynamics, in acoustooptics, and in many-body theory in atomic, nuclear, plasma, and condensed matter physics. *See* ACOUSTOOPTICS; ELEMENTARY PARTICLE; FUNDAMENTAL INTERACTIONS; QUANTUM CHROMODYNAMICS; QUANTUM ELECTRODYNAMICS; QUANTUM FIELD THEORY; WEAK NUCLEAR INTERACTIONS.   [P.M.]

Feynman integral   A technique, also called the sum over histories, which is basic to understanding and analyzing the dynamics of quantum systems. It is named after fundamental work of Richard Feynman. The crucial formula gives the quantum probability density for transition from a point q_0 to a point q_1 in time t as the expression below, where $S(\text{path})$ is the classical mechanical action of a trial path, and $\hbar$ is the

$$\int \exp[i\,S(\text{path})/\hbar]\,d(\text{path})$$

rationalized Planck's constant. The integral is a formal one over the infinite-dimensional space of all paths which go from q_0 to q_1 in time t. Feynman defines it by a limiting procedure using approximation by piecewise linear paths.

Feynman integral ideas are especially important in quantum field theory, where they not only are a useful device in analyzing perturbation series but are also one of the few nonperturbative tools available. *See* QUANTUM FIELD THEORY.

An especially attractive element of the Feynman integral formulation of quantum dynamics is the classical limit, $\hbar \to 0$. Formal application of the method of stationary phase to the above expression says that the significant paths for small $\hbar$ will be the paths of stationary action. One thereby recovers classical mechanics in the hamiltonian stationary action formulation. *See* LEAST-ACTION PRINCIPLE.   [B.Si.]

Fiber-optic sensor   A sensor that uses thin optical fibers to carry light to and from a location to be probed. In performing the sensing, light can be lost from the fibers or modified in velocity by the action of the phenomena on the fiber. Fiber-optic sensors are ideal for probing in remote or hostile locations, where miniature sensors are required such as in the body, or where extreme sensitivity is required. Two classes of fiber sensors have evolved: intensity sensors, in which the amplitude of light in the fiber is changed during sensing, and interferometric sensors, in which the velocity of light or its phase is modified during sensing. The latter class has proved to be extremely sensitive; intensity sensors are used where moderate performance is acceptable and lower cost is important. Depending on their design, intensity sensors can respond to pressure, temperature, liquid level, position, flow, smoke, displacement, electric and magnetic fields, chemical composition, and numerous other conditions.

Optical interferometry is one of the most sensitive means of detecting displacements as small as 10^{-13} m. Interferometric fiber sensors apply this technology to sense many physical phenomena. A fiber gyro based on the Sagnac effect is formed by making a fiber loop which, when rotated, causes the light traveling in both directions in the loop to experience different velocities with or against the rotation. A second type of interferometric sensor is constructed by using the Mach-Zehnder interferometer. Sensitivities equal to or surpassing the best conventional technologies have been achieved in these sensors. *See* INTERFEROMETRY.   [T.G.G.]

Field emission   The emission of electrons from a metal or semiconductor into vacuum (or a dielectric) under the influence of a strong electric field. In field emission, electrons tunnel through a potential barrier, rather than escaping over it as in thermionic or photoemission. The effect is purely quantum-mechanical, with no classical analog. It occurs because the wave function of an electron does not vanish at the classical turning point, but decays exponentially into the barrier (where the electron's total energy is less than the potential energy). Thus there is a finite probability that the electron will be found on the outside of the barrier. *See* PHOTOEMISSION; QUANTUM MECHANICS; THERMIONIC EMISSION.

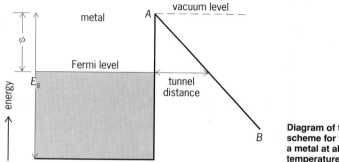

Diagram of the energy-level scheme for field emission from a metal at absolute zero temperature.

For a metal at low temperature, the process can be understood in terms of the illustration. The metal can be considered a potential box, filled with electrons to the Fermi level, which lies below the vacuum level by several electronvolts. The distance from Fermi to vacuum level is called the work function, ϕ. The vacuum level represents the potential energy of an electron at rest outside the metal, in the absence of an external field. In the presence of a strong field, the potential outside the metal will be deformed along the line AB, so that a triangular barrier is formed, through which electrons can tunnel. Most of the emission will occur from the vicinity of the Fermi level where the barrier is thinnest. *See* ELECTRON EMISSION.   [R.Gom.]

Fine structure (spectral lines)   The closely spaced groups of lines observed in the spectra of the lightest elements, notably hydrogen and helium. The components of any one such group are characterized by identical values of the principal quantum number n, but different values of the azimuthal quantum number l and the angular momentum quantum number j.

In atoms having several electrons, this fine structure becomes the multiplet structure resulting from spin-orbit coupling. This gives splittings of the terms and the spectral lines that are "fine" for the lightest elements but that are very large, of the order of an electronvolt, for the heavy elements. [F.A.J./W.W.W.]

FitzGerald-Lorentz contraction   The contraction of a moving body in the direction of its motion. In 1892 G. F. FitzGerald and H. A. Lorentz proposed independently that the failure of the Michelson-Morley experiment to detect an absolute motion of the Earth in space arose from a physical contraction of the interferometer in the direction of the Earth's motion. According to this hypothesis, as formulated more exactly by Albert Einstein in the special theory of relativity, a body in motion with speed v is contracted by a factor $\sqrt{1 - v^2/c^2}$ in the direction of motion, where c is the speed of light. *See* LIGHT; RELATIVITY. [E.L.Hi.]

Flavor   Any of the six different varieties of quarks. All hadronic matter is composed of quarks, the most elementary constituents of matter. The six different flavors are labeled u, d, s, c, b, and t, corresponding to up, down, strange, charmed, bottom, and top. Quarks are all spin-$1/2$ fermions. The u, c, and t flavors carry a positive electric charge equal in magnitude to two-thirds that of the electron; the d, s, and b flavors have a negative charge one-third that of the electron. Different flavored quarks have vastly different masses ranging from the lightest, the u quark, with a mass around 5 MeV/c^2 (where c is the speed of light), equal to the mass of about 10 electrons, to the top quark, with a mass 35,000 times greater, or 175 GeV/c^2, about the mass of a gold atom. Quarks of any flavor are further characterized by three additional quantum numbers called color: red, green, and blue. Each quark has an antiquark counterpart carrying the corresponding anticolor. *See* ANTIMATTER; COLOR (QUANTUM MECHANICS); ELEMENTARY PARTICLE; QUARKS. [V.F.]

Flow-induced vibration   The dynamic response of structures immersed in or conveying fluid flow. Fluid flow is a source of energy that can induce structural and mechanical oscillations. Flow-induced vibrations best describe the interaction that occurs between the fluid's dynamic forces and a structure's inertial, damping, and elastic forces. *See* FLUID MECHANICS; VIBRATION.

The study of flow-induced vibrations has rapidly developed in aeronautical and nonaeronautical engineering. In aeronautics, flow-induced vibration is often referred to as flutter, a topic of aeroelasticity concerning the mutual interactions of aerodynamic, elastic, and inertial forces in a flying object, its components, or its propulsion systems. Flow-induced vibration also covers classical flutter of an airfoil in a low-speed flow, stall flutter associated with a separated flow, and buffeting flutter related to turbulent wakes. Nonaeronautical flow-induced vibrations are often found in blood vessels, smokestacks, suspension bridges, oil pipe lines, power transmission lines, telephone wires, television antennas, heat exchanger tubes, nuclear fuel assemblies, and submarine periscopes and hulls. All nonaeronautical structures are unstreamlined and susceptible to both stall flutter and buffeting flutter caused by flow separation. The interaction of these structures with a fluid stream usually is more complicated than that of aeronautical structures and offers more possibilities for the flow to trigger unstable oscillations in the structures.

Aircraft wing flutter.   The fluid-elastic instability of an airplane wing or its control surfaces in a smooth flow without shock waves is called classical flutter. Flight tests show that the lift on an airfoil increases with increasing Mach number for a fixed angle of attack. This lifting force reaches a maximum at a critical Mach number, then drops sharply, and never increases no matter how high the flight speed is. This drop is due to flow separation or shock wave formation. Either of these two flow mechanisms can cause the airfoil to stall or can damage it. In these cases, the airfoil is said to be stall-fluttered or shock-stalled, depending on the flow process. *See* SHOCK WAVE.

When a flow separates from an airplane wing, the flow behind the wing is turbulent and random in nature. The airplane's tail is therefore subject to random excitations from the wing's turbulent wake. The wings and tails that oscillate randomly can lose stability. This type of dynamic aeroelastic instability is called buffeting flutter because the oscillations are random. *See* BOUNDARY-LAYER FLOW; TURBULENT FLOW; WAKE FLOW.

Vibrations of cylinder arrays. Among the topics of flow-induced vibration, cylindrical structures play very important and vital roles. For instance, the slender bodies of aircraft fuselages, missiles, and rockets, and the main bodies of industrial smokestacks, power transmission cables, telephone wires, oil pipelines, reactor fuel rods, heat-exchanger tubes, and offshore structures, as well as the blood vessels are primarily made up of cylindrical structures. These structures, when in operation, are subject to unsteady fluid-dynamic forces and prone to vibrations. Such vibrations can be classified as axial-flow-induced or cross-flow-induced vibrations, depending on the incident angle of the incoming flow with respect to the cylinders' axes. Stall and buffeting flutter, also called fluid-elastic instabilities, including fluid-damping- and fluid-stiffness-controlled instabilities, as well as vortex shedding are all possible during flow-induced vibration.   [S.S.C.; W.W.Li.]

Fluid flow   Motion of a fluid subjected to unbalanced forces or stresses. The motion continues as long as unbalanced forces are applied. For example, in the pouring of water from a pitcher the water velocity is very high over the lip, moderately high approaching the lip, and very low near the bottom of the pitcher. The unbalanced force is gravity, that is, the weight of the tilted water particles near the surface. The flow continues as long as water is available and the pitcher remains tilted. *See* FLUIDS.

A fluid may be a liquid, vapor, or gas. The term vapor denotes a gaseous substance interacting with its own liquid phase, for example, steam above water. If this phase interaction is not important, the vapor is simply termed a gas.

Gases have weak intermolecular forces and expand to fill any container. Left free, gases expand and form the atmosphere of the Earth. Gases are highly compressible; doubling the pressure at constant temperature doubles the density.

Liquids, in contrast, have strong intermolecular forces and tend to retain constant volume. Placed in a container, a liquid occupies only its own volume and forms a free surface which is at the same pressure as any gas touching it. Liquids are nearly incompressible; doubling the pressure of water at room temperature, for example, increases its density by only 0.005%.

Liquids and vapors can flow together as a mixture, such as steam condensing in a pipe flow with cold walls. This constitutes a special branch of fluid mechanics, covering two-phase-flow.

The physical properties of a fluid are essential to formulating theories and developing designs for fluid flow. Especially important are pressure, density, and temperature.

Since shear stresses cause motion in a fluid and result in differences in normal stresses at a point, it follows that a fluid at rest must have zero shear and uniform pressure at a point. This is the hydrostatic condition. The fluid pressure increases deeper in the fluid to balance the increased weight of fluid above. For liquids, and for gases over short vertical distances, the fluid density can be assumed to be constant. *See* HYDROSTATICS.

When a fluid is subjected to shear stress, it flows and resists the shear through molecular momentum transfer. The macroscopic effect of this molecular action, for most common fluids, is the physical property called viscosity. Shear stress results in a gradient in fluid velocity; the converse is also true.

The common fluids for which the linear relationship of flow velocity and sheer stress holds are called newtonian viscous fluids. More complex fluids, such as paints, pastes, greases, and slurries, exhibit nonlinear or non-newtonian behavior and require more complex theories to account for their behavior. *See* NEWTONIAN FLUID; NON-NEWTONIAN FLUID; VISCOSITY.

A common characteristic of all fluids, whether newtonian or not, is that they do not slip at a solid boundary. No matter how fast they flow away from the boundary, fluid particles at a solid surface become entrapped by the surface structure. The macroscopic effect is that the fluid velocity equals the solid velocity at a boundary. This is called the no-slip condition where the solid is fixed, so that the fluid velocity drops to zero there. No-slip sets up a slow-moving shear layer or boundary layer when fluid flows near a solid surface. The theory of boundary-layer flow is well developed and explains many effects involving viscous flow past immersed bodies or within passages. *See* BOUNDARY-LAYER FLOW.

All fluids are at least slightly compressible, that is, their density increases as pressure is applied. In many flows, however, compressibility effects may be neglected. A very important parameter in determining compressibility effects is the Mach number Ma, or ratio of flow velocity V to fluid speed of sound. For subsonic flow, Ma < 1, whereas for supersonic flow, Ma > 1. The flow is essentially incompressible if Ma < 0.3; hence for air the flow velocity is less than about 100 m/s (330 ft/s). Almost all liquid flows and many gas flows are thus treated as incompressible. Even a supersonic airplane lands and takes off in the incompressible regime. *See* COMPRESSIBLE FLOW; MACH NUMBER.

For various types of fluid flow *see* FLUID MECHANICS; HYDRODYNAMICS; ISENTROPIC FLOW; LAMINAR FLOW; RAREFIED GAS FLOW; TURBULENT FLOW; WAKE FLOW.   [F.M.Wh.]

Fluid mechanics   The engineering science concerned with the relation between the forces acting on fluids (liquids and gases) and their motions, and with the forces caused by fluids on their surroundings. It is distinct from solid mechanics by virtue of the different responses of fluids and solids to applied forces. In an ideal elastic solid, the deflection or deformation is proportional to the applied stress, whereas a fluid cannot support an applied shear stress unless it is in motion. In most fluids, called simple or newtonian fluids, it is the rate of deformation of the fluid, as opposed to the amount of deformation in a solid, that is proportional to the applied stress. *See* FLUID FLOW; NEWTONIAN FLUID.

Many substances of everyday experience and of engineering importance are found naturally in the fluid state. These include water (liquid and vapor), air (gaseous and liquid), as well as other liquids and gases of natural and industrial importance. The most common fluids are newtonian under most flow conditions.

Fluid mechanics treats the fluid as a continuum, ignoring the fact that it actually consists of individual molecules that may be, in the case of gases, widely spaced compared to molecular dimensions. Nevertheless, the continuum assumption is valid for almost all applications down to the size of bacteria. An exception occurs with gases at very low densities, such as exist in the uppermost regions of the atmosphere. At extremely high altitudes the mean free paths of air molecules—that is, the distances they travel between collisions in random thermal motion—can become as large, or even larger than, the dimensions of a space vehicle, making the assumption of a continuum invalid. *See* RAREFIED GAS FLOW.

Fluid mechanics is of fundamental importance to a number of disciplines, including aerospace, chemical, civil, environmental, mechanical, and ocean engineering, as well as to climatology, geology, meteorology, and oceanography. Applications in these fields include, but are not limited to, the study of fluid forces acting on vehicles; flows in natural rivers and artificial channels and the flow of ground water; the dispersion of pollutants in the atmosphere, lakes, rivers, and oceans; the flows in the circulatory and pulmonary systems of humans and animals; the flows in pipelines that carry crude oil and natural gas over many hundreds, or even thousands, of miles from the petroleum fields of their origin to deep-water ports or refineries; the flow of molten plastics or metals filling molds in the manufacture of numerous solid parts; the flow in pumps for water distribution systems; and both hydraulic and gas turbines for power generation and propulsion. Fluid mechanics forms the basis for much of chemistry and physics, and is sometimes applied to such apparently remote fields as cosmology. The fluid mechanical behavior of gases and liquids plays an important role in the dispersion of dissolved or entrained substances. *See* AERODYNAMIC FORCE; AERODYNAMICS; HYDRODYNAMICS.

[D.A.Ca.; J.A.Li.]

Fluids   Substances that flow under shear stress and have a density range from essentially zero (gases) to solidlike values (liquids). Fluids are one of the two major forms of matter. Solids, the other form, generally deform very little when shear forces are applied, and their densities do not change significantly with pressure or temperature.

The distinction between solids and fluids is easily seen in substances and mixtures which show a well-defined melting process. For substances with large molecules, such as polymers, ceramics, and biologicals, this distinction is less clear. Instead, there is a slow evolution of structure and of resistance to flow as temperature or some other variable is changed. *See* GLASS TRANSITION.

Molecular density varies greatly in fluids and is their most important characteristic. The distinction between vapors (or gases) and liquids is most clear for substances and mixtures that show well-defined vaporizing (boiling) and condensing processes. The high-density liquid boils to make a low-density gaseous vapor. The illustration shows the pressures and temperatures for which pure substances are single phases. At the conditions of the lines between the single-phase regions, two phases can be observed to coexist. At the state of intersection of the lines (the triple point), three phases can coexist. For most substances, the triple-point pressure is well below atmospheric. However, for carbon dioxide, it is very high, so that dry ice sublimes rather than melts, as water ice does. Beyond the end of the liquid-vapor (saturation or vapor-pressure) line, vaporization and condensation cannot be observed. The state at the end of this line is called the critical point, where all the properties of the vapor and liquid become the same. There is no such end to the solid-liquid (or melting) line, because the solid and fluid structures cannot become the same. *See* CRITICAL PHENOMENA; PHASE TRANSITIONS.

Mixtures of fluids show the same general density and multiphase behavior as pure fluids, but the composition is an extra variable to be considered. For example, the density differences between the vapor and the liquid phases cause them to have different relative amounts of the components. This difference in composition is the basis of the separation process of distillation, where the vapor will be richer in some components while the liquid will be richer in others. It is also possible for mixtures of liquids to be partially or nearly wholly immiscible, as are water and oil. The separation process of liquid extraction, used in some metal-purification systems and chemical-pollution-abatement

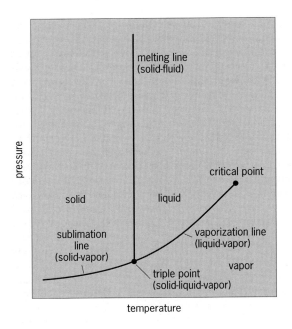

melting line
(solid-fluid)

pressure

critical point

solid   liquid

sublimation
line
(solid-vapor)

vaporization line
(liquid-vapor)

vapor

triple point
(solid-liquid-vapor)

temperature

**Conditions of pure-component
phase behavior.**

processes, depends on different preferences of chemical solutes for one liquid phase or the other. *See* PHASE RULE.

The usual observation of the presence of more than one fluid phase is the appearance of the boundary or interface between them. This is seen because the density or composition (or both) changes over a distance of a few molecular diameters, and this variation bends or scatters light in a detectable way. At the interface, the molecules feel different forces than in the bulk phases and thus have special properties. Energy is always required to create interface from bulk, the amount per unit area being called the interfacial tension. Water is a fluid with an extremely high vapor-liquid (or surface) tension; this surface tension allows insects to crawl on ponds and causes sprinkler streams to break up into sprays of droplets.

In mixtures, the molecules respond differently to the interfacial forces, so the interfacial composition is generally different from that of the bulk. This has also been the basis of a separation process. If the difference of composition is great enough and it varies with time and position because of evaporation of one or more of the components, the interfacial forces can push the fluid into motion, as can be observed on the walls of a glass of brandy (the Marangoni effect). Some substances strongly adsorb at the interface because their chemical structure has one part that prefers to be in one phase, such as water, and another part that prefers the other phase, such as oil or air. Such surfactants or detergents help solubilize dirt into wash water, keep cosmetics and other immiscible mixtures together, and form foams when air and soapy water are whipped together.

Besides the relations among pressure, density, temperature, and composition of static or equilibrium fluids, there are also characteristics associated with fluid flow, heat transfer, and material transport. For example, when a liquid or gas flows through a tube, energy must be supplied by a pump, and there is a drop in pressure from the beginning to the end of the tube that matches the rise in pressure in the pump. The pump work and pressure drop depend on the flow rate, the tube size and shape, the density, and a property of the molecules called the viscosity. The effect arises because the fluid molecules at the solid tube wall do not move and there are velocity gradients and shear in the flow. The molecules that collide with one another transfer momentum to the wall and work against one another, in a sort of friction which dissipates mechanical energy into internal energy or heat. The greater the viscosity, the greater the amount of energy dissipated by the collisions and the greater the pressure drop. If only chemical constitution and physical state are needed to characterize the viscosity, and if shear stress is directly proportional to velocity gradient, the fluid is called newtonian and the relation for pressure drop is relatively simple. If the molecules are large or the attractive forces are very strong over long ranges, as in polymers, gels, and foods such as bread dough and cornstarch, the resistance to flow can also depend on the rate of flow and even the recent deformations of the substance. These fluids are called non-newtonian, and the relationship of flow resistance to the applied forces can be very complex. *See* FLUID FLOW; NEWTONIAN FLUID; NON-NEWTONIAN FLUID; VISCOSITY.

Another fluid-transport property, thermal conductivity, indicates the ability of a static fluid to pass heat from higher to lower temperature. This characteristic is a function of chemical constitution and physical state, as is the viscosity. In mixtures, these properties may involve simple or complex dependence on composition, the variation becoming extreme if the unlike species strongly attract each other. The values of both properties increase rapidly near a critical point. *See* CONDUCTION (HEAT); HEAT TRANSFER.

Finally, the ability of molecules to change their relative position in a static fluid is called the diffusivity. This is a particularly important characteristic for separation processes whose efficiency depends on molecular motion from one phase to another through a relatively static interface, or on the ability of some molecules to move faster than others in a static fluid under an applied force. *See* DIFFUSION.   [J.P.O'C.]

Fluorescence   Fluorescence is generally defined as a luminescence emission that is caused by the flow of some form of energy into the emitting body, this emission ceasing abruptly when the exciting energy is shut off. In attempts to make this definition more meaningful it is often stated, somewhat arbitrarily, that the decay time, or afterglow, of the emission must be of the order of the natural lifetime for allowed radiative transitions in an atom or a molecule, which is about 10^{-8} s for transitions involving visible light. Perhaps a better distinction between fluorescence and its counterpart, phosphorescence, rests not on the magnitude of the decay time per se, but on the criterion that the fluorescence decay is temperature-independent.

In the literature of organic luminescence, the term fluorescence is used exclusively to denote a luminescence which occurs when a molecule makes an allowed optical transition. Luminescence with a longer exponential decay time, corresponding to an optically forbidden transition, is called phosphorescence, and it has a different special distribution from the fluorescence. *See* PHOSPHORESCENCE.

The decay time of fluorescent materials varies widely, from the order of 5×10^{-9} s for many organic crystalline materials up to 2 s for the europium-activated strontium silicate phosphor. Fluorescent materials with decay times between 10^{-9} and 10^{-7} s are used to detect and measure high-energy radiations, such as x-rays and gamma rays, and high-energy particles such as alpha particles, beta particles, and neutrons. These agents produce light flashes (scintillations) in certain crystalline solids, in solutions of many polynuclear aromatic hydrocarbons, or in plastics impregnated with these hydrocarbons. The so-called fluorescent lamps employ the luminescence of gases and solids in combination to produce visible light. *See* LUMINESCENCE.   [J.H.S.; C.C.K.]

Focal length   A measure of the collecting or diverging power of a lens or an optical system. Focal length, usually designated f' in formulas, is measured by the distance of the focal point (the point where the image of a parallel entering bundle of light rays is formed) from the lens, or more exactly by the distance from the principal point to the focal point. *See* GEOMETRICAL OPTICS.

The power of a lens system is equal to n'/f', where n' is the refractive index in the image space (n' is usually equal to unity). A lens of zero power is said to be afocal. Telescopes are afocal lens systems. *See* DIOPTER; LENS (OPTICS); TELESCOPE.   [M.J.H.]

Force   Force may be briefly described as that influence on a body which causes it to accelerate. In this way, force is defined through Newton's second law of motion.

This law states in part that the acceleration of a body is proportional to the resultant force exerted on the body and is inversely proportional to the mass of the body. An alternative procedure is to try to formulate a definition in terms of a standard force, for example, that necessary to stretch a particular spring a certain amount, or the gravitational attraction which the Earth exerts on a standard object. Even so, Newton's second law inextricably links mass and force. *See* ACCELERATION; MASS.

One may choose either the absolute or the gravitational approach in selecting a standard particle or object. In the so-called absolute systems of units, it is said that the standard object has a mass of one unit. Then the second law of Newton defines unit force as that force which gives unit acceleration to the unit mass. Any other mass may in principle be compared with the standard mass (m) by subjecting it to unit force and measuring the acceleration (**a**), with which it varies inversely. By suitable appeal to experiment, it is possible to conclude that masses are scalar quantities and that forces are vector quantities which may be superimposed or resolved by the rules of vector addition and resolution.

In the absolute scheme, then, the equation $\mathbf{F} = m\mathbf{a}$ is written for nonrelativistic mechanics; boldface type denotes vector quantities. This statement of the second law of Newton is in fact the definition of force. In the absolute system, mass is taken as a fundamental quantity and force is a derived unit of dimensions MLT^{-2} (M = mass, L = length, T = time).

The gravitational system of units uses the attraction of the Earth for the standard object as the standard force. Newton's second law still couples force and mass, but since force is here taken as the fundamental quantity, mass becomes the derived factor of proportionality between force and the acceleration it produces. In particular, the standard force (the Earth's attraction for the standard object) produces in free fall what one measures as the gravitational acceleration, a vector quantity proportional to the standard force (weight) for any object. It follows from the use of Newton's second law as a defining relation that the mass of that object is $m = w/g$, with g the magnitude of the gravitational acceleration and w the magnitude of the weight. The derived quantity mass has dimensions $FT^2 L^{-1}$. *See* FREE FALL.   [G.E.P.]

Forced oscillation   A response of a mechanical or electrical system in reaction to an external signal.

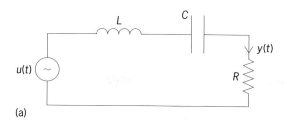

(a)

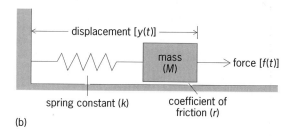

(b)

Examples of forced oscillation. (*a*) Electrical system, an *RLC* circuit. (*b*) Analogous mechanical system, a spring-mass-damper system.

A simple electrical *RLC* circuit (illus. *a*) consists of a resistor with resistance *R* (measured in ohms), an inductor with inductance *L* (measured in henrys), and a capacitor with capacitance *C* (measured in farads). The dynamics relating the input voltage, $u(t)$, to the current, $y(t)$, passing through the resistor are described by Eq. (1). Equation (1)

$$L\frac{dy}{dt} + Ry + \frac{1}{C}\int y\,dt = u(t) \tag{1}$$

states that the input voltage is equal to the sum of the voltage across the inductor, the voltage across the resistor, and the voltage across the capacitor, where the voltage across the inductor is the product of its inductance (L) and the rate of change of the current through the inductor; the voltage across the resistor is the product of its resistance (R) and the current passing through it; and the voltage across the capacitor is the integral over time of the current through the capacitor (that is, the charge on the capacitor plates) divided by the capacitance (C).

A fundamental property of differential equations states that the response of a differential equation to a periodic input can be decomposed as a sum of two responses. The first one, called the zero-input response or free oscillation, is due to initial energy stored in the circuit and decays eventually to zero. The second one, due to the voltage input $u(t)$, converges to a periodic signal with the same frequency as $u(t)$. The latter is referred to as the forced oscillation or the steady-state response. The decaying rate of the free oscillation depends on the time constant of the circuit which is determined by the values of R, L, and C and the structure of the circuit. *See* TIME CONSTANT.

Similarly, an analogous mechanical system, a simple spring-mass-damper system (illus. *b*), consists of a body with mass M, which is attached to a wall by a spring with spring constant k, and rests on a horizontal surface over which it moves with friction coefficient r. The dynamic equation that relates the force applied to the body, $f(t)$, to the body's displacement, $y(t)$, is given by Eq. (2). Equation (2) states that the force

$$M\frac{d^2 y(t)}{dt^2} + r\frac{dy(t)}{dt} + ky(t) = f(t) \tag{2}$$

applied to the body equals the sum of the three quantities: the product of the body's mass and its acceleration, the negative of the frictional force, and the negative of the force exerted by the spring. Here, the negative of the frictional force is the product of the coefficient of friction and the body's velocity, and the negative of the force exerted by the spring is the product of the spring constant and the body's displacement. Moreover, the body's velocity is the first derivative of its displacement with respect to time, and its acceleration is the second derivative of its displacement with respect to time.

Analogous to the *RLC* circuit case, application of a sinusoidal force $f(t)$ results eventually in a forced oscillation of the displacement $y(t)$ that is also a sinusoidal function. The magnitude and the phase of the displacement $y(t)$ depends on the complex mechanical impedance that is a function of the mass (M), the spring constant (k), and the friction coefficient (r). The exact evaluation is similar to the *RLC* circuit case. *See* MECHANICAL IMPEDANCE; OSCILLATION; VIBRATION.   [E.W.Ba.]

Foucault pendulum   A pendulum or swinging weight, supported by a long wire, by which J. B. L. Foucault demonstrated in 1851 the rotation of Earth on its axis. Foucault used a 62-lb (28-kg) iron ball suspended on about a 200-ft (60-m) wire in the Pantheon in Paris. The upper support of the wire restrains the wire only in the vertical direction. The bob is set swinging along a meridian in pure translation (no lateral or circular motion). In the Northern Hemisphere the plane of swing appears to turn clockwise; in the Southern Hemisphere it appears to turn counterclockwise, the rate being 15 degrees times the sine of the local latitude per sidereal hour. Thus, at the Equator the plane of swing is carried around by Earth and the pendulum shows no apparent rotation; at either pole the plane of swing remains fixed in space while Earth completes one rotation each sidereal day. *See* PENDULUM.   [F.H.R.]

Fractals   Geometrical objects that are self-similar under a change of scale, for example, magnification. The concept is helpful in many disciplines to allow order to be perceived in apparent disorder. For instance, in the case of a river and its tributaries, every tributary has its own tributaries so that it has the same structure organization as the entire river except that it covers a smaller area. The branching of trees and their roots as well as that of blood vessels, nerves, and bronchioles in the human body follows the same pattern. Other examples include a landscape with peaks and valleys of all sizes, a coastline with its multitude of inlets and peninsulas, the mass distribution within a galaxy, the distribution of galaxies in the universe, and the structure of vortices in a turbulent flow. The rise and fall of economic indices has a self-similar structure when plotted as a function of time.

The triadic Koch curve, shown in the illustration, is a good example of how a fractal may be constructed. The procedure begins with a straight segment. This segment is divided into three equal parts, and the (single) central piece is replaced by two similar pieces (illus. *a*). The same procedure is now applied to each of the four new segments (illus. *b*), and this is repeated an infinite number of times. The curve is self-similar, because a magnification by 3 of any portion will look the same as the original curve.

Fractals came into natural sciences when it was recognized that natural objects are random versions of mathematical fractals. They are self-similar in a statistical sense; that is, given a sufficiently large number of samples, a suitable magnification of a part of one sample can be matched closely with some member of the ensemble. Unlike the Koch curve which must be magnified by an integral power of 3 to

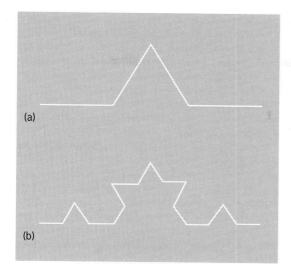

Koch curve, (*a***) first and
(***b***) second stages.**

achieve self-similarity, natural fractal objects are usually self-similar under arbitrary magnification.

Physicists have used the concept of fractals to study the properties of amorphous solids and rough interfaces and the dynamics of turbulence. It has also been found useful in physiology to analyze the heart rhythm and to model blood circulation, and in ecology to understand population dynamics. In computer graphics it has been shown that the vast amount of information contained in a natural scene can be compressed very effectively by identifying the basic set of fractals therein together with their rules of construction. When the fractals are reconstructed, a close approximation of the original scene is reproduced. *See* AMORPHOUS SOLID.   [S.H.L.]

Frame of reference   A base to which to refer physical events. A physical event occurs at a point in space and at an instant of time. Each reference frame must have an observer to record events, as well as a coordinate system for the purpose of assigning locations to each event. The latter is usually a three-dimensional space coordinate system and a set of standardized clocks to give the local time of each event. For a discussion of the geometrical properties of space-time coordinate systems *see* SPACE-TIME. *See also* RELATIVITY.

In the ordinary range of experience, where light signals, for all practical purposes, propagate instantaneously, the time of an event is quite distinct from its space coordinates, since a single clock suffices for all observers, regardless of their state of relative motion. The set of reference frames which have a common clock or time is called newtonian, since Isaac Newton regarded time as having invariable significance for all observers.

For discussion of other types of reference frames.   [B.G.]

Franck-Condon principle   The generalization that the transition from one energy state of a molecular system to another occurs so nearly instantaneously that the nuclei of the atoms involved can be considered as stationary during the process. The Franck-Condon principle is closely related to the Born-Oppenheimer approximation, in which the various motions (electronic, nuclear vibrations and rotations) are considered to be separable, and in which the electrons respond to the instantaneous vibrations of the system whereas the system responds only to the average position of the electrons. The principle, proposed by J. Franck in 1925 and developed quantum-mechanically by E. U. Condon in 1928, is important in discussing systems of more than one atom. It is therefore valuable in molecular spectroscopy and in the interpretation of the optical properties of liquids and solids. *See* MOLECULAR STRUCTURE AND SPECTRA. [J.H.S.; C.C.K.]

Free-electron theory of metals

Free-electron theory of metals   The treatment of a metal as containing a gas of electrons completely free to move within it. The theory was originally proposed in 1900 to describe and correlate the electrical and thermal properties of metals. Later, quantum mechanics became the basis for the theory of most of the general properties of simple metals such as sodium, with one free electron per atom, magnesium with two, and aluminum with three. Transition metals, such as iron, have partially filled electronic *d* states and are not treated by the free-electron model.

Three years after J. J. Thomson's 1897 discovery of the electron, P. Drude suggested that the transport properties of metals might be understood by assuming that their electrons are free and in thermal equilibrium with their atoms. This theory was made more quantitative by H. A. Lorentz. Assuming that the mean free path of electrons was limited by collisions, he was able to derive Ohm's law for the electrical conductivity and obtain the ratio of thermal to electrical conductivity in excellent agreement with experiment. This ratio, divided by the absolute temperature, is called the Wiedemann-Franz ratio and had been observed to be universal 50 years earlier. *See* CONDUCTION (ELECTRICITY); CONDUCTION (HEAT); KINETIC THEORY OF MATTER; MEAN FREE PATH; OHM'S LAW; THERMAL CONDUCTION IN SOLIDS; WIEDEMANN-FRANZ LAW.

The theory, however, had two major shortcomings. First, it predicted a large component of the specific heat of a metal, not present in insulators, which was not observed. Second, comparison of the theory with experiment indicated that the mean free path of the electrons became extremely large at low temperatures; the model offered no justification.

In 1928 A. Sommerfeld revised Lorentz's treatment by using quantum statistics, which removed the difficulty of the specific heat without losing the successful description of transport properties. The resulting theory remains the basis for the understanding of most transport properties of metals and semiconductors. At about the same time, W. V. Houston and F. Bloch solved the quantum-mechanical wave equation for electrons in a regular periodic structure, finding that they could indeed have arbitrarily large mean free paths if there were no defects in the periodicity, thereby putting the free-electron theory on a firm basis. *See* BAND THEORY OF SOLIDS; BLOCH THEOREM; FERMI-DIRAC STATISTICS; STATISTICAL MECHANICS.

Even in the context of a free-electron gas, there are strong Coulomb interactions between electrons which are frequently neglected in the free-electron theory of metals. This neglect was justified in the late 1950s by L. D. Landau, who asserted that, even with strong electron–electron interactions, there is a one-to-one correspondence between the excited states, called quasiparticle states, of the real system and the one-electron excitations from the ground state of the noninteracting electron gas. Thus, the formulations for free-electron theory still follow, but perhaps with modifications of parameters such as mass. Subsequent theory indicates that indeed these modification due to the electron-electron interaction are extremely small for the low-energy excitations present in thermal equilibrium, and so again the simplest theory succeeds for many properties, although substantial modifications are required for the higher-energy excitations caused by light. There are additional corrections, which are much larger than those from the electron-electron interaction, arising from the interaction between electrons and phonons, the quantum-mechanical term for lattice vibrations. In many metals these vibrations reduce the electron velocities by factors of as much as 2, increasing the electronic specific heat although they turn out not to modify the conductivity itself. *See* PHONON.

Another feature of the electron–phonon interaction is a resulting interaction among electrons, which is attractive and tends to cancel or exceed the repulsive electron-electron interaction. At low temperatures the net attraction binds electrons in pairs in a superconducting state. The theory of J. Bardeen, J. R. Schrieffer, and L. N. Cooper (the BCS theory of superconductivity), which first explained this phenomenon, is also a free-electron theory, but assumes that the free electrons have such a net attractive interaction. In contrast, it is generally believed that the high-temperature superconductors discovered in 1986 are very far from free-electron in character, and most workers do not believe that phonons are primarily responsible for the attractive interaction. *See* SOLID-STATE PHYSICS; SUPERCONDUCTIVITY.   [W.A.H.]

Free energy   A term in thermodynamics which in different treatments may designate either of two functions defined in terms of the internal energy E or enthalpy H, and the temperature-entropy product TS.

The function $(E - TS)$ is the Helmholtz free energy and is the function ordinarily meant by free energy in European references. The Gibbs free energy is the function $(H - TS)$. For the Lewis and Randall school of American chemical thermodynamics, this is the function meant by the free energy F. To avoid confusion with the symbol F as applied elsewhere to the Helmholtz free energy, the symbol G has also been used. Another development was the introduction of the name free enthalpy, with symbol G, for the Gibbs function. *See* WORK FUNCTION (THERMODYNAMICS).

For a closed system (no transfer of matter across its boundaries), the work which can be donein a reversible isothermal process is given by the series shown in Eq. (1).

$$W_{rev} = -\Delta A = -\Delta(E - TS) = -(\Delta E - T\Delta S) \qquad (1)$$

For these conditions, $T\Delta S$ represents the heat given up to the surroundings. Should the process be exothermal, $T\Delta S < 0$, then actual work done on the surroundings is less than the decrease in the internal energy of the system. The quantity $(\Delta E - T\Delta S)$ can then be thought of as a change in free energy, that is, as that part of the internal energy change which can be converted into work under the specified conditions. This then is the origin of the name free energy. Such an interpretation of thermodynamic quantities can be misleading, however; for the case in which $T\Delta S$ is positive, Eq. (1) shows that the decrease in "free" energy is greater than the decrease in internal energy.

For constant temperature and pressure in a reversible process the decrease in the Gibbs function G for the system again corresponds to a free-energy change in the above sense, since it is equal to the work which can be done by the closed system other than that associated with its change in volume ΔV under the given constant pressure P. The relations shown in Eq. (2) can be formed since $\Delta H = \Delta E + P\Delta V$.

$$\Delta G = -(\Delta H - T\Delta S) = W_{net} = W_{rev} - P\Delta V \tag{2}$$

Each of these free-energy functions is an extensive property of the state of the thermodynamic system. For a specified change in state, both ΔA and ΔG are independent of the path by which the change is accomplished. Only changes in these functions can be measured, not values for a single state.

The thermodynamic criteria for reversibility, irreversibility, and equilibrium for processes in closed systems at constant temperature and pressure are expressed naturally in terms of the function G. For any infinitesimal process at constant temperature and pressure, $-dG \geqq \delta w_{net}$. If δw_{net} is never negative, that is, if the surroundings do no net work on the system, then the change dG must be negative or zero. For a reversible differential process, $-dG > \delta w_{net}$; for an irreversible process, $-dG > \delta w_{net}$. The free energy G thus decreases to a minimum value characteristic of the equilibrium state at the given temperature and pressure. At equilibrium, $dG = 0$ for any differential process taking place, for example, an infinitesimal change in the degree of completion of a chemical reaction. A parallel role is played by the work function A for conditions of constant temperature and volume. Because temperature and pressure constitute more convenient working variables than temperature and volume, it is the Gibbs free energy which is the more commonly used in thermodynamics. *See* ENTROPY; THERMODYNAMIC PRINCIPLES.   [P.J.B.]

Free fall   The accelerated motion toward the center of the Earth of a body acted on by the Earth's gravitational attraction and by no other force. If a body falls freely from rest near the surface of the Earth, it gains a velocity of approximately 9.8 m/s every second. Thus, the acceleration of gravity g equals 9.8 m/s² or 32.16 ft/s². This acceleration is independent of the mass or nature of the falling body. For short distances of free fall, the value of g may be considered constant. After t seconds the velocity v_t of a body failing from rest near the Earth is given by Eq. (1).

$$v_t = gt \tag{1}$$

If a falling body has an initial constant velocity in any direction, it retains that velocity if no other forces are present. If other forces are present, they may change the observed direction and rate of fall of the body, but they do not change the Earth's gravitational pull; therefore a body may still be thought of as freely "failing" even though the resultant observed motion is upward.

For a body failing a very large distance from the Earth, the acceleration of gravity can no longer be considered constant. According to Newton's law of gravitation, the force between any two bodies varies inversely with the square of the distance between them; therefore with increasing distance between any body and the Earth, the acceleration of the body toward the Earth decreases rapidly. The final velocity v_f, attained when a body falls freely from an infinite distance to the surface of the Earth, is given by Eq. (2), where R is the radius of the Earth, which gives a numerical value of 11.3 km/s or

$$v_f = \sqrt{2gR} \tag{2}$$

7 mi/s. This is consequently the "escape velocity," the initial upward velocity for a rising body to completely overcome the Earth's attraction.

Because of the independent action of the forces involved, a ball thrown horizontally or a projectile fired horizontally with velocity v will be accelerated downward at the same rate as a body falling from rest, regardless of the horizontal motion.

At a sufficiently large horizontal velocity, a projectile would fall from the horizontal only at the same rate that the surface of the Earth curves away beneath it. The projectile would thus remain at the same elevation above the Earth and in effect become an Earth satellite. *See* BALLISTICS; GRAVITATION.

[R.D.Ru.]

Frequency (wave motion)   The number of times which sound pressure, electrical intensity, or other quantities specifying a wave vary from their equilibrium value through a complete cycle in unit time. The most common unit of frequency is the hertz (Hz), which is equal to 1 cycle per second. In one cycle there is a positive variation from equilibrium, a return to equilibrium, then a negative variation, and return to equilibrium. This relationship is often described in terms of the sine wave, and the frequency referred to is that of an equivalent sine-wave variation in the parameter under discussion. *See* FREQUENCY MEASUREMENT; SINE WAVE; WAVE MOTION.   [W.J.G.]

Frequency measurement   The determination of the number of cycles of a periodically varying quantity occurring in unit time. Many physical systems demonstrate cyclic behavior; that is, one or more of their properties vary in a characteristic fashion before returning to the initial value and then repeating the cycle. Examples are the angular positions of the planets and satellites in the solar system, the pressure in a cylinder in a reciprocating engine, and the heights and fields associated with surface, acoustic, and electromagnetic waves. The duration of a single cycle, the period, may vary widely, from 10^{-27} s for the electromagnetic field associated with a cosmic gamma ray to 10^8 years for the rotation of a galaxy in space. The frequency, which is the inverse of the period, is the number of cycles, including fractions, occurring in unit time. The unit of frequency is the hertz (Hz), named after Heinrich Hertz, who investigated the nature of electromagnetic radiation. Measurement of the characteristic frequencies of a system, and their variation with time or under changing conditions, yields valuable information on its properties and behavior. Together with temperature and voltage, frequency ranks as one of the quantities most often measured in modern science and technology. *See* CERENKOV RADIATION; ELECTROMAGNETIC RADIATION; WAVE MOTION.

The measurement of an unknown frequency requires a standard producing a fixed, stable, and known frequency, and a system or technique for the comparison of the unknown frequency with the standard. In the past, a wide variety of analog techniques and material standards have been employed. An example is the use of a tuning fork to adjust a musical instrument, usually a piano. Analog frequency measurement techniques possessed two major disadvantages: The frequency of the standards depended upon the material properties and dimensions of critical components, which meant that they were prone to drift and affected by variations in the ambient temperature. In addition, optimum accuracy was achieved only when the unknown and standard frequencies were close or harmonically related.

Developments in electron-tube and, later, solid-state electronics improved matters. These included the quartz crystal oscillator, in which a thin slice of crystalline quartz acts as the resonant element in an electronic feedback circuit. As a result of the sharpness of the resonance and the stability of the properties of the quartz, this device provides a stable frequency in the range from 10 kHz to 100 MHz and remains the most common secondary frequency standard in use. In addition, a range of circuits were developed to generate more complex harmonic and subharmonic frequencies from a standard source. This led ultimately to the frequency synthesizer which, with an array of phase-locked loops, could be set to produce one of a very wide range of output frequencies. In use, however, it was still necessary to measure the beat or heterodyne frequency from the unknown frequency. *See* PIEZOELECTRICITY.

Fast, inexpensive solid-state digital circuits have replaced analog frequency measurement techniques and many of their associated standards. The underlying principle of the digital technique is simple: the electrical signal from the sensor or transducer observing the physical system under test generally contains, from Fourier analysis, the fundamental frequency and components at integral harmonics of this frequency. It is filtered to select the fundamental and converted into a rectangular waveform, representing transitions between the binary logic levels 0 and 1. A frequency measurement then consists of counting the number of positive- or negative-going transitions between the two levels in a known time.

In parallel with the production of counters capable of operating at frequencies up to around 1 GHz, frequency standards based upon selected atomic transitions rather than the properties of bulk materials have been developed. These have the advantage that the frequency produced from a particular transition is in principle universal; that is, it is largely independent of the design of the standard and the materials used in its construction, and of changes in the ambient conditions. The combination of high-speed digital counters and of very stable atomic reference sources allows a wide range of frequencies to be determined simply, inexpensively, and very accurately.

As a result, much work has been carried out on the definition and measurement of other physical quantities in terms of frequency. Clearly, time and frequency are closely related; not only are the measurement, calibration, and dissemination techniques largely interchangeable, but any frequency standard may be converted into a standard of time, that is, a clock, by adding an appropriately designed counter. The unit of time, the second, is itself defined as the duration of 9 192 631 770 cycles of the electromagnetic radiation corresponding to the transition between the two hyperfine levels of the ground state of the cesium-133 atom. The primary standards of voltage and resistance are now also realized in terms of frequency using, respectively, the superconducting Josephson effect and the von Klitzing (quantum Hall) effect. *See* ELECTRICAL UNITS AND STANDARDS; RESISTANCE MEASUREMENT; TIME; VOLTAGE MEASUREMENT.

To calibrate the internal quartz oscillators in frequency counters, and to enable frequency measurements to be made at the highest accuracies, up to and occasionally beyond 10^{-12}, standards laboratories require a selection of very stable frequency standards. The four types in common use are the temperature-stabilized or ovened quartz crystal oscillator, the rubidium gas cell, the cesium atomic beam standard, and the hydrogen maser. Their performance depends essentially upon the quality factor Q—the ratio of the resonant or transition frequency f_T to its half-bandwidth—and the sensitivity of f_T to changes in the properties of materials or in the ambient and operating conditions.

Quartz oscillators are employed in most of the atomic standards to reduce the short-term noise and to provide a convenient output frequency (usually 10 MHz). In these, f_T is set by atomic transitions whose properties are in principle fixed and universal. In practice, small interactions with the containment system and the operating conditions mean that this ideal is not completely realized. In the rubidium gas cell, the transition is perturbed by collisions with other buffer gas atoms whose temperature and composition may change in time; in the hydrogen maser, collisions of the hydrogen atoms with the inert coating inside the containing bulb produce the so-called wall shift, which depends upon the condition of the coating. Atoms in the cesium beam standard are very well isolated from each other and the container, and this is reflected in the low drift rates and temperature coefficients observed. *See* ATOMIC CLOCK.   [P.B.Co.]

Fringe (optics)   One of the light or dark bands produced by interference or diffraction of light. Distances between fringes are usually very small, because of the short wavelength of light. Fringes are clearer and more numerous when produced with light of a single color.

Diffraction fringes are formed when light from a point source, or from a narrow slit, passes by an opaque object of any shape. Interference fringes are obtained by bringing together two or more beams of light that have originated from a common source. This is usually accomplished by means of an apparatus especially designed for the purpose called an interferometer, although interference fringes may also be seen in nature. Examples are the colors in a soap film and in an oil film on water. *See* DIFFRACTION; INTERFERENCE OF WAVES; INTERFEROMETRY; RESOLVING POWER (OPTICS).   [F.A.J./W.W.W.]

Froude number   The dimensionless quantity $U(gL)^{-1/2}$, where U is a characteristic velocity of flow, g is the acceleration of gravity, and L is a characteristic length. The Froude number can be interpreted as the ratio of the inertial to gravity forces in the flow. This ratio may also be interpreted physically as the ratio between the mean flow velocity and the speed of an elementary gravity (surface or disturbance) wave traveling over the water surface.

When the Froude number is equal to one, the speed of the surface wave and that of the flow is the same. The flow is in the critical state. When the Froude number is less than one, the flow velocity is smaller than the speed of a disturbance wave traveling on the surface. Flow is considered to be subcritical (tranquil flow). Gravitational forces are dominant. The surface wave will propagate upstream and, therefore, flow profiles are calculated in the upstream direction. When the Froude number is greater than one, the flow is supercritical (rapid flow) and inertial forces are dominant. The surface wave will not propagate upstream, and flow profiles are calculated in the downstream direction.

The Froude number is useful in calculations of hydraulic jump, design of hydraulic structures, and ship design, where forces due to gravity and inertial forces are governing. In these cases, geometric similitude and the same value of the Froude number in model and prototype produce a good approximation to dynamic similitude. *See* DIMENSIONAL ANALYSIS; DIMENSIONLESS GROUPS.   [R.M.Wr.]

Fundamental constants   That group of physical constants which play a fundamental role in the basic theories of physics. These constants include the speed of light in vacuum, c; the magnitude of the charge on the electron, e, which is the fundamental unit of electric charge; the mass of the electron, m_e; Planck's constant, $\hbar$; and the fine-structure constant, α.

These five quantities typify the different origins of the fundamental constants: c and $\hbar$ are examples of quantities which appear naturally in the mathematical formulation of certain physical theories—Einstein's theories of relativity, and quantum theory, respectively; e and m_e are examples of quantities which characterize the elementary particles of which all matter is constituted; and α, the fundamental constant of quantum electrodynamics (QED), is an example of quantities which are combinations of other fundamental constants, but are actually constants in their own right since the same combination always appears together in the basic equations of physics.

Reliable numerical values for the fundamental physical constants are required for two main reasons. First, they are necessary if quantitative predictions from physical theory are to be obtained. Second, and even more important, the self-consistency of the basic theories of physics can be critically tested by a careful intercomparison of the

Recommended values (1986) of selected fundamental physical constants

Quality	Symbol	Numerical value[*]	Units[†]	Relative uncertainty, ppm
Speed of light in vacuum	c	299792458	m/s	(defined)
Constant of gravitation	G	6.67259(85)	10^{-11} m^3/(kg · s^2)	128
Planck constant	h	6.6260755(40)	10^{-34} J · s	0.60
Elementary charge	e	1.60217733(49)	10^{-19} C	0.30
Magnetic flux quantum, $h/(2e)$	Φ_0	2.06783461(61)	10^{-15} Wb	0.30
Fine-structure constant,	α	7.29735308(33)	10^{-3}	0.045
$\mu_0 c e^2/(2h)$	α^{-1}	137.0359895(61)		0.045
Electron mass	m_e	9.1093897(54)	10^{-31} kg	0.59
Proton mass	m_p	1.6726231(10)	10^{-27} kg	0.59
Neutron mass	m_n	1.6749286(10)	10^{-27} kg	0.59
Proton-electron mass ratio	m_p/m_e	1836.152701(37)		0.020
Rydberg constant, $m_e c \alpha^2/(2h)$	R_∞	1.0973731534(13)	m^{-1}	0.0012
Bohr radius, $\alpha/(4\pi R_\infty)$	a_0	5.29177249(24)	10^{-11} m	0.045
Compton wavelength of the electron, $h/(m_e c) = \alpha^2/(2R_\infty)$	λ_c	2.42631058(22)	10^{-12} m	0.089
Classical electron radius, $\mu_0 e^2/(4\pi m_e) = \alpha^3/(4\pi R_\infty)$	r_e	2.81794092(38)	10^{-15} m	0.13
Bohr magneton, $eh/(4\pi m_e)$	μ_B	9.2740154(31)	10^{-24} J/T	0.34
Electron magnetic moment in Bohr magnetons	μ_e/μ_B	1.001159652193(10)		10^{-5}
Nuclear magneton, $eh/(4\pi m_p)$	μ_N	5.0507866(17)	10^{-27} J/T	0.34
Proton magnetic moment in nuclear magnetons	μ_p/μ_N	2.792847386(63)		0.023
Boltzmann constant	k	1.380658(12)	10^{-23} J/K	8.5
Avogadro constant	N_A	6.0221367(36)	10^{-23}/mol	0.59
Fataday constant, $N_A e$	F	96485.309(29)	C/mol	0.30
Molar gas constant, $N_A k$	R	8.314510(70)	J/(mol-K)	8.4

[*]The digits in parentheses represent the one-standard-deviation uncertainties in the last digits of the quoted value.
[†]C = coulomb, J = joule, kg = kilogram, m = meter, mol = mole, s = second, T = tesla, Wb = weber.

numerical values of fundamental constants obtained from experiments in the different fields of physics. In general, the accuracy of fundamental constants determinations has continually improved over the years. Whereas in the past, 100 ppm (0.01%) and even 1000 ppm (0.1%) measurements were commonplace, today 0.01 ppm and better determinations are not unusual (ppm = parts per million).

Complex relationships can exist among groups of constants and conversion factors, and a particular constant may be determined either directly by measurement or indirectly by an appropriate combination of other directly measured constants. If the direct and indirect values have comparable accuracy, then both must be taken into account in order to arrive at a best value for that quantity. (By best value is meant that value believed to be closest to the true but unknown value.) Generally, each of the several routes which can be followed to a particular constant, both direct and indirect, will give a slightly different numerical value. Such a situation may be satisfactorily handled by the mathematical method known as least-squares. This technique provides a self-consistent procedure for calculating best "compromise" values of the constants from all of the available data. It automatically takes into account all possible routes and determines a single final value for each constant being calculated. It does this by weighting the different routes according to their relative uncertainties. The appropriate weights follow from the uncertainties assigned the individual measurements constituting the original set of data.

The 1986 least-squares adjustment, carried out under the auspices of the CODATA Task Group on Fundamental Constants, succeeded a CODATA adjustment in 1973 by E. R. Cohen and B. N. Taylor; CODATA, the Committee on Data for Science and Technology, is an interdisciplinary committee of the International Council of Scientific Unions. Recommended values are shown in the table.   [E.R.Co.]

Fundamental interactions   Fundamental forces that act between elementary particles, of which all matter is assumed to be composed.

At present, four fundamental interactions are distinguished. Their properties are summarized in the table.

Properties of the four fundamental interactions

Interaction	Range	Exchanged quanta
Gravitational	Long-range	Gravitons (g)
Electromagnetic	Long-range	Photons (γ)
Weak nuclear	Short-range $\approx 10^{-18}$ m	W^+, Z^0, W^-
Strong nuclear	Short-range $\approx 10^{-15}$ m	Gluons (G)

The gravitational interaction manifests itself as a long-range force of attraction between all elementary particles.

The electromagnetic interaction is responsible for the long-range force of repulsion of like, and attraction of unlike, electric charges. At comparable distances, the ratio of gravitational to electromagnetic interactions (as determined by the strength of respective forces between an electron and a proton) is approximately 4×10^{-37}. See COULOMB'S LAW; ELECTROSTATICS; GRAVITATION.

In modern quantum field theory, the electromagnetic interaction and the forces of attraction or repulsion between charged particles are pictured as arising secondarily as a consequence of the primary process of emission of one or more photons (particles or quanta of light) emitted by an accelerating electric charge (in accordance with Maxwell's equations) and the subsequent reabsorption of these quanta by a second charged particle. A similar picture may also be valid for the gravitational interaction.

The third fundamental interaction is the weak nuclear interaction, which is responsible for the decay of a neutron into a proton, an electron, and an antineutrino. Unlike electromagnetism and gravitation, weak interactions are short-range, the range of the force being of the order of 10^{-18} m.

An important question was finally answered in 1983: Is the weak interaction similar to electromagnetism in being mediated primarily by intermediate objects, the W^+ and W^- particles. The experimental answer, discovered at the CERN laboratory at Geneva, is that W^+ and W^- do exist, with a mass of 80.4 GeV/c^2. Each carries a spin of magnitude $\hbar$, where $\hbar$ is Planck's constant divided by 2π, just as does the photon (γ). The mass of these particles gives the range of the weak interaction. See INTERMEDIATE VECTOR BOSON.

Another crucial discovery in weak interaction physics was the neutral current phenomenon in 1973, that is, the discovery of new types of weak interactions where (as in the case of electromagnetism or gravity) the nature of the interacting particles is not changed during the interaction. The 1983 experiments at CERN also gave evidence for the existence of an intermediate particle Z^0, with a mass of 91.2 GeV/c^2, which is believed to mediate such reactions. See NEUTRAL CURRENTS; WEAK NUCLEAR INTERACTIONS.

The fourth fundamental interaction is the strong nuclear interaction between protons and neutrons, which resembles the weak nuclear interaction in being short-range, although the range is of the order of 10^{-15} m rather than 10^{-18} m. Within this range of distances the strong force overshadows all other forces between protons and neutrons.

Protons and neutrons are themselves believed to be made up of yet more fundamental entities, the up (u) and down (d) quarks ($P = uud$, $N = udd$). Each quark is assumed to be endowed with one of three color quantum numbers [conventionally labeled red (r), yellow (y), and blue (b)]. The strong nuclear force can be pictured as ultimately arising through an exchange of zero rest-mass color-carrying quanta of spin $\hbar$ called gluons (G) [analogous to photons in electromagnetism], which are exchanged between quarks (contained inside protons and neutrons). Since neutrinos, electrons, and muons (the so-called leptons) do not contain quarks, their interactions among themselves or with protons and neutrinos do not exhibit the strong nuclear force. *See* COLOR (QUANTUM MECHANICS); GLUONS; LEPTON; QUANTUM CHROMODYNAMICS; QUARKS; STRONG NUCLEAR INTERACTIONS.

Three of the four fundamental interactions (electromagnetic, weak nuclear, and strong nuclear) appear to be mediated by intermediate quanta (photons γ; W^+, Z^0, and W^-; and gluons G, respectively), each carrying spin of magnitude $\hbar$. This is characteristic of the gauge interactions, whose general theory was given by H. Weyl, C. N. Yang, R. Mills, and R. Shaw. This class of interactions is further characterized by the fact that the force between any two particles (produced by the mediation of an intermediate gauge particle) is universal in the sense that its strength is (essentially) proportional to the product of the intrinsic charges (electric, or weak-nuclear, or strong-color) carried by the two interacting particles concerned.

The fourth interaction (the gravitational) can also be considered as a gauge interaction, with the intrinsic charge in this case being the mass; the gravitational force between any two particles is proportional to the product of their masses. The only difference between gravitation and the other three interactions is that the gravitational gauge quantum (the graviton) carries spin $2\hbar$ rather than $\hbar$. It is an open question whether all fundamental interactions are gauge interactions.

Ever since the discovery and clear classification of these four interactions, particle physicists have attempted to unify these interactions as aspects of one basic interaction between all matter. A unification of weak and electromagnetic interactions, employing the gauge ideas was suggested by S. Glashow and by A. Salam and J. C. Ward in 1959. Following this initial attempt, Glashow (and independently Salam and Ward) noted that such a unification could be effected only if neutral current weak interactions were postulated to exist.

There were two major problems with this unified electroweak gauge theory considered as a fundamental theory. Yang and Mills had shown that masslessness of gauge quanta is the hallmark of unbroken gauge theories. The origin of the masses of the weak interaction quanta W^+, W^-, and Z^0 (or equivalently the short-range of weak interactions), as contrasted with the masslessness of the photon (or equivalently the long-range character of electromagnetism), therefore required explanation. The second problem concerned the possibility of reliably calculating higher-order quantum effects with the new unified electroweak theory, on the lines of similar calculations for the "renormalized" theory of electromagnetism elaborated by S. Tomonaga, Schwinger, Feynman, and F. J. Dyson around 1949. The first problem was solved by S. Weinberg and Salam and the second by G. t'Hooft and by B. W. Lee and J. Zinn-Justin. *See* RENORMALIZATION.

Weinberg and Salam considered the possibility of the electroweak interaction being a "spontaneously broken" gauge theory. By introducing an additional self-interacting

Higgs-Englert-Brout-Kibble particle into the theory, they were able to show that the W^+, W^-, and Z^0 would acquire well-defined masses through the so-called Higgs mechanism. The predicted theoretical mass values of the W and Z particles are in good accord with the experimental values found by the CERN 1983 experiments.

The Weinberg-Salam electroweak theory contains an additional neutral particle (the Higgs) but does not predict its mass. A search for this particle will be undertaken when the large hadron collider (LHC) at CERN comes into commission. *See* ELECTROWEAK INTERACTION; HIGGS BOSON; PARTICLE ACCELERATOR; SYMMETRY BREAKING.

The gauge unification of weak and electromagnetic interactions, which started with the observation that the relevant mediating quanta (W^+, W^-, Z^0, and γ) possess intrinsic spin $\hbar$, can be carried further to include strong nuclear interactions as well, if these strong interactions are also mediated through quanta (gluons) carrying spin $\hbar$. The resulting theory, which appears to explain all known low-energy phenomena, is called the standard model. (It is a model based on three similarly constituted generations of quarks and leptons plus the mediating quanta W^+, W^-, Z^0, photons, and gluons plus the Higgs particle.) A complete gauge unification of all three forces (electromagnetic, weak nuclear, and strong nuclear) into a single electronuclear interaction seems plausible. Such a (so-called grand) unification necessarily means that the distinction between quarks on the one hand and neutrinos, electrons, and muons (leptons) on the other, must disappear at sufficiently high energies, with all interactions (weak, electromagnetic, and strong) clearly manifesting themselves then as facets of one universal gauge force. The fact that at low energies presently available, these interactions exhibit vastly different effective strengths is ascribed to differing renormalizations due to successive spontaneous symmetry breakings. A startling consequence of the eventual universality and the disappearance of distinction between quarks and leptons is the possibility of protons transforming into leptons and pions. Contrary to the older view, protons would therefore decay into leptons and pions and not live forever. *See* GRAND UNIFICATION THEORIES; PROTON; STANDARD MODEL.

Research in unification theories of fundamental interactions is now concerned with uniting the gauge theories of gravity and of the electronuclear interactions. The most promising approach appears to be that of superstring theories. Such theories appear to describe the only possible theory of gravity which is finite and suffers from no ultraviolet infinities. A closed string is a (one-dimensional) loop which may exist in a d-dimensional space-time (where d must equal 10 to completely eliminate all ultraviolet infinities). The quantum oscillations of the string correspond to particles of higher spins and higher masses. The theory has a unique built-in gauge symmetry. *See* SUPERSTRING THEORY.   [A.S.]

Fuzzy-structure acoustics   Large structures such as ships and airplanes can undergo a variety of complicated vibrations. Such structures typically consist of an outer body made of metal plating (for example, the hull of a ship) or perhaps a massive metallic frame (for example, the chassis of a truck), and a large variety of internal objects that are connected to either the plating or the frame. In designing such structures, it is highly desirable to have some method for predicting how they will vibrate under various conditions. The radiation of sound caused by these vibrations, either into the environment or into the empty portions of the structure, is also of interest because this sound is often either unwanted noise or a means of inferring information about the details of the structure or the excitation. Fuzzy-structure acoustics refers to a class of conceptual viewpoints in which precise, computationally intensive models of the overall structure are replaced by nonprecise analytical models, for which the initial information is said to be fuzzy.

Fuzzy-structure theories divide the overall structure into a master structure and one or more attached structures, the latter being referred to as the fuzzy substructures, the internal structures, or the internals. (An example of a master structure is the hull and major framework of a ship.) The master structure is presumed to be sufficiently well known at the outset that its vibrations or dynamical response could be predicted if the forces that were exerted on it were known. Some of the forces are exerted on it by the substructures at the points at which they are attached. Such forces can be very complicated; nevertheless, there is some hope that a satisfactory approximate prediction of the vibrations of the master structure itself can be achieved with a highly simplified model.

The fuzzy substructures can be regarded as structures that are not known precisely. Recently developed theories of fuzzy structures lead, after various plausible ideal-izations, to a formulation that requires only a single function, this being the modal mass per unit frequency bandwidth. The influence of fuzzy substructures attached to the master structure tends to resemble that of an added frequency-dependent mass attached to the master structure in parallel with a frequency-dependent dashpot connecting the master structure to a hypothetical rigid wall. The added mass is a frequency-weighted integral over the modal mass per unit natural frequency, the weight-ing being such that the natural modes whose natural frequencies are less than the driving frequency have a positive contribution, while those for which the natural fre-quencies are greater than the driving frequency have a negative contribution. The master structure can seem to be less massive than it actually is when the bulk of the substructure mass is associated with resonant frequencies less than the excitation frequency.

One implication of the newly emerging fuzzy-structure theories is that, insofar as there is concern with the vibrations of only the master structure, it is possible to drastically curtail the estimation or measurement of any parameters within the substructures that are associated with internal damping. *See* VIBRATION.   [A.D.P.]

G

Galilean transformations The family of mathematical transformations used in newtonian mechanics to relate the space and time variables of uniformly moving (inertial) reference systems. In the simple case of two similarly oriented cartesian reference frames, moving along their common (x, x') axis, the transformation equations can be put in the form

$$x' = x - vt \qquad y' = y \qquad z' = z \qquad t' = t$$

where x, y, z and x', y', z' are the space coordinates of a given particle, and v is the speed of one system relative to the other. *See* FRAME OF REFERENCE.   [E.L.Hi.]

Galvanomagnetic effects Electrical and thermal phenomena occurring when a current-carrying conductor or semiconductor is placed in a magnetic field. The galvanomagnetic effects are closely related to the thermomagnetic effects. *See* THERMOMAGNETIC EFFECTS.

Let the electric current density j be transverse to the magnetic field H_z, for example, along x. Then the following transverse-transverse effects are observed: (1) Hall effect, an electric field along y. (2) Ettingshausen effect, a temperature gradient along y. Also the following transverse-longitudinal effects are observed: (3) Transverse magnetoresistance, an electrical potential change along x. (4) Nernst effect, a temperature gradient along x. *See* HALL EFFECT; MAGNETORESISTANCE.

Let the electric current density j be along H. Then, the most important effect is longitudinal magnetoresistance, or an electrical potential change along H.   [E.A.; F.Ke.]

Galvanometer A device for indicating very small electric currents. Although the deflection of a galvanometer results from current in the moving coil, the voltage in a closed circuit producing this current is frequently the quantity of interest to the user. In this mode, galvanometers are used to detect a null or an unbalanced condition in a bridge or potentiometer circuit. Electronic instruments that employ amplifying circuits to achieve sensitivities approaching the nanovolt level are also used as balance detectors in bridge circuits. However, in applications where extreme sensitivity and high rejection of ac signals are required, a galvanometer in combination with a photocell amplifier is to be preferred.

Galvanometers may also be used ballistically to integrate a transient current, as from the discharge of a capacitor, or a transient voltage, as produced when a coil moves relative to a magnetic field. *See* CURRENT MEASUREMENT; VOLTAGE MEASUREMENT.

The d'Arsonval galvanometer is the most common type and is widely used. Its indicating system consists of a light coil of wire suspended from a copper or gold ribbon a few thousandths of an inch wide and less than 0.001 in. (0.025 mm) thick. This coil, free to rotate in the radial field between the shaped pole pieces of a permanent

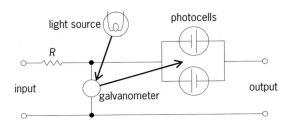

Simplified schematic of
photocell galvanometer
amplifier.

magnet, carries a small mirror which serves as an optical pointer and indicates the coil
position by reflecting a light beam onto a fixed scale. Current is conducted to and from
the coil by the suspension ribbons. The torque which deflects the indicating element
is produced by the reaction of the coil current with the magnetic field in which it is
suspended.

Voltage sensitivity is of importance in applications where the galvanometer serves as
the detector of unbalance in a bridge or potentiometer network. The energy of motion
in the indicating system must be dissipated and the system brought to rest in order
for the deflection to be evaluated. This process, common to all indicating systems in
which equilibrium must be achieved between a driving torque and a restoring torque,
is called damping. *See* DAMPING.

The sensitivity of modern galvanometers ranges up to 0.04 in. (1 mm) of deflection,
on a scale 40 in. (1 m) distant from the mirror, for a current of 0.00001 microampere. Such a galvanometer may have a coil resistance of 800 ohms and a critical
damping resistance of 100,000 Ω. The voltage response of this instrument amounts to
0.04 in./microvolt (1 mm/μV) at critical damping. A galvanometer designed for voltage sensitivity, has a coil resistance of 20 Ω, a critical damping resistance of 30 Ω, and
a response of 0.04 in. (1 mm) for 0.05 μV in the critically damped circuit. It will be
seen from these examples that a large response to current is associated with large coil
resistance and high critical damping resistance, whereas voltage response is associated
with low coil resistance and low critical damping resistance.

The limitations reached in improving a galvanometer's resolution by increasing its
scale length or using multiple reflections can be overcome by operating the galvanometer in its equilibrium position by using a negative-feedback network. In this system the
beam of light reflected from the galvanometer mirror is directed onto a pair of photocells connected in series opposition (see illustration). As the galvanometer coil rotates,
the light received on one photocell increases while the other decreases, and an amplified output current is produced whose sign and magnitude depend on the direction
and amount of the coil deflection, respectively. The photocell output current is monitored by an external indicator and is approximately 1800 times the galvanometer
input current. This photocell output also provides a negative-feedback current to servo
the galvanometer at its equilibrium position, so that the linearity and input resistance
of the system are increased. In such an arrangement it is frequently possible to approach rather closely the theoretical limit of resolution of angular motion imposed by
the brownian motion of the coil, or by the Johnson noise of the circuit connected to
the galvanometer. In a low-resistance galvanometer this theoretical resolution may be
about 0.001 μV. *See* BROWNIAN MOVEMENT; KINETIC THEORY OF MATTER.

A galvanometer is an extremely efficient low-pass filter, and it is still the most sensitive
low-level detector at input impedances of 1 kΩ or less. At higher input impedances,
chopper-stabilized amplifiers, called nanovoltmeters, exhibit better resolution, noise,
and drift characteristics than galvanometers. These instruments modulate the input dc
signal by using a chopper (mechanical contact, transistor, or photocell switching), then

amplify the signal by traditional ac techniques to achieve sensitivities of 10 nV or less. State-of-the-art digital voltmeters are now available with nanovolt resolution and are useful as low-level detectors. *See* AMMETER; ELECTRICAL MEASUREMENTS; ELECTROMETER; VOLTMETER.

[F.K.H.; R.F.Dz.]

Gamma-ray detectors Instruments that register the presence of gamma (γ) radiation. Such detectors convert some or all of the energy of gamma radiation into an electrical signal. Most instruments are capable of detecting individual gamma-ray photons (or quanta), each of which produces a short (0.1–5-microsecond) current pulse at the detector output. The output pulses may be made visible on an oscilloscope, made audible through a speaker (such as the familiar Geiger counter), or be electronically processed further, depending on the application. *See* GAMMA RAYS; OSCILLOSCOPE.

In common with most radiation detectors, gamma-ray detectors respond not to the radiation but to its secondary effects, in this case energetic electrons. Photons have neither mass nor charge and pass through matter rather easily. In so doing, they lose energy by (1) elastic scattering of electrons (Compton effect), (2) electron-positron ($\beta^+\beta^-$) pair production, and (3) at lower energies by photoabsorption. In these processes the energy of the photon is converted to the kinetic energy of the few electrons with which it interacts. Since electrons are much less penetrating than gamma-ray photons, their energy is largely trapped within the detector, where their ionizing effect creates a response convertible to an electrical output. In a gas-ionization device, such as a Geiger counter, this occurs by the production of ion-electron pairs and in a solid-state device, such as a germanium detector, by production of electron-hole pairs. In a scintillation device, for example, a sodium iodide (NaI) detector, the response is caused by the emission of optical photons from atoms excited by the passage of energetic electrons. *See* COMPTON EFFECT; ELECTRON-POSITRON PAIR PRODUCTION; IONIZATION CHAMBER; SCINTILLATION COUNTER.

In accurate instruments the magnitude of the current pulse created by a single gamma-ray photon is closely proportional to the energy within the detector volume. However, gamma radiation is so penetrating that any particular event may involve only partial absorption of the photon. For example, a single Compton scattering may be followed by the escape of the scattered photon (now reduced in energy) from the detector, leaving behind only the energy of the scattered electron.

Gamma-ray detectors range from hand-held devices capable of giving some indication of the intensity of a radiation field to devices that accurately measure the energy and event time of individual photons reaching detectors assembled into a single complex instrument. These diverse detectors are widely used in industry, medicine, and research.

[D.Wa.]

Gamma rays Electromagnetic radiation emitted from excited atomic nuclei as an integral part of the process whereby the nucleus rearranges itself into a state of lower excitation (that is, energy content). *See* NUCLEAR STRUCTURE; RADIOACTIVITY.

The gamma ray is an electromagnetic radiation pulse—a photon—of very short wavelength. The electric (E) and magnetic (H) fields associated with the individual radiations oscillate in planes mutually perpendicular to each other and also the direction of propagation with a frequency ν which characterizes the energy of the radiation. The E and H fields exhibit various specified phase-and-amplitude relations, which define the character of the radiation as either electric (EL) or magnetic (ML). The second term in the designation indicates the order of the radiation as 2^L-pole, where the orders are monopole (2^0), dipole (2^1), quadrupole (2^2), and so on. The most common radiations are dipole and quadrupole. Gamma rays range in energy from a few kiloelectronvolts to 100 MeV, although most radiations are in the range 50–6000 keV. As such, they lie at the very upper high-frequency end of the family of electromagnetic radiations, which include also radio waves, light rays, and x-rays. *See* ELECTROMAGNETIC RADIATION; MULTIPOLE RADIATION; PHOTON.

The dual nature of gamma rays is well understood in terms of the wavelike and particlelike behavior of the radiations. For a gamma ray of intrinsic frequency v, the wavelength is $\lambda = c/v$, where c is the velocity of light; energy is $E = hv$, where h is Planck's constant. The photon has no rest mass or electric charge but, following the concept of mass-energy equivalence set forth by Einstein, has associated with it a momentum given by $p = hv/c = E/c$. *See* LIGHT; QUANTUM MECHANICS; RELATIVITY.

Various nuclear species exhibit distinctly different nuclear configurations; the excited states, and thus the γ-rays which they produce, are also different. Precise measurements of the γ-ray energies resulting from nuclear decays may therefore be used to identify the γ-emitting nucleus. This has ramifications for nuclear research and also for a wide variety of more practical applications. One of the most useful studies of the nucleus involves the bombardment of target nuclei by energetic nuclear projectiles to form final nuclei in various excited states. Measurements of the decay γ-rays are routinely used to identify the various final nuclei according to their characteristic γ-rays.

In practical applications, the presence of γ-rays is used to detect the location or presence of radioactive atoms which have been deliberately introduced into the sample. In irradiation studies, for example, the sample is activated by placing it in the neutron flux from a reactor. The resultant γ-rays are identified according to isotope, and thus the composition of the original sample can be inferred. Such studies have been used to identify trace elements found as impurities in industrial production, or in ecological studies of the environment, such as minute quantities of tin or arsenic in plant and animal tissue.

In tracer studies, a small quantity of radioactive atoms is introduced into fluid systems (such as the human blood stream), and the flow rate and diffusion can be mapped out by following the radioactivity. Local concentrations, as in tumors, can also be determined.

For the three types of interaction with matter which together are responsible for the observable absorption of γ-rays, namely, Compton scattering, the photoelectric effect, and pair production, *see* COMPTON EFFECT; ELECTRON-POSITRON PAIR PRODUCTION; PHOTOEMISSION. [J.W.Ol.]

Gas constant   The universal constant R that appears in the ideal gas law, Eq. (1),

$$PV = nRT \tag{1}$$

where P is the pressure, V the volume, n the amount of substance, and T the thermodynamic (absolute) temperature. The gas constant is universal in that it applies to all gases, providing they are behaving ideally (in the limit of zero pressure). The gas constant is related to the more fundamental Boltzmann constant, k, by Eq. (2), where N_A

$$R = N_A k \tag{2}$$

is the Avogadro constant (the number of entities per mole). The best modern value in SI units is $R = 8.314\ 472\ (15)\ \text{J/K} \cdot \text{mol}$, where the number in parentheses represents the uncertainty in the last two digits. *See* BOLTZMANN CONSTANT.

According to the equipartition principle, at a temperature T, the average molar energy of each quadratic term in the expression for the energy is $(1/2)RT$; as a consequence, the translational contribution to the molar heat capacity of a gas at constant volume is $(3/2)R$; the rotational contribution of a linear molecule is R. *See* KINETIC THEORY OF MATTER.

Largely because R is related to the Boltzmann constant, it appears in a wide variety of contexts, including properties unrelated to gases. Thus, it occurs in Boltzmann's formula for the molar entropy of any substance, Eq. (3), where W is the number of arrangements

$$S = R \ln W \tag{3}$$

of the system that are consistent with the same energy; and in the Nernst equation for the potential of an electrochemical cell, Eq. (4), where $E°$ is a standard potential, F

$$E = E° - (RT/nF) \ln Q \tag{4}$$

is the Faraday constant, and Q is a function of the composition of the cell. The gas constant also appears in the Boltzmann distribution for the population of energy levels when the energy of a level is expressed as a molar quantity. *See* BOLTZMANN STATISTICS; ENTROPY.

[P.W.A.]

Gas discharge   A system made up of a gas, electrodes, and an enclosing wall in which an electric current is carried by charged particles in response to an electric field, the gradient of the electric potential, or the voltage between two surfaces. The gas discharge is manifested in a variety of modes (including Townsend, glow, arc, and corona discharges) depending on parameters such as the gas composition and density, the external circuit or source of the voltage, electrode geometry, and electrode material. *See* ELECTRICAL BREAKDOWN.

Gas discharges are useful both as tools to study the physics existing under various conditions and in technological applications such as in the lighting industry and in electrically excited gas lasers. New applications in gas insulation, in high-power electrical switching, and in materials reclamation and processing will assure a continuing effort to better understand all aspects of gas discharges. *See* LASER.

Electrons, rather than ions, are the main current carriers in gas discharges because their mass is smaller and their mobility is correspondingly much higher than that of ions. Electrons are produced by ionization of the gas itself, or they originate at the electrodes present in the system. Gas ionization can be accomplished in several ways, including electron impact ionization, photoionization, and associative ionization. Bombardment by photons, energetic ions or electrons, and excited neutral particles can cause secondary emission from the electrodes. A high-energy-per-unit electrode surface area can induce thermionic or field emission of electrons. Each of these means of producing electrons leads to a different response of the gas discharge as a circuit element. *See* ELECTRICAL CONDUCTION IN GASES; ELECTRON EMISSION; FIELD EMISSION; IONIZATION; PHOTOEMISSION; SECONDARY EMISSION; THERMIONIC EMISSION.

[L.C.P.]

Gas dynamics   The study of gases in motion. In general, matter exists in any of three states: solid, liquid, or gas. Liquids are incompressible under normal conditions; water is a typical example. In contrast, gases are compressible fluids; that is, their density varies depending on the pressure and temperature. The air surrounding a high-speed aircraft is an example.

Gas dynamics can be treated in a variety of ways. One such way deals with gases as a continuum. The structure of gases on the particle level is called rarefied gas dynamics. *See* AEROTHERMODYNAMICS; COMPRESSIBLE FLOW; FLUID FLOW.

Gases in motion are subject to certain fundamental laws. These are the laws of the conservation of mass, momentum, and energy. In the case of the dynamics of incompressible fluids, it is usually sufficient to satisfy only the laws of conservation of mass and momentum. This distinction constitutes the fundamental difference between high-speed gas dynamics and hydrodynamics. If irreversibilities are involved, a fourth equation called the entropy balance equation may be considered. Whereas mass, momentum, and energy are conserved, the entropy is not. Real problems are irreversible; that is, losses such as friction are involved. However, as a first approximation such effects are generally not considered. See CONSERVATION LAWS (PHYSICS); CONSERVATION OF ENERGY; CONSERVATION OF MASS; CONSERVATION OF MOMENTUM.

The mass, momentum, and energy equations are higher-order, nonlinear equations that have no general solution, and only a limited number of special cases can be solved. Another approach is to resort to numerical solutions using high-speed digital computers. While this approach has proven to be very useful, it limits the degree to which flow phenomena can be conceptualized. Accordingly, it is frequently permissible to write the equations in one-dimensional form. By one-dimensional flow is meant that the properties of gas such as its velocity and density are assumed to be constant in the direction perpendicular to the direction of the gas flow. Generally, the one-dimensional approach gives excellent insights into understanding the physical behavior of the flow. It is also very useful in setting up the computer algorithm for numerical solutions. See COMPUTATIONAL FLUID DYNAMICS.

One other matter must be considered, namely whether the flow is steady or unsteady. In steady flow, the flow characteristics do not vary with time, whereas unsteady flow implies that the flow assumes different configurations over time. Thus, unsteady flow is broader in scope. In this case the continuity equations for conservation of mass may be written as Eq. (1). In this equation, the first term defines the mass-flow changes with

$$\frac{\partial(\rho V)}{\partial x} + \frac{\partial \rho}{\partial t} = 0 \tag{1}$$

respect to the space coordinates, whereas the second term indicates the changes with time. Here, ∂ is the partial differential operator; x denotes the space coordinate, in this case the direction of flow; ρ is the gas density; V is the gas velocity; and t is the time.

If the flow is steady, there is no time-dependent term, and hence the continuity equation can be written in integrated form as Eq. (2), where A denotes the area in the direction perpendicular to the flow direction.

$$\rho V A = \text{constant} \tag{2}$$

The momentum equation is the mathematical representation of the law of conservation of momentum. It is a statement of the forces acting on the gas. Different types of forces must be recognized. Body forces, such as gravitation and electromagnetic forces, act at a distance. The so-called surface forces may assume different forms, such as normal stresses and viscosity. The simple form of the momentum equation is Eq. (3), which, in spite of its simplicity, is very powerful. Called Bernoulli's theorem,

$$\frac{p}{\rho} + \frac{V^2}{2} = \text{constant} \tag{3}$$

this equation makes a crucial statement that when the velocity increases, the pressure p decreases. See BERNOULLI'S THEOREM.

The energy equation expresses the first law of thermodynamics and accounts for the changes in energy as the gas moves about its path. It can also take into consideration energy exchanges between the gas and its environment, such as radiation. Its simplest form is Eq. (4), where c_p denotes the specific heat at constant pressure.

$$c_p \frac{p}{\rho} + \frac{V^2}{2} = \text{constant} \qquad (4)$$

The speed of sound or the acoustic velocity is a very important term in gas dynamics because it serves as a criterion to identify flow regimes. Being able to do so is crucial because the designer must know the conditions that the gas will generate or, conversely, experience. In prescribing flow regimes, the flow velocity of the gas is compared with the acoustic velocity. This ratio, called the Mach number (M), is defined by Eq. (5).

$$M = \frac{V}{a} \qquad (5)$$

Using the Mach number the following flow regimes are described:

$$
\begin{array}{ll}
M < 1 & \text{subsonic flow} \\
M = 1 & \text{sonic flow} \\
0.9 < M < 1.1 & \text{transonic flow} \\
M > 1 & \text{supersonic flow} \\
M > 5 & \text{hypersonic flow}
\end{array}
$$

High-speed aircraft are categorized by the Mach number. *See* MACH NUMBER; SOUND.

Flows can be classified as internal flow and external flow. Internal flow refers to the cases where the gas is constrained by a duct of some sort. Characteristically external flow is flow over an airplane or missile. Internal flows are conveniently characterized by (1) the shape of the duct and its variation, (2) the heat transfer through the walls of the duct and internal heat sources, and (3) frictional effects. By varying one of these characteristics at a time, the essential features of internal flow can be discussed most simply.

Boundary layers and wakes are the centers of interest in external flows. Here the effects of compressibility are substantially more difficult to analyze than in internal flows, if for no other reason than the inapplicability of a one-dimensional approach. *See* BOUNDARY-LAYER FLOW; WAKE FLOW. [J.Men.; A.B.C.]

Rarefied gas dynamics is that branch of gas dynamics dealing with the flow of gases under conditions where the molecular mean free path is not negligibly small compared to some characteristic dimension of the flow field. Rarefied flows occur when the gas density is extremely low, as in the cases of vacuum systems and high-altitude flight, but also when gases are at normal densities if the characteristic dimension is sufficiently small, as in the case of very small particles suspended in the atmosphere.

The dimensionless parameter which describes the degree of rarefaction existing in a flow is the Knudsen number, $Kn = \lambda/L$, defined as the ratio of the mean free path λ to some characteristic dimension L of the flow field. Depending on the situation, L might be chosen, for example, as the diameter of a duct in a vacuum system, the wavelength of a high-frequency sound wave, the diameter of a suspended submicrometer-size particle, the length of a high-altitude rocket, or the thickness of a boundary layer or a shock wave. The mean free path λ, which is the average distance traveled by a gas molecule between successive collisions with other molecules, is equal to the molecular mean speed, given by Eq. (6) [where R is the gas constant and T is the gas temperature],

$$\bar{C} = \sqrt{\frac{8}{\pi} RT} \qquad (6)$$

divided by the collision frequency v_c: thus, Eq. (7) is satisfied. However, it is often more

$$\lambda = \frac{\bar{C}}{v_c} \tag{7}$$

convenient in evaluating the Knudsen number to use the viscosity-based mean free path given by Eq. (8), where v is the kinematic viscosity. *See* VISCOSITY.

$$\lambda \simeq \frac{2v}{\bar{C}} \tag{8}$$

It is convenient to divide rarefied flows into three flow regimes, according to the range of values of the appropriate Knudsen numbers. The regime of highly rarefied flow, which obtains for Kn much greater than 1 (typically greater than 10), is called collisionless or free-molecule flow, while the regime of slight rarefaction, where Kn is much less than 1 (typically less than 0.1), is called slip flow. Flows at Knudsen numbers intermediate to these limiting values are termed transition flows. The phenomena and methods of analysis associated with the three regimes are in general quite dissimilar.

[L.T.]

Gas thermometry   A method of measuring temperatures with gas as the thermometric fluid. Gas thermometry is the primary source of information about a fundamental physical parameter, temperature, over the range from about 3 to 900 K (−454 to 1160°F).

In principle, gas thermometry consists of using the ideal gas law, Eq. (1), where P is

$$PV = nRT \tag{1}$$

the pressure, V the volume, n the number of moles of gas, R the molar gas constant, and T the thermodynamic temperature, to evaluate an unknown temperature by reference to the single defining temperature of the Kelvin thermodynamic temperature scale, namely the triple-point temperature of water. This reference temperature, achievable within about one part in 10^7 by standard laboratory practice, has the value 273.16 K (0.01°C or 32.018°F). Determination of the unknown temperature requires two sets of measurements of the pressure P and the volume V of n moles of an ideal gas. One set of measurements usually is performed while the gas is maintained at the water triple-point temperature; the second set is obtained with the gas at the unknown temperature. The unknown temperature can readily be evaluated by rewriting Eq. (1) as $n = (PV)/(RT)$, which has the same value for both sets of measurements. The desired result is given by Eq. (2), where the primed quantities refer to the unknown temperature. In Eq. (2)

$$T' = \frac{273.16(P'V')}{(PV)} \tag{2}$$

the precise values of the gas constant R and of n need not be considered.

Constant-volume gas thermometry is by far the most commonly used of the gas-thermometry methods. The name is somewhat misleading because no gas bulb truly exhibits constant volume over any substantial temperature range. Several steps are involved: inserting a fixed mass of a working gas into a rigid container or bulb; determining the pressure of the gas at the triple-point temperature of water; heating or cooling the container to a new temperature whose value is to be determined; and measuring the gas pressure at the new temperature.

A typical constant-volume gas thermometer (see illustration) includes a mercury manometer, used for the measurement of pressure; a gas-bulb system; and a

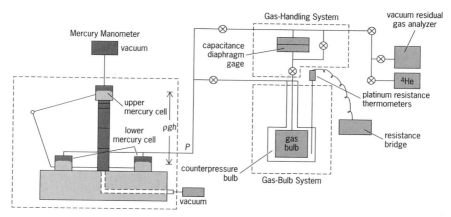

Typical constant-volume gas thermometer. (*After J. F. Schooley, Thermometry, CRC Press, 1986*)

gas-handling system. The part-per-million accuracy achievable in manometric pressure measurements is central to the thermometric accuracy of this instrument. The entire manometer is operated in a temperature-controlled environment. The distance between the surface of the mercury in an upper cell and mercury surfaces in two lower cells is measured by the use of wrung stacks of calibrated end gage blocks. Axial holes through the gage blocks permit detection of the quality of the wringing process by measurement of the internal vacuum of the stack. The pressure exerted by the column of mercury is given by the product $\rho g h$, where ρ is the density of the mercury, g is the acceleration due to gravity at the manometer, and h is the height of the gage-block stack.

The gas bulb is completely enclosed by a second bulb in which a so-called counterpressure of helium gas equal to the gas-bulb pressure is maintained at all times. The counterpressure gas minimizes pressure-induced changes in the gas-bulb volume and helps to reduce contamination of the working gas from impurities in the gas-bulb thermostat.

It is possible to determine the thermodynamic temperature of a gas bulb by repeatedly adding measured quantities of a working gas to it and measuring the resulting pressures. The repeated measurements at a single temperature are known as an isotherm. The virial equation (3) can be used to obtain both the unknown gas-bulb temperature and

$$PV = nRT(1 + BP + CP^2 + \cdots) \qquad (3)$$

the value of the second virial coefficient. *See* VIRIAL EQUATION.

Two variations on the technique can be used, the absolute isotherm and the relative isotherm. In absolute isothermal gas thermometry, measured quantities of working gas are introduced into a gas bulb at the unknown temperature from a known volume that is maintained at 273.16 K. In the relative isotherm method, the working gas is added stepwise to a gas bulb while it is maintained at another, more convenient reference temperature. *See* LOW-TEMPERATURE THERMOMETRY; PHYSICAL MEASUREMENT; TEMPERATURE MEASUREMENT.   [J.F.Sc.]

Gauge theory   The theoretical foundation of the four fundamental forces of nature, the electromagnetic, weak, strong, and gravitational interactions. Gauge symmetry lies at the heart of gauge theory. A gauge symmetry differs from an ordinary symmetry in two important respects:

1. Gauge symmetry is a local symmetry rather than a global symmetry. For a local symmetry, the element of the symmetry group (G) that acts on the fields of a theory at a space-time point $(\mathbf{x}, t)$ depends on the position $\mathbf{x}$ and time t, whereas for a global symmetry a fixed group element acts on fields at different space-time points.

2. A gauge transformation leaves a physical state invariant. Gauge symmetry reflects a redundancy in the variables used to describe a physical state. By contrast, a global symmetry acting on a physical state in general produces a new, distinct physical state. *See* FUNDAMENTAL INTERACTIONS; SPACE-TIME.

The simplest example of a gauge theory is electromagnetism. In classical electrodynamics, gauge invariance reflects the arbitrariness that exists in choosing the potentials $\mathbf{A}(\mathbf{x}, t)$ and $\phi(\mathbf{x}, t)$ to represent the electric and magnetic fields, $\mathbf{E}$ and $\mathbf{B}$, according to Eqs. (1), where c is the speed of light. If $\Lambda(\mathbf{x}, t)$ is an arbitrary scalar field, Eqs. (2)

$$\mathbf{E}(\mathbf{x}, t) = -\nabla\phi - \frac{1}{c}\frac{\partial \mathbf{A}}{\partial t} \tag{1}$$
$$\mathbf{B}(\mathbf{x}, t) = \nabla \times \mathbf{A}$$

$$\phi \to \phi' = \phi - \frac{1}{c}\frac{\partial \Lambda}{\partial t} \tag{2}$$
$$\mathbf{A} \to \mathbf{A}' = \mathbf{A} + \nabla\Lambda$$

define a gauge transformation. The potentials $\mathbf{A}'$ and ϕ' may equally well be used to represent the electromagnetic fields $\mathbf{E}$ and $\mathbf{B}$. *See* MAXWELL'S EQUATIONS; POTENTIALS; RELATIVISTIC ELECTRODYNAMICS.

In nonrelativistic quantum mechanics, gauge invariance is realized as follows. The Schrödinger equation for a particle with an electromagnetic charge q and mass m is Eq. (3), where $\hat{H}$ is the hamiltonian operator, $\hat{\mathbf{p}}$ is the momentum operator, Ψ is the

$$\hat{H}\Psi = \frac{1}{2m}\left(\hat{\mathbf{p}} - \frac{q}{c}\mathbf{A}\right)^2 \Psi + q\phi\Psi$$

$$= \frac{1}{2m}\left(-i\hbar\nabla - \frac{q}{c}\mathbf{A}\right)^2 \Psi + q\phi\Psi$$

$$= i\hbar\frac{\partial}{\partial t}\Psi \tag{3}$$

wave function of the particle, and $\hbar$ is Planck's constant divided by 2π. Equation (3) is invariant under the gauge transformation given by Eqs. (4). Electromagnetism is

$$\Psi \to \Psi' = \exp\left(\frac{iq\Lambda}{\hbar c}\right) \cdot \Psi$$

$$\phi \to \phi' = \phi - \frac{1}{c}\frac{\partial \Lambda}{\partial t} \tag{4}$$

$$\mathbf{A} \to \mathbf{A}' = \mathbf{A} + \nabla\Lambda$$

a U(1) gauge symmetry [where U(1) is the one-dimensional unitary group, which is represented by the complex numbers $e^{i\varphi}$ with $0 \leq \varphi < 2\pi$] because a gauge transformation rotates the phase of the wave function in a space-time–dependent manner and adjusts the potentials A and ϕ accordingly. *See* QUANTUM MECHANICS; SCHRÖDINGER'S WAVE EQUATION; UNITARY SYMMETRY.

The concept of a gauge theory may be generalized to larger, nonabelian Lie groups, such as $G = $ SU(2), SU(3), SU(5), or SO(10). In 1954 C. N. Yang and R. L. Mills suggested gauging the SU(2) isospin symmetry, thus developing the first nonabelian gauge theory, also known as Yang-Mills theory. In 1971 G. 't Hooft demonstrated the renor- malizability of nonabelian gauge theory. Nonabelian gauge theory is the foundation of the electroweak and strong interactions. In the electroweak theory, formulated by S. Weinberg and A. Salam in 1967, the gauge group $G = SU(2)_L \times U(1)_Y$ is spontaneously broken to $U(1)_Q$, the gauge group of ordinary electromagnetism, by the condensation of a Higgs field. There are four kinds of particles that mediate the gauge interactions, called gauge bosons: two massive charged weak gauge bosons, the W^+ and W^-, with a mass of 80.4 GeV; a neutral gauge boson, the Z, with a mass of 91.2 GeV; and finally a massless photon (γ). The $W^\pm$ and the Z are responsible for the charged and neutral weak interactions, respectively, and the photon is the gauge particle responsible for the electromagnetic interaction. In quantum chromodynamics (QCD) the symmetry group is $SU(3)_{color}$, which remains unbroken. There are eight gauge bosons called gluons. Quarks, which come in three colors, carry a color charge, and the quark field forms an SU(3) triplet. *See* ELECTROWEAK INTERACTION; SYMMETRY BREAKING.

One remarkable property of quantum chromodynamics is that color charge is anti- screened rather than screened, a feature arising from the nonabelian character of the gauge symmetry. This situation is opposite to that in QED. Consequently, as shorter distance scales are probed, the coupling constant α_{QCD} decreases, so that at very short distances QCD approaches a free (noninteracting) field theory. This feature, pointed out in 1973 by D. Gross and F. Wilczek and by H. D. Politzer, is known as asymptotic freedom. Although on large scales QCD is a strongly coupled theory, at very small scales, which may be probed by scattering at very high energies, the constituents of hadrons (quarks and gluons) behave almost as if they were free particles, originally called partons. Similarly, at larger distances scales α_{QCD} becomes larger, and pertur- bation theory breaks down. This increase in the coupling constant at large distances leads to a phenomenon known as confinement, which prevents colored objects (such as quarks or gluons) from being isolated. *See* GLUONS; QUANTUM CHROMODYNAMICS; QUARKS.

[M.Bu.]

Geometric phase   A unifying mathematical concept that describes the relation between the history of internal states of a system and the system's resulting orien- tation in space. Under various aspects, this concept occurs in geometry, astronomy, classical mechanics, and quantum theory. In geometry it is known as holonomy. In quantum theory it is known as Berry's phase, after M. Berry, who isolated the concept (which was already known in special cases) and explained its wide-ranging signi- ficance.

A system is envisioned whose possible states can be visualized as points in a suitable abstract space. At the same time, the system has some position or orientation in another space. A history of internal states can be represented by a curve in the first space; and the effect of this history on the disposition of the system, by a curve in the second space. The mapping between these two curves is described by the geometric phase. Especially interesting is the case when a closed curve (cycle) in the first space maps onto an open curve in the second, for then there is no net change in internal state, yet the disposition of the system with respect to the outside world is altered.

The power of the geometric phase ideas is that they make it possible, in complex dynamical problems, to find some simple universal regularities without having to solve the complete equations. Significant uses of these ideas include demonstrations of the fractional electric charge and quantum statistics of the quasiparticles in the quantum Hall effect, and of the occurrence of anomalies in quantum field theory. *See* ANYONS; HALL EFFECT; QUANTUM FIELD THEORY.

[F.Wil.]

Geometrical optics The geometry of light rays and their images, through optical systems. In the modern view of the wave nature of light, geometrical optics as a model is simply wrong. In spite of this geometrical optics is remarkably robust, remaining as a most practical tool in the solution of optical problems where at first glance it would seem to be totally inappropriate. The principal application of geometrical optics remains in the field of optical design.

Light is a form of energy which flows from a source to a receiver. It consists of particles (corpuscles) called photons. The speed with which the particles travel depends on the medium. In a vacuum this speed is 3×10^8 m $\cdot$ s^{-1} (1.86×10^5 mi $\cdot$ s^{-1}) for all colors. In a material medium, whether gas, liquid, or solid, light travels more slowly. Moreover, different colors travel at different rates. The ratio of the speed in a vacuum to the speed in the medium is called the refractive index of the medium. The variation in refractive index with color is called dispersion. *See* COLOR; DISPERSION (RADIATION); REFRACTION OF WAVES.

The paths that particles take in going from the source to the receiver are called rays. The product of the refractive index and the path length is called the optical path length along the ray. The optical path length is equal to the distance that the particle would have traveled in a vacuum in the same time interval.

A point source is an infinitesimal region of space which emits photons. An extended source is a dense array of point sources. Each point source emits photons along a family of rays associated with it. For each such family of rays there is also a family of surfaces, each of which is a surface of constant transit time from the source for all the particles, or alternatively, a surface of constant optical path length from the source. These surfaces are called geometrical wavefronts, because they are often good approximations to the wavefronts predicted by a wave theory.

The ray path which any particle takes as it propagates is determined by Fermat's principle, which states that the ray path between any two points in space is that path along which the optical path length is stationary (usually a minimum) among all neighboring paths. In a homogeneous medium (one with a constant refractive index) the ray paths are straight lines.

In a system that consists of a sequence of separately homogeneous media with different refractive indices and with smooth boundaries between them (such as a lens system), the ray paths are straight lines in each medium, but the directions of the ray paths will change in passing through a boundary surface. This change in direction is called refraction, and is governed by Snell's law, which states that the product of the refractive index and the sine of the angle between the normal to the surface and the ray is the same on both sides of a surface separating two media. The normal in question is at the point where the ray intersects the surface.

The primary area of application of geometrical optics is in the analysis and design of image-forming systems. An optical image-forming system consists of one or more optical elements (lenses or mirrors) which when directed at a luminous (light-emitting) object will produce a spatial distribution of the light emerging from it which more or less resembles the object. The latter is called an image. *See* OPTICAL IMAGE.

In order to judge the performance of the system, it is first necessary to have a clear idea of what constitutes ideal behavior. Departures from this ideal behavior are called aberrations, and the purpose of optical design is to produce a system in which the aberrations are small enough to be tolerable. *See* ABERRATION (OPTICS).

In an ideal optical system the rays from *every* point in the object space pass through the system so that they converge to or diverge from a corresponding point in the image space. This corresponding point is the image of the object point, and the two are said to be conjugate to each other (object and image functions are interchangeable).

The geometry of the object and image spaces must be connected by some mapping transformation. The one generally used to represent ideal behavior is the collinear transformation. If three object points lie on the same straight line, they are said to be collinear. If the corresponding three image points are also collinear, and if this relationship is true for all sets of three conjugate pairs of points, then the two spaces are connected by a collinear transformation. In this case, not only are points conjugate to points, but straight lines and planes are conjugate to corresponding straight lines and planes.

Another feature usually incorporated in the ideal behavior is the assumption that all refracting or reflecting surfaces in the system are figures of revolution about a common axis, and this axis of symmetry applies to the object-image mapping as well. Every object plane containing the axis, called a meridional plane, has a conjugate which is a meridional plane coinciding with the object plane. In addition, every object plane perpendicular to the axis must have a conjugate image plane which is also perpendicular to the axis, because of axial symmetry. In the discussion below, the terms object plane and image plane refer to planes perpendicular to the axis.

An object line parallel to the axis will have a conjugate line which either intersects the axis in image space or is parallel to it. The first case is called a focal system, and the second an afocal system.

The point of intersection of the image-space conjugate line of a focal system with the axis is called the rear focal point (see illustration). It is conjugate to an object point on axis at infinity. The image plane passing through the rear focal point is the rear focal plane, and it is conjugate to an object plane at infinity. Every other object plane has a conjugate located at a finite distance from the rear focal point, except for one which will

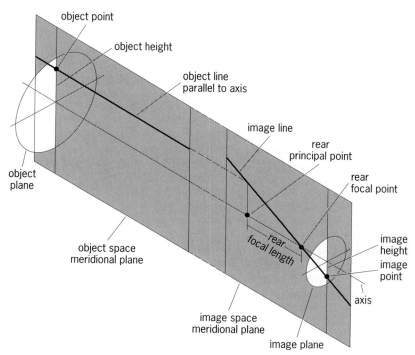

Focal system.

have its image at infinity. This object plane is the front focal plane, and its intersection with the axis is the front focal point.

Now take an arbitrary object plane and its conjugate image plane. Select a point off axis in the object plane and construct a line parallel to the axis passing through the off-axis point. The conjugate line in image space will intersect the axis at the rear focal point and the image plane at some off-axis point. The distance of the object point from the axis is called the object height, and the corresponding distance for the image point is the image height. The ratio of the image height to the object height is called the transverse magnification, and is positive or negative according to whether the image point is on the same or the opposite side of the axis relative to the object point.

Every pair of conjugate points has associated with it a unique transverse magnification. The conjugate pair which have a transverse magnification of +1 are called the (front and rear) principal planes. The intersections of the principal plane with the axis are called the principal points. The distance from the rear principal point to the rear focal point is called the rear focal length, and likewise for the front focal length. *See* FOCAL LENGTH.

The focal points and principal points are four of the six gaussian cardinal points. The remaining two are a conjugate pair, also on axis, called the nodal points. They are distinguished by the fact that any conjugate pair of lines passing through them make equal angles with the axis. The function of the cardinal points and their associated planes is to simplify the mapping of the object space into the image space.

In addition to the transverse magnification, the concept of longitudinal magnification is useful. If two planes are separated axially, their conjugate planes are also separated axially. The longitudinal magnification is defined as the ratio of the image plane separation to the object plane separation.

In the case of afocal systems, any line parallel to the axis in object space has a conjugate which is also parallel to the axis. The transverse and longitudinal magnifications are constant for the system. Cardinal points do not exist for afocal systems.

The most common use of an afocal system is as a telescope, where both the object and the image are at infinity. The angular magnification is defined as the ratio of the transverse magnification to the longitudinal magnification. The power of a telescope or a pair of binoculars is the magnitude of the angular magnification. *See* MAGNIFICATION; TELESCOPE.

Real optical systems cannot obey the laws of the collinear transformation, and departures from this ideal behavior are identified as aberrations. However, if the system is examined in a region restricted to the neighborhood of the axis, the so-called paraxial region, where angles and their sines are indistinguishable from their tangents, a behavior is found which is exactly congruent with the collinear transformation. Paraxial ray tracing can therefore be used to determine the ideal collinear properties of the system.

The above discussion does not take into account the fact that the sizes of the elements of any optical system are finite, and the light that can get through the system to form the image is limited. A circular aperture which limits the cone of rays from an axial object point that gets through the system and participates in image formation is called the aperture stop of the system. An observer who looks into the front of the system from the axial object point sees not the aperture stop itself (unless it is in front of the system), but the image of it formed by the elements preceding it. This image of the aperture stop is called the entrance pupil, and is situated in the object space on the system. The image of the aperture stop formed by the rear elements is the exit pupil, and is situated in the image space of the system.   [R.V.S.]

Ghost image (optics)   An undesired image appearing at the image plane of an optical system. Each surface of an optical system divides the incoming light into two parts: (1) the reflected light, which returns into the first medium, and (2) the refracted light. The reflected light is again divided into two parts when it in turn strikes another dividing surface. The light thus reflected twice forms an image which may be near the plane of the primary image. This may be a false image of the object or an out-of-focus image of a bright source of light in the field of the optical system. Thus a large number of undesired or ghost images may appear. *See* OPTICAL IMAGE; REFLECTION OF ELECTROMAGNETIC RADIATION; REFRACTION OF WAVES.

If the ghost images are far out of focus, they only diminish the contrast in the primary image, a condition known as flare. But if the ghost images are near the focal plane, they are very disturbing. This effect is especially noticeable if there is a bright light source in the field of the instrument, since the ghost image of the light source may have an even greater brightness than the image of the desired object. The coating of lenses with layers of fluorite and other materials has nearly eliminated ghost images from modern optical systems.

[M.J.H.]

Giant nuclear resonances   Elementary modes of oscillation of the whole nucleus, closely related to the normal modes of oscillation of coupled mechanical systems. Giant nuclear resonances occur systematically in most, if not all, nuclei, with oscillation energies typically in the range of 10–30 MeV. Among the best-known examples is the giant electric dipole (E1) resonance, in which all the protons and all the neutrons oscillate with opposite phase, producing a large time-varying electric dipole moment which acts as an effective antenna for radiating gamma rays. *See* GAMMA RAYS.

Giant resonances are usually classified in terms of three characteristic quantum numbers: L, S, and T, where L is the orbital angular momentum, S is the (intrinsic) spin, and T is the isospin carried by the resonance oscillation. The number L is also the multipole order, with possible values $L = 0$ (monopole), $L = 1$ (dipole), $L = 2$ (quadrupole), $L = 3$ (octupole), and so on. The spin quantum number S is either 0 or 1. The $S = 0$ resonances are often called electric, and the $S = 1$ ones magnetic (EL or ML, where L is the multipole order), stemming from the fact that these giant resonances have strong decay modes involving the emission of either electric (for EL resonances) or magnetic (for ML resonances) multipole photons of the same multipole order as the resonance. A giant resonance with $S = 0$ corresponds to a purely spatial oscillation of the nuclear mass (or charge density), while one with $S = 1$ corresponds to a spin oscillation. The isospin quantum number T, which is also either 0 or 1, determines the relative behavior of neutrons versus protons; in a $T = 0$ or isoscalar giant resonance, the neutrons and protons oscillate in phase, whereas in a $T = 1$ or isovector resonance the neutrons and protons oscillate with opposite phase. *See* MULTIPOLE RADIATION; NUCLEAR MOMENTS.

These resonances are called giant because of their great strength, 50–100% of the theoretical limit, concentrated in a compact energy region. The oscillation energy is characteristic of the type of giant resonance and is determined by the restoring force and the nuclear mass; the force is due to the nuclear attraction between nucleons, the most important part being the component of the same multipole order as the giant resonance.

The giant electric dipole (E1) resonance is the oldest and best known of the nuclear giant resonances. It is the dominant feature in reactions initiated by gamma rays. The absorption of a gamma ray induces the giant E1 oscillation, which breaks up, in this case, by emitting neutrons. This resonance is also the dominant feature in the reverse process, in which gamma rays are produced by proton and neutron bombardments of nuclei. The resonance is isovector ($L = 1$, $S = 0$, $T = 1$).

The isoscalar giant E0 (electric monopole; $L = 0$, $S = 0$, $T = 0$) resonance lies very close in energy to the giant E1 resonance, whereas the isoscalar giant E2 (electric quadrupole; $L = 2$, $S = 0$, $T = 0$) resonance lies somewhat lower. Both are strongly excited in forward-angle inelastic scattering of energetic alpha particles.

The isoscalar E0 resonance is called the breathing mode, as the whole nucleus undergoes a purely radial oscillation, alternately expanding and contracting. The isoscalar E0 resonance energy is important in determining the nuclear compressibility.

In ordinary nuclear beta decay, a neutron inside a nucleus is transformed into a proton, and an electron and an antineutrino are produced. In one of the simplest types of beta decay, called Gamow-Teller decay, the transformed neutron is otherwise undisturbed, except that its spin may be reversed. As a result, the nucleus usually gains a small amount of energy. If beta decay involved a higher energy transfer to the nucleus, it would drive the giant Gamow-Teller resonance, which is a pure spin oscillation where the neutron spin and the proton spin oscillate out of phase ($L = 0$, $S = 1$, $T = 1$). A giant Gamow-Teller resonance is a strong feature in the (p, n) reaction in which neutrons emerge at $0°$ from nuclei struck by energetic protons. This reaction substitutes a proton for a neutron in the nucleus via a spin-dependent interaction, in a manner analogous to beta decay but with a much larger energy transfer.

The properties of the giant Gamow-Teller resonance are important in certain problems in nuclear astrophysics.

Studies of the giant electric dipole resonance have been extended to highly excited hot nuclei. These studies provide unique information about the properties of such nuclei, in particular their shape. The shape sensitivity arises from the resonance splitting in a deformed nucleus. The size of the splitting gives the magnitude of the deformation, whereas the relative strength of the components determines the sense of the deformation: prolate (football-shaped) or oblate (doorknob-shaped).

Giant resonances play an important role in energetic nuclear reactions occurring in nature. Among the best examples are supernovae explosions. The rate of electron capture reactions, which cool the core of the massive star involved in the explosion and accelerate its gravitational collapse, depends on the properties of the giant Gamow-Teller resonance. The strength of the shock wave created by the collapse is directly related to the nuclear compressibility discussed above in the context of the giant isoscalar E0 resonance. Higher-energy neutrinos from the central region of the star travel outward and heat the nuclei in the mantle via inelastic scattering reactions which excite various giant resonances. Certain elements found in nature may have been produced primarily as giant resonance decay products in these reactions. *See* NEUTRINO; NUCLEAR REACTION; NUCLEAR SPECTRA; NUCLEAR STRUCTURE; RESONANCE (QUANTUM MECHANICS).

[K.A.Sn.]

Glass transition   The transition that occurs when a liquid is cooled to an amorphous or glassy solid. This can occur only if the cooling rate is fast enough to prevent crystallization which would otherwise occur if time had been sufficient for the sample to reach true equilibrium at each temperature. Since the crystal is invariably the thermodynamically stable low-temperature phase, the glass transition corresponds to a transition from a high-temperature liquid into a nonequilibrium meta-stable low-temperature solid. *See* AMORPHOUS SOLID; CRYSTAL; VISCOSITY.

For many organic and polymeric systems, the difficulty of molecular packing and the steric hindrances are sufficient to prevent crystallization, and glass formation in these systems is relatively easy. In other systems, for example, metallic systems, rapid quench rates on the order of 10^6 K/s (2×10^6 °F/s) may be necessary to avoid crystallization, suggesting that any system can be quenched from the liquid state to an amorphous glassy state assuming that the system can be cooled rapidly enough.

[G.S.G.]

Gluons The hypothetical force particles believed to bind quarks into "elementary" particles. Although theoretical models in which the strong interactions of quarks are mediated by gluons have been successful in predicting, interpreting, and explaining many phenomena in particle physics, free gluons remain undetected in experiments (as do free quarks). According to prevailing opinion, an individual gluon cannot be isolated.

According to quantum chromodynamics (QCD), the mediators of the strong interaction are eight massless vector bosons, which are named gluons because they make up the "glue" that binds quarks together. It is hoped that the infinite range of the forces mediated by the gluons may help to explain why free quarks have not been isolated. The gluons themselves carry color. Hence, strong interactions among gluons will also occur through the exchange of gluons. It is therefore believed that gluons, as well as quarks, may be permanently confined. According to this view, only colorless objects may exist in isolation. *See* ELEMENTARY PARTICLE; QUANTUM CHROMODYNAMICS; QUARKS.

[C.Q.]

Goldhaber triangle The phase space triangle, or Goldhaber triangle, corresponds to the kinematically allowed boundary for a high-energy reaction leading to four or more particles. In a high-energy reaction between two particles a and b yielding four particles 1, 2, 3, and 4 in the final state ($a + b \rightarrow 1 + 2 + 3 + 4$), it is convenient to consider the reaction in terms of the production of two intermediate-state quasiparticle composites x and y, which then decay into two particles each, as in expression (1).

$$a + b \rightarrow x \qquad + y$$
$$\quad \downarrow_{\rightarrow 1+2} \quad \downarrow_{\rightarrow 3+4} \tag{1}$$

Most high-energy interactions indeed proceed through such intermediate steps, in which, for specific values of the invariant masses $m_x = M_x^*$ and $m_y = M_y^*$, the quasiparticle composites may form resonances. However, the description in terms of the

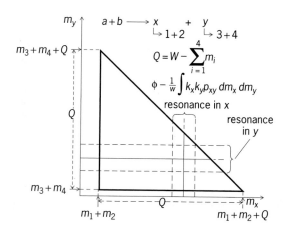

Definition of the kinematical boundary of the Goldhaber triangle for four particles.

composites x and y, with the invariant masses m_x and m_y as variables, is valid irrespective of whether or not these composites form resonances.

The kinematical limits in this representation are particularly simple, namely, they form a right-angle isosceles triangle. A Goldhaber triangle plot corresponds to plotting a point (m_x, m_y) for each event occurring in the above high-energy reaction. Because of the kinematical constraints, these points must all lie inside the triangle.

If one considers the general reaction given in Eq. (1), then the length of each of the two equal sides of the triangle is Q, defined in Eq. (2). Here W is the total energy in the

$$Q = W - \sum_{i=1}^{4} m_i \tag{2}$$

center of mass of particles a and b, and m_i, for $i = 1$ to 4, is the mass of the particles 1 to 4. Hence Q corresponds to the total kinetic energy available in the reaction.

In the triangle corresponding to the general reaction (see illustration), the vertical and horizontal bands indicate resonances at masses M_x^* and M_y^* with full width at half-maximum height or Γ_x and Γ_y, respectively. The bands shown of width 2Γ represent the regions usually chosen if the events corresponding to a given resonance are selected.   [G.G.]

Grain boundaries   The internal interfaces that separate neighboring misoriented single crystals in a polycrystalline solid. Most solids such as metals, ceramics, and semiconductors have a crystalline structure, which means that they are made of atoms which are arranged in a three-dimensional periodic manner within the constituent crystals. Most engineering materials are polycrystalline in nature in that they are made of many small single crystals which are misoriented with respect to each other and meet at internal interfaces called grain boundaries. These interfaces, which are frequently planar, have a two-dimensionally periodic atomic structure. A polycrystalline cube 1 cm on edge, with grains 0.0001 cm in diameter, would contain 10^{12} crystals with a grain boundary area of several square meters. Thus, grain boundaries play an important role in controlling the electrical and mechanical properties of the polycrystalline solid. It is believed that the properties are influenced by the detailed atomic structure of the grain boundaries, as well as by the defects that are present, such as dislocations and ledges. Grain boundaries generally have very different atomic configurations and local atomic densities than those of the perfect crystal, and so they act as sinks for impurity atoms which tend to segregate to interfaces. See CRYSTAL DEFECTS; CRYSTAL STRUCTURE.

Using electron microscopy and x-ray diffraction, it was determined that the grain boundary structure is frequently periodic in two dimensions. The geometry of a grain boundary is described by the rotation axis and angle, θ, that relate the orientations of the two crystals neighboring the interface, and the interface plane (or plane of contact) between the two crystals. Grain boundaries are typically divided into categories characterized by the magnitude of θ and the orientation of the rotation axis with respect to the interface plane. When θ is less than (arbitrarily) $15°$, the boundary is called small-angle, and when θ is greater than $15°$, the boundary is large-angle. See X-RAY DIFFRACTION.

Because of the large differences in atomic structure and density between the grain boundary region and the bulk solid, the properties of the boundary are also quite different from those of the bulk, and have a strong influence on the bulk properties of the polycrystalline solid. The mechanical behavior of a solid, that is, its response to an applied stress, often involves the movement of dislocations in the bulk, and the presence of boundaries impedes their motion since, in order for deformation to be transmitted from one crystal to its neighbor, the dislocations must transfer across the boundary and change direction. The detailed structure at the interface influences the ease or difficulty with which the dislocations accomplish this change in direction.

Since grain boundaries in engineering materials are not in a high-purity environment, the presence of impurities dissolved in the solid may have a strong influence on their behavior. The presence of one-half of a monolayer of impurity atoms, such as sulfur or antimony in iron, at the grain boundary can have a drastic effect on mechanical properties, making iron, which is ductile in the high-purity state, extremely brittle, so that it fractures along grain boundaries. The segregation of the impurity atoms to the boundaries has been well documented by the use of Auger electron spectroscopy, and studies have led to the suggestion that the change in properties may be related to a change in the dislocation structure of the grain boundary induced by the presence of these impurities.

Since modern electronic devices are fabricated from semiconductors, which may be polycrystalline, the presence of grain boundaries and their effect on electrical properties is of great technological interest. In a semiconductor such as silicon the local change in structure at the interface gives rise to disruption of the normal crystal bonding, or sharing of valence electrons. One consequence can be the charging of the grain boundaries, which produces a barrier to current flowing across them and thus raises the overall resistance of the sample. This polycrystalline effect is exploited in devices such as zinc oxide varistors, which are used as voltage regulators and surge protectors.

[S.L.S.]

Grand unification theories

Attempts to unify three fundamental interactions—strong, electromagnetic, and weak—with a postulate that the three forces, with the exception of gravity, can be unified into one at some very high energy. The basic idea is motivated by the incompleteness of the electroweak theory of S. Weinberg, A. Salam, and S. Glashow, which has been extremely successful in the energy region presently accessible with the use of accelerators, and by the observation that the coupling constant for strong nuclear forces becomes smaller as energy increases, whereas the fine-structure constant ($\alpha = 1/137$) for electromagnetic interactions is expected to increase with energy. *See* GRAVITATION; STRONG NUCLEAR INTERACTIONS; WEAK NUCLEAR INTERACTIONS.

The simplest grand unification theory (GUT), proposed by H. Georgi and Glashow, is based on the assumption that the new symmetry that emerges when the three forces are unified is given by a special unitary group SU(5) of dimension 24. This symmetry is not observable in the low-energy region since it is badly broken. In this model, as in most GUTs, the coupling constants for the three interactions merge into one at an energy of about 10^{14} GeV. Quarks and leptons belong to the same multiplets, implying that distinctions between them disappear at the energy of 10^{14} GeV or above. In addition to the known 12 quanta of strong, electromagnetic, and weak interactions, there appear, in this model, 12 new quanta with the mass of 10^{14} GeV. These generate new but extremely weak interactions that violate baryon- and lepton-number conservation. The most spectacular prediction of GUTs is the instability of the proton, which is a consequence of baryon-number (and lepton-number) violation. *See* LEPTON; PROTON; QUARKS; SYMMETRY BREAKING; SYMMETRY LAWS (PHYSICS).

GUTs, in general, explain why the charge of the electron is precisely that of the proton with the opposite sign. Massive neutrinos are a distinct possibility in GUTs, and the smallness of their mass can also be understood. *See* NEUTRINO.

According to the scenario based on the GUTs, the universe underwent a phase transition when its temperature cooled to 10^{27} K, which corresponds to 10^{14} GeV in energy and to the first 10^{-35} s after the big bang. The phase transition caused an exponential expansion (10^{30}-fold in 10^{-32} s) of the universe, which explains why the observed 3 K microwave background radiation is uniform (the horizon problem), and why the universe behaves as if space is practically flat (the flatness problem). *See* PHASE TRANSITIONS.

In spite of its theoretical triumph and spectacular predictions, the simple SU(5) model is practically untested by experiment and appears to be incomplete or even incorrect. No experimental evidence of proton decay has been established, and the problems which GUTs leave unsolved are numerous. *See* ELEMENTARY PARTICLE; FUNDAMENTAL INTERACTIONS; SUPERGRAVITY; SUPERSYMMETRY.   [C.W.K.]

Gravitation   The mutual attraction between all masses and particles of matter in the universe. In a sense this is one of the best-known physical phenomena. During the eighteenth and nineteenth centuries gravitational astronomy, based on Newton's laws, attracted many of the leading mathematicians and was brought to such a pitch that it seemed that only extra numerical refinements would be needed in order to account in detail for the motions of all celestial bodies. In the twentieth century, however, A. Einstein shattered this complacency, and the subject is currently in a healthy state of flux.

Newton's law of gravitation. Newton's law of universal gravitation states that every two particles of matter in the universe attract each other with a force that acts in the line joining them, the intensity of which varies as the product of their masses and inversely as the square of the distance between them. Or, the gravitational force F exerted between two particles with masses m_1 and m_2 separated by a distance d is given by the equation below, where G is called the constant of gravitation.

$$F = \frac{Gm_1m_2}{d^2}$$

Gravitational constant. In 1774, G was determined by measuring the deflection of the vertical by the attraction of a mountain. This method is much inferior to the laboratory method in which the gravitational force between known masses is measured. In the torsion balance two small spheres, each of mass m, are connected by a light rod, suspended in the middle by a thin wire. The deflection caused by bringing two large spheres each of mass M near the small ones on opposite sides of the rod is measured, and the force is evaluated by observing the period of oscillation of the rod under the influence of the torsion of the wire (see illustration). This is known as the Cavendish experiment, in honor of H. Cavendish, who achieved the first reliable results by this method in 1797–1798. More recent determinations using various refinements yield the results: constant of gravitation $G = 6.67 \times 10^{-11}$ SI (mks) units; mass of Earth $= 5.98 \times 10^{24}$ kg. The result of the best available laboratory measurements, announced in 2002, is $G = (6.6742 \pm 0.0010) \times 10^{-11}$ in SI (mks) units.

In newtonian gravitation, G is an absolute constant, independent of time, place, and the chemical composition of the masses involved. Partial confirmation of this was provided before Newton's time by the experiment attributed to Galileo in which different weights released simultaneously from the top of the Tower of Pisa reached the ground at the same time. Newton found further confirmation, experimenting with pendulums made out of different materials. Early in this century, R. Eötvös found that different materials fall with the same acceleration to within 1 part in 10^7. The accuracy of this figure has been extended to 1 part in 10^{11}, using aluminum and gold, and to 0.9×10^{-12} with a confidence of 95%, using aluminum and platinum.

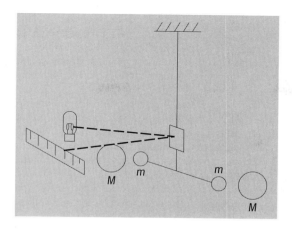

Diagram of the torsion balance.

Mass and weight. In the equations of motion of newtonian mechanics, the mass of a body appears as inertial mass, a measure of resistance to acceleration, and as gravitational mass in the expression of the gravitational force. The equality of these masses is confirmed by the Eötvös experiment. It justifies the assumption that the motion of a particle in a gravitational field does not depend on its physical composition. In Newton's theory the equality can be said to be a coincidence, but not in Einstein's theory, where this equivalence becomes a cornerstone of relativistic gravitation.

While mass in newtonian mechanics is an intrinsic property of a body, its weight depends on certain forces acting on it. For example, the weight of a body on the Earth depends on the gravitational attraction of the Earth on the body and also on the centrifugal forces due to the Earth's rotation. The body would have lower weight on the Moon, even though its mass would remain the same. *See* CENTRIFUGAL FORCE.

Gravity. This should not be confused with the term gravitation. Gravity is the older term, meaning the quality of having weight, and so came to be applied to the tendency of downward motion on the Earth. Gravity or the force of gravity is today used to describe the intensity of gravitational forces, usually on the surface of the Earth or another celestial body. So gravitation refers to a universal phenomenon, while gravity refers to its local manifestation.

Accuracy of newtonian gravitation. A discrepancy in newtonian gravitation was discovered by U.J.J. Leverrier in the orbit of Mercury. Because of the action of the other planets, the perihelion of Mercury's orbit advances. But allowance for all known gravitational effects still left an observed motion of about 43 seconds of arc per century unaccounted for by Newton's theory. Attempts to account for this by adding an unknown planet or by drag with an interplanetary medium were unsatisfactory, and a very small change was suggested in the exponent of the inverse square of force. This particular discordance was accounted for by A. Einstein's general theory of relativity in 1916, but the final word on the subject has yet to be said. *See* RELATIVITY.

Gravitational lens. Light is deflected when it passes through a gravitational field, and an analogy can be made to the refraction of light passing through a lens. It has been suggested that a galaxy situated between an observer and a more distant source might have a focusing effect, and that this might account for some of the observed properties of quasistellar objects. The multiple images of the quasar (Q0957 + 561 A,B) are almost certainly caused by the light from a single body passing through a gravitational lens. While this is the best-studied gravitational lens, many other examples of this phenomenon have been discovered.

Relativistic theories. In spite of his success and the absence of a reasonable alternative, Newton's theory was heavily criticized, not least with regard to its requirement of "action at a distance" (that is, through a vacuum). Newton himself considered this to be "an absurdity," and he recognized the weaknesses in postulating in his system of mechanics the existence of preferred reference systems (that is, inertial reference systems) and an absolute time.

The theory of relativity grew from attempts to describe electromagnetic phenomena in moving systems. No physical effect can propagate with a speed exceeding that of light in vacuum; therefore, Newton's theory must be the limiting case of a field theory in which the speed of propagation approaches infinity. Einstein's field theory of gravitation (general relativity) is based on the identification of the gravitational field with the curvature of space-time. The geometry of space-time is affected by the presence of matter and radiation. The relationship between mass-energy and the space-time curvature is therefore a relativistic generalization of the newtonian law of gravitation. The relativistic theory is mathematically far more complicated than Newton's. Instead of the single newtonian potential described above, Einstein worked with 10 quantities that form a tensor.

An important step in Einstein's reasoning is his "principle of equivalence": A uniformly accelerated reference system imitates completely the behavior of a uniform gravitational field. This principle requires that all bodies fall in a gravitational field with precisely the same acceleration, a result that is confirmed by the Eötvös experiment mentioned earlier. Also, if matter and antimatter were to repel one another, it would be a violation of the principle. *See* FREE FALL.

Gravitational waves. The existence of gravitational waves, or gravitational "radiation," was predicted by Einstein shortly after he formulated his general theory of relativity. They are now a feature of any relativity theory. Gravitational waves are "ripples in the curvature of space-time." In other words, they are propagating gravitational fields, or propagating patterns of strain, traveling at the speed of light. They carry energy and can exert forces on matter in their path, producing, for instance, very small vibrations in elastic bodies. The gravitational wave is produced by change in the distribution of some matter. It is not produced by a rotating sphere, but would result from a rotating body not having symmetry about its axis of rotation: a pulsar, perhaps. In spite of the relatively weak interaction between gravitational radiation and matter, the measurement of this radiation is now technically possible.   [J.M.A.D.; B.Ma.]

Graviton   A theoretically deduced particle postulated as the quantum of the gravitational field. According to Einstein's theory of general relativity, accelerated masses (or other distributions of energy) should emit gravitational waves, just as accelerated charges emit electromagnetic waves. And according to quantum field theory, such a radiation field should be quantized; that is, its energy should appear in discrete quanta, called gravitons, just as the energy of light appears in discrete quanta, namely photons. *See* ELEMENTARY PARTICLE; GRAVITATION; QUANTUM FIELD THEORY; RELATIVITY.   [C.J.G.]

Gravity   The gravitational attraction at the surface of a planet or other celestial body. The quantity g is often referred to simply as "gravity" or "the force of gravity" of Earth, both of which are incorrect. The force of gravity means the force with which a celestial body attracts an object, that is, the weight of the object. The letter g represents the acceleration caused by the gravitational force and, of course, has the dimensions of acceleration. *See* GRAVITATION.   [D.Br./G.M.C.]

Graybody   An energy radiator which has a blackbody energy distribution, reduced by a constant factor, throughout the radiation spectrum or within a certain wavelength interval. The designation "gray" has no relation to the visual appearance of a body but only to its similarity in energy distribution to a blackbody. Most metals, for example, have a constant emissivity within the visible region of the spectrum and thus are graybodies in that region. The graybody concept allows the calculation of the total radiation intensity of certain substances by multiplying the total radiated energy (as given by the Stefan-Boltzmann law) by the emissivity. The concept is also quite useful in determining the true temperatures of bodies by measuring the color temperature. For a discussion of the Stefan-Boltzmann law and color temperature *see* BLACKBODY; HEAT RADIATION.
[H.G.S.; P.J.W.]

Green's function   A solution of a partial differential equation for the case of a point source of unit strength within the region under examination. The Green's function is an important mathematical tool that has application in many areas of theoretical physics including mechanics, electromagnetism, acoustics, solid-state physics, thermal physics, and the theory of elementary particles. The underlying physics in each of these areas is generally described by some linear partial differential equation which relates the physical variable of interest (electrostatic potential or pressure amplitude in a sound wave, for example) to a source function. For present purposes the source may be regarded as an independent entity, although in some applications (for example, particle physics) this view masks an inherent nonlinearity. The source may be physically located within the region of interest, it may be simulated by certain boundary conditions on the surface of that region, or it may consist of both possibilities. A Green's function is a solution to the relevant partial differential equation for the particular case of a point source of unit strength in the interior of the region and some designated boundary condition on the surface of the region. Solutions to the partial differential equation for a general source function and appropriate boundary condition can then be written in terms of certain volume and surface integrals of the Green's function.
[P.Sh.]

Ground state   In quantum mechanics, the stationary state of lowest energy of a particle or a system of particles. The ground state may be bound or unbound; when bound, its energy generally is a finite amount less than the energy of the next higher or first excited state. In the typical circumstance that the potential energy is zero at infinite separation, the magnitude of the negative ground-state energy is the binding energy, that is, the energy required to separate all the particles infinitely. *See* ENERGY LEVEL (QUANTUM MECHANICS); EXCITED STATE; NUCLEAR BINDING ENERGY.
[E.G.]

Group velocity   The velocity of propagation of a group of waves forming a wave packet; also, the velocity of energy flow in a traveling wave or wave packet. The pure sine waves used to define phase velocity v_p do not ever really exist, for they would require infinite extent. What do exist are groups of waves, wave packets, which are combined disturbances of a group of sine waves having a range of frequencies and wavelengths. Good approximations to pure sine waves exist, provided the extent of the media is very large in comparison with the wavelength of the sine wave. In nondispersive media, pure sine waves of different frequencies all travel at the same speed v_p, and any wave packet retains its shape as it propagates. In this case, the group velocity v_g is the same as v_p. But if there is dispersion, the wave packet changes shape as it moves, because each different frequency which makes up the packet moves with a different phase velocity. If v_p is frequency-dependent, then v_g is not equal to v_p. *See* PHASE VELOCITY; SINE WAVE; WAVE MOTION.   [S.A.Wi.]

Gyromagnetic effect   An effect arising from the relation between the angular momentum and the magnetization of a magnetic substance. It is the effect which is exploited in the measurement of the gyromagnetic ratio of magnetic materials. The gyromagnetic effect is demonstrated by a simple experiment in which a freely suspended magnetic substance is subjected to a magnetic field. Upon a change in direction of the magnetic field, the magnetization of the substance must change. In order for this to happen, the atoms must change their angular momentum. Since there are no external torques acting on the system, the total angular momentum must remain constant. Thus the sample must acquire a mass rotation which may be measured. In this way, the gyromagnetic ratio may be determined. *See* GYROMAGNETIC RATIO.   [E.A.; F.Ke.]

Gyromagnetic ratio   The ratio of angular momentum to magnetic moment for atomic systems. This ratio is usually expressed in terms of the magnetomechanical factor g', as in Eq. (1). The ratio is written here in electromagnetic units; thus, e/c and

$$\frac{\text{Angular momentum}}{\text{Magnetic moment}} = \frac{2mc}{g'e} \tag{1}$$

m are the charge and mass of the electron. The factor g' is sometimes loosely called the gyromagnetic ratio.

The magnetomechanical ratio is the inverse of the gyromagnetic ratio. It is usually denoted by γ and is equal to $g'e/2mc$. The magnetomechanical ratio of a substance identifies the origin of the magnetic moment. For example, for electron spin the angular momentum is $\frac{1}{2}\hbar$, where $\hbar$ is Planck's constant divided by 2π. The magnetic moment is the Bohr magneton $e\hbar/2mc$. Thus, the magnetomechanical ratio is given by Eq. (2).

$$\gamma = \frac{e\hbar/2mc}{\hbar2} = \frac{e}{mc} \tag{2}$$

Since $\gamma = g'e/2mc$, for electron spin $g' = 2$. For orbital angular momentum, $\gamma = e/mc$ and $g' = 1$. The experimental values of g' for most ferromagnetic materials are in the neighborhood of 2, showing that the major contribution to the magnetization comes from the electron spin. In superconductors, on the other hand, the fact that $g' = 1$ shows that the diamagnetic currents which cause the Meissner effect are caused by electrons. *See* MEISSNER EFFECT; SUPERCONDUCTIVITY.   [E.A.; F.Ke.]

Hadron   The generic name of a class of particles which interact strongly with one another. Examples of hadrons are protons, neutrons, the π, K, and D mesons, and their antiparticles. Protons and neutrons, which are the constituents of ordinary nuclei, are members of a hadronic subclass called baryons, as are strange and charmed baryons. Baryons have half-integral spin, obey Fermi-Dirac statistics, and are known as fermions. Mesons, the other subclass of hadrons, have zero or integral spin, obey Bose-Einstein statistics, and are known as bosons. The electric charges of baryons and mesons are either zero or ± 1 times the charge on the electron. Masses of the known mesons and baryons cover a wide range, extending from the pi meson, with a mass approximately one-seventh that of the proton, to values of the order of 10 times the proton mass. The spectrum of meson and baryon masses is not understood. *See* BARYON; BOSE-EINSTEIN STATISTICS; FERMI-DIRAC STATISTICS; MESON; NEUTRON; PROTON.

Based on an enormous body of data, hadrons are now thought to consist of elementary fermion constituents known as quarks which have electric charges of $+\frac{2}{3}|e|$ and $\frac{1}{3}|e|$, where $|e|$ is the absolute value of the electron charge. For example, a quark-antiquark pair makes up a meson, while three quarks constitute a baryon. *See* ELEMENTARY PARTICLE; QUARKS.

[A.K.M.]

Hadronic atom   A hydrogenlike system that consists of a strongly interacting particle (hadron) bound in the Coulomb field and in orbit around any ordinary nucleus. The kinds of hadronic atoms that have been made and the years in which they were first identified include pionic (1952), kaonic (1966), Σ^- hyperonic (1968), and antiprotonic (1970). They were made by stopping beams of negatively charged hadrons in suitable targets of various elements, for example, potassium, zinc, or lead. The lifetime of these atoms is of the order of 10^{-12} s, but this is long enough to identify them and study their characteristics by means of their x-ray spectra. They are available for study only in the beams of particle accelerators. Pionic atoms can be made by synchrocyclotrons and linear accelerators in the 500-MeV range. The others can be generated only at accelerators where the energies are greater than about 6 GeV. *See* ELEMENTARY PARTICLE; HADRON; PARTICLE ACCELERATOR.

The hadronic atoms are smaller in size than their electronic counterparts by the ratio of electron to hadron mass. For example, in pionic calcium, atomic number $Z = 20$, the Bohr radius of the ground state is about 10 fermis (1 fermi $= 10^{-15}$ m), and in ordinary calcium it is about 2500 fermis. Thus the atomic electrons are practically not involved in the hadronic atoms, and the equations of the hydrogen atom are applicable. The close approach of the hadrons to their host nuclei suggests that hadron-nucleon and hadron-nucleus forces will be in evidence, and this is one of the motivations for studying these relatively new types of atoms.

Antiproton atoms are the latest in the series of hadronic atoms to be observed. The main research effort involving antiproton atoms has been dedicated to the investigation of the x-ray spectra of the antiprotonic hydrogen. The transitions to the ground state depend directly on the elementary antiproton-proton interaction at the threshold. If this interaction turns out to be simple enough, the antiprotonic atoms will be a future tool for measuring the matter distribution of the nuclear surface. Another source of low-energy antiprotons—the Low Energy Antiproton Ring (LEAR), which makes precision measurements on antiprotonic atoms feasible—was put into operation at CERN near Geneva, Switzerland.

There are two additional hadrons with lifetimes that are long enough to be candidates for hadronic atom formation: the negative xi (Ξ^-) and the negative omega (Ω^-), but even at the largest accelerators, these particles are too scarce for their atoms to be detected.   [C.E.W.; B.Po.]

Hall effect   An effect whereby a conductor carrying an electric current perpendicular to an applied magnetic field develops a voltage gradient which is transverse to both the current and the magnetic field. It was discovered by E. H. Hall in 1879. Important information about the nature of the conduction process in semiconductors and metals may be obtained through analysis of this effect.

A simple model which accounts for the phenomenon is the following. For a magnetic field of strength B in the z direction (see illustration), particles flowing with speed v in the x direction suffer a Lorentz force F_L in the y direction given by Eq. (1), where q is

$$F_L = -qvB \qquad (1)$$

the charge of the particles. This force deflects the particles so that a charge imbalance develops between opposite sides of the conductor. Deflection continues until the electric field E_y resulting from this charge imbalance produces a force $F_y = qE_y$ which cancels the Lorentz force. In practice, the equilibrium condition $F_L + F_y = 0$ is achieved almost instantaneously, giving a steady-state Hall field as in Eq. (2). The current density is $J_x = nqv$, where n is the carrier density. The Hall resistivity, defined by Eq. (3), is thus given by Eq. (4). The Hall coefficient, defined by Eq. (5), satisfies Eq. (6) and thus

$$E_y = vB \qquad (2)$$

$$\rho_{yx} = \frac{E_y}{J_x} \qquad (3)$$

$$\rho_{yx} = \frac{B}{nq} \qquad (4)$$

$$R_0 = \frac{\rho_{yx}}{B} \qquad (5)$$

$$R_0 = \frac{1}{nq} \qquad (6)$$

R_0 provides a measure of the sign and magnitude of the mobile charge density in a conductor. Within the free-electron theory of simple metals, q is expected to be the electron charge $-e$, and n is taken to be $n = Zn_A$, where Z is the valence of the metal and n_A is the density of the atoms. This yields Eq. (7).

$$R_0 = \frac{-1}{n_A Z e} \qquad (7)$$

See FREE-ELECTRON THEORY OF METALS.

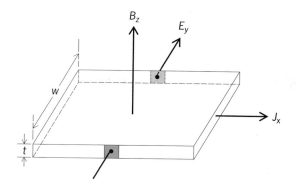

Configuration of fields and currents in the Hall effect experiment.

Equation (7) is approximately valid in simple monovalent metals but fails drastically for other materials, often even giving the wrong sign. The explanation of the failures of Eq. (7) was one of the great early triumphs of the quantum theory of solids. The theory of band structure shows how collisions with the periodic array of atoms in a crystal can cause the current carriers to be holes which have an effective positive charge which changes the sign of the Hall coefficient. Band structure theory also accounts for the observed dependence of R_0 on the orientation of the current and the magnetic field relative to the crystal axes, an effect which is very useful for studying the topology of the Fermi surface. *See* BAND THEORY OF SOLIDS; FERMI SURFACE; HOLE STATES IN SOLIDS.

In certain special field-effect transistors, it is possible to create an electron gas which is effectively two-dimensional. The Hall resistance for an idealized system in two dimensions is given by Eq. (8), where n_S is the density of electrons per unit area (rather

$$\rho_{xy} = -\rho_{yx} = \frac{B}{n_S e} \qquad (8)$$

than volume). However, if the measured value of ρ_{xy} for a high-quality (low-disorder) device is plotted as a function of B, the linear behavior predicted by Eq. (8) is observed only at low fields. At high fields the Hall resistance exhibits plateau regions in which it is a constant independent of B. Furthermore, the values of ρ_{xy} on these plateaus are given quite accurately by the universal relation of Eq. (9), where h is Planck's con-

$$\rho_{xy} = \frac{h}{e^2 \nu} \qquad (9)$$

stant and ν is an integer or simple rational fraction. The absolute accuracy with which Eq. (9) has been verified is better than 1 part in 10^6.

This extremely accurate quantization of ρ_{xy} allows the realization of a new standard of resistance based solely on fundamental constants of nature. In addition, the quantum unit of Hall resistance, $h/e^2 \simeq 25,812.80$ ohms, determines the fine-structure constant. *See* ELECTRICAL UNITS AND STANDARDS; FUNDAMENTAL CONSTANTS.

The explanation of this remarkable phenomenon involves several subtle quantum-mechanical effects. In the quantum regime (small ν), ρ_{xx}, which is the dissipative (longitudinal) resistivity, approaches zero on the Hall plateaus. The quantization of the Hall resistance is intimately connected with this fact. It is speculated that at zero temperature the dissipation is zero and that Eq. (9) is then obeyed exactly. *See* QUANTUM MECHANICS.

The nearly complete lack of dissipation in the quantum Hall regime is reminiscent of superconductivity. In both effects the ability of the current to flow without dissipation has its origin in the existence of a quantum-mechanical excitation gap, that is, a minimum threshold energy needed to disturb the special microscopic order in the system. *See* ENTROPY; SUPERCONDUCTIVITY.

In the integer quantum Hall effect [where ν in Eq. (9) in an integer], this excitation gap is a single-particle effect associated with the quantization by the strong magnetic field of the kinetic energy of the individual electrons into discrete states called Landau levels. In the fractional effect, the gap is associated with the highly collective, many-body ordering of the electrons into a quantum state which minimizes the strong Coulomb repulsion and hence lowers the overall energy. Thus, while the integer and fractional quantum Hall effects look superficially similar on a plot of resistivities versus magnetic field, their physical origins are actually quite different. *See* DE HAAS-VAN ALPHEN EFFECT; GALVANOMAGNETIC EFFECTS. [S.M.G.]

Hamilton's equations of motion

The motion of a mechanical system may be described by a set of first-order ordinary differential equations known as Hamilton's equations. Because of their remarkably symmetrical form, they are often referred to as the canonical equations of motion of a system. They are equivalent to Lagrange's equations, but the fact that they are of first order and highly symmetrical makes them advantageous for general discussions of the motion of systems. *See* LAGRANGE'S EQUATIONS.

Hamilton's equations can be derived from Lagrange's equations. Let the coordinates of the system be q_j ($j = 1, 2, \ldots, f$), and let the dynamical description of the system be given by the lagrangian $L(q, \dot{q}, t)$, where q denotes all the coordinates and a dot denotes total time derivative. Lagrange's equations are then given by Eq. (1). The momentum p_j canonically conjugate to q_j is defined by Eq. (2).

$$\frac{d}{dt}\frac{\partial L}{\partial \dot{q}_j} - \frac{\partial L}{\partial q_j} = 0 \tag{1}$$

$$p_j = \frac{\partial L}{\partial \dot{q}_j} \tag{2}$$

The hamiltonian H is defined by Eq. (3). Then Hamilton's canonical equations are Eqs. (4).

$$H = \sum_{j=1}^{f} p_j \dot{q}_j - L(q, \dot{q}, t) \tag{3}$$

$$\dot{q}_j = \frac{\partial H(q, p, t)}{\partial p_j} \quad \dot{p}_j = -\frac{\partial H(q, p, t)}{\partial q_j} \tag{4}$$

As they stand, Hamilton's equations are no easier to integrate directly than Lagrange's. Hamilton's equations are of great advantage in more general discussions, and they permit the making of canonical transformations which can lead to simplifications. *See* CANONICAL TRANSFORMATIONS.

The hamiltonian function H of classical mechanics is used to form the quantum-mechanical hamiltonian operator. [P.M.S.]

Hamilton's principle

A variational principle from which can be derived the equations of motion of a classical dynamical system in which friction or other forms of dissipation of energy do not occur. In the original formulation of Newton's laws of motion, the position of each particle of the system of interest is specified by the cartesian coordinates of that particle. In many cases, these coordinates are not all independent of each other or do not reflect the structure of the system in a convenient way. It is then advantageous to introduce a system of generalized coordinates which are independent of each other and do reflect any special features of the system such as its symmetry about some center. The number of degrees of freedom of the system, f, is the number of such coordinates required to specify the configuration of the system at any time. *See* DEGREE OF FREEDOM (MECHANICS).

The problem of determining how a system moves may be formulated in the following way: If the configuration of the system at time t_1 is specified by the generalized coordinates $q_1(t_1), \ldots, q_f(t_1)$ and at the time t_2 by $q_1(t_2), \ldots, q_f(t_2)$, then it is required to find the trajectory along which the system travels from the initial to the final configuration. Hamilton's principle addresses this problem similarly to the way that a geometer addresses the problem of finding the shortest path lying in a curved surface between two given points on the surface. The geometer specifies the distance ds between any two close-lying points in terms of the coordinates q_i of the two points and their differences, the coordinate differentials dq_i, as in Eq. (1). The path length D between the

$$(ds)^2 = \sum_{i,j=1}^{f} g_{ij}(q_1, \ldots, q_f) dq_i dq_j \tag{1}$$

$$D = \int_{t_1}^{t_2} ds \tag{2}$$

two specified points, given by the integral in Eq. (2), is then required to be a minimum. Hamilton defined a characteristic function Φ, analogous to D, by Eq. (3), using the

$$\Phi = \int_{t_1}^{t_2} L(q, \dot{q}, t) \, dt \tag{3}$$

lagrangian function $L(q, \dot{q}, t)$ of the system in a way analogous to the geometer's g. Hamilton's principle states that the system follows the trajectory that makes the integral in Eq. (3) have a minimum value, provided the time interval between times t_1 and t_2 is not too great. It can be shown that this principle implies Lagrange's equations of motion for the system, and that it follows from Lagrange's equations. *See* LAGRANGE'S EQUATIONS; LAGRANGIAN FUNCTION; LEAST-ACTION PRINCIPLE; MINIMAL PRINCIPLES; VARIATIONAL METHODS (PHYSICS). [P.St.]

Harmonic (periodic phenomena)
A sinusoidal quantity having a frequency that is an integral multiple of the frequency of a periodic quantity to which it is related. *See* MODE OF VIBRATION.

A harmonic series of sounds is one in which the basic frequency of each sound is an integral multiple of some fundamental frequency. The name exists for historical reasons, even though according to the usual mathematical definition such frequencies form an arithmetic series. An ideal string (or air column) can vibrate as a whole or in a number of equal parts, and the respective periods of vibration are proportional to the lengths. These increasingly shorter lengths or periods form a harmonic series. The name came from the harmonious relation of such sounds, and the science of musical acoustics was once called harmonics. Nowadays, it is customary to deal with ratios of frequency rather than ratios of length and, because frequency is the reciprocal of

period, the definition of harmonic in acoustics becomes that given here. *See* Musical
Acoustics. [R.W.Y.]

Harmonic motion A periodic motion that is a sinusoidal function of time. It is
often called simple harmonic motion (SHM). It is the simplest possible type of vibratory
motion. The motion is symmetric about its midpoint, at which the velocity is greatest
and the acceleration is zero. At the extreme displacements or turning points, the velocity
is zero, and the acceleration is a maximum. The motion is characterized by a unique
frequency (without overtones).

Harmonic motion may be present in very simple mechanisms. For example, if a
wheel is rotating at a constant speed about a fixed axis, the projection on any fixed line
of the motion of a point on the wheel is simple harmonic. Harmonic motion may also
result from the response of a vibrating system to a periodic—in particular a sinusoidal—
force. Harmonic motion is the typical motion of most simple systems that have been
displaced from a position of stable equilibrium and then released, provided that the
damping is negligible. The motion of a pendulum is approximately simple harmonic
for small amplitudes. *See* Pendulum.

The realization that atoms are continually vibrating in motions that are nearly harmonic is essential for understanding many properties of matter, including molecular
spectra, heat capacity, and heat conduction. *See* Damping; Forced oscillation; Harmonic oscillator; Lattice vibrations; Molecular structure and spectra; Periodic
motion; Vibration; Wave motion. [J.M.Ke.]

Harmonic oscillator Any physical system that is bound to a position of stable
equilibrium by a restoring force or torque proportional to the linear or angular displacement from this position. If such a body is disturbed from its equilibrium position and
released, and if damping can be neglected, the resulting vibration will be simple harmonic motion, with no overtones. The frequency of vibration is the natural frequency
of the oscillator, determined by its inertia (mass) and the stiffness of its restoring force.

The harmonic oscillator is not restricted to a mechanical system, but might, for
example, be electric. Typical electronic oscillators, however, are only approximately
harmonic.

If a harmonic oscillator, instead of vibrating freely, is driven by a periodic force, it
will vibrate harmonically with the period of the force; initially the natural frequency
will also be present, but any damping will eventually remove the natural motion. *See*
Damping; Forced oscillation; Harmonic motion.

In both quantum mechanics and classical mechanics, the harmonic oscillator is an
important problem. It is one of the few rigorously soluble problems of quantum mechanics. The quantum-mechanical description of electromagnetic, electronic, mesonic,
and other fields is usually carried out in terms of a (time) Fourier analysis. The individual Fourier components of noninteracting fields are independent harmonic oscillators.
See Anharmonic oscillator. [J.M.Ke.]

Heat For the purposes of thermodynamics, it is convenient to define all energy
while in transit, but unassociated with matter, as either heat or work. Heat is that form
of energy in transit due to a temperature difference between the source from which
the energy is coming and the sink toward which the energy is going. The energy is
not called heat before it starts to flow or after it has ceased to flow. A hot object does
contain energy, but calling this energy heat as it resides in the hot object can lead to
widespread confusion. *See* Energy; Internal energy; Temperature; Thermodynamic
principles. [H.C.W.; W.A.S.]

Heat balance An application of the first law of thermodynamics to a process in which any work terms are negligible.

For a closed system, one that always consists of the same material, the first law is $Q + W = \Delta E$, where Q is the heat supplied to the system, W is the work done on the system, and ΔE is the increase in energy of the material forming the system. It is convenient to treat ΔE as the sum of changes in mechanical energy, such as kinetic energy and potential energy in a gravitational field, and of internal energy ΔU that depends on changes in the thermodynamic state of the material. Because the rates at which any changes occur are usually of interest, heat balances are often written in terms of heat flow rates (heat per unit time), sometimes denoted by a dot over the symbol, $\dot{Q}$, so that for a process with negligible work, kinetic energy and potential energy terms, $\dot{Q} = \dot{Q}_{IN} - \dot{Q}_{OUT} = dU/dt$, the rate of change of internal energy with time.

Often it is more convenient to apply the first law or a heat balance to an open system, a fixed region or control volume across the boundaries of which materials may travel and inside which they may accumulate, such as a building, an aircraft engine, or a section of a chemical process plant. Then the first law is expressed by the equation below, where $\dot{W}_S$ is the rate of doing shaft work on the system; $\dot{m}$ is the mass flow

$$\dot{Q} + \dot{W}_S = \sum \dot{m}\left(h + c^2/2 + gz\right)_{OUT} - \sum \dot{m}\left(h + c^2/2 + gz\right)_{IN} + dE/dt$$

rate of any stream entering or leaving the control volume; h is the enthalpy per unit mass; c is the velocity; gz is the gravitational potential for each stream at the point of crossing the boundary of the control volume; and E is the energy of all material inside the control volume. When conditions inside the control volume do not change with time, although they need not be spatially uniform, $dE/dt = 0$, and the balance equation is known as the steady-flow energy equation.

Enthalpy is a thermodynamic property defined by $h = u + pv$, where u is the specific internal energy (enthalpy per unit mass), p the pressure, and v the specific volume. It is used, along with shaft work, because the derivation of the first-law equation for a control volume from the more fundamental equation for a closed system involves work terms pv that are not available for use outside the control volume. Changes in enthalpy occur because of changes in temperature, pressure, physical state (for example, from liquid to vapor), and changes in chemical state. See ENTHALPY; THERMODYNAMIC PRINCIPLES; THERMODYNAMIC PROCESSES.   [D.B.R.K.]

Heat capacity The quantity of heat required to raise a unit mass of homogeneous material one unit in temperature along a specified path, provided that during the process no phase or chemical changes occur, is known as the heat capacity of the material in question. Moreover, the path is so restricted that the only work effects are those necessarily done on the surroundings to cause the change to conform to the specified path. The path is usually at either constant pressure or constant volume.

In accordance with the first law of thermodynamics, heat capacity at constant pressure C_p is equal to the rate of change of enthalpy with temperature at constant pressure $(\partial H/\partial T)_p$. Heat capacity at constant volume C_v is the rate of change of internal energy with temperature at constant volume $(\partial U/\partial T)_v$. Moreover, for any material, the first law yields the relation

$$C_p - C_v = \left[P + \left(\frac{\partial U}{\partial V}\right)_T\right]\left(\frac{\partial U}{\partial T}\right)_P$$

See ENTHALPY; INTERNAL ENERGY; THERMODYNAMIC PRINCIPLES.   [H.C.W.]

Heat radiation   The energy radiated by solids, liquids, and gases as a result of their temperature. Such radiant energy is in the form of electromagnetic waves and covers the entire electromagnetic spectrum, extending from the radio-wave portion of the spectrum through the infrared, visible, ultraviolet, x-ray, and gamma-ray portions. From most hot bodies on Earth this radiant energy lies largely in the infrared region. *See* ELECTROMAGNETIC RADIATION; INFRARED RADIATION.

Radiation is one of the three basic methods of heat transfer, the other two methods being conduction and convection. *See* CONDUCTION (HEAT); CONVECTION (HEAT); HEAT TRANSFER.

A hot plate at 260°F (400 K) may show no visible glow; but a hand which is held over it senses the warming rays emitted by the plate. A temperature of more than 1300°F (1000 K) is required to produce a perceptible amount of visible light. At this temperature a hot plate glows red and the sensation of warmth increases considerably, demonstrating that the higher the temperature of the hot plate the greater the amount of radiated energy. Part of this energy is visible radiation, and the amount of this visible radiation increases with increasing temperature. A steel furnace at 2800°F (1800 K) shows a strong yellow glow. If a tungsten wire (used as the filament in incandescent lamps) is raised by resistance heating to a temperature of 4600°F (2800 K), it emits a bright white light. As the temperature of a substance increases, additional colors of the visible portion of the spectrum appear, the sequence being first red, then yellow, green, blue, and finally violet. The violet radiation is of shorter wavelength than the red radiation, and it is also of higher quantum energy. In order to produce strong violet radiation, a temperature of almost 5000°F (3000 K) is required. Ultraviolet radiation necessitates even higher temperatures. The Sun emits considerable ultraviolet radiation; its temperature is about 10,000°F (6000 K). Such temperatures have been produced on Earth in gases ionized by electrical discharges. The mercury-vapor lamp and the fluorescent lamp emit large amounts of ultraviolet radiation. Temperatures up to 36,000°F (20,000 K), however, are still much too low to produce x-rays or gamma radiation. A gas maintained at temperatures above $2 \times 10^{6\circ}$F (1×10^6 K), encountered in nuclear fusion experiments, emits x-rays and gamma rays. *See* NUCLEAR FUSION; ULTRAVIOLET RADIATION.

A blackbody is defined as a body which emits the maximum amount of heat radiation. Although there exists no perfect blackbody radiator in nature, it is possible to construct one on the principle of cavity radiation. *See* BLACKBODY.

A cavity radiator is usually understood to be a heated enclosure with a small opening which allows some radiation to escape or enter. The escaping radiation from such a cavity has the same characteristics as blackbody radiation.

Kirchhoff's law correlates mathematically the heat radiation properties of materials at thermal equilibrium. It is often called the second law of thermodynamics for radiating systems. Kirchhoff's law can be expressed as follows: The ratio of the emissivity of a heat radiator to the absorptivity of the same radiator is a function of frequency and temperature alone. This function is the same for all bodies, and it is equal to the emissivity of a blackbody. A consequence of Kirchhoff's law is the postulate that a blackbody has an emissivity which is greater than that of any other body. *See* KIRCHHOFF'S LAWS OF ELECTRIC CIRCUITS.

Planck's radiation law represents mathematically the energy distribution of the heat radiation from 1 cm² of surface area of a blackbody at any temperature. Formulated by Max Planck early in the twentieth century, it laid the foundation for the advance of modern physics and the advent of quantum theory. Equation (1) is the mathematical

$$R_\lambda = 37,418/\lambda^5[e^{14.388(\lambda T)} - 1] \tag{1}$$

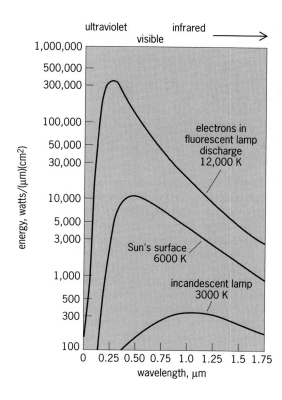

Graphs of Planck's law for various temperatures.

expression of Planck's radiation law, where R_λ is the total energy radiated from the body measured in watts per square centimeter per unit wavelength, at the wavelength λ. The wavelength in this formula is measured in micrometers. The quantity T is the temperature in kelvins, and e is the base of the natural logarithms. The illustration presents graphs of Planck's law for various temperatures and shows substances which attain these temperatures. It should be noted that these substances will not radiate as predicted by Planck's law since they are not blackbodies themselves.

The Stefan-Boltzmann law states that the total energy radiated from a hot body increases with the fourth power of the temperature of the body. This law can be derived from Planck's law by the process of integration and is expressed mathematically as Eq. (2), where R_T is the total amount of energy radiated from a blackbody in watts

$$R_T = 5.670 \times 10^{-10} T^4 \tag{2}$$

per square centimeter. When R_T is multiplied by the total emissivity, the total energy radiated from a real heat radiator is obtained.   [H.G.S.; P.J.W.]

Heat transfer Heat, a form of kinetic energy, is transferred in three ways: conduction, convection, and radiation. Heat transfer (also called thermal transfer) can occur only if a temperature difference exists, and then only in the direction of decreasing temperature. Beyond this, the mechanisms and laws governing each of these ways are quite different. *See* CONDUCTION (HEAT); CONVECTION (HEAT); HEAT RADIATION.

By utilizing a knowledge of the principles governing the three methods of heat transfer and by a proper selection and fabrication of materials, the designer attempts to obtain the required heat flow. This may involve the flow of large amounts of heat to some point in a process or the reduction in flow in others. All three methods operate in processes that are commonplace.

In industry, for example, it is generally desired to extract heat from one fluid stream and add it to another. Devices used for this purpose have passages for each of the two streams separated by a heat-exchange surface in the form of plates or tubes and are known as heat exchangers. The automobile radiator, the hot-water heater, the steam or hot-water radiator in a house, the steam boiler, the condenser and evaporator on the household refrigerator or air conditioner, and even the ordinary cooking utensils in everyday use are all heat exchangers. *See* HEAT.   [R.H.L.]

Helicity (quantum mechanics)   A fundamental quantized variable used in quantum mechanics to specify the relative orientations of spin and linear momentum of massless particles. It is a requirement of fundamental Dirac quantum mechanics that such particles have their spins aligned either parallel or antiparallel to their linear momentum. Particles having parallel alignment are arbitrarily assigned helicity +1; those having antiparallel alignment, −1. *See* MOMENTUM; SPIN (QUANTUM MECHANICS).

In a classic experiment on K electron capture by ^{152}Eu, M. Goldhaber, L. Grodzins, and A. Sunyar first showed that the neutrino emitted in the weak nuclear interaction had negative helicity—that its spin was aligned antiparallel to its momentum. An equivalent description of this situation is that these neutrinos are left-handed. Symmetry requires that antineutrinos be right-handed and have positive helicity. *See* ELECTRON CAPTURE; ELEMENTARY PARTICLE; NEUTRINO; QUANTUM MECHANICS; SYMMETRY LAWS (PHYSICS).   [D.A.B.]

Helimagnetism   A property possessed by some metals, alloys, and salts of transition elements in which the atomic magnetic moments, at sufficiently low temperatures, are arranged in a spiral or helix. Simple antiferromagnets and ferromagnets can be

Some representative helimagnets		
Substance	Magnetic structure	Temperature, K
MnO_2	Nonconical helix	$0 < T < 84$
$MnAu_2$	Nonconical helix	$0 < T < 363$
Dy	Nonconical helix	$85 < T < 179$
	Ferromagnet	$0 < T < 85$
$MnCr_2O_4$	Simple ferrimagnet	$18 < T < 43$
	Complex conical helix	$0 < T < 18$
Er	Conical helix	$0 < T < 20$
	Complex oscillation	$20 < T < 53$
	Sinusoidal antiferromagnet	$53 < T < 85$

considered as nonconical helimagnets with helical angles of 180 and 0°, respectively. In the same way, nonconical helimagnets may be considered as conical helimagnets with cone angle of 0°. Some typical helimagnets are listed in the table. The magnetic structures have been detected by neutron diffraction.   [F.Ke.]

Helmholtz coils   A pair of flat circular coils with equal numbers of turns and equal diameters, arranged with a common axis, and connected in series to have a common current (see illustration). The purpose of the arrangement is to obtain a magnetic field that is more nearly uniform than that of a single coil without the use of a long

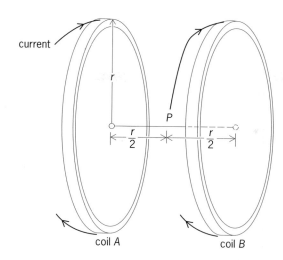

current

$\dfrac{r}{2}$

$\dfrac{r}{2}$

P

coil A

coil B

Helmholtz coils separated by distance *r*, resulting in a nearly uniform field of point *P*. (*After L. B. Loeb, Fundamentals of Electricity and Magnetism*, 3d ed., 1947)

solenoid. The optimum arrangement is that in which the distance between the two coils is equal to the radius of one of the coils.   [K.V.M.]

Hidden variables   Additional variables or parameters that would supplement quantum mechanics so as to make it like classical mechanics. Hidden variables would make it possible to unambiguously predict (as in classical mechanics) the result of a specific measurement on a single microscopic system. In contrast, quantum mechanics can give only probabilities for the various possible results of that measurement. Hidden variables would thus provide deeper insights into the quantum-mechanical probabilities. In this sense the relationship between quantum mechanics and hidden variables could be analogous to the relationship between thermodynamics (for example, temperature) and statistical mechanics (the motions of the individual molecules). *See* STATISTICAL MECHANICS.

F. J. Belinfante formulated a three-section classification scheme for hidden variable theories—zeroth kind, first kind, and second kind. Most interest, both theoretically and experimentally, has been focused on hidden variable theories of the second kind, also known as local hidden variable theories.

In 1932 J. von Neumann provided an axiomatic basis for the mathematical methods of quantum mechanics. As a sidelight to this work, he rigorously proved from the axioms that any hidden variable theory was inconsistent with quantum mechanics. This was the most famous of a number of proofs, appearing as recently as 1980 and purporting to show the impossibility of any hidden variable theory. In 1966 J. S. Bell pinpointed the difficulty with von Neumann's proof—one of his axioms was fine for a pure quantum theory which makes statistical predictions, but the axiom was inherently incompatible with any hidden variable theory. The other impossibility proofs have also been found to be based on self-contradictory theories. Such theories are called hidden variable theories of the zeroth kind.

Hidden variable theories of the first kind are constructed so as to be self-consistent and to reproduce all the statistical predictions of quantum mechanics when the hidden variables are in an "equilibrium" distribution. Hidden variables of the second kind predict deviations from the statistical predictions of quantum mechanics, even for the "equilibrium" situations for which theories of the first kind agree with quantum mechanics. They are generally called local hidden variable theories because they are required to satisfy a locality condition. Intuitively, this seems to be a very natural condition. Locality requires that an apparatus at one location should operate independently of any settings or actions of a second apparatus at a spatially separated location. In the strict Einstein sense of locality, the two apparatus must be independent during any time interval less than the time required for a light signal to travel from one apparatus to the other.

The focus for much of the discussion of local hidden variable theories is provided by a famous thought experiment (a hypothetical, idealized experiment in which the experimental results are deduced) that was introduced in 1935 by A. Einstein, B. Podolsky, and N. Rosen. Their thought experiment involves an examination of the correlation between measurements on two parts of a single system after the parts have become spatially separated. They used this thought experiment to argue that quantum mechanics was not a complete theory. Although they did not refer to hidden variables as such, these would presumably provide the desired completeness. The Einstein-Podolsky-Rosen (EPR) experiment led to a long-standing philosophical controversy; it has also provided the framework for a great deal of research on hidden variable theories.

New efforts were stimulated in 1952 when Bohm did the "impossible" by designing a hidden variable theory of the first kind. Bohm's theory was explicitly nonlocal, and this fact led Bell to reexamine the Einstein-Podolsky-Rosen experiment. He came to the remarkable conclusion that any hidden variable theory that satisfies the condition of locality cannot possibly reproduce all the statistical predictions of quantum mechanics. Specifically, in Einstein-Podolsky-Rosen type experiments, quantum mechanics predicts a very strong correlation between measurements on the spatially separated parts. Bell showed that there is an upper limit on the strength of these correlations in the statistical prediction of any local hidden variable theory. Bell's result can be put in the form of inequalities which must be satisfied by any local hidden variable theory but which may be violated by the statistical predictions of quantum mechanics under appropriate experimental conditions.

Experiments performed under conditions in which the statistical predictions of quantum mechanics violate Bell's inequalities can test the entire class of local hidden variable theories. However, all existing experiments have required supplementary assumptions regarding detector efficiencies. Due to the supplementary assumptions, small loopholes still remain, and experiments have been proposed to eliminate them. The overwhelming experimental evidence is against any theory that would supplement quantum mechanics with hidden variables and still retain the locality condition; that is, any hidden variable theory that reproduces all the statistical predictions of quantum mechanics must be nonlocal. The remarkable Einstein-Podolsky-Rosen correlations have defied any reasonable classical kind of explanation. *See* QUANTUM MECHANICS.   [E.S.F.]

Higgs boson   An elementary scalar particle in the Glashow-Weinberg-Salam theory of electromagnetic and weak interactions. At present, there is no direct experimental evidence for its existence. It is closely associated with the origin of mass for all known elementary particles, as described in the Glashow-Weinberg-Salam theory. This gives the Higgs boson distinctive properties which shape the search for it. *See* ELECTROWEAK INTERACTION; STANDARD MODEL.

The mass of the Higgs boson is uncertain theoretically, being determined by the parameters of the scalar self-interactions. It is strongly suspected, however, that the Higgs mass is between about 114 and 200 GeV/c^2 (where c is the speed of light).

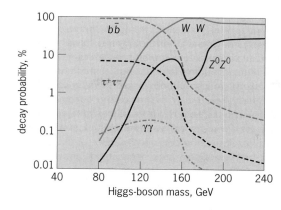

Dependence of the decay probabilities of a standard-model Higgs boson on its mass.

The fact that the Higgs particle is closely related to the origin of mass endows it with special properties crucial in its production and detection. The illustration shows how the decay pattern of a standard-model Higgs boson depends on its mass.

The first extensive searches for the Higgs boson were carried out at the Large Electron Positron Storage Ring (LEP) and Stanford Linear Collider (SLC), e^+e^- machines with sufficient center-of-mass energy to operate at the Z resonance. The LEP was subsequently upgraded, and there were signs that experiments at center-of-mass energies up to 208 GeV/c^2 might be seeing the Higgs boson, but these were not sufficient to delay the machine's shutdown in November 2000 to make way for the construction of the Large Hardon Collider (LHC) in the same tunnel.

The next particle accelerators to extend the Higgs boson search will probably be hadron colliders (proton-proton or proton-antiproton), particularly the LHC. *See* Elementary particle; Particle accelerator.   [A.Ch.; M.Sol.]

High magnetic fields   Magnetic fields that are large enough to significantly alter the properties of objects that are placed in them. Valuable research is conducted at high magnetic fields. *See* Magnetism.

High-field magnets. Research and development efforts in magnets and magnet materials have led to gradual increases in the fields available for scientific research to fields near 20 tesla from superconducting magnets, 33 T in copper-core (resistive) magnets, and 45 T for hybrid magnets. Superconducting magnets have the advantage that they use no electrical power once the field is established and the temperature is maintained at liquid-helium temperatures of 4.2 K (−452°F) or below. The disadvantage is that there is a critical magnetic field, H_{c2}, determined by the type of conductor, that limits the attainable field to about 22 T in superconducting materials currently available. Resistive magnets, which consume enormous amounts of power and are very expensive to build and operate, are confined to a few central facilities worldwide. *See* Superconductivity.

Advanced pulsed magnets that are not self-destructing provide fields beyond 70 T for about 0.1 s. Pulsed magnets using explosive magnetic flux compression have achieved fields above 500 T for periods of 10 microseconds. *See* Magnet.

Materials research. Research at very high magnetic fields spans a wide spectrum of experimental techniques for studies of materials. These techniques include nuclear magnetic resonance (NMR) in biological molecules utilizing the highest-field superconducting magnets, while the resistive magnet research is primarily in the investigation of semiconducting, magnetic, superconducting, and low-dimensional conducting materials.

Much of the progress in semiconductor physics and technology has come from high-field studies. For example, standard techniques for mapping the allowed electronic states (the Fermi surface) of semiconductors and metals are to measure the resistance (in the Shubnikov-de Haas effect) or magnetic susceptibility (in the de Haas-van Alphen effect) as a function of magnetic field and to observe the oscillatory behavior arising from the Landau levels of the electron orbits. Measurements at low fields are limited to low impurity concentrations since the orbits are large and impurity scattering wipes out the oscillations. At high fields of 20–200 T, the orbits are smaller, and higher impurity concentrations (higher carrier concentrations) have been studied. Another area in which very high magnetic fields have an important role is in high-temperature superconductors, which have great potential for high-field applications, from magnetic resonance imaging, to magnetically levitated trains, to basic science. *See* De Haas-van Alphen effect; Fermi surface; Semiconductor.

Studies at high magnetic fields have played an important role in advancing understanding of magnetic materials. For example, in many organic conductors the conduction electrons (or holes) are confined to one or two dimensions, leading to very rich magnetic phase diagrams. High-field phases above 20 T include spin-density waves, a modulation of the electron magnetic moments that can propagate through the crystal, modifying the conduction and magnetic properties. Another area of interest is the magnetic levitation of diamagnetic materials (the most common materials). *See* Magnetic materials; Phase transitions; Spin-density wave.   [W.G.M.]

High-pressure physics   The study of the effects of high pressure on the properties of matter. Since most properties of matter are modified by pressure, the field of high-pressure physics encompasses virtually all branches of physics.

The "high" of high-pressure physics connotes experimental difficulty. At liquid-helium temperatures, pressures of several hundred bars are considered high. In general, however, the high-pressure range may be arbitrarily regarded as extending from about 1 kbar (100 MPa or 14,500 lb/in.^{2}) upward to the present experimental limit. Prolonged static pressures in excess of 1 megabar (100 gigapascals or 1.45×10^7 lb/in.^{2}) can be achieved in very small samples weighing about 1 microgram.

Transient pressures as high as about 10^7 bars (1000 GPa or 1.45×10^8 lb/in.^{2}) have been attained in shock waves produced by high explosives or by projectile impact.

The major effects of high pressure on matter include diminution of volume, phase transitions, changes in electrical, optical, magnetic, and chemical properties, increases in viscosity of liquids, and increases in the strength of most solids. In general solids are less compressible than liquids, and the compressibility of both solids and liquids decreases with increasing pressure.

At high pressure many solids exhibit polymorphic phase changes, that is, a rearrangement of the atoms or molecules in the solid. There are no universally applicable rules governing the number of phase changes or the kind of phase change to be expected at high pressure, but there is a thermodynamic requirement that the phase that is stable at high pressure must have a smaller volume than the phase that is stable at low pressure. *See* Thermodynamic principles.

Frequently, dramatic changes in physical properties result from phase changes. Ferromagnetic iron transforms to a paramagnetic form at pressures somewhat above 100 kbar (10 GPa or 1.45×10^6 lb/in.^{2}). In the same pressure range, the semiconducting element germanium transforms into a metallic phase that has an electrical conductivity greater than a million times that of the semiconductor. Similar semiconductor-to-metal transitions at high pressure have been observed in the cases of silicon, indium arsenide, gallium antimonide, indium phosphide, aluminum antimonide, and gallium arsenide. *See* SEMICONDUCTOR.

Many phases that form at high pressure transform back to low-pressure phases as the pressure is released. However, some high-pressure phases may be retained in a metastable condition at low pressures, and some low-pressure phases can persist metastably at high pressure. Diamond, the high-pressure form of carbon, is thermodynamically unstable at room temperature and pressures below about 12 kbar (1.2 GPa or 1.74×10^5 lb/in.^{2}). Nonetheless diamond persists indefinitely as a metastable phase at low temperatures; it transforms to the stable form, graphite, only when heated to temperature in excess of 1800°F (1000°C) at low pressure.   [R.K.L.; P.S.DeC.]

Hole states in solids
Vacant electron energy states near the top of an energy band in a solid are called holes. A full band cannot carry electric current; a band nearly full with only a few unoccupied states near its maximum energy can carry current, but the current behaves as though the charge carriers are positively charged. *See* BAND THEORY OF SOLIDS.

The process of conduction in such a system may be visualized in the following way. An electron moves against an applied electric field by jumping into a vacant state. This transfers the position of the vacant state, or propagates the hole, in the direction of the field.

Hole conduction is important in many semiconductors, notably germanium and silicon. The occurrence of hole conduction in semiconductors can be favored by alloying with a material of lower valence than the "host." Semiconductors in which the conduction is primarily due to holes are called *p* type. *See* SEMICONDUCTOR.   [J.C.]

Holography
A technique for recording, and later reconstructing, the amplitude and phase distributions of a coherent wave disturbance. Invented by Dennis Gabor in 1948, the process was originally envisioned as a possible method for improving the resolution of electron microscopes. While this original application has not proved feasible, the technique is widely used as a method for optical image formation, and in addition has been successfully used with acoustical and radio waves. *See* ACOUSTICAL HOLOGRAPHY.

The technique is accomplished by recording the pattern of interference between the unknown "object" wave of interest and a known "reference" wave (Fig. 1). In general, the object wave is generated by illuminating the (possibly three-dimensional) subject of concern with a highly coherent beam of light, such as supplied by a laser source. The waves reflected from the object strike a light-sensitive recording medium, such as photographic film or plate. Simultaneously a portion of the light is allowed to bypass the object, and is sent directly to the recording plane, typically by means of a mirror placed next to the object. Thus incident on the recording medium is the sum of the light from the object and a mutually coherent "reference" wave. *See* LASER.

The photographic recording obtained is known as a hologram (meaning a "total recording"); this record generally bears no resemblance to the original object, but rather is a collection of many fine fringes which appear in rather irregular patterns. Nonetheless, when this photographic transparency is illuminated by coherent light, one of the transmitted wave components is an exact duplication of the original object wave (Fig. 2). This wave component therefore appears to originate from the object (although the object has long since been removed) and accordingly generates a virtual image of it, which appears to an observer to exist in three-dimensional space behind the transparency. The image is truly three-dimensional in the sense that the observer's

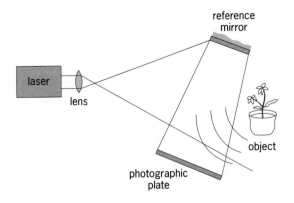

Fig. 1. Recording a hologram.

eyes must refocus to examine foreground and background, and indeed can "look behind" objects in the foreground simply by moving his or her head laterally.

Holography has been demonstrated to offer the capability of several unique kinds of interferometry. This capability is a consequence of the fact that holographic images are coherent; that is, they have well-defined amplitude and phase distributions. Any use of holography to achieve the superposition of two coherent images will result in a potential method of interferometry. *See* INTERFEROMETRY.

Optical memories for storing large volumes of binary data in the form of holograms have been developed for commercial use. Such a memory consists of an array of small holograms, each capable of reconstructing a different "page" of binary data. When one of these holograms is illuminated by coherent light, it generates a real image consisting of an array of bright or dark spots, each spot representing a binary digit.

There has been interest in the use of holography for purposes of display of three-dimensional images. Applications have been found in the field of advertising, and there is increased use of holography as a medium for artistic expression.   [J.W.Goo.]

Microwave holography is microwave imaging by means of coherent continuous-wave electromagnetic radiation in the wavelength range from 1 mm to 1 m. As a long-wavelength imaging modality, it differs from techniques which employ echo timing (for example, conventional radar) by its requirement for phase information. In this respect it resembles optical holography, from which it has departed significantly. The technique usually involves small-scale systems, that is, systems in which the effective data acquisition aperture is of the order of tens or hundreds of wavelengths. Microwave holographic imaging is characterized by high lateral-resolution capability in comparison with images obtained from echo timing. The natural image format of the data

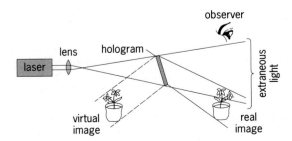

Fig. 2. Obtaining images from a hologram.

it presents to the human observer enhances its diagnostic potential. In particular, it conveniently produces phase imagery which increases further its diagnostic capability.

Microwave holography is useful in applications where images of concealed structure are required. Microwave radiation penetrates a variety of dielectric media to a depth depending on the attenuation of a given wavelength in a particular medium. One such application is the mapping of subsurface pipes and cables. Plastic pipes as well as metal pipes can be imaged. Hence this noninvasive microwave technique has a diagnostic power greater than the normal metal detectors.

The major limitation of the microwave holographic techniques is that the images produced are essentially two-dimensional. The reason is that the microwave wavelength is so long (10^4–10^6 times that of light) that the depth of focus of the microwave hologram is prohibitive. This disadvantage is overcome by employing a tomographic mode of imaging which exploits the ability of microwaves to penetrate many materials and thereby characterize their three-dimensional structure more accurately. Microwave holographic tomography requires holograms to be recorded from different views of the object and synthesized.

[A.P.An.]

Huygens' principle An assumption regarding the behavior of light waves, originally proposed by C. Huygens in the seventeenth century to explain the fact that light travels in straight lines and casts sharp shadows. Large-scale waves, such as sound waves or water waves, bend appreciably into the shadow. The special behavior of light may be explained by Huygens' principle, which states that "each point on a wavefront may be regarded as a source of secondary waves, and the position of the wavefront at a later time is determined by the envelope of these secondary waves at that time." Thus a wave WW originating at S is shown in the illustration at the instant it passes through an aperture. If a large number of circular secondary waves, originating at various points on WW, are drawn with the radius r representing the distance the wave would travel in time t, the envelope of these secondary waves is the heavily drawn circular arc $W'W'$. This represents the wave after t. If, as Huygens' principle requires, the disturbance is confined to the envelope, it will be 0 outside the limits indicated by points W'.

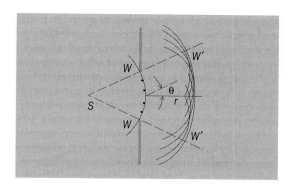

Huygens' principle: the construction for a spherical wave.

Careful observation shows that there is a small amount of light beyond these points, decreasing rapidly with distance into the geometrical shadow. This is called diffraction. *See* DIFFRACTION.

[F.A.J./W.W.W.]

Hydraulic jump An abrupt increase of depth in a free-surface liquid flow. A hydraulic jump is characterized by rapid flow and small depths on the upstream side, and by larger depths and smaller velocities on the downstream side. A jump can form only when the upstream flow is supercritical, that is, when the fluid velocity is greater than the propagation velocity c of a small, shallow-water gravity wave ($c = \sqrt{gh}$, where g is the acceleration of gravity and h is the depth). A considerable amount of energy is dissipated in the conversion from supercritical to subcritical flow. *See* OPEN CHANNEL.

[D.R.F.H.]

Hydrodynamics   The study of fluids in motion. The study is based upon the physical conservation laws of mass, momentum, and energy. The mathematical statements of these laws may be written in either integral or differential form. The integral form is useful for large-scale analyses and provides answers that are sometimes very good and sometimes not, but that are always useful, particularly for engineering applications. The differential form of the equations is used for small-scale analyses. In principle, the differential forms may be used for any problem, but exact solutions can be found only for a small number of specialized flows. Solutions for most problems must be obtained by using numerical techniques, and these are limited by the computer's inability to model small-scale processes. *See* CONSERVATION OF ENERGY; CONSERVATION OF MASS; CONSERVATION OF MOMENTUM; FLUID FLOW; FLUID MECHANICS.

Applications of hydrodynamics include the study of closed-conduit and open-channel flow, and the calculation of forces on submerged bodies.

Flow in closed conduits, or pipes, has been extensively studied both experimentally and theoretically. If the pipe Reynolds number, given by the equation below, where

$$\mathrm{Re}_D = \frac{VD\rho}{\mu}$$

V is the average velocity and D is the pipe diameter, is less than about 2000, the flow in the pipe is laminar. In this case, the solution to the continuity, momentum, and energy equations is readily obtained, particularly in the case of steady flows. If Re_D is greater than about 4000, the flow in the pipe is turbulent, and the solution to the continuity, momentum, and energy equations can be obtained only by employing empirical correlations and other approximate modeling tools. The Re_D region between 2000 and 4000 is the transition region in which the flow is intermittently laminar and turbulent. *See* LAMINAR FLOW; REYNOLDS NUMBER; TURBULENT FLOW.

Confined flows that have a liquid surface exposed to the atmosphere (a free surface) are called open-channel flows. Flows in rivers, canals, partially full pipes, and irrigation ditches are examples. The difficulty with these flows is that the shape of the free surface is one of the unknowns to be calculated.

In most open-channel flows the bottom slope and the water depth change with position, and the free surface is not parallel to the channel bottom. If the slopes are small and the changes are not too sudden, the flow is called a gradually varied flow. An energy balance between two sections of the channel yields a differential equation for the rate of change of the water depth with respect to the distance along the channel. The solution of this equation, which must be accomplished by using one of many available numerical techniques, gives the shape of the water surface.

Flow over spillways and weirs and flow through a hydraulic jump are examples of rapidly varying flows. In these cases, changes of water depth with distance along the channel are large. Here, because of large accelerations, the pressure distribution with depth may not be hydrostatic as it is in the cases of gradually varied and uniform flows. Solutions for rapidly varying flows are accomplished by using approximation techniques. *See* HYDRAULIC JUMP; OPEN CHANNEL.

The force exerted by a fluid flowing past a submerged body is calculated by integrating the pressure distribution over the surface of the body. The pressure distribution is determined from the simultaneous solution of the continuity and momentum equations along with the appropriate boundary conditions. In almost all cases, this solution must be accomplished by using an appropriate approximation. *See* BOUNDARY-LAYER FLOW.

Usually the force exerted on the body is resolved into two components, the lift and the drag. The drag force is the component parallel to the velocity of the undisturbed stream (flow far away from the body), and the lift force is the component perpendicular to the undisturbed stream.

[W.M.H.]

Hydrostatics   The study of liquids at rest. In the absence of motion, there are no shear stresses; the internal state of stress at any point is determined by pressure alone. Hence, the pressure at a point is the same in all directions. Pressure acts normally to all boundary surfaces. For equilibrium under gravity, regardless of the shape of the containing vessel, the pressure is uniform over any horizontal cross section. Pressure varies with height or depth. Two different reference levels are used in measuring pressure. For many engineering purposes, gage pressure is used with pressure measured relative to atmospheric pressure as zero. For most scientific purposes, pressure is referred to true zero. Normal atmospheric pressure at sea level caused by the weight of the air above is approximately 101 kilopascals or 14.7 pounds per square inch absolute.

The buoyant force is the force exerted vertically upward by a fluid on a body wholly or partly immersed in it. Its magnitude is equal to the weight of the fluid displaced by the body. This value is also the vertical component of the fluid pressure force acting upward against the bottom of the body minus the fluid pressure force component (if any) acting vertically downward against the top of the body. If this buoyant force equals the weight of the body, the body will remain at the given level. If it exceeds the weight of the body, the latter will rise, and vice versa. The buoyant force as a single magnitude acts vertically upward through the center of buoyancy which is the center of gravity of the displaced fluid. *See* ARCHIMEDES' PRINCIPLE.

Pressure applied to a confined liquid is transmitted with equal intensity throughout the liquid and by it to all surfaces of the confining vessel or piping. Hence, a small force applied to a small area of a confined liquid can create a large force against a large area. If the small and large areas are pistons, the device may be a hydraulic press or jack. Because the transmitting liquid is practically incompressible and its volume virtually constant, the linear movement of the large piston will be to that of the small piston in inverse proportion to their areas. The principle of multiplying a force by means of liquid pressure applies also to hydraulic brakes, power steering, control systems, and the like; the actuating force may be a pump instead of a small piston.

[W.A.]

Hypercharge   A quantized attribute, analogous to electric charge, introduced in the classification of a subset of elementary particles—the so-called baryons—including the proton and neutron as its lightest members. As far as is known, electric charge is absolutely conserved in all physical processes. Hypercharge was introduced to formalize the observation that certain decay modes of baryons expected to proceed by means of the strong nuclear force simply were not observed. *See* ELECTRIC CHARGE.

Unlike electric charge, however, the postulated hypercharge was found not to be conserved absolutely; the weak nuclear interactions do not conserve hypercharge—and indeed can change hypercharge by ± 1 or 0 units.

When the known baryons are classified according to their electric charge and their hypercharge, they naturally group into octets in the scheme first proposed by M. Gell-Mann and K. Nishijima. The quarks, hypothesized as the fundamental building blocks of matter, must have fractional hypercharge as well as electrical charge; the simplest quark model suggests values of 1/3 and 2/3, respectively. *See* BARYON; ELEMENTARY PARTICLE; QUANTUM MECHANICS; QUARKS; UNITARY SYMMETRY.   [D.A.B.]

Hyperfine structure   A closely spaced structure of the spectrum lines forming a multiplet component in the spectrum of an atom or molecule, or of a liquid or solid. In the emission spectrum for an atom, when a multiplet component is examined at the highest resolution, this component may be seen to be resolved, or split, into a group of spectrum lines which are extremely close together. This hyperfine structure may be due to a nuclear isotope effect, to effects related to nuclear spin, or to both. *See* ISOTOPE SHIFT; SPIN (QUANTUM MECHANICS).

The measurement of a hyperfine structure spectrum for a gaseous atomic or molecular system can lead to information about the nuclear magnetic and quadrupole moments, and about the atomic or molecular electron configuration. Important methods for the measurement of hyperfine structure for gaseous systems may employ an interferometer, or use atomic beams, electron spin resonance, or nuclear spin resonance. *See* INTERFEROMETRY; MAGNETIC RESONANCE.   [L.D.R.]

Hypernuclei   Nuclei that consist of protons, neutrons, and one or more strange particles such as lambda particles. The lambda particle is the lightest strange baryon (hyperon); its lifetime is 2.6×10^{-10} s. Because strangeness is conserved in strong interactions, the lifetime of the lambda particle remains essentially unchanged in the nucleus also. Lambda hypernuclei live long enough to permit detailed study of their properties. *See* BARYON; ELEMENTARY PARTICLE; HYPERON; NUCLEAR STRUCTURE; STRONG NUCLEAR INTERACTIONS.   [B.Po.]

Hyperon   A collective name for any baryon with nonzero strangeness number s. The name hyperon has generally been limited to particles which are semistable, that is, which have long lifetimes relative to 10^{-22} s and which decay by photon emission or through weaker decay interactions. Hyperonic particles which are unstable (that is, with lifetimes shorter than 10^{-22} s) are commonly referred to as excited hyperons. The known hyperons with spin $^{1}/_{2} \hbar$ (where $\hbar$ is Planck's constant divided by 2π) are Λ, Σ^{-}, Σ^{0}, and Σ^{+} with $s = -1$, and Ξ^{-} and Ξ^{0}, with $s = -2$, together with the Ω^{-} particle, which has spin $^{3}/_{2} \hbar$ and $s = -3$. The corresponding antihyperons have baryon number $B = -1$, opposite strangenesse s, and charge Q; they are all known empirically.

There is no deep distinction between hyperons and excited hyperons, beyond the phenomenological definition above. Indeed, the hyperon $\Omega(1672)^{-}$ and the excited hyperons $\Xi(1530)$ and $\Sigma(1385)$, together with the unstable nucleonic states $\Delta(1236)$, are known to form a unitary decuplet of states with spin $^{3}/_{2} \hbar$. *See* BARYON; ELEMENTARY PARTICLE; SYMMETRY LAWS (PHYSICS); UNITARY SYMMETRY.   [R.H.D.]

Hysteresis   A phenomenon wherein two (or more) physical quantities bear a relationship which depends on prior history. More specifically, the response Y takes on different values for an increasing input X than for a decreasing X.

If one cycles X over an appropriate range, the plot of Y versus X gives a closed curve which is referred to as the hysteresis loop. The response Y appears to be lagging the input X.

Hysteresis occurs in many fields of science. Perhaps the primary example is of magnetic materials where the input variable H (magnetic field) and response variable B (magnetic induction) are traditionally chosen. For such a choice of conjugate variables, the area of the hysteresis loop takes on a special significance, namely the conversion of energy per unit volume to heat per cycle. For mechanical hysteresis, it is customary to take the variables stress and strain, where the energy density loss per cycle is related to the internal friction. Thermal hysteresis is characteristic of many systems, particularly those involving phase changes, but here the hysteresis loops are not usually related to energy loss. *See* FERROELECTRICS.   [H.B.H.; R.K.MacC.]

I-spin A quantum-mechanical variable or quantum number applied to quarks and their compounds, the strongly interacting fundamental hadrons, and the compounds of those hadrons (such as nuclear states) to facilitate consideration of the consequences of the charge independence of the strong (nuclear) forces. This variable is also known as isotopic spin, isobaric spin, and isospin.

The many strongly interacting particles (hadrons) and the compounds of these particles, such as nuclei, are observed to form sets or multiplets such that the members of the multiplet differ in their electric charge and magnetic moments, and other electromagnetic properties but are otherwise almost identical. For example, the neutron and proton, with electric charges that are zero and plus one (in units of the magnitude of the electron charge), form a set of two such states. The pions, one with a unit of positive charge, one with zero charge, and one with a unit of negative charge, form a set of three. It appears that if the effects of electromagnetic forces and the closely related weak nuclear forces (responsible for beta decay) are neglected, leaving only the strong forces effective, the different members of such a multiplet are equivalent and cannot be distinguished in their strong interactions. The strong interactions are thus independent of the different electric charges held by different members of the set; they are charge-independent. *See* ELEMENTARY PARTICLE; FUNDAMENTAL INTERACTIONS; HADRON; STRONG NUCLEAR INTERACTIONS.

The I-spin (I) of such a set or multiplet of equivalent states is defined such that Eq. (1) is satisfied, where N is the number of states in the set. Another quantum number

$$N = 2I + 1 \tag{1}$$

I_3, called the third component of I-spin, is used to differentiate the numbers of a multiplet where the values of I_3 vary from $+I$ to $-I$ in units of one. The charge Q of a state and the value of I_3 for this state are connected by the Gell-Mann–Okubo relation (2), where

$$Q = I_3 + \frac{Y}{2} \tag{2}$$

Y, the charge offset, is called hypercharge. For nuclear states, Y is simply the number of nucleons. Electric charge is conserved in all interactions; Y is observed to be conserved by the strong forces so that I_3 is conserved in the course of interactions mediated by the strong forces. *See* HYPERCHARGE.

This description of a multiplet of states with I-spin is similar to the quantum-mechanical description of a particle with a total angular momentum or spin of j (in units of $\hbar$, Planck's constant divided by 2π). Such a particle can be considered as a set of states which differ in their orientation or component of spin j_z in a z direction of quantization. There are $2j + 1$ such states, where j_z varies from $-j$ to $+j$ in steps of one unit. To the extent that the local universe is isotropic (or there are no external forces on the states that depend upon direction), the components of angular momentum in any direction are conserved, and states with different values of j_z are dynamically equivalent.

There is then a logical or mathematical equivalence between the descriptions of (1) a multiplet of states of definite I and different values of I_3 with respect to charge-independent forces and (2) a multiplet of states of a particle with a definite spin j and different values of j_z with respect to direction-independent forces. In each case, the members of the multiplet with different values of the conserved quantity I_3 on the one hand and j_z on the other are dynamically equivalent; that is, they are indistinguishable by any application of the forces in question. *See* ANGULAR MOMENTUM; SPIN (QUANTUM MECHANICS).

The charge independence of the strong interactions has important consequences, defining the intensity ratios of different charge states produced in those particle reactions and decays which are mediated by the strong interactions. I-spin considerations also provide insight into the total energies or masses of nuclear and particle states. The basis for the symmetry described by I-spin is to be found in the quark structure of the strongly interacting particles and the character of the interactions between quarks. *See* QUARKS; SELECTION RULES (PHYSICS); SYMMETRY LAWS (PHYSICS).   [R.K.A.]

Illuminance   A term expressing the density of luminous flux incident on a surface. This word has been proposed by the Colorimetry Committee of the Optical Society of America to replace the term illumination. The definitions are the same. The symbol of illumination is E, and the equation is $E = dF/dA$, where A is the area of the illuminated surface and F is the luminous flux. *See* LUMINOUS FLUX; PHOTOMETRY.   [R.C.Pu.]

Immittance   The impedance or admittance of an alternating-current circuit. It is sometimes convenient to use the term immittance when referring to a complex number which may be either the impedance (ratio of voltage to current) or the admittance (ratio of current to voltage) of an electrical circuit. The units of impedance and admittance are, of course, different and so units cannot be assigned to an immittance. However, in certain theoretical work it may be necessary to deal with general functions which afterward will be specialized to become either an impedance or an admittance by the assignment of suitable units; in such cases it is convenient to refer to the functions as immittances. *See* ADMITTANCE; ELECTRICAL IMPEDANCE.   [J.O.S.]

Impact   A force, also known as impulsive force, which acts only during a short time interval but which is sufficiently large to cause an appreciable change in the momentum of the system on which it acts. The momentum change produced by the impulsive force is described by the momentum-impulse relation. *See* COLLISION (PHYSICS); IMPULSE (MECHANICS).   [P.W.S.]

Impulse (mechanics)   The integral of a force over an interval of time. For a force $\mathbf{F}$, the impulse $\mathbf{J}$ over the interval from t_0 to t_1 can be written as Eq. (1). The

$$\mathbf{J} = \int_{t_0}^{t_1} \mathbf{F}\, dt \qquad (1)$$

impulse thus represents the product of the time interval and the average force acting during the interval. Impulse is a vector quantity with the units of momentum.

The momentum-impulse relation states that the change in momentum of a mass m over a given time interval equals the impulse of the resultant force acting during that interval. The momentum change can be expressed in terms of the velocities $\mathbf{v}_1$ and $\mathbf{v}_0$

at times t_1 and t_0, respectively, giving Eq. (2).

$$\mathbf{J} = m(\mathbf{v}_1 - \mathbf{v}_0) \tag{2}$$

See IMPACT; MOMENTUM.

[P.W.S.]

Incandescence   The emission of visible radiation by a hot body. A theoretically perfect radiator, called a blackbody, will emit radiant energy according to Planck's radiation law at any temperature. Prediction of the visual brightness requires additional consideration of the sensitivity of the eye, and the radiation will be visible only for temperatures of the blackbody which are above some minimum. The relation between brightness and temperature is plotted in the illustration. As shown, the minimum tem-

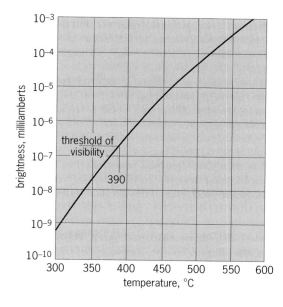

Relation between brightness of blackbody and temperature.

perature for incandescence for the dark-adapted eye is about 390°C (730°F). Under these ideal observing conditions, the incandescence appears as a colorless glow. The dull red light commonly associated with incandescence of objects in a lighted room requires a temperature of about 500°C (930°F). See BLACKBODY; HEAT RADIATION.

[H.W.Ru./G.R.H.]

Incompressible flow   Fluid motion with negligible changes in density. No fluid is truly incompressible, since even liquids can have their density increased through application of sufficient pressure. But density changes in a flow will be negligible if the Mach number, Ma, of the flow is small. This condition for incompressible flow is given by the equation below, where V is the fluid velocity and a is the speed of sound of the

$$\text{Ma} = \frac{V}{a} < 0.3$$

fluid. It is nearly impossible to attain Ma = 0.3 in liquid flow because of the very high pressures required. Thus liquid flow is incompressible. See MACH NUMBER.

Gases may easily move at compressible speeds. Doubling the pressure of air—from, say, 1 to 2 atm—may accelerate it to supersonic velocity. In principle, practically any large Mach number may be achieved in gas flow. As Mach number increases above 0.3, the four compressible speed ranges occur: subsonic, transonic, supersonic, and hypersonic flow. Each of these has special characteristics and methods of analysis.

Air at 68°F (20°C) has a speed of sound of 760 mi/h (340 m/s). Thus inequality indicates that air flow will be incompressible at velocities up to 228 mi/h (102 m/s). This includes a wide variety of practical air flows: ventilation ducts, fans, automobiles, baseball pitches, light aircraft, and wind forces. The result is a wide variety of useful incompressible flow relations applicable to both liquids and gases. *See* COMPRESSIBLE FLOW; FLUID FLOW.   [F.M.Wh.]

Inductance   That property of an electric circuit or of two neighboring circuits whereby an electromotive force is induced (by the process of electromagnetic induction) in one of the circuits by a change of current in either of them. The term inductance coil is sometimes used as a synonym for inductor, a device possessing the property of inductance. *See* ELECTROMAGNETIC INDUCTION; ELECTROMOTIVE FORCE (EMF); INDUCTOR.

For a given coil, the ratio of the electromotive force of induction to the rate of change of current in the coil is called the self-inductance of the coil. An alternative definition of self-inductance is the number of flux linkages per unit current. Flux linkage is the product of the flux and the number of turns in the coil. Self-inductance does not affect a circuit in which the current is unchanging; however, it is of great importance when there is a changing current, since there is an induced emf during the time that the change takes place. For example, in an alternating-current circuit, the current is constantly changing and the inductance is an important factor.

The mutual inductance of two neighboring circuits is defined as the ratio of the emf induced in one circuit to the rate of change of current in the other circuit.

The International System (SI) unit of mutual inductance is the henry, the same as the unit of self-inductance. The same value is obtained for a pair of coils, regardless of which coil is the starting point.

The mutual inductance of two circuits may also be expressed as the ratio of the flux linkages produced in a circuit by the current in a second circuit to the current in the second circuit. *See* INDUCTANCE MEASUREMENT.   [K.V.M.]

Inductance measurement   The measurement of self- or mutual inductance. An electrical reactance such as the angular frequency ($2\pi f$, where f is the frequency) times self- or mutual inductance is the ratio of the alternating voltage having the appropriate phase, which appears across specified terminals, to the current through the device. Commercial instruments often measure inductance from this ratio by comparing it with the voltage-to-current ratio associated with a noninductive resistor. *See* ELECTRICAL IMPEDANCE.

Some practical precautions must be taken if accurate results are to be obtained. Any magnetic field associated with the inductor must not interact significantly with magnetic or conducting material in the vicinity of the inductor, since the field, and therefore the inductance, would be altered. The varying magnetic field of an inductor will induce eddy currents in any nearby conducting material, which will in turn produce a magnetic field which interacts with the inductor and measuring system. Errors in a measurement of inductance may also arise from the interaction of the magnetic field of an inductor with the rest of the measuring circuit. Capacitance to other parts or to the surroundings of an inductor arising from its associated electric field will inevitably affect the impedance or apparent inductance of an inductor by a frequency-dependent amount but capacitive currents associated with screening of the measuring circuit can be routed in such a way as not to affect the measurement. *See* EDDY CURRENT; INDUCTOR.

If the magnetic circuit of an inductor includes magnetic material whose permeability depends on its previous magnetic history, or the magnetic flux caused by a direct current flowing simultaneously in the coil, its inductance will also be current- or history-dependent, and these conditions must be specified if the measurement is to be meaningful.

The electrical property of self- or mutual inductance is only defined for complete circuits. Since a measuring device or network forms part of the complete circuit when it is connected to an inductor to perform a measurement, it is necessary to ensure that either the inductance associated with the measuring circuit is negligible or that the measured quantity is defined as the change in inductance occurring when the unknown is replaced by a short circuit. The former procedure is usual for mutual inductors, and the latter for self-inductors. *See* ELECTRICAL MEASUREMENTS; INDUCTANCE.   [B.P.K.]

Inductive voltage divider   An autotransformer that has its winding divided into a number of equal-turn sections (usually 10) so that when an alternating voltage V is applied to the whole winding, the voltage across each section is nominally V/10. The progressive voltages from one end to the section junctions are thus V/10, 2V/10, 3V/10, . . . , and 9V/10. This voltage division can be realized with errors considerably less than 1 part per million of V, and such units therefore find wide use as standards of ac voltage ratio in the discipline of electrical measurements.

The division of voltage will be in error if there are differences of resistance and leakage inductance from section to section, and these errors will be significant if the differences are significant in relation to the input impedance of the winding. Leakage inductance is caused by that very small fraction of the flux from one section's winding which fails to thread the rest of the windings. The most commonly used constructional technique for minimizing such errors is to take 10 equal lengths of insulated copper wire and twist them into a "rope." The rope is wound onto a toroidal core made of thin, high-permeability, low-iron-loss magnetic material. The strands of the winding are then connected in series so that each strand forms the winding of one section of the 10-section divider (see illustration). The resistances of the sections are very nearly equal since the strands are the same length, and the leakage inductances are also closely equal and small because of the close flux coupling of this type of winding. The low-reluctance magnetic path of the core ensures a very high value of input impedance.

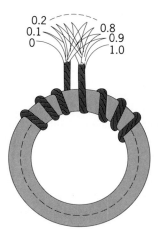

Rope winding on toroidal core making single-decade inductive voltage divider.

Thus, voltage division at low audio frequencies can be accurate to a few parts in 10^8 of V.

Inductive voltage dividers operate most accurately in the frequency range 20–1592 Hz but can be constructed to operate at frequencies up to 1 MHz. They are usually designed to accept an input voltage of up to about 0.25 times the frequency in hertz. *See* VOLTAGE MEASUREMENT.   [T.A.D.; B.P.K.]

Inductor   A device for introducing inductance into a circuit. The term covers devices with a wide range of uses, sizes, and types, including components for electric-wave filters, tuned circuits, electrical measuring circuits, and energy storage devices.

Inductors are classified as fixed, adjustable, and variable. All are made either with or without magnetic cores. Inductors without magnetic cores are called air-core coils, although the actual core material may be a ceramic, a plastic, or some other nonmagnetic material. Inductors with magnetic cores are called iron-core coils. A wide variety of magnetic materials are used, and some of these contain very little iron.

In fixed inductors coils are wound so that the turns remain fixed in position with respect to each other. Adjustable inductors have either taps for changing the number of turns desired, or consist of several fixed inductors which may be switched into various series or parallel combinations. Variable inductors are constructed so that the effective inductance can be changed. Means for doing this include (1) changing the permeability of a magnetic core; (2) moving the magnetic core, or part of it, with respect to the coil or the remainder of the core; and (3) moving one or more coils of the inductor with respect to one or more of the other coils, thereby changing mutual inductance. *See* INDUCTANCE.   [B.L.R.; W.S.P.]

Inertia   That property of matter which manifests itself as a resistance to any change in the motion of a body. Thus when no external force is acting, a body at rest remains at rest and a body in motion continues moving in a straight line with a uniform speed (Newton's first law of motion). The mass of a body is a measure of its inertia. *See* MASS.   [L.N.]

Infrared radiation   Electromagnetic radiation in which wavelengths lie in the range from about 1 micrometer to 1 millimeter. This radiation therefore has wavelengths just a little longer than those of visible light and cannot be seen with the unaided eye. The radiation was discovered in 1800 by William Herschel.

An infrared source can be described by the spectral distribution of power emitted by an ideal body (a blackbody curve). This distribution is characteristic of the temperature of the body. A real body is related to it by a radiation efficiency factor or emissivity which is the ratio at every wavelength of the emission of a real body to that of a blackbody under identical conditions. The illustration shows curves for these ideal blackbodies radiating at a number of different temperatures. The higher the temperature, the greater the total amount of radiation. *See* EMISSIVITY.

Infrared detectors are based either on the generation of a change in voltage due to a change in the detector temperature resulting from the power focused on it, or on the generation of a change in voltage due to some photon-electron interaction in the detector material. This latter effect is sometimes called the internal photoelectric effect.

Infrared techniques have been applied in military, medical, industrial, meteorological, ecological, forestry, agricultural, chemical, and other disciplines. Weather satellites use infrared imaging devices to map cloud patterns and provide the imagery seen in

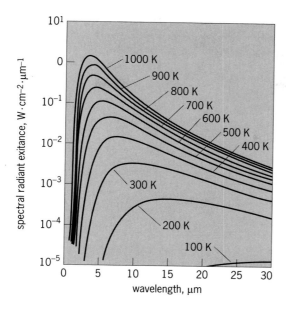

Radiation from blackbodies at different temperatures, shown on a logarithmic scale.

many weather reports. Infrared imaging devices have also been used for breast cancer screening and other medical diagnostic applications. In most of these applications, the underlying principle is that pathology produces inflammation, and these locations of increased temperature can be found with an infrared imager. Airborne infrared imagers have been used to locate the edge of burning areas in forest fires.   [W.L.Wo.]

Infrared spectroscopy   The spectroscopic study of the interaction of matter with infrared radiation. Electromagnetic waves from the long-wavelength limit of visible light at 800 nanometers to the shortest microwaves at 1 mm are used. In the wave-number units usually employed (oscillations per centimeter, read as reciprocal centimeters), this corresponds to 12,500–10 cm^{-1}. *See* INFRARED RADIATION.

The broad wavelength range of infrared radiation, and the few transparent optical materials available, require that infrared instruments be designed with reflecting optics: radiation is focused with front-surface aluminized mirrors rather than lenses. Because of strong infrared absorption by water vapor and carbon dioxide, operation takes place in a vacuum or the optical path is purged with dry nitrogen. Absorption spectroscopy is the principal method, where attention centers on the frequencies absorbed by the sample from a broadly emitting source. However, spectrometers and interferometers can easily be adapted to emission spectroscopy.

In a dispersive spectrometer (see illustration), infrared radiation from the source Q is focused by a spherical mirror M_2 onto the entrance slit S_1 of the monochromator, after passing through the sample cell SC. The beam is collimated by the off-axis paraboloid mirror M_3, dispersed by refraction through the prism P, and focused in the plane of the exit slit S_2 by a second reflection from M_3. A narrow spectral region of the dispersed radiation passes the exit slit and is focused by the ellipsoidal mirror M_7 onto the detector D, which converts the radiant energy into an electrical signal. Since the beam has been chopped at constant frequency by a rotating mechanical chopper C, this signal is an alternating current that is amplified by the lock-in amplifier A, controlling the pen of

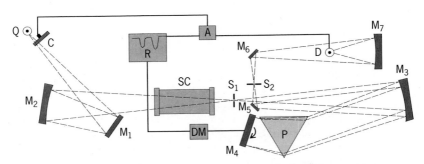

Basic single-beam recording infrared prism spectrometer. The monochromator (the portion from the entrance slit S_1 to the exit slit S_2) is a Littrow mounting, a common arrangement for infrared instruments.

the chart recorder R. To scan the spectrum, M_4 is rotated by the drive mechanism DM, which also drives the recorder. Successive frequencies are thus moved across the exit slit, producing a record of signal intensity as a function of mirror position; with a proper mechanical linkage, this record is made linear in wavelength or wave number. The same arrangement can be used with a diffraction grating as the dispersive element; the prism is removed and mirror M_4 is replaced by the grating.

In the near-infrared, quartz prisms are used, and in the mid-infrared, alkali-metal halide or calcium fluoride (CaF_2) prisms, but no prism material is suitable beyond about 50 μm. Diffraction gratings can be used in all regions with the advantage, for equivalent optical configurations, of significantly higher resolving power than prisms. Prism instruments have resolutions of little better than 1 cm^{-1} near wavelengths of maximum dispersion, and much poorer than this elsewhere. Grating resolution can be several tenths of a reciprocal centimeter for commercial spectrometers; some specially built instruments can resolve a few hundredths of a reciprocal centimeter. In most laboratories these instruments are being replaced by other techniques, although inexpensive double-beam grating spectrometers are still manufactured. *See* DIFFRACTION GRATING; OPTICAL PRISM; RESOLVING POWER (OPTICS).

In a Michelson interferometer, light from the source strikes a thin, partially reflecting beam splitter at an angle of 45° and is divided into two beams that are returned by mirrors and recombined at the beam splitter. The intensity of the recombined and interfering beams, recorded as a function of the optical path difference (or retardation) as one mirror is moved, yields an interferogram. From this, the desired spectrum (that is, the source intensity as a function of wave number) can be recovered by the mathematical procedure known as a Fourier transform. *See* INTERFEROMETRY.

Fourier-transform spectroscopy offers several advantages over dispersive methods; these are especially important for effective use of the limited radiant power available from most infrared sources. Whereas a spectrometer samples only one small frequency range at any given instant and must scan these frequencies sequentially to produce a spectrum, an interferometer processes information from all frequencies simultaneously (multiplex or Fellgett advantage). Furthermore, an interferometer passes a much greater light flux for a given resolving power than a spectrometer, which can accept only a very limited solid angle of source radiation because of the narrow slits required (throughput or Jacquinot advantage). These advantages can be translated into improvements of orders of magnitude in any one of the three interrelated important parameters of resolution, signal-to-noise ratio, and scan time. Another advantage is that the mirror movement is monitored to high precision with a fixed-frequency laser, so that the wave-number scale of the transformed spectrum is highly accurate compared with spectrometer readings.

Commercial Fourier-transform spectrometers are marketed with resolutions from a few reciprocal centimeters for routine qualitative analyses to research instruments that can resolve better than 0.002 cm^{-1}. Typically, they operate from 4000 to 400 cm^{-1} on one beam splitter such as germanium-coated potassium bromide (KBr); this range can be extended broadly in either direction with different beam-splitter materials. For the far-infrared, Mylar films in various thicknesses are used. These instruments are controlled by microprocessors and include a digital computer to handle the Fourier transform. This computing power allows data manipulation such as repetitive scanning and signal averaging; background subtraction; spectral smoothing, fitting, and scaling; and searching digitized spectral libraries and databases to identify unknowns. Although backgrounds can be subtracted with software, some Fourier-transform instruments are designed for true optical double-beam operation. Many offer rapid-scan capability for the study of short-lived species and can be adapted to recording Fourier-transform Raman spectra. *See* RAMAN EFFECT.

Infrared spectra are usually plotted as percent transmittance T or absorbance A on a scale linear in wave number ν (less commonly, in wavelength λ). Transmittance is the ratio of the intensity of radiation transmitted by the sample (I) to that incident on the sample (I_0), expressed as a percentage, so that $T = 100I/I_0$.

Infrared spectra are ideal for identifying chemical compounds because every molecule [except homonuclear diatomics such as nitrogen (N_2), oxygen (O_2), and chlorine (Cl_2)] has an infrared spectrum. Since the vibrational frequencies depend upon the masses of the constituent atoms and the strengths and geometry of the chemical bonds, the spectrum of every molecule is unique (except for optical isomers). Pure unknowns can thus be identified by comparing their spectra with recorded spectra; catalogs are available in digitized versions, and searches can be made rapidly by computer.

Simple mixtures can be identified with the help of computer software that subtracts the spectrum of a pure compound from that of the unknown mixture. More complex mixtures may require fractionation first. This has led to the development of combinations of analytical techniques, such as gas chromatography used together with Fourier-transform infrared spectroscopy; instruments that combine these functions are available commercially.

Many functional groups have characteristic infrared frequencies that are relatively independent of the molecular environment. Often, specific conclusions can be drawn from frequencies, and it may be possible to identify even a new compound from its spectrum. Group frequencies are most useful above about 1500 cm^{-1}; below this the absorptions are due to, or are influenced more by, the skeletal vibrations of the molecule. This is the "fingerprint" region, where even similar molecules may have quite different spectra.

Many details of molecular structure and dynamics can be extracted from an infrared spectrum, especially for light molecules that can be examined in the gas phase and therefore exhibit rotational structure.

Other applications include calculation of thermodynamic quantities from vibrational and rotational energy levels; studies of intermolecular forces in condensed phases; distinguishing between free or hindered internal rotation (as of methyl groups); quantitative intensity measurements to obtain bond dipole moments; studies of molecular interactions in adsorption, surface chemistry, and catalytic processes; time-resolved monitoring of transient species and chemical reaction kinetics; characterization of reactive molecules isolated at cryogenic temperatures in rare-gas lattices (matrix isolation);

electronic energy states in semiconductors and superconductors; studies of biological molecules and membranes (effects of hydrogen bonding); and analysis of energy levels in laser systems. Double-resonance techniques, in which molecules are excited by an intense pulse of ultraviolet, infrared, or microwave energy, and simultaneously probed by infrared or microwave spectroscopy, are useful in monitoring population changes in molecular energy levels, energy transfer and relaxation, and multiphoton absorption processes. *See* INTERMOLECULAR FORCES; SEMICONDUCTOR; SPECTROSCOPY; SUPERCONDUCTIVITY. [R.S.McD.]

Infrasound   Sound waves, particularly in the atmosphere, whose frequencies of pressure variation and of vibration are below the audible range, that is, lower than about 20 Hz. Earthquake and seismic waves are elastic waves which occur at infrasonic frequencies in the Earth's crust and in the oceans and seas. The physical laws of propagation in the atmosphere are essentially the same as for audible sound. The local speed of infrasound in air at ambient temperatures near 20°C (68°F) is about 340 m/s (1115 ft/s), the same as for audible sound.

At frequencies less than about 1.0 Hz, infrasound propagates through the atmosphere for distances of thousands of kilometers without substantial loss of energy. Sounds at these frequencies are almost always present at measurable intensities. Those of natural origin have many causes, including tornadoes, volcanic explosions, earthquakes, the aurora borealis, waves on the seas, large meteorites, and lightning discharges. When the wind blows, turbulent pressure fluctuations in the atmosphere occur at amplitudes up to tens of pascals, at infrasonic frequencies. People are unaware of these pressures via the sensation of hearing.

Sufficiently strong infrasound is "audible," contrary to simple acoustic tradition. The threshold sound pressure level (the least intensity for audibility) is about 92 dB at 16 Hz, and increases 12 dB per octave to about 140 dB at 1.0 Hz. However, there is no sensation of tone. Listeners variously describe audible infrasound as pumping, popping effect, or chugging. For vibration at very low frequencies, motion sickness of people in boats must have been one of the earliest noticeable effects. The human body is particularly sensitive to vibrations and infrasound near 7 Hz, at which frequency there is an overall mechanical resonance of organs in the abdominal and chest cavities. *See* ATMOSPHERIC ACOUSTICS; SOUND. [R.K.C.]

Intercalation compounds   Crystalline or partially crystalline solids consisting of a host lattice containing voids into which guest atoms or molecules are inserted. Candidate hosts for intercalation reactions may be classified by the number of directions (0 to 3) along which the lattice is strongly bonded and thus unaffected by the intercalation reaction. Isotropic, three-dimensional lattices (including many oxides and zeolites) contain large voids that can accept multiple guest atoms or molecules. Layer-type, two-dimensional lattices (graphite and clays) swell up perpendicular to the layers when the guest atoms enter. The chains in one-dimensional structures (polymers such as polyacetylene) rotate cooperatively about their axes during the intercalation reaction to form channels that are occupied by the guest atoms. In the intercalation family based on solid C_{60} (buckminsterfullerene), the zero-dimensional host lattice consists of 60-atom carbon clusters with strong internal bonding but weak intercluster bonding. These clusters pack together like hard 1-nm-diameter spheres, creating interstitial voids which are large enough to accept most elements in the periodic table. The proportions of guest and host atoms may be varied continuously in many of these materials, which are therefore not true compounds. Many ternary and quaternary substances, containing two or three distinct guest species, are known. The guest may be an atom or inorganic molecule (such as an alkali metal, halogen, or metal halide), an organic molecule (for example, an aromatic such as benzene, pyridine, or ammonia), or both. *See* CRYSTAL STRUCTURE.

Many applications of intercalation compounds derive from the reversibility of the intercalation reaction. The best-known example is pottery: Water intercalated between the silicate sheets makes wet clay plastic, while driving the water out during firing results in a dense, hard, durable material. Many intercalation compounds are good ionic conductors and are thus useful as electrodes in batteries and fuel cells. A technology for lightweight rechargeable batteries employs lithium ions which shuttle back and forth between two different intercalation electrodes as the battery is charged and discharged: vanadium oxide (three-dimensional) and graphite (two-dimensional). Zeolites containing metal atoms remain sufficiently porous to serve as catalysts for gas-phase reactions. Many compounds can be used as convenient storage media, releasing the guest molecules in a controlled manner by mild heating.

[J.E.Fi.]

Interference filter

Interference filter   An optical filter in which the wavelengths that are not transmitted are removed by interference phenomena rather than by absorption or scattering. In addition to being able to duplicate most of the spectral characteristics of absorption color filters, these devices can be made to transmit a very narrow band of wavelengths. They can thus be used as monochromators to examine a radiation source at the wavelength of a single spectrum line. For example, the solar disk can be observed in light of the hydrogen line Hα and thus the distribution of excited hydrogen over the disk can be determined. Most narrow-band interference filters are based on the Fabry-Perot interferometer. *See* INTERFERENCE OF WAVES; INTERFEROMETRY.   [B.H.Bi.]

Interference of waves   The process whereby two or more waves of the same frequency or wavelength combine to form a wave whose amplitude is the sum of the amplitudes of the interfering waves. The interfering waves can be electromagnetic, acoustic, or water waves, or in fact any periodic disturbance.

The most striking feature of interference is the effect of adding two waves in which the trough of one wave coincides with the peak of another. If the two waves are of equal amplitude, they can cancel each other out so that the resulting amplitude is zero. This is perhaps most dramatic in sound waves; it is possible to generate acoustic waves to arrive at a person's ear so as to cancel out disturbing noise. In optics, this cancellation can occur for particular wavelengths in a situation where white light is a source. The resulting light will appear colored. This gives rise to the iridescent colors of beetles' wings and mother-of-pearl, where the substances involved are actually colorless or transparent.

To observe interference with waves generated by atomic or molecular transitions, it is necessary to use a single source and to split the light from the source into parts which can then be recombined. In this case, the amplitude and phase changes occur simultaneously in each of the parts at the same time.

The simplest technique for producing a splitting from a single source was done by T. Young in 1801 and was one of the first demonstrations of the wave nature of light. In this experiment, a narrow slit is illuminated by a source, and the light from this slit is caused to illuminate two adjacent slits. The light from these two parallel slits can interfere, and the interference can be seen by letting the light from the two slits fall on a white screen. The screen will be covered with a series of parallel fringes. The location of these fringes can be derived approximately as follows: In the illustration, S_1 and S_2 are the two slits separated by a distance d. Their plane is a distance l from the screen.

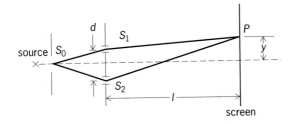

Since the slit S_0 is equidistant from S_1 and S_2, the intensity and phase of the light at each slit will be the same. The light falling on P from slit S_1 can be represented by Eq. (1) and from S_2 by Eq. (2), where f is the frequency, t the time, c the velocity of

$$A = A_0 \sin 2\pi f \left(t - \frac{x_1}{c} \right) \tag{1}$$

$$B = A_0 \sin 2\pi f \left(t - \frac{x_2}{c} \right) \tag{2}$$

light; x_1 and x_2 are the distances of P from S_1 and S_2, and A_0 is the amplitude. This amplitude is assumed to be the same for each wave since the slits are close together, and x_1 and x_2 are thus nearly the same. The square of the amplitude or the intensity at P can then be written as Eq. (3).

$$I = 4A_0^2 \cos^2 \frac{2\pi f}{c} (x_1 - x_2) \tag{3}$$

In general, I is very much larger than y so that Eq. (3) can be simplified to Eq. (4).

$$I = 4A_0^2 \cos^2 \pi \left(\frac{yd}{l\lambda} \right) \tag{4}$$

Equation (4) is a maximum when Eq. (5) holds and a minimum when Eq. (6) holds, where n is an integer.

$$y = n\lambda \frac{l}{d} \tag{5}$$

$$y = (n + 1/2)\lambda \frac{l}{d} \tag{6}$$

Accordingly, the screen is covered with a series of light and dark bands called interference fringes. If the source behind slit S_0 is white light and thus has wavelengths varying perhaps from 400 to 700 nanometers, the fringes are visible only where $x_1 - x_2$ is a few wavelengths, that is, where n is small. At large values of n, the position of the nth fringe for red light will be very different from the position for blue light, and the fringes will blend together and be washed out.

The energy carried by a wave is measured by the intensity, which is equal to the square of the amplitude. The total energy failing on the screen is not changed by the presence of interference. The energy density at a particular point is, however, drastically changed. This fact is most important for those waves of the electromagnetic spectrum which can be generated by vacuum-tube oscillators. The sources of radiation or antennas can be made to emit coherent waves which will undergo interference. This makes possible a redistribution of the radiated energy. Quite narrow beams of radiation can be produced by the proper phasing of a linear antenna array. *See* INTERFEROMETRY.   [B.H.Bi.]

Interferometry The design and use of optical interferometers. Optical interferometers based on both two-beam interference and multiple-beam interference of light are extremely powerful tools for metrology and spectroscopy. A wide variety of measurements can be performed, ranging from determining the shape of a surface to an accuracy of less than a millionth of an inch (25 nanometers) to determining the separation, by millions of kilometers, of binary stars. In spectroscopy, interferometry can be used to determine the hyperfine structure of spectrum lines. By using lasers in classical interferometers as well as holographic interferometers and speckle interferometers, it is possible to perform deformation, vibration, and contour measurements of diffuse objects that could not previously be performed. There are two basic classes of interferometers: division of wavefront and division of amplitude.

Michelson interferometer. The Michelson interferometer (Fig. 1) is based on division of amplitude. Light from an extended source S is incident on a partially reflecting plate (beam splitter) P_1. The light transmitted through P_1 reflects off mirror M_1 back to plate P_1. The light which is reflected proceeds to M_2 which reflects it back to P_1. At P_1, the two waves are again partially reflected and partially transmitted, and a portion of each wave proceeds to the receiver R, which may be a screen, a photocell, or a human eye. Depending on the difference between the distances from the beam splitter to the mirrors M_1 and M_2, the two beams will interfere constructively or destructively. Plate P_2 compensates for the thickness of P_1.

The function of the beam splitter is to superimpose (image) one mirror onto the other. When the mirrors' images are completely parallel, the interference fringes appear circular. If the mirrors are slightly inclined about a vertical axis, vertical fringes are formed across the field of view. These fringes can be formed in white light if the path difference in part of the field of view is made zero. Just as in other interference experiments, only a few fringes will appear in white light.

Twyman-Green interferometer. If the Michelson interferometer is used with a point source instead of an extended source, it is called a Twyman-Green interferometer. The use of the laser as the light source for the Twyman-Green interferometer has made it an extremely useful instrument for testing optical components. The great advantage of a laser source is that it makes it possible to obtain bright, good-contrast, interference fringes even if the path lengths for the two arms of the interferometer are quite different. See LASER.

The Twyman-Green interferometer can be used to test a flat mirror. In this case, M_1 in Fig. 1 is a reference surface and M_2 is the flat surface being tested. If the test surface is perfectly flat, then straight, equally spaced fringes are obtained. Departure from the

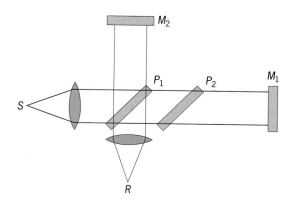

Fig. 1. Michelson interferometer.

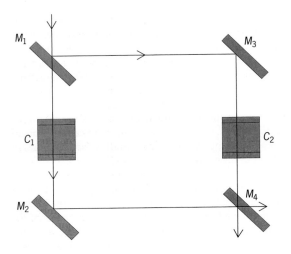

Fig. 2. Mach-Zehnder interferometer.

straight, equally spaced condition shows directly how the surface differs from being perfectly flat. A height change of half a wavelength will cause an optical path change of one wavelength and a deviation from fringe straightness of one fringe. Thus, the fringes give surface height information, just as a topographical map gives height or contour information.

The basic Twyman-Green interferometer can be modified to test concave-spherical mirrors. In the interferometer, the center of curvature of the surface under test is placed at the focus of a high-quality diverger lens so that the wavefront is reflected back onto itself. Likewise, a convex-spherical mirror can be tested. Also, if a high-quality spherical mirror is used, the high-quality diverger lens can be replaced with the lens to be tested.

Fizeau interferometer. One of the most commonly used interferometers in optical metrology is the Fizeau interferometer, which can be thought of as a folded Twyman-Green interferometer. In the Fizeau, the two surfaces being compared, which can be flat, spherical, or aspherical, are placed in close contact. The light reflected off these two surfaces produces interference fringes. For each fringe, the separation between the two surfaces is a constant. If the two surfaces match, straight, equally spaced fringes result. Surface height variations between the two surfaces cause the fringes to deviate from straightness or equal separation.

Mach-Zehnder interferometer. The Mach-Zehnder interferometer (Fig. 2) is a variation of the Michelson interferometer and, like the Michelson interferometer, depends on amplitude splitting of the wavefront. Light enters the instrument and is reflected and transmitted by the semitransparent mirror M_1. The reflected portion proceeds to M_3, where it is reflected through the cell C_2 to the semitransparent mirror M_4. Here it combines with the light transmitted by M_1 to produce interference. The light transmitted by M_1 passes through a cell C_1, similar to C_2, and is used to compensate for the windows of C_2. The major application of this instrument is in studying airflow around models of aircraft, missiles, or projectiles.

Shearing interferometers. In a lateral-shear interferometer a wavefront is interfered with a shifted version of itself. A bright fringe is obtained at the points where the slope of the wavefront times the shift between the two wavefronts is equal to an integer number of wavelengths. That is, for a given fringe the slope or derivative of the wavefront is a constant. For this reason a lateral-shear interferometer is often called a differential interferometer. Another type of shearing interferometer is a radial-shear interferometer. Here, a wavefront is interfered with an expanded version of itself. This interferometer is sensitive to radial slopes.

Michelson stellar interferometer. A Michelson stellar interferometer can be used to measure the diameter of stars which are as small as 0.01 second of arc. This task is impossible with a ground-based optical telescope since the atmosphere limits the resolution of the largest telescope to not much better than 1 second of arc.

Fabry-Perot interferometer. All the interferometers discussed above are two-beam interferometers. The Fabry-Perot interferometer is a multiple-beam interferometer since the two glass plates are partially silvered on the inner surfaces, and the incoming wave is multiply reflected between the two surfaces. The position of the fringe maxima is the same for multiple beam interference as two-beam interference; however, as the reflectivity of the two surfaces increases and the number of interfering beams increases, the fringes become sharper.

Holographic interferometry. A wave recorded in a hologram is effectively stored for future reconstruction and use. Holographic interferometry is concerned with the formation and interpretation of the fringe pattern which appears when a wave, generated at some earlier time and stored in a hologram, is later reconstructed and caused to interfere with a comparison wave. It is the storage or time-delay aspect which gives the holographic method a unique advantage over conventional optical interferometry. *See* HOLOGRAPHY.

Speckle interferometry. A random intensity distribution, called a speckle pattern, is generated when light from a highly coherent source, such as a laser, is scattered by a rough surface. The use of speckle patterns in the study of object displacements, vibration, and distortion is becoming of more importance in the nondestructive testing of mechanical components. *See* SPECKLE.

Phase-shifting interferometry. Electronic phase-measurement techniques can be used in interferometers such as the Twyman-Green, where the phase distribution across the interferogram is being measured. Phase-shifting interferometry is often used for these measurements since it provides for rapid precise measurement of the phase distribution. In phase-shifting interferometry, the phase of the reference beam in the interferometer is made to vary in a known manner. This can be achieved, for example, by mounting the reference mirror on a piezoelectric transducer. By varying the voltage on the transducer, the reference mirror is moved a known amount to change the phase of the reference beam a known amount. A solid-state detector array is used to detect the intensity distribution across the interference pattern. This intensity distribution is read into computer memory three or more times, and between each intensity measurement the phase of the reference beam is changed a known amount. From these three or more intensity measurements, the phase across the interference pattern can be determined to within a fraction of a degree.

[J.C.Wy.]

Intermediate vector boson   One of the three fundamental particles that transmit the weak force. (An example of a weak interaction process is nuclear beta decay.) These elementary particles—the W^+, W^-, and Z^0 particles—were discovered in 1983 in very high-energy proton-antiproton collisions. It is through the exchange of W and Z bosons that two particles interact weakly, just as it is through the exchange of photons that two charged particles interact electromagnetically. The intermediate vector bosons were postulated to exist in the 1960s; however, their large masses prevented their production and study at accelerators until 1983. Their discovery was a key step toward unification of the weak and electromagnetic interactions.

See ELECTROWEAK INTERACTION; ELEMENTARY PARTICLE; FUNDAMENTAL INTERACTIONS; WEAK NUCLEAR INTERACTIONS.

The W and Z particles are roughly 100 times the mass of a proton. Therefore, the experiment to search for the W and the Z demanded collisions of elementary particles at the highest available center-of-mass energy. Such very high center-of-mass energies capable of producing the massive W and Z particles were achieved with collisions of protons and antiprotons at the laboratory of the European Organization for Nuclear Research (CERN) near Geneva, Switzerland. *See* PARTICLE ACCELERATOR; PARTICLE DETECTOR.

Striking features of both the charged W and the Z^0 particles are their large masses. The charged boson (W^+ and W^-) mass is measured to be about 80 GeV/c^2, and the neutral boson (Z^0) mass is measured to be about 91 GeV/c^2. (For comparison, the proton has a mass of about 1 GeV/c^2.) Prior to the discovery of the W and the Z, particle theorists had met with some success in the unification of the weak and electromagnetic interactions. The electroweak theory as it is understood today is due largely to the work of S. Glashow, S. Weinberg, and A. Salam. Based on low-energy neutrino scattering data, which in this theory involves the exchange of virtual W and Z particles, theorists made predictions for the W and Z masses. The actual measured values are in agreement (within errors) with predictions. The discovery of the W and the Z particles at the predicted masses is an essential confirmation of the electroweak theory.

Only a few intermediate vector bosons are produced from 10^9 proton-antiproton collisions at a center-of-mass energy of 540 GeV. This small production probability per $p\bar{p}$ collision is understood to be due to the fact that the bosons are produced by a single quark-antiquark annihilation. The other production characteristics of the intermediate vector bosons, such as longitudinal and transverse momentum distributions (with respect to the $p\bar{p}$ colliding beam axis), all provide support for this theoretical picture. *See* QUARKS.   [J.W.R.]

Intermolecular forces   Attractive or repulsive interactions that occur between all atoms and molecules. Intermolecular forces become significant at molecular separations of about 1 nanometer or less, but are much weaker than the forces associated with chemical bonding. They are important, however, because they are responsible for many of the physical properties of solids, liquids, and gases. These forces are also largely responsible for the three-dimensional arrangements of biological molecules and polymers.

Intermolecular forces can be classified into several types, of which two are universal. The attractive force known as dispersion arises from the quantum-mechanical fluctuation of the electron density around the nucleus of each atom. At distances greater than 1 nm or so, the electrons of each atom move independently of the other, and the charge distribution is spherically symmetric. At shorter distances, an instantaneous fluctuation of the charge density in one atom can affect the other. If the electrons of one atom move briefly to the side nearer the other, the electrons of the other atom are repelled to the far side. In this configuration, both atoms have a small dipole moment, and they attract each other electrostatically. At another moment, the electrons may move the other way, but their motions are correlated so that an attractive force is maintained on average. Molecular orbital theory shows that the electrons of each atom are slightly more likely to be on the side nearer to the other atom, so that each atomic nucleus is attracted by its own electrons in the direction of the other atom.

At small separations the electron clouds can overlap, and repulsive forces arise. These forces are described as exchange-repulsion, and are a consequence of the Pauli exclusion principle, a quantum-mechanical effect which prevents electrons from occupying the same region of space simultaneously. To accommodate it, electrons are squeezed out from the region between the nuclei, which repel each other as a result. Each element can be assigned, approximately, a characteristic van der Waals radius; that is, when atoms in different molecules approach more closely than the sum of their radii, the repulsion ennergy increases sharply. It is this effect that gives molecules their characteristic shape, leading to steric effects in chemical reactions. *See* EXCLUSION PRINCIPLE.

The other important source of intermolecular forces is the electrostatic interaction. When molecules are formed from atoms, electrons flow from electropositive atoms to electronegative ones, so that the atoms become somewhat positively or negatively charged. In addition, the charge distribution of each atom may be distorted by the process of bond formation, leading to atomic dipole and quadrupole moments. The electrostatic interaction between these is an important source of intermolecular forces, especially in polar molecules, but also in molecules that are not normally thought of as highly polar. The electrostatic field of a molecule may cause polarization of its neighbors, and this leads to a further induction contribution to the intermolecular interaction. An induction interaction can often polarize both molecules in such a way as to favor interactions with further molecules, leading to a cooperative network of intermolecular attractions. This effect is important in the network structure of water and ice.

Intermolecular forces are responsible for many of the bulk properties of matter in all its phases. A realistic description of the relationship between pressure, volume, and temperature of a gas must include the effects of attractive and repulsive forces between molecules. The viscosity, diffusion, and surface tension of liquids are examples of physical properties which depend strongly on intermolecular forces. Intermolecular forces are also responsible for the ordered arrangement of molecules in solids, and account for their elasticity and properties (such as the velocity of sound in materials).   [A.J.S.]

Internal energy   A characteristic property of the state of a thermodynamic system, introduced in the first law of thermodynamics. For a static, closed system (no bulk motion, no transfer of matter across its boundaries), the change in internal energy for a process is equal to the heat absorbed by the system from its surroundings minus the work done by the system on its surroundings. Only a change in internal energy can be measured, not its value for any single state. For a given process, the change in internal energy is fixed by the initial and final states and is independent of the path by which the change in state is accomplished. *See* THERMODYNAMIC PRINCIPLES.   [P.J.B.]

Inverse scattering theory   A theory whose objective is to determine the scattering object, or an interaction potential energy, from the knowledge of a scattered field. This is the opposite problem from direct scattering theory, where the scattering amplitude is determined from the equations of motion, including the potential. The equations of motion are usually linear (operator-valued) equations. *See* SCATTERING EXPERIMENTS (ATOMS AND MOLECULES); SCATTERING EXPERIMENTS (NUCLEI).

Inverse scattering theories can be divided into two types: (1) pure inverse problems, when the data consist of complete, noise-free information of the scattering amplitude; and (2) applied inverse problems, when incomplete data which are corrupted by noise are given. Many different applied inverse problems can be obtained from any pure inverse problem by using different band-limiting procedures and different noise spectra.

The difficulty of determining the exact object which produced a scattering amplitude is evident. It is often a priori information about the scatterer that makes the inversion possible.

Much of the basic knowledge of systems of atoms, molecules, and nuclear particles is obtained from inverse scattering studies using beams of different particles as probes. For the Schrödinger equation with spherical symmetry or in one dimension, there is an exact solution of the inverse problem.

A number of high-technology areas (nondestructive evaluation, medical diagnostics including acoustic and ultrasonic imaging, x-ray absorption and nuclear magnetic resonance tomography, radar scattering and geophysical exploration) use inverse scattering theory. Several classical waves including acoustic, electromagnetic, ultrasonic, x-rays, and others are used.

All of the inverse scattering technologies require the solution to ill-posed or improperly posed problems. A model equation is well posed if it has a unique solution which depends continuously on the initial data. It is ill posed otherwise. The ill-posed problems which are amenable to analysis, called regularizable ill-posed problems, are those which depend discontinuously upon the data. This destroys uniqueness, although solutions (in fact, many solutions) exist.   [B.DeF.]

Ion sources   Devices which produce positive or negative electrically charged atoms or molecules.

In general, ion sources fall into three major categories: those designed for positive-ion generation, those for negative-ion generation, and a highly specialized type of source designed to produce a polarized ion beam. The positive-ion source category may further be subdivided into sources specifically designed to generate singly charged ions and those designed to produce very highly charged ions.

Ion sources have acquired a wide variety of applications. They are used in a variety of different types of accelerators for nuclear research; have application in the field of fusion research; and are used for ion implantation, in isotope separators, in ion microprobes, as a means of rocket propulsion, in mass spectrometers, and for ion milling. *See* NUCLEAR FUSION; PARTICLE ACCELERATOR.   [R.M.]

Ionic crystals   A class of crystals in which the lattice-site occupants are charged ions held together primarily by their electrostatic interaction. Such binding is called ionic binding. Empirically, ionic crystals are distinguished by strong absorption of infrared radiation, good ionic conductivity at high temperatures, and the existence of planes along which the crystals cleave easily. *See* CRYSTAL STRUCTURE.

Compounds of strongly electropositive and strongly electronegative elements form solids which are ionic crystals, for example, the alkali halides, other monovalent metal halides, and the alkaline-earth halides, oxides, and sulfides. Crystals in which some of the ions are complex, such as metal carbonates, metal nitrates, and ammonium salts, may also be classed as ionic crystals.

As a crystal type, ionic crystals are to be distinguished from other types such as molecular crystals, valence crystals, or metals. The ideal ionic crystal as defined is approached most closely by the alkali halides (see illustration). Other crystals often classed as ionic have binding which is not exclusively ionic but includes a certain admixture of covalent binding. Thus the term ionic crystal refers to an idealization to which real crystals correspond to a greater or lesser degree, and crystals exist having characteristics of more than one crystal type.

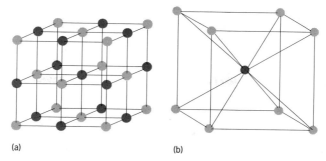

Lattices of (*a*) sodium chloride and (*b*) cesium chloride. The darker circles represent positive ions and the lighter circles negative ions. (*After F. Seitz, The Modern Theory of Solids, Dover, 1987*)

Ionic crystals, especially alkali halides, have played a very prominent role in the development of solid-state physics. They are relatively easy to produce as large, quite pure, single crystals suitable for accurate and reproducible experimental investigations. In addition, they are relatively easy to subject to theoretical treatment since they have simple structures and are bound by the well-understood Coulomb force between the ions. This is in contrast to metals and covalent crystals, which are bound by more complicated forces, and to molecular crystals, which either have complicated structures or are difficult to produce as single crystals. Being readily available and among the simplest known solids, they have thus been a frequent and profitable meeting place between theory and experiment. These same features of ionic crystals have made them attractive as host crystals for the study of crystal defects: deliberately introduced impurities, vacancies, interstitials, and color centers. *See* COLOR CENTERS; CRYSTAL DEFECTS.

[B.G.D.]

Most ionic crystals have large band gaps, and are therefore generally good electronic insulators. However, electrical conduction occurs by the motion of ions through these crystals. The presence of point defects, that is, deviations from ideal order in the crystalline lattice, facilitates this motion, thus giving rise to transport of electric charge. In an otherwise perfect lattice where all lattice sites are fully occupied, ions cannot be mobile.

Many so-called normal ionic crystals possess conductivities of about 10^{-10} (ohm $\cdot$ cm)$^{-1}$ or lower at room temperature. However, a relatively small number of ionic materials, called superionic conductors or fast ionic conductors, display conductivities of the order of 10^{-1} to 10^{-2} (ohm $\cdot$ cm)$^{-1}$, which imply ionic liquidlike behavior. In most of these crystals, only one kind of ionic species is mobile, and its diffusion coefficient and mobility attain values such as found otherwise only in liquids. Due to their high value of ionic conductivity as well as their ability to selectively transport ionic species, superionic conductors have successfully been employed as solid electrolytes in many applications. *See* DIFFUSION.

[T.M.Gü.]

Ionization   The process by which an electron is removed from an atom, molecule, or ion. It is of basic importance to electrical conduction in gases and liquids. In the simplest case, ionization may be thought of as a transition between an initial state consisting of a neutral atom and a final state consisting of a positive ion and a free electron. In more complicated cases, a molecule may be converted to a heavy positive ion and a heavy negative ion which are separated.

[G.H.M.]

Ionization chamber   An instrument for detecting ionizing radiation by measuring the amount of charge liberated by the interaction of ionizing radiation with suitable gases, liquids, or solids.

While the gold leaf electroscope is the oldest form of ionization chamber, instruments of this type are still widely used as monitors of radiations by workers in the nuclear or radiomedical professions. However, for many purposes it is useful to measure the ionization pulse produced by a single ionizing particle. *See* ELECTROSCOPE.

The simplest form of a pulse ionization chamber consists of two conducting electrodes in a container filled with gas (see illustration). A battery, or other power supply, maintains an electric field between the positive anode and the negative cathode. When ionizing radiation penetrates the gas in the chamber—entering, for example, through a thin gas-tight window—this radiation liberates electrons from the gas atoms leaving positively charged ions. The electric field present in the gas sweeps these electrons and ions out of the gas, the electrons going to the anode and the positive ions to the cathode.

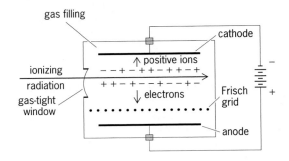

Parallel-plate ionization chamber.

In a chamber, such as that represented in the illustration, the current begins to flow as soon as the electrons and ions begin to separate under the influence of the applied electric field. The time it takes for the full current pulse to be observed depends on the drift velocity of the electrons and ions in the gas. Because the ions are thousands of times more massive than the electrons, the electrons always travel several orders of magnitude faster than the ions. As a result, virtually all pulse ionization chambers make use of only the relatively fast electron signal.

One of the most important uses of an ionization chamber is to measure the total energy of a particle or, if the particle does not stop in the ionization chamber, the energy lost by the particle in the chamber. In addition to energy information, ionization chambers are now routinely built to give information about the position within the gas volume where the initial ionization event occurred. This information can be important not only in experiments in nuclear and high-energy physics where these position-sensitive detectors were first developed, but also in medical and industrial applications.

Foremost among the other applications is the use of gas ionization chambers for radiation monitoring. Portable instruments of this type usually employ a detector containing approximately 60 in.^{3} (1 liter) of gas, and operate by integrating the current produced by the ambient radiation. Another application of ionization chambers is the use of air-filled chambers as domestic fire alarms. Yet another development in ion chamber usage is that of two-dimensional imaging in x-ray medical applications to replace the use of photographic plates.

Gaseous ionization chambers have also found application as total-energy monitors for high-energy accelerators. Such applications involve the use of a very large number of interleaved thin parallel metal plates immersed in a gas inside a large container.

Ionization chambers can be made where the initial ionization occurs, not in gases, but in suitable liquids or solids. In the solid-state ionization chamber (or solid-state detector) the gas filling is replaced by a large single crystal of suitably chosen solid material. In this case the incident radiation creates electron-hole pairs in the crystal, and this constitutes the signal charge. Silicon and germanium detectors have proved to be highly successful and have led to detectors that have revolutionized low-energy nuclear spectroscopy. The use of a liquid in an ionization chamber combines many of the advantages of both solid and gas-filled ionization chambers; most importantly, such devices have the flexibility in design of gas chambers with the high density of solid chambers. During the 1970s a number of groups built liquid argon ionization chambers and demonstrated their feasibility.

[W.A.L.]

Ionization gage   An instrument for measuring vacuum by ionizing the gas present and measuring the ion current. There are two types of ionization gages.

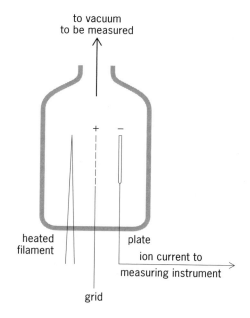

Hot-filament ionization gage.

In the hot-filament ionization gage (see illustration), electrons emitted by a filament are attracted toward a positively charged grid electrode. Collisions of electrons with gas molecules produce ions, which are then attracted to a negatively charged electrode. The current measured at this electrode is directly proportional to the pressure or gas density.

In the cold-cathode (Philips or Penning) ionization gage, a high voltage is applied between two electrodes. Fewer electrons are emitted, but a strong magnetic field deflects the electron stream, increasing the length of the electron path which increases the chance for ionizing collisions of electrons with gas molecules. *See* VACUUM MEASUREMENT.

[R.C.]

Ionization potential   The potential difference through which a bound electron must be raised to free it from the atom or molecule to which it is attached. In particular, the ionization potential is the difference in potential between the initial state, in which the electron is bound, and the final state, in which it is at rest at infinity.

The ionization potential for the removal of an electron from a neutral atom other than hydrogen is more correctly designated as the first ionization potential. The potential associated with the removal of a second electron from a singly ionized atom or molecule is then the second ionization potential, and so on.   [G.H.M.]

Isentropic flow   Compressible flow in which entropy remains constant throughout the flowfield. A slight distinction is sometimes made, especially in Europe, as follows. If the entropy of a fluid element moving along a streamline in a flow remains constant, the flow is isentropic along a streamline. However, the value of the entropy may be different along different streamlines, thus allowing entropy changes normal to the streamlines. An example is the flowfield behind a curved shock wave; here, streamlines that pass through different locations along the curved shock wave experience different increases in entropy. Hence, downstream from this shock, the entropy can be constant along a given streamline but differs from one streamline to another. This type of flow, with entropy constant along streamlines, is sometimes defined as isentropic. Flow with entropy constant everywhere is then called homentropic. *See* COMPRESSIBLE FLOW; ENTROPY; ISENTROPIC PROCESS.

Because of the second law of thermodynamics, an isentropic flow does not strictly exist. From the definition of entropy, an isentropic flow is both adiabatic and reversible. However, all real flows experience to some extent the irreversible phenomena of friction, thermal conduction, and diffusion. Any nonequilibrium, chemically reacting flow is also irreversible. However, there are a large number of gas dynamic problems with entropy increase negligibly slight, which for the purpose of analysis are assumed to be isentropic. Examples are flow through subsonic and supersonic nozzles, as in wind tunnels and rocket engines; and shock-free flow over a wing, fuselage, or other aerodynamic shape. For these flows, except for the thin boundary-layer region adjacent to the surface where friction and thermal conduction effects can be strong, the outer inviscid flow can be considered isentropic. If shock waves exist in the flow, the entropy increase across these shocks destroys the assumption of isentropic flow, although the flow along streamlines between shocks may be isentropic. *See* ADIABATIC PROCESS; BOUNDARY-LAYER FLOW; SHOCK WAVE; THERMODYNAMIC PRINCIPLES; THERMODYNAMIC PROCESSES.   [J.D.A.]

Isentropic process   In thermodynamics, a process involving change without any increase or decrease of entropy. Since the entropy always increases in a spontaneous process, one must consider reversible or quasistatic processes. During a reversible process the quantity of heat transferred is directly proportional to the system's entropy change. Systems which are thermally insulated from their surroundings undergo processes without any heat transfer; such processes are called adiabatic. Thus during an isentropic process there are no dissipative effects and the system neither absorbs nor gives off heat. For this reason the isentropic process is sometimes called the reversible adiabatic process. *See* ADIABATIC PROCESS; ENTROPY; THERMODYNAMIC PROCESSES.   [P.E.Bl.]

Ising model   A model which consists of a lattice of "spin" variables with two characteristic properties: (1) each of the spin variables independently takes on either the value $+1$ or the value -1; and (2) only pairs of nearest-neighboring spins can interact. The study of this model in two dimensions forms the basis of the modern theory of phase transitions and, more generally, of cooperative phenomena.

A macroscopic piece of material consists of a large number of atoms, the number being of the order of the Avogadro number (approximately 6×10^{23}). Thermodynamic phenomena all depend on the participation of such a large number of atoms. Even though the fundamental interaction between atoms is short-ranged, the presence of this large number of atoms can, under suitable conditions, lead to an effective interaction between widely separated atoms. Phenomena due to such effective long-range interactions are referred to as cooperative phenomena. The simplest examples of cooperative phenomena are phase transitions. The most familiar phase transition is either the condensation of steam into water or the freezing of water into ice. Only slightly less familiar is the ferromagnetic phase transition that takes place at the Curie temperature, which, for example, is roughly 1043 K for iron. *See* CURIE TEMPERATURE; FERROMAGNETISM; PHASE TRANSITIONS.

[B.M.McC.; T.T.W.]

Isobar (nuclear physics)   One of two or more atoms which have a common mass number A but which differ in atomic number Z. Thus, although isobars possess approximately equal masses, they differ in chemical properties; they are atoms of different elements. Isobars whose atomic numbers differ by unity cannot both be stable; one will inevitably decay into the other. *See* ELECTRON CAPTURE; RADIOACTIVITY.   [H.E.D.]

Isobaric process   A thermodynamic process during which the pressure remains constant. When heat is transferred to or from a gaseous system, a volume change occurs at constant pressure. This thermodynamic process can be illustrated by the expansion of a substance when it is heated. The system is then capable of doing an amount of work on its surroundings. The maximum work is done when the external pressure of the surroundings on the system is equal to the pressure of the system. *See* ISOMETRIC PROCESS; POLYTROPIC PROCESS.

[P.E.Bl.]

Isoelectronic sequence   A term used in spectroscopy to designate the set of spectra produced by different chemical elements ionized in such a way that their atoms or ions contain the same number of electrons. The sequence in the table is an example.

Example of isoelectronic sequence		
Designation of spectrum	Emitting atom or ion	Atomic number, Z
CaI	Ca	20
ScII	Sc^+	21
TiIII	Ti^{2+}	22
VIV	V^{3+}	23
CrV	Cr^{4+}	24
MnVI	Mn^{5+}	25

Since the neutral atoms of these elements each contain Z electrons, removal of one electron from scandium, two from titanium, and so forth, yields a series of ions all of which have 20 electrons. Isoelectronic sequences are useful in predicting unknown spectra of ions belonging to a sequence in which other spectra are known. *See* ATOMIC STRUCTURE AND SPECTRA.

[F.A.J./W.W.W.]

Isometric process   A constant-volume thermodynamic process in which the system is confined by mechanically rigid boundaries. No direct mechanical work can be done on the surroundings by a system with rigid boundaries; therefore the heat transferred into or out of the system equals the change of internal energy stored in the system. This change in the internal energy, in turn, is a function of the specific heat and the temperature change in the system. *See* POLYTROPIC PROCESS.   [P.E.Bl.]

Isothermal process   A thermodynamic process which occurs with a heat addition or removal rate just adequate to maintain constant temperature. The change in the internal energy per mole U accompanying a change in volume in an isothermal process is given by the equation below, where T is the temperature, P is the pressure,

$$U_2 - U_1 = \int_{V_1}^{V_2} \left[T \left(\frac{\partial P}{\partial T} \right)_V - P \right] dV$$

and V is the volume per mole. The integral in the equation is zero for an ideal gas, and approximately zero for a condensed phase (solid or liquid) for which the volume changes vary little with pressure. Thus, in both these cases, $U_2 = U_1$. For real gases, the integral is nonzero, and the internal energy change is computed using the equation of state of the gas in the equation. *See* THERMODYNAMIC PROCESSES.   [S.I.S.]

Isotone   One of two or more atoms which display a constant difference $A - Z$ between their mass number A and their atomic number Z. Thus, despite differences in the total number of nuclear constituents, the numbers of neutrons in the nuclei of isotones are the same. The numbers of naturally occurring isotones provide useful evidence concerning the stability of particular neutron configurations. For example, the relatively large number (six and seven, respectively) of naturally occurring 50- and 82-neutron isotones suggests that these nuclear configurations are especially stable. On the other hand, from the fact that most atoms with odd numbers of neutrons are anisotonic, one may conclude that odd-neutron configurations are relatively unstable. *See* NUCLEAR STRUCTURE.   [H.E.D.]

Isotope   One member of a (chemical-element) family of atomic species which has two or more nuclides with the same number of protons (Z) but a different number of neutrons (N). Because the atomic mass is determined by the sum of the number of protons and neutrons contained in the nucleus, isotopes differ in mass. Since they contain the same number of protons (and hence electrons), isotopes have the same chemical properties. However, the nuclear and atomic properties of isotopes can be different. The electronic energy levels of an atom depend upon the nuclear mass. Thus, corresponding atomic levels of isotopes are slightly shifted relative to each other. A nucleus can have a magnetic moment which can interact with the magnetic field generated by the electrons and lead to a splitting of the electronic levels. The number of resulting states of nearly the same energy depends upon the spin of the nucleus and the characteristics of the specific electronic level. *See* ATOMIC STRUCTURE AND SPECTRA; HYPERFINE STRUCTURE; ISOTOPE SHIFT.

Of the 12 elements onfirmed thus far, 81 have at least one stable isotope whereas the others exist only in the form of radioactive nuclides. Some radioactive nuclides (for example, ^{115}In, ^{232}Th, ^{235}U, ^{238}U) have survived from the time of formation of the elements. Several thousand radioactive nuclides produced through natural or artificial means have been identified. *See* RADIOISOTOPE.

Of the 83 elements which occur naturally in significant quantities on Earth, 20 are found as a single isotope (mononuclidic), and the others as admixtures containing from

2 to 10 isotopes. Isotopic composition is mainly determined by mass spectroscopy.

Nuclides with identical mass number (that is, $A = N + Z$) but differing in the number of protons in the nucleus are called isobars. Nuclides having different mass number but the same number of neutrons are called isotones. *See* ISOBAR (NUCLEAR PHYSICS); ISOTONE.

Isotopic abundance refers to the isotopic composition of an element found in its natural terrestrial state. The isotopic composition for most elements does not vary much from sample to sample. This is true even for samples of extraterrestrial origin such as meteorites and lunar materials brought back to Earth by space missions. However, there are a few exceptional cases for which variations of up to several percent have been observed. There are several phenomena that can account for such variations, the most likely being some type of nuclear process which changes the abundance of one isotope relative to the others. For some of the lighter elements, the processes of distillation or chemical exchange between different chemical compounds could be responsible for isotopic differences. *See* NUCLEAR REACTION; RADIOACTIVITY.

The areas in which separated (or enriched) isotopes are utilized have become fairly extensive, and a partial list includes nuclear research, nuclear power generation, nuclear weapons, nuclear medicine, and agricultural research. For many applications there is a need for separated radioactive isotopes. These are usually obtained through chemical separations of the desired element following production by means of a suitable nuclear reaction. Separated radioactive isotopes are used for a number of diagnostic studies in nuclear medicine, including the technique of positron tomography.

Studies of metabolism, drug utilization, and other reactions in living organisms can be done with stable isotopes such as ^{13}C, ^{15}N, ^{18}O, and ^{2}H. Molecular compounds are "spiked" with these isotopes, and the metabolized products are analyzed by using a mass spectrometer to measure the altered isotopic ratios.   [D.J.Ho.]

Isotope shift   A small difference between the different isotopes of an element in the transition energies corresponding to a given spectral line transition. For a spectral line transition between two energy levels a and b in an atom or ion with atomic number Z, the small difference $\Delta E_{ab} = E_{ab}(A') - E_{ab}(A)$ in the transition energy between isotopes with mass numbers A', and A is the isotope shift. It consists largely of the sum of two contributions, the mass shift (MS) and the field shift, also called the volume shift. The mass shift is customarily divided into a normal mass shift and a specific mass shift; each is proportional to the fractional mass difference $(A' - A)/A'A$. The normal mass shift is a reduced mass correction that is easily calculated for all transitions. The specific mass shift is produced by the correlated motion of different pairs of atomic electrons and is, therefore, absent in one-electron systems. The field shift is produced by the change in the finite size and shape of the nuclear charge distribution when neutrons are added to the nucleus. *See* ATOMIC STRUCTURE AND SPECTRA; NUCLEAR STRUCTURE.

[P.M.K.]

J

J/psi particle An elementary particle with an unusually long lifetime or, from the Heisenberg uncertainty principle, with an extremely narrow width $\Gamma = 91.0 \pm 3.2$ keV, and a large mass $m = 3096.916 \pm 0.011$ MeV. It is a bound state containing a charm quark and an anticharm quark. The discovery of the J/psi particle is one of the cornerstones of the standard model.

Since its discovery in 1974, more than 10^9 J/psi particles have been produced. More than 100 different decay modes and new particles radiating from the J/psi particle have been observed. The J/psi particle has been shown to be a bound state of charm quarks. The long lifetime of the J/psi results from its mass being less than the masses of particles which separately contain a charm and an anticharm quark. This situation permits the J/psi to decay only into noncharm quarks, and empirically this restriction was found to lead to a suppression of the decay rate resulting in a long lifetime and narrow width. The subsequent discovery of the b quark and the intermediate vector bosons Z^0 and $W^{\pm}$, and studies of Z^0 decays into charm, b, and other quarks, show that the theory of the standard model is in complete agreement with experimental data to an accuracy of better than 1%. *See* CHARM; ELEMENTARY PARTICLE; INTERMEDIATE VECTOR BOSON; QUARKS; STANDARD MODEL; UPSILON PARTICLES. [S.C.C.T.]

Jahn-Teller effect A distortion of a highly symmetrical molecule, which reduces its symmetry and lowers its energy. The effect occurs for all nonlinear molecules in degenerate electronic states, the degeneracy of the state being removed by the effect. It was first predicted in 1937 by H. A. Jahn and E. Teller. In early experimental work, the effect often "disappeared" or was masked by other molecular interactions. This has surrounded the Jahn-Teller effect with a certain mystery and allure, rarely found in science today. However, there are now a number of clear-cut experimental examples which correlate well with theoretical predictions. These examples range from the excited states of the most simple polyatomic molecule, H_3, through moderate-sized organic molecules, like the ions of substituted benzene, to complex solid state phenomena involving crystals or localized impurity centers. *See* CRYSTAL DEFECTS; DEGENERACY (QUANTUM MECHANICS); MOLECULAR STRUCTURE AND SPECTRA; QUANTUM MECHANICS.

With the exception of linear molecules which suffer Renner-Teller effects, all polyatomic molecules of sufficiently high symmetry to possess orbitally degenerate electronic states will be subject to the Jahn-Teller instability. However, in cases other than molecules with fourfold symmetry, the proof is somewhat involved and requires the use of the principles of group theory. *See* RENNER-TELLER EFFECT. [V.E.B.; T.A.M.]

Jet flow A fluid flow in which a stream of one fluid mixes with a surrounding medium, at rest or in motion. Such flows occur in a wide variety of situations, and the geometries, sizes, and flow conditions cover a large range. Jet flows vary greatly, depending on the values of two numbers. The first is the Reynolds number, defined in Eq. (1), where ρ is the density, V is a characteristic velocity (for example, the jet

$$\text{Re} \equiv \frac{\rho V L}{\mu} \tag{1}$$

exit velocity), L is a characteristic length (for example, the jet diameter), and μ is the viscosity. The second is the Mach number, defined in Eq. (2), where a is the speed of

$$\text{M} \equiv \frac{V}{a} \tag{2}$$

sound. *See* MACH NUMBER; REYNOLDS NUMBER.

For conditions where Re < 2300 and M $\ll$ 1, jet flows take on a simple character. An example is the water jet formed by a household tap when the valve is partially opened to produce a low flow. If the flow or the diameter is increased or the viscosity is decreased so that Re > 2300, the jet will change dramatically. For example, a water jet exiting into water at rest with Re $\approx$ 2300 is initially in the simple laminar state, but at this Reynolds number that state is unstable and the flow undergoes a transition to the more chaotic turbulent state. Turbulent structures called eddies are formed with a large range of sizes. The large-scale structures are responsible for capturing fluid from the surroundings and entraining it into the jet. However, the jet and external fluids are not thoroughly mixed until diffusion is completed by the small-scale structures. *See* DIFFUSION; LAMINAR FLOW; TURBULENT FLOW.

When the velocities in the jet are greater than the speed of sound (M > 1) the flow is said to be supersonic, and important qualitative changes in the flow occur. The most prominent change is the occurrence of shock waves. For example, a supersonic air jet exhausting from a nozzle at low pressure into higher-pressure air at rest is said to be overexpanded. As the jet leaves the nozzle, it senses the higher pressure around it and adjusts through oblique shock waves emanating from the edges of the nozzle. *See* SHOCK WAVE; SUPERSONIC FLOW.

Another class of jet flows is identified by the fact that the motion of the jet is induced primarily by buoyancy forces. A common example is a hot gas exhaust rising in the atmosphere. Such jet flows are called buoyant plumes, or simply plumes, as distinct from the momentum jets, or simply jets, discussed above. *See* FLUID FLOW.   [J.A.Sc.]

Josephson effect

The passage of paired electrons (Cooper pairs) through a weak connection (Josephson junction) between superconductors, as in the tunnel passage of paired electrons through a thin dielectric layer separating two superconductors.

Quantum-mechanical tunneling of Cooper pairs through a thin insulating barrier (on the order of a few nanometers thick) between two superconductors was theoretically predicted by Brian D. Josephson in 1962. Josephson found that a current of paired electrons (supercurrent) would flow in addition to the usual current that results from the tunneling of single electrons. Josephson predicted that if the current did not exceed a limiting value (the critical current), there would be no voltage drop across the tunnel barrier. This zero-voltage current flow is known as the dc Josephson effect. Josephson also predicted that if a constant nonzero voltage were maintained across the tunnel barrier, an alternating supercurrent would flow through the barrier in addition to the dc current produced by the tunneling of unpaired electrons. This phenomenon is known as the ac Josephson effect. *See* TUNNELING IN SOLIDS.

Josephson pointed out that the magnitude of the maximum zero-voltage supercurrent would be reduced by a magnetic field. In fact, the magnetic field dependence of the magnitude of the critical current is one of the more striking features of the Josephson effect. Circulating supercurrents flow through the tunnel barrier to screen an applied magnetic field from the interior of the Josephson junction just as if the tunnel barrier itself were weakly superconducting. The screening effect produces a spatial variation of the transport current, and the critical current goes through a series of maxima and minima as the field is increased.

Josephson junctions, and instruments incorporating Josephson junctions, are used in applications for metrology at dc and microwave frequencies, frequency metrology, magnetometry, measurement of absolute temperatures below about 1 K, detection and amplification of electromagnetic signals, and other superconducting electronics such as high-speed analog-to-digital converters and computers. A Josephson junction, like a vacuum tube or a transistor, is capable of switching signals from one circuit to another; a Josephson tunnel junction is the fastest switch known. Josephson junction circuits are capable of storing information. Finally, because a Josephson junction is a super-conducting device, its power dissipation is extremely small, so that Josephson junction circuits can be packed together as tightly as fabrication techniques permit. All the basic circuit elements required for a Josephson junction computer have been developed. *See* LOW-TEMPERATURE THERMOMETRY; SUPERCONDUCTING DEVICES; SUPERCONDUCTIVITY.

[L.B.H.]

Joule's law   A quantitative relationship between the quantity of heat produced in a conductor and an electric current flowing through it. As experimentally determined and announced by J. P. Joule, the law states that when a current of voltaic electricity is propagated along a metallic conductor, the heat evolved in a given time is proportional to the resistance of the conductor multiplied by the square of the electric intensity. Today the law would be stated as $H = RI^2$, where H is rate of evolution of heat in watts, the unit of heat being the joule; R is resistance in ohms; and I is current in amperes. This statement is more general than the one sometimes given that specifies that R be independent of I. Also, it is now known that the application of the law is not limited to metallic conductors.

[L.G.H./J.W.St.]

Junction detector   A device in which detection of radiation takes place in or near the depletion region of a reverse-biased semiconductor junction. The electrical output pulse is linearly proportional to the energy deposited in the junction depletion layer by the incident ionizing radiation. *See* IONIZATION CHAMBER.

Introduced into nuclear studies in 1958, the junction detector, or more generally, the nuclear semiconductor detector, revolutionized the field. In the detection of both charged particles and gamma radiation, these devices typically improved experimentally attainable energy resolutions by about two orders of magnitude over that previously attainable. To this they added unprecedented flexibility of utilization, speed of response, miniaturization, freedom from deleterious effects of extraneous electromagnetic (and often nuclear) radiation fields, low-voltage requirements, and effectively perfect linearity of output response. They are now used for a wide variety of diverse applications. They are used for general analytical applications, giving both qualitative and quantitative analysis in the microprobe and the scanning transmission electron microscopes. They are used in medicine, biology, environmental studies, and the space program. In the last category they continue to play a very fundamental role, ranging from studies of the radiation fields in the solar system to the composition of extraterrestrial surfaces. *See* PARTICLE DETECTOR; SEMICONDUCTOR.

[J.M.McK.]

K

Kapitza resistance A resistance to the flow of heat across the interface between liquid helium and a solid. A temperature difference is required to drive heat from a solid into liquid helium, or vice versa; the temperature discontinuity occurs right at the interface. The Kapitza resistance, discovered by P. L. Kapitza, is defined in the equation below, where T_S and T_H are the solid and helium temperatures and $\dot{Q}/A$ is

$$R_K = \frac{T_S - T_H}{\dot{Q}/A}$$

the heat flow per unit area across the interface. *See* CONDUCTION (HEAT).

In principle, the measured Kapitza resistance should be easily understood. In liquid helium and solids (such as copper), heat is carried by phonons, which are thermal-equilibrium sound waves with frequencies in the gigahertz to terahertz region. The acoustic impedance of helium and solids can differ by up to 1000 times, which means that the phonons mostly reflect at the boundary, like an echo from a cliff face. This property together with the fact that the number of phonons dies away very rapidly at low temperatures means that at about 1 K there are few phonons to carry heat and even fewer get across the interface. The prediction is that the Kapitza resistance at the interface is comparable to the thermal resistance of a 10-m (30-ft) length of copper with the same cross section. *See* ACOUSTIC IMPEDANCE; PHONON; QUANTUM ACOUSTICS.

The reality is that above 0.1 K and below 0.01 K (10 mK) more heat is driven by a temperature difference than is predicted. Above 0.1 K this is now understood to be a result of imperfections such as defects and impurities at the interface, which scatter the phonons and allow greater transmission. *See* CRYSTAL DEFECTS.

The enormous interest in ultralow-temperature (below 10 mK) research generated by the invention of the dilution refrigerator and the discovery of superfluidity in liquid helium-3 (^{3}He) below 0.9 mK also regenerated interest in Kapitza resistance, because heat exchange between liquid helium and solids was important for both the dilution refrigerator and superfluidity research. An ingenious technique was invented to overcome the enormous Kapitza resistance at 1 mK: The solid is powdered, and the powder is packed and sintered to a spongelike structure to enhance the surface area. In this way a 1-cm^3 (0.06-in.^{3}) chamber can contain up to 1 m^2 (10 ft^2) of interface area between the solid and the liquid helium.

It was found that at 1 mK the Kapitza resistance is 100 times smaller than predicted by the phonon model. There have been two explanations for the anomaly, and probably both are relevant. One is that energy is transferred by magnetic coupling between the magnetic ^{3}He atoms and magnetic impurities in the solid or at the surface of the solid; the other is that the spongelike structure has quite different, and many more, phonons than a bulk solid and that these can transfer heat directly to the ^{3}He atoms. Whatever its cause, this anomaly has had a major impact on ultralow-temperature physics. *See* ADIABATIC DEMAGNETIZATION; LIQUID HELIUM; LOW-TEMPERATURE PHYSICS; SUPERFLUIDITY.   [J.P.Ha.]

Kármán vortex street   A double row of line vortices in a fluid. Under certain conditions a Kármán vortex street is shed in the wake of bluff cylindrical bodies when the relative fluid velocity is perpendicular to the generators of the cylinder, as illustrated.

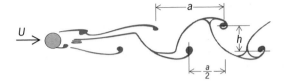

Kármán vortex street. **U** = stream speed; **a** = spacing between vortices; **h** = distance between two rows of vortices.

This periodic shedding of eddies occurs first from one side of the body and then from the other, an unusual phenomenon because the oncoming flow may be perfectly steady. Vortex streets can often be seen, for example, in rivers downstream of the columns supporting a bridge. They can be created by steady winds blowing past smokestacks, transmission lines, bridges, missiles about to be launched vertically, and pipelines aboveground in the desert. *See* VORTEX.   [A.E.Br.]

Kelvin bridge   A specialized version of the Wheatstone bridge network designed to eliminate, or greatly reduce, the effect of lead and contact resistance and thus permit accurate measurement of low resistance. The circuit shown in the illustration accomplishes this by effectively placing relatively high-resistance-ratio arms in series with the potential leads and contacts of the low-resistance standards and the unknown resistance. In this circuit R_A and R_B are the main ratio resistors, R_a and R_b the auxiliary ratio, R_x the unknown, R_s the standard, and R_y a heavy copper yoke of low resistance connected between the unknown and standard resistors.

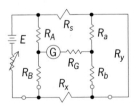

A Kelvin bridge used to measure an unknown low resistance.

As with the Wheatstone bridge, the Kelvin bridge for routine engineering measurements is constructed using both adjustable-ratio arms and adjustable standards. However, the ratio is usually continuously adjustable, over a short span, and the standard is adjustable in appropriate steps to cover the required range. *See* BRIDGE CIRCUIT; RESISTANCE MEASUREMENT; WHEATSTONE BRIDGE.   [C.E.A.]

Kelvin's circulation theorem   A theorem in fluid dynamics that pertains to the dynamics of vortices and the use of ideal-fluid potential-flow equations. The theorem states that the circulation (defined as the line integral of the component of velocity tangential to the closed contour) in an inviscid and incompressible fluid subject to only conservative forces is constant. By using Stokes' theorem of integral calculus, it may be shown that the circulation is also related to the flux of vorticity (defined as the curl of the velocity field) normal to the area transcribed by the contour.

The principal use of Kelvin's theorem is in the study of incompressible, inviscid fluid flows. If a body is moving through such a fluid, the vorticity far from the body is, by definition, zero. Then according to Kelvin's theorem, the vorticity in the fluid will everywhere be zero and the flow will be irrotational. This permits the reduction of the governing equations from the Euler equations to the Laplace equation and presents the many mathematical techniques of potential theory for solving fluid-flow problems. *See* LAPLACE'S IRROTATIONAL MOTION; VORTEX.

[E.Pa.; F.Ste.]

Kelvin's minimum-energy theorem   A theorem in fluid dynamics that pertains to the kinetic energy of an ideal fluid (that is, inviscid, incompressible, and irrotational) and provides uniqueness statements concerning the solution of potential-flow problems. The theorem states that the irrotational motion of a liquid occupying a simply connected region has less kinetic energy than any other motion consistent with the same normal motion of the boundary S. *See* POTENTIALS.

The implications of the minimum-energy of irrotational motion are: (1) irrotational motion is impossible in a simply connected region bounded by fixed walls since in this case the normal derivative of the velocity potential vanishes at all points on the boundary, and therefore, according to the energy theorem, the kinetic energy is zero, or the system is at rest; (2) irrotational motion is impossible in a fluid in which the velocity at infinity vanishes if the internal boundaries are also at rest; (3) if the velocity at infinity vanishes, then the irrotational motion due to prescribed motion of an internal boundary is unique; and (4) if a fluid is in motion with uniform velocity at infinity, then the irrotational motion due to prescribed motion of an internal boundary is unique. *See* LAPLACE'S IRROTATIONAL MOTION.

[E.Pa.; F.Ste.]

Kerr effect   Electrically induced birefringence that is proportional to the square of the electric field. When a substance (especially a liquid or a gas) is placed in an electric field, its molecules may become partly oriented. This renders the substance anisotropic and gives it birefringence, that is, the ability to refract light differently in two directions. This effect, which was discovered in 1875 by John Kerr, is called the electrooptical Kerr effect, or simply the Kerr effect.

When a liquid is placed in an electric field, it behaves optically like a uniaxial crystal with the optical axis parallel to the electric lines of force. The Kerr effect is usually observed by passing light between two capacitor plates inserted in a glass cell containing the liquid. Such a device is known as a Kerr cell or optical Kerr shutter. Light passing through the medium normal to the electric lines of force (that is, parallel to the capacitor plates) is split into two linearly polarized waves.

In certain crystals there may be an electrically induced birefringence that is proportional to the first power of the electric field. This is called the Pockels effect. In these crystals the Pockels effect usually overshadows the Kerr effect, which is nonetheless present. In crystals of cubic symmetry and in isotropic solids (such as glass) only the Kerr effect is present. *See* ELECTROOPTICS.

[M.A.D.]

Kinematics   That branch of mechanics which deals with the motion of a system of material particles without reference to the forces which act on the system. Kinematics differs from dynamics in that the latter takes these forces into account. *See* DYNAMICS.

For a single particle moving in a straight line (rectilinear motion), the motion is prescribed when the position of the particle is known as a function of the time. Plane kinematics of a particle is concerned with the specification of the position of a particle moving in a plane by means of two independent variables. The kinematics of a particle in space is concerned with the ways in which three independent coordinates may be chosen to specify the position of the particle at a given time, and with the relations between the first and second time derivatives of these coordinates and the components of velocity and acceleration of the particle.

Among the coordinate systems studied in kinematics are those used by observers who are in relative motion. In nonrelativistic kinematics the time coordinate for each such observer is assumed to be the same, but in relativistic kinematics proper account must be taken of the fact that lengths and time intervals appear different to observers moving relative to each other. *See* RELATIVITY.   [H.C.Co./B.G.]

Kinetic theory of matter   A theory which states that the particles of matter in all states of aggregation are in vigorous motion. In computations involving kinetic theory, the methods of statistical mechanics are applied to specific physical systems. The atomistic or molecular structure of the system involved is assumed, and the system is then described in terms of appropriate distribution functions. The main purpose of kinetic theory is to deduce, from the statistical description, results valid for the whole system. The distinction between kinetic theory and statistical mechanics is thus of necessity arbitrary and vague. Historically, kinetic theory is the oldest statistical discipline. Today a kinetic calculation refers to any calculation in which probability methods, models, or distribution functions are involved. *See* STATISTICAL MECHANICS.

Kinetic calculations are not restricted to gases, but occur in chemical problems, solid-state problems, and problems in radiation theory. Even though the general procedures in these different areas are similar, there are a sufficient number of important differences to make a general classification useful.

In classical ideal equilibrium problems there are no interactions between the constituents of the system. The system is in equilibrium, and the mechanical laws governing the system are classical. The basic information is contained in the Boltzmann distribution f (also called Maxwell or Maxwell-Boltzmann distribution) which gives the number of particles in a given momentum and positional range ($d^3x = dxdydz$, $d^3v = dv_xdv_ydv_z$, where x, y, and z are coordinates of position, and v_x, v_y, and v_z are coordinates of velocity). In Eq. (1) ϵ is the energy, $\beta = 1/kT$ (where k is the Boltzmann constant and T is the

$$f(xyz, v_xv_yv_z) = Ae^{-\beta\epsilon} \tag{1}$$

absolute temperature), and A is a constant determined from Eq. (2). The calculations

$$\int\int d^3x \, d^3v f = N \qquad \text{total number of particles} \tag{2}$$

of gas pressure, specific heat, and the classical equipartition theorem are all based on these relations. *See* BOLTZMANN STATISTICS.

Many important physical properties refer not to equilibrium but to nonequilibrium states. Phenomena such as thermal conductivity, viscosity, and electrical conductivity all require a discussion starting from the Boltzmann transport equation. If one deals with states that are near equilibrium, the exact Boltzmann equation need not be solved; then it is sufficient to describe the nonstationary situation as a small perturbation superimposed on an equilibrium state.

The basic classical procedure for arbitrary systems (systems with interactions taken into account) that allows the calculation of macroscopic entities is that using the partition function.

Classical nonideal nonequilibrium theory is the most general situation that classical statistics can describe. In general, very little is known about such systems.

There are quantum counterparts to the classifications just described. In a quantum treatment a distribution function is also used for an ideal system in equilibrium to describe its general properties. For systems of particles which must be described by symmetrical wave functions, such as helium atoms and photons, one has the Bose distribution, Eq. (3), where $\beta = 1/kT$, and A is determined by Eq. (2).

$$f(v_x v_y v_z) = \frac{1}{(1/A)e^{\beta\epsilon} - 1} \tag{3}$$

See BOSE-EINSTEIN STATISTICS.

For systems of particles which must be described by antisymmetrical wave functions, such as electrons, protons, and neutrons, one has the Fermi distribution, Eq. (4). The

$$f(v_x v_y v_z) = \frac{1}{(1/A)e^{\beta\epsilon} + 1} \tag{4}$$

application to electrons as an (ideal) Fermi-Dirac gas in a metal is the basis of the Sommerfeld theory of metals. *See* FERMI-DIRAC STATISTICS; FREE-ELECTRON THEORY OF METALS; QUANTUM STATISTICS.   [M.Dr.]

Kinetics (classical mechanics)

That part of classical mechanics which deals with the relation between the motions of material bodies and the forces acting upon them. It is synonymous with dynamics of material bodies. *See* DYNAMICS.

Kinetics proceeds by adopting certain intuitively acceptable concepts which are associated with measurable quantities. These essential concepts and the measurable quantities used for their specification are as follows:

1. Space configuration refers to the positions and orientations of bodies in a reference frame adopted by the observer. It is expressed quantitatively by an arbitrarily chosen set of space coordinates, of which cartesian and polar coordinates are examples. All space coordinates rest on the notion of distance measurement.

2. Duration is expressed quantitatively by time measured by a clock or comparable mechanism.

3. Motion refers to change of configuration with time and is expressed by time rates of coordinate change called velocities and time rates of velocity change called accelerations. The classical assumption that coordinates behave as analytic functions of time permits representation of velocities and accelerations as first and second derivatives, respectively, of the space coordinates with respect to time.

4. Inertia is an attribute of bodies implying their capacity to resist changes of motion. A body's inertia with respect to linear motion is denoted by its mass.

5. Momentum is an attribute proportional to both the mass and velocity of a body. Momentum of linear motion is expressed as the product of mass and linear velocity.

6. Force serves to designate the influence exercised upon the motion of a particular body by other bodies, not necessarily specified. A quantitative connection between the motion of a body and the force applied to it is expressed by Newton's second law of motion, which is discussed later.

Distance, time, and mass are commonly regarded as fundamental, all other dynamical quantities being definable in terms of them.

A primary objective of classical kinetics is the prediction of the behavior of bodies which are subject to known forces when only initial values of the coordinates and momenta are available. This is accomplished by use of a principle first recognized by

Isaac Newton. Newton's statement of the principle was restricted to the linear motion of an idealized body called a mass particle, having negligible extension in space.

The basic dynamical law set forth by Newton and known as his second law states that the time rate of change of a particle's linear momentum is proportional to and in the direction of the force applied to the particle. Stated analytically, Newton's second law becomes the differential equation (1), in which m represents the particle's mass,

$$\frac{d(mv)}{dt} = F \tag{1}$$

v its velocity, F the applied force, and t the time. Equation (1) provides a definition of force and of its units if units of mass, distance, and time have previously been adopted. The classical assumption of constancy of mass permits Eq. (1) to be expressed as Eq. (2), where a represents the linear acceleration.

$$ma = F \tag{2}$$

The behavior of systems composed of two or more interacting particles is treated by Newtonian dynamics augmented by Newton's third law of motion which states that when two bodies interact, the forces they exert on one another are equal and oppositely directed. The important laws of momentum and energy conservation are derivable for such systems (the latter only for forces of special type) and useful in solution of problems. *See* ACCELERATION; FORCE; GRAVITATION; HARMONIC MOTION; MASS; MOMENTUM; RIGID-BODY DYNAMICS; VELOCITY.   [R.A.Fi.]

Kirchhoff's laws of electric circuits   Fundamental natural laws dealing with the relation of currents at a junction and the voltages around a loop. These laws are commonly used in the analysis and solution of networks. They may be used directly to solve circuit problems, and they form the basis for network theorems used with more complex networks.

One way of stating Kirchhoff's voltage law is: "At each instant of time, the algebraic sum of the voltage rise is equal to the algebraic sum of the voltage drops, both being taken in the same direction around the closed loop."

Kirchhoff's current law may be expressed as follows: "At any given instant, the sum of the instantaneous values of all the currents flowing toward a point is equal to the sum of the instantaneous values of all the currents flowing away from the point."   [K.Y.T./R.T.W.]

Knudsen number   In gas dynamics, the ratio of the molecular mean free path λ to some characteristic length L: $\mathrm{Kn} = \lambda/L$. The length chosen will depend on the problem under consideration. It may be, for example, the diameter of a pipe or an object immersed in a flow, or the thickness of a boundary layer or a shock wave. *See* MEAN FREE PATH.

The magnitude of the Knudsen number determines the appropriate gas dynamic regime. When the Knudsen number is small compared to unity, of the order of $\mathrm{Kn} \leq 0.1$, the fluid can be treated as a continuous medium and described in terms of the macroscopic variables: velocity, density, pressure, and temperature. In the transition flow regime, for Knudsen numbers of the order of unity or greater, a microscopic approach is required, wherein the trajectories of individual representative molecules are considered, and macroscopic variables are obtained from the statistical properties of their motions. In both internal and external flows, for $\mathrm{Kn} \geq 10$, intermolecular collisions in the region of interest are much less frequent than molecular interactions with solid boundaries, and can be ignored. Flows under such conditions are termed collisionless or free molecular. In the range $0.1 \leq Kn \leq 1.0$, termed the slip flow regime, it is sometimes possible to obtain useful results by treating the gas as a continuum, but allowing for discontinuities in velocity and temperature at solid boundaries. *See* GAS DYNAMICS; KINETIC THEORY OF MATTER; RAREFIED GAS FLOW.

[L.T.]

Kondo effect   An unusual, temperature-dependent effect displayed in the thermal, electrical, and magnetic properties of nonmagnetic metals containing very small quantities of magnetic impurities. A striking example is the anomalous, logarithmic increase in the electrical resistivity with decreasing temperature. Other properties, such as heat capacity, magnetic susceptibility, and thermoelectric power, also display anomalous behavior because of the Kondo effect. For these properties, the temperature dependence of a typical dilute magnetic metal (Kondo alloy) differs greatly from the behavior expected of an ordinary metal containing no magnetic impurities.

The Kondo effect has been observed in a wide variety of dilute magnetic alloys. Usually these alloys are made from a nonmagnetic host such as copper, silver, gold, magnesium, or zinc and a small amount of a magnetic metal impurity such as chromium, manganese, iron, cobalt, nickel, vanadium, or titanium. Typical concentrations range from about one to a few hundred magnetic atoms per million host atoms. At higher concentrations, the dilute magnetic alloys may display spin-glass behavior. *See* SPIN GLASS.

The Kondo effect is used in thermometry applications, especially thermocouple thermometers at very low temperatures (that is, millikelvin temperatures). In other applications where the properties of pure metals are studied, the Kondo effect serves as a useful indicator of the metal's magnetic-impurity level.

The problem of understanding the Kondo effect is considered important since it is recognized to be a simpler version of the more complex problem of understanding ferromagnetism in magnetic materials, which is one of the great challenges in physics. Basically the Kondo effect is an example of the most simple possible magnetic system— a single magnetic atom in a nonmagnetic environment. (The alloys used are so dilute that the interaction between different magnetic impurities can be safely ignored.) Although this involves a simple physical model, the problem has required some of the most sophisticated mathematical techniques known to advance its understanding.

An important step in this direction was the development of a partial mathematical solution of the Kondo problem using renormalization field theory techniques. Information gained in this step helped with the final development of a mathematically exact solution of the Kondo problem. The exact solution permits a systematic calculation of all properties (resistivity, thermal conductivity, thermopower, specific heat, magnetic susceptibility, neutron scattering behavior, and so forth) and provides a physical understanding of these properties. The theoretical work on the Kondo problem has been connected with new understanding in a variety of other scientific disciplines such as condensed-matter physics, surface physics, critical phenomena, elementary particle physics, magnetism, molecular physics, and chemistry, where parallels and analogs to the Kondo problem can be identified and utilized. *See* CRITICAL PHENOMENA; FERROMAGNETISM; RENORMALIZATION.

[W.P.K.]

Kronig-Penney model   An idealized, one-dimensional model of a crystal which exhibits many of the basic features of the electronic structure of real crystals. Consider the potential energy $V(x)$ of an electron shown in the illustration with an infinite sequence of potential wells of depth $-V_0$ and width a, arranged with a spacing b. The width and the curvatures of the allowed bands increase with energy. The

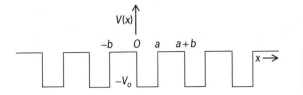

Potential energy which is assumed for the one-dimensional Kronig-Penney model.

Kronig-Penney model has been extended to include the effects of impurity atoms. *See* BAND THEORY OF SOLIDS.

[J.C.]

L

Lagrange's equations   Equations of motion of a mechanical system for which a classical (non-quantum-mechanical) description is suitable, and which relate the kinetic energy of the system to the generalized coordinates, the generalized forces, and the time. If the configuration of the system is specified by giving the values of f independent quantities $q_1, \ldots, q_f$, there are f such equations of motion.

In their usual form, these equations are equivalent to Newton's second law of motion and are differential equations of the second order for the q's as functions of the time t.

[P.M.S.]

Lagrangian function   A function of the generalized coordinates and velocities of a dynamical system from which the equations of motion in Lagrange's form can be derived. The Lagrangian function is denoted by $L(q_1, \ldots, q_f; \dot{q}_1, \ldots, \dot{q}_f; t)$. For systems in which the forces are derivable from a potential energy V, if the kinetic energy is T, the equation below holds.

$$L = T - V$$

See Lagrange's equations.

[P.M.S.]

Laminar flow   A smooth, streamline type of viscous fluid motion characteristic of flow at low-to-moderate deformation rates. The name derives from the fluid's moving in orderly layers or laminae.

The chief criterion for laminar flow is a relatively small value for the Reynolds number, $\mathrm{Re} = \rho V L / \mu$, where ρ is fluid density, V is flow velocity, L is body size, and μ is fluid viscosity. Laminar flow may be achieved in many ways: low-density flows as in rarefied gases; low-velocity or "creeping" motions; small-size bodies such as microorganisms swimming in the ocean; or high-viscosity fluids such as lubricating oils. At higher values of the Reynolds number, the flow becomes disorderly or turbulent, with many small eddies, random fluctuations, and streamlines intertwining. *See* Creeping flow; Reynolds number; Turbulent flow; Viscosity.

Nearly all of the many known exact solutions of the equations of motion of a viscous fluid are for the case of laminar flow. These mathematically accurate descriptions can be used to give insight into the more complex turbulent and transitional flow patterns for which no exact analyses are known. *See* Navier-Stokes equation.

The theory of viscous lubricating fluids in bearings is a highly developed area of laminar flow analysis. Even large Reynolds number flows, such as aircraft in flight, have regions of laminar flow near their leading edges, so that laminar flow analysis can be useful in a variety of practical and scientifically relevant flows. *See* Boundary-layer flow; Fluid flow.

[F.M.W.]

Langevin function   A mathematical function which is important in the theory of paramagnetism and in the theory of the dielectric properties of insulators. The analytical expression for the Langevin function is shown in the equation below. If $x \ll 1$,

$$L(x) = \coth x - (1/x)$$

$L(x) \simeq x/3$. The paramagnetic susceptibility of a classical (non-quantum-mechanical) collection of magnetic dipoles is given by the Langevin function, as is the polarizability of molecules having a permanent electric dipole moment. *See* PARAMAGNETISM.

[E.A.; F.Ke.]

Laplace's irrotational motion   Laplace's equation for irrotational motion of an inviscid, incompressible fluid is partial differential equation (1), where x_1, x_2, x_3 are rectangular cartesian coordinates in an inertial reference frame, and Eq. (2) gives

$$\partial^2 \phi / \partial x_1{}^2 + \partial^2 \phi / \partial x_2{}^2 + 2^2 \phi / 2 x_3{}^2 = 0 \tag{1}$$

$$\phi = \phi(x_1, \ x_2, \ x_3, \ t) \tag{2}$$

the velocity potential. The fluid velocity components, u_1, u_2, u_3 in the three respective rectangular coordinate directions are given by $u_i = \partial\phi/\partial x_i$, $i = 1, 2, 3$. More generally, in any inertial coordinate system, the equation is div (grad ϕ) = 0 and the velocity vector is **v** = grad ϕ. Irrotational motion implies that the fluid particles translate without rotation (like the cars on a ferris wheel). *See* FLUID FLOW.   [A.E.Br.]

Larmor precession   A precession in a magnetic field of the motion of charged particles or of particles possessing magnetic moments.

The Larmor theorem states that, for electrons moving in a single central field of force, the motion in a uniform magnetic field H is, to first order in H, the same as a possible motion in the absence of H except for the superposition of a common precession of angular frequency given by Eq. (1). Here e/c is the magnitude of the

$$\omega_L = \frac{eH}{2mc} \tag{1}$$

electronic charge in electromagnetic units, and m is the electronic mass. The frequency ω_L is called the Larmor frequency and is numerically equal to 2π times 1.40 MHz per oersted or 2π times 111 MHz per SI unit of magnetic field strength (ampere-turn per meter). *See* PRECESSION.

In stating the Larmor theorem, use was made of the phrase "a possible motion." If H is applied sufficiently slowly, it can be proved that the motion is the same as in the absence of H, except for the superposition of the Larmor precession. However, a sudden application of H may change, for example, a circular orbit into an elliptical one.

According to elementary electromagnetic theory, a current loop of area A and of current I possesses a magnetic moment μ of magnitude IA and of direction normal to the loop. Thus an electron moving in a circular orbit has an orbital magnetic moment.

The electron also has orbital angular momentum, which by quantum theory must equal $\hbar J$, where J is an integer and $\hbar$ is Planck's constant h divided by 2π. In terms of the equivalent magnetic moment, Eq. (1) may be written in the form of Eq. (2).

$$\omega_L = -\frac{\mu}{\hbar J} H \tag{2}$$

In this form the Larmor precession is exhibited by any magnetic moment μ including magnetic moments associated with spin angular momentum as well as those associated with orbital angular momentum. In this form the Larmor precession applies to experiments in molecular beams, electron paramagnetic resonance (EPR), and nuclear magnetic resonance (NMR). *See* ANGULAR MOMENTUM; ELECTRON SPIN; MAGNETIC RESONANCE.

[E.A.; F.Ke.]

Laser A device that uses the principle of amplification of electromagnetic waves by stimulated emission of radiation and operates in the infrared, visible, or ultraviolet region. The term laser is an acronym for light amplification by stimulated emission of radiation, or a light amplifier. However, just as an electronic amplifier can be made into an oscillator by feeding appropriately phased output back into the input, so the laser light amplifier can be made into a laser oscillator, which is really a light source. Laser oscillators are so much more common than laser amplifiers that the unmodified word "laser" has come to mean the oscillator, while the modifier "amplifier" is generally used when the oscillator is not intended. *See* MASER.

The process of stimulated emission can be described as follows: When atoms, ions, or molecules absorb energy, they can emit light spontaneously (as with an incandescent lamp) or they can be stimulated to emit by a light wave. This stimulated emission is the opposite of (stimulated) absorption, where unexcited matter is stimulated into an excited state by a light wave. If a collection of atoms is prepared (pumped) so that more are initially excited than unexcited (population inversion), then an incident light wave will stimulate more emission than absorption, and there is net amplification of the incident light beam. This is the way the laser amplifier works.

A laser amplifier can be made into a laser oscillator by arranging suitable mirrors on either end of the amplifier. These are called the resonator. Thus the essential parts of a laser oscillator are an amplifying medium, a source of pump power, and a resonator. Radiation that is directed straight along the axis bounces back and forth between the mirrors and can remain in the resonator long enough to build up a strong oscillation. (Waves oriented in other directions soon pass off the edge of the mirrors and are lost before they are much amplified.) Radiation may be coupled out by making one mirror partially transparent so that part of the amplified light can emerge through it (see illustration). The output wave, like most of the waves being amplified between the mirrors, travels along the axis and is thus very nearly a plane wave. *See* OPTICAL PUMPING.

Continuous-wave gas lasers. Perhaps the best-known gas laser is the neutral-atom helium-neon (HeNe) laser, which is an electric-discharge-excited laser involving the noble gases helium and neon. The lasing atom is neon. The wavelength of the transition most used is 632.8 nanometers; however, many helium-neon lasers operate at longer and shorter wavelengths including 3390, 1152, 612, 594, and 543 nm. Output powers are mostly around 1 milliwatt.

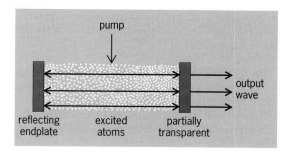

Structure of a parallel-plate laser.

A useful gas laser for the near-ultraviolet region is the helium-cadmium (HeCd) laser, where lasing takes place from singly ionized cadmium. Wavelengths are 325 and 442 nm, with powers up to 150 mW.

The argon ion laser provides continuous-wave (CW) powers up to about 50 W, with principal wavelengths of 514.5 and 488 nm, and a number of weaker transitions at nearby wavelengths. The argon laser is often used to pump other lasers, most importantly tunable dye lasers and titanium:sapphire lasers. For applications requiring continuous-wave power in the red, the krypton ion laser can provide continuous-wave lasing at 647.1 and 676.4 nm (as well as 521, 568, and other wavelengths), with powers somewhat less than those of the argon ion laser.

The carbon dioxide (CO_2) molecular laser has become the laser of choice for many industrial applications, such as cutting and welding.

Short-pulsed gas lasers. Some lasers can be made to operate only in a pulsed mode. Examples of self-terminating gas lasers are the nitrogen laser (337 nm) and excimer lasers (200–400 nm). The nitrogen laser pulse duration is limited because the lower level becomes populated because of stimulated transitions from the upper lasing level, thus introducing absorption at the lasing wavelength. Peak powers as large as 1 MW are possible with pulse durations of 1–10 nanoseconds. Excimer lasers are self-terminating because lasing transitions tear apart the excimer molecules and time is required for fresh molecules to replace them.

Solid-state lasers. The term solid-state laser should logically cover all lasers other than gaseous or liquid. Nevertheless, current terminology treats semiconductor (diode) lasers separately from solid-state lasers because the physical mechanisms are somewhat different. With that reservation, virtually all solid-state lasers are optically pumped.

Historically, the first laser was a single crystal of synthetic ruby, which is aluminum oxide (Al_2O_3 or sapphire), doped with about 0.05% (by weight) chromium oxide (Cr_2O_3). Three important rare-earth laser systems in current use are neodymium:YAG, that is, yttrium aluminum garnet ($Y_3Al_5O_{12}$) doped with neodymium; neodymium:glass; and erbium:glass. Other rare earths and other host materials also find application.

Semiconductor (diode) lasers. The semiconductor laser is the most important of all lasers, both by economic standards and by the degree of its applications. Its main features include rugged structure, small size, high efficiency, direct pumping by low-power electric current, ability to modulate its output by direct modulation of the pumping current at rates exceeding 20 GHz, compatibility of its output beam dimensions with those of optical fibers, feasibility of integrating it monolithically with other semiconductor optoelectronic devices to form integrated circuits, and a manufacturing technology that lends itself to mass production.

Most semiconductor lasers are based on III–V semiconductors. The laser can be a simple sandwich of *p*- and *n*-type material such as gallium arsenide (GaAs). The active region is at the junction of the *p* and *n* regions. Electrons and holes are injected into the active region from the *p* and *n* regions respectively. Light is amplified by stimulating electron-hole recombination. The mirrors comprise the cleaved end facets of the chip (either uncoated or with enhanced reflective coatings). *See* ELECTRON-HOLE RECOMBINATION; SEMICONDUCTOR; SEMICONDUCTOR DIODE.

Monochromaticity. When lasers were first developed, they were widely noted for their extreme monochromaticity. They provided far more optical power per spectral range (as well as per angular range) than was previously possible. It has since proven useful to relate laser frequencies to the international time standard (defined by an energy-level difference in the cesium atom), and this was done so precisely, through the use of optical heterodyne techniques, that the standard of length was redefined in such a way that the speed of light is fixed. In addition, extremely stable and monochromatic lasers have been developed, which can be used, for example, for optical communication between remote and moving frames, such as the Moon and the Earth. *See* FREQUENCY MEASUREMENT; LASER SPECTROSCOPY; LIGHT.

Tunable lasers. Having achieved lasers whose frequencies can be monochromatic, stable, and absolute (traceable to the time standard), the next goal is tunability. Most lasers allow modest tuning over the gain bandwidth of their amplifying medium. However, the laser most widely used for wide tunability has been the (liquid) dye laser. This laser must be optically pumped, either by a flash lamp or by another laser, such as the argon ion laser. Considerable engineering has gone into the development of systems to rapidly flow the dye and to provide wavelength tunability. About 20 different dyes are required to cover the region from 270 to 1000 nm.

Free-election lasers. The purpose of the free-electron laser is to convert the kinetic energy in an electron beam to electromagnetic radiation. Since it is relatively simple to generate electron beams with peak powers of 10^{10} W, the free-electron laser has the potential for providing high optical power, and since there are no prescribed energy levels, as in the conventional laser, the free-electron laser can operate over a broad spectral range.   [S.F.J.; A.L.S.; R.Pan.]

Laser cooling   Reducing the thermal motion of atoms with the force exerted by a laser beam. Typically, such cooling is used to reduce the temperature of a gas of atoms, or the velocity spread of atoms in an atomic beam.

Light affects atomic motion when the atoms absorb or emit photons, the particles or quanta that make up light. Photons carry momentum $p = h/\lambda$, where h is Planck's constant and λ is the light's wavelength. By conservation of momentum, when an atom absorbs or emits a photon, the atom's momentum must change by an amount equal to the photon momentum. Each absorption or emission thus gives the atom a tiny kick, changing its velocity. For most atoms this change is only a few millimeters to a few centimeters per second, while atoms in a gas at room temperature have velocities of a few hundred to a few thousand meters per second. Nevertheless, repeated absorption and emission of photons can have a significant effect on even hot atomic gases or beams. *See* CONSERVATION OF MOMENTUM; LIGHT; MOMENTUM; PHOTON.

The keys to using such repeated kicks to reduce the random, thermal motion of a gas of atoms are the monochromatic nature of laser light, the selectivity of absorption of light by atoms, and the Doppler effect. Light is an oscillating electromagnetic wave whose frequency of oscillation determines its color. The energy of each photon is $E = h\nu$, where ν is the frequency. Laser light can have nearly a single frequency or color, so that all the photons have almost identical energies. Atoms absorb only photons whose energy is equal, within a small range, to the difference in energy between two of its quantum states or energy levels. For sodium atoms this resonance frequency is $\nu_0 \equiv 5 \times 10^{14}$ Hz (wavelength $\lambda \equiv 589$ nanometers), but the absorption is efficient only over a range $\Delta\nu = 10^7$ Hz. Moving atoms, however, experience a Doppler shift so that, depending on their speed and whether they are moving along the direction of the laser beam or against it, the light appears to room-temperature atoms to have a frequency shifted up or down by a hundred or more times the natural absorption width $\Delta\nu$. *See* DOPPLER EFFECT; LASER.

If the frequency ν of the laser is tuned to be slightly lower than ν_0, those atoms moving against the laser beam see the laser upshifted, closer to ν_0. These atoms are more likely to absorb photons, receive kicks opposite to the direction of their velocity, and slow down. After absorbing a photon, the atoms are in an excited state and return to the original state by spontaneously emitting a photon. Such photons are radiated in random directions, so the effect of their kicks averages to zero. For atoms held in a trap, as ions generally are, any trapped atom will at some time be traveling against the laser beam and be cooled. Laser cooling was first demonstrated in 1978 with such trapped ions. For free atoms, another, similarly tuned laser beam is added, aimed in the opposite sense, to cool those atoms moving in the opposite direction. More generally, one uses three pairs of mutually perpendicular, counterpropagating laser beams, all tuned below ν_0. Then, no matter the direction of an atom's velocity, there are one or more laser beams that oppose the velocity and slow the atom.

Improving atomic clocks, where the thermal motion of atoms reduces the precision and accuracy, was a major motivation to developing laser cooling. Laser cooling is also used in atom optics, where well-collimated, monoenergetic atomic beams are more easily and effectively manipulated. In addition, laser cooling has been used to study collisions between very slow atoms. See ATOMIC CLOCK; SCATTERING EXPERIMENTS (ATOMS AND MOLECULES).

Laser cooling is intimately connected with trapping of atoms, because atoms must often be slowed down before they can be held in a trap and because atoms must often be trapped in order to observe laser cooling or its effects. Such effects include cold, trapped ions arranging themselves into a crystal because of the electric repulsion between the charged ions. Neutral atoms can become arrayed on an optical lattice of tiny traps formed by interference between the laser beams used to cool them. In both cases, the spacing between atoms is thousands of times larger than the spacing in solid crystals. Another effect is Bose-Einstein condensation, wherein a gas of atoms whose de Broglie wavelength is comparable to the spacing between atoms has a transition to a state where a significant fraction of the atoms are in the lowest kinetic energy state possible. See BOSE-EINSTEIN CONDENSATION; PARTICLE TRAP; QUANTUM MECHANICS.

[W.D.P.]

Laser-solid interactions   Interactions of laser light with solids. The term usually refers to the thermal effects of absorption of high-intensity laser beams. For nonthermal laser interactions with matter see LASER SPECTROSCOPY; NONLINEAR OPTICS.

The high power densities attainable with lasers allow melting and even vaporization of any solid material that is sufficiently opaque at a given wavelength or photon energy. This has led to a number of applications involving cutting and drilling of ceramics and other brittle materials, even diamonds. Welding of components from the smallest wires to huge steel plates is done commercially with high-power lasers. Metal alloying in surface regions is also a domain of lasers.

Ion implantation has become a dominant method of introducing controlled quantities of impurities near the surface of silicon and other semiconductors. The implanted layers need a heat treatment to repair the displacement damage caused by bombardment with energetic ions and to move the implanted impurity ions into lattice locations where they replace host atoms and become electrically active. Laser heating is particularly suitable for annealing since only the implanted regions are heated.

Thin films of single-crystalline silicon over an insulating substrate are very attractive for high-speed integrated circuits. An important approach to the formation of such films is the controlled melting of thin polycrystalline layers deposited over fused silica substrates or over oxidized silicon wafers. Through a careful control of temperature gradients around the molten spot, by shaping the laser beam or patterning the film, single-crystalline regions can be obtained. The formation of silicon-on-insulator structures will lead in the future to three-dimensional circuits, with several levels of transistors on the same chip. See LASER.

[G.K.C.]

Laser spectroscopy   Spectroscopy with laser light or, more generally, studies of the interaction between laser radiation and matter. Lasers have led to a rejuvenescence of classical spectroscopy, because laser light can far surpass the light from other sources in brightness, spectral purity, and directionality, and if required, laser light can be produced in extremely intense and short pulses. The use of lasers can greatly increase the resolution and sensitivity of conventional spectroscopic techniques, such as absorption spectroscopy, fluorescence spectroscopy, or Raman spectroscopy. Moreover, interesting new phenomena have become observable in the resonant interaction of intense coherent laser light with matter. Laser spectroscopy has become a wide and diverse field, with applications in numerous areas of physics, chemistry, and biology. *See* LASER; SPECTROSCOPY.

[T.W.Ha.]

Lattice vibrations   The oscillations of atoms in a solid about their equilibrium positions. In a crystal, these positions form a regular lattice. Because the atoms are bound not to their average positions but to the neighboring atoms, vibrations of neighbors are not independent of each other. In a regular lattice with harmonic forces between atoms, the normal modes of vibrations are lattice waves. These are progressive waves, and at low frequencies they are the elastic waves in the corresponding anisotropic continuum. The spectrum of lattice waves ranges from these low frequencies to frequencies of the order of 10^{13} Hz, and sometimes even higher. The wavelengths at these highest frequencies are of the order of interatomic spacings. *See* CRYSTAL STRUCTURE; VIBRATION; WAVE MOTION.

At room temperature and above, most of the thermal energy resides in the waves of highest frequency. Because of the short wavelength, the motion of neighboring atoms is essentially uncorrelated, so that for many purposes the vibrations can be regarded as those of independently vibrating atoms, each moving about its average position in three dimensions with average vibrational energy of $3kT$, where k is the Boltzmann constant and T the absolute temperature. The wave character of the vibrations is needed, however, to describe heat transport by lattice waves. Also, lattice vibrations interact with free electrons in a conducting solid and give rise to electrical resistance. The temperature variation at low temperatures provides evidence that this interaction is with waves. *See* ELECTRICAL RESISTIVITY.

Scattering of lattice waves by defects increases with increasing frequency (f); its variation depends on the nature of the defect. Scattering by external and internal boundaries is almost independent of frequency, thus dominating at low frequencies and hence at low temperatures. A study of the thermal conductivity of nonmetallic crystals as function of temperature yields information about the defects present, and about the anharmonic nature of the interatomic forces in the crystal lattice. *See* CRYSTAL DEFECTS; THERMAL CONDUCTION IN SOLIDS.

[P.G.Kl.]

Lawson criterion   A necessary but not sufficient condition for the achievement of a net release of energy from nuclear fusion reactions in a fusion reactor. As originally formulated by J. D. Lawson, this condition simply stated that a minimum requirement for net energy release is that the fusion fuel charge must combust for at least enough time for the recovered fusion energy release to equal the sum of energy invested in heating that charge to fusion temperatures, plus other energy losses occurring during combustion. The Lawson criterion is to be thought of as only a rule of thumb for measuring fusion progress; detailed evaluation of all energy dissipative and energy recovery processes is required in order properly to evaluate any specific system. *See* NUCLEAR FUSION.

[R.F.P.]

Least-action principle   Like Hamilton's principle, the principle of least action is a variational statement that forms a basis from which the equations of motion of a classical dynamical system may be deduced. Consider a mechanical system described by coordinates $q_1, \ldots, q_f$ and their canonically conjugate momenta $p_1, \ldots, p_f$. The action S associated with a segment of the trajectory of the system is defined by the equation below, where the integral is evaluated along the given segment c of the

$$S = \int_c \sum_j p_j \, dq_j$$

trajectory. The action is of interest only when the total energy E is conserved. The principle of least action states that the trajectory of the system is that path which makes the value of S stationary relative to nearby paths between the same configurations and for which the energy has the same constant value. The principle is misnamed, as only the stationary property is required. It is a minimum principle for sufficiently short but finite segments of the trajectory. *See* HAMILTON'S EQUATIONS OF MOTION; HAMILTON'S PRINCIPLE; MINIMAL PRINCIPLES.   [P.M.S.]

Length   A one-dimensional extension in space. Length is one of the three fundamental physical quantities, the others being mass and time. It can be measured by comparison with an arbitrary standard; the specific one in most common usage is the international meter. In 1983, at the meeting of the Conférence Général des Poids et Mésures, the meter was redefined in terms of time and the speed of light: "The meter is the length of the path traveled by light in vacuum during a time interval of 1/299 792 458 of a second." This definition defines the speed of light to be exactly 299 792 458 m/s, and defines the meter in terms of the most accurately known quantity, the second. *See* LIGHT; MASS; TIME.   [D.A.J.]

Lens (optics)   A curved piece of ground and polished or molded material, usually glass, used for the refraction of light. Its two surfaces have the same axis. Usually this is an axis of rotation symmetry for both surfaces; however, one or both of the surfaces can be toric, cylindrical, or a general surface with double symmetry (see illustration). The intersection points of the symmetry axis with the two surfaces are called the front and back vertices and their separation is called the thickness of the lens. There are three lens types, namely, compound, single, and cemented. A group of lenses used together is a lens system. Such systems may be divided into four classes: telescopes, oculars (eyepieces), photographic objectives, and enlarging lenses.

Lens types. A compound lens is a combination of two or more lenses in which the second surface of one lens has the same radius as the first surface of the following lens and the two lenses are cemented together. Compound lenses are used instead of single lenses for color correction, or to introduce a surface which has no effect on the aperture rays but large effects on the principal rays, or vice versa. Sometimes the term

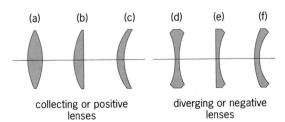

(a)   (b)   (c)   (d)   (e)   (f)

collecting or positive lenses   diverging or negative lenses

Common lenses. (*a*) Biconvex. (*b*) Plano-convex. (*c*) Positive meniscus. (*d*) Biconcave. (*e*) Plano-concave. (*f*) Negative meniscus. (*After F. A. Jenkins and H. E. White, Fundamentals of Optics, 4th ed., McGraw-Hill, 1976*)

compound lens is applied to any optical system consisting of more than one element, even when they are not in contact.

The diameter of a simple lens is called the linear aperture, and the ratio of this aperture to the focal length is called the relative aperture. This latter quantity is more often specified by its reciprocal, called the *f*-number. Thus, if the focal length is 50 mm and the linear aperture 25 mm, the relative aperture is 0.5 and the *f*-number is *f*/2. *See* FOCAL LENGTH.

A compound lens made of two or more simple thin lenses cemented together is called a cemented lens.

Lens systems. A lens system consisting of two systems combined so that the back focal point of the first (the objective) coincides with the front focal point of the second (the ocular) is called a telescope. Parallel entering rays leave the system as parallel rays. The magnification is equal to the ratio of the focal length of the first system to that of the second. *See* TELESCOPE.

A photographic objective images a distant object onto a photographic plate or film. The amount of light reaching the light-sensitive layer depends on the aperture of the optical system, which is equivalent to the ratio of the lens diameter to the focal length. The larger the aperture (the smaller the *f*-number), the less adequate may be the scene luminance required to expose the film. Therefore, if pictures of objects in dim light are desired, the *f*-number must be small. On the other hand, for a lens of given focal length, the depth of field is inversely proportional to the aperture.

In general, photographic objectives with large fields have small apertures; those with large apertures have small fields.

The basic type of enlarger lens is a holosymmetric system consisting of two systems of which one is symmetrical with the first system except that all the data are multiplied by the enlarging factor *m*. When the object is in the focus of the first system, the combination is free from all lateral errors even before correction. A magnifier in optics is a lens that enables an object to be viewed so that it appears larger than its natural size. The magnifying power is usually given as equal to one-quarter of the power of the lens expressed in diopters. *See* DIOPTER; MAGNIFICATION.   [M.J.H.]

Lenz's law   A law of electromagnetism which states that, whenever there is an induced electromotive force (emf) in a conductor, it is always in such a direction that the current it would produce would oppose the change which causes the induced emf. If the change is the motion of a conductor through a magnetic field, the induced current must be in such a direction as to produce a force opposing the motion. If the change causing the emf is a change of flux threading a coil, the induced current must produce a flux in such a direction as to oppose the change. Lenz's law is a form of the law of conservation of energy, since it states that a change cannot propagate itself. *See* CONSERVATION OF ENERGY; ELECTROMAGNETIC INDUCTION.   [K.V.M.]

Lepton   An elementary particle having no internal constituents which interacts through the electromagnetic, weak, and gravitational forces, but does not interact through the strong (nuclear) force. Leptons are very small, less than 10^{-18} m in size. This is less than 10^{-3} the size of a nucleus and less than 10^{-8} the size of an atom. Indeed, existing measurements are consistent with leptons being point particles.

These properties of the lepton family of particles are to be contrasted with the properties of the quark family of particles. Quarks interact through the strong force as well as through the electromagnetic, weak, and gravitational forces. By means of the strong force, quark-antiquark pairs bind together to form hadrons such as the π meson, and the quarks bind together to form hadrons such as the proton. In contrast, leptons act as individual particles and can be studied as isolated particles whereas, as far as is known, quarks are always inside hadrons and cannot be studied as isolated particles. *See* FUNDAMENTAL INTERACTIONS; HADRON; QUARKS.

Six leptons are known. There are three known charged leptons: the electron (e), muon (μ), and tau (τ). Associated with each charged lepton is a neutral lepton called a neutrino. A charged lepton and its associated neutrino is said to form a lepton generation. Thus there are three known lepton generations. *See* ELECTRON; NEUTRINO.   [M.L.P.]

Light   The term light, as commonly used, refers to the kind of radiant electromagnetic energy that is associated with vision. In a broader sense, light includes the entire range of radiation known as the electromagnetic spectrum. The branch of science dealing with light, its origin and propagation, its effects, and other phenomena associated with it is called optics. Spectroscopy is the branch of optics that pertains to the production and investigation of spectra. *See* OPTICS; SPECTROSCOPY.

Principal effects. The electromagnetic spectrum is a broad band of radiant energy that extends over a range of wavelengths running from trillionths of inches to hundreds of miles; wavelengths of visible light are measured in hundreds of thousandths of an inch. Arranged in order of increasing wavelength, the radiation making up the electromagnetic spectrum is termed gamma rays, x-rays, ultraviolet rays, visible light, infrared waves, microwaves, radio waves, and very long electromagnetic waves. *See* ELECTROMAGNETIC RADIATION.

The fact that light travels at a finite speed or velocity is well established. In round numbers, the speed of light in vacuum or air may be said to be 186,000 mi/s or 300,000 km/s. Measurements of the speed of light, c, which had attracted physicists for 308 years, came to an end in 1983 when the new definition of the meter fixed the value of the speed of light. Highly precise values of c were obtained by extending absolute frequency measurements into a region of the electromagnetic spectrum where wavelengths can be most accurately measured. These advances were facilitated by the use of stabilized lasers and high-speed tungsten-nickel diodes which were used to measure the lasers' frequencies. The measurements of the speed of light and of the frequency of lasers yielded a value of the speed of light limited only by the standard of length which was then in use. This permitted a redefinition of the meter in which the value of the speed of light assumed an exact value, 299,792,458 m/s. The meter is defined as the length of the path traveled by light in vacuum during a time interval of 1/299 792 458 of a second. *See* LASER.   [K.M.E.]

One of the most easily observed facts about light is its tendency to travel in straight lines. Careful observation shows, however, that a light ray spreads slightly when passing the edges of an obstacle. This phenomenon is called diffraction. The reflection of light is also well known. Reflection of light from smooth optical surfaces occurs so that the angle of reflection equals the angle of incidence, a fact that is most readily observed with a plane mirror. When light is reflected irregularly and diffusely, the phenomenon is termed scattering. The scattering of light by gas particles in the atmosphere causes the blue color of the sky. *See* DIFFRACTION; REFLECTION OF ELECTROMAGNETIC RADIATION.

The type of bending of light rays called refraction is caused by the fact that light travels at different speeds in different media—faster, for example, in air than in either glass or water. Refraction occurs when light passes from one medium to another in which it moves at a different speed. Familiar examples include the change in direction of light rays in going through a prism, and the bent appearance of a slick partially immersed in water. *See* REFRACTION OF WAVES.

In the phenomenon called interference, rays of light emerging from two parallel slits combine on a screen to produce alternating light and dark bands. This effect can be obtained quite easily in the laboratory, and is observed in the colors produced by a thin film of oil on the surface of a pool of water. Polarization of light is usually shown with Polaroid disks. Such disks are quite transparent individually. When two of them are placed together, however, the degree of transparency of the combination depends upon the relative orientation of the disks. It can be varied from ready transmission of light to almost total opacity, simply by rotating one disk with respect to the other. *See* INTERFERENCE OF WAVES; POLARIZED LIGHT.

When light is absorbed by certain substances, chemical changes take place. This fact forms the basis for the science of photochemistry.

Theory. Phenomena involving light may be classed into three groups: electromagnetic wave phenomena, corpuscular or quantum phenomena, and relativistic effects. The relativistic effects appear to influence similarly the observation of both corpuscular and wave phenomena. *See* RELATIVITY.

Wave phenomena. Interference and diffraction are the most striking manifestations of the wave character of light. Their fundamental similarity can be demonstrated in a number of experiments. The wave aspect of the entire spectrum of electromagnetic radiation is most convincingly shown by the similarity of diffraction pictures produced on a photographic plate, placed at some distance behind a diffraction grating, by radiations of different frequencies, such as x-rays and visible light. The interference phenomena of light are, moreover, very similar to interference of electronically produced microwaves and radio waves.

Polarization demonstrates the transverse character of light waves. Further proof of the electromagnetic character of light is found in the possibility of inducing, in a transparent body that is being traversed by a beam of plane-polarized light, the property of rotating the plane of polarization of the beam when the body is placed in a magnetic field. *See* FARADAY EFFECT.

The fact that the velocity of light had been calculated from electric and magnetic parameters (permittivity and permeability) was at the root of Maxwell's conclusion in 1865 that "light, including heat and other radiations if any, is a disturbance in the form of waves propagated. . . according to electromagnetic laws." Finally, the observation that electrons and neutrons can give rise to diffraction patterns quite similar to those produced by visible light has made it necessary to ascribe a wave character to particles. *See* ELECTRON DIFFRACTION; NEUTRON DIFFRACTION.

Corpuscular phenomena. In its interactions with matter, light exchanges energy only in discrete amounts, called quanta. This fact is difficult to reconcile with the idea that light energy is spread out in a wave, but is easily visualized in terms of corpuscles, or photons, of light.

The radiation from theoretically perfect heat radiators, called blackbodies, involves the exchange of energy between radiation and matter in an enclosed cavity. The observed frequency distribution of the radiation emitted by the enclosure at a given temperature of the cavity can be correctly described by theory only if one assumes that light of frequency ν is absorbed in integral multiples of a quantum of energy equal to $h\nu$, where h is a fundamental physical constant called Planck's constant.

When a monochromatic beam of electromagnetic radiation illuminates the surface of a solid (or less commonly, a liquid), electrons are ejected from the surface in the phenomenon known as photoemission or the external photoelectric effect. It is found that the emission of these photoelectrons, as they are called, is immediate, and independent of the intensity of the light beam, even at very low light intensities. This fact excludes the possibility of accumulation of energy from the light beam until an amount corresponding to the kinetic energy of the ejected electron has been reached.

The scattering of x-rays of frequency ν_0 by the lighter elements is caused by the collision of x-ray photons with electrons. Under such circumstances, both a scattered x-ray photon and a scattered electron are observed, and the scattered x-ray has a lower frequency than the impinging x-ray. The kinetic energy of the impinging x-ray, the scattered x-ray, and the scattered electron, as well as their relative directions, are in agreement with calculations involving the conservation of energy and momentum. *See* COMPTON EFFECT; HEAT RADIATION; PHOTON.

Quantum theories. The need for reconciling Maxwell's theory of the electromagnetic field, which describes the electromagnetic wave character of light, with the particle nature of photons, which demonstrates the equally important corpuscular character of light, has resulted in the formulation of several theories which go a long way toward giving a satisfactory unified treatment of the wave and the corpuscular picture. These theories incorporate, on one hand, the theory of quantum electrodynamics, first set forth by P. A. M. Dirac, P. Jordan, W. Heisenberg, and W. Pauli, and on the other, the earlier quantum mechanics of L. de Broglie, Heisenberg, and E. Schrödinger. Unresolved theoretical difficulties persist, however, in the higher-than-first approximations of the interactions between light and elementary particles.

Dirac's synthesis of the wave and corpuscular theories of light is based on rewriting Maxwell's equations in a Hamiltonian form resembling the Hamiltonian equations of classical mechanics. Using the same formalism involved in the transformation of classical into wave-mechanical equations by the introduction of the quantum of action $h\nu$, Dirac obtained a new equation of the electromagnetic field. The solutions of this equation require quantized waves, corresponding to photons. The superposition of these solutions represents the electromagnetic field. The quantized waves are subject to Heisenberg's uncertainty principle. The quantized description of radiation cannot be taken literally in terms of either photons or waves, but rather is a description of the probability of occurrence in a given region of a given interaction or observation. *See* HAMILTON'S EQUATIONS OF MOTION; QUANTUM ELECTRODYNAMICS; QUANTUM FIELD THEORY; QUANTUM MECHANICS; RELATIVISTIC QUANTUM THEORY; UNCERTAINTY PRINCIPLE.   [G.W.S.]

Light-emitting diode

Light-emitting diode   A rectifying semiconductor device which converts electrical energy into electromagnetic radiation. The wavelength of the emitted radiation ranges from the near-ultraviolet to the near-infrared, that is, from about 400 to over 1500 nanometers.

Most commercial light-emitting diodes (LEDs), both visible and infrared, are fabricated from III–V semiconductors. These compounds contain elements such as gallium, indium, and aluminum from column III (or group 13) of the periodic table, as well as arsenic, phosphorus, and nitrogen from column V (or group 15) of the periodic table. There are also LED products made of II–VI (or group 12–16) semiconductors, for example ZnSe and related compounds. Taken together, these semiconductors possess the proper band-gap energies to produce radiation at all wavelengths of interest. Most of these compounds have direct band gaps and, as a consequence, are efficient in the conversion of electrical energy into radiation. With the addition of appropriate chemical impurities, called dopants, both III–V and II–VI compounds can be made p- or n-type, for the purpose of forming pn junctions. All modern-day LEDs contain pn junctions. Most of them also have heterostructures, in which the pn junctions are surrounded by semiconductor materials with larger band-gap energies. *See* ACCEPTOR ATOM; DONOR ATOM; ELECTROLUMINESCENCE; ELECTRON-HOLE RECOMBINATION; LASER; SEMICONDUCTOR; SEMICONDUCTOR DIODE.

Conventional low-power, visible LEDs are used as solid-state indicator lights in instrument panels, telephone dials, cameras, appliances, dashboards, and computer terminals, and as light sources for numeric and alphanumeric displays. Modern high-brightness, visible LED lamps are used in outdoor applications such as traffic signals, changeable message signs, large-area video displays, and automotive exterior lighting. General-purpose white lighting and multielement array printers are applications in which high-power visible LEDs may soon displace present-day technology. Infrared LEDs, when combined in a hybrid package with solid-state photodetectors, provide a unique electrically isolated optical interface in electronic circuits. Infrared LEDs are also used in optical-fiber communication systems as a low-cost, high-reliability alternative to semiconductor lasers.

[J.M.Woo.; L.J.G.]

Line spectrum   A discontinuous spectrum characteristic of excited atoms, ions, and certain molecules in the gaseous phase at low pressures. If an electric arc or spark between metallic electrodes, or an electric discharge through a low-pressure gas, is viewed through a spectroscope, images of the spectroscope slit are seen in the characteristic colors emitted by the atoms or ions present. See ATOMIC STRUCTURE AND SPECTRA; SPECTROSCOPY.

[G.R.H.]

Lines of force   Imaginary lines in fields of force whose tangents at any point give the direction of the field at that point and whose number through unit area perpendicular to the field represents the intensity of the field. The concept of lines of force is perhaps most common when dealing with electric or magnetic fields.

Electric lines of force are drawn to represent, or map, an electric field graphically in the space around a charged body. They are of great help in visualizing an electric field and in quantitative thinking about such a field. A magnetic field may also be represented by lines of force. Magnetic lines of force due to magnets originate on north poles and terminate on south poles, both inside and outside the magnet. See ELECTRIC FIELD.

[R.P.Wi.]

Linewidth   A measure of the width of the band of frequencies of radiation emitted or absorbed in an atomic or molecular transition. One of the dominant sources of electromagnetic radiation of all frequencies is transitions between two energy levels of an atomic or molecular system. The frequency of the radiation is related to the difference in the energy of the two levels by the Bohr relation (1), where ν_0 is the frequency of

$$\nu_0 = (E_1 - E_2)/h \tag{1}$$

the radiation, h is Planck's constant, and E_1 and E_2 are the energies of the levels. This radiation is not monochromatic, but consists of a band of frequencies centered about ν_0 whose intensity $I(\nu)$ can be characterized by the linewidth. The linewidth is the full width at half height of the distribution function $I(\nu)$. The simplest case is for a transition from an excited state to the ground state for an atom or molecule at rest. For this case, the normalized distribution function is the lorentzian line profile given by Eq. (2). Here

$$I(\nu) = \frac{1}{\pi} \frac{\Delta\nu/2}{(\nu - \nu_0)^2 + (\Delta\nu/2)^2} \tag{2}$$

$\Delta\nu$ is the full width at half maximum (FWHM). The FWHM is related to the lifetime τ of the excited level through Eq. (3). This is a manifestation of the quantum-mechanical

$$(\Delta\nu)(\tau) = \frac{1}{2\pi} \tag{3}$$

uncertainty principle, and the linewidth $\Delta\nu$ is referred to as the natural linewidth. See ENERGY LEVEL (QUANTUM MECHANICS); QUANTUM MECHANICS; UNCERTAINTY PRINCIPLE.

Another major source of line broadening for atomic and molecular transitions is the Doppler shift due to thermal motion. For most situations the Doppler width is greater than the natural linewidth. *See* DOPPLER EFFECT.

A third major source of line broadening is collisions of the radiating molecule with other molecules. This broadens the line, shifts the center of the line, and shortens the lifetime of the radiating state.

For radiating atoms in a liquid or solid the width is usually dominated by the strong interaction of the radiator with the surrounding molecules. The net result is a broad line profile with a complex structure. *See* BAND THEORY OF SOLIDS.   [F.M.P.]

Liquefaction of gases   The process of refrigerating a gas to a temperature below its critical temperature so that liquid can be formed at some suitable pressure, also below the critical pressure.

Gas liquefaction is a special case of gas refrigeration. The gas is first compressed to an elevated pressure in an ambient-temperature compressor. This high-pressure gas is passed through a countercurrent heat exchanger to a throttling valve or expansion engine. Upon expanding to the lower pressure, cooling may take place, and some liquid may be formed. The cool, low-pressure gas returns to the compressor inlet to repeat the cycle. The purpose of the countercurrent heat exchanger is to warm the low-pressure gas prior to recompression, and simultaneously to cool the high-pressure gas to the lowest temperature possible prior to expansion. Both refrigerators and liquefiers operate on this same basic principle. *See* CRITICAL PHENOMENA.

An important distinction between refrigerators and liquefiers is that in a continuous refrigeration process, there is no accumulation of refrigerant in any part of the system. This contrasts with a gas-liquefying system, where liquid accumulates and is withdrawn. Thus, in a liquefying system, the total mass of gas that is warmed in the countercurrent heat exchanger is less than the gas to be cooled by the amount that is liquefied, creating an unbalanced flow in the heat exchanger. In a refrigerator, the warm and cool gas flows are equal in the heat exchanger. This results in balanced flow condition. The thermodynamic principles of refrigeration and liquefaction are identical. However, the analysis and design of the two systems are quite different due to the condition of balanced flow in the refrigerator and unbalanced flow in liquefier systems.

The prerequisite refrigeration for gas liquefaction is accomplished in a thermodynamic process when the process gas absorbs heat at temperatures below that of the environment. A process for producing refrigeration at liquefied gas temperatures usually involves equipment at ambient temperature in which the gas is compressed and heat is rejected to a coolant. During the ambient-temperature compression process, the enthalpy and entropy, but usually not the temperature of the gas, are decreased. The reduction in temperature of the gas is usually accomplished by heat exchange between the cooling and warming gas streams followed by an expansion of the high-pressure stream. This expansion may take place either through a throttling device (isenthalpic expansion) where there is a reduction in temperature only (when the Joule-Thomson coefficient is positive) or in a work-producing device (isentropic expansion) where both temperature and enthalpy are decreased. *See* ENTHALPY; ENTROPY; ISENTROPIC PROCESS; THERMODYNAMIC PRINCIPLES; THERMODYNAMIC PROCESSES.   [T.M.F.]

Liquid helium   Helium boils at a substantially lower temperature, 4.2 K ($-452°$F or $-269°$C), than any other substance; and below 2.172 K ($-455.76°$F) the liquid exhibits the extraordinary properties of superfluidity, notably the ability to flow through narrow channels with complete absence of friction. In addition to the common isotope

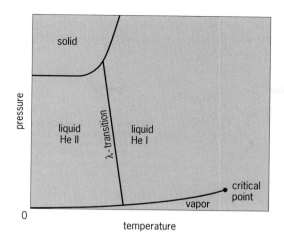

solid

pressure

liquid
He II

λ-transition

liquid
He I

critical
point

vapor

0

temperature

**Phase diagram for ⁴He. The
critical point is at $T_c = 5.20$ K
(−450.3°F), $P_c = 229$ kPa
(2.26 atm or 33.2 lb/in.²).**

of atomic weight 4, helium has a rare isotope of atomic weight 3 with a normal boiling point of 3.2 K (−454°F) and a superfluid transition at a very much lower temperature near 0.001 K. Both forms of helium remain in a liquid state at absolute zero. All of these characteristics are due to the weakness of the attractive force between two helium atoms and to the small atomic mass, which according to the laws of quantum mechanics makes the atoms difficult to localize.

At 4.2 K (−452°F) liquid ⁴He is colorless and of low refractive index ($n = 1.024$), with a density of 0.125 g/cm³ (0.125 times that of water). The latent heat of vaporization, 5 cal/g (21 J/g), is very small, and so care must be taken to reduce the heat input by conduction and radiation into the storage container. The classical container consists of two vacuum-insulated vessels of silvered glass (Dewar flask) or metal, with the inner vessel containing the liquid helium immersed in a larger outer vessel filled with liquid nitrogen. Modern superinsulated Dewars are able to dispense with the liquid nitrogen.

The phase diagram of ⁴He (see illustration) shows several remarkable characteristics. Helium remains a liquid down to absolute zero unless a pressure greater than 2.53 megapascals (25.0 atm or 367 lb/in.²) is applied. A more subtle feature is a transition between two different liquid phases. This λ-transition is so named because the specific heat has a singularity resembling the Greek letter lambda. There is no latent heat; such a transition is called second-order. The high-temperature liquid phase, called helium I, is a rather ordinary liquid. The λ-transition at 2.172 K or −455.76°F (at vapor pressure) marks the onset of superfluidity, which is the characteristic property of the low-temperature phase, helium II. *See* SUPERFLUIDITY.   [B.S.S.]

Lissajous figures   Plane curves traced by a point which executes two independent harmonic motions in perpendicular directions, the frequencies of the motion being in the ratio of two integers. Such figures are widely used in frequency and phase measurements (see illustration). *See* HARMONIC MOTION.

The cathode-ray oscilloscope furnishes the most important and practical means for the generation of the figures. The x-deflection plates of the tube are supplied with one alternating voltage, and the y-deflection plates with another. If the frequencies are incommensurable, the figure is not a closed curve and, except for very low frequencies,

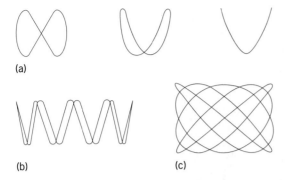

(a)

(b)   (c)

Typical Lissajous figures for
ratios of vertical frequency to
horizontal frequency. (*a*) 2:1,
with various phase relations.
(*b*) 8:1. (*c*) 5:4. (*After F. E.
Terman, Radio Engineers'
Handbook, McGraw-Hill, 1943*)

will appear as a patch of light because of the persistence of the screen. On the other
hand, if the frequencies are commensurable, the figure is closed and strictly periodic; it
is a true Lissajous figure, stationary on the screen and, if the persistence is sufficient, visible continuously as a complete pattern. *See* FREQUENCY MEASUREMENT; OSCILLOSCOPE;
PHASE-ANGLE MEASUREMENT.   [M.Gr.]

Low-temperature acoustics

The application of acoustics to research on
the properties of condensed matter at low temperatures. Acoustic techniques are readily
adaptable to the cryogenic environment and make possible many measurements of
the structural and thermodynamic properties of materials at temperatures approaching
absolute zero (0 K, which is $-273°$C). The study of sound propagation has also yielded
major insights into the low-temperature phenomena of superconductivity in metals and
superfluidity in liquid helium.

Solid materials. Acoustic measurements have been used to characterize the properties of a wide variety of solid-state materials, such as metals, dielectric crystals, amorphous solids, and magnetic materials. A measurement of the velocity of sound in a
substance gives information on its elastic properties, while the attenuation of the sound
characterizes the interaction of the lattice vibrations with the electronic and structural
properties of the material. Ultrasonic frequencies, in the range from 20 kHz to 100 MHz
and above, are commonly employed in these measurements because of the ease of
generating and detecting the sound with piezoelectric quartz crystals.

Because the sound velocity effectively measures elastic constants, such measurements are used to characterize phase transitions in crystals where the structure of the
lattice changes. The attenuation of sound in many crystals is due to defects and impurities in the crystal lattice and provides information on such structures. In a metal at very
low temperatures, the dominant source of attenuation is the interaction of the sound
with the conduction electrons. *See* CRYSTAL; LATTICE VIBRATIONS; PHASE TRANSITIONS;
PHONON; SOUND ABSORPTION.

A large variety of magnetoacoustic effects are observed in metals and crystals. In
these measurements, changes in the sound attenuation occur as the strength of a
magnetic field applied to the sample is increased. One example is the phenomenon
of nuclear acoustic resonance, resulting from the interaction of the nuclear spins in a
crystal with vibrations of the lattice. There are also a number of other magnetoacoustic
effects in metals which are useful in determining the orbits followed by the conduction
electrons in the metal. *See* DE HAAS-VAN ALPHEN EFFECT.

Sound propagation is useful for studying amorphous materials. In materials such
as silica glass (amorphous silicon dioxide, SiO_2), only two quantum energy levels are

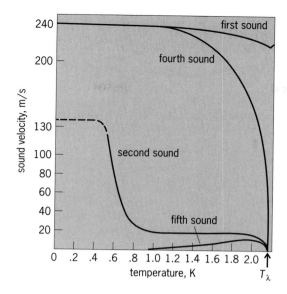

Velocity of the various types of sound in superfluid ^{4}He as a function of temperature. 1 m/s = 3.28 ft/s. $^\circ$F = (K × 1.8) − 459.67.

found to be important at low temperatures. These levels correspond to two nearly equivalent arrangements of the atoms, with one arrangement having slightly higher energy. An imposed sound field can cause a transition from one arrangement to the other. If the relaxation rate back to the original configuration is comparable to the sound frequency, there will be a net absorption of energy from the sound wave. A peak in the attenuation in silica glass near 50 K (−370°F) has been identified as being due to this process, and measurements as a function of frequency allow a determination of the relaxation rate. *See* AMORPHOUS SOLID.

When a metal is cooled below its superconducting transition temperature, there are striking changes in the attenuation of sound. At the transition some of the electrons near the Fermi surface begin to pair together, due to the attractive electron-phonon coupling. Once this occurs, the electrons can no longer exchange momentum with the lattice, and hence have zero resistance. This also means that the paired electrons no longer absorb energy from the sound wave, and the attenuation is from the remaining unpaired normal electrons. As the temperature is lowered well below the transition, the density of the unpaired electrons drops rapidly, and the attenuation becomes very small. *See* SUPERCONDUCTIVITY.

Superfluid helium. Sound propagation has been extensively used to probe many of the unusual properties of superfluid helium. The novel features of the superfluid (zero viscosity and entropy) give rise to a rich variety of different types of sound which can propagate in the superfluid helium. Five distinct sound modes have been identified and observed experimentally. The sound velocities of a number of these modes are shown in the illustration as a function of temperature.

First sound is a pressure wave which propagates in the bulk liquid. It is quite similar to sound in ordinary fluids.

Second sound is an unusual type of wave: it is a temperature wave in the bulk superfluid. In this mode the normal fluid and superfluid move in opposite directions. This keeps the density constant, and hence there are no pressure oscillations in the wave (as in first sound); but because only the normal fluid carries entropy, there are oscillations in the entropy and thus in the temperature of the liquid. *See* SECOND SOUND.

Third sound is a wave which propagates in very thin films of helium. The third sound is a wave in which the thickness of the film varies, somewhat like waves in a tank of water. Because the films are so thin, only the superfluid can move, the normal fluid being immobilized by its viscosity.

Fourth sound is a pressure wave which propagates in superfluid helium when it is confined in a porous material such as a tightly packed powder. In such a situation the normal fluid is immobilized, and only the superfluid can flow freely because of its zero viscosity (the porous materials are often called superleaks for this reason). The fourth sound is analogous to first sound because it involves density and pressure oscillations.

Fifth sound is a temperature wave which can propagate in helium confined in a superleak. It is analogous to second sound, except that again only the superfluid component can flow. *See* LIQUID HELIUM; SUPERFLUIDITY.   [G.A.W.]

Low-temperature physics   A branch of physics dealing with physical properties of matter at temperatures such that thermal fluctuations are greatly reduced and effects of interactions at the quantum-mechanical level can be observed. As the temperature is lowered, order sets in (either in space or in motion), and quantum-mechanical phenomena can be observed on a macroscopic scale.

Some of the most interesting manifestations of low temperatures have been investigated in the temperature range from 4 K ($-452°$F) down to less than a nanokelvin above absolute zero. (1 K is equal to $1.8°$F above absolute zero, or $-459.67°$F.) Certain metals become superconducting, losing their electrical resistance entirely; hence persistent currents can flow indefinitely in a superconducting ring or coil, displaying quantum-mechanical coherence over large distances. The liquids helium-3 (^{3}He) and helium-4 (^{4}He) remain liquid down to absolute zero under their own vapor pressure due to the large zero-point energy of these light atoms. (To overcome the large zero-point energy in liquid ^{3}He and liquid ^{4}He, a large pressure, approximately 30 atm or 3 megapascals, must be applied to cause these systems to solidify.) Liquid ^{4}He becomes superfluid, exhibiting no resistance to flow under certain conditions; when set in circulation, the fluid current persists indefinitely. Liquid ^{3}He also becomes superfluid at a much lower temperature with interesting magnetic and orbital effects. At sufficiently low temperatures, nuclear magnetic ordering has been observed in solid ^{3}He, in magnetic insulators, and in metallic systems. Silver becomes a nuclear antiferromagnet in the nanokelvin range as a result of quantum-mechanical exchange interactions. Considerable attention has been addressed to the general problem of ordering in disordered systems leading to studies of spin glasses, localization, and lower dimensionality. Quantum statistics are investigated in atomic hydrogen and deuterium, stabilized in states known as spin-polarized hydrogen (H↓) and spin-polarized deuterium (D↓). Because of its light mass and weak interactions, spin-polarized hydrogen is expected to remain gaseous down to absolute zero, whereas spin-polarized deuterium might liquefy at low temperatures. *See* LIQUID HELIUM; SUPERFLUIDITY.

Low-temperature research also deals with problems of thermometry and heat transfer between systems and within systems. Many practical applications have emerged, including the use of superconductivity for large magnets, ultrafast electronics for computers, and low-noise and high-sensitivity instrumentation. This type of instrumentation has opened new areas of research in biophysics, and in fundamental problems such as the search for magnetic monopoles, gravity waves, and quarks. *See* LOW-TEMPERATURE THERMOMETRY; SUPERCONDUCTING DEVICES; SUPERCONDUCTIVITY.

The development of low-temperature techniques has revealed a wide range of other phenomena. The behavior of oriented nuclei is studied by observing the distribution of gamma-ray emission of radioactive nuclei oriented in a magnetic field. Other areas of study include surfaces of liquid ^{3}He and liquid ^{4}He, ^{3}He–^{4}He mixtures, cryogenics, acoustic microscopy, phonon spectroscopy, monolayer helium films, molecular hydrogen, determination of the voltage standard, and phase transitions. *See* CRYOGENICS; ELECTRICAL UNITS AND STANDARDS; NUCLEAR ORIENTATION; PHASE TRANSITIONS.

[O.G.S.]

Low-temperature thermometry   The measurement of temperature below 32°F (0°C). Very few thermometers are truly wide-range, and hence most of the conventional methods of thermometry tend to fail the further one moves below room temperature (see table). The defining instrument for the lower regions of the International Practical Temperature Scale, the platinum resistance thermometer, rapidly loses sensitivity below −405°F (30 K), and its official limit is set at −434.81°F (13.81 K), the triple point of equilibrium hydrogen. This scale is based upon measurements of thermodynamic temperature made with the gas thermometer; the gas thermometer may be used down to about −456°F (2 K). *See* GAS THERMOMETRY.

Ranges and sensitivities of low-temperature thermometers	
Thermometer	Temperature range, K
Thermocouples	
300 to 700 ppm Fe in Au/Ag + 0.37 at. % Au	1–25
Chromel/300 to 700 ppm Fe in Au	1–300
Chromel/Constantan	20–1100
Resistance thermometers	
Platinum (capsule)	4–500
Rhodium + 0.5 at. % Fe	0.5–300
Carbon	0.01–300
Germanium	0.01–30
Saturation vapor pressure thermometers	
Hydrogen	14–21
Helium-4	1.0–5.2
Helium-3	0.5–3.3
Noise thermometers*	0.002–0.1
Magnetic thermometers	
Gadolinium metaphosphate, Gd(PO$_3$)$_3$	2–100
Cerous magnesium nitrate (CMN),	
Ce$_2$Mg$_3$(NO$_3$)$_{12}$ · 24H$_2$O; single crystal	0.003–4
CMN powder sphere or cylinder	0.002–4
Copper (and other nuclear paramagnets)	0.001–0.01
Gamma-ray anisotropy thermometers*	0.002–0.05
^{60}Co in hexagonal close-packed cobalt single crystal	0.002–0.04
^{54}Mn in iron	0.003–0.03
^{54}Mn in nickel	0.004–0.045
^{3}He melting-curve thermometer	0.001–1
Nuclear resonance thermometer	310 nK–2K

*Primary thermometer.

The low-temperature region is unique in having available several different types of primary thermometers, all of which are quite practical. The least practical, perhaps, is the acoustic thermometer, which uses the property that, extrapolated to zero pressure, the speed of sound in a gas is proportional to $T^{1/2}$. This has been used in the range −456 to −424°F (2–20 K) as an alternative to, and check upon, the gas thermometer. The Johnson noise in a resistor can be used with particular advantage at low temperatures when allied with SQUID detector technology. *See* SOUND.

In suitable systems it is possible to spatially orient atomic nuclei at very low temperatures, and if these nuclei are emitters of gamma rays, the emission pattern is anisotropic to a degree which is a measure of the thermodynamic temperature, Finally the magnetic susceptibility of suitable atomic nuclei may also be employed via the Curie law. Nuclear magnetic resonance or static SQUID-based techniques may be employed, but nuclear magnetic resonance is preferable in being unaffected by magnetic impurities, to which the second method falls hostage. *See* SQUID; TEMPERATURE; TEMPERATURE MEASUREMENT; THERMOCOUPLE; THERMOMETER.   [R.P.Hu.]

Luminance   The luminous intensity of any surface in a given direction per unit of projected area of the surface viewed from that direction. The International Commission on Illumination defines it as the quotient of the luminous intensity in the given direction of an infinitesimal element of the surface containing the point under consideration, by the orthogonally projected area of the element on a plane perpendicular to the given direction. Simply, it is the luminous intensity per unit area. Luminance is also called photometric brightness.

Since the candela is the unit of luminous intensity, the luminance, or photometric brightness, of a surface may be expressed in candelas/cm^2, candelas/in.^{2}, and so forth.

The stilb is a unit of luminance (photometric brightness) equal to 1 candela/cm^2. It is often used in Europe, but the practice in America is to use the term candela/cm^2 in its place.

The apostilb is another unit of luminance sometimes used in Europe. It is equal to the luminance of a perfectly diffusing surface emitting or diffusing light at the rate of 1 lumen/m^2. *See* LUMINOUS INTENSITY; PHOTOMETRY.   [R.C.Pu.]

Luminescence   Light emission that cannot be attributed merely to the temperature of the emitting body. Various types of luminescence are often distinguished according to the source of the energy which excites the emission. When the light energy emitted results from a chemical reaction, such as in the slow oxidation of phosphorus at ordinary temperatures, the emission is called chemiluminescence. When the luminescent chemical reaction occurs in a living system, such as in the glow of the firefly, the emission is called bioluminescence. In the foregoing two examples part of the energy of a chemical reaction is converted into light. There are also types of luminescence that are initiated by the flow of some form of energy into the body from the outside. According to the source of the exciting energy, these luminescences are designated as cathodoluminescence if the energy comes from electron bombardment; radioluminescence or roentgenoluminescence if the energy comes from x-rays or from γ-rays; photoluminescence if the energy comes from ultraviolet, visible, or infrared radiation; and electroluminescence if the energy comes from the application of an electric field. By attaching a suitable prefix to the word luminescence, similar designations may be coined to characterize luminescence excited by other agents. Since a given substance can frequently be made to luminesce by a number of different external exciting agents, and since the atomic and electronic phenomena that cause luminescence are basically the same regardless of the mode of excitation, the classification of luminescence phenomena into the foregoing categories is only a matter of convenience, not of fundamental distinction.

When a luminescent system provided with a special configuration is excited, or "pumped," with sufficient intensity of excitation to cause an excess of excited atoms over unexcited atoms (a so-called population inversion), it can produce laser action. (Laser is an acronym for light amplification by stimulated emission of radiation.) This laser emission is a coherent stimulated luminescence, in contrast to the incoherent spontaneous emission from most luminescent systems as they are ordinarily excited and used. *See* LASER; OPTICAL PUMPING.

A second basis frequently used for characterizing luminescence is its persistence after the source of exciting energy is removed. Many substances continue to luminesce for extended periods after the exciting energy is shut off. The delayed light emission (afterglow) is generally called phosphorescence; the light emitted during the period of excitation is generally called fluorescence. In an exact sense, this classification, based on persistence of the afterglow, is not meaningful because it depends on the properties of the detector used to observe the luminescence. With appropriate instruments one can detect afterglows lasting on the order of a few thousandths of a microsecond, which would be imperceptible to the human eye. The characterization of such a luminescence, based on its persistence, as either fluorescence or phosphorescence would therefore depend upon whether the observation was made by eye or by instrumental means. These terms are nevertheless commonly used in the approximate sense defined here, and are convenient for many practical purposes. However, they can be given a more precise meaning. For example, fluorescence may be defined as a luminescence emission having an afterglow duration which is temperature-independent, while phosphorescence may be defined as a luminescence with an afterglow duration which becomes shorter with increasing temperature. *See* CATHODOLUMINESCENCE; ELECTROLUMINESCENCE; FLUORESCENCE; PHOSPHORESCENCE; PHOTOLUMINESCENCE; THERMOLUMINESCENCE.

[C.C.K.; J.H.S.]

Luminous efficacy

There are three ways this term can be used: (1) The luminous efficacy of a source of light is the quotient of the total luminous flux emitted divided by the total lamp power input. Light is visually evaluated radiant energy. Luminous flux is the time rate of flow of light. Luminous efficacy is expressed in lumens per watt. (2) The luminous efficacy of radiant power is the quotient of the total luminous flux emitted divided by the total radiant power emitted. This is always somewhat larger for a particular lamp than the previous measure, since not all the input power is transformed into radiant power. (3) The spectral luminous efficacy of radiant power is the quotient of the luminous flux at a given wavelength of light divided by the radiant power at that wavelength. A plot of this quotient versus wavelength displays the spectral response of the human visual system. It is, of course, zero for all wavelengths outside the range from 380 to 760 nanometers. It rises to a maximum near the center of this range. Both the value and the wavelength of this maximum depend on the degree of dark adaptation present. However, an accepted value of 683 lumens per watt maximum at 555 nanometers represents a standard observer in a light-adapted condition. *See* LUMINOUS EFFICIENCY; LUMINOUS FLUX; PHOTOMETRY.

[G.A.Ho.]

Luminous efficiency

Visual efficacy of visible radiation, a function of the spectral distribution of the source radiation in accordance with the "spectral luminous efficiency curve," usually for the light-adapted eye or photopic vision, or in some instances for the dark-adapted eye or scotopic vision.

The spectral luminous efficiency of radiant flux is the ratio of luminous efficacy for a given wavelength to the value of maximum luminous efficacy. It is a dimensionless ratio. *See* LUMINOUS EFFICACY; PHOTOMETRY.

[G.A.Ho.]

Luminous energy

The radiant energy in the visible region or quantity of light. It is in the form of electromagnetic waves, and since the visible region is commonly taken as extending 380–760 nanometers in wavelength, the luminous energy is contained within that region. It is equal to the time integral of the production of the luminous flux. *See* PHOTOMETRY.   [R.C.Pu.]

Luminous flux   The time rate of flow of light. It is radiant flux in the form of electromagnetic waves which affects the eye or, more strictly, the time rate of flow of radiant energy evaluated according to its capacity to produce visual sensation. The visible spectrum is ordinarily considered to extend from 380 to 760 nanometers in wavelength; therefore, luminous flux is radiant flux in that region of the electromagnetic spectrum. The unit of measure of luminous flux is the lumen. *See* PHOTOMETRY.   [R.C.Pu.]

Luminous intensity   The solid angular luminous flux density in a given direction from a light source. It may be considered as the luminous flux on a small surface normal to the given direction, divided by the solid angle (in steradians) which the surface subtends at the source of light. Since the apex of a solid angle is a point, this concept applies exactly only to a point source. The size of the source, however, is often extremely small when compared with the distance from which it is observed, so in practice the luminous flux coming from such a source may be taken as coming from a point. *See* CANDLEPOWER; PHOTOMETRY.   [R.C.Pu.]

M

Mach number In the flow of a fluid, the ratio of the flow velocity, V, at a given point in the flow to the local speed of sound, a, at that same point. That is, the Mach number, M, is defined as V/a. In a flowfield where the properties vary in time and/or space, the local value of M will also vary in time and/or space. In aeronautics, Mach number is frequently used to denote the ratio of the airspeed of an aircraft to the speed of sound in the freestream far ahead of the aircraft; this is called the freestream Mach number. The Mach number is a convenient index used to define the following flow regimes: (1) subsonic, where M is less than 1 everywhere throughout the flow; (2) supersonic, where M is greater than 1 everywhere throughout the flow; (3) transonic, where the flow is composed of mixed regions of locally subsonic and supersonic flows, all with local Mach numbers near 1, typically between 0.8 and 1.2; and (4) hypersonic, where (by arbitrary definition) M is 5 or greater.

Perhaps the most important physical aspect of Mach number is in the completely different ways that disturbances propagate in subsonic flow compared to that in a supersonic flow. Shock waves are a ubiquitous aspect of supersonic flows. *See* COMPRESSIBLE FLOW; SHOCK WAVE; SONIC BOOM.
[J.D.A.]

Madelung constant A numerical constant α_M in terms of which the electrostatic energy U of a three-dimensional periodic crystal lattice of positive and negative point charges q_+, $-q_-$, N in number, is given by the equation below, where d is the

$$U = -\frac{1}{2}\frac{Nq_+q_-}{d}\alpha_M$$

nearest-neighbor distance between positive and negative charges and N is large. Knowledge of such electrostatic energies as given by the Madelung constant is of importance in the calculation of the cohesive energies of ionic crystals and in many other problems in the physics of solids. *See* IONIC CRYSTALS.

Madelung constants for some common ionic crystals	
Crystal structure	Madelung constant, α_M
Sodium chloride, NaCl	1.7476
Cesium chloride, CsCl	1.7627
Zinc blende, α-ZnS	1.6381
Wurtzite, β-ZnS	1.641
Fluorite, CaF_2	5.0388
Cuprite, CuO_2	4.1155
Rutile, TiO_2	4.816
Anatase, TiO_2	4.800
Corundum, Al_2O_3	25.0312

The Madelung constants for a number of common ionic crystal structures are given in the table. For these cases d is chosen as the nearest-neighbor distance. *See* CRYSTAL.

[B.G.D.]

Magic numbers   The number of neutrons or protons in nuclei which are required to fill major quantum shells. They occur at particle numbers 2, 8, 20, 50, and 82.

In atoms, the electrons that orbit the nucleus fill quantum electron shells at atomic numbers of 2 (helium), 10 (neon), 18 (argon), 36 (krypton), and 54 (xenon). These elements are chemically inert and difficult to ionize because the energies of orbits are grouped in bunches or shells with large gaps between them. In nuclei, an analogous behavior is found; quantum orbits completely filled with neutrons or protons result in extra stability. The neutrons and protons fill their quantum states independently, so that both full neutron and full proton shells can occur as magic nuclei. In a few cases, for example oxygen-16 ($^{16}_{8}O_8$) and calcium-40 ($^{40}_{20}Ca_{20}$), doubly magic nuclei have full neutron and proton shells. Between the major shell gaps, smaller subshell gaps cause some extra stabilization and semimagic behavior is found at particle numbers 14, 28, 40, and 64. *See* ATOMIC STRUCTURE AND SPECTRA; ELECTRON CONFIGURATION.

In very heavy nuclei the Coulomb repulsion between the protons results in a different sequence of states for neutrons and protons and different major shell gaps. For neutrons the magic sequence continues at $N = 126$; the next shell gap is predicted at $N = 184$. For protons the next major shell gap is anticipated at $Z = 114$. The latter shell gaps lie beyond the heaviest nuclei known, but calculations indicate that the extra stability gained by producing nuclei with these particle numbers may result in an island of long-lived superheavy nuclei.

The closing of nuclear quantum shells has many observable consequences. The nuclei are more tightly bound than average, and the extra stability leads to anomalously high abundances of magic nuclei in nature. The full shells require unusually high energies to remove the least bound neutron or proton, and the probability of capturing extra particles is lower than expected. Furthermore, the full shells are spherically symmetric, and the nuclei have very small electric quadrupole moments. Many of these properties were known before the nuclear shell model was developed to account for quantum-level ordering and gaps between major shells. The different shell closures for atomic and nuclear systems reflect the differences between the Coulomb force that binds electrons to nuclei and the strong force that holds the nucleus together. An important component of the strong force in nuclei is the spin-orbit term, which makes the energy of a state strongly dependent on the relative orientation of spin and orbital angular momentum. *See* ANGULAR MOMENTUM; ISOTOPE; NUCLEAR MOMENTS; NUCLEAR STRUCTURE; STRONG NUCLEAR INTERACTIONS.   [C.J.Li.]

Magnet   An object or device that produces a magnetic field. Magnets are essential for the generation of electric power and are used in motors, generators, labor-saving electromechanical devices, information storage, recording, and numerous specialized applications, for example, seals of refrigerator doors. The magnetic fields produced by magnets apply a force at a distance on other magnets, charged particles, electric currents, and magnetic materials.

Magnets may be classified as either permanent or excited. Permanent magnets are composed of so-called hard magnetic material, which retains an alignment of the magnetization in the presence of ambient fields. Excited magnets use controllable energizing currents to generate magnetic fields in either electromagnets or air-cored magnets. *See* FERROMAGNETISM; SUPERCONDUCTIVITY.

The essential characteristic of permanent-magnet materials is an inherent resistance to change in magnetization over a wide range of field strength. Resistance to change in magnetization in this type of material is due to two factors: (1) the material consists of particles smaller than the size of a domain, a circumstance which prevents the gradual change in magnetization which would otherwise take place through the movement of domain wall boundaries; and (2) the particles exhibit a marked magnetocrystalline anisotropy. During manufacture the particles are aligned in a magnetic field before being sintered or bonded in a soft metal or polyester resin. Compounds of neodymium, iron, and boron are used.

Electromagnets rely on magnetically soft or permeable materials which are well annealed and homogeneous so as to allow easy motion of domain wall boundaries. Ideally the coercive force should be zero, permeability should be high, and the flux density saturation level should be high. Coincidentally the hysteresis energy loss represented by the area of the hysteresis curve is small. This property and high electrical resistance (for the reduction of eddy currents) are required where the magnetic field is to vary rapidly. This is accomplished by laminating the core and using iron alloyed with a few percent silicon that increases the resistivity.

Electromagnets usually have an energizing winding made of copper and a permeable iron core. Applications include relays, motors, generators, magnetic clutches, switches, scanning magnets for electron beams (for example, in television receivers), lifting magnets for handling scrap, and magnetic recording heads.

Special iron-cored electromagnets designed with highly homogeneous fields are used for special analytical applications in, for example, electron or nuclear magnetic resonance, or as bending magnets for particle accelerators. *See* MAGNETIC RESONANCE; PARTICLE ACCELERATOR.

Air-cored electromagnets are usually employed above the saturation flux density of iron (about 2 T); at lower fields, iron-cored magnets require much less power because the excitation currents needed then are required only to generate a small field to magnetize the iron. The air-cored magnets are usually in the form of a solenoid with an axial hole allowing access to the high field in the center. The conductor, usually copper or a copper alloy, must be cooled to dissipate the heat generated by resistive losses. In addition, the conductor and supporting structure must be sufficiently strong to support the forces generated in the magnet.

In pulsed magnets, higher fields can be generated by limiting the excitation to short pulses (usually furnished by the energy stored in a capacitor bank) and cooling the magnet between pulses. The highest fields are generally achieved in small volumes. A field of 75 T has been generated for 120 microseconds.

Large-volume or high-field magnets are often fabricated with superconducting wire in order to avoid the large resistive power losses of normal conductors. The two commercially available superconducting wire materials are (1) alloys of niobium-titanium, a ductile material which is used for generating fields up to about 9 T; and (2) a brittle alloy of niobium and tin (Nb_3Sn) for fields above 9 T. Practical superconducting wires use complex structures of fine filaments of superconductor that are twisted together and embedded in a copper matrix. The conductors are supported against the electromagnetic forces and cooled by liquid helium at 4.2 K ($-452°F$). A surrounding thermal insulating enclosure such as a dewar minimizes the heat flow from the surroundings.

Superconducting magnets operating over 20 T have been made with niobium-titanium outer sections and niobium-tin inner sections. Niobium-titanium is used in whole-body nuclear magnetic resonance imaging magnets for medical diagnostics. Other applications of superconducting magnets include their use in nuclear magnetic resonance for chemical analysis, particle accelerators, containment of plasma in fusion reactors, magnetic separation, and magnetic levitation. *See* Nuclear fusion; Superconducting devices.

The highest continuous fields are generated by hybrid magnets. A large-volume (lower-field) superconducting magnet that has no resistive power losses surrounds a water-cooled inner magnet that operates at the highest field. The fields of the two magnets add. Over 35 T has been generated continuously.   [S.Fo.]

Magnetic ferroelectrics   Materials that display both magnetic order and spontaneous electric polarization. Research on these materials has enabled considerable advances to be made in understanding the interplay between magnetism and ferroelectricity. The existence of both linear and higher-order coupling terms has been confirmed, and their consequences studied. They have given rise, in particular, to a number of magnetically induced polar anomalies and have even provided an example of a ferromagnet whose magnetic moment per unit volume is totally induced by its coupling via linear terms to a spontaneous electric dipole moment.

Most known ferromagnetic materials are metals or alloys. Ferroelectric materials, on the other hand, are nonmetals by definition. It therefore comes as no surprise to find that there are no known room-temperature ferromagnetic ferroelectrics. In fact, there are no well-characterized materials which are known to be both strongly ferromagnetic and ferroelectric at any temperature.

Somewhat unaccountably, antiferromagnetic ferroelectrics are also comparative rarities in nature. Nevertheless, a few are known, and among them the barium-transition-metal fluorides are virtually unique in providing a complete series of isostructural examples. They have the chemical composition $BaXF_4$ in which X is a divalent ion of one of the $3d$ transition metals, manganese, iron, cobalt, or nickel. These materials are orthorhombic and all spontaneously polar (that is, pyroelectric) at room temperature. For all except the iron and manganese materials, which have a higher electrical conductivity than the others, the polarization has been reversed by the application of an electric field, so that they are correctly classified as ferroelectric. Long-range antiferromagnetic ordering sets in at temperatures somewhat below 100 K ($-280°F$). Structurally the materials consist of XF_6 octahedra which share corners to form puckered xy sheets which are linked in the third dimension z by the barium atoms. *See* Crystal.

The importance of these magnetic ferroelectrics is the opportunity they provide to study and to separate the effects of a variety of magnetic and nonmagnetic excitations upon the ferroelectric properties and particularly upon the spontaneous polarization. Measurements are often made via the pyroelectric effect, which is the variation of polarization with temperature. This effect is an extremely sensitive indicator of electronic and ionic charge perturbations in polar materials. Through these perturbations the effects of propagating lattice vibrations (phonons), magnetic excitations (magnons), electronic excitations (excitons), and even subtle structural transitions can all be probed with precision. *See* Pyroelectricity.

Of all the X ions present in the series $BaXF_4$, the largest is Mn^{2+}. As the temperature is reduced from room temperature, the fluorine cages contract and eventually the divalent manganous ion becomes too big for its cage, precipitating a complicated structural transition at 250 K ($-10°F$). One interesting effect of this phase transition is that it produces a lower-temperature phase with a crystal symmetry low enough to support the existence of the linear magnetoelectric effect, a linear coupling between magnetization and polarization. Below the antiferromagnetic transition at 26 K in $BaMnF_4$ this linear coupling produces a canting of the antiferromagnetic sublattices through a very small angle (of order 0.2 degree of arc). The result is a spontaneous, polarization-induced magnetic moment. At low temperatures $BaMnF_4$ is therefore technically a weak ferromagnet, although the resultant magnetic moment is extremely small, and it is more usually referred to as a canted antiferromagnet. This is the only well-categorized example of pyroelectrically driven ferromagnetism. *See* ANTIFERROMAGNETISM; FERROELECTRICS; FERROMAGNETISM; MAGNETISM.

[M.E.L.]

Magnetic instruments
Instruments designed for the measurement of magnetic field strength or magnetic flux density, depending on their principle of operation.

Hall-effect instruments. Often called gaussmeters, these instruments measure magnetic field strength. They have a useful working range from 10 A/m to 2.4 MA/m (0.125 oersted to 30 kilooersteds). When a magnetic field, H_z, is applied in a direction at right angles to the current flowing in a conductor (or semiconductor), a voltage proportional to H_z is produced across the conductor in a direction mutually perpendicular to the current and the applied magnetic field. This phenomenon is called the Hall effect. The output voltage of the Hall probe is proportional to the Hall coefficient, which is a characteristic of the Hall-element material, and is inversely proportional to the thickness of this material. For a sensitive Hall probe, the material is thin with a large Hall coefficient. The semiconducting materials indium arsenide and indium antimonide are particularly suitable. *See* HALL EFFECT.

Fluxgate magnetometer. This instrument is used to measure low magnetic field strengths. It is usually calibrated as a gaussmeter with a useful range of 0.2 millitesla to 0.1 nanotesla (2 gauss to 1 microgauss).

Fluxmeter. This instrument is designed to measure magnetic flux. A fluxmeter is a form of galvanometer in which the torsional control is very small and heavy damping is produced by currents induced in the coil by its motion. This enables a fluxmeter to accurately integrate an emf produced in a search coil when the latter is withdrawn from a magnetic field, almost independently of the time taken for the search coil to be moved. *See* GALVANOMETER.

Electronic charge intergrators. Often termed an integrator or gaussmeter, an electronic charge integrator, in conjunction with a search coil of known effective area, is used for the measurement of magnetic flux density. Integrators have almost exclusively replaced fluxmeters because of their independence of level and vibration. The instrument (see illustration) consists of a high-open-loop-gain (10^7 or more) operational amplifier with a capacitive feedback and resistive input.

Rotating-coil gaussmeter. This instrument measures low magnetic field strengths and flux densities. It comprises a coil mounted on a nonmagnetic shaft remote from a motor mounted at the other end. The motor causes the coil to rotate at a constant speed, and in the presence of a magnetic field or magnetic flux density a voltage is induced in the search coil. The magnitude of the voltage is proportional to the effective area

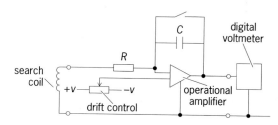

Arrangement of an electronic charge integrator.

of the search coil and the speed of rotation. [A.E.D.]

Magnetic lens   A magnetic field with axial symmetry capable of converging beams of charged particles of uniform velocity and of forming images of objects placed in the path of such beams. Magnetic lenses are employed as condensers, objectives, and projection lenses in magnetic electron microscopes, as final focusing lenses in the electron guns of cathode-ray tubes, and for the selection of groups of charged particles of specific velocity in velocity spectrographs.

Magnetic lenses may be formed by solenoids or helical coils of wire traversed by electric current, by axially symmetric pole pieces excited by a coil encased in a high-permeability material such as soft iron, or by similar pole pieces excited by permanent magnets. In the last two instances the armatures and pole pieces serve to concentrate the magnetic field in a narrow region about the axis.

Magnetic lenses are always converging lenses. Their action differs from that of electrostatic lenses and glass lenses in that they produce a rotation of the image in addition to the focusing action. For the simple uniform magnetic field within a long solenoid the image rotation is exactly $180°$. Thus a uniform magnetic field forms an erect real image of an object on its axis. [E.G.R.]

Magnetic materials   Materials exhibiting ferromagnetism. The magnetic properties of all materials make them respond in some way to a magnetic field, but most materials are diamagnetic or paramagnetic and show almost no response. The materials that are most important to magnetic technology are ferromagnetic and ferrimagnetic materials. Their response to a field H is to create an internal contribution to the magnetic induction B proportional to H, expressed as $B = \mu H$, where μ, the permeability, varies with H for ferromagnetic materials. Ferromagnetic materials are the elements iron, cobalt, nickel, and their alloys, some manganese compounds, and some rare earths. Ferrimagnetic materials are spinels of the general composition MFe_2O_4, and garnets, $M_3Fe_5O_{12}$, where M represents a metal. *See* FERRIMAGNETISM; FERROMAGNETISM; MAGNETISM; MAGNETIZATION.

Ferromagnetic materials are characterized by a Curie temperature, above which thermal agitation destroys the magnetic coupling giving rise to the alignment of the elementary magnets (electron spins) of adjacent atoms in a crystal lattice. Below the Curie temperature, ferromagnetism appears spontaneously in small volumes called domains. In the absence of a magnetic field, the domain arrangement minimizes the external energy, and the bulk material appears unmagnetized. *See* CURIE TEMPERATURE.

Magnetic materials are further classified as soft or hard according to the ease of magnetization. Soft materials are used in devices in which change in the magnetization during operation is desirable, sometimes rapidly, as in ac generators and transformers. Hard materials are used to supply a fixed field either to act alone, as in a magnetic separator, or to interact with others, as in loudspeakers and instruments. *See* ELECTRICAL MEASUREMENTS; INDUCTOR. [F.E.Lu.]

Magnetic monopoles   Magnetically charged particles. Such particles are predicted by various physical theories, but so far all experimental searches have failed to demonstrate their existence.

The fundamental laws governing electricity and magnetism become symmetric if particles exist that carry magnetic charge. Current understanding of electromagnetic physical phenomena is based on the existence of electric monopoles, which are sources

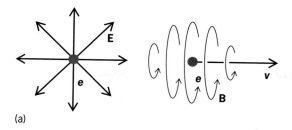

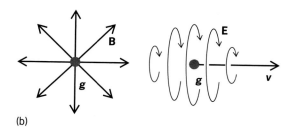

Electric (E) and magnetic
(B) field lines generated by
monopoles and by their motion
with velocity *v*. (*a*) Electric
monopole with electric charge *e*.
(*b*) Magnetic monopole with
magnetic charge *g*.

or sinks of electric field lines (illus. *a*), and which when set into motion generate magnetic fields. The magnetic field lines produced by such a current have no beginning or end and form closed loops. All magnetic fields occurring in nature can be explained as arising from currents. However, theories of electromagnetism become symmetric if magnetic charges also exist. These would be sources or sinks of magnetic field and when set into motion would generate electric fields whose lines would be closed without ends (illus. *b*). See ELECTRIC FIELD.

In 1931 P. A. M. Dirac found a more fundamental reason for hypothesizing magnetic charges, when he showed that this would explain the observed quantization of electric charge. He showed that all electric and magnetic charges *e* and *g* must obey Eq. (1),

$$eg = k(^1/_2 \hbar c) \tag{1}$$

where k must be an integer and $\hbar$ is Planck's constant divided by 2π. Equation (1) can be satisfied only if all electric and magnetic charges are integer multiples of an elementary electric charge e_0 and an elementary magnetic charge g_0. Since the size of the elementary electric charge, the charge carried by an electron or proton, is known experimentally, Dirac's equation predicts the size of the elementary magnetic charge to be given by Eq. (2). Since the fine structure constant α is given by Eq. (3), the

$$g_0 = \frac{1}{2} \frac{\hbar c}{e_0} \tag{2}$$

$$\alpha = \frac{e_0^2}{\hbar c} \approx \frac{1}{137} \tag{3}$$

elementary magnetic charge g_0 is about 68.5 times larger than the elementary electric charge e_0. See FINE STRUCTURE (SPECTRAL LINES); FUNDAMENTAL CONSTANTS.

In 1983 a successful theoretical unification of the electromagnetic and weak forces culminated in the detection of the W^+, the W^-, and the Z^0 particles predicted by the theory. This success has encouraged the search for a grand unification theory that would include the electroweak force and the nuclear or color force under one consistent description. In 1974 G. 't Hooft and independently A. M. Polyakov showed that magnetically charged particles are necessarily present in all true unification theories (those based on simple or semisimple compact groups). These theories predict the same long-range field and thus the same charge g_0 as the Dirac solution; now, however, the near field is also specified, leading to a calculable mass. The SU(5) model predicts a monopole mass of 10^{16} GeV/c^2, while theories based on supersymmetry or Kaluza-Klein models yield even higher masses up to the Planck mass of 10^{19} GeV/c^2. *See* ELECTROWEAK INTERACTION; FUNDAMENTAL INTERACTIONS; GRAND UNIFICATION THEORIES; QUANTUM GRAVITATION; SUPERGRAVITY; SUPERSYMMETRY.

There are two classes of magnetic monopole detectors, superconducting and conventional. On February 14, 1982, a prototype superconducting detector operating at Stanford University observed a single candidate event. Since then a number of groups have operated larger second- and third-generation detectors, and their combined data have placed a limit on the monopole flux more than 3000 times lower than the value from the data set that included the original event. Thus the possibility that this event was caused by the passage of a magnetic monopole has been largely discounted.   [B.Ca.]

Magnetic relaxation   The relaxation or approach of a magnetic system to an equilibrium or steady-state condition as the magnetic field is changed. This relaxation is not instantaneous but requires time. The characteristic times involved in magnetic relaxation are known as relaxation times. Relaxation has been studied for nuclear magnetism, electron paramagnetism, and ferromagnetism.

Magnetism is associated with angular momentum called spin, because it usually arises from spin of nuclei or electrons. The spins may interact with applied magnetic fields, the so-called Zeeman energy; with electric fields, usually atomic in origin; and with one another through magnetic dipole or exchange coupling, the so-called spin-spin energy. Relaxation which changes the total energy of these interactions is called spin-lattice relaxation; that which does not is called spin-spin relaxation. (As used here, the term lattice does not refer to an ordered crystal but rather signifies degrees of freedom other than spin orientation, for example, translational motion of molecules in a liquid.) Spin-lattice relaxation is associated with the approach of the spin system to thermal equilibrium with the host material; spin-spin relaxation is associated with an internal equilibrium of the spins among themselves. *See* MAGNETISM; SPIN (QUANTUM MECHANICS).   [C.P.S.]

Magnetic resonance   A phenomenon exhibited by the magnetic spin systems of certain atoms whereby the spin systems absorb energy at specific (resonant) frequencies when subjected to alternating magnetic fields. The magnetic fields must alternate in synchronism with natural frequencies of the magnetic system. In most cases the natural frequency is that of precession of the bulk magnetic moment of constituent atoms or nuclei about some magnetic field. Because the natural frequencies are highly specific as to their origin (nuclear magnetism, electron spin magnetism, and so on), the resonant method makes possible the selective study of particular features of interest. For example, it is possible to study weak nuclear magnetism unmasked by the much larger electronic paramagnetism or diamagnetism which usually accompanies it.

Nuclear magnetic resonance (that is, resonance exhibited by nuclei) reveals not only the presence of a nucleus such as hydrogen, which possesses a magnetic moment, but also its interaction with nearby nuclei. It has therefore become a most powerful method of determining molecular structure. The detection of resonance displayed by unpaired electrons, called electron paramagnetic resonance, is also an important application. *See* MAGNETIC RESONANCE; MAGNETISM.

[C.P.S.]

Magnetic susceptibility

The magnetization of a material per unit applied field. It describes the magnetic response of a substance to an applied magnetic field. *See* MAGNETISM; MAGNETIZATION.

All ferromagnetic materials exhibit paramagnetic behavior above their ferromagnetic Curie points. The general behavior of the susceptibility of ferromagnetic materials at temperatures well above the ferromagnetic Curie temperature follows the Curie-Weiss law. The paramagnetic Curie temperature is usually slightly greater than the temperature of transition. *See* CURIE TEMPERATURE; CURIE-WEISS LAW; FERROMAGNETISM.

Most paramagnetic substances at room temperature have a static susceptibility which follows a Langevin-Debye law. Saturation of the paramagnetic susceptibility occurs when a further increase of the applied magnetic field fails to increase the magnetization, because practically all the magnetic dipoles are already oriented parallel to the field. *See* PARAMAGNETISM.

The susceptibility of diamagnetic materials is negative, since a diamagnetic substance is magnetized in a direction opposite to that of the applied magnetic field. The diamagnetic susceptibility is independent of temperature. Diamagnetic susceptibility depends upon the distribution of electronic charge in an atom and upon the energy levels. *See* DIAMAGNETISM.

The susceptibility of antiferromagnetic materials above the Néel point, which marks the transition from antiferromagnetic to paramagnetic behavior, follows a Curie-Weiss law with a negative paramagnetic Curie temperature.

[E.A.; F.Ke.]

Magnetic thermometer

A thermometer whose operation is based on Curie's law, which states that the magnetic susceptibility of noninteracting (that is, paramagnetic) dipole moments is inversely proportional to absolute temperature. Magnetic thermometers are typically used at temperatures below 1 K ($-458°$F). The magnetic moments in the thermometric material may be of either electronic or nuclear origin. Generally the magnetic thermometer must be calibrated at one or more reference temperatures. *See* ELECTRON; NUCLEAR MOMENTS; PARAMAGNETISM.

At temperatures from a few millikelvins upward, the thermometric material is preferably an electronic paramagnet, typically a nonconducting hydrous rare-earth salt. For higher temperatures, an ion is selected with a large magnetic moment in a crystalline environment with a high density of magnetic ions. In contrast, for low temperature use the magnetic exchange interactions between the magnetic ions should be small, which is accomplished by selecting an ion with a well-localized moment and by maintaining a large separation between the magnetic ions by means of diamagnetic atoms. This is the case in cerium magnesium nitrate (CMN) [$2Ce(NO_3)_3 \cdot 3Mg(NO_3)_3 \cdot 24H_2O$]. Here, the Ce^{3+} ion is responsible for the magnetic moment, which is well localized within the incompletely filled $4f$ shell relatively deep below the outer valence electrons. To reduce the magnetic interactions between the Ce^{3+} ions further, Ce^{3+} may be partly substituted with diamagnetic La^{3+} ions. Lanthanum-diluted CMN has been used for thermometry to below 1 mK. *See* EXCHANGE INTERACTION.

A mutual-inductance bridge, originally known as the Hartshorn bridge, has been the most widely employed measuring circuit for precision thermometry. The bridge is driven by a low-frequency alternating-current source. The inductance at low temperatures consists of two coils, which are as identical as possible. The voltages induced across them by the drive current are compared by means of a high-input-impedance ratio transformer. The output level of this voltage divider is adjusted to equal that of the midpoint between the two coils, using as null indicator a narrow-band preamplifier and a phase-sensitive (lock-in) detector. Thus, without a paramagnetic specimen, the bridge is balanced with the decade divider adjusted at its midpoint, while with the specimen inside one of the coils the change in the divider reading at bridge balance is proportional to the sample magnetization. For high-resolution thermometry it has become standard practice to replace the room-temperature zero detector with a SQUID magnetometer circuit. This also allows the mass of the sample to be reduced from several grams to the 1-mg level. *See* INDUCTANCE MEASUREMENT; SQUID.

Nuclear magnetic moments are smaller by a factor of 10^3 and are used for thermometry only in the ultralow-temperature region. For this the Curie-law behavior is generally sufficient down to the lowest temperatures. The nuclear paramagnetic thermometer loses adequate sensitivity for calibration purposes above 50–100 millikelvins, unless it is operated in a high polarizing field (H greater than 0.1 tesla). It can be utilized as a self-calibrating primary thermometer if the spin-lattice relaxation time is measured in parallel with the nuclear Curie susceptibility. Pulsed NMR measurement on the ^{195}Pt isotope in natural platinum metal provides presently the most widely used thermometry at temperatures below 1 mK. In the Curie-susceptibility measuring mode, it has been extended down to 10 μK. *See* LOW-TEMPERATURE THERMOMETRY; MAGNETIC RELAXATION. [M.K.]

Magnetism   The branch of science that describes the effects of the interactions between charges due to their motion and spin. These interactions may appear in various forms, including electric currents and permanent magnets. They are described in terms of the magnetic field, although the field hypothesis cannot be tested independently of the electrokinetic effects by which it is defined. The magnetic field complements the concept of the electrostatic field used to describe the potential energy between charges due to their relative positions. Special relativity theory relates the two, showing that magnetism is a relativistic modification of the electrostatic forces. The two together form the electromagnetic interactions which are propagated as electromagnetic waves, including light. They control the structure of materials at distances between the long-range gravitational actions and the short-range "strong" and "weak" forces most evident within the atomic nucleus. *See* ELECTROMAGNETIC RADIATION; RELATIVITY.

The magnetic field can be visualized as a set of lines (Fig. 1) illustrated by iron filings scattered on a suitable surface. The intensity of the field is indicated by the line spacing,

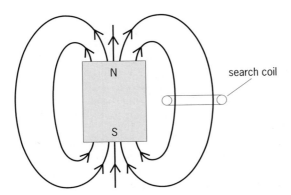

Fig. 1. Magnetic lines of a bar magnet.

and the direction by arrows pointing along the lines. The sign convention is chosen so that the Earth's magnetic field is directed from the north magnetic pole toward the south magnetic pole. The field can be defined and measured in various ways, including the forces on the equivalent magnetic poles, and on currents or moving charges. Bringing a coil of wire into the field, or removing it, induces an electromotive force (emf) which depends on the rate at which the number of field lines, referred to as lines of magnetic flux, linking the coil changes in time. This provides a definition of flux, Φ, in terms of the emf, e, given by Eq. (1) for a coil of N turns wound sufficiently closely to make

$$e = -N \, d\Phi/dt \qquad \text{volts} \tag{1}$$

the number of lines linking each the same. The International System (SI) unit of Φ, the weber (Wb), is defined accordingly as the volt-second. The symbol B is used to denote the flux, or line, density, as in Eq. (2), when the area of the coil is sufficiently

$$B = \Phi/\text{area} \tag{2}$$

small to sample conditions at a point, and the coil is oriented so that the induced emf is a maximum. The SI unit of B, the tesla (T), is the Wb/m^2. The sign of the emf, e, is measured positively in the direction of a right-hand screw pointing in the direction of the flux lines. It is often convenient, particularly when calculating induced emfs, to describe the field in terms of a magnetic vector potential function instead of flux.

Magnetic circuits. The magnetic circuit provides a useful method of analyzing devices with ferromagnetic parts, and introduces various quantities used in magnetism. It describes the use of ferromagnetic materials to control the flux paths in a manner analogous to the role of conductors in carrying currents around electrical circuits. For example, pieces of iron may be used to guide the flux which is produced by a magnet along a path which includes an air gap (Fig. 2), giving an increase in the flux density, B, if the cross-sectional area of the gap is less than that of the magnet. See MAGNET; MAGNETIC MATERIALS.

The magnet may be replaced by a coil of N turns carrying a current, i, wound over a piece of iron, or ferromagnetic material, in the form of a ring of uniform cross section. The flux linking each turn of the coil, and each turn of a secondary coil wound separately from the first, is then approximately the same, giving the same induced emf per turn [according to Eq. (1)] when the supply current, i, and hence the flux, Φ, changes in time. The arrangement is typical of many different devices. It provides, for example, an electrical transformer whose input and output voltages are directly proportional to the numbers of turns in the windings. Emf's also appear within the iron, and tend to produce circulating currents and losses. These are commonly reduced by dividing the material into thin laminations. See EDDY CURRENT.

The amount of flux produced by a given supply current is reduced by the presence of any air gaps which may be introduced to contribute constructional convenience or to allow a part to move. The effects of the gaps, and of different magnetic materials, can

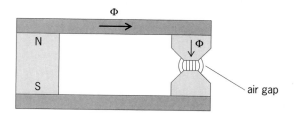

Fig. 2. Magnetic circuit with an air gap.

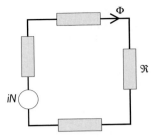

Fig. 3. Circuit analogy. Components of a magnetic circuit carrying a flux Φ analogous to current. The reluctances of the components are analogous to resistance.

be predicted by utilizing the analogy between flux, Φ, and the flow of electric current through a circuit consisting of resistors connected in series (Fig. 3). Since Φ depends on the product, iN, of the winding current and number of turns, as in Eq. (3), the ratio

$$iN = \Phi\Re \tag{3}$$

between them, termed the reluctance, $\Re$, is the analog of electrical resistance. It may be constant or may vary with Φ. The quantity iN is the magnetomotive force (mmf), analogous to voltage or emf in the equivalent electrical circuit. The relationship between the two exchanges the potental and flow quantities, since the magnetic mmf depends on current, i, and the electrical emf on $d\Phi/dt$. Electric and magnetic equivalent circuits are referred to as duals. *See* RELUCTANCE.

Any part of the magnetic circuit of length l, in which the cross section, a, and flux density, B, are uniform has a reluctance given by Eq. (4). This equation parallels Eq. (5) for the resistance, R, of a conduct of the same dimensions. The permeability, μ,

$$\Re = l/(a\mu) \tag{4}$$

$$R = l/(a\sigma) \tag{5}$$

is the magnetic equivalent of the conductivity, σ, of the conducting material. Using a magnet as a flux source (Fig. 2) gives an mmf which varies with the air gap reluctance. In the absence of any magnetizable materials, as in the air gaps, the permeability is given by Eq. (6) in SI units (Wb/A-m). The quantity μ_0 is sometimes referred to as the

$$\mu = \mu_0 = 4\pi \times 10^{-7} \tag{6}$$

permeability of free space. Material properties are described by the relative permeability, μ_r in accordance with Eq. (7). The materials which are important in magnetic circuits

$$\mu = \mu_r \mu_0 \tag{7}$$

are the ferromagnetics and ferrites characterized by large value of μ_r, sometimes in excess of 10,000 at low flux densities.

Magnetic field strength. It is convenient to introduce two different measures of the magnetic field: the flux density, B, and the field strength, or field intensity, H. The field strength, H, can be defined as the mmf per meter. It provides a measure of the currents and other magnetic field sources, excluding those representing polarizable materials. It may also be defined in terms of the force on a unit pole.

A straight wire carrying a current I sets up a field (Fig. 4) whose intensity at a point at distance r is given by Eq. (8). The field strength, H, like B, is a vector quantity pointing

$$H = \frac{I}{2\pi r} \tag{8}$$

in the direction of rotation of a right-hand screw advancing in the direction of current

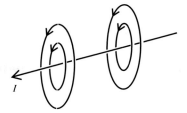

Fig. 4. Magnetic field of a straight wire.

flow. The intensity of the field is shown by the number of field lines intersecting a unit area. The straight wire provides one example of the circuital law, known as Ampère's law, given by Eq. (9). Here, θ is the angle between H and the element dl of any closed

$$\oint H \cos \theta \, dl = I \tag{9}$$

path of summation, or integration, and I is the current which links this path. Choosing a circular path, centered on a straight wire, reduces the integral to $H\,(2\pi r)$.

A long, straight, uniformly wound coil (Fig. 5), for example, produces a field which is uniform in the interior and zero outside. The interior magnetic field, H, points in the direction parallel to the coil axis. Applying Eq. (14) to the rectangle $pqrs$ of unit length in the axial direction shows that the only contribution is from pq, giving Eq. (10), where

$$H = In \tag{10}$$

n is the number of turns, per unit length, carrying the current, I. The magnetic field strength, H, remains the same, by definition, whether the interior of the coil is empty or is filled with ferromagnetic material of uniform properties. The interior forms part of a magnetic circuit in which In is the mmf per unit length, where mmf is the magnetic analog of electric voltage, or scalar potential, in an electric circuit. The magnetic field strength, H, is the analog of the electric field vector, E, as a measure of potential gradient, pointing down the gradient. The flux density, B, describes the effect of the field, in the sense of the voltage which is induced in a search coil by changes in time [Eq. (1)]. The ratio of H to B is the reluctance of a volume element of unit length and unit cross section in which the field is uniform, so that, from Eq. (4), the two quantities are related by Eq. (11). The permeability, μ, is defined by Eq. (11). The relative permeability, μ_r,

$$B = \mu H \tag{11}$$

of polarizable materials is measured accordingly by subjecting a sample to a uniform

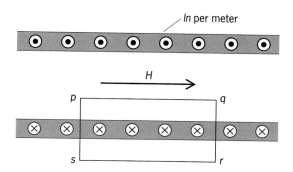

Fig. 5. Cross section of part of a long, straight uniformly wound coil. Black circles indicate current out of the plane of the page, and crosses indicate current into the plane of the page. Rectangle *pqrs* is used to calculate magnetic field strength, *H*, within the coil.

field inside a long coil such as that shown in Fig. 5 and using the emf induced in a search coil wound around the specimen to observe the flux in it.

Magnetic flux and flux density. Magnetic flux is defined in terms of the forces exerted by the magnetic field on electric charge. The forces can be described in terms of changes in flux with time [Eq. (1)], caused either by motion relative to the source or by changes in the source current, describing the effect of charge acceleration.

Since the magnetic, or electrokinetic, energy of current flowing in parallel wires depends on their spacing, the wires are subject to forces tending to change the configuration. The force, dF, on an element of wire carrying a current, i, is given by Eq. (12),

$$dF = Bi\, dl \qquad \text{newtons} \qquad (12)$$

and this provides a definition of the flux density, B, due to the wires which exert the force. The SI unit of B, called the tesla, or Wb/m^2, is the N/A-m. The flux density, B, equals $\mu_0 H$ in empty space, or in any material which is not magnetizable [Eq. (11)]. An example is the force, F, per meter (length) which is exerted by a long straight wire on another which is parallel to it, at distance r. From Eq. (8), this force is given by Eq. (13), when the wires carry currents I and i. The force, F, is accounted for by the

$$F = \frac{\mu_0 I i}{2\pi r} = 2 \times 10^{-7}\, Ii/r \qquad \text{newtons} \qquad (13)$$

electrokinetic interactions between the conduction charges, and describes the relativistic modification of the electric forces between them due to their relative motion.

In general, any charge, q, moving at velocity u is subject to a force given by Eq. (14), where $\mathbf{u} \times \mathbf{B}$ denotes the cross-product between vector quantities. That

$$\mathbf{f} = q\,\mathbf{u} \times \mathbf{B} \qquad \text{newtons} \qquad (14)$$

is, the magnitude of $\mathbf{f}$ depends on the sine of the angle θ between the vectors $\mathbf{u}$ and $\mathbf{B}$, of magnitudes u and B, according to Eq. (15). The force on a positive charge is at

$$f = quB \sin\theta \qquad (15)$$

right angles to the plane containing $\mathbf{u}$ and $\mathbf{B}$ and points in the direction of a right-hand screw turned from $\mathbf{u}$ to $\mathbf{B}$.

The same force also acts in the axial direction on the conduction electrons in a wire moving in a magnetic field, and this force generates an emf in the wire. The emf in an element of wire of length dl is greatest when the wire is at right angles to the $\mathbf{B}$ vector, and the motion is at right angles to both. The emf is then given by Eq. (16). More

$$\text{emf} = uB\, dl \qquad (16)$$

generally, u is the component of velocity normal to $\mathbf{B}$, and the emf depends on the sine of the angle between $\mathbf{dl}$ and the plane containing the velocity and the $\mathbf{B}$ vectors. The sign is given by the right-hand screw rule, as applied to Eq. (15).

Magnetic flux linkage. The magnetic flux linking any closed path is obtained by counting the number of flux lines passing through any surface, s, which is bounded by the path. Stated more formally, the linkage depends on the sum given in Eq. (17),

$$\Phi = \int\int B_n\, ds \qquad \text{webers} \qquad (17)$$

where B_n denotes the component of B in the direction normal to the area element, ds. The rate of change of linkage gives the emf induced in any conducting wire which follows the path [Eq. (1)].

The flux linkage with a coil (Fig. 6) is usually calculated by assuming that each turn of the coil closes on itself, giving a flux pattern which likewise consists of a large number of

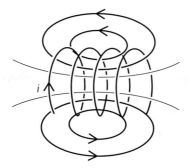

Fig. 6. *B* (magnetic flux) field of a short coil.

separate closed loops. Each links some of the turns, so that the two cannot be separated without breaking, or "tearing," either the loop or the turn. The total linkage with the coil is then obtained by adding the contributions from each turn.

The inductance, L, is a property of a circuit defined by the emf which is induced by changes of current in time, as in Eq. (18). The SI unit of inductance is the henry

$$e = -L \, di/dt \qquad (18)$$

(H), or V-s/A. The negative sign shows that e opposes an increase in current (Lenz's law). From Eq. (1) the inductance of a coil of N turns, each linking the same flux, Φ, is given by Eq. (19), so that the henry is also the Wb/A. When different turns, or different

$$L = N \, \Phi/i \qquad \text{henrys} \qquad (19)$$

parts of a circuit, do not link the same flux, the product $N\Phi$ is replaced by the total flux linkage, Φ, with the circuit as a whole.

The mutual inductance, M, between any two coils, or circuit parts, is defined by emf which is induced in one by a change of current in the other. Using 1 and 2 to distinguish between them, the emf induced in coil 1 is given by Eq. (20a), where the

$$e_1 = -M_{12} \, di_2/dt \qquad (20a)$$

$$e_2 = -M_{21} \, di_1/dt \qquad (20b)$$

sign convention is consistent with that used for L, referred to as the self-inductance. Likewise, the emf induced in coil 2 when the roles of the windings are reversed is given by Eq. (20b). The interaction satisfies the reciprocity condition of Eq. (21), so that the suffixes may be omitted.

$$M_{21} = M_{12} \qquad (21)$$

Magnetostatics. The term "magnetostatics" is usually interpreted as the magnet equivalent of the electrostatic interactions between electric charges. The equivalence is described most directly in terms of the magnetic pole, since the forces between poles, like those between charges, vary inversely with the square of the separation distance. Although no isolated poles, or monopoles, have yet been observed, the forces which act on both magnets and on coils are consistent with the assumption that the end surfaces are equivalent to magnetic poles.

Magnetic moment. The magnetic moment of a small current loop, or magnet, can be defined in terms of the torque which acts on it when placed in a magnetic flux density, B, which is sufficiently uniform in the region of the loop. For a rectangular loop with dimensions a and b and with N turns, carrying a current, i, equal but opposite forces act on the opposite sides of length a (Fig. 7). The force is $iNBa$ [Eq. (12)], and

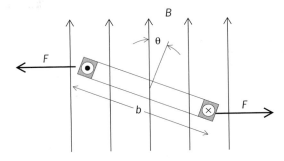

Fig. 7. Cross section of a rectangular current loop placed in magnetic flux density B. Equal but opposite forces, F, act on opposite sides of the loop carrying current into the plane of the page (indicated by a cross) and out of the plane of the page (indicated by black circle). These forces give rise to a torque on the loop.

the torque, given by Eq. (22), depends on the effective distance, $b \sin \theta$, between the

$$T = iNBab \sin \theta \quad \text{N-m} \tag{22}$$

wires. It is proportional to the area ab, and is a maximum when the angle θ between B and the axis of the loop is $90°$. A current loop of any other shape can be replaced by a set of smaller rectangles placed edge to edge, and the torques of these added to give the total on the loop. The magnetic moment of any loop of area s is defined as the ratio of the maximum torque to the flux density, so its magnitude is given by Eq. 23). It is a

$$m = iNs \tag{23}$$

vector quantity pointing in the direction of a right-hand screw turned in the direction of current flow. It is expressed in vector cross-product notation by Eq. (24). *See* Torque.

$$\mathbf{T} = \mathbf{m} \times \mathbf{B} \tag{24}$$

An electron orbiting at frequency f is the equivalent of a current $i = qf$, giving Eq. (25)

$$m_0 = qfs \tag{25}$$

for the moment, where s is the area of the orbit. The permissible values are determined by the quantum energy levels. The electron spin is a quantum state which can likewise be visualized as a small current loop. Atomic nuclei also possess magnetic moments. *See* Electron spin; Magneton; Nuclear moments.

Magnetic polarization. Materials are described as magnetic when their response to the magnetic field controls the ratio of B to H. The behavior is accounted for by the magnetic moments produced mainly by the electron spins and orbital motions. These respond to the field and contribute to it in a process referred to as magnetic polarization. The effects are greatest in ferromagnetics and in ferrites, in which the action is described as ferrimagnetic. *See* Ferrimagnetism; Ferromagnetism.

The sources are the equivalent of miniature "Ampèrean currents" whose sum, in any volume element, is equivalent to a loop of current flowing along the surface of the element. The flux density, B, depends on the field intensity, H, which is defined so that its value inside a long ferromagnetic rod of uniform cross section placed inside a long coil (Fig. 5) is the same as in the annular gap between the rod and the coil, in accordance with Eq. (10). If the field is not sufficiently uniform, H can be measured by using a search coil to observe the flux density, $\mu_0 H$, in the gap. The flux density inside the rod is given by Eq. (26), where B_0 denotes $\mu_0 H$, and μ_r is the relative permeability

$$B = \mu_r B_0 \tag{26}$$

[Eqs. (7) and (11)]. The same flux, B, is obtained by replacing the material by a coil in which the current in amperes per unit length is given by Eq. (27). The magnetic

$$J_s = (B - B_0)/\mu_0$$
$$= (1 - 1/\mu_r)B/\mu_0 \tag{27}$$

moment, dm, of a volume element of length dz is due to the current flowing over the surface enclosing the area, $dydz$; from Eq. (23), it is given by Eq. (28). The moment per unit volume defines the magnetic polarization, as in Eq. (29). The polarization, **M**, is a

$$dm = (J_s \, dx) \, dy \, dz \tag{28}$$

$$M = dm/dx \, dy \, dz \tag{29}$$

vector pointing in the direction of **dm** with magnitude J_s. The surface current produces an **H**-like, or B/μ_0, field which is entirely different from **H** in the material. Substituting

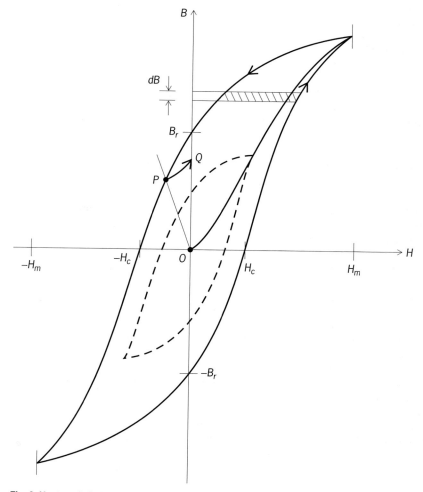

Fig. 8. Hysteresis behavior of a ferromagnetic material.

from Eq. (27) gives Eq. (30). This model of the material accounts for the flux field,

$$\mathbf{H} = \frac{\mathbf{B}}{\mu_0} - \mathbf{M} \qquad (30)$$

B, as observed by the voltage induced in a search coil wound around the specimen, and **H**, becomes an auxiliary quantity representing the sum of the polarization, **M**, and the magnetizing field, $\mathbf{B}\mu_0$, to which **M** responds. The polarization, **M**, also makes the largest contribution to that field, since the equivalent surface current is in the same direction as the current in the magnetizing coil.

Magnetic hysteresis. The relationship between the flux density, B, and the field intensity, H, in ferromagnetic materials depends on the past history of magnetization. The effect is known as hysteresis. It is demonstrated by subjecting the material to a symmetrical cycle of change during which H is varied continuously between the positive and negative limits $+H_m$ and $-H_m$ (Fig. 8). The path that is traced by repeating the cycle a sufficient number of times is the hysteresis loop. The sequence is counterclockwise, so that B is larger when H is diminishing than when it is increasing, in the region of positive H. The flux density, B_r, which is left when H falls to zero is called the remanence, or retentivity. The magnetically "hard" materials used for permanent magnets are characterised by a high B_r, together with a high value of the field strength, $-H_c$, which is needed to reduce B to zero. The field strength, H_c, is known as the coercive force, or coercivity. Cycling the material over a reduced range in H gives the path in Fig. 8 traced by the broken line, lying inside the larger loop. The locus of the tips of such loops is known as the normal magnetization curve. The initial magnetization curve is the B-H relationship which is followed when H is progressively increased in one direction after the material has first been demagnetized ($B = H = 0$).   [C.J.C.]

Magnetization   The process of becoming magnetized; also the property and in particular the extent of being magnetized. Magnetization has an effect on many of the physical properties of a substance. Among these are electrical resistance, specific heat, and elastic strain. *See* MAGNETOCALORIC EFFECT; MAGNETORESISTANCE; MAGNETOSTRICTION.

The magnetization **M** of a body is caused by circulating electric currents or by elementary atomic magnetic moments, and is defined as the magnetic moment per unit volume of such currents or moments. In the mks (SI) system of units, **M** is measured in webers per square meter. For **M**, 1 weber/m^2 = $10^4/4\pi$ gauss.

The magnetic induction or magnetic flux density **B** is given by the equation below,

$$\mathbf{B} = \mu_0\mathbf{H} + \mathbf{M} \text{ (mks)}$$

where **B** and **M** are in webers/m^2, **H**, the applied magnetic field, is in ampere-turns/m, and μ_0, the permeability of free space, is defined as $4\pi \times 10^{-7}$ henry/m, that is, webers/(ampere-turn)(m). *See* ELECTRICAL UNITS AND STANDARDS.

The topic of magnetization is generally restricted to materials exhibiting spontaneous magnetization, that is, magnetization in the absence of **H**. All such materials will be referred to as ferromagnets, including the special category of ferrimagnets. A ferromagnet is composed of an assemblage of spontaneously magnetized regions called domains. Within each domain, the elementary atomic magnetic moments are essentially aligned, that is, each domain may be envisioned as a small magnet. An unmagnetized ferromagnet is composed of numerous domains, oriented in some fashion.

The process of magnetization in an applied field **H** consists of growth of those domains oriented most nearly in the direction of **H** at the expense of others, followed

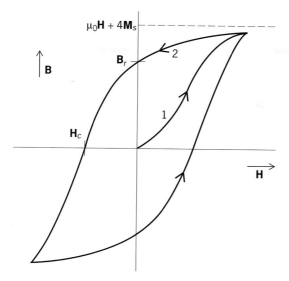

$\mu_0\mathbf{H} + 4\mathbf{M}_s$

B

$\mathbf{B}_r$

$\mathbf{H}_c$

2

1

$\overrightarrow{\mathbf{H}}$

Magnetization or *B-H* curves.

by rotation of the direction of magnetization against anisotropy forces. *See* FERROMAG-NETISM.

On removal of the field **H**, some magnetization will remain, called the remanence **M**$_r$.

Curves, sometimes called *B-H* curves, are used to describe magnetic materials. They are plotted with **H** as abscissa and with either **M** or **B** as ordinate. In the illustration, **B**$_r$ is the remanent induction (**B**$_r$ = **M**$_r$); **H**$_c$ is the coercive force, or reverse field required to bring the induction **B** back to zero; and **M**$_s$ is the saturation magnetization, or magnetization when all domains are aligned. The saturation magnetization is equal to the spontaneous magnetization of a single domain, except that it is possible to increase this magnetization slightly by application of an extremely large field. Saturation magnetization is temperature dependent, and disappears completely above the Curie temperature T_c where a ferromagnet changes into a paramagnet. *See* CURIE TEMPERATURE.

The irreversible nature of magnetization is shown most strikingly by the fact that the path of demagnetization does not retrace the path of magnetization—path 2 of the illustration does not retrace path 1. There is a tendency for the magnetization to show hysteresis, that is, to lag behind the applied field, and the loop of the illustration is called a hysteresis loop.

[E.A.; F.Ke.; K.V.M.]

Magnetocaloric effect   The reversible change of temperature accompanying the change of magnetization of a ferromagnetic or paramagnetic material. This change in temperature may be of the order of 1°C (2°F), and is not to be confused with the much smaller hysteresis heating effect, which is irreversible. *See* THERMAL HYSTERESIS.

[E.A.; F.Ke.]

Magnetohydrodynamics   The interaction of electrically conducting fluids with magnetic fields. The fluids can be ionized gases (commonly called plasmas) or liquid metals. Magnetohydrodynamic (MHD) phenomena occur naturally in the Earth's interior, constituting the dynamo that produces the Earth's magnetic field; in the magnetosphere that surrounds the Earth; and in the Sun and throughout astrophysics. In the laboratory, magnetohydrodynamics is important in the magnetic confinement of plasmas in experiments on controlled thermonuclear fusion. Magnetohydrodynamic principles are also used in plasma accelerators for ion thrusters for spacecraft propulsion, for light-ion-beam powered inertial confinement, and for magnetohydrodynamic power generation. *See* NUCLEAR FUSION; PLASMA (PHYSICS).

The conducting fluid and magnetic field interact through electric currents that flow in the fluid. The currents are induced as the conducting fluid moves across the magnetic field lines. In turn, the currents influence both the magnetic field and the motion of the fluid. Qualitatively, the magnetohydrodynamic interactions tend to link the fluid and the field lines so as to make them move together. *See* ELECTRIC CURRENT.

The generation of the currents and their subsequent effects are governed by the familiar laws of electricity and magnetism. The motion of a conductor across magnetic lines of force causes a voltage drop or electric field at right angles to the direction of the motion and the field lines; the induced voltage drop causes a current to flow as in the armature of a generator.

The currents themselves create magnetic fields which tend to loop around each current element. The currents heat the conductor and also give rise to mechanical ponderomotive forces when flowing across a magnetic field. (These are the forces which cause the armature of an electric motor to turn.) In a fluid, the ponderomotive forces combine with the pressure forces to determine the fluid motion. *See* ELECTRICITY; MAGNETISM.

Magnetohydrodynamic phenomena involve two well-known branches of physics, electrodynamics and hydrodynamics, with some modifications to account for their interplay. The basic laws of electrodynamics as formulated by J. C. Maxwell apply without any change. However, Ohm's law, which relates the current flow to the induced voltage, has to be modified for a moving conductor. *See* ELECTRODYNAMICS; HYDRODYNAMICS; MAXWELL'S EQUATIONS; OHM'S LAW.

It is useful to consider first the extreme case of a fluid with a very large electrical conductivity. Maxwell's equations predict, according to H. Alfvén, that for a fluid of this kind the lines of the magnetic field move with the material. The picture of moving lines of force is convenient but must be used with care because such a motion is not observable. It may be defined, however, in terms of observable consequences by either of the following statements: (1) a line moving with the fluid, which is initially a line of force, will remain one; or (2) the magnetic flux through a closed loop moving with the fluid remains unchanged.

If the conductivity is low, this is not true and the fluid and the field lines slip across each other. This is similar to a diffusion of two gases across one another and is governed by similar mathematical laws.

As in ordinary hydrodynamics, the dynamics of the fluid obeys theorems expressing the conservation of mass, momentum, and energy. These theorems treat the fluid as a continuum. This is justified if the mean free path of the individual particles is much shorter than the distances that characterize the structure of the flow. Although this assumption does not generally hold for plasmas, one can gain much insight into magnetohydrodynamics from the continuum approximation. The ordinary laws of hydrodynamics can then easily be extended to cover the effect of magnetic and electric fields on the fluid by adding a magnetic force to the momentum-conservation equation and electric heating and work to the energy-conservation equation.   [M.G.H.]

Magneton   A unit of magnetic moment used to describe atomic, molecular, or nuclear magnets. More precisely, one unit, the Bohr magneton, is used at the atomic and molecular levels, and another unit, the nuclear magneton, is used at the nuclear level. Still another unit (which might be called the muon magneton, but is usually not named) is used to describe the magnetic moment of the muon.

The Bohr magneton μ_B is defined and its value given in Eq. (1), where $-e$ and m are

$$\mu_B = \frac{e\hbar}{2m} = (9.274\ 009\ 49 \pm 0.000\ 000\ 80) \times 10^{-24}\ \text{joule/tesla} \qquad (1)$$

the charge and mass of the electron respectively and $\hbar$ is Planck's constant divided by 2π. In Dirac's theory the magnetic moment of the electron is exactly $-\mu_B$, but according to the theory of quantum electrodynamics the electron has a small anomalous magnetic moment. The experimental value of the electron magnetic moment μ_e is given by Eq. (2), in agreement with the prediction of quantum electrodynamics within the errors.

$$\mu_e = -(1.001\ 159\ 652\ 1883 \pm 0.000\ 000\ 000\ 0042)\mu_B \qquad (2)$$

The unit of magnetic moment to describe the muon is obtained from the Bohr magneton by replacing m in Eq. (1) by the muon mass m_μ. The experimental value of the muon magnetic moment is given in Eq. (3). The deviation of the muon magnetic

$$\mu_\mu = (1.001\ 165\ 9203 \pm 0.000\ 000\ 0007)\frac{e\hbar}{2m_\mu} \qquad (3)$$

moment from its Dirac value can also be accounted for by the theory of quantum electrodynamics. *See* LEPTON.

The nuclear magneton is obtained from the Bohr magneton by replacing m by the proton mass m_p. The value of the nuclear magneton is given in Eq. (4). The nuclear

$$\mu_N = (5.050\ 783\ 43 \pm 0.000\ 000\ 43) \times 10^{-27}\ \text{joule/tesla} \qquad (4)$$

magneton is used not only as the unit for the magnetic moment of the proton but also for the neutron and other hadrons and for atomic nuclei. If the proton and neutron were Dirac particles, the proton's magnetic moment would be one nuclear magneton (except for a small correction arising from quantum electrodynamics) and the neutron's magnetic moment would be zero (because the neutron is uncharged). However, the proton and neutron have large anomalous magnetic moments, given in Eqs. (5).

$$\mu_p = (2.792\ 847\ 351 \pm 0.000\ 000\ 028)\mu_N$$
$$\mu_n = (-1.913\ 042\ 73 \pm 0.000\ 000\ 45)\mu_N \qquad (5)$$

See NEUTRON; NUCLEAR MOMENTS; PROTON.

According to present theory, the proton, neutron, and other hadrons have large anomalous magnetic moments because these particles are not elementary but composite. In the theory of quantum chromodynamics, the principal constituents of a baryon, such as the proton or neutron, are three quarks. *See* BARYON; ELEMENTARY PARTICLE; FUNDAMENTAL CONSTANTS; QUANTUM CHROMODYNAMICS; QUARKS.   [D.B.L.]

Magnetooptics   That branch of physics which deals with the influence of a magnetic field on optical phenomena. Considering the fact that light is electromagnetic radiation, an interaction between light and a magnetic field would seem quite plausible. It is, however, not the direct interaction of the magnetic field and light that produces the known magnetooptic effects, but the influence of the magnetic field upon matter which is in the process of emitting or absorbing light. [G.H.Di.; W.W.W.]

Magnetoresistance The change of electrical resistance produced in a current-carrying conductor or semiconductor on application of a magnetic field H. Magnetoresistance is one of the galvanomagnetic effects. It is observed with H both parallel to and transverse to the current flow. The change of resistance usually is proportional to H^2 for small fields, but at high fields it can rise faster than H^2, increase linearly with H, or tend to a constant (that is, saturate), depending on the material. In most nonmagnetic solids the magnetoresistance is positive. *See* GALVANOMAGNETIC EFFECTS.

In semiconductors, the magnetoresistance is unusually large and is highly anisotropic with respect to the angle between the field direction and the current flow in single crystals. When the magnetoresistance is measured as a function of field, it is the basis for the Shubnikov–de Haas effect, much as the field dependence of the magnetization gives rise to the de Haas–van Alphen effect. Measurement of either effect as the field direction changes with respect to the crystal axes serves as a powerful probe of the Fermi surface. Magnetoresistance measurements also yield information about current carrier mobilities. Important to practical applications is the fact that the geometry of a semiconductor sample can generate very large magnetoresistance, as in the Corbino disk. *See* DE HAAS-VAN ALPHEN EFFECT; FERMI SURFACE; SEMICONDUCTOR.

Multilayered structures composed of alternating layers of magnetic and nonmagnetic metals, such as iron/chromium or cobalt/copper, can feature very large, negative values of magnetoresistance. This effect, called giant magnetoresistance, arises from the spin dependence of the electron scattering which causes resistance. When consecutive magnetic layers have their magnetizations antiparallel (antiferromagnetic alignment), the resistance of the structure is larger than when they are parallel (ferromagnetic alignment). Since the magnetic alignment can be changed with an applied magnetic field, the resistance of the structure is sensitive to the field. Giant magnetoresistance can also be observed in a simpler structure known as a spin valve, which consists of a nonmagnetic layer (for example, copper) sandwiched between two ferromagnetic layers (for example, cobalt). The magnetization direction in one of the ferromagnetic layers is fixed by an antiferromagnetic coating on the outside, while the magnetization direction in the other layer, and hence the resistance of the structure, can be changed by an external magnetic field. Films of nonmagnetic metals containing ferromagnetic granules, such as cobalt precipitates in copper, have been found to exhibit giant magnetoresistance as well. *See* ANTIFERROMAGNETISM; FERROMAGNETISM; MAGNETIZATION.

Magnetoresistors, especially those consisting of semiconductors such as indium antimonide or ferromagnets such as permalloy, are important to a variety of devices which detect magnetic fields. These include magnetic recording heads and position and speed sensors. *See* MAGNETIC MATERIALS. [J.F.He.]

Magnetostriction The change of length of a ferromagnetic substance when it is magnetized. More generally, magnetostriction is the phenomenon that the state of strain of a ferromagnetic sample depends on the direction and extent of magnetization. The phenomenon has an important application in devices known as magnetostriction transducers. *See* FERROMAGNETISM.

The magnetostrictive effect is exploited in transducers used for the reception and transmission of high-frequency sound vibrations. Nickel is often used for this application. *See* ULTRASONICS. [E.A.; F.Ke.]

Magnification   A measure of the effectiveness of an optical system in enlarging or reducing an image. For an optical system that forms a real image, such a measure is the lateral magnification m, which is the ratio of the size of the image to the size of the object. If the magnification is greater than unity, it is an enlargement; if less than unity, it is a reduction.

The angular magnification is the ratio of the angles formed by the image and the object at the eye. In telescopes the angular magnification (or, better, the ratio of the tangents of the angles under which the object is seen with and without the lens, respectively) can be taken as a measure of the effectiveness of the instrument.

Magnifying power is the measure of the effectiveness of an optical system used in connection with the eye. The magnifying power of a spectacle lens is the ratio of the tangents of the angles under which the object is seen with and without the lens, respectively. The magnifying power of a magnifier or an ocular is the ratio of the size under which an object would appear when seen through the instrument at a distance of 10 in. or 250 mm (the distance of distinct vision) divided by the object size. *See* Lens (optics); Optical image.
                                                                                    [M.J.H.]

Magnon   A quantum of a spin wave; an elementary excitation of a magnetic system which is usually long-range-ordered, such as a ferromagnet. *See* Antiferromagnetism; Ferrimagnetism; Ferromagnetism.

In the lowest energy state of a simple ferromagnet, all the magnetic moments of the individual atoms are parallel (say, to the z axis). Each atomic moment derives mainly from the electron spin angular momentum of the atom. In the next-to-lowest

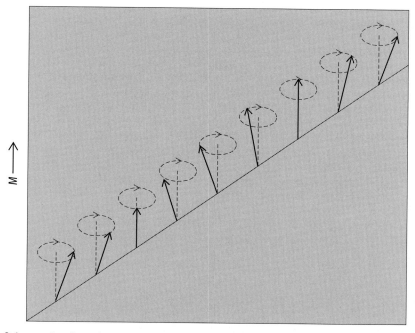

Spin wave in a linear ferromagnetic array of precessing atomic spins of equal magnitude, represented as arrows (vectors) in perspective. The axis of precession is along the vertical direction of the total magnetization, *M*.

energy state (first excited state), the total z component of spin angular momentum, S_z, is reduced by one unit of $\hbar = h/2\pi$, where h is Planck's constant. In the case of a crystalline material, this unit is shared equally by all the spins, each of which lies on a cone (see illustration), precessing at an angular rate ω. These spins form a wave, known as a spin wave, having a repeat distance or wavelength, λ. The wave amplitude (that is, the cone angle) is extremely small, because of the sharing among all the spins whose number N is very large, roughly 10^{23}. Thus, each atom's share of the reduction in S_z, labeled Δ, is only $\hbar/N$, whereas the z component of the atomic spin in the fully aligned state is typically 1–10 times $\hbar$. It follows from simple geometry that the cone half-angle is of order 10^{-11} to 10^{-12} radian. The state with this value of the amplitude is said to be a one-magnon state with wave number $k = 1/\lambda$. If Δ is doubled to $2\hbar/N$, the state is a two-magnon state, and so forth. The integer values of $N\Delta/\hbar$ correspond to the possible changes in S_z being integral multiples of $\hbar$. See ELECTRON SPIN; WAVE MOTION.

While the spin waves associated with energy states, that is, stationary states, in crystals vary sinusoidally in space (see illustration), magnons can be associated, instead, with nonstationary states (wave packets) in some situations. Closely analogous to magnons are phonons and photons, quanta of mass-density waves and electromagnetic waves, respectively. See ELECTROMAGNETIC WAVE; PHONON; PHOTON; QUANTUM MECHANICS.

[T.A.K.]

Manometer   A double-leg liquid-column gage used to measure the difference between two fluid pressures. Micromanometers are precision instruments which typically measure from very low pressures to 50 mm of mercury (6.7 kilopascals). The barometer is a special case of manometer with one pressure at zero absolute. See BAROMETER.

The various types of manometers have much in common with the U-tube manometer, which consists of a hollow tube, usually glass, a liquid partially filling the tube, and a scale to measure the height of one liquid surface with respect to the other (see illustration). If the legs of this manometer are connected to separate sources of pressure, the liquid will rise in the leg with the lower pressure and drop in the other leg. The difference between the levels is a function of the applied pressure and the specific gravity of the pressurizing and fill fluids.

A well-type manometer has one leg with a relatively small diameter, and the second leg is a reservoir. The cross-sectional area of the reservoir may be as much as 1500 times that of the vertical leg, so that the level of the reservoir does not change appreciably

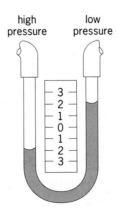

U-tube manometer.

with a change of pressure. Mercurial barometers are commonly made as well-type manometers.

The inclined-tube manometer is used for gage pressures below 10 in. (250 mm) of water differential. The leg of the well-type manometer is inclined from the vertical to elongate the scale. Inclined double-leg U-tube manometers are also used to measure very low differential pressures. *See* PRESSURE MEASUREMENT.   [J.H.Z.]

Maser   A device for coherent amplification or generation of electromagnetic waves by use of excitation energy in resonant atomic or molecular systems. "Maser" is an acronym for microwave amplification by stimulated emission of radiation. The device uses an unstable ensemble of atoms or molecules that may be stimulated by an electromagnetic wave to radiate energy at the same frequency and phase as the stimulating wave, thus providing coherent amplification. Amplifiers and oscillators operating on the same principle as the maser exist in many regions of the electromagnetic spectrum. Those operating in the optical region were once called optical masers, but they are now universally called lasers (the "l" stands for "light"). Amplification by maser action is also observed arising naturally from interstellar gases. *See* COHERENCE; LASER.

Maser amplifiers can have exceptionally low internally generated noise, approaching the limiting effective input power of one-half quantum of energy per unit bandwidth. Their inherently low noise makes maser oscillators that use a narrow atomic or molecular resonance extremely monochromatic, providing a basis for frequency standards. The hydrogen maser, which uses a hyperfine resonance of a gas of hydrogen atoms as the amplification source, is the prime example of this use. Also, because of their low noise and consequent high sensitivity, maser amplifiers are particularly useful for reception and detection of very weak signals in radio astronomy, microwave radiometry, and the like. A maser amplifier was used in the experiments that detected the cosmic microwave radiation left over from the big bang that created the universe. *See* FREQUENCY MEASUREMENT; UNCERTAINTY PRINCIPLE.

The quantum theory describes discrete particles such as atoms or molecules as existing in one or more members of a discrete set of energy levels, corresponding to the various possible internal motions of the particle (vibrations, rotations, and so forth). Thermal equilibrium of an ensemble of such particles requires that the number of particles n_1 in a lower energy level 1 be related to the number of particles n_2 in a higher energy level 2 by the Boltzmann distribution, given by the equation below, where E_1

$$\frac{n_1}{n_2} = \exp \frac{(E_2 - E_1)}{kT}$$

and E_2 are the respective energies of the two levels, k is Boltzmann's constant, and T is the absolute (Kelvin) temperature. *See* BOLTZMANN STATISTICS; QUANTUM MECHANICS.

Particles may be stimulated by an electromagnetic wave to make transitions from a lower energy level to a higher one, thereby absorbing energy from the wave and decreasing its amplitude, or from a higher energy level to a lower one, thereby giving energy to the wave and increasing its amplitude. These two processes are inverses of each other, and their effects on the stimulating wave add together. The upward and downward transition rates are the same, so that, for example, if the number of particles in the upper and lower energy states is the same, the stimulated emission and absorption processes just cancel. For any substance in thermal equilibrium at a positive (ordinary) temperature, the Boltzmann distribution requires that n_1 be greater than n_2, resulting in net absorption of the wave. If n_2 is greater than n_1, however, there are more particles that emit than those that absorb, so that the particles amplify the wave. In such a case, the ensemble of particles is said to have a negative temperature T, to be consistent with the Boltzmann condition. If there are not too many counterbalancing losses from other sources, this condition allows net amplification. This is the basic description of how a maser amplifies an electromagnetic wave. An energy source is required to create the negative temperature distribution of particles needed for a maser. This source is called the pump.

Gas masers. In the first known maser of any kind, the amplifying medium was a beam of ammonia (NH_3) molecules, and the molecular resonance used was the strongest of the rotation-inversion lines, at a frequency near 23.87 GHz (1.26-cm wavelength). Molecules from a pressurized tank of ammonia issued through an array of small orifices to form a molecular beam in a meter-long vacuum chamber. Spatially varying electric fields in the vacuum chamber created by a cylindrical array of electrodes formed a focusing device, which ejected from the beam the molecules in the lower energy level and directed the molecules in the upper energy level into a metal-walled electromagnetic cavity resonator. When the cavity resonator was tuned to the molecular transition frequency, the number of molecules was sufficiently large to produce net amplification and self-sustained oscillation. This type of maser is particularly useful as a frequency or time standard because of the relative sharpness and invariance of the resonance frequencies of molecules in a dilute gas. *See* MOLECULAR BEAMS.

Solid-state masers. Solid-state masers usually involve the electrons of paramagnetic ions in crystalline media immersed in a magnetic field. At least three energy levels are needed for continuous maser action. The energy levels are determined both by the interaction of the electrons with the internal electric fields of the crystal and by the interaction of the magnetic moments of the electrons with the externally applied magnetic field. The resonant frequencies of these materials can be tuned to a desired condition by changing the strength of the applied magnetic field and the orientation of the crystal in the field. An external oscillator, the pump, excites the transition between levels 1 and 3 [at the frequency $\nu_{31} = (E_3 - E_1)/h$], equalizing their populations. Then, depending on other conditions, the population of the intermediate level 2 may be greater or less than that of levels 1 and 3. If greater, maser amplification can occur at the frequency ν_{21}, or if less, at the frequency ν_{32}. Favorable conditions for this type of maser are obtained only at very low temperature, as in a liquid-helium cryostat. A typical material is synthetic ruby, which contains paramagnetic chromium ions (Cr^{3+}), and has four pertinent energy levels. The important feature of solid-state masers is their sensitivity when used as amplifiers. *See* PARAMAGNETISM.

Astronomical masers. Powerful, naturally occurring masers have probably existed since the earliest stages of the universe, though that was not realized until a few years after masers were invented and built on Earth. Their existence was first proven by discovery of rather intense 18-cm-wavelength microwave radiation of the free radical hydroxyl (OH) molecule coming from very localized regions of the Milky Way Galaxy.

Masers in astronomical objects differ from those generally used on Earth in that they involve no resonators or slow-wave structures to contain the radiation and so increase its interaction with the amplifying medium. Instead, the electromagnetic waves in astronomical masers simply travel a very long distance through astronomical clouds of gas, far enough to amplify the waves enormously even on a single pass through the cloud. It is believed that usually these clouds are large enough in all directions that a wave passing through them in any direction can be strongly amplified, and hence astronomical maser radiation emerges from them in all directions.

Naturally occurring masers have been important tools for obtaining information about astronomical objects. Since they are very intense localized sources of microwave radiation, their positions around stars or other objects can be determined very accurately with microwave antennas separated by long distances and used as interferometers. This provides information about the location of stars themselves as well as that of the masers often closely surrounding them. The masers' velocity of motion can also be determined by Doppler shifts in their wavelengths. The location and motion of masers surrounding black holes at the centers of galaxies have also provided information on the impressively large mass of these black holes. Astronomical masers often vary in power on time scales of days to years, indicating changing conditions in the regions where they are located. Such masers also give information on likely gas densities, temperature, motions, or other conditions in the rarefied gas of which they are a part. *See* DOPPLER EFFECT.

[C.H.T.; J.P.Go.]

Masking of sound   Interference with the audibility of a sound caused by the presence of another sound. More specifically, the number of decibels (dB) by which the intensity level of a sound (signal) must be raised above its threshold of audibility, to be heard in the presence of a second sound (masker), is called the masking produced by the masker on the signal. The masker and the signal may be identical or may differ in frequency, complexity, or time.

When both the masker and signal are pure tones and the tonal signal and masker have the same frequency, a very low level masker is required to mask the signal, indicating significant masking. As the difference in the frequency between the signal and masker increases, the signal is easier to detect, requiring a higher-level masker to mask the signal. Results from these psychophysical tuning curve measures of masking agree very well with data obtained from single auditory neurons in the auditory periphery, suggesting that tonal masking is mediated by the activity of these peripheral neurons.

The most widely studied complex masking sound is random noise which has energy at all frequencies and is said to be flat if the level for each 1-Hz bandwidth of the noise is the same. When random-flat noise is used to mask a pure tone, only a narrow frequency band (critical band) of the noise centered at the tonal frequency causes masking. If the bandwidth of the noise is narrower than this critical bandwidth, the tone's intensity can be lowered before the tone is masked. If the bandwidth is wider than this critical bandwidth, further widening of the bandwidth causes no changes in the detectability of the signal. The width of the critical band increases proportionally to its center frequency, that is, to the signal frequency. When noise masks speech, either the detectability of speech or speech intelligibility can be measured. The level for speech intelligibility is about 10–14 dB higher than for speech detectability.

Masking can occur when the signal either precedes or follows the masker in time. In backward masking the signal precedes the masker, while in forward masking the signal follows the masker. The physiological basis for this effect, as well as its implications for auditory processing of complex stimuli, is of great interest to the auditory scientist. *See* ACOUSTIC NOISE; SOUND.

[W.A.Y.]

Mass   The quantitative or numerical measure of a body's inertia, that is, of its resistance to being accelerated.

Because it is often necessary to compare masses of such dissimilar bodies as a sample of sugar, a sample of air, an electron, and the Moon, it is necessary to define mass in terms of a property that not only is inherent and permanent but is also universal in that it is possessed by all known forms of matter. All matter possesses two properties, gravitation and inertia. The property of gravitation is that every material body attracts every other material body. The property of inertia is that every material body resists any attempt to change its motion. A body's motion is said to change if the body is accelerated, that is, if it increases or decreases its speed or changes the direction of its motion. Because of its inertia a body cannot be accelerated unless a force is exerted on it. The greater the inertia of a body, the less will be the acceleration produced by a given force. *See* GRAVITATION; INERTIA.

The present definition of mass is in terms of inertia. The masses of two bodies are compared by applying equal forces to the bodies and measuring their accelerations. For example, the two bodies may be allowed to collide. According to Newton's third law, each body will then experience an equally strong force. If there are no external forces, and if a_1 and a_2 are the measured accelerations of the two bodies, the ratio of the masses of the two bodies is by definition given by the equation

$$\frac{m_1}{m_2} = \frac{a_2}{a_1}$$

This equation gives only ratios of masses; it is therefore necessary to designate the mass of some one body as the standard mass to which the masses of all other bodies can be compared. The body that has been chosen for this purpose is a cylinder of platinum-iridium alloy. It is known as the international standard of mass; its mass is called 1 kilogram (kg), and it is kept at the International Bureau of Weights and Measures near Paris, France. Replicas of the standard mass, kept at various national laboratories, are periodically compared with this standard.

Einstein's special theory of relativity predicts that the inertia of a body should increase if the energy of the body increases. This prediction has been conclusively verified experimentally. It follows that the mass of a body will increase if, for example, the body gains speed (addition of kinetic energy), or its temperature rises (addition of heat energy), or the body is compressed (addition of elastic energy). *See* CONSERVATION OF MASS.   [L.N.]

Mass defect   The difference between the mass of an atom and the sum of the masses of its individual components in the free (unbound) state. The mass of an atom is always less than the total mass of its constituent particles; this means, according to Albert Einstein's well-known formula, that an energy of $E = mc^2$ has been released in the process of combination, where m is the difference between the total mass of the constituent particles and the mass of the atom, and c is the velocity of light. The mass defect, when expressed in energy units, is called the binding energy, a term which is perhaps more commonly used. *See* NUCLEAR BINDING ENERGY.   [W.W.W.]

Mass number   The mass number A of an atom is the total number of its nuclear constituents, or nucleons, as the protons and neutrons are collectively called. The mass number is placed before and above the elemental symbol, thus ^{238}U. The mass number gives a useful rough figure for the atomic mass; for example, $^1H = 1.00814$ atomic mass units (amu), $^{238}U = 238.124$ amu, and so on. *See* ATOMIC NUMBER.   [H.E.D.]

Mass spectroscope   An instrument used for determining the masses of atoms or molecules found in a sample of gas, liquid, or solid. It is analogous to the optical spectroscope, in which a beam of light containing various colors (white light) is sent through a prism to separate it into the spectrum of colors present. In a mass spectroscope, a beam of ions (electrically charged atoms or molecules) is sent through a combination of electric and magnetic fields so arranged that a mass spectrum is

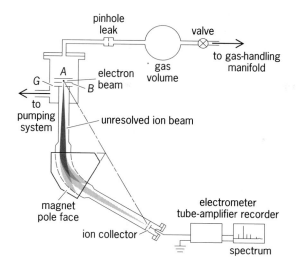

pinhole leak

valve

to gas-handling manifold

gas volume

electron beam

A
G
B

to pumping system

unresolved ion beam

magnet pole face

ion collector

electrometer tube-amplifier recorder

spectrum

Schematic drawing of mass spectrometer tube. Ion currents are in the range 10^{-10} to 10^{-15} ampere and require special electrometer tube amplifiers for their detection. In actual instruments the radius of curvature of ions in a magnetic field is 4–6 in. (10–15 cm).

produced (see illustration). If the ions fall on a photographic plate which after development shows the mass spectrum, the instrument is called a mass spectrograph; if the spectrum is allowed to sweep across a slit in front of an electrical detector which records the current, it is called a mass spectrometer.

Mass spectroscopes are used in both pure and applied science. Atomic masses can be measured very precisely. Because of the equivalence of mass and energy, knowledge of nuclear structure and binding energy of nuclei is thus gained. The relative abundances of the isotopes in naturally occurring or artificially produced elements can be determined.

Empirical and theoretical studies have led to an understanding of the relation between molecular structure and the relative abundances of the fragments observed when a complex molecule, such as a heavy organic compound, is ionized. When a high-resolution instrument is employed, the masses of the molecular or fragment ions can be determined so accurately that identification of the ion can frequently be made from the mass alone.

Because chemical compounds may have mass spectra as unique as fingerprints, mass spectroscopes are widely used in industries such as oil refineries, where analyses of complex hydrocarbon mixtures are required. See BETA PARTICLES.   [A.O.N.]

Mathematical physics   An area of science concerned with the application of mathematical concepts to the physical sciences and the development of mathematical ideas in response to the needs of physics. Historically, the concept of mathematical physics was synonymous with that of theoretical physics. In present-day terminology, however, a distinction is made between the two. Whereas most of theoretical physics uses a large amount of mathematics as a tool and as a language, mathematical physics places greater emphasis on mathematical rigor, and devotes attention to the development of areas of mathematics that are, or show promise to be, useful to physics. The results obtained by pure mathematicians, with no thought to applications, are almost always found to be both useful and effective in formulating physical theories.

Mathematical physics forms the bridge between physics as the description of nature and its structure on the one hand, and mathematics as a construction of pure logical thought on the other. This bridge between the two disciplines benefits and strengthens both fields enormously. *See* PHYSICS; THEORETICAL PHYSICS.

The methods employed in mathematical physics range over most of mathematics, the areas of analysis and algebra being the most commonly used. Partial differential equations and differential geometry, with heavy use of vector and tensor methods, are of particular importance in the formulation of field theories, and functional analysis as well as operator theory in quantum mechanics. Group theory has become an especially valuable tool in the construction of quantum field theories and in elementary-particle physics. There has also been an increase in the use of general geometrical approaches and of topology. For solution methods and the calculation of quantities that are amenable to experimental tests, of particular prominence are Fourier analysis, complex analysis, variational methods, the theory of integral equations, and perturbation theory. *See* VARIATIONAL METHODS (PHYSICS); VECTOR METHODS (PHYSICS).   [R.G.Ne.]

Matrix mechanics   A formulation of quantum theory in which the operators are represented by time-dependent matrices. Matrix mechanics is not useful for obtaining quantitative solutions to actual problems; on the other hand, because it is concisely expressed in a form independent of special coordinate systems, matrix mechanics is useful for proving general theorems.   [E.G.]

Matter (physics)   A term that traditionally refers to the substance of which all bodies consist. Matter in classical mechanics is closely identified with mass. Modern analyses distinguish two types of mass: inertial mass, by which matter retains its state of rest or uniform rectilinear motion in the absence of external forces; and gravitational mass, by which a body exerts forces of attraction on other bodies, and by which it reacts to those forces. Expressed in appropriate units, these two properties are numerically equal—a purely experimental fact, unexplained by theory. Albert Einstein made the equality of inertial and gravitational mass a fundamental principle (principle of equivalence), as one of the two postulates of the theory of general relativity. *See* GRAVITATION; INERTIA; MASS; RELATIVITY; WEIGHT.

In quantum mechanics, mass is only one among many properties (quantum numbers) that a particle can have, for example, electric charge, spin, and parity. The nearest quantum-mechanical analogs of traditional matter are fermions, having half-integral values of spin. Forces are mediated by exchange of bosons, particles having integral spins. Fermions correspond to classical matter in exhibiting impenetrability (a consequence of the exclusion principle), but the correspondence is only rough. For example, fermions can also be exchanged in interactions (a photon and an electron can exchange an electron), and they also exhibit wavelike (nonlocalized) behavior. States of classical matter-particles were given by their positions and momenta, but in quantum mechanics it is impossible to assign simultaneous precise positions and momenta to particles. *See* EXCLUSION PRINCIPLE; QUANTUM ELECTRODYNAMICS; QUANTUM MECHANICS; QUANTUM STATISTICS.

The primary constituents of ordinary matter are baryonic, consisting of quarks. However it is possible that as much as 99% (by mass) of the matter in the universe consists of nonbaryonic "dark matter" whose nature is yet to be discovered. *See* BARYON; QUARKS.
   [D.Sha.]

Matthiessen's rule An empirical rule which states that the total resistivity of a crystalline metallic specimen is the sum of the resistivity due to thermal agitation of the metal ions of the lattice and the resistivity due to the presence of imperfections in the crystal. This rule is a basis for understanding the resistivity behavior of metals and alloys at low temperatures.

The resistivity of a metal results from the scattering of conduction electrons. Lattice vibrations scatter electrons because the vibrations distort the crystal. Imperfections such as impurity atoms, interstitials, dislocations, and grain boundaries scatter conduction electrons because in their immediate vicinity the electrostatic potential differs from that of the perfect crystal.

[F.J.B.]

Maxwell's demon An imaginary being whose action appears to contradict the second law of thermodynamics, which identifies the natural direction of change with the direction of increasing entropy. There has always been a certain degree of discomfort associated with the acceptance of the law, particularly in relation to the time reversibility of physical laws and the role of molecular fluctuations. In 1867, J. C. Maxwell considered, in this connection, the action of "a finite being who knows the paths and velocities of all the molecules by inspection." This being was later referred to as a demon by Lord Kelvin, and the usage has been generally adopted. *See* ENTROPY; THERMODYNAMIC PRINCIPLES; TIME, ARROW OF.

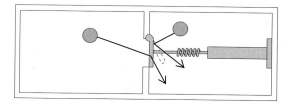

Type of device that emulates mechanically the actions of Maxwell's demon. Molecules traveling to the right can open the trapdoor and enter the right-hand compartment, but those striking it from the right cannot open it, so do not move into the left-hand compartment.

The activity of Maxwell's demon can be modeled by a trapdoor in a partition between two regions full of gas at the same pressure and temperature. The trapdoor needs to be restrained by a light spring to ensure that it is closed unless it is struck by molecules traveling from the left (see illustration). Its hinging is such that molecules traveling from the right cannot open it. The essential point of Maxwell's vision was that molecules striking the trapdoor from the left would be able to penetrate into the right-hand region but those present on the right would not be able to escape back into the left-hand region. Therefore, the initial equilibrium state of the two regions, that of equal pressures, would be slowly replaced by a state in which the two regions acquired different pressures as molecules accumulated in the right-hand region at the expense of the left-hand region. Only a slightly more elaborate mechanical arrangement is needed to change the apparatus to one in which the temperatures of the two regions move apart. In each case, the demonic trapdoor appears to be contriving a change that is contrary to the second law, for an implication of that law is that systems in either mechanical equilibrium (at the same pressure) or thermal equilibrium (at the same temperature) cannot spontaneously diverge from equilibrium.

As frequently occurs in science, the resolution of a paradox or the elimination of an apparent conflict with a firmly based law depends on a detailed analysis of the proposed arrangement. Numerous analyses of this kind have shown that the activities of Maxwell's demon do not in fact result in the overthrow of the second law. [P.W.A.]

Maxwell's equations   Four differential equations proposed by James Clerk Maxwell in 1864 as the basis of the theory of electromagnetic waves. They may be written, in vector notation, as Eqs. (1)–(4), where **D** is the electric displacement, **B**

$$\nabla \cdot \mathbf{D} = \rho \tag{1}$$

$$\nabla \cdot \mathbf{B} = 0 \tag{2}$$

$$\nabla \times \mathbf{E} = -\frac{\delta \mathbf{B}}{\delta t} \tag{3}$$

$$\nabla \times \mathbf{H} = \mathbf{i} + \frac{\delta \mathbf{D}}{\delta t} \tag{4}$$

the magnetic flux density, **E** the electric field strength or intensity, **H** the magnetic field strength or intensity, ρ the charge density, and **i** the current density.

The first equation states that electric flux lines, if they end at all, will do so on electric charges. The second states that magnetic flux lines never terminate. The third is a form of Faraday's law of induction, which states that the rate of change of the magnetic flux threading a circuit equals the electromotive force or line integral of **E** around the circuit. The fourth integral is based partially on A. M. Ampère's experiments on steady currents which show that the line integral of the magnetic intensity **H** (or $\mathbf{B}/\mu$, where μ is the permeability) around a closed curve equals the current encircled. *See* DISPLACEMENT CURRENT.   [W.R.Sm.]

McLeod gage   A type of instrument used to measure vacuum by application of the principle of Boyle's law. A known volume of a gas whose pressure is to be measured is trapped by raising the level of a fluid (mercury or oil) by means of a plunger, by lifting a reservoir, by using pressure, or by tipping the apparatus. As the fluid level is further raised, the gas is compressed into the capillary tube (see illustration). Obeying Boyle's

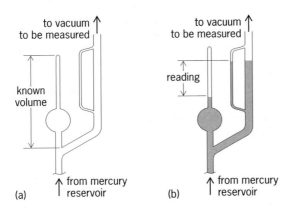

McLeod gage. (*a*) Filling (charging) position. (*b*) Measuring position.

law, the compressed gas now exerts enough pressure to support a column of fluid high enough to read. Readings are somewhat independent of the composition of the gas under pressure. *See* VACUUM MEASUREMENT.   [R.C.]

Mean free path   The average distance traveled between two similar events. The concept of mean free path is met in all fields of science and is classified by the events which take place. The concept is most useful in systems which can be treated statistically, and is most frequently used in the theoretical interpretation of transport phenomena in gases and solids, such as diffusion, viscosity, heat conduction, and electrical conduction. The types of mean free paths which are used most frequently are for elastic collisions of molecules in a gas, of electrons in a crystal, of phonons in a crystal, and of neutrons in a moderator. *See* KINETIC THEORY OF MATTER.   [W.D.W.]

Measure   A reference sample used in comparing lengths, areas, volumes, masses, and the like. The measures employed in scientific work are based on the international units of length, mass, and time—the meter, the kilogram, and the second—but decimal multiples and submultiples are commonly employed. Prior to the development of the international metric system, many special-purpose systems of measures had evolved and many still survive, especially in the United Kingdom and the United States. *See* METRIC SYSTEM; PHYSICAL MEASUREMENT; TIME; UNITS OF MEASUREMENT; WEIGHT.   [D.Wi.]

Mechanical impedance   For a system executing simple harmonic motion, the mechanical impedance is the ratio of force to particle velocity. If the force is that which drives the system and the velocity is that of the point of application of the force, the ratio is the input or driving-point impedance. If the velocity is that at some other point, the ratio is the transfer impedance corresponding to the two points.

Mechanical impedance is a complex quantity. The real part, the mechanical resistance, is independent of frequency if the dissipative forces are proportional to velocity; the imaginary part, the mechanical reactance, varies with frequency, becoming zero at the resonant and infinite at the antiresonant frequencies of the system. *See* FORCED OSCILLATION; HARMONIC MOTION.   [M.Gr.]

Mechanics   In its original sense, mechanics refers to the study of the behavior of systems under the action of forces. Mechanics is subdivided according to the types of systems and phenomena involved.

An important distinction is based on the size of the system. Those systems that are large enough can be adequately described by the newtonian laws of classical mechanics; in this category, for example, are celestial mechanics and fluid mechanics. On the other hand, the behavior of microscopic systems such as molecules, atoms, and nuclei can be interpreted only by the concepts and mathematical methods of quantum mechanics.

Mechanics may also be classified as nonrelativistic or relativistic mechanics, the latter applying to systems with material velocities comparable to the velocity of light. This distinction pertains to both classical and quantum mechanics.

Finally, statistical mechanics uses the methods of statistics for both classical and quantum systems containing very large numbers of similar subsystems to obtain their large-scale properties. *See* CLASSICAL FIELD THEORY; CLASSICAL MECHANICS; DYNAMICS; FLUID MECHANICS; QUANTUM MECHANICS; STATICS; STATISTICAL MECHANICS.   [B.G.]

Meissner effect   The expulsion of magnetic flux from the interior of a superconducting metal when it is cooled in a magnetic field to below the critical temperature, near absolute zero, at which the transition to superconductivity takes place. It was discovered by Walther Meissner in 1933, when he measured the magnetic field surrounding two adjacent long cylindrical single crystals of tin and observed that at $-452.97°F$ (3.72 K) the Earth's magnetic field was expelled from their interior. This indicated that at the onset of superconductivity they became perfect diamagnets. This discovery showed that the transition to superconductivity is reversible, and that the laws of thermodynamics apply to it. The Meissner effect forms one of the cornerstones in the understanding of superconductivity, and its discovery led F. London and H. London to develop their phenomenological electrodynamics of superconductivity. *See* DIAMAGNETISM; THERMODYNAMIC PRINCIPLES.

The magnetic field is actually not completely expelled, but penetrates a very thin surface layer where currents flow, screening the interior from the magnetic field.

The Meissner effect is subject to limitations. Full diamagnetism is not observed in polycrystalline samples, and the effect is not observed in impure samples or samples with certain geometrics, such as a round flat disk, with the magnetic field parallel to the axis of rotation. *See* SUPERCONDUCTIVITY.   [H.W.M.]

Meson   The generic name for any hadronic particle with baryon number zero. Such particles were first envisaged in 1935 by H. Yukawa, who pointed out that the main features of nuclear forces would be explained if these forces were transmitted between nucleons through an intermediate field coupled with nucleons, provided that its quanta (nuclear force mesons) were massive [200–300 electron masses (m_e)] and could carry electric charge between the nucleons. *See* BARYON; HADRON; NUCLEAR STRUCTURE; QUANTUM FIELD THEORY.

All mesons are unstable. Those with relatively long lifetimes are referred to as semistable. Nearly 200 highly unstable mesons are established, with lifetimes shorter than 10^{-22} s, and more continue to be discovered. These mesons decay to lighter mesons through the strong nuclear (hadronic) interactions, whereas the hadronic decays of the semistable mesons are forbidden or strongly suppressed. Alternative decay modes involve the weak interactions or the electromagnetic interactions, which are much weaker than the strong interactions and therefore lead to much smaller decay rates and longer lifetimes. The longest-lived mesons are those that decay only through the weak interactions; these include the charged π mesons (pions) and the K mesons (kaons), with lifetimes of about 10^{-8} to 10^{-10} s. *See* FUNDAMENTAL INTERACTIONS; WEAK NUCLEAR INTERACTIONS.

Hadrons are now considered to be composite, consisting of spin-1/2 quarks (q), corresponding antiquarks (q), and some number of gluons (g), the last being the quanta of the intermediate field which binds the quarks and antiquarks to form hadrons. Baryon number $B = +1/3$ holds for a quark q, $B = -1/3$ for antiquark q, while $B = 0$ holds for a gluon. In this view, the simplest possibility is that each meson is a quark-antiquark ($q\bar{q}$) pair bound together by the gluon field, and this model does account quite well for most of the known mesons and their properties. However, more complicated systems (for example, consisting of two quarks with two antiquarks) can be considered and may even be required by some of the present data. The quarks must be assigned fractional charge values, relative to the proton charge. *See* GLUONS; QUARKS.   [R.H.D.; C.G.]

Mesoscopic physics   A subdiscipline of condensed-matter physics that focuses on the properties of solids in a size range intermediate between bulk matter and individual atoms or molecules. The size scale of interest is determined by the appearance of novel physical phenomena absent in bulk solids and has no rigid definition; however, the systems studied are normally in the range of 100 nanometers (the size of a typical virus) to 1000 nm (the size of a typical bacterium). Other branches of science, such as chemistry and molecular biology, also deal with objects in this size range, but mesoscopic physics has dealt primarily with artificial structures of metal or semiconducting material which have been fabricated by the techniques employed for producing microelectronic circuits. Thus mesoscopic physics has a close connection to

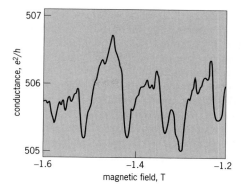

Conductance of 2000-nm gold wire as a function of magnetic field measured at a temperature of 0.04 K. The pattern observed is reproducible over a period of days. (*After R. A. Webb and S. Washburn, Quantum interference fluctuations in disordered materials, Phys. Today, 41(12):46–55, December 1988*)

the fields of nanofabrication and nanotechnology. Three categories of new phenomena in such systems are interference effects, quantum size effects, and charging effects. *See* Artificially layered structures; Quantized electronic structure (QUEST); Semiconductor heterostructures.

Interference effects. In the mesoscopic regime, scattering from defects induces interference effects which modulate the flow of electrons. The experimental signature of mesoscopic interference effects is the appearance of reproducible fluctuations in physical quantities. For example, the conductance of a given specimen oscillates in an apparently random manner as a function of experimental parameters (see illustration). However, the same pattern may be retraced if the experimental parameters are cycled back to their original values; in fact, the patterns observed are reproducible over a period of days.

Quantum size effects. Another prediction of quantum mechanics is that electrons confined to a particular region of space may exist only in a certain set of allowed energy levels. The spacing between these levels increases as the confining region becomes smaller. One striking phenomenon which arises from these quantum size effects is the steplike increase of the conductance of electrons flowing through a constriction of several hundred nanometers' width. *See* Energy level (quantum mechanics).

Another mesoscopic system that shows quantum size effects consists of isolated islands of electrons that may be formed at the appropriately patterned interface between two different semiconducting materials. The electrons typically are confined to disk-shaped regions termed quantum dots. The confinement of the electrons in these systems changes their interaction with electromagnetic radiation significantly.

Charging effects. Isolated mesoscopic solids such as quantum dots or metallic grains on an insulating substrate also show novel effects associated with the discreteness of the charge on the electron. Devices known as single-electron transistors (SETs) are by far the most sensitive electrometers (instruments for measuring electrical charge) presently known. *See* Electrometer; Transistor.   [A.D.S.]

Metastable state   In quantum mechanics, a state that is not truly stationary but is almost stationary.

In practice, especially in atomic and nuclear physics applications, the designation metastable state usually is reserved for states whose lifetimes are unusually long. For example, the excited states of atoms usually decay with the emission of a single photon, in a time of the order of 10^{-8} s. However, the necessity for angular momentum and parity conservation forces the second excited state ($2S_{1/2}$) of atomic hydrogen to decay by simultaneous emission of two photons; consequently, the lifetime is increased to an estimated value of 0.15 s. Thus, the $2S_{1/2}$ state of atomic hydrogen is usually termed metastable, but most other hydrogenic states are not. Similarly, emission of a gamma-ray photon by an excited nucleus usually occurs in 10^{-13} s or less; however, the lifetime of one excited state of the ^{113}In nucleus, the state that customarily is termed metastable, is about 100 min. Since radiative transition probabilities for emission of photons generally decrease rapidly with decreasing frequency, a low-lying atomic or nuclear excited state may have a lifetime longer than most excited states of atoms and nuclei and yet not be metastable in the practical sense just described, because photon emission from the state may not be hindered by any general requirement or selection rule, such as is invoked for the $2S_{1/2}$ state of hydrogen. *See* EXCITED STATE; NUCLEAR ISOMERISM; RADIOACTIVITY.   [E.G.]

Metric system   A system of units used in scientific work throughout the world and employed in general commercial transactions and engineering applications in most of the developed nations of the world except for the United Kingdom and the United States. The basic units of the metric system define length (meter), mass (kilogram), and time (second).

The chief advantage of the metric system is that it is based on standards that have been accepted by international agreement, and it therefore provides a common basis for all scientific measurements. A second advantage of the metric system lies in the fact that only decimal multiples and submultiples of the fundamental length and mass units and of other derived units are employed. *See* PHYSICAL MEASUREMENT; TIME; UNITS OF MEASUREMENT.   [D.Wi.]

Microscope   An instrument used to obtain an enlarged image of a small object. The image may be seen, photographed, or sensed by photocells or other receivers, depending upon the nature of the image and the use to be made of the information of the image.

A simple microscope, hand lens, or magnifier usually is a round piece of transparent material, ground thinner at the edge than at the center, which can form an

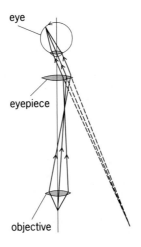

Compound microscope diagram. (*After F. A. Jenkins and H. E. White, Fundamentals of Optics, 4th ed., McGraw-Hill, 1976*)

enlarged image of a small object. Commonly, simple microscopes are double convex or planoconvex lenses, or systems of lenses acting together to form the image.

The compound microscope utilizes two lenses or lens systems. One lens system forms an enlarged image of the object and the second magnifies the image formed by the first. The total magnification is then the product of the magnifications of both lens systems (see illustration).

The typical compound microscope consists of a stand, a stage to hold the specimen, a movable body-tube containing the two lens systems, and mechanical controls for easy movement of the body and the specimen. The lens system nearest the specimen is called the objective; the one nearest the eye is called the eyepiece or ocular. A mirror is placed under the stage to reflect light into the instrument when the illumination is not built into the stand. For objectives of higher numerical aperture than 0.4, a condenser is provided under the stage to increase the illumination of the specimen. Various optical and mechanical attachments may be added to facilitate the analysis of the information in the doubly enlarged image. *See* LENS (OPTICS). [O.W.R.]

Microwave measurements   A collection of techniques particularly suited for development of devices and monitoring of systems where physical size of components varies from a significant fraction of an electromagnetic wavelength to many wavelengths.

Virtually all microwave devices are coupled together with a transmission line having a uniform cross section. The concept of traveling electromagnetic waves on that transmission line is fundamental to the understanding of microwave measurements.

At any reference plane in a transmission line there are considered to exist two independent traveling electromagnetic waves moving in opposite directions. One is called the forward or incident wave, and the other the reverse or reflected wave. The electromagnetic wave is guided by the transmission line and is composed of electric and magnetic fields with associated electric currents and voltages. Any one of these parameters can be used in considering the traveling waves, but the measurements in the early development of microwave technology made principally on the voltage waves led to the custom of referring only to voltage. One parameter in very common use is the voltage reflection coefficient Γ, which is related to the incident, V_i, and reflected, V_r voltage waves by Eq. (1).

$$\Gamma = \frac{V_r}{V_i} \tag{1}$$

Impedance. The voltage reflection coefficient Γ is related to the impedance terminating the transmission line and to the impedance of the line itself. If a wave is launched to travel in only one direction on a uniform reflectionless transmission line of infinite length, there will be no reflected wave. The input impedance of this infinitely long transmission line is defined as its characteristic impedance Z_0. An arbitrary length of transmission line terminated in an impedance Z_0 will also have an input impedance Z_0. *See* ELECTRICAL IMPEDANCE.

If the transmission line is terminated in the arbitrary complex impedance load Z_L, the complex voltage reflection coefficient Γ_L at the termination is given by Eq. (2).

$$\Gamma = \frac{Z_L - Z_0}{Z_L + Z_0} \tag{2}$$

Even when there is no unique expression for Z_L and Z_0 such as in the case of hollow uniconductor waveguides, the voltage reflection coefficient Γ has a value because it is simply a voltage ratio. In general, the measurement of microwave impedance is the measurement of Γ. Both amplitude and phase of Γ can be measured by direct probing of the voltage standing wave set up along a transmission line by the two opposed traveling waves, but this is a slow technique. Directional couplers have been used for many years to perform much faster swept frequency measurement of the magnitude of Γ, and more recently the use of automatic network analyzers under computer control has made possible rapid, accurate measurements of amplitude and phase of Γ over very broad frequency ranges.

Power. A required increase in microwave power is expensive whether it be the output from a laboratory signal generator, the power output from a power amplifier on a satellite, or the cooking energy from a microwave oven. To minimize this expense, absolute power must be measured. Most techniques involve conversion of the microwave energy to heat energy which, in turn, causes a temperature rise in a physical body. This temperature rise is measured and is approximately proportional to the power dissipated. The whole device can be calibrated by reference to low-frequency electrical standards and application of appropriate corrections. *See* RADIOMETRY.

The power sensors are simple and can be made to have a very broad frequency response. A power meter can be connected directly to the output of a generator to measure available power P_A, or a directional coupler may be used to permit measurement of a small fraction of the power actually delivered to the load.

Scattering coefficients. While the measurement of absolute power is important, there are many more occasions which require the measurement of relative power which is equivalent to the magnitude of voltage ratio and is related to attenuation. Also there arises frequently the need to measure the relative phase of two voltages. Measurement systems having this capability are referred to as vector network analyzers, and they are used to measure scattering coefficients of multi-port devices. The concept of scattering coefficients is an extension of the voltage reflection coefficient applied to devices having more than one port. The most simple is a two-port. Its characteristics can be specified completely in terms of a 2×2 scattering matrix, the coefficients of which are indicated in the illustration. The incident voltage at the reference plane of each port is defined as a, and the reflected voltage is b. Voltages a and b are related by matrix equation (3), where (S_{nm}) is the scattering matrix of the junction. Writing Eq. (3) out for a two-port

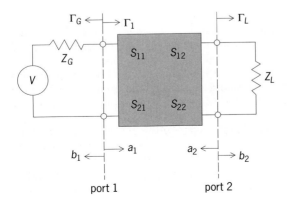

A two-port inserted between a load and a generator. S_{nm} are the scattering coefficients of the two-port.

device gives Eqs. (4) and (5). Examination of Eq. (4) shows, for example, that S_{11} is

$$(b_n) = (S_{nm})(a_m) \tag{3}$$

$$b_1 = S_{11}a_1 + S_{12}a_2 \tag{4}$$

$$b_2 = S_{21}a_1 + S_{22}a_2 \tag{5}$$

the voltage reflection coefficient looking into port 1 if port 2 is terminated with a Z_0 load ($a_2 = 0$).

Heterodyne. The heterodyne principle is used for scalar attenuation measurements because of its large dynamic range and for vector network analysis because of its phase coherence. The microwave signal at frequency f_s is mixed with a microwave local oscillator at frequency f_{LO}, in a nonlinear mixer. The mixer output signal at frequency $f_s - f_{LO}$ is a faithful amplitude and phase reproduction of the original microwave signal but is at a low, fixed frequency so that it can be measured simply with low-frequency techniques. One disadvantage of the heterodyne technique at the highest microwave frequencies is its cost. Consequently, significant effort has been expended in development of multiport network analyzers which use several simple power detectors and a computer analysis approach which allows measurement of both relative voltage amplitude and phase with reduced hardware cost.

Noise. Microwave noise measurement is important for the communications field and radio astronomy. The measurement of thermal noise at microwave frequencies is essentially the same as low-frequency noise measurement, except that there will be impedance mismatch factors which must be carefully evaluated. The availability of broadband semiconductor noise sources having a stable, high, noise power output has greatly reduced the problems of source impedance mismatch because an impedance-matching attenuator can be inserted between the noise source and the amplifier under test.

Use of computers. The need to apply calculated corrections to obtain the best accuracy in microwave measurement has stimulated the adoption of computers and computer-controlled instruments. An additional benefit of this development is that measurement techniques that are superior in accuracy but too tedious to perform manually can now be considered.

[R.F.Cl.]

Microwave optics The study of those properties of microwaves which are analogous to the properties of light waves in optics. The fact that microwaves and light waves are both electromagnetic waves, the major difference being that of frequency, already suggests that their properties should be alike in many respects. But the reason microwaves behave more like light waves than, for instance, very low-frequency waves for electrical power (50 or 60 Hz) is primarily that the microwave wavelengths are usually comparable to or smaller than the ordinary physical dimensions of objects interacting with the waves.

As is the case with light, a beam of microwaves propagates along a straight line in a perfectly homogeneous infinite medium. This phenomenon follows directly from a general solution of the wave equation in which the direction of a wave normal does not change in a homogeneous medium. *See* WAVE EQUATION.

With some modification the laws of reflection and refraction can be applied to the propagation of microwaves inside a dielectric-filled metallic waveguide. Another interesting application is associated with the microwave analog of total internal reflection in optics. A properly designed dielectric rod (without metal walls) can serve as a waveguide by totally reflecting the elementary plane waves. Still another case of interest is that of a microwave lens. *See* REFLECTION OF ELECTROMAGNETIC RADIATION; REFRACTION OF WAVES.

In an analogous manner to light, a microwave undergoes diffraction when it encounters an obstacle or an opening which is comparable to or somewhat smaller than its wavelength.   [C.K.J.]

Microwave spectroscopy

Microwave spectroscopy   The study of the interaction of matter and electromagnetic radiation in the microwave region of the spectrum. *See* SPECTROSCOPY.

The interaction of microwaves with matter can be detected by observing the attenuation or phase shift of a microwave field as it passes through matter. These are determined by the imaginary or real parts of the microwave susceptibility (the index of refraction). The absorption of microwaves may also trigger a much more easily observed event like the emission of an optical photon in an optical double-resonance experiment or the deflection of a radioactive atom in an atomic beam. *See* MOLECULAR BEAMS.

At room temperature, the relative population difference between the states involved in a microwave transition is a few percent or less. The population difference can be close to 100% at liquid helium temperatures, and microwave spectroscopic experiments are often performed at low temperatures to enhance population differences and to eliminate certain line-broadening mechanisms. The population differences between the states involved in a microwave transition can also be enhanced by artificial means. When the molecules or atoms with inverted populations are placed in an appropriate microwave cavity, the cavity will oscillate spontaneously as a maser (microwave amplification by stimulated emission of radiation). *See* MASER.

The magnetic dipole and electric quadrupole interactions between the nuclei and electrons in atoms and molecules can lead to energy splittings in the microwave region of the spectrum. Thus, microwave spectroscopy has been used extensively for precision determinations of spins and moments of nuclei. *See* HYPERFINE STRUCTURE; NUCLEAR MOMENTS.

The rotational frequencies of molecules often fall within the microwave range, and microwave spectroscopy has contributed a great deal of information about the moments of inertia, the spin-rotation coupling mechanisms, and other physical properties of rotating molecules. *See* MOLECULAR STRUCTURE AND SPECTRA.

The magnetic resonance frequencies of electrons in fields of a few thousand gauss (a few tenths of a tesla) lie in the microwave region. Thus, microwave spectroscopy is used in the study of electron-spin resonance or paramagnetic resonance. *See* MAGNETIC RESONANCE.

The cyclotron resonance frequencies of electrons in solids at magnetic fields of a few thousand gauss (a few tenths of a tesla) lie within the microwave region of the spectrum. Microwave spectroscopy has been used to map out the dependence of the effective mass on the electron momentum.

For other applications *see* ATOMIC CLOCK.   [W.Hap.]

Minimal principles

Minimal principles   In the treatment of physical phenomena, it can sometimes be shown that, of all the processes or conditions which might occur, the ones actually occurring are those for which some characteristic physical quantity assumes a minimum value. These processes or conditions are known as minimal principles.

The application of minimal principles provides a powerful method of attacking certain problems that would otherwise prove formidable if approached directly from first principles.

One simple minimal principle asserts that the state of stable equilibrium of any mechanical system is the state for which the potential energy is a minimum. Other general theorems of classical dynamics that are related to minimal principles are Hamilton's principle and the principle of least action. *See* HAMILTON'S PRINCIPLE; LEAST-ACTION PRINCIPLE.

[D.Wi.]

Mirror optics   The use of plane or curved reflecting surfaces for the purpose of reverting, directing, or forming images. An optical surface which specularly reflects the largest fraction of the incident light is called a reflecting surface. Such surfaces are commonly fabricated by polishing of glass, metal, or plastic substrates, and then coating the surface of the substrate with a thin layer of metal, which may be covered in addition by a single or multiple layers of thin dielectric films. The law of reflection states that the incident and reflected rays will lie in the plane containing the local normal to the reflecting surface and that the angle of the reflected ray from the normal will be equal to the angle of the incident ray from the normal. *See* GEOMETRICAL OPTICS.

The formation of images in the plane mirrors is easily understood by applying the law of reflection. The illustration shows the formation of the image of a point formed by a plane mirror. Each of the reflected rays appears to come from a point image located a distance behind the mirror equal to the distance of the object point in front of the mirror. The face of the observer can be considered as a set of points, each of which is imaged by the plane mirror. Since the observer is viewing the facial image from the object side of the mirror, the face will appear to be reversed left for right in the virtual image formed by the mirror. The illustration also indicates the redirection of light by a plane mirror, in that a viewer who cannot observe the object point directly can observe the virtual image of the point formed by the mirror. A simple optical device which is

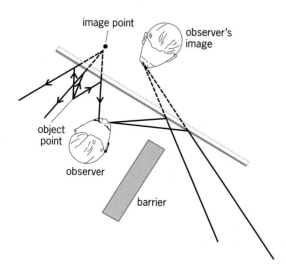

Formation of images by a plane mirror.

based on this principle is the simple mirror periscope, which uses two mirrors to permit viewing of scenes around an obstacle.

A curved mirror, either spherical or conic in form, will produce a real or virtual image in much the same manner as a lens, but generally with reduced aberrations. There will be no chromatic aberrations since the law of reflection is independent of the color or wavelength of the incident light. *See* ABERRATION (OPTICS); OPTICAL IMAGE.

Both concave and convex spherical mirrors are commonly encountered. Convex mirrors are commonly used as wide-angle rearview mirrors in automobiles or on trucks. A common application of concave mirrors is the magnifying shaving mirror frequently found in bathrooms.

A spherical mirror will form an image which is not perfect, except for particular conjugate distances. The use of a mirror which has the shape of a rotated conic section, such as a parabola, ellipsoid, or hyperboloid, will form a perfect image for a particular set of object-image conjugate distances and will have reduced aberrations for some range of conjugate relations. The most familiar applications for conic mirrors are in reflecting telescopes. *See* OPTICAL PRISM; OPTICAL SURFACES; REFLECTION OF ELECTROMAGNETIC RADIATION; TELESCOPE.   [R.R.S.]

Mode of vibration   A characteristic manner in which vibration occurs. In a freely vibrating system, oscillation is restricted to certain characteristic frequencies; these motions are called normal modes of vibration.

An ideal string, for example, can vibrate as a whole with a characteristic frequency $f = (1/2L)\sqrt{T/m}$, where L is the length of string between rigid supports, T the tension, and m the mass per unit length of the string. The displacements of different parts of the string are governed by a characteristic shape function. The frequency of the second mode of vibration is twice that of the first mode. Similarly, modes of higher order have frequencies that are integral multiples of the fundamental frequency.

Because the frequencies are in the ratios 1:2:3. . ., the modes of vibration of an ideal string are properly called harmonics. Not all vibrating bodies have harmonic modes of vibration, however. *See* HARMONIC (PERIODIC PHENOMENA); VIBRATION.   [R.W.Y.]

Moiré pattern   When one family of curves is superposed on another family of curves, a new family called the moiré pattern appears.

To produce moiré patterns, the lines of the overlapping figures must cross at an angle of less than about 45°. The moiré lines are then the locus of points of intersection. The illustration shows the case of two identical figures of simple gratings of alternate black and white bars of equal spacing. When the figures are crossed at 90°, a checkerboard pattern with no moiré effect is seen. At crossing angles of less than 45°, however, one sees a moiré pattern of equispaced lines, the moiré fringes. The spacing of the fringes increases with decreasing crossing angle. This provides one with a simple method for measuring extremely small angles (down to 1 second of arc). As the angle of crossing approaches zero, the moiré fringes approach 90° with respect to the original figures.

Even when the spacings of the original figures are far below the resolution of the eye, the moiré fringes will still be readily seen. This phenomenon provides a means of checking the fidelity of a replica of a diffraction grating. *See* DIFFRACTION GRATING.

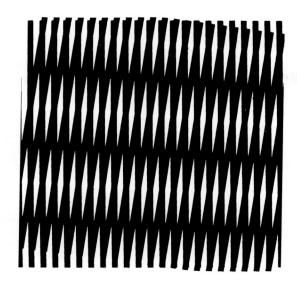

Two simple gratings crossed at a small angle.

Moiré techniques are widely used in the stress analysis of metals, in the examination of large optical surfaces, in investigating aberrations of lenses, and in determining a refractive index gradient (for example, that of sugar molecules diffusing into water).   [G.O.]

Molecular beams   Well-directed streams of atoms or molecules in vacuum. Utilization of molecular beams is a cornerstone technique in the investigation of molecular structure and interactions. Molecular beams are usually formed at sufficiently low particle density for the interaction of one beam molecule with another to be negligible. This ensemble of truly isolated molecules is available for the spectroscopic study of molecular energy levels using photon probes from the radio-frequency to optical portions of the electromagnetic spectrum. Some of the best-determined fundamental knowledge of physics comes from spectroscopic molecular-beam experiments. Beyond this, beams can be applied as probes of the multifaceted nature of gases, plasmas, surfaces, and even the structure of solids. An application intermediate in complexity is the study of molecular interactions by means of two colliding beams, where one might be a beam of charged particles such as ions or electrons. *See* SCATTERING EXPERIMENTS (ATOMS AND MOLECULES).

One simple means of forming a beam is to permit gas from an enclosed chamber to escape through a small orifice into a second chamber maintained at high vacuum by means of large pumps (illus. *a*). A useful number of molecules passes forward along the horizontal axis of the apparatus. A well-collimated beam is then formed by requiring that those molecules entering the test chamber where an experiment is to be performed pass not only through the orifice but also through a second small hole separating the collimating and test chambers.

If higher velocities are desired, a charge-exchange beam system can be used. In this scheme (illus. *b*), ions are produced by some ionizing process such as electron impact on atoms within a gas discharge. Since the ions are electrically charged, they can be accelerated to the desired velocity and focused into a beam using electric or magnetic fields. The last step in neutrally charged beam formation is to pass the ions through

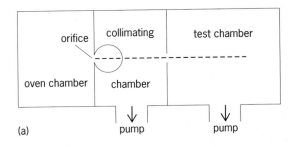

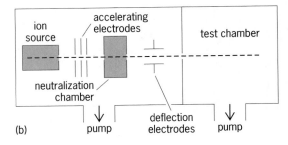

Schematic diagrams of systems for producing molecular beams. (*a*) Conventional oven-beam system. (*b*) Charge-exchange beam system.

a neutralizing gas where electrons from the gas molecules are transferred to the beam ions in charge-exchange molecular collisions. *See* ION SOURCES.

Much of molecular spectroscopy involves the absorption or emission of light by molecules in a gas sample. The frequency of the light photon is proportional to the separation of molecular energy levels involved in the spectroscopic transition. However, the molecule density in typical gas samples is so high that the energy levels are slightly altered by collisions between molecules, with the transition frequency no longer characteristic of the free molecule. The use of low-density molecular beams with their sensitive detection techniques can reduce this collision alteration problem, with the result that atomic properties can be measured to accuracies of parts per million or even better. If the very simplest atoms or molecules are employed, the basic electromagnetic interactions holding the component electrons and nuclei together can be precisely studied. This is of great importance to fundamental physics, since theoretical understanding of electromagnetic interactions through quantum electrodynamics represents the most successful application of quantum field theory to elementary particle physics problems. *See* QUANTUM ELECTRODYNAMICS; QUANTUM FIELD THEORY.

The development of tunable, strong laser sources of single-frequency light beams has added another dimension to molecular-beam experiments. With laser radiation resonantly tuned to excite a molecule from its normal ground state to one of its infinite number of vibrationally, rotationally, and electronically excited states, the number of possible studies and applications of excited molecular beams becomes enormous. *See* LASER SPECTROSCOPY; MOLECULAR STRUCTURE AND SPECTRA; NUCLEAR STRUCTURE.   [J.E.B.]

Molecular physics   The study of the physical properties of molecules. Molecules possess a far richer variety of physical and chemical properties than do isolated atoms. This is attributable primarily to the greater complexity of molecular structure, as compared to that of the constituent atoms. Molecules also possess additional energy modes because they can vibrate; that is, the constituent nuclei oscillate about their equilibrium positions and rotate when unhindered. These modes give rise to additional spectroscopic properties, as compared to those of an atom; molecular spectroscopy in the optical, infrared, and microwave regions is one of the physical chemist's most powerful means of identifying and understanding molecular structure. Molecular spectroscopy has also given rise to the rapidly growing field of molecular astronomy.

Molecular physics is primarily concerned with the study of properties of isolated molecules, as contrasted to the more general study of molecular reactions, which is the domain of physical chemistry. Such properties, in addition to the broad field of spectroscopy, include electron affinities (for the formation of molecular negative ions); polarizabilities (the "distortability" of the molecule along its various symmetry axes by external electric fields); magnetic and electric multipole moments, attributable to the distributions of electric charge; currents and spins of the molecule; and the (nonreactive) interactions of molecules with other molecules, atoms, and ions. *See* INFRARED SPECTROSCOPY; INTERMOLECULAR FORCES; MICROWAVE SPECTROSCOPY; MOLECULAR BEAMS; MOLECULAR STRUCTURE AND SPECTRA; SPECTROSCOPY.   [B.B.]

Molecular structure and spectra

Until the advent of quantum theory, ideas about the structure of molecules evolved gradually from analysis and interpretation of the facts of chemistry. Chemists developed the concept of molecules as built from atoms in definite proportions, and identified and constructed (synthesized) a great variety of molecules. Later, when the structure of atoms as built from nuclei and electrons began to be understood with the help of quantum theory, a beginning was made in seeing why atoms can combine in definite ways to form molecules; also, infrared spectra began to be used to obtain information about the dimensions and the nuclear motions (vibrations) in molecules. However, a fundamental understanding of chemical bonding and molecular structure became possible only by application of the present form of quantum theory, called quantum mechanics. This theory makes it possible to obtain from the spectra of molecules a great deal of information about the nature of molecules in their normal as well as excited states, and about dissociation energies and other characteristics of molecules.

Molecular sizes. The size of a molecule varies approximately in proportion to the numbers and sizes of the atoms in the molecule. Simplest are diatomic molecules. These may be thought of as built of two spherical atoms of radii r and r', flattened where they are joined. The equilibrium value R_e of the distance R between their nuclei is then smaller than the sum of the atomic radii. However, the nuclei of atoms in two different molecules cannot normally approach more closely than a distance $r + r'$; r and r' are called the van der Waals radii of the atoms.

To describe a polyatomic molecule, one must specify not merely its size but also its shape or configuration. For example, carbon dioxide (CO_2) is a linear symmetrical molecule, the O—C—O angle being 180°. The H—O—H angle in the nonlinear water (H_2O) molecule is 105°. Many molecules which are essential for life contain thousands or even millions of atoms. Proteins are often coiled or twisted and cross-linked in ways which are important for their biological functioning.

Dipole moments. Most molecules have an electric dipole moment. In atoms, the electron cloud surrounds the nucleus so symmetrically that its electrical center coincides with the nucleus, giving zero dipole moment; in a molecule, however, these coincidences are disturbed, and a dipole moment usually results.

Thus, when the atoms of HCl come together, there is some shifting of the H-atom electron toward the Cl. A complete shift would give H^+Cl^-, which would constitute an electric dipole of magnitude eR_e, where e is the electronic charge. But in fact the dipole moment is only 0. 17 eR_e. This is because the actual electronic shift is only fractional.

Molecular polarizability. In the preceding consideration of dipole moments, the discussion has been in terms of atoms and molecules free from external forces. An electric field pulls the electrons of an atom or molecule toward it and pushes the nuclei away, or vice versa. This action creates a small induced dipole moment, whose magnitude per unit strength of the field is called the polarizability.

Molecular energy levels. The states of motion of nuclei and electrons in a molecule, or of electrons in an atom, are restricted by quantum mechanics to special forms with definite energies. The state of lowest energy is called the ground state; all others are excited states. In analogy to water levels, one speaks of energy levels. Excited states exist only momentarily, following an electrical or other stimulus.

Excitation of an atom consists of a change in the state of motion of its electrons. Electronic excitation of molecules can also occur, but alternatively or additionally, molecules can be excited to discrete states of vibration and rotation.

The total energy of any molecule can be written as Eq. (1). Both the electronic energy

$$E = E_{el} + E_v + (E_r + E_{fs} + E_{hfs} + E_{ext}) \tag{1}$$

E_{el} and vibration energy E_v can be discrete or continuous. The quantities E_r, break E_{fs}, and E_{hfs} denote rotational, fine-structure, and hyperfine-structure energies. The last two appear as small or minute splittings of the rotation levels. The spacings ΔE of adjacent discrete levels of each type are usually in the order given in notation (2). The E_{ext} term

$$\Delta E_{el} \gg \Delta E_v \gg \Delta E_r \gg \Delta E_{fs} \gg \Delta E_{hfs} \tag{2}$$

in Eq. (1) refers to additional fine structure which appears on subjecting molecules to external magnetic fields (Zeeman effect) or electric fields (Stark effect). *See* FINE STRUCTURE (SPECTRAL LINES); HYPERFINE STRUCTURE; STARK EFFECT; ZEEMAN EFFECT.

Polyatomic molecules have much more complicated patterns of vibrational and (usually) rotational energy levels than diatomic molecules.

Molecular spectra. The frequencies cv (c = speed of light) of electromagnetic spectra obey the Einstein-Bohr equation (3), where h is Planck's constant. Molecular

$$hcv = E' - E'' \tag{3}$$

emission spectra accompany jumps in energy from higher to lower levels; absorption spectra accompany jumps from lower to higher levels.

Molecular spectra can be classified as fine-structure or low-frequency spectra, rotation spectra, vibration-rotation spectra, and electronic spectra. *See* MAGNETIC RESONANCE; MICROWAVE SPECTROSCOPY; MOLECULAR BEAMS; SPECTROSCOPY.

Transitions between energy levels differing only in rotational state give rise to pure rotation spectra. These typically consist of a sequence of lines spaced almost equidistantly, and lying in the far infrared or the microwave region.

Spectra involving only vibrational and rotational state changes consist of bands which lie mainly in the infrared. Each band consists of two sets of closely spaced rotational lines, one on each side of a central frequency. Vibration-rotation absorption bands of liquids and solutions are widely used in chemical analysis. Here the rotational structure is blurred out, and only an "envelope" is seen. *See* INFRARED SPECTROSCOPY.

Electronic band spectra are the most general type of molecular spectra. For any one electronic transition, the spectrum consists typically of many bands. *See* ATOMIC

Molecular weight   The sum of the atomic weights of all atoms making up a molecule. Actually, what is meant by molecular weight is molecular mass. The use of this expression is historical, however, and will be maintained. The atomic weight is the mass, in atomic mass units, of an atom. It is approximately equal to the total number of nucleons, protons, and neutrons composing the nucleus. Since 1961 the official definition of the atomic mass unit (amu) has been that it is 1/12 the mass of the carbon-12 isotope, which is assigned the value 12.000 exactly. *See* ATOMIC MASS; ATOMIC MASS UNIT.

A mole is an amount of substance containing the Avogadro number, N_A, approximately 6.022×10^{23}, of molecules or atoms. Molecule, in this definition, is understood to be the smallest unit making up the characteristic compound. Originally, the mole was interpreted as that number of particles whose total mass in grams was numerically equivalent to the atomic or molecular weight in atomic mass units, referred to as gram-atomic or gram-molecular weight. This is how the above value for N_A was calculated. As the ability to make measurements of the absolute masses of single atoms and molecules has improved, however, modern metrology is tending to alter its approach and define the Avogadro number as an exact quantity, thereby changing slightly the definition of the atomic mass unit and removing the need to define atomic weight with respect to a particular isotopic species. The latest and most accurate value for the Avogadro number is $6.0221415(10) \times 10^{23} \, \mathrm{mol}^{-1}$.

As the masses of all the atomic species are now well known, masses of molecules can be determined once the composition of the molecule has been ascertained. Alternatively, if the molecular weight of the molecule is known and enough additional information about composition is available, such as the basic atomic constituents, it is possible to begin to assemble structural information about the molecule. Thus, the determination of the molecular weight is one of the first steps in the analysis of an unknown species. Given the increasing emphasis on the study of biologically important molecules, particular attention has been focused on the determination of molecular weights of larger and larger units. There are a number of methods available, and the one chosen will depend on the size and physical state of the molecule. All processes are physical macroscopic measurements and determine the molecular weight directly. Connection to the absolute mass scale is straightforward by using the Avogadro number, although, for extremely large molecules, this connection is often unnecessary or impossible, as the accuracy of the measurements is not that good. The main function of molecular weight determination of large molecules is elucidation of structure.

Molecular weight determination of materials which are solid or liquid at room temperature is best achieved by taking advantage of one of the colligative properties of solutions, boiling-point elevation, freezing-point lowering, or osmotic pressure, which depend on the number of particles in solution, not on the nature of the particle. The choice of which to use will depend on a number of properties of the substance, the most important of which will be the size. All require that the molecule be small enough to dissolve in the solution but large enough not to participate in the phase change or pass through a semipermeable membrane. Freezing-point lowering is an excellent method for determining molecular weights of smaller organic molecules, and osmometry, as the osmotic pressure determination is called, for determining molecular weights of larger organic molecules, particularly polymeric species. Boiling-point elevation is used less frequently.

The basis of all the methods involving colligative properties of solutions is that the chemical potentials of all phases must be the same. (Chemical potential is the partial change in energy of a system as matter is transferred into or out of it. For two systems in contact at equilibrium, the chemical potentials for each must be equal.)

Another measurement from which molecular weights can be obtained is based on the scattering of light from the molecule. A beam of light falling on a molecule will induce in the molecule a dipole moment which in its turn will radiate. The interference between the radiated beam and the incoming beam produces an angular dependence of the scattered radiation which depends on the molecular weight of the molecule. This occurs whether the molecule is free or in solution. While the theory for this effect is complicated and varies according to the size of the molecule, the general result for molecules whose size is considerably less than that of the wavelength λ of the radiation (less than $\lambda/50$) is given by the equation below; $I(\theta)$ is the intensity of radiation at

$$\frac{I(\theta)}{I_0} = \text{constant } (1 + \cos^2\theta) \, Mc$$

angle θ, I_0 the intensity of the incoming beam, M the molecular weight, and c the concentration in grams per cubic centimeter of the molecule. If the molecules are much larger than $\lambda/50$ (about 9 nanometers for visible light), this relationship in this simple form is no longer valid, but the method is still viable with appropriate adjustments to the theory. In fact, it can be used in its extended version even for large aggregates. *See* SCATTERING OF ELECTROMAGNETIC RADIATION. [C.D.C.]

Molecule A molecule may be thought of either as a structure built of atoms bound together by chemical forces or as a structure in which two or more nuclei are maintained in some definite geometrical configuration by attractive forces from a surrounding swarm of negative electrons. Besides chemically stable molecules, short-lived molecular fragments called free radicals can be observed under special circumstances. *See* MOLECULAR STRUCTURE AND SPECTRA. [R.S.M.]

Moment of inertia A relation between the area of a surface or the mass of a body to the position of a line. The analogous positive number quantities, moment of inertia of area and moment of inertia of mass, are involved in the analysis of problems of statics and dynamics respectively.

The moment of inertia of a figure (area or mass) about a line is the sum of the products formed by multiplying the magnitude of each element (of area or of mass) by the square of its distance from the line. The moment of inertia of a figure is the sum of moments of inertia of its parts.

For a body of mass distributed continuously within volume V, the movement of inertia of the mass about the X axis is given by either $I_X = \int r_x^2 \, dm$ or $I_X = \int r_x^2 \rho \, dV$, where dm is the mass included in volume element dV at whose position the mass per unit volume is ρ (see illustration). Similarly $I_y = \int r_y^2 \rho \, dV$ and $I_z = \int r_z^2 \rho \, dV$.

The moments of inertia of a figure about lines which intersect at a common point are generally unequal. The moment is greatest about one line and least about another line perpendicular to the first one. A set of three orthogonal lines consisting of these two and a line perpendicular to both are the principal axes of inertia of the figure relative to that point. If the point is the figure's centroid, the axes are the central principal axes of

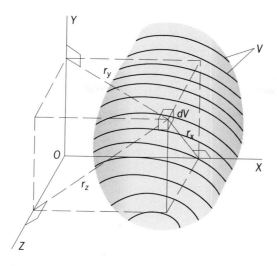

Moment of inertia of a volume.

inertia. The moments of inertia about principal axes are principal moments of inertia. *See* PRODUCT OF INERTIA; RADIUS OF GYRATION.   [N.S.F.]

Momentum   Linear momentum is the product of the mass and the linear velocity of a body. It is defined by Eq. (1), where m is the mass and $\mathbf{v}$ is the linear velocity.

$$\mathbf{P} = m\mathbf{v} \tag{1}$$

Since linear momentum is the product of a scalar and a vector quantity, it is a vector and hence has both magnitude and direction.

According to the general statement of Newton's second law, for a force $\mathbf{F}$, a momentum $\mathbf{P}$, and a time t, Eq. (2) holds. Thus Newton's second law involves the time rate

$$\mathbf{F} = d\mathbf{P}/dt \tag{2}$$

of change of momentum. Changes of momentum are important in collision processes. *See* COLLISION (PHYSICS).

When a group of bodies is subject only to forces that members of the group exert on one another, the total momentum of the group remains constant. *See* ANGULAR MOMENTUM; CONSERVATION OF MOMENTUM; IMPULSE (MECHANICS).   [P.W.S.]

Mössbauer effect   Recoil-free gamma-ray resonance absorption. The Mössbauer effect, also called nuclear gamma resonance fluorescence, has become the basis for a type of spectroscopy which has found wide application in nuclear physics, structural and inorganic chemistry, biological sciences, the study of the solid state, and many related areas of science.

The fundamental physics of this effect involves the transition (decay) of a nucleus from an excited state of energy E_e to a ground state of energy E_g with the emission of a gamma ray of energy E_γ. If the emitting nucleus is free to recoil, so as to conserve momentum, the emitted gamma ray energy is $E_\gamma = (E_e - E_g) - E_r$, where E_r is the recoil energy of the nucleus. Tlie magnitude of E_r is given classically by the relationship $E_r = E_\gamma^2/2mc^2$, where m is the mass of the recoiling atom and c is the speed of light. Since E_r is a positive number, the E_γ will always be less than the difference $E_e - E_g$,

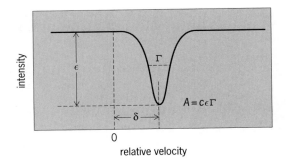

intensity

relative velocity

$A = c_{\epsilon} \Gamma$

0

Mössbauer spectrum of an absorber which gives an unsplit resonance line. The spectrum is characterized by a position δ, a line width Γ, and an area A related to the effect magnitude ε.

and if the gamma ray is now absorbed by another nucleus, its energy is insufficient to promote the transition from E_g to E_e.

In 1957 R. L. Mössbauer discovered that if the emitting nucleus is held by strong bonding forces in the lattice of a solid, the whole lattice takes up the recoil energy, and the mass in the recoil energy equation given above becomes the mass of the whole lattice. Since this mass typically corresponds to that of 10^{10} to 10^{20} atoms, the recoil energy is reduced by a factor of 10^{-10} to 10^{-20}, with the important result that $E_r \approx 0$ so that $E_{\gamma} = E_e - E_g$; that is, the emitted gamma-ray energy is exactly equal to the difference between the nuclear ground-state energy and the excited-state energy. Consequently, absorption of this gamma ray by a nucleus which is also firmly bound to a solid lattice can result in the "pumping" of the absorber nucleus from the ground state to the excited state. *See* ENERGY LEVEL (QUANTUM MECHANICS); EXCITED STATE; GAMMA RAYS.

In a typical Mössbauer experiment the radioactive source is mounted on a velocity transducer which imparts a smoothly varying motion (relative to the absorber, which is held stationary), up to a maximum of several centimeters per second, to the source of the gamma rays. These gamma rays are incident on the material to be examined (the absorber). Some of the gamma rays are absorbed and reemitted in all directions, while the remainder of the gamma rays traverse the absorber and are registered in an appropriate detector.

A typical display of a Mössbauer spectrum, which is the result of many repetitive scans through the velocity range of the transducer, is shown in the illustration. In certain nuclides the Mössbauer resonance line displays splitting that arises from the coupling of the nuclear electric quadrupole moment with the electric field gradient or of the nuclear magnetic dipole moment with the magnetic field at the nucleus, providing information on the magnitude of these interactions.

Mössbauer effect experiments have been used to elucidate problems in a very wide range of scientific disciplines. Applications include the measurement of nuclear magnetic and quadrupole moments and of excited-state lifetimes involved in the nuclear decay process; study of the chemical consequences of nuclear decay; study of the nature of magnetic interactions in iron-containing alloys and of the dependence of the magnetic field in these alloys on various parameters; study of the effects of high pressure on chemical properties of materials; investigation of the relationship between chemical composition and structure on the one hand and the superconductive transition on the other; investigation of the structure of compounds; and study of the structure and bonding properties of metal atoms in complex biological molecules.   [R.H.He.]

Motion   If the position of a material system as measured by a particular observer changes with respect to time, that system is said to be in motion with respect to the observer. Absolute motion, then, has no significance, and only relative motion may be defined; what one observer measures to be at rest, another observer in a different frame of reference may regard as being in motion. *See* FRAME OF REFERENCE; RELATIVE MOTION.

The time derivatives of the various coordinates used to specify the system may be used to prescribe the motion at any instant of time. How the motion develops in subsequent instants is then determined by the laws of motion. In classical dynamics it is supposed that in principle the motion and configuration of the system may be specified to an arbitrary precision, although in quantum mechanics it is recognized that the measurement of the one disturbs the other.

The most general theory of motion that has yet been developed is quantum field theory, which combines both quantum mechanics and relativity theory, as well as the experimentally observed fact that elementary particles can be created and annihilated. *See* DEGREE OF FREEDOM (MECHANICS); DYNAMICS; HAMILTON'S EQUATIONS OF MOTION; HARMONIC MOTION; KINEMATICS; KINETICS (CLASSICAL MECHANICS); LAGRANGE'S EQUATIONS; NEWTON'S LAWS OF MOTION; OSCILLATION; PERIODIC MOTION; QUANTUM FIELD THEORY; QUANTUM MECHANICS; RECTILINEAR MOTION; RELATIVITY; ROTATIONAL MOTION.

[H.C.Co./B.G.]

Multimeter   An instrument designed to measure electrical quantities. A typical multimeter can measure alternating- and direct-current potential differences (voltages), current, and resistance, with several full-scale ranges provided for each quantity. Sometimes referred to as a volt-ohm meter (VOM), it is a logical development of the electrical meter, providing a general-purpose instrument. Many kinds of special-purpose multimeters are manufactured to meet the needs of such specialists as telephone engineers and automobile mechanics testing ignition circuits. *See* AMMETER; CURRENT MEASUREMENT; OHMMETER; RESISTANCE MEASUREMENT; VOLTAGE MEASUREMENT; VOLTMETER.

Multimeters originated when all electrical measuring instruments used analog techniques. They were generally based on a moving-coil indicator, in which a pointer moves across a graduated scale. Accuracy was typically limited to about 2%, although models achieving 0.1% were available. Analog multimeters are still preferred for some applications. For most purposes, digital instruments are now used. In these, the measured value is presented as a row of numbers in a window. Inexpensive hand-held models perform at least as well as a good analog design. High-resolution multimeters have short-term errors as low as 0.1 part per million (ppm) and drift less than 5 ppm in one year. Many digital multimeters can be commanded by, and send their indications to, computers or control equipment.

[R.B.D.K.]

Multipole radiation   Standard patterns of radiation distribution about their source. The term radiation applies primarily to the transport of energy by acoustic, elastic, electromagnetic, or gravitational waves, and extends to the transport of atomic or subatomic particles (as represented by quantum-mechanical wave functions). *See* ELECTROMAGNETIC RADIATION; QUANTUM MECHANICS; SOUND; WAVE MOTION.

Each multipole pattern reflects the source's geometrical shape (or the shape of a source component). These geometrical features stand out clearly for the static electric potentials generated by fixed charges as shown by the small set of monopole, dipole, and quadrupole charges (see illustration), elements of all multipoles being named (in terms of powers of 2) 2^l-poles, with l equal to any nonnegative integer. A monopole ($l = 0$) acoustic wave radiates from a perfectly spherical bubble with oscillating radius; higher multipoles would arise from bubble distortions. So-called transverse waves, elastic or

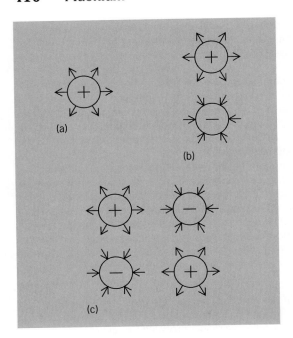

Static electric potentials generated by fixed multipoles. (a) Monopole (l = 0). (b) Dipole (l = 1). (c) Quadrupole (l = 2).

electromagnetic (including light), have only $l \geq 1$ components, gravitational waves only $l \geq 2$. The angular distributions, in azimuth (φ) and colatitude (θ), of 2^l-pole waves have amplitudes distributed in directions (θ, φ) in proportion to the spherical harmonic functions $Y_l^m(\theta, \varphi)$. The index m is a positive or negative integer whose absolute value is equal to less than l. See DIPOLE.

The multipolarity index l also represents the number of angular momentum quanta $\hbar$ (Planck's constant divided by 2π) radiated together with each energy quantum $h\nu$ (phonon, photon, graviton, and so forth). Detection and measurement of received energy quanta, together with measurement of their detection rate and mapping of their directional distribution, generally serve to diagnose the mechanics of the radiation source. Energy and momentum conservation underlie this analysis; so does the conservation of angular momentum which states that the initial angular momentum of the source equals the vector sum of the final angular momentum of the source and the angular momentum of the radiation. The quantitative implications of this vector relation are studied by the branch of quantum theory called angular momentum algebra. The balancing of parity, that is, of each variable's sign reversal (or persistence) under reflection through the source's center, also contributes to the analysis of experimental data. Further, more complex angular-momentum considerations play a role in the analysis of the behavior of spin-carrying particles. See ANGULAR MOMENTUM; CONSERVATION LAWS (PHYSICS); GRAVITON; PHONON; SELECTION RULES (PHYSICS); SPIN (QUANTUM MECHANICS).

[U.F.]

Muonium   An exotic atom, Mu or $(\mu^+ e^-)$, formed when a positively charged muon (μ^+) and an electron are bound by their mutual electrical attraction. It is a light, unstable isotope of hydrogen, with a muon replacing the proton. Muonium has a mass 0.11 times that of a hydrogen atom due to the lighter mass of the muon, and a mean lifetime of 2.2 microseconds, determined by the spontaneous decay of the muon $(\mu^+ \rightarrow e^+ \nu_e \overline{\nu}_\mu)$.

Muonium is formed when beams of μ^+ produced in particle accelerators are stopped in certain nonmetallic targets.

Since muonium is a system consisting only of leptons, it serves as a testing ground for the theory of quantum electrodynamics (QED), which describes the electromagnetic interaction between particles. Muonium chemistry and muonium spin rotation (MSR) are two developing subfields which seek to understand the chemical and physical behavior of a light hydrogen isotope in matter and to probe the structure of materials. *See* POSITRONIUM; QUANTUM ELECTRODYNAMICS.

[P.O.E.]

Musical acoustics   The branch of acoustics that deals with the generation of sound by musical instruments, the transmission of sound to the listener, and the perception of musical sound. A main research activity in musical acoustics is the study of the way in which musical instruments vibrate and produce sound. The most common way of classifying musical instruments is according to the nature of the primary vibrator, into string instruments, wind instruments, and percussion instruments. The vibrations of a plucked string, a struck membrane, or a blown pipe can be described in terms of normal modes of vibration. Determining the normal modes of a complex vibrator is often termed modal analysis. Much of the progress in understanding how musical instruments generate sound is due to new methods of modal analysis, such as holographic interferometry and experimental modal testing. *See* INTERFEROMETRY; MODE OF VIBRATION; VIBRATION.

In the case of most percussion, plucked string, and struck string instruments, the player delivers energy to the primary vibrator (string, membrane, bar, or plate) and thereafter has little control over the way it vibrates. In the case of wind and bowed string instruments, however, the continuing flow of energy is controlled by feedback from the vibrating system. In brass and reed woodwinds, pressure feedback opens or closes the input valve. In flutes or flue organ pipes, however, the input valve is flow-controlled. In bowed string instruments, pulses on the string control the stick-slip action of the bow on the string.

Four attributes are frequently used to describe musical sound: loudness, pitch, timbre, and duration. Each of the subjective qualities depends on one or more physical parameters that can be measured. Loudness, for example, depends mainly on sound pressure but also on the spectrum of the partials and the physical duration. Pitch depends mainly on frequency, but also shows lesser dependence on sound pressure and envelope. Timbre includes all the attributes by which sounds with the same pitch and loudness are distinguished. Relating the subjective qualities of sound to the physical parameters is a central problem in psychoacoustics, and musical acousticians are concerned with this same problem as it applies to musical sound.

Sound pressure level is measured with a sound level meter and is generally expressed on a logarithmic scale of decibels (dB) using an appropriate reference level and weighting network. From measurements of the sound pressure level at different frequencies, it is possible to calculate a subjective loudness, expressed in sones, which describes the sensation of loudness heard by an average listener. Musicians prefer to use dynamic markings ranging from *ppp* (very soft) to *fff* (very loud). *See* DECIBEL; SOUND; SOUND PRESSURE.

Pitch is defined as that attribute of auditory sensation in terms of which sounds may be ordered on a scale extending from low to high. Pitch is generally related to a musical scale where the octave, rather than the critical bandwidth, is the "natural" pitch interval.

Timbre is defined as that attribute of auditory sensation in terms of which a listener can judge two sounds similarly presented and having the same loudness and pitch as dissimilar. Timbre depends primarily on the spectrum of the sound, but it also depends upon the waveform, the sound pressure, the frequency location of the spectrum, and the temporal characteristics of the sound. It has been found impossible to construct a single subjective scale of timbre (such as the sone scale of loudness); multidimensional scales have been constructed. The term "tone color" is often used to refer to that part of timbre that is attributable to the steady-state part of the tone, but the time envelope (and especially the attack) has been found to be very important in determining timbre as well.

Another subject relating to the perception of music is combination tones. When two tones that are close together in frequency are sounded at the same time, beats generally are heard, at a rate that is equal to their frequency difference. When the frequency difference Δf exceeds 15 Hz or so, the beat sensation disappears, and a roughness appears. As Δf increases still further, a point is reached at which the "fused" tone at the average frequency gives way to two tones, still with roughness. The respective resonance regions on the basilar membrane are now separated sufficiently to give two distinct pitches, but the excitations overlap to give a sense of roughness. When the separation Δf exceeds the width of the critical band, the roughness disappears, and the two tones begin to blend.

Pythagoras of ancient Greece is considered to have discovered that the tones produced by a string vibrating in two parts with simple ratios such as 2:1, 3:2, or 4:3 sound harmonious. These ratios define the so-called perfect intervals of music, which are considered to have the greatest consonance. Other consonant intervals in music are the major sixth ($f_2/f_1 = 5/3$), the major third ($f_2/f_1 = 5/4$), the major sixth ($f_2/f_1 = 8/5$), and the minor third ($f_2/f_1 = 6/5$). Why are some intervals more consonant than others? H. Helmholtz concluded that dissonance (the opposite of consonance) is greatest when partials of the two tones produce 30 to 40 beats per second (which are not heard as beats but produce roughness). The more the partials of one tone coincide in frequency with the partials of the other, the less chance of roughness. This explains why simple frequency ratios define the most consonant intervals. More recent research has concluded that consonance is related to the critical band. If the frequency difference between two pure tones is greater than a critical band, they sound consonant; if it is less than a critical band, they sound dissonant. The maximum dissonance occurs when Δf is approximately 1/4 of a critical band, which agrees reasonably well with Helmholtz's criterion for tones around 500 Hz.   [T.D.R.]

N

Navier-Stokes equation A partial differential equation which describes the conservation of linear momentum for a linearly viscous (newtonian), incompressible fluid flow. In vector form, this relation is written as Eq. (1), where ρ is fluid density, $\mathbf{V}$ is

$$\rho\left[\frac{\partial \mathbf{V}}{\partial t} + (\mathbf{V} \cdot \nabla)\mathbf{V}\right] = -\nabla p + \rho\mathbf{g} + \mu\nabla^2\mathbf{V} \tag{1}$$

fluid velocity, p is fluid pressure, $\mathbf{g}$ is the gravitational acceleration, μ is fluid viscosity, ∇ is the del or grad operator, and ∇^2 is the laplacian operator. The equation is named after its two principal developers, French engineer C. L. M. H. Navier (1823) and Irish scientist George G. Stokes (1845). When coupled with the conservation of mass relation, $\mathbf{V} \cdot \mathbf{V} = 0$, Eq. (1) can be solved for the space-time distribution of $\mathbf{V}$ and p in a given region of viscous fluid flow. Typical boundary conditions are (1) the knowledge of the velocity and pressure in the far field, and (2) the no-slip condition at solid surfaces (fluid velocity equals solid velocity). See NEWTONIAN FLUID; VISCOSITY.

Equation (1) correctly models the continuum behavior of all newtonian fluids, including the disorderly fluctuating motion known as turbulence. However, since the left-hand side is highly nonlinear, only a few score of exact solutions are known, mostly for simple geometries. The primary dimensionless parameter which governs Eq. (1) is the Reynolds number, given by Eq. (2), where L is a characteristic body dimension.

$$\mathrm{Re} = \frac{\rho V L}{\mu} \tag{2}$$

For small $\mathrm{Re} \ll 1$, Eq. (1) can be simplified by neglecting the left-hand side, resulting in a linear approximation called Stokes flow, or creeping flow, for which many solutions are known. See CREEPING FLOW; REYNOLDS NUMBER.

For large $\mathrm{Re} \gg 1$, viscous effects are often confined to a thin boundary layer near solid surfaces, with the remaining flow being nearly inviscid. See BOUNDARY-LAYER FLOW.

[F.M.Wh.; A.E.Br.]

Negative ion An atomic or molecular system with an excess of negative charge. Negative ions, also called anions, are formed in attachment processes in which an additional electron is captured by an atom or molecule. Negative ions were first reported in the early days of mass spectrometry. It was soon learned that even a small concentration of such weakly bound, negatively charged systems had an appreciable effect on the electrical conductivity of gaseous discharges. Negative ions now play a major role in a number of areas of physics and chemistry involving weakly ionized gases and plasmas. Applications include accelerator technology, injection heating of thermonuclear plasmas, material processing, and the development of tailor-made gaseous dielectrics.

In nature, negative ions are known to be present in tenuous plasmas such as those found in astrophysical and aeronomical environments. The absorption of radiation by negative hydrogen ions in the solar photosphere, for example, determines the Sun's spectral distribution. *See* ION SOURCES.   [D.J.Pe.]

Negative temperature   The property of a thermodynamical system which satisfies certain conditions and whose thermodynamically defined absolute temperature is negative. The essential requirements for a thermodynamical system to be capable of negative temperature are: (1) the elements of the thermodynamical system must be in thermodynamical equilibrium among themselves in order for the system to be described by a temperature at all; (2) there must be an upper limit to the possible energy of the allowed states of the system; and (3) the system must be thermally isolated from all systems which do not satisfy both requirements (1) and (2); that is, the internal thermal equilibrium time among the elements of the system must be short compared to the time during which appreciable energy is lost to or gained from other systems.

The second condition must be satisfied if negative temperatures are to be achieved with a finite energy. Most systems do not satisfy this condition; for example, there is no upper limit to the possible kinetic energy of a gas molecule. Systems of interacting nuclear spins, however, have the characteristic that under suitable circumstances they can satisfy all three of the conditions, in which case the nuclear spin system can be at negative absolute temperature. *See* KINETIC THEORY OF MATTER; STATISTICAL MECHANICS.

The transition between positive and negative temperatures is through infinite temperature, not absolute zero; negative absolute temperatures should therefore not be thought of as colder than absolute zero, but as hotter than infinite temperature. *See* ABSOLUTE ZERO; TEMPERATURE.   [N.F.R.]

Neutral currents   Exchange currents which carry no electric charge and mediate certain types of electroweak interactions. The discovery of the neutral-current weak interactions and the agreement of their experimentally measured properties with the theoretical predictions were of great significance in establishing the validity of the Weinberg-Salam model of the electroweak forces.

The electroweak forces come in three subclasses: the electromagnetic interactions, the charged-current weak interactions, and the neutral-current weak interactions. The electromagnetic interaction is mediated by an exchanged photon γ. Since the photon carries no electric charge, there is no change in charge between the incoming and the outgoing particles. The charged-current weak interaction is mediated by the exchange of a charged intermediate boson, the W^+, and thus, for example, an incoming neutral lepton such as the ν_μ is changed into a charged lepton, the μ^-. In the neutral-current weak interactions, the exchanged intermediate boson, the Z^0, carries no electric charge (hence the name neutral-current interaction), and thus, for example, an incident neutral lepton, such as the ν_μ, remains an outgoing neutral ν_μ. *See* ELECTRON; INTERMEDIATE VECTOR BOSON; LEPTON; NEUTRINO; PHOTON.

The neutral-current interactions were experimentally discovered in 1973, and have since been extensively studied, in neutrino scattering processes. Very important information about the properties of the neutral currents have been obtained by studying the interference effects between the electromagnetic and the neutral-current weak interactions in the scattering of polarized electrons on deuterium. Parity-violating effects in atomic physics processes due to the neutral weak currents have been observed, and predicted parity-violating nuclear effects have been searched for. *See* ELEMENTARY PARTICLE; FUNDAMENTAL INTERACTIONS; SYMMETRY LAWS (PHYSICS); WEAK NUCLEAR INTERACTIONS.   [C.B.]

Neutrino   An elusive elementary particle that interacts with matter principally through the weak nuclear force. Neutrinos are electrically neutral spin-$\frac{1}{2}$ fermions with left-handed helicity. Many weak interaction processes (interactions that involve the weak force), such as radioactive nuclear beta decay and thermonuclear fusion, involve neutrinos. Present experimental knowledge is consistent with neutrinos being point particles that have no internal constituents. Neutrinos are classified as neutral leptons, where leptons are defined as elementary particles that interact with the electroweak (electromagnetic and weak nuclear) and gravitational forces but not with the strong nuclear force. *See* ELEMENTARY PARTICLE; FUNDAMENTAL INTERACTIONS; HELICITY (QUANTUM MECHANICS); LEPTON; SPIN (QUANTUM MECHANICS); WEAK NUCLEAR INTERACTIONS.

Because the role of gravitational forces is negligible in nuclear and particle interactions and because neutrinos have zero electric charge, neutrinos have the unique property that they interact almost completely via the weak nuclear force. Consequently, neutrinos can be used as sensitive probes of the weak force. As such, neutrino beams at particle accelerators have been employed to study charge-changing (charged current) and charge-preserving (neutral current) weak interactions. However, the extreme weakness (compared to the electromagnetic and strong forces) and short range (of the order of 10^{-18} m) of the weak interaction have made determination of many neutrino properties extremely difficult.

Currently, three distinct flavors (or types) of neutrinos are known to exist: the electron neutrino (ν_e), the muon neutrino (ν_μ), and the tau neutrino (ν_τ). Each neutrino flavor is associated with a corresponding charged lepton, the electron (e), muon (μ), and tau (τ) particle. The electron, muon, and tau neutrinos (or their antiparticles) have been observed in experiments. Based on present measurements, the lepton flavor families, which comprise the charged and neutral leptons and their antiparticles (e^-, ν_e, e^+, $\bar{\nu}_e$; μ^-, ν_μ, μ^+, $\bar{\nu}_\mu$; τ^-, ν_τ, τ^+, $\bar{\nu}_\tau$), obey laws of conservation of lepton number. These empirical laws state that the number of leptons minus antileptons does not change, both within a flavor family and overall. *See* ELECTRON; SYMMETRY LAWS (PHYSICS).

The existence of neutrino oscillations (a phenomenon whereby neutrinos change their flavors during the flight from a neutrino source to a detector), seen clearly in observations of atmospheric neutrinos, shows that neutrinos have tiny finite masses which are many orders of magnitude smaller than the masses of their charged lepton counterparts, and also shows that the physical neutrinos do not have pure flavors (quantum-mechanical states) but contain mixtures of two or more neutrino states. This mixing indicates that the empirical laws of lepton number conservation are not exact and that they are violated in some physical processes. It is not known whether neutrinos have magnetic or electric dipole moments.   [Y.S.]

Neutron   An elementary particle having approximately the same mass as the proton, but lacking a net electric charge. It is indispensable in the structure of the elements, and in the free state it is an important reactant in nuclear research and the propagating agent of fission chain reactions. Neutrons, in the form of highly condensed matter, constitute the substance of neutron stars.

Neutrons and protons are the constituents of atomic nuclei. The number of protons in the nucleus determines the chemical nature of an atom, but without neutrons it would be impossible for two or more protons to exist stably together within nuclear dimensions, which are of the order of 10^{-13} cm. The protons, being positively charged, repel one another by virtue of their electrostatic interactions. The presence of neutrons weakens the electrostatic repulsion, without weakening the nuclear forces of cohesion. In light nuclei the resulting balanced, stable configurations contain protons and neutrons in almost equal numbers, but in heavier elements the neutrons outnumber the protons;

in ^{238}U, for example, 146 neutrons are joined with 92 protons. Only one nucleus, ^{1}H, contains no neutrons. For a given number of protons, neutrons in several different numbers within a restricted range often yield nuclear stability—and hence the isotopes of an element. *See* ISOTOPE; NUCLEAR STRUCTURE; PROTON.

Free neutrons have to be generated from nuclei, and since they are bound therein by cohesive forces, an amount of energy equal to the binding energy must be expended to get them out. Nuclear machines, such as cyclotrons and electrostatic generators, induce many nuclear reactions when their ion beams strike target material. Some of these reactions release neutrons, and these machines are sources of high neutron flux. Neutrons are released in the act of fission, and nuclear reactors are unexcelled as intense neutron sources. *See* NUCLEAR BINDING ENERGY; NUCLEAR FISSION.

Neutrons occur in cosmic rays, being liberated from atomic nuclei in the atmosphere by collisions of the high-energy primary or secondary charged particles. They do not themselves come from outer space.

Having no electric charge, neutrons interact so slightly with atomic electrons in matter that energy loss by ionization and atomic excitation is essentially absent. Consequently they are vastly more penetrating than charged particles of the same energy. The main energy-loss mechanism occurs when they strike nuclei. The most efficient slowing-down occurs when the bodies that are struck in an elastic collision have the same mass as the moving bodies; hence the most efficient neutron moderator is hydrogen, followed by other light elements: deuterium, beryllium, and carbon. The great penetrating power of neutrons imposes severe shielding problems for reactors and other nuclear machines, and it is necessary to provide walls, usually of concrete, several feet in thickness to protect personnel. The currently accepted health tolerance levels for an 8-h day correspond for fast neutrons to a flux of 20 neutrons/(cm^2)(s) or 130 neutrons/(in.^{2})(s); for slow neutrons, 700/(cm^2)(s) or 4500/(in.^{2})(s). On the other hand, fast neutrons are useful in some kinds of cancer therapy.

Free neutrons are radioactive, each transforming spontaneously into a proton, an electron (β^- particle), and an antineutrino. This instability is a reflection of the fact that neutrons are slightly heavier than hydrogen atoms. The neutron's rest mass is 1.0086652 atomic mass units on the unified mass scale (1.67495 $\times$ 10^{-24} g), as compared with 1.0078252 atomic mass units for the hydrogen atom.

Neutrons are, individually, small magnets. This property permits the production of beams of polarized neutrons, that is, beams of neutrons whose magnetic dipoles are aligned predominantly parallel to one direction in space. The magnetic moment is -1.913042 nuclear magnetons. *See* MAGNETON; NUCLEAR MOMENTS; NUCLEAR ORIENTATION; SPIN (QUANTUM MECHANICS).

Despite its overall neutrality, the neutron does have an internal distribution of electric charge, as has been revealed by scattering experiments. On a still finer scale, the neutron can also be presumed to have a quark structure in analogy of that of the proton. *See* QUANTUM CHROMODYNAMICS; QUARKS.

When neutrons are completely slowed down in matter, they have a maxwellian distribution in energy that corresponds to the temperature of the moderator with which they are in equilibrium. The de Broglie wavelength of these ultracold neutrons is greater than 50 nm, which is so much larger than interatomic distances in solids that they interact with regions of a surface rather than with individual atoms, and as a result they are reflected from polished surfaces at all angles of incidence. Ultracold neutrons are important in basic physics and have applications in studies of surfaces and of the structure of inhomogeneities and magnetic domains in solids. *See* ELEMENTARY PARTICLE; NEUTRON DIFFRACTION. [A.H.Sn.]

Neutron diffraction   The phenomenon associated with the interference processes which occur when neutrons are scattered by the atoms within solids, liquids, and gases. The use of neutron diffraction as an experimental technique is relatively new compared to electron and x-ray diffraction, since successful application requires high thermal-neutron fluxes, which can be obtained only from nuclear reactors. These diffraction investigations are possible because thermal neutrons have energies with equivalent wavelengths near 0.1 nanometer and are therefore ideally suited for interatomic interference studies.

In the scattering of neutrons by atoms, there are two important interactions. One is the short-range, nuclear interaction of the neutron with the atomic nucleus. This interaction produces isotropic scattering because the atomic nucleus is essentially a point scatterer relative to the wavelengths of thermal neutrons. Strong resonances associated with the scattering process prevent any regular variation of the nuclear scattering amplitudes with atomic number. The other important process for the scattering of neutrons by atoms is the interaction of the magnetic moment of the neutron with the spin and orbital magnetic moments of the atom. *See* SCATTERING EXPERIMENTS (ATOMS AND MOLECULES); SCATTERING EXPERIMENTS (NUCLEI).

Since the nuclear scattering amplitudes for neutrons do not vary uniformly with atomic number, there are certain types of chemical structures which can be investigated more readily by neutron diffraction than by x-ray diffraction. Moreover, since neutron scattering is a nuclear process, when the scattering amplitude of an element is not favorable for a particular investigation, it is frequently possible to substitute an enriched isotope which has scattering characteristics that are markedly different. The most significant application of neutron diffraction in chemical crystallography is the structure determination of composite crystals which contain both heavy and light atoms, and the most important compounds in this general classification are the hydrogen-containing substances.

The interaction of the magnetic moment of the neutron with the orbital and spin moments in magnetic atoms makes neutron scattering a unique tool for the study of a wide variety of magnetic phenomena, because information is obtained on the magnetic properties of the individual atoms in a material. This interaction depends on the size of the atomic magnetic moment and also on the relative orientation of the neutron spin and of the atomic magnetic moment with respect to the scattering vector and with respect to each other. Consequently, detailed information can be obtained on both the magnitude and orientation of magnetic moments in any substance which displays magnetic properties.

The investigation of antiferromagnetic and ferrimagnetic substances is one of the most important applications of the neutron diffraction technique, because detailed information on the magnetic configuration in these systems cannot be obtained by other methods.

One of the most important uses of inelastic neutron scattering is the study of thermal vibrations of atoms about their equilibrium positions, because lattice vibration quanta, or phonons, can be excited or annihilated in their interactions with low-energy neutrons. The measurements provide a direct determination of the dispersion relations for the normal vibrational modes of the crystal and do not require the large corrections necessary in similar x-ray investigations. These measured dispersion relations furnish the best experimental information available on interatomic forces that exist in crystals. *See* ELECTRON DIFFRACTION; MAGNON; NEUTRON SPECTROMETRY.   [M.K.W.]

Neutron optics   The general class of experiments designed to emphasize the wavelike character of neutrons. Like all elementary particles, neutrons can be made to display wavelike, as well as particlelike, behavior. They can be reflected and refracted, and they can scatter, diffract, and interfere, like light or any other type of wave. Many classical optical effects, such as Fresnel diffraction, have been performed with neutrons, including even those involving the construction of Fresnel zone plates. *See* DIFFRACTION; INTERFERENCE OF WAVES; REFLECTION OF ELECTROMAGNETIC RADIATION; REFRACTION OF WAVES; SCATTERING OF ELECTROMAGNETIC RADIATION; WAVE (PHYSICS).

The typical energy of a neutron produced by a moderated nuclear reactor is about 0.02 eV, which is approximately equal to the kinetic energy of a particle at about room temperature (80°F or 300 K), and which corresponds to a wavelength of about 10^{-10} m. This is also the typical spacing of atoms in a crystal, so that solids form natural diffraction gratings for the scattering of neutrons, and much information about crystal structure can be obtained in this way. However, the wavelike properties of neutrons have been confirmed over a vast energy range from 10^{-7} eV to over 100 MeV. *See* NEUTRON DIFFRACTION.

Neutrons, being uncharged, can be made to interfere over large spatial distances, since they are relatively unaffected by the stray fields in the laboratory that deflect charged particles. This property has been exploited by using the neutron interferometer. This device is made possible by the ability to grow essentially perfect crystals of up to 4 in. (10 cm). The typical interferometer is made from a single perfect crystal cut so that three parallel "ears" are presented to the neutron beam. This allows the incident beam to be split and subsequently recombined coherently. *See* COHERENCE; INTERFEROMETRY; SINGLE CRYSTAL.

One of the most significant experiments performed with the interferometer involved rotating the interferometer about the incident beam so that one neutron path was higher than the other, creating a minute gravitational potential difference (of 10^{-9} eV) between the paths. This was sufficient to cause a path difference of 20 or so wavelengths between the beams. This remains the only type of experiment that has ever seen a quantum-mechanical interference effect due to gravity. It also verifies the extension of the equivalence principle to quantum theory (although in a form more subtle than its classical counterpart). *See* GRAVITATION; RELATIVITY.

Many noninterferometer experiments have also been done with neutrons. In one experiment, resonances were produced in transmitting ultracold neutrons (energy about 10^{-7} eV) through several sheets of material. This is theoretically similar to seeing the few lowest states in a square-well potential in the Schrödinger equation. *See* NEUTRON; QUANTUM MECHANICS.   [D.M.Gre.]

Neutron spectrometry   A generic term applied to experiments in which neutrons are used as the probe for measuring excited states of nuclides and for determining the properties of these states. The term neutron spectroscopy is also used. The strength of the interaction between a neutron and a target nuclide can vary rapidly as a function of the energy of the incident neutron, and it is different for every nuclide. At particular neutron energies the interaction strength for a specific nuclide can be very strong; these narrow energy regions of strong interactions are called resonances (see illustration). The strength of the interaction, expressing the probability that an interaction of a given kind will take place, can be considered as the effective cross-sectional area σ presented by a nucleus to an incident neutron.

Neutron spectroscopy can be carried out by two different techniques (or a combination): (1) by the use of a time-pulsed neutron source which emits neutrons of many energies simultaneously, combined with the time-of-flight technique to measure the velocities of the neutrons; this time-of-flight technique can be used for neutron measurements from 10^{-3} eV to about 200 MeV; (2) by the use of a beam of nearly

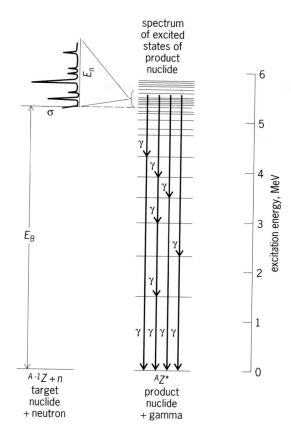

Energy-level diagram for the product nucleus $^AZ^*$ with mass number A and charge number Z. The asterisk emphasizes that the product nucleus is in an excited state, from which it returns to ground state by emitting gamma (γ) rays. Excitation energy is the sum of the energy of the neutron E_n, and the binding energy E_B of the neutron which has been added to the target nuclide.

monoenergetic neutrons whose energy can be varied in small steps approximately equal to the energy spread of the neutron beam; however, useful "monoenergetic" neutron sources are not available from about 10 eV to about 10 keV.

Neutron spectroscopy has yielded a mass of valuable information on nuclear systematics for almost all nuclides. The distribution of the spacings between nuclear levels and the average of these spacings have provided valuable tests for various nuclear theories. The properties of these levels, that is, the probabilities that they decay by neutron or gamma-ray emission, or by fission, and the averages and distribution of these probabilities have stimulated much theoretical effort.

In addition, knowledge of neutron cross sections is fundamental for the optimum design of thermal fission power reactors and fast neutron breeder reactors, as well as fusion power reactors now in the conceptual stage. Cross sections are needed for nuclear fuel materials such as ^{235}U or ^{239}Pu, for fertile materials such as ^{238}U, for structural materials such as iron and chromium, for coolants such as sodium, for moderators such as beryllium, and for shielding materials such as concrete. *See* NUCLEAR STRUCTURE.

[J.A.H.]

Newtonian fluid A fluid whose stress at each point is linearly proportional to its strain rate at that point. The concept was first deduced by Isaac Newton and is directly analogous to Hooke's law for a solid. All gases are newtonian, as are most common liquids such as water, hydrocarbons, and oils.

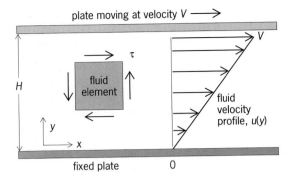

plate moving at velocity $V \longrightarrow$

H

τ

fluid element

fluid velocity profile, $u(y)$

V

y

x

fixed plate   0

A fluid sheared between two plates. The resulting strain rate equals V/H.

A simple example, often used for measuring fluid deformation properties, is the steady one-dimensional flow $u(y)$ between a fixed and a moving wall (see illustration). The no-slip condition at each wall forces the fluid into a uniform shear strain rate ϵ, given by Eq. (1), which is induced by a uniform shear stress τ. Here V is the speed of

$$\varepsilon = \frac{\partial u}{\partial y} = \frac{V}{H} \tag{1}$$

the moving wall, H is the perpendicular distance between the walls, and u is the fluid velocity at distance y from the fixed wall.

If the fluid is newtonian, the experimental plot of τ versus ε will be a straight line. The constant of proportionality is called the viscosity μ of the fluid, as stated in Eq. (2).

$$\tau = \mu\varepsilon \tag{2}$$

The viscosity coefficients of common fluids vary by several orders of magnitude. *See* FLUID FLOW; FLUIDS; VISCOSITY.   [F.M.Wh.]

Newton's laws of motion   Three fundamental principles which form the basis of classical, or newtonian, mechanics. They are stated as follows:

First law: A particle not subjected to external forces remains at rest or moves with constant speed in a straight line.

Second law: The acceleration of a particle is directly proportional to the resultant external force acting on the particle and is inversely proportional to the mass of the particle.

Third law: If two particles interact, the force exerted by the first particle on the second particle (called the action force) is equal in magnitude and opposite in direction to the force exerted by the second particle on the first particle (called the reaction force).

The newtonian laws have proved valid for all mechanical problems not involving speeds comparable with the speed of light and not involving atomic or subatomic particles. *See* DYNAMICS; FORCE; KINETICS (CLASSICAL MECHANICS).   [D.Wi.]

Nonlinear acoustics   The study of amplitude-dependent acoustical phenomena. The amplitude dependence is due to the nonlinear response of the medium in which the sound propagates, and not to the nonlinear behavior of the sound source. According to the linear theory of acoustics, increasing the level of a source by 10 dB results in precisely the same sound field as before, just 10 dB more intense. Linear theory also predicts that only frequency components radiated directly by the source can be present in the sound field. These principles do not hold in nonlinear acoustics. *See* NONLINEAR PHYSICS.

The extent to which nonlinear acoustical effects are strong or even significant depends on the competing influences of energy loss, frequency dispersion, geometric spreading, and diffraction. When conditions are such that nonlinear effects are strong, acoustic signals may experience substantial waveform distortion and changes in frequency content as they propagate, and shock waves may be present. Nonlinear acoustical effects occur in gases, liquids, and solids, and they are observed over a broad range of frequencies. Shock waves present in sonic booms and thunder claps are in the audio frequency range. Principles of nonlinear acoustics form the basis for procedures at megahertz frequencies used in medical ultrasound and nondestructive evaluation of materials. Nonlinearity can also induce changes in nonfluctuating properties of the medium. These include acoustic streaming, which is the steady fluid flow produced by the absorption of sound, and radiation pressure, which results in a steady force exerted by sound on its surroundings. *See* ACOUSTIC RADIATION PRESSURE; SHOCK WAVE; SONIC BOOM.

The principal feature that distinguishes nonlinear acoustics from nonlinear optics is that most acoustical media exhibit only weak dispersion, whereas media in which nonlinear optical effects arise exhibit strong dispersion. Dispersion is the dependence of propagation speed on frequency. In optical media, strong nonlinear wave interactions require that phase-matching conditions be satisfied, which can be accomplished only for several frequency components at one time. In contrast, all frequency components in a sound wave propagate at the same speed and are automatically phase-matched, which permits strong nonlinear interactions to occur among all components in the frequency spectrum. *See* NONLINEAR OPTICS.

Acoustic streaming is a nonlinear effect because the velocity of the flow depends quadratically on the amplitude of the sound, and the flow is not predicted by linear theory. Absorption due to viscosity and heat conduction results in a transfer of momentum from the sound field to the fluid. This momentum transfer manifests itself as steady fluid flow.

Acoustic streaming produced in sound beams is enhanced considerably when shocks develop. Shock formation generates a frequency spectrum rich in higher harmonics. Because thermoviscous absorption increases quadratically with frequency, attenuation of the wave, and therefore the streaming velocity, increases markedly following shock formation. Streaming is also generated in acoustic boundary layers formed by standing waves in contact with surfaces. Measurements of acoustic streaming have been used to determine the bulk viscosity coefficients of fluids. Thermoacoustic engines and refrigerators are adversely affected by heat transport associated with streaming. *See* THERMOACOUSTICS.

Phase conjugation refers to wavefront reversal, also called time reversal, at a single frequency. The latter terminologies more clearly describe this procedure. A waveform is captured by a phase conjugation device and reversed in such a way that it propagates back toward the source in the same way that it propagated toward the conjugator. Sound that is radiated from a point source and propagates through an inhomogeneous medium that introduces phase distortion in the wave field is thus retransmitted by the conjugator in such a way as to compensate for the phase distortion and to focus the wave back on the point source.

Phase conjugation is used to compensate for phase distortion in applications involving imaging and retargeting of waves on sources. The most successful techniques for acoustical phase conjugation are based on modulation of acoustical properties of a material that captures the incident sound wave. The modulation is twice the frequency of the incident sound wave, and it is induced by an electric field applied to piezoelectric material, or a magnetic field applied to magnetostrictive material. Often the modulated property of interest is the sound speed in the material. When the incident wave at frequency f propagates through a medium in which the sound speed fluctuates at frequency $2f$, parametric interaction generates a wave at the difference frequency f that propagates backward as though reversed in time. *See* MAGNETOSTRICTION; OPTICAL PHASE CONJUGATION; PIEZOELECTRICITY.

Phenomena associated with nonlinear acoustics have proved useful in both diagnostic and therapeutic applications of biomedical ultrasound. A very significant breakthrough in diagnostic imaging, especially for echocardiography and abdominal ultrasound imaging, is based on second-harmonic generation. Medical ultrasound imaging is performed at frequencies of several megahertz. Images constructed from the backscattered second-harmonic component have substantially reduced clutter and haze associated with the propagation of ultrasound through the outer layers of skin, which is the primary cause of phase aberrations. In another technique, microbubbles are injected into the bloodstream to enhance echoes backscattered from blood flow. The microbubbles are fabricated to make them resonant at diagnostic imaging frequencies, and they become strongly nonlinear oscillators when excited by ultrasound. Imaging is based on echoes at harmonics of the transmitted signal. Frequencies backscattered from the microbubbles differ from those in echoes coming from the surrounding tissue, which highlights the locations of the microbubbles and therefore of the blood flow itself.

A notable therapeutic application is lithotripsy, which refers to the noninvasive disintegration of kidney stones and gallstones with focused shock waves. Nonlinear acoustical effects in lithotripsy are associated not only with propagation of the shock wave but also with the generation of cavitation activity near the stones. Radiation of shock waves due to the collapse of cavitation bubbles is believed to be the dominant cause of stone breakup. An emerging therapeutic application, high-intensity focused ultrasound (HIFU), utilizes the heat dissipated by shock waves that develop in beams of focused ultrasound. The heating is so intense and localized that the potential exists for noninvasive cauterization of internal wounds and removal of tumors and scar tissue. *See* CAVITATION; ULTRASONICS. [M.F.H.]

Nonlinear optical devices Devices that use the fact that the polarization in any real medium is a nonlinear function of the optical field strength to implement various useful functions. The nonlinearities themselves can be grouped roughly into second-order and third-order. Materials that possess inversion symmetry typically exhibit only third-order nonlinearities, whereas materials without inversion symmetry can exhibit both second- and third-order nonlinearities. *See* CRYSTALLOGRAPHY; ELECTRIC SUSCEPTIBILITY; ELECTROMAGNETIC RADIATION; POLARIZATION OF DIELECTRICS.

Second-order devices. Devices based on the second-order nonlinearity involve three-photon (or three-wave) mixing. In this process, two photons are mixed together to create a third photon, subject to energy- and momentum-conservation constraints. Different names are ascribed to this mixing process, depending upon the relative magnitudes of the energies of the three photons. *See* CONSERVATION OF ENERGY; CONSERVATION OF MOMENTUM.

When the two beginning photons are of equal energy or frequency, the mixing process gives a single photon with twice the energy or frequency of the original ones. This mixing process is called second-harmonic generation. Second-harmonic generation is used often in devices where photons of visible frequency are desired but the available underlying laser system is capable of producing only infrared photons. For example, the neodymium-doped yttrium-aluminum-garnet (Nd:YAG) laser produces photons in the infrared with a wavelength of 1.06 micrometers. These photons are mixed in a crystal with a large second-order nonlinearity and proper momentum-conservation characteristics to yield green second-harmonic photons of 0.532-μm wavelength. Under different momentum-conservation constraints, a similar interaction can take place between two photon fields of different frequency, resulting in photons whose energy or frequency is the sum of those of the original photons. This process is called sum-frequency mixing. *See* LASER.

Optical parametric oscillation/amplification occurs when one of the two initial photons has the largest energy and frequency of the three. A high-energy photon and a low-energy photon mix to give a third photon with an energy equal to the difference between the two initial photons. If initially the third field amplitude is zero, it is possible to generate a third field from nothing; in this case the process is called optical parametric oscillation. If the third field exists but at a low level, it can be amplified through the optical parametric amplification process.

Third-order devices. Devices based on the third-order nonlinearity involve a process called four-photon (or four-wave) mixing. In this process, three photons are mixed together to create a fourth photon, subject to energy- and momentum-conservation constraints. The four-photon mixing nonlinearity is responsible for the existence of so-called self-action effects where the refractive index and absorption coefficient of a light field are modified by the light field's own presence, for third-harmonic generation and related processes, and for phase-conjugation processes.

In a medium with a third-order nonlinearity, the refractive index and absorption coefficient of a light field present in the medium are modified by the strength of the light intensity. Because the field effectively acts on itself, this interaction is termed a self-action effect. The momentum-conservation constraints are automatically satisfied because of the degenerate frequencies involved in the interaction. Such an interaction manifests itself by changing the total absorption experienced by the light field as well as by changing the velocity of propagation of the light field. *See* ABSORPTION OF ELECTROMAGNETIC RADIATION; REFRACTION OF WAVES.

There are many devices based on the self-action effects. A reverse saturable absorber becomes more opaque because of the nonlinear absorption (also called two-photon adsorption) that it manifests. Refractive-index changes can be used to change the transmission characteristics of resonant cavities and other structures by modifying the effective optical path length (the product of actual structure length times the effective refractive index for the structure) and shifting the cavity resonances to other frequencies. Several nonlinear optical switches have been proposed based upon this resonance-shifting phenomenon. *See* OPTICAL BISTABILITY.

In a third-harmonic generation process, three photons of like energy and frequency are mixed to yield a single photon with three times the energy and frequency of the initial photons. Applications of third-harmonic generation are typically in the areas of frequency upconversion.

Phase-conjugation devices make use of a property that third-order media possess whereby energy- and frequency-degenerate photons from two counterpropagating fields are mixed with an incoming photon to yield a photon with exactly the opposite propagation direction and conjugate phase. This phase-conjugate field will pass out of the nonlinear optical device in exactly the direction opposite to the incoming field. Such devices are used in phase-conjugate mirrors, mirrors which have the ability to cancel phase variation in a beam due to, for example, atmospheric turbulence. *See* ADAPTIVE OPTICS; OPTICAL PHASE CONJUGATION.

The suitability of available nonlinear optical materials is a critical factor in the development of nonlinear optical devices. For certain applications, silica glass fibers may be used. Because of the long propagation distances involved in intercontinental transmission systems, the small size of the optical nonlinearity in silica is not a drawback.

Other key materials are semiconductors [such as gallium arsenide (GaAs), zinc selenide (ZnSe), and indium gallium arsenide phosphide (InGaAsP)], certain organic polymeric films, hybrid materials such as semiconductor-doped glasses, and liquid crystals. *See* NONLINEAR OPTICS; OPTICAL MATERIALS.   [D.R.A.]

Nonlinear optics   A field of study concerned with the interaction of electromagnetic radiation and matter in which the matter responds in a nonlinear manner to the incident radiation fields. The nonlinear response can result in intensity-dependent variation of the propagation characteristics of the radiation fields or in the creation of radiation fields that propagate at new frequencies or in new directions. Nonlinear effects can take place in solids, liquids, gases, and plasmas, and may involve one or more electromagnetic fields as well as internal excitations of the medium. Most of the work done in the field has made use of the high powers available from lasers. The wavelength range of interest generally extends from the far-infrared to the vacuum ultraviolet, but some nonlinear interactions have been observed at wavelengths extending from the microwave to the x-ray ranges. *See* LASER.

Nonlinear materials. Nonlinear effects of various types are observed at sufficiently high light intensities in all materials. It is convenient to characterize the response of the medium mathematically by expanding it in a power series in the electric and magnetic fields of the incident optical waves. The linear terms in such an expansion give rise to the linear index of refraction, linear absorption, and the magnetic permeability of the medium, while the higher-order terms give rise to nonlinear effects. *See* ABSORPTION OF ELECTROMAGNETIC RADIATION; REFRACTION OF WAVES.

In general, nonlinear effects associated with the electric field of the incident radiation dominate over magnetic interactions. The even-order dipole susceptibilities are zero except in media which lack a center of symmetry, such as certain classes of crystals, certain symmetric media to which external forces have been applied, or at boundaries between certain dissimilar materials. Odd-order terms can be nonzero in all materials regardless of symmetry. Generally the magnitudes of the nonlinear susceptibilities decrease rapidly as the order of the interaction increases. Second- and third-order effects have been the most extensively studied of the nonlinear interactions, although effects up to order 30 have been observed in a single process. In some situations, multiple low-order interactions occur, resulting in a very high effective order for the overall nonlinear process. For example, ionization through absorption of effectively 100 photons has been observed. In other situations, such as dielectric breakdown or saturation of absorption, effects of different order cannot be separated, and all orders must be included in the response. *See* ELECTRIC SUSCEPTIBILITY; POLARIZATION OF DIELECTRICS.

Stimulated scattering. Light can scatter inelastically from fundamental excitations in the medium, resulting in the production of radiation at a frequency that is shifted from that of the incident light by the frequency of the excitation involved. The difference in photon energy between the incident and scattered light is accounted for by excitation or deexcitation of the medium. Some examples are Brillouin scattering from acoustic vibrations; various forms of Raman scattering involving molecular rotations or vibrations, electronic states in atoms or molecules, lattice vibrations or spin waves in solids, spin flips in semiconductors, and electron plasma waves in plasmas; Rayleigh scattering involving density or entropy fluctuations; and scattering from concentration fluctuations in gases. *See* SCATTERING OF ELECTROMAGNETIC RADIATION.

At the power levels available from pulsed lasers, the scattered light experiences exponential gain, and the process is then termed stimulated, in analogy to the process of stimulated emission in lasers. In stimulated scattering, the incident light can be almost completely converted to the scattered radiation. Stimulated scattering has been observed for all of the internal excitations listed above. The most widely used of these processes are stimulated Raman scattering and stimulated Brillouin scattering.

Self-action and related effects. Nonlinear polarization components at the same frequencies as those in the incident waves can result in effects that change the index of refraction or the absorption coefficient, quantities that are constants in linear optical theory. For example, propagation through optical fibers can involve several nonlinear optical interactions. Self-phase modulation resulting from the nonlinear index can be used to spread the spectrum, and subsequent compression with diffraction gratings and prisms can be used to reduce the pulse duration. The shortest optical pulses, with durations of the order of 6 femtoseconds, have been produced in this manner. Linear dispersion in fibers causes pulses to spread in duration and is one of the major limitations on data transmission through fibers. Dispersive pulse spreading can be minimized with solitons, which are specially shaped pulses that propagate long distances without spreading. They are formed by a combined interaction of spectral broadening due to the nonlinear refractive index and anomalous dispersion found in certain parts of the spectrum. *See* SOLITON.

Coherent effects. Another class of effects involves a coherent interaction between the optical field and an atom in which the phase of the atomic wave functions is preserved during the interaction. These interactions involve the transfer of a significant fraction of the atomic population to an excited state. As a result, they cannot be described with the simple perturbation expansion used for the other nonlinear optical effects. Rather they require that the response be described by using all powers of the incident fields. These effects are generally observed only for short light pulses, of the order of several nanoseconds or less. In one interaction, termed self-induced transparency, a pulse of light of the proper shape, magnitude, and duration can propagate unattenuated in a medium which is otherwise absorbing.

Other coherent effects involve changes of the propagation speed of a light pulse or production of a coherent pulse of light, termed a photon echo, at a characteristic time after two pulses of light spaced apart by a time interval have entered the medium. Still other coherent interactions involve oscillations of the atomic polarization, giving rise to effects known as optical nutation and free induction decay. Two-photon coherent effects are also possible.

Nonlinear spectroscopy. The variation of the nonlinear susceptibility near the resonances that correspond to sum- and difference-frequency combinations of the input frequencies forms the basis for various types of nonlinear spectroscopy which allow study of energy levels that are not normally accessible with linear optical spectroscopy.

Nonlinear spectroscopy can be performed with many of the interactions discussed earlier. Multiphoton absorption spectroscopy can be performed by using two strong laser beams, or a strong laser beam and a weak broadband light source. If two counter-propagating laser beams are used, spectroscopic studies can be made of energy levels in gases with spectral resolutions much smaller than the Doppler limit. Nonlinear optical spectroscopy has been used to identify many new energy levels with principal quantum numbers as high as 150 in several elements. *See* RESONANCE IONIZATION SPECTROSCOPY; RYDBERG ATOM.

Many types of four-wave mixing interactions can also be used in nonlinear spectroscopy. The most widespread of these processes, termed coherent anti-Stokes Raman spectroscopy (CARS), offers the advantage of greatly increased signal levels over linear Raman spectroscopy for the study of certain classes of materials.

Phase conjugation. Optical phase conjugation is an interaction that generates a wave that propagates in the direction opposite to a reference, or signal, wave, and has the same spatial variations in intensity and phase as the original signal wave, but with the sense of the phase variations reversed. Several nonlinear interactions are used to produce phase conjugation.

Optical phase conjugation allows correction of optical distortions that occur because of propagation through a distorting medium. This process can be used for improvement of laser-beam quality, optical beam combining, correction of distortion because of mode dispersion in fibers, and stabilized aiming. It can also be used for neural networks that exhibit learning properties. *See* OPTICAL PHASE CONJUGATION.   [J.F.R.]

Photorefractive effect. The photorefractive effect occurs in many electrooptic materials. A change in the index of refraction in a photorefractive medium arises from the redistribution of charge that is induced by the presence of light. Charge carriers that are trapped in impurity sites in a photorefractive medium are excited into the material's conduction band when exposed to light. The charges migrate in the conduction band until they become retrapped at other sites. The charge redistribution produces an electric field that in turn produces a spatially varying index change through the electrooptic effect in the material. Unlike most other nonlinear effects, the index change of the photorefractive effect is retained for a time in the absence of the light and thus may be used as an optical storage mechanism. Storage times range from milliseconds to months or years, depending upon the material and the methods employed. *See* TRAPS IN SOLIDS.

Photorefractive materials are often used for holographic storage. In this case, the index change mimics the intensity interference pattern of two beams of light. Over 500 holograms have been stored in the volume of a single crystal of iron-doped lithium niobate. *See* HOLOGRAPHY.

Photorefractive materials are typically sensitive to very low light levels. The photorefractive effect is, however, extremely slow by the standards of optical nonlinearity. Because of their sensitivity, photorefractive materials are increasingly used for image and optical-signal processing applications. *See* NONLINEAR OPTICAL DEVICES.   [D.Z.A.]

Nonlinear physics   The study of situations where, in a general sense, cause and effect are not proportional to each other; in other words, if the measure of what is considered to be the cause is doubled, the measure of its effect is not simply twice as large. Many examples have been known in physics for a long time, and they seemed well understood. Over the last few decades, however, physicists have noticed that this lack of proportionality in some of the basic laws of physics often leads to unexpected complications, if not to outright contradictions. Thus, the term nonlinear physics refers more narrowly to these developments in the understanding of physical reality.

Linearity in nonlinear systems. When a large number of particles starts out in a condition of stable equilibrium, the result of small external forces is well-coordinated vibrations of the whole collection, for example, the vibrations of a violin string, or of the electric current in an antenna. Each collective motion acts like an independent oscillator, each with its own frequency. In more complicated systems, many vibrational modes can be active simultaneously without mutual interference. A large dynamical system is, therefore, described in terms of its significant degrees of freedom, thought to be only loosely coupled. The motion of any part of the whole becomes multiperiodic; for example, a water molecule has bending and stretching vibrations with different frequencies, and both are coupled with the rotational motion of the whole molecule. *See* DEGREE OF FREEDOM (MECHANICS); MOLECULAR STRUCTURE AND SPECTRA; VIBRATION.

Failure of perturbation theory. H. Poincaré discovered at the end of the nineteenth century that for many problems this perturbation theory is not entirely satisfactory. He showed, in the case of the Moon's motion around the Earth, that the disturbance by the Sun is strong enough that this standard mathematical procedure fails. The main culprits are resonances, which occur when the frequencies of different degrees of freedom are combined through their nonlinear coupling. A key nonperturbative phenomenon is known to engineers as phase lock: When different frequencies arise in simple multiples of one another, the whole dynamical system falls into a dynamical trap; and for a continuous range of initial conditions, the interaction changes the frequencies of the individual degrees of freedom sufficiently to "lock" the motion into the resonance. *See* RESONANCE (ACOUSTICS AND MECHANICS).

KAM theorem. In the 1950s, A. N. Kolmogoroff provided a first account of how the addition of a weak coupling generates chaotic regions and islands in phase space. This problem was later worked out in detail by V. Arnold and J. Moser to yield the KAM theorem. This theorem gives detailed information about the loss of the regular structure as the strength of the coupling increases. It does not say anything, however, about the trajectories in the newly created areas of chaotic behavior. These further investigations are the main goal of such fields as chaos or complexity. The impact of Poincaré's general arguments and the KAM theorem reaches into every area of nonlinear physics. The oldest among the areas is hydrodynamics, where the phenomenon of turbulent flow has so far resisted any effective control. This is what makes weather prediction so difficult. Signal propagation along the nerves and transmission of pulses through synaptic connections are other well-known nonlinear processes. *See* CHAOS; NONLINEAR ACOUSTICS; NONLINEAR OPTICS.

[M.C.G.]

Non-newtonian fluid   A fluid that departs from the classic linear newtonian relation between stress and shear rate. In a strict sense, a fluid is any state of matter that is not a solid, and a solid is a state of matter that has a unique stress-free state. A conceptually simpler definition is that a fluid is capable of attaining the shape of its container and retaining that shape for all time in the absence of external forces. Therefore, fluids encompass a wide variety of states of matter including gases and liquids as well as many more esoteric states (for example, plasmas, liquid crystals, and foams). *See* FLUIDS; PLASMA (PHYSICS).

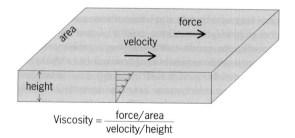

$$\text{Viscosity} = \frac{\text{force}/\text{area}}{\text{velocity}/\text{height}}$$

Fig. 1. Steady shear flow of a fluid between a fixed plate and a parallel plate, illustrating the concept of viscosity.

A newtonian fluid is one whose mechanical behavior is characterized by a single function of temperature, the viscosity, a measure of the "slipperiness" of the fluid. For the example of Fig. 1, where a fluid is sheared between a fixed plate and a moving plate, the viscosity is given by Eq. (1). Thus, as the viscosity of a fluid increases, it requires a

$$\text{Viscosity} = \frac{\text{force}/\text{area}}{\text{velocity}/\text{height}} \tag{1}$$

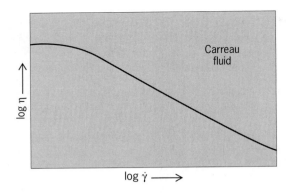

Fig. 2. Typical dependence of the viscosity (η) on shear rate ($\dot{\gamma}$) for a non-newtonian fluid (Carreau model).

larger force to move the top plate at a given velocity. For simple, newtonian fluids, the viscosity is a constant dependent on only temperature; but for non-newtonian fluids, the viscosity can change by many orders of magnitude as the shear rate (velocity/height in Fig. 1) changes. Typically, the viscosity (η) of these fluids is given as a function of the shear rate ($\dot{\gamma}$). A common dependence for this function is given in Fig. 2. For other non-newtonian fluids, the viscosity might increase as the shear rate increases (shear-thickening fluids). *See* NEWTONIAN FLUID; VISCOSITY.

Many of the fluids encountered in everyday life (such as water, air, gasoline, and honey) are adequately described as being newtonian, but there are even more that are not. Common examples include mayonnaise, peanut butter, toothpaste, egg whites, liquid soaps, and multigrade engine oils. Other examples such as molten polymers and slurries are of considerable technological importance. A distinguishing feature of many non-newtonian fluids is that they have microscopic or molecular-level structures that can be rearranged substantially in flow.

Our intuitive understanding of how fluids behave and flow is built primarily from observations and experiences with newtonian fluids. However, non-newtonian fluids display a rich variety of behavior that is often in dramatic contrast to these expectations. For example, an intuitive feel for the slipperiness of fluids can be gained from rubbing them between the fingers. Furthermore, the slipperiness of water, experienced in this way, is expected to be the same as the slipperiness of automobile tires on a wet road. However, the slipperiness (viscosity) of many non-newtonian fluids changes a great deal depending on how fast they move or the forces applied to them.

Intuitive expectations for how the surface of a fluid will deform when the fluid is stirred (with the fluid bunching up at the wall of the container) are also in marked contrast to the behavior of non-newtonian fluids. When a cylindrical rod is rotated inside a container of a newtonian fluid, centrifugal forces cause the fluid to be higher at the wall. However, for non-newtonian fluids, the normal stress differences cause the fluid to climb the rod; this is called the Weissenberg effect. Intuitive understanding about the motion of material when the flow of a fluid is suddenly stopped, for example, by turning off a water tap, is also notably at odds with the behavior of non-newtonian fluids. *See* CENTRIFUGAL FORCE.

A non-newtonian fluid also displays counterintuitive behavior when it is extruded from an opening. A newtonian fluid tapers to a smaller cross section as it leaves the opening, but the cross section for a non-newtonian fluid first increases before it eventually tapers. This phenomenon is called die swell.

When a newtonian fluid is siphoned and the fluid level goes below the entrance to the siphon tube, the siphoning action stops. For many non-newtonian fluids, however, the siphoning action continues as the fluid climbs from the surface and continues to enter the tube. This phenomenon is called the tubeless siphon.

Perhaps the most striking behavior of non-newtonian fluids is a consequence of their viscoelasticity. Solids can be thought of as having perfect memory. If they are deformed through the action of a force, they return to their original shape when the force is removed. This happens when a rubber ball bounces; the ball is deformed as it hits a surface, but the rubber remembers its undeformed spherical shape. Recovery of the shape causes the ball to bounce back. In contrast, newtonian fluids have no memory; when a force is removed, they retain their condition at the time the force is removed (or continue moving as the result of inertia). When a newtonian fluid is dropped onto a surface, it does not bounce. Non-newtonian fluids are viscoelastic in the sense that they have fading memory. If a force is removed shortly after it is applied, the fluid will remember its undeformed shape and return toward it. However, if the force is applied on the fluid for a long time, the fluid will eventually forget its undeformed shape. If a sample of a non-newtonian fluid is dropped onto a surface, it will bounce like a ball. However, if the fluid is simply placed on the surface, it will flow smoothly. Viscoelasticity is frequently the cause of many of the secondary flows that are observed for non-newtonian fluids. These are fluid motions that are small for newtonian fluids (for example, swirling motions) but can become dominant for non-newtonian fluids.

Analysis of fluid flow operations is typically performed by examining local conservation relations—conservation of mass, momentum (Newton's second law), and energy. This analysis requires material-specific information (for example, the relation between density, pressure, and temperature) that is collectively known as constitutive relations. The science devoted to obtaining suitable constitutive equations for description of the behavior of non-newtonian fluids is called rheology. The most important constitutive equation for fluid mechanics is that relating the stress in the fluid to the kinematics of the motion (that is, the velocity, the derivatives of the velocity with respect to position, and the time history of the velocity).

Although the non-newtonian behavior of many fluids has been recognized for a long time, the science of rheology is, in many respects, still in its infancy, and new phenomena are constantly being discovered and new theories proposed. Advancements in computational techniques are making possible much more detailed analyses of complex flows and more sophisticated simulations of the structural and molecular behavior that gives rise to non-newtonian behavior. Engineers, chemists, physicists, and mathematicians are actively pursuing research in rheology, particularly as more technologically important materials are found to display non-newtonian behavior. *See* RHEOLOGY.

[J.M.Wie.]

Nuclear binding energy

Nuclear binding energy   The amount by which the mass of an atom is less than the sum of the masses of its constituent protons, neutrons, and electrons expressed in units of energy. This energy difference accounts for the stability of the atom. In principle, the binding energy is the amount of energy which was released when the several atomic constituents came together to form the atom. Most of the binding energy is associated with the nuclear constituents (protons and neutrons) or nucleons, and it is customary to regard this quantity as a measure of the stability of the nucleus alone. *See* NUCLEAR STRUCTURE.

A widely used term, the binding energy (BE) per nucleon, is defined by the equation below, where $_ZM^A$ represents the mass of an atom of mass number A and atomic number

$$BE/nucleon = \frac{[ZH + (A - Z)n - _ZM^A]c^2}{A}$$

Z, H and n are the masses of the hydrogen atom and neutron, respectively, and c is the velocity of light. The binding energies of the orbital electrons, here practically neglected, are not only small, but increase with Z in a gradual manner; thus the BE/nucleon gives an accurate picture of the variations and tends in nuclear stability.

The binding energy, when expressed in mass units, is known as the mass defect, a term sometimes incorrectly applied to quantity $M - A$, where M is the mass of the atom. *See* MASS DEFECT.

The term binding energy is sometimes also used to describe the energy which must be supplied to a nucleus in order to remove a specified particle to infinity, for example, a neutron, proton, or alpha particle. A more appropriate term for this energy is the separation energy. This quantity varies greatly from nucleus to nucleus and from particle to particle. For example, the binding energies for a neutron, a proton, and a deuteron in ^{16}O are 15.67, 12.13, and 20.74 MeV, respectively, while the corresponding energies in ^{17}O are 4.14, 13.78, and 14.04 MeV, respectively. The usual order of neutron or proton separation energy is 7–9 MeV for most of the periodic table. [H.E.D.; D.H.W.]

Nuclear fission An extremely complex nuclear reaction representing a cataclysmic division of an atomic nucleus into two nuclei of comparable mass. This rearrangement or division of a heavy nucleus may take place naturally (spontaneous fission) or under bombardment with neutrons, charged particles, gamma rays, or other carriers of energy (induced fission). Although nuclei with mass number A of approximately 100 or greater are energetically unstable against division into two lighter nuclei, the fission process has a small probability of occurring, except with the very heavy elements. Even for these elements, in which the energy release is of the order of 200 megaelectronvolts, the lifetimes against spontaneous fission are reasonably long. *See* NUCLEAR REACTION.

Liquid-drop model. The stability of a nucleus against fission is most readily interpreted when the nucleus is viewed as being analogous to an incompressible and charged liquid drop with a surface tension. Long-range Coulomb forces between protons act to disrupt the nucleus, whereas short-range nuclear forces, idealized as a surface tension, act to stabilize it. The degree of stability is then the result of a delicate balance between the relatively weak electromagnetic forces and the strong nuclear forces. Although each of these forces results in potentials of several hundred megaelectronvolts, the height of a typical barrier against fission for a heavy nucleus, because they are of opposite sign but do not quite cancel, is only 5 or 6 MeV. Investigators have used this charged liquid-drop model with great success in describing the general features of nuclear fission and also in reproducing the total nuclear binding energies. *See* NUCLEAR BINDING ENERGY; NUCLEAR STRUCTURE.

Shell corrections. The general dependence of the potential energy on the fission coordinate representing nuclear elongation or deformation for a heavy nucleus such as ^{240}Pu is shown in Fig. 1. The expanded scale used in this figure shows the large decrease in energy of about 200 MeV as the fragments separate to infinity. It is known that ^{240}Pu is deformed in its ground state, which is represented by the lowest minimum of 7–1813 MeV near zero deformation. This energy represents the total nuclear binding energy when zero of potential energy is the energy of the individual nucleons at a separation of infinity. The second minimum to the right of zero deformation illustrates

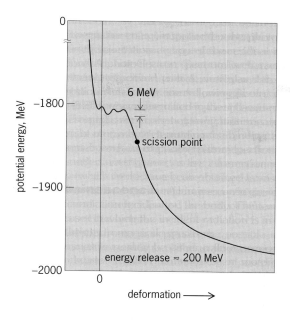

Fig. 1. Plot of the potential energy in MeV as a function of deformation for the nucleus ^{240}Pu. (*After M. Bolsteli et al., New calculations of fission barriers for heavy and superheavy nuclei, Phys. Rev., 5C:1050–1077, 1972*)

structure introduced in the fission barrier by shell corrections, that is, corrections dependent upon microscopic behavior of the individual nucleons, to the liquid-drop mass. Although shell corrections introduce small wiggles in the potential-energy surface as a function of deformation, the gross features of the surface are reproduced by the liquid-drop model. Since the typical fission barrier is only a few megaelectronvolts, the magnitude of the shell correction need only be small for irregularities to be introduced into the barrier. This structure is schematically illustrated for a heavy nucleus by the double-humped fission barrier in Fig. 2, which represents the region to the right of zero

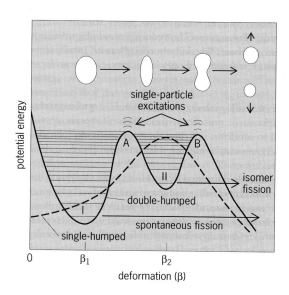

Fig. 2. Schematic plots of single-humped fission barrier of liquid-drop model and double-humped barrier introduced by shell corrections. Humps at *A* and *B* result in minima in potential energy at deformation of β_1 and β_2. States in these wells are designated class I and class II, respectively. (*After J. R. Huizenga, Nuclear fission revisited, Science, 168:1405–1413, 1979*)

deformation in Fig. 1 on an expanded scale. The fission barrier has two maxima and a rather deep minimum in between. For comparison, the single-humped liquid-drop barrier is also schematically illustrated. The transition in the shape of the nucleus as a function of deformation is schematically represented in the upper part of the figure.

Experimental consequences. The observable consequences of the double-humped barrier have been reported in numerous experimental studies. In the actinide region more than 30 spontaneously fissionable isomers have been discovered between uranium and berkelium, with half-lives ranging from 10^{-11} to 10^{-2} s. These decay rates are faster by 20 to 30 orders of magnitude than the fission half-lives of the ground states, because of the increased barrier tunneling probability (Fig. 2). Several cases in which excited states in the second minimum decay by fission are also known. Normally these states decay within the well by gamma decay; however, if there is a hindrance in gamma decay due to spin, the state (known as a spin isomer) may undergo fission instead.

Fission probability. The cross section for particle-induced fission $\sigma(y, f)$ represents the cross section for a projectile y to react with a nucleus and produce fission, as shown by the equation below. The quantities $\sigma_R(y)$, Γ_f and Γ_t are the total reaction

$$\sigma(y, f) = \sigma_R(y)(\Gamma_f / \Gamma_t)$$

across sections for the incident particle y, the fission width, and the total level width, respectively, where $\Gamma_t = \Gamma_f + \Gamma_n + \Gamma_y + \cdots$ is the sum of all partial-level widths. All the quantities in the above equation are energy-dependent.

When the incoming neutron has low energy, the likelihood of reaction is substantial only when the energy of the neutron is such as to form a compound nucleus in one or another of its resonance levels. The requisite sharpness of the "tuning" of the energy is specified by the total level width Γ. The nuclei ^{233}U, ^{235}U, and ^{239}Pu have a very large cross section to take up a slow neutron and undergo fission because both their absorption cross section and their probability for decay by fission are large. The probability for fission decay is high because the binding energy of the incident neutron is sufficient to raise the energy of the compound nucleus above the fission barrier. The very large, slow neutron fission cross sections of these isotopes make them important fissile materials in a chain reactor. *See* CHAIN REACTION (PHYSICS).

Postscission phenomena. After the nuclear fragments are separated, they are further accelerated as the result of the large Coulomb repulsion. The initially deformed fragments collapse to their equilibrium shapes, and the excited primary fragments lose energy by evaporating neutrons. After neutron emission, the fragments lose the remainder of their energy by gamma radiation, with a lifetime of about 10^{-11} s. The variation of neutron yield with fragment mass is directly related to the fragment excitation energy. Minimum neutron yields are observed for nuclei near closed shells because of the resistance to deformation of nuclei with closed shells. Maximum neutron yields occur for fragments that are "soft" toward nuclear deformation.

After the emission of the prompt neutrons and gamma rays, the resulting fission products are unstable against β-decay. For example, in the case of thermal neutron fission of ^{235}U, each fragment undergoes on the average about three β-decays before it settles down to a stable nucleus. For selected fission products (for example, ^{87}Br and ^{137}I) β-decay leaves the daughter nucleus with excitation energy exceeding its neutron binding energy. The resulting delayed neutrons amount, for thermal neutron fission of ^{235}U, to about 0.7% of all the neutrons given off in fission. Though small in number, they are quite important in stabilizing nuclear chain reactions against sudden minor fluctuations in reactivity. *See* NEUTRON.   [J.R.Hu.]

Nuclear fusion   One of the primary nuclear reactions, the name usually designating an energy-releasing rearrangement collision which can occur between various isotopes of low atomic number. *See* NUCLEAR REACTION.

Interest in the nuclear fusion reaction arises from the expectation that it may someday be used to produce useful power, from its role in energy generation in stars, and from its use in the fusion bomb. Since a primary fusion fuel, deuterium, occurs naturally and is therefore obtainable in virtually inexhaustible supply, solution of the fusion power problem would permanently solve the problem of the present rapid depletion of chemically valuable fossil fuels. As a power source, the lack of radioactive waste products from the fusion reaction is another argument in its favor as opposed to the fission of uranium. *See* NUCLEAR FISSION.

In a nuclear fusion reaction the close collision of two energy-rich nuclei results in a mutual rearrangement of their nucleons (protons and neutrons) to produce two or more reaction products, together with a release of energy. The energy usually appears in the form of kinetic energy of the reaction products, although when energetically allowed, part may be taken up as energy of an excited state of a product nucleus. In contrast to neutron-produced nuclear reactions, colliding nuclei, because they are positively charged, require a substantial initial relative kinetic energy to overcome their mutual electrostatic repulsion so that reaction can occur. This required relative energy increases with the nuclear charge Z, so that reactions between low-Z nuclei are the easiest to produce. The best known of these are the reactions between the heavy isotopes of hydrogen, deuterium, and tritium.

Nuclear fusion reactions can be self-sustaining if they are carried out at a very high temperature. That is to say, if the fusion fuel exists in the form of a very hot ionized gas of stripped nuclei and free electrons termed a plasma, the agitation energy of the nuclei can overcome their mutual repulsion, causing reactions to occur. This is the mechanism of energy generation in the stars and in the fusion bomb. It is also the method envisaged for the controlled generation of fusion energy.

The cross sections (effective collisional areas) for many of the simple nuclear fusion reactions have been measured with high precision. It is found that the cross sections generally show broad maxima as a function of energy and have peak values in the general range of 0.01 barn (1 barn = 10^{-24} cm^2) to a maximum value of 5 barns, for the deuterium-tritium (D-T) reaction. The energy releases of these reactions can be readily calculated from the mass difference between the initial and final nuclei or determined by direct measurement.

Some of the important simple fusion reactions, their reaction products, and their energy releases are:

$$D + D \rightarrow He^3 + n + 3.25 \text{ MeV}$$
$$D + D \rightarrow T + p + 4.0 \text{ MeV}$$
$$T + D \rightarrow He^4 + n + 17.6 \text{ MeV}$$
$$He^3 + D \rightarrow He^4 + p + 18.3 \text{ MeV}$$
$$Li^6 + D \rightarrow 2He^4 + 22.4 \text{ MeV}$$
$$Li^7 + p \rightarrow 2He^4 + 17.3 \text{ MeV}$$

If it is remembered that the energy release in the chemical reaction in which hydrogen and oxygen combine to produce a water molecule is about 1 eV per reaction, it will be seen that, gram for gram, fusion fuel releases more than 1,000,000 times as much energy as typical chemical fuels.   [R.F.P.]

Nuclear isomerism   The existence of excited states of atomic nuclei with un-usually long lifetimes. If the lifetime of a specific excited state is unusually long, com-pared with the lifetimes of other excited states in the same nucleus, the state is said to be isomeric. The definition of the boundary between isomeric and normal decays is arbitrary, and the term is therefore used loosely. *See* EXCITED STATE; PARITY (QUANTUM MECHANICS); SPIN (QUANTUM MECHANICS).

The predominant decay mode of excited nuclear states is by γ-ray emission. The rate at which this process occurs is determined largely by the spins, parities, and excitation energies of the decaying state and of those to which it is decaying. In particular, the rate is extremely sensitive to the difference in the spins of initial and final states and to the difference in excitation energies. Both extremely large spin differences and extremely small energy differences can result in a slowing of the γ-ray emission by many orders of magnitude, resulting in some excited states having unusually long lifetimes and therefore being termed isomeric.

In addition to spin isomers, two other types of isomers have been identified. The first of these arises from the fact that some excited nuclear states represent a drastic change in shape of the nucleus from the shape of the ground state. In many cases this extremely deformed shape displays unusual stability, and states with this shape are therefore isomeric. A particularly important class of these shape isomers is observed in the decay of heavy nuclei by fission, and the study of such fission isomers has been the subject of intensive effort. *See* NUCLEAR FISSION.

A more esoteric form of isomer has also been observed, the so-called pairing isomer which results from differences in the microscopic motions of the constituent nucleons in the nucleus. A state of this type has a quite different character from the ground state of the nucleus, and is therefore also termed isomeric. *See* NUCLEAR STRUCTURE.   [R.Bet.]

Nuclear molecule   A quasistable entity of nuclear dimensions formed in nu-clear collisions and comprising two or more discrete nuclei that retain their identities and are bound together by strong nuclear forces. Whereas the stable molecules of chemistry and biology consist of atoms bound through various electronic mechanisms, nuclear molecules do not form in nature except possibly in the hearts of giant stars; this simply reflects the fact that all nuclei carry positive electrical charges, and that under all natural conditions the long-range electrostatic repulsion prevents nuclear components from coming within the grasp of the short-range attractive nuclear force which could provide molecular binding. But in energetic collisions this electrostatic repulsion can be overcome. *See* NUCLEAR STRUCTURE.   [D.A.B.]

Nuclear moments   Intrinsic properties of atomic nuclei: electric moments result from deviations of the nuclear charge distribution from spherical symmetry; magnetic moments are a consequence of the intrinsic spin and the rotational motion of nucleons within the nucleus. The classical definitions of the magnetic and electric multipole mo-ments are written in general in terms of multipole expansions. *See* NUCLEAR STRUCTURE; SPIN (QUANTUM MECHANICS).

In special cases nuclear moments can be measured by direct methods involving the interaction of the nucleus with an external magnetic field or with an electric field gradi-ent produced by the scattering of high-energy charged particles. In general, however, nuclear moments manifest themselves through the hyperfine interaction between the nuclear moments and the fields or field gradients produced by either the atomic elec-trons' currents and spins, or the molecular or crystalline electronic and lattice structures. *See* HYPERFINE STRUCTURE.   [N.Ko.]

Nuclear orientation   The directional ordering of an assembly of nuclear spins *I* with respect to some axis in space. Under normal conditions nuclei are not oriented; that is, all directions in space are equally probable. For a system of nuclear spins with rotational symmetry about an axis, the degree of orientation is completely characterized by the relative populations a_m of the $2I + 1$ magnetic sublevels $m \ (= I, I - 1, \ldots, -I)$.

Nuclear orientation can be achieved in various ways. The most obvious way is to modify the energies of the $2I + 1$ magnetic sublevels so as to remove their degeneracy and thereby change the populations of these sublevels. The spin degeneracy can be removed by a magnetic field interacting with the nuclear magnetic dipole moment, or by an inhomogeneous electric field interacting with the nuclear electric quadrupole moment. Significant differences in the populations of the sublevels can be established by cooling the nuclear sample to low temperatures. This means of producing nuclear orientation is called the static method. In contrast, there is the dynamic method, which is related to optical pumping in gases. There are other ways to produce oriented nuclei; for example, in a nuclear reaction such as the capture of polarized neutrons (produced by magnetic scattering) by unoriented nuclei. *See* Dynamic nuclear polarization; Optical pumping.

Oriented nuclei have been used to measure nuclear properties, for example, magnetic dipole and electric quadrupole moments, spins, parities, and mixing ratios of nuclear states. Oriented nuclei have been used to examine some of the fundamental properties of nuclear forces, for example, nonconservation of parity in the weak interaction. Measurement of hyperfine fields, electric-field gradients, and other properties relating to the environment of the nucleus have been made by using oriented nuclei. Nuclear orientation thermometry is one of the few sources of a primary temperature scale at low temperatures. Oriented nuclear targets used in conjunction with beams of polarized and unpolarized particles have proved very useful in examining certain aspects of the nuclear force. *See* Low-temperature thermometry; Nuclear moments; Nuclear structure; Parity (quantum mechanics).   [H.Mar.]

Nuclear physics   The discipline involving the structure of atomic nuclei and their interactions with each other, with their constituent particles, and with the whole spectrum of elementary particles that is provided by very large accelerators. The nuclear domain occupies a central position between the atomic range of forces and sizes and those of elementary-particle physics, characteristically within the nucleons themselves. As the only system in which all the known natural forces can be studied simultaneously, it provides a natural laboratory for the testing and extending of many fundamental symmetries and laws of nature. Containing a reasonably large, yet manageable number of strongly interacting components, the nucleus also occupies a central position in the universal many-body problem of physics. *See* Atomic nucleus; Atomic structure and spectra; Elementary particle; Symmetry laws (physics).

Nuclear physics is unique in the extent to which it merges the most fundamental and the most applied topics. Its instrumentation has found broad applicability throughout science, technology, and medicine; nuclear engineering and nuclear medicine are two very important areas of applied specialization.

Nuclear chemistry, certain aspects of condensed matter and materials science, and nuclear physics together constitute the broad field of nuclear science; outside the United States and Canada elementary particle physics is frequently included in this more general classification. *See* Analog states; Fundamental interactions; Isotope; Nuclear fission; Nuclear fusion; Nuclear isomerism; Nuclear moments; Nuclear reaction; Nuclear spectra; Nuclear

STRUCTURE; PARTICLE ACCELERATOR; PARTICLE DETECTOR; RADIOACTIVITY; SCATTERING EX-
PERIMENTS (NUCLEI); WEAK NUCLEAR INTERACTIONS. [D.A.B.]

Nuclear quadrupole resonance
A selective absorption phenomenon ob-
servable in a wide variety of polycrystalline compounds containing nonspherical atomic
nuclei when placed in a magnetic radio-frequency field. Nuclear quadrupole resonance
(NQR) is very similar to nuclear magnetic resonance (NMR), and was originated as an
inexpensive (no stable homogeneous large magnetic field is required) alternative way
to study nuclear moments. It later gained a modest popularity. *See* MAGNETIC RESO-
NANCE.

In the simplest case, for example, ^{35}Cl in solid Cl_2, NQR is associated with the
precession of the angular momentum I (and the nuclear magnetic dipole moment μ)
of the nucleus, depicted in the illustration as a flat ellipsoid of rotation, around the
symmetry axis (taken as the z axis) of the Cl_2 molecule fixed in the crystalline solid.
The precession, with constant angle θ between the nuclear axis and symmetry axis
of the molecule, is due to the torque which the inhomogeneous molecular electric
field exerts on the nucleus of electric quadrupole moment eQ. The absorption occurs
classically when the frequency of the rf field and that of the precessing motion of the
angular momentum coincide.

NQR spectra have been observed in the approximate range 1–1000 MHz. Most of
the NQR work has been on molecular crystals. For such crystals the coupling constants
found do not differ very much from those measured for the isolated molecules in
microwave spectroscopy. The most precise nuclear information which may be extracted
from NQR data are quadrupole moment ratios of isotopes of the same element. If values
for the axial gradient of the molecular electric field can be estimated from atomic fine
structure data, then fair values of the quadrupole moment may be obtained. However,
it has also proved very productive to use the quadrupole nucleus as a probe of bond

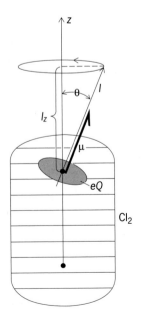

Interaction of ^{35}Cl nucleus with
the electric field of a Cl_2
molecule.

character and orientation and crystalline electric fields and lattice sites, and extensive data have been accumulated in this area. *See* MICROWAVE SPECTROSCOPY.   [H.De.]

Nuclear radiation   All particles and radiations emanating from an atomic nucleus due to radioactive decay and nuclear reactions. Thus the criterion for nuclear radiations is that a nuclear process is involved in their production. The term was originally used to denote the ionizing radiations observed from naturally occurring radioactive materials. These radiations were alpha rays (energetic helium nuclei), beta rays (negative electrons), and gamma rays (electromagnetic radiation with wavelength much shorter than visible light). *See* ALPHA PARTICLES; BETA PARTICLES; GAMMA RAYS.

Nuclear radiations have traditionally been considered to be of three types based on the manner in which they interact with matter as they pass through it. These are the charged heavy particles with masses comparable to that of the nuclear mass (for example, protons, alpha particles, and heavier nuclei), electrons (both negatively and positively charged), and electromagnetic radiation. For all of these, the interactions with matter are considered to be primarily electromagnetic. The behavior of mesons and other particles is intermediate between that of the electron and heavy charged particles.

A striking difference in the absorption of the three types of radiations is that only heavy charged particles have a range. That is, a monoenergetic beam of heavy charged particles, in passing through a certain amount of matter, will lose energy without changing the number of particles in the beam. Ultimately, they will be stopped after crossing practically the same thickness of absorber. For electromagnetic radiation (gamma rays) and neutrons, on the other hand, the absorption is exponential. The difference in behavior reflects the fact that charged particles are not removed from the beam by individual interactions, whereas gamma radiation photons (and neutrons) are removed. Electrons exhibit a more complex behavior. *See* ELECTRON; NUCLEAR REACTION.
     [D.G.K.]

Nuclear reaction   A process that occurs as a result of interactions between atomic nuclei when the interacting particles approach each other to within distances of the order of nuclear dimensions ($\simeq 10^{-12}$ cm). While nuclear reactions occur in nature, understanding of them and use of them as tools have taken place primarily in the controlled laboratory environment. In the usual experimental situation, nuclear reactions are initiated by bombarding one of the interacting particles, the stationary target nucleus, with nuclear projectiles of some type, and the reaction products and their behaviors are studied.

Types of nuclear interaction. As a generalized nuclear process, consider a collision in which an incident particle strikes a previously stationary particle, to produce an unspecified number of final products. If the final products are the same as the two initial particles, the process is called scattering. The scattering is said to be elastic or inelastic, depending on whether some of the kinetic energy of the incident particle is used to raise either of the particles to an excited state. If the product particles are different from the initial pair, the process is referred to as a reaction.

The most common type of nuclear reaction, and the one which has been most extensively studied, involves the production of two final products. Such reactions can be observed, for example, when deuterons with a kinetic energy of a few megaelectronvolts are allowed to strike a carbon nucleus of mass 12. Protons, neutrons, deuterons, and alpha particles are observed to be emitted, and reactions (1)–(4) are responsible. In

$$\ce{^2_1H} + \ce{^{12}_6C} \rightarrow \ce{^2_1H} + \ce{^{12}_6C} \tag{1}$$

$$\ce{^2_1H} + \ce{^{12}_6C} \rightarrow \ce{^1_1H} + \ce{^{13}_6C} \tag{2}$$

$$\ce{^2_1H} + \ce{^{12}_6C} \rightarrow \ce{^0_1n} + \ce{^{13}_7N} \tag{3}$$

$$\ce{^2_1H} + \ce{^{12}_6C} \rightarrow \ce{^4_2He} + \ce{^{10}_5B} \tag{4}$$

these equations the nuclei are indicated by the usual chemical symbols; the subscripts indicate the atomic number (nuclear charge) of the nucleus, and the superscripts the mass number of the particular isotope. These reactions are conventionally written in the compact notation $^{12}C(d,d)^{12}C$, $^{12}C(d,p)^{13}C$, $^{12}C(d,n)^{13}N$, and $^{12}C(d,\alpha)^{10}B$, where d represents deuteron, p proton, n neutron, and α alpha particle. In each of these cases the reaction results in the production of an emitted light particle and a heavy residual nucleus. If the residual nucleus is formed in an excited state, it will subsequently emit this excitation energy in the form of gamma rays or, in special cases, electrons. The residual nucleus may also be a radioactive species, in which case it will undergo further transformation in accordance with its characteristic radioactive decay scheme. *See* RADIOACTIVITY.

Nuclear cross section. In general one is interested in the probability of occurrence of the various reactions as a function of the bombarding energy of the incident particle. The measure of probability for a nuclear reaction is its cross section. Consider a reaction initiated by a beam of particles incident on a region which contains N atoms per unit area (uniformly distributed), and where I particles per second striking the area result in R reactions of a particular type per second. The fraction of the area bombarded which is effective in producing the reaction products is R/I. If this is divided by the number of nuclei per unit area, the effective area or cross section $\sigma = R/IN$. This is referred to as the total cross section for the specific reaction, since it involves all the occurrences of the reaction. The dimensions are those of an area, and total cross sections are expressed in either square centimeters or barns (1 barn $= 10^{-24}$ cm^2). The differential cross section refers to the probability that a particular reaction product will be observed at a given angle with respect to the beam direction. Its dimensions are those of an area per unit solid angle (for example, barns per steradian).

Reaction mechanism. Various reaction models have been extremely successful in describing certain classes or types of nuclear reaction processes. In general, all reactions can be classified according to the time scale on which they occur, and the degree to which the kinetic energy of the incident particle is converted into internal excitation of the final products. A large fraction of the reactions observed has properties consistent with those predicted by two reaction mechanisms which represent the extremes in this general classification. These are the mechanisms of compound nucleus formation and direct interaction.

Compound nucleus formation is envisioned to take place in two distinct steps. In the first step the incident particle is captured by (or fuses with) the target nucleus, forming an intermediate or compound nucleus which lives a long time ($\simeq 10^{-16}$ s) compared to the approximately 10^{-22} s it takes the incident particle to travel past the target. During this time the kinetic energy of the incident particle is shared among all the nucleons, and all memory of the incident particle and target is lost. The compound nucleus is always formed in a highly excited unstable state, is assumed to approach thermodynamic equilibrium involving all or most of the available degrees of freedom, and will decay, as the second step, into different reaction products, or through so-called exit channels.

The essential feature of the compound nucleus formation or fusion reaction is that the probability for a specific reaction depends on two independent probabilities: the probability for forming the compound nucleus, and the probability for decaying into that specific exit channel.

Some reactions have properties which are in striking conflict with the predictions of the compound nucleus hypothesis. Many of these are consistent with the picture of a mechanism where no long-lived intermediate system is formed, but rather a fast mechanism where the incident particle, or some portion of it, interacts with the surface, or some nucleons on the surface, of the target nucleus. These direct reactions are assumed to involve only a very small number of the available degrees of freedom. Most direct reactions are of the transfer type, where one or more nucleons are transferred to or from the incident particle as it passes the target, leaving the two final partners either in their ground states or in one of their many excited states. Such transfer reactions are generally referred to as stripping or pickup reactions, depending on whether the incident particle has lost or acquired nucleons in the reaction.

Inelastic scattering is also a direct reaction. Whereas the states preferentially populated in transfer reactions are those of specific single-particle or shell-model structure, the states preferentially excited in inelastic scattering are collective in nature. See NU-CLEAR STRUCTURE; SCATTERING EXPERIMENTS (NUCLEI).   [D.G.K.]

Nuclear spectra   The distribution of the intensity of particles (or radiation) emitted in a nuclear process as a function of energy. The nuclear spectrum is a unique signature of the process.

For example, when very slow neutrons (with speeds less than 0.5% of the speed of light) hit nitrogen nuclei, there is a high probability that they will be captured and that the nuclear system which is formed will emit a set of gamma rays (electromagnetic radiation) of very precise energies. The 24 gamma rays have energies ranging from 1.68 to 10.83 MeV, and their relative intensities are well known. A spectrum of these gamma rays, that is, the number of gamma rays having a particular energy, versus that energy can provide a unique signature of the presence of nitrogen. An application is the passing of a beam of slow neutrons through luggage at an airport: the presence of unusual amounts of nitrogen indicates that a plastic explosive may be present. This testing is nondestructive: relatively few neutrons are needed to produce the characteristic spectrum, and the luggage and its contents are not harmed. See GAMMA RAYS.

Measurements. The methods used to measure nuclear spectra depend on the nature of the particles (radiation) involved. The most accurate energy measurements are those of gamma rays. Gamma-ray spectra can be measured by determining the energy deposited by the gamma rays in a crystal, often made of sodium iodide, containing thallium impurities [NaI(Tl)], or of germanium, containing lithium impurities [Ge(Li)]. In a NaI(Tl) detector, the gamma-ray energy is transferred to electrons within the crystal, and these charged particles in turn produce electromagnetic radiation with frequencies in the visible range. The crystal is surrounded by detectors (photomultipliers) that are sensitive to the visible light. The intensity of the signal in the photomultipliers is proportional to the energy of the gamma rays that entered the NaI(Tl) crystal. The signal pulse is amplified electronically, and the pulse heights (pulse sizes) are displayed in a pulse-height multichannel analyzer in a histogram. Usually the number of pulses having a certain height (strength) is plotted versus the height. What results is a plot showing the number of gamma rays having a certain energy versus the energy of the gamma rays, a spectrum.

Neutron spectra are often determined by measuring their velocities. This is done by a time-of-flight technique in which an electronic timer measures the time interval between the emission of the neutron from a nucleus and its arrival at a detector a known distance away. This measurement uniquely determines the velocity, and thus the kinetic energy, of the neutrons. *See* NEUTRON SPECTROMETRY.

Measurements of nuclear spectra involving charged particles, such as pions, protons and alpha particles, are often made by determining their momenta (mass × velocity) and then calculating the corresponding kinetic energy. Momentum measurements are made by passing the beam of charged particles through a region in which a magnetic field exists. A magnetic field that is constant in time will not cause a change in a charged particle's speed, but it will cause a charged particle to deviate in its path. *See* PARTICLE ACCELERATOR.

Modern magnetic spectrometers use sophisticated counter telescopes and multiwire proportional counters, which permit not only the registering of the particles characterized by a certain value of the radius of curvature (and therefore of momentum) but enable the particular particle (proton, alpha particle, or whatever) that caused the signal to be identified. Contemporary magnetic spectrometer systems not only utilize complex arrangements of magnetic fields, detectors, and electronics but also generally require powerful computers to monitor and analyze the results. *See* PARTICLE DETECTOR.

[F.A.S.]

Nuclear structure   At the center of every atom lies a small, dense nucleus, which carries more than 99.97% of the atomic mass in less than 10^{-12} of its volume. The nucleus is a tightly bound system of protons and neutrons which is held together by strong forces that are not normally perceptible in nature because of their extremely short range. The small size, strong forces, and many particles in the nucleus result in a highly complex and unique quantal system that at present defies exact analysis. The study of the nucleus and the forces that hold it together constitute the field of nuclear structure physics. *See* ATOMIC STRUCTURE AND SPECTRA; NEUTRON; PROTON; QUANTUM MECHANICS; STRONG NUCLEAR INTERACTIONS.

The protons of the nucleus, being positively charged, generate a spherically symmetric electric field in which the atomic electrons orbit. The cloud of negatively charged atomic electrons normally balances the positive nuclear charge, making the atom electrically neutral. The atomic number of protons is usually denoted by Z and the number of neutrons, which are electrically neutral, by N. The total number of protons and neutrons (or nucleons) is the mass number $A = Z + N$. Isotopes have the same atomic number, Z, and hence are forms of the same chemical element, having the same chemical properties, but they differ in neutron number; isotones have a common number of neutrons, N, and isobars have the same mass number, A. *See* ISOBAR (NUCLEAR PHYSICS); ISOTONE; ISOTOPE.

Nuclei have masses less than the sum of the constituents, the missing mass ΔM being accounted for by the binding energy $\Delta M c^2$ (where c is the speed of light), which holds the nuclear system together. The characteristic energy scale is in megaelectronvolts (1 MeV = 1.6×10^{-13} joule). The internuclear forces generate an attractive potential field which holds the nucleus together and in which the nucleons orbit in highly correlated patterns. The volume of nuclei increases approximately linearly with mass number A, and the radius is roughly $R = 1.2 \times 10^{-15} \cdot A^{1/3}$ m. *See* NUCLEAR BINDING ENERGY.

Size, shape, and density distributions. A variety of sophisticated techniques have been developed for precise estimates of the nuclear charge distribution, including electron scattering, the study of muonic atoms, and the laser spectroscopy of hyperfine atomic structure. An overall picture of the nuclear charge distributions emerges. The nuclear charge density saturates in the interior and has a roughly constant value in all but the lightest nuclei. The nucleus has a diffuse skin which is of nearly constant thickness.

Many nuclei are found to have nonspherical shapes. Unlike the atom, which has a spherically symmetric Coulomb field generated by the nucleus, the nuclear field is composed of a complicated superposition of short-range interactions between nucleons, and the most stable nuclear shape is the one that minimizes the energy of the system. In general, it is not spherical, and the nuclear shape is most simply described by a multipole power series, the most important term of which is the nuclear quadrupole moment. A positive quadrupole moment reflects the elongation of nuclei into a prolate or football-like shape, while a negative value reflects an oblate shape like that of Earth. *See* NUCLEAR MOMENTS.

An accurate determination of nuclear matter distributions, that is, the distribution of both protons and neutrons in nuclei, is harder to precisely ascertain.

Nuclear masses and binding energies. The variation of average binding energy with mass number is approximated by the Bethe-Weizsacker mass formula, which is noteworthy for its simplicity in reproducing the overall binding energy systematics. The formula is developed by modeling the nucleus on a liquid drop. By analogy with a drop of liquid, there is an attractive volume term, which depends on the number of particles; a repulsive surface-tension term; and a term due to the mutual Coulomb repulsion of protons, which is responsible for the decrease in binding energy for heavy nuclei. The model is spectacularly successful in reproducing the overall trends in nuclear binding energies, masses, and the energetics of nuclear fission, and in predicting the limits of stability where neutrons and protons become unbound. As in the case of predicting a mean nuclear shape, a comparison of the prediction of the Bethe-Weizsacker mass formula to measured masses shows periodic fluctuations with both N and Z, which are due to the quantum shell effects. *See* NUCLEAR FISSION.

Nuclear excited states. The small nuclear size and tightly bound nature impose very restrictive constraints on the orbits that protons and neutrons can undergo inside the system. Thus, each nucleus has a series of quantum states that particles can occupy. The Pauli principle requires that each particle have a unique set of quantum labels. Each nuclear state can then be filled with four particles: protons with internal angular momentum "up" and "down," and likewise two neutrons. *See* ANGULAR MOMENTUM; ENERGY LEVEL (QUANTUM MECHANICS); EXCLUSION PRINCIPLE; PARITY (QUANTUM MECHANICS); QUANTUM NUMBERS; QUARKS; SPIN (QUANTUM MECHANICS).

A nucleus is most stable when all of its nucleons occupy the lowest possible states without violating this occupancy rule. This is called the nuclear ground state. During nuclear collisions the protons and neutrons can be excited from their most bound states and promoted to higher-lying unoccupied states. The process is usually very short-lived and the particles deexcite to their most stable configuration on a time scale of the order of 10^{-12} s. The energy is usually released in the form of gamma rays of well-defined energy corresponding to the difference in energy of the initial and final nuclear states. Occasionally, gamma decay is not favored because of angular momentum selection rules, and long-lived nuclear isomers result. *See* GAMMA RAYS; NUCLEAR ISOMERISM; NUCLEAR SPECTRA; SELECTION RULES (PHYSICS).

Nuclear models. The detailed categorization of the excitation of protons and neutrons allows a mapping of the excited states of each nucleus and determination of its quantum numbers. These data are the essential information required for development of detailed models that can describe the motion of nucleons inside nuclei. Unlike atomic molecules, where rotational, vibrational, and single-particle degrees of freedom involve different time scales and energies, the nucleus is highly complex, with rotation, vibration, and single-particle degrees of freedom being excited at similar energies and often strongly mixed. *See* MOLECULAR STRUCTURE AND SPECTRA.

The measurement of static electric and magnetic moments of nuclear states and of dynamic transition moments has provided a great deal of information. Electric moments have revealed a variety of enhanced collective modes of excitation, including elongated, flattened, and pear-shaped nuclear shapes. Magnetic moments have provided detailed information on the differences between excitations involving neutrons (negative moments) and protons (positive moments).

For atoms, the solutions of the Schrödinger equation with a Coulomb potential lead to a reasonable prediction of the energies of quantized atomic states, as well as their spins, parities, and moments. Attempts to make the same progress for nuclei, using a variety of spherically symmetric geometric potentials of nuclear dimensions, failed to reproduce known properties until it was realized by M. G. Meyer and independently by J. H. Jensen in 1949 that an additional spin-orbit potential was required to reproduce the known sequence of nuclear states. A potential of this form binds states having the internal spin of the nucleons parallel to its orbital angular momentum more tightly than when they are antiparallel. The ensuing sequence of quantum shell gaps and subgaps was then correctly reproduced. The shell model has evolved rapidly, and its domain of applicability has widened from the limited regions of sphericity near doubly magic nuclei to encompass most light nuclei with $A < 60$ as well as enlarged regions around shell closures.

As valence particles are added to a closed core of nucleons, the mutual residual interactions can act coherently and polarize the nuclear system away from sphericity. The polarization effects are strongest when both valence protons and neutrons are involved. The deformed nuclear potential can then undergo collective rotation, which generally involves less energy than other degrees of freedom and thus dominates the spectrum of strongly deformed nuclei.

Nuclei undergo collective vibrations about both spherical and deformed shapes. The degree of softness of these vibrations is characterized by the excitation energy required to populate states. The distinguishing feature of vibrational excited states is that they are grouped in nearly degenerate angular momentum multiplets, each group being separated by a characteristic phonon energy.

It has been a goal of nuclear structure studies to develop models that incorporate all of the features described above in order to produce a unified nuclear picture. The development of generalized nuclear models has relevance to other fields of physics. There are many isotopes that will never be accessible in the laboratory but may exist in stars or may have existed earlier in cosmological time. The evolution of generalized models greatly increases the power to predict nuclear behavior and provides information that is required for cosmological calculations.

Nuclei at high excitation energies. As nuclei are excited to ever higher excitation energies, it is anticipated that shell effects will be slowly replaced by statistical, or chaotic, behavior. The number of states per megaelectronvolt of excitation energy with each spin and parity rise exponentially with increasing excitation energy until the levels become sufficiently close that they overlap and mix strongly and so become a continuum of states.

Toward the top of this energy regime, new modes of nuclear collectivity become accessible. Giant resonance states can be excited that involve compression and oscillation of the nuclear medium with vibrations of the protons and neutrons in phase (isoscalar) or beating against each other (isovector). The excitation and decay of these giant resonances can provide information about shapes of nuclei at high excitation and about the compressibility of nuclear matter. Results from giant resonance studies indicate that the shell effect persists high into the region previously thought to be statistical. *See* GIANT NUCLEAR RESONANCES.

The semiclassical statistical and hydrodynamic behavior of hot nuclear matter and its experimental, theoretical, and astrophysical aspects are of great interest at the highest nuclear energies. The influence of compression and heat on the density of nuclear matter is being investigated in order to measure a nuclear equation of state in analogy with the properties of a classical fluid. It has been suggested that the nuclear matter may undergo phase changes under compression, with high-density condensates possibly providing a new metastable state. At the highest densities and temperatures, the nucleons themselves are forced to overlap and merge, leading to a plasma of quarks and gluons that are the nucleonic constituents. *See* QUARK-GLUON PLASMA; RELATIVISTIC HEAVY-ION COLLISIONS.
[C.J.Li.]

Nucleon   The collective name for a proton or a neutron. These subatomic particles are the principal constituents of atomic nuclei and therefore of most matter in the universe. The proton and neutron share many characteristics. They have the same intrinsic spin, nearly the same mass, and similar interactions with other subatomic particles, and they can transform into one another by means of the weak interactions. Hence it is often useful to view them as two different states or configurations of the same particle, the nucleon. Nucleons are small compared to atomic dimensions and relatively heavy. Their characteristic size is of order 1/10,000 the size of a typical atom, and their mass is of order 2000 times the mass of the electron.

The proton and neutron differ chiefly in their electromagnetic properties. The proton has electric charge $+1$, the opposite of the electron, while the neutron is electrically neutral. They have significantly different intrinsic magnetic moments. Because the neutron is slightly heavier than the proton, roughly 1 part in 1000, the neutron is unstable, decaying into a proton, an electron, and an antineutrino with a characteristic lifetime of approximately 900 s. Although some unified field theories predict that the proton is unstable, no experiment has detected proton decay.

The complex forces between nucleons and the discovery during the 1950s of many similar subatomic particles led physicists to suggest that nucleons might not be fundamental particles. During the late 1960s and 1970s, inelastic electron and neutrino scattering experiments indicated that nucleons are composed of pointlike particles with spin $1/2$ and electric charges that are fractions of the charge on the electron. Particles with similar properties, named quarks, had been hypothesized in the early 1960s to explain other regularities among the properties of hadrons. In the early 1970s, it became clear that nucleons and other hadrons are indeed bound states of quarks. *See* HADRON; NUCLEAR STRUCTURE.

Quarks are believed to be fundamental particles without internal structure. The proton consists of two up-type quarks and one down-type quark (*uud*), while the neutron consists of *ddu*. Quarks are bound into nucleons by strong forces carried by gluons. The nucleon contains ambient gluon fields in somewhat the same way that the atom contains ambient electromagnetic fields. Because quarks and gluons are much less massive than the nucleon itself, their motion inside the nucleon is relativistic, making quark-antiquark pair creation a significant factor. Thus the nucleon contains fluctuating quark-antiquark pairs in addition to quarks and gluons. The theory of quark-gluon interactions is known as quantum chromodynamics (QCD), in analogy to the quantum theory of electrodynamics (QED). *See* ELEMENTARY PARTICLE; GLUONS; NEUTRON; PROTON; QUANTUM CHROMODYNAMICS; QUANTUM ELECTRODYNAMICS; QUARKS.   [R.L.Ja.]

Nuclide   A species of atom that is characterized by the constitution of its nucleus, in particular by its atomic number Z and its neutron number $A - Z$, where A is the mass number. The total number of stable nuclides is approximately 275. About a dozen radioactive nuclides are found in nature, and hundreds of others have been created artificially.   [H.E.D.]

Nutation (astronomy and mechanics)   In mechanics, a bobbing motion that accompanies the precession of a spinning rigid body, such as a top. In simple precession, the axis of a top with a fixed point of contact sweeps out a cone, whose axis is the vertical direction. In the general motion, the angle between the axis of the top and the vertical varies with time. This motion of the top's axis, bobbing up and down as it precesses, is known as nutation. *See* RIGID-BODY DYNAMICS.

Astronomical nutation refers to irregularities in the precessional motion of the equinoxes caused by the varying torque applied to the Earth by the Sun and Moon. Astronomical nutation should not be confused with nutation as defined in mechanics; the latter is present even if the source of the torques is unvarying.   [V.D.B.]

Ohmmeter A portable instrument for measuring relatively low values of electrical resistance. The range of resistance measured is typically from 0.1 microhm to 1999 ohms (Ω). The ohmmeter solves quickly and easily a variety of measurement problems, including measuring the resistance of cladding and tracks on printed circuit boards, electrical connectors, and switch and relay contacts, as well as determining the quality of ground-conductor continuity and bonding, cables, bus-bar joints, and welded connector tags. *See* RESISTANCE MEASUREMENT.
[A.D.Sk.]

Ohm's law The direct current flowing in an electrical circuit is directly proportional to the voltage applied to the circuit. The constant of proportionality R, called the electrical resistance, is given by the equation below, in which V is the applied voltage

$$V = RI$$

and I is the current. Numerous deviations from this simple, linear relationship have been discovered. *See* ELECTRICAL RESISTANCE.
[C.E.A.]

Open channel A covered or uncovered conduit in which liquid (usually water) flows with its top surface bounded by the atmosphere. Typical open channels are rivers, streams, canals, flumes, reclamation or drainage ditches, sewers, and water-supply or hydropower aqueducts.

Open-channel flow is classified according to steadiness, a condition in relation to time, and to uniformity, a condition in relation to distance. Flow is steady when the velocity at any point of observation does not change with time; if it changes from instant to instant, flow is unsteady. At every instant, if the velocity is the same at all points along the channel, flow is uniform; if it is not the same, flow is nonuniform. Nonuniform flow which is steady is called varied; nonuniform flow which is unsteady is called variable.

Flow occurs from a higher to a lower elevation by action of gravity. If the phenomenon is short, wall friction is small or negligible, and gravity shapes the flow behavior. Gravity phenomena are local; they include the hydraulic jump, flow over weirs, spillways, or sills, flow under sluices, and flow into culvert entrances.

If the phenomenon is long, friction shapes the flow behavior. Friction phenomena include flows in rivers, streams, canals, flumes, and sewers.
[W.A.]

Open circuit A condition in an electric circuit in which there is no path for current between two points; examples are a broken wire and a switch in the open, or off, position. Open-circuit voltage is the potential difference between two points in a circuit when a branch (current path) between the points is open-circuited. Open-circuit voltage is measured by a voltmeter which has a very high resistance (theoretically infinite).
[C.F.G.]

Optical bistability   A phenomenon exhibited by certain resonant optical structures whereby it is possible to have two stable steady transmission states for the device, depending upon the history of the input. Such a bistable device may be useful for optical computing elements because of its memory characteristics. The bistability can result from the intrinsic properties of the optical device or from some external feedback such as an electrical voltage supplied by another device. This second type, extrinsic or hybrid optical bistability, is not true optical bistability.

Optical bistability is an inherently steady-state phenomenon, and typically any cycling of the device through its hysteresis cycle must be done adiabatically; that is, changes in the propagating light amplitude, envelope phase, and profile must occur sufficiently slowly that their impact on the evolution of the system may be neglected. This requirement imposes some rather severe frequency-response limitations on the use of intrinsically bistable devices in optical circuits. The two primary types of intrinsic optical bistability, each arising from a distinct physical mechanism, are absorptive bistability and refractive bistability. *See* ADIABATIC PROCESS; HYSTERESIS.

Absorptive optical bistability is based upon coupling the feedback mechanism inherent in an optical cavity with an absorbing nonlinear optical medium in which the absorption coefficient decreases with increasing light intensity (a saturable absorber). The basic theory of operation is: the saturable absorber is placed in the cavity, and the cavity is resonantly pumped. For low light intensities, the transmission coefficient for the cavity is small because of the presence of the highly absorbing medium inside the cavity. As the pump intensity is increased, the absorption of the nonlinear medium decreases. Finally, for some threshold pump intensity, the cavity switches into a high transmission state, because the absorption coefficient is reduced sufficiently that the intrinsic cavity feedback mechanism dominates. The threshold is very sharp because, when the cavity is in a highly transmittive state, the builtup intensity inside the cavity becomes very large compared to the pump intensity (due to the feedback) and effectively bleaches virtually all of the absorption in the nonlinear medium. The intense pump is then largely transmitted, although some energy is stored in the cavity to bleach the absorber. *See* ABSORPTION OF ELECTROMAGNETIC RADIATION; LASER; OPTICAL PUMPING.

This device exhibits two characteristics that constrain its usefulness in particular applications. (1) The device is based on an absorption mechanism, so the energy absorbed from the pump light must be dissipated in the bistable element or heat-sinked elsewhere. (2) It is highly frequency-sensitive because its operation is based on the switching characteristics of a resonant cavity.

Refractive optical bistability is based on coupling the feedback mechanism inherent in an optical cavity with a nonlinear optical medium that exhibits a change in the refractive index as a function of light intensity. The nonlinear refractive medium is placed inside the optical cavity, and the cavity is pumped slightly off-resonance so that the transmission coefficient is small compared to unity. However, a small amount of light intensity does exist inside the cavity, and changes the effective optical path length inside the cavity by inducing change in the refractive index of the nonlinear medium. As the pump intensity is increased, this change in the effective path length becomes larger, until at some point the cavity switches into, and possibly past, resonance. The transmission coefficient switches abruptly to a value close to unity, and the builtup intensity inside the cavity increases abruptly. If the pump intensity is increased further, it is possible to switch the cavity through a second resonance, with an additional threshold in the transmission coefficient. *See* REFRACTION OF WAVES.

The most common implementation scheme for a bistable optical device is the nonlinear Fabry-Perot etalon. The device is typically fabricated from a semiconductor, and consists of a slab of material of approximately 1 micrometer thickness. On each surface of the semiconductor, a highly reflective coating may be deposited to increase the bandwidth of the Fabry-Perot cavity. The choice of a proper nonlinear material is based upon the operating wavelength and the temporal response time desired, and possibly other considerations. Typically for applications in the far-infrared, near-infrared, and visible wavelengths, the proper materials are indium antimonide (InSb), gallium arsenide (Ga As), and zinc selenide (ZnSe), respectively. *See* INTERFEROMETRY.   [D.R.A.]

Optical fibers

Flexible transparent fiber devices, sometimes called lightguides, used for either image or information transmission, in which light is propagated by total internal reflection. In simplest form, the optical fiber or lightguide consists of a core of material with a refractive index higher than the surrounding cladding. The optical fiber properties and requirements for image transfer, in which information is continuously transmitted over relatively short distances, are quite different than those for information transmission, where typically digital encoding of information into on-off pulses of light (on = 1, off = 0) is used to transmit audio, video, or data over much longer distances at high bit rates. Another application for optical fibers is in sensors, where a change in light transmission properties is used to sense or detect a change in some property, such as temperature, pressure, or magnetic field. *See* REFLECTION OF ELECTROMAGNETIC RADIATION; REFRACTION OF WAVES.

There are three basic types of optical fibers. Propagation in these lightguides is most easily understood by ray optics, although the wave or modal description must be used for an exact description. In a multimode, stepped-refractive-index-profile fiber (illus. *a*), the number of rays or modes of light which are guided, and thus the amount of light power coupled into the lightguide, is determined by the core size and the core-cladding refractive index difference. Such fibers, used for conventional image transfer, are limited to short distances for information transmission due to pulse broadening. An initially sharp pulse made up of many modes broadens as it travels long distances in the fiber, since high-angle modes have a longer distance to travel relative to the low-angle modes. This limits the bit rate and distance because it determines how closely input pulses can be spaced without overlap at the output end.

A graded-index multimode fiber (illus. *b*), where the core refractive index varies across the core diameter, is used to minimize pulse broadening due to intermodal dispersion. Since light travels more slowly in the high-index region of the fiber relative to the

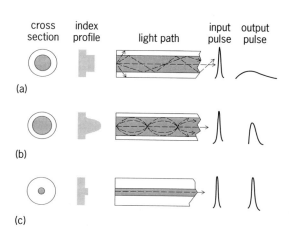

| cross section | index profile | light path | input pulse | output pulse |

(a)

(b)

(c)

Types of optical fiber designs.
(*a*) Multimode, stepped-refractive-index-profile.
(*b*) Multimode, graded-index-profile. (*c*) Single-mode, stepped-index. Graded-index is possible.

low-index region, significant equalization of the transit time for the various modes can be achieved to reduce pulse broadening. This type of fiber is suitable for intermediate-distance, intermediate-bit-rate transmission systems. For both fiber types, light from a laser or light-emitting diode can be effectively coupled into the fiber. *See* LASER; LIGHT-EMITTING DIODE.

A single-mode fiber (illus. *c*) is designed with a core diameter and refractive index distribution such that only one fundamental mode is guided, thus eliminating intermodal pulse-broadening effects. Material and waveguide dispersion effects cause some pulse broadening, which increases with the spectral width of the light source. These fibers are best suited for use with a laser source in order to efficiently couple light into the small core of the lightguide and to enable information transmission over long distances at very high bit rates.

A special class of single-mode fibers comprises polarization-preserving fibers. In an ideal, perfectly circular single-mode fiber core, the polarization state of the propagating light is preserved, but in a real fiber various imperfections can cause birefringence; that is, the two orthogonally polarized modes of the fundamental mode travel at different speeds. For applications such as sensors, where controlling the polarization is important, polarization-maintaining fibers can be designed that deliberately introduce a polarization. This is typically accomplished by using noncircular cores (shape birefringence) or by introducing asymmetric stresses (stress-induced birefringence) on the core. *See* BIREFRINGENCE; POLARIZED LIGHT.

The attenuation or loss of light intensity is an important property of the lightguide since it limits the achievable transmission distance, and is caused by light absorption and scattering. Optical fibers based on silica glass have an intrinsic transmission window at near-infrared wavelengths with extremely low losses. Glass fibers, intrinsically brittle, are coated with a protective plastic to preserve their strength. *See* OPTICAL MATERIALS.
[S.R.N.]

Optical image   The image formed by the light rays from a self-luminous or an illuminated object that traverse an optical system. The image is said to be real if the light rays converge to a focus on the image side and virtual if the rays seem to come from a point within the instrument (see illustration).

The optical image of an object is given by the light distribution coming from each point of the object at the image plane of an optical system. The ideal image of a point

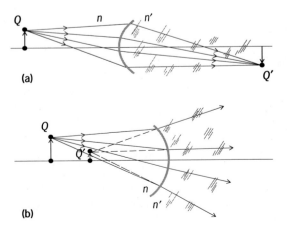

Optical images. (*a*) Real image. Rays leaving object point *Q* and passing through the refracting surface separating media *n* and *n′* are brought to a focus at the image point *Q′*. (*b*) Virtual image. Rays leaving *A* and refracted by the concave surface separating *n* and *n′* appear to be coming from the virtual image point *Q′*. As the rays are diverging, they cannot be focused at any point. (*Modified from F. A. Jenkins and H. E. White, Fundamentals of Optics, 4th ed., McGraw-Hill, 1976*)

according to geometrical optics is obtained when all rays from an object point unite in a single image point. However, diffraction theory teaches that even in this case the image is not a point but a minute disk. *See* DIFFRACTION.

From the standpoint of geometrical optics, if this most desirable type of image formation cannot be achieved, the next best objective is to have the image free from all but aperture errors (spherical aberration). In this case the light distribution in the image plane is still circular, resembling the point image; there is a true coordination of object point and image, although the image may be slightly unsharp. If the aperture errors are small, or if the image is viewed from a distance, such an image formation may be very satisfactory. *See* ABERRATION (OPTICS).

Asymmetry and deformation errors may be very disturbing if not held in check, because the light distribution of the image of a point in this case has a decidedly undesirable shape.

[M.J.H.]

Optical materials All substances used in the construction of devices or instruments whose function is to alter or control electromagnetic radiation in the ultraviolet, visible, or infrared spectral regions. Optical materials are fabricated into optical elements such as lenses, mirrors, windows, prisms, polarizers, detectors, and modulators. These materials serve to refract, reflect, transmit, disperse, polarize, detect, and transform light. The term "light" refers here not only to visible light but also to radiation in the adjoining ultraviolet and infrared spectral regions. At the microscopic level, atoms and their electronic configurations in the material interact with the electromagnetic radiation (photons) to determine the material's macroscopic optical properties such as transmission and refraction. These optical properties are functions of the wavelength of the incident light, the temperature of the material, the applied pressure on the material, and in certain instances the external electric and magnetic fields applied to the material. *See* ATOMIC STRUCTURE AND SPECTRA; DISPERSION (RADIATION); ELECTROMAGNETIC RADIATION; ELECTROOPTICS; INFRARED RADIATION; LENS (OPTICS); LIGHT; MAGNETOOPTICS; MIRROR OPTICS; OPTICAL MODULATORS; OPTICAL PRISM; POLARIZED LIGHT; REFLECTION OF ELECTROMAGNETIC RADIATION; REFRACTION OF WAVES; ULTRAVIOLET RADIATION.

There is a wide range of substances that are useful as optical materials. Most optical elements are fabricated from glass, crystalline materials, polymers, or plastic materials. In the choice of a material, the most important properties are often the degree of transparency and the refractive index, along with each property's spectral dependency. The uniformity of the material, the strength and hardness, temperature limits, hygroscopicity, chemical resistivity, and availability of suitable coatings may also need to be considered.

Glass technology provided the foundation for classical optical elements, such as lenses, prisms, and filters. Glasses developed for use in the visible region have internal transmittances of over 99% throughout the wavelength range of 380–780 nanometers. However, the silicate structure in glasses limits their transmission to about 2.5 micrometers in the infrared. Chalcogenide glasses, heavy-metal fluoride glasses, and heavy-metal oxide glasses extend this transmission to 8–12 μm. *See* COLOR FILTER.

Advances in the process for manufacturing optical fibers led to the present fiber-optic communication systems that operate in the near-infrared region with windows at wavelengths of 850, 1310, 1550, and 1625 nm. An advanced fiber-optic system, LEAF (Large Effective Area Fiber), was designed to minimize nonlinearities by spreading the optical power over large areas. *See* OPTICAL FIBERS.

The use of photolithography for printing integrated circuits has necessitated the improvement in the transmission of glasses for the ultraviolet region. Fused silica, which transmits to about 180 nm, is well suited for the lithography in the ultraviolet region. However, the crystalline material calcium fluoride, which transmits into the ultraviolet region to about 140 nm, outperforms any glass in printing microchips using fluorine excimer lasers. Deep-ultraviolet applications of fused-silica glasses include high-energy lasers, spacecraft windows, blanks for large astronomical mirrors, optical imaging, and cancer detection using ultraviolet-laser-induced autofluorescence. *See* FLUORESCENCE; TELESCOPE.

The need for an inexpensive, unbreakable lens that could be easily mass-produced precipitated the introduction of plastic optics in the mid-1930s. Although the variety of plastics suitable for precision optics is limited compared to glass or crystalline materials, plastics are often preferred when difficult or unusual shapes, lightweight elements, or economical mass-production techniques are required.

The softness, inhomogeneity, and susceptibility to abrasion intrinsic to plastics often restrict their application. Haze (which is the light scattering due to microscopic defects) and birefringence (resulting from stresses) are inherent to plastics. Plastics also exhibit large variations in the refractive index with changes in temperature. Shrinkage resulting during the processing must be considered. *See* BIREFRINGENCE.

Organic synthetic polymers are emerging as key materials for information technologies. Polymers often have an advantage over inorganic materials because they can be designed and synthesized into compositions and architectures not possible with crystals, glasses, or plastics. They are manufactured to be durable, optically efficient, reliable, and inexpensive. Many uses of polymers in photonic and optoelectronic devices have emerged, including light-emitting diodes, liquid-crystal–polymer photodetectors, polymer-dispersed liquid-crystal devices (for projection television), optical-fiber amplifiers doped with organic dyes (rhodamine), organic thin-film optics, and electrooptic modulators. *See* LIGHT-EMITTING DIODE.

Although most of the early improvements in optical devices were due to advancements in the production of glasses, the crystalline state has taken on increasing importance. Historically, the naturally occurring crystals such as rock salt, quartz, and fluorite plus suitable detectors permitted the first extension of visible optical techniques to harness the invisible ultraviolet and infrared rays. Synthetic crystal-growing techniques have made available single crystals such as lithium fluoride (of special value in the ultraviolet region, since it transmits at wavelengths down to about 120 nm), calcium fluoride, and potassium bromide (useful as a prism at wavelengths up to about 25 μm in the infrared). Many alkali-halide crystals are important because they transmit into the far-infrared. *See* CRYSTAL GROWTH; CRYSTAL STRUCTURE; SINGLE CRYSTAL.

Following the invention of the transistor, germanium and silicon ushered in the use of semiconductors as infrared optical elements or detectors. Polycrystalline forms of these semiconductors could be fabricated into windows, prisms, lenses, and domes by casting, grinding, and polishing. Compound semiconductors such as gallium arsenide (GaAs), ternary compounds such as gallium aluminum arsenide ($Ga_{1-x}Al_xAs$), and quaternary compounds such as indium gallium arsenide phosphide (InGaAsP) now serve as lasers, light-emitting diodes, and photodetectors. *See* SEMICONDUCTOR.

Single crystals are indispensable for transforming, amplifying, and modulating light. Birefringent crystals serve as retarders, or wave plates, which are used to convert the polarization state of the light. In many cases, it is desirable that the crystals not only be birefringent, but also behave nonlinearly when exposed to very large fields such as those generated by intense laser beams. A few examples of such nonlinear crystals are ammonium dihydrogen phosphate (ADP), potassium dihydrogen phosphate (KDP), beta barium borate (BBO), lithium borate (LBO), and potassium titanyl phosphate (KTP). *See* CRYSTAL OPTICS; NONLINEAR OPTICS.

Other optical materials are the liquid crystals used in displays as light valves, materials used in erasable optical disks for computers and in liquid cells (Kerr cells), laser dyes, dielectric multilayer films, filter materials, and the many metals (aluminum, gold, beryllium, and so forth) and alloys that are important as coating materials. *See* KERR EFFECT.

[J.S.Bro.]

Optical modulators

Optical modulators   Devices that serve to vary some property of a light beam. The direction of the beam may be scanned as in an optical deflector, or the phase or frequency of an optical wave may be modulated. Most often, however, the intensity of the light is modulated.

Rotating or oscillating mirrors and mechanical shutters can be used at relatively low frequencies (less than 10^5 Hz). However, these devices have too much inertia to operate at much higher frequencies. At higher frequencies it is necessary to take advantage of the motions of the low-mass electrons and atoms in liquids or solids. These motions are controlled by modulating the applied electric fields, magnetic fields, or acoustic waves in phenomena known as the electrooptic, magnetooptic, or acoustooptic effect, respectively. *See* ACOUSTOOPTICS; ELECTROOPTICS; KERR EFFECT; MAGNETOOPTICS.   [I.P.K.]

Optical phase conjugation   A process that involves the use of nonlinear optical effects to precisely reverse the direction of propagation of each plane wave in an arbitrary beam of light, thereby causing the return beam to exactly retrace the path of the incident beam. The process is also known as wavefront reversal or time-reversal reflection. The unique features of this phenomenon suggest widespread application to the problems of optical beam transport through distorting or inhomogeneous media. Although closely related, the field of adaptive optics will not be discussed here. *See* ADAPTIVE OPTICS.

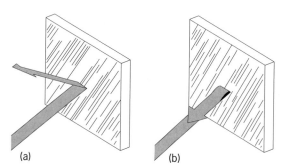

Comparison of reflections (*a*) from a conventional mirror and (*b*) from an optical phase conjugator. (*After V. J. Corcoran, ed., Proceedings for the International Conference on Laser '78 for Optical and Quantum Electronics, STS Press, McLean, Virginia, 1979*)

Optical phase conjugation is a process by which a light beam interacting in a nonlinear material is reflected in such a manner as to retrace its optical path. As the illustration shows, the image-transformation properties of this reflection are radically different from those of a conventional mirror. The incoming rays and those reflected by a conventional mirror (illus. *a*) are related by reversal of the component of the wave vector $\bar{k}$ which is normal to the mirror surface. Thus a light beam can be arbitrarily redirected by adjusting the orientation of a conventional mirror. In contrast, a phase-conjugate reflector (illus. *b*) inverts the vector quantity $\bar{k}$ so that, regardless of the orientation of the device, the reflected conjugate light beam exactly retraces the path of the incident beam. This retracing occurs even though an aberrator (such as a piece of broken glass) may be in the path of the incident beam. Looking into a conventional mirror, one would see one's own face, whereas looking into a phase-conjugate mirror, one would see only the pupil of the eye.

These new and remarkable image-transformation properties (even in the presence of a distorting optical element) open the door to many potential applications in areas such as laser fusion, atmospheric propagation, fiber-optic propagation, image restoration, real-time holography, optical data processing, nonlinear microscopy, laser resonator design, and high-resolution nonlinear spectroscopy. *See* HOLOGRAPHY; LASER; MIRROR OPTICS; NONLINEAR OPTICS; OPTICAL FIBERS. [R.A.F.; B.J.F.]

Optical prism   A simple component, made of a light-refracting and transparent material such as glass and bounded by two or more plane surfaces at an angle, that is used in optical devices, especially to change the direction of light travel, to accomplish image rotation or inversion, and to disperse light into its constituent colors. Once light enters a prism, it can be reflected one or more times before it exits the prism.

A variety of prisms can be classified according to their function. Some prisms, such as the dove prism, can be used to rotate an image and to change its parity. Image inversion by prisms in traditional binoculars is a typical application. Some prisms take advantage of the phenomenon of total internal reflection to deviate light, such as the right-angle prism and the pentaprism used in single lens reflex cameras. A thin prism is known as an optical wedge; it can be used to change slightly the direction of light travel, and therefore it can be used in pairs as an alignment device. Optical wedges are also use in stereoscopic instruments to allow the viewer to observe the three-dimensional effect without forcing the eyes to point in different directions. A variable wedge can be integrated into a commercial pair of binoculars to stabilize the line of sight in the presence of the user's slight hand movements. Other prisms such as corner-cubes can be used to reflect light backward, and are fabricated in arrays for car and bicycle retroreflectors. *See* MIRROR OPTICS; REFLECTION OF ELECTROMAGNETIC RADIATION.

An important application of a prism is to disperse light. When light enters at an angle to the face of a prism, it is refracted. Since the index of refraction depends on the wavelength, the light is refracted at different angles and therefore it is dispersed into a spectrum of colors. The blue color is refracted more than the red. When light reaches the second face of the prism, it is refracted again and the initial dispersion can be added to or canceled, depending on the prism angle. A combination of prisms in tandem can increase the amount of light dispersion. Dispersing prisms have been used in monochromators and spectroscopic instruments. With two prisms of different materials, it is possible to obtain light deviation without dispersion (an achromatic prism) or dispersion without deviation. *See* DISPERSION (RADIATION); REFRACTION OF WAVES; SPECTROSCOPY. [J.M.Sa.]

Optical pulses   Bursts of electromagnetic radiation of finite duration. Optical pulses are used to transmit information or to record the chronology of physical events. The simplest example is the photographic flash. This was probably first developed by early photographers who used flash powder that, when ignited, produced a short burst of intense light. This was followed by the flash lamp, in which a tube filled with an inert gas such as xenon is excited by a brief electrical pulse. A great advance in the creation of short optical pulses came with the invention of the laser. Lasers are now the most common and effective way of generating a variety of short optical pulses, of different durations, energies, and wavelengths. *See* LASER.

Pulses of millisecond (10^{-3} s) duration are very simply generated by mechanically modulating a constant light source such as a lamp or a continuous-wave laser. This can be done, for example, by placing a rotating disk with holes in it in front of the light source. Shorter laser pulses, of microsecond (10^{-6} s) or nanosecond (10^{-9} s) duration, are generated by using a technique known as Q-switching. A modulating device is incorporated inside the laser cavity that allows the buildup of the laser radiation inside the cavity and then switches it out in an instant. The modulating device is usually controlled by external electrical pulses. Semiconductor diode lasers, which are used to transmit information (voice or data) over a fiber-optic cable, are pumped by electricity and can be directly pulsed by applying to them a pulsed electrical signal. *See* OPTICAL FIBERS.

Ultrashort laser pulses, with durations of the order of picoseconds (1 ps = 10^{-12} s) or femtoseconds (1 fs = 10^{-15} s), are generated by using a general principle known as mode locking, whereby several frequency modes of the laser structure are made to resonate simultaneously and with a well-orchestrated relationship so as to form a short-duration pulse at the laser output.

Pulses as short as 11 fs have been produced directly by a passively mode-locked titanium:sapphire laser. The titanium:sapphire laser has also allowed the extension of ultrashort optical pulses to other wavelength ranges, such as the near-infrared (2–10 μm). Dye lasers, based on organic dyes in solution, have achieved durations as short as 27 fs. Ultrashort diode laser pulses have been obtained by active and passive mode locking and produce pulses as short as a few hundred femtoseconds. They are more commonly operated so as to give rise to pulses in the picosecond range, appropriate for optical communication systems.

The generation of ultrashort laser pulses has been motivated by the quest for ever better resolution in the study of the temporal evolution and dynamics of physical systems, events, and processes. Such laser pulses are capable of creating snapshots in time of many events that occur on the atomic or molecular scale, a technique known as time-resolved spectroscopy. This stroboscopic aspect of ultrashort laser pulses is their most important scientific application and is used in physics, engineering, chemistry, and biology. For example, ultrashort pulses can excite and take snapshots of molecular vibrations and deformations. They can track the passage of charge carriers through a microscopic semiconductor device. This ability to understand the dynamics of the more elemental building blocks of nature can in turn make it possible to build ever faster devices for use in information processing and information transmission, in addition to providing a better understanding of the physical world. *See* LASER SPECTROSCOPY.

[P.C.Be.]

Optical pumping

The process of causing strong deviations from thermal equilibrium populations of selected quantized states of different energy in atomic or molecular systems by the use of optical radiation (that is, light of wavelengths in or near the visible spectrum), called the pumping radiation.

Optical pumping is vital for light amplification by stimulated emission in an important class of lasers. For example, the action of the ruby laser involves the fluorescent emission of red light by a transition from an excited level E_2 to the ground level E_1. In this case E_2 is relatively high above E_1 and the equilibrium population of E_2 is practically zero. Amplification of the red light by laser action requires that number of atoms N_2 exceed N_1 (population inversion). The inversion is accomplished by intense green and violet light from an external source which excites the chromium ion in the ruby to a band of levels, E_3 above E_2. From E_3 the ion rapidly drops without radiation to E_2, in which its lifetime is relatively long for an excited state. Sufficiently intense pumping forces more luminescent ions into E_2 by way of the E_3 levels than remain in the ground state E_1, and amplification of the red emission of the ruby by stimulated emission can then occur. *See* LASER.

[W.W.]

Optical surfaces   Interfaces between different optical media at which light is refracted or reflected. From a physical point of view, the basic elements of an optical system are such things as lenses and mirrors. However, from a conceptual point of view, the basic elements of an optical system are the refracting or reflecting surfaces of such components. Surfaces are the basic elements of an optical system because they are the elements that affect the light passing through the system. Every wavefront has its curvature changed on passing through each surface so that the final set of wavefronts in the image space may converge on the appropriate image points. Also, the aberrations of the system depend on each surface, the total aberrations of the system being the sum of the aberrations generated at the individual surfaces. *See* ABERRATION (OPTICS); REFLECTION OF ELECTROMAGNETIC RADIATION; REFRACTION OF WAVES.

Optical systems are designed by ray tracing, and refraction at an optical surface separating two media of different refractive index is the fundamental operation in the process. The transfer between two surfaces is along a straight line if, as is usually the case, the optical media are homogeneous. The refraction of the ray at a surface results in a change in the direction of the ray. This change is governed by Snell's law.

The vast majority of optical surfaces are spherical in form. This is so primarily because spherical surfaces are much easier to generate than nonspherical, or aspheric, surfaces. Moreover, lens systems seldom need aspherics because the aberrations can be controlled by changing the shape of the component lenses without changing their function in the system, apart from modifying the aberrations. Also, many lens components can be included in a lens system in order to control the aberrations. *See* LENS (OPTICS).

On the other hand, mirror systems usually require aspheric surfaces. Unlike lenses, where the shape can be changed to modify the aberrations, mirrors cannot be changed except by introducing aspheric surfaces. Mirror systems are further constrained by the fact that only a few mirrors, usually two, are used in a system because each successive mirror occludes part of the beam going to the mirror preceding it. *See* MIRROR OPTICS.

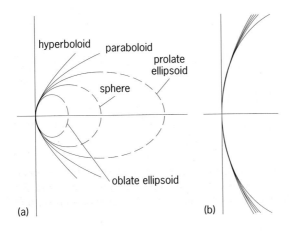

Conics of revolution. (*a*) Cross sections of entire surfaces. (*b*) Cross sections of portions near the optical axis.

The most common form of rotationally symmetric surface is the conic of revolution. The departure of conic surfaces from spherical form is shown in the illustration. The classical virtue of the conics of revolution for mirrors is the fact that light from a point located at one focus of the conic is perfectly imaged at the other focus. If these conic foci are located on the axis of revolution, the mirror is free of spherical aberration for such conjugate points.

[R.V.S.]

Optics   Narrowly, the science of light and vision; broadly, the study of the phenomena associated with the generation, transmission, and detection of electromagnetic radiation in the spectral range extending from the long-wave edge of the x-ray region to the short-wave edge of the radio region. This range, often called the optical region or the optical spectrum, extends in wavelength from about 1 nanometer to about 1 millimeter. *See* GEOMETRICAL OPTICS; PHYSICAL OPTICS.

The discoveries of the experimentalists of the early seventeenth century formed the basis of the science of optics. The statement of the law of refraction, the development of the astronomical telescope, observations of diffraction, and the principles of the propagation of light all came in this relatively short period. The publication of Isaac Newton's *Opticks* in 1704, with its comprehensive and original studies of refraction, dispersion, interference, diffraction, and polarization, established the science.

In the early 19th century many productive investigators established the transverse-wave nature of light. The relationship between optical and magnetic phenomena led to the crowning achievement of classical optics—the electromagnetic theory of J. C. Maxwell. Maxwell's theory, which holds that light consists of electric and magnetic fields propagated together through space as transverse waves, provided a general basis for the treatment of optical phenomena. In particular, it served as the basis for understanding the interaction of light with matter and, hence, as the basis for treatment of the phenomena of physical optics. *See* ELECTROMAGNETIC RADIATION; LIGHT; MAXWELL'S EQUATIONS.

In the twentieth century optics has been in the forefront of the revolution in physical thinking caused by the theory of relativity and especially by the quantum theory.

The science of optics finds itself in a position that is satisfactory for practical purposes but less so from a theoretical standpoint. The theory of Maxwell is sufficiently valid for treating the interaction of high-intensity radiation with systems considerably larger than those of atomic dimensions. The modern quantum theory is adequate for an understanding of the spectra of atoms and molecules and for the interpretation of phenomena involving low-intensity radiation, provided one does not insist on a very detailed description of the process of emission or absorption of radiation. However, a general theory of relativistic quantum electrodynamics valid for all conditions and systems has not been worked out.

The development of the laser has been an outstanding event in the history of optics. The theory of electromagnetic radiation from its beginnings was able to comprehend and treat the properties of coherent radiation, but the controlled generation of coherent monochromatic radiation of high power was not achieved in the optical region until the work of C. H. Townes and A. L. Schawlow in 1958 pointed the way. Many achievements in optics, such as holography and interferometry over long paths, have resulted from the laser. *See* HOLOGRAPHY; INTERFEROMETRY; LASER.

[R.C.L.]

Oscillation   Any effect that varies in a back-and-forth or reciprocating manner. Examples of oscillation include the variations of pressure in a sound wave and the fluctuations in a mathematical function whose value repeatedly alternates above and below some mean value.

The term oscillation is for most purposes synonymous with vibration, although the latter sometimes implies primarily a mechanical motion. The alternating current and the associated electric and magnetic fields are referred to as electric (or electromagnetic) oscillations.

If a system is set into oscillation by some initial disturbance and then left alone, the effect is called a free oscillation. A forced oscillation is one in which the oscillation is in response to a steadily applied periodic disturbance.

Any oscillation that continually decreases in amplitude, usually because the oscillating system is sending out energy, is spoken of as a damped oscillation. An oscillation that maintains a steady amplitude, usually because of an outside source of energy, is undamped. *See* ANHARMONIC OSCILLATOR; DAMPING; FORCED OSCILLATION; HARMONIC OSCILLATOR; VIBRATION.   [J.M.Ke.]

Oscilloscope   An electronic measuring instrument which produces a display showing the relationship of two or more variables. In most cases it is an orthogonal (x,y) plot with the horizontal axis being a linear function of time. The vertical axis is normally a linear function of voltage at the signal input terminal of the instrument. Because transducers of many types are available to convert almost any physical phenomenon into a corresponding voltage, the oscilloscope is a very versatile tool that is useful for many forms of physical investigation.

The oscillograph is an instrument that performs a similar function but provides a permanent record. The light-beam oscillograph used a beam of light reflected from a mirror galvanometer which was focused onto a moving light-sensitive paper. These instruments are obsolete. The mechanical version, in which the galvanometer drives a pen which writes on a moving paper chart, is still in use, particularly for process control. *See* GALVANOMETER.

Oscilloscopes are one of the most widely used electronic instruments because they provide easily understood displays of electrical waveforms and are capable of making measurements over an extremely wide range of voltage and time. Although a very large number of analog oscilloscopes are in use, digitizing oscilloscopes (also known as digital oscilloscopes or digital storage oscilloscopes) are preferred, and analog instruments are likely to be superseded.

An analog oscilloscope, in its simplest form, uses a linear vertical amplifier and a time base to display a replica of the input signal waveform on the screen of a cathode-ray tube (CRT). The screen is typically divided into 8 vertical divisions and 10 horizontal divisions. Analog oscilloscopes may be classified into nonstorage oscilloscopes, storage oscilloscopes, and sampling oscilloscopes.

Analog nonstorage oscilloscopes are the oldest and most widely used type. Except for the cathode-ray tube, the circuit descriptions also apply to analog storage oscilloscopes. A typical oscilloscope might have a bandwidth of 150 MHz, two main vertical channels plus two auxiliary channels, two time bases (one usable for delay), and a cathode-ray-tube display area; and it might include on-screen readout of some control settings and measurement results. A typical oscilloscope is composed of five basic elements: (1) the cathode-ray tube and associated controls; (2) the vertical or signal amplifier system with input terminal and controls; (3) the time base, which includes sweep generator, triggering circuit, horizontal or x-amplifier, and unblanking circuit; (4) auxiliary facilities such as a calibrator and on-screen readout; and (5) power supplies.

Digital techniques are applied to both timing and voltage measurement in digitizing oscilloscopes. A digital clock determines sampling instants at which analog-to-digital converters obtain digital values for the input signals. The resulting data can be stored indefinitely or transferred to other equipment for analysis or plotting. *See* VOLTAGE MEASUREMENT; WAVEFORM DETERMINATION.

In its simplest form a digitizing oscilloscope comprises six basic elements: (1) analog vertical input amplifier; (2) high-speed analog-to-digital converter and digital waveform memory; (3) time base, including triggering and clock drive for the analog-to-digital converter and waveform memory; (4) waveform reconstruction and display circuits; (5) display, generally, but not restricted to, a cathode-ray tube; (6) power supplies and ancillary functions. In addition, most digitizing oscilloscopes provide facilities for further manipulation of waveforms prior to display, for direct measurements of waveform parameters, and for connection to external devices such as computers and hard-copy units.

Higher measurement accuracy is available from digitizing oscilloscopes. The first decision to be made in choosing an oscilloscope is whether this or any of the other properties exclusive to the digitizing type are essential. If not, the option of an analog design remains. The selected instrument must be appropriate for the signal under examination. It must have enough sensitivity to give an adequate deflection from the applied signal, sufficient bandwidth, adequately short rise time, and time-base facilities capable of providing a steady display of the waveform. An analog oscilloscope needs to be able to produce a visible trace at the sweep speed and repetition rate likely. A digitizing oscilloscope must have an adequate maximum digitizing rate and a sufficiently long waveform memory.   [R.B.D.K.]

Parallel circuit   An electric circuit in which the elements, branches (elements in series), or components are connected between two points with one of the two ends of each component connected to each point. The illustration shows a simple parallel circuit. In more complicated electric networks one or more branches of the network may be made up of various combinations of series or series-parallel elements.

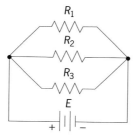

Schematic of a parallel circuit. E is a battery; R_1, R_2, and R_3 are resistors.

In a parallel circuit the potential difference (voltage) across each component is the same. However, the current through each branch of the parallel circuit may be different. For example, the lights and outlets in a house are connected in parallel so that each load will have the same voltage (120 volts) but each load may draw a different current (0.5 ampere in a 60-watt lamp and 10 amperes in a toaster).   [C.F.G.]

Paramagnetism   A property exhibited by substances which, when placed in a magnetic field, are magnetized parallel to the field to an extent proportional to the field (except at very low temperatures or in extremely large magnetic fields). Paramagnetic materials always have permeabilities greater than 1, but the values are in general not nearly so great as those of ferromagnetic materials. Paramagnetism is of two types, electronic and nuclear.

The following types of substances are paramagnetic:

1. All atoms and molecules which have an odd number of electrons. According to quantum mechanics, such a system cannot have a total spin equal to zero; therefore, each atom or molecule has a net magnetic moment which arises from the electron spin angular momentum. Examples are organic free radicals and gaseous nitric oxide.

2. All free atoms and ions with unfilled inner electron shells and many of these ions when in solids or in solution. Examples are transition, rare-earth, and actinide elements and many of their salts. This includes ferromagnetic and antiferromagnetic

materials above their transition temperatures. For a discussion of these materials *see* ANTIFERROMAGNETISM; FERRIMAGNETISM; FERROMAGNETISM.

3. Several miscellaneous compounds including molecular oxygen and organic biradicals.

4. Metals. In this case, the paramagnetism arises from the magnetic moments associated with the spins of the conduction electrons and is called Pauli paramagnetism.

Relatively few substances are paramagnetic. Aside from the Pauli paramagnetism found in metals, the most important paramagnetic effects are found in the compounds of the transition and rare-earth elements which have partially filled 3d and 4f electron shells respectively.

Electronic paramagnetism arises in a substance if its atoms or molecules possess a net electronic magnetic moment. The magnetization arises because of the tendency of a magnetic field to orient the electronic magnetic moments parallel to itself.

Nuclear paramagnetism arises when there is a net magnetic moment due to the magnetic moments of the nuclei in a substance. Nuclear magnetic moments are about 10^3 times smaller than electron magnetic moments. As a result, nuclear paramagnetism produces effects 10^6 times smaller than electron paramagnetic or diamagnetic effects. *See* DIAMAGNETISM; MAGNETIC RESONANCE; NUCLEAR MOMENTS.   [E.Ad.; F.Ke.]

Parametric arrays   Arrays of sources (or receivers) of sound formed by variation of appropriate parameters of the propagation medium. Normally, these parameters are the local sound speed and the particle velocity which vary because of the presence of large-amplitude pump, or primary, sound waves.

The usual parametric source configuration simply consists of a directional transducer (often a plane piston or planar array) driven at two frequencies near the transducer resonance, forming a dual-frequency sound beam called the primary beam. Because sound-wave propagation is not a completely linear process, signals at new frequencies are formed effectively through the interaction of sound with sound as the beam progresses and are generated along the length of the primary beam. The lowest of these new frequencies is the difference of the two primary frequencies, and so the primary beam acts as an end-fire array of sources at the difference frequency. The effective length of the array will be determined by the attenuation of the primary beam, which occurs either as a result of small-signal absorption or, for sufficiently high primary amplitudes, as a result of nonlinear losses due to the generation of harmonics of the primary frequencies and other intermodulation components, such as the sum-frequency component.

Most applications of parametric sources have been to underwater acoustics, but their use in air, as well as in other media, may be expected. Because the effective length of a parametric source can be made quite long in practice, it is possible to generate highly directional difference-frequency beams, and because the primary amplitude is shaded very gradually along the length of the array, these beams can be made practically side-lobe-free, in contrast to the beams from conventional acoustic sources. As a result, echoes from a parametric source exhibit practically no reverberation, whereas conventional echoes may be obscured by reverberation from reflection of the side lobes. Thus, the parametric source may be expected to be useful in reverberation-limited situations where one desires a narrow beam from a small projector. Such applications include precision fathometry, subbottom profiling, echo ranging, communications, and Doppler navigation logs. In order to obtain the advantages of a parametric source, however, one must be willing to tolerate low efficiency and low search rate. *See* SOUND; UNDERWATER SOUND.   [M.B.M.]

Parity (quantum mechanics)   A physical property of a wave function which specifies its behavior under simultaneous reflection of all spatial coordinates through the origin, that is, when x is replaced by $-x$, y by $-y$, and z by $-z$. If the single-particle wave function ψ satisfies Eq. (1), it is said to have even parity. If, on the other hand, Eq. (2) holds, the wave function is said to have odd parity. These two expressions can be combined in Eq. (3), where $P = \pm 1$ is a quantum number, parity,

$$\psi(x,y,z) = \psi(-x,-y,-z) \tag{1}$$

$$\psi(x,y,z) = -\psi(-x,-y,-z) \tag{2}$$

$$\psi(x,y,z) = P\psi(-x,-y,-z) \tag{3}$$

having only the two values $+1$ (designated as even parity) and -1 (odd parity). More precisely, parity is defined as the eigenvalue of the operation of space inversion. Parity is a concept that has meaning only for fields or waves and therefore has meaning only in classical field theory or in quantum mechanics. *See* QUANTUM MECHANICS.

The conservation of parity follows from the inversion symmetry of space, that is, the invariance of the Schrödinger equation $H\psi = E\psi$ (the wave equation satisfied by the wave function ψ) to the inversion of space coordinates, $\mathbf{r} \rightarrow -\mathbf{r}$. The parity (or inversion) operator, which changes $\mathbf{r}$ to $-\mathbf{r}$, has the alternative interpretation that the coordinate values remain unchanged but the coordinate axes are inverted; that is, the positive x axis of the new frame points along the old negative x axis, and similarly for y and z. If the original frame was right-handed, then the new frame is left-handed. [A cartesian coordinate system (frame, for short) is called right-handed if it is possible to place the right hand at the origin and point the thumb and first and second fingers along the positive x, y, and z axes, respectively.] Thus, parity would be conserved if the statement of physical laws were independent of the handedness of the coordinate system that was being used. Of course, the fact that most people are right-handed is not a physical law but an accident of evolution; there is nothing in the relevant laws of physics which favors a right-handed over a left-handed human. The same holds for optically active organic compounds, such as the amino acids. However, the statement that the neutrino is left-handed is a physical law. *See* NEUTRINO.

All the strong interactions between hadrons (for example, nuclear forces) and the electromagnetic interactions are symmetrical to inversion, so that parity is conserved by these interactions. As far as is known, only the weak interactions fail to conserve parity. Thus parity is not conserved in the weak decays of elementary particles (including beta decay of nuclei); in all other processes the weak interactions play a small role, and parity is very nearly conserved. Likewise, in energy eigenstates, weak interactions can be neglected to a very good approximation, and parity is very nearly a good quantum number, so that each atomic, nuclear, or hadronic state is characterized by a definite value of parity, and its conservation in reactions is an important principle. *See* FUNDAMENTAL INTERACTIONS; WEAK NUCLEAR INTERACTIONS.

One of the selection rules which follows from parity conservation is the following: A spin zero boson cannot decay sometimes into two π mesons and sometimes into three π mesons, because these final states have different parities, even and odd respectively. But the positive K meson is observed to have both these decay modes, originally called the θ and the τ mesons, respectively, but later shown by the identity of masses and lifetimes to be decay modes of the same particle. This τ-θ puzzle was the first observation of parity nonconservation. In 1956, T. D. Lee and C. N. Yang made the bold hypothesis that parity also is not conserved in beta decay. They reasoned that the magnitude of the beta-decay coupling is about the same as the coupling which leads to decay of the K meson, and so these decay processes may be manifestations of a single kind of coupling. Also, there is a very natural way to introduce parity nonconservation in beta decay, namely, by assuming a restriction on the possible states of the neutrino (two-component theory). They pointed out that no beta-decay experiment had ever looked for the spin-momentum correlations that would indicate parity nonconservation; they urged that these correlations be sought.

In the first experiment to show parity nonconservation in beta decay, the spins of the beta-active nuclei cobalt-60 were polarized with a magnetic field at low temperature; the decay electrons were observed to be emitted preferentially in directions opposite to the direction of the ^{60}Co spin. The magnitude of this correlation shows that the parity-nonconserving and parity-conserving parts of the beta interaction are of equal size, substantiating the two-component neutrino theory.

It was at first somewhat disconcerting to find parity not conserved, for that seemed to imply a handedness of space. But this is not really the situation; the saving thing is that anti-^{60}Co decays in the opposite direction. Thus, after all, there is nothing intrinsically left-handed about the world, just as there is nothing intrinsically positively charged about nuclei. What really exists here is a correlation between handedness and sign of charge.   [C.J.G.]

Particle accelerator   An electrical device which accelerates charged atomic or subatomic particles to high energies. The particles may be charged either positively or negatively. If subatomic, the particles are usually electrons or protons and, if atomic, they are charged ions of various elements and their isotopes throughout the entire periodic table of the elements.

Accelerators that produce various subatomic particles at high intensity have many practical applications in industry and medicine as well as in basic research. Electrostatic generators, pulse transformer sets, cyclotrons, and electron linear accelerators are used to produce high levels of various kinds of radiation that in turn can be used to polymerize plastics, provide bacterial sterilization without heating, and manufacture radioisotopes which are utilized in industry and medicine for direct treatment of some illnesses as well as research. They can also be used to provide high-intensity beams of protons, neutrons, heavy ions, pi mesons, or x-rays that are used for cancer therapy and research. The x-rays used in industry are usually produced by arranging for accelerated electrons to strike a solid target. However, with the advent of electron synchrotron storage rings that produce x-rays in the form of synchrotron radiation, many new industrial applications of these x-rays have been realized, especially in the field of solid-state microchip fabrication and medical diagnostics. See RADIOISOTOPE; SYNCHROTRON RADIATION.

Particle accelerators fall into two general classes—electrostatic accelerators that provide a steady dc potential, and varieties of accelerators that employ various combinations of time-varying electric and magnetic fields.

Electrostatic accelerators. Electrostatic accelerators in the simplest form accelerate the charged particle either from the source of high voltage to ground potential or from ground potential to the source of high voltage. All particle accelerations are carried out inside an evacuated tube so that the accelerated particles do not collide with air molecules or atoms and may follow trajectories characterized specifically by the electric fields utilized for the acceleration. The maximum energy available from this kind of accelerator is limited by the ability of the evacuated tube to withstand some maximum high voltage.

Time-varying field accelerators. In contrast to the high-voltage-type accelerator which accelerates particles in a continuous stream through a continuously maintained increasing potential, the time-varying accelerators must necessarily accelerate particles in small discrete groups or bunches.

An accelerator that varies only in electric field and does not use any magnetic guide or turning field is customarily referred to as a linear accelerator or linac. In the simplest version of this kind of accelerator, the electrodes that are used to attract and accelerate the particles are connected to a radio-frequency (rf) power supply or oscillator so that alternate electrodes are of opposite polarity. In this way, each successive gap between adjacent electrodes is alternately accelerating and decelerating. If these acceleration gaps are appropriately spaced to accommodate the increasing velocity of the accelerated particle, the frequency can be adjusted so that the particle bunches are always experiencing an accelerating electric field as they cross each successive gap. In this way, modest voltages can be used to accelerate bunches of particles indefinitely, limited only by the physical length of the accelerator construction.

All conventional (but not superconducting) research linacs usually are operated in a pulsed mode because of the extremely high rf power necessary for their operation. The pulsed operation can then be adjusted so that the duty cycle or amount of time actually on at full power averages to a value that is reasonable in cost and practical for cooling. This necessarily limited duty cycle in turn limits the kinds of research that are possible with linacs; however, they are extremely useful (and universally used) as pulsed high-current injectors for all electron and proton synchrotron ring accelerators. Superconducting linear accelerators have been constructed that are used to accelerate electrons and also to boost the energy of heavy ions injected from electrostatic machines. These linacs can easily operate in the continuous-wave (cw) rather than pulsed mode, because the rf power losses are only a few watts.

The Continuous Electron Beam Accelerator Facility (CEBAF) uses two 400-MeV superconducting linacs to repeatedly accelerate electrons around a racetrack-like arrangement where the two linacs are on the opposite straight sides of the racetrack and the circular ends are a series of recirculation bending magnets, a different set for each of five passes through the two linacs in succession. The continuous electron beam then receives a 400-MeV acceleration on each straight side or 0.8 GeV per turn, and is accelerated to a final energy of 4 GeV in five turns and extracted for use in experiments. The superconducting linacs allow for continuous acceleration and hence a continuous beam rather than a pulsed beam. This makes possible many fundamental nuclear and quark structure measurements that are impossible with the pulsed electron beams from conventional electron linacs. *See* SUPERCONDUCTING DEVICES.

As accelerators are carried to higher energy, a linac eventually reaches some practical construction limit because of length. This problem of extreme length can be circumvented conveniently by accelerating the particles in a circular path maintained by either static or time-varying magnetic fields. Accelerators utilizing steady magnetic fields as guide paths are usually referred to as cyclotrons or synchrocyclotrons, and are arranged to provide a steady magnetic field over relatively large areas that allow the particles to travel in an increasing spiral orbit of gradually increasing size as they increase in energy.

Practical limitations of magnet construction and cost have kept the size of circular proton accelerators with static magnetic fields to the vicinity of 100 to 1000 MeV. For even higher energies, up to 400 GeV per nucleon in the largest conventional (not superconducting) proton synchrotron in operation, it is necessary to vary the magnetic field as well as the electric field in time. In this way the magnetic field can be of a minimal practical size, which is still quite extensive for a 980-GeV accelerator (6500 ft or 2000 m in diameter). This circular magnetic containment region, or "racetrack," is injected with relatively low-energy particles that can coast around the magnetic ring when it is at minimum field strength. The magnetic field is then gradually increased to stay in step with the higher magnetic rigidity of the particles as they are gradually accelerated with a time-varying electric field.

Superconducting magnets. The study of the fundamental structure of nature and all associated basic research require an ever increasing energy in order to allow finer and finer measurements on the basic structure of matter. Since the voltage-varying and magnetic-field-varying accelerators also have limits to their maximum size in terms of cost and practical construction problems, the only way to increase particle energies even further is to provide higher-varying magnetic fields through superconducting magnet technology, which can extend electromagnetic capability by a factor of 4 to 5. Large superconducting cyclotrons and superconducting synchrotrons are in operation. *See* MAGNET.

Storage rings. Beyond the limit just described, the only other possibility is to accelerate particles in opposite directions and arrange for them to collide at certain selected intersection regions around the accelerator. The main technical problem is to provide adequate numbers of particles in the two colliding beams so that the probability of a collision is moderately high. Such storage ring facilities are in operation for both electrons and protons. Besides storing the particles in circular orbits, the rings can operate initially as synchrotrons and accelerate lower-energy injected particles to much higher energies and then store them for interaction studies at the beam interaction points.

Large proton synchrotrons have been used as storage-ring colliders by accelerating and storing protons in one direction around the ring while accelerating and storing antiprotons (negative charge) in the opposite direction. The proton and antiproton beams are carefully programmed to be in different orbits as they circulate in opposite directions and to collide only when their orbits cross at selected points around the ring where experiments are located. The antiprotons are produced by high-energy proton collisions with a target, collected, stored, cooled, and eventually injected back into the synchrotron as an antiproton beam.

Electron-positron synchrotron accelerator storage rings have been in operation for many years in the basic study of particle physics, with energies ranging from 2 GeV + 2 GeV to 104 GeV + 104 GeV. The by-product synchrotron radiation from many of these machines is used in numerous applications. However, the synchrotron radiation loss forces the machine design to larger and larger diameters, characterized by the Large Electron Positron Storage Ring (LEP) at CERN, near Geneva, Switzerland (closed down in 2000), which was 17 mi (27 km) in circumference. Conventional rf cavities enable electron-positron acceleration only up to 50–70 GeV (limited by synchrotron radiation loss) while higher energies of 100–150 GeV require superconducting cavities. *See* SYNCHROTRON RADIATION.

Advanced linacs. Although circular machines with varying magnetic fields have been developed because linacs of comparable performance would be too long (many miles), developments in linac design and utilization of powerful laser properties may result in a return to linacs that will outperform present ring machines at much lower cost. As a first example, the 20-GeV electron linac at Stanford University, Palo Alto, California, has been modified to provide simultaneous acceleration of positrons and electrons to energies as high as 50 GeV, while operating in what is called the SLED mode. After acceleration the electrons and positrons are separated by a magnet, and the two beams are magnetically directed around the opposite sides of a circle so that they collide at one intersection point approximately along a diameter extending from the end of the linac across the circle. This collider arrangement is much less expensive than the

17-mi (27-km) ring at CERN and provides electron-positron collisions of comparable energies but at lower intensities.
[H.E.W.]

Particle detector   A device used to detect and measure radiation characteristically emitted in nuclear processes, including gamma rays or x-rays, lightweight charged particles (electrons or positrons), nuclear constituents (neutrons, protons, and heavier ions), and subnuclear constituents such as mesons. The device is also known as a radiation detector. Since human senses do not respond to these types of radiation, detectors are essential tools for the discovery of radioactive minerals, for all studies of the structure of matter at the atomic, nuclear, and subnuclear levels, and for protection from the effects of radiation. They have also become important practical tools in the analysis of materials using the techniques of neutron activation and x-ray fluorescence analysis. *See* ELEMENTARY PARTICLE; NUCLEAR REACTION; NUCLEAR SPECTRA; PARTICLE ACCELERATOR; RADIOACTIVITY.

A convenient way to classify radiation detectors is according to their mode of use: (1) For detailed observation of individual photons or particles, a pulse detector is used to convert each such event (that is, photon or particle) into an electrical signal. (2) To measure the average rate of events, a mean-current detector, such as an ion chamber, is often used. Radiation monitoring and neutron flux measurements in reactors generally fall in this category. Sometimes, when the total number of events in a known time is to be determined, an integrating version of this detector is used. (3) Position-sensitive detectors are used to provide information on the location of particles or photons in the plane of the detector. (4) Track-imaging detectors image the whole three-dimensional structure of a particle's track. The output may be recorded by immediate electrical readout or by photographing tracks as in the bubble chamber. (5) The time when a particle passes through a detector or a photon interacts in it is measured by a timing detector. Such information is used to determine the velocity of particles and when observing the time relationship between events in more than one detector.

The ionization produced by a charged particle is the effect commonly employed in a particle detector. In the basic type of gas ionization detector, an electric field applied between two electrodes separates and collects the electrons and positive ions produced in the gas by the radiation to be measured. Multiwire proportional chambers and spark chambers are position-sensitive adaptations of gas detectors. The signal division or time delay that occurs between the ends of an electrode made of resistive material is sometimes used to provide position sensitivity in gas and semiconductor detectors. Track-imaging detectors rely on a secondary effect of the ionization along a particle's track to reveal its structure. *See* IONIZATION CHAMBER.

In a semiconductor detector, a solid replaces the gas. The "insulating" region (depletion layer) of a reverse-biased pn junction in a semiconductor is employed. Since solids are approximately 1000 times denser than gases, absorption of radiation can be accomplished in relatively small volumes. A less obvious but fundamental advantage of semiconductor detectors is the fact that much less energy is required ($\sim$3 eV) to produce a hole-electron pair than that required ($\sim$30 eV) to produce an ion electron pair in gases. *See* CRYSTAL COUNTER; JUNCTION DETECTOR.

In addition to producing free electrons and ions, the passage of a charged particle through matter temporarily raises electrons in the material into excited states. When these electrons fall back into their normal state, light may be emitted and detected as in the scintillation detector. *See* SCINTILLATION COUNTER.

Neutral particles, such as neutrons, cannot be detected directly by ionization. Consequently, they must be converted into charged particles by a suitable process and then observed by detecting the ionization caused by these particles.

Although ionization detectors dominate the field, a number of detector types based on other radiation-induced effects are used. Examples are (1) transition radiation detectors, which depend on the x-rays and light emitted when a particle passes through the interface between two media of different refractive indices; (2) track detectors, in which the damage caused by charged particles in plastic films and in minerals is revealed by etching procedures; (3) thermoluminescent and radiophotoluminescent detectors, which rely on the latent effects of radiation in creating traps in a material or in creating trapped charge; and (4) Cerenkov detectors, which depend on measurement of the light produced by passage of a particle whose velocity is greater than the velocity of light in the detector medium. See CERENKOV RADIATION; TRANSITION RADIATION DETECTORS.

The very large detector systems used in relativistic heavy-ion experiments and in the detection of the products of collisions of charged particles at very high energies, typically at the intersection region of storage rings, deserve special consideration. These detectors are frequently composites of several of the basic types of detectors discussed above and are designed to provide a detailed picture of the multiple products of collisions at high energies. The complete detector system may occupy a space tens of feet in extent and involve tens or hundreds of thousands of individual signal processing channels, together with large computer recording and analysis facilities.   [F.S.G.]

Particle trap   A device used to confine charged or neutral particles where their interaction with the wall of a container must be avoided. Electrons or protons accelerated to energies as high as 1 teraelectronvolt (10^{12} electronvolts) are trapped in magnetic storage rings in high-energy collision studies. Other forms of magnetic bottles are designed to hold dense hot plasmas of hydrogen isotopes for nuclear fusion. At the other end of the energy spectrum, ion and atom traps can store isolated atomic systems at temperatures below 1 millikelvin. Other applications of particle traps include the storage of antimatter such as antiprotons and positrons (antielectrons) for high-energy collision studies or low-energy experiments. See ANTIMATTER; NUCLEAR FUSION; PARTICLE ACCELERATOR; PLASMA (PHYSICS); POSITRON.

Charged-particle traps. Charged particles can be trapped in a variety of ways. An electrostatic (Kingdon) trap is formed from a thin charged wire. The ion is attracted to the wire, but its angular momentum causes it to spiral around the wire in a path with a low probability of hitting the wire.

A magnetostatic trap (magnetic bottle) is based on the fact that a charged particle with velocity perpendicular to the magnetic field lines travels in a circle, whereas a particle moving parallel to the field is unaffected by it. In general, the particle has velocity components both parallel and perpendicular to the field lines and moves in a helical spiral. In high-energy physics, accelerators and storage rings also use magnetic forces to guide and confine charged particles. A tokamak has magnetic field lines configured in the shape of a torus, confining particles in spiral orbits. This type of bottle is used to contain hot plasmas in nuclear fusion studies. Another type of bottle uses a magnetic mirror.

The radio-frequency Paul trap uses inhomogeneous radio-frequency electric fields to confine particles, forcing them to oscillate rapidly in the alternating field (see illustration). If the amplitude of oscillation (micromotion) is small compared to the trap dimensions, the trap may be thought of as increasing the (kinetic) energy of the

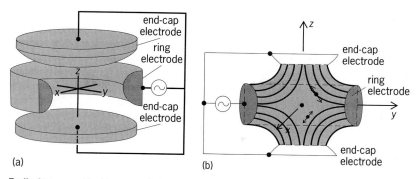

Radio-frequency Paul trap consisting of two end caps and a ring electrode. (*a*) Cutaway view (*after G. Kamas, ed., Time and Frequency Users's Manual, National Bureau of Standards Technical Note 695, 1977*). (*b*) Cross section, showing the amplitude of the instantaneous oscillations for several locations in the trap.

particle in a manner that is a function of the particle position. The particle moves to the position of minimal energy and is therefore attracted to the center of the trap where the oscillating electric fields are weakest. At the center of the trap, the fields are exactly zero, and a single, cold ion or electron trapped there is essentially at rest with almost no micromotion.

The Penning trap, with the same electrode configuration as the Paul trap, uses a combination of static electric and magnetic fields instead of oscillating electric fields.

Neutral-particle traps. Uncharged particles such as neutrons or atoms are manipulated by higher-order moments of the charge distribution such as the magnetic or electric dipole moments.

Magnetic traps of neutral particles use the fact that atoms usually have a magnetic dipole moment on which the gradient of a magnetic field exerts a force. The atom can be in a state whose magnetic energy increases or decreases with the field strength, depending on whether the moment is antiparallel or parallel to the field. A magnetic field cannot be constructed with a local maximum in a current-free region, but a local minimum is possible, allowing particles seeking a weak field to be trapped.

Laser traps use the strong electric fields of the laser beam to induce an electric dipole moment on the atom. A laser field tuned below the atomic resonance polarizes the atom in phase with the driving field; the instantaneous dipole moment points in the same direction as the field. Thus the energy of the atom is lowered if it is in a region of high laser intensity. The high-intensity trapping region is formed simply by focusing the beam of a laser. *See* LASER.

Magnetooptic hybrid traps use, instead of the dipole forces induced by the laser field, the scattering force that arises when an atom absorbs photons. An inhomogeneous magnetic field separates the magnetic substates of an atom in a position-dependent manner. These states interact differently with circularly polarized light. It is possible to arrange a combination of laser beams with proper polarizations to create net scattering forces that drive the atom into the region of zero magnetic field. Such a trap requires much lower laser intensities and weaker magnetic fields. *See* LASER COOLING.   [S.Ch.]

Pascal's law   A law of physics which states that a confined fluid transmits externally applied pressure uniformly in all directions. More exactly, in a static fluid, force is transmitted at the velocity of sound throughout the fluid. The force acts normal to any surface. This natural phenomenon is the basis of the pneumatic fire, balloon, hydraulic jack, and related devices. *See* HYDROSTATICS.   [K.Arn.; R.S.R.]

Paschen-Back effect   An effect on spectral lines obtained when the light source is placed in a very strong magnetic field, first explained by F. Paschen and E. Back in 1921. In such a field the anomalous Zeeman effect, which is obtained with weaker fields, changes over to what is, in a first approximation, the normal Zeeman effect. The term "very strong field" is a relative one, since the field strength required depends on the particular lines being investigated. It must be strong enough to produce a magnetic splitting that is large compared to the separation of the components of the spin-orbit multiplet. *See* ATOMIC STRUCTURE AND SPECTRA; ZEEMAN EFFECT.   [F.A.J./W.W.W.]

Peltier effect   A phenomenon discovered in 1834 by J. C. A. Peltier, who found that at the junction of two dissimilar metals carrying a small current the temperature rises or falls, depending upon the direction of the current. In view of experiments, which establish that the rate of intake or output of heat is proportional to the magnitude of the current, it can be shown that an electromotive force resides at a junction. Electromotive forces of this type are called Peltier emf's. *See* SEEBECK EFFECT; THERMOELECTRICITY; THOMSON EFFECT.   [J.W.St.]

Pendulum   A rigid body mounted on a fixed horizontal axis, about which it is free to rotate under the influence of gravity. The period of the motion of a pendulum is virtually independent of its amplitude and depends primarily on the geometry of the pendulum and on the local value of g, the acceleration of gravity. Pendulums have therefore been used as the control elements in clocks, or inversely as instruments to measure g.

　　Motion. In the schematic representation of a pendulum shown in the illustration, O represents the axis and C the center of mass. The line OC makes an instantaneous angle θ with the vertical. In rotary motion of any rigid body about a fixed axis, the angular acceleration is equal to the torque about the axis divided by the moment of inertia I about the axis. If m represents the mass of the pendulum, the force of gravity can be considered as the weight mg acting at the center of mass C.

　　If the amplitude of motion is small, the motion is simple harmonic. The period T, time for a complete vibration (for example, from the extreme displacement right to the

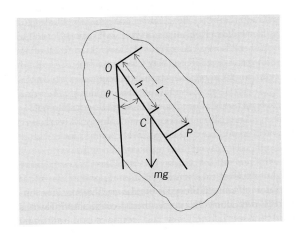

Schematic diagram of a pendulum. *O* represents the axis, *C* is the center of mass, and *P* the center of oscillation.

next extreme displacement right), is given by Eq. (1). *See* HARMONIC MOTION.

$$T = 2\pi\sqrt{I/mgh} \tag{1}$$

The actual form of a pendulum often consists of a long, light bar or a cord that serves as a support for a small, massive bob. The idealization of this form into a point mass on the end of a weightless rod of length L is known as a simple pendulum. An actual pendulum is sometimes called a physical or compound pendulum. In a simple pendulum the lengths h and L become identical, and the moment of inertia I equals mL^2. Equation (1) for the period becomes Eq. (2).

$$T = 2\pi\sqrt{L/g} \tag{2}$$

Center of oscillation. Equation (2) can be used to define the equivalent length of a physcal pendulum. Comparison with Eq. (1) shows that Eq. (3) holds. The point P

$$L = I/mh \tag{3}$$

on line OC of the illustration, whose distance from the axis O equals L, is called the center of oscillation. Points O and P are reciprocally related to each other in the sense that if the pendulum were suspended at P, O would be the center of oscillation.

Types. Kater's reversible pendulum is designed to measure g, the acceleration of gravity. It consists of a body with two knife-edge supports on opposite sides of the center of mass as at O and P (and with at least one adjustable knife-edge). If the pendulum has the same period when suspended from either knife-edge, then each is located at the center of oscillation of the other, and the distance between them must be L, the length of the equivalent simple pendulum. The value for g follows from Eq. (2).

The ballistic pendulum is a device to measure the momentum of a bullet. The pendulum bob is a block of wood into which the bullet is fired. The bullet is stopped within the block and its momentum transferred to the pendulum. This momentum is determined from the amplitude of the pendulum swing. *See* BALLISTICS.

The spherical pendulum is a simple pendulum mounted on a pivot so that its motion is not confined to a plane. The bob then moves over a spherical surface. A Foucault pendulum is a spherical pendulum suspended so that its plane of oscillation is free to rotate. Its purpose is to demonstrate the rotation of the Earth. *See* FOUCAULT PENDULUM.

[J.M.Ke.]

Period doubling A scenario for the transition of a natural process from regular motion to chaos. Various natural processes develop in time in a way that depends upon prevailing environmental details. A quantity that specifies the particular state of the environment of a process is called a parameter, and is taken as a fixed value over the course of development of the process.

It is a frequent natural occurrence for a process to have a regular and easily describable motion for some range of parameters, but to have complex, irregular, and difficult-to-describe motions for other ranges of parameters. In the context of fluid flow, the latter circumstance is termed turbulence. In a more general context it is called chaos (which includes fluid turbulence but presages an underlying generality). *See* FLUID FLOW; TURBULENT FLOW.

Sometimes, as the environmental parameters are varied, a process may systematically exhibit more irregular motions, turning over into chaotic motion beyond some parameter value. In analogy to the phenomenology of phase transitions, this circumstance is termed a transition to chaos. There are a variety of qualitatively different transitions to chaos, each termed a scenario. Period doubling is one frequently encountered scenario leading to chaos for which a full theoretical account exists. Since it occurs in a wide variety of processes of significantly divergent physical characters (for example, fluid-flow, chemical reactions, and electronic devices), it is sensible to consider it as a phenomenon in its own right. *See* PHASE TRANSITIONS.

In order to observe this scenario, it is sufficient that all but one parameter is held fixed. Over some range of this varied parameter (it shall be defined to increase over the range of investigation) the motion is observed to be periodic. Above a certain value of the parameter the motion grows more complicated (a bifurcation has occurred): after the amount of time T for which the motion exactly repeated itself just prior to the bifurcation, the motion now slightly fails to do so, exactly repeating, however, after another T seconds. That is, the period has doubled from T to $2T$. As the parameter is further increased, the error to repeat after the first half of the new period systematically increases. A still further increase of parameter produces another bifurcation resulting in a new doubling of the period: the motion slightly fails to repeat after two roughly periodic cycles, exactly doing so after four. As the parameter is further increased, there are successive period-doubling bifurcations, more and more closely spaced in parameter value until at a critical value the doubling has occurred an infinite number of times, so that the motion is now no longer periodic and hence of a more complex character than had yet been encountered. Unpredictably complex motions occur for values of the parameter above its critical value, although ranges of parameter still exist for which the system exhibits new periodic motions. Indeed any period-doubling system exhibits the same sequence of truly chaotic motion and interspersed periodicities as its parameter increases. Thus there is a strong degree of qualitatively universal behavior for all systems experiencing this scenario.

However, there is also a precise quantitative universality. That is, without knowing the system (or its equations) essentially all measurable quantities can be predicted: By looking at the data alone, it would not be possible to guess the physical system responsible for that data. Thus, reminiscent of thermodynamics, questions can be posed and answered in a general manner that bypasses the specific mechanisms governing any particular system.   [M.J.F.]

Periodic motion   Any motion that repeats itself identically at regular intervals. If $x(t)$ represents the displacement of any coordinate of the system at time t, a periodic motion has the property defined by the equation below for every value of the variable

$$x(t + T) = x(t)$$

time t. The fixed time interval T between repetitions, or the duration of a cycle, is known as the period of the motion.

The motion of the escapement mechanism of a watch, the motion of the Earth about the Sun, and the more complicated motion of the crankshaft, piston rods, and pistons in an engine running at uniform speed are all examples of periodic motion.

The vibration of a piano string after it is struck is a damped periodic motion, not strictly periodic according to the definition. Although the motion very nearly repeats itself, and with a fixed repetition time, each successive cycle has a slightly smaller amplitude. *See* DAMPING; HARMONIC MOTION; VIBRATION; WAVE MOTION.   [J.M.Ke.]

Permeance   The reciprocal of reluctance in a magnetic circuit. It is the analog of conductance (the reciprocal of resistance) in an electric circuit, and is given by Eq. (1),

$$P_m = \frac{\text{magnetic flux}}{\text{magnetomotive force}} = \frac{\iint \mathbf{B} \cdot d\mathbf{S}}{\oint \mathbf{H} \cdot d\mathbf{l}} \tag{1}$$

where **B** is the magnetic flux density, **H** is the magnetic field strength, and the integrals are respectively over a cross section of the circuit and around a path within it.

From Eq. (1), it can be shown that Eq. (2) is valid, where A is the cross-sectional

$$P_m = \mu A / l \qquad (2)$$

area of the magnetic circuit, l its length, and μ the permeability. If the material is ferromagnetic, as is often the case, then μ is not constant but varies with the flux density and the complete magnetization curve of B against H may have to be used to determine the permeance. *See* MAGNETIC MATERIALS; RELUCTANCE.   [A.E.Ba.]

Permittivity   A property of a dielectric medium that determines the forces that electric charges placed in the medium exert on each other. If two charges of q_1 and q_2 coulombs in free space are separated by a distance r meters, the electrostatic force F newtons acting upon each of them is proportional to the product of the charges and inversely proportional to the square of the distance between them. Thus, F is given by Eq. (1), where $1/(4\pi\epsilon_0)$ is the constant of proportionality, having the magnitude and

$$F = \frac{q_1 q_2}{4\pi\epsilon_0 r^2} \qquad \text{newtons} \qquad (1)$$

dimensions necessary to satisfy Eq. (1). This condition leads to a value for ϵ_0, termed the permittivity of free space, given by Eq. (2), where c is the velocity of light in vacuum.

$$\epsilon_0 = \frac{1}{4\pi 10^{-7}c^2} \simeq 8.8542 \times 10^{-12} \qquad \text{farads/meter} \qquad (2)$$

If now the charges are placed in a dielectric medium that is homogeneous and isotropic, the force on each of them is reduced by a factor ϵ_r, where ϵ_r is greater than 1. This dimensionless scalar quantity is termed the relative permittivity of the medium, and the product $\epsilon_0\epsilon_r$ is termed the absolute permittivity ϵ of the medium.

A consequence is that if two equal charges of opposite sign are placed on two separate conductors, then the potential difference between the conductors will be reduced by a factor ϵ_r when the conductors are immersed in a dielectric medium compared to the potential difference when they are in vacuum. Hence a capacitor filled with a dielectric material has a capacitance ϵ_r times greater than a capacitor with the same electrodes in vacuum would have. Except for exceedingly high applied fields, unlikely normally to be reached, ϵ_r is independent of the magnitude of the applied electric field for all dielectric materials used in practice, excluding ferroelectrics. *See* CAPACITANCE; CAPACITOR; FERROELECTRICS.   [J.H.Ca.]

Perpetual motion   The expression perpetual motion, or perpetuum mobile, arose historically in connection with the quest for a mechanism which, once set in motion, would continue to do useful work without an external source of energy or which would produce more energy than it absorbed in a cycle of operation. This type of motion, now called perpetual motion of the first kind, involves only one of the three distinct concepts presently associated with the idea of perpetual motion.

Perpetual motion of the first kind refers to a mechanism whose efficiency exceeds 100%. Clearly such a mechanism violates the now firmly established principle of conservation of energy, in particular that statement of the principle of conservation of energy embodied in the first law of thermodynamics. (Indeed, the first law of thermodynamics is sometimes stated as "A perpetuum mobile of the first kind cannot exist."). *See* CONSERVATION OF ENERGY.

Perpetual motion of the second kind refers to a device that extracts heat from a source and then converts this heat completely into other forms of energy, a process which satisfies the principle of conservation of energy. A dramatic scheme of this type would be an ocean liner, which extracts heat from the nearly limitless oceanic source and then uses this heat for propulsion. This type of perpetual motion is, however, precluded by the second law of thermodynamics which is sometimes stated as "A perpetuum mobile of the second kind cannot exist."

The third type of perpetual motion is, in contrast to the two types described above wherein useful output was the goal, merely a device which can continue moving forever. It could result in actual systems if all mechanisms by which energy is dissipated could be eliminated. Since experience indicates that dissipative effects in mechanical systems can be reduced, by lubrication in the case of friction, for example, but not eliminated, mechanical perpetual motion of the third kind can be approximated but never achieved. An example of a genuine case of this kind occurs in a superconductor. If a direct current is caused to flow in a superconducting ring, this current will continue to flow undiminished in time without application of any external force. *See* SUPERCONDUCTIVITY; THERMODYNAMIC PRINCIPLES.   [K.L.K.]

Perturbation (quantum mechanics)

Perturbation (quantum mechanics)   An expansion technique useful for solving complicated quantum-mechanical problems in terms of solutions for simple problems. Perturbation theory in quantum mechanics provides an approximation scheme whereby the physical properties of a system, modeled mathematically by a quantum-mechanical description, can be estimated to a required degree of accuracy. Such a scheme is useful because very few problems occurring in quantum mechanics can be solved analytically. Consequently an approximation technique must be employed in order to give an approximate analytic solution or to provide suitable algorithms for a numerical solution. Even for problems which admit an exact analytic solution, the exact solution may be of such mathematical complexity that its physical interpretation is not apparent. For these situations, perturbation techniques are also desirable.

Here the discussion of the application of perturbation techniques to quantum mechanics is limited to the domain of nonrelativistic quantum theory. Applications of a similar but mathematically more intricate nature have also been made in quantum electrodynamics and quantum field theory. *See* QUANTUM ELECTRODYNAMICS; QUANTUM FIELD THEORY; QUANTUM MECHANICS.

Perturbation theory is applied to the Schrödinger equation, $H\Psi = (H_0 + \lambda V)\Psi = i\hbar(\partial/\partial t)\Psi$ [where $\hbar$ is Planck's constant h divided by 2π, and $(\partial/\partial t)$ represents partial differentiation with respect to the time variable t], for which the exact hamiltonian H is split into two parts: the approximate (unperturbed) time-independent hamiltonian H_0 whose solutions of the corresponding Schrödinger equation are known analytically, and the perturbing potential λV. The basic idea is to expand the exact solution Ψ in terms of the solution set of the unperturbed hamiltonian H_0 by means of a power series in the coupling constant λ. Such a procedure is expected to be successful if the system characterized by the unperturbed hamiltonian closely resembles that characterized by the exact hamiltonian. Supposedly the differences are not singular in character, but change as a continuous function of the parameter λ.

Perturbation theory is used in two contexts to provide information about the state of the system, which in quantum mechanics is determined by the wave function Ψ. If λV is time-independent, an objective may be to find the stationary states of the system Ψ_n whose time dependence is given by $\exp(-iE_n t/\hbar)$, where $i = \sqrt{-1}$ and E_n represents the energy of the stationary state labeled by n. If λV is either time-independent or time-dependent, an objective may be to find the time evolution of a state which at some specified time was a stationary state of the unperturbed hamiltonian. The perturbing potential is then considered as causing transitions from the original state to other states of the unperturbed hamiltonian, and application of time-dependent perturbation theory provides the probability of such transitions.   [D.M.Fr.]

Phase (periodic phenomena)   The fractional part of a period through which the time variable of a periodic quantity (alternating electric current, vibration) has moved, as measured at any point in time from an arbitrary time origin. In the case of a sinusoidally varying quantity, the time origin is usually assumed to be the last point at which the quantity passed through a zero position from a negative to a positive direction.

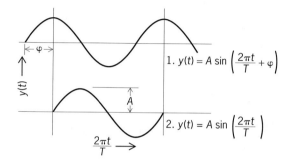

1. $y(t) = A \sin \left(\dfrac{2\pi t}{T} + \varphi \right)$

2. $y(t) = A \sin \left(\dfrac{2\pi t}{T} \right)$

An illustration of the meaning of phase for a sinusoidal wave. The difference in phase between waves 1 and 2 is φ and is called the phase angle. For each wave, A is the amplitude and T is the period.

In comparing the phase relationships at a given instant between two time-varying quantities, the phase of one is usually assumed to be zero, and the phase of the other is described, with respect to the first, as the fractional part of a period through which the second quantity must vary to achieve a zero of its own (see illustration). In this case, the fractional part of the period is usually expressed in terms of angular measure, with one period being equal to $360°$ or 2π radians. See PHASE-ANGLE MEASUREMENT; SINE WAVE.   [W.J.G.]

Phase-angle measurement   Measurement of the time delay between two periodic signals. The phase difference between two sinusoidal waveforms that have the same frequency and are free of a dc component can be conveniently described as shown in the illustration. It can be seen that the phase angle can be considered as a measure of the time delay between two periodic signals expressed as a fraction of the wave period. This fraction is normally expressed in units of angle, with a full cycle corresponding to $360°$. For example, in the illustration, where the voltage v_1 passes through zero $\frac{1}{8}$ cycle before a second voltage v_2, it leads by $360°/8$ or $45°$. Phase angle is usually defined from the fundamental component of each waveform; therefore distortion of either or both signals can give rise to errors, the extent of which depends on the nature of the distortion and the method of measurement.

The majority of modern phase-measuring devices are based on the use of zero-crossing detectors. The time at which each signal crosses the zero-voltage axis is determined, usually by means of a squaring-up circuit (for example, an overdriven amplifier) followed by a high-speed comparator. This produces, in each channel, a trigger pulse that is used to drive a bistable flip-flop. The output from the bistable is a rectangular wave, the duty cycle of which is proportional to the phase difference between the input

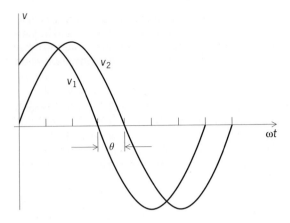

Phase angle θ between voltages v_1 and v_2.

signals. If this signal is integrated by means of a suitable filter, a dc voltage is produced that is an analog representation of the phase angle. This voltage is then displayed on a panel meter (analog or digital) suitably scaled in degrees or radians. Instrumentation using this principle is capable of measuring phase differences to approximately $\pm 0.05°$ over a wide range of amplitudes and frequencies.

Conventional phase meters have an upper frequency limit of a few hundred kilohertz. This limit is imposed mainly by the ability of the arrangement consisting of a comparator and a flip-flop to maintain a clean and precise rectangular waveform under conditions of high-speed operation. In order to measure phase angle at frequencies between about 100 kHz and several gigahertz, it is necessary to down-convert the radio-frequency signals to a frequency that can be handled correctly by the phase meter. At microwave frequencies, instruments such as slotted lines, air lines, and vector network analyzers are also used for phase-angle measurements. *See* MICROWAVE MEASUREMENTS.   [J.Hur.]

Phase rule   A relationship used to determine the number of state variables F, usually chosen from among temperature, pressure, and species compositions in each phase, which must be specified to fix the thermodynamic state of a system in equilibrium. It was derived by J. Willard Gibbs. The phase rule (in the absence of electric, magnetic, and gravitational phenomena) is given by the equation below, where C is the

$$F = C - P - M + 2$$

number of chemical species present at equilibrium, P is the number of phases, and M is the number of independent chemical reactions. Here phase is used to indicate a homogeneous, mechanically separable portion of the system, and the term independent reactions refers to the smallest number of chemical reactions which, upon forming various linear combinations, includes all reactions which occur among the species present. The number of independent state variables F is referred to as the degrees of freedom or variance of the system. *See* THERMODYNAMIC PROCESSES.   [S.I.S.]

Phase transitions   Changes of state brought about by a change in an intensive variable (for example, temperature or pressure) of a system. Some familiar examples of phase transitions are the gas-liquid transition (condensation), the liquid-solid transition (freezing), the normal-to-superconducting transition in electrical conductors, the paramagnet-to-ferromagnet transition in magnetic materials, and the superfluid transition in liquid helium. Further examples include transitions involving amorphous or glassy structures, spin glasses, charge-density waves, and spin-density waves. *See* AMORPHOUS SOLID; CHARGE-DENSITY WAVE; SPIN-DENSITY WAVE; SPIN GLASS; SUPERCONDUCTIVITY; SUPERFLUIDITY.

Typically the phase transition is brought about by a change in the temperature of the system. The temperature at which the change of state occurs is called transition temperature (usually denoted by T_c). For example, the liquid-solid transition occurs at the freezing point.

The two phases above and below the phase transition can be distinguished from each other in terms of some ordering that takes place in the phase below the transition temperature. For example, in the liquid-solid transition, the molecules of the liquid get "ordered" in space when they form the solid phase. In a paramagnet, the magnetic moments on the individual atoms can point in any direction (in the absence of an internal magnetic field), but in the ferromagnetic phase the moments are lined up along a particular direction, which is then the direction of ordering. Thus in the phase above the transition, the degree of ordering is smaller than in the phase below the transition. One measure of the amount of disorder in a system is its entropy, which is the negative of the first derivative of the thermodynamic free energy with respect to temperature. When a system possesses more order, the entropy is lower. Thus at the transition temperature the entropy of the system changes from a higher value above the transition to some lower value below the transition. *See* ENTROPY; FERROMAGNETISM; PARAMAGNETISM.

This change in entropy can be continuous or discontinuous at the transition temperature. In other words, the development of order in the system at the transition temperature can be gradual or abrupt. This leads to a convenient classification of phase transitions into two types, namely, discontinuous and continuous.

Discontinuous transitions involve a discontinuous change in the entropy at the transition temperature. A familiar example of this type of transition is the freezing of water into ice. As water reaches the freezing point, order develops without any change in temperature. Thus there is a discontinuous decrease in the entropy at the freezing point. This is characterized by the amount of latent heat that must be extracted from the water for it to be "ordered" into the solid phase (ice). Discontinuous transitions are also called first-order transitions.

In a continuous transition, entropy changes continuously, and hence the growth of order below T_c is also continuous. There is no latent heat involved in a continuous transition. Continuous transitions are also called second-order transitions. The paramagnet-to-ferromagnet transition in magnetic materials is an example of such a transition.

The degree of ordering in a system undergoing a phase transition can be made quantitative in terms of an order parameter. At temperatures above the transition temperature the order parameter has a value zero, and below the transition it acquires some nonzero value. For example, in a ferromagnet the order parameter is the magnetic moment per unit volume (in the absence of an externally applied magnetic field). It is zero in the paramagnetic state since the individual magnetic moments in the solid may point in any random direction. Below the transition temperature, however, there exists a preferred direction of ordering, and as the temperature is lowered below T_c, more and more individual magnetic moments start to align along the preferred direction of ordering, leading to a continuous growth of the magnetization or the macroscopic magnetic moment per unit volume in the ferromagnetic state. Thus the order parameter changes continuously from zero above to some nonzero value below the transition

temperature. In a first-order transition, the order parameter would change discontinuously at the transition temperature.   [D.J.S.; S.J.]

Phase velocity   The velocity of propagation of a pure sine wave of infinite
extent. In one dimension, for example, the form of the disturbance for such a wave is
given by $y(x, t) = A \sin [2\pi (x/\lambda - t/T)]$. Here x is the position at which the disturbance
$y(x, t)$ exists at time t, λ is the wavelength, T is the period which is related to the wave
frequency by $T = 1/f$, and A is the disturbance amplitude. The argument of the sine
function is called the phase. The phase velocity is the speed with which a point of
constant phase can be said to move. Thus $x/\lambda - ft = $ constant, so the phase velocity
v_p is given by $dx/dt = v_p = \lambda f$. This is the basic relationship connecting phase velocity,
wavelength, and frequency. See PHASE (PERIODIC PHENOMENA); SINE WAVE; WAVE MOTION.
   The phase velocity for waves in a medium is determined in part by intrinsic properties
of the medium. For all mechanical waves in elastic media, the square of the phase
velocity is proportional to the ratio of the appropriate elastic property of the medium to
the appropriate inertia property. The phase velocity of electromagnetic waves depends
upon the medium as well. In vacuum, the phase velocity c is given by $c^2 = 1/\epsilon_0\mu_0 \approx$
9×10^{16} m^2/s^2, where ϵ_0 and μ_0 are respectively the permittivity and permeability of
the vacuum. Phase velocity may also depend upon the mode of wave propagation—in
general, upon the frequency of the wave. Waves of different frequencies will travel at
different speeds, resulting in a phenomenon called dispersion. See ELECTROMAGNETIC
RADIATION; LIGHT; WAVE EQUATION.   [S.A.Wi.]

Phonon   A quantum of vibrational energy in a solid or other elastic medium. This
vibrational energy can be transported by elastic waves. The energy content of each
wave is quantized. For a wave of frequency f, the energy is $(N + \frac{1}{2})hf$, where N is an
integer and h is Planck's constant. Apart from the zero-point energy, $\frac{1}{2}hf$, there are
N quanta of energy hf. In elastic or lattice waves, these quanta are called phonons.
Quantization of energy is not related to the discreteness of the lattice, and also applies
to waves in a continuum. See QUANTUM MECHANICS; WAVE MOTION.
   The concept of phonons closely parallels that of photons, quanta of electromagnetic
wave energy. The indirect consequences of quantization were established for phonons
just as for photons in the early days of quantum mechanics—for example, the decrease
of the specific heat of solids at low temperatures. Direct evidence that the energy of
vibrational modes is changed one phonon at a time came much later than that for
photons—for example, the photoelectric effect—because phonons exist only within a
solid, are subject to strong attenuation and scattering, and have much lower quantum
energy than optical or x-ray photons. See PHOTOEMISSION.
   Like photons, phonons can be regarded as particles, each of energy hf and momentum proportional to the wave vector of the elastic or lattice wave. Such a particle can
be said to transport energy, thus moving with a velocity equal to the group velocity of
the underlying wave. See LATTICE VIBRATIONS; PHOTON.   [P.G.Kl.]

Phosphorescence   A delayed luminescence, that is, a luminescence that persists after removal of the exciting source. It is sometimes called afterglow.
   This original definition is rather imprecise, because the properties of the detector
used will determine whether or not there is an observable persistence. There is no generally accepted rigorous definition or uniform usage of the term phosphorescence. In
the literature of inorganic luminescent systems, some authors define phosphorescence
as delayed luminescence whose persistence time decreases with increasing temperature. According to this usage, luminescence whose persistence time is independent of

temperature is called fluorescence regardless of the length of the afterglow; a temperature-independent afterglow of long duration is called simply a slow fluorescence, which implies that the atomic or molecular transition involved is forbidden to a greater or lesser degree by the spectroscopic selection rules. The most common mechanism of phosphorescence in photoconductive inorganic systems, however, occurs when electrons or holes, set free by the excitation process and trapped at lattice defects, are expelled from their traps by the thermal energy in the system and recombine with oppositely charged carriers with the emission of light. *See* HOLE STATES IN SOLIDS; SELECTION RULES (PHYSICS).

In the organic literature the term phosphorescence is reserved for the forbidden luminescent transition from a metastable energy state M to the ground state G, while the afterglow corresponding to the M→E→G process (where E is a higher energy state) is called delayed fluorescence. *See* FLUORESCENCE; LIGHT; LUMINESCENCE.   [C.C.K.; J.H.S.]

Photoconductivity   The increase in electrical conductivity caused by the excitation of additional free charge carriers by light of sufficiently high energy in semiconductors and insulators. Effectively a radiation-controlled electrical resistance, a photoconductor can be used for a variety of light- and particle-detection applications, as well as a light-controlled switch. Other major applications in which photoconductivity plays a central role are television cameras (vidicons), normal silver halide emulsion photography, and the very large field of electrophotographic reproduction. *See* OPTICAL MODULATORS; PARTICLE DETECTOR.

Although all insulators and semiconductors may be said to be photoconductive, that is, they show some increase in electrical conductivity when illuminated by light of sufficiently high energy to create free carriers, only a few materials show a large enough change, that is, show a large enough photosensitivity, to be practically useful in applications of photoconductors.

Since the electrical conductivity σ of a material is given by the product of the carrier density, its charge, and its mobility, an increase in the conductivity can be formally due to either an increase in carrier density or an increase in mobility. Although cases are found in which both types of effects are observable, photoconductivity in single-crystal materials is due primarily to an increase in earner density. In polycrystalline materials, on the other hand, where transport may be limited by potential barriers between the crystalline grains, an increase in mobility due to photoexcitation effects on these intergrain barriers may dominate the photoconductivity.

The variation of photoconductivity with photon energy is called the spectral response of the photoconductor. Spectral response curves typically show a fairly well-defined maximum at a photon energy close to that of the bandgap of the material, that is, the minimum energy required to excite an electron from a bond in the material into a higher-lying conduction band where it is free to contribute to the conductivity. This energy ranges from 3.7 eV, in the ultraviolet, for zinc sulfide (ZnS) to 0.2 eV, in the infrared, for cooled lead selenide (PbSe).

Another major characteristic of a photoconductor of practical concern is the rate at which the conductivity changes with changes in photoexcitation intensity. If a steady photoexcitation is turned off at some time, for example, the length of time required for the current to decrease to $1/e$ of its initial value is called the decay time of photoconductivity, t_d. The magnitude of the decay time is determined by the lifetime π and by the density of carriers trapped in imperfections as a result of the previous photoexcitation, which must now also be released in order to return to the thermal equilibrium situation.
                                                                    [R.H.Bu.]

Photoemission   The ejection of electrons from a solid (or less commonly, a liquid) by incident electromagnetic radiation. Photoemission is also called the external photoelectric effect. The visible and ultraviolet regions of the electromagnetic spectrum are most often involved, although the infrared and x-ray regions are also of interest.

The salient experimental features of photoemission are the following: (1) There is no detectable time lag between irradiation of an emitter and the ejection of photoelectrons. (2) At a given frequency the number of photoelectrons ejected per second is proportional to the intensity of the incident radiation. (3) The photoelectrons have kinetic energies ranging from zero up to a well-defined maximum, which is proportional to the frequency of the incident radiation and independent of the intensity.

In 1905 Albert Einstein made the clarifying assumption that electromagnetic radiation had characteristics like those of particles when it delivered energy to electrons in the emitter. In Einstein's approach the light beam behaves like a stream of photons, each of energy $h\nu$, where h is Planck's constant, and ν is the frequency of the photon. The energy required to eject an electron from the emitter has a well-defined minimum value ϕ called the photoelectric threshold energy. When a photon interacts with an electron, the latter absorbs the entire photon energy. *See* PHOTON.

For $h\nu$ values below the threshold, photoelectrons are not ejected. Even though the electrons absorb photon energy, they do not receive enough to surmount the potential barrier at the surface, which normally holds the electrons in the solid. For photon energies above ϕ, the kinetic energies of photo-electrons range from zero up to a maximum value, $E = h\nu - \phi$. This is the Einstein photoelectric law, and E is commonly termed the Einstein maximum energy. *See* HEAT RADIATION; SCHOTTKY EFFECT.   [L.Ap.]

Photoionization   The ejection of one or more electrons from an atom, molecule, or positive ion following the absorption of one or more photons. The process of electron ejection from matter following the absorption of electromagnetic radiation has been under investigation for over a century. The earliest measurements involved the ultraviolet irradiation of metal surfaces. The theoretical interpretation of this phenomenon, known as the photoelectric effect, played an important role in establishing quantum mechanics. It was shown that, contrary to classical ideas, energy exchanges between radiation and matter are mediated by integral numbers of photons. In the gas phase the photoeffect is called either photoionization (atoms, molecules, and their positive ions) or photodetachment (atomic and molecular negative ions). *See* PHOTOEMISSION.

Photoionization involves a radiative bound-free transition from an initial state consisting of n photons and an atom, molecule, or ion in a bound state to a final continuum state consisting of a residual ion (or an atom in the case of photodetachment) and m free electrons; that is,

$$n h\nu + X \rightarrow X^{m+} + me^-$$

In the simplest atomic photoionization process a single electron is ejected from an atom following the absorption of a single photon. Each mode of fragmentation defines a final-state channel that is characterized by the energy and angular momentum of the outgoing electron as well as the excitation state of the residual ion. Since the photoionization process is endoergic, each channel has a well-defined threshold energy below which the channel is energetically closed. The threshold photon energy for a particular channel is equal to the binding energy of the electron that is to be ejected plus the excitation energy, if any, of the residual ion.

Above threshold, the energy carried off by the outgoing electron represents the balance between the energy supplied by the photon and the binding energy of the electron plus the excitation energy of the residual ion (neglecting the small recoil of the heavy ion). A photoelectron spectrum is characterized by a discrete set of peaks, each peak being associated with a particular state of the residual ion. Information on the excitation state of the ion following photoionization can also be obtained by monitoring the fluorescence emitted in the subsequent radiative decay of the state. One of the earliest applications of photoionization measurements was the investigation of the structure of atoms by determining the binding energies of both outer- and inner-shell electrons by means of photoelectron spectroscopy. *See* ATOMIC STRUCTURE AND SPECTRA.

[D.J.Pe.]

Photoluminescence A luminescence excited in a body by some form of electromagnetic radiation incident on the body. The term photoluminescence is generally limited to cases in which the incident radiation is in the ultraviolet, visible, or infrared regions of the electromagnetic spectrum.

Photoluminescence may be either a fluorescence or a phosphorescence, or both. Energy can be stored in certain luminescent materials by subjecting them to light or some other exciting agent, and can be released by subsequent illumination of the material with light of certain wavelengths. This type of photoluminescence is called stimulated photoluminescence. *See* FLUORESCENCE; LUMINESCENCE; PHOSPHORESCENCE.

[C.C.K.; J.H.S.]

Photometer An instrument used for making measurements of light, or electromagnetic radiation, in the visible range. In general, photometers may be divided into two classifications: laboratory photometers, which are usually fixed in position and yield results of high accuracy; and portable photometers, which are used in the field or outside the laboratory and yield results of lower accuracy. Each class may be subdivided into visual (subjective) photometers and photoelectric (objective or physical) photometers. These in turn may be grouped according to function, such as photometers to measure luminous intensity (candelas or candlepower), luminous flux, illumination (illuminance), luminance (photometric brightness), light distribution, light reflectance and transmittance, color, spectral distribution, and visibility. Visual photometric methods have largely been supplanted commercially by physical methods, but because of their simplicity, visual methods are still used in educational laboratories to demonstrate photometric principles. *See* ILLUMINANCE; LUMINANCE; LUMINOUS FLUX; LUMINOUS INTENSITY.

[G.A.Ho.]

Photometry That branch of science which deals with measurements of light (visible electromagnetic radiation) according to its ability to produce visual sensation. Specifically, photometry deals with the attribute of light that is perceived as intensity, while the related attribute of light that is perceived as color is treated in colorimetry. *See* COLOR.

The purely physical attributes of light such as energy content and spectral distribution are treated in radiometry. Sometimes the word photometry is used to denote measurements that have nothing to do with human vision, but this is a mistake according to modern usage. Such measurements are properly referred to as radiometry, even if they are performed in the visible spectral region. *See* RADIOMETRY.

The relative visibility of a fixed power level of monochromatic electromagnetic radiation varies with wavelength over the visible spectral region (400–700 nanometers). The relative visibility of radiation also depends upon the illumination level that is being observed. The cone cells in the retina determine the visual response at high levels of illumination, while the rod cells dominate in the dark-adapted eye at very low levels (such as starlight). Cone-controlled vision is called photopic, and rod-controlled vision is called scotopic, while the intermediate region where both rods and cones play a role is called mesopic.

Originally, photometry was carried out by using the human visual sense as the detector of light. As a result, photometric measurements were subjective. In order to put photometric measurements on an objective basis, and to allow convenient electronic detectors to replace the eye in photometric measurements, the Commission Internationale de l'Eclairage (CIE; International Commission on Illumination) has adopted two relative visibility functions as standards. These internationally accepted functions are called the spectral luminous efficiency functions for photopic and scotopic vision, and are denoted by $V(\lambda)$ and $V'(\lambda)$, respectively. See LUMINOUS EFFICIENCY.

Thus photopic and scotopic (but not mesopic) photometric quantities have objective definitions, just as do the purely physical quantities. However, there is a difference. The purely physical quantities are defined in terms of physical laws, whereas the photometric quantities are defined by convention. In recognition of this difference the photometric quantities are called psychophysical quantities.

According to the International System of Units, SI, the photometric units are related to the purely physical units through a defined constant called the maximum spectral luminous efficacy. This quantity, which is denoted by K_m, is the number of lumens per watt at the maximum of the $V(\lambda)$ function. K_m is defined in SI to be 683 lm/W for monochromatic radiation whose wavelength is 555 nanometers, and this defines the photometric units with which the photometric quantities are to be measured.

At various times, the photometric units have been defined in terms of the light from different standard sources, such as candles made according to specified procedures, and blackbodies at the freezing point of platinum. According to these definitions, K_m was a derived, rather than defined, quantity. See LIGHT; PHYSICAL MEASUREMENT; UNITS OF MEASUREMENT.   [J.Gei.]

Photon   An entity that can be loosely described as a quantum of energy of electromagnetic radiation. According to classical electromagnetic theory, an electromagnetic wave can transfer arbitrarily small amounts of energy to matter. According to the quantum theory of radiation, however, the energy is transferred in discrete amounts. The energy of a photon is the product of Planck's constant and the frequency of the electromagnetic field. In addition to energy, the photon possesses momentum and also possesses angular momentum corresponding to a spin of unity. The interaction of radiation with matter involves the absorption, scattering, and emission of photons. Consequently, the energy interchange is inherently quantized. See ANGULAR MOMENTUM; ENERGY; MOMENTUM; SPIN (QUANTUM MECHANICS).

For many purposes, the photon behaves like a particle of zero rest mass moving at the speed of light. The particlelike nature of the photon is vividly exhibited by the photoelectric effect, predicted by A. Einstein, in which light is absorbed in a metal, causing electrons to be ejected. An electron absorbs a photon, gaining its energy. In leaving the metal, it loses energy because of interactions with the surface; the energy loss equals the product of the so-called work function of the surface and the charge of the electron. The final kinetic energy of the electron therefore equals the energy of the incident photon minus this energy loss. See PHOTOEMISSION.

A second demonstration of the particlelike behavior of photons is provided by the scattering of an x-ray photon from an electron bound in an atom. The electron recoils because of the momentum of the photon, thereby gaining energy. As a result, the frequency, and hence the wavelength of the scattered x-ray, is altered. If the x-ray is scattered through a certain angle, the wavelength is shifted by an amount determined by this scattering angle and the mass of an electron, according to the laws of conservation of energy and momentum. *See* COMPTON EFFECT.

From a more fundamental view, the photon is the quantum of excitation of a single mode of a radiation field. The dynamical equations for the electric and magnetic energy in such a field are identical to those of a harmonic oscillator. According to quantum theory, the allowed energies of a harmonic oscillator are given by $E = (j + \frac{1}{2})hf$, where h is Planck's constant, f is the frequency of the oscillator, and the quantum number $j = 0, 1, 2, \ldots$, describes the state of excitation of the oscillator. This quantum relation was first postulated by M. Planck for the material oscillators in the walls of a thermal enclosure in order to obtain the correct form for the density of radiation in a thermal field, but it was quickly applied by Einstein to describe the state of the radiation field itself. In this picture, j describes the number of photons in the field. *See* HARMONIC OSCILLATOR; QUANTUM ELECTRODYNAMICS; QUANTUM MECHANICS.   [D.Kl.]

Photovoltaic effect   The conversion of electromagnetic radiation into electric power through absorption by a semiconducting material. Devices based on this effect serve as power sources in remote terrestrial locations and for satellites and other space applications. Photovoltaic–powered calculators and other consumer electronic products are widely available, and solar photovoltaic automobiles and aircraft have been demonstrated.

The basic requirements for the photovoltaic effect are (1) the absorption of photons through the creation of electron-hole pairs in a semiconductor; (2) the separation of the electron and hole so that their recombination is inhibited and the electric field within the semiconductor is altered; and (3) the collection of the electrons and holes, separately, by each of two current-collecting electrodes so that current can be induced to flow in a circuit external to the semiconductor itself.

There are many approaches to achieving these three requirements simultaneously. A very common approach for separating the electrons from the holes is to use a single-crystal semiconductor, for example, silicon, into which a *pn* junction has been diffused. Silicon is often chosen because its optical band gap permits the absorption of a substantial portion of solar photons via the generation of electron-hole pairs. The fabrication of such a device structure causes a local transfer of negative charges from the *n* layer into the *p* layer, bending the conduction and valence bands in the vicinity of the *p-n* boundary, and thereby creating a rectifying junction. Electrons generated in the *p* region can lower their energy by migrating into the *n* region, which they will do by a random walk process in the electric-field-free region far from the junction, or by drift induced by the electric field in the junction region. Holes created in the *n* region, conversely, lose energy by migrating into the *p* region. Thus the presence of such a junction leads to the spontaneous spatial separation of the photogenerated carriers, thereby inducing a voltage difference between current-carrying electrodes connected to the *p* and *n* regions. This process will continue until the difference in potential between the two electrodes is large enough to flatten the bands in the vicinity of the junction, canceling out the internal electric field existing there and so eliminating the source of carrier separation. The resulting voltage is termed the open-circuit voltage, and approximates the built-in voltage associated with the *pn* junction in the dark, a value which cannot exceed the band gap of the semiconductor. *See* HOLE STATES IN SOLIDS; SEMICONDUCTOR; SEMICONDUCTOR DIODE.

In the limit when the device is short-circuited by the external circuit, no such buildup of potential can occur. In this case, one electron flows in the external circuit for each electron or hole which crosses the junction, that is, for each optically generated electron-hole pair which is successfully separated by the junction. The resulting current is termed the short-circuit current and, in most practical photovoltaic devices, approaches numerically the rate at which photons are being absorbed within the device. Losses can arise from the recombination of minority carriers (for example, electrons in the p-type region, holes in the n-type region) with majority carriers. *See* ELECTRON-HOLE RECOMBINATION.

For a photovoltaic device to generate power, it is necessary to provide a load in the external circuit which is sufficiently resistive to avoid short-circuiting the device. In this case, the voltage will be reduced compared to the open-circuit voltage because a continuing requirement exists for carrier separation at the junction; thus some band bending and its associated internal field must be retained.

Various multiple-layered device configurations based on doped and undoped alloys of amorphous silicon have been developed for photovoltaic devices used in applications ranging from solar watches and calculators to remote power generators. The photovoltaic effect in these devices is particularly intriguing since it is possible to build up so-called tandem devices by stacking one device electrically and optically in series above another. In addition to the increased voltage and concomitant reduction in the required current-carrying capability of electrode grid structures, such devices permit, in principle, an increased efficiency of solar photovoltaic energy conversion.

[J.P.deN.]

Physical law   A term that designates four different concepts: (1) objective pattern (or natural regularity), (2) formula purporting to represent an objective pattern, (3) law-based rule (or uniform procedure), and (4) principle concerning any of the preceding.

For example, Newton's second law of motion, $ma = F$, is a law of type 2. It represents, to a good approximation, the actual behavior (law of type 1) of medium-size particles moving slowly relative to the speed of light. Alternative laws of motion, such as the relativistic and quantum-mechanical ones, are different laws of type 2 representing the same objective pattern or law of type 1 to even better approximations. One of the rules (laws of type 3) associated with Newton's second law of motion is: In order to set in motion a stationary particle, exert a force on it. Another is: In order to stop a moving particle, exert on it a force in the opposite direction. An example of a law of type 4 is: Newton's laws of motion are invariant under a Galileo transformation. *See* NEWTON'S LAWS OF MOTION.

A physical law of type 1, or objective pattern, is a constant relation among two or more properties of a physical entity. In principle, any such pattern can be conceptualized in different ways, that is, as alternative laws of type 2. The history of theoretical physics is to a large extent a sequence of laws of type 2. Every one of these is hoped to constitute a more accurate representation of the corresponding objective pattern or law of type 1, which is assumed to be constant and, in particular, untouched by human efforts to grasp it. Likewise, the history of engineering is to some extent a sequence of laws of type 3, or law-based rules of action, of which there are least two for every law of type 2. As for the laws of type 4, or laws of laws, they are of two kinds: scientific and philosophical. The general covariance principle is of the first kind, whereas the hypothesis that all events are lawful is a philosophical thesis. Unlike the former, whose truth can be checked, the principle of lawfulness is irrefutable. *See* THEORETICAL PHYSICS.

Not all formulas are called physical laws. For example, the regularities found by curve fitting are called empirical formulas. In physics a formula is called a law if and only if it meets the following conditions: it is part of a theory, and it has been satisfactorily confirmed by measurement or experiment at least within a certain domain (for example, for small mass densities or high field intensities). Thus, the basic assumptions of all the standard physical theories are laws, and so are their logical consequences. In particular, the usual variational principles, such as Hamilton's, are basic laws. However, the equations of motion and field equations entailed by such principles are derived laws (theorems); so are the conservation laws entailed by the equations of motion and field equations. However, the distinction between basic and derived laws is contextual: what is a principle in one theory may be a theorem in another. For example, Newton's second law of motion is a theorem in analytical dynamics, and the first principle of thermodynamics is a theorem of statistical mechanics. *See* CONSERVATION LAWS (PHYSICS); HAMILTON'S PRINCIPLE; PHYSICAL THEORY; STATISTICAL MECHANICS; THERMODYNAMIC PRINCIPLES; VARIATIONAL METHODS (PHYSICS).

[M.Bun.]

Physical measurement

Physical measurement Quantitative information on physical conditions, properties, or relations essential for coordination of activities, efficiency of communication, and understanding of the nature of things in science and engineering and in much of everyday life. Time, distance, mass, temperature, force, power, and all other physical quantities (or parameters or variables), as well as the properties of matter, materials, and devices, must be described and measured in terms which have the same meaning for everyone. The measuring device or instrument is calibrated (that is, the functional relationship between its indication and the magnitude of the measured quantity is determined) by direct or indirect comparison with a standard which embodies, possesses, or generates a fixed or reproducible magnitude of the physical quantity which is taken as the unit or some multiple or fraction of the unit. Any measured quantity may thus be expressed by a number (the magnitude ratio) and the name of the unit, for example, a length of 1.54 meters. The general area of scientific activity relating to standards and units and the accuracy of measurement is called metrology. *See* UNITS OF MEASUREMENT.

Metric system. The basic unit of length in the decimal metric system was defined as one ten-millionth of the Earth's polar quadrant (as determined from latitude surveys), and is termed the meter. The basic unit for mass was defined as the mass of a cubic decimeter of water, to be called the kilogram.

The United States has adopted the Metric Conversion Act, declaring that "the policy of the U.S. shall be to coordinate and plan the increasing use of the metric system in the United States," and established the U.S. Metric Board "to coordinate the voluntary conversion to the metric system." However, English units have become almost universal in some worldwide industries—for example, dimensions of oil-drilling equipment, or altitude measurement in aviation. Thus it is likely that there will always be exceptions to uniformity, requiring special knowledge of special units for at least some people even as the whole world "goes metric" in principle.

International System of Units (SI). At present the International System of Units (abbreviated SI, from the French Système International d'Unités) is constructed from seven base units for independent quantities (Table 1). Units for all other quantities are derived from these seven units. In Table 2 are listed 22 SI derived units with special

Table 1. SI base units

Quantity*	Unit name	Symbol
Length	meter	m
Mass	kilogram	kg
Time	second	s
Electric current	ampere	A
Thermodynamic temperature	kelvin	K
Amount of substance	mole	mol
Luminous intensity	candela	cd

*Quantity here and in Table 2 means a measurable attribute.

names. These units are derived from the base units in a coherent manner, which means they are expressed as products and quotients of the seven base units without numerical factors. All other SI derived units are similarly derived in a coherent manner from the 29 base and special-name SI units. For use with the SI units, there is a set of 20 prefixes (Table 3) to form multiples and submultiples of these units. For mass, the prefixes are to be applied to the gram instead of to the SI unit, the kilogram. *See* DIMENSIONAL ANALYSIS.

The SI units together with the SI prefixes provide a logical and interconnected framework for measurements in science, industry, and commerce.

In some cases, quantities are commonly expressed in terms of fundamental constants of nature, and use of these constants or "natural units" is acceptable. *See* FUNDAMENTAL CONSTANTS.

Table 2. SI derived units with special names

Quantity	Unit name	Symbol	Expression in terms of other units	Expression in terms of SI base units
Plane angle	radian	rad		$m \cdot m^{-1} = 1$
Solid angle	steradian	sr		$m^2 \cdot m^{-2} = 1$
Frequency	hertz	Hz		s^{-1}
Force	newton	N		$m \cdot kg \cdot s^{-2}$
Pressure, stress	pascal	Pa	N/m^2	$m^{-1} \cdot kg \cdot s^{-2}$
Energy, work, quantity of heat	joule	J	$N \cdot m$	$m^2 \cdot kg \cdot s^{-2}$
Power, radiant flux	watt	W	J/s	$m^2 \cdot kg \cdot s^{-3}$
Quantity of electricity, electric charge	coulomb	C	$A \cdot s$	$s \cdot A$
Electric potential difference, electromotive force, voltage	volt	V	W/A	$m^2 \cdot kg \cdot s^{-3} \cdot A^{-1}$
Capacitance	farad	F	C/V	$m^{-2} \cdot kg^{-1} \cdot s^4 \cdot A^2$
Electric resistance	ohm	Ω	V/A	$m^2 \cdot kg \cdot s^{-3} \cdot A^{-2}$
Electric conductance	siemens	S	A/V	$m^{-2} \cdot kg^{-1} \cdot s^3 \cdot A^2$
Magnetic flux	weber	Wb	$V \cdot s$	$m^2 \cdot kg \cdot s^{-2} \cdot A^{-1}$
Magnetic flux density	tesla	T	Wb/m^2	$kg \cdot s^{-2} \cdot A^{-1}$
Inductance	henry	H	Wb/A	$m^2 \cdot kg \cdot s^{-2} \cdot A^{-2}$
Celsius temperature	degree Celsius	°C		K
Luminous flux	lumen	lm	$cd \cdot sr$	$m^2 \cdot m^{-2} \cdot cd = cd$
Illuminance	lux	lx	lm/m^2	$m^2 \cdot m^{-4} \cdot cd = m^{-2} \cdot cd$
Activity (of a radionuclide)	becquerel	Bq		s^{-1}
Absorbed dose, specific energy imparted, kerma	gray	Gy	J/kg	$m^2 \cdot s^{-2}$
Dose equivalent	sievert	Sv	J/kg	$m^2 \cdot s^{-2}$
Catalytic activity	katal	kat		$s^{-1} \cdot mol$

Table 3. SI prefixes

Factor	Prefix	Symbol	Factor	Prefix	Symbol
10^{24}	yotta	Y	10^{-1}	deci	d
10^{21}	zetta	Z	10^{-2}	centi	c
10^{18}	exa	E	10^{-3}	milli	m
10^{15}	peta	P	10^{-6}	micro	μ
10^{12}	tera	T	10^{-9}	nano	n
10^{9}	giga	G	10^{-12}	pico	p
10^{6}	mega	M	10^{-15}	femto	f
10^{3}	kilo	k	10^{-18}	atto	a
10^{2}	hecto	h	10^{-21}	zepto	z
10^{1}	deka	da	10^{-24}	yocto	y

Typical examples of natural units, with their symbols, are:

elementary charge	e
electron mass	m_e
proton mass	m_p
Bohr radius	a_0
electron radius	r_e
Compton wavelength of electron	λ_c
Bohr magneton	μ_B
nuclear magneton	μ_N
speed of light	c
Planck constant	h

Certain units which are not part of the SI are used so widely that it is impractical to abandon them. The units that are accepted for continued use with the International System are listed in Table 4. It is likewise necessary to recognize, outside the International System, the following units which are used in specialized fields:

electronvolt	eV
unified atomic mass unit	u
astronomical unit	AU
parsec	pc

Logarithmic measures such as pH, dB (decibel), and Np (neper) are acceptable. *See* ATOMIC MASS UNIT; DECIBEL; ELECTRONVOLT.

The internationally accepted definitions for the seven base units follow:

Mass. The kilogram (kg) is equal to the mass of the International Prototype Kilogram. The International Prototype is a platinum-iridium cylinder preserved at the International Bureau of Weights and Measures at Sèvres, France.

Mass is the only one of the base quantities for which the standard is an arbitrarily defined object. No basic property of matter involving mass can be measured with more precision than is possible in comparing kilogram masses by weighing, about 1 part in 10^8. *See* MASS.

[W.A.Wi.]

Length. The meter is defined in terms of time and the speed of light: "The meter is the length of the path traveled by light in a vacuum during a time interval of 1/299 792 458 of a second." This definition defines the speed of light to be exactly 299 792 458 m/s and defines the meter in terms of the most accurately known quantity, the second. *See* LIGHT.

Table 4. Units in use with the International System

Name	Symbol	Value in SI unit
Minute	min	1 min = 60 s
Hour	h	1 h = 60 min = 3600 s
Day	d	1 d = 24 h = 86,400 s
Degree	°	$1° = (\pi/180)$ rad
Minute	′	$1′ = (1/60)° = (\pi/10{,}800)$ rad
Second	″	$1″ = (1/60)′ = (\pi/648{,}000)$ rad
Liter	L*	$1 \text{ L} = 1 \text{ dm}^3 = 10^{-3} \text{ m}^3$
Metric ton	t	$1 \text{ t} = 10^3 \text{ kg}$
Neper[a]	Np	1 Np = 1
Bel[b]	B	1 B = (1/2) ln 10 (Np)

*An alternate symbol for liter is "l." Since "l" can be easily confused with the numeral 1, the symbol "L" is recommended for United States use.
[a]The neper is used to express values of various logarithmic quantities. Natural logarithms are used to obtain the numerical values of quantities expressed in nepers. The neper is coherent with the SI, but is not yet adopted as an SI unit.
[b]The bel is used to express values of various logarithmic quantities. Logarithms to base ten are used to obtain the numerical values of quantities expressed in bels.

The most accurate method of realizing the meter is by means of an interferometrically measured distance by fringe counting in which each vacuum fringe is a half wavelength from the next one. This wavelength, λ, is obtained from the measured frequency, f, using the relation $\lambda = c/f$, where c is the value of the speed of light in vacuum. To this end, major standards laboratories have measured the frequencies of several lasers stabilized to narrow molecular absorptions in the visible and near-infrared spectral regions. These stabilized lasers now serve as standards of length. *See* INTERFEROMETRY; WAVELENGTH STANDARDS. [D.A.J]

Time interval. The second (s) is the duration of 9 192 631 770 periods of the radiation corresponding to the transition between the two hyperfine levels of the ground state of the cesium-133 atom. In the best equipments the stability and accuracy of the cesium frequency generator correspond to an uncertainty of a few parts in 10^{15}.

The second was long defined, for physical measurements as well as for civil affairs, as 1/86,400 of the time required for an average complete rotation of the Earth on its axis with respect to the Sun. Because of the slight slowing of the Earth's rotation rate, now averaging about 1 second per year (that is, 3 parts in 10^8) but with erratic and unexplained fluctuations, the universal second thus defined is not a constant. A time scale called Coordinated Universal Time (UTC) recommended by the General Conference of Weights and Measures (CGPM) in 1975 is defined in such a manner that it differs from international atomic time (TAI) by an exact whole number of seconds. This difference is adjusted occasionally by the use of a positive or negative leap second at the end of certain months to keep UTC in agreement with the time defined by the rotation of the Earth with an approximation better than 9/10 second. *See* ATOMIC CLOCK; FREQUENCY MEASUREMENT; TIME. [W.A.Wi.]

Temperature. The kelvin (K), the unit of thermodynamic temperature, is the fraction 1/273.16 of the thermodynamic temperature of the triple point of water. The unit kelvin and its symbol K should also be used to express an interval or differences of temperature.

To provide convenient and adequately accurate means for practical realization and measurement of temperature, the International Temperature Scale is used, based on the assigned values of the temperatures of a number of reproducible equilibrium states (defining fixed points), on standard instruments calibrated at those temperatures, and on vapor-pressure temperature relationships. Interpolation between the fixed-point temperatures is provided by formulas used to establish the relation between indications of the standard instruments and values of International Temperature. An extensive revision, which came into effect in 1990, is called the ITS-90. *See* TEMPERATURE; TEMPERATURE MEASUREMENT. [B.W.M.]

Electric current. The ampere (A) is that constant current which, if maintained in two straight parallel conductors of infinite length and of negligible circular sections, and placed 1 meter apart in a vacuum, would produce between these conductors a force equal to 2×10^{-7} newton per meter of length. *See* ELECTRICAL UNITS AND STANDARDS.

Luminous intensity. The CGPM, in 1979, redefined the base SI unit candela as the luminous intensity, in a given direction, of a source that emits monochromatic radiation of frequency 540×10^{12} hertz and of which the radiant intensity in that direction is 1/683 watt per steradian. *See* LIGHT; LUMINOUS EFFICACY; LUMINOUS EFFICIENCY; LUMINOUS INTENSITY; PHOTOMETRY; RADIOMETRY.

Amount of substance. The mole is the amount of substance of a system which contains as many elementary entities as there are atoms in 0.012 kilogram of carbon-12. When the mole is used, the elementary entities must be specified, and may be atoms, molecules, ions, electrons, other particles, or specified groups of such particles. [W.A.Wi.]

Physical optics   The study of the interaction of electromagnetic waves in the optical range with material systems. The optical range of wavelengths may be taken as the range from about 1 nanometer to about 1 millimeter.

The explanation of the absorption, reflection, scattering, polarization, and dispersion of light by a material medium in terms of the properties of the atoms and molecules making up the medium is the objective of physical optics. In the course of seeking this objective, physicists have found that optical investigations are powerful methods of determining the structures of atoms and molecules and of large systems composed thereof. *See* ATOMIC STRUCTURE AND SPECTRA; CRYSTAL OPTICS; DIFFRACTION; DISPERSION (RADIATION); ELECTROMAGNETIC RADIATION; ELECTROOPTICS; FARADAY EFFECT; FLUORESCENCE; INTERFERENCE OF WAVES; LASER; LIGHT; MAGNETOOPTICS; MOLECULAR STRUCTURE AND SPECTRA; POLARIZED LIGHT; REFLECTION OF ELECTROMAGNETIC RADIATION; REFRACTION OF WAVES; SCATTERING OF ELECTROMAGNETIC RADIATION; SPECTROSCOPY. [R.C.L.]

Physical theory   A physical theory usually involves the attempt to explain a certain class of physical phenomena by deducing them as necessary consequences of other phenomena regarded as more primitive and less in need of explanation. The value of a theory depends on both the success with which it coordinates a wide range of presently known facts and its fertility in suggesting places to look for presently unknown phenomena. [P.W.Br./G.Ho.]

Physics   Formerly called natural philosophy, physics is concerned with those aspects of nature which can be understood in a fundamental way in terms of elementary principles and laws. In the course of time, various specialized sciences broke away from physics to form autonomous fields of investigation. In this process physics retained its original aim of understanding the structure of the natural world and explaining natural phenomena.

The most basic parts of physics are mechanics and field theory. Mechanics is concerned with the motion of particles or bodies under the action of given forces. The physics of fields is concerned with the origin, nature, and properties of gravitational, electromagnetic, nuclear, and other force fields. Taken together, mechanics and field theory constitute the most fundamental approach to an understanding of natural phenomena which science offers. The ultimate aim is to understand all natural phenomena in these terms. *See* CLASSICAL FIELD THEORY; MECHANICS; QUANTUM FIELD THEORY.

The older, or classical, divisions of physics were based on certain general classes of natural phenomena to which the methods of physics had been found particularly applicable. The divisions are all still current, but many of them tend more and more to designate branches of applied physics or technology, and less and less inherent divisions in physics itself. The divisions or branches of modern physics are made in accordance with particular types of structures in nature with which each branch is concerned.

In every area physics is characterized not so much by its subject-matter content as by the precision and depth of understanding which it seeks. The aim of physics is the construction of a unified theoretical scheme in mathematical terms whose structure and behavior duplicates that of the whole natural world in the most comprehensive manner possible. Where other sciences are content to describe and relate phenomena in terms of restricted concepts peculiar to their own disciplines, physics always seeks to understand the same phenomena as a special manifestation of the underlying uniform structure of nature as a whole. In line with this objective, physics is characterized by accurate instrumentation, precision of measurement, and the expression of its results in mathematical terms.

For the major areas of physics and for additional listings of articles in physics *see* ACOUSTICS; ATOMIC PHYSICS; CLASSICAL MECHANICS; ELECTRICITY; ELECTROMAGNETISM; ELEMENTARY PARTICLE; FLUID MECHANICS; HEAT; LOW-TEMPERATURE PHYSICS; MOLECULAR PHYSICS; NUCLEAR PHYSICS; OPTICS; SOLID-STATE PHYSICS; STATISTICAL MECHANICS; THEORETICAL PHYSICS. [W.G.P.]

Piezoelectricity Electricity, or electric polarity, resulting from the application of mechanical pressure on a dielectric crystal. The application of a mechanical stress produces in certain dielectric (electrically nonconducting) crystals an electric polarization (electric dipole moment per cubic meter) which is proportional to this stress. If the crystal is isolated, this polarization manifests itself as a voltage across the crystal, and if the crystal is short-circuited, a flow of charge can be observed during loading. Conversely, application of a voltage between certain faces of the crystal produces a mechanical distortion of the material. This reciprocal relationship is referred to as the piezoelectric effect. The phenomenon of generation of a voltage under mechanical stress is referred to as the direct piezoelectric effect, and the mechanical strain produced in the crystal under electric stress is called the converse piezoelectric effect. *See* POLARIZATION OF DIELECTRICS.

The necessary condition for the piezoelectric effect is the absence of a center of symmetry in the crystal structure. Of the 32 crystal classes, 21 lack a center of symmetry, and with the exception of one class, all of these are piezoelectric. Hydrostatic pressure produces a piezoelectric polarization in the crystals of those 10 classes that show pyroelectricity in addition to piezoelectricity. *See* CRYSTALLOGRAPHY; PYROELECTRICITY.

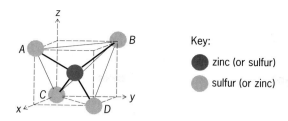

Key:

● zinc (or sulfur)

● sulfur (or zinc)

Tetrahedral structure of zincblende, ZnS. Only part of unit cell is shown. Size of circles has no relation to size of ions.

Molecular theory. Quantitative theories based on the detailed crystal structure are very involved. Qualitatively, however, the piezoelectric effect is readily understood for simple crystal structures. The illustration shows this for a particular cubic crystal, zincblende (ZnS). Every Zn ion is positively charged and is located in the center of a regular tetrahedron $ABCD$, the corners of which are the centers of sulfur ions, which are negatively charged. When this system is subjected to a shear stress in the xy plane, the edge AB, for example, is elongated, and the edge CD of the tetrahedron becomes shorter. Consequently, these edges are no longer equivalent, and the Zn ion will be displaced along the z axis, thus giving rise to an electric dipole moment. The dipole moments arising from different octahedrons sum up because they all have the same orientation with respect to the axes x, y, and z.

Applications. The sharp resonance curve of a piezoelectric resonator makes it useful in the stabilization of the frequency of radio oscillators. Quartz crystals are used almost exclusively in this application. In vacuum-tube oscillators, the crystal generally is part of the feedback circuit. Selective band-pass filters with low losses can be built by using piezoelectric resonators as circuit elements. A synthetic piezoelectric crystal which is often substituted for quartz in this application is ethylene diamine tartrate.

Piezoelectric materials are used extensively in transducers for converting a mechanical strain into an electrical signal. Such devices include microphones, phonograph pickups, vibration-sensing elements, and the like. The converse effect, in which a mechanical output is derived from an electrical signal input, is also widely used in such devices as sonic and ultrasonic transducers, headphones, loudspeakers, and cutting heads for disk recording. *See* ULTRASONICS. [H.G.]

Piezoelectric materials. The principal piezoelectric materials used commercially are crystalline quartz and rochelle salt, although the latter is being superseded by other materials, such as barium titanate. Quartz has the important qualities of being a completely oxidized compound (silicon dioxide), and is almost insoluble in water. Therefore, it is chemically stable against changes occurring with time. It also has low internal losses when used as a vibrator. Rochelle salt has a large piezoelectric effect, and is thus useful in acoustical and vibrational devices where sensitivity is necessary, but it decomposes at high temperatures (131°F or 55°C) and requires protection against moisture. Barium titanate provides lower sensitivity, but greater immunity to temperature and humidity effects. Other crystals that have been used for piezoelectric devices include tourmaline, ammonium dihydrogen phosphate (ADP), and ethylenediamine tartrate (EDT). [F.D.L.]

Pinch effect

A name given to manifestations of the magnetic self-attraction of parallel electric currents having the same direction. The effect at modest current levels of a few amperes can usually be neglected, but when current levels approach a million amperes such as occur in electrochemistry, the effect can be damaging and must be

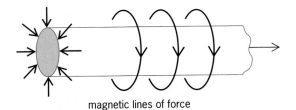

magnetic lines of force

Fig. 1. Pinch pressure on a current-carrying conductor. Arrows at left show direction of pinch pressure.

taken into account by electrical engineers. The pinch effect in a gas discharge has been the subject of intensive study, since it presents a possible way of achieving the magnetic confinement of a hot plasma (a highly ionized gas) necessary for the successful operation of a thermonuclear or fusion reactor.

The law of attraction which describes the interaction between parallel electric currents was discovered by A. M. Ampère in 1820. For a cylindrical wire of radius r meters carrying a total surface current of I amperes, it manifests itself as an inward pressure on the surface (Fig. 1) given by $I^2/2 \times 10^7 \pi r^2$ pascals. For the electric currents of normal experience, this force is small and passes unnoticed, but it is significant that the pressure increases with the square of the current, I^2. For example, at 25,000 amperes the pressure amounts to about 1 atm (100 kilopascals) for a wire of 1-cm radius, but at 10^6 amperes the pressure is about 1600 atm or about 12 tons in.$^{-2}$ (160 megapascals).

There are a number of ways in which the magnetic field of a fusion reactor can be arranged around the plasma to hold it together, and one of these methods is the pinch effect. A fusion reactor using this type of confinement would ideally be a toroidal tube in which the confined plasma would carry a large electric current induced in it by magnetic induction from a transformer core passing through the major axis of the torus. The current would have the double function of ohmically heating the plasma and compressing the plasma toward the center of the tube.

Characteristically, as can be shown by high-speed photography, the pinch forms at the inner surface of a discharge tube wall and contracts radially inward, forming an intense line, the pinch, on the axis; the pinch rebounds slightly; the contracted discharge rapidly develops necks and kinks; and in a few microseconds all structure is lost in an apparently turbulent glowing gas which fills the tube. Thus, the pinch turns out to be unstable, and plasma confinement is soon lost by contact with the wall. The cause of the instability is easily seen qualitatively: The pinch confinement can be described as being caused by the magnetic field lines encircling the pinch which are stretched longitudinally but which are in compression transversely (Fig. 2). For a uniform cylindrical pinch, the magnetic pinch pressure is everywhere equal to the outward plasma pressure, but at a neck or on the inward side of a kink, the magnetic field lines crowd together, creating a higher magnetic pressure than the outward gas pressure. Consequently, the neck contracts still further, the kink cuts in on the concave side and bulges out on the convex side, and both perturbations grow. The instability has a disastrous effect on the confinement time.

The term theta pinch has come into wide usage to denote an important plasma confinement system which relies on the repulsion of oppositely directed currents and which is thus not in accord with the original definition of the pinch effect (self-attraction of currents in the same direction). Plasma confinement systems based on the original pinch effect are known as Z pinches.

Tokamak is essentially a low-density, slow Z pinch in a torus with a very strong longitudinal field. The helical magnetic field lines, resultant from the externally applied

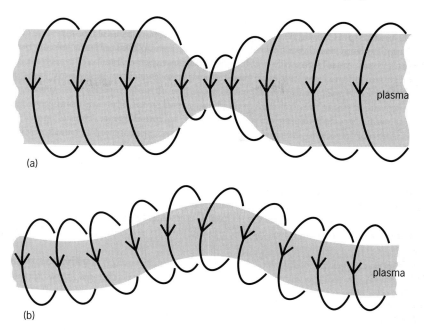

Fig. 2. Instability. (*a*) **Sausage type.** (*b*) **Kink type.**

field and that of the pinch, do not close, that is, do not complete one revolution of the minor axis in going around the major axis of the torus once. This is known theoretically to prevent the growth of certain helical distortions of the plasma. The performance of tokamak experiments has raised the possibility of achieving a net power balance. *See* NUCLEAR FUSION.

[J.A.Ph.]

Pionium   An exotic atom, also called the pi-mu atom, which is similar in structure to the hydrogen atom but with the proton replaced by a pion and the electron replaced by a muon. Pionium is unique among atoms that have been observed in the laboratory in that all of its constituents are unstable particles not found in ordinary matter. Pionium is formed during the decay of a certain heavier particle called the neutral kaon. A kaon has many modes of decay, one of which results in the formation of a pion, a muon, and a neutrino. *See* ELEMENTARY PARTICLE.

The nomenclature of exotic atoms is not well established, and the name pionium may also refer to the pion-electron atom.

[P.A.S.]

Pirani gage   A type of instrument used to measure vacuum by utilizing a resistance change due to a temperature change in a filament. This fine-wire filament, one of the four electrical resistances forming a Wheatstone bridge circuit, is exposed to the vacuum to be measured. Electric current heats the wire; the surrounding gas (in the vacuum) conducts heat away from the wire. At a stable vacuum, the wire quickly reaches equilibrium temperature. If the pressure rises, the gas carries away more heat, and the temperature of the wire decreases. Since the resistance of the filament is a function of temperature, the electrical balance of the Wheatstone bridge is changed. Pressure measurement range of this type of gage is usually 1 to 10^{-4} torr (10^2 to 10^{-2} pascals). *See* VACUUM MEASUREMENT.

[R.C.]

Planck's constant   A fundamental physical constant which represents the elementary quantum of action, action being defined as energy multiplied by time. Introduced by Max Planck in 1900, it has the value $h = 6.6261 \times 10^{-27}$ erg-second or 6.6261×10^{-34} joule-second. The symbol $\hbar$, sometimes called the Dirac h, is often used for convenience in physics to denote the quantity $h/2\pi$, where $\pi = 3.1416.\ldots$

As used by Planck in deriving his radiation law, h multiplied by the frequency of radiation represented a bundle of energy, that is, a quantum of energy. Radiant energy at any wavelength can occur only as multiples of this energy; thus energy is quantized. *See* COMPTON EFFECT; FUNDAMENTAL CONSTANTS; HEAT RADIATION; QUANTUM MECHANICS.

[H.G.S./P.J.W.]

Planck's radiation law   A law of physics which gives the spectral energy distribution of the heat radiation emitted from a so-called blackbody at any temperature. Discovered by Max Planck, this law laid the foundation for the advent of the quantum theory because it was the first physical law to postulate that electromagnetic energy exists in discrete bundles, or quanta. *See* HEAT RADIATION; QUANTUM MECHANICS.

[H.G.S.; P.J.W.]

Plasma (physics)   The field of physics that studies highly ionized gases. Plasma is a gas of charged and neutral particles which exhibits collective behavior. All gases become ionized at sufficiently high temperatures, creating what has been called a fourth state of matter, together with solids, liquids, and gases. It has been estimated that more than 99% of the universe is in the plasma state. On the Earth, plasmas are much less common. Lightning is a familiar natural manifestation, and fluorescent lights are a practical application. Plasma applications and studies make use of an enormous range of plasma temperatures, densities, and neutral pressures. They extend from plasma processing applications at relatively low temperatures (such as plasma etching of semiconductor chips at low pressure, or plasma cutting torches at atmospheric pressure) to studies of controlled fusion at very high temperatures.

Plasma physics is a many-body problem that can be described by a combination of Newton's laws and Maxwell's equations. The charged particles in plasmas are usually ions, both positive and negative, and electrons. Plasmas are normally quasineutral; that is, the net positive ion charge density approximately equals the net negative charge density everywhere in the bulk of the plasma. Quasineutrality refers to charge density and does not imply equal densities of electrons and ions since ions can be multiply charged and can also have negative charge. In space and fusion plasmas, plasmas are normally magnetized, while in application plasmas on Earth, such as plasma processing, both magnetized and unmagnetized plasmas are employed. *See* MAXWELL'S EQUATIONS; NEWTON'S LAWS OF MOTION.

It is convenient to keep track of plasma properties in terms of characteristic lengths, frequencies, and velocities. Among these are the Debye length, the electron and ion plasma frequencies, the electron and ion gyrofrequencies and gyroradii, the electron and ion thermal velocities, the ion sound velocity, the Alfvén velocity, and various collision lengths. The definition of a plasma depends on several of these characteristic parameters, and the magnitude of ratios of these parameters to system size or applied frequencies determines most plasma behavior.

The simplest plasma is a collisionless, unmagnetized collection of ions and electrons with no significant currents. Such plasmas have quasineutral regions and nonneutral regions. The nonneutral regions are highly localized. They are usually located near boundaries (where they are known as sheaths), but are sometimes located within the plasma (where they are known as double layers).

Collective behavior refers to the plasma properties not present in single-particle motion. Collective behavior is a distinguishing characteristic of a plasma. It consists of flows, waves, instabilities, and so forth. Common examples are fluctuations in the aurora, generation of microwaves in devices such as magnetrons and klystrons, and reflection of electromagnetic waves from the ionosphere.

Curiously, very high density collections of equal numbers of ions and electrons are not plasmas. Such systems are referred to as strongly coupled plasmas (even though, strictly speaking, they are not plasmas at all).

A collection of either electrons or ions can exhibit properties similar to those of an electrically neutral plasma if the charged-particle density is sufficiently large. For such so-called plasmas, the Debye length and the characteristic frequency of electrons or ions can still be defined, and collective behavior is still exhibited when the Debye length is less than the system's characteristic dimension. So-called pure electron plasmas or pure ion plasmas are unconfined in an unmagnetized system. However, particle traps consisting of a combination of electric and magnetic fields can be used to confine the charges. *See* PARTICLE TRAP.

The visual appearance of a plasma depends on the kind of ion present, the electron temperature, and the plasma density. Some plasmas are invisible. Curiously, if a plasma is present and not glowing, it is either very hot or very cold. For example, an H^+ plasma, or any other relatively hot plasma with fully stripped ions, contains atomic nuclei with no electrons, so there is no atomic physics and no optical emission or absorption. If plasma electrons and ions are very cold, there is insufficient energy to excite optical transitions. The glow often associated with plasmas indicates only where visible energy transitions are excited by energetic electrons or perhaps absorption of ultraviolet radiation, and may have little to do with the presence of bulk plasma. In fusion plasmas, the edges are often copious sources of emission associated with the dissociation and ionization of hydrogen and edge-generated impurities, while much of the hotter core plasma is fully ionized and invisible.

Direct-current glow-discharge plasmas originate from electrons created by secondary electron emission due to ion bombardment of a negatively biased cathode. The secondary electrons are accelerated through the cathode sheath potential (called the cathode fall) to energies the order of 1 keV, and partially ionize the neutral gas, releasing additional energetic electrons in a multiplicative process. The energetic electrons also undergo inelastic collisions with neutrals which result in optical emission that contributes to the so-called glow. *See* SECONDARY EMISSION.

The understanding of plasma physics begins with an understanding of the motion of single charged particles in a combination of electric and magnetic fields (E and B), produced by a combination of external fields and the motion of the charged particles themselves. The motion of a single particle, with mass m, charge q, and velocity $\mathbf{v}$, is governed by the Lorentz force, as given in Eq. (1). From the perpendic-

$$m\frac{d\mathbf{v}}{dt} = q(\mathbf{E} + \mathbf{v} \times \mathbf{B}) \tag{1}$$

ular component of Eq. (1), it can be shown that the charged particles gyrate about magnetic field lines with a characteristic frequency (the cyclotron frequency). Ions rotate about the magnetic field in the clockwise direction, while electrons rotate counterclockwise with the magnetic field pointing outward. *See* ELECTRIC FIELD; PARTICLE ACCELERATOR.

In addition to the motion parallel to the magnetic field and the gyromotion about the magnetic field, there are drifts perpendicular to the magnetic field. For a general force, **F**, in the presence of a magnetic field, the perpendicular drift velocity is given by Eq. (2).

$$\mathbf{v}_D = \frac{\mathbf{F} \times \mathbf{B}}{qB^2} \tag{2}$$

Given a perpendicular electric field, particles can walk across a magnetic field. Forces associated with magnetic-field curvature give rise to a curvature drift in the direction orthogonal to the magnetic field, and to the radius of curvature of the magnetic field lines.

For gyro motion in a slowly changing magnetic field, which is approximately periodic, it can be shown that the ratio of the perpendicular energy to the magnetic field is approximately constant. This means that a charged particle moving parallel to a magnetic field and gyrating about the field will gyrate faster as the magnetic field increases. If the magnetic field changes in space and is constant in time, the total energy is conserved. For a sufficiently large magnetic field, a point is reached where the total energy equals the perpendicular energy, so that the parallel energy goes to zero and the particle reflects. This is known as magnetic mirroring.

Magnetic mirroring is the chief natural mechanism of charged-particle confinement. For example, this process confines charged particles in the ionosphere and magnetosphere. The magnetic field lines that connect the north and south magnetic poles of the Earth provide a mirror magnetic field which increases as either pole is approached. In the absence of collisions, a particle moving along and gyrating about such a magnetic field is magnetically confined, if it has a sufficiently large velocity perpendicular to the magnetic field. The Van Allen belts are composed of such mirror-trapped charged particles. The source of these particles is the solar wind, a stream of charged particles continuously emitted by the Sun.

For fully ionized plasmas, it is convenient to describe the plasma as a single fluid together with Maxwell's equations. This gives the magnetohydrodynamic (MHD) equations, which are used to describe plasma equilibria and plasma waves and instabilities. Their relative simplicity has made them ideal for solutions of fusion problems in complicated geometries, and they have been widely used to describe astrophysical plasmas and magnetohydrodynamic energy conversion. *See* MAGNETOHYDRODYNAMICS.

Plasmas can support an impressive variety of electrostatic and electromagnetic waves not present in the absence of plasma. The waves are distinguished by their frequency, the presence or absence of dc magnetic fields, and the plasma temperature and density.

Ionization is the key to plasma production and can be accomplished in many different ways. The most common approach is to employ energetic electrons with energies greater than the ionization potential of the gas being ionized. In dc glow discharges, electrons produced by ion secondary electron emission are accelerated by the cathode sheath potential, as are electrons created by thermionic emission in hot-cathode plasmas. Electrons can also pick up energy by reflecting from oscillating radio-frequency sheath electric fields, or by cyclotron resonance in magnetic fields, or from collisions with other energetic electrons. *See* ELECTRICAL CONDUCTION IN GASES; GAS DISCHARGE; IONIZATION POTENTIAL; THERMIONIC EMISSION.

Several other approaches involving collisions, which do not require energetic electrons, also exist. These techniques include photoionization, ion-neutral charge exchange, surface ionization, and Penning ionization. Ions can also be produced in the dissociation of molecules. Yet another mechanism, called critical ionization velocity, is instability driven, and occurs when the kinetic energy of the neutral gas atoms streaming perpendicular to a magnetic field exceeds their ionization potential. *See* ION SOURCES; IONIZATION; PHOTOIONIZATION.

A vacuum chamber provides the simplest approach to confinement. In an unmagnetized plasma, electrons are lost more rapidly than ions, and the plasma acquires a net positive charge. The excess positive charge appears in a sheath at the plasma boundary with the bulk plasma potential more positive than the boundary potential. The decrease in potential at the boundary provides plasma electron confinement, reducing their loss rate to balance the ion loss rate.

Addition of a uniform magnetic field reduces the loss rate of ions and electrons transverse to the magnetic field, but has no effect on losses parallel to the magnetic field because the Lorentz force has no components along this field. Effective confinement by magnetic fields requires that the ion and electron gyroradii be small compared to device dimensions. Plasma transport across the magnetic field can still occur as a result of collisions or of perpendicular drifts.

In the absence of magnetic fields (both inside and outside the plasma), an equilibrium can be achieved by establishing a pressure balance between plasma and edge walls or edge gas. The existence of an equilibrium does not guarantee that a particular configuration is stable.

Plasma processing can be defined as the collection of techniques which make use of plasmas to create new materials or to modify properties of existing materials. It is used in a large variety of applications including semiconductor etching, preparing plastic surfaces to accept ink, depositing polymers, depositing diamond films, and hardening artificial hip joints. The technique has its foundations in plasma physics, chemistry, electrical and chemical engineering, and materials science.

Controlled fusion aims at taking advantage of nuclear fusion reactions to generate net power. Advances in fusion studies have been tied to the techniques developed for plasma confinement and heating. Fusion experiments employ either magnetic confinement or inertial confinement, in which fusion reactions take place before the plasma has a chance to expand to chamber boundaries. Magnetic mirrors are an example of open systems, while tokamaks, stellarators, and reversed-field pinches are examples of closed toroidal systems. Most magnetic confinement research experiments are done on tokamaks. *See* NUCLEAR FUSION.

[N.He.]

Naturally occurring plasmas exist throughout the solar system and beyond. Above the atmosphere, most matter is ionized. The lower-density ionized materials are considered to be plasmas, and they behave in manners very different from the behavior of nonplasmas. Some dense materials, such as stellar matter or electrolytic solutions, are often not considered to be plasmas even though they are ionized; they behave, for the most part, as ordinary fluids.

Some of the major plasma-physics issues that are under study with naturally occurring plasmas are the energization of charged particles, the reconnection of magnetic fields (temporal changes in magnetic-field topology), the production of magnetic fields by dynamos, the production of electromagnetic waves, the interaction between waves and particles, and the transport of mass, momentum, and energy across magnetic fields.

Naturally occurring plasmas are in general difficult to measure. The solar-wind, ionospheric, and magnetospheric plasmas are diagnosed by single-point measurements by rockets and satellites; the solar atmosphere and all astrophysical plasmas are unreachable and must be diagnosed by the light and radio waves that they emit; and lightning is unpredictable and inhospitable to instruments and must be diagnosed primarily by the light that it emits. As a consequence of limited diagnostics, theoretical analysis and laboratory-plasma experiments play supporting roles in the investigations of naturally occurring plasmas.

[J.E.Bo.]

Plasmon   The quanta of waves produced by collective effects of large numbers of electrons in matter when the electrons are disturbed from equilibrium. Metals provide the best evidence of plasmons, because they have a high density of electrons free to move. The results of plasmon stimulation are seen in the illustration. The graph shows the probability of energy losses by fast electrons transmitted through a thin aluminum foil. The number of detected electrons in a beam is plotted against their energy loss during transit through the foil. Each energy-loss peak corresponds to excitation of one or more plasmons. Within experimental error, the peaks occur at integral multiples of a fundamental loss. Further evidence is the fact that the areas under the peaks (a measure of the energy-loss probability) follow a Poisson distribution.

The name plasmon derives from the physical plasma as a state of matter in which the atoms are ionized. At the lowest densities this means an ionized gas, or classical plasma; but densities are much higher in a metal, or quantum plasma, the atoms of a solid metal being in the form of ions. In both types of physical plasma, the frequency of plasma-wave oscillation is determined by the electronic density. In a quantum plasma the energy of the plasmon is its frequency multiplied by Planck's constant, a basic relationship of quantum mechanics. *See* FREE-ELECTRON THEORY OF METALS; QUANTUM MECHANICS.

The plasmon energy for most metals corresponds to that of an ultraviolet photon. However, for silver, gold, the alkali metals, and a few other materials, the plasmon energy is sufficiently low to correspond to that of a visible or near-ultraviolet photon. This means there is a possibility of exciting plasmons by light. If plasmons are confined upon a surface, optical effects can be easily observed. In this case, the quanta are called surface plasmons, and they have the bulk plasmon energy as an upper energy limit.

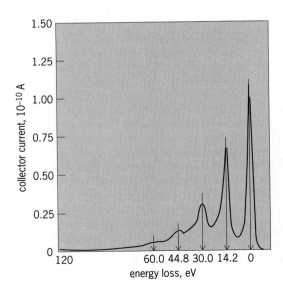

Number of detected electrons in a beam versus their energy loss during transit through a thin aluminum foil. (Number of electrons is expressed as a current; 10^{-14} A $= 6.7 \times 10^{6}$ electrons per second.) Peaks at approximate multiples of 14.2 eV correspond to energy donated to plasmons in the aluminum. (*After T. L. Ferrell, T. A. Calicott, and R. J. Warmack, Plasmons and surfaces, Amer. Sci., 73:344–353, 1985*)

Surface plasmons were first proposed to explain energy losses by electrons reflected from metal surfaces. Since then, numerous experiments have involved coupling photons to surface plasmons. Potential applications extend to new light sources, solar cells, holography, Raman spectroscopy, and microscopy.   [T.L.F.]

Pleochroism   In some colored transparent crystals, the effect wherein the color is quite different in different directions through the crystals. In such a crystal the absorption of light is different for different polarization directions. In colored transparent tourmaline the effect may be so strong that one polarized component of a light beam is wholly absorbed, and the crystal can be used as a polarizer. *See* DICHROISM.   [B.H.Bi.]

Polar molecule   A molecule possessing a permanent electric dipole moment. Molecules containing atoms of more than one element are polar except where forbidden by symmetry; molecules formed from atoms of a single element are nonpolar (except ozone). The dipole moments of polar molecules result in stronger intermolecular attraction, increased viscosities, higher melting and boiling points, and greater solubility in polar solvents than in nonpolar molecules.   [R.D.W.]

Polarimetry   The science of determining the polarization state of electromagnetic radiation (x-rays, light or radio waves). Radiation is said to be linearly polarized when the electric vector oscillates in only one plane. It is circularly polarized when the x-plane component of the electric vector oscillates $90°$ out of phase with the y-plane component.

To completely specify the polarization state, it is necessary to make six intensity measurements of the light passed by a quarter-wave retarder and a rotatable linear polarizer, such as a Polaroid or a Nicol prism. The retarder converts circular light into linear light.

Most starlight is unpolarized. However, atoms in the presence of a magnetic field align themselves at fixed, quantized angles to the field direction. Then the spectral lines they emit are circularly polarized when the magnetic field is parallel to the line of sight, and linearly polarized when the field is perpendicular. The light from sunspots is polarized because the magnetic fields impose some direction in the emitting gas. Other phenomena also remove isotropy and produce polarization. *See* ZEEMAN EFFECT.

Electrooptical devices are rapidly replacing rotating polarizers and fixed retarders. The magnetograph consists of a spectrograph to isolate the atomic spectral line for study; a Pockels cell, an electrooptic crystal whose retardance depends on an applied voltage; a polarizing prism to isolate the polarization state passed by the retarder; a pair of photocells to detect the transmitted light; and a scanning mechanism to sweep the solar image across the spectrograph entrance slit. Two photocells are needed to simultaneously measure left- and right-circular polarization. *See* SPECTROGRAPH.

A magnetograph can be made sensitive to linear polarization, but the signal levels are about 100 times weaker for the inferred transverse fields than for longitudinal fields of comparable strength. To improve signal-to-noise levels, the spectrograph can be replaced with an optical filter having a narrow passband, and the photocells can be replaced with an array of photosensitive picture elements (pixels).   [D.M.R.]

Polarization of dielectrics   A vector quantity representing the electric dipole moment per unit volume of a dielectric material. *See* DIELECTRIC MATERIALS.

Dielectric polarization arises from the electrical response of individual molecules of a medium and may be classified as electronic, atomic, orientation, and space-charge or interfacial polarization, according to the mechanism involved.

Electronic polarization represents the distortion of the electron distribution or motion about the nuclei in an electric field.

Atomic polarization arises from the change in dipole moment accompanying the stretching of chemical bonds between unlike atoms in molecules. *See* MOLECULAR STRUCTURE AND SPECTRA.

Orientation polarization is caused by the partial alignment of polar molecules, that is, molecules possessing permanent dipole moments, in an electric field. This mechanism leads to a temperature-dependent component of polarization at lower frequencies.

Space-charge or interfacial polarization occurs when charge carriers are present which can migrate an appreciable distance through a dielectric but which become trapped or cannot discharge at an electrode. This process always results in a distortion of the macroscopic field and is important only at low frequencies. *See* ELECTRIC FIELD; ELECTRIC SUSCEPTIBILITY.   [R.D.W.]

Polarization of waves   The directional dependence of certain wave phenomena, at right angles to the propagation direction of the wave. In particular, ordinary light may be regarded as composed of two such asymmetrical components, referred to as its two states of linear polarization.

These two components are refracted differently by doubly refracting crystals, such as calcite, or Iceland spar. Each state of linear polarization is refracted according to its own separate refractive index. On a subsequent refraction by the same crystal, but now rotated through an angle θ about the direction of the beam, each component appears as a mixture of the original two polarization components, according to the proportions $\cos^2 \theta : \sin^2 \theta$. *See* BIREFRINGENCE; CRYSTAL OPTICS; REFRACTION OF WAVES.

In the early nineteenth century, T. Young suggested that light polarization arises from transverse oscillations. In J. C. Maxwell's theory of light as electromagnetic waves, visible light—and also other types of electromagnetic radiation such as radio waves, microwaves, and x-rays (distinguished from visible light only by wavelength)—consists of electric and magnetic fields, each oscillating in directions perpendicular to the propagation direction, the electric and magnetic field vectors being perpendicular to each other. The plane of polarization of the wave contains the electric vector (or magnetic vector; there is no general agreement which) and the propagation direction. *See* ELECTROMAGNETIC RADIATION; LIGHT; MAXWELL'S EQUATIONS.

If the plane of polarization remains constant along the wave (as in the case of each light component in a doubly refracting medium), the wave has linear (or plane) polarization. However, the plane of polarization can also rotate. If the rotation rate is constant, the intensity of the wave being also constant, a circularly polarized wave results. These are of two types: right-handed and left-handed.

Any electromagnetic wave can be considered to be composed of monochromatic components, and each monochromatic component can be decomposed into a left-handed and a right-handed circularly polarized part. The states of linear polarization are each made up of equal magnitudes of the two circularly polarized parts, with differing phase relations to provide the different possible directions of plane polarization. Monochromatic waves composed of unequal magnitudes of the two circularly polarized parts are called elliptically polarized. This refers to the fact that the electric and magnetic vectors trace out ellipses in the plane perpendicular to the direction of motion.

Photons have quantum-mechanical spin, which refers to the angular momentum of the photon, necessarily about its direction of motion. A photon's spin has magnitude 1, in fundamental units. This spin can point along the direction of motion (positive helicity, right-handed spin) or opposite to it (negative helicity, left-handed spin), and this corresponds (depending on conventions used) to a classical electromagnetic wave of right- or left-handed circular polarization. *See* HELICITY (QUANTUM MECHANICS); PHOTON.

Electromagnetic and gravitational waves both have the specific property that they are entirely transverse in character, which is a consequence of their speed of propagation being the absolute speed of relativity theory (the speed of light). This corresponds to the fact that their respective quanta, namely photons and gravitons, are massless particles. In the case of waves that travel at a smaller speed, as with fields whose quanta are massive rather than massless, there can be (unpolarized) longitudinal as well as transverse effects. Seismic waves traveling through the Earth's material, for example, can be transverse (polarized sideways oscillations) or longitudinal (unpolarized pressure waves). *See* SOUND.

In most situations encountered in practice, light (or gravitational waves) consists of an incoherent mixture of different polarization states, and is referred to as unpolarized. However, light reflected off a refracting surface (for example, glass or water) is polarized to some extent; that is, there is a certain preponderance of one state of linear polarization over the orthogonal possibility. Complete polarization occurs for a particular angle of incidence, known as the Brewster angle. *See* POLARIZED LIGHT; REFLECTION OF ELECTROMAGNETIC RADIATION; WAVE MOTION.   [R.Pe.]

Polarized light

Polarized light   Light which has its electric vector oriented in a predictable fashion with respect to the propagation direction. In unpolarized light, the vector is oriented in a random, unpredictable fashion. Even in short time intervals, it appears to be oriented in all directions with equal probability. Most light sources seem to be partially polarized so that some fraction of the light is polarized and the remainder unpolarized.

According to all available theoretical and experimental evidence, it is the electric vector rather than the magnetic vector of a light wave that is responsible for all the effects of polarization and other observed phenomena associated with light. Therefore, the electric vector of a light wave, for all practical purposes, can be identified as the light vector. *See* CRYSTAL OPTICS; ELECTROMAGNETIC RADIATION; LIGHT; POLARIZATION OF WAVES.

One of the simplest ways of producing linearly polarized light is by reflection from a dielectric surface. At a particular angle of incidence, known as Brewster's angle, the reflectivity for light whose electric vector is in the plane of incidence becomes zero. The reflected light is thus linearly polarized at right angles to the plane of incidence.

Linear polarizing devices. The first polarizers were glass plates inclined so that the incident light was at Brewster's angle. Such polarizers are quite inefficient since only a small percentage of the incident light is reflected as polarized light.

Certain natural materials absorb linearly polarized light of one vibration direction much more strongly than light vibrating at right angles. Such materials are termed dichroic. Tourmaline is one of the best-known dichroic crystals, and tourmaline plates were used as polarizers for many years. *See* DICHROISM.

Other natural materials exist in which the velocity of light depends on the vibration direction. These materials are called birefringent. One of the best-known of these birefringent crystals is transparent calcite (Iceland spar). The Nicol prism is made of two pieces of calcite cemented together (Fig. 1). The cement is Canada balsam, in which

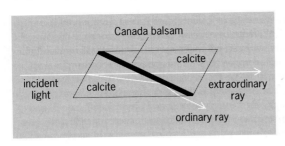

Fig. 1. Nicol prism. The ray for which Snell's law holds is called the ordinary ray.

the wave velocity is intermediate between the velocity in calcite for the fast and the slow ray. The angle at which the light strikes the boundary is such that for one ray the angle of incidence is greater than the critical angle for total reflection. Thus the rhomb is transparent for only one polarization direction. *See* BIREFRINGENCE; OPTICS.

A different type of polarizer, made of quartz, is shown in Fig. 2. Here the vibration directions are different in the two pieces so that the two rays are deviated as they pass through the material. The incoming light beam is thus separated into two oppositely linearly polarized beams which have an angular separation between them, and it is possible to select either beam.

A third mechanism for obtaining polarized light is the Polaroid sheet polarizer, of which there are three types. The first is a microcrystalline polarizer in which small crystals of a dichroic material are oriented parallel to each other in a plastic medium. The second type depends for its dichroism on a property of an iodine-in-water solution. The iodine appears to form a linear high polymer. If the iodine is put on a transparent oriented sheet of material such as polyvinyl alcohol (PVA), the iodine chains apparently line themselves parallel to the PVA molecules and the resulting dyed sheet is strongly dichroic. A third type of sheet polarizer depends for its dichroism directly on the molecules of the plastic itself. This plastic consists of oriented polyvinylene.

Polarization by scattering. When an unpolarized light beam is scattered by molecules or small particles, the light observed at right angles to the original beam is polarized. The best-known example of polarization by scattering is the light of the north sky. *See* SCATTERING OF ELECTROMAGNETIC RADIATION.

Types. Polarized light is classified according to the orientation of the electric vector. In linearly polarized light, the electric vector remains in a plane containing the propagation direction. For monochromatic light, the amplitude of the vector changes sinusoidally with time. In circularly polarized light, the tip of the electric vector describes a circular helix about the propagation direction. The amplitude of the vector is constant. The frequency of rotation is equal to the frequency of the light. In elliptically polarized light, the vector also rotates about the propagation direction, but the amplitude of the vector changes so that the projection of the vector on a plane at right angles to the propagation direction describes an ellipse.

Circular and elliptical polarizing devices. Circularly and elliptically polarized light are normally produced by combining a linear polarizer with a wave plate. A Fresnel rhomb can be used to produce circularly polarized light.

A plate of material (quartz, calcite, or other birefringent crystals) which is linearly birefringent is called a wave plate or retardation sheet. Wave plates have a pair of orthogonal axes which are designated fast and slow. Polarized light with its electric vector parallel to the fast axis travels faster than light polarized parallel to the slow axis. The thickness of the material can be chosen so that for light traversing the plate, there

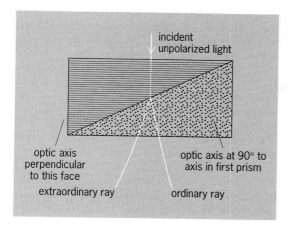

incident
unpolarized light

optic axis
perpendicular
to this face

optic axis at 90° to
axis in first prism

extraordinary ray

ordinary ray

Fig. 2. Wollaston prism.

is a definite phase shift between the fast component and the slow component. A plate with a 90° phase shift is termed a quarter-wave plate.

If linearly polarized light is incident normally on a quarter-wave plate and oriented at 45° to the fast axis, the transmitted light will be circularly polarized. If the linearly polarized light is at an angle other than 45° to the fast axis, the transmitted radiation will be elliptically polarized.

Analyzing devices. Polarized light is one of the most useful tools for studying the characteristics of materials. The absorption constant and refractive index of a metal can be calculated by measuring the effect of the metal on polarized light reflected from its surface. *See* REFLECTION OF ELECTROMAGNETIC RADIATION.

The analysis of polarized light can be performed with a variety of different devices. If the light is linearly polarized, it can be extinguished by a linear polarizer and the direction of polarization of the light determined directly from the orientation of the polarizer. If the light is elliptically polarized, it can be analyzed with the combination of a quarter-wave plate and a linear polarizer. Any such combination of polarizer and analyzer is called a polariscope.   [B.H.Bi.]

Polaron   The object that results when an electron in the conduction band of a crystalline insulator or semiconductor polarizes or otherwise deforms the lattice in its vicinity. The polaron comprises the electron plus its surrounding lattice deformation. (Polarons can also be formed from holes in the valence band.) If the deformation extends over many lattice sites, the polaron is "large," and the lattice can be treated as a continuum. Charge carriers inducing strongly localized lattice distortions form "small" polarons. *See* BAND THEORY OF SOLIDS; HOLE STATES IN SOLIDS; SEMICONDUCTOR.   [D.M.L.]

Polymorphism (crystallography)   The existence of different crystal structures with the same chemical composition. If only one chemical element is present, the forms are called allotropes. Graphite and diamond are allotropes of carbon, whereas quartz and cristobalite are polymorphs of silica (silicon dioxide, SiO_2). Although properties are different in these forms, reversible transformations, which involve small shifts in atom positions and no bulk transport of material, are common. The quartz transformation at 1063°F (573°C) is a reversible, atom-displacement transformation.

In metals and ceramics, similar transformations are called martensitic. Advantage is taken of the localized nature of reversible transformation in steel by controlling the melting atmosphere, temperature, composition, mechanical working (alloying), and tempering and quenching operations.

Control over transformations to achieve desirable properties as either devices or structural materials in extreme environments is a frequent objective. In the case of tin, reversibility on the atomic scale can have devastating consequences for bulk properties. Similar transformations may be beneficial in the right place and in the desired degree. Such transformation is attempted with metals and ceramics. *See* CRYSTAL STRUCTURE.

[D.Ev.]

Polytropic process   A process which occurs with an interchange of both heat and work between the system and its surroundings. The nonadiabatic expansion or compression of a fluid is an example of a polytropic process. The interrelationships between the pressure (P) and volume (V) and pressure and temperature (T) for a gas undergoing a polytropic process are given by Eqs. (1) and (2), where a and b

$$PV^a = \text{constant} \tag{1}$$

$$P^b/T = \text{constant} \tag{2}$$

are the polytropic constants for the process of interest. These constants, which are usually determined from experiment, depend upon the equation of state of the gas, the amount of heat transferred, and the extent of irreversibility in the process. *See* ISENTROPIC PROCESS; ISOTHERMAL PROCESS; THERMODYNAMIC PROCESSES.   [S.I.S.]

Positron   An elementary particle with mass equal to that of the electron, and positive charge equal in magnitude to the electron's negative charge. The positron is thus the antiparticle (charge-conjugate particle) to the electron. The positron has the same spin and statistics as the electron. Positrons, like electrons, appear as decay products of many heavier particles; electron-positron pairs are produced by high-energy photons in matter. *See* ANTIMATTER; ELECTRON; ELECTRON-POSITRON PAIR PRODUCTION; ELEMENTARY PARTICLE.

A positron is, in itself, stable, but cannot exist indefinitely in the presence of matter, for it will ultimately collide with an electron. The two particles will be annihilated as a result of this collision, and photons will be created. However, a positron can first become bound to an electron to form a short-lived "atom" termed positronium. *See* POSITRONIUM.

Quantum field theory predicts the occurrence of a fundamental positron creation process in the presence of strong, static electric fields. For a bare nucleus with atomic number $Z > 173$, it becomes energetically favorable to transform the electron binding energy of larger than $2m_0c^2$, where m_0 is the electron rest mass and c is the speed of light, into simultaneously creating an electron bound to the nucleus and a positron that escapes from the nucleus. This process of spontaneous positron emission has not been observed since atoms with $Z > 173$ are not available in nature. However, with the introduction of heavy-ion accelerators, it has become possible to simulate such an atom for a short period in a high-energy collision between two stable heavy atoms such as uranium. Experiments have utilized a variety of such collision systems with total Z ranging from 180 to 188 to search for spontaneous positron emission. A number of these experiments reproduce the salient features expected for this process. However, some inconsistencies with the predictions of the theory have yet to be resolved before spontaneous positron emission is established experimentally. *See* Nuclear molecule; Quasiatom; Supercritical fields.

[C.G.; J.Gre.]

Positronium   An atomic-like system consisting of an electron and positron. Just as in the hydrogen atom, the energy levels of positronium are quantized, with the deepest levels bound by about 6.8 eV. The electron and positron spins can be aligned in the same direction (singlet states) or in opposite directions (triplet states). Annihilation of the positron and electron destroys the lowest-energy singlet state (parapositronium) in about 10^{-10} s, but the lowest triplet state (orthopositronium) survives longer, about 10^{-7} s. This allows sufficient time for precise measurement of the energy levels of triplet states. Because of the absence of nuclei in positronium, these measurements provide an accurate test of theories of the electromagnetic force (quantum electrodynamics) without interference from the strong force. *See* Atomic structure and spectra; Electron; Fundamental interactions; Positron.

Since the formation of positronium requires the close approach of a positron and an electron, beams of slow positrons can be used as probes of the electron density in gases, in insulating solids, or near surfaces. Since the singlet and triplet forms of positronium have very different lifetimes, and transitions between the two states can be induced by neighboring electrons, study of the decay of positronium can also provide information about electron densities on a microscopic scale. This is especially useful in the study of density fluctuations in gases near the critical point for condensation into liquids or solids. *See* Critical phenomena.

Annihilation radiation from positronium forms a component of the gamma-ray spectrum observed by astronomers, in particular from the galactic center.

[J.N.Ba.]

Potential flow   A fluid flow that is isentropic and that, if incompressible, can be mathematically described by Laplace's equation. For an ideal fluid, or a flow in which viscous effects are ignored, vorticity (defined as the curl of the velocity) cannot be produced, and any initial vorticity existing in the flow simply moves unchanged with the fluid. Ideal fluids, of course, do not exist since any actual fluid has some viscosity, and the effects of this viscosity will be important near a solid wall, in the region known as the boundary layer. Nevertheless, the study of potential flow is important in hydrodynamics, where the fluid is considered incompressible, and even in aerodynamics, where the fluid is considered compressible, as long as shock waves are not present. *See* Boundary-layer flow; Compressible flow; Isentropic flow.

In the absence of viscous effects, a flow starting from rest will be irrotational for all subsequent time. For an irrotational flow, the curl of the velocity is zero ($\nabla \times V = 0$). The curl of the gradient of any scalar function is zero ($\nabla \times \nabla\phi = 0$). It then follows mathematically that the condition of irrotationality can be satisfied identically by choosing the scalar function, ϕ, such that the velocity is the gradient of ϕ ($V = \nabla\phi$). For this reason, this scalar function ϕ has been traditionally referred to as the velocity potential, and the flow as a potential flow. *See* Potentials.

By applying the continuity equation to the definition of the potential function, it becomes possible to represent the flow by the well-known Laplace equation ($\nabla^2\phi = 0$), instead of the coupled system of the continuity and nonlinear Euler equations. The linearity of the Laplace equation, which also governs other important physical phenomena such as electricity and magnetism, makes it possible to use the principle of superposition to combine elementary solutions in solving more complex problems. *See* Fluid flow.

[P.E.R.]

Potentials   Functions or sets of functions from whose first derivatives a vector can be formed. A vector is a quantity which has a magnitude and a direction, such as force.

A single function, the scalar potential, is used in gravitation theory, electricity and magnetism, fluid mechanics, and other areas. The vectors obtained from it by partial differentiation are in these cases the gravitational, electric and magnetic field strengths, and the velocity, respectively. The vector potential is a set of three functions whose first derivatives give the magnetic induction. *See* ELECTRIC FIELD; GRAVITATION; LAPLACE'S IRROTATIONAL MOTION.

[F.Ro.]

Potentiometer   An instrument that precisely measures an electromotive force (emf) or a voltage by opposing to it a known potential drop established by passing a definite current through a resistor of known characteristics. (A three-terminal resistive voltage divider is sometimes also called a potentiometer.) There are two ways of accomplishing this balance: (1) the current I may be held at a fixed value and the resistance R across which the IR drop is opposed to the unknown may be varied; (2) current may be varied across a fixed resistance to achieve the needed IR drop. *See* ELECTROMOTIVE FORCE (EMF); RESISTOR.

The essential features of a general-purpose constant-current instrument are shown in the illustration. The value of the current is first fixed to match an IR drop to the emf of a reference standard cell. With the standard-cell dial set to read the emf of the reference cell, and the galvanometer (balance detector) in position G_1, the resistance of the supply branch of the circuit is adjusted until the IR drop in 10 steps of the coarse dial plus the set portion of the standard-cell dial balances the known reference emf, indicated by a null reading of the galvanometer. This adjustment permits the potentiometer

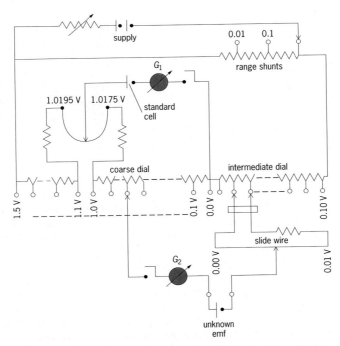

Circuit diagram of a general-purpose constant-current potentiometer, showing essential features.

to be read directly in volts. Then, with the galvanometer in position G_2, the coarse, intermediate, and slide-wire dials are adjusted until the galvanometer again reads null. If the potentiometer current has not changed, the emf of the unknown can be read directly from the dial settings. There is usually a switching arrangement so that the galvanometer can be quickly shifted between positions 1 and 2 to check that the current has not drifted from its set value. See GALVANOMETER.

Potentiometer techniques may also be used for current measurement, the unknown current being sent through a known resistance and the IR drop opposed by balancing it at the voltage terminals of the potentiometer. Here, of course, internal heating and consequent resistance change of the current-carrying resistor (shunt) may be a critical factor in measurement accuracy; and the shunt design may require attention to dissipation of heat resulting from its I^2R power consumption. See CURRENT MEASUREMENT; JOULE'S LAW.

Potentiometer techniques have been extended to alternating-voltage measurements, but generally at a reduced accuracy level (usually 0.1% or so). Current is set on an ammeter which must have the same response on ac as on dc, where it may be calibrated with a potentiometer and shunt combination. Balance in opposing an unknown voltage is achieved in one of two ways: (1) a slide-wire and phase-adjustable supply; (2) separate in-phase and quadrature adjustments on slide wires supplied from sources that have a 90° phase difference. Such potentiometers have limited use in magnetic testing. See ELECTRICAL MEASUREMENTS; VOLTAGE MEASUREMENT.   [F.K.H.; R.F.Dz.]

Power   The time rate of doing work. Like work, power is a scalar quantity, that is, a quantity which has magnitude but no direction. Some units often used for the measurement of power are the watt (1 joule of work per second) and the horsepower (550 foot-pounds of work per second). See WORK.

Power is a concept which can be used to describe the operation of any system or device in which a flow of energy occurs. In many problems of apparatus design, the power, rather than the total work to be done, determines the size of the component used. Any device can do a large amount of work by performing for a long time at a low rate of power, that is, by doing work slowly. However, if a large amount of work must be done rapidly, a high-power device is needed. High-power machines are usually larger, more complicated, and more expensive than equipment which need operate only at low power. A motor which must lift a certain weight will have to be larger and more powerful if it lifts the weight rapidly than if it raises it slowly. An electrical resistor must be large in size if it is to convert electrical energy into heat at a high rate without being damaged.
                                                                            [P.W.S.]

Poynting's vector   A vector, the outward normal component of which, when integrated over a closed surface in an electromagnetic field, represents the outward flow of energy through that surface. It is given by the equation below, where $\mathbf{E}$ is the

$$\Pi = \mathbf{E} \times \mathbf{H} = \mu^{-1}\mathbf{E} \times \mathbf{B}$$

electric field strength, $\mathbf{H}$ the magnetic field strength, $\mathbf{B}$ the magnetic flux density, and μ the permeability.

When an electromagnetic wave is incident on a conducting or absorbing surface, theory predicts that it should exert a force on the surface in the direction of the difference between the incident and the reflected Poynting's vector. See ELECTRIC FIELD; ELECTROMAGNETIC RADIATION; MAXWELL'S EQUATIONS; RADIATION PRESSURE.   [W.R.Sm.]

Precession   The motion of an axis fixed in a body around a direction fixed in space. If the angle between the two is constant so that the axis sweeps out a circular cone, the motion is pure precession; oscillation of the angle is called nutation. An example of precession is the motion of the Earth's polar axis around the normal to the plane of the ecliptic; this is the precession of the equinoxes. A fast-spinning top, with nonvertical axis, which precesses slowly around the vertical direction, is another example. In both examples the precession is due to torque acting on the body. Another kind of precession, called free or fast precession, with a rate which is comparable to the rotation rate of the body, is seen, for instance, in a coin spun into the air.

As a simple example of gyroscopic motion, consider a rapidly spinning wheel with a horizontal axis supported at a distance d from the plane of the wheel (see illustration).

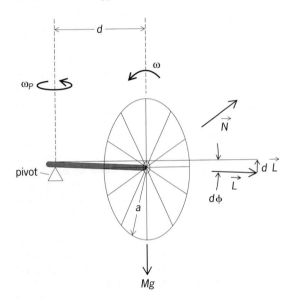

Simple precession of a rapidly spinning wheel with a horizontal axis supported by a pivot.

The angular momentum $\vec{L}$ is along the wheel symmetry axis and is approximately given by the angular momentum of the wheel about this axis; in the simple precession approximation to the motion, the angular momentum associated with precessional motion is neglected. The external torque $\vec{N}$ due to the gravitational force is perpendicular to the wheel axis in the horizontal plane. The change in the angular momentum $\vec{L}$ in an infinitesimal time interval dt is given by the rotational equation of motion in Eq. (1).

$$d\vec{L} = \vec{N}\, dt \qquad (1)$$

See Rigid-body dynamics.

Since $d\vec{L}$ and $\vec{L}$ are perpendicular, the length $\vec{L}$ is unchanged to first order in dt. The direction of $\vec{L}$ is rotated counterclockwise in the horizontal plane. The angular velocity of precession ω_P about the vertical axis is then given by Eq. (2), where g is

$$\omega_P = \frac{gd}{\omega a^2} \qquad (2)$$

the gravitational acceleration, ω is the spin angular velocity, and a is the radius of the wheel. Thus ω_P is independent of the mass of the wheel and inversely proportional to ω. For ω very large, the precession rate ω_P is quite slow.   [V.D.B.]

Pressure   The ratio of force to area. Atmospheric pressure at the surface of Earth is in the vicinity of 15 lbf/in.^{2} (1.0×10^5 Pa). Pressures in enclosed containers less than this value are spoken of as vacuum pressures; for example, the vacuum pressure inside a cathode-ray tube is 10^{-8} mmHg, meaning that the pressure is equal to the pressure that would be produced by a column of mercury, with no force acting above it, that is 10^{-8} mm high. This is absolute pressure measured above zero pressure as a reference level. Inside a steam boiler, the pressure may be 800 lbf/in.^{2} (5.5×10^6 Pa) or higher. Such pressure, measured above atmospheric pressure as a reference level, is gage pressure, designated psig. *See* PRESSURE MEASUREMENT.   [F.H.R.]

Pressure measurement   The determination of the magnitude of a fluid force applied to a unit area. Pressure measurements are generally classified as gage pressure, absolute pressure, or differential pressure. *See* PRESSURE.

Pressure gages generally fall in one of three categories, based on the principle of operation: liquid columns, expansible-element gages, and electrical pressure transducers.

Liquid-column gages include barometers and manometers. They consist of a U-shaped tube partly filled with a nonvolatile liquid. Water and mercury are the two most common liquids used in this type of gage. *See* BAROMETER; MANOMETER.

There are three classes of expansible metallic-element gages: bourdon, diaphragm, and bellows. Bourdon-spring gages, in which pressure acts on a shaped, flattened, elastic tube, are by far the most widely used type of instrument. These gages are simple, rugged, and inexpensive. In diaphragm-element gages, pressure applied to one or more contoured diaphragm disks acts against a spring or against the spring rate of the diaphragms, producing a measurable motion. In bellows-element gages, pressure in or around the bellows moves the end plate of the bellows against a calibrated spring, producing a measurable motion.

Electrical pressure transducers convert a pressure to an electrical signal which may be used to indicate a pressure or to control a process. Such devices as strain gages and resistive, magnetic, crystal, and capacitive pressure transducers are commonly used to convert the measured pressure to an electrical signal. *See* PRESSURE TRANSDUCER.
                    [J.H.Z.]

Pressure transducer   An instrument component which detects a fluid pressure and produces an electrical, mechanical, or pneumatic signal related to the pressure.

In general, the complete instrument system comprises a pressure-sensing element such as a bourdon tube, bellows, or diaphragm element; a device which converts motion or force produced by the sensing element to a change of an electrical, mechanical, or pneumatic parameter; and an indicating or recording instrument. Frequently the instrument is used in an autocontrol loop to maintain a desired pressure.

Although pneumatic and mechanical transducers are commonly used, electrical measurement of pressure is often preferred because of a need for long-distance transmission, higher accuracy requirements, more favorable economics, or quicker response. Electrical pressure transducers may be classified by the operating principle as resistive transducers, strain gages, magnetic transducers, crystal transducers, capacitive transducers, and resonant transducers.

In resistive pressure transducers, pressure is measured by an element that changes its electrical resistance as a function of pressure. Many types of resistive pressure transducers use a movable contact, positioned by the pressure-sensing element. One form is a contact sliding along a continuous resistor, which may be straight-wire, wire-wound, or nonmetallic such as carbon.

Strain-gage pressure transducers might be considered to be resistive transducers, but are usually classified separately, They convert a physical displacement into an electrical signal. When a wire is placed in tension, its electrical resistance increases. The change in resistance is a measure of the displacement, hence of the pressure. Another variety of strain gage transducer uses integrated circuit technology. Resistors are diffused onto the surface of a silicon crystal within the boundaries of an area which is etched to form a thin diaphragm.

In magnetic pressure transducers, a change of pressure is converted into change of magnetic reluctance or inductance when one part of a magnetic circuit is moved by a pressure-sensing element—bourdon tube, bellows, or diaphragm.

Piezoelectric crystals produce an electric potential when placed under stress by a pressure-sensing element. Crystal transducers offer a high speed of response and are widely used for dynamic pressure measurements in such applications as ballistics and engine pressures. *See* Piezoelectricity.

Capacitive pressure transducers almost invariably sense pressure by means of a metallic diaphragm, which is also used as one plate of a capacitor.

The resonant transducer consists of a wire or tube fixed at one end and attached at the other (under tension) to a pressure-sensing element. The wire is placed in a magnetic field and allowed to oscillate. As the pressure is increased, the element increases the tension in the wire or tube, thus raising its resonant frequency. *See* Pressure Measurement.   [J.H.Z.]

Product of inertia   The product of inertia of area A relative to the indicated XY rectangular axes is $I_{XY} = \int xy \, dA$ (see illustration). The product of inertia of the mass contained in volume V relative to the XY axes is $I_{XY} = \int xy\rho \, dV$—similarly for I_{YZ} and I_{ZX}.

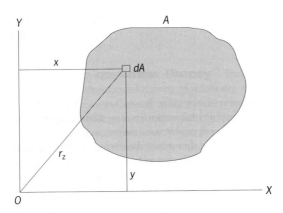

Product of inertia of an area.

Relative to principal axes of inertia, the product of inertia of a figure is zero. If a figure is mirror symmetrical about a YZ plane, $I_{ZX} = I_{XY} = 0$. *See* Moment of Inertia.   [N.S.F.]

Propagator (field theory)   The probability amplitude for a particle to move or propagate to some new point of space and time when its amplitude at some point of origination is known. The propagator occurs as an important part of the probability in reactions and interactions in all branches of modern physics. Its properties are best

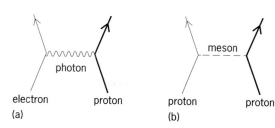

Feynman diagrams for scattering processes. (*a*) Electron-proton scattering via photon exchange. (*b*) Proton-proton scattering via meson exchange.

described in the framework of quantum field theory for relativistic particles, where it is written in terms of energy and momentum. Concrete examples for electron-proton and proton-proton scattering are provided in the illustration. The amplitude for these processes contains the propagators for the exchanged proton and meson, which actually specify the dominant part of the probability of each process when the scattering occurs at small angles. In similar fashion, for any electromagnetic process, a propagator for each internal line of the Feynman diagram (each line not connected directly to the outside world) enters the probability amplitude. See FEYNMAN DIAGRAM; QUANTUM ELECTRODYNAMICS; QUANTUM FIELD THEORY; QUANTUM MECHANICS.   [K.E.L.]

Proton   A positively charged particle that is the nucleus of the lightest chemical element, hydrogen. The hydrogen atom consists of a proton as the nucleus, to which a single negatively charged electron is bound by an attractive electrical force (since opposite charges attract). The proton is about 1836 times heavier than the electron, so that the proton constitutes almost the entire mass of the hydrogen atom. Most of the interior of the atom is empty space, since the sizes of the proton and the electron are very small compared to the size of the atom. See ATOMIC STRUCTURE AND SPECTRA.

For chemical elements heavier than hydrogen, the nucleus can be thought of as a tightly bound system of Z protons and N neutrons. An electrically neutral atom will then have Z electrons bound comparatively loosely in orbits outside the nucleus. See NEUTRON; NUCLEAR STRUCTURE.

The numerical values of some overall properties of the proton can be summarized as follows: charge, 1.602×10^{-19} coulomb; mass, 1.673×10^{-27} kg; spin, $(^1/_2)\hbar$ (where $\hbar$ is Planck's constant h divided by 2π); magnetic dipole moment, 1.411×10^{-26} joule/tesla; radius, about 10^{-15} m. See FUNDAMENTAL CONSTANTS; NUCLEAR MOMENTS; SPIN (QUANTUM MECHANICS).

It is instructive to contrast the proton's properties with those of the electron. All of the electron's properties have been found to be those expected of a spin-$^1/_2$ particle which is described by the Dirac equation of quantum mechanics. Such a Dirac particle has no internal size or structure. See ELECTRON; RELATIVISTIC QUANTUM THEORY.

By contrast, although it also has a spin of $^1/_2$, the proton's magnetic moment, which is different from that for a Dirac particle, and its binding with neutrons into nuclei strongly suggest that it has some kind of internal structure, rather than being a point particle. Two different kinds of high-energy physics experiments have been used to study the internal structure of the proton. An example of the first type of experiment is the scattering of high-energy electrons, above say 1 GeV, from a target of protons. The angular pattern and energy distribution of the scattered electrons give direct information about the size and structure of the proton. The second type of high-energy experiment involves the production and study of excited states of the proton, often called baryonic resonances.

It has been found that the spectrum of higher-mass states which are produced in high-energy collisions follows a definite pattern. *See* BARYON.

In 1963, M. Gell-Mann and, independently, G. Zweig pointed out that this pattern is what would be expected if the proton were composed of three spin-$\frac{1}{2}$ particles, quarks, with two of the quarks (labeled *u*) each having a positive electric charge of magnitude equal to $\frac{2}{3}$ of the electron's charge (*e*), and the other quark (labeled *d*) having a negative charge of magnitude of $\frac{1}{3}e$. Subsequently, the fractionally charged quark concept was developed much further, and has become central to understanding every aspect of the behavior and structure of the proton. *See* QUARKS.

An important class of fundamental theories, called grand unification theories (GUTs), makes the prediction that the proton will decay. The predicted lifetime of the proton is very long, about 10^{30} years or more—which is some 10^{20} times longer than the age of the universe—but this predicted rate of proton decay may be detectable in practical experiments. *See* GRAND UNIFICATION THEORIES.

If the proton is observed to decay, this new interaction will also have profound consequences for understanding of cosmology. The very early times of the big bang (about 10^{-30} s) are characterized by energies so high that the same grand unified interaction which would allow proton decay would also completely determine the subsequent evolution of the universe. This could then explain the remarkable astrophysical observation that the universe appears to contain only matter and not an equal amount of antimatter. *See* ELEMENTARY PARTICLE. [T.H.F.]

Pyroelectricity　The property of certain crystals to produce a state of electric polarity by a change of temperature. Certain dielectric (electrically nonconducting) crystals develop an electric polarization (dipole moment per unit volume) when they are subjected to a uniform temperature change. This pyroelectric effect occurs only in crystals which lack a center of symmetry and also have polar directions (that is, a polar axis). These conditions are fulfilled for 10 of the 32 crystal classes. Typical examples of pyroelectric crystals are tourmaline, lithium sulfate monohydrate, cane sugar, and ferroelectric barium titanate.

Pyroelectric crystals can be regarded as having a built-in or permanent electric polarization. When the crystal is held at constant temperature, this polarization does not manifest itself because it is compensated by free charge carriers that have reached the surface of the crystal by conduction through the crystal and from the surroundings. However, when the temperature of the crystal is raised or lowered, the permanent polarization changes, and this change manifests itself as pyroelectricity.

The magnitude of the pyroelectric effect depends upon whether the thermal expansion of the crystal is prevented by clamping or whether the crystal is mechanically unconstrained. In the clamped crystal, the primary pyroelectric effect is observed, whereas in the free crystal, a secondary pyroelectric effect is superposed upon the primary effect. The secondary effect may be regarded as the piezoelectric polarization arising from thermal expansion, and is generally much larger than the primary effect. *See* PIEZOELECTRICITY. [H.Gr.]

Pyroelectrics have a broad spectrum of potential scientific and technical applications. The most developed is the detection of infrared radiation. In addition, pyroelectric detectors can be used to measure the power generated by a radiation source (in radiometry), or the temperature of a remote hot body (in pyrometry, with corrections due to deviations from the blackbody emission). *See* PYROMETER; RADIOMETRY.

An infrared image can be projected on a pyroelectric plate and transformed into a relief of polarization on the surface. Other potential applications of pyroelectricity include solar energy conversion, refrigeration, information storage, and solid-state science.

[A.Had.]

Pyrometer   A temperature-measuring device, originally an instrument that measures temperatures beyond the range of thermometers, but now in addition a device that measures thermal radiation in any temperature range. This article discusses radiation pyrometers; for other temperature-measuring devices *see* BOLOMETER; THERMISTOR; THERMOCOUPLE.

The illustration shows a very simple type of radiation pyrometer. Part of the thermal radiation emitted by a hot object is intercepted by a lens and focused onto a thermopile. The resultant heating of the thermopile causes it to generate an electrical signal (proportional to the thermal radiation) which can be displayed on a recorder.

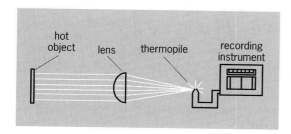

Elementary radiation pyrometer. **(After D. M. Considine and S. D. Ross, Process Instruments and Controls Handbook, 2d ed., McGraw-Hill, 1974)**

Unfortunately, the thermal radiation emitted by the object depends not only on its temperature but also on its surface characteristics. The radiation existing inside hot, opaque objects is so-called blackbody radiation, which is a unique function of temperature and wavelength and is the same for all opaque materials. However, such radiation, when it attempts to escape from the object, is partly reflected at the surface. In order to use the output of the pyrometer as a measure of target temperature, the effect of the surface characteristics must be eliminated. A cavity can be formed in an opaque material and the pyrometer sighted on a small opening extending from the cavity to the surface. The opening has no surface reflection, since the surface has been eliminated. Such a source is called a blackbody source, and is said to have an emittance of 1.00. By attaching thermocouples to the blackbody source, a curve of pyrometer output voltage versus blackbody temperature can be constructed. *See* BLACKBODY; HEAT RADIATION.

Pyrometers can be classified generally into types requiring that the field of view be filled, such as narrow-band and total-radiation pyrometers; and types not requiring that the field of view be filled, such as optical and ratio pyrometers. The latter depend upon making some sort of comparison between two or more signals.

The optical pyrometer should more strictly be called the disappearing-filament pyrometer. In operation, an image of the target is focused in the plane of a wire that can be heated electrically. A rheostat is used to adjust the current through the wire until the wire blends into the image of the target (equal brightness condition), and the temperature is then read from a calibrated dial on the rheostat.

The ratio, or "two-color," pyrometer makes measurements in two wavelength regions and electronically takes the ratio of these measurements. If the emittance is the same for both wavelengths, the emittance cancels out of the result, and the true temperature of the target is obtained. This so-called graybody assumption is sufficiently valid in some cases so that the "color temperature" measured by a ratio pyrometer is close to the true temperature. *See* TEMPERATURE MEASUREMENT; THERMOMETER.   [T.P.M.]

Q

Q (electricity)   Often called the quality factor of a circuit, Q is defined in various ways, depending upon the particular application. In the simple RL and RC series circuits, Q is the ratio of reactance to resistance, as in Eqs. (1), where X_L is the inductive

$$Q = \frac{X_L}{R} \qquad Q = \frac{X_C}{R} \quad \text{(a numerical value)} \tag{1}$$

reactance, X_C is the capacitive reactance, and R is the resistance. An important application lies in the dissipation factor or loss angle when the constants of a coil or capacitor are measured by means of the alternating-current bridge.

   Q has greater practical significance with respect to the resonant circuit, and a basic definition is given by Eq. (2), where Q_0 means evaluation at resonance. For certain

$$Q_0 = 2\pi \frac{\text{max stored energy per cycle}}{\text{energy lost per cycle}} \tag{2}$$

circuits, such as cavity resonators, this is the only meaning Q can have.

   For the RLC series resonant circuit with resonant frequency f_0, Eq. (3) holds, where

$$Q_0 = \frac{2\pi f_0 L}{R} = \frac{1}{2\pi f_0 CR} \tag{3}$$

R is the total circuit resistance, L is the inductance, and C is the capacitance. Q_0 is the Q of the coil if it contains practically the total resistance R. The greater the value of Q_0, the sharper will be the resonance peak.

   The practical case of a coil of high Q_0 in parallel with a capacitor also leads to $Q_0 = 2\pi f_0 L/R$. R is the total series resistance of the loop, although the capacitor branch usually has negligible resistance.

   In terms of the resonance curve, Eq. (4) holds, where f_0 is the frequency at resonance,

$$Q_0 = \frac{f_0}{f_2 - f_1} \tag{4}$$

and f_1 and f_2 are the frequencies at the half-power points. [B.L.R.]

Q meter   A direct-reading instrument widely used for measuring the Q of an electric circuit at radio frequencies. Originally designed to measure the Q of coils, the Q meter has been developed into a flexible, general-purpose instrument for determining many other quantities such as (1) the distributed capacity, effective inductance, and self-resonant frequency of coils; (2) the capacitance, Q or power factor, and self-resonant frequency of capacitors; (3) the effective resistance, inductance or capacitance, and the

Q of resistors; (4) characteristics of intermediate- and radio-frequency transformers; and (5) the dielectric constant, dissipation factor, and power factor of insulating materials. *See* ELECTRICAL MEASUREMENTS; Q (ELECTRICITY).   [I.F.K./E.C.St.]

Quantized electronic structure (QUEST)   A material that confines electrons in such a small space that their wavelike behavior becomes important and their properties are strongly modified by quantum-mechanical effects. Such structures occur in nature, as in the case of atoms, but can be synthesized artificially with great flexibility of design and applications. They have been fabricated most frequently with layered semiconductor materials. Generally, the confinement regions for electrons in these structures are 1–100 nanometers in size. The allowable energy levels, motion, and optical properties of the electrons are strongly affected by the quantum-mechanical effects. The structures are referred to as quantum wells, wires, and dots, depending on whether electrons are confined with respect to motion in one, two, or three dimensions. Multiple closely spaced wells between which electrons can move by quantum-mechanical tunneling through intervening thin barrier-material layers are referred to as superlattices. *See* QUANTUM MECHANICS.

The most frequently used fabrication technique for quantized electronic structures is epitaxial growth of thin single-crystal semiconductor layers by molecular-beam epitaxy or by chemical vapor growth techniques. These artificially synthesized quantum structures find major application in high-performance transistors such as the microwave high-electron-mobility transistor (HEMT), and in high-performance solid-state lasers such as the semiconductor quantum-well laser. They also have important scientific applications for the study of fundamental two-dimensional, one-dimensional, and zero-dimensional physics problems in which particles are confined so that they have free motion in only two, one, or zero directions. Chemically formed nanocrystals, carbon nanotubes, zeolite cage compounds, and carbon buckyball C_{60} molecules are also important quantized electronic structures.

The optical applications are based on the interactions between light and electrons in the quantum structures. The absorption of a photon by an electron in a quantum well raises the electron from occupied quantum states to unoccupied quantum states. Electrons and holes in quantum wells may also recombine, with the resultant emission of photons from the quantized electronic structure as the electron drops from a higher state to a lower state. *See* ELECTRON-HOLE RECOMBINATION.

The photon emission is the basis for quantum-well semiconductor lasers, which have widespread applications in optical fiber communications and compact disk and laser disk optical recording. Quantum-well lasers operate by electrically injecting or pumping electrons into the lowest-conduction-band ($n = 1$) quantum-well state, where they recombine with holes in the highest-valence-band ($n = 1$) quantum-well state (that is, the electrons drop to an empty $n = 1$ valence-band state; illus. *a*), producing the emission of photons. These photons stimulate further photon emission and produce high-efficiency lasing. *See* LASER.

The photon absorption is the basis for quantum-well photodetectors and light modulators. In the quantum-well infrared photodetector an electron is promoted from lower (say, $n = 1$) to higher (say, $n = 2$) conduction band quantum-well states (illus. *b*) by absorption of an infrared photon. An electron in the higher state can travel more freely across the barriers, enabling it to escape from the well and be collected in a detector circuit. Changes in quantum-well shapes produced by externally applied elec-

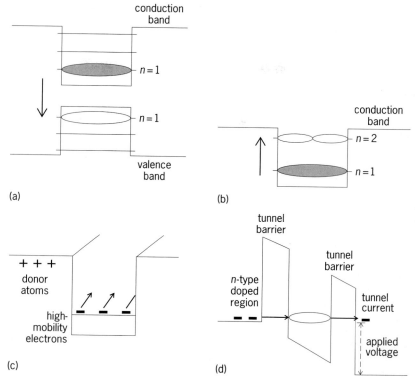

Principles of operation of quantum-well devices. (*a*) **Quantum-well laser.** (*b*) **Quantum-well infrared detector.** (*c*) **High-electron-mobility transistor (HEMT or MODFET). Electrons in the quantum well that came from donor atoms in the barrier are free to move with high mobility in the direction perpendicular to the page.** (*d*) **Resonant tunneling device.**

tric fields can change the absorption wavelengths for light in a quantized electronic structure. The shift in optical absorption wavelength with electric field is known as the quantum-confined Stark effect. It forms the basis for semiconductor light modulators and semiconductor optical logic devices. *See* OPTICAL MODULATORS; STARK EFFECT.

Modulation doping is a special way of introducing electrons into quantum wells for electrical applications. The electrons come from donor atoms lying in adjacent barrier layers (illus. *c*). Modulation doping is distinguished from conventional uniform doping in that it produces carriers in the quantum well without introducing impurity dopant atoms into the well. Since there are no impurity atoms to collide with in the well, electrons there are free to move with high mobility along the quantum-well layer. Resistance to electric current flow is thus much reduced relative to electrical resistance in conventional semiconductors. This enhances the low-noise and high-speed applications of quantum wells and is the basis of the high-electron-mobility transistor (HEMT), which is also known as the modulation-doped field-effect transistor (MODFET). HEMTs are widely used in microwave receivers for direct reception of satellite television broadcasts. *See* TRANSISTOR.

Electrical conductivity in carbon nanotubes occurs without doping and results from the absence of any energy gap in the electronic energy band structure of the nanotubes and the presence of allowed states at the Fermi energy. Individual nanotubes can be electrically contacted. Simple quantum wire transistors displaying quantized electron motion have been formed from single nanotubes.

Quantum-mechanical tunneling is another important property of quantized electronic structures. Tunneling of electrons through thin barrier layers between quantum wells is a purely quantum-mechanical effect without any real analog in classical physics or classical mechanics. It results from the fact that electrons have wavelike properties and that the particle waves can penetrate into the barrier layers. This produces a substantial probability that the particle wave can penetrate entirely through a barrier layer and emerge as a propagating particle on the opposite side of the barrier. The penetration probability has an exponential drop-off with barrier thickness. The tunneling is greatest for low barriers and thin barriers.

This effect finds application in resonant tunnel devices, which can show strong negative resistance in their electrical properties. In such a device (illus. *d*), electrons from an *n*-type doped region penetrate the barrier layers of a quantum well by tunneling. The tunneling current is greatest when the tunneling electrons are at the same energy as the quantum-well energy. The tunneling current actually drops at higher applied voltages, where the incident electrons are no longer at the same energy as the quantum-state energy, thus producing the negative resistance characteristic of the resonant tunneling diode. *See* ARTIFICIALLY LAYERED STRUCTURES; RESONANCE (QUANTUM MECHANICS); TUNNELING IN SOLIDS.   [A.C.Go.]

Quantized vortices   A type of flow pattern exhibited by superfluids, such as liquid ^{4}He below 2.17 K ($-455.76°$F). The term vortex designates the familiar whirlpool pattern where the fluid moves circularly around a central line and the velocity diminishes inversely proportionally to the distance from the center. The strength of a vortex is determined by the circulation, which is the line integral of the velocity around any path enclosing the central line. *See* VORTEX.

A superfluid is believed to be characterized by a macroscopic (that is, large-scale) quantum-mechanical wave function ψ. This wave function locks the superfluid into a coherent state. Since the velocity around the vortex increases without limit as the center is approached, the superfluid density and thus ψ must vanish at the center in order to avoid an infinite energy. Thus the central core of the vortex marks the zeros, or nodal lines, in the macroscopic wave function. *See* QUANTUM MECHANICS.

Quantized vortex lines are usually produced by rotating a vessel containing superfluid helium. At very low rotation speeds, no vortices exist: the superfluid remains at rest while the vessel rotates. At a certain speed the first vortex appears and corresponds to the first excited rotational state of the system. If the container continues to accelerate, additional quantized vortices will appear. At any given speed the vortices form a regular array which rotates with the vessel.

Quantized vortex lines were first detected in the mid-1950s by their influences on superfluid thermal waves traveling across the lines. In the late 1950s it was discovered that electrons in liquid helium form tiny charged bubbles which can become trapped on the vortex core but can move quite freely along the line. These electron bubbles (often referred to as ions) have been one of the most useful probes of quantized vortices. Researchers have been able to use ions to detect single quantized vortex lines. In one experiment the trapped ions are pulled out at the top of the vortex lines, accelerated, and focused onto a phosphor screen. The pattern of light thus produced on the phosphor

1 mm

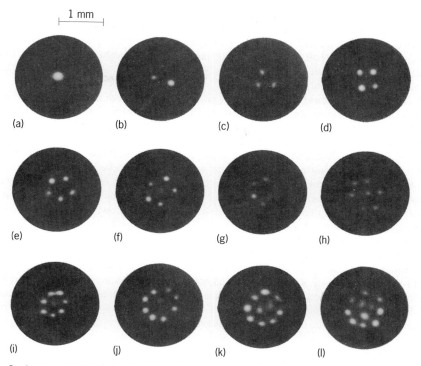

(a)   (b)   (c)   (d)

(e)   (f)   (g)   (h)

(i)   (j)   (k)   (l)

Stationary configurations of vortices which appear when a cylindrical container of superfluid ⁴He is rotated about its axis. As the rotation speed increases from a to f, more vortices appear and the patterns become more complex. (*From E. J. Yarmchuk, M. J. V. Gordon, and R. E. Packard, Observation of stationary vortex arrays in rotating superfluid helium, Phys. Rev. Lett., 43:214–217, 1979*)

is a map of the position of the vortices where they contact the liquid meniscus (*see* illustration). *See* LIQUID HELIUM; SUPERFLUIDITY.   [R.E.P.]

Quantum (physics)   A term characterizing an excitation in a wave or field, connoting fundamental particlelike properties such as energy or mass, momentum, and angular momentum for this excitation. In general, any field or wave equation that is quantized, including systems already treated in quantum mechanics that are second-quantized, leads to a particle interpretation for the excitations which are called quanta of the field. This term historically was first applied to indivisible amounts of electromagnetic, or light, energy usually referred to as photons. The photon, or quantum of the electromagnetic field, is a massless particle, best interpreted as such by quantizing Maxwell's equations. Analogously, the electron can be said to be the quantum of the Dirac field through second quantization of the Dirac equation, which also leads to the prediction of the existence of the positron as another quantum of this field with the same mass but with a charge opposite to that of the electron. In similar fashion, quantization of the gravitational field equations suggests the existence of the graviton. The pi meson or pion was theoretically predicted as the quantum of the nuclear force field. Another quantum is the quantized lattice vibration, or phonon, which can be interpreted as a quantized sound wave since it travels through a quantum solid or fluid, or through nuclear matter, in the same manner as sound goes through air.

The use of quantum as an adjective (quantum mechanics, quantum electrodynamics) implies that the particular subject is to be treated according to the modern rules that have evolved for quantized systems. *See* ELEMENTARY PARTICLE; GRAVITATION; GRAVITON; MAXWELL'S EQUATIONS; MESON; PHONON; PHOTON; QUANTUM ELECTRODYNAMICS; QUANTUM FIELD THEORY; QUANTUM MECHANICS.

[K.E.L.]

Quantum acoustics

The investigation of the effects of the laws of quantum mechanics on the propagation and absorption of sound. At the present stage of development of physical science, quantum mechanics is the most fundamental theory of physical phenomena. However, for many applications in the everyday world, a sufficiently accurate description of nature is provided by classical mechanics. Quantum acoustics refers to acoustic experiments that are carried out under conditions such that the results can be understood only in terms of quantum theory. As a general tendency, quantum effects become more important in acoustic experiments that are performed with higher-frequency sound waves or that are carried out at lower temperatures. *See* QUANTUM MECHANICS.

Sound and phonons. To understand the quantum nature of sound, it is valuable to consider the origins of the quantum theory of light. Experiments in the latter part of the nineteenth century showed that the classical theory of electromagnetism combined with the laws of statistical mechanics could not explain the spectrum of light emitted by a heated surface. To reconcile theory and experiment, Max Planck in 1900 proposed that the energy of a light wave is quantized. This means that only certain values of the energy of the wave are allowed. These allowed values are given by

$$E = n\hbar\omega$$

where $\hbar$ is Planck's constant divided by 2π, ω is the angular frequency of the light wave, and n is an integer. The modern interpretation of this formula describes this energy in terms of elementary quanta called photons. Each photon has an energy $\hbar\omega$, and the wave is made up of n photons. It was later realized that this same quantization of energy should apply to sound waves. These fundamental units of sound energy were given the name phonons. *See* HEAT RADIATION; LIGHT; PHONON; PHOTON.

Phonoatomic effect. One of the earliest experiments that confirmed the idea that light consists of photons was the photoelectric effect. In this effect the energy of light quanta is used to eject electrons from the surface of a metal. Only if the frequency of the light is sufficiently high do the photons have enough energy to knock the electrons out of the metal. The detection of this threshold frequency is thus strong evidence for quantization of energy. An analogous experiment has been performed with sound. In the experiment, sound is generated by a source in a vessel of liquid helium maintained at very low temperature, less than 0.1 K above absolute zero. In liquid helium, each atom is bound to the other atoms in the liquid by an energy that is unusually small compared to that of other liquids. For sound of sufficiently high frequency, the sound quanta when they arrive at the surface of the liquid have sufficient energy to knock helium atoms out of the liquid. These atoms can be detected by a suitable receiver placed above the surface of the liquid. The energy of an ejected atom, which determines its velocity, is equal to the energy of a phonon minus the binding energy. The time of arrival of the ejected atoms at the receiver is, in turn, determined by this velocity and hence is dependent on the frequency of the sound wave. Experiments have confirmed that the helium atoms arrive at the time expected based on this theory. *See* PHOTOEMISSION.

Phonon-phonon interactions. Sound waves in solids are attenuated during their propagation by a wide variety of physical processes. In many materials the most important mechanisms are related to impurities or defects in the solid, such as cracks, grain boundaries, or dislocations. Even when sound travels through a perfect crystal containing no defects, it is found that a measurable attenuation still occurs. In insulating crystals, where there are no free electrons, this attenuation is due to an interaction between the sound wave and the random thermal vibrations of the atoms in the solid. These random vibrations, which constitute the heat energy of the solid, are also quantized and are called thermal phonons. The attenuation of the sound wave can be attributed to collisions with the thermal phonons in which some of the sound quanta are scattered out of the sound beam. This mechanism is referred to as phonon-phonon scattering. *See* CRYSTAL DEFECTS; LATTICE VIBRATIONS; SOUND ABSORPTION.

In a linear elastic solid the elastic stress is exactly proportional to the strain, a phenomenon that is called Hooke's law. If Hooke's law holds, the presence of one wave does not affect the propagation of another, and so there are no interactions between phonons. Phonon-phonon scattering occurs because real solids always exhibit some deviations from linear elastic behavior. This nonlinearity is called anharmonicity. *See* NONLINEAR ACOUSTICS.

Even when the temperature is very low and there are very few thermal phonons, anharmonicity can still give rise to an attenuation of a sound wave. This is because the sound phonons can spontaneously decay into phonons of lower frequencies. The rate at which this decay occurs is proportional to the frequency of the sound wave to the fifth power, and so the attenuation is important only for sound waves of very high frequency. If the number of phonons is sufficiently large, it is possible under some circumstances for there to be a large buildup in the population of some of the decay phonons. This process is called parametric amplification.

[H.J.Ma.]

Quantum anomalies
Phenomena that arise when a quantity that vanishes according to the dynamical rules of classical physics acquires a finite value when quantum rules are used. For example, the classical Poisson bracket for some entities may vanish; yet the corresponding quantum commutator may be nonzero—this is a commutator anomaly. Alternatively, the flow of some material current may satisfy a continuity equation by virtue of the classical equations of motion, indicating conservative flow; but upon quantization the continuity equation may fail and the flow may no longer be conservative in the quantum theory—this is an anomalous divergence (of the current in question). Since the forms of Poisson brackets and quantum commutators as well as the occurrence of continuity equations for currents are related to symmetries and conservation laws of the theory, quantum anomalies serve to break some symmetries and destroy some conservation laws of classical models. This violation of symmetry is not driven by explicit symmetry-breaking terms in the dynamical equations—rather the quantization procedure itself violates the classical symmetry. The mathematical reason for this phenomenon is that classical dynamics, involving a finite number of degrees of freedom, usually leads to a quantum theory on an infinite-dimensional vector space (Hilbert space), and this "infinity" gives rise to novel effects. *See* CANONICAL TRANSFORMATIONS; CONSERVATION LAWS (PHYSICS); SYMMETRY LAWS (PHYSICS).

The physically interesting setting for these phenomena is in quantum field theory, especially as applied to elementary particle physics, where the mechanism serves as an important source for symmetry breaking. Quantum anomalies also play a role in various other branches of physics, in which quantum field theory finds application, including condensed matter, supersymmetry, string theory, and motion in curved space-time. *See* ELEMENTARY PARTICLE; QUANTUM FIELD THEORY; SPACE-TIME; SUPERSTRING THEORY; SUPERSYMMETRY; SYMMETRY BREAKING.

[R.J.]

Quantum chromodynamics   A theory of the strong ("nuclear") interactions among quarks, which are regarded as fundamental constituents of matter. Quantum chromodynamics (QCD) seeks to explain why quarks combine in certain configurations to form the observed patterns of subnuclear particles, such as the proton and pi meson. According to this picture, the strong interactions among quarks are mediated by a set of force particles known as gluons. Strong interactions among gluons may lead to new structures that correspond to as-yet-undiscovered particles. The long-studied nuclear force that binds protons and neutrons together in atomic nuclei is regarded as a collective effect of the elementary interactions among constituents of the composite protons and neutrons. *See* NUCLEAR STRUCTURE.

Quantum chromodynamics has not yet been subjected to precise experimental tests. Several qualitative predictions of quantum chromodynamics do seem to have been borne out. Part of the esthetic appeal of the theory is due to the fact that quantum chromodynamics is nearly identical in mathematical structure to quantum electrodynamics (QED) and to the unified theory of weak and electromagnetic interactions. This resemblance encourages the hope that a unified description of the strong, weak, and electromagnetic interactions may be at hand. *See* ELECTROWEAK INTERACTION; QUANTUM ELECTRODYNAMICS; WEAK NUCLEAR INTERACTIONS.

Gauge theories. At the heart of current theories of the fundamental interactions is the idea of gauge invariance. It is widely believed that gauge theories constructed to embody various symmetry principles represent the correct quantum-mechanical descriptions of the strong, weak, and electromagnetic interactions. *See* SYMMETRY LAWS (PHYSICS).

Color. Although the idea that the strongly interacting particles are built up of quarks brought new order to hadron spectroscopy and suggested new relations among mesons and baryons, the constituent description brought with it a number of puzzles. According to the Pauli exclusion principle, identical spin-$\frac{1}{2}$ particles cannot occupy the same quantum state. As a consequence, the observed baryons such as Δ^{++} (uuu) and Ω^- (sss), which would be composed of three identical quarks in the same state, would seem to be forbidden configurations. To comply with the Pauli principle, it is necessary to make the three otherwise identical quarks distinguishable by supposing that *every* flavor of quark exists in three varieties, fancifully labeled by the colors red, green, and blue. Color may be regarded as the strong-interaction analog of electric charge. Color cannot be created or destroyed by any of the known interactions. Like electric charge, it is said to be conserved. *See* COLOR (QUANTUM MECHANICS); EXCLUSION PRINCIPLE.

In the face of evidence that color could be regarded as the conserved charge of the strong interactions, it was natural to seek a gauge symmetry that would have color conservation as its consequence. An obvious candidate for the gauge symmetry group is the unitary group SU(3), now to be applied to color rather than flavor. The theory of strong interactions among quarks that is prescribed by local color gauge symmetry is known as quantum chromodynamics. The mediators of the strong interactions are eight massless spin-1 bosons, one for each generator of the symmetry group. These strong-force particles are named gluons because they make up the "glue" that binds quarks together into hadrons. Gluons also carry color and hence have strong interactions among themselves.

Asymptotic freedom. The theoretical description of the strong interactions has historically been inhibited by the very strength of the interaction, which renders low-order perturbative calculations untrustworthy. However, in 1973 it was found that in many circumstances the effective strength of the interaction in Yang-Mills theories becomes increasingly feeble at short distances, a property known as asymptotic freedom. For quantum chromodynamics, this remarkable observation implies that the interaction between quarks becomes weak at small separations. This discovery raises the hope that some aspects of the strong interactions might be treated by using familiar computational techniques that are predicated upon the smallness of the interaction strength.

Quarkonium. It was suggested in 1974 that the bound system of an extremely massive quark with its antiquark would be so small that the strong force would be extremely feeble. In this case, the binding between quark and antiquark is mediated by the exchange of a single massless gluon, and the spectrum of bound states resembles that of an exotic atom composed of an electron and an antielectron (positron) bound electromagnetically in a Coulomb potential generated by the exchange of a massless photon. Since the electron-positron atom is known as positronium, the heavy quark-antiquark atom has been called quarkonium. Two families of heavy quark-antiquark bound states, the ψ/J system composed of charmed quarks and the Υ system made up of b quarks, have been discovered. Both have level schemes characteristic of atomic spectra, which have been analyzed by using tools of nonrelativistic quantum mechanics developed for ordinary atoms. The atomic analogy has proved extremely fruitful for studying the strong interaction. *See* CHARM; J/PSI PARTICLE; POSITRONIUM.

Lattice models. To deal with the existence and properties of the hadrons themselves, it is necessary to devise a new computational approach that does not break down when the interaction becomes strong. The most promising method has been the crystal lattice formulation of the theory. By considering the values of the color field only on individual lattice sites, it is possible to use many of the techniques developed in statistical physics for the study of spin systems such as magnetic substances. *See* ELEMENTARY PARTICLE; FUNDAMENTAL INTERACTIONS; GLUONS; ISING MODEL; QUANTUM FIELD THEORY; QUARKS; STANDARD MODEL; STATISTICAL MECHANICS; STRONG NUCLEAR INTERACTIONS.

[C.Q.]

Quantum electrodynamics

The field of physics that studies the interaction of electromagnetic radiation with electrically charged matter within the framework of relativity and quantum mechanics. It is the fundamental theory underlying all disciplines of science concerned with electromagnetism, such as atomic physics, chemistry, biology, the theory of bulk matter, and electromagnetic radiation.

Efforts to formulate quantum electrodynamics (QED) were initiated by P. A. M. Dirac, W. Heisenberg, and W. Pauli soon after quantum mechanics was established. The first step was to remedy the obvious shortcoming of quantum mechanics: that it applies only to the case where particle speeds are small compared with that of light, c. This led to Dirac's discovery of a relativistic wave equation, in which the wave function has four components and is multiplied by certain 4×4 matrices. His equation incorporates in a natural manner the observed electron-spin angular momentum, which implies that the electron is a tiny magnet. The strength of this magnet (magnetic moment) was predicted by Dirac and agreed with observation. A detailed prediction of the hydrogen spectrum was also in good agreement with experiment. *See* ATOMIC STRUCTURE AND SPECTRA; ELECTRON SPIN.

In order to go beyond this initial success and calculate higher-order effects, however, the interaction of charge and electromagnetic field had to be treated dynamically. To begin with, a good theoretical framework had to be found for describing the wave-particle duality of light, that is, the experimentally well-established fact that light behaves like a particle (photon) in some cases but like a wave in others. Similarly, the electron manifests wave-particle duality, another observed fact. Once this problem was settled, the next question was how to deal with the interaction of charge and electromagnetic field. It is here that the theory ran into severe difficulties. Its predictions often diverged when attempts were made to calculate beyond lowest-order approximations. This inhibited the further development of the theory for nearly 20 years. Stimulated by spectroscopic experiments vastly refined by microwave technology developed during World War II, however, S. Tomonaga, R. P. Feynman, and J. Schwinger discovered that the difficulties disappear if all observable quantities are expressed in terms of the experimentally measured charge and mass of the electron. With the discovery of this procedure, called renormalization, quantum electrodynamics became a theory in which all higher-order corrections are finite and well defined. *See* PHOTON; QUANTUM MECHANICS; RELATIVISTIC QUANTUM THEORY; RELATIVITY; WAVE MECHANICS.

Quantum electrodynamics is the first physical theory ever developed that has no obvious intrinsic limitation and describes physical quantities from first principles. Nature accommodates forces other than the electromagnetic force, such as those responsible for radioactive disintegration of heavy nuclei (called the weak force) and the force that binds the nucleus together (called the strong force). A theory called the standard model has been developed which unifies the three forces and accounts for all experimental data from very low to extremely high energies. This does not mean, however, that quantum electrodynamics fails at high energies. It simply means that the real world has forces other than electromagnetism.

High-precision tests have provided excellent confirmation for the validity of the renormalization theory of quantum electrodynamics. In the high-energy regime, tests using electron-positron colliding-beam facilities at various high-energy physics laboratories have confirmed the predictions of quantum electrodynamics at center-of-mass energies up to 1.8×10^{11} electronvolts (180 GeV). The uncertainty principle implies that this is equivalent to saying that quantum electrodynamics is valid down to about 10^{-17} meter, a distance 100 times shorter than the radius of the proton.

High-precision tests of quantum electrodynamics have also been carried out at low energies by using various simple atomic systems. The most accurate is that of the measurement of the magnetic moment of the electron, or the gyromagnetic ratio g, the ratio of spin and rotation frequencies, which is correctly predicted by quantum electrodynamics to 12 significant figures. This is the most precise confirmation of any theory ever carried out. *See* QUANTUM FIELD THEORY.   [T.K.]

Quantum field theory   The quantum-mechanical theory of physical systems whose dynamical variables are local functions of space and time. As distinguished from the quantum mechanics of atoms, quantum field theories describe systems with an infinite number of degrees of freedom. Such theories provide the natural language for describing the interactions and properties of elementary particles, and have proved to be successful in providing the basis for the fundamental theories of the interactions of matter. The present understanding of the basic forces of nature is based on quantum field theories of the strong, weak, electromagnetic, and gravitational interactions. Quantum field theory is also useful in the study of many-body systems, especially in situations where the characteristic length of a system is large compared to its microscopic scale. *See* QUANTUM MECHANICS.

Quantum field theory originated in the attempt, in the late 1920s, to unify P. A. M. Dirac's relativistic electron theory and J. C. Maxwell's classical electrodynamics in a quantum theory of interacting photon and electron fields. This effort was completed in the 1950s and was extremely successful. At present the quantitative predictions of the theory are largely limited to perturbative expansions (believed to be asymptotic) in powers of the fine-structure constant. However, because of the extremely small value of this parameter, $\alpha = e^2/\hbar c \approx 1/137$ (where e is the electron charge, $\hbar$ is Planck's constant divided by 2π, and c is the speed of light), such an expansion is quite adequate for most purposes. The remarkable agreement of the predictions of quantum electrodynamics with high-precision experiments (sometimes to an accuracy of 1 part in 10^{12}) provides strong evidence for the validity of the basic tenets of relativistic quantum field theory. See CLASSICAL FIELD THEORY; ELECTROMAGNETIC RADIATION; MAXWELL'S EQUATIONS; PERTURBATION (QUANTUM MECHANICS); PHOTON; QUANTUM ELECTRODYNAMICS; RELATIVISTIC ELECTRODYNAMICS; RELATIVISTIC QUANTUM THEORY.

Quantum field theory also provides the natural framework for the treatment of the weak, strong, and gravitational interactions.

The first of such applications was Fermi's theory of the weak interactions, responsible for radioactivity, in which a hamiltonian was constructed to describe beta decay as a product of four fermion fields, one for each lepton or nucleon. This theory has been superseded by the modern electroweak theory that unifies the weak and the electromagnetic interactions into a common framework. This theory is a generalization of Maxwell's electrodynamics which was the first example of a gauge theory, based on a continuous local symmetry. In the case of electromagnetism the local gauge symmetry is the space-time-dependent change of the phase of a charged field. The existence of massless spin-1 particles, photons, is one of the consequences of the gauge symmetry. The electroweak theory is based on generalizing this symmetry to space-time-dependent transformations of the labels of the fields, based on the group SU(2) × U(1). However, unlike electromagnetism, part of this extended symmetry is not shared by the ground state of the system. This phenomenon of spontaneous symmetry breaking produces masses for all the elementary fermions and for the gauge bosons that are the carriers of the weak interactions, the $W^{\pm}$ and Z bosons. (This is known as the Higgs mechanism.) The electroweak theory has been confirmed by many precision tests, and almost all of its essential ingredients have been verified. See ELECTROWEAK INTERACTION; INTERMEDIATE VECTOR BOSON; SYMMETRY BREAKING; SYMMETRY LAWS (PHYSICS); WEAK NUCLEAR INTERACTIONS.

The application of quantum field theory to the strong or nuclear interactions dates from H. Yukawa's hypothesis that the short-range nuclear forces arise from the exchange of massive particles that are the quanta of local fields coupled to the nucleons, much as the electromagnetic interactions arise from the exchange of massless photons that are the quanta of the electromagnetic field. The modern theory of the strong interactions, quantum chromodynamics, completed in the early 1970s, is also based on a local gauge theory. This is a theory of spin-$\frac{1}{2}$ quarks invariant under an internal local SU(3) (color) gauge group. The observed hadrons (such as the proton and neutron) are SU(3) color-neutral bound states of the quarks whose interactions are dictated by the gauge fields (gluons). This theory exhibits almost-free-field behavior of quarks and gluons over distances and times short compared to the size of a hadron (asymptotic freedom), and a strong binding of quarks at large separations that results in the absence of colored states (confinement). See ELEMENTARY PARTICLE; GLUONS; MESON; QUANTUM CHROMODYNAMICS; QUARKS.

Quantum field theory has been tested down to distances of 10^{-20} m. There appears to be no reason why it should not continue to work down to Planck's length, $(G\hbar/c^3)^{1/2} \approx 10^{-35}$ m (where G is the gravitational constant), where the quantum effects of gravity become important. In the case of gravity, A. Einstein's theory of general relativity already provides a very successful classical field theory. However, the union of quantum mechanics and general relativity raises conceptual problems that seem to call for a radical reexamination of the foundations of quantum field theory. See FUNDAMENTAL INTERACTIONS; GRAVITATION; QUANTUM GRAVITATION; RELATIVITY.   [D.Gr.]

Quantum gravitation   The quantum theory of the gravitational field; also, the study of quantum fields in a curved space-time. In classical general relativity, the gravitational field is represented by the metric tensor $g_{\mu\nu}$ of space-time. This tensor satisfies Einstein's field equation, with the energy-momentum tensor of matter and radiation as a source. However, the equations of motion for the matter and radiation fields also depend on the metric.

Classical field theories such as Maxwell's electromagnetism or the classical description of particle dynamics are approximations valid only at the level of large-scale macroscopic observations. At a fundamental level, elementary interactions of particles and fields must be described by relativistic quantum mechanics, in terms of quantum fields. Because the geometry of space-time in general relativity is inextricably connected to the dynamics of matter and radiation, a consistent theory of the metric in interaction with quantum fields is possible only if the metric itself is quantized. *See* MAXWELL'S EQUATIONS; QUANTUM FIELD THEORY; RELATIVISTIC QUANTUM THEORY; RELATIVITY.

Under ordinary laboratory conditions the curvature of space-time is so extremely small that in most quantum experiments gravitational effects are completely negligible. Quantization in Minkowski space is then justified. Gravity is expected to play a significant role in quantum physics only at rather extreme conditions of strongly time-dependent fields, near or inside very dense matter. The scale of energies at which quantization of the metric itself becomes essential is given by $(\hbar c^5/G)^{1/2} \approx 10^{19}$ GeV, where G is the gravitational constant, $\hbar$ is Planck's constant divided by 2π, and c is the velocity of light. Energies that can be reached in the laboratory or found in cosmic radiation are far below this order of magnitude. Only in the very early stages of the universe, within a proper time of the order of $(G\hbar/c^5)^{1/2} \approx 10^{-43}$ s after the big bang, would such energies have been produced.

In most physical systems the metric is quasistationary over macroscopic distances so that its fluctuations can be ignored. A quantum description of fields in a curved space-time can then be given by treating the metric as a classical external field in interaction with the quantum fields.

Quantum effects of black holes. The most striking quantum effect in curved space-time is the emission of radiation by black holes. A black hole is an object that has undergone gravitational collapse. Classically this means that it becomes confined to a space-time region in which the metric has a singularity (the curvature becomes infinite). This region is bounded by a surface, called the horizon, such that any matter or radiation falling inside becomes trapped. Therefore, classically the mass of a black hole can only increase. However, this is no longer the case if quantum effects are taken into account. When, because of fluctuations of the quantum field, particle-antiparticle or photon pairs are created near the horizon of a black hole, one of the particles carrying negative energy may move toward the hole, being absorbed by it, while the other moves out with positive energy.

It is found that the total rate of emission is inversely proportional to the square of the mass. For stellar black holes whose masses are of the order of a solar mass, the emission rate is negligibly small and unobservable. Only primordial black holes, of mass less than 10^{13} kg, formed very early in the quantum era of the universe, would have been small enough to produce quantum effects that could play any significant role in astrophysics or in cosmology.

Quantization of the metric. There are basically two approaches to the quantization of the metric, the canonical and the covariant quantization. A third method, which can be derived from the first and is now most widely used, is based on the Feynman path integral representation for the vacuum-to-vacuum amplitude, which is the generator of Green's functions for the quantum theory. One important feature of this method is that since the topology of the manifold is not specified at the outset it is possible to include a sum over paths in different topologies. The outcome of this idea is that the vacuum would, at the level of the Planck length, $(G\hbar/c^3)^{1/2} \approx 10^{-35}$ m, acquire a foamlike structure. *See* FEYNMAN INTEGRAL; GREEN'S FUNCTION.

At present a complete, consistent theory of quantum gravity is still lacking. The formal theory fails to satisfy the power-counting criterion for renormalizability. In every order of the perturbation expansion, new divergences appear which could only be canceled by counterterms that do not exist in the original lagrangian. This may not be just a technical problem but the reflection of a conceptual difficulty stemming from the dual role, geometric and dynamic, played by the metric. *See* RENORMALIZATION.

Supergravity and superstrings. Supergravity is a geometric extension of general relativity which incorporates the principle of supersymmetry. Supersymmetry is a kind of symmetry, discovered in the 1970s, that allows for the transformation of fermions and bosons into each other. (Fermions carry half-integer spin while bosons carry integer spin; they also obey different statistics.) Supergravity can be formulated in space-time manifolds with a total of $D = d + 1$ dimensions, where d, the number of space dimensions, can be as large as 10. They constitute truly unified theories of all interactions including gravity.

In the early 1980s, some encouraging results were found with a theory based on the idea that the basic objects of nature are not pointlike but actually one-dimensional objects like strings, which can be open or closed. Incorporating supersymmetry into the theory leads to a critical dimension $D = 10$.

In the approximation of neglecting string excitations, certain superstring models may be described in terms of local fields as a $D = 10$ supergravity theory. At present, these are the only theories that both include gravity and can be consistently quantized. Although a superstring theory may eventually become the ultimate theory of all the interactions, there is still a very long way to go in making the connection between its fundamental fields and the fields representing the particles and their interactions as observed at low energies. *See* FUNDAMENTAL INTERACTIONS; GRAVITATION; SUPERGRAVITY; SUPERSTRING THEORY; SUPERSYMMETRY.

[S.W.MacD.]

Quantum mechanics

The modern theory of matter, of electromagnetic radiation, and of the interaction between matter and radiation; also, the mechanics of phenomena to which this theory may be applied. Quantum mechanics, also termed wave mechanics, generalizes and supersedes the older classical mechanics and Maxwell's electromagnetic theory. Atomic and subatomic phenomena provide the most striking evidence for the correctness of quantum mechanics and best illustrate the differences between quantum mechanics and the older classical physical theories. Quantum mechanics is needed to explain many properties of bulk matter, for instance, the temperature dependence of the specific heats of solids.

The formalism of quantum mechanics is not the same in all domains of applicability. In approximate order of increasing conceptual difficulty, mathematical complexity, and likelihood of future fundamental revision, these domains are the following: (i) Nonrelativistic quantum mechanics, applicable to systems in which particles are neither created nor destroyed, and in which the particles are moving slowly compared to the velocity of light. Here a particle is defined as a material entity having mass, whose internal structure either does not change or is irrelevant to the description of the system. (ii) Relativistic quantum mechanics, applicable in practice to a single relativistic particle (one whose speed equals or nearly equals c); here the particle may have zero rest mass, in which event, its speed must equal c. (iii) Quantum field theory, applicable to systems in which particle creation and destruction can occur; the particles may have zero or nonzero rest mass. This article is concerned mainly with nonrelativistic quantum mechanics, which apparently applies to all atomic and molecular phenomena, with the exception of the finer details of atomic spectra. Nonrelativistic quantum mechanics also is well established in the realm of low-energy nuclear physics. *See* ATOMIC STRUCTURE AND SPECTRA; NUCLEAR PHYSICS; QUANTUM FIELD THEORY; RELATIVISTIC QUANTUM THEORY.

Planck's constant. The quantity 6.626×10^{-34} joule-second, first introduced into physical theory by Max Planck in 1901, is a basic ingredient of the formalism of quantum mechanics. Planck's constant commonly is denoted by the letter h; the notation $\hbar = h/2\pi$ also is standard.

Uncertainty principle. In classical physics the observables characterizing a given system are assumed to be simultaneously measurable (in principle) with arbitrarily small error. For instance, it is thought possible to observe the initial position and velocity of a particle and therewith, using Newton's laws, to predict exactly its future path in any assigned force field. According to the uncertainty principle, accurate measurement of an observable quantity necessarily produces uncertainties in one's knowledge of the values of other observables. In particular, for a single particle relation (1a) holds, where

$$\Delta x \Delta p_x > \hbar \qquad (1a)$$

$$\Delta t \, \Delta E \gtrsim \hbar \qquad (1b)$$

Δx represents the uncertainty (error) in the location of the x coordinate of the particle at any instant, and Δp_x is the simultaneous uncertainty in the x component of the particle momentum. Relation (1a) asserts that under the best circumstances, the product $\Delta x \Delta p_x$ of the uncertainties cannot be less than about 10^{-34} joule second.

The uncertainty relation (1b) is derived and interpreted somewhat differently than relation (1a); it asserts that for any system, an energy measurement with error ΔE must be performed in a time not less than $\Delta t \sim \hbar/\Delta E$. If a system endures for only Δt seconds, any measurement of its energy must be uncertain by at least $\Delta E \sim \hbar \Delta t$. *See* UNCERTAINTY PRINCIPLE.

Wave-particle duality. It is natural to identify such fundamental constituents of matter as protons and electrons with the mass points or particles of classical mechanics. According to quantum mechanics, however, these particles, in fact all material systems, necessarily have wavelike properties. Conversely, the propagation of light, which, by Maxwell's electromagnetic theory, is understood to be a wave phenomenon, is associated in quantum mechanics with massless energetic and momentum-transporting particles called photons. The quantum-mechanical synthesis of wave and particle concepts is embodied in the de Broglie relations, given by Eqs. (2a) and (2b). These give

$$\lambda = h/p \qquad (2a)$$

$$f = E/h \qquad (2b)$$

the wavelength λ and wave frequency f associated with a free particle (a particle moving freely under no forces) whose momentum is p and energy is E; the same relations give the photon momentum p and energy E associated with an electromagnetic wave in free space (that is, in a vacuum) whose wavelength is λ and frequency is f. *See* PHOTON.

The wave properties of matter have been demonstrated conclusively for beams of electrons, neutrons, atoms (hydrogen, H, and helium, He), and molecules (H_2). When incident upon crystals, these beams are reflected into certain directions, forming diffraction patterns. Diffraction patterns are difficult to explain on a particle picture; they are readily understood on a wave picture, in which wavelets scattered from regularly spaced atoms in the crystal lattice interfere constructively along certain directions only. *See* ELECTRON DIFFRACTION; NEUTRON DIFFRACTION.

The particle properties of light waves are observed in the photoelectric effect and the compton effect. *See* COMPTON EFFECT; PHOTOEMISSION.

Complementarity. Wave-particle duality and the uncertainty principle are thought to be examples of the more profound principle of complementarity, first enunciated by Niels Bohr (1928). According to the principle of complementarity, nature has "complementary" aspects; an experiment which illuminates one of these aspects necessarily simultaneously obscures the complementary aspect. To put it differently, each experiment or sequence of experiments yields only a limited amount of information about the system under investigation; as this information is gained, other equally interesting information (which could have been obtained from another sequence of experiments) is lost. Of course, the experimenter does not forget the results of previous experiments, but at any instant, only a limited amount of information is usable for predicting the future course of the system.

Quantization. In classical physics the possible numerical values of each observable, meaning the possible results of exact measurement of the observable, generally form a continuous set. For example, the x coordinate of the position of a particle may have any value between $-\infty$ and $+\infty$. In quantum mechanics the possible numerical values of an observable need not form a continuous set, however. For some observables, the possible results of exact measurement form a discrete set; for other observables, the possible numerical values are partly discrete, partly continuous; for example, the total energy of an electron in the field of a proton may have any positive value between 0 and $+\infty$, but may have only a discrete set of negative values, namely, -13.6, $-13.6/4$, $-13.6/9$, $-13.6/16$ eV,.... Such observables are said to be quantized; often there are simple quantization rules determining the quantum numbers which specify the allowable discrete values. Spectroscopy, especially the study of atomic spectra, probably provides the most detailed quantitative confirmation of quantization.

Probability considerations. The uncertainty and complementarity principles, which limit the experimenter's ability to describe a physical system, must limit equally the experimenter's ability to predict the results of measurement on that system. Suppose, for instance, that a very careful measurement determines that the x coordinate of a particle is precisely $x = x_0$. This is permissible in nonrelativistic quantum mechanics. Then, formally, the particle is known to be in the eigenstate corresponding to the eigenvalue $x = x_0$ of the x operator. Under these circumstances, an immediate repetition of the position measurement again will indicate that the particle lies at $x = x_0$. Knowing that the particle lies at $x = x_0$ makes the momentum p_x of the particle completely uncertain, however, according to relation (1a). A measurement of p_x immediately after the particle is located at $x = x_0$ could yield any value of p_x from $-\infty$ to $+\infty$.

More generally, suppose the system is known to be in the eigenstate corresponding to the eigenvalue α of the observable A. Then for any observable B, which is to some extent complementary to A, that is, for which an uncertainty relation of the form of relations (1) limits the accuracy with which A and B can simultaneously be measured, it is not possible to predict which of the many possible values $B = \beta$ will be observed. However, it is possible to predict the relative probabilities $P_\alpha(\beta)$ of immediately thereafter finding the observable B equal to β, that is, of finding the system in the eigenstate corresponding to the eigenvalue $B = \beta$.

To the eigenvalues correspond eigenfunctions, in terms of which $P_\alpha \lesssim \beta$ can be computed. In particular, when α is a discrete eigenvalue of A, and the operators depend only on x and p_x, the probability $P_\alpha(\beta)$ is postulated as in Eq. (3), where $u(x,\alpha)$ is the

$$P\alpha(\beta) = \left| \int_{-\infty}^{\infty} dx \ v^*(x, \beta)u(x, \alpha) \right|^2 \tag{3}$$

eigenfunction corresponding to $A = \alpha$; $v(x,\beta)$ is the eigenfunction corresponding to $B = \beta$; and the * denotes the complex conjugate. The integral in Eq. (3) is called the projection of $u(x,\alpha)$ on $u(x,\beta)$. The quantity $|u(x,\alpha)|^2 \ dx$ is the probability that the system, known to be in the eigenstate $A = \alpha$, will be found in the interval x to $x + dx$. *See* EIGENVALUE (QUANTUM MECHANICS).

Wave function. When the system is known to be in the eigenstate corresponding to $A = \alpha$, the eigenfunction $u(x,\alpha)$ is the wave function; that is, it is the function whose projection on an eigenfunction $v(x,\beta)$ of any observable B gives the probability of measuring $B = \beta$. The wave function $\psi(x)$ may be known exactly; in other words, the state of the system may be known as exactly as possible (within the limitations of uncertainty and complementarity), even though $\psi(x)$ is not the eigenfunction of a known operator. This circumstance arises because the wave function obeys Schrödinger's wave equation. Knowing the value of $\psi(x)$ at time $t = 0$, the wave equation completely determines $\psi(x)$ at all future times. In general, however, if $\psi(x,0) = u(x,\alpha)$, that is, if $\psi(x,t)$ is an eigenfunction of A at $t = 0$, then $\psi(x,t)$ will not be an eigenfunction of A at later times $t > 0$.

A system described by a wave function is said to be in a pure state. Not all systems are described by wave functions, however. For example, a beam of hydrogen atoms streaming out of a small hole in a hydrogen discharge tube can be regarded as a statistical ensemble or mixture of pure states oriented with equal probability in all directions.

Schrödinger equation. Equation (4) describes a plane wave of frequency f, wave-

$$\psi(x, t) = A(\lambda) \exp\left[2\pi i \left(\frac{x}{\lambda} - ft \right) \right] \tag{4}$$

length λ, and amplitude $A(\lambda)$, propagating in the positive x direction. The previous discussion concerning wave-particle duality suggests that this is the form of the wave function for a beam of free particles moving in the x direction with momentum $p = p_x$, with Eq. (2) specifying the connections between f, λ, and E, p. Differentiating Eq. (4), it is seen that Eqs. (5) hold. Since for a free particle $E = p^2/2m$, it follows also that Eq. (6) is valid.

$$p_x \psi = \frac{h}{\lambda} \psi = \frac{\hbar}{i} \frac{\partial \psi}{\partial x} \tag{5a}$$

$$E\psi = hf \psi = -\frac{\hbar}{i} \frac{\partial \psi}{\partial t} \tag{5b}$$

$$\frac{-\hbar^2}{2m} \frac{\partial^2 \psi}{\partial x^2} = -\frac{\hbar}{i} \frac{\partial \psi}{\partial t} \tag{6}$$

See WAVE MOTION

Equation (6) holds for a plane wave of arbitrary λ, and therefore for any superposition of waves of arbitrary λ, that is, arbitrary p_x. Consequently, Eq. (6) should be the wave equation obeyed by the wave function of any particle moving under no forces, whatever the projections of the wave function on the eigenfunctions of p_x. Equations (5) and (6) further suggest that for a particle whose potential energy $V(x)$ changes, in other words, for a particle in a conservative force field, $\psi(x,t)$ obeys Eq. (7).

$$\frac{-\hbar^2}{2m} \frac{\partial^2 \psi}{\partial x^2} + V(x)\psi = -\frac{\hbar}{i} \frac{\partial \psi}{\partial t} \tag{7}$$

Equation 7 is the time-dependent Schrödinger equation for a one-dimensional (along x), spinless particle. Noting Eq. (5b), and observing that Eq. (7) has a solution for the form of Eq. (8), it is inferred that $\psi(x)$ of Eq. (8) obeys the time-dependent Schrödinger equation, Eq. (9).

$$\psi(x, t) = \psi(x) \exp(-i\,Et/\hbar) \tag{8}$$

$$\frac{-\hbar^2}{2m}\frac{\partial^2\psi}{\partial x^2} + V(x)\psi = E\psi \tag{9}$$

See FORCE

Equation (9) is solved subject to reasonable boundary conditions, for example, that ψ must be continuous and must not become infinite as x approaches $\pm\infty$. These boundary conditions restrict the values of E for which there exist acceptable solutions $\psi(x)$ to Eq. (9), the allowed values of E depending on $V(x)$. In this manner, the allowed energies of atomic hydrogen listed in the earlier discussion of quantization are obtained.

The forms of Eqs. (5a), (7), and (9) suggest that the classical observable p_x, must be replaced by the operator $(\hbar/i)\,(\partial/\partial x)$. With this replacement, Eq. (10) holds. In other

$$(xp_x - p_x x)\psi = i\hbar\psi \tag{10}$$

words, whereas the classical canonically conjugate variables x and p_x are numbers, obeying the commutative law in Eq. (11a), the quantum-mechanical quantities x and p_x are noncommuting operators, obeying Eq. (11b).

$$xp_x - p_x x = 0 \tag{11a}$$
$$xp_x - p_x x = i\hbar \tag{11b}$$

Correspondence principle. Since classical mechanics and Maxwell's electromagnetic theory accurately describe macroscopic phenomena, quantum mechanics must have a classical limit in which it is equivalent to the older classical theories. Although there is no rigorous proof of this principle for arbitrarily complicated quantum-mechanical systems, its validity is well established by numerous illustrations.   [E.G.]

Quantum numbers   The quantities, usually discrete with integer or half-integer values, which are needed to characterize a physical system of one or more atomic or subatomic particles. Specification of the set of quantum numbers serves to define such a system or, in other words, to label the possible states the system may have. In general, quantum numbers are obtained from conserved quantities determinable by performing symmetry transformations consisting of arbitrary variations of the system which leave the system unchanged. For example, since the behavior of a set of particles should be independent of the location of the origin in space and time (that is, the symmetry operation is translation in space-time), it follows that momentum and energy are rigorously conserved. *See* SYMMETRY LAWS (PHYSICS).

In general, each physical system must be studied individually to find the symmetry transformations, and thus the conserved quantities and possible quantum numbers. The quantum numbers themselves, that is, the actual state labels, are usually the eigenvalues of the physical operators corresponding to the conserved quantities for the system in question. *See* EIGENVALUE (QUANTUM MECHANICS); ELEMENTARY PARTICLE; PARITY (QUANTUM MECHANICS).

It is not necessary that the conserved quantity be "quantized" in order to be regarded as a quantum number; for example, a free particle possesses energy and momentum, both of which can have values from a continuum but which are used to specify the state of the particle.   [K.E.L.]

Quantum solids   A class of solids whose atoms or molecules undergo large zero-point motion even in the quantum ground state (at temperature $T = 0\,K = -459.67°F$) as a result of their small mass and the weak attractive part of their interaction potential. The most striking examples are the isotopes of helium, ^3He and ^4He, which have a root-mean-square displacement from their lattice sites of approximately 25%. Further examples are the molecular hydrogens, H_2, D_2, and HD, as well as some heavier molecular solids. *See* INTERMOLECULAR FORCES; QUANTUM MECHANICS.

These materials display quantum effects in their bulk properties when cooled to temperatures near absolute zero so that the chaotic thermal motion is reduced. Both of the helium isotopes remain liquid all the way to absolute zero, unless external pressure (~ 3 megapascals ≈ 30 atm) is applied. This is because the atoms are not at rest at 0 K; the zero-point motion acts as an internal pressure which must be overcome in order to bring the atoms close enough together for solidification. All other substances, including the hydrogens, freeze under their own vapor pressure above 10 K.

The two melting curves are quite different in detail because of the different types of quantum statistics which the particles obey. There is a pronounced minimum in the ^3He melting pressure which is unique, and can be understood by considering the entropies of the liquid and the solid. The pressure minimum leads to the bizarre situation of the addition of heat causing freezing. The inverse of this process, the adiabatic formation of the solid by compression, is an important process which has been used extensively to cool liquid and solid ^3He to temperatures of approximately 1 mK. *See* CRYOGENICS; LIQUID HELIUM.   [E.D.A.]

Quantum statistics   The statistical description of particles or systems of particles whose behavior must be described by quantum mechanics rather than by classical mechanics. As in classical, that is, Boltzmann statistics, the interest centers on the construction of appropriate distribution functions. However, whereas these distribution functions in classical statistical mechanics describe the number of particles in given (in fact, finite) momentum and positional ranges, in quantum statistics the distribution functions give the number of particles in a group of discrete energy levels. In an individual energy level there may be, according to quantum mechanics, either a single particle or any number of particles. This is determined by the symmetry character of the wave functions. For antisymmetric wave functions only one particle (without spin) may occupy a state; for symmetric wave functions, any number is possible. Based on this distinction, there are two separate distributions, the Fermi-Dirac distribution for systems described by antisymmetric wave functions and the Bose-Einstein distribution for systems described by symmetric wave functions. *See* BOLTZMANN STATISTICS; BOSE-EINSTEIN STATISTICS; EXCLUSION PRINCIPLE; FERMI-DIRAC STATISTICS; KINETIC THEORY OF MATTER; QUANTUM MECHANICS; STATISTICAL MECHANICS.   [M.Dr.]

Quantum theory of matter   The microscopic explanation of the properties of condensed matter, that is, solids and liquids, based on the fundamental laws of quantum mechanics. Without the quantum theory, some properties of matter such as magnetism and superconductivity have no explanation at all, while for others only a phenomenological description can be obtained. With the theory, it is at least possible to comprehend what is needed to approach a complete understanding.

The theoretical problem of condensed matter—large aggregates of elementary particles with mutual interactions—is the quantum-mechanical many-body problem: an enormous number, of order 10^{23}, of constituent particles in the presence of a heat bath and interacting with each other according to quantum-mechanical laws. What makes the quantum physics of matter different from the traditional quantum theory of elementary particles is that the fundamental constituents (electrons and ions) and their interactions (Coulomb interactions) are known but the solutions of the appropriate quantum-mechanical equations are not. This situation is not due to the lack of a sufficiently large computer, but is caused by the fact that totally new structures, such as crystals, magnets, ferroelectrics, superconductors, liquid crystals, and glasses, appear out of the complexity of the interactions among the many constituents. The consequence is that entirely new conceptual approaches are required to construct predictive theories of matter. The usual technique for approaching the quantum many-body problem for a condensed-matter system is to try to reduce the huge number of variables (degrees of freedom) to a number which is more manageable but still can describe the essential physics of the phenomena being studied. *See* Crystal; Ferroelectrics; Magnetic materials; Quantum mechanics; Solid-state physics; Superconductivity.

[E.A.]

Quantum theory of measurement
The attempt to reconcile the counter-intuitive features of quantum mechanics with the hypothesis that quantum mechanics is in principle a complete description of the physical world, even at the level of everyday objects. A paradox arises because, at the atomic level where the quantum formalism has been directly tested, the most natural interpretation implies that where two or more different outcomes are possible it is not necessarily true that one or the other is actually realized, whereas at the everyday level such a state of affairs seems to conflict with direct experience.

The resolution of this paradox that is probably most favored by practicing physicists proceeds in two stages. At stage 1, it is pointed out that, quite generically, whenever the quantum formalism appears to generate a superposition of macroscopically distinct states it is impossible to demonstrate the effects of interference between them. The reasons for this claim include the facts that the initial state of a macroscopic system is likely to be unknown in detail; the initial state has extreme sensitivity to random external noise; and most important, merely by virtue of its macroscopic nature any such system will rapidly have its quantum-mechanical state correlated (entangled) with that of its environment in such a way that no measurement on the system alone (without a simultaneous measurement of the complete state of the environment) can demonstrate any interference between the two states in question—a result often known as decoherence. Thus, it is argued, the outcome of any possible experiment on the ensemble of macroscopic systems prepared in this way will be indistinguishable from that expected if each system had actually realized one or the other of the two macroscopically distinct states in question. Stage 2 of the argument (often not stated explicitly) is to conclude that if this is indeed true, then it may be legitimately asserted that such realization of a definite macroscopic outcome has indeed taken place by this stage.

Most physicists agree with stage 1 of the argument. However, not all agree that the radical reinterpretation of the meaning of the quantum formalism which is implicit at stage 2 is legitimate; that is, an interpretation in terms of realization, by each individual system, of one alternative or the other, forbidden at the atomic level by the observed phenomenon of interference, is allowed once, on going to the macroscopic level, the phenomenon disappears. Consequently, various alternative interpretations have been developed. *See* Quantum mechanics.

[A.J.L.]

Quark-gluon plasma
A predicted state of matter containing deconfined quarks and gluons. According to the theory of strong interactions, called quantum chromodynamics, hadrons such as mesons and nucleons (the generic name for protons and neutrons) are bound states of more fundamental objects called quarks. The

quarks are confined within the individual hadrons by the exchange of particles called gluons. However, calculations indicate that at sufficiently high temperatures or densities, hadronic matter should evolve into a new phase of matter containing deconfined quarks and gluons, called a quark-gluon plasma or quark matter. Such a state of matter is thought to have existed briefly in the period about 1–10 microseconds after the big bang, and might also exist inside the cores of dense neutron stars. See HADRON; QUANTUM CHROMODYNAMICS.

The study of such a new state of matter requires a means for producing it under controlled laboratory conditions. Experimentally the transition from the hadronic to the quark-gluon phase requires collisions of beams of heavy ions such as nuclei of gold or uranium (although lighter nuclei can be used) with other heavy nuclei at high enough energies to produce the necessary extreme conditions of heat and compression. Quantum chromodynamics calculations using the lattice gauge model indicate that energy densities of at least 1–2 GeV/fm^3 (1 femtometer $= 10^{-15}$ m), about 10 times that found in ordinary nuclear matter, must be produced in the collision for plasma formation to occur. See NUCLEAR REACTION; RELATIVISTIC HEAVY-ION COLLISIONS.

Accelerator experiments using beams of nuclei with energies of 10–200 GeV/nucleon bombarding stationary nuclear targets have found interesting phenomena such as nuclear stopping. In such cases, the colliding nucleons of the target and projectile are observed to pile up on each other, achieving large nuclear matter densities (two to four times normal nuclear density, or higher) corresponding to energy densities near the threshold for quark matter production. Other results of these experiments suggest that conditions favorable to thermal and chemical equilibrium may be present in some of these collisions. Such experiments can provide critical tests of the theory of the strong interaction and illuminate the earliest moments of the universe. See ELEMENTARY PARTICLE; GLUONS; QUARKS.   [L.S.S.]

Quarks   The basic constituent particles of which elementary particles are understood to be composed. Theoretical models built on the quark concept have been very successful in understanding and predicting many phenomena in the physics of elementary particles.

The study of the elastic scattering of electrons on protons demonstrated that the proton has a finite form factor, that is, a finite radial extent of its electric charge and magnetic moment distributions. It was plausible that the charge cloud which constitutes the proton is a probability distribution of some smaller, perhaps pointlike constituents, just as the charge cloud of an atom was learned to be the probability distribution of electrons. Subsequent high-energy, deep inelastic scattering experiments of electrons on protons, leading to meson production, revealed form factors corresponding to pointlike constituents of the proton. These proton constituents, first referred to as partons, are now understood to include the constituent quarks of the proton.

These high-energy collisions also produced an abundance of resonance states, equivalent to short-lived particles. The spectroscopy of these hadronic states revealed an order and symmetry among the observed hadrons that could be interpreted in terms of representations of the SU(3) symmetry group. This in turn is interpreted as a consequence of the grouping of elementary constituents of fractional electric charge in pairs and triplets to form the observed particles. The general features of the quark model of hadrons have withstood the tests of time, and the static properties of hadrons are consistent with predictions of this model. See SYMMETRY LAWS (PHYSICS); UNITARY SYMMETRY.

Thus, the proton and neutron are not fundamental constituents of matter, but each is composed of three quarks, very much as the nuclei of ^{3}H and ^{3}He are made of protons

Properties of quarks

Flavor	Mass[†], GeV/c^2	Electric charge[‡]	Baryon number	Spin[§]	Isotopic spin	Strangeness	Charm
u	0.0015–0.004	+2/3	+1/3	1/2	1/2	0	0
d	0.004–0.008	−1/3	+1/3	1/2	1/2	0	0
c	1.15–1.35	+2/3	+1/3	1/2	0	0	+1
s	0.080–0.130	−1/3	+1/3	1/2	0	−1	0
t	174.3±5.1[¶]	+2/3	+1/3	1/2	0	0	0
b	4.1–4.4	−1/3	+1/3	1/2	0	0	0

[†]As the mass of baryons composed of quarks is strongly influenced by the gluons binding the quarks, and as free quarks are not observed, the masses are theoretical estimates.
[‡]Charge is in units of the magnitude of the charge of an electron, 1.6×10^{-19} coulomb.
[§]Spin is in units of Planck's constant divided by 2π, written as $\hbar$.
[¶]The top quark mass is deduced from experimental measurements of its decay dynamics.

and neutrons, and the molecules of NO_2 and N_2O are made of oxygen and nitrogen atoms.

There are two kinds (or "flavors") of quarks of very low mass of which the proton, neutron, and pions are composed, and a third, more massive quark which is a constituent of "strange" particles, including the K mesons and hyperons such as the Λ^0. These are known as the up quark (u), the down quark (d), and the strange quark (s). Baryons are composed of three quarks, for example the proton (uud), neutron (udd), Λ^0 (uds), and Ξ^- (dss). Antiparticles such as antiprotons are formed by the antiquarks of those forming the particle, for example, the antiproton $\bar{p}(\bar{u}\bar{u}\bar{d})$. Mesons are composed of a quark-antiquark pair, such as the $\pi^+(u\bar{d})$, $\pi^-(\bar{u}d)$, $K^+(u\bar{s})$, and $K^-(\bar{u}s)$. See BARYON; HYPERON; MESON; STRANGE PARTICLES.

The quantum numbers of quarks are added to give the quantum numbers of the elementary particle which they form on combination. The unit of electrical charge of a quark is +2/3 or −1/3 of the charge on a proton (1.6×10^{-19} coulomb), and the baryon number of each quark is +1/3 (see table). The charge, baryon number, and so forth, of each antiquark are just the negative of that for each quark.

During the 1970s, experiments at electron-positron colliders and proton accelerators detected a relatively long-lived (that is, very narrow, in energy) resonant state of about 3.1 GeV total energy. This was interpreted as evidence for a new quark, the charm (c) quark, produced as a quark-antiquark resonance analogous to the ϕ. The discovery of this J/ψ resonance was followed by the observation and study of meson systems, now labeled D mesons, containing a single c or quark (paired with an antiquark of another flavor), as well as baryon states containing these quarks. See CHARM; J/PSI PARTICLE; PARTICLE ACCELERATOR.

A few years later, experiments with higher-energy proton beams, studying the spectra of muon-antimuon pairs at the Fermi National Accelerator Laboratory, discovered a more massive, narrow resonant state at about 9.4 GeV, which was labeled the Υ (upsilon). This was interpreted as evidence for a more massive quark, the b (bottom) quark. Subsequent experiments at proton and electron accelerators confirmed the existence of the b quark and also observed a corresponding family of meson resonant states, now referred to as B mesons.

During the 1990s, experiments observing collisions of protons and antiprotons at an energy of 1.8 TeV in the center of mass established the existence of the t (top) quark, primarily through analysis of its decay to a B meson and a W intermediate vector boson. The t mass of 174.3 ± 5.1 GeV/c^2 (about the mass of a tungsten atom) is so great that its weak decay through this channel is very fast, and mesonic states of the t

and $\bar{t}$ quark (analogous to the Υ, the J/ψ, and the ϕ) are not observed, although the observed t's are from the production of pairs. *See* INTERMEDIATE VECTOR BOSON.

Quarks are understood to have a spin of 1/2; that is, their intrinsic angular momentum is $\hbar/2$ (where $\hbar$ is Planck's constant h divided by 2π), just as for the electron and muon. A problem arose when the structure of observed baryons required two or, in some cases, three quarks of the same flavor in the same quantum state, a situation forbidden for spin-1/2 particles by the Pauli exclusion principle. In order to accommodate this contradiction, a new quantum variable, arbitrarily labeled color, was introduced; the idea is that each quark is red, green, or blue (and the antiquarks, antired, and so forth). The color quantum number then breaks the degeneracy and allows up to three quarks of the same flavor to occupy a single quantum state. Confirmation of the color concept has been obtained from experiments with electron-positron storage rings, and the theory of quantum chromodynamics (QCD), based on this concept, has been developed. According to quantum chromodynamics, hadrons must be colorless; for example, baryons must consist of a red, a green, and a blue quark, and mesons of a quark-antiquark pair of the same color (for example, a red quark and an antired antiquark). *See* COLOR (QUANTUM MECHANICS); EXCLUSION PRINCIPLE; SPIN (QUANTUM MECHANICS).

The field quanta of quantum chromodynamics are gluons, massless, spin-1 quanta which interact with quarks. This is very analogous to the manner in which photons, the quanta of electromagnetic interaction, interact with particles containing electric charge and are responsible for electromagnetic forces. The QCD theory is part of the now widely accepted standard model of elementary particle interactions, together with the electroweak theory. Experiments have increasingly confirmed details of the standard model to the extent that most physicists are confident that it is fundamentally correct. *See* ELECTROWEAK INTERACTION; GLUONS; QUANTUM CHROMODYNAMICS; STANDARD MODEL.

There are three sets, or "generations," of quarks and leptons. Each generation contains a charged lepton (electron, muon, or tau lepton); a correspoding neutrino; a charge $-1/3$ quark color triad; and a charge $+2/3$ quark triad. *See* LEPTON; NEUTRINO.

Quarks and the theory of quantum chromodynamics are now firmly established as cornerstones of the standard model of elementary particles (together with the electroweak theory, charged leptons, neutrinos, and so forth). However, unanswered questions remain.

The advanced string and M theories have the property of supersymmetry, which demands that every spin-1/2 quark and lepton must have a partner with integral spin. As of 2004, no experimental evidence for any of these supersymmetric (SUSY) particles had been found. *See* SUPERSTRING THEORY; SUPERSYMMETRY.

Contemporary theories also predict that there exists one or more massive particles of integral spin, the Higgs paticles, responsible for the rest masses of the quarks and charged leptons. Again, the lack of evidence suggests that the Higgs particles must also have a rest mass of over $100 \text{ GeV}/c^2$, if they exist. *See* HIGGS BOSON.

Quarks may be permanently stable against decay via the weak interaction; however, it is also possible that quarks spontaneously decay to leptons. Intensive searches for the decay of the proton (into a neutral pion and a positron, for example) have been negative, setting a lower limit of over 10^{32} years for the proton lifetime. However, the apparent asymmetry of the universe between matter and antimatter (there is, at present, no evidence for primordial antimatter) suggests that antiprotons, for example, may spontaneously decay (or transform) more readily. *See* ANTIMATTER; PROTON.

Some theories have postulated that quarks are composed of smaller constituents, just as other objects that were originally believed to be fundamental subsequently were found to have internal structure. So far, all observations are compatible with the quarks as point objects, like the electron.

[L.W.J.]

Quasiatom   A structure in which the nuclei of two atoms approach each other closely and their electrons are arranged in atomic orbitals characteristic of a single atom of atomic number equal to the sum of the nuclear charges. Quasiatoms can be formed for short times in atom-atom and ion-atom collisions when the nuclei are much closer than the mean orbital radius of the innermost K-shell electrons. The electrons are then bound in the electric field of both nuclear charges Z_1 and Z_2, which resembles the spherically symmetric $1/r^2$ Coulomb field of a single united atom having charge $Z_{ua} = Z_1 + Z_2$. See ATOMIC STRUCTURE AND SPECTRA.

An interesting effect is associated with quasiatoms with $Z > 173$, in which the $1s$ binding energy is more than twice the electron rest mass, $E_{1s} > 2mc^2$. If a vacancy exists in this orbital, it is energetically favorable to create an electron-positron pair with the electron bound in this state. The positron would be repelled from the nucleus with kinetic energy equal to $E_{e^+} = |E_{1s}| - 2mc^2$. In the Dirac hole picture, in which the vacuum consists of a negative energy continuum ($E < -mc^2$) filled with electrons, the $1s$ level is said to fall into the negative-energy Dirac sea as Z increases above the critical value, $Z_{cr} = 173$. A $1s$ hole (vacancy) becomes embedded in the negative continuum as an unstable resonance state that decays in a time of $\sim 10^{-19}$ s to a bound electron and a spontaneously emitted monoenergetic positron.

The quantum electrodynamic vacuum in the presence of a bare supercritical nuclear charge is therefore unstable and decays to a fundamentally new charged vacuum, which consists of the nucleus with two $1s$ electrons (from the two spin orientations). At higher values of Z_{ua}, as additional quasiatomic levels enter the negative continuum, the charge of the quantum electrodynamic vacuum increases accordingly. If detected, spontaneous positron emission would represent the first observation of a phase transition in a gauge field theory. See ANTIMATTER; ELECTRON-POSITRON PAIR PRODUCTION; PHASE TRANSITIONS; POSITRON; QUANTUM ELECTRODYNAMICS; SUPERCRITICAL FIELDS.

[T.E.C.]

Quasicrystal   A solid with conventional crystalline properties but exhibiting a point-group symmetry inconsistent with translational periodicity. Like crystals, quasicrystals display discrete diffraction patterns, crystallize into polyhedral forms, and have long-range orientational order, all of which indicate that their structure is not random. But the unusual symmetry and the finding that the discrete diffraction pattern does not fall on a reciprocal periodic lattice suggest a solid that is quasiperiodic. Their discovery in 1982 contradicted a long-held belief that all crystals would be periodic arrangements of atoms or molecules.

It is easily shown that in two and three dimensions the possible rotations that superimpose an infinitely repeating periodic structure on itself are limited to angles that are $360°/n$, where n can be only 1, 2, 3, 4, or 6. Various combinations of these rotations lead to only 32 point groups in three dimensions, and 230 space groups which are combinations of the 14 Bravais lattices that describe the periodic translations with the allowed rotations. Until the 1980s, all known crystals could be classified according to this limited set of symmetries allowed by periodicity. Periodic structures diffract only at discrete angles (Bragg's law) that can be described by a reciprocal lattice, in which the diffraction intensities fall on lattice points that, like all lattices, are by definition periodic, and which has a symmetry closely related to that of the structure. See CRYSTAL; CRYSTALLOGRAPHY; X-RAY CRYSTALLOGRAPHY; X-RAY DIFFRACTION.

Quasicrystals of an alloy of aluminum, copper, and iron, displaying an external form consistent with their icosahedral symmetry.

Icosahedral quasicrystals were discovered in 1982 during a study of rapid solidification of molten alloys of aluminum with one or more transition elements, such as manganese, iron, and chromium. Since then, many different alloys of two or more metallic elements have led to quasicrystals with a variety of symmetries and structures. The illustration shows the external polyhedral form of an icosahedral aluminum-copper-iron alloy.

The diffraction patterns of quasicrystals violate several predictions resulting from periodicity. Quasicrystals have been found in which the quantity n is 5, 8, 10, and 12. In addition, most quasicrystals exhibit icosahedral symmetry in which there are six intersecting fivefold rotation axes. Furthermore, in the electron diffraction pattern the diffraction spots do not fall on a (periodic) lattice but on what has been called a quasilattice. *See* ELECTRON DIFFRACTION.   [J.W.C.; D.S.]

Radiance The physical quantity that corresponds closely to the visual brightness of a surface. A simple radiometer for measuring the (average) radiance of an incident beam of optical radiation (light, including invisible infrared and ultraviolet radiation) consists of a cylindrical tube, with a hole in each end cap to define the beam cross section there, and with a photocell against one end to measure the total radiated power in the beam of all rays that reach it through both holes (see illustration). If A_1 and A_2

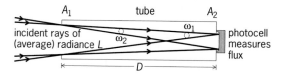

A simple radiometer.

are the respective areas of the two holes, D is the length of the tube (distance between holes), and Φ is the radiant flux or power measured by the photocell, then the (average) radiance is approximately given by the equation $L = \Phi/(A_1 \cdot A_2/D^2) \ \mathrm{W \cdot m^{-2} \cdot sr^{-1}}$.

[F.E.N.]

Radiation The emission and propagation of energy; also, the emitted energy itself. The etymology of the word implies that the energy propagates rectilinearly, and in a limited sense, this holds for the many different types of radiation encountered.

The major types of radiation may be described as electromagnetic, acoustic, and particle, and within these major divisions there are many subdivisions. Electromagnetic radiation is classified roughly in order of decreasing wavelength as radio, microwave, visible, ultraviolet, x-rays, and γ-rays. Acoustic or sound radiation may be classified by frequency as infrasonic, sonic, or ultrasonic in order of increasing frequency, with sonic being between about 16 and 20,000 Hz. The traditional examples of particle radiation are the α- and β-rays of radioactivity. *See* ELECTROMAGNETIC RADIATION; RADIOACTIVITY; SOUND.

[McA.H.H.]

Radiation pressure Pressure exerted by electromagnetic radiation on objects on which it impinges. This pressure is caused by the fact that electromagnetic radiation transmits energy and possesses momentum. These pressures are very small. The effect is conspicuous in the case of a comet near the Sun, where the radiation pressure from the Sun forces the lighter cometary constituents away from the Sun. *See* ELECTROMAGNETIC RADIATION.

[W.R.Sm.]

Radiative transfer The study of the propagation of energy by radiative processes; it is also called radiation transport. Radiation is one of the three mechanisms by which energy moves from one place to another, the other two being conduction and convection. *See* ELECTROMAGNETIC RADIATION; HEAT TRANSFER.

The kinds of problems requiring an understanding of radiative transfer can be characterized by looking at meteorology, astronomy, and nuclear reactor design. In meteorology, the energy budget of the atmosphere is determined in large part by energy gained and lost by radiation. In astronomy, almost all that is known about the abundance of elements in space and the structure of stars comes from modeling radiative transfer processes. Since neutrons moving in a reactor obey the same laws as radiation being scattered by atmospheric particles, radiative transfer plays an important part in nuclear reactor design.

Each of these three fields—meteorology, astronomy, and nuclear engineering—concentrates on a different aspect of radiative transfer. In meteorology, situations are studied in which scattering dominates the interaction between radiation and matter; in astronomy, there is more interest in the ways in which radiation and the distribution of electrons in atoms affect each other; and in nuclear engineering, problems relate to complicated, three-dimensional geometry.

Radiative transfer is a complicated process because matter interacts with the radiation. This interaction occurs when the photons that make up radiation exchange energy with matter. These processes can be understood by considering the transfer of visible light through a gas made up of atoms. Similar processes occur when radiation interacts with solid dust particles or when it is transmitted through solids or liquids. *See* PHOTON.

If a gas is hot, collisions between atoms can convert the kinetic energy of motion to potential energy by raising atoms to an excited state. Emission is the process which releases this energy in the form of photons and cools the gas by converting the kinetic energy of atoms to energy in the form of radiation. The reverse process, absorption, occurs when a photon raises an atom to an excited state, and the energy is converted to kinetic energy in a collision with another atom. Absorption heats the gas by converting energy from radiation to kinetic energy. Occasionally an atom will absorb a photon and reemit another photon of the same energy in a random direction. If the photon is reradiated before the atom undergoes a collision, the photon is said to be scattered. Scattering has no net effect on the temperature of the gas. *See* ATOMIC STRUCTURE AND SPECTRA; SCATTERING OF ELECTROMAGNETIC RADIATION.   [A.H.K.]

Radioactive beams   Beams of radioactive (unstable) nuclei. In several nuclear physics laboratories, a capability exists to produce such beams and, before these nuclei spontaneously decay, use them to gain insight into the reactions on and structure of nuclei never before accessible. Radioactive beams are particularly useful to study stellar explosions such as novae, supernovae, and x-ray bursts. These explosions are some of the most catastrophic events in the universe, generating enormous amounts of energy while synthesizing the elements that make up lifeforms and the world. These spectacular explosions involve, and in some cases are driven by, reactions where the atomic nuclei of hydrogen (protons) and helium (alpha particles) fuse with (are captured by) radioactive isotopes of heavier elements to form new elements. The capability to produce beams of radioactive nuclei allows direct measurements of these reactions, providing crucial information needed to theoretically model cataclysmic stellar events and to understand the origin of many chemical elements.

One approach to radioactive beam production is the isotope separator on-line (ISOL) technique. One accelerator bombards a target with a beam of stable nuclei, and a small number of the radioactive atoms of interest are produced through nuclear reactions. These atoms are transported, by various techniques, including thermal diffusion, to an ion source where they are ionized (removing or adding electrons to give atoms an electrical charge) and extracted. The radioactive ions are then mass-separated from other ions and accelerated to energies needed for nuclear physics experiments by a second accelerator. The ISOL technique can produce very high beam qualities, purities, and intensities; the disadvantages are that only a few radioactive beam species can be generated from each combination of production target and primary beam, and that beams with short lifetimes (less than 1 s) are difficult to produce. *See* ION SOURCES.

A complementary radioactive beam production technique is projectile fragmentation. When a high-energy beam of stable heavy ions passes through a thin target, the beam particles (projectiles) can break up into fragments—some of which are the radioactive isotope of interest. The desired fragments are then mass-separated from other ions and steered toward a target to undergo the reaction of interest. The projectile fragmentation technique can produce beams of very short lifetimes (10^{-6} s or less), and the same setup can be used to produce many different beam species; the disadvantages are that high beam quality, purity, and intensity are difficult to obtain. *See* NUCLEAR FUSION; NUCLEAR REACTION; PARTICLE ACCELERATOR.   [M.S.S.]

Radioactivity   A phenomenon resulting from an instability of the atomic nucleus in certain atoms whereby the nucleus experiences a spontaneous but measurably delayed nuclear transition or transformation with the resulting emission of radiation. The discovery of radioactivity by H. Becquerel in 1896 marked the birth of nuclear physics.

All chemical elements may be rendered radioactive by adding or by subtracting (except for hydrogen and helium) neutrons from the nucleus of the stable ones. Studies of the radioactive decays of new isotopes far from the stable ones in nature continue as a major frontier in nuclear research. The availability of this wide variety of radioactive isotopes has stimulated their use in a wide variety of fields, including chemistry, biology, medicine, industry, artifact dating, agriculture, and space exploration. *See* ALPHA PARTICLES; BETA PARTICLES; GAMMA RAYS; ISOTOPE.

A particular radioactive transition may be delayed by less than a microsecond or by more than a billion years, but the existence of a measurable delay or lifetime distinguishes a radioactive nuclear transition from a so-called prompt nuclear transition, such as is involved in the emission of most gamma rays. The delay is expressed quantitatively by the radioactive decay constant, or by the mean life, or by the half-period for each type of radioactive atom, discussed below.

The most commonly found types of radioactivity are alpha, beta negatron, beta positron, electron capture, and isomeric transition. Each is characterized by the particular type of nuclear radiation which is emitted by the transforming parent nucleus. In addition, there are several other decay modes that are observed more rarely in specific regions of the periodic table.

Transition rates and decay laws. The rate of radioactive transformation, or the activity, of a source equals the number A of identical radioactive atoms present in the source, multiplied by their characteristic radioactive decay constant λ. Thus Eq. (1)

$$\text{Activity} = A\lambda \text{ disintegrations per second} \tag{1}$$

holds, where the decay constant λ has dimensions of s^{-1}. The numerical value of λ expresses the statistical probability of decay of each radioactive atom in a group of identical atoms, per unit time. For example, if $\lambda = 0.01 \text{ s}^{-1}$ for a particular radioactive species, then each atom has a chance of 0.01 (1%) of decaying in 1 s, and a chance of 0.99 (99%) of not decaying in any given 1-s interval. The constant λ is one of the most important characteristics of each radioactive nuclide: λ is essentially independent of all physical and chemical conditions such as temperature, pressure, concentration, chemical combination, or age of the radioactive atoms.

Many radioactive nuclides have two or more independent and alternative modes of decay. For example, ^{238}U can decay either by alpha-particle emission or by spontaneous fission. When two or more independent modes of decay are possible, the nuclide is said to exhibit dual decay. The competing modes of decay of any nuclide have independent partial decay constants given by the probabilities $\lambda_1, \lambda_2, \lambda_3, \ldots$, per second, and the total probability of decay is represented by the total decay constant λ, defined by Eq. (2).

$$\lambda = \lambda_1 + \lambda_2 + \lambda_3 + \cdots \tag{2}$$

The actual life of any particular atom can have any value between zero and infinity. The average or mean life of a large number of identical radioactive atoms is, however, a definite and important quantity. The total L of the life-spans of all the A_0 atoms initially present is given by Eq. (3). Then the average lifetime L/A_0, which is called the mean life τ, is given by Eq. (4).

$$L = \frac{A_0}{\lambda} \tag{3}$$

$$\tau = 1/\lambda \tag{4}$$

The time interval over which the chance of survival of a particular radioactive atom is exactly one-half is called half-period T (also called the half-life, written $T_{1/2}$). The half-period T is related to the total radioactive decay constant λ, and to the mean life τ, by Eq. (5). For mnemonic reasons, the half-period T is much more frequently employed

$$T = 0.693/\lambda = 0.693\tau \tag{5}$$

than the total decay constant λ or the mean life τ.

Radioactive series decay. In a number of cases a radioactive nuclide A decays into a nuclide B which is also radioactive; the nuclide B decays into C which is also radioactive, and so on. For example, $^{232}_{90}$Th decays into a series of 10 successive radioactive nuclides. Substantially all the primary products of nuclear fission are negatron beta-particle emitters which decay through a chain or series of two to six successive beta-particle emitters before a stable nuclide is reached as an end product.

Alpha-particle decay. Alpha-particle decay is that type of radioactivity in which the parent nucleus expels an alpha particle (a helium nucleus), which contains two protons and two neutrons. Thus, the atomic number, or nuclear charge, of the decay product is 2 units less than that of the parent, and the nuclear mass of the product is 4 atomic mass units less than that of the parent, because the emitted alpha particle carries away this amount of nuclear charge and mass. This decrease of 2 units of atomic number or nuclear charge between parent and product means that the decay product will be a different chemical element, displaced by 2 units to the left in a periodic table of the elements.

In the simplest case of alpha decay, every alpha particle would be emitted with exactly the same velocity and hence the same kinetic energy. However, in most cases there are two or more discrete energy groups called lines. For example, in the alpha decay of a large group of ^{238}U atoms, 77% of the alpha decays will be by emission of alpha particles whose kinetic energy is 4.20 MeV, while 23% will be by emission of 4.15-MeV alpha particles. When the 4.20-MeV alpha particle is emitted, the decay product nucleus is formed in its ground (lowest energy) level. When a 4.15-MeV alpha particle is emitted, the decay product is produced in an excited level, 0.05 MeV above the ground level.

This nucleus promptly transforms to its ground level by the emission of a 0.05-MeV gamma ray or alternatively by the emission of the same amount of energy in the form of a conversion electron and the associated spectrum of characteristic x-rays. Thus in all alpha-particle spectra, the alpha particles are emitted in one or more discrete and homogeneous energy groups, and alpha-particle spectra are accompanied by gamma-ray and conversion electron spectra whenever there are two or more alpha-particle groups in the spectrum.

Among all the known alpha-particle emitters, most alpha-particle energy spectra lie in the domain of 4–6 MeV, although a few extend as low as 2 MeV and as high as 10 MeV. There is a systematic relationship between the kinetic energy of the emitted alpha particles and the half-period of the alpha emitter. The highest-energy alpha particles are emitted by short-lived nuclides, and the lowest-energy alpha particles are emitted by the very-long-lived alpha-particle emitters. H. Geiger and J. M. Nuttall showed that there is a linear relationship between log λ and the energy of the alpha particle.

The Geiger-Nuttall rule is inexplicable by classical physics, but emerges clearly from quantum, or wave, mechanics. In 1928 the hypothesis of transmission through nuclear potential barriers was shown to give a satisfactory account of the alpha-decay data, and it has been altered subsequently only in details.

Beta-particle decay. Beta-particle decay is a type of radioactivity in which the parent nucleus emits a beta particle. There are two types of beta decay established: in negatron beta decay (β^-) the emitted beta particle is a negatively charged electron (negatron); in positron beta decay (β^+) the emitted beta particle is a positively charged electron (positron). In beta decay the atomic number shifts by one unit of charge, while the mass number remains unchanged. In contrast to alpha decay, when beta decay takes place between two nuclei which have a definite energy difference, the beta particles from a large number of atoms will have a continuous distribution of energy.

For each beta-particle emitter, there is a definite maximum or upper limit to the energy spectrum of beta particles. This maximum energy, E_{max}, corresponds to the change in nuclear energy in the beta decay. As in the case of alpha decay, most beta-particle spectra include additional continuous spectra which have less maximum energy and which leave the product nucleus in an excited level from which gamma rays are then emitted.

For nuclei very far from stability, the energies of these excited states populated in beta decay are so large that the excited states may decay by proton, two-proton, neutron, two-neutron, three-neutron, or alpha emission, or spontaneous fission.

The continuous spectrum of beta-particle energies implies the simultaneous emission of a second particle besides the beta particle, in order to conserve energy and angular momentum for each decaying nucleus. This particle is the neutrino. The neutrino has zero charge and extremely small rest mass, travels at nearly the same speed as light (3×10^{10} cm/s), and is emitted as a companion particle with each beta ray. By postulating the simultaneous emission of a beta particle and a neutrino, E. Fermi developed in 1934 a quantum-mechanical theory which satisfactorily gives the shape of beta-particle spectra, and the relative half-periods of beta-particle emitters for allowed beta decays. *See* NEUTRINO.

When the ground state of a nucleus differing by two units of charge from nucleus A has lower energy than A, then it is theoretically possible for A to emit two beta particles, either $\beta^+\beta^+$ or $\beta^-\beta^-$ as the case may be, and two neutrinos or antineutrinos, and go from Z to $Z \pm 2$. Here two protons decay into two neutrons, or vice versa. This is a second-order process and so should go much slower than beta decay. There are a number of cases where such decays should occur, but their half-lives are of the order of 10^{20} years or greater. Such decay processes are obviously very difficult to detect.

The first direct evidence for two-neutrino double-beta-minus decay of selenium-82, was found only in 1987.

Whenever it is energetically allowed by the mass difference between neighboring isobars, a nucleus Z may capture one of its own atomic electrons and transform to the isobar of atomic number $Z - 1$. Usually the electron-capture (EC) transition involves an electron from the K shell of atomic electrons, because these innermost electrons have the greatest probability density of being in or near the nucleus. See ELECTRON CAPTURE.

Gamma-ray decay. Gamma-ray decay involves a transition between two excited levels of a nucleus, or between an excited level and the ground level. A nucleus in its ground level cannot emit any gamma radiation. Therefore gamma-ray decay occurs only as a sequel of another radioactive decay process or of some other process whereby the product nucleus is left in an excited state. Such additional processes include the fusion of two nuclei, Coulomb excitation, and induced nuclear fission. See COULOMB EXCITATION; NUCLEAR FISSION; NUCLEAR FUSION.

A gamma ray is high-frequency electromagnetic radiation (a photon) in the same family with radio waves, visible light, and x-rays. The energy of a gamma ray is given by $h\nu$, where h is Planck's constant and ν is the frequency of oscillation of the wave in hertz. The gamma-ray or photon energy $h\nu$ lies between 0.05 and 3 MeV for the majority of known nuclear transitions. Higher-energy gamma rays are seen in neutron capture and some reactions. See ELECTROMAGNETIC RADIATION.

Gamma rays carry away energy, linear momentum, and angular momentum, and account for changes of angular momentum, parity, and energy between excited levels in a given nucleus. This leads to a set of gamma-ray selection rules for nuclear decay and a classification of gamma-ray transitions as "electric" or as "magnetic" multipole radiation of multipole order 2^l, where $l = 1$ is called dipole radiation, $l = 2$ is quadrupole radiation, and $l = 3$ is octupole, l being the vector change in nuclear angular momentum. The most common type of gamma-ray transition in nuclei is the electric quadrupole (E2). There are cases where several hundred gamma rays with different energies are emitted in the decays of atoms of only one isotope. See MULTIPOLE RADIATION.

An alternative type of deexcitation which always competes with gamma-ray emission is known as internal conversion. Instead of the emission of a gamma ray, the nuclear excitation energy can be transferred directly to a bound electron of the same atom. Then the nuclear energy difference is converted to energy of an atomic electron, which is ejected from the atom.

When the energy between two states in the same nucleus exceeds 1.022 MeV, twice the rest mass energy of an electron, it is also possible for the nucleus to give up its excess energy to an electron-positron pair—a pair creation process. See ELECTRON-POSITRON PAIR PRODUCTION.

Spontaneous fission. This involves the spontaneous breakup of a nucleus into two heavy fragments and neutrons. Spontaneous fission can occur when the sum of the masses of the two heavy fragments and the neutrons is less than the mass of the parent undergoing decay. After the discovery of fission in 1939, it was subsequently discovered that isotopes like ^{238}U had very weak decay branches for spontaneous fission, with branching ratios on the order of 10^{-6}. Some isotopes with relatively long half-lives such as ^{252}Cf have large (3.1%) spontaneous fission branching.   [J.H.H.]

Heavy cluster decays. Alpha-particle decay and spontaneous fission are two natural phenomena in which an atomic nucleus spontaneously breaks into two fragments, but the fragments are of very different size in one case and almost equal size in the other. On the basis of fragmentation theory and the two-center shell model, new kinds of radioactivities that are intermediate between alpha-particle decay and fission were predicted in 1980. Subsequently, it was shown theoretically that the new processes should occur throughout a very broad region of the nuclear chart, including elements with atomic numbers higher than 40. However, experimentally observable emission rates could be expected only for nuclei heavier than lead, in a breakup leading to a very stable heavy fragment with proton and neutron numbers equal or very close to $Z = 82, N = 126$ ($^{208}_{82}$Pb or its neighborhood). The main competitor is always alpha-particle decay. In 1984, a series of experimental confirmations began with the discovery of $^{14}_{6}$C radioactivity of $^{223}_{88}$Ra. A very promising technique uses solid-state track-recording detectors with special plastic films and glasses that are sensitive to heavier clusters but not to alpha particles.

[W.Gr.; J.H.H.]

Proton radioactivity. Proton radioactivity is a mode of radioactive decay that is generally expected to arise in proton-rich nuclei far from the stable isotopes, in which the parent nucleus changes its chemical identity by emission of a proton in a single-step process. Its physical interpretation parallels almost exactly the quantum-mechanical treatment of alpha-particle decay. For many years only a few examples of this decay mode were observed, because of the narrow range of half-lives and decay energies where this mode can compete with other modes. However, in the late 1990s, experimental techniques using new recoil mass spectrometers, which can separate rare reaction products, and new double-sided silicon strip detectors became available and opened up the discovery of many new proton radioactivities. Two-proton radioactivity (the simultaneous emission of two protons) was first observed in 2001 in an excited state of neon-18 and in 2002 in the ground state of iron-45.

Delayed particle emissions. Twelve types of beta-delayed particle emissions have been observed. Beta-delayed deuteron (^{2}H) emission, which is not shown there, also can be expected. Over 100 beta-delayed particle radioactivities are now known. Theoretically, the number of isotopes which can undergo beta-delayed particle emission could exceed 1000. Thus, this mode, which was observed in only a few cases prior to 1965, is among the important ones in nuclei very far from the stable ones in nature. Studies of these decays can provide insights into the nucleus which can be gained in no other way.

[J.H.H.]

Radio-frequency impedance measurements
Measurements of electrical impedance at frequencies ranging from a few tens of kilohertz to about 1 gigahertz. In the electrical context, impedance is defined as the ratio of voltage to current (or electrical field strength to magnetic field strength), and it is measured in units of ohms (Ω). *See* ELECTRICAL IMPEDANCE.

At zero frequency, that is, when the current involved is a direct current, both voltage and current are expressible as real numbers. Their ratio, the resistance, is a scalar (real) number. However, at nonzero frequencies, the voltage is not necessarily in phase with the current, and both are represented by vectors, and therefore are conveniently described by using complex numbers. To distinguish between the scalar quantity of resistance at zero frequency and the vectorial quantity at nonzero frequencies, the word impedance is used for the complex ratio of voltage to current. *See* ELECTRICAL RESISTANCE.

The measurement of impedance at radio frequencies cannot always be performed directly by measuring an rf voltage and dividing it by the corresponding rf current, for the following reasons: (1) it may be difficult to measure rf voltages and currents without loading the circuit by the sensing probes; (2) the distributed parasitic reactances (stray capacitances to neighboring objects, and lead inductances) may be altered by the sensing probes; and (3) the spatial voltage and current distributions may prevent unambiguous measurements (in waveguides, for instance).

At low frequencies, impedance measurements are often carried out by measuring separately the resistive and reactive parts, using either Q-meter instruments (for resonance methods) or reconfigurable bridges, which are sometimes called universal LCR (inductance-capacitance-resistance) bridges. In one such bridge the resistive part of the impedance is measured at dc with a Wheatstone bridge. Capacitive reactance is measured with a series-resistance-capacitance bridge, and inductive reactance is measured with a Maxwell bridge, using alternating-current (ac) excitation and a standard capacitance. *See* WHEATSTONE BRIDGE.

Transformer bridges are capable of operating up to 100 MHz. The use of transformers offers the following advantages: (1) only two bridge arms are needed, the standard and the unknown arms, and (2) both the detector and the source may be grounded at one of their terminals, minimizing ground-loop problems and leakage.

A coaxial line admittance bridge is usable from 20 MHz to 1.5 GHz. The currents flowing in three coaxial branch lines are driven from a common junction, and are sampled by three independently rotatable, electrostatically shielded loops, whose outputs are connected in parallel.

A quantity related to impedance is the complex (voltage) reflection coefficient, defined as the ratio of the reflected voltage to the incident voltage, when waves propagate along a uniform transmission line in both directions. Usually, uppercase gamma (Γ) or lowercase rho (ρ) is used to represent the reflection coefficient. When a transmission line of characteristic impedance Z_0 is terminated in impedance Z_T, the reflection coefficient at the load is given by Eq. (1), and the voltage standing-wave ratio (VSWR) is related to the magnitude of Γ by Eq. (2).

$$\Gamma = \frac{Z_T - Z_0}{Z_T + Z_0} \tag{1}$$

$$\text{VSWR} = \frac{1 + |\Gamma|}{1 - |\Gamma|} \tag{2}$$

When it is sufficient to measure only the voltage standing-wave ratio, resistive bridges may be used. Resistive bridges employed as reflectometers use a matched source and detector, and therefore differ from the Wheatstone bridge, which aims to use a zero-impedance voltage source and an infinite-impedance detector.

Some specialized electronic instruments make use of the basic definition of impedance, and effectively measure voltage and current. One such instrument is called an rf vector impedance meter. Instead of measuring both the voltage and the current, it drives a constant current into the unknown impedance, and the resultant voltage is measured.

Vector voltmeters (VVM) are instruments with two (high-impedance) voltmeter probes, which display the voltages at either probe (relative to ground) as well as the phase difference between them. One type operates from 1 MHz to 1 GHz, and linearly converts to a 20-kHz intermediate frequency by sampling.

When the magnitude of the reactive part of the impedance is much greater than the resistive part at a given frequency, resonance methods may be employed to measure impedance. The most commonly used instrument for this purpose is the Q meter. *See* Q METER.

At the upper end of the rf range, microwave methods of impedance measurement may also be used, employing slotted lines and six-port junctions. *See* MICROWAVE MEASUREMENTS.

[P.I.S.]

Radioisotope   A radioactive isotope (as distinguished from a stable isotope) of an element. Atomic nuclei are of two types, unstable and stable. Those in the former category are said to be radioactive and eventually are transformed, by radioactive decay, into the latter. One of the three types of particles or radiation (alpha particles, beta particles, and gamma rays) is emitted during each stage of the decay. *See* ISOTOPE; RADIOACTIVITY.

The term radioisotope is also loosely used to refer to any radioactive atomic species. Whereas approximately a dozen radioisotopes are found in nature in appreciable amounts, hundreds of different radioisotopes have been artificially produced by bombarding stable nuclei with various atomic projectiles.

A very wide variety of radioisotopes are produced in particle accelerators, such as the cyclotron. Charged particles, such as deuterons (D^+) and protons (H^+), are accelerated to great speeds by high-voltage electrical fields and allowed to strike targets in which nuclear reactions take place; for example, proton in, neutron out (p,n), increasing the target-atom atomic number by one without changing the atomic mass; and deuteron in, proton out (d,p), increasing the atomic mass by one without changing the atomic number. The target elements become radioactive because the nuclei of the atoms are unbalanced, having an excess or deficit of neutrons or protons. Although the particle-accelerating machines are most versatile in producing radioisotopes, the amount of radioactive material that can be produced is relatively smaller than that made in a nuclear reactor [less than curie amounts; a curie (abbreviated Ci) is that quantity of a radioisotope required to supply 3.7×10^{10} disintegrations per second or 3.7×10^{10} becquerels (Bq)]. For large-scale production, nuclear reactors with neutron fluxes of 1×10^{10} to 5×10^{15} neutrons per square centimeter per second are required. *See* NUCLEAR REACTION; PARTICLE ACCELERATOR; UNITS OF MEASUREMENT.

[A.F.Ru.]

Radiometry   A branch of science that deals with the measurement or detection of radiant electromagnetic energy. Radiometry is divided according to regions of the spectrum in which the same experimental techniques can be used. Thus, vacuum ultraviolet radiometry, intermediate-infrared radiometry, far-infrared radiometry, and microwave radiometry are considered separate fields, and all of these are to be distinguished from radiometry in the visible spectral region. Curiously, radiometry in the visible is called radiometry, optical radiation measurement science, or photometry, but it is not called visible radiometry. *See* ELECTROMAGNETIC RADIATION; INFRARED RADIATION; LIGHT; ULTRAVIOLET RADIATION.

Any radiation detector (such as a thermometer) that responds to an increase in temperature caused by the absorption of radiant energy is known as a thermal detector. Similarly, any detector (such as a photochemical reaction) that responds to the excitation of a bound electron is called a photon or quantum detector.

Liquid-in-glass thermometers are sluggish and relatively insensitive. The key to developing thermal detectors with better performance than liquid-in-glass thermometers has been to secure a large and rapid rise in temperature associated with a high sensitivity to temperature changes.

Thermal detectors have been based upon a number of different principles. Radiation thermocouples produce a voltage, bolometers undergo a change in resistance, pyroelectric detectors undergo a change in spontaneous electric polarization, and the gas in pneumatic detectors (Golay cells) and photoacoustic detectors expands in response to incident radiation. The periodic expansion and contraction of the gas in response to high-frequency modulated radiation is detected by a sensitive microphone in the case of the photoacoustic detector. The Golay cell, on the other hand, uses a sensitive photomultiplier and a reference beam of light to detect distortion of a flexible membrane mirror caused by the expansion and contraction of the gas. *See* BOLOMETER; PYROELECTRICITY; THERMOCOUPLE.

The main problem with thermal detectors is that they respond not only to electromagnetic radiation but to any source of heat. This makes their design, construction, and use rather difficult, because they must be made sensitive to the radiation of interest while remaining insensitive to all other sources of heat, such as conduction, convection, and background radiation, that are of no interest in the particular measurement.

Photon detectors respond only to photons of electromagnetic radiation that have energies greater than some minimum value determined by the quantum-mechanical properties of the detector material. Since heat radiation from the environment at room temperature consists of infrared photons, photon detectors for use in the visible can be built so that they do not respond to any source of heat except the radiation of interest.

Following the introduction of planar silicon technology for microelectronics, the same technology was quickly exploited to make planar photodiodes based on the internal photoelectric effect in silicon. In these devices, the separation of a photogenerated electron-hole pair by the built-in field surrounding the p^+n junction induces the flow of one electron in an external short circuit (such as the inputs to an operational amplifier) across the electrodes. The number of electrons flowing in an external short circuit per absorbed photon is called the quantum efficiency. The use of these diodes has grown to the point where they are the most widely used detector for the visible and nearby spectral regions. Their behavior as a radiation detector in the visible is so nearly ideal that they can be used as a standard, their cost is so low that they can be used for the most mundane of applications, and their sensitivity is so high that they can be used to measure all but the weakest radiation (which requires the most sensitive photomultipliers). *See* SEMICONDUCTOR DIODE.

Research efforts have been directed at producing photon detectors based on more exotic semiconductors, and more complicated structures to extend the sensitivity, time response, and spectral coverage.   [J.Gei.]

Radius of gyration   A relation of the area or mass of a figure to its moment of inertia. If I is the moment of inertia about a line of a figure whose area is A, the figure's radius of gyration with respect to that line is $k = +\sqrt{I/A}$. Accordingly, $I = k^2A$. For a figure of mass M, $k = +\sqrt{I/M}$; $I = k^2M$. In these equations, k is measured in length units such as feet. Geometrically similar figures have equal radii of gyration about corresponding centroidal axes. *See* MOMENT OF INERTIA.   [N.S.F.]

Raman effect   A phenomenon observed in the scattering of light as it passes through a material medium, whereby the light suffers a change in frequency and a random alteration in phase. Raman scattering differs in both these respects from Rayleigh and Tyndall scattering, in which the scattered light has the same frequency as the unscattered and bears a definite phase relation to it. The intensity of normal Raman scattering is roughly one-thousandth that of Rayleigh scattering in liquids and smaller still in gases. *See* SCATTERING OF ELECTROMAGNETIC RADIATION.

Because of its low intensity, the Raman effect was not discovered until 1928, although the scattering of light by transparent solids, liquids, and gases had been investigated for many years before. The development of the laser has led to a resurgence of interest in the Raman effect and to the discovery of a number of related phenomena. *See* LASER.

When the exciting radiation falls within the frequency range of a molecule's absorption band in the visible or ultraviolet spectrum, the radiation may be scattered by two different processes, resonance fluorescence or the resonance Raman effect. Both these processes give much more intense scattering than the normal nonresonant Raman effect. The absolute frequencies of the resonance Raman effect shift by exactly the amount of any shift in the exciting frequency, just as do those of the normal Raman effect. Thus the main characteristic of the resonance as compared to the normal Raman effect is its intensity, which may be greater by two or three orders of magnitude. *See* FLUORESCENCE.

Raman scattering is analyzed by spectroscopic means. The collection of new frequencies in the spectrum of monochromatic radiation scattered by a substance is characteristic of the substance and is called its Raman spectrum. Although the Raman effect can be made to occur in the scattering of radiation by atoms, it is of greatest interest in the spectroscopy of molecules and crystals. In a typical experiment monochromatic radiation from a laser impinges on the sample in an appropriate transparent cell. Raman scattering is approximately uniform in all directions and is usually studied at right angles. In this way the intense radiation of the laser beam interferes least with the observation of the weak scattered light.

Raman spectroscopy is of considerable value in determining molecular structure and in chemical analysis. Molecular rotational and vibrational frequencies can be determined directly, and from these frequencies it is sometimes possible to evaluate the molecular geometry, or at least to find the molecular symmetry. Even when a precise determination of structure is not possible, much can often be said about the arrangement of atoms in a molecule from empirical information about the characteristic Raman frequencies of groups of atoms. This kind of information is closely similar to that provided by infrared spectroscopy; in fact, Raman and infrared spectra often provide complementary data about molecular structure. Raman spectra also provide information for solid-state physicists, particularly with respect to lattice dynamics but also concerning the electronic structures of solids. *See* INFRARED SPECTROSCOPY; LATTICE VIBRATIONS; MOLECULAR STRUCTURE AND SPECTRA.

[R.C.L.]

Random matrices   Collections of large matrices, chosen at random from some ensemble. Random-matrix theory is a branch of mathematics which emerged from the study of complex physical problems, for which a statistical analysis is often more enlightening than a hopeless attempt to control every degree of freedom, or every detail of the dynamics. Although the connections to various parts of mathematics are very rich, the relevance of this approach to physics is also significant.

Random matrices were introduced by Eugene Wigner in nuclear physics in 1950. In quantum mechanics the discrete energy levels of a system of particles, bound together, are given by the eigenvalues of a hamiltonian operator, which embodies the interactions between the constituents. This leads to the Schrödinger equation which, in most cases of interest in the physics of nuclei, cannot be solved exactly, even with the most advanced computers. For a complex nucleus, instead of finding the location of the nuclear energy levels through untrustworthy approximate solutions, Wigner proposed to study the statistics of eigenvalues of large matrices, drawn at random from some ensemble. The only constraint is to choose an ensemble which respects the symmetries that are present in the forces between the nucleons in the original problem, and to select a sequence of levels corresponding to the quantum numbers that are conserved as a consequence of these symmetries, such as angular momentum and parity. The statistical theory does not attempt to predict the detailed sequence of energy levels of a given nucleus, but only the general properties of those sequences and, for instance, the presence of hidden symmetries. In many cases this is more important than knowing the exact location of a particular energy level. This program became the starting point of a new field, which is now widely used in mathematics and physics for the understanding of quantum chaos, disordered systems, fluctuations in mesoscopic systems, random surfaces, zeros of analytic functions, and so forth. *See* CONSERVATION LAWS (PHYSICS); EIGENVALUE (QUANTUM MECHANICS); QUANTUM MECHANICS.

The mathematical theory underlying the properties of random matrices overlaps with several active fields of contemporary mathematics, such as the asymptotic behavior of orthogonal polynomials at large-order, integrable hierachies, tau functions, semiclassical expansions, combinatorics, and group theory; and it is the subject of active research and collaboration between physics and mathematics.   [E.B.]

Rarefied gas flow   Flow of gases below standard atmospheric pressure, sometimes called low-pressure gas flow. The flow may be confined to pipes between a chamber or vessel to be evacuated and a pump, or it may be the beam of molecules issuing from an orifice into a large evacuated chamber or the plume of exhaust gases from a rocket launched into the upper atmosphere, for example. The flow velocity is measured with respect to a fixed boundary such as the wall of a pipe, the surface of a rocket or jet plane, or a model in a wind tunnel. *See* FLUID FLOW; MOLECULAR BEAMS.

For flow through ducts, the gases concerned are initially those of the original atmosphere inside a chamber that must be evacuated. However, even after the bulk of the original gas has been removed, the pumps must continue to remove gas evolved from surfaces and leaking in through imperfections in the walls. In some cases, gas is introduced through valves at a controlled rate as part of the process being carried out at a low pressure.

Since the flow through pipes involves an interaction or drag at the walls, a pressure drop is generated across the entrance and exit of the pipe. Also, gaseous impurities from the pump may flow toward the chamber when the pressure is very low. Proper design of the duct system therefore involves selecting pipes and valves of adequate internal diameter to ensure a minimal pressure drop and the insertion of baffles or traps to prevent impurities from the pumps from entering the process chamber.

The resistance due to the walls depends on the mass flow velocity, and may depend on the gas viscosity and the pressure or density of the gas. The mean free path of molecules is the distance between collisions with other molecules in the gas. *See* KINETIC THEORY OF MATTER; MEAN FREE PATH; VISCOSITY.

The analysis of low-pressure flow is divided into three or four flow regimes depending on the value of the Knudsen number Kn, defined as the ratio of the mean free path to a characteristic length, and the dimensionless Reynolds number. The characteristic length may be chosen as the mean pipe diameter in the case of confined flow or as some length associated with a test model suspended in a wind tunnel, for example. *See* GAS DYNAMICS; KNUDSEN NUMBER; REYNOLDS NUMBER.

Another dimensionless number used in gas flow dynamics is the Mach number (Ma), defined as the ratio of the mass flow velocity to the local velocity of sound in the gas. *See* MACH NUMBER.

When the mean free path is much smaller than the pipe diameter (Kn < 0.01), the gas flows as a continuous viscous fluid with velocity near the axis of the pipe at locations well beyond the pipe entrance much higher than the velocity in gas layers near the wall. The velocity profile as a function of radial distance from the axis depends on the distance from the entrance and the viscosity of the gas. When the Reynolds number is less than 2000, the profile is a simple curved surface so that the flow is laminar (laminar flow regime). When the mean free path becomes greater than about 0.01 times the diameter, the profile is distorted by boundary-layer effects, and the velocity near the wall does not approach zero (sometimes referred to as slip flow). *See* LAMINAR FLOW.

For Reynolds numbers above the critical value (approximately 2100), the flow is subject to instabilities depending on the geometry of the boundary and at high Reynolds numbers becomes turbulent (turbulent flow regime). *See* TURBULENT FLOW.

When the mean free path is about equal to or greater than the pipe diameter (Kn ≥ 1), the gas molecules seldom collide with each other, but can either pass through the pipe without striking the wall or scatter randomly back and forth between various points on the wall and eventually escape through the exit or pass back through the entrance. This type of gas flow is known as free-molecule flow (molecular flow regime). The transition region (0.01 < Kn < 1) between the laminar flow regime and the molecular flow regime is referred to as the Knudsen or transition flow regime.

The flow may also be classified by the boundary conditions or by the Mach number. For example, Couette flow involves the flow of rarefied gas between two surfaces that are moving with respect to each other with different parallel tangential velocities. For hypersonic flow, Ma ≥ 5.

[B.B.Da.]

Reactance   The imaginary part of the impedance of an alternating-current circuit.

The impedance Z of an alternating current circuit is a complex number given by Eq. (1). The imaginary part X is the reactance. The units of reactance, like those

$$Z = R + jX \tag{1}$$

$$Z = jL\omega = jX \tag{2}$$

$$Z = -\frac{j}{C\omega} = jX \tag{3}$$

of impedance, are ohms. Reactance may be positive or negative. For example, the impedance of an inductor L at frequency ω is given by Eq. (2), so X is positive. The impedance of a capacitor C is given by Eq. (3), so X is negative.

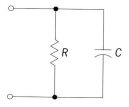

Circuit with a resistor and capacitor in parallel.

The reactance of a circuit may depend on both the resistors and the inductors or capacitors in the circuit. For example, the circuit in the illustration has admittance [Eq. (4)] and impedance [Eq. (5)], so that the reactance [Eq. (6)], depends on both the capacitor C and the resistor R.

$$Y = \frac{1}{R} + jC\omega \tag{4}$$

$$Z = \frac{R}{1 + jRC\omega} \tag{5}$$

$$X = -\frac{R^2 C\omega}{1 + R^2 C^2 \omega^2} \tag{6}$$

See ADMITTANCE; ELECTRICAL IMPEDANCE.   [J.O.S.]

Reciprocity principle

In the scientific sense, a theory that expresses various reciprocal relations for the behavior of some physical systems. Reciprocity applies to a physical system whose input and output can be interchanged without altering the response of the system to a given excitation. Optical, acoustical, electrical, and mechanical devices that operate equally well in either direction are reciprocal systems, whereas unidirectional devices violate reciprocity. The theory of reciprocity facilitates the evaluation of the performance of a physical system. If a system must operate equally well in two directions, there is no need to consider any nonreciprocal components when designing it.

Some systems that obey the reciprocity principle are any electrical network composed of resistances, inductances, capacitances, and ideal transformers; systems of antennas, which obey certain restrictions; mechanical gear systems; and light sources, lenses, and reflectors.

Devices that violate the theory of reciprocity are transistors, vacuum tubes, gyrators, and gyroscopic couplers. Any system that contains the above devices as components must also violate the reciprocity theory. The gyrator differs from the transistor and vacuum tube in that it is linear and passive, as opposed to the active and nonlinear character of the other two devices. *See* TRANSISTOR.   [H.S.La.]

Rectilinear motion

Motion is defined as continuous change of position of a body. If the body moves so that every particle of the body follows a straight-line path, then the motion of the body is said to be rectilinear. *See* MOTION.

When a body moves from one position to another, the effect may be described in terms of motion of the center of mass of the body from a point A to a point B (see illustration). If the center of mass of the body moves along a straight line connecting points A and B, then the motion of the center of mass of the body is rectilinear. If the body as a whole does not rotate while it is moving, then the path of every particle of which the body is composed is a straight line parallel to or coinciding with the path of the center of mass, and the body as a whole executes rectilinear motion. This is shown by the straight line connecting points P_1 and P_2 in the illustration. *See* CENTER OF MASS.

Rectilinear motion is an idealized form of motion which rarely, if ever, occurs in actual experience, but it is the simplest imaginable type of motion and thus forms the basis for the analysis of more complicated motions. However, many actual motions are approximately rectilinear and may be treated as such without appreciable error. For

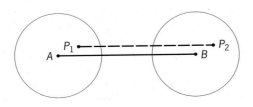

Rectilinear motion. All points move parallel to the center of mass.

example, a ball thrown directly upward may follow, for all practical purposes, a straight-line path. The motion of a high-speed rifle bullet fired horizontally may be essentially rectilinear for a short length of path, even though in its larger aspects the ideal path is a parabola. The motion of an automobile traveling over a straight section of roadway is essentially rectilinear if minor variations of path are neglected. The motion of a single wheel of the car is not rectilinear, although the motion of the center of mass of the wheel may be essentially so. *See* BALLISTICS.

[R.D.Ru.]

Reflection and transmission coefficients

When an electromagnetic wave passes from a medium of permeability μ_1 and dielectric constant ϵ_1 to one with values μ_2 and ϵ_2, part of the wave is reflected at the boundary and part transmitted. The ratios of the amplitudes in the reflected wave and the transmitted wave to that in the incident wave are called the reflection and transmission coefficients, respectively. For oblique incidence, the reflection and refraction formulas of optics are most convenient, but for normal incidence of plane waves on plane boundaries, such as occur with transmission lines, waveguides, and some free waves, the concept of wave impedance and characteristic impedance is useful. *See* ELECTROMAGNETIC RADIATION.   [W.R.Sm.]

Reflection of electromagnetic radiation

The returning or throwing back of electromagnetic radiation such as light, ultraviolet rays, radio waves, or microwaves by a surface upon which the radiation is incident. In general, a reflecting surface is the boundary between two materials of different electromagnetic properties, such as the boundary between air and glass, air and water, or air and metal. Devices designed to reflect radiation are called reflectors or mirrors.

The simplest reflection laws are those that govern plane waves of radiation. The law of reflection concerns the incident and reflected rays (as in the case of a beam from a flashlight striking a mirror) or, more precisely, the wave normals of the incident and reflected waves. The law states that the incident and reflected rays and the normal to the reflecting surface all lie in one plane, called the plane of incidence, and that the reflection angle equals the angle of incidence as in Eq. (1) [see illustration]. The

$$\theta_{\text{refl}} = \theta_{\text{inc}} \tag{1}$$

angles θ_{inc} and θ_{refl} are measured between the surface normal and the incident and reflected rays, respectively. The surface (in the above example, that of the mirror) is assumed to be smooth, with surface irregularities small compared to the wavelength of the radiation. This results in so-called specular reflection. In contrast, when the surface is rough, the reflection is diffuse. An example of this is the diffuse scattering of light from a screen or from a white wall where light is returned through a whole range of different angles.

The reflectivity of a surface is a measure of the amount of reflected radiation. It is defined as the ratio of the intensities of the reflected and incident radiation. The reflectivity depends on the angle of incidence, the polarization of the radiation, and the electromagnetic properties of the materials forming the boundary surface. These properties usually change with the wavelength of the radiation. Reflecting materials are divided into two groups: transparent materials also called dielectrics, and opaque conducting materials, usually metals.

The reflectivity of polished metal surfaces is usually quite high. Silver and aluminum, for example, reflect more than 90% of visible light. In ordinary mirrors the reflecting surface is the interface between metal and glass, which is thus protected from oxidation, dirt, and other forms of deterioration. When it is not permissible to use this protection

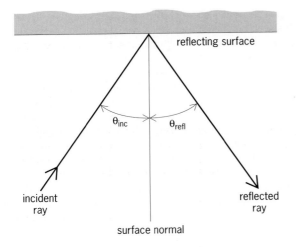

reflecting surface

θ_{inc} θ_{refl}

incident
ray

reflected
ray

surface normal

**Reflection of electromagnetic
radiation from a smooth surface.**

for technical reasons, one uses "front-surface" mirrors, which are usually coated with evaporated aluminum.

The material property that determines the amount of radiation reflected from an interface between two dielectric media is the phase velocity v of the electromagnetic radiation in the two materials. In optics one uses as a measure for this velocity the refractive index n of the material, which is defined by Eq. (2) as the ratio of the velocity

$$n = c/v \tag{2}$$

of light c in vacuum and the phase velocity in the material. For visible light, for example, the refractive index of air is about $n = 1$, the index of water is about $n = 1.33$, and the index of glass is about $n = 1.5$. *See* PHASE VELOCITY; REFRACTION OF WAVES.

For normal incidence ($\theta_{\text{inc}} = 0$) the reflectivity R of the interface is given by Eq. (3),

$$R = \left(\frac{v_1 - v_2}{v_1 + v_2}\right)^2 = \left(\frac{n_2 - n_1}{n_2 + n_1}\right)^2 \tag{3}$$

in which the material constants are labeled 1 and 2, where the radiation is incident in material 1. The reflectivity of an air-water interface is about 2% ($R = 0.02$) and that of an air-glass interface about 4% ($R = 0.04$); the other 98% or 96% are transmitted through the water or glass, respectively. *See* GEOMETRICAL OPTICS; MIRROR OPTICS; REFLECTION OF SOUND.   [H.K.]

Reflection of sound   The return of sound waves from surfaces on which they are incident. The geometrical laws for reflection of sound waves are the same as those for light waves. The apparent differences involve only questions of scale, because the average wavelength of sound is about 100,000 times that of light. For example, a mirror or lens used to produce a beam of sound waves must be enormously large compared to mirrors and lenses used in optical systems. *See* REFLECTION OF ELECTROMAGNETIC RADIATION.

A concave surface tends to concentrate the reflected sound waves. Convex reflectors tend to spread the reflected waves. Therefore, when placed at the boundaries of a room, they tend to diffuse the sound throughout the room. For this reason, some radio-broadcasting studios employ cylindrical convex panels as part of their wall construction to promote diffusion. *See* ECHO; SOUND.   [C.M.H.]

Refraction of waves   The change of direction of propagation of any wave phenomenon which occurs when the wave velocity changes. The term is most frequently applied to visible light, but it also applies to all other electromagnetic waves, as well as to sound and water waves.

The physical basis for refraction can be readily understood with the aid of the illustration. Consider a succession of equally spaced wavefronts approaching a boundary surface obliquely. The direction of propagation is in ordinary cases perpendicular to the wavefronts. In the case shown, the velocity of propagation is less in medium 2 than in medium 1, so that the waves are slowed down as they enter the second medium. Thus, the direction of travel is bent toward the perpendicular to the boundary surface (that is, $\theta_2 < \theta_1$). If the waves enter a medium in which the velocity of propagation is faster than in their original medium, they are refracted away from the normal.

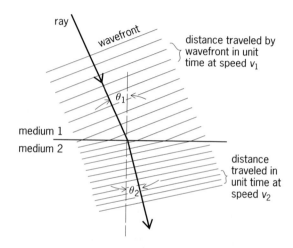

Physical basis for Snell's law.

Snell's law. The simple mathematical relation governing refraction is known as Snell's law. If waves traveling through a medium at speed v_1 are incident on a boundary surface at angle θ_1 (with respect to the normal), and after refraction enter the second medium at angle θ_2 (with the normal) while traveling at speed v_2, then Eq. (1) holds.

$$\frac{v_1}{v_2} = \frac{\sin \theta_1}{\sin \theta_2} \qquad (1)$$

The index of refraction n of a medium is defined as the ratio of the speed of waves in vacuum c to their speed in the medium. Thus $c = n_1 v_1 = n_2 v_2$, and therefore Eq. (2)

$$n_1 \sin \theta_1 = n_2 \sin \theta_2 \qquad (2)$$

holds. The refracted ray, the normal to the surface, and the incident ray always lie in the same place.

The relative index of refraction of medium 2 with respect to that of medium 1 may be defined as $n = n_1/n_2$. Snell's law then becomes Eq. (3). For sound and other

$$\sin \theta_1 = n \sin \theta_2 \qquad (3)$$

elastic waves which require a medium in which to propagate, only this last form has meaning. Equation (3) is frequently used for light when one medium is air, whose index of refraction is very nearly unity.

When the wave travels from a region of low velocity (high index) to one of high velocity (low index), refraction occurs only if $(n_1/n_2) \sin \theta_1 \leqq 1$. If θ_1 is too large for this relation to hold, then $\sin \theta_2 > 1$, which is meaningless. In this case the waves are totally reflected from the surface back into the first medium. The largest value that θ_1 can have without total internal reflection taking place is known as the critical angle θ_c. Thus $\sin \theta_c = n_2/n_1$.

Atmospheric refraction. The index of refraction of the Earth's atmosphere increases continuously from 1.000000 at the edge of space to 1.000293 (yellow light) at $0°C$ and 760 mmHg (101.3 kilopascals) pressure. Thus celestial bodies as seen in the sky are actually nearer to the horizon than they appear to be. The effect decreases from a maximum of about 35 minutes of arc for an object on the horizon to zero at the zenith, where the light enters the atmosphere at perpendicular incidence.

Other manifestations of atmospheric refraction are the mirages and "looming" of distant objects which occur over oceans or deserts, where the vertical density gradient of the air is quite uniform over a large area.

Sound waves. The velocity of sound in a gas is proportional to the square root of the absolute temperature. Because of the vertical temperature gradients in the atmosphere, refraction of sound can be quite pronounced. As in mirage formation, to allow large-scale refraction the temperature at a given height must be uniform over a rather large horizontal area. *See* ATMOSPHERIC ACOUSTICS; SOUND.

Seismic waves. The velocity of elastic waves in a solid depends upon the modulus of elasticity and upon the density of the material. Waves propagating through solid earth are refracted by changes of material or changes of density. Worldwide observations of earthquake waves enable scientists to draw conclusions on the distribution of density within the Earth.

Water waves. As the waves enter shallower water they travel more slowly. As a train of waves approaches a coastline obliquely, its direction of travel becomes more nearly perpendicular to the shore because of refraction. *See* WAVE MOTION.   [J.W.St.]

Relative motion   All motion is relative to some frame of reference. The simplest laboratory frame of reference is three mutually perpendicular axes at rest with respect to an observer. In terms of the frame of reference of an observer some distance from Earth, the laboratory frame of reference would be moving with Earth as it rotates on its axis and as it revolves about the Sun. What would be a simple form of motion in the laboratory frame of reference would appear to be a much more complicated motion in the frame of reference of the distant observer. *See* FRAME OF REFERENCE.

Motion means continuous change of position of an object with respect to an observer. To another observer in a different frame of reference the object may not be moving at all, or it may be moving in an entirely different manner. The motions of the planets were found in ancient times to appear quite complicated in the laboratory frame of reference of an observer on Earth. By transferring to the frame of reference of an imaginary observer on the Sun, Johannes Kepler showed that the relative motion of the planets could be simply described in terms of elliptical orbits. The validity of one description is no greater than the other, but the latter description is far more convenient.   [R.D.Ru.]

Relativistic electrodynamics   The study of the interaction between electrically charged particles and electromagnetic fields when the velocities involved are comparable to that of light.

A group of charged particles in motion can be represented by a distribution in charge and distribution in current. During the latter part of the eighteenth century and the early part of the nineteenth century, experiments by C. A. Coulomb, M. Faraday, A. M.

Ampère, and others showed that electric and magnetic fields are produced by charge and current distributions. These fields, in turn, act on other charges and currents. The interaction between charges and currents on the one hand and electric and magnetic fields on the other is the topic of study of electrodynamics. This field of study was established as a quantitative and self-contained subject in 1864 when J. C. Maxwell formulated his equations for the electromagnetic field. Maxwell conjectured that a time-varying electric field is equivalent to an electric current in its effect of producing a magnetic field, and named it the displacement current. The inclusion of this displacement current enabled Maxwell to combine all the previously established laws of electromagnetism into a coherent whole in his equations. *See* CLASSICAL FIELD THEORY; DISPLACEMENT CURRENT; MAXWELL'S EQUATIONS.

With the inclusion of the displacement current, the Maxwell equations are relativistically covariant, meaning that they are valid for all velocities, even those approaching the velocity of light. However, the implications of the covariance of the equations were not fully appreciated until A. Einstein formulated the special theory of relativity in 1905. Relativistic electrodynamics was then rapidly developed into a powerful and precise field of physics. It describes and predicts all macroscopic electrodynamic phenomena to the minutest detail and with perfect accuracy, and now forms the foundation on which the entire electrical industry is based. However, its limitations soon became evident when attempts were made to apply it to atomic phenomena: Straightforwardly applied, relativistic electrodynamics failed to explain many of these phenomena, and its predictions frequently disagreed with experimental observations. For these microscopic phenomena, quantum electrodynamics (QED) was developed in the 1930s to replace classical relativistic electrodynamics. In 1967 quantum electrodynamics was further unified by S. Weinberg and A. Salam with the theory of weak interactions to form the electroweak theory. *See* QUANTUM ELECTRODYNAMICS; RELATIVITY; WEAK NUCLEAR INTERACTIONS.

Electrodynamic problems generally fall into one of two categories:

1. Finding the electromagnetic field produced by prescribed charge and current distributions. For example, one may want to determine the electromagnetic field produced or radiated by a given oscillatory electric current in a transmitting antenna, or the field radiated by an accelerating electron.

2. Finding the effect of a predetermined electromagnetic field on the motion of charges and currents. This is the inverse problem corresponding to that of the receiving antenna or of the motion of charged particles in an accelerator.

All other electrodynamic problems are combinations or iterations of these two basic types. For instance, the scattering of light (electromagnetic radiation) by a charged particle is composed of, first, the incident light shaking the charge and, second, the subsequent emission of the scattered light by the shaken charge. *See* SCATTERING OF ELECTROMAGNETIC RADIATION.

[L.C.T.]

Relativistic heavy-ion collisions
Collisions between heavy atomic nuclei at relative velocities close to the speed of light. These high-energy nuclear collisions are usually divided into two different domains, relativistic and ultrarelativistic collisions, depending on whether the kinetic energy per nucleon (the generic name for protons and neutrons) is either close to the rest mass of the nucleon (relativistic collisions) or much larger than the nucleon rest mass (ultrarelativistic collisions).

By utilizing high-energy nuclear collisions, it is possible to study nuclear matter under conditions of very high temperatures and densities. The most common form of nuclear matter, at least under terrestrial conditions, is found in the atomic nucleus, which consists of protons and neutrons bound together by the strong nuclear force. If nuclear matter is heated up to temperatures comparable to the rest mass of the pion, it becomes a mixture of nucleons, pions, and various other particles, collectively denoted hadrons. Under these circumstances, nuclear matter is referred to as hadronic matter. *See* HADRON; NEUTRON; NUCLEAR STRUCTURE; PROTON.

According to the quantum chromodynamics (QCD) theory, all hadrons are bound states of a set of more fundamental entities called quarks. The quarks are confined within the hadrons by the exchange of gluons. Quantum chromodynamics calculations using the most powerful computers available show that if hadronic matter is further heated or compressed to very high densities it will undergo a phase transition into a new phase of matter, called the quark-gluon plasma. In this phase the hadrons will lose their identity, and the quarks and gluons will be deconfined within volumes much larger than the typical hadron volume of 0.1–0.5 cubic femtometer. Quantum chromodynamics calculations indicate that the phase transition will occur at a critical density around 5–10 times the normal nuclear matter density of approximately 0.2 nucleon/fm^3, or at a critical temperature around 150 MeV. *See* GLUONS; QUANTUM CHROMODYNAMICS; QUARK-GLUON PLASMA; QUARKS.

When two nuclei collide at high energies, some of the nucleons in each nucleus, called spectators, will continue their motion unaffected, while other nucleons, called participants, will strike one or several nucleons in the other nucleus. In the overlap volume a hot and dense fireball will develop. If the temperature or density of the fireball becomes larger than the critical values, a quark-gluon plasma will be created with an estimated lifetime of $1–5 \times 10^{-23}$ s. The fireball will start to expand and cool, and the quarks in the plasma will eventually be reconfined into a large number of hadrons (hadronization). After further expansion the hadrons will cease interacting with each other (freeze out) and leave the collision zone without further mutual interactions.

In the search for the quark-gluon plasma, a fundamental problem is that even if the plasma is created in the early phases of the collisions, the subsequent hadronization and scattering of the hadrons before freeze-out might mask any traces of the plasma. In order to circumvent this problem, many plasma signatures have been proposed. [S.P.S.]

Relativistic quantum theory The quantum theory of particles which is consistent with the special theory of relativity, and thus can describe particles moving arbitrarily close to the speed of light. It is now realized that the only satisfactory relativistic quantum theory is quantum field theory; the attempt to relativize the Schrödinger equation for the wave function of a single particle fails [Eq. (1)]. However, with a change of interpretation, relativistic wave equations do correctly describe some aspects of the motions of particles in an electromagnetic field. *See* QUANTUM FIELD THEORY; QUANTUM MECHANICS; RELATIVITY.

The Schrödinger equation for the wave function $\psi(\mathbf{r},t)$ of a particle is Eq. (1), where

$$E\psi = H(\mathbf{p},\mathbf{r})\psi \tag{1}$$

E is the energy operator $i\hbar(\partial/\partial t)$, $\mathbf{p}$ is the momentum operator $-i\hbar\Delta$, $H(\mathbf{p},\mathbf{r})$ is the classical hamiltonian, and $\hbar$ is Planck's constant divided by 2π. For a nonrelativistic free particle, $H = \mathbf{p}^2/2m$. The naive way to relativize Eq. (1) would be to use the relativistic hamiltonian, Eq. (2). However, this equation is not relativistically invariant.

$$H = \sqrt{(mc^2)^2 + \mathbf{p}^2 c^2} \tag{2}$$

The so-called Klein-Gordon equation (3) is relativistically invariant. However, the

$$E^2\varphi = [(mc^2)^2 + \mathbf{p}^2c^2]\varphi \tag{3}$$

only possible density of a conserved quantity formed from φ is of the form shown in (4). But this cannot be a probability density, because it is not positive definite (it changes sign when φ is replaced by φ^*).

$$\rho \propto \varphi^* E\varphi - \varphi E\varphi^* \tag{4}$$

But ρ, in relation (4), can be interpreted as a charge density (when multiplied by a unit charge e); φ is then to be interpreted as a matrix element of a field operator Φ of a quantized field whose quanta are particles with mass m and charge e or $-e$ and zero spin.

P. A. M. Dirac found a relativized form of Eq. (1), Eq. (5), which is both linear in E and has a positive definite density form ρ, where β and α are constants which obey Eqs. (6). Obviously the four constants β and α_i cannot be numbers; however, they can

$$E\psi = [\beta mc^2 + \alpha \cdot pc]\psi (\rho \propto \psi^*\psi) \tag{5}$$

$$\alpha_i\alpha_j + \alpha_j\alpha_i = 0, \quad i \neq j$$
$$\alpha\beta + \beta\alpha = 0 \qquad \alpha_i^2 = 1 \qquad \beta^2 = 1 \qquad i, j = 1, 2, 3 \tag{6}$$

be 4×4 matrices, and Ψ is then a four-component object called a Dirac spinor.

If plane wave solutions of Dirac's equation (5) are considered, then $\mathbf{P}$ is now a number. Taking Eq. (5) as an eigenequation for E, four eigenstates are found (because H is a 4×4 matrix): two with

$$E = -\sqrt{(mc^2)^2 + p^2c^2}$$

and two with

$$E = +\sqrt{(mc^2)^2 + p^2c^2}$$

The interpretation of the two positive energy states is that they are the two spin states of a particle with spin $1/2[\hbar]$. But the two negative energy states are an embarrassment; even a particle that was initially in a positive energy state would quickly make radiative transitions down through the negative energy states. Dirac's solution was to observe that if the particle described by ψ obeyed the Pauli principle, then one can suppose that all the negative energy states are already filled with particles, thus excluding any more. There are still four single-particle states for a given momentum $\mathbf{p}$: the two spin states of a particle with positive energy, and the two states obtained by removing a negative energy particle (of momentum $-\mathbf{p}$). These last states ("hole states") have positive energy and a charge opposite the charge of the particle. The hole is in fact the antiparticle; if the particle is an electron, the hole is a positron. *See* ANTIMATTER; POSITRON.

With the filling up of the negative energy states, one no longer has a single-particle system, and ψ, just as in the Klein-Gordon case, no longer can be interpreted as a wave function but must be interpreted as a matrix element of a field operator Ψ.

[C.J.G.]

Relativity   A general theory of physics, primarily conceived by Albert Einstein, which involves a profound analysis of time and space, leading to a generalization of physical laws, with far-reaching implications in important branches of physics and in cosmology. Historically, the theory developed in two stages. Einstein's initial formulation in 1905 (now known as the special, or restricted, theory of relativity) does not treat gravitation; and one of the two principles on which it is based, the principle of relativity (the other being the principle of the constancy of the speed of light), stipulates the form invariance of physical laws only for inertial reference systems. Both restrictions were removed by Einstein in his general theory of relativity developed in 1915, which exploits a deep-seated equivalence between inertial and gravitational effects, and leads to a successful "relativistic" generalization of Isaac Newton's theory of gravitation.

Special theory. The key feature of the theory of special relativity is the elimination of an absolute notion of simultaneity in favor of the notion that all observers always measure light to have the same velocity, in vacuum, c, independently of their own motion. The impetus for the development of the theory arose from the theory of electricity and magnetism developed by J. C. Maxwell. This theory accounted for all observed phenomena involving electric and magnetic fields and also predicted that disturbances in these fields would propagate as waves with a definite speed, c, in vacuum. These electromagnetic waves predicted in Maxwell's theory successfully accounted for the existence of light and other forms of electromagnetic radiation. However, the presence of a definite speed, c, posed a difficulty, since if one inertial observer measures light to have velocity c, it would be expected that another inertial observer, moving toward the light ray with velocity v with respect to the first, would measure the light to have velocity $c + v$. Hence, it initially was taken for granted that there must be a preferred rest frame (often referred to as the ether) in which Maxwell's equations would be valid, and only in that frame would light be seen to travel with velocity c. However, this viewpoint was greatly shaken by the 1887 experiment of A. A. Michelson and E. W. Morley, which failed to detect any motion of the Earth through the ether. By radically altering some previously held beliefs concerning the structure of space and time, the theory of special relativity allows Maxwell's equations to hold, and light to propagate with velocity c, in all frames of reference, thereby making Maxwell's theory consistent with the null result of Michelson and Morley. *See* ELECTROMAGNETIC RADIATION; LIGHT; MAXWELL'S EQUATIONS.

Simultaneity in prerelativity physics. The most dramatic aspect of the theory of special relativity is its overthrowing of the notion that there is a well-defined, observer-independent meaning to the notion of simultaneity. The following terminology will be introduced: An event is a point of space at an instant of time. Since it takes four numbers to specify an event—one for the time at which the event occurred and three for its spatial position—it follows that the set of all events constitutes a four-dimensional continuum, which is referred to as space-time.

A space-time diagram (Fig. 1) is a plot of events in space-time, with time, t, represented by the vertical axis and two spatial directions (x, y) represented by the horizontal axes. (The z direction is not shown.) For any event A shown in the diagram, there are many other events in this diagram—say, an event B—having the property that an observer or material body starting at event B can, in principle, be present at event A. The collection of all such events constitutes the past of event A. Similarly, there are many events—say, an event C—having the property that an observer or material body starting at event A can, in principle, be present at event C. These events constitute the future of A. Finally, there remain some events in space-time which lie neither to past nor future of A. In prerelativity physics, these events are assumed to make a three-dimensional set, and they are referred to as the events which are simultaneous with event A.

In both prerelativity physics and special relativity, an inertial observer is one who is not acted upon by any external forces. In both theories it is assumed that any inertial

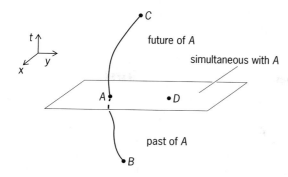

Fig. 1. Space-time diagram illustrating the causal relationships with respect to an event, *A*, in prerelativity physics. Event *B* lies to the past of *A*, event *C* lies to the future of *A*, and event *D* is simultaneous with *A*.

observer, $\mathcal{O}$, can build a rigid cartesian grid of meter sticks, all of which intersect each other at right angles. Observer $\mathcal{O}$ may then label the points on this cartesian grid by the coordinates (x, y, z) representing the distance of the point from $\mathcal{O}$ along the three orthogonal directions of the grid. A clock may then be placed at each grid point. In prerelativity physics, these clocks may be synchronized by requiring that they start simultaneously with each other. Any event in space-time may then be labeled by the four numbers t, x, y, z, where t is the time of the event as determined by the synchronized clock situated at that grid point. *See* FRAME OF REFERENCE.

It is of interest to compare the coordinate labelings given to events in space-time by two inertial observers, $\mathcal{O}$ and $\mathcal{O}'$, who are in relative motion. The relationship occurring in prerelativity physics is called a galilean transformation. In the simple case where $\mathcal{O}'$ moves with velocity v in the x direction with respect to $\mathcal{O}$, and these observers meet at the event A labeled by $(t, x, y, z) = (t', x', y', z') = (0, 0, 0, 0)$, with the axes of the grid of meter sticks carried by $\mathcal{O}'$ aligned (that is, not rotated) with respect to those of $\mathcal{O}$, the transformation is given by Eqs. (1). The galilean transformation displays in an

$$t' = t \tag{1a}$$

$$x' = x - vt \tag{1b}$$

$$y' = y \tag{1c}$$

$$z' = z \tag{1d}$$

explicit manner that the two inertial observers, $\mathcal{O}$ and $\mathcal{O}'$, agree upon the time labeling of events and, in particular, agree upon which events are simultaneous with a given event.

Causal structure in special relativity. In special relativity there is a different causal relationship between an arbitrary event A and other events in space-time (Fig. 2). As in prerelativity physics, there are many events, B, which lie to the past of A. There also are many events which lie to the future of A. However, there is now a much larger class of events which lie neither to the past nor to the future of A. These events are referred to as being spacelike-related to A.

The most striking feature of this causal structure (Fig. 2) is the absence of any three-dimensional surface of simultaneity. Indeed, the closest analog to the surface of simultaneity in prerelativity physics is the double-cone-shaped surface that marks the boundaries of the past and future of event A. This surface comprises the paths in space-time of all light rays which pass through event A, and for this reason it is referred to as the light cone of A. Thus, the statement that the events lying to the future of A are contained within the light cone of A is equivalent to the statement that a material

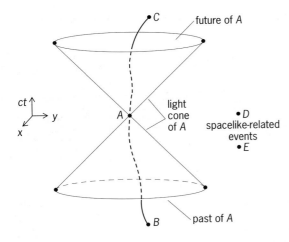

light cone of A

*D spacelike-related events

*E

Fig. 2. Space-time diagram illustrating the causal relationships with respect to an event, A, in special relativity. Event B lies to the past of A, event C lies to the future of A, and events D and E are spacelike-related to A.

body present at event A can never overtake a light ray emitted at event A. In special relativity, the light cone of an event A replaces the surface of simultaneity with event A as the absolute, observer-independent structure of space-time related to causality.

As in prerelativity physics, it is assumed in special relativity that an inertial observer, $\mathcal{O}$, can build a rigid grid of meter sticks, place clocks at the grid points, and label events in space-time by global inertial coordinates (t, x, y, z). The only difference from the procedure used for the construction of the similar coordinates in prerelativity physics is that the synchronization of clocks is now nontrivial, since the causal structure of space-time no longer defines an absolute notion of simultaneity. Nevertheless, any pair of clocks in $\mathcal{O}$'s system can be synchronized—and thereby all of $\mathcal{O}$'s clocks can be synchronized—by having an assistant stationed half-way between the clocks send a signal to the two clocks in a symmetrical manner. This synchronization of clocks allows $\mathcal{O}$ to define a notion of simultaneity; that is, $\mathcal{O}$ may declare events A_1 and A_2 to be simultaneous if time readings t_1 and t_2 of the synchronized clocks at events A_1 and A_2 satisfy $t_1 = t_2$. However, events judged by $\mathcal{O}$ to be simultaneous will, in general, be judged by $\mathcal{O}'$ to be nonsimultaneous.

The key assumptions of special relativity are encapsulated by the following two postulates.

Postulate 1: The laws of physics do not distinguish between inertial observers; in particular, no inertial observer can be said to be at rest in an absolute sense. Thus, if observer $\mathcal{O}$ writes down equations describing laws of physics obeyed by physically measurable quantities in her global inertial coordinate system (t, x, y, z), then the form of these equations must be identical when written down by observer $\mathcal{O}'$ in his global inertial coordinates (t', x', y', z').

Postulate 2: All inertial observers (independent of their relative motion) must always obtain the same value, c, when they measure the velocity of light in vacuum. In particular, the path of a light ray in space-time must be independent of the motion of the emitter of the light ray. Furthermore, no material body can have a velocity greater than c.

The precise relationship between the labeling of events in space-time by the coordinate systems of two inertial observers, $\mathcal{O}$ and $\mathcal{O}'$, in special relativity is given by the Lorentz transformation formulas. In the simple case where $\mathcal{O}'$ moves with velocity v in the x direction with respect to $\mathcal{O}$ and crosses $\mathcal{O}$'s world line at the event labeled by $(t, x, y, z) = (t', x', y', z') = (0, 0, 0, 0)$ with spatial axes aligned, the Lorentz transformation is given by Eqs. (2a)–(2d). Equation (2a) shows explicitly that $\mathcal{O}$ and $\mathcal{O}'$ disagree over simultaneity.

$$t' = \frac{t - \dfrac{xv}{c^2}}{\sqrt{1 - \dfrac{v^2}{c^2}}} \tag{2a}$$

$$x' = \frac{x - vt}{\sqrt{1 - \dfrac{v^2}{c^2}}} \tag{2b}$$

$$y' = y \tag{2c}$$

$$z' = z \tag{2d}$$

Space-time geometry. A key question both in prerelativity physics and in special relativity is what quantities, describing the space-time relationships between events, are observer independent. Such quantities having observer-independent status may be viewed as describing the fundamental, intrinsic structure of space-time.

It has already been seen that in special relativity the time interval, Δt, between two events is no longer observer independent. Furthermore, since different inertial observers disagree over simultaneity, the spatial interval between two simultaneous events is not even a well-defined concept, and cannot be observer independent. Remarkably, however, in special relativity, all inertial observers agree upon the value of the space-time interval, I, between any two events, where I is defined by Eq. (3). What is most

$$I = (\Delta x)^2 + (\Delta y)^2 + (\Delta z)^2 - c^2(\Delta t)^2 \tag{3}$$

remarkable about this formula for I is that it is very closely analogous to the formula for squared distance in euclidean geometry. The minus sign occurring in the last term in Eq. (3) is of considerable importance, since it distinguishes between the notions of time and space in special relativity. Nevertheless, this minus sign turns out not to be a serious obstacle to the mathematical development of the theory of lorentzian geometry based upon the space-time interval, I, in a manner which parallels closely the development of euclidean geometry. In particular, notions such as geodesics (straightest possible lines) can be introduced in lorentzian geometry in complete analogy with euclidean geometry. The Lorentz transformation between the two inertial observers is seen from this perspective to be the mathematical analog of a rotation between two cartesian frames in euclidean geometry.

The formulation of special relativity as a theory of the lorentzian geometry of space-time is of great importance for the further development of the theory, since it makes possible the generalization which describes gravitation. The lorentzian geometry defined by Eq. (3) is a flat geometry, wherein initially parallel geodesics remain parallel forever. The theory of general relativity accounts for the effects of gravitation by allowing the lorentzian geometry of space-time to be curved. *See* SPACE-TIME.

Consequences. The theory of special relativity makes many important predictions, the most striking of which concern properties of time. One such effect, known as time dilation, is predicted directly by the Lorentz transformation formula (2a). If observer $\mathcal{O}$ carries a clock, then the event at which $\mathcal{O}$'s clock reads time τ would be labeled by her as $(\tau, 0, 0, 0)$. According to Eq. (2a), the observer $\mathcal{O}'$ would label the event as

$(t', x', 0, 0)$ where t' is given by Eq. (4). Thus, $\mathcal{O}'$ could say that a clock carried by $\mathcal{O}$

$$t' = \frac{\tau}{\sqrt{1 - \dfrac{v^2}{c^2}}} > \tau \tag{4}$$

slows down on account of $\mathcal{O}$'s motion relative to $\mathcal{O}'$. Similarly, $\mathcal{O}$ would find that a clock carried by $\mathcal{O}'$ slows down with respect to hers. This apparent disagreement between $\mathcal{O}$ and $\mathcal{O}'$ as to whose clock runs slower is resolved by noting that $\mathcal{O}$ and $\mathcal{O}'$ use different notions as to simultaneity in comparing the readings of their clocks.

The decay of unstable elementary particles provides an important direct application of the time dilation effect. If a particle is observed to have a decay lifetime T when it is at rest, special relativity predicts that its observed lifetime will increase according to Eq. (6) when it is moving. Exactly such an increase is routinely observed in experiments using particle accelerators, where particle velocities can be made to be extremely close to c. *See* PARTICLE ACCELERATOR.

An even more striking prediction of special relativity is the clock paradox: Two identical clocks which start together at an event A, undergo different motions, and then rejoin at event B will, in general, register different total elapsed time in going from A to B. This effect is the lorentzian geometry analog of the mundane fact in euclidean geometry that different paths between two points can have different total lengths. *See* CLOCK PARADOX.   [R.H.Wa.]

General theory. One of the basic tenets of special relativity is that no physical effect can propagate with a velocity greater than the speed of light, c, which represents a universal speed limit. On the other hand, classical gravitational theory describes the gravitational field of a body throughout space as a function of its instantaneous position, which is equivalent to the assumption that gravitational effects propagate with an infinite velocity. Thus, special relativity and classical gravitational theory are inconsistent, and a modified theory of gravity is necessary.

Principle of equivalence. It had long been considered a fundamental question why bodies of different mass fall with the same acceleration in a gravitational field. This situation was explained by Newton with the statement that both the gravitational force on a body and its inertial resistance to acceleration are proportional to its mass.

Newton's explanation is more in the nature of an ad hoc description. A deeper and more natural explanation occurred to Einstein. There are numerous forces other than gravity which are mass-proportional. These generally arise due to the use of accelerated coordinate systems to describe the motion, for example, the centrifugal force encountered in a rotating coordinate system. If an observer in the gravitational field of the Earth and another in an accelerating elevator or rocket in free space both drop a test body, they will both observe it to accelerate relative to the floor. According to classical theory, the Earth-based observer would attribute this to a gravitational force and the elevator-based observer would attribute it to the accelerated floor overtaking the uniformly moving body. In both cases the motion is identical, and in particular the acceleration is independent of the mass of the test body. Einstein elevated this fact to a general principle, the principle of equivalence; the principle states that on a local scale all physical effects of a gravitational field are indistinguishable from the physical effects of an accelerated coordinate system. This profound principle is the physical cornerstone of the theory of general relativity. From the point of view of the principle of equivalence, it is obvious why the motion of a test body in a gravitational field is independent of its mass. But the principle applies not only to mechanics but to all physical phenomena and thereby has profound consequences for electromagnetic and other nonmechanical phenomena. *See* CENTRIFUGAL FORCE.

Tensor field equations. The close connection between gravity and accelerating coordinate systems convinced Einstein that gravity is fundamentally a geometric phenomenon. Because of this, it is naturally described by the mathematics of higher-dimensional abstract geometry. This geometry involves systems of equations, called tensor equations, that are manifestly independent of the coordinate system. Tensors are a simple generalization of vectors.

The space-time of relativity contains one covariant second-rank tensor of particularly great importance, called the metric tensor $g_{\mu\nu}$, which is a generalization of the Lorentz metric of special relativity, introduced in Eq. (3). Nearby points in space-time, called events, which are separated by coordinate distances dx^μ have an invariant physical separation whose square, called the line element, is defined by Eq. (5). This quantity

$$ds^2 = g_{\mu\nu}dx^\mu dx^\nu \tag{5}$$

is a generalization of the space-time interval, I, in special relativity.

Tensor equations are equations in which one tensor of a given type is set equal to another of the same type. The field equations of general relativity are tensor equations for the metric tensor, which completely describes the geometry of the space. The Riemann tensor (or curvature tensor), $R^\alpha_{\mu\beta\nu}$, plays a central role in the geometric structure of a space; if it is zero, the space is termed flat and has no gravitational field; if nonzero, the space is termed curved, and a gravitational field is present. In terms of the contracted Riemann tensor, that is, a Riemann tensor summed over $\alpha = \beta$, the Einstein field equations for empty space are given by Eq. (6).

$$\sum_\alpha R^\alpha_{\mu\alpha\nu} \equiv R_{\mu\nu} = 0 \qquad \text{(empty space)} \tag{6}$$

The field equations are a set of 10 second-order partial differential equations since the four-by-four symmetric tensor $R^{\mu\nu}$ has 10 independent components; they are to be solved for the metric tensor. A solution in a given coordinate system defines an Einstein space-time. The curvature of this space corresponds to the intrinsic presence of a gravitational field. Thus the concept of a field of mechanical force in classical gravitational theory is replaced by the geometric concept of curved space in relativity theory.

In a nonempty region of space the field equations (6) must be modified to include a tensor representing the matter or energy content of space, the energy-momentum tensor $T_{\mu\nu}$. The modified equations are Eq. (7), where G is the gravitational constant,

$$G_{\mu\nu} \equiv R_{\mu\nu} - {}^1\!/_2 g_{\mu\nu} \sum_\alpha R^\alpha_\alpha = -\frac{8\pi G}{c^4} T_{\mu\nu} \qquad \text{(nonempty space)} \tag{7}$$

equal to 6.67×10^{-11} N · m² · kg⁻². On the left is the tensor $G_{\mu\nu}$ representing the geometry of space, and on the right is the tensor $T_{\mu\nu}$ representing the mass or energy content of space. The tensor $G_{\mu\nu}$ defined in Eq. (7) is called the Einstein tensor. These equations automatically imply the conservation of energy and momentum, which is an extremely important result. Moreover, in the limiting case when the mass densities of all the gravitating bodies are small and their velocities are small compared to c, the equations reduce to the classical newtonian equations of gravity.

Cosmological term. The field equations were given in the form of Eq. (7) by Einstein in 1916. However, they can be consistently generalized by the addition of another term on the left side, which he called the cosmological term, $\Lambda g_{\mu\nu}$. The more general equations are (8). The constant Λ is called the cosmological constant.

$$G_{\mu\nu} + \Lambda g_{\mu\nu} = -\frac{8\pi G}{c^4} T_{\mu\nu} \tag{8}$$

Einstein introduced the cosmological term in 1917 in order to obtain mathematical models of the universe that were independent of time, since it was then believed that the universe was static. When it was discovered in 1929 that the universe is expanding, as evidenced by the Doppler shifts of distant galaxies, Einstein abandoned the cosmological term. However, interest in the cosmological constant has revived in connection with a serious inconsistency between relativity and quantum theory involving the quantum energy of the vacuum, and with observations since 1998 of type Ia supernovae which suggest that the expansion of the universe is accelerating.

Motion of test bodies. The path of a test body is a generalization of a straight line in euclidean space; it is the shortest "distance" (in terms of intervals ds) between points in space-time, known as a geodesic. General relativity theory possesses an extraordinary property: because the field equations are nonlinear, unlike those of newtonian theory, the motion of a test body in a gravitational field is not arbitrary since the body itself has mass and contributes to the field. Indeed, the field equations are so restrictive that the geodesic equation of motion is a necessary consequence and need not be treated as a separate postulate.

Schwarzschild solution. A very important solution of the field equations was obtained by K. Schwarzschild in 1916, surprisingly soon after the inception of general relativity. This solution represents the field in free space around a spherically symmetric body such as the Sun. It is the basis for a relativistic description of the solar system and most of the experimental tests of general relativity which have been carried out.

Gravitational redshift. Electromagnetic radiation of a given frequency emitted in a gravitational field will appear to an outside observer to have a lower frequency; that is, it will be redshifted. The redshift can be derived from the principle of equivalence. The most accurate test of the redshift to date was performed using a hydrogen maser on a rocket. Comparison of the maser frequency with Earth-based masers gave a measured redshift in agreement with theory to about 1 part in 10^4.

Perihelion shift. The equations of motion can be solved for a planet considered as a test body in the Schwarzschild field of the Sun. As should be expected, the orbits obtained are very similar to the ellipses of classical theory. However, the ellipse rotates very slowly in the plane of the orbit so that the perihelion, the point of closest approach of the planet to the Sun, is at a slightly different angular position for each orbit. This shift is extremely small. It is greatest in the case of the planet Mercury, whose perihelion advance is predicted to be 43 seconds of arc in a century. This agrees with the discrepancy between classical theory and observation, which was well known for many years before the discovery of general relativity.

Deflection of light. The principle of equivalence suggests an extraordinary phenomenon of gravity. Light or other electromagnetic radiation crossing the Einstein elevator horizontally will appear to be deflected downward in a parabolic arc because of the upward acceleration of the elevator. The same phenomenon must occur for light in the gravitational field of the Sun; it must be deflected toward the Sun. A calculation of this deflection gives 1.75 seconds for the net deflection of starlight grazing the edge of the Sun. Modern measurements, made by tracking quasars as they pass near or behind the Sun, find the deflection to be within 1% of the value predicted by general relativity.

In 1936 Einstein observed that if two stars were exactly lined up with the Earth, the more distant star would appear as a ring of light, distorted from its point appearance by the lens effect of the gravitational field of the nearer star. It was soon pointed out that a very similar phenomenon was much more likely to occur for entire galaxies instead of individual stars. Many candidates for such gravitational lens systems have been found.

Radio time delay. In the curved space around the Sun the distance between points in space, for example between two planets, is not the same as it would be in flat space. In particular, the round-trip travel time of a radar signal sent between the Earth and the planet will be measurably increased by the curvature effect when the Earth, the Sun, and the planet are approximately lined up. Using a transponder on the Viking spacecraft, the time delay was found to agree with the predictions of general relativity to an accuracy of about one-half of 1%.

[R.J.A.]

Relaxation time of electrons   The characteristic time for a distribution of electrons in a solid to approach or "relax" to equilibrium after a disturbance is removed. A familiar example is the property of electrical conductivity, in which an applied electric field generates an electron current which relaxes to an equilibrium zero current after the field is turned off. The conductivity of a material is directly proportional to this relaxation time; highly conductive materials have relatively long relaxation times. The closely related concept of a lifetime is the mean time that an electron will reside in a given quantum state before changing state as a result of collision with another particle or intrinsic excitation. This lifetime is related to equilibrium properties of the material, whereas the relaxation time relates to the thermal and electrical transport properties. The average distance that an electron travels before a collision is called the mean free path. Although typical collision times in metals are quite short (on the order of 10^{-14} s at room temperature), mean free paths range from about 100 atomic distances at room temperature to 10^6 atomic distances in pure metals near absolute zero temperature. Considering the very dense packing of atoms in a solid, these surprisingly long electron path lengths are analogous to the unlikely event that a rifle bullet might travel for miles through a dense forest without hitting a tree. The detailed explanation of the electron mean free path in metals is a major success of the modern theory of solids.

A relaxation time appears in the simplest expression for the transport property of electrical conductivity, which states that the electrical conductivity equals the product of the relaxation time, the density of conduction electrons, and the square of the electron charge, divided by the electron effective mass in the solid. *See* BAND THEORY OF SOLIDS.

The conduction process is a steady-state balance between the accelerating force of an electric field and the decelerating friction of electron collisions which occur on the time scale of the relaxation time. This process may be described in terms of the probability distribution function for the electrons, which depends on the electron momentum (proportional to the wave vector, **k**, which labels the quantum state of the electrons), the position, and the time. Viewed in **k**-space, the entire distribution will shift from an equilibrium state under the influence of a perturbation such as an electric field (see illustration). For example, in the ground state, the collection of occupied electron states in **k**-space is bounded by the Fermi surface centered at the origin, while in an electric field this region is shifted. Because of electron collisions with impurities, lattice imperfections, and vibrations (also called phonons), the displaced surface may be maintained in a steady state in an electric field. These collisions also restore the equilibrium distribution

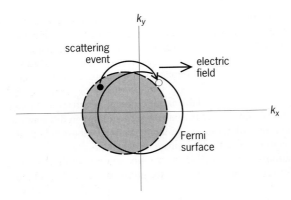

k_y

scattering
event

electric
field

k_x

Fermi
surface

**Effect of the electric field on
electron distribution in a solid,
viewed in k-space, where k is
the electron wave vector. The
shaded area indicates occupied
states in distribution which
result when the field is applied.**

after the field is turned off, and the relaxation time is determined by the rate at which the shifted distribution returns to equilibrium. Specifically, the contribution of collisions to the rate of change of the shifted distribution after the field is turned off equals the difference between the shifted and equilibrium distributions divided by the relaxation time. This statement is referred to as the relaxation-time approximation, and is a simple way of expressing the role of collisions in the maintenance of thermodynamic equilibrium. The details of the various collision mechanisms are lumped into the parameter of the relaxation time. For example, in the case of mixed scattering by impurities and phonons, the inverse of the relaxation time can be determined from the sum of collision rates as the sum of the inverses of the electron-impurity and electron-phonon scattering times. *See* CRYSTAL DEFECTS; FERMI SURFACE; LATTICE VIBRATIONS; PHONON.

In pure metals at low temperatures, the long mean free path of conduction electrons results from their large velocity (on the order of 10^6 m/s near the Fermi surface) and relatively long relaxation time, on the order of 10^{-9} s. From a practical standpoint, this is what makes metals useful as electrical conductors even at room temperature, where a relaxation time on the order of 10^{-14} s and a mean free path (equal to the product of the velocity and the relaxation time) on the order of 10^{-8} m (or about 100 atomic distances) is typical.

As compared to poorly conducting solids or insulators, an excess of so-called free electrons in a metal contributes to the long mean free paths. From a quantum mechanical standpoint, the free-electron wave function readjusts in a perfectly periodic atomic lattice to avoid the atomic ion cores and spend most of the time in the spaces between. In the analogy of the rifle fired into a forest, the bullet will not travel far, but the sound of the gunshot can, because the sound waves bend around the trees in a constructive manner. In a perfectly periodic lattice, the electron waves scatter constructively from the atomic ion cores, resulting in screening of the cores and coherent transmission of the waves over large distances. Any disturbance to the lattice periodicity tends to destroy this wave phenomenon, resulting in lower transmission or energy loss. At room temperature, the conductivity is usually limited by scattering from lattice vibrations (phonons). At lower temperatures, vibrations are greatly reduced, but the conductivity is still limited by scattering from impurities and imperfections. *See* ELECTRICAL RESISTIVITY; FREE-ELECTRON THEORY OF METALS; QUANTUM MECHANICS.

In the superconductive state, observed in many metals and certain complex compounds at sufficiently low temperatures, impurities, imperfections, and lattice vibrations become completely ineffective in retarding current flow, leading to persistent electric currents even after the driving electric field is removed. This state of resistanceless conduction is described as an ordered quantum state of pairs of electrons resulting from lattice vibrations which deform the ion core potential in such a way as to provide an attractive electron-electron interaction. The relaxation time thus becomes infinite (as nearly as can be measured), a rare macroscopic manifestation of a quantum effect. *See* SUPERCONDUCTIVITY.

[G.L.Ee.]

Reluctance   A property of a magnetic circuit analogous to resistance in an electric circuit.

Every line of magnetic flux is a closed path. Whenever the flux is largely confined to a well-defined closed path, there is a magnetic circuit. That part of the flux that departs from the path is called flux leakage.

For any closed path of length l in a magnetic field H, the line integral of $H \cos \alpha\, dl$ around the path is the magnetomotive force (mmf) of the path, as in Eq. (1), where

$$\mathrm{mmf} = \oint H \cos \alpha\, dl \qquad (1)$$

α is the angle between H and the path. If the path encloses N conductors, each with current I, Eq. (2) holds.

$$\mathrm{mmf} = \oint H \cos \alpha\, dl = NI \qquad (2)$$

Consider the closely wound toroid shown in the illustration. For this arrangement of currents, the magnetic field is almost entirely within the toroidal coil, and there the

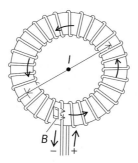

A toroidal coil.

flux density or magnetic induction B is given by Eq. (3), where l is the mean circumference of the toroid and μ is the permeability. The flux Φ within the toroid of cross-sectional area A is given by either form of Eqs. (4), which is similar in form to the

$$B = \mu \frac{NI}{l} \qquad (3)$$

$$\Phi = BA = \frac{\mu A}{l} NI$$
$$\Phi = \frac{NI}{l/\mu A} = \frac{\mathrm{mmf}}{l/\mu A} = \frac{\mathrm{mmf}}{\mathscr{R}} \qquad (4)$$

equation for the electric circuit, although nothing actually flows in the magnetic circuit. The factor $l/\mu A$ is called the reluctance $\mathscr{R}$ of the magnetic circuit. The reluctance is not constant because the permeability μ varies with changing flux density.   [K.V.M.]

Renner-Teller effect   The splitting, into two, of the potential function along the bending coordinate in degenerate electronic states of linear triatomic or polyatomic molecules. Most of the areas and methods of molecular physics and spectroscopy assume the validity of the Born-Oppenheimer approximation. The nuclei generally move much more slowly than the electrons, the frequencies associated with electronic transitions are much higher than vibrational frequencies, and one can consider separately the three types of molecular motion: electronic, vibrational, and rotational. These statements are no longer necessarily valid for electronic states which are degenerate or at least close to degeneracy, and the Born-Oppenheimer approximation breaks down.

Degenerate electronic states usually occur in molecules having a high degree of symmetry. The symmetric equilibrium geometry which causes the electronic degeneracy is, in general, lowered in the course of molecular vibrations, and this may lead to splitting of the potential. The molecular potential is usually expressed in terms of a polynomial expansion in displacements r, and, in nonlinear molecules, the linear terms may lead to coupling of the electronic and vibrational degrees of freedom. The resulting breakdown of the Born-Oppenheimer approximation is in this case known as the Jahn-Teller effect. In linear molecules the symmetry is lowered during bending vibrations. In the bending potential the linear (and other odd) terms are zero by symmetry. The first nonvanishing terms which can couple the degenerate electronic states are quadratic in the bending coordinate. The results of this coupling in linear molecules are referred to as the Renner-Teller effect, or simply the Renner effect. *See* JAHN-TELLER EFFECT; MOLECULAR STRUCTURE AND SPECTRA.   [V.E.B.; T.A.M.]

Renormalization   A program in quantum field theory consisting of a set of rules for calculating S-matrix amplitudes which are free of ultraviolet (or short-distance) divergences, order by order in perturbative calculations in an expansion with respect to coupling constants. *See* SCATTERING MATRIX.

So far the only field theories known to be renormalizable in four dimensions are those which include spin-0, spin-$1/2$, and spin-1 fields such that no term in the lagrangian exceeds operator dimension 4. The operator dimension of any term is calculated by assigning dimension 1 to bosons and derivatives ∂_μ, and dimension $3/2$ to fermions. Spin-1 fields are allowed only if they correspond to the massless gauge potentials of a locally gauge-invariant Yang-Mills-type theory associated with any compact Lie group. The gauge invariance can remain exact or can be allowed to break via spontaneous breakdown without spoiling the renormalizability of the theory. In the latter case the spin-1 field develops a mass. The successful quantum chromodynamics theory describing the strong forces and the SU(2) × U(1) Weinberg-Salam-Glashow gauge model of unified electroweak particle interactions are such renormalizable gauge models containing spin 0, $1/2$, and 1 fields. *See* ELECTROWEAK INTERACTION; FUNDAMENTAL INTERACTIONS; QUANTUM CHROMODYNAMICS; QUANTUM ELECTRODYNAMICS; WEAK NUCLEAR INTERACTIONS.
                                                                              [I.Ba.]

Effective field theory is a general and powerful method for analyzing quantum field theories over a wide range of length scales. Together with a closely related idea, the Wilson renormalization group, it places renormalization theory on a more general, physical, and rigorous basis. This method is most naturally developed in the Feynman path integral formulation of quantum field theory, where amplitudes are given by an integral over all histories. Each history is weighted by a phase equal to the classical action divided by Planck's constant. *See* ACTION.   [J.Pol.]

Resistance measurement   The quantitative determination of that property of an electrically conductive material, component, or circuit called electrical resistance. The ohm, which is the International System (SI) unit of resistance, is defined through the application of Ohm's law as the electric resistance between two points of a conductor when a constant potential difference of 1 volt applied to these points produces in the conductor a current of 1 ampere. Ohm's law can thus be taken to define resistance R as the ratio of dc voltage V to current I, Eq. (1). For bulk metallic conductors, for

$$R = \frac{V}{I} \tag{1}$$

example, bars, sheets, wires, and foils, this ratio is constant. For most other substances, such as semiconductors, ceramics, and composite materials, it may vary with voltage, and many electronic devices depend on this fact. The resistance of any conductor is given by the integral of expression (2), where l is the length, A the cross-sectional area,

$$\int_0^l \frac{\rho \, dl}{A} \tag{2}$$

and ρ the resistivity. See ELECTRICAL RESISTANCE; ELECTRICAL RESISTIVITY; OHM'S LAW; SEMICONDUCTOR.

Since January 1, 1990, all resistance measurements worldwide have been referred to the quantized Hall resistance standard, which is used to maintain the ohm in all national standards laboratories. Conventional wire-wound working standards are measured in terms of the quantized Hall resistance and then used to disseminate the ohm through the normal calibration chain. These working standards can be measured in terms of the quantized Hall resistance with a one-standard-deviation uncertainty of about 1 part in 10^8. See HALL EFFECT.

The value of an unknown resistance is determined by comparison with a standard resistor. The Wheatstone bridge is perhaps the most basic and widely used resistance- or impedance-comparing device. Its principal advantage is that its operation and balance are independent of variations in the supply. The greatest sensitivity is obtained when all resistances are similar in value, and the comparison of standard resistors can then be made with a repeatability of about 3 parts in 10^8, the limit arising from thermal noise in the resistors. In use, the direction of supply is reversed periodically to eliminate effects of thermal or contact emf's.

The bridge is normally arranged for two-terminal measurements, and so is not suitable for the most accurate measurement at values below about 100 Ω, although still very convenient for lower resistances if the loss of accuracy does not matter. However, a Wheatstone bridge has also been developed for the measurement of four-terminal resistors. This involves the use of auxiliary balances, and resistors of the same value can be compared with uncertainties of a few parts in 10^8.

Typically a bridge will have two decade-ratio arms, for example, of 1, 10, 100, 1000, and 10,000 Ω, and a variable switched decade arm of 1–100,000 Ω, although many variations are encountered. For the measurement of resistors of values close to the decade values, a considerable increase in accuracy can be obtained by substitution measurement, in which the bridge is used only as an indicating instrument. The resistors being compared can be brought to the same value by connecting a much higher variable resistance across the larger of them, and the accuracy of this high-resistance shunt can be much less than that of the resistance being compared. See WHEATSTONE BRIDGE.

The Kelvin double bridge is a double bridge for four-terminal measurements, and so can be used for very low resistances. The addition to its use for accurate laboratory measurement of resistances below 100 Ω, it is very valuable for finding the resistance of conducting rods or bars, or for the calibration in the field of air-cooled resistors used for measurement of large currents. See KELVIN BRIDGE.

Measurements of resistances from 10 megohms to 1 terohm (10^{12} Ω) or even higher with a Wheatstone bridge present additional problems. The resistance to be measured will usually be voltage-dependent, and so the measurement voltage must be specified. The resistors in the ratio arms must be sufficiently high in value that they are not overloaded. If a guard electrode is fitted, it is necessary to eliminate any current flowing to the guard from the measurement circuit. The power dissipated in the 1-MΩ resistor is then 10 mW, and the bridge ratio is 10^6. The guard is connected to a subsidiary divider of the same ratio, so that any current flowing to it does not pass through the detector. Automated measurements can be made by replacing the ratio arms of the Wheatstone bridge by programmable voltage sources. An alternative method that can also be automated is to measure the RC time constant of the unknown resistor R combined with a capacitor of known value C.

An obvious and direct way of measuring resistance is by the simultaneous measurement of voltage and current, and this is usual in very many indicating ohmmeters and multirange meters. In most digital instruments, which are usually also digital voltage meters, the resistor is supplied from a constant-current circuit and the voltage across it is measured by the digital voltage meter. This is a convenient arrangement for a four-terminal measurement, so that long leads can be used from the instrument to the resistor without introducing errors. The simplest systems, used in passive pointer instruments, measure directly the current through the meter which is adjusted to give full-scale deflection by an additional resistor in series with the battery. This gives a nonlinear scale of limited accuracy, but sufficient for many practical applications. *See* CURRENT MEASUREMENT; VOLTAGE MEASUREMENT.   [C.H.Di.; R.G.Jon.]

Resistor   One of the three basic passive components of an electric circuit that displays a voltage drop across its terminals and produces heat when an electric current passes through it. The electrical resistance, measured in ohms, is equal to the ratio of the voltage drop across the resistor terminals measured in volts divided by the current measured in amperes. *See* OHM'S LAW.

Resistors are described by stating their total resistance in ohms along with their safe power-dissipating ability in watts. The tolerance and temperature coefficient of the resistance value may also be given. *See* ELECTRICAL RESISTANCE; ELECTRICAL RESISTIVITY.

All resistors possess a finite shunt capacitance across their terminals, leading to a reduced impedance at high frequencies. Resistors also possess inductance, the magnitude of which depends greatly on the construction and is largest for wire-wound types. *See* CAPACITANCE; ELECTRICAL IMPEDANCE; INDUCTANCE.

Resistors may be classified according to the general field of engineering in which they are used. Power resistors range in size from about 5 W to many kilowatts and may be cooled by air convection, air blast, or water. The smaller sizes, up to several hundred watts, are used in both the power and electronics fields of engineering.

Direct-current (dc) ammeters employ resistors as meter shunts to bypass the major portion of the current around the low-current elements. These high-accuracy, four-terminal resistors are commonly designed to provide a voltage drop of 50–100 mV when a stated current passes through the shunt. *See* AMMETER.

Voltmeters of both the dc and the ac types employ scale-multiplying resistors designed for accuracy and stability. The arc-over voltage rating of these resistors is of importance in the case of high-voltage voltmeters. *See* VOLTMETER.

Standard resistors are used for calibration purposes in resistance measurements and are made to be as stable as possible, in value, with time, temperature, and other influences. Resistors with values from 1 ohm to 10 megohms are wound by using wire made from special alloys. The best performance is obtained from quaternary alloys, which contain four metals. The proportions are chosen to give a shallow parabolic variation of resistance with temperature, with a peak, and therefore the slowest rate of change, near room temperature. *See* ELECTRICAL UNITS AND STANDARDS.

By far the greatest number of resistors manufactured are intended for use in the electronics field. The major application of these resistors is in transistor analog and digital circuits which operate at voltage levels between 0.1 and 200 V, currents between 1 μA and 100 mA, and frequencies from dc to 100 MHz. Their power-dissipating ability is small, as is their physical size.

Since their exact value is rarely important, resistors are supplied in decade values (0.1, 1, 10, 100 ohms, and so forth) with the interval between these divided into a geometric series, thus having a constant percentage increase. For noncritical applications, values from a series with intervals of 20% (12 per decade) are appropriate. A series with 10% intervals (24 per decade) is often used for resistors having a tolerance of 1%. Where the precise value of a resistor is important, a series with 2.5% intervals (96 per decade) may be used.

Resistors are also classified according to their construction, which may be composition, film-type, wire-wound, or integrated circuit.

The composition resistor is in wide use because of its low cost, high reliability, and small size. Basically it is a mixture of resistive materials, usually carbon, and a suitable binder molded into a cylinder. Copper wire leads are attached to the ends of the cylinder, and the entire resistor is molded into a plastic or ceramic jacket. Composition resistors are commonly used in the range from several ohms to 10–20 MΩ, and are available with tolerances of 20, 10, or 5%.

The film-type resistor is now the preferred type for most electronic applications because its performance has surpassed that of composition resistors and mass-production techniques have reduced the cost to a comparable level. Basically this resistor consists of a thin conducting film of carbon, metal, or metal oxide deposited on a cylindrical ceramic or glass former. The resistance is controlled by cutting a helical groove through the conducting film. This helical groove increases the length and decreases the width of the conducting path, thereby determining its ohmic value. By controlling the conductivity, thickness of the film, and pitch of the helix, resistors over a wide range of values can be manufactured. Film construction is used for very high value resistors, up to and even beyond 1 TΩ (10^{12} ohms).

Wire remains the most stable form of resistance material available; therefore, all high-precision instruments rely upon wire-wound resistors. Wire also will tolerate operation at high temperatures, and so compact high-power resistors use this construction. Power resistors are available in resistance values from a fraction of an ohm to several hundred thousand ohms, at power ratings from one to several thousand watts, and at tolerances from 10 to 0.1%. The usual design of a power resistor is a helical winding of wire on a cylindrical ceramic former. After winding, the entire resistor is coated in vitreous enamel. Alternatively, the wound element may be fitted inside a ceramic or metal package, which will assist in heat dissipation. The helical winding results in the resistor having significant inductance, which may become objectionable at the higher audio frequencies and all radio frequencies. Precision wire-wound resistors are usually wound in several sections on ceramic or plastic bobbins and are available in the range from 0.1 Ω to 10 MΩ.

Integrated circuit resistors must be capable of fabrication on a silicon integrated circuit chip along with transistors and capacitors. There are two major types: thin-film resistors and diffused resistors. Thin-film resistors are formed by vacuum deposition or sputtering of nichrome, tantalum, or Cermet (Cr-SiO). Such resistors are stable, and the resistance may be adjusted to close tolerances by trimming the film by using a laser. Typical resistor values lie in the range from 100 Ω to 10 kΩ with a matching tolerance of $\pm 0.2\%$ and a temperature coefficient of resistance of ± 10 to ± 200 ppm/°C.

Diffused resistors are based upon the same fabrication geometry and techniques used to produce the active transistors on the silicon chip or die. A diffused base, emitter, or epitaxial layer may be formed as a bar with contacts at its extremities. The resistance of such a semiconductor resistor depends upon the impurity doping and the length and cross section of the resistor region. In the case of the base-diffused resistor, the emitter and collector regions may be formed so as to pinch the base region to a very small cross-sectional area, thereby appreciably increasing the resistance. The relatively large impurity carrier concentration in n- and p-type regions limits the resistance value. Resistor values between $100\ \Omega$ and $10\ \mathrm{k}\Omega$ are common.

The deposited-film and wire-wound resistors lend themselves to the design of adjustable resistors or rheostats and potentiometers. Adjustable-slider power resistors are constructed in the same manner as any wire-wound resistor on a cylindrical form except that when the vitreous outer coating is applied an uncovered strip is provided. The resistance wire is exposed along this strip, and a suitable slider contact can be used to adjust the overall resistance, or the slider can be used as the tap on a potentiometer. *See* POTENTIOMETER.   [R.B.D.K.]

Resolving power (optics)   A quantitative measure of the ability of an optical instrument to produce separable images. The images to be resolved may differ in position because they represent (1) different points on the object, as in telescopes and microscopes, or (2) images of the same object in light of two different wavelengths, as in prism and grating spectroscopes. For the former class of instruments, the resolving limit is usually quoted as the smallest angular or linear separation of two object points, and for the latter class, as the smallest difference in wavelength or wave number that will produce separate images. Since these quantities are inversely proportional to the power of the instrument to resolve, the term resolving power has generally fallen into disfavor. It is still commonly applied to spectroscopes, however, for which the term chromatic resolving power is used, signifying the ratio of the wavelength itself to the smallest wavelength interval resolved. The figure quoted as the resolving power or resolving limit of an instrument may be the theoretical value that would be obtained if all optical parts were perfect, or it may be the actual value found experimentally. Aberrations of lenses or defects in the ruling of gratings usually cause the actual resolution to fall below the theoretical value, which therefore represents the maximum that could be obtained with the given dimensions of the instrument in question. This maximum is fixed by the wave nature of light and may be calculated for given conditions by diffraction theory. *See* DIFFRACTION; OPTICAL IMAGE.   [F.A.J.; G.R.H.]

Resonance (acoustics and mechanics)   When a mechanical or acoustical system is acted upon by an external periodic driving force whose frequency equals a natural free oscillation frequency of the system, the amplitude of oscillation becomes large and the system is said to be in a state of resonance.

A knowledge of both the resonance frequency and the sharpness of resonance is essential to any discussion of driven vibrating systems. When a vibrating system is sharply resonant, careful tuning is required to obtain the resonance condition. Mechanical standards of frequency must be sharply resonant so that their peak response can easily be determined. In other circumstances, resonance is undesirable. For example, in the faithful recording and reproduction of musical sounds, it is necessary either to have all vibrational resonances of the system outside the band of frequencies being reproduced or to employ heavily damped systems. *See* ACOUSTIC RESONATOR; SYMPATHETIC VIBRATION; VIBRATION.

[L.E.K.]

Resonance (quantum mechanics)

An enhanced coupling between quantum states with the same energy. The concept of resonance in quantum mechanics is closely related to resonances in classical physics. *See* RESONANCE (ACOUSTICS AND MECHANICS).

The matching of frequencies is central to the concept of resonance. An example is provided by waves, acoustic or electromagnetic, of a spectrum of frequencies propagating down a tube or waveguide. If a closed side tube is attached, its characteristic natural frequencies will couple and resonate with waves of those same frequencies propagating down the main tube. This simple illustration provides a description of all resonances, including those in quantum mechanics. The propagation of all quantum entities, whether electrons, nucleons, or other elementary particles, is represented through wave functions and thus is subject to resonant effects. *See* ACOUSTIC RESONATOR; HARMONIC (PERIODIC PHENOMENA).

An important allied element of quantum mechanics lies in its correspondence between frequency and energy. Instead of frequencies, differences between allowed energy levels of a system are considered. In the presence of degeneracy, that is, of different states of the system with the same energy, even the slightest influence results in the system resonating back and forth between the degenerate states. These states may differ in their internal motions or in divisions of the system into subsystems. The above example of wave flow suggests the terminology of channels, each channel being a family of energy levels similar in other respects. These energies are discretely distributed for a closed channel, whereas a continuum of energy levels occurs in open channels whose subsystems can separate to infinity. If all channels are closed, that is, within the realm of bound states, resonance between degenerate states leads to a theme of central importance to quantum chemistry, namely, stabilization by resonance and the resulting formation of resonant bonds. *See* DEGENERACY (QUANTUM MECHANICS); ENERGY LEVEL (QUANTUM MECHANICS).

Resonances occur in scattering when at least one channel is closed and one open. Typically, a system is divided into two parts: projectile + target, such as electron + atom or nucleon + nucleus. One channel consists of continuum states with their two parts separated to infinity. The other, closed channel consists of bound states. In the atomic example, a bound state of the full system would be a state of the negative ion and, in the nuclear example, a state of the larger nucleus formed by incorporating one extra nucleon in the target nucleus. *See* QUANTUM MECHANICS; SCATTERING EXPERIMENTS (ATOMS AND MOLECULES); SCATTERING EXPERIMENTS (NUCLEI).

[A.R.P.R.]

Resonance ionization spectroscopy

A form of atomic and molecular spectroscopy in which wavelength-tunable lasers are used to remove electrons from (that is, ionize) a given kind of atom or molecule. Laser-based resonance ionization spectroscopy (RIS) methods have been developed and used with ionization detectors, such as proportional counters, to detect single atoms. Resonance ionization spectroscopy is combined with mass spectrometers to provide analytical systems for a wide range of applications, including physics, chemistry, materials sciences, medicine, and the environmental sciences.

When an atom or molecule is irradiated with a light source of frequency ν, photons at this selected frequency are absorbed only when the energy $h\nu$ (h is Planck's constant) is almost exactly the same as the difference in energy between some excited state and the ground state of the atom or molecule. If a laser source is tuned to a very

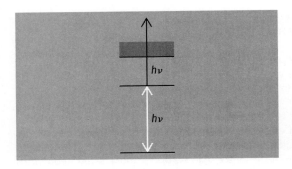

Basic laser scheme for resonance ionization spectroscopy. The atom or molecule is irradiated by a light source with frequency ν and photons of energy $h\nu$, where h is Planck's constant.

narrow bandwidth at a frequency that excites a given kind of atom (see illustration), it is highly unlikely that any other kind of atom will be excited. An atom in an excited state can be ionized by photons of the specified frequency ν, provided that $2h\nu$ is greater than the ionization potential of the atom. While the final ionization step can occur with any energy above a threshold, the entire process of ionization is a resonance one. Resonance ionization spectroscopy is a selective process in which only those atoms that are in resonance with the light source are ionized. Modern pulsed lasers have made resonance ionization spectroscopy a practical method for the sensitive (and highly selective) detection of nearly every type of atom in the periodic table. See ATOMIC STRUCTURE AND SPECTRA; IONIZATION POTENTIAL; LASER; LASER SPECTROSCOPY; PHOTOIONIZATION; RESONANCE (QUANTUM MECHANICS).

Resonance ionization spectroscopy is used to analyze very low levels of trace elements in extremely pure materials, for example, semiconductors in the electronics industry. A sputter-initiated resonance ionization spectroscopy (SIRIS) apparatus uses an argon ion beam to sputter a tiny cloud of atoms from a sample placed in a high-vacuum system and a pulsed laser tuned to detect the specified impurity atom.

The sputter-initiated resonance ionization spectroscopy method is also used for chemical and materials research, geophysical research and explorations, medical diagnostics, biological research, and environment analysis. Thermal-atomization resonance ionization spectroscopy (TARIS) may be used for the bulk analysis of materials. By simply using resonance ionization spectroscopy with ionization chambers or proportional counters, gas-phase work can be done to study the diffusion of atoms, measure chemical reaction rates, and investigate the statistical behavior of atoms and molecules. See DIFFUSION.

Resonance ionization spectroscopy is used in sophisticated nuclear physics studies involving high-energy accelerators. It is used as an on-line detector to record the hyperfine structure of nuclei with short lifetimes and hence to determine several nuclear properties such as nuclear spin and the shape of nuclei. See FINE STRUCTURE (SPECTRAL LINES); NUCLEAR STRUCTURE.

Resonance ionization spectroscopy is used for measurements of krypton-81 in the natural environment to determine the ages of polar ice caps and old ground-water deposits. Oceanic circulation and the mixing of oceans could also be studied by measuring the concentrations of noble-gas isotopes by resonance ionization spectroscopy.   [G.S.H.]

Rest mass   A constant intrinsic to a body which determines its inertial and energy-momentum properties. It is a fundamental concept of special relativity, and in particular it determines the internal energy content of a body. It is the same as the inertial mass of classical mechanics. According to the principle of equivalence, the basic physical principle of general relativity, the inertial mass of a body is also equal to its gravitational mass. *See* CLASSICAL MECHANICS; GRAVITATION; RELATIVITY.

The rest mass or inertial mass of a body, m, is a measure of its resistance to being accelerated at a by **a** force **F**; in classical mechanics the relation between inertial mass, acceleration, and force is given by Newton's law, Eq. (1). In special relativity Newton's

$$\mathbf{F} = m\mathbf{a} \tag{1}$$

law holds exactly only in the body's rest frame, that is, the frame in which the body is instantaneously at rest. *See* NEWTON'S LAWS OF MOTION.

Associated with the rest mass of a body, there is an internal or rest energy. In the system where the body is at rest, the energy of the body is given by Eq. (2).

$$E = mc^2 \qquad \text{(body at rest)} \tag{2}$$

The experimental realization of the interconversion of mass and energy is accomplished in the reactions of nuclei and elementary particles. In particular, the energy source of nuclear bombs and nuclear fission reactors is a small decrease in the total mass of the interacting nuclei, which gives rise to a large energy release because of the large numerical value of c^2. *See* ELEMENTARY PARTICLE; NUCLEAR FISSION.   [R.J.A.]

Resultant of forces

A system of at most a single force and a single couple whose external effects on a rigid body are identical with the effects of the several actual forces that act on the body. For analytic purposes, forces are grouped and replaced by their resultant. Forces can be added graphically (see illustration) or analytically. The sum of more than two vector forces can be found by extending the method of illus. *c* to a three-dimensional vector polygon in which one force is drawn from the tip of the previous one until all are laid out.

The resultant force is the force vector required to close the polygon directed from the tail of the first force vector to the tip of the last. A force system has a zero force resultant if its vector polygon closes.

Two force systems are equivalent if their resultant forces, as described above, are equal and if their total vector moments about the same point are also equal. Vector moments are combined in the same manner as forces, that is, by parallelograms,

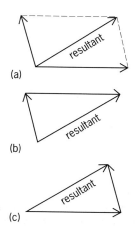

Resultant of two forces acting through a common center. (*a*) Diagonal of parallelogram. (*b*, *c*) Hypotenuse of triangle.

triangles, or polygons. A resultant is the equivalent force system having the fewest possible forces and couples. *See* COUPLE; FORCE; STATICS.   [N.S.F.]

Reverberation   After sound has been produced in, or enters, an enclosed space, it is reflected repeatedly by the boundaries of the enclosure, even after the source ceases to emit sound. This prolongation of sound after the original source has stopped is called reverberation. A certain amount of reverberation adds a pleasing characteristic to the acoustical qualities of a room. However, excessive reverberation can ruin the acoustical properties of an otherwise well-designed room. A typical record representing the sound-pressure level at a given point in a room plotted against time, after a sound source has

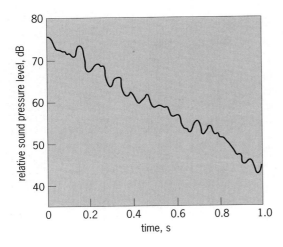

Typical decay curve illustrating reverberation.

been turned off, is given in the decay curve shown in the illustration. The rate of sound decay is not uniform but fluctuates about an average slope. *See* SOUND.
                                                                                          [C.M.H.]

Reynolds number   In fluid mechanics, the ratio $\rho v d / \mu$, where ρ is fluid density, v is velocity, d is a characteristic length, and μ is fluid viscosity. The Reynolds number is significant in the design of a model of any system in which the effect of viscosity is important in controlling the velocities or the flow pattern. In the evaluation of drag on a body submerged in a fluid and moving with respect to the fluids, the Reynolds number is important.

The Reynolds number also serves as a criterion of type of fluid motion. In a pipe, for example, laminar flow normally exists at Reynolds numbers less than 2000, and turbulent flow at Reynolds numbers above about 3000. *See* DYNAMIC SIMILARITY; FLUID MECHANICS; LAMINAR FLOW; TURBULENT FLOW.   [G.Mu.]

Rheology   In the broadest sense of the term, that part of mechanics which deals with the relation between force and deformation in material bodies. The nature of this relation depends on the material of which the body is constituted. It is customary to represent the deformation behavior of metals and other solids by a model called the linear or hookean elastic solid (displaying the property known as elasticity) and that of fluids by the model of the linear viscous or newtonian fluid (displaying the property known as viscosity). These classical models are, however, inadequate to depict certain

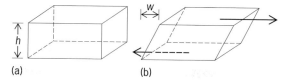

Fig. 1. Simple shear. (*a*) Undeformed block of height *h*. (*b*) Deformed block after top has moved a distance *w* parallel to itself. The arrows indicate the net forces acting on the top and bottom faces. The forces which must be applied to left and right faces to maintain a steady state are not indicated.

nonlinear and time-dependent deformation behavior that is sometimes observed. It is these nonclassical behaviors which are the chief interest of rheologists and hence referred to as rheological behavior. *See* VISCOSITY.

Rheological behavior is particularly readily observed in materials containing polymer molecules which typically contain thousands of atoms per molecule, although such properties are also exhibited in some experiments on metals, glasses, and gases. Thus rheology is of interest not only to mathematicians and physicists, who consider it to be a part of continuum mechanics, but also to chemists and engineers who have to deal with these materials. It is of special importance in the plastics, rubber, film, and coatings industries. *See* FLUID MECHANICS.

Models and properties. Consider a block of material of height h deformed in the manner indicated in Fig. 1; the bottom surface is fixed and the top moves a distance w parallel to itself. A measure of the deformation is the shear strain γ given by Eq. (1).

$$\gamma = \frac{w}{h} \tag{1}$$

To achieve such a deformation if the block is a linear elastic material, it is necessary to apply uniformly distributed tangential forces on the top and bottom of the block as shown in Fig. 1*b*. The intensity of these forces, that is, the magnitude of the net force per unit area, is called the shear stress S. For a linear elastic material, γ is much less than unity and is related to S by Eq. (2), where the proportionality constant G is a

$$S = G\gamma \tag{2}$$

property of the material known as the shear modulus.

If the material in the block is a newtonian fluid and a similar set of forces is imposed, the result is a simple shearing flow, a deformation as pictured in Fig. 1*b* with the top surface moving with a velocity dw/dt. This type of motion is characterized by a rate of shear $\dot{\gamma} = (dw/dt)/h$, which is proportional to the shear stress S as given by Eq. (3), where η is a property of the material called the viscosity.

$$S = \eta\dot{\gamma} \tag{3}$$

Linear viscoelasticity. If the imposed forces are small enough, time-dependent deformation behavior can often be described by the model of linear viscoelasticity. The material properties in this model are most easily specified in terms of simple experiments.

In a creep experiment a stress is suddenly applied and then held constant; the deformation is then followed as a function of time. This stress history is indicated in the

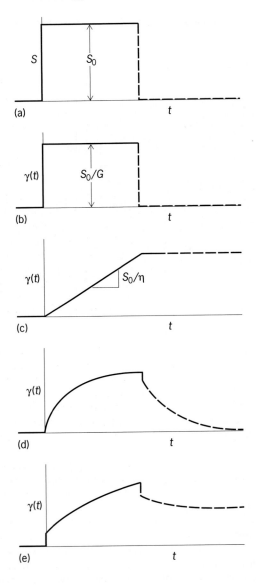

Fig. 2. Creep and recovery; solid lines indicate creep; broken lines indicate recovery. (a) Applied stress history. (b) Corresponding strain history for linear elastic solid, (c) linear viscous fluid, (d) viscoelastic solid, and (e) viscoelastic fluid.

solid line of Fig. 2a for the case of an applied constant shear stress S_0. If such an experiment is performed on a linear elastic solid, the resultant deformation is indicated by the full line in Fig. 2b and for the linear viscous fluid in Fig. 2c. In the case of elasticity, the result is an instantly achieved constant strain; in the case of the fluid, an instantly achieved constant rate of strain. In the case of viscoelastic materials, there are some which eventually attain a constant equilibrium strain (Fig. 2d) and hence are called viscoelastic solids. Others eventually achieve constant rate of strain (Fig. 2e) and are called viscoelastic fluids. If the material is linear viscoelastic, the deformation $\gamma(S_0, t)$ is a function of the time t since the stress was applied and also a linear function of S_0;

that is, Eq. (4) is satisfied, where $J(t)$ is independent of S_0. The function $J(t)$ is a property

$$\gamma(S_0, t) = S_0 J(t) \qquad (4)$$

of the material known as the shear creep compliance.

Nonlinear viscoelasticity. If stresses become too high, linear viscoelasticity is no longer an adequate model for materials which exhibit time-dependent behavior. In a creep experiment, for example, the ratio of the strain to stress, $\gamma(t, S_0)/S_0$, is no longer independent of S_0; this ratio generally decreases with increasing S_0. Two examples of nonlinear viscoelasticity are shear thinning and thixotropy.

For polymer melts, solutions, and suspensions, generally speaking, the viscosity decreases as the shear rate increases. This type of behavior, called shear thinning, is of considerable industrial significance. For example, paints are formulated to be shear-thinning. A high viscosity at low flow rates keeps the paint from dripping from the brush or roller and prevents sagging of the paint film newly applied to a vertical wall. The lower viscosity at the high deformation rates while brushing or rolling means that less energy is required, and hence the painter's arm does not become overly tired.

Thixotropy is a property of suspensions (for example, bentonite clay in water) which, after remaining at rest for a long time, act as solids; for example, they cannot be poured. However, if it is stirred, such a suspension can be poured quite freely. If the suspension is then allowed to rest, the viscosity increases with time and finally sets again. This whole process is reversible; it can be repeated again and again. *See* NON-NEWTONIAN FLUID. [H.Mark.]

Rigid body An idealized extended solid whose size and shape are definitely fixed and remain unaltered when forces are applied. Treatment of the motion of a rigid body in terms of Newton's laws of motion leads to an understanding of certain important aspects of the translational and rotational motion of real bodies without the necessity of considering the complications involved when changes in size and shape occur. Many of the principles used to treat the motion of rigid bodies apply in good approximation to the motion of real elastic solids. *See* RIGID-BODY DYNAMICS. [D.Wi.]

Rigid-body dynamics The study of the motion of a rigid body under the influence of forces. A rigid body is a system of particles whose distances from one another are fixed. The general motion of a rigid body consists of a combination of translations (parallel motion of all particles in the body) and rotations (circular motion of all particles in the body about an axis). Its equations of motion can be derived from the equations of motion of its constituent particles. *See* RECTILINEAR MOTION; ROTATIONAL MOTION.

The location of a mass point m_i can be specified relative to a fixed-coordinate system by a position vector $\vec{r}_i$ with cartesian components (x_i, y_i, z_i). The vector force $\vec{F}_i$ which acts on the mass points has corresponding components (F_{ix}, F_{iy}, F_{iz}). Newton's second law for the motion of m_i is stated in Eq. (1). Here $\ddot{\vec{r}}_i \equiv d^2\vec{r}_i/dt^2 = \vec{a}$ is the acceleration

$$\vec{F}_i = m_i \ddot{\vec{r}}_i \qquad (1)$$

of m_i. *See* ACCELERATION; FORCE; NEWTON'S LAWS OF MOTION.

Translational motion. If Eq. (1) is summed over all particles in the rigid body, the left-hand side becomes the total force

$$\vec{F} = \sum_i \vec{F}_i$$

acting on the rigid body. If the internal forces satisfy Newton's third law (to each action there is an equal but opposite reaction), the contributions of the internal forces cancel in pairs and $\vec{F}$ is the total external force on the rigid body, $\vec{F}^{\text{ext}}$. The right-hand side can be expressed in terms of the center-of-mass (CM) position vector defined by Eq. (2), where M is the total mass of the body. Then the sum of the equations of motion,

$$\vec{R} = \frac{1}{M} \sum_i m_i \vec{r}_i \qquad (2)$$

$$M = \sum_i m_i$$

Eq. (1), takes the form of Eq. (3). This equation of motion for the center of mass of the

$$\vec{F}^{\text{ext}} = M\ddot{\vec{R}} \qquad (3)$$

rigid body has exactly the form of the equation of motion for a particle of mass M and position $\vec{R}$, under the influence of an external force $\vec{F}^{\text{ext}}$. Consequently, the second law of motion holds, not just for a particle, but for an arbitrary rigid body, if the position of the body is interpreted to mean the position of its center of mass. *See* CENTER OF MASS.

The momentum of a mass point is given by the product of the mass and the velocity, $\vec{p}_i = m_i \dot{\vec{r}}_i$, where $\dot{\vec{r}}_i \equiv d\vec{r}_i/dt$. The total momentum of the center of mass of the rigid body, obtained by summing over the momenta $\vec{P}_i$ of its constituent masses, is given by Eq. (4). In terms of the center-of-mass momentum $\vec{P}$, the equation of motion for the center of mass is expressed by Eq. (5). For an isolated rigid body, the external force

$$\vec{P} = M\dot{\vec{R}} \qquad (4)$$

$$\dot{\vec{P}} = \vec{F}^{\text{ext}} \qquad (5)$$

is zero and therefore $\vec{P}$ is constant. According to Eq. (4), this implies that the center of mass moves with constant velocity $\vec{V} = \vec{P}/M$. *See* CONSERVATION OF MOMENTUM; MOMENTUM.

In fact, the preceding equations for translational motion hold for any body, rigid or nonrigid.

Rotational motion. The total angular momentum of a rigid body about a point O with coordinate $\vec{r}_O$ is the sum of the angular momenta of its constituent masses, and is given by Eq. (6). Here × denotes the cross-product of the coordinate vector $(\vec{r}_i - \vec{r}_O)$

$$\vec{L}_O = \sum_i (\vec{r}_i - \vec{r}_O) \times m_i(\dot{\vec{r}}_i - \dot{\vec{r}}_O) \qquad (6)$$

with the momentum vector $m_i(\dot{\vec{r}}_i - \dot{\vec{r}}_O)$. The time derivative of $\vec{L}$ is given in Eq. (7).

$$\dot{\vec{L}}_O = \sum_i (\vec{r}_i - \vec{r}_O) \times m_i(\ddot{\vec{r}}_i - \ddot{\vec{r}}_O) \qquad (7)$$

Hereafter the point O is taken to be either a fixed point (in which case $\dot{\vec{r}}_O = \ddot{\vec{r}}_O = 0$) or the center-of-mass point. Using the equation of motion (1), $m_i\ddot{\vec{r}}_i$ can be replaced by $\vec{F}_i$. Thus the rotational equation of motion (8) is obtained. The right-hand side of Eq. (8) is

$$\dot{\vec{L}}_O = \sum_i (\vec{r}_i - \vec{r}_O) \times \vec{F}_i \qquad (8)$$

known as the torque, $\vec{N}$. The contribution of the internal forces to the torque vanishes if the "extended third law" holds; namely, action equals reaction and is directed along a line between the particles. In this circumstance the rotational equation of motion is given by Eqs. (9), where $\vec{N}^{\text{ext}}$ is the total torque associated with external forces that act

$$\dot{\vec{L}}_0 = \vec{N}_0^{\text{ext}} \tag{9a}$$

$$\vec{N}_0^{\text{ext}} = \sum_i (\vec{r}_i - \vec{r}_O) \times \vec{F}_i^{\text{ext}} \tag{9b}$$

on the rigid body. See ANGULAR MOMENTUM; TORQUE.

It is straightforward to show from Eq. (6) that the angular momentum about an arbitrary point O is related to the angular momentum about the center of mass by Eq. (10). The torque about an arbitrary point O can also be easily related to the torque about the center of mass by Eq. (11).

$$\vec{L}_O = \vec{L}_{\text{CM}} + (\vec{R} - \vec{r}_O) \times \vec{P} \tag{10}$$

$$\vec{N}_O^{\text{ext}} = \vec{N}_{\text{CM}}^{\text{ext}} + (\vec{R} - \vec{r}_O) \times \vec{F}^{\text{ext}} \tag{11}$$

Six coordinates determine the positions of all particles in a rigid body, and the motion of a rigid body is described by six equations of motion. The translational motion of the center of mass is determined by Eq. (5), and the rotational motion about the center of mass, or a fixed point, is determined by Eq. (9). These six equations, which hold for any system of particles, completely describe the motion of a rigid body.

Motion of an isolated system. The equation $\dot{\vec{L}} = \vec{N}^{\text{ext}}$ has the same form as $\dot{\vec{P}} = \vec{F}^{\text{ext}}$. Both $\vec{L}$ and $\vec{p}$ are constants for an isolated system since $\vec{N}^{\text{ext}} = 0$ and $\vec{F}^{\text{ext}} = 0$. Even though the two conditions $\vec{N}^{\text{ext}} = 0$ and $\vec{F}^{\text{ext}} = 0$ appear similar, there are some important differences for systems in which internal motion is possible. If $\vec{F}^{\text{ext}} = 0$, a center of mass which is at rest will remain so, regardless of internal forces or internal motion. If $\vec{N}^{\text{ext}} = 0$, the total angular momentum is constant, and if initially zero, will remain zero. However, $\vec{L} = 0$ does not exclude changes in orientation of the system by the use of merely internal forces. There is no rotational analog to the equation $\vec{r}(t) = (\vec{P}/M)t + \vec{r}(0)$ for linear motion of the center of mass.

Static equilibrium. In the design of permanent structures, the conditions under which a rigid body remains in steady motion under the action of a set of forces are of great importance. The six conditions for complete equilibrium of a rigid body are given in Eqs. (12). However, in many circumstances equilibrium is desired only for a

$$\vec{F}^{\text{ext}} = \sum_i \vec{F}_i^{\text{ext}} = 0$$
$$\vec{N}_{\text{CM}}^{\text{ext}} = \sum_i (\vec{r}_i - \vec{R}) \times \vec{F}_i^{\text{ext}} = 0 \tag{12}$$

subset of the six independent directions of motion. To illustrate, the external force in the direction of motion of an accelerating automobile is nonzero, but equilibrium is maintained in all other directions. See STATICS.   [V.D.B.]

Ritz's combination principle   The empirical rule, formulated by W. Ritz in 1905, that sums and differences of the frequencies of spectral lines often equal other observed frequencies. The rule is an immediate consequence of the quantum-mechanical formula $hf = E_i - E_f$ relating the energy hf of an emitted photon to the initial energy E_i and final energy E_f, of the radiating system; h is Planck's constant and f is the frequency of the emitted light. See ATOMIC STRUCTURE AND SPECTRA; ENERGY LEVEL (QUANTUM MECHANICS); QUANTUM MECHANICS.   [E.G.]

Rotational motion   The motion of a rigid body which takes place in such a way that all of its particles move in circles about an axis with a common angular velocity; also, the rotation of a particle about a fixed point in space. Rotational motion is illustrated by (1) the fixed speed of rotation of the Earth about its axis; (2) the varying speed of rotation of the flywheel of a sewing machine; (3) the rotation of a satellite about a planet; (4) the motion of an ion in a cyclotron; and (5) the motion of a pendulum. Circular motion is a rotational motion in which each particle of the rotating body moves in a circular path about an axis. Such motion is exhibited by the first and second examples. For information concerning the other examples see HARMONIC MOTION; PARTICLE ACCELERATOR; PENDULUM.

The speed of rotation, or angular velocity, remains constant in uniform circular motion. In this case, the angular displacement θ experienced by the particle or rotating body in a time t is $\theta = \omega t$, where ω is the constant angular velocity.

A special case of circular motion occurs when the rotating body moves with constant angular acceleration. If a body is moving in a circle with an angular acceleration of α radians/s^2, and if at a certain instant it has an angular velocity ω_0, then at a time t seconds later, the angular velocity may be expressed as $\omega = \omega_0 + \alpha t$, and the angular displacement as $\theta = \omega_0 t + \frac{1}{2}\alpha t^2$. See ACCELERATION; VELOCITY.

A rotating body possesses kinetic energy of rotation which may be expressed as $T_{\text{rot}} = \frac{1}{2}I\omega^2$, where ω is the magnitude of the angular velocity of the rotating body and I is the moment of inertia, which is a measure of the opposition of the body to angular acceleration. The moment of inertia of a body depends on the mass of a body and the distribution of the mass relative to the axis of rotation. For example, the moment of inertia of a solid cylinder of mass M and radius R about its axis of symmetry is $\frac{1}{2}MR^2$.

The action of a torque L is to produce an angular acceleration α according to the equation below, where $I\omega$, the product of moment of inertia and angular velocity, is

$$L = I\alpha = I\frac{d\omega}{dt} = \frac{d}{dt}(I\omega)$$

called the angular momentum of the rotating body. This equation points out that the angular momentum $I\omega$ of a rotating body, and hence its angular velocity ω, remains constant unless the rotating body is acted upon by a torque. Both L and $I\omega$ may be represented by vectors.

It is readily shown that the work done by the torque L acting through an angle θ on a rotating body originally at rest is exactly equal to the kinetic energy of rotation. See ANGULAR MOMENTUM; MOMENT OF INERTIA; RIGID-BODY DYNAMICS; TORQUE; WORK.

[C.E.H./R.J.S.]

Runge vector   The Runge vector describes certain unchanging features of a non-relativistic two-body interaction for which the potential energy is inversely proportional to the distance r between the bodies or, alternatively, in which each body exerts a force on the other that is directed along the line between them and proportional to r^{-2}. Two basic interactions in nature are of this type: the gravitational interaction between two masses (called the classical Kepler problem), and the Coulomb interaction between like or unlike charges (as in the hydrogen atom). Both at the classical level and the quantum-mechanical level, the existence of a Runge vector is a reflection of the symmetry inherent in the interaction. See COULOMB'S LAW; QUANTUM MECHANICS; SYMMETRY LAWS (PHYSICS).

[D.M.Fr.]

Rydberg atom   An atom which possesses one valence electron orbiting about an atomic nucleus within an electron shell well outside all the other electrons in the atom. Such an atom approximates the hydrogen atom in that a single electron is interacting with a positively charged core. Early observations of atomic electrons in such Rydberg quantum states involved studies of the Rydberg series in optical spectra. Electrons jumping between Rydberg states with adjacent principal quantum numbers, n and

$n - 1$, with n near 80 produce microwave radiation. Microwave spectral lines due to such electronic transitions in Rydberg atoms have been observed both in laboratory experiments and in the emissions originating from certain low-density partially ionized portions of the universe called HII regions. *See* ELECTRON CONFIGURATION.

The advent of the laser has made possible the production of sizable numbers of Rydberg atoms within a bulb containing gas at low pressures, 10^{-2} torr (1.3 pascals) or less. The rapid energy-resonance absorption of several laser light photons by an atom in its normal or ground state results in a Rydberg atom in a state with a selected principal quantum number. Aggregates of Rydberg atoms have been used as sensitive detectors of infrared radiation, including thermal radiation. They have also been observed to collectively participate in spontaneous photon emission, called superradiance. Such aggregates form the active medium for infrared lasers that operate through the usual laser mechanism of collective stimulated photon emission. All these developments are based upon the great sensitivity of Rydberg atoms to external electromagnetic radiation fields. Atoms with n near 40 can absorb almost instantaneously over a hundred microwave photons and become ionized at easily achievable microwave power levels. Isotope separation techniques have been developed that combine the selectivity of laser excitation of Rydberg states with the ready ionizability of Rydberg atoms. Such applications have been pursued for atoms ranging from deuterium through uranium. *See* INFRARED RADIATION; LASER.

[J.E.B.]

Rydberg constant   The most accurately measured of the fundamental constants; it is a universal scaling factor for any spectroscopic transition and an important cornerstone in the determination of other constants.

This constant was introduced empirically. J. Balmer's formula described the visible spectral lines of atomic hydrogen, while J. Rydberg's formula applied to the spectra of many elements. Their results may be summarized by Eq. (1), where λ is the wavelength

$$\frac{1}{\lambda} = R\left(\frac{1}{n_1^2} - \frac{1}{n_2^2}\right) \tag{1}$$

of the spectral line and R is a constant. In Balmer's account of the visible hydrogen spectrum, n_1 was equal to 2, while n_2 took on the integer values 3, 4, 5, and so forth. In Rydberg's more general work, n_1 and n_2 differed slightly from integer values. A remarkable result of Rydberg's work was that the constant R was the same for all spectral series he studied, regardless of the element. This constant R has come to be known as the Rydberg constant.

Applied to hydrogen, Niels Bohr's atomic model leads to Balmer's formula with a predicted value for the Rydberg constant given by Eq. (2), where m_e is the electron

$$R_\infty = \frac{m_e e^4}{8h^3 \epsilon_0^2 c} \tag{2}$$

mass, e is the electron charge, h is Planck's constant, ϵ_0 is the permittivity of vacuum, and c is the speed of light. The equation expresses the Rydberg constant in SI units. To express it in cgs units, the right-hand side must be multiplied by $(4\pi\epsilon_0)^2$. The subscript ∞ means that this is the Rydberg constant corresponding to an infinitely massive nucleus.

E. Schrödinger's wave mechanics predicts the same energy levels as the simple Bohr model, but the relativistic quantum theory of P. A. M. Dirac introduces small corrections or fine-structure splittings. The modern theory of quantum electrodynamics predicts further corrections. Additional small hyperfine-structure corrections account for the interaction of the electron and nuclear magnetic moments. *See* Fine structure (spectral lines); Hyperfine structure.

The Rydberg constant is determined by measuring the wavelength or frequency of a spectral line of a hydrogenlike atom or ion. The highest resolution and accuracy has been achieved by the method of Doppler-free two-photon spectroscopy, which permits the observation of very sharp resonance transitions between long-living states. The 2002 adjustment of the fundamental constants, taking into account different measurements, adopted the value $R_{\infty} = 10{,}973{,}731.568{,}525 \pm 0.000{,}073$ m^{-1} for the Rydberg constant. The measurements provide an important cornerstone for fundamental tests of basic laws of physics. *See* Atomic structure and spectra; Fundamental constants; Laser; Laser spectroscopy.   [T.W.Ha.; M.We.]

S

Scanning tunneling microscope An instrument for producing surface images with atomic-scale lateral resolution, in which a fine probe tip is scanned over the surface at a distance of 0.5–1 nanometer, and the resulting tunneling current, or the position of the tip required to maintain a constant tunneling current, is monitored.

Scanning tunneling microscopes have pointed electrodes that are scanned over the surface of a conducting specimen, with help from a piezoelectric crystal whose dimensions can be altered electronically. They normally generate images by holding the current between the tip of the electrode and the specimen at some constant (set-point) value by using a piezoelectric crystal to adjust the distance between the tip and the specimen surface, while the tip is piezoelectrically scanned in a raster pattern over the region of specimen surface being imaged. By holding the force, rather than the electric current, between tip and specimen at a set-point value, atomic force microscopes similarly allow the exploration of nonconducting specimens. In either case, when the height of the tip is plotted as a function of its lateral position over the specimen, an image that looks very much like the surface topography results.

It is becoming increasingly possible to record other signals (such as lateral force, capacitance, scan-related tip displacement, temperature, light intensity, or magnetic resonance) as the tip scans. For example, modern atomic force microscopes can map lateral force and conductivity along with height, while image pairs from scanning tunneling microscopes scanning to and fro can provide information about friction as well as topography.

Scanning tunneling microscopes make it possible not just to view atoms but to push them and even to rearrange them in unlikely combinations (sometimes whether or not these rearrangements are desirable). A few considerations of scale are important in understanding this process. Atoms comprise a positive nucleus and a surrounding cloud of negative electrons. These charges rearrange when another atom approaches, with unlike charges shifting to give rise to the van der Waals force of attraction between neutral atoms. This force makes gravity (and most accelerations) ignorable when contact between solid objects in the micrometer size range and smaller is involved, since surface-to-volume ratios are inversely proportional to object size.

The electric field in the scanning tunneling microscope allows plucking as well, in which adsorbed or substrate atoms are removed and transferred to the electrode tip with a suitable voltage pulse. Because the electric field from the tip falls off less rapidly with separation than do van der Waals forces, the most weakly attached nearby atom rather than the nearest may end up being removed. One solution to this problem is a hybrid approach. By invoking the tip electric field for bond breaking only when the tip is sufficiently close to the target atom that the van der Waals forces contribute as well, atoms on silicon could be singly removed and redeposited at will.

A third kind of selective bond breaking was also demonstrated. It involved the selective breaking of silicon-hydrogen bonds using electron energies (that is, pulse voltages) below those necessary to break bonds directly. Since the desorption probability was observed to vary exponentially with the tip-specimen current, it is believed that vibrational heating from inelastic electron tunneling mediated the chemical transition in this work. This work involves bond alteration at the level of signal atoms, the ultimate frontier for lithographic miniaturization.   [P.B.F.]

Scattering experiments (atoms and molecules)   Experiments in which a beam of incident electrons, atoms, or molecules is deflected by collisions with an atom or molecule. Such experiments provide tests of the theory of scattering as well as information about atomic and molecular forces. Scattering experiments can be designed to simulate conditions in planetary atmospheres, electrical discharges, gas lasers, fusion reactors, stars, and planetary nebulae. *See* ELECTRICAL CONDUCTION IN GASES; GAS DISCHARGE; LASER; NUCLEAR FUSION.

In general, in any type of collision, scattering occurs, which causes the direction of relative motion of the two systems to be rotated to a new direction after the collision. More than two systems may also result from such an impact. A complete description of a collision event requires measurement of the directions, speeds, and internal states of all the products. *See* COLLISION (PHYSICS).

There are two basic types of scattering experiments. The simpler involves passing a collimated beam of particles (electrons, atoms, molecules, or ions) through a dilute target gas (in a cell or a jet) and measuring the fraction of incident particles that are deflected into a certain angle relative to the incident beam direction. In the second method, a collimated beam of particles intersects a second beam. The scattering events are usually registered by measuring the deflection or internal-state change of the beam particles. *See* MOLECULAR BEAMS.

Scattering in a particular type of collision is specified in terms of a differential cross section. The probability that, in a particular type of collision, the direction of motion of the electron is turned through a specified scattering angle into a specified solid angle is proportional to the corresponding differential scattering cross section. Collision cross sections can be measured with appropriately designed experimental apparatus. Depending on the type of collision process, that apparatus may measure the scattering angle, energy, charge, or mass of the scattered systems.

For the simplest case, the scattering of a beam of structureless particles of specified mass and speed by a structureless scattering center, the differential cross section may be calculated exactly by using the quantum theory. In the special case where the Coulomb force fully describes the interaction, both the quantum and classical theory give the same exact value for the differential cross section at all values of the scattering angle.

For scattering of systems with internal structure (for example, molecules, and their ions), no exact theoretical calculation of the cross section is possible. Methods of approximation specific to different types of collisions have been developed. The power of modern high-speed computers has greatly increased their scope and effectiveness, with scattering experiments serving as benchmarks. *See* ATOMIC STRUCTURE AND SPECTRA.   [R.A.Ph.]

Scattering experiments (nuclei)   Experiments in which beams of particles such as electrons, nucleons, alpha particles and other atomic nuclei, and mesons are deflected by elastic collisions with atomic nuclei. Much is learned from such experiments about the nature of the scattered particle, the scattering center, and the forces acting between them. Scattering experiments, made possible by the construction of high-energy particle accelerators and the development of specialized techniques for detecting the scattered particles, are one of the main sources of information regarding the structure of matter. *See* NUCLEAR STRUCTURE; PARTICLE ACCELERATOR; PARTICLE DETECTOR; SCATTERING MATRIX.

[K.A.Er.]

Scattering matrix An infinite-dimensional matrix or operator that expresses the state of a scattering system consisting of waves or particles or both in the far future in terms of its state in the remote past; also called the S matrix. In the case of electromagnetic (or acoustic) waves, it connects the intensity, phase, and polarization of the outgoing waves in the far field at various angles to the direction and polarization of the beam pointed toward an obstacle. It is used most prominently in the quantum-mechanical description of particle scattering, in which context it was invented in 1937 by J. A. Wheeler to describe nuclear reactions. Because an analog of the Schrödinger equation for the description of particle dynamics is lacking in the relativistic domain, W. Heisenberg proposed in 1943 that the S matrix rather than the hamiltonian or the lagrangian be regarded as the fundamental dynamical entity of quantum mechanics. This program played an important role in high-energy physics during the 1960s but is now largely abandoned. The physics of fundamental particles is now described primarily in terms of quantum gauge fields, and these are used to determine the S matrix and its elements for the collision and reaction processes observed in the laboratory. *See* ELEMENTARY PARTICLE; NUCLEAR REACTION; QUANTUM MECHANICS; RELATIVISTIC QUANTUM THEORY; SCATTERING EXPERIMENTS (ATOMS AND MOLECULES); SCATTERING EXPERIMENTS (NUCLEI).

The mathematical properties of the S matrix in nonrelativistic quantum mechanics have been thoroughly studied and are, for the most part, well understood. If the potential energy in the Schrödinger equation, or the scattering obstacle, is spherically symmetric, the eigenfunctions of the S matrix are spherical harmonics and its eigenvalues are of the form $\exp (2i\delta_l)$, where the real number δ_l is the phase shift of angular momentum l. In the nonspherically symmetric case, analogous quantities are called the eigenphase shifts, and the eigenfunctions depend on both the energy and the dynamics. In the relativistic regime, without an underlying Schrödinger equation for the particles, the mathematical properties are not as well known. Causality arguments (no signal should propagate faster than light) lead to dispersion relations, which constitute experimentally verifiable consequences of very general assumptions on the properties of nature that are independent of the detailed dynamics. *See* ANGULAR MOMENTUM; CAUSALITY; DISPERSION RELATIONS; EIGENFUNCTION.

[R.G.Ne.]

Scattering of electromagnetic radiation The process in which energy is removed from a beam of electromagnetic radiation and reemitted with a change in direction, phase, or wavelength. All electromagnetic radiation is subject to scattering by the medium (gas, liquid, or solid) through which it passes.

It has been known since the work of J. Maxwell in the nineteenth century that accelerating electric charges radiate energy and, conversely, that electromagnetic radiation consists of fields which accelerate charged particles. Light in the visible, infrared, or ultraviolet region interacts primarily with the electrons in gases, liquids, and solids—not the nuclei. The scattering process in these wavelength regions consists of acceleration of the electrons by the incident beam, followed by reradiation from the accelerating charges. *See* ELECTROMAGNETIC RADIATION.

Scattering processes may be divided according to the time between the absorption of energy from the incident beam and the subsequent reradiation. True "scattering" refers only to those processes which are essentially instantaneous. Mechanisms in which there is a measurable delay between absorption and reemission are usually termed luminescence. *See* LUMINESCENCE.

Instantaneous scattering processes may be further categorized according to the wavelength shifts involved. Some scattering is "elastic"; there is no wavelength change, only a phase shift. In 1928 C. V. Raman discovered the process in which light was inelastically scattered and its energy was shifted by an amount equal to the vibrational energy of a molecule or crystal.

In liquids or gases two distinct processes generate inelastic scattering with small wavelength shifts. The first is Brillouin scattering from pressure waves. When a sound wave propagates through a medium, it produces alternate regions of high compression (high density) and low compression (or rarefaction). Brillouin scattering of light to higher (or lower) frequencies occurs because the medium is moving toward (or away from) the light source. This is an optical Doppler effect. *See* DOPPLER EFFECT.

The second kind of inelastic scattering studied in fluids is due to entropy and temperature fluctuations, and is known as Rayleigh scattering. These entropy fluctuations produce a broadening in the scattered radiation centered about the exciting wavelength, rather than sharp, well-defined wavelength shifts. Under the assumption that the scattering in fluids is from particles much smaller than the wavelength of the exciting light, Lord Rayleigh derived in 1871 an equation for such scattering. The dependence of scattering intensity upon the inverse fourth power of the wavelength given in Rayleigh's equation is responsible for the fact that daytime sky looks blue and sunsets red: blue light is scattered out of the sunlight by the air molecules more strongly than red; at sunset, more red light passes directly to the eyes without being scattered. *See* ENTROPY.

Rayleigh's derivation of his scattering equation relies on the assumption of small, independent particles. Under some circumstances of interest, both of these assumptions fail. Colloidal suspensions provide systems in which the scattering particles are comparable to or larger than the exciting wavelengths. Such scattering is called the Tyndall effect and results in a nearly wavelength-independent (that is, white) scattering spectrum. The Tyndall effect is the reason clouds are white (the waterdroplets become larger than the wavelengths of visible light).

The breakdown of Rayleigh's second assumption—that of independent particles— occurs in all liquids. There is strong correlation between the motion of neighboring particles. This leads to fixed phase relations and destructive interference for most of the scattered light. The remaining scattering arises from fluctuations in particle density discussed above.   [J.F.S.]

Schlieren photography
Any technique for the photographic recording of schlieren, which are regions or stria in a medium that is surrounded by a medium of different refractive index. Refractive index gradients in transparent media cause light rays to bend (refract) in the direction of increasing refractive index. This is a result of the reduced light velocity in a higher-refractive-index material. This phenomenon is exploited in viewing the schlieren, with schlieren photographs as the result. Electronic video recorders, scanning diode array cameras, and holography are widely used as supplements. *See* HOLOGRAPHY; REFRACTION OF WAVES.

There are many techniques for optically enhancing the appearance of the schlieren in an image of the field of interest. In the oldest of these, called the knife-edge method (see illustration), a point or slit source of light is collimated by a mirror and passed through a field of interest, after which a second mirror focuses the light, reimaging the point or slit where it is intercepted by an adjustable knife edge (commonly a razor blade).

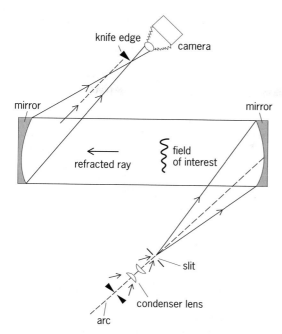

Knife-edge method of viewing schlieren, employing the "z" configuration.

The illustration shows the "z" configuration which minimizes the coma aberration in the focus. Mirrors are most often used because of the absence of chromatic aberration. *See* ABERRATION (OPTICS).

Rays of light that are bent by the schlieren in the direction of the knife edge are intercepted and removed from the final image of the region of interest, causing those regions to appear dark. Consequently, the system is most sensitive to the density gradients that are perpendicular to the knife edge. The knife edge is commonly mounted on a rotatable mount so that it can be adjusted during a measurement to optimally observe different gradients in the same field of interest. The intensity in the processed image is proportional to the refractive index gradient. A gradient in the same direction as the knife edge appears dark. Gradients in the opposite direction appear bright. This method, employed with arc light sources, is still one of the simplest ways to view refractive index changes in transparent solids, liquids, and gases. A well-designed schlieren system can easily detect the presence of a refractive index gradient that causes 1 arc-second deviation of a light ray.

Except for locating and identifying schlieren-causing events such as turbulent eddies, shock waves, and density gradients, schlieren systems are usually considered to be qualitative instruments. Quantitative techniques for determining density are possible but are much more difficult to employ. The most common of these is color schlieren. The knife edge is replaced with a multicolored filter. Rays of light refracted through different angles appear in different colors in the final image.

The availability of lasers and new optical components has expanded the method considerably. When a coherent light source such as a laser is used, the knife edge can be replaced by a variety of phase-, amplitude-, or polarization-modulating filters to produce useful transformations in the image intensity. *See* INTERFEROMETRY; LASER; POLARIZED LIGHT.

[J.D.Tr.]

Schottky anomaly   A contribution to the heat capacity of a solid arising from the thermal population of discrete energy levels as the temperature is raised. The effect is particularly prominent at low temperatures, where other contributions to the heat capacity are generally small. *See* SPECIFIC HEAT.

Discrete energy levels may arise from a variety of causes, including the removal of orbital or spin degeneracy by magnetic fields, crystalline electric fields, and spin orbit coupling, or from the magnetic hyperfine interaction. Such effects commonly occur in paramagnetic ions. *See* LOW-TEMPERATURE THERMOMETRY.

Corresponding to the Schottky heat capacity, there is a contribution to the entropy. This can act as a barrier to the attainment of low temperatures if the substance is to be cooled either by adiabatic demagnetization or by contact with another cooled substance. Conversely, a substance with a Schottky anomaly can be used as a heat sink in experiments at low temperatures (generally below 1 K or −457.9°F) to reduce temperature changes resulting from the influx or generation of heat. *See* ADIABATIC DEMAGNETIZATION; LOW-TEMPERATURE PHYSICS.   [W.P.W.]

Schottky effect   The enhancement of the thermionic emission of a conductor resulting from an electric field at the conductor surface. Since the thermionic emission current is given by the Richardson formula, an increase in the current at a given temperature implies a reduction in the work function of the emitter. The reduction in work function can be calculated by considering the effect of a constant externally applied field on the potential energy of an electron near the conductor surface, and is found to be proportional to the square root of the field. *See* THERMIONIC EMISSION; WORK FUNCTION (ELECTRONICS).

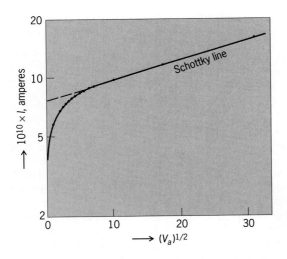

Logarithm of thermionic emission current I of tungsten as function of square root of anode voltage V_a. (*After W. B. Nottingham, Phys. Rev., 58:927–928, 1940*)

A plot of the logarithm of the current versus the square root of the anode voltage should yield a straight line. An example is given in the illustration for tungsten; the deviation from the straight line for low anode voltages is due to space-charge effects. *See* SPACE CHARGE.   [A.J.D.]

Schrödinger's wave equation   A linear, homogeneous partial differential equation that determines the evolution with time of a quantum-mechanical wave function.

Quantum mechanics was developed in the 1920s along two different lines, by W. Heisenberg and by E. Schrödinger. Schrödinger's approach can be traced to the notion of wave-particle duality that flowed from A. Einstein's association of particle-like energy bundles (photons, as they were later called) with electromagnetic radiation, which, classically, is a wavelike phenomenon. For radiation of definite frequency f, each bundle carries energy hf. The proportionality factor, $h = 6.626 \times 10^{-34}$ joule-second, is a fundamental constant of nature, introduced by M. Planck in his empirical fit to the spectrum of blackbody radiation. This notion of wave-particle duality was extended in 1923 by L. de Broglie, who postulated the existence of wavelike phenomena associated with material particles such as electrons. See PHOTON; WAVE MECHANICS.

There are certain purely mathematical similarities between classical particle dynamics and the so-called geometric optics approximation to propagation of electromagnetic signals in material media. For the case of a single (nonrelativistic) particle moving in a potential $V(\mathbf{r})$, this analogy leads to the association with the system of a wave function, $\Psi(\mathbf{r})$, which obeys Eq. (1). Here m is the mass of the particle, E its energy, $\hbar = h/(2\pi)$,

$$-\frac{\hbar^2}{2m}\nabla^2\Psi + V\Psi = E\Psi \tag{1}$$

and ∇^2 is the laplacian operator. See GEOMETRICAL OPTICS.

It is possible to ask what more general equation a time- as well as space-dependent wave function, $\Psi(\mathbf{r}, t)$, might obey. What suggests itself is Eq. (2), which is now called

$$i\hbar\frac{\partial\Psi}{\partial t} = -\frac{\hbar^2}{2m}\nabla^2\Psi + V\Psi \tag{2}$$

the Schrödinger equation.

The wave function can be generalized to a system of more than one particle, say N of them. A separate wave function is not assigned to each particle. Instead, there is a single wave function, $\Psi(\mathbf{r}_1, \mathbf{r}_2, \ldots, \mathbf{r}_N, t)$, which depends at once on all the position coordinates as well as time. This space of position variables is the so-called configuration space. The generalized Schrödinger equation is Eq. (3), where the potential V may now

$$ih\frac{\partial\Psi}{\partial t} = -\sum_{i=1}^{N}\frac{\hbar^2}{2m_i}\nabla_i^2\Psi + V\Psi \tag{3}$$

depend on all the position variables. Three striking features of this equation are to be noted:

1. The complex number i (the square root of minus one) appears in the equation. Thus Ψ is in general complex.

2. The time derivative is of first order. Thus, if the wave function is known as a function of the position variables at any one instant, it is fully determined for all later times.

3. The Schrödinger equation is linear and homogeneous in Ψ, which means that if Ψ is a solution so is $c\Psi$, where c is an arbitrary complex constant. More generally, if Ψ_1 and Ψ_2 are solutions, so too is the linear combination $c_1\Psi_1 + c_2\Psi_2$, where c_1 and c_2 are arbitrary complex constants. This is the superposition principle of quantum mechanics. See SUPERPOSITION PRINCIPLE.

The Schrödinger equation suggests an interpretation in terms of probabilities. Provided that the wave function is square integrable over configuration space, it follows from Eq. (3) that the norm, $\langle \Psi | \Psi \rangle$, is independent of time, where the norm is defined by Eq. (4).

$$\langle \Psi \mid \Psi \rangle \; = \int d^3x_1 d^3x_2 \ldots d^3x_N \Psi^* \Psi \tag{4}$$

It is possible to normalize Ψ (multiply it by a suitable constant) to arrange that this norm is equal to unity. With that done, the Schrödinger equation itself suggests that expression (5) is the joint probability distribution at time t for finding particle 1 in the

$$\Psi^* \Psi d^3x_1 d^3x_2 \ldots d^3x_N \tag{5}$$

volume element d^3x_1, particle 2 in d^3x_2, and so forth.   [S.Tr.]

Scintillation counter   A particle or radiation detector which operates through emission of light flashes that are detected by a photosensitive device, usually a photomultiplier or a silicon PIN diode. The scintillation counter not only can detect the presence of a particle, gamma ray, or x-ray, but can measure the energy, or the energy loss, of the particle or radiation in the scintillating medium. The sensitive medium may be solid, liquid, or gaseous, but is usually one of the first two. The scintillation counter is one of the most versatile particle detectors, and is widely used in industry, scientific research, medical diagnosis, and radiation monitoring, as well as in exploration for petroleum and radioactive minerals that emit gamma rays. Many low-level radioactivity measurements are made with scintillation counters. *See* PARTICLE DETECTOR.

Scintillation counters are made of transparent crystalline materials, liquids, plastics, or glasses. In order to be an efficient detector, the bulk scintillating medium must be transparent to its own luminescent radiation, and since some detectors are quite extensive, covering meters in length, the transparency must be of a high order. One face of the scintillator is placed in optical contact with the photosensitive surface of the photomultiplier or PIN diode (see illustration). In order to direct as much as possible of the light flash to the photosensitive surface, reflecting material is placed between the scintillator and the inside surface of the container.

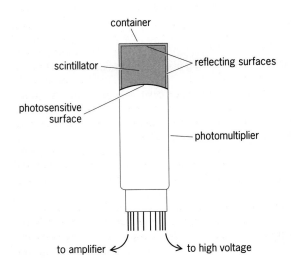

Diagram of a scintillation counter.

In many cases it is necessary to collect the light from a large area and transmit it to the small surface of a photomultiplier. In this case, a "light pipe" leads the light signal from the scintillator surface to the photomultiplier with only small loss. The best light guides and light fibers are made of glass, plastic, or quartz. It is also possible to use lenses and mirrors in conjunction with scintillators and photomultipliers. *See* OPTICAL FIBERS.

A charged particle, moving through the scintillator, loses energy and leaves a trail of ions and excited atoms and molecules. Rapid interatomic or intermolecular transfer of electronic excitation energy follows, leading eventually to a burst of luminescence characteristic of the scintillator material. When a particle stops in the scintillator, the integral of the resulting light output, called the scintillation response, provides a measure of the particle energy, and can be calibrated by reference to particle sources of known energy. Photomultipliers or PIN diodes may be operated so as to generate an output pulse of amplitude proportional to the scintillation response.

When a particle passes completely through a scintillator, the energy loss of the particle is measured. When a gamma ray converts to charged particles in a scintillator, its energy may also be determined. When the scintillator is made of dense material and of very large dimensions, the entire energy of a very energetic particle or gamma ray may be contained within the scintillator, and again the original energy may be measured. Such is the case for energetic electrons, positrons, or gamma rays which produce electromagnetic showers in the scintillator. Energy spectra can be determined in these various cases by using electronic equipment to convert amplitudes of the output pulses from the photomultiplier or PIN diode to digital form, for further processing by computers or pulse-height analyzers.   [F.D.B.; R.Ho.]

Second sound   A type of wave propagated in the superfluid phase of liquid helium (helium II) and in certain other substances under special conditions. The name is misleading since second sound is not in any sense a sound wave, but a temperature or entropy wave. In ordinary or first sound, pressure and density variations propagate with very small accompanying variations in temperature; in second sound, temperature variations propagate with no appreciable variation in density or pressure. *See* LIQUID HELIUM; SUPERFLUIDITY.

The two-fluid model of helium II provides further insight into the nature of second sound. In this model the liquid can be described as consisting of superfluid and normal components of densities ρ_s and ρ_n, respectively, such that the total density $\rho = \rho_s + \rho_n$. The superfluid component is frictionless and devoid of entropy; the normal component has a normal viscosity and contains the entropy and thermal energy of the system. In a temperature or second-sound wave, the normal and superfluid flows are oppositely directed so that $\rho_s \mathbf{V}_s + \rho_n \mathbf{V}_n = 0$, where $\mathbf{V}_s$ and $\mathbf{V}_n$ are the superfluid and normal flow velocities. Thus a variation in relative densities of the two components, and hence a temperature fluctuation, propagates with no change in total density or pressure. In a first-sound wave, the two components move in phase, that is, $\mathbf{V}_n \cong \mathbf{V}_s$.

Theoretical predictions that second sound should exist in certain solid dielectric crystals under suitable conditions have been confirmed experimentally for solid helium single crystals at temperatures between 0.4 and 1.0 K (-459.0 and $-457.9°F$). *See* DIELECTRIC MATERIALS.

Another quite different class of materials can exhibit second sound. In smectic A liquid crystals, when the wave vector is oblique with respect to the layers of these ordered structures, a modulation of the interlayer spacing can propagate at nearly constant density.   [H.A.F.]

Secondary emission   The emission of electrons from the surface of a solid into vacuum caused by bombardment with charged particles, in particular with electrons. The mechanism of secondary emission under ion bombardment is quite different from that under electron bombardment; it is only in the latter case that the term secondary emission is generally used.

The bombarding electrons and the emitted electrons are referred to, respectively, as primaries and secondaries. Secondary emission has important practical applications because the secondary yield, that is, the number of secondaries emitted per incident primary, may exceed unity. Thus, secondary emitters are used in electron multipliers, especially in photomultipliers, and in other electronic devices such as television pickup tubes, storage tubes for electronic computers, and so on.

The emission of secondary electrons can be described as the result of three processes: (1) excitation of electrons in the solid into high-energy states by the impact of high-energy primary electrons, (2) transport of these secondary electrons to the solid-vacuum interface, and (3) escape of the electrons over the surface barrier into the vacuum. The efficiency of each of these three processes, and hence the magnitude of the secondary emission yield δ, varies greatly for different materials.

Most of the materials used in practical devices are semiconductors or insulators whose band-gap energies are much larger than their electron affinities. Examples are magnesium oxide (MgO), beryllium oxide (BeO), cesium antimonide (Cs_3Sb), and potassium chloride (KCl). Maximum δ values in the 8–15 range are typically obtained at primary energies of several hundred volts.

In certain semiconductors the bands are bent downward to such an extent that the vacuum level lies below the bottom of the conduction band in the bulk. A material with this characteristic is said to have negative effective electron affinity. The most

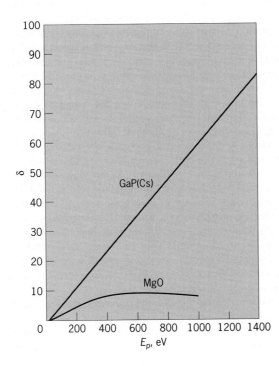

Secondary emission yield versus primary energy for GaP(Cs). The curve for MgO is shown for comparison.

important material in this category is cesium-activated gallium phosphide, GaP(Cs). The illustration shows the curve of yield δ versus primary energy E_p for GaP(Cs) by comparison with MgO. Values of δ exceeding 100 are readily obtained, with maximum yields at energies in the 5–10-kV region. *See* BAND THEORY OF SOLIDS; SEMICONDUCTOR.

[A.H.So.]

Seebeck effect   The generation of a temperature-dependent electromotive force (emf) at the junction of two dissimilar metals. This phenomenon provides the physical basis for the thermocouple. In 1821, T. J. Seebeck discovered that near a closed circuit composed of two linear conductors of two different metals a magnetic needle would be deflected if, and only if, the two junctions were at different temperatures, and that if the temperatures of the two junctions were reversed the direction of deflection would also be reversed. He investigated 35 different metals and arranged them in a series such that at a hot junction, current flows from a metal earlier in the series to a later one. *See* ELECTROMOTIVE FORCE (EMF).

A thermocouple consists of a pair of wires of dissimilar metals, joined at the ends. One junction is kept at an accurately known cold temperature, usually that of melting ice, and the other is used for the measurement of an unknown temperature, by measuring the emf generated as a result of the Seebeck effect. *See* THERMOCOUPLE; THERMOELECTRICITY.

[A.E.Ba.]

Selection rules (physics)   General rules concerning the transitions which may occur between the states of a quantum-mechanical physical system. They derive in almost all cases from the symmetry properties of the states and of the interaction which gives rise to the transitions. The system may have a classical (nonquantum) counterpart, and in this case the selection rules may often be related to the classical conserved quantities. A first use of selection rules is in determining the symmetry classes of the states; but in a great variety of ways they may yield other information about the system and the conservation laws. *See* QUANTUM MECHANICS; SYMMETRY LAWS (PHYSICS).

For an isolated system the total angular momentum is a conserved quantity; this fact derives from a fundamental fact of nature, namely, that space is isotropic. Each state is then classifiable by angular momentum J and its z component M ($= -J, -J + 1, \ldots, +J$). Angular momenta combine in a vectorial fashion. Thus, if the system makes a particle-emitting transition $J_1, M_1 \rightarrow J_2, M_2$, the emitted particles must carry away angular momentum (j, μ), where $\mathbf{j} = \mathbf{j}_1 - \mathbf{j}_2$. This implies that $\mu = M_1 - M_2$ and that j takes on values $J_1 - J_2, J_1 - J_2 + 1, \ldots, (J_1 + J_2)$. Thus in transitions ($J = 4 \leftrightarrow J = 2$) the possible j values comprise only 2, 3, 4, 5, 6, and, if it is also specified that $M_1 - M_2 = \pm 4$, only 4, 5, 6. Observe that J_2 is additive. *See* ANGULAR MOMENTUM; QUANTUM NUMBERS.

Another fundamental symmetry, the parity, which determines the behavior of a system (or of its description) under inversion of the coordinate axes, is conserved by the strong and electromagnetic interactions, and gives a classification of systems as even ($\pi = +1$) or odd ($\pi = -1$). Under combination the parity combines multiplicatively. Thus, if the transition above is $4^\pm \rightarrow 2^\pm$, it follows that $j^\pi = 2^-, \ldots, 6^-$, while $4^\pm \rightarrow 2^\pm$ would give $j^\pi = 2^+, \ldots, 6^+$. The angular momentum $\mathbf{j}$ may be a combination of intrinsic spin $\mathbf{s}$ and orbital angular momentum $\mathbf{l}$. Scalar, pseudoscalar, vector, and pseudovector particles are respectively characterized by $s^{\pi s} = 0^+, 0^-, 1^-, 1^+$, where π_s is the "intrinsic" parity, while l always carries $\pi_l = (-1)^l$. *See* PARITY (QUANTUM MECHANICS); SPIN (QUANTUM MECHANICS).

The isospin symmetry of the elementary particles is almost conserved, being broken by electromagnetic and weak interactions. It is described by the group SU(2), of unimodular unitary transformations in two dimensions. Since the SU(2) algebra is identical with that of the angular momentum SO(3), isospin behaves like angular momentum with its three generators **T** replacing **J**.

The isospin group is a subgroup of SU(3) which defines a more complex fundamental symmetry of the elementary particles. Two of its eight generators commute, giving two additive quantum numbers, T_z and strangeness S' (or, equivalently, charge and hypercharge). The strangeness is conserved ($\Delta S' = 0$) for strong and electromagnetic, but not for weak, interactions. The selection rules and combination laws for SU(3) and its many extensions, and the quark-structure underlying them, correlate an enormous amount of information and make many predictions about the elementary particles. *See* BARYON; ELEMENTARY PARTICLE; MESON; QUARKS; UNITARY SYMMETRY.

A great variety of other groups have been introduced to define relevant symmetries for atoms, molecules, nuclei, and elementary particles. They all have their own selection rules, representing one aspect of the symmetries of nature.   [J.B.Fr.]

Semiconductor   A solid crystalline material whose electrical conductivity is intermediate between that of a metal and an insulator. Semiconductors exhibit conduction properties that may be temperature-dependent, permitting their use as thermistors (temperature-dependent resistors), or voltage-dependent, as in varistors. By making suitable contacts to a semiconductor or by making the material suitably inhomogeneous, electrical rectification and amplification can be obtained. Semiconductor devices, rectifiers, and transistors have replaced vacuum tubes almost completely in low-power electronics, making it possible to save volume and power consumption by orders of magnitude. In the form of integrated circuits, they are vital for complicated systems. The optical properties of a semiconductor are important for the understanding and application of the material. Photodiodes, photoconductive detectors of radiation, injection lasers, light-emitting diodes, solar-energy conversion cells, and so forth are examples of the wide variety of optoelectronic devices. *See* LASER; LIGHT-EMITTING DIODE; SEMICONDUCTOR DIODE; THERMISTOR; TRANSISTOR.

Conduction in semiconductors. The electrical conductivity of semiconductors ranges from about 10^3 to 10^{-9} ohm^{-1} cm^{-1}, as compared with a maximum conductivity of 10^7 for good conductors and a minimum conductivity of 10^{-17} ohm^{-1} cm^{-1} for good insulators. *See* ELECTRIC INSULATOR.

The electric current is usually due only to the motion of electrons, although under some conditions, such as very high temperatures, the motion of ions may be important. The basic distinction between conduction in metals and in semiconductors is made by considering the energy bands occupied by the conduction electrons. *See* BAND THEORY OF SOLIDS; IONIC CRYSTALS.

At absolute zero temperature, the electrons occupy the lowest possible energy levels, with the restriction that at most two electrons with opposite spin may be in the same energy level. In semiconductors and insulators, there are just enough electrons to fill completely a number of energy bands, leaving the rest of the energy bands empty. The highest filled energy band is called the valence band. The next higher band, which is empty at absolute zero temperature, is called the conduction band. The conduction band is separated from the valence band by an energy gap, which is an important characteristic of the semiconductor. In metals, the highest energy band that is occupied by the electrons is only partially filled. This condition exists either because the number of electrons is not just right to fill an integral number of energy bands or because the highest occupied energy band overlaps the next higher band without an intervening energy gap. The electrons in a partially filled band may acquire a small amount of energy from an applied electric field by going to the higher levels in the same band. The electrons are accelerated in a direction opposite to the field and thereby constitute an electric current. In semiconductors and insulators, the electrons are found only in completely filled bands at low temperatures. In order to increase the energy of the electrons, it is necessary to raise electrons from the valence band to the conduction band across the energy gap. The electric fields normally encountered are not large enough to accomplish this with appreciable probability. At sufficiently high temperatures, depending on the magnitude of the energy gap, a significant number of valence electrons gain enough energy thermally to be raised to the conduction band. These electrons in an unfilled band can easily participate in conduction. Furthermore, there is now a corresponding number of vacancies in the electron population of the valence band. These vacancies, or holes as they are called, have the effect of carriers of positive charge, by means of which the valence band makes a contribution to the conduction of the crystal. *See* HOLE STATES IN SOLIDS.

The type of charge carrier, electron or hole, that is in largest concentration in a material is sometimes called the majority carrier and the type in smallest concentration the minority carrier. The majority carriers are primarily responsible for the conduction properties of the material. Although the minority carriers play a minor role in electrical conductivity, they can be important in rectification and transistor actions in a semiconductor.

Intrinsic semiconductors. A semiconductor in which the concentration of charge carriers is characteristic of the material itself rather than of the content of impurities and structural defects of the crystal is called an intrinsic semiconductor. Electrons in the conduction band and holes in the valence band are created by thermal excitation of electrons from the valence to the conduction band. Thus an intrinsic semiconductor has equal concentrations of electrons and holes. The carrier concentration, and hence the conductivity, is very sensitive to temperature and depends strongly on the energy gap. The energy gap ranges from a fraction of 1 eV to several electronvolts. A material must have a large energy gap to be an insulator.

Extrinsic semiconductors. Typical semiconductor crystals such as germanium and silicon are formed by an ordered bonding of the individual atoms to form the crystal structure. The bonding is attributed to the valence electrons which pair up with valence electrons of adjacent atoms to form so-called shared pair or covalent bonds. These materials are all of the quadrivalent type; that is, each atom contains four valence electrons, all of which are used in forming the crystal bonds. *See* CRYSTAL STRUCTURE.

Atoms having a valence of +3 or +5 can be added to a pure or intrinsic semiconductor material with the result that the +3 atoms will give rise to an unsatisfied bond with one of the valence electrons of the semiconductor atoms, and +5 atoms will result in an extra or free electron that is not required in the bond structure. Electrically, the +3 impurities add holes and the +5 impurities add electrons. They are called acceptor and donor impurities, respectively. Typical valence +3 impurities used are boron, aluminum, indium, and gallium. Valence +5 impurities used are arsenic, antimony, and phosphorus.

Semiconductor material "doped" or "poisoned" by valence +3 acceptor impurities is termed p-type, whereas material doped by valence +5 donor material is termed n-type. The names are derived from the fact that the holes introduced are considered to carry positive charges and the electrons negative charges. The number of electrons in the energy bands of the crystal is increased by the presence of donor impurities and decreased by the presence of acceptor impurities. *See* ACCEPTOR ATOM; DONOR ATOM.

At sufficiently high temperatures, the intrinsic carrier concentration becomes so large that the effect of a fixed amount of impurity atoms in the crystal is comparatively small and the semiconductor becomes intrinsic. When the carrier concentration is predominantly determined by the impurity content, the conduction of the material is said to be extrinsic. Physical defects in the crystal structure may have similar effects as donor or acceptor impurities. They can also give rise to extrinsic conductivity.

Materials. The group of chemical elements which are semiconductors includes germanium, silicon, gray (crystalline) tin, selenium, tellurium, and boron. Germanium, silicon, and gray tin belong to group 14 of the periodic table and have crystal structures similar to that of diamond. Germanium and silicon are two of the best-known semiconductors. They are used extensively in devices such as rectifiers and transistors.

A large number of compounds are known to be semiconductors. A group of semiconducting compounds of the simple type AB consists of elements from columns symmetrically placed with respect to column 14 of the periodic table. Indium antimonide (InSb), cadmium telluride (CdTe), and silver iodide (AgI) are examples of III–V, II–IV, and I–VI compounds, respectively. The various III–V compounds are being studied extensively, and many practical applications have been found for these materials. Some of these compounds have the highest carrier mobilities known for semiconductors. The compounds have zincblende crystal structure which is geometrically similar to the diamond structure possessed by the elemental semiconductors, germanium and silicon, of column 14, except that the four nearest neighbors of each atom are atoms of the other kind. The II–VI compounds, zinc sulfide (ZnS) and cadmium sulfide (CdS), are used in photoconductive devices. Zinc sulfide is also used as a luminescent material. *See* LUMINESCENCE; PHOTOCONDUCTIVITY.

The properties of semiconductors are extremely sensitive to the presence of impurities. It is therefore desirable to start with the purest available materials and to introduce a controlled amount of the desired impurity. The zone-refining method is often used for further purification of obtainable materials. The floating zone technique can be used, if feasible, to prevent any contamination of molten material by contact with the crucible.

For basic studies as well as for many practical applications, it is desirable to use single crystals. Various methods are used for growing crystals of different materials. For many semiconductors, including germanium, silicon, and the III–V compounds, the Czochralski method is commonly used. The method of condensation from the vapor phase is used to grow crystals of a number of semiconductors, for instance, selenium and zinc sulfide. *See* CRYSTAL GROWTH.

The introduction of impurities, or doping, can be accomplished by simply adding the desired quantity to the melt from which the crystal is grown. When the amount to be added is very small, a preliminary ingot is often made with a larger content of the doping agent; a small slice of the ingot is then used to dope the next melt accurately. Impurities which have large diffusion constants in the material can be introduced directly by holding the solid material at an elevated temperature while this material is in contact with the doping agent in the solid or the vapor phase.

A doping technique, ion implantation, has been developed and used extensively. The impurity is introduced into a layer of semiconductor by causing a controlled dose of highly accelerated impurity ions to impinge on the semiconductor.

An important subject of scientific and technological interest is amorphous semiconductors. In an amorphous substance the atomic arrangement has some short-range but no long-range order. The representative amorphous semiconductors are selenium, germanium, and silicon in their amorphous states, and arsenic and germanium chalco-genides, including such ternary systems as Ge-As-Te. Some amorphous semiconductors can be prepared by a suitable quenching procedure from the melt. Amorphous films can be obtained by vapor deposition.

Rectification in semiconductors. In semiconductors, narrow layers can be pro-duced which have abnormally high resistances. The resistance of such a layer is nonohmic; it may depend on the direction of current, thus giving rise to rectification. Rectification can also be obtained by putting a thin layer of semiconductor or insulator material between two conductors of different material.

A narrow region in a semiconductor which has an abnormally high resistance is called a barrier layer. A barrier may exist at the contact of the semiconductor with another material, at a crystal boundary in the semiconductor, or at a free surface of the semiconductor. In the bulk of a semiconductor, even in a single crystal, barriers may be found as the result of a nonuniform distribution of impurities. The thickness of a barrier layer is small, usually 10^{-3} to 10^{-5} cm.

A barrier is usually associated with the existence of a space charge. In an intrinsic semiconductor, a region is electrically neutral if the concentration n of conduction electrons is equal to the concentration p of holes. Any deviation in the balance gives a space charge equal to $e(p - n)$, where e is the charge on an electron. In an extrinsic semiconductor, ionized donor atoms give a positive space charge and ionized acceptor atoms give a negative space charge.

Surface electronics. The surface of a semiconductor plays an important role tech-nologically, for example, in field-effect transistors and charge-coupled devices. Also, it presents an interesting case of two-dimensional systems where the electric field in the surface layer is strong enough to produce a potential wall which is narrower than the wavelengths of charge carriers. In such a case, the electronic energy levels are grouped into subbands, each of which corresponds to a quantized motion normal to the surface, with a continuum for motion parallel to the surface. Consequently, various properties cannot be trivially deduced from those of the bulk semiconductor. See SURFACE PHYSICS.

[H.Y.F]

Semiconductor diode   A two-terminal electronic device that utilizes the prop-erties of the semiconductor from which it is constructed. In a semiconductor diode without a *pn* junction, the bulk properties of the semiconductor itself are used to make a device whose characteristics may be sensitive to light, temperature, or electric field. In a diode with a *pn* junction, the properties of the *pn* junction are used. The most important property of a *pn* junction is that, under ordinary conditions, it will allow electric current to flow in only one direction. Under the proper circumstances, how-ever, a *pn* junction may also be used as a voltage-variable capacitance, a switch, a light source, a voltage regulator, or a means to convert light into electrical power. See SEMICONDUCTOR.

The conductivity of a semiconductor is proportional to the number of electrical carriers (electrons and holes) it contains. In a temperature-compensating diode, or thermistor, the number of carriers changes with temperature. See THERMISTOR.

In a photoconductor the semiconductor is packaged so that it may be exposed to light. Light photons whose energies are greater than the band gap can excite electrons from the valence band to the conduction band, increasing the number of electrical carriers in the semiconductor. See PHOTOCONDUCTIVITY.

In some semiconductors the conduction band has more than one minimum. This results in a region of negative differential conductivity, and a device operated in this region is unstable. The current pulsates at microwave frequencies, and the device, a

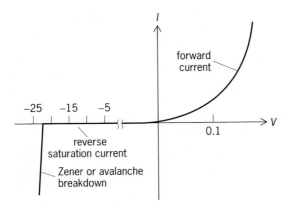

Current-voltage characteristic of a *pn* junction.

Gunn diode, may be used as a microwave power source.

A rectifying junction is formed whenever two materials of different conductivity types are brought into contact. Most commonly, the two materials are an *n*-type and a *p*-type semiconductor, and the device is called a junction diode. However, rectifying action also occurs at a boundary between a metal and a semiconductor of either type. If the metal contacts a large area of semiconductor, the device is known as a Schottky barrier diode; if the contact is a metal point, a point-contact diode is formed. *See* SCHOTTKY EFFECT.

The contact potential between the two materials in a diode creates a potential barrier which tends to keep electrons on the *n* side of the junction and holes on the *p* side. When the *p* side is made positive with respect to the *n* side by an applied field, the barrier height is lowered and the diode is forward biased. Majority electrons from the *n* side may flow easily to the *p* side, and majority holes from the *p* side may flow easily to the *n* side. When the *p* side is made negative, the barrier height is increased and the diode is reverse-biased. Then, only a small leakage current flows: Minority electrons from the *p* side flow into the *n* side, and minority holes from the *n* side flow into the *p* side. The current-voltage characteristic of a typical diode is shown in the illustration. Rectifying diodes can be made in a variety of sizes, and much practical use can be made of the fact that such a diode allows current to flow in essentially one direction only. [S.N.]

Semiconductor heterostructures Structures consisting of two different semiconductor materials in junction contact, with unique electrical or electrooptical characteristics. A heterojunction is a junction in a single crystal between two dissimilar semiconductors. The most important differences between the two semiconductors are generally in the energy gap and the refractive index. In semiconductor heterostructures, differences in energy gap permit spatial confinement of injected electrons and holes, while the differences in refractive index can be used to form optical waveguides. Semiconductor heterostructures have been used for diode lasers, light-emitting diodes, optical detector diodes, and solar cells. In fact, heterostructures must be used to obtain continuous operation of diode lasers at room temperature. Heterostructures also exhibit other interesting properties such as the quantization of confined carrier motion in ultrathin heterostructures and enhanced carrier mobility in modulation-doped heterostructures. Structures of current interest utilize III–V and IV–VI compounds having similar crystal structures and closely matched lattice constants. *See* BAND THEORY OF SOLIDS; LASER; LIGHT-EMITTING DIODE; REFRACTION OF WAVES.

The most intensively studied and thoroughly documented materials for heterostructures are GaAs and $Al_xGa_{1-x}As$. Several other III–V and IV–VI systems also are used for semiconductor heterostructures. A close lattice match is necessary in heterostructures in order to obtain high-quality crystal layers by epitaxial growth and thereby to prevent excessive carrier recombination at the heterojunction interface.

When the narrow energy gap layer in heterostructures becomes a few tens of nanometers or less in thickness, new effects that are associated with the quantization of confined carriers are observed. These ultrathin heterostructures are referred to as superlattices or quantum well structures, and they consist of alternating layers of GaAs and $Al_xGa_{1-x}As$. These structures are generally prepared by molecular-beam epitaxy. Each layer is 5 to 40 nanometers thick.

In the GaAs layers, the motion of the carriers is restricted in the direction perpendicular to the heterojunction interfaces, while they are free to move in the other two directions. The carriers can therefore be considered as a two-dimensional gas. The Schrödinger wave equation shows that the carriers moving in the confining direction can have only discrete bound states. *See* QUANTUM MECHANICS.

Another property of semiconductor heterostructures is illustrated by a modulation doping technique that spatially separates conduction electrons in the GaAs layer and their parent donor impurity atoms in the $Al_xGa_{1-x}As$ layer. Since the carrier mobility in semiconductors is decreased by the presence of ionized and neutral impurities, the carrier mobility in the modulation-doped GaAs is larger than for a GaAs layer doped with impurities to give the same free electron concentration. Higher carrier mobilities should permit preparation of devices that operate at higher frequencies than are possible with doped layers. *See* SEMICONDUCTOR.   [H.C.C.]

Series circuit   An electric circuit in which the principal circuit elements have their terminals joined in sequence so that a common current flows through all the elements. The circuit may consist of any number of passive and active elements, such as resistors, inductors, capacitors, electron tubes, and transistors.   [R.L.R.]

Shadow   A region of darkness caused by the presence of an opaque object interposed between such a region and a source of light. A shadow can be totally dark only in that part called the umbra, in which all parts of the source are screened off. With a point source, the entire shadow consists of an umbra, since there can be no region in which only part of the source is eclipsed. If the source has an appreciable extent, however, there exists a transition surrounding the umbra, called the penumbra, which is illuminated by only part of the source. Depending on what fraction of the source is exposed, the illumination in the penumbra varies from zero at the edge of the full shadow to the maximum where the entire source is exposed. The edge of the umbra is not perfectly sharp, even with an ideal point source, because of the wave character of light. *See* DIFFRACTION.   [F.A.J./W.W.W.]

Shadowgraph   An optical method of rendering fluid flow patterns visible by using index-of-refraction differences in the flow. The method relies on the fact that rays of light bend toward regions of higher refractive index while passing through a transparent material. The fluid is usually illuminated by a parallel beam of light. The illustration depicts the method as it might be applied to a fluid sample undergoing thermal convection between two parallel plates, with the lower plate being kept warmer

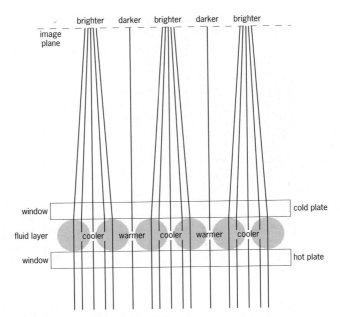

brighter   darker   brighter   darker   brighter

image plane

window .... cold plate

fluid layer   cooler   warmer   cooler   warmer   cooler

window .... hot plate

Schematic of the shadowgraph method applied to reveal convection patterns in a fluid layer. A cross section of the apparatus perpendicular to the convection rolls is shown.

than the upper one. As illustrated, the rays bend toward the cooler down-flowing regions, where the refractive index is higher, and away from the warmer up-flowing ones. After they have passed through the fluid layer, the rays tend to focus above the cooler regions and defocus above the warmer regions. If an image of the light beam is recorded not too far from the sample, brighter areas of the image will lie above regions of down flow, where the rays have been concentrated, and darker areas will lie above regions of up flow. Because the light passes completely through the sample, the bending effect for each ray is averaged over the sample thickness. *See* CONVECTION (HEAT); REFRACTION OF WAVES.

In convection experiments the refractive index varies because of thermal expansion of the fluid, but the method is not restricted regarding the mechanism responsible for disturbing the refractive index. Thus the same method may be used to visualize denser and less dense regions in a gas flowing in a wind tunnel, including Mach waves and shock waves, where the denser regions have a higher-than-average refractive index. *See* SHOCK WAVE; SUPERSONIC FLOW.

Images are usually recorded by means of a charge-coupled-device (CCD) camera, digitized, and stored in a computer. Such a digitized image consists of an array of numbers, each number being proportional to the brightness at a particular point in the image. The image points (pixels) form a closely spaced rectangular grid. A reference image may be taken in the absence of any fluid flow, and the reference image may be divided point by point into images taken with the fluid moving.   [D.S.Ca.]

Shock wave   A mechanical wave of large amplitude, propagating at supersonic velocity, across which pressure or stress, density, particle velocity, temperature, and related properties change in a nearly discontinuous manner. Unlike acoustic waves, shock waves are characterized by an amplitude-dependent wave velocity. Shock waves

Schlieren photograph of supersonic flow over blunt object. Shock wave is approximately parabolic, and detached from object. (*Avco Everett Research Laboratory, Inc.*)

arise from sharp and violent disturbances generated from a lightning stroke, bomb blast, or other form of intense explosion, and from steady supersonic flow over bodies.

The abrupt nature of a shock wave in a gas can best be visualized from a schlieren photograph or shadow graph of supersonic flow over objects. Such photographs show well-defined surfaces in the flow field across which the density changes rapidly, in contrast to waves within the range of linear dynamic behavior of the fluid. Measurements of fluid density, pressure, and temperature across the surfaces show that these quantities always increase along the direction of flow, and that the rates of change are usually so rapid as to be beyond the spatial resolution of most instruments. These surfaces of abrupt change in fluid properties are called shock waves or shock fronts. *See* SCHLIEREN PHOTOGRAPHY.

Shock waves in supersonic flow may be classified as normal or oblique according to whether the orientation of the surface of abrupt change is perpendicular or at an angle to the direction of flow. A schlieren photograph of a supersonic flow over a blunt object is shown in the illustration. Although this photograph was obtained from a supersonic flow over a stationary model in a shock tube, the general shape of the shock wave around the object is quite typical of those observed in a supersonic wind tunnel, or of similar objects (or projectiles) flying at supersonic speeds in a stationary atmosphere. The shock wave in this case assumes an approximately parabolic shape and is clearly detached from the blunt object. The central part of the wave, just in front of the object, may be considered an approximate model of the normal shock; the outer part of the wave is an oblique shock wave of gradually changing obliqueness and strength.

[S.C.Li.]

Some common examples of shock waves in condensed materials are encountered in the study of underground or underwater explosions, meteorite impacts, and ballistics problems. The field of shock waves in condensed materials (solids and liquids) has grown into an important interdisciplinary area of research involving condensed matter physics, geophysics, materials science, applied mechanics, and chemical physics. The nonlinear aspect of shock waves is an important area of applied mathematics.

Experimentally, shock waves are produced by rapidly imparting momentum over a large flat surface. This can be accomplished in many different ways: rapid deposition of radiation using electron or photon beams (lasers or x-rays), detonation of a high explosive in contact with the material, or high-speed impact of a plate on the sample surface. The impacting plate itself can be accelerated by using explosives, electrical discharge, underground nuclear explosions, and compressed gases. The use of compressed gas to accelerate projectiles with appropriate flyer plates provides the highest precision and control as well as convenience in laboratory experiments.

Large-amplitude one-dimensional compression and shear waves have been studied in solids. In these experiments, a macroscopic volume element is subjected to both a compression and shear deformation. The combined deformation state is produced by impacting two parallel flyer plates that are inclined at an angle to the direction of the plate motion. Momentum conservation coupled with different wave velocities for compression and shear waves leads to a separation of these waves in the sample interior. These experiments provide direct information about the shear response of shocked solids, and subject samples to more general loading states than the uniaxial strain state.

Shock waves subject matter to unusual conditions and therefore provide a good test of understanding of fundamental processes. The majority of the studies on condensed materials have concentrated on mechanical and thermodynamic properties. These are obtained from measurements of shock velocity, stress, and particle velocity in well-controlled experiments. Advanced techniques using electromagnetic gages, laser interferometry, piezoelectric gages, and piezoresistance gages have given continuous, time-resolved measurements at different sample thicknesses.

The study of residual effects, that is, the postshock examination of samples subjected to a known pulse amplitude and duration, is of considerable importance to materials science and metallurgy. The conversion of graphite to diamonds is noteworthy. Other effects that have been observed are microstructural changes, enhanced chemical activity, changes in material hardness and strength, and changes in electrical and magnetic properties. The generation of shock-induced lattice defects is thought to be important for explaining these changes in material properties. There has been growing interest in using shock methods for material synthesis and powder compaction.   [Y.M.G.]

Short circuit   An abnormal condition (including an arc) of relatively low impedance, whether made accidentally or intentionally, between two points of different potential in an electric network or system. *See* ELECTRICAL IMPEDANCE.

Common usage of the term implies an undesirable condition arising from failure of electrical insulation, from natural causes (lightning, wind, and so forth), or from human causes (accidents, intrusion, and so forth). From an analytical viewpoint, however, short circuits represent a severe condition that the circuit designer must consider in designing an electric system that must withstand all possible operating conditions.

In circuit theory the short-circuit condition represents a basic condition that is used analytically to derive important information concerning the network behavior and operating capability. Thus, along with the open-circuit voltage, the short-circuit current provides important basic information about the network at a given point.

The short-circuit condition is also used in network theory to describe a general condition of zero voltage. Thus the term short-circuit admittance (or impedance) is used to describe a network condition in which certain terminals have had their voltage reduced to zero for the purpose of analysis. This leads to the terms short-circuit driving point admittance, short-circuit transfer admittance, and similar references to the zero voltage condition. *See* ADMITTANCE.

Short-circuit protection is a separate discipline dedicated to the study, analysis, application, and design of protective apparatus that are intended to minimize the effect of unintentional short circuits in power supply systems. For these analyses the short circuit is an important limiting (worst) case, and is used to compute the coordination of fuses, circuit reclosers, circuit breakers, and other devices designed to recognize and isolate short circuits. The short circuit is also an important parameter in the specification of these protective devices, which must have adequate capability for interrupting the high short-circuit current.

Short circuits are also important on high-frequency transmission lines where shorted stub lines, one-quarter wavelength long and shorted at the remote end, are used to design matching sections of the transmission lines which also act as tuning elements.

[P.M.A.]

Sine wave   A wave having a form which, if plotted, would be the same as that of a trigonometric sine or cosine function. The sine wave may be thought of as the projection on a plane of the path of a point moving around a circle at uniform speed. It is characteristic of one-dimensional vibrations and one-dimensional waves having no dissipation. *See* HARMONIC MOTION.

The sine wave is the basic function employed in harmonic analysis. It can be shown that any complex motion in a one-dimensional system can be described as the superposition of sine waves having certain amplitude and phase relationships. The technique for determining these relationships is known as Fourier analysis. *See* WAVE EQUATION; WAVE MOTION.

[W.J.G.]

Single crystal   In crystalline solids the atoms or molecules are stacked in a regular manner, forming a three-dimensional pattern which may be obtained by a three-dimensional repetition of a certain pattern unit called a unit cell. When the periodicity of the pattern extends throughout a certain piece of material, one speaks of a single crystal. A single crystal is formed by the growth of a crystal nucleus without secondary nucleation or impingement on other crystals. *See* CRYSTAL STRUCTURE; CRYSTALLOGRAPHY.

When grown from a melt, single crystals usually take the form of their container. Crystals grown from solution (gas, liquid, or solid) often have a well-defined form which reflects the symmetry of the unit cell. *See* CRYSTAL GROWTH.

Ideally, single crystals are free from internal boundaries. They give rise to a characteristic x-ray diffraction pattern.

Many types of single crystal exhibit anisotropy, that is, a variation of some of their physical properties according to the direction along which they are measured. For example, the electrical resistivity of a randomly oriented aggregate of graphite crystallites is the same in all directions. This anisotropy exists both for structure-sensitive properties, which are strongly affected by crystal imperfections (such as cleavage and crystal growth rate), and for structure-insensitive properties, which are not affected by imperfections (such as elastic coefficients).

The structure-sensitive properties of crystals (for example, strength and diffusion coefficients) seem governed by internal defects, often on an atomic scale. *See* CRYSTAL DEFECTS.

[D.T.]

Skin effect (electricity)   The tendency for an alternating current to concentrate near the outer part or "skin" of a conductor. For a steady unidirectional current through a homogeneous conductor, the current distribution is uniform over the cross section; that is, the current density is the same at all points in the cross section. With an alternating current, the current is displaced more and more to the surface as the frequency increases. The conductor's effective cross section is therefore reduced so the resistance and energy dissipation are increased compared with the values for a uniformly distributed current. The effective resistance of a wire rises significantly with frequency; for example, for a copper wire of 1-mm (0.04-in.) diameter, the resistance at a frequency of 1 MHz is almost four times the dc value. *See* ELECTRICAL RESISTANCE.

A skin depth or penetration depth δ is frequently used in assessing the results of skin effect; it is the depth below the conductor surface at which the current density has decreased to $1/e$ (approximately 37%) of its value at the surface. This concept applies strictly only to plane solids, but can be extended to other shapes provided the radius of curvature of the conductor surface is appreciably greater than δ.

At a frequency of 60 Hz the penetration depth in copper is 8.5 mm (0.33 in.); at 10 GHz it is only 6.6×10^{-7} m. Waveguide and resonant cavity internal surfaces for use at microwave frequencies are therefore frequently plated with a high-conductivity material, such as silver, to reduce the energy losses since nearly all the current is concentrated at the surface. Provided the plating material is thick compared to δ, the conductor is as good as a solid conductor of the coating material.   [F.A.Be.]

Slow neutron spectroscopy   The use of beams of slow neutrons, from nuclear reactors or nuclear accelerators, in studies of the structure or structural dynamics of solid, liquid, or gaseous matter. Studies of the chemical or magnetic structure of substances are usually referred to under the term neutron diffraction, while studies of atomic and magnetic dynamics go under the terms slow neutron spectroscopy, inelastic neutron scattering, or simply neutron spectroscopy. *See* NEUTRON DIFFRACTION; PARTICLE ACCELERATOR; SPECTROSCOPY; X-RAY DIFFRACTION.

In a neutron spectroscopy experiment, a beam of neutrons is scattered by a specimen and the scattered neutrons are detected at various angles to the initial beam. From these measurements, the linear momenta of the incoming and outgoing neutrons (and the vector momentum changes experienced by individual neutrons) can be computed. In general, just those neutrons which have been scattered once only by the specimen are useful for analysis; the specimen must be "thin" with respect to neutron scattering power as well as to neutron absorption. In practice, the experiments are usually intensity-limited, since even the most powerful reactors or accelerators are sources of weak luminosity when, as here, individual slow neutrons are to be considered as quanta of radiation. *See* NEUTRON.

Neutron spectroscopy requires slow neutrons, with energies of the order of neutrons in equilibrium with matter at room temperature, or approximately 0.025 eV. The corresponding de Broglie wavelengths are approximately 0.2 nanometer, of the order of interatomic spacings in solids or liquids. The fast neutrons emitted in nuclear or slow fission reactions can be slowed down to thermal velocities in matter which is transparent to neutrons and which contains light elements, such as hydrogen, carbon, and beryllium, by a process of diffusion and elastic (billiard-ball) scattering known as neutron moderation. By selection of those diffusing neutrons which travel in a certain restricted range of directions (collimation), a beam of thermal and near-thermal neutrons can be obtained. *See* QUANTUM MECHANICS.

The bulk of the observations can be accounted for in terms of scattering of semiclassical neutron waves by massive, moving-point scatterers in the forms of atomic nuclei and their bound electron clouds. The spatial structure of the scatterers, time-averaged over the duration of the experiment, gives rise to the elastic scattering from the specimen that is studied in neutron diffraction; the spatial motions of the scatterers give rise to the Doppler-shifted inelastic scattering involved in slow neutron spectroscopy.

Just as (slow) neutron diffraction is the most powerful available scientific tool for study of the magnetic structure of matter on an atomic scale, so slow neutron spectroscopy is the most powerful tool for study of the atomic magnetic and nuclear dynamics of matter in all its phases. The direct nature of the analysis has in some cases added considerable support to the conceptual structure of solid-state and liquid-state physics and thus to the confidence with which the physics is applied. For example, neutron spectroscopy has confirmed the existence of phonons, magnons, and the quasiparticles (rotons) of liquid helium II. Detailed information has been obtained on the lattice vibrations of most of the crystalline elements and numerous simple compounds, on the atomic dynamics of many simple liquids, on the dynamics of liquid helium in different phases, and on the atomic magnetic dynamics of a great variety of ferromagnetic, ferrimagnetic, antiferromagnetic, and modulated magnetic substances.   [B.N.B.]

Solid-state physics
The study of the physical properties of solids, such as electrical, dielectric, elastic, and thermal properties, and their understanding in terms of fundamental physical laws. Most problems in solid-state physics would be called solid-state chemistry if studied by scientists with chemical training, and vice versa. Solid-state physics emphasizes the properties common to large classes of compounds rather than the dependence of properties upon compositions, the latter receiving greater emphasis in solid-state chemistry. In addition, solid-state chemistry tends to be more descriptive, while solid-state physics focuses upon quantitative relationships between properties and the underlying electronic structure.

Many of the scientists who study the physics of liquids identify with solid-state physics, and the term "condensed-matter physics" has been used by some researchers to replace "solid-state physics" as a division of physics. It includes noncrystalline solids such as glass as well as crystalline solids. *See* AMORPHOUS SOLID.

In solid-state physics it is generally assumed that the electronic states can be described as wavelike. The individual electronic states, called Bloch states, have energies which depend upon the wave number (a vector equal to the momentum divided by $\hbar$, which is Planck's constant divided by 2π), and the wave number is restricted to a domain called the Brillouin zone. This energy given as a function of the wave number is called the band structure. There are several curves, called bands, for each line in the Brillouin zone. *See* BRILLOUIN ZONE.

The total energy of a solid includes a sum of the energies of the occupied electronic states. Since the energy bands depend upon the positions of the atoms, so does the total energy, and the stable crystal structure is that which minimizes this energy. The theory has not proved adequate to really predict the crystal structure of various solids, but it is possible to predict the changes in energy under various distortions of the lattice. There are in fact three times as many independent distortions, called normal modes, as there are atoms in the solid. Each has a wave number, and the frequencies of the normal vibrational modes, as a function of wave number in the Brillouin zone, form vibrational bands in direct analogy with the electronic energy bands. These can be directly calculated from quantum theory or measured by using neutron or x-ray diffraction. *See* CRYSTAL; LATTICE VIBRATIONS; NEUTRON DIFFRACTION; X-RAY DIFFRACTION.

[W.A.H.]

Soliton   An isolated wave that propagates without dispersing its energy over larger and larger regions of space. In most of the scientific literature, the requirement that two solitons emerge unchanged from a collision is also added to the definition; otherwise the disturbance is termed a solitary wave.

There are many equations of mathematical physics which have solutions of the soliton type. Correspondingly, the phenomena which they describe, be it the motion of waves in shallow water or in an ionized plasma, exhibit solitons. The first observation of this kind of wave was made in 1834 by John Scott Russell, who followed on horseback a soliton propagating in the windings of a channel. In 1895, D. J. Korteweg and H. de Vries proposed an equation for the motion of waves in shallow waters which possesses soliton solutions, and thus established a mathematical basis for the study of the phenomenon. Interest in the subject, however, lay dormant for many years, and the major body of investigations began only in the 1950s. Researches done by analytical methods and by numerical methods made possible with the advent of computers gradually led to a complete understanding of solitons.

Eventually, the fact that solitons exhibit particlelike properties, because the energy is at any instant confined to a limited region of space, received attention, and solitons were proposed as models for elementary particles. However, it is difficult to account for all of the properties of known particles in terms of solitons. More recently it has been realized that some of the quantum fields which are used to describe particles and their interactions also have solutions of the soliton type. The solitons would then appear as additional particles, and may have escaped experimental detection because their masses are much larger than those of known particles. In this context the requirement that solitons emerge unchanged from a collision has been found too restrictive, and particle theorists have used the term soliton where traditionally the term solitary wave would be used. *See* ELEMENTARY PARTICLE; QUANTUM FIELD THEORY.   [C.R.]

A hydrodynamic soliton is simply described by the equation of Korteweg and de Vries, which includes a dispersive term and a term to represent nonlinear effects. Easily observed in a wave tank, a bell-shaped solution of this equation balances the effects of dispersion and nonlinearity, and it is this balance that is the essential feature of the soliton phenomenon. Tidal waves in the Firth of Forth were found by Scott Russell to be solitons, as are internal ocean waves and tsunamis. At an even greater level of energy, it has been suggested that the Great Red Spot of the planet Jupiter is a hydrodynamic soliton.

The most significant technical application of the soliton is as a carrier of digital information along an optical fiber. The optical soliton is governed by the nonlinear Schrödinger equation, and again expresses a balance between the effects of optical dispersion and nonlinearity that is due to electric field dependence of the refractive index in the fiber core. If the power is too low, nonlinear effects become negligible, and the information spreads (or disperses) over an ever increasing length of the fiber. At a pulse power level of about 5 milliwatts, however, a robustly stable soliton appears and maintains its size and shape in the presence of disturbing influences. Present designs for data transmission systems based on the optical soliton have a data rate of 4×10^9 bits per second.

A carefully studied soliton system is the transverse electromagnetic (TEM) wave that travels between two strips of superconducting metal separated by an insulating layer thin enough (about 2.5 nanometers) to permit transverse Josephson tunneling. Since each soliton carries one quantum of magnetic flux, it is also called a fluxon if the magnetic flux points in one direction, and an antifluxon if the flux points in the opposite direction. Oscillators based on this system reach into the submillimeter wave region of the electromagnetic spectrum (frequencies greater than 10^{11} Hz). *See* JOSEPHSON EFFECT.

The all-or-nothing action potential or nerve impulse that carries a bit of biological information along the axon of a nerve cell shares many properties with the soliton. Both are solutions of nonlinear equations that travel with fixed shape at constant speed, but the soliton conserves energy, while the nerve impulse balances the rate at which electrostatic energy is released from the nerve membrane to the rate at which it is consumed by the dissipative effects of circulating ionic currents. The nerve process is much like the flame of a candle.

[A.Sco.]

Sonic boom   An audible sound wave generated by an object that moves faster than the speed of sound (supersonic object). The sonic boom forms because the air is pushed away faster than the air molecules can move. The displaced air becomes highly compressed and creates a very strong sound wave, referred to as a compressional head shock or bow shock. At the back of the supersonic object the air has to fill the void left as the object moves forward; in this case, the gas becomes rarefied and a rarefractional tail shock develops. These shock waves are the main components of a sonic boom, and they are generated the entire time that an object flies faster than the speed of sound, not just when it breaks the sonic barrier. *See* SHOCK WAVE.

Sonic booms may be natural or generated by human activity. A natural sonic boom is thunder, created when lightning ionizes air, which expands supersonically. Meteors can create sonic booms if they enter the atmosphere at supersonic speeds. Human sources of sonic booms include aircraft, rockets, the space shuttle during reentry, and bullets.

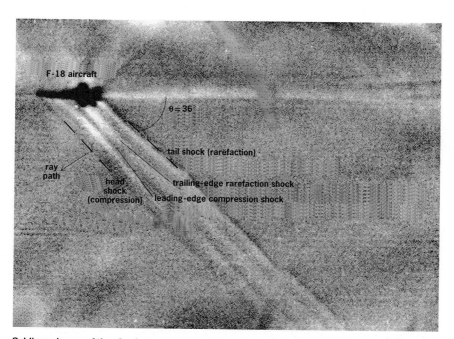

Schlieren image of the shock waves generated by an F-18 aircraft flying at Mach 1.4. The Mach cone generated by the head and tail shocks can be seen as well as the shocks generated by the leading and trailing edges of the wings. (*NASA Dryden Flight Research Center*)

Sonic booms are commonly associated with supersonic aircraft. The shock waves associated with sonic booms propagate away from the aircraft in a unique fashion. These waves form a cone, called the Mach cone, that is dragged behind the aircraft. The illustration shows the outline of the Mach cone generated by an F-18 fighter aircraft flying at Mach 1.4. The schlieren photographic technique was used to display the sonic boom, which is normally invisible. The half-angle of the cone is determined solely by the Mach number of the aircraft, $\theta = \arctan(1/M)$, $36°$ for $M = 1.4$. The shock waves travel along rays that are perpendicular to the Mach cone (see illustration). As the Mach number increases, θ becomes smaller and the sound travels almost directly downward. *See* MACH NUMBER; SCHLIEREN PHOTOGRAPHY.

The typical peak pressure amplitude (or overpressure) of a sonic boom on the ground is about 50–100 pascals. A sonic boom with 50 Pa (1 lbf/ft^2 or 0.007 psi) overpressure will produce no damage to buildings. Booms in the range of 75–100 Pa are considered disturbing by some people. Occasionally there is minor damage to buildings from sonic booms in the range of 100–250 Pa; however, buildings in good condition will be undamaged by overpressures up to 550 Pa. Very low flying aircraft (30 m or 100 ft) can produce sonic booms of 1000–7000 Pa. These pressures are about five times less than that needed to injure the human ear, but can lead to damage to buildings, such as the breaking of glass windows and the cracking of plaster. Although sonic booms are not dangerous, they can evoke a strong startle response in people and animals. [R.Cl.]

Sound The mechanical excitation of an elastic medium. Originally, sound was considered to be only that which is heard. This admitted questions such as whether or not sound was generated by trees falling where no one could hear. A more mechanistic approach avoids these questions and also allows acoustic disturbances too high in frequency (ultrasonic) to be heard or too low (infrasonic) to be classed as extensions of those events that can be heard.

A source of sound undergoes rapid changes of shape, size, or position that disturb adjacent elements of the surrounding medium, causing them to move about their equilibrium positions. These disturbances in turn are transmitted elastically to neighboring elements. This chain of events propagates to larger and larger distances, constituting a wave traveling through the medium. If the wave contains the appropriate range of frequencies and impinges on the ear, it generates the nerve impulses that are perceived as hearing.

Acoustic pressure. A sound wave compresses and dilates the material elements it passes through, generating associated pressure fluctuations. An appropriate sensor (a microphone, for example) placed in the sound field will record a time-varying deviation from the equilibrium pressure found at that point within the fluid. The changing total pressure P measured will vary about the equilibrium pressure P_0 by a small amount called the acoustic pressure, $p = P - P_0$. The SI unit of pressure is the pascal (Pa), equal to 1 newton per square meter (N/m^2). Standard atmospheric pressure (14.7 lb/in.^{2}) is approximately 1 bar $= 10^6$ dyne/cm$^2 = 10^5$ Pa. For a typical sound in air, the amplitude of the acoustic pressure may be about 0.1 Pa (one-millionth of an atmosphere); most sounds cause relatively slight perturbations of the total pressure. *See* PRESSURE; PRESSURE MEASUREMENT; PRESSURE TRANSDUCER; SOUND PRESSURE.

Plane waves. One of the more basic sound waves is the traveling plane wave. This is a pressure wave progressing through the medium in one direction, say the $+x$ direction, with infinite extent in the y and z directions. A two-dimensional analog is ocean surf advancing toward a very long, straight, and even beach. *See* WAVE (PHYSICS); WAVE EQUATION; WAVE MOTION.

A most important plane wave, called harmonie, is the smoothly oscillating monofrequency plane wave described by Eq. (1). The amplitude of this wave is P. The phase

$$p = P \cos \left[2\pi f \left(t - \frac{x}{c} \right) \right] \tag{1}$$

(argument of the cosine) increases with time, and at a point in space the cosine will pass through one full cycle for each increase in phase of 2π. The period T required for each cycle must therefore be such that $2\pi fT = 2\pi$, or $T = 1/f$, so that $f = 1/T$ can be identified as the frequency of oscillation of the pressure wave. During this period T, each portion of the waveform has advanced through a distance $\lambda = cT$, and this distance λ must be the wavelength. This gives the fundamental relation (2) between

$$\lambda f = c \tag{2}$$

the frequency, wavelength, and speed of sound c in any medium. For example, in air at room temperature the speed of sound is 343 m/s (1125 ft/s). A sound of frequency 1 kHz (1000 cycles per second) will have a wavelength of $\lambda = c/f = 343/1000$ m = 0.34 m (1.1 ft). Lower frequencies will have longer wavelengths: a sound of 100 Hz in air has a wavelength of 3.4 m (11 ft). For comparison, in fresh water at room temperature the speed of sound is 1480 m/s (4856 ft/s), and the wavelength of 1-kHz sound is nearly 1.5 m (5 ft), almost five times greater than the wavelength for the same frequency in air.

Description of sound. The characterization of a sound is based primarily on human psychological responses to it. Because of the nature of human perceptions, the correlations between basically subjective evaluations such as loudness, pitch, and timbre and more physical qualities such as energy, frequency, and frequency spectrum are subtle and not necessarily universal.

The strength of a sound wave is described by its intensity. From basic physical principles, the instantaneous rate at which energy is transmitted by a sound wave through unit area is given by the product of acoustic pressure and the component of particle velocity perpendicular to the area. The time average of this quantity is the acoustic intensity. If all quantities are expressed in SI units (pressure amplitude or effective pressure amplitude in Pa, speed of sound in m/s, and density in kg/m^3), then the intensity will be in watts per square meter (W/m^2). *See* SOUND INTENSITY.

Because of the way the strength of a sound is perceived, it has become conventional to specify the intensity of sound in terms of a logarithmic scale with the (dimensionless) unit of the decibel (dB). An individual with unimpaired hearing has a threshold of perception near 10^{-12} W/m^2 between about 2 and 4 kHz, the frequency range of greatest sensitivity. As the intensity of a sound of fixed frequency is increased, the subjective evaluation of loudness also increases, but not proportionally. Rather, the listener tends to judge that *every* successive doubling of the acoustic intensity corresponds to the same increase in loudness. For sounds lying higher than 4 kHz or lower than 500 Hz, the sensitivity of the ear is appreciably lessened. Sounds at these frequency extremes must have higher threshold intensity levels before they can be perceived, and doubling of the loudness requires smaller changes in the intensity with the result that at higher levels sounds of equal intensities tend to have more similar loudnesses. It is because of this characteristic that reducing the volume of recorded music causes it to sound thin or tinny, lacking both highs and lows of frequency. Since most sound-measuring equipment detects acoustic pressure rather than intensity, it is convenient to define an equivalent scale in terms of the sound pressure level. The intensity level and sound-pressure level are usually taken as identical, but this is not always true. *See* DECIBEL.

How "high" sound of a particular frequency appears to be is described by the sense of pitch. A few minutes with a frequency generator and a loudspeaker show that pitch is closely related to the frequency. Higher pitch corresponds to higher frequency, with small influences depending on loudness, duration, and the complexity of the waveform. For the pure tones (monofrequency sounds) encountered mainly in the laboratory, pitch and frequency are not found to be proportional. Doubling the frequency less than doubles the pitch. For the more complex waveforms usually encountered, however, the presence of harmonics favors a proportional relationship between pitch and frequency.

Propagation of sound. Plane waves are a considerable simplification of an actual sound field. The sound radiated from a source (such as a loudspeaker, a hand clap, or a voice) must spread outward much like the widening circles from a pebble thrown into a lake. A simple model of this more realistic case is a spherical source vibrating uniformly in all directions with a single frequency of motion. The sound field must be spherically symmetric with an amplitude that decreases with increasing distance from the source, and the fluid elements must have particle velocities that are directed radially.

Not all sources radiate their sound uniformly in all directions. When someone is speaking in an unconfined space, for example an open field, a listener circling the speaker hears the voice most well defined when the speaker is facing the listener. The voice loses definition when the speaker is facing away from the listener. Higher frequencies tend to be more pronounced in front of the speaker, whereas lower frequencies are perceived more or less uniformly around the speaker.

Diffraction. It is possible to hear but not see around the corner of a tall building. However, higher-frequency sound (with shorter wavelength) tends to bend or "spill" less around edges and corners than does sound of lower frequency. The ability of a wave to spread out after traveling through an opening and to bend around obstacles is termed diffraction. This is why it is often difficult to shield a listener from an undesired source of noise, like blocking aircraft or traffic noise from nearby residences. Simply erecting a brick or concrete wall between source and receiver is often an insufficient remedy, because the sounds may diffract around the top of the wall and reach the listeners with sufficient intensity to be distracting or bothersome. *See* ACOUSTIC NOISE; DIFFRACTION.

Rays. Since the speed of sound varies with the local temperature (and pressure, in other than perfect gases), the speed of a sound wave can be a function of position. Different portions of a sound wave may travel with different speeds of sound.

Each small element of a surface of constant phase traces a line in space, defining a ray along which acoustic energy travels. The sound beam can then be viewed as a ray bundle, like a sheaf of wheat, with the rays distributed over the cross-sectional area of the surface of constant phase. As the major lobe spreads with distance, this area increases and the rays are less densely concentrated. The number of rays per unit area transverse to the propagation path measures the energy density of the sound at that point.

It is possible to use the concept of rays to study the propagation of a sound field. The ray paths define the trajectories over which acoustic energy is transported by the traveling wave, and the flux density of the rays measures the intensity to be found at each point in space. This approach, an alternative way to study the propagation of sound, is approximate in nature but has the advantage of being very easy to visualize.

Reflection and transmission. If a sound wave traveling in one fluid strikes a boundary between the first fluid and a second, then there may be reflection and transmission of sound. For most cases, it is sufficient to consider the waves to be planar. The first fluid contains the incident wave of intensity I_i and reflected wave of intensity I_r; the second fluid, from which the sound is reflected, contains the transmitted wave of intensity I_t. The directions of the incident, reflected, and transmitted plane sound waves may be specified by the grazing angles θ_i, θ_r, and θ_t (measured between the respective directions of propagation and the plane of the reflecting surface). *See* REFLECTION OF SOUND.

Absorption. When sound propagates through a medium, there are a number of mechanisms by which the acoustic energy is converted to heat and the sound wave weakened until it is entirely dissipated. This absorption of acoustic energy is characterized by a spatial absorption coefficient for traveling waves. *See* SOUND ABSORPTION.

[A.B.Co.]

Sound absorption   The process by which the intensity of sound is diminished by the conversion of the energy of the sound wave into heat. The absorption of sound is an important case of sound attenuation. Regardless of the material through which sound passes, its intensity, measured by the average flow of energy in the wave per unit time per unit area perpendicular to the direction of propagation, decreases with distance from the source. This decrease is called attenuation. In the simple case of a point source of sound radiating into an ideal medium (having no boundaries, turbulent fluctuations, and the like), the intensity decreases inversely as the square of the distance from the source. This relationship exists because the spherical area through which the energy propagates per unit time increases as the square of the propagation distance. This attenuation or loss may be called geometrical attenuation.

In addition to this attenuation due to spreading, there is effective attenuation caused by scattering within the medium. Sound can be reflected and refracted when incident on media of different physical properties, and can be diffracted and scattered as it bends around obstacles. These processes lead to effective attenuation, for example, in a turbulent atmosphere; this is easily observed in practice and can be measured, but is difficult to calculate theoretically with precision. *See* DIFFRACTION; REFLECTION OF SOUND; REFRACTION OF WAVES.

In actual material media, geometrical attenuation and effective attenuation are supplemented by absorption due to the interaction between the sound wave and the physical properties of the propagation medium itself. This interaction dissipates the sound energy by transforming it into heat and hence decreases the intensity of the wave. In all practical cases, the attenuation due to such absorption is exponential in character. *See* SOUND INTENSITY.

The four classical mechanisms of sound absorption in material media are shear viscosity, heat conduction, heat radiation, and diffusion. These attenuation mechanisms are generally grouped together and referred to as classical attenuation or thermoviscous attenuation. *See* CONDUCTION (HEAT); DIFFUSION; HEAT RADIATION; VISCOSITY.

Sound absorption in fluids can be measured in a variety of ways, referred to as mechanical, optical, electrical, and thermal methods. All these methods reduce essentially to a measurement of sound intensity as a function of distance from the source.

The amount of sound that air absorbs increases with audio frequency and decreases with air density, but also depends on temperature and humidity. Sound absorption in air depends heavily on relative humidity. The reason for the strong dependence on relative humidity is molecular relaxation. One can note the presence of two transition regimes in most of the actual absorption curves, representing the relaxation effects of nitrogen and oxygen, the dominant constituents of the atmosphere. *See* ATMOSPHERIC ACOUSTICS.

Sound absorption in water is generally much less than in air. It also rises with frequency, and it strongly depends on the amount of dissolved materials (in particular, salts in seawater), due to chemical relaxation. *See* UNDERWATER SOUND.

The theory of sound attenuation in solids is complicated because of the presence of many mechanisms responsible for it. These include heat conductivity, scattering due to anisotropic material properties, scattering due to grain boundaries, magnetic domain losses in ferromagnetic materials, interstitial diffusion of atoms, and dislocation relaxation processes in metals. In addition, in metals at very low temperature the interaction between the lattice vibrations (phonons) due to sound propagation and the valence of electrons plays an important role, particularly in the superconducting domain. *See* CRYSTAL DEFECTS; DIFFUSION; FERROMAGNETISM; SUPERCONDUCTIVITY.

[H.E.Ba.; A.J.Ca.; J.P.C.; R.B.L.]

Sound intensity   A fundamental acoustic quantity which describes the rate of flow of acoustic energy through a unit of area perpendicular to the flow direction. The unit of sound intensity is watt per square meter. The intensity is calculated at a field point (x) as a product of acoustic pressure p and particle velocity u. Generally, both p and u are functions of time, and therefore an instantaneous intensity vector is defined by the equation below. The time-variable instantaneous intensity, $\vec{I}_i(x, t)$, which has the

$$\vec{I}_i(x, t) = p(x, t) \cdot \vec{u}(x, t)$$

same direction as $\vec{u}(x, t)$, is a nonmoving static vector representing the instantaneous power flow through a point (x). *See* POWER; SOUND PRESSURE.

Many acoustic sources are stable at least over some time interval so that both the sound pressure and the particle velocity in the field of such a source can be represented in terms of their frequency spectra.

The applications of sound intensity were fully developed after a reliable technique for intensity measurement was perfected. Sound intensity measurement requires measuring both the sound pressure and the particle velocity. Very precise microphones for sound-pressure measurements are available.

An application of the intensity technique is the measurement of sound power radiated from sources. The knowledge of the radiated power makes it possible to classify, label, and compare the noise emissions from various pieces of equipment and products and to provide a reliable input into environmental design. *See* SOUND.   [J.Ti.]

Sound pressure   The incremental variation in the static pressure of a medium when a sound wave is propagated through it. Sound refers to small-amplitude, propagating pressure perturbations in a compressible medium. These pressure disturbances are related to the corresponding density perturbation via the material equation of state, and the manner in which these disturbances propagate is governed by a wave equation. Since a pressure variation with time is easily observed, the science of sound is concerned with small fluctuating pressures and their spectral characteristics. The unit of pressure commonly used in acoustics is the micropascal (1 μPa = 1 μN/m^2 = 10^{-5} dyne/cm^2 = 10^{-5} μbar). One micropascal is approximately 10^{-11} times the normal atmospheric pressure. *See* PRESSURE; PRESSURE MEASUREMENT; WAVE MOTION.

The instantaneous sound pressure at a point can be harmonic, transient, or a random collection of waves. This pressure is usually measured with an instrument that is sensitive to a particular band of frequencies. A concept widely used in acoustics is "level," which refers to the logarithm of the ratio of any two field quantities. When the ratio is proportional to a power ratio, the unit for measuring the logarithm of the ratio is called a bel, and the unit for measuring this logarithm multiplied by 10 is called a decibel (dB). The sound intensity, which describes the rate of flow of acoustic energy (acoustic power flow) per unit area, is given by the mean square pressure divided by the acoustic impedance, defined as the product of the medium density and compressional wave speed. *See* DECIBEL; SOUND; SOUND INTENSITY.

[W.M.C.]

Space charge   The net electric charge within a given volume. If both positive and negative charges are present, the space charge represents the excess of the total positive charge diffused through the volume in question over the total negative charge.

[E.G.R.]

Space-time   A term used to denote the geometry of the physical universe as suggested by the theory of relativity. It is also called space-time continuum. Whereas in Newtonian physics space and time had been considered quite separate entities, A. Einstein and H. Minkowski showed that they are actually intimately intertwined.

Einstein showed that in general two observers, each using the same techniques of observation but being in motion relative to each other, will disagree concerning the simultaneity of distant events. But if they do disagree, they are also unable to compare unequivocally the rates of clocks moving in different ways, or the lengths of scales and measuring rods. Instead, clock rates and scale lengths of different observers and different frames of reference must be established so as to assure the principal observed fact. Each observer, using his or her own clocks and scales, must measure the same speed of propagation of light. This requirement leads to a set of relationships known as the Lorentz transformations.

In accordance with the Lorentz transformations, both the time interval and the spatial distance between two events are relative quantities, depending on the state of motion of the observer who carries out the measurements. There is, however, a new absolute quantity that takes the place of the two former quantities. It is known as the invariant, or proper, space-time interval τ and is defined by Eq. (1), where T is the ordinary

$$\tau^2 = T^2 - \frac{1}{c^2} R^2 \tag{1}$$

time interval, R the distance between the two events, and c the speed of light in empty space. Whereas T and R are different for different observers, τ has the same value. In the event that Eq. (1) would render τ imaginary, its place may be taken by σ, defined by Eq. (2). If both τ and σ are zero, then a light signal leaving the location of one event

$$\sigma^2 = R^2 - c^2 T^2 \tag{2}$$

while it is taking place will reach the location of the other event precisely at the instant the signal from the latter is coming forth.

The existence of a single invariant interval led the mathematician Minkowski to conceive of the totality of space and time as a single four-dimensional continuum, which is often referred to as the Minkowski universe. In this universe, the history of a single space point in the course of time must be considered as a curve (or line), whereas an event, limited both in space and time, represents a point. So that these geometric concepts in the Minkowski universe may be distinguished from their analogs in ordinary three-dimensional space, they are referred to as world curves (world lines) and world points, respectively. *See* GRAVITATION; RELATIVITY.

[P.G.B.]

Spallation reaction   A nuclear reaction that can take place when two nuclei collide at very high energy (typically 500 MeV per nucleon and up), in which the involved nuclei are either disintegrated into their constituents (protons and neutrons), light nuclei, and elementary particles, or a large number of nucleons are expelled from the colliding system resulting in a nucleus with a smaller atomic number. This mechanism is clearly different from fusion reactions induced by heavy or light ions with modest kinetic energy (typically 5 MeV per nucleon) where, after formation of a compound nucleus, only a few nucleons are evaporated. A spallation reaction can be compared to a glass that shatters in many pieces when it falls on the ground. The way that the kinetic energy is distributed over the different particles involved in a spallation reaction and the process whereby this results in residues and fluxes of outgoing particles are not well understood. *See* NUCLEAR FUSION.

Spallation reactions take place in interstellar space when energetic cosmic rays (such as high-energy protons) collide with interstellar gas, which contains atoms such as carbon, nitrogen, and oxygen. This leads to the synthesis of light isotopes, such as ^{6}Li, ^{9}Be, ^{10}Be, and ^{11}B, that cannot be produced abundantly in nucleosynthesis scenarios in the big bang or stellar interiors.

In terrestrial laboratories spallation reactions are initiated by bombarding targets with accelerated light- or heavy-ion beams, and they are used extensively in basic and applied research, such as the study of the equation of state of nuclear matter, production of energetic neutron beams, and radioactive isotope research. *See* NEUTRON DIFFRACTION; RELATIVISTIC HEAVY-ION COLLISIONS; SLOW NEUTRON SPECTROSCOPY.   [P.V.D.]

Specific charge   The ratio of charge to mass expressed as e/m, of a particle. The acceleration of a particle in electromagnetic fields is proportional to its specific charge. Specific charge can be determined by measuring the velocity which the particle acquires in falling through an electric potential; by measuring the frequency of revolution in a magnetic field; or by observing the orbit of the particles in combined electric and magnetic fields. *See* ELEMENTARY PARTICLE.   [C.J.G.]

Specific gravity   The specific gravity of a material is defined as the ratio of its density to the density of some standard material, such as water at a specified temperature, for example, $60°F$ ($15°C$), or (for gases) air at standard conditions of temperature and pressure. Specific gravity is a convenient concept because it is usually easier to measure than density, and its value is the same in all systems of units. *See* DENSITY.   [L.N.]

Specific heat   A measure of the heat required to raise the temperature of a substance. When the heat ΔQ is added to a body of mass m, raising its temperature by ΔT, the ratio C given in Eq. (1) is defined as the heat capacity of the body. The quantity c defined in Eq. (2) is called the specific heat capacity or specific heat. A commonly

$$C = \frac{\Delta Q}{\Delta T} \tag{1}$$

$$c = \frac{C}{m} = \frac{1}{m}\frac{\Delta Q}{\Delta T} \tag{2}$$

used unit for heat capacity is joule $\cdot$ kelvin^{-1} ($J \cdot K^{-1}$); for specific heat capacity, the unit joule $\cdot$ gram$^{-1} \cdot K^{-1}$ ($J \cdot g^{-1} \cdot K^{-1}$) is often used. Joule should be preferred over the unit calorie = 4.18 J. As a unit of specific heat capacity, Btu $\cdot$ lb$^{-1} \cdot °F^{-1}$ = 4.21 J $\cdot$ g$^{-1} \cdot$ K^{-1} is also still in use in English-language engineering literature. If the heat capacity is referred to the amount of substance in the body, the molar heat capacity c_m results, with the unit J $\cdot$ mol$^{-1} \cdot$ K^{-1}.

If the volume of the body is kept constant as the energy ΔQ is added, the entire energy will go into raising its temperature. If, however, the body is kept at a constant pressure, it will change its volume, usually expanding as it is heated, thus converting some of the heat ΔQ into mechanical energy. Consequently, its temperature increase will be less than if the volume is kept constant. It is therefore necessary to distinguish between these two processes, which are identified with the subscripts V (constant volume) and p (constant pressure): C_V, c_V, and C_p, c_p. For gases at low pressures, which obey the ideal gas law, the molar heat capacities differ by R, the molar gas constant, as given in Eq. (3), where $R = 8.31 \text{ J} \cdot \text{mol}^{-1} \cdot \text{K}^{-1}$; that is, the expanding gas heats up less.

$$c_p - c_V = R \tag{3}$$

For solids, the difference between c_p and c_V is of the order of 1% of the specific heat capacities at room temperature. This small difference can often be ignored. *See* HEAT CAPACITY; THERMODYNAMIC PROCESSES.

[R.O.P.]

Speckle   The generation of a random intensity distribution, called a speckle pattern, when light from a highly coherent source, such as a laser, is scattered by a rough surface or inhomogeneous medium. *See* LASER.

The surfaces of most materials are extremely rough on the scale of an optical wavelength (approximately 5×10^{-7} m). When nearly monochromatic light is reflected from such a surface, the optical wave resulting at any moderately distant point consists of many coherent wavelets, each arising from a different microscopic element of the surface. Since the distances traveled by these various wavelets may differ by several wavelengths if the surface is truly rough, the interference of the wavelets of various phases results in the granular pattern of intensity called speckle. If a surface is imaged with a perfectly corrected optical system, diffraction causes a spread of the light at an image point, so that the intensity at a given image point results from the coherent addition of contributions from many independent surface areas. As long as the diffraction-limited point-spread function of the imaging system is broad by comparison with the microscopic surface variations, many dephased coherent contributions add at each image point to give a speckle pattern.

The basic random interference phenomenon underlying laser speckle exists for sources other than lasers. For example, it explains radar "clutter," results for scattering of x-rays by liquids, and electron scattering by amorphous carbon films. Speckle theory also explains why twinkling may be observed for stars, but not for planets. *See* COHERENCE; DIFFRACTION; INTERFERENCE OF WAVES.

In metrology, the most obvious application of speckle is to the measurement of surface roughness. If a speckle pattern is produced by coherent light incident on a rough surface, then surely the speckle pattern, or at least the statistics of the speckle pattern, must depend upon the detailed surface properties. An application of growing importance in engineering is the use of speckle patterns in the study of object displacements, vibration, and distortion that arise in nondestructive testing of mechanical components.

[J.C.Wy.]

Astronomical speckle interferometry is a technique for obtaining spatial information on astronomical objects at the diffraction-limited resolution of a telescope, despite the presence of atmospheric turbulence. Speckle interferometry techniques have proven to be an invaluable tool for astronomical research, allowing studies of a wide range of scientifically interesting problems. They have been widely used to determine the separation and position angle of binary stars, and for accurate diameter measurements of a large number of stars, planets, and asteroids. Speckle imaging techniques have successfully uncovered details in the morphology of a range of astronomical objects, including the Sun, planets, asteroids, cool giants and supergiants, young stellar objects, the supernova SN1987A in the Large Magellanic Cloud, Seyfert galaxies, and quasars. *See* INTERFEROMETRY.   [M.Ka.]

Spectrograph   An optical instrument that consists of an entrance slit, collimator, disperser, camera, and detector and that produces and records a spectrum. A spectrograph is used to extract a variety of information about the conditions that exist where light originates and along the paths of light. It reveals the details that are stored in the light's spectral distribution, whether this light is from a source in the laboratory or a quasistellar object a billion light-years away.

Spectrograph design takes into account the type of light source to be measured, and the circumstances under which these measurements will be made. Since observational astronomy presents unusual problems in these areas, the design of astronomical spectrographs may also be unique.

Astronomical spectrographs have the same general features as laboratory spectrographs (see illustration). The width of the entrance slit influences both spectral resolution and the amount of light entering the spectrograph, two of the most important variables in spectroscopy. The collimator makes this light parallel so that the disperser (a grating or prism) may properly disperse it. The camera then focuses the dispersed spectrum onto a detector, which records it for further study.

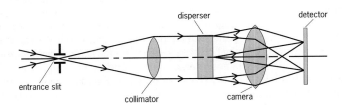

Basic optical components of a spectrograph.

Laboratory spectrographs usually function properly only in a fixed orientation under controlled environmental conditions. By contrast, most astronomical spectrographs are used on a moving telescope operating at local temperature. Thus, their structures must be mechanically and optically insensitive to orientation and temperature.

The brightness, spectral characteristics, and geometry of laboratory sources may be tailored to experimental requirements and to the capabilities of a spectrograph. Astronomical sources, in the form of images at the focus of a telescope, cannot be manipulated, and their faintness and spectral diversity make unusual and difficult demands on spectrograph performance.

Typical laboratory spectrographs use either concave gratings, which effectively combine the functions of collimator, grating, and camera in one optical element, or plane reflection gratings with spherical reflectors for collimators and cameras.   [R.Hi.]

Spectroscopy   An analytic technique concerned with the measurement of the interaction (usually the absorption or the emission) of radiant energy with matter, with the instruments necessary to make such measurements, and with the interpretation of the interaction both at the fundamental level and for practical analysis.

A display of such data is called a spectrum, that is, a plot of the intensity of emitted or transmitted radiant energy (or some function of the intensity) versus the energy of that light. Spectra due to the emission of radiant energy are produced as energy is emitted from matter, after some form of excitation, then collimated by passage through a slit, then separated into components of different energy by transmission through a prism (refraction) or by reflection from a ruled grating or a crystalline solid (diffraction), and finally detected. Spectra due to the absorption of radiant energy are produced when radiant energy from a stable source, collimated and separated into its components in a monochromator, passes through the sample whose absorption spectrum is to be measured, and is detected. Instruments which produce spectra are variously called spectroscopes, spectrometers, spectrographs, and spectrophotometers. *See* SPECTRUM.

Interpretation of spectra provides fundamental information on atomic and molecular energy levels, the distribution of species within those levels, the nature of processes involving change from one level to another, molecular geometries, chemical bonding, and interaction of molecules in solution. At the practical level, comparisons of spectra provide a basis for the determination of qualitative chemical composition and chemical structure, and for quantitative chemical analysis.

Origin of spectra. Atoms, ions, and molecules emit or absorb characteristically; only certain energies of these species are possible; the energy of the photon (quantum of radiant energy) emitted or absorbed corresponds to the difference between two permitted values of the energy of the species, or energy levels. (If the flux of photons incident upon the species is great enough, simultaneous absorption of two or more photons may occur.) Thus the energy levels may be studied by observing the differences between them. The absorption of radiant energy is accompanied by the promotion of the species from a lower to a higher energy level; the emission of radiant energy is accompanied by falling from a higher to a lower state; and if both processes occur together, the condition is called resonance.

Instruments. Spectroscopic methods involve a number of instruments designed for specialized applications.

An optical instrument consisting of a slit, collimator lens, prism or grating, and a telescope or objective lens which produces a spectrum for visual observation is called a spectroscope.

If a spectroscope is provided with a photographic camera or other device for recording the spectrum, the instrument is called a spectrograph.

A spectroscope that is provided with a calibrated scale either for measurement of wavelength or for measurement of refractive indices of transparent prism materials is called a spectrometer.

A spectrophotometer consists basically of a radiant-energy source, monochromator, sample holder, and detector. It is used for measurement of radiant flux as a function of wavelength and for measurement of absorption spectra.

An interferometer is an optical device that measures differences of geometric path when two beams travel in the same medium, or the difference of refractive index when the geometric paths are equal. Interferometers are employed for high-resolution measurements and for precise determination of relative wavelengths. *See* INTERFEROMETRY.

Methods and applications. Since the early methods of spectroscopy there has been a proliferation of techniques, often incorporating sophisticated technology.

Acoustic spectroscopy uses modulated radiant energy that is absorbed by a sample. The loss of that excess produces a temperature increase that can be monitored around the sample by using a microphone transducer. This is the optoacoustic effect.

In astronomical spectroscopy, the radiant energy emitted by celestial objects is studied by combined spectroscopic and telescopic techniques to obtain information about their chemical composition, temperature, pressure, density, magnetic fields, electric forces, and radial velocity.

Atomic absorption and fluorescence spectroscopy is a branch of electronic spectroscopy that uses line spectra from atomized samples to give quantitative analysis for selected elements at levels down to parts per million, on the average.

Attenuated total reflectance spectroscopy is the study of spectra of substances in thin films or on surfaces obtained by the technique of attenuated total reflectance or by a closely related technique called frustrated multiple internal reflection. In either method the radiant-energy beam penetrates only a few micrometers of the sample. The technique is employed primarily in infrared spectroscopy for qualitative analysis of coatings and of opaque liquids.

Electron spectroscopy includes a number of subdivisions, all of which are associated with electronic energy levels. The outermost or valence levels are studied in photoelectron spectroscopy. Electron impact spectroscopy uses low-energy electrons (0–100 eV).

X-ray photoelectron spectroscopy (XPS), also called electron spectroscopy for chemical analysis (ESCA), and Auger spectroscopy use x-ray photons to remove inner-shell electrons. Ion neutralization spectroscopy uses protons or other charged particles instead of photons. *See* AUGER EFFECT; SURFACE PHYSICS.

Fourier transform spectroscopy is a technique that has been applied to infrared spectrometry and nuclear magnetic resonance spectrometry to allow the acquisition of spectra from smaller samples in less time, with high resolution and wavelength accuracy.

Gamma-ray spectroscopy employs the techniques of activation analysis and Mössbauer spectroscopy. *See* MÖSSBAUER EFFECT; NEUTRON SPECTROMETRY.

Information on processes which occur on a picosecond time scale can be obtained by making use of the coherent properties of laser radiation, as in coherent anti-Stokes-Raman spectroscopy. Laser fluorescence spectroscopy provides the lowest detection limits for many materials of interest in biochemistry and biotechnology. Ultrafast laser spectroscopy may be used to study some aspects of chemical reactions, such as transition states of elementary reactions and orientations in bimolecular reactions. *See* LASER SPECTROSCOPY.

In mass spectrometry, the source of the spectrometer produces ions, often from a gas, but also in some instruments from a liquid, a solid, or a material absorbed on a surface. The dispersive unit provides either temporal or spatial dispersion of ions according to their mass-to-charge ratio.

In multiplex or frequency-modulated spectroscopy, each optical wavelength exiting the spectrometer output is encoded or modulated with an audio frequency that contains the optical wavelength information. Use of a wavelength analyzer then allows recovery of the original optical spectrum.

When a beam of light passes through a sample, a small fraction of the light exits the sample at a different angle. If the wavelength of the scattered light is different than the original wavelength, it is called Raman scattering. Raman spectroscopy is used in structural chemistry and is a valuable tool for surface analysis. A related process, resonance Raman spectroscopy, makes use of the fact that Raman probabilities are greatly increased when the exciting radiation has an energy which approaches the energy of an allowed electronic absorption. *See* RAMAN EFFECT.

In x-ray spectroscopy, the excitation of inner electrons in atoms is manifested as x-ray absorption; emission of a photon as an electron falls from a higher level into the vacancy thus created is x-ray fluorescence. The techniques are used for chemical analysis.

[M.M.Bu.]

Spectrum   The term spectrum is applied to any class of similar entities or properties strictly arrayed in order of increasing or decreasing magnitude. In general, a spectrum is a display or plot of intensity of radiation (particles, photons, or acoustic radiation) as a function of mass, momentum, wavelength, frequency, or some other related quantity. For example, a β-ray spectrum represents the distribution in energy or momentum of negative electrons emitted spontaneously by certain radioactive nuclides, and when radionuclides emit α-particles, they produce an α-particle spectrum of one or more characteristic energies. A mass spectrum is produced when charged particles (ionized atoms or molecules) are passed through a mass spectrograph in which electric and magnetic fields deflect the particles according to their charge-to-mass ratios. The distribution of sound-wave energy over a given range of frequencies is also called a spectrum. *See* SOUND.

In the domain of electromagnetic radiation, a spectrum is a series of radiant energies arranged in order of wavelength or of frequency. The entire range of frequencies is subdivided into wide intervals in which the waves have some common characteristic of generation or detection, such as the radio-frequency spectrum, infrared spectrum, visible spectrum, ultraviolet spectrum, and x-ray spectrum.

Spectra are also classified according to their origin or mechanism of excitation, as emission, absorption, continuous, line, and band spectra. An emission spectrum is produced whenever the radiations from an excited light source are dispersed. An absorption spectrum is produced against a background of continuous radiation by interposing matter that reduces the intensity of radiation at certain wavelengths or spectral regions. The energies removed from the continuous spectrum by the interposed absorbing medium are precisely those that would be emitted by the medium if properly excited. A continuous spectrum contains an unbroken sequence of waves or frequencies over a long range. Line spectra are discontinuous spectra characteristic of excited atoms and ions, whereas band spectra are characteristic of molecular gases or chemical compounds. *See* ATOMIC STRUCTURE AND SPECTRA; ELECTROMAGNETIC RADIATION; LINE SPECTRUM; MOLECULAR STRUCTURE AND SPECTRA; SPECTROSCOPY.   [W.F.M./W.W.W.]

Speed   The time rate of change of position of a body without regard to direction. It is the numerical magnitude only of a velocity and hence is a scalar quantity. Linear speed is commonly measured in such units as meters per second, miles per hour, or feet per second.

Average linear speed is the ratio of the length of the path traversed by a body to the elapsed time during which the body moved through that path. Instantaneous speed is the limiting value of the foregoing ratio as the elapsed time approaches zero. *See* VELOCITY.

[R.D.Ru.]

Spin (quantum mechanics)   The intrinsic angular momentum of a particle. It is that part of the angular momentum of a particle which exists even when the particle is at rest, as distinguished from the orbital angular momentum. The total angular momentum of a particle is the sum of its spin and its orbital angular momentum resulting from its translational motion. The general properties of angular momentum in quantum mechanics imply that spin is quantized in half integral multiples of $\hbar$ ($=h/2\pi$, where h is Planck's constant); orbital angular momentum is restricted to half even integral multiples of $\hbar$. A particle is said to have spin $^3/_2$, meaning that its spin angular momentum is $^3/_2$. *See* ANGULAR MOMENTUM.

A nucleus, atom, or molecule in a particular energy level, or a particular elementary particle, has a definite spin. The spin is an intrinsic or internal characteristic of a particle, along with its mass, charge, and isotopic spin. *See* QUANTUM MECHANICS; SYMMETRY LAWS (PHYSICS).   [C.J.G.]

Spin-density wave   The ground state of a metal in which the conduction-electron-spin density has a sinusoidal variation in space, with a wavelength usually incommensurate with the crystal structure. This antiferromagnetic state normally occurs in metals, alloys, and compounds with a transition-metal component. It occurs also, however, in quasi-one-dimensional organic conductors. *See* ANTIFERROMAGNETISM; CRYSTAL STRUCTURE; ELECTRON SPIN.

There are well over 100 materials which, over a temperature range, support a spin-density wave. These include some of the rare-earth elements of the lanthanide series and the 3d transition metals, manganese and chromium, the latter being the prototypical itinerant electron antiferromagnet. The occurrence of inelastic neutron-scattering peaks at incommensurate points indicates the existence of spin-density-wave fluctuations in some metals thought to be nonmagnetic (for example, copper and yttrium) when doped with magnetic impurities (manganese and gadolinium, respectively). This behavior suggests that the spin-density-wave instability may be common, even in non-transition metals.   [E.F.]

Spin glass   One of a wide variety of materials which contain interacting atomic magnetic moments and also possess some form of disorder, in which the temperature variation of the magnetic susceptibility undergoes an abrupt change in slope, that is, a cusp, at a temperature generally referred to as the freezing temperature. At lower temperatures the spins have no long-range magnetic order, but instead are found to have static or quasistatic orientations which vary randomly over macroscopic distances. The latter state is referred to as spin-glass magnetic order. Spin-glass ordering is usually detected by means of magnetic susceptibility measurements, although additional data are required to demonstrate the absence of long-range order. Closely related susceptibility cusps can also be observed by using neutron diffraction. It is not generally agreed whether spin glasses undergo a phase transition or not. *See* MAGNETIC SUSCEPTIBILITY; NEUTRON DIFFRACTION; PHASE TRANSITIONS.   [R.E.W.]

Spray flow   A special case of a two-phase (gas and liquid) flow in which the liquid phase is the dispersed phase and exists in the form of many droplets. The gas phase is the continuous phase, so abstract continuous lines (or surfaces) can be constructed through the gas at any instant without intersection of the droplets. The droplets and the gas have velocities that can be different, so both phases can move through some fixed volume or chamber and the droplets can move relative to the surrounding gas.

Spray flows have many applications. Sprays are used to introduce liquid fuel into the combustion chambers of diesel engines, turbojet engines, liquid-propellant rocket engines, and oil-burning furnaces. They are used in agricultural and household applications of insecticides and pesticides, for materials and chemicals processing, for fire extinguishing, for cooling in heat exchangers, for application of medicines, and for application of coatings (including paint and various other types of layered coatings). Common liquids (such as water, fuels, and paints) are used in sprays. It is sometimes useful to spray uncommon liquids such as molten metals. In the various applications, the approximately spherical droplets typically have submillimeter diameters that can be as small as a few micrometers.

Sprays are formed for industrial, commercial, agricultural, and power generation purposes by injection of a liquid stream into a gaseous environment. In addition, sprays can form naturally in a falling or splashing liquid. Injected streams of liquid tend to become unstable when the dynamic pressure (one-half of the gas density times the square of the liquid velocity) is much larger than the coefficient of surface tension divided by the transverse dimension. Typically, the liquid stream disintegrates into ligaments (coarse droplets) and then into many smaller spherical droplets. The breakup (or atomization) process is faster at higher stream velocity, and the final droplet sizes are smaller for higher stream velocities. Spray droplet sizes vary and typically are represented statistically by a distribution function. The number of droplets in a spray can be as high as a few million in a volume smaller than a liter. *See* JET FLOW.   [W.A.Si.]

Squeezed quantum states

Squeezed quantum states   Quantum states for which certain variables can be measured more accurately than is normally possible.

All matter and radiation fluctuate. Much random fluctuation derives from environmental influence, but even if all these influences are removed, there remains the intrinsic uncertainty prescribed by the laws of quantum physics. The position and momentum of a particle, or the electric and magnetic components of an electromagnetic field, are conjugate variables that cannot simultaneously possess definite values (Heisenberg uncertainty principle). It is possible, however, to have the position of a particle more and more accurately specified at the expense of increasing momentum uncertainty; the same applies to electromagnetic field amplitudes. This freedom underlies the phenomenon of squeezing or the possibility of having squeezed quantum states. With squeezed states, the inherent quantum fluctuation may be partly circumvented by focusing on the less noisy variable, thus permitting more precise measurement or information transfer than is otherwise possible. *See* UNCERTAINTY PRINCIPLE.

According to quantum electrodynamics, the vacuum is filled with a free electromagnetic field in its ground state that consists of fluctuating field components with significant noise energy. If ϕ is a phase angle and $a(\phi)$ and $a[\phi + (\pi/2)]$ are two quadrature components of the field (for example, the electric and magnetic field amplitudes), the vacuum mean-square field fluctuation is given by Eq. (1), independently of the phase angle.

$$\langle \Delta a^2(\phi) \rangle = \frac{1}{4} \tag{1}$$

Equation (1) is normalized to a photon; the corresponding equivalent noise temperature at optical frequencies is thousands of kelvins. Equation (1) also gives the general fluctuation of an arbitrary coherent state, which is the quantum state of ordinary lasers. Further environment-induced randomness is introduced in addition to Eq. (1) for other conventional light sources, including light-emitting diodes. *See* COHERENCE; LASER; QUANTUM ELECTRODYNAMICS.

In a squeezed state, the quadrature fluctuation is reduced below Eq. (1) for some ϕ, as given in Eq. (2). At that point, squeezing, that is, reduction of field fluctuation

$$\langle \Delta a^2(\phi) \rangle < \frac{1}{4} \tag{2}$$

below the coherent state level, occurs. The fluctuation of the conjugate quadrature is

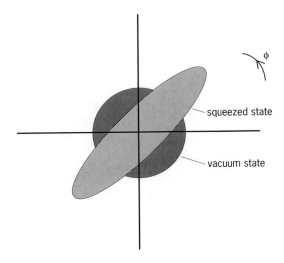

squeezed state

vacuum state

Field-amplitude fluctuation $\langle \Delta a^2(\phi) \rangle$ as a function of phase angle ϕ. The noise circle of the vacuum state, Eq. (1) in the text, is squeezed to an ellipse according to Eq. (3).

correspondingly increased to preserve the uncertainty relation, Eq. (3). In a two-photon

$$\langle \Delta a^2(\phi) \rangle \left\langle \Delta a^2 \left(\phi + \frac{\pi}{2} \right) \right\rangle \geq \frac{1}{16} \tag{3}$$

coherent state, or squeezed state in the narrow sense, Eq. (3) is satisfied with equality. As seen in the illustration, the designation "squeezed state" is partly derived from the fact that the noise circle of Eq. (1) is squeezed to an ellipse when Eq. (3) is satisfied with equality.

Squeezed light can be generated by a variety of processes, especially nonlinear optical processes. The first successful experimental demonstration of squeezing, in 1985, involved a four-wave mixing process in an atomic beam of sodium atoms.

Squeezing was first studied in connection with optical communication, although it is evident that reduced quantum fluctuation might find applications in precision measurements. *See* NONLINEAR OPTICS. [H.P.Yu.]

SQUID An acronym for superconducting quantum interference device, which actually refers to two different types of device, the dc SQUID and the rf SQUID.

The dc SQUID consists of two Josephson tunnel junctions connected in parallel on a superconducting loop (see illustration). A small applied current flows through the junctions as a supercurrent, without developing a voltage, by means of Cooper pairs of electrons tunneling through the barriers. However, when the applied current exceeds a certain critical value, a voltage is generated. When a magnetic field is applied so that a magnetic flux threads the loop, the critical value oscillates as the magnetic flux is changed, with a period of one flux quantum, weber, where h is Planck's constant and e is the electron charge. The oscillations arise from the interference of the two waves describing the Cooper pairs at the two junctions, in a way that is closely analogous to the interference between two coherent electromagnetic waves. *See* INTERFERENCE OF WAVES; JOSEPHSON EFFECT; SUPERCONDUCTIVITY.

The rf SQUID consists of a single junction interrupting a superconducting loop. In operation, it is coupled to the inductor of an *LC*-tank circuit excited at its resonant frequency by a radio-frequency (rf) current. The rf voltage across the tank circuit oscillates as a function of the magnetic flux in the loop, again with a period of one flux

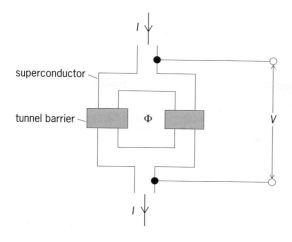

Direct-current (dc) SQUID with enclosed magnetic flux Φ. $I=$ **applied current;** $V=$ **generated voltage.**

quantum. Although SQUIDs were for many years operated while immersed in liquid helium, ceramic superconductors with high transition temperatures make possible devices operating in liquid nitrogen at 77 K.

SQUIDs have important device applications. Usually with the addition of a superconducting input circuit known as a flux transformer, both dc and rf SQUIDs are used as magnetometers to detect tiny changes in magnetic field. The output of the SQUID is amplified by electronic circuitry at room temperature and fed back to the SQUID so as to cancel any applied flux. This makes it possible to detect changes in flux as small as 10^{-6} of one flux quantum with SQUIDs based on low-transition-temperature superconductors, corresponding to magnetic field changes of the order of 1 femtotesla in a 1-hertz bandwidth. Suitable modifications to the input circuit enable the SQUID to measure other physical quantities, including voltages, displacement, or magnetic susceptibility. SQUIDs are also used for logic and switching elements in experimental digital circuits and high-speed analog-to-digital converters. *See* SUPERCONDUCTING DEVICES.

[J.Cl.]

Standard An accepted reference sample which is used for establishing a unit for the measurement of physical quantities. A physical quantity is specified by a numerical factor and a unit; for example, a mass might be expressed as 8 g, a length as 6 cm, and a time interval as 2 min. Here the gram is a mass unit defined in terms of the international kilogram, which serves as the primary standard of mass. The centimeter is defined in terms of the international meter, which is the primary standard of length and is defined as the length of path traveled by light in a vacuum during a time interval of 1/299,792,458 of a second. In similar fashion, the minute is a time interval defined as 60 s, where the second is the international standard of time and is defined as the duration of 9 192 631 770 periods of the radiation corresponding to the transition between the two hyperfine energy levels of the ground state of the cesium-133 atom.

The National Institute of Standards and Technology in the United States and comparable laboratories in other countries are responsible for maintaining accurate secondary standards for various physical quantities. *See* ELECTRICAL UNITS AND STANDARDS; LIGHT; METRIC SYSTEM; PHYSICAL MEASUREMENT; TIME.

[D.Wi.]

Standard model   The theory that explains the three major interactions of elementary particle physics—the strong interaction responsible for nuclear forces, the weak interaction responsible for radioactive decay, and the electromagnetic interaction—in terms of a common physical picture. The model for this picture is quantum electrodynamics, the fundamental theory underlying electromagnetism. In that theory, electrons, viewed as structureless elementary constituents of matter, interact with photons, structureless elementary particles of light. The standard model extends quantum electrodynamics to explain all three interactions of subnuclear physics in terms of similar basic constituents. *See* ELECTRON; ELECTROWEAK INTERACTION; ELEMENTARY PARTICLE; LIGHT; PHOTON; QUANTUM CHROMODYNAMICS; QUANTUM ELECTRODYNAMICS; STRONG NUCLEAR INTERACTIONS; WEAK NUCLEAR INTERACTIONS.   [M.E.Pe.]

Stark effect   The effect of an electric field on spectrum lines. The electric field may be externally applied; but in many cases it is an internal field caused by the presence of neighboring ions or atoms in a gas, liquid, or solid. Discovered in 1913 by J. Stark, the effect is most easily studied in the spectra of hydrogen and helium, by observing the light from the cathode dark space of an electric discharge. Because of the large potential drop across this region, the lines are split into several components. For observation perpendicular to the field, the light of these components is linearly polarized.

The linear Stark effect exhibits large, nearly symmetrical patterns. The interpretation of the linear Stark effect was one of the first successes of the quantum theory. According to this theory, the effect of the electric field on the electron orbit is to split each energy level of the principal quantum number n into $2n - 1$ equidistant levels, of separation proportional to the field strength. *See* ATOMIC STRUCTURE AND SPECTRA.

The quadratic Stark effect occurs in lines resulting from the lower energy states of many-electron atoms. The quadratic Stark effect is basic to the explanation of the formation of molecules from atoms, of dielectric constants, and of the broadening of spectral lines.

The intermolecular Stark effect is produced by the action of the electric field from surrounding atoms or ions on the emitting atom. The intermolecular effect causes a shifting and broadening of spectrum lines. The molecules being in motion, these fields are inhomogeneous in space and also in time. Hence the line is not split into resolved components but is merely widened.   [F.A.J./W.W.W.]

The quantum-confined Stark effect is the Stark effect observed in structures in which the hydrogenic system is confined in a layer of thickness much less than its normal diameter. This is not practical with atoms, but the effect is observed with excitons in semiconductor quantum-well heterostructures. It is important that quantum-confined Stark shifts can be much larger than the binding energy of the hydrogenic system. The resulting shifts of the exciton optical absorption lines can be used to make optical beam modulators and self-electrooptic-effect optical switching devices. *See* ARTIFICIALLY LAYERED STRUCTURES; ELECTROOPTICS; EXCITON; OPTICAL MODULATORS; SEMICONDUCTOR HETEROSTRUCTURES.   [D.A.B.M.]

Static electricity   Electric charge at rest, generally produced by friction or electrostatic induction. Triboelectrification is the process whereby charge transfer between dissimilar materials, at least one of which must have a high electrical resistivity, occurs due to rubbing or mere contact. *See* ELECTRIC CHARGE; ELECTRICAL RESISTIVITY.

In modern industry, highly insulating synthetic materials, such as plastic powders and insulating liquids, are used in large quantities in an ever-increasing number of applications. Such materials charge up readily, and large quantities of electrical energy may develop with an attendant risk of incendiary discharges. When, for example, powder is pneumatically transported along pipes, charge levels of up to about 100 microcoulombs per kilogram can develop and potentials of thousands of volts are generated within powder layers and the powder cloud. Energetic sparking from charged powder may initiate an explosion of the powder cloud. Similar problems occur when insulating liquids, such as certain fuels, are pumped along pipes, and it is essential that strict grounding procedures are followed during the refueling of aircraft, ships, and other large vehicles.

The capacity of a person for retaining charge depends upon stature, but is typically about 150 picofarads. Even the simple operations of removing items of clothing or sliding off a chair can lead to body discharges to ground of about 0.1 μC, which are energetic enough to ignite a mixture of natural gas and air. Human body capacitance is sufficiently high that, if poorly conducting shoes are worn, body potential may rise to 15,000 V or so above ground during industrial operations such as emptying bags of powder. Sparking may then occur with energy exceeding the minimum ignition energy of powder or fumes, so initiating a fire or explosion. Conducting footware should be used to prevent charge accumulation on personnel in industrial situations where triboelectrification may occur. *See* CAPACITANCE.

In the microelectronics industry, extremely low-energy discharges, arising from body potentials of only a few tens of volts, can damage microelectronics systems or corrupt computer data. During the handling of some sensitive semiconductor devices, it is imperative that operators work on metallic grounded surfaces and are themselves permanently attached to ground by conducting wrist straps. *See* ELECTROSTATICS.   [A.G.B.]

Statics   The branch of mechanics that describes bodies which are acted upon by balanced forces and torques so that they remain at rest or in uniform motion. This includes point particles, rigid bodies, fluids, and deformable solids in general. Static point particles, however, are not very interesting, and special branches of mechanics are devoted to fluids and deformable solids. For example, hydrostatics is the study of static fluids, and elasticity and plasticity are two branches devoted to deformable bodies. Therefore this article will be limited to the discussion of the statics of rigid bodies in two- and three-space dimensions. *See* BUOYANCY; HYDROSTATICS; MECHANICS.

In statics the bodies being studied are in equilibrium. The equilibrium conditions are very similar in the planar, or two-dimensional, and the three-dimensional rigid-body statics. These are that the vector sum of all forces acting upon the body must be zero; and the resultant of all torques about any point must be zero. Thus it is necessary to understand the vector sums of forces and torques.

In studying statics problems, two principles, superposition and transmissibility, are used repeatedly on force vectors. They are applicable to all vectors, but specifically to forces and torques (first moments of forces). The principle of superposition of vectors is that the sum of any two vectors is another vector. The principle of transmissibility of a force applied to a rigid body is that the same mechanical effect is produced by any shift of the application of the force along its line of action. To use the superposition principle to add two vectors, the principle of transmissibility is used to move some vectors along their line of action in order to add to their components.

The moment of a force about a directed line is a signed number whose value can be obtained by applying these two rules: (1) The moment of a force about a line parallel to the force is zero. (2) The moment of a force about a line normal to a plane containing the force is the product of the magnitude of the force and the least distance from the line to the line of the force. *See* EQUILIBRIUM OF FORCES; FORCE; TORQUE.   [B.DeF.]

Statistical mechanics   That branch of physics which endeavors to explain the macroscopic properties of a system on the basis of the properties of the microscopic constituents of the system. Usually the number of constituents is very large. All the characteristics of the constituents and their interactions are presumed known; it is the task of statistical mechanics (often called statistical physics) to deduce from this information the behavior of the system as a whole.

Scope. Elements of statistical mechanical methods are present in many widely separated areas in physics. For instance, the classical Boltzmann problem is an attempt to explain the thermodynamic behavior of gases on the basis of classical mechanics applied to the system of molecules.

Statistical mechanics gives more than an explanation of already known phenomena. By using statistical methods, it often becomes possible to obtain expressions for empirically observed parameters, such as viscosity coefficients, heat conduction coefficients, and virial coefficients, in terms of the forces between molecules. Statistical considerations also play a significant role in the description of the electric and magnetic properties of materials. *See* BOLTZMANN STATISTICS; INTERMOLECULAR FORCES; KINETIC THEORY OF MATTER.

If the problem of molecular structure is attacked by statistical methods, the contributions of internal rotation and vibration to thermodynamic properties, such as heat capacity and entropy, can be calculated for models of various proposed structures. Comparison with the known properties often permits the selection of the correct molecular structure.

Perhaps the most dramatic examples of phenomena requiring statistical treatment are the cooperative phenomena or phase transitions. In these processes, such as the condensation of a gas, the transition from a paramagnetic to a ferromagnetic state, or the change from one crystallographic form to another, a sudden and marked change of the whole system takes place. *See* PHASE TRANSITIONS.

Statistical considerations of quite a different kind occur in the discussion of problems such as the diffusion of neutrons through matter. In this case, the probability of the various events which affect the neutron are known, such as the capture probability and scattering cross section. The problem here is to describe the physical situation after a large number of these individual events. The procedures used in the solution of these problems are very similar to, and in some instances taken over from, kinetic considerations. Similar problems occur in the theory of cosmic-ray showers.

It happens in both low-energy and high-energy nuclear physics that a considerable amount of energy is suddenly liberated. An incident particle may be captured by a nucleus, or a high-energy proton may collide with another proton. In either case, there is a large number of ways (a large number of degrees of freedom) in which this energy may be utilized. To survey the resulting processes, one can again invoke statistical considerations. *See* SCATTERING EXPERIMENTS (NUCLEI).

Of considerable importance in statistical physics are the random processes, also called stochastic processes or sometimes fluctuation phenomena. The brownian motion, the motion of a particle moving in an irregular manner under the influence of molecular bombardment, affords a typical example. The stochastic processes are in a sense intermediate between purely statistical processes, where the existence of fluctuations may safely be neglected, and the purely atomistic phenomena, where each particle requires its individual description. *See* BROWNIAN MOVEMENT.

All statistical considerations involve, directly or indirectly, ideas from the theory of probability of widely different levels of sophistication. The use of probability notions is, in fact, the distinguishing feature of all statistical considerations.

Methods. For a system of N particles, each of the mass m, contained in a volume V, the positions of the particles may be labeled $x_1, y_1, z_1, \ldots, x_N, y_N, z_N$, their cartesian velocities $v_{x1}, \ldots, v_{zN}$, and their momenta $P_{x1}, \ldots, P_{zN}$. This simplest statistical description concentrates on a discussion of the distribution function $f(x,y,z;v_x,v_y,v_z;t)$. The quantity $f(x,y,z;v_x,v_y,v_z;t) \cdot (dxdydzdv_xdv_ydv_z)$ gives the (probable) number of particles of the system in those positional and velocity ranges where x lies between x and $x + dx$; v_x between v_x and $v_x + dv_x$, and so on. These ranges are finite.

Observations made on a system always require a finite time; during this time the microscopic details of the system will generally change considerably as the phase point moves. The result of a measurement of a quantity Q will therefore yield the time average, as in Eq. (1). The integral is along the trajectory in phase space; Q depends on the

$$\overline{Q}_t = \frac{1}{t} \int_0^t Q \, dt \tag{1}$$

variables $x_1, \ldots, P_{zN}$, and t. To evaluate the integral, the trajectory must be known, which requires the solution of the complete mechanical problem.

Ensembles. J. Willard Gibbs first suggested that instead of calculating a time average for a single dynamical system, a collection of systems, all similar to the original one, should instead be considered. Such an ensemble of systems is to be constructed in harmony with the available knowledge of the single system, and may be represented by an assembly of points in the phase space, each point representing a single system. If, for example, the energy of a system is precisely known, but nothing else, the appropriate representative example would be a uniform distribution of ensemble points over the energy surface, and no ensemble points elsewhere. An ensemble is characterized by a density function $\rho(x_1, \ldots, z_N; p_{x1}, \ldots, p_{zN}; t) \equiv p(x,p,t)$. The significance of this function is that the number of ensemble systems dN_e contained in the volume element $dx_1 \ldots dz_N; dp_x \ldots dp_{zN}$ of the phase space (this volume element will be called $d\Gamma$) at time t is as given in Eq. (2).

$$\rho(x,p,t) \, d\Gamma = dN_e \tag{2}$$

The ensemble average of any quantity Q is given by Eq. (3). The basic idea now

$$\overline{Q}_{\text{ens}} = \frac{\int Q\rho \, d\Gamma}{\int \rho \, d\Gamma} \tag{3}$$

is to replace the time average of an individual system by the ensemble average, at a fixed time, of the representative ensemble. Stated formally, the quantity $\overline{Q}_t$ defined by Eq. (1), in which no statistics is involved, is identified with $\overline{Q}_{\text{ens}}$ defined by Eq. (3), in which probability assumptions are explicitly made.

Relation to thermodynamics. It is certainly reasonable to assume that the appropriate ensemble for a thermodynamic equilibrium state must be described by a density function which is independent of the time, since all the macroscopic averages which are to be computed as ensemble averages are time-independent.

The so-called microcanonical ensemble is defined by Eq. (4a), where c is a constant, for the energy E between E_0 and $E_0 + \Delta E$; for other energies Eq. (4b) holds. By

$$\rho(p,x) = c \tag{4a}$$

$$\rho(p,x) = 0 \tag{4b}$$

using Eq. (3), any microcanonical average may be calculated. The calculations, which involve integrations over volumes bounded by two energy surfaces, are not trivial. Still, many of the results of classical Boltzmann statistics may be obtained in this way. For applications and for the interpretation of thermodynamics, the canonical ensembles is much more preferable. This ensemble describes a system which is not isolated but which is in thermal contact with a heat reservoir.

There is yet another ensemble which is extremely useful and which is particularly suitable for quantum-mechanical applications. Much work in statistical mechanics is based on the use of this so-called grand canonical ensemble. The grand ensemble describes a collection of systems; the number of particles in each system is no longer the same, but varies from system to system. The density function $p(N,p,x)\, d\Gamma_N$ gives the probability that there will be in the ensemble a system having N particles, and that this system, in its $6N$-dimensional phase space Γ_N, will be in the region of phase space $d\Gamma_N$.   [M.Dr.]

Strange particles   Bound states of quarks, in which at least one of these constituents is of the strange (s) type. Strange quarks are heavier than the up (u) and down (d) quarks, which form the neutrons and protons in the atomic nucleus. Neutrons (udd) and protons (uud) are the lightest examples of a family of particles composed of three quarks, known as baryons. These and other composite particles which interact dominantly through the strong (nuclear) force are known as hadrons. The first strange hadron discovered (in cosmic rays in 1947) was named the lambda baryon, Λ; it is made of the three-quark combination uds. A baryon containing a strange quark is also called a hyperon. Although strange particles interact through the strong (nuclear) force, the strange quark itself can decay only by conversion to a quark of different type (such as u or d) through the weak interaction. For this reason, strange particles have very long lifetimes, of the order of 10^{-10} s, compared to the lifetimes of the order of 10^{-23} s for particles which decay directly through the strong interaction. This long lifetime was the origin of the term strange particles. *See* BARYON; HADRON; NEUTRON; PROTON; STRONG NUCLEAR INTERACTIONS.

In addition to strange baryons, strange mesons occur. The lightest of these are the kaons ($K^+ = u\bar{s}$ and $K^0 = d\bar{s}$) and the antikaons ($\bar{K}^0 = s\bar{d}$ and $\bar{K}^- = s\bar{u}$). Kaons and their antiparticles have been very important in the study of the weak interaction and in the detection of the very weak *CP* violation, which causes a slow transition between neutral kaons and neutral antikaons. *See* ELEMENTARY PARTICLE; MESON; QUARKS.   [T.Bar.]

Stream function   In fluid mechanics, a mathematical idea which satisfies identically, and therefore eliminates completely, the equation of mass conservation. If the flow field consists of only two space coordinates, for example, x and y, a single and very useful stream function $\psi(x, y)$ will arise. If there are three space coordinates, such as (x, y, z), multiple stream functions are needed, and the idea becomes much less useful and is much less widely employed.

The stream function not only is mathematically useful but also has a vivid physical meaning. Lines of constant ψ are streamlines of the flow; that is, they are everywhere parallel to the local velocity vector. No flow can exist normal to a streamline; thus, selected ψ lines can be interpreted as solid boundaries of the flow.

Further, ψ is also quantitatively useful. In plane flow, for any two points in the flow field, the difference in their stream function values represents the volume flow between the points. *See* CREEPING FLOW; FLUID FLOW.   [F.M.W.]

Streamlining   The contouring of a body to reduce its resistance (drag) to motion through a fluid.

For fluids with relatively low viscosity such as water and air, effects of viscous friction are confined to a thin layer of fluid on the surface termed the boundary layer. Under the influence of an increasing pressure, the flow within the boundary layer tends to reverse and flow in an upstream direction. Viscosity tends to cause the flow to separate from the body surface with consequent formation of a region of swirling or eddy flow (termed the body wake; illus. a). This eddy formation leads to a reduction in the downstream pressure on the body and hence gives rise to a force opposite to the body motion, known as pressure drag. *See* WAKE FLOW.

In general, streamlining in subsonic flow involves the contouring of the body in such a manner that the wake is reduced and hence the pressure drag is reduced. The contouring should provide for gradual deceleration to avoid flow separation, that is, reduced adverse pressure gradients. These considerations lead to the following general rules for subsonic streamlining: The forward portion of the body should be well rounded, and the body should curve back gradually from the forward section to a tapering aftersection with the avoidance of sharp corners along the body surface. These conditions are well illustrated by teardrop shapes (illus. b).

At supersonic speeds the airflow can accommodate sudden changes in direction by being compressed or expanded. Where this change in direction occurs at the nose of the body, a compression wave is created, the strength of which depends upon the magnitude of the change in flow direction. Lowering the body-induced flow angle weakens this compression shock wave. When the flow changes direction again at the midpoint of the body, the air will expand to follow the shape of the body. This change in direction creates expansion waves. At the tail of the body the direction changes again, creating another compression or shock wave. At each of these shock waves, changes in pressure, density, and velocity occur, and in this process energy is lost. This loss results in a retarding force known as wave drag. *See* SHOCK WAVE.

Bodies which are streamlined for supersonic speeds are characterized by a sharp nose and small flow deflection angles. Because the intensity of the shock wave and the

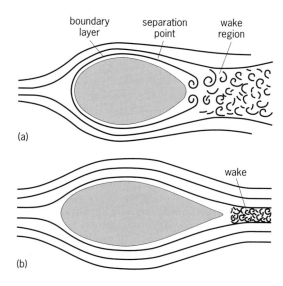

(a)

(b)

Flow about bodies in uniform subsonic flow. (*a*) Blunt body. (*b*) Streamlined body.

drag level is dependent upon the magnitude of the change in flow direction, the width or thickness of the body should be minimal. *See* BOUNDARY-LAYER FLOW.   [A.G.H.; D.M.Bu.]

Strong nuclear interactions   One of the fundamental physical interactions, which acts between a pair of hadrons. Hadrons include the nucleons, that is, neutrons and protons; the strange baryons, such as lambda (Λ) and sigma (Σ); the mesons, such as pion (π) and rho (ρ); and the strange meson, kaon (K). The nature of the interaction is determined principally through observations of the collision of a hadron pair. From this it is found that the interaction has a short range of about 10^{-15} m (10^{-13} in.) and is by far the dominant force within this range, being much larger than the electromagnetic interaction, which is next in magnitude. The strong interaction conserves parity and is time-reversal-invariant. *See* BARYON; HADRON; MESON; NUCLEON; PARITY (QUANTUM MECHANICS); STRANGE PARTICLES; SYMMETRY LAWS (PHYSICS).

The interaction between baryons for distances greater than 10^{-15} m arises from the exchange of mesons. At relatively large distances, single-pion exchange dominates (illus. *a*). At shorter separation distances, the two-pion systems such as the ρ become important (illus. *b*). The interaction between the strange baryons, and between the strange baryons and the nucleons, is moderated by kaon exchange. To summarize this description, the interaction between baryons in the SU(3) multiplet is the consequence of the exchange of SU(3) spin 0 and spin 1 bosons. In a second approximation, the exchange of two pions (illus. *c, d*), or more generally, the exchange of two members of the SU(3) spin 0 and spin 1 multiplet, is responsible for a component of the strong nuclear interaction. *See* UNITARY SYMMETRY.

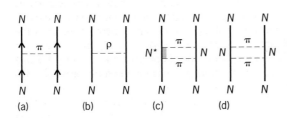

Interaction between nucleons (*a*) from exchange of single pion, (*b*) from exchange of ρ-meson, a two-pion system, (*c*) from exchange of two separate pions with formation of excited state of nucleon, and (*d*) without formation of excited state.

The range of the interaction generated by these exchanges can be calculated by using the formula below, where m is the mass of the exchanged particles, $\hbar$ is Planck's

$$\text{Range} = \frac{\hbar}{mc}$$

constant divided by 2π, and c is the speed of light. According to the above equation, the range of the interaction developed when a single pion is exchanged (illus. *a*) is equal to 1.4×10^{-15} m (5.5×10^{-14} in.), while that due to two-pion exchange (illus. *d*) is 0.7×10^{-15} m (2.8×10^{-14} in.).

At short separation distances the quark-gluon structure of the baryons must be taken into account. The interaction must be considered as a property of the six-quark-plus-gluon system. The decisive elements are the Pauli principle obeyed by the quarks, and the mismatch between the six-quark wave function and the two-baryon wave function. Thus, at short distances the interaction is effectively repulsive or more generally independent of the kinetic energy of the baryons at infinite separations. *See* ELEMENTARY PARTICLE; EXCLUSION PRINCIPLE; FUNDAMENTAL INTERACTIONS; GLUONS; QUARKS.   [H.F.]

Sum rules Formulas in quantum mechanics for transitions between energy levels, in which the sum of the transition strengths is expressed in a simple form. Sum rules are used to describe the properties of many physical systems, including solids, atoms, atomic nuclei, and nuclear constituents such as protons and neutrons. The sum rules are derived from quite general principles, and are useful in situations where the behavior of individual energy levels is too complex to describe by a precise quantum-mechanical theory. See ENERGY LEVEL (QUANTUM MECHANICS).

In general, sum rules are derived by using Heisenberg's quantum-mechanical algebra to construct operator equalities, which are then applied to particles or the energy levels of a system. See QUANTUM MECHANICS. [G.F.Be.]

Superconducting devices Devices that perform functions in the superconducting state that would be difficult or impossible to perform at room temperature, or that contain components which perform such functions. The superconducting state involves a loss of electrical resistance and occurs in many metals and alloys at temperatures near absolute zero. An enormous impetus was provided by the discovery in 1986 of a new class of ceramic, high-transition-temperature (T_c) superconductors, which has resulted in a new superconducting technology at liquid nitrogen temperature. Superconducting devices may be conveniently divided into two categories: small-scale thin-film devices, and large-scale devices which employ zero-resistance superconducting windings made of type II superconducting materials. See SUPERCONDUCTIVITY.

Small-scale devices. A variety of thin-film devices offer higher performance than their nonsuperconducting counterparts. The prediction and discovery in the early 1960s of the Josephson effects introduced novel opportunities for ultrasensitive detectors, high-speed switching elements, and new physical standards. Niobium-based devices, patterned on silicon wafers using photolithographic techniques taken over from the semiconductor industry, have reached a high level of development, and a variety of such devices are commercially available. These devices operate at or below 4.2 K ($-452°$F), the temperature of liquid helium boiling under atmospheric pressure. See LIQUID HELIUM.

The discovery of the high-transition-temperature superconductors has enabled the operation of devices in liquid nitrogen at 77 K ($-321°$F). Not only is liquid nitrogen much cheaper and more readily available than liquid helium, but it also boils away much more slowly, enabling the use of simpler and more compact dewars or simpler, relatively inexpensive refrigerators. Of the new ceramic superconductors, only $YBa_2Cu_3O_{7-x}$ (YBCO) has been developed in thin-film form to the point of practical applications, and several devices are available. Intensive materials research has resulted in techniques, notably laser-ablation and radio-frequency sputtering, for the epitaxial growth of high-quality films with their crystalline planes parallel to the surface of the substrate. Most of the successful Josephson-junction devices have been formed at the interface between two grains of YBCO. These so-called grain-boundary junctions are made by depositing the film either on a bicrystal in which the two halves of the substrate have a carefully engineered in-plane misalignment of the crystal axes, or across a step-edge patterned in the substrate. See CRYOGENICS; GRAIN BOUNDARIES.

Two types of superconducting quantum interference device (SQUID) detect changes in magnetic flux: the dc SQUID and the rf SQUID. The dc SQUID, which operates with a dc bias current, consists of two Josephson junctions incorporated into a superconducting loop. The maximum dc supercurrent, known as the critical current, and the current-voltage (I-V) characteristic of the SQUID oscillate when the magnetic field applied to the device is changed. The oscillations are periodic in the magnetic flux Φ threading the loop with a period of one flux quantum, $\Phi_0 = h/2e \approx 2.07 \times 10^{-15}$ weber, where h is Planck's constant and e is the magnitude of the charge of the electron. Thus, when the SQUID is biased with a constant current, the voltage is periodic in the flux. The SQUID is almost invariably operated in a flux-locked loop. A change in the applied flux gives rise to a corresponding current in the coil that produces an equal and opposite flux in the SQUID. The SQUID is thus the null detector in a feedback circuit, and the output voltage is linearly proportional to the applied flux. *See* JOSEPHSON EFFECT; SQUID.

The rf SQUID consists of a single Josephson junction incorporated into a superconducting loop and operates with an rf bias. The SQUID is coupled to the inductor of an LC-resonant circuit excited at its resonant frequency, typically 30 MHz. The characteristics of rf voltage across the tank circuit versus the rf current depends on applied flux. With proper adjustment of the rf current, the amplitude of the rf voltage across the tank circuit oscillates as a function of applied flux. The rf SQUID is also usually operated in a feedback mode.

SQUIDs are mostly used in conjunction with an input circuit. For example, magnetometers are made by connecting a superconducting pickup loop to the input coil to form a flux transformer. A magnetic field applied to the pickup loop induces a persistent current in the transformer and hence a magnetic flux in the SQUID. These magnetometers have found application in geophysics, for example, in magnetotellurics.

Low-transition-temperature SQUIDs are widely used to measure the magnetic susceptibility of tiny samples over a wide temperature range. Another application is a highly sensitive voltmeter, used in measurements of the Hall effect and of thermoelectricity. Low-transition-temperature SQUIDs are used as ultrasensitive detectors of nuclear magnetic and nuclear quadrupole resonance, and as transducers for gravitational-wave antennas. So-called scanning SQUIDs are used to obtain magnetic images of objects ranging from single-flux quanta trapped in superconductors to subsurface damage in two metallic sheets riveted together. *See* HALL EFFECT; MAGNETIC SUSCEPTIBILITY; THERMOELECTRICITY; VOLTMETER.

Perhaps the single largest area of application is biomagnetism, notably to image magnetic sources in the human brain or heart. In these studies an array of magnetometers or gradiometers is placed close to the subject, both generally being in a magnetically shielded room. The fluctuating magnetic signals recorded by the various channels are analyzed to locate their source. These techniques have been used, for example, to pinpoint the origin of focal epilepsy and to determine the function of the brain surrounding a tumor prior to its surgical removal.

The most sensitive detector available for millimeter and submillimeter electromagnetic radiation is the superconductor-insulator-superconductor (SIS) quasiparticle mixer. In this tunnel junction, usually niobium–aluminum oxide–niobium, the Josephson supercurrent is quenched and only single-electron tunneling occurs. The current-voltage characteristic exhibits a very sharp onset of current at a voltage $2\Delta/e$, where Δ is the superconducting energy gap. The mixer is biased near this onset where the characteristics are highly nonlinear and used to mix the signal frequency with the frequency of a local oscillator to produce an intermediate frequency that is coupled out into a low-noise amplifier. These mixers are useful at frequencies up to about 750 GHz (wavelengths down to 400 micrometers). Such receivers are of great importance in radio astronomy, notably for airborne, balloon-based, or high-altitude, ground-based telescopes operating above most of the atmospheric water vapor.

The advent of high-transition-temperature superconductors stimulated major efforts to develop passive radio-frequency and microwave components that take advantage of the low electrical losses offered by these materials compared with normal conductors in liquid nitrogen. The implementation of thin-film YBCO receiver coils has improved the signal-to-noise ratio of nuclear magnetic resonance (NMR) spectrometers by a factor of 3 compared to that achievable with conventional coils. This improvement enables the data acquisition time to be reduced by an order of magnitude. These coils also have potential applications in low-frequency magnetic resonance imaging (MRI). High-transition-temperature bandpass filters have application in cellular communications.

[J.Cl.]

Large-scale devices. Large-scale applications of superconductivity comprise medical, energy, transportation, high-energy physics, and other miscellaneous applications such as high-gradient magnetic separation. When strong magnetic fields are needed, superconducting magnets offer several advantages over conventional copper or aluminum electromagnets. Most important is lower electric power costs because once the system is energized only the refrigeration requires power input, generally only 5–10% that of an equivalent-field resistive magnet. Relatively high magnetic fields achievable in unusual configurations and in smaller total volumes reduce the costs of expensive force-containment structures. *See* MAGNET.

Niobium-titanium (NbTi) has been used most widely for large-scale applications, followed by the A15 compounds, which include niobium-tin (Nb_3Sn), niobium-aluminum (Nb-Al), niobium-germanium (Nb-Ge), and vanadium-gallium (Va_3Ga). Niobium-germanium held the record for the highest critical field (23 K; $-418.5°F$) until the announcement of high-temperature ceramic superconductors. *See* A15 PHASES.

Significant advances have been made in high-temperature superconducting wire development. Small coils have been wound that operate at 20 K ($-410°F$). Current leads are in limited commercial use. Considerable development remains necessary to use these materials in very large applications.

MRI dominates superconducting magnet systems applications. Most of the MRI systems are in use in hospitals and clinics, and incorporate superconducting magnets.

Some of the largest-scale superconducting magnet systems are those considered for energy-related applications. These include magnetic confinement fusion, superconducting magnetic energy storage, magnetohydrodynamic electrical power generation, and superconducting generators. *See* NUCLEAR FUSION.

In superconducting magnetic energy storage superconducting magnets are charged during off-peak hours when electricity demand is low, and then discharged to add electricity to the grid at times of peak demand. The largest systems would require large land areas, for example, an 1100-m-diameter (3600-ft) site for a 5000-MWh system. However, intermediate-size systems are viable. A 6-T peak-field solenoidal magnet system designed for the Alaskan power network stores 1800 megajoules (0.5 MWh). High-purity-aluminum-stabilized niobium-titanium alloy conductor carrying 16 kiloamperes current is used for the magnet winding.

Superconducting magnets have potential applications for transportation, such as magnetically levitated vehicles. In addition, superconducting magnets are used in particle accelerators and particle detectors. *See* PARTICLE ACCELERATOR; PARTICLE DETECTOR.

[A.M.Da.]

Superconductivity A phenomenon occurring in many electrical conductors, in which the electrons responsible for conduction undergo a collective transition into an ordered state with many unique and remarkable properties. These include the vanishing of resistance to the flow of electric current, the appearance of a large diamagnetism and other unusual magnetic effects, substantial alteration of many thermal properties, and the occurrence of quantum effects otherwise observable only at the atomic and subatomic level.

Superconductivity was discovered by H. Kamerlingh Onnes in Leiden in 1911 while studying the temperature dependence of the electrical resistance of mercury within a few degrees of absolute zero. He observed that the resistance dropped sharply to an unmeasurably small value at a temperature of 4.2 K ($-452°$F). The temperature at which the transition occurs is called the transition or critical temperature, T_c. The vanishingly small resistance (very high conductivity) below T_c suggested the name given the phenomenon.

In 1933 W. Meissner and R. Ochsenfeld discovered that a metal cooled into the superconducting state in a moderate magnetic field expels the field from its interior. This discovery demonstrated that superconductivity involves more than simply very high or infinite electrical conductivity, remarkable as that alone is. *See* MEISSNER EFFECT.

In 1957, J. Bardeen, L. N. Cooper, and J. R. Schrieffer reported the first successful microscopic theory of superconductivity. The Bardeen-Cooper-Schrieffer (BCS) theory describes how the electrons in a conductor form the ordered superconducting state. The BCS theory still stands as the basic explanation of superconductivity, even though extensive theoretical work has embellished it.

There are a number of practical applications of superconductivity. Powerful superconducting electromagnets guide elementary particles in particle accelerators, and they also provide the magnetic field needed for magnetic resonance imaging. Ultrasensitive superconducting circuits are used in medical studies of the human heart and brain and for a wide variety of physical science experiments. A completely superconducting prototype computer has even been built. *See* PARTICLE ACCELERATOR; SUPERCONDUCTING DEVICES.

Transition temperatures. It was realized from the start that practical applications of superconductivity could become much more widespread if a high-temperature superconductor, that is, one with a high T_c, could be found. For instance, the only practical way to cool superconductors with transition temperatures below 20 K ($-424°$F) is to use liquid helium, which boils at a temperature of 4.2 K ($-452°$F) and which is rather expensive. On the other hand, a superconductor with a transition temperature of 100 K ($-280°$F) could be cooled with liquid nitrogen, which boils at 77 K ($-321°$F) and which is roughly 500 times less expensive than liquid helium. Another advantage of a high-T_c material is that, since many of the other superconducting properties are proportional to T_c, such a material would have enhanced properties. In 1986 the discovery of transition temperatures possibly as high as 30 K ($-406°$F) was reported in a compound containing barium, lanthanum, copper, and oxygen. In 1987 a compound of yttrium, barium, copper, and oxygen was shown to be superconducting above 90 K ($-298°$F). In 1988 researchers showed that a bismuth, strontium, calcium, copper, and oxygen compound was superconducting below 110 K ($-262°$F), and transition temperatures as high as 135 K ($-216°$F) were found in a mercury, thallium, barium, calcium, copper, and oxygen compound.

Occurrence. Some 29 metallic elements are known to be superconductors in their normal form, and another 17 become superconducting under pressure or when prepared in the form of thin films. The number of known superconducting compounds and alloys runs into the thousands. Superconductivity is thus a rather common characteristic of metallic conductors. The phenomenon also spans an extremely large temperature range. Rhodium is the element with the lowest transition temperature (370 μK), while $Hg_{0.2}Tl_{0.8}Ca_2Ba_2Cu_3O$ is the compound with the highest (135 K or $-216°F$).

Despite the existence of a successful microscopic theory of superconductivity, there are no completely reliable rules for predicting whether a metal will be a superconductor. Certain trends and correlations are apparent among the known superconductors, however—some with obvious bases in the theory—and these provide empirical guidelines in the search for new superconductors. Superconductors with relatively high transition temperatures tend to be rather poor conductors in the normal state.

The ordered superconducting state appears to be incompatible with any long-range-ordered magnetic state: Usually the ferromagnetic or antiferromagnetic metals are not superconducting. The presence of nonmagnetic impurities in a superconductor usually has very little effect on the superconductivity, but the presence of impurity atoms which have localized magnetic moments can markedly depress the transition temperature even in concentrations as low as a few parts per million. *See* ANTIFERROMAGNETISM; FERROMAGNETISM.

Some semiconductors with very high densities of charge carriers are superconducting, and others such as silicon and germanium have high-pressure metallic phases which are superconducting. Many elements which are not themselves superconducting form compounds which are.

Certain organic conductors are superconducting. For instance, brominated polymeric chains of sulfur and nitrogen, known as $(SNBr_{0.4})_x$, are superconducting below 0.36 K. Other more complicated organic materials have T_c values near 10 K ($-442°F$).

Although nearly all the classes of crystal structure are represented among superconductors, certain structures appear to be especially conducive to high-temperature superconductivity. The so-called A15 structure, shared by a series of intermetallic compounds based on niobium, produced several superconductors with T_c values above 15 K ($-433°F$) as well as the record holder, NbGe, at 23 K ($-418°F$). Indeed, the robust applications of superconductivity that depend on the ability to carry high current in the presence of high magnetic fields still exclusively use two members of this class: NbTi with $T_c = 8$ K ($-445°F$), and Nb_3Sn with $T_c = 18.1$ K ($-427°F$). *See* A15 PHASES.

After 1986 the focus of superconductivity research abruptly shifted to the copper-oxide-based planar structures, due to their significantly higher transition temperatures. Basically there are three classes of these superconductors, all of which share the common feature that they contain one or more conducting planes of copper and oxygen atoms. The first class is designated by the chemical formula $La_{2-x}A_xCuO_4$, where the A atom can be barium, strontium, or calcium. Superconductivity was originally discovered in the barium-doped system, and systematic study of the substitutions of strontium, calcium, and so forth have produced transition temperatures as high as 40 K ($-388°F$).

The second class of copper-oxide superconductor is designated by the chemical formula $Y_1Ba_2Cu_3O_{7-\delta}$, with $\delta < 1.0$. Here, single sheets of copper and oxygen atoms straddle the rare-earth yttrium ion and chains of copper and oxygen atoms thread among the barium ions. The transition temperature, 92 K ($-294°F$), is quite insensitive to replacement of yttrium by many other rare-earth ions.

The third class is the most complicated. These compounds contain either single thallium-oxygen layers, represented by the chemical formula $Tl_1Ca_{n-1}Ba_2Cu_nO_{2n+3}$, where n refers to the number of copper-oxygen planes, or double thallium-oxygen layers, represented by the chemical formula $Tl_2Ca_{n-1}Ba_2Cu_nO_{2n+4}$. The number of copper-oxygen planes may be varied, and as many as three planes have been included in the structure. Thallium may be replaced by bismuth, thus generating a second family of superconductors. In all of these compounds, the transition temperature appears to increase with the number of planes, but T_c decreases for larger values of n.

The spherical molecule comprising 60 carbon atoms (C_{60}), known as a buckyball, can be alloyed with various alkaline atoms which contribute electrons for conduction. By varying the number of conductors in C_{60}, it is possible to boost T_c to a maximum value of 52 K ($-366°F$).

Superconductivity was discovered in magnesium diboride (MgB_2) in January 2001 in Japan. This material may be a good alternative for some of the applications envisioned for high-T_c superconductivity, since this compound has T_c of 39 K ($-389°F$), is relatively easy to make, and consists of only two elements.

Magnetic properties. The existence of the Meissner-Ochsenfeld effect, the exclusion of a magnetic field from the interior of a superconductor, is direct evidence that the superconducting state is not simply one of infinite electrical conductivity. Instead, it is a true thermodynamic equilibrium state, a new phase which has lower free energy than the normal state at temperatures below the transition temperature and which somehow requires the absence of magnetic flux.

The exclusion of magnetic flux by a superconductor costs some magnetic energy. So long as this cost is less than the condensation energy gained by going from the normal to the superconducting phase, the superconductor will remain completely superconducting in an applied magnetic field. If the applied field becomes too large, the cost in magnetic energy will outweigh the gain in condensation energy, and the superconductor will become partially or totally normal. The manner in which this occurs depends on the geometry and the material of the superconductor. The geometry which produces the simplest behavior is that of a very long cylinder with field applied parallel to its axis. Two distinct types of behavior may then occur, depending on the type of superconductor—type I or type II.

Below a critical field H_c which increases as the temperature decreases below T_c, the magnetic flux is excluded from a type I superconductor, which is said to be perfectly diamagnetic. For a type II superconductor, there are two critical fields, the lower critical field H_{c1} and the upper critical field H_{c2}. In applied fields less than H_{c1}, the superconductor completely excludes the field, just as a type I superconductor does below H_c. At fields just above H_{c1}, however, flux begins to penetrate the superconductor, not in a uniform way, but as individual, isolated microscopic filaments called fluxoids or vortices. Each fluxoid consists of a normal core in which the magnetic field is large, surrounded by a superconducting region in which flows a vortex of persistent supercurrent which maintains the field in the core. *See* DIAMAGNETISM.

Thermal properties. The appearance of the superconducting state is accompanied by rather drastic changes in both the thermodynamic equilibrium and thermal transport properties of a superconductor.

The heat capacity of a superconducting material is quite different in the normal and superconducting states. In the normal state (produced at temperatures below the transition temperature by applying a magnetic field greater than the critical field), the heat capacity is determined primarily by the normal electrons (with a small contribution from the thermal vibrations of the crystal lattice) and is nearly proportional to the temperature. In zero applied magnetic field, there appears a discontinuity in the heat capacity at the transition temperature. At temperatures just below the transition temperature, the heat capacity is larger than in the normal state. It decreases more rapidly with decreasing temperature, however, and at temperatures well below the transition temperature varies exponentially as $e^{-\Delta/kT}$, where Δ is a constant and k is Boltzmann's constant. Such an exponential temperature dependence is a hallmark of a system with a gap Δ in the spectrum of allowed energy states. Heat capacity measurements provided the first indications of such a gap in superconductors, and one of the key features of the macroscopic BCS theory is its prediction of just such a gap.

Ordinarily a large electrical conductivity is accompanied by a large thermal conductivity, as in the case of copper, used in electrical wiring and cooking pans. However, the thermal conductivity of a pure superconductor is less in the superconducting state than in the normal state, and at very low temperatures approaches zero. Crudely speaking, the explanation for the association of infinite electrical conductivity with vanishing thermal conductivity is that the transport of heat requires the transport of disorder (entropy). The superconducting state is one of perfect order (zero entropy), and so there is no disorder to transport and therefore no thermal conductivity. *See* ENTROPY; THERMAL CONDUCTION IN SOLIDS.

Two-fluid model. C. J. Gorter and H. B. G. Casimir introduced in 1934 a phenomenological theory of superconductivity based on the assumption that in the superconducting state there are two components of the conduction electron "fluid" (hence the name given this theory, the two-fluid model). One, called the superfluid component, is an ordered condensed state with zero entropy; hence it is incapable of transporting heat. It does not interact with the background crystal lattice, its imperfections, or the other conduction electron component and exhibits no resistance to flow. The other component, the normal component, is composed of electrons which behave exactly as they do in the normal state. It is further assumed that the superconducting transition is a reversible thermodynamic phase transition between two thermodynamically stable phases, the normal state and the superconducting state, similar to the transition between the liquid and vapor phases of any substance. The validity of this assumption is strongly supported by the existence of the Meissner-Ochsenfeld effect and by other experimental evidence. This assumption permits the application of all the powerful and general machinery of the theory of equilibrium thermodynamics. The results tie together the observed thermodynamic properties of superconductors in a very satisfying way.

Microscopic (BCS) theory. The key to the basic interaction between electrons which gives rise to superconductivity was provided by the isotope effect. It is an interaction mediated by the background crystal lattice and can crudely be pictured as follows: An electron tends to create a slight distortion of the elastic lattice as it moves, because of the Coulomb attraction between the negatively charged electron and the positively charged lattice. If the distortion persists for a brief time (the lattice may ring like a struck bell), a second passing electron will see the distortion and be affected by it. Under certain circumstances, this can give rise to a weak indirect attractive interaction between the two electrons which may more than compensate their Coulomb repulsion.

The first forward step was taken by Cooper in 1956, when he showed that two electrons with an attractive interaction can bind together to form a "bound pair" (often called a Cooper pair) if they are in the presence of a high-density fluid of other electrons, no matter how weak the interaction is. The two partners of a Cooper pair have opposite momenta and spin angular momenta. Then, in 1957, Bardeen, Cooper, and Schrieffer showed how to construct a wave function in which all of the electrons (at least, all of the important ones) are paired. Once this wave function is adjusted to minimize the free energy, it can be used as the basis for a complete microscopic theory of superconductivity.

The successes of the BCS theory and its subsequent elaborations are manifold. One of its key features is the prediction of an energy gap. Excitations called quasiparticles (which are something like normal electrons) can be created out of the superconducting ground state by breaking up pairs, but only at the expense of a minimum energy of

Δ per excitation; Δ is called the gap parameter. The original BCS theory predicted that Δ is related to T_c by $\Delta = 1.76kT_c$ at $T = 0$ for all superconductors. This turns out to be nearly true, and where deviations occur they are understood in terms of modifications of the BCS theory. The manifestations of the energy gap in the low-temperature heat capacity and in electromagnetic absorption provide strong confirmation of the theory.

[D.N.L.; R.J.So.; M.O.]

Supercritical fields   Static fields that are strong enough to cause the normal vacuum, which is devoid of real particles, to break down into a new vacuum in which real particles exist. This phenomenon has not yet been observed for electric fields, but it is predicted for these fields as well as others such as gravitational fields and the gluon field of quantum chromodynamics.

Vacuum decay in quantum electrodynamics. The original motivation for developing the new concept of a charged vacuum arose in the late 1960s in connection with attempts to understand the atomic structure of superheavy nuclei expected to be produced by heavy-ion linear accelerators. See PARTICLE ACCELERATOR.

The best starting point for discussing this concept is to consider the binding energy of atomic electrons as the charge Z of a heavy nucleus is increased. If the nucleus is assumed to be a point charge, the total energy E of the $1s_{1/2}$ level drops to 0 when $Z = 137$. This so-called $Z = 137$ catastrophe had been well known, but it was argued loosely that it disappears when the finite size of the nucleus is taken into account. However, in 1969 it was shown that the problem is not removed but merely postponed, and reappears around $Z = 173$. Any level $E(nj)$ can be traced down to a binding energy of twice the electronic rest mass if the nuclear charge is further increased. At the corresponding charge number, called Z_{cr}, the state dives into the negative-energy continuum of the Dirac equation (the so-called Dirac sea). The overcritical state acquires a width and is spread over the continuum. See ANTIMATTER; RELATIVISTIC QUANTUM THEORY.

When Z exceeds Z_{cr} a K-shell electron is bound by more than twice its rest mass, so that it becomes energetically favorable to create an electron-positron pair. The electron becomes bound in the $1s_{1/2}$ orbital and the positron escapes. The overcritical vacuum state is therefore said to be charged. See POSITRON.

Clearly, the charged vacuum is a new ground state of space and matter. The normal, undercritical, electrically neutral vacuum is no longer stable in overcritical fields: it decays spontaneously into the new stable but charged vacuum. Thus the standard definition of the vacuum, as a region of space without real particles, is no longer valid in very strong external fields. The vacuum is better defined as the energetically deepest and most stable state that a region of space can have while being penetrated by certain fields.

Superheavy quasimolecules. Inasmuch as the formation of a superheavy atom of $Z > 173$ is very unlikely, a new idea is necessary to test these predictions experimentally. That idea, based on the concept of nuclear molecules, was put forward in 1969: a superheavy quasimolecule forms temporarily during the slow collision of two heavy ions. It is sufficient to form the quasimolecule for a very short instant of time, comparable to the time scale for atomic processes to evolve in a heavy atom, which is typically of the order 10^{-18} to 10^{-20}. Suppose a uranium ion is shot at another uranium ion at an energy corresponding to their Coulomb barrier, and the two, moving slowly (compared to the K-shell electron velocity) on Rutherford hyperbolic trajectories, are close to each other (compared to the K-shell electron orbit radius). Then the atomic electrons move in the combined Coulomb potential of the two nuclei, thereby experiencing a field corresponding to their combined charge of 184. This happens because the ionic velocity (of the order of $c/10$) is much smaller than the orbital electron velocity (of the order of c), so that there is time for the electronic molecular orbits to be established, that is, to adjust to the varying distance between the charge centers, while the two ions are in the vicinity of each other. *See* QUASIATOM.

Giant nuclear systems. The energy spectrum for positrons created in, for example, a uranium-curium collision consists of three components: the induced, the direct, and the spontaneous, which add up to a smooth spectrum. The presence of the spontaneous component leads only to 5–10% deviations for normal nuclear collisions along Rutherford trajectories. This situation raises the question as to whether there is any way to get a clear qualitative signature for spontaneous positron production. Suppose that the two colliding ions, when they come close to each other, stick together for a certain time Δt before separating again. The longer the sticking, the better is the static approximation. For Δt very long, a very sharp line should be observed in the positron spectrum with a width corresponding to the natural lifetime of the resonant positron-emitting state. The observation of such a sharp line will indicate not only the spontaneous decay of the vacuum but also the formation of giant nuclear systems ($Z > 180$). *See* LINEWIDTH; NUCLEAR MOLECULE.

Search for spontaneous positron emission. The search for spontaneous positron emission in heavy-ion collisions began in 1976. Of special interest are peak structures in the positron energy distribution. However, the issue of spontaneous positron production in strong fields remains open. If line structures have been observed at all, they are most likely due to nuclear conversion processes. The observation of vacuum decay very much depends on the existence of sufficiently long-lived (at least 10^{-20} s) giant nuclear molecular systems. Therefore, the investigation of nuclear properties of heavy nuclei encountering heavy nuclei at the Coulomb barrier is a primary task.

Other field theories. The idea of overcriticality also has applications in other field theories, such as those of pion fields, gluon fields (quantum chromodynamics), and gravitational fields (general relativity).

A heavy meson may be modeled as an ellipsoidal bag, with a heavy quark Q and antiquark $\bar{Q}$ located at the foci of the ellipsoid. The color-electric or glue-electric field lines do not penetrate the bag surface. The Dirac equation may be solved for light quarks q and antiquarks $\bar{q}$ in this field of force. In the spherical case the potential is zero, and the solutions with different charges degenerate. As the source charges Q and $\bar{Q}$ are pulled apart, the wave functions start to localize. At a critical deformation of the bag, positive and negative energy states cross; that is, overrcriticality is reached and the color field is strong enough that the so-called perturbative vacuum inside the bag rearranges so that the wave functions are pulled to opposite sides and the color charges of the heavy quarks are completely shielded. Hence two new mesons of types $\bar{Q}q$ and $Q\bar{q}$ appear; the original meson fissions. *See* GLUONS; QUANTUM CHROMODYNAMICS; QUARKS.

[W.Gr.]

Superfluidity   The frictionless flow of liquid helium at low temperature; also, the flow of electric current without resistance in certain solids at low temperature (superconductivity).

Both helium isotopes have a superfluid transition, but the detailed properties of their superfluid states differ considerably because they obey different statistics. ^{4}He, with an intrinsic spin of 0, is subject to Bose-Einstein statistics, and ^{3}He, with a spin of $^1/_2$, to Fermi-Dirac statistics. There are two distinct superfluid states in ^{3}He called A and B.

The term "superfluidity" usually implies He II or the A and B phases of ^{3}He, but the basic similarity between these and the "fluid" consisting of pairs of electrons in superconductors is sufficiently strong to designate the latter as a charged superfluid. Besides flow without resistance, superfluid helium and superconducting electrons display quantized circulating flow patterns in the form of microscopic vortices. *See* BOSE-EINSTEIN STATISTICS; LIQUID HELIUM; QUANTIZED VORTICES; SECOND SOUND; SUPERCONDUCTIVITY.   [L.J.C.]

Supergravity A theory that attempts to unify gravitation with the other fundamental interactions. The first, and only, completely successful unified theory was constructed by James Clerk Maxwell, in which the up-to-then unrelated electric and magnetic phenomena were unified in his electrodynamics. *See* FUNDAMENTAL INTERACTIONS; MAXWELL'S EQUATIONS.

Electroweak theory. The second stage of unification concerns the unification of electromagnetic and weak interactions, using Maxwell's theory as a guide. This was accomplished making use of the nonabelian gauge theories invented by C. N. Yang and R. L. Mills, and of spontaneous symmetry breaking. The symmetry of Maxwell's theory is very similar to spatial rotations about an axis, rotating the vector potentials while leaving the electric and magnetic fields unchanged. It is a local invariance because the rotations about a fixed axis can be made by different amounts at different points in space-time. Thus, Maxwell's theory is invariant under a one-parameter group of transformations U(1). In Yang-Mills theory this local invariance was generalized to theories with larger symmetry groups such as the three-dimensional rotation group SO(3) $\simeq$ SU(2) which has three parameters. The number of parameters of the local symmetry (gauge) group is also equal to the number of 4-vector potentials in the gauge theory based on that group. A detailed analysis of weak and electromagnetic forces shows that their description requires four 4-vector potentials (gauge fields), so that the gauge group must be a four-parameter group. In fact, it is the product SU(2) · U(1). *See* ELECTROWEAK INTERACTION; SYMMETRY BREAKING.

Grand unified theories. In the third stage of unification, electroweak and strong forces are regarded as different components of a more general force which mediates the interactions of particles in a grand unified model. Strong forces are responsible for the interactions of hadrons and for keeping quarks confined inside hadrons. They are described by eight massless 4-vector potentials (gluons), the corresponding eight-parameter group being SU(3). This local symmetry is called color, and the corresponding theory quantum chromodynamics (QCD). The combination SU(3) · SU(2) · SU(1) has strong experimental support, and has come to be known as the standard model. Thus the gauge group of any grand unified model must include the standard model as a subsymmetry. The most dramatic prediction of these theories is the decay of protons. *See* GLUONS; GRAND UNIFICATION THEORIES; PROTON; QUANTUM CHROMODYNAMICS; QUARKS; STANDARD MODEL.

Supersymmetry and supergravity theories. A still higher and more ambitious stage of unification deals with the possibility of combining grand unified and gravity theories into a superunified theory, also known as supergravity. To achieve this, use is made of the dual role played by local internal symmetry groups. On the one hand, they describe the behavior of forces. On the other hand, they classify the elementary particles (fields) of the theory into multiplets: spin-zero fields in one multiplet, spin-1/2 fields in another multiplet, and so forth, but never fermions and bosons in a single irreducible multiplet of internal symmetry. This last restriction used to be a major obstacle on the way to superunification. This is because, of all the elementary particles, only the quanta of the gravitational field (gravitons) have spin 2, so that a multiplet of elementary particles including the graviton must of necessity involve particles of different spin. But then by an internal symmetry transformation, which is by definition distinct from space-time (Lorentz) transformations, it is possible to "rotate" particles of different spin into one another, thus altering their space-time transformation properties. This apparent paradox can be circumvented if both the internal symmetry and Lorentz transformations are part of a larger (supersymmetric) transformation group which also includes the spin-changing transformations. The irreducible multiplets of such supergroups naturally contain both fermions and bosons. This is how supersymmetry makes its appearance in supergravity theories. *See* GRAVITON; RELATIVITY; SUPERSYMMETRY; SYMMETRY LAWS (PHYSICS).

Effective theory. If supergravity models are regarded not as fundamental theories but as effective theories describing the low-energy behavior of superstring theories, it is possible to make a strong case for their usefulness. In that case, since supergravity is no longer a fundamental theory, it is no longer crucial that supergravity satisfy very stringent physical requirements such as renormalizability. In its role as an effective theory, supergravity has been used in a number of problems in particle physics. *See* ELEMENTARY PARTICLE; QUANTUM FIELD THEORY; SUPERSTRING THEORY.   [F.M.]

Supermultiplet   A generalization of the concept of a multiplet. A multiplet is a set of quantum-mechanical states, each of which has the same value of some fundamental quantum number and differs from the other members of the set by another quantum number which takes values from a range of numbers dictated by the fundamental quantum number. The number of states in the set is called the multiplicity or dimension of the multiplet. The concept was originally introduced to describe the set of states in a nonrelativistic quantum-mechanical system with the same value of the orbital angular momentum, L, and different values of the projection of the angular momentum on an axis, M. The values that M can take are the integers between $-L$ and L, $2L + 1$ in all. This is the dimension of the multiplet. If the hamiltonian operator describing the system is rotationally invariant, all states of the multiplet have the same energy. A supermultiplet is a generalization of the concept of multiplet to the case when there are several quantum numbers that describe the quantum-mechanical states. *See* ANGULAR MOMENTUM; SYMMETRY LAWS (PHYSICS).

Both concepts, multiplet and supermultiplet, acquire a precise mathematical meaning by the use of the theory of group transformations. A multiplet is an irreducible representation of a group, G. The quantum number called fundamental described above labels the representation of the group. The other quantum number labels the representation of a subgroup G' of G. For angular momentum, the group G is the rotation group, called special orthogonal group in three dimensions, SO(3), and its subgroup G' is the group of special orthogonal transformations in two dimensions, SO(2). A supermultiplet is a generalization to the case in which the group G is not a group of rank one but has larger rank. A group of rank one has only one quantum number to label its representations. The concept of a multiplet or supermultiplet is particularly useful in the classification of states of physical systems. *See* QUANTUM MECHANICS; QUANTUM NUMBERS.

The term supermultiplet was first used by E. P. Wigner in 1932 in order to classify the quantum-mechanical states of light atomic nuclei. The constituents of these are protons, p, and neutrons, n. Each proton and neutron has an intrinsic spin, S, of $1/2$ in units of $\hbar$, which is Planck's constant divided by 2π. The projection of the intrinsic spin on an axis, S_z, is then $S_z = 1/2$ or $-1/2$ (spin up or down). In addition to having the same spin, the proton and neutron have essentially the same mass but differ in that the proton is charged whereas the neutron is not. They can thus be regarded as different charge states of the same particle, a nucleon. The distinction can be made formal by introducing a quantum number called isotopic spin, T, which has the value $1/2$. The two charge orientations, T_z, are taken to be $1/2$ for the proton and $-1/2$ for the neutron. There are thus four constituents of nuclei, protons and neutrons with spin up and down, that is, $p \uparrow$, $p \downarrow$, $n \uparrow$, and $n \downarrow$. The set of transformations among these constituents forms a group called SU(4), the special unitary group in four dimensions. This is the group G for Wigner's theory. The representations of SU(4), that is, Wigner supermultiplets, are characterized by three quantum numbers $(\lambda_1, \lambda_2, \lambda_3)$ with $\lambda_1 \geq \lambda_2 \geq \lambda_3$. *See* I-SPIN; NUCLEAR STRUCTURE.   [F.I.]

Superposition principle   The principle, obeyed by many equations describing physical phenomena, that a linear combination of the solutions of the equation is also a solution.

An effect is proportional to a cause in a variety of phenomena encountered at the level of fundamental physical laws as well as in practical applications. When this is true, equations which describe such a phenomenon are known as linear, and their solutions obey the superposition principle. Thus, when f, g, h, $\cdots$, solve the linear equation, then s ($s = \alpha f + \beta g + \gamma h + \cdots$, where α, β, γ, $\cdots$, are coefficients) also satisfies the same equation.

For example, an electric field is proportional to the charge that generates it. Consequently, an electric force caused by a collection of charges is given by a superposition—a vector sum—of the forces caused by the individual charges. The same is true for the magnetic field and its cause—electric currents. Each of these facts is connected with the linearity of Maxwell's equations, which describe electricity and magnetism. *See* ELECTRIC FIELD; MAXWELL'S EQUATIONS.

The superposition principle is important both because it simplifies finding solutions to complicated linear problems (they can be decomposed into sums of solutions of simpler problems) and because many of the fundamental laws of physics are linear. Quantum mechanics is an especially important example of a fundamental theory in which the superposition principle is valid and of profound significance. This property has proved most useful in studying implications of quantum theory, but it is also a source of the key conundrum associated with its interpretation.

Its effects are best illustrated in the double-slit superposition experiment, in which the wave function representing a quantum object such as a photon or electron can propagate toward a detector plate through two openings (slits). As a consequence of the superposition principle, the wave will be a sum of two wave functions, each radiating from its respective slit. These two waves interfere with each other, creating a pattern of peaks and troughs, as would the waves propagating on the surface of water in an analogous experimental setting. However, while this pattern can be easily understood for the normal (for example, water or sound) waves resulting from the collective motion of vast numbers of atoms, it is harder to understand its origin in quantum mechanics, where the wave describes an individual quantum, which can be detected, as a single particle, in just one spot along the detector (for example, photographic) plate. The interference pattern will eventually emerge as a result of many such individual quanta, each of which apparently exhibits both wave (interference-pattern) and particle (one-by-one detection) characteristics. This ambivalent nature of quantum phenomena is known as the wave-particle duality. *See* INTERFERENCE OF WAVES; QUANTUM MECHANICS.
                                                                                   [W.H.Z.]

Supersonic flow   Fluid motion in which the Mach number M, defined as the speed of the fluid relative to the sonic speed in the same medium, is more than unity.

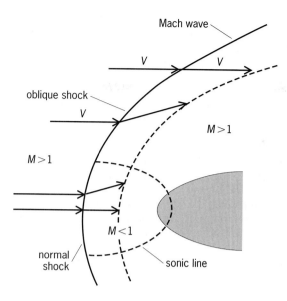

Typical normal shock, oblique shock, and Mach wave pattern in supersonic flow past a blunt body. *M* is the Mach number and *V* is the particle speed. The curved line parallel to normal and oblique shock waves indicates the end of the velocity vectors.

It is, however, common to call the flow transonic when $0.8 < M < 1.4$, and hypersonic when $M > 5$. *See* MACH NUMBER.

Mach waves. A particle moving in a compressible medium, such as air, emits acoustic disturbances in the form of spherical waves. These waves propagate at the speed of sound ($M = 1$). If the particle moves at a supersonic speed, the generated waves cannot propagate upstream of the particle. The spherical waves are enveloped in a circular cone called the Mach cone. The generators of the Mach cone are called Mach lines or Mach waves.

Shock waves. When a fluid at a supersonic speed approaches an airfoil (or a high-pressure region), no information is communicated ahead of the airfoil, and the flow adjusts to the downstream conditions through a shock wave. Shock waves propagate faster than Mach waves, and the flow speed changes abruptly from supersonic to less supersonic or subsonic across the wave. Similarly, other properties change discontinuously across the wave. A Mach wave is a shock wave of minimum strength. A normal shock is a plane shock normal to the direction of flow, and an oblique shock is inclined at an angle to the direction of flow. The velocity upstream of a shock wave is always supersonic. Downstream of an oblique shock, the velocity may be subsonic resulting in a strong shock, or supersonic resulting in a weak shock. The downstream velocity component normal to any shock wave is always subsonic. There is no change in the tangential velocity component across the shock.

In a two-dimensional supersonic flow around a blunt body (see illustration), a normal shock is formed directly in front of the body, and extends around the body as a curved oblique shock. At a sufficient distance away, the flow field is unaffected by the presence of the body, and no discontinuity in velocity occurs. The shock then reduces to a Mach wave. *See* COMPRESSIBLE FLOW; FLUID FLOW. [M.A.S.]

Superstring theory

A proposal for a unified theory of all interactions, including gravity. At present, the strong, weak, and electromagnetic interactions are accounted for within the framework of the standard model. This model correctly describes experiments up to the highest energies performed so far, and gives a complete description of the elementary particles and their interactions down to distances of the order of 10^{-18} m. Nevertheless, it has serious limitations, and attempts to overcome them and to unify the forces of nature have been only partly successful. Moreover, these attempts have left standing fundamental difficulties in reconciling gravitation and the laws of quantum mechanics. Superstring theory represents an ambitious program to unify all of the interactions observed in nature, including gravitation, in a theory with no unexplained parameters. In other words, this theory, if successful, should be able to account for all of the particles observed in nature and their interactions. *See* ELEMENTARY PARTICLE; FUNDAMENTAL INTERACTIONS.

String concept. In string theory, the fundamental objects are not point particles, as in standard theories of elementary particles, but one-dimensional extended objects, the open and closed strings. In such a theory, what are usually called the elementary particles are simply particular quantum states of the string. In superstring theories, space-time is ten-dimensional (space is nine-dimensional). If such theories are to describe nature, six dimensions must be "curled up" or "compact." The main consequence of such extra dimensions is the existence of certain very massive particles. *See* SPACE-TIME.

The essential features of string theories can be understood by analogy with the strings of a musical instrument. Such strings vibrate at a characteristic frequency, as well as any integer multiple of that frequency. Each of these modes of vibration (so-called normal modes) can be excited by plucking or striking the string. In classical physics, the amplitudes of vibration of each mode can take on a continuum of values. If there were a string of atomic dimensions, subject to the laws of quantum mechanics, the energies of this quantum string could take on only discrete values, corresponding to particular quantum states. *See* QUANTUM MECHANICS; VIBRATION.

The strings of superstring theory are quite similar. The main difference is that they obey Einstein's principles of special relativity. As a result, since each quantum state has a particular energy, it has a definite mass. Thus, each state of the string behaves as a particle of definite mass. Because it is possible, in principle, to pump an arbitrarily large amount of energy into the string, the theory contains an infinity of different types of particles of arbitrarily large mass. The interactions of these particles are governed by the ways in which the strings themselves interact. To be consistent with the principles of relativity, a string can interact only by splitting into two strings or by joining together with another string to form a third string. As a result, the interactions of strings are nearly unique. This geometric picture of string interactions translates into a precise set of rules for calculating the interaction of individual string states, that is, particles. *See* RELATIVITY.

Classical solutions. Obtaining a description of superstring theory analogous to quantum field theory is an active topic of research. However, even though the equations that describe this field theory are not completely known at present, it is known how to find classical solutions of these equations, and by various techniques, an enormous number of such solutions have been found. These include states in which space-time has any dimension between one and ten, and states with many bizarre symmetries and spectra. Each of these solutions then corresponds to a possible ground state of the system. The theories built around some of these states look very much like the real world. Not only are four dimensions flat while six are compact, but they possess gauge symmetries close to that of the standard model. Some have three or four generations of quarks and leptons, as well as light Higgs particles, which are of crucial importance in the standard model. Many of these solutions possess space-time supersymmetry. *See* HIGGS BOSON; LEPTON; QUARKS.

However, if the theory does describe nature, it must have some mechanism that chooses one of the possible ground states. Because the masses and couplings of the elementary particles depend only on the choice of ground state, determining this true ground state will yield a set of predictions for these quantities. If string theory is a correct theory, these predictions must agree with the experimental values. [M.Di.]

Supersymmetry A conjectured enhanced symmetry of the laws of nature that would relate two fundamental observed classes of particles, bosons and fermions.

All particles can be classified as fermions, such as the electron and quarks, or bosons, such as the photon and graviton. A fundamental characteristic distinguishing these two classes is that they carry different quantum-mechanical spin. If the amount of spin of an elementary particle is measured in terms of the fundamental quantum unit of angular momentum—$\hbar$, equal to Planck's constant divided by 2π—then bosons always have integer amounts of spin (that is, 0, 1, 2 ...), while fermions have odd half-integer amounts of spin (that is, 1/2, 3/2, 5/2, ...). *See* SPIN (QUANTUM MECHANICS).

There is seemingly a fundamental distinction between particles with differing amounts of spin. For example, bosons like to act collectively (Bose-Einstein statistics), producing such distinctive behavior as the laser, while, conversely, fermions obey the Pauli exclusion principle (and the Pauli-Dirac statistics), which disallows two identical fermions to be in the same state, and explains the stability of matter. Moreover, all the symmetries that are observed in the world relate different particles of the same spin. *See* BOSE-EINSTEIN STATISTICS; FERMI-DIRAC STATISTICS; QUANTUM STATISTICS; SYMMETRY LAWS (PHYSICS).

In contrast, supersymmetry would relate bosons and fermions. This would be a remarkable step forward in understanding the physical world. However, if supersymmetry were realized as an exact symmetry, the particles so related should have almost all their characteristics, such as mass and charge, preserved. Explicitly, any fermion of spin 1/2 should have a boson superpartner of spin 0, while any gauge boson of spin 1 should have a fermion superpartner of spin 1/2. This is apparently a disaster for the idea of supersymmetry since it predicts, for instance, that there should exist a spin-0 boson partner of the electron, the selectron, with electric charge and mass equal to that of the electron. Such a particle would be easy to detect and is certainly ruled out by very many experiments.

The crucial caveat to this negative result is the condition that supersymmetry be realized as an exact symmetry. A fundamental concept of modern physics is spontaneously broken symmetry. Physics displays many examples of symmetries that are exact symmetries of the fundamental equations describing a system, but not of their solutions. In particle physics the spontaneous breaking of a symmetry usually results in a difference in the masses of the particles related by the symmetry; the amount of breaking can be quantified by this mass difference. *See* SYMMETRY BREAKING.

If supersymmetry is broken by a large amount, then all the superpartners have masses much greater than the particles that are currently observed, and there is little hope of seeing evidence for supersymmetry. However, evidence that supersymmetry is broken by only a moderate amount comes from examination of the properties of the fundamental forces at high energy.

Of the four fundamental forces, the three excluding gravity are very similar in their basic formulation; they are all described by gauge theories, generalizations of the quantum theory of electromagnetism, and quantum electrodynamics (QED). The strength of electrical interaction between two electrons can be quantified in terms of a number, the coupling constant α_1. However, the quantity α_1 is actually not a constant, but depends on the energies at which the interaction strength is measured. In fact, the interaction strengths, α_1, α_2, and α_3, of the three forces (excluding gravity) all depend on energy, μ.

The couplings $\alpha_{1,2,3}$ satisfy differential equations—renormalization group equations—that depend on the types of elementary particles that exist with mass at or below the energy scale μ and that are charged with respect to each of the three interactions. If the fundamental particles include not only the observed particles but also their superpartners, taken to have masses not greater than 1000 GeV heavier than their (observed) partners, then from the renormalization group equations, the couplings α_i are predicted to meet (unify) at a huge energy of 2×10^{16} GeV. In contrast, if either supersymmetry is not an underlying symmetry of the world, or it is very badly broken so that the superpartners are very massive, the couplings fail to unify at a single point. *See* FUNDAMENTAL INTERACTIONS; QUANTUM ELECTRODYNAMICS; RENORMALIZATION.

Although the unification of couplings is the most significant indication that supersymmetry is a new law of nature, there are a number of other hints in this same direction. By observing the large-scale motions of the galaxies, the average density of large volumes of the universe can be deduced, resulting in a value that is substantially greater than that directly observed in luminous matter such as stars and hot gas. Therefore, a substantial fraction of the mass of the universe must be composed of some form of nonluminous or dark matter. Remarkably, many attractive models of supersymmetry predict that the lightest of all the superpartners is a weakly interacting massive particle with just the right characteristics to be this dark matter. *See* WEAKLY INTERACTING MASSIVE PARTICLE (WIMP). [J.M.R.]

Surface physics   The study of the structure and dynamics of atoms and their associated electron clouds in the vicinity of a surface, usually at the boundary between a solid and a low-density gas. Surface physics deals with those regions of large and rapid variations of atomic and electron density that occur in the vicinity of an interface between the two "bulk" components of a two-phase system. In conventional usage, surface physics is distinguished from interface physics by the restriction of the scope of the former to interfaces between a solid (or liquid) and a low-density gas, often at ultrahigh-vacuum pressures $p = 10^{-10}$ torr (1.33×10^{-8} newton/m^2 or 10^{-13} atm). *See* SOLID-STATE PHYSICS.

Surface physics is concerned with two separate but complementary areas of investigation into the properties of such solid-"vacuum" interfaces. Interest centers on the experimental determination and theoretical prediction of surface composition and structure (that is, the masses, charges, and positions of surface species), of the dynamics of surface atoms (such as surface diffusion and vibrational motion), and of the energetics and dynamics of electrons in the vicinity of a surface (such as electron density profiles and localized electronic surface states). As a practical matter, however, the nature and dynamics of surface species are determined experimentally by scattering and emission measurements involving particles or electromagnetic fields (or both) external to the surface itself. Thus, a second major interest in surface physics is the study of the interaction of external entities (that is, atoms, ions, electrons, electromagnetic fields, or mechanical probes) with solids at their vacuum interfaces. It is this aspect of surface physics that most clearly distinguishes it from conventional solid-state physics, because quite different scattering, emission, and local probe experiments are utilized to examine surface as opposed to bulk properties.

Techniques for characterizing the solid-vacuum interface are based on one of three simple physical mechanisms for achieving surface sensitivity. The first, which is the basis for field emission, field ionization, and scanning tunneling microscopy (STM), is the achievement of surface sensitivity by utilizing electron tunneling through the potential-energy barrier at a surface. This concept provides the basis for the development of STM to directly examine the atomic structure of surfaces by measuring with atomic resolution the tunneling current at various positions along a surface. It also has been utilized for direct determinations of the energies of individual electronic orbitals of adsorbed complexes via the measurement of the energy distributions either of emitted electrons or of Auger electrons emitted in the process of neutralizing a slow (energy $E \sim 10$ eV) external ion. *See* SCANNING TUNNELING MICROSCOPE; TUNNELING IN SOLIDS.

The second mechanism for achieving surface sensitivity is the examination of the elastic scattering or emission of particles which interact strongly with the constituents of matter, for example, "low energy" ($E \lesssim 10^3$ eV) electrons, thermal atoms and molecules, or "slow" (300 eV $\lesssim E \lesssim 10^3$ eV) ions. Since such entities lose appreciable ($\Delta E \sim 10$ eV) energy in distances of the order of tenths of a nanometer, typical electron analyzers with resolutions of tenths of an electronvolt are readily capable of identifying scattering and emission processes which occur in the upper few atomic layers of a solid. This second mechanism is responsible for the surface sensitivity of photoemission, Auger electron, electron characteristic loss, low-energy electron diffraction (LEED), and ion scattering spectroscopy techniques. The strong particle-solid interaction criterion that renders these measurements surface-sensitive is precisely the opposite of that used in selecting bulk solid-state spectroscopies. In this case, weak particle-solid interactions (that is, penetrating radiation) are desired in order to sample the bulk of the specimen via, for example, x-rays, thermal neutrons, or fast ($E \gtrsim 10^4$ eV) electrons. These probes, however, can sometimes be used to study surface properties by virtue of special geometry, for example, the use of glancing-angle x-ray diffraction to determine surface atomic structure. *See* AUGER EFFECT; ELECTRON DIFFRACTION; PHOTOEMISSION; X-RAY CRYSTALLOGRAPHY.

The third mechanism for achieving surface sensitivity is the direct measurement of the force on a probe in mechanical contact or near contact with the surface. At near contact, the van der Waals force can be measured directly by probes of suitable sensitivity. After contact is made, a variety of other forces dominate, for example, the capillary force for solid surfaces covered with thin layers of adsorbed liquid (that is, most solid surfaces in air at atmospheric pressure). When this mechanism is utilized via measuring the deflection of a sharp tip mounted on a cantilever near a surface, the experiment is referred to as atomic force microscopy (AFM) and results in maps of the force across the surface. Under suitable circumstances, atomic resolution can be achieved by this method as well as by STM. Atomic force microscopy opens the arena of microscopic surface characterization of insulating samples as well as electrochemical and biochemical interfaces at atmospheric pressure. Thus, its development is a major driving force for techniques based on surface physics. *See* INTERMOLECULAR FORCES.

Another reason for the renaissance in surface physics is the capability to generate in a vacuum chamber special surfaces that approximate the ideal of being atomically flat. These surfaces may be prepared by cycles of fast-ion bombardment, thermal outgassing, and thermal annealing for bulk samples (for example, platelets with sizes of the order of 1 cm × 1 cm × 1 mm), molecular beam epitaxy of a thin surface layer on a suitably prepared substrate, or field evaporation of etched tips for field-ion microscopes. Alternatively, the sample may be cleaved in a vacuum chamber. In such a fashion, reasonable facsimiles of uncontaminated, atomically flat solid-vacuum interfaces of many metals and semiconductors have been prepared and subsequently characterized by various spectroscopic techniques. Such characterizations must be carried out in an ultrahigh vacuum ($p \sim 10^{-8}$ N/m^2) so that the surface composition and structure are not altered by gas adsorption during the course of the measurements.

[C.B.D.]

Susceptance   The imaginary part of the admittance of an alternating-current circuit.

The admittance, Y, of an alternating current circuit is a complex number given by Eq. (1). The imaginary part, B, is the susceptance. The units of susceptance like those of

$$Y = G + jB \tag{1}$$

admittance are called siemens or mhos. Susceptance may be either positive or negative. For example, the admittance of a capacitor C at frequency ω is given by Eq. (2), and so B is positive. For an inductor L, the admittance is given by Eq. (3), and so B is negative.

$$Y = jC\omega = jB \tag{2}$$

$$Y = -\frac{j}{L\omega} = jB \tag{3}$$

In general, the susceptance of a circuit may depend on the resistors as well as the capacitors and inductors. For example, the circuit in the illustration has impedance

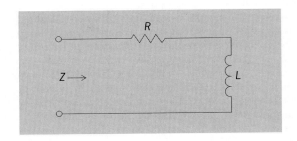

Circuit with a resistor and inductor in series.

given by Eq. (4) and admittance given by Eq. (5), so that the susceptance, given by Eq. (6), depends on the resistor R as well as the inductor L.

$$Z = R + jL\omega \tag{4}$$

$$Y = \frac{1}{R + jL\omega} \tag{5}$$

$$B = \frac{-L\omega}{R^2 + L^2\omega^2} \tag{6}$$

See ADMITTANCE; ELECTRICAL IMPEDANCE.   [J.O.S.]

Symmetry breaking   A deviation from exact symmetry. According to modern physical theory the fundamental laws of physics possess a very high degree of symmetry. Several deep insights into nature arise in understanding why specific physical systems, or even the universe as a whole, exhibit less symmetry than the laws themselves.

Spontaneous symmetry breaking. This mechanism occurs in quite diverse circumstances. The most symmetrical solutions of the fundamental equations governing a given system may be unstable, so that in practice the system is found to be in a less symmetrical, but stable, state. When this occurs, the symmetry is said to have been broken spontaneously.

For example, the laws of physics are unchanged by any translation in space, but a crystalline lattice is unchanged only by special classes of translations. A crystal does retain a large amount of symmetry, for it is unchanged by those finite translations, but this falls far short of the full symmetry of the underlying laws. *See* CRYSTAL STRUCTURE.

Another example is provided by ferromagnetic materials. The spins of electrons within such materials are preferentially aligned in some particular direction, the axis of the poles of the magnet. The laws of physics governing the interactions among these spins are unchanged by any rotation in space, but the aligned configuration of spins has less symmetry. Indeed, it is left unchanged only by rotations about the polar axis. *See* FERROMAGNETISM.

In both these examples, the loss of symmetry is associated with the appearance of order. This is a general characteristic of spontaneous symmetry breaking.

Consequences. There is a cluster of important observable consequences associated with spontaneous symmetry breaking.

Nambu-Goldstone bosons are a class of low-energy excitations associated with gentle variations of the order. Thus, there is a class of excitations of the ferromagnet, the magnons, that exist as a consequence of the spontaneous symmetry breaking, and that have very low energy. Similarly, in the case of crystals, phonons are associated with gentle distortions of the lattice structure. *See* MAGNON; PHONON.

At high temperatures the energy gained by assuming an ordered structure is increasingly outweighed by the entropy loss associated with the constraints it imposes, and at some point it will no longer be favorable to have spontaneous symmetry breaking in thermal equilibrium. Changes from broken symmetry to unbroken symmetry are marked by phase transitions. For a magnet, the transition occurs at the Curie temperature. For a crystal, it is melting into a liquid or sublimation into a gas. *See* CURIE TEMPERATURE; ENTROPY; PHASE TRANSITIONS; THERMODYNAMIC PRINCIPLES.

Defects are imperfections in the ordering. The most familiar examples are domain walls in magnets. *See* CRYSTAL DEFECTS; DOMAIN (ELECTRICITY AND MAGNETISM).

In systems with long-range forces as well as spontaneous symmetry breaking, it need no longer be true that gradual changes require only a small input of energy, because even distant regions interact significantly. Thus, the Nambu-Goldstone bosons no longer have very low energies, and they are not easily excited. Conversely, the system will exhibit a special rigidity, with strong correlations between distant points. These ideas are central to modern theories of superconductivity and of particle physics (the Higgs mechanism). *See* ELECTROWEAK INTERACTION; HIGGS BOSON; SUPERCONDUCTIVITY.

[F.Wil.]

Symmetry laws (physics)

Symmetry laws (physics) The physical laws which are expressions of symmetries. The term symmetry, as it is used in mathematics and the exact sciences, refers to a special property of bodies or of physical laws, namely that they are left unchanged by transformations which, in general, might have changed them. For example, the geometric form of a sphere is not changed by any rotation of the sphere around its center, and so a sphere can be said to be symmetric under rotations. Symmetry can be very powerful in constraining form. Indeed, referring to the same example, the only sort of surface which is symmetric under arbitrary rotations is a sphere.

The concept that physical laws exhibit symmetry is more subtle. A naive formulation would be that a physical law exhibits symmetry if there is some transformation of the universe that might have changed the form of the law but in reality does not. However, the comparison of different universes is generally not feasible or desirable. A more fruitful definition of the symmetry of physical law exploits locality, the principle that the behavior of a given system is only slightly affected by the behavior of other bodies far removed from it in space or time. Because of locality, it is possible to define symmetry by using transformations that do not involve the universe as a whole but only a suitably isolated portion of it. Thus the statement that the laws of physics are symmetric under rotations means that (say) astronauts in space would not be able to orient themselves—to determine a preferred direction—by experiments internal to their space station. They could do this only by referring to weak effects from distant objects, such as the light of distant stars or the small residual gravity of Earth.

Symmetries of space and time. Perhaps the most basic and profound symmetries of physical laws are symmetry under translation in time and under translations in space.

The statement that fundamental physical laws are symmetric under translation in time is equivalent to the statement that these laws do not change or evolve. Time-translation symmetry is supposed to apply, fundamentally, to simple isolated systems. Large complicated systems, and in particular the universe as a whole, do of course age and evolve. Thus in constructing the big-bang model of cosmology, it is assumed that the properties of individual electrons or protons do not change in time, although of course the state of the universe as a whole, according to the model, has changed quite drastically.

The statement that fundamental laws are symmetric under translations in space is another way of formulating the homogeneity of space. It is the statement that the laws are the same throughout the universe. It says that the astronauts in the previous example cannot infer their location by local experiments within their space station. The power of this symmetry is that it makes it possible to infer, from observations in laboratories on Earth, the behavior of matter anywhere in the universe.

The symmetry of physical law under rotations, mentioned above, embodies the isotropy of space.

In the mathematical formulation of dynamics, there is an intimate connection between symmetries and conservation laws. Symmetry under time translation implies conservation of energy; symmetry under spatial translations implies conservation of momentum; and symmetry under rotation implies conservation of angular momentum. *See* ANGULAR MOMENTUM; CONSERVATION OF ENERGY; CONSERVATION OF MOMENTUM.

The fundamental postulate of the special theory of relativity, that the laws of physics take the same form for observers moving with respect to one another at a fixed velocity, is clearly another statement about the symmetry of physical law. The idea that physical laws should be unchanged by such transformations was discussed by Galileo, who illustrated it by an observer's inability to infer motion while on a calm sea voyage in an enclosed cabin. The novelty of Einstein's theory arises from combining this velocity symmetry with a second postulate, deduced from experiments, that the speed of light is a universal constant and must take the same value for both stationary and uniformly moving observers. *See* GALILEAN TRANSFORMATIONS; RELATIVITY.

Discrete symmetries. Before 1956, it was believed that all physical laws obeyed an additional set of fundamental symmetries, denoted P, C, and T, for parity, charge conjugation, and time reversal, respectively. Experiments involving particles known as K mesons led to the suggestion that P might be violated in the weak interactions, and violations were indeed observed. This discovery led to questioning—and in some cases overthrow—of other cherished symmetry principles.

Parity, P, roughly speaking, transforms objects into the shapes of their mirror images. If P were a symmetry, the apparent behavior of the images of objects reflected in a mirror would also be the actual behavior of corresponding real objects. *See* PARITY (QUANTUM MECHANICS).

Charge conjugation, C, changes particles into their antiparticles. It is a purely internal transformation; that is, it does not involve space and time. If the laws of physics were symmetric under charge conjugation, the result of an experiment involving antiparticles could be inferred from the corresponding experiment involving particles.

Remarkably, by combining the transformations P and C, a result is obtained, CP, which is much more nearly a valid symmetry than either of its components separately. However, in 1964 it was discovered experimentally that even CP is not quite a valid symmetry.

Although the preceding discussion has emphasized the failure of P, C, and CP to be precise symmetries of physical law, both the strong force responsible for nuclear structure and reactions and the electromagnetic force responsible for atomic structure and chemistry do obey these symmetries. Only the weak force, responsible for beta radioactivity and some relatively slow decays of exotic elementary particles, violates them. Thus these symmetries, while approximate, are quite useful and powerful in nuclear and atomic physics. *See* ELECTROWEAK INTERACTION; FUNDAMENTAL INTERACTIONS; WEAK NUCLEAR INTERACTIONS.

The operation of time-reversal symmetry, T, involves changing the direction of motion of all particles. For example, it relates reactions of the type $A + B \rightarrow C + D$ to their reverse $C + D \rightarrow A + B$. No direct violation of T has been detected.

Time-reversal symmetry, even if valid, applies in a straightforward way only to elementary processes. It does not, for example, contradict the one-way character of the second law of thermodynamics, which states that entropy can only increase with time. *See* THERMODYNAMIC PRINCIPLES; TIME, ARROW OF.

Fundamental principles of quantum field theory suggest that the combined operation PCT, which involves simultaneously reflecting space, changing particles into antiparticles, and reversing the direction of time, must be a symmetry of physical law. Existing evidence is consistent with this prediction. *See* CPT THEOREM.

Internal symmetry. Internal symmetries, like C, do not involve transformations in space-time but change one type of particle into another. An important, although approximate, symmetry of this kind is isospin or i-spin symmetry. It is observed experimentally that the strong interactions of the proton and neutron are essentially the same. *See* I-SPIN.

There have been several successful predictions of the existence and properties of new particles, based on postulates of internal symmetries. Perhaps the most notable was the prediction of the mass and properties of the Ω^- baryon, based on an extension of the symmetry group SU(2) of isospin to a larger approximate SU(3) symmetry acting on strange particles as well. These symmetries were an important hint that the fundamental strong interactions are at some level universal, that is, act on all quarks in the same way, and thus paved the way toward modern quantum chromodynamics, which does implement such universality. *See* BARYON; QUANTUM CHROMODYNAMICS; UNITARY SYMMETRY.

Much simpler mathematically than SU(2) internal symmetry, but quite profound physically, is U(1) internal symmetry. The important case of the electric charge quantum number will be considered. The action of the U(1) internal symmetry transformation with parameter λ is to multiply the wave function of a state of electric charge q by the factor $e^{i\lambda q}$. An amplitude between two states with electric charges q_r and q_s will therefore be multiplied by a factor $e^{i\lambda(q_s - q_r)}$. Since the physical predictions of quantum mechanics depend on such amplitudes, these predictions will be unchanged only if the phase factors multiplying all nonvanishing amplitudes are trivial. This will be true, in turn, only if the amplitudes between states of unequal charge vanish; that is, if charge-changing amplitudes are forbidden, which is just a backhanded way of expressing the conservation of charge. *See* QUANTUM MECHANICS; QUANTUM NUMBERS.

Localization of symmetry. The concept of local gauge invariance, which is central to the standard model of fundamental particle interactions, and in a slightly different form to general relativity, may be approached as a generalization of the U(1) internal symmetry transformation, where a parameter λ independent of space and time appears. Such a parameter goes against the spirit of locality, according to which each point in space-time has a certain independence. There is therefore reason to consider a more general symmetry, involving a space-time-dependent transformation in which the wave function is multiplied by $e^{i\lambda(x,t)q(x,t)}$, where $q(x,t)$ is the density of charge at the space-time point (x,t). These transformations are much more general than those discussed above, and invariance under them leads to much more powerful and specific consequences.

For electromagnetism, the required interactions of matter with the electromagnetic field are predicted precisely. Thus, the theory of the electromagnetic field—Maxwell's equations and quantum electrodynamics—can be said to be the unique ideal embodiment of the abstract concept of a space-time-dependent symmetry, that is, of local gauge symmetry. *See* MAXWELL'S EQUATIONS; QUANTUM ELECTRODYNAMICS. [F.Wil.]

Sympathetic vibration The driving of a mechanical or acoustical system at its resonant frequency by energy from an adjacent system vibrating at this same frequency. Examples include the vibration of wall panels by sounds issuing from a loudspeaker, vibration of machinery components at specific frequencies as the speed of a motor increases, and the use of tuned air resonators under the bars of a xylophone to enhance the acoustic output. Increasing the damping of a vibrating system will decrease the amplitude of its sympathetic vibration but at the same time widen the band of frequencies over which it will partake of sympathetic vibration. *See* RESONANCE (ACOUSTICS AND MECHANICS); VIBRATION. [L.E.K.]

Synchrotron radiation Electromagnetic radiation emitted by relativistic charged particles curving in magnetic or electric fields. With the development of electron storage rings, radiation with increasingly high flux, brightness, and coherent power levels has become available for a wide variety of basic and applied research in biology, chemistry, and physics, as well as for applications in medicine and technology. *See* ELECTROMAGNETIC RADIATION; PARTICLE ACCELERATOR; RELATIVISTIC ELECTRODYNAMICS.

Electron storage rings provide radiation from the infrared through the visible, near-ultraviolet, vacuum-ultraviolet, soft-x-ray, and hard-x-ray parts of the electromagnetic spectrum extending to 100 keV and beyond. The flux [photons/(second, unit bandwidth)], brightness (or brilliance) [flux/(unit source size, unit solid angle)], and coherent power (important for imaging applications and proportional to brightness) available for experiments, particularly in the vacuum-ultraviolet, soft-x-ray, and hard-x-ray parts of the spectrum, are many orders of magnitude higher than is available from other sources.

The radiation has many features (natural collimation, high intensity and brightness, broad spectral bandwidth, high polarization, pulsed time structure, small source size, and high-vacuum environment) that make it ideal for a wide variety of applications in experimental science and technology. Very powerful sources of synchrotron radiation in the ultraviolet and x-ray parts of the spectrum became available when high-energy physicists began operating electron synchrotrons in the 1950s. Although synchrotrons produce large amounts of radiation, their cyclic nature results in pulse-to-pulse intensity

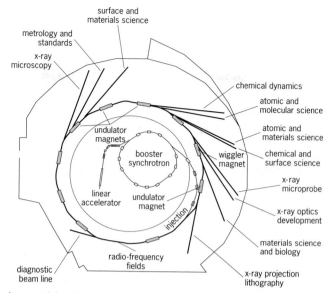

**Layout of the 1.5-GeV Advanced Light Source at Lawrence Berkeley
National Laboratory, a low-energy, third-generation synchrotron
radiation source. Applications of experimental stations on beam lines
are indicated.**

changes and variations in spectrum and source shape during each cycle. By contrast,
the electron-positron storage rings developed for colliding-beam experiments starting
in the 1960s offered a constant spectrum and much better stability. Beam lines were
constructed on both synchrotrons and storage rings to allow the radiation produced in
the bending magnets of these machines to leave the ring vacuum system and reach
experimental stations. In most cases the research programs were pursued on a parasitic
basis, secondary to the high-energy physics programs.

Since about 1980, fully dedicated storage ring sources have been completed in
several countries. They are called second-generation facilities to distinguish them from
the first-generation rings that were built for research in high-energy physics.

Special magnets may be inserted into the straight sections between ring bending
magnets to produce beams with extended spectral range or with higher flux and brightness than is possible with the ring bending magnets. These devices, called wiggler and
undulator magnets, utilize periodic transverse magnetic fields to produce transverse
oscillations of the electron beam with no net deflection or displacement. They provide another order-of-magnitude or more improvement in flux and brightness over
ring bending magnets, again opening up new research opportunities. However, their
potential goes well beyond their performance levels, in first- and second-generation
sources.

Third-generation sources are storage rings with many straight sections for wiggler
and undulator insertion device sources and with a smaller transverse size and angular
divergence of the circulating electron beam. The product of the transverse size and divergence is called the emittance. The lower the electron-beam emittance, the higher the
photon-beam brightness and coherent power level. With smaller horizontal emittances
and with straight sections that can accommodate longer undulators, third-generation

rings provide two or more orders of magnitude higher brightness and coherent power level than earlier sources.

One consequence of the extraordinary brilliance of these sources is that the x-ray beam is partially coherent. By aperturing the beam, a fully coherent beam can be obtained, but at the expense of flux. Nonetheless, there is still sufficient flux remaining to explore the use and application of coherent x-ray beams. *See* COHERENCE.

Several third-generation rings are in operation. Low-energy (typically 1–2-GeV) third-generation rings (see illustration) are optimized to produce high-brightness radiation in the vacuum ultraviolet (VUV) and soft x-ray spectral range, up to photon energies of about 2–3 keV. High-energy rings (typically 6–8 GeV) aim at harder x-rays with energies of 10–20 keV and above.

The radiation produced by an electron in circular motion at low energy (speed much less than the speed of light) is weak and rather nondirectional. At relativistic energies (speed close to the speed of light) the radiated power increases markedly, and the emission pattern is folded forward into a cone with a half-opening angle in radians given approximately by $\gamma - 1 = mc^2/E$, where mc^2 is the rest-mass energy of the electron (0.51 MeV) and E is the total energy. Thus, at electron energies of the order of 1 GeV, much of the very strong radiation produced is confined to a forward cone with an instantaneous opening angle of about 1 mrad (0.06°). At higher electron energies this cone is even smaller. The large amount of radiation produced combined with the natural collimation gives synchrotron radiation its intrinsic high brightness. Brightness is further enhanced by the small cross-sectional area of the electron beam, which is as low as 0.01 mm^2 in the third-generation rings. [A.Bi.; D.Mi.; G.She.; H.Win.]

Tachyon   A hypothetical faster-than-light particle consistent with the special theory of relativity. According to this theory, a free particle has an energy E and a momentum $\mathbf{p}$ which form a Lorentz four-vector. The length of this vector is a scalar, having the same value in all inertial reference frames. One writes Eq. (1), where c is the speed of

$$E^2 - c^2\mathbf{p}^2 = m^2c^4 \qquad (1)$$

light and the parameter m^2 is a property of the particle, independent of its momentum and energy. Three cases may be considered: m^2 may be positive, zero, or negative. The case $m^2 > 0$ applies for atoms, nuclei, and the macroscopic objects of everyday experience. The positive root m is called the rest-mass. If $m^2 = 0$, the particle is called massless. A few of these are known: the electron neutrino, the muon neutrino, the photon, and the graviton. The name tachyons (after a Greek word for swift) has been given to particles with $m^2 < 0$.

In general, the particle speed is given by Eq. (2). If $m^2 < 0$, Eq. (1) implies $E > c\mathbf{p}$

$$v = \frac{c\mathbf{p}}{E}\,c \qquad (2)$$

and Eq. (2) gives $v < c$. If $m^2 = 0$, then $E = c\mathbf{p}$ and $v = c$. In case $m^2 < 0$, one finds $E < c\mathbf{p}$ and $v > c$. Tachyons exist only at faster-than-light speeds. *See* ELEMENTARY PARTICLE; RELATIVITY.

[R.H.Go.]

Telescope   An instrument used to collect, measure, or analyze electromagnetic radiation from distant objects. A telescope overcomes the limitations of the eye by increasing the ability to see faint objects and discern fine details. In addition, when used in conjunction with modern detectors, a telescope can "see" light that is otherwise invisible. The wavelength of the light of interest can have a profound effect on the design of a telescope. *See* ELECTROMAGNETIC RADIATION; LIGHT.

For many applications, the Earth's atmosphere limits the effectiveness of larger telescopes. The most obvious deleterious effect is image scintillation and motion, collectively known as poor seeing. Atmospheric turbulence produces an extremely rapid motion of the image resulting in a smearing. On the very best nights at ideal observing sites, the image of a star will be spread out over a 0.25-arcsecond seeing disk; on an average night, the seeing disk may be between 0.5 and 2.0 arcseconds.

The upper atmosphere glows faintly because of the constant influx of charged particles from the Sun. The combination of the finite size of the seeing disk of stars and the presence of airglow limits the telescope's ability to see faint objects. One solution is placing a large telescope in orbit above the atmosphere. In practice, the effects of air and light pollution outweigh those of airglow at most observatories in the United States.

There are basically three types of optical systems in use in astronomical telescopes: refracting systems whose main optical elements are lenses which focus light by refraction; reflecting systems, whose main imaging elements are mirrors which focus light by reflection; and catadioptric systems, whose main elements are a combination of a lens and a mirror. The most notable example of the last type is the Schmidt camera.

Astronomers seldom use large telescopes for visual observations. Instead, they record their data for future study. Modern developments in photoelectric imaging devices are supplanting photographic techniques for many applications. The great advantages of detectors such as charge-coupled devices is their high sensitivity, and the images can be read out in a computer-compatible format for immediate analysis.

Light received from most astronomical objects is made up of radiation of all wavelengths. The spectral characteristics of this radiation may be extracted by special instruments called spectrographs.

As collectors of radiation from a specific direction, telescopes may be classified as focusing and nonfocusing. Nonfocusing telescopes are used for radiation with energies of x-rays and above (x-ray, gamma-ray, cosmic-ray, and neutrino telescopes). Focusing telescopes, intended for nonvisible wavelengths, are similar to optical ones (solar, radio, infrared, and ultraviolet telescopes), but they differ in the details of construction. *See* CERENKOV RADIATION.

The 5-m (200-in.) Hale telescope at Palomar Mountain, California, was completed in 1950. The primary mirror is 5 m in diameter with a 1.02-m (40-in.) hole in the center.

The 4-m (158-in.) Mayall reflector at the Kitt Peak National Observatory was dedicated in 1973. The prime focus has a field of view six times greater than that of the Hale reflector. An identical telescope was subsequently installed at Cerro Tololo Inter-American Observatory, in Chile.

The mirrors for these traditional large telescopes were all produced using the same general methodology. A large, thick glass mirror blank was first cast; then the top surface of the mirror was laboriously ground and polished to the requisite shape. The practical and economical limit to the size of traditional mirror designs was nearly reached by the 6-m (236-in.) telescope in the Caucasus Mountains, Russia. Newer telescopes have been designed and built that use either a number of mirrors mounted such that the light collection by them is brought to a common focus, or lightweight mirrors in computer-controlled mounts.

The Keck Telescope on Mauna Kea, Hawaii, completed in 1993, is the largest of the segmented mirror telescopes to be put into operation. The telescope itself is a fairly traditional design. However, its primary mirror is made up of 36 individual hexagonal segments mosaiced together to form a single 10-m (386-in.) mirror. Electronic sensors built into the edges of the segments monitor the relative positions of the segments, and feed the results to a computer-controlled actuator system.

In 1989, the European Southern Observatory put into operation their New Technology Telescope. The 3.58-m (141-in.) mirror was produced by a technique known as spin-casting, where molten glass is poured into a rotating mold.

Worldwide efforts are under way on a new generation of large, ground-based telescopes, using both the spin-casting method and the segmented method to produce large mirrors. The Gemini project of the National Optical Astronomy Observatories included twin 8.1-m (319-in.) telescopes, Gemini North (see illustration) on Mauna Kea, Hawaii, and Gemini South on Cerro Pachon in Chile.

Gemini North telescope in its dome. (*Copyright Gemini Observatory/AURA/NOAO/NSF, all rights reserved*)

The Very Large Telescope (VLT), operated by the European Southern Observatory on Cerro Paranel, Chile, consists of four 8-m (315-in.) "unit" telescopes with spin-cast mirrors. The light from the four telescopes is combined to give the equivalent light-gathering power of a 16-m (630-in.) telescope. The last of the four telescopes began collecting scientific data in September 2000.

The ability of large telescopes to resolve fine detail is limited by a number of factors. Distortion due to the mirror's own weight causes problems in addition to those of atmospheric seeing. The Earth-orbiting Hubble Space Telescope (HST), with an aperture of 2.4 m (94 in.), was designed to eliminate these problems. The telescope operates in ultraviolet as well as visible light, resulting in a great improvement in resolution not only by the elimination of the aforementioned terrestrial effects but by the reduced blurring by diffraction in the ultraviolet. *See* DIFFRACTION; RESOLVING POWER (OPTICS).

Soon after the telescope was launched in 1990, it was discovered that the optical system was plagued with spherical aberration, which severely limited its spatial resolution. After space-shuttle astronauts serviced and repaired the telescope in 1993, adding what amounted to eyeglasses for the scientific instruments, the telescope exceeded its prelaunch specifications for spatial resolution. Subsequent servicing missions replaced instruments with newer technology.   [R.D.Ch.; W.M.S.]

Temperature   A concept related to the flow of heat from one object or region of space to another. The term refers not only to the senses of hot and cold but to numerical scales and thermometers as well. Fundamental to the concept are the absolute scale and absolute zero and the relation of absolute temperatures to atomic and molecular motions.

Thermometers do not measure a special physical quantity. They measure length (as of a mercury column) or pressure or volume (with the gas thermometer at the National Institute of Standards and Technology) or electrical voltage (with a thermocouple). The basic fact is that if a mercury column has the same length when touching two different, separated objects when the objects are placed in contact, no heat will flow from one to the other. *See* THERMOMETER.

The numbers on the thermometer scales are merely historical choices; they are not scientifically fundamental. The most widely used scales are the Fahrenheit (°F) and the Celsius (°C). The centigrade scale with 0° assigned to ice water (ice point) and 100° assigned to water boiling under one atmosphere pressure (steam point) was formerly used, but it has been succeeded by the Celsius scale, defined in a different way than the centigrade scale. However, on the Celsius scale the temperatures of the ice and steam points differ by only a few hundredths of a degree from 0° and 100°, respectively. The illustration shows how the Celsius and Fahrenheit scales compare and how they fit onto the absolute scales.

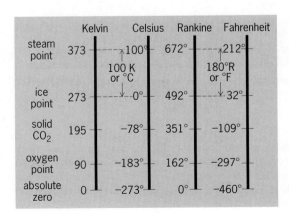

Comparisons of Kelvin, Celsius, Rankine, and Fahrenheit temperature scales. Temperatures are rounded off to nearest degree. (*After M. W. Zemansky, Temperatures Very Low and Very High, Van Nostrand, 1964*)

In 1848 William Thomson (Lord Kelvin), following ideas of Sadi Carnot, stated the concept of an absolute scale of temperature in terms of measuring amounts of heat flowing between objects. Most important, Kelvin conceived of a body which would not give up any heat and which was at an absolute zero of temperature. Experiments have shown that absolute zero corresponds to $-273.15°C$ or $-459.7°F$. Two absolute scales, shown in the illustration, are the Kelvin (K) and the Rankine (°R).

In practice, absolute temperatures are measured by using low-density helium gas and dilute paramagnetic crystals, the most nearly ideal of real materials. The measurement of a single temperature with a gas or magnetic thermometer is a major scientific event done at a national standards laboratory. Only a few temperatures have been measured, including the freezing point of gold (1337.91 K or 1948.57°F), and the boiling points of sulfur (717.85 K or 832.46°F), oxygen (90.18 K or $-297.35°F$), and helium (4.22 K or $-452.07°F$). Various types of thermometers (platinum, carbon, and doped germanium resistors; thermocouples) are calibrated at these temperatures and used to measure intermediate temperatures. *See* TEMPERATURE MEASUREMENT; THERMODYNAMIC PRINCIPLES.

[R.A.Hu.]

Temperature measurement

Measurement of the hotness of a body relative to a standard scale. The fundamental scale of temperature is the thermodynamic scale, which can be derived from any equation expressing the second law of thermodynamics. Efforts to approximate the thermodynamic scale as closely as possible depend on relating measurements of temperature-dependent physical properties of systems to thermodynamic relations expressed by statistical thermodynamic equations, thus in general linking temperature to the average kinetic energy of the measured system. Temperature-measuring devices, thermometers, are systems with properties that change with temperature in a simple, predictable, reproducible manner. *See* TEMPERATURE; THERMODYNAMIC PRINCIPLES.

In the establishment of a useful standard scale, assigned temperature values of thermodynamic equilibrium fixed points are agreed upon by an international body (General Conference of Weights and Measures), which updates the scale about once every 20 years. Thermometers for interpolating between fixed points and methods for realizing the fixed points are prescribed, providing a scheme for calibrating thermometers used in science and industry.

The scale now in use is the International Temperature Scale of 1990 (ITS-90). Its unit is the kelvin, K, arbitrarily defined as 1/273.16 of the thermodynamic temperature T of the triple point of water (where liquid, solid, and vapor coexist). For temperatures above 273.15 K, it is common to use International Celsius Temperatures, t_{90} (rather than International Kelvin Temperatures, T_{90}), having the unit degree Celsius, with symbol °C. The degree Celsius has the same magnitude as the kelvin. Temperatures, t_{90}, are defined as $t_{90}/°C = T_{90}/K - 273.15$, that is, as differences from the ice-point temperature at 273.15 K. The ice point is the state in which the liquid and solid phases of water coexist at a pressure of 1 atm (101,325 pascals). [The Fahrenheit scale, with symbol °F, still in common use in the United States, is given by $t_F/°F = (t_{90}/°C \times 1.8) + 32$, or $t_F/°F = (T_{90}/K \times 1.8) - 459.67$.] The ITS-90 is defined by 17 fixed points.

Primary thermometers are devices which relate the thermodynamic temperature to statistical mechanical formulation. The fixed points of ITS-90 are all based on one or more types of gas thermometry or on spectral radiation pyrometry referenced to gas thermometry. Secondary thermometers are used as reference standards in the laboratory because primary thermometers are often too cumbersome. It is necessary to establish standard secondary thermometers referenced to one or more fixed points for interpolation between fixed points. Lower-order thermometers are used for most practical purposes and, when high accuracy is required, can usually be calibrated against reference standards maintained at laboratories, such as the U.S. National Institute of Standards and Technology, or against portable reference devices (sealed boiling or melting point cells). *See* GAS THERMOMETRY; LOW-TEMPERATURE THERMOMETRY; PYROMETER; THERMISTOR; THERMOCOUPLE; THERMOMETER. [B.W.M.]

Theoretical physics The description of natural phenomena in mathematical form. It is impossible to separate theoretical physics from experimental physics, since a complete understanding of nature can be obtained only by the application of both theory and experiment. There are two main purposes of theoretical physics: the discovery of the fundamental laws of nature and the derivation of conclusions from these fundamental laws.

Physicists aim to reduce the number of laws to a minimum to have as far as possible a unified theory. When the laws are known, it is possible from any given initial conditions of a physical system to derive the subsequent events in the system. Sometimes, especially in quantum theory, only the probability of various events can be predicted. *See* DETERMINISM; QUANTUM MECHANICS.

The conclusions to be derived from the fundamental laws of nature may be of several different types.

1. Conclusions may be derived in order to test a given theory, particularly a new theory. An example is the derivation of the spectrum of the hydrogen atom from quantum mechanics; the verification of the predictions by accurate measurements is a good test of quantum mechanics. On rather rare occasions an experiment has been found to contradict the predictions of an existing theory, and this has then led to the discovery of important new physical laws. An example is the Michelson-Morley experiment on the constancy of the velocity of light, an experiment which led to special relativity theory. *See* ATOMIC STRUCTURE AND SPECTRA; LIGHT; RELATIVITY.

2. Theory may be required for experiments designed to determine physical constants. Most fundamental physical constants cannot be accurately measured directly. Elaborate theories may be required to deduce the constant from indirect experiments. *See* FUNDAMENTAL CONSTANTS.

3. Predictions of physical phenomena may be made in order to gain understanding of the structure of the physical world. In this category fall theories of the structure of the atom leading to an understanding of the periodic system of elements, or of the structure of the nucleus in which various models are tested (for example, shell model or collective model). In the same category fall applications of theoretical physics to other sciences, for example, to chemistry (theory of the chemical bond and of the rate of chemical reactions), astronomy (theory of planetary motion, internal constitution, and energy production of stars), or biology.

4. Engineering applications may be drawn from fundamental laws. All of engineering may be considered an application of physics, and much of it is an application of mathematical physics, such as elasticity theory, aerodynamics, electricity, and magnetism. The generation and propagation of radio waves of all frequencies are examples of application of theoretical physics to direct practice. *See* AERODYNAMICS; ELECTRICITY; MAGNETISM.

Apart from the classification of the fields of theoretical physics according to purpose, a classification can also be made according to content. Here one may perhaps distinguish three classification principles: type of force, scale of physical phenomena, and type of phenomena. *See* MATHEMATICAL PHYSICS; PHYSICS. [H.A.Be.]

Thermal conduction in solids   Thermal conduction in a solid is generally measured by stating the thermal conductivity K, which is the ratio of the steady-state heat flow (heat transfer per unit area per unit time) along a long rod to the temperature gradient along the rod. Thermal conductivity varies widely among different types of solids, and depends markedly on temperature and on the purity and physical state of the solids, particularly at low temperatures.

From the kinetic theory of gases the thermal conductivity can be written as $K =$ (constant) Svl, where S is the specific heat per unit volume, v is the average particle velocity, and l is the mean free path. In solids, thermal conduction results from conduction by lattice vibrations and from conduction by electrons. In insulating materials, the conduction is by lattice waves; in pure metals, the lattice contribution is negligible and the heat conduction is primarily due to electrons. In many alloys, impure metals, and semiconductors, both conduction mechanisms contribute. *See* CONDUCTION (HEAT); KINETIC THEORY OF MATTER; LATTICE VIBRATIONS; SPECIFIC HEAT.

In superconductors at temperatures below the critical temperature, the electronic conduction is reduced; at sufficiently low temperatures, the thermal conductivity becomes entirely due to lattice waves and is similar to the form of the thermal conductivity of an insulating material. *See* SUPERCONDUCTIVITY.   [K.A.McC.]

Thermal converters   Devices consisting of a conductor heated by an electric current, with one or more hot junctions of a thermocouple attached to it, so that the output emf responds to the temperature rise, and hence the current. Thermal converters are used with external resistors for alternating-current (ac) and voltage measurements over wide ranges and generally form the basis for calibration of ac voltmeters and the ac ranges of instruments providing known voltages and currents.

In the most common form, the conductor is a thin straight wire less than 0.4 in. (1 cm) long, in an evacuated glass bulb, with a single thermocouple junction fastened to the midpoint by a tiny electrically insulating bead. Thermal inertia keeps the temperature of the heater wire constant at frequencies above a few hertz, so that the constant-output emf is a true measure of the root-mean-square (rms) heating value of the current. The reactance of the short wire is so small that the emf can be independent of frequency up

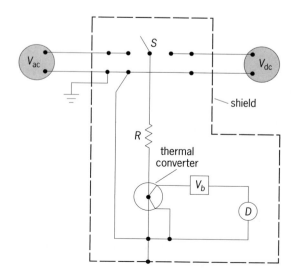

Basic circuit for ac-dc transfer measurements of ac voltages.

to 10 MHz or more. An emf of 10 mV can be obtained at a rated current less than 5 mA, so that resistors of reasonable power dissipation, in series or in shunt with the heater, can provide voltage ranges up to 1000 V and current ranges up to 20 A. However, the flow of heat energy cannot be controlled precisely, so the temperature, and hence the emf, generally changes with time and other factors. Thus an ordinary thermocouple instrument, consisting of a thermal converter and a millivoltmeter to measure the emf, is accurate only to about 1–3%. *See* THERMOCOUPLE; VOLTMETER.

To overcome this, a thermal converter is normally used as an ac-dc transfer instrument (ac-dc comparator) to measure an unknown alternating current or voltage by comparison with a known nearly equal dc quantity (see illustration). By replacing the millivoltmeter with an adjustable, stable, opposing voltage V_b in series with a microvoltmeter D, very small changes in emf can be detected. The switch S is connected to the unknown ac voltage V_{ac}, and V_b is adjusted for a null (zero) reading of D. Then S is immediately connected to the dc voltage V_{dc}, which is adjusted to give a null again, without changing V_b. Thus $V_{ac} = V_{dc}(1 + d)$, where d is the ac-dc difference of the transfer instrument, which can be as small as a few parts per million (ppm).

In many commercial instruments, all of the components are conveniently packaged in the shield, shown with a broken line, and several ranges are available by taps on R. Accuracies of 0.001% are attainable at audio frequencies.   [F.L.H.; J.R.K.]

Thermal expansion   Solids, liquids, and gases all exhibit dimensional changes for changes in temperature while pressure is held constant. The molecular mechanisms at work and the methods of data presentation are quite different for the three cases.

The temperature coefficient of linear expansion α_l is defined by Eq. (1), where l is the

$$\alpha_l = \frac{1}{l}\left(\frac{\partial l}{\partial t}\right)_{p=\text{const}} \tag{1}$$

length of the specimen, t is the temperature, and p is the pressure. For each solid there is a Debye characteristic temperature Θ, below which α_l is strongly dependent upon temperature and above which α_l is practically constant. Many common substances are near or above Θ at room temperature and follow approximate equation (2), where l_0 is

$$l = l_0(1 + \alpha_l t) \tag{2}$$

the length at 0°C and t is the temperature in °C. The total change in length from absolute zero to the melting point has a range of approximately 2% for most substances.

So-called perfect gases follow the relation in Eq. (3), where p is absolute pressure, v

$$\frac{pv}{T} = \frac{R}{\text{molecular weight}} \tag{3}$$

is specific volume, T is absolute temperature, and R is the so-called gas constant. Real gases often follow this equation closely. *See* GAS CONSTANT.

The coefficient of cubic expansion α_v is defined by Eq. (4), and for a perfect gas this

$$\alpha_v = \frac{1}{v}\left(\frac{\partial v}{\partial t}\right)_{p=\text{const}} \tag{4}$$

is found to be $1/T$. The behavior of real gases is largely accounted for by the van der Waals equation. *See* KINETIC THEORY OF MATTER.

For liquids, α_v is somewhat a function of pressure but is largely determined by temperature. Though α_v may often be taken as constant over a sizable range of temperature (as in the liquid expansion thermometer), generally some variation must be accounted for. For example, water contracts with temperature rise from 32 to 39°F (0 to 4°C), above which it expands at an increasing rate. *See* THERMOMETER. [R.A.Bu.]

Thermal hysteresis A phenomenon in which a physical quantity depends not only on the temperature but also on the preceding thermal history. It is usual to compare the behavior of the physical quantity while heating and the behavior while cooling through the same temperature range. The illustration shows the thermal hysteresis which has been observed in the behavior of the dielectric constant of single crystals of barium titanate. On heating, the dielectric constant was observed to follow the path *ABCD*, and on cooling the path *DCEFG*. *See* FERROELECTRICS.

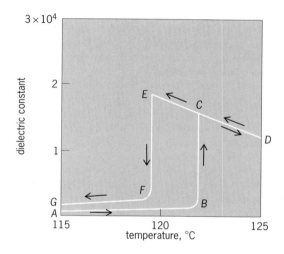

Plot of dielectric constant versus temperature for a single crystal of barium titanate. (*After M. E. Drougard and D. R. Young, Phys. Rev., 95:1152–1153, 1954*)

Perhaps the most common example of thermal hysteresis involves a phase change such as solidification from the liquid phase. In many cases these liquids can be dramatically supercooled. Elaborate precautions to eliminate impurities and outside disturbances can be instrumental in supercooling 60 to 80°C. On raising the temperature after freezing, however, the system follows a completely different path, with melting coming at the prescribed temperature for the phase change. *See* CRYSTAL; PHASE TRANSITIONS.
[H.B.H.; R.K.Mac.C.]

Thermionic emission The emission of electrons into vacuum by a heated electronic conductor. In its broadest meaning, thermionic emission includes the emission of ions, but this process is quite different from that normally understood by the term. Thermionic emitters are used as cathodes in electron tubes and hence are of great technical and scientific importance. Although in principle all conductors are thermionic emitters, only a few materials satisfy the requirements set by practical applications. Of the metals, tungsten is an important practical thermionic emitter; in most electron tubes, however, the oxide-coated cathode is used to great advantage.

The thermionic emission of a material may be measured by using the material as the cathode in a vacuum tube and collecting the emitted electrons on a positive anode. If the anode is sufficiently positive relative to the cathode, space charge (a concentration of electrons near the cathode) can be avoided and all electrons emitted can be collected; the saturation thermionic current is then measured. *See* SCHOTTKY EFFECT.

The emission current density J increases rapidly with increasing temperature; this is illustrated by the following approximate values for tungsten:

$T(K)$	1000	2000	2500	3000
$J(amperes/cm^2)$	10^{-15}	10^{-3}	0.3	15

The temperature dependence of J is given by the Richards (or Dushman-Richardson) equation below. Here A is a constant, k is Boltzmann's constant ($=1.38 \times 10^{-23}$ joule/

$$J = AT^2 e^{-(\phi - kT)}$$

degree), and ϕ is the work function of the emitter. The work function has the dimensions of energy and is a few electronvolts for thermionic emitters. *See* WORK FUNCTION (ELECTRONICS).   [A.J.D.]

Thermistor   An electrical resistor with a relatively large negative temperature coefficient of resistance. Thermistors are useful for measuring temperature and gas flow or wind velocity. Often they are employed as bolometer elements to measure radiofrequency, microwave, and optical power. They also are used as electrical circuit components for temperature compensation, voltage regulation, circuit protection, time delay, and volume control. Thermistors are semiconducting ceramics composed of mixtures of several metal oxides. Metal electrodes or wires are attached to the ceramic material so that the thermistor resistance can be measured conveniently. *See* BOLOMETER; ELECTRICAL RESISTIVITY.

At room temperature the resistance of a thermistor may typically change by several percent for a variation of $1°$ of temperature, but the resistance does not change linearly with temperature. The temperature coefficient of resistance of a thermistor is approximately equal to a constant divided by the square of the temperature in kelvins. The constant is equal to several thousand kelvins and is specified for a given thermistor and the temperature range of intended use.

The resistance and heat capacity of a thermistor depend upon the material composition, the physical dimensions, and the environment provided by the thermistor enclosure. Thermistors range in form from small beads and flakes less than 10^{-3} in. (25 micrometers) thick to disks, rods, and washers with inch dimensions. The small beads are often coated with glass to prevent changes in composition or encased in glass probes or cartridges to prevent damage. Beads are available with room-temperature resistances ranging from less than 100 Ω to tens of megohms, with heat capacities as low as tens of microwatts per degree celsius, and with time constants of less than a second. Large disks and washers have heat capacities as high as a few watts per degree Celsius and time constants of minutes. *See* TEMPERATURE MEASUREMENT; TIME CONSTANT.
                                                                                    [R.Pow.]

Thermoacoustics   The study of phenomena that involve both thermodynamics and acoustics. A sound wave in a gas is usually regarded as consisting of coupled pressure and displacement oscillations, but temperature oscillations accompany the pressure oscillations. When there are spatial gradients in the oscillating temperature, oscillating heat flow also occurs. The combination of these four oscillations produces a rich variety of thermoacoustic effects. *See* ACOUSTICS; OSCILLATION; SOUND; THERMODYNAMIC PRINCIPLES.

Although the oscillating heat transfer at solid boundaries does contribute significantly to the dissipation of sound in enclosures such as buildings, thermoacoustic effects are usually too small to be obviously noticeable in everyday life. However, thermoacoustic effects in intense sound waves inside suitable cavities can be harnessed to produce extremely powerful pulsating combustion, thermoacoustic refrigerators, and thermoacoustic engines.

Pulsating combustion. Oscillations can occur whenever combustion takes place in a cavity. In industrial equipment and residential appliances, these oscillations are sometimes encouraged in order to stir or pump the combustion ingredients, while in rocket engines such oscillations must usually be suppressed because they can damage the rocket structure. The oscillations occur spontaneously if the combustion progresses more rapidly or efficiently during the compression phase of the pressure oscillation than during the rarefaction (expansion) phase—the Rayleigh criterion. *See* GAS DYNAMICS.

Thermoacoustic refrigerators. Thermoacoustic refrigerators use acoustic power to pump heat from a low temperature to ambient temperature (see illustration). The heat-pumping mechanism takes place in the pores of a structure called a stack. As a typical parcel of the gas oscillates along a pore, it experiences changes in temperature. Most of the temperature change comes from adiabatic compression and expansion of the gas by the sound pressure, and the rest is a consequence of the local temperature of the solid wall of the pore. A thermodynamic cycle results from the coupled pressure,

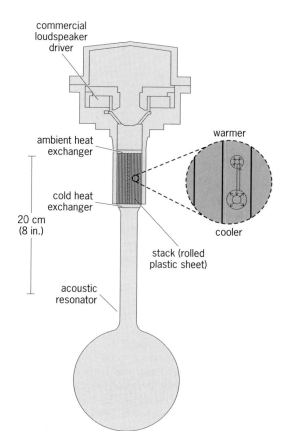

An early standing-wave thermoacoustic refrigerator that cooled to −60°C (−76°F). Heat is carried up the temperature gradient in the stack. At the right is a magnified view of the oscillating motion of a typical parcel of gas. The volume of the parcel depends on its pressure and temperature. (*After T. J. Hofler, Thermoacoustic Refrigerator Design and Performance, Ph.D. thesis, University of California at San Diego, 1996*)

temperature, position, and heat oscillations. The overall effect, much as in a bucket brigade, is the net transport of heat from the cold heat exchanger to room temperature. *See* ADIABATIC PROCESS; SOUND PRESSURE; THERMODYNAMIC PROCESSES.

Thermoacoustic engines. While standing-wave thermoacoustic systems have matured only recently, Stirling engines and refrigerators have a long, rich history. New insights have resulted from applying thermoacoustics to Stirling systems, treating them as traveling-wave thermoacoustic systems in which the extrema in pressure and gas motion are approximately 90° out of phase in time. In the thermoacoustic-Stirling engine, the thermodynamic cycle is accomplished in a traveling-wave acoustic network, and acoustic power is produced from heat with an efficiency of 30%.   [G.Swi.]

Thermocouple   A device in which the temperature difference between the ends of a pair of dissimilar metal wires is deduced from a measurement of the difference in the thermoelectric potentials developed along the wires. The presence of a temperature gradient in a metal or alloy leads to an electric potential gradient being set up along the temperature gradient. This thermoelectric potential gradient is proportional to the temperature gradient and varies from metal to metal. It is the fact that the thermoelectric emf is different in different metals and alloys for the same temperature gradient that allows the effect to be used for the measurement of temperature.

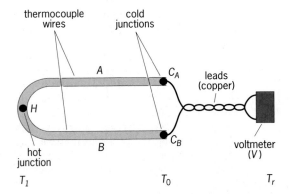

Basic circuit of a thermocouple.

The basic circuit of a thermocouple is shown in the illustration. The thermocouple wires, made of different metals or alloys A and B, are joined together at one end H, called the hot (or measuring) junction, at a temperature T_1. The other ends, C_A and C_B (the cold or reference junctions), are maintained at a constant reference temperature T_0, usually but not necessarily 32°F (0°C). From the cold junctions, wires, usually of copper, lead to a voltmeter V at room temperature T_r. Due to the thermoelectric potential gradients being different along the wires A and B, there exists a potential difference between C_A and C_B. This can be measured by the voltmeter, provided that C_A and C_B are at the same temperature and that the lead wires between C_A and V and C_B and V are identical (or that V is at the temperature T_0, which is unusual). Such a thermocouple will produce a thermoelectric emf between C_A and C_B which depends only upon the temperature difference $T_1 - T_0$. *See* TEMPERATURE MEASUREMENT; THERMOELECTRICITY.

A large number of pure metal and alloy combinations have been studied as thermocouples, and the seven most widely used are listed in the table. The thermocouples in the table together cover the temperature range from about −420°F (−250°C or 20 K) to about 3300°F (1800°C). The most accurate and reproducible are the platinum/rhodium

Letter designations and compositions for standardized thermocouples*	
Type designation	Materials
B	Platinum-30% rhodium/platinum-6% rhodium
E	Nickel-chromium alloy/a copper-nickel alloy
J	Iron/another slightly different copper-nickel alloy
K	Nickel-chromium alloy/nickel-aluminum alloy
R	Platinum-13% rhodium/platinum
S	Platinum-10% rhodium/platinum
T	Copper/a copper-nickel alloy

*After T. J. Quinn, *Temperature*, Academic Press, 1983.

thermocouples, types R and S, while the most widely used industrial thermocouples are probably types K, T, and E.   [T.J.Q.]

Thermodynamic principles   Laws governing the transformation of energy. Thermodynamics is the science of the transformation of energy. It differs from the dynamics of Newton by taking into account the concept of temperature, which is outside the scope of classical mechanics. In practice, thermodynamics is useful for assessing the efficiencies of heat engines (devices that transform heat into work) and refrigerators (devices that use external sources of work to transfer heat from a hot system to cooler sinks), and for discussing the spontaneity of chemical reactions (their tendency to occur naturally) and the work that they can be used to generate.

The subject of thermodynamics is founded on four generalizations of experience, which are called the laws of thermodynamics. Each law embodies a particular constraint on the properties of the world. The connection between phenomenological thermodynamics and the properties of the constituent particles of a system is established by statistical thermodynamics, also called statistical mechanics. Classical thermodynamics consists of a collection of mathematical relations between observables, and as such is independent of any underlying model of matter (in terms, for instance, of atoms). However, interpretations in terms of the statistical behavior of large assemblies of particles greatly enriches the understanding of the relations established by thermodynamics. *See* STATISTICAL MECHANICS.

Zeroth law of thermodynamics. The zeroth law of thermodynamics establishes the existence of a property called temperature. This law is based on the observation that if a system A is in thermal equilibrium with a system B (that is, no change in the properties of B take places when the two are in contact), and if system B is in thermal equilibrium with a system C, then it is invariably the case that A will be found to be in equilibrium with C if the two systems are placed in mutual contact. This law suggests that a numerical scale can be established for the common property, and if A, B, and C have the same numerical values of this property, then they will be in mutual thermal equilibrium if they were placed in contact. This property is now called the temperature. *See* TEMPERATURE.

First law of thermodynamics. The first law of thermodynamics establishes the existence of a property called the internal energy of a system. It also brings into the discussion the concept of heat.

The first law is based on the observation that a change in the state of a system can be brought about by a variety of techniques. Indeed, if attention is confined to an adiabatic system, one that is thermally insulated from its surroundings, then the work of J. P. Joule shows that same change of state is brought about by a given quantity of work regardless of the manner in which the work is done. This observation suggests that, just as the height through which a mountaineer climbs can be calculated from the difference in altitudes regardless of the path the climber takes between two fixed points, so the work, w, can be calculated from the difference between the final and initial properties of a system. The relevant property is called the internal energy, U. However, if the transformation of the system is taken along a path that is not adiabatic, a different quantity of work may be required. The difference between the work of adiabatic change and the work of nonadiabatic change is called heat, q. In general, Eq. (1) is satisfied, where ΔU is the change in internal energy between the final and

$$\Delta U = w + q \tag{1}$$

initial states of the system. *See* ADIABATIC PROCESS; ENERGY; HEAT.

The implication of this argument is that there are two modes of transferring energy between a system and its surroundings. One is by doing work; the other is by heating the system. Work and heat are modes of transferring energy. They are not forms of energy in their own right. Work is a mode of transfer that is equivalent (if not the case in actuality) to raising a weight in the surroundings. Heat is a mode of transfer that arises from a difference in temperature between the system and its surroundings. What is commonly called heat is more correctly called the thermal motion of the molecules of a system.

The first law of thermodynamics states that the internal energy of an isolated system is conserved. That is, for a system to which no energy can be transferred by the agency of work or of heat, the internal energy remains constant. This law is a cousin of the law of the conservation of energy in mechanics, but it is richer, for it implies the equivalence of heat and work for bringing about changes in the internal energy of a system (and heat is foreign to classical mechanics).

Second law of thermodynamics. The second law of thermodynamics deals with the distinction between spontaneous and nonspontaneous processes. A process is spontaneous if it occurs without needing to be driven. In other words, spontaneous changes are natural changes, like the cooling of hot metal and the free expansion of a gas. Many conceivable changes occur with the conservation of energy globally, and hence are not in conflict with the first law; but many of those changes turn out to be nonspontaneous, and hence occur only if they are driven.

The second law was formulated by Lord Kelvin and by R. Clausius in a manner relating to observation: "no cyclic engine operates without a heat sink" and "heat does not transfer spontaneously from a cool to a hotter body," respectively (see illustration). The two statements are logically equivalent in the sense that failure of one implies failure of the other. However, both may be absorbed into a single statement: the entropy of

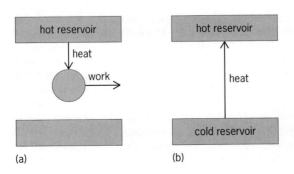

Representation of the statements of the second law of thermodynamics by (*a*) Lord Kelvin and (*b*) R. Clausius. In each case, the law states that the device shown cannot operate as shown.

an isolated system increases when a spontaneous change occurs. The property of entropy is introduced to formulate the law quantitatively in exactly the same way that the properties of temperature and internal energy are introduced to render the zeroth and first laws quantitative and precise.

The entropy, S, of a system is a measure of the quality of the energy that it stores. The formal definition is based on Eq. (2), where dS is the change in entropy of a system, dq

$$dS = \frac{dq_{\text{reversible}}}{T} \qquad (2)$$

is the energy transferred to the system as heat, T is the temperature, and the subscript "reversible" signifies that the transfer must be carried out reversibly (without entropy production other than in the system). When a given quantity of energy is transferred as heat, the change in entropy is large if the transfer occurs at a low temperature and small if the temperature is high.

This definition of entropy is illuminated by L. Boltzmann's interpretation of entropy as a measure of the disorder of a system. The connection can be appreciated qualitatively at least by noting that if the temperature is high, the transfer of a given quantity of energy as heat stimulates a relatively small additional disorder in the thermal motion of the molecules of a system; in contrast, if the temperature is low, the same transfer could stimulate a relatively large additional disorder.

The illumination of the second law brought about by the association of entropy and disorder is that in an isolated system the only changes that may occur are those in which there is no increase in order. Thus, energy and matter tend to disperse in disorder (that is, entropy tends to increase), and this dispersal is the driving force of spontaneous change. *See* ENTROPY; TIME, ARROW OF.

Third law of thermodynamics. The practical significance of the second law is that it limits the extent to which the internal energy may be extracted from a system as work. In order for a process to generate work, it must be spontaneous. For the process to be spontaneous, it is necessary to discard some energy as heat in a sink of lower temperature. In other words, nature in effect exacts a tax on the extraction of energy as work. There is therefore a fundamental limit on the efficiency of engines that convert heat into work.

The quantitative limit on the efficiency, ϵ, which is defined as the work produced divided by the heat absorbed from the hot source, was first derived by S. Carnot. He found that, regardless of the details of the construction of the engine, the maximum efficiency (that is, the work obtained after payment of the minimum allowable tax to ensure spontaneity) is given by Eq. (3), where T_{hot} is the temperature of the hot source

$$\epsilon = 1 - \frac{T_{\text{cold}}}{T_{\text{hot}}} \qquad (3)$$

and T_{cold} is the temperature of the cold sink. The greatest efficiencies are obtained with the coldest sinks and the hottest sources, and these are the design requirements of modern power plants.

Perfect efficiency ($\epsilon = 1$) would be obtained if the cold sink were at absolute zero ($T_{\text{cold}} = 0$). However, the third law of thermodynamics, which is another summary of observations, asserts that absolute zero is unattainable in a finite number of steps for any process. Therefore, heat can never be completely converted into work in a heat engine. The implication of the third law in this form is that the entropy change accompanying any process approaches zero as the temperature approaches zero. That implication in turn implies that all substances tend toward the same entropy as the temperature is reduced to zero. It is therefore sensible to take the entropy of all perfect crystalline substances (substances in which there is no residual disorder arising from the location of atoms) as equal to zero. A common short statement of the third law is therefore that all perfect crystalline substances have zero entropy at absolute zero ($T = 0$). This statement is consistent with the interpretation of entropy as a measure of disorder, since at absolute zero all thermal motion has been quenched. *See* ABSOLUTE ZERO.   [P.W.A.]

Thermodynamic processes   Changes of any property of an aggregation of matter and energy, accompanied by thermal effects.

Systems and processes. To evaluate the results of a process, it is necessary to know the participants that undergo the process, and their mass and energy. A region, or a system, is selected for study, and its contents determined. This region may have both mass and energy entering or leaving during a particular change of conditions, and these mass and energy transfers may result in changes both within the system and within the surroundings which envelop the system.

To establish the exact path of a process, the initial state of the system must be determined, specifying the values of variables such as temperature, pressure, volume, and quantity of material. The number of properties required to specify the state of a system depends upon the complexity of the system. Whenever a system changes from one state to another, a process occurs.

The path of a change of state is the locus of the whole series of states through which the system passes when going from an initial to a final state. For example, suppose a gas expands to twice its volume and that its initial and final temperatures are the same. An extremely large number of paths connect these initial and final states. The detailed path must be specified if the heat or work is to be a known quantity; however, changes in the thermodynamic properties depend only on the initial and final states and not

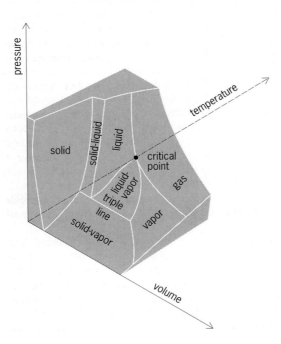

Fig. 1. Portion of pressure-volume-temperature (*P-V-T*) surface for a typical substance.

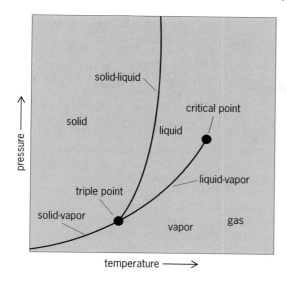

Fig. 2. Portion of equilibrium surface projected on pressure-temperature (*P-T*) plane.

upon the path. A quantity whose change is fixed by the end states and is independent of the path is a point function or a property.

Pressure-volume-temperature diagram. Whereas the state of a system is a point function, the change of state of a system, or a process, is a path function. Various processes or methods of change of a system from one state to another may be depicted graphically as a path on a plot using thermodynamic properties as coordinates.

The variable properties most frequently and conveniently measured are pressure, volume, and temperature. If any two of these are held fixed (independent variables), the third is determined (dependent variable). To depict the relationship among these physical properties of the particular working substance, these three variables may be used as the coordinates of a three-dimensional space. The resulting surface is a graphic presentation of the equation of state for this working substance, and all possible equilibrium states of the substance lie on this *P-V-T* surface.

Because a *P-V-T* surface represents all equilibrium conditions of the working substance, any line on the surface represents a possible reversible process, or a succession of equilibrium states.

The portion of the *P-V-T* surface shown in Fig. 1 typifies most real substances; it is characterized by contraction of the substance on freezing. Going from the liquid surface to the liquid-solid surface onto the solid surface involves a decrease in both temperature and volume. Water is one of the few exceptions to this condition; it expands upon freezing, and its resultant *P-V-T* surface is somewhat modified where the solid and liquid phases abut.

One can project the three-dimensional surface onto the *P-T* plane as in Fig. 2. The triple point is the point where the three phases are in equilibrium. When the temperature exceeds the critical temperature (at the critical point), only the gaseous phase is possible.

Temperature-entropy diagram. Energy quantities may be depicted as the product of two factors: an intensive property and an extensive one. Examples of intensive properties are pressure, temperature, and magnetic field; extensive ones are volume, magnetization, and mass. Thus, in differential form, work is the product of a pressure exerted against an area which sweeps through an infinitesimal volume, as in Eq. (1). As

$$dW = P\,dV \qquad (1)$$

a gas expands, it is doing work on its environment. However, a number of different kinds of work are known. For example, one could have work of polarization of a dielectric, of magnetization, of stretching a wire, or of making new surface area. In all cases, the infinitesimal work is given by Eq. (2), where X is a generalized applied force which is

$$dW = X\,dx \qquad (2)$$

an intensive quantity, and dx is a generalized displacement of the system and is thus extensive.

By extending this approach, one can depict transferred heat as the product of an intensive property, temperature, and a distributed or extensive property defined as entropy, for which the symbol is S. See ENTROPY.

Reversible and irreversible processes. Not all energy contained in or associated with a mass can be converted into useful work. Under ideal conditions only a fraction of the total energy present can be converted into work. The ideal conversions which retain the maximum available useful energy are reversible processes.

Characteristics of a reversible process are that the working substance is always in thermodynamic equilibrium and the process involves no dissipative effects such as viscosity, friction, inelasticity, electrical resistance, or magnetic hysteresis. Thus, reversible processes proceed quasistatically so that the system passes through a series of states of thermodynamic equilibrium, both internally and with its surroundings. This series of states may be traversed just as well in one direction as in the other.

Actual changes of a system deviate from the idealized situation of a quasistatic process devoid of dissipative effects. The extent of the deviation from ideality is correspondingly the extent of the irreversibility of the process. See THERMODYNAMIC PRINCIPLES.

[P.E.Bl.; W.A.S.]

Thermoelectricity   The direct conversion of heat into electrical energy, or the reverse, in solid or liquid conductors by means of three interrelated phenomena—the Seebeck effect, the Peltier effect, and the Thomson effect—including the influence of magnetic fields upon each. The Seebeck effect concerns the electromotive force (emf) generated in a circuit composed of two different conductors whose junctions are maintained at different temperatures. The Peltier effect refers to the reversible heat generated at the junction between two different conductors when a current passes through the junction. The Thomson effect involves the reversible generation of heat in a single current-carrying conductor along which a temperature gradient is maintained. Specifically excluded from the definition of thermoelectricity are the phenomena of Joule heating and thermionic emission. See ELECTROMOTIVE FORCE (EMF); JOULE'S LAW; SEEBECK EFFECT; THERMIONIC EMISSION; THOMSON EFFECT.

The three thermoelectric effects are described in terms of three coefficients: the absolute thermoelectric power (or thermopower) S, the Peltier coefficient II, and the Thomson coefficient μ, each of which is defined for a homogeneous conductor at a given temperature. These coefficients are connected by the Kelvin relations, which convert complete information about one into complete information about all three. It is therefore necessary to measure only one of the three coefficients; usally the thermopower S is chosen.

The most important practical application of thermoelectric phenomena is in the accurate measurement of temperature. The phenomenon involved is the Seebeck effect. Of less importance are the direct generation of electrical power by application of heat

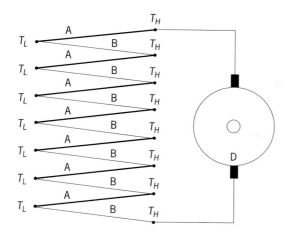

Thermopile, a battery of thermocouples connected in series. *D* is a device appropriate to the particular application; *A* and *B* are the two different conductors.

(also involving the Seebeck effect) and thermoelectric cooling and heating (involving the Peltier effect).

A basic system suitable for all four applications is shown schematically in the illustration. Several thermocouples are connected in series to form a thermopile, a device with increased output (for power generation or cooling and heating) or sensitivity (for temperature measurement) relative to a single thermocouple. The junctions forming one end of the thermopile are all at the same low temperature T_L, and the junctions forming the other end are at the high temperature T_H. The thermopile is connected to a device D which is different for each application. For temperature measurement, the temperature T_L is fixed, for example, by means of a bath; the temperature T_H becomes the running temperature T, which is to be measured; and the device is a potentiometer for measuring the thermoelectric emf generated by the thermopile. For power generation, the temperature T_L is fixed by connection to a heat sink; the temperature T_H is fixed at a value determined by the output of the heat source and the thermal conductance of the thermopile; and the device is whatever is to be run by the electricity that is generated. For heating or cooling, the device is a current generator that passes current through the thermopile. If the current flows in the proper direction, the junctions at T_H will heat up, and those at T_L will cool down. If T_H is fixed by connection to a heat sink, thermoelectric cooling will be provided by T_L. Alternatively, if T_L is fixed, thermoelectric heating will be provided at T_H. Such a system has the advantage that at any given location it can be converted from a cooler to a heater merely by reversing the direction of the current.

Thermoelectric power generators, heaters, or coolers made from even the best presently available materials have the disadvantages of relatively low efficiencies and concomitant high cost per unit of output. Their use has therefore been largely restricted to situations in which these disadvantages are outweighed by such advantages as small size, low maintenance due to lack of moving parts, quiet and vibration-free performance, light weight, and long life. [J.B.]

Thermoluminescence The emission of light when certain solids are warmed, generally to a temperature lower than that needed to provoke visible incandescence. Two characteristics of thermoluminescence distinguish it from incandescence. First, the intensity of thermoluminescent emission does not remain constant at constant temperature, but decreases with time and eventually ceases altogether. Second, the spectrum of the thermoluminescence is highly dependent on the composition of the material and is only slightly affected by the temperature of heating. If a thermoluminescent material emits both thermoluminescence and incandescent light at some temperature of observation, the transient light emission is the thermoluminescence and the remaining steady-state emission is the incandescence. The transient nature of the thermoluminescent emission suggests that heating merely triggers the release of stored energy previously imparted to the material. Supporting this interpretation is the fact that after the thermoluminescence has been reduced to zero by heating, the sample can be made thermoluminescent again by exposure to one of a number of energy sources: x-rays and gamma rays, electron beams, nuclear particles, ultraviolet light, and, in some cases, even short-wave visible light (violet and blue). A thermoluminescent material, therefore, has a memory of its earlier exposure to an energizing source, and this memory is utilized in a number of applications. Many natural minerals are thermoluminescent, but the most efficient materials of this type are specially formulated synthetic solids (phosphors). See LUMINESCENCE.

In addition to special sites capable of emitting light (luminescent centers), thermoluminescent phosphors have centers that can trap electrons or holes when these are produced in the solid by ionizing radiation. The luminescent center itself is often the hole trap, and the electron is trapped at another center, although the reverse situation can also occur. In the former case, if the temperature is low and the energy required to release an electron from a trap (the trap depth) is large, electrons will remain trapped and no luminescence will occur. If, however, the temperature of the phosphor is progressively raised, electrons will receive increasing amounts of thermal energy and will have an increased probability of escape from the traps. Freed electrons may then go over to luminescent centers and recombine with holes trapped at or near these centers. The energy liberated by the recombination can excite the luminescent centers, causing them to emit light. See HOLE STATES IN SOLIDS; TRAPS IN SOLIDS.

Radiation dosimeters based on thermoluminescence are widely used for monitoring integrated radiation exposure in nuclear power plants, hospitals, and other installations where high-energy radiations are likely to be encountered. The key elements of the dosimeters, thermoluminescent phosphors with deep traps, can store some of the energy absorbed from these radiations for very long periods of time at normal temperatures and release it as luminescence on demand when appropriately heated. The brightness (or light sum) of the luminescence is a measure of the original radiation dose.   [J.H.S.]

Thermomagnetic effects   Electrical and thermal phenomena occurring when a conductor or semiconductor which is carrying a thermal current (that is, is in a temperature gradient) is placed in a magnetic field. See SEMICONDUCTOR.

Let the temperature gradient be transverse to the magnetic field H_z, for example, along x. Then the following transverse-transverse effects are observed:

1. Ettingshausen-Nernst effect, an electric field along y.

2. Righi-Leduc effect, a temperature gradient along y.

3. An electric potential change along x, amounting to a change of thermoelectric power.

4. A temperature gradient change along x, amounting to a change of thermal resistance.

Let the temperature gradient be along H. Then changes in thermoelectric power and in thermal conductivity are observed in the direction of H.

For related phenomena see HALL EFFECT; MAGNETORESISTANCE.   [E.A.; F.Ke.]

Thermometer An instrument that measures temperature. Although this broad definition includes all temperature-measuring devices, they are not all called thermometers. Other names have been generally adopted. For a discussion of two such devices *see* PYROMETER; THERMOCOUPLE. *See also* TEMPERATURE MEASUREMENT.

Liquid-in-glass thermometer. This thermometer consists of a liquid-filled glass bulb and a connecting partially filled capillary tube. When the temperature of the thermometer increases, the differential expansion between the glass and the liquid causes the liquid to rise in the capillary. A variety of liquids, such as mercury, alcohol, toluene, and pentane, and a number of different glasses are used in thermometer construction, so that various designs cover diverse ranges between about $-300°F$ and $+1200°F$ ($-184°C$ and $+649°C$).

Bimetallic thermometer. In this thermometer the differential expansion of thin dissimilar metals, bonded together into a narrow strip and coiled into the shape of a helix or spiral, is used to actuate a pointer. In some designs the pointer is replaced with low-voltage contacts to control, through relays, operations which depend upon temperature, such as furnace controls.

Filled-system thermometer. This type of thermometer has a bourdon tube connected by a capillary tube to a hollow bulb. When the system is designed for and filled with a gas (usually nitrogen or helium) the pressure in the system substantially follows the gas law, and a temperature indication is obtained from the bourdon tube. The temperature-pressure-motion relationship is nearly linear. Atmospheric pressure effects are minimized by filling the system to a high pressure. When the system is designed for and filled with a liquid, the volume change of the liquid actuates the bourdon tube.

Vapor-pressure thermal system. This filled-system thermometer utilizes the vapor pressure of certain stable liquids to measure temperature. The useful portion of any liquid-vapor pressure curve is between approximately 15 psia (100 kilopascals absolute) and the critical pressure, that is, the vapor pressure at the critical temperature, which is the highest temperature for a particular liquid-vapor system. A nonlinear relationship exists between the temperature and the vapor pressure, so the motion of the bourdon tube is greater at the upper end of the vapor-pressure curve. Therefore, these thermal systems are normally used near the upper end of their range, and an accuracy of 1% or better can be expected.

Resistance thermometer. In this type of thermometer the change in resistance of conductors or semiconductors with temperature change is used to measure temperature. Usually, the temperature-sensitive resistance element is incorporated in a bridge network which has a reasonably constant power supply. Although a deflection circuit is occasionally used, almost all instruments of this class use a null-balance system, in which the resistance change is balanced and measured by adjusting at least one other resistance in the bridge. Metals commonly used as the sensitive element in resistance thermometers are platinum, nickel, and copper.

Thermistor. This device is made of a solid semiconductor with a high temperature coefficient of resistance. The thermistor has a high resistance, in comparison with metallic resistors, and is used as one element in a resistance bridge. Since thermistors are more sensitive to temperature changes than metallic resistors, accurate readings of small changes are possible. *See* THERMISTOR. [H.S.B.]

Thermonuclear reaction A nuclear fusion reaction which occurs between various nuclei of the light elements when they are constituents of a gas at very high temperatures. Thermonuclear reactions, the source of energy generation in the Sun and the stable stars, are utilized in the fusion bomb. *See* NUCLEAR FUSION.

Thermonuclear reactions occur most readily between isotopes of hydrogen (deuterium and tritium) and less readily among a few other nuclei of higher atomic number. At the temperatures and densities required to produce an appreciable rate of thermonuclear reactions, all matter is completely ionized; that is, it exists only in the plasma state. Thermonuclearer fusion reactions may then occur within such an ionized gas when the agitation energy of the stripped nuclei is sufficient to overcome their mutual electrostatic repulsions, allowing the colliding nuclei to approach each other closely enough to react. For this reason, reactions tend to occur much more readily between energy-rich nuclei of low atomic number (small charge) and particularly between those nuclei of the hot gas which have the greatest relative kinetic energy. This latter fact leads to the result that, at the lower fringe of temperatures where thermonuclear reactions may take place, the rate of reactions varies exceedingly rapidly with temperature. *See* KINETIC THEORY OF MATTER; MAGNETOHYDRODYNAMICS; NUCLEAR REACTION; PINCH EFFECT.   [R.F.P.]

Thomson effect   A phenomenon discovered in 1854 by William Thomson, later Lord Kelvin. He found that there occurs a reversible transverse heat flow into or out of a conductor of a particular metal, the direction depending upon whether a longitudinal electric current flows from colder to warmer metal or from warmer to colder. Any temperature gradient previously existing in the conductor is thus modified if a current is turned on. The Thomson effect does not occur in a current-carrying conductor which is initially at uniform temperature. *See* THERMOELECTRICITY.   [J.W.St.]

Time   The dimension of the physical universe which orders the sequence of events at a given place; also, a designated instant in this sequence, such as the time of day, technically known as an epoch, or sometimes as an instant.

Measurement.   Time measurement consists of counting the repetitions of any recurring phenomenon and possibly subdividing the interval between repetitions. Two aspects to be considered in the measurement of time are frequency, or the rate at which the recurring phenomena occur, and epoch, or the designation to be applied to each instant.

Time units are the intervals between successive recurrences of phenomena, such as the period of rotation of the Earth or a specified number of periods of radiation derived from an atomic energy-level transition. Other units are arbitrary multiples and subdivisions of these intervals, such as the hour being 1/24 of a day, and the minute being 1/60 of an hour. *See* TIME-INTERVAL MEASUREMENT.

Time bases.   Several phenomena are used as bases with which to determine time. The phenomenon traditionally used has been the rotation of the Earth, where the counting is by days. Days are measured by observing the meridian passages of stars and are subdivided with the aid of precision clocks. The day, however, is subject to variations in duration. Thus, when a more uniform time scale is required, other bases for time must be used.

The angle measured along the celestial equator between the observer's local meridian and the vernal equinox, known as the hour angle of the vernal equinox, is the measure of sidereal time. It is reckoned from 0 to 24 hours, each hour being subdivided into 60 sidereal minutes and the minutes into 60 sidereal seconds. Sidereal clocks are used for convenience in most astronomical observatories because a star or other object outside the solar system comes to the same place in the sky at virtually the same sidereal time.

The hour angle of the Sun is the apparent solar time. The only true indicator of local apparent solar time is a sundial. Mean solar time has been devised to eliminate the irregularities in apparent solar time that arise from the obliquity of the ecliptic and the varying speed of the Earth in its orbit around the Sun. It is the hour angle of a fictitious point moving uniformly along the celestial equator at the same rate as the average rate of the Sun along the ecliptic. Both sidereal and solar time depend on the rotation of the Earth for their time base.

The mean solar time determined for the meridian of $0°$ longitude from the rotation of the Earth by using astronomical observations is referred to as UT1. Observations are made at a number of observatories around the world. The International Earth Rotation Service (IERS) receives these data and maintains a UT1 time scale.

Because the Earth has a nonuniform rate of rotation and since a uniform time scale is required for many timing applications, a different definition of a second was adopted in 1967. The international agreement calls for the second to be defined as 9,192,631,770 periods of the radiation derived from an energy-level transition in the cesium atom. This second is referred to as the international or SI (International System) second and is independent of astronomical observations. International Atomic Time (TAI) is maintained by the International Bureau of Weights and Measures (BIPM) from data contributed by time-keeping laboratories around the world.

Coordinated Universal Time (UTC) uses the SI second as its time base. However, the designation of the epoch may be changed at certain times so that UTC does not differ from UT1 by more than 0.9 s. UTC forms the basis for civil time in most countries and may sometimes be referred to as Greenwich mean time. The adjustments to UTC to bring this time scale into closer accord with UT1 consist of the insertion or deletion of integral seconds. These "leap seconds" may be applied at 23 h 59 m 59 s of June 30 or December 31 of each year according to decisions made by the IERS. UTC differs from TAI by an integral number of atomic seconds.

Civil and standard times. Because rotational time scales are defined as hour angles, at any instant they vary from place to place on the Earth. Persons traveling westward around the Earth must advance their time 1 day, and those traveling eastward must retard their time 1 day in order to be in agreement with their neighbors when they return home. The International Date Line is the name given to a line where the change of date is made. It follows approximately the 180th meridian but avoids inhabited land. To avoid the inconvenience of the continuous change of mean solar time with longitude, zone time or civil time is generally used. The Earth is divided into 24 time zones, each approximately $15°$ wide and centered on standard longitudes of $0°$, $15°$, $30°$, and so on. Within each of these zones the time kept is the mean solar time of the standard meridian.

Many countries, including the United States, advance their time 1 hour, particularly during the summer months, into "daylight saving time."   [D.D.McC.]

Time, arrow of   The uniform and unique direction associated with the apparent inevitable flow of time into the future. There appears to be a fundamental asymmetry in the universe. Herein lies a paradox, for all the laws of physics, whether they are the equations of classical mechanics, classical electromagnetism, general relativity, or quantum mechanics, are time reversible in that they admit solutions in either direction of time. This reversibility raises the question of how these fundamentally time-symmetrical equations can result in the perceived asymmetry of temporally ordered events.

The symmetry breaking of temporal order has not yet been fully explained. There are certain indications that an intrinsic asymmetry exists in temporal evolution. Thus it may be that the fundamental laws of physics are not really time symmetric and that the currently known laws are only symmetrized approximations to the truth. Indeed, the decay of the K^0 meson is not time reversible. However, it is not clear how such a rare and exotic instance of time asymmetry could emerge into the world of essentially macroscopic, electromagnetic phenomena as an everyday observable.

Another, more ubiquitous example of a fundamentally time-asymmetric process is the expansion of the universe. It has been speculated that this expansion is the true basis of time asymmetry.

Alternatively, even a time-symmetrical universe will have a statistical behavior in which configurations of molecules and localizations of energy have significant probabilities of recurring only after enormously long time intervals. Indeed, such time intervals are longer than the times required for the ceaseless expansion of the universe and the evolution of its component particles. Time's arrow is destined, either by the nature of space-time or the statistics of large assemblies, to fly into the future. *See* STATISTICAL MECHANICS.   [P.W.A.]

Time constant   A characteristic time that governs the approach of an exponential function to a steady-state value. When a physical quantity is varying as a decreasing exponential function of time as in Eq. (1), or as an increasing exponential function as in Eq. (2), the approach to the steady-state value achieved after a long time is governed by a characteristic time T as given in Eq. (3). This time T is called the time constant.

$$f(t) = e^{-kt} \tag{1}$$

$$f(t) = 1 - e^{-kt} \tag{2}$$

$$t = \frac{1}{k} = T \tag{3}$$

When time t is zero, $f(t)$ in Eq. (1) has the magnitude 1, and when t equals T the magnirude is $1/e$. Here e is the transcendental number whose value is approximately 2.71828, and the change in magnitude is $1 - (1/e) = 0.63212$. The function has moved 63.2% of the way to its final value. The same factor also holds for Eq. (2).

The initial rate of change of both the increasing and decreasing functions is equal to the maximum amplitude of the function divided by the time constant.

The concept of time constant is useful when evaluating the presence of transient phenomena.   [R.L.R.]

Time-interval measurement   A determination of the duration between two instants of time (epochs). Time intervals are measured with high precision with a digital display counter. An electronic oscillator generates pulses; the count begins with a start signal and ends with a second signal. Two atomic clocks can be compared in epoch to 1 picosecond (1 ps = 10^{-12} s) by electronic interpolation. *See* ATOMIC CLOCK; OSCILLOSCOPE.

Rapid motions can be studied at short intervals by means of a large variety of high-speed cameras, including stroboscopic, rotating film-drum, rotating mirror, streak, and image converter cameras. An electronic streak camera can separate two pulses 1 ps apart.

Ultrashort laser pulses are used to study rapid processes caused by the interaction of photons with an atom or molecule. Pulses as short as three wavelengths of 620-nm light, with $\tau = 6$ femtoseconds (1 fs = 10^{-15} s), have been formed. *See* LASER; OPTICAL PULSES.

Radioactive decay is used to measure long time intervals, to about 5×10^9 years, concerning human history, the Earth, and the solar system.   [W.M.]

Time-of-flight spectrometer   Any of a general class of instruments in which the speed of a particle is determined directly by measuring the time that it takes to travel a measured distance. By knowing the particle's mass, its energy can be calculated. If the particles are uncharged (for example, neutrons), difficulties arise because standard methods of measurement (such as deflection in electric and magnetic fields) are not possible. The time-of-flight method is a powerful alternative, suitable for both uncharged and charged particles.

The time intervals are best measured by counting the number of oscillations of a stable oscillator that occur between the instants that the particle begins and ends its journey. Oscillators operating at 100 MHz are in common use. See MASS SPECTROSCOPE; NEUTRON SPECTROMETRY; TIME-INTERVAL MEASUREMENT.   [F.W.K.F.]

Time-projection chamber   An advanced particle detector for the study of ultra-high-energy collisions of positrons and electrons. The underlying physics of the scattering process can be studied through precise measurements of the momenta, directions, particle species, and correlations of the collision products. The time-projection chamber (TPC) provides a unique combination of capabilities for these studies and other problems in elementary particle physics by offering particle identification over a wide momentum range, and by offering high resolution of intrinsically three-dimensional spatial information for accurate event reconstruction.

The time-projection chamber concept is based on the maximum utilization of ionization information, which is deposited by high-energy charged particles traversing a gas. The ionization trail, a precise image of the particle trajectory, also contains information about the particle velocity. A strong, uniform magnetic field and a uniform electric field are generated within the time-projection chamber active volume in an exactly parallel orientation. The parallel configuration of the fields permits electrons, products of the ionization processes, to drift through the time-projection chamber gas over great distances without distortion; the parallel configuration offers a further advantage in that the diffusion of the electrons during drift can be greatly suppressed by the magnetic field, thus preserving the quality of track information. See PARTICLE DETECTOR.   [D.R.N.]

Time reversal invariance   A symmetry of the fundamental (microscopic) equations of motion of a system; if it holds, the time reversal of any motion of the system is also a motion of the system. With one exception (K_L meson decay), all observations are consistent with time reversal invariance (T invariance).

Time reversal invariance is not evident from casual observation of everyday phenomena. If a movie is taken of a phenomenon, the corresponding time-reversed motion can be exhibited by running the movie backward. The result is usually strange. For instance, water in the ground is not ordinarily observed to collect itself into drops and shoot up into the air. However, if the system is sufficiently well observed, the direction of time is not obvious. For instance, a movie which showed the motion of the planets would look just as right run backward or forward. The apparent irreversibility of everyday phenomena results from the combination of imprecise observation and starting from an improbable situation (a state of low entropy, to use the terminology of statistical mechanics). See ENTROPY; STATISTICAL MECHANICS; TIME, ARROW OF.

If time reversal invariance holds, no particle (a physical system with a definite mass and spin) can have an electric dipole moment. A polar body, for example, a water (H_2O) molecule, has an electric dipole moment, but its energy and spin eigenstates (which are particles) do not. No particle has been observed to have an electric dipole moment; for instance, the present experimental upper limit on the electric moment of the neutron is approximately 10^{-25} cm times e, where e is the charge of the proton. Even smaller upper limits have been reported for the electric moment of some nuclei. *See* DIPOLE MOMENT; NEUTRON; POLAR MOLECULE; SPIN (QUANTUM MECHANICS).

Another test of time reversal invariance is to compare the cross sections for reactions which are inverse to one another. The present experimental upper limit on the relative size of the time reversal invariance-violating amplitude of such reactions is approximately 3×10^{-3}; unfortunately, this is far larger than any expected violation. *See* NUCLEAR REACTION.

If time reversibility holds, then by the *CPT* theorem *CP* invariance must hold, that is, invariance of the fundamental equations under the combined operations of charge conjugation *C* and space inversion *P*. Conversely, violation of *CP* invariance implies violation of *T* (time reversal invariance). In 1964, *CP* violation was observed in the decay of the long-lived neutral *K* meson, the K_L. For many years, no other evidence for *T* or *CP* violation was seen. From this it was deduced that the interactions which violate *CP* are very weak and are evident in K_L decay only because there are two neutral *K* mesons that have practically the same mass and are therefore easily mixed. *See* CPT THEOREM; MESON.

Within the current understanding of particle physics, namely the so-called standard model, *CP* violation comes from the Kobayashi-Maskawa matrix of coefficients that relate the quark weak *i*-spin eigenstates with the quark mass eigenstates. It turns out that because the number of flavors is greater than four, the Kobayashi-Maskawa matrix can be nonreal, resulting in nonconservation of *CP* and *T*. *See* ELECTROWEAK INTERACTION; FLAVOR; QUARKS; STANDARD MODEL; WEAK NUCLEAR INTERACTIONS.

It is also possible that *CP* violation comes from yet unknown interactions. More is being learned from observations since 2001 of *CP* violation in the decay of the neutral *B* mesons at the so-called *B* factories, particle accelerators built for the purpose of copiously producing *B* mesons. Initial results are consistent with the standard model. *See* ELEMENTARY PARTICLE; PARTICLE ACCELERATOR; SYMMETRY LAWS (PHYSICS).   [C.J.G.]

Torque   The product of a force and its perpendicular distance to a point of turning; also called the moment of the force. Torque produces torsion and tends to produce rotation. Torque arises from a force or forces acting tangentially to a cylinder or from any force or force system acting about a point. A couple, consisting of two equal, parallel, and oppositely directed forces, produces a torque or moment about the central point. A prime mover such as a turbine exerts a twisting effort on its output shaft, measured as torque. In structures, torque appears as the sum of moments of torsional shear forces acting on a transverse section of a shaft or beam. *See* COUPLE.   [N.S.F.]

Torricelli's theorem   The speed of efflux of a liquid from an opening in a reservoir equals the speed that the liquid would acquire if allowed to fall from rest from the surface of the reservoir to the opening.

Torricelli, a student of Galileo, observed this relationship in 1643. In equation form, $v^2 = 2gh$, in which v is the speed of efflux, h the head (or elevation difference between reservoir surface and center line of opening if in a vertical plane), and g the acceleration due to gravity. (The equation is the same as that for a solid particle dropped a distance *h* in a vacuum.) The relationship can be derived from the energy equation for flow along a streamline, if energy losses are neglected.

[V.L.S.]

Trajectory   The curve described by a body moving through space, as of a meteor through the atmosphere, a planet around the Sun, a projectile fired from a gun, or a rocket in flight. In general, the trajectory of a body in a gravitational field is a conic section—ellipse, hyperbola, or parabola—depending on the energy of motion. The trajectory of a shell or rocket fired from the ground is a portion of an ellipse with the Earth's center as one focus; however, if the altitude reached is not great, the effect of gravity is essentially constant, and the parabola is a good approximation. *See* BALLISTICS.

[J.P.H.]

Transistor   A solid-state device involved in amplifying small electrical signals and in processing of digital information. Transistors act as the key element in amplification, detection, and switching of electrical voltages and currents. They are the active electronic component in all electronic systems which convert battery power to signal power. Almost every type of transistor is produced in some form of semiconductor, often single-crystal materials, with silicon being the most prevalent. There are several different types of transistors, classified by how the internal mobile charges (electrons and holes) function. The main categories are bipolar junction transistors (BJTs) and field-effect transistors (FETs).

Single-crystal semiconductors, such as silicon from column 14 of the periodic table of chemical elements, can be produced with two different conduction species, majority and minority carriers. When made with, for example, 1 part per million of phosphorus (from column 15), the silicon is called *n*-type because it adds conduction electrons (negative charge) to form the majority carrier. When doped with boron (from column 13), it is called *p*-type because it has added positive mobile carriers called holes. For *n*-type doping, electrons are the majority carrier while holes become the minority carrier. For *p*-type doping holes are in larger numbers, hence they are the majority carriers, while electrons are the minority carriers. All transistors are made up of regions of *n*-type and *p*-type semiconducting material. *See* SEMICONDUCTOR; SINGLE CRYSTAL.

The bipolar transistor has two conducting species, electrons and holes. Field-effect transistors can be called unipolar because their main conduction is by one carrier type, the majority carrier. Therefore, field-effect transistors are either *n*-channel (majority electrons) or *p*-channel (majority holes). For the bipolar transistor, there are two forms, n^+pn and p^+np, depending on which carrier is majority and which is the minority in a given region. As a result the bipolar transistor conducts by majority as well as by minority carriers. The n^+pn version is by far the most used as it has several distinct performance advantages, as does the *n*-channel for the field-effect transistors. (The n^+ indicates that the region is more heavily doped than the other two regions.)

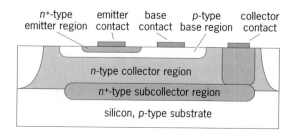

Fig. 1. Isolated n^+pn bipolar junction transistor for integrated-circuit operation.

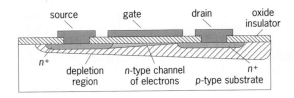

Fig. 2. An *n*-channel enhancement-mode metal-oxide-semiconductor field-effect transistor (MOSFET).

Bipolar transistors. Bipolar transistors have additional categories: the homojunction for one type of semiconductor (all silicon), and heterojunction for more than one (particularly silicon and silicon-germanium, $Si/Si_{1-x}Ge_x/Si$). At present the silicon homojunction, usually called the BJT, is by far the most common. However, the highest performance (frequency and speed) is a result of the heterojunction bipolar transistor (HBT).

Bipolar transistors are manufactured in several different forms, each appropriate for a particular application. They are used at high frequencies, for switching circuits, in high-power applications, and under extreme environmental stress. The bipolar junction transistor may appear in discrete form as an individually encapsulated component, in monolithic form (made in and from a common material) in integrated circuits, or as a so-called chip in a thick-film or thin-film hybrid integrated circuit. In the *pn*-junction isolated integrated-circuit n^+pn bipolar transistor, an n^+ subcollector, or buried layer, serves as a low-resistance contact which is made on the top surface (Fig. 1).

Field-effect transistors. Majority-carrier field-effect transistors are classified as metal-oxide-semiconductor field-effect transistor (MOSFET), junction "gate" field-effect transistor (JFET), and metal "gate" on semiconductor field-effect transistor (MESFET) devices. MOSFETs are the most used in almost all computers and system applications. However, the MESFET has high-frequency applications in gallium arsenide (GaAs), and the silicon JFET has low-electrical noise performance for audio components and instruments. In general, the *n*-channel field-effect transistors are preferred because of larger electron mobilities, which translate into higher speed and frequency of operation.

An *n*-channel MOSFET (Fig. 2) has a so-called source, which supplies electrons to the channel. These electrons travel through the channel and are removed by a drain electrode into the external circuit. A gate electrode is used to produce the channel or to remove the channel; hence it acts like a gate for the electrons, either providing a channel for them to flow from the source to the drain or blocking their flow (no channel). With a large enough voltage on the gate, the channel is formed, while at a low gate voltage it is not formed and blocks the electron flow to the drain. This type of MOSFET is called enhancement mode because the gate must have sufficiently large voltages to create a channel through which the electrons can flow. Another way of saying the same idea is that the device is normally "off" in an nonconducting state until the gate enhances the channel.

In the JFET (Fig. 3), a conducting majority-carrier *n* channel exists between the source and drain. When a negative voltage is applied to the p^+ gate, the depletion regions widen with reverse bias and begin to restrict the flow of electrons between the source and drain. At a large enough negative gate voltage (symbolized V_P), the channel pinches off.

The MESFET is quite similar to the JFET in its mode of operation. A conduction channel is reduced and finally pinched off by a metal Schottky barrier placed directly on the semiconductor. Metal on gallium arsenide is extensively used for high-frequency

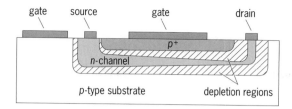

gate   source   gate   drain

p^+

n-channel

p-type substrate   depletion regions

Fig. 3. An *n*-channel junction field-effect transistor (JFET).

communications because of the large mobility of electrons, good gain, and low noise characteristics. Its cross section is similar to that of the JFET (Fig. 3), with a metal used as the gate.   [G.W.N.]

Transition radiation detectors   Detectors of energetic charged particles that make use of radiation emitted as the particle crosses boundaries between regions with different indices of refraction. An energetic charged particle moving through matter momentarily polarizes the material nearby. If the particle crosses a boundary where the index of refraction changes, the change in polarization gives rise to the emission of electromagnetic transition radiation. About one photon is emitted for every 100 boundaries crossed, for transitions between air and matter of ordinary density. Transition radiation is emitted even if the velocity of the particle is less than the light velocity of a given wavelength, in contrast to Cerenkov radiation. Consequently, this radiation can take place in the x-ray region of the spectrum where there is no Cerenkov radiation, because the index of refraction is less than one. *See* CERENKOV RADIATION; PARTICLE DETECTOR; REFRACTION OF WAVES.   [W.J.W.]

Translucent medium   A medium which transmits rays of light so diffused that objects cannot be seen distinctly; that is, the medium is only partially transparent. Familiar examples are various forms of glass which admit considerable light but impede vision. Inasmuch as the term translucent seems to imply seeing, usage of the term is ordinarily limited to the visible region of the spectrum.   [M.G.M.]

Traps in solids   Localized regions in a material that can capture and localize an electron or hole, thus preventing the electron or hole from moving through the material until supplied with sufficient thermal or optical energy. Traps in solids are associated with imperfections in the material caused by either impurities or crystal defects. *See* BAND THEORY OF SOLIDS; CRYSTAL DEFECTS; HOLE STATES IN SOLIDS.

   Imperfections that behave as traps are commonly distinguished from imperfections that behave as recombination centers. If the probability for a captured electron (or hole) at the imperfection to be thermally reexcited to the conduction (or valence) band before recombination with a free hole (or free electron) is greater than the probability for such recombination, then the imperfection is said to behave like an electron (or hole) trap. If the probability for a captured electron (or hole) at the imperfection to recombine with a free hole (or free electron) is greater than the probability for being thermally reexcited to the band, the imperfection is said to behave like a recombination center. It is possible for a specific chemical or structural imperfection in the material to behave like a trap under one set of conditions of temperature and light intensity, and as a recombination center under another.

   Traps play a significant role in many phenomena involving photoconductivity and luminescence. In photoconductors, for example, the presence of traps decreases the sensitivity and increases the response time. Their effect is detectable through changes in the rise and decay transients of photoconductivity and luminescence, thermally stimulated conductivity and luminescence in which the traps are filled at a low temperature and then emptied by increasing the temperature in a controlled way, electron spin responance associated with trapped electrons with unpaired spins, and a variety of techniques involving the capacitance of a semiconductor junction such as photocapacitance and deep-level transient spectroscopy. *See* LUMINESCENCE; PHOTOCONDUCTIVITY; THERMOLUMINESCENCE.   [R.H.Bu.]

Trichroism   When certain optically anisotropic transparent crystals are subjected to white light, a cube of the material is found to transmit a different color through each of the three pairs of parallel faces. Such crystals are sometimes termed trichroic, and the phenomenon is called trichroism. This expression is used only rarely today since the colors in a particular crystal can appear quite different if the cube is cut with a different orientation with respect to the crystal axes. Accordingly, the term is frequently replaced by the more general term pleochroism. Even this term is being replaced by the phrase linear dichroism or circular dichroism to correspond with linear birefringence or circular birefringence. *See* BIREFRINGENCE; CRYSTAL OPTICS; DICHROISM.   [B.H.Bi.]

Triton   The nucleus of $^{2}_{1}$H (tritium); it is the only known radioactive nuclide belonging to hydrogen. The triton is produced in nuclear reactors by neutron absorption in deuterium ($^{2}_{1}$H + $^{1}_{0}n \rightarrow +\gamma$), and decays by β^{-} emission to $^{3}_{2}$H with a half-life of 12.4 years. Much of the interest in producing $^{3}_{1}$H arises from the fact that the fusion reaction $^{3}_{1}$H + $^{1}_{1}$H $\rightarrow$ $^{4}_{2}$H releases about 20 MeV of energy. Tritons are also used as projectiles in nuclear bombardment experiments. *See* NUCLEAR REACTION.
[H.E.D.]

Tuning fork   A steel instrument consisting of two prongs and a handle which, when struck, emits a tone of fixed pitch. Because of their simple mechanical structure, purity of tone, and constant frequency, tuning forks are widely used as standards of

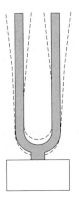

A tuning fork vibrating at its fundamental frequency.

frequency in musical acoustics. In its electrically driven form, a tuning fork serves to control electric circuits by producing frequency standards of high accuracy and stability. A tuning fork is essentially a transverse vibrator (see illustration). *See* VIBRATION.   [L.E.K.]

Tunneling in solids   A quantum-mechanical process which permits electrons to penetrate from one side to the other through an extremely thin potential barrier to electron flow. The barrier would be a forbidden region if the electron were treated as a classical particle. A two-terminal electronic device in which such a barrier exists and primarily governs the transport characteristic (current-voltage curve) is called a tunnel junction. See QUANTUM MECHANICS.

During the infancy of the quantum theory, L. de Broglie introduced the fundamental hypothesis that matter may be endowed with a dualistic nature—particles such as electrons, alpha particles, and so on, may also have the characteristics of waves. This hypothesis found expression in the definite form now known as the Schrödinger wave equation, whereby an electron or an alpha particle is represented by a solution to this equation. The nature of such solutions implies an ability to penetrate classically forbidden regions of negative kinetic energy and a probability of tunneling from one classically allowed region to another. The concept of tunneling, indeed, arises from this quantum-mechanical result. The subsequent experimental manifestations of this concept, such as high-field electron emission from cold metals, alpha decay, and so on, in the 1920s, can be regarded as one of the early triumphs of the quantum theory. See FIELD EMISSION; RADIOACTIVITY; SCHRÖDINGER'S WAVE EQUATION.

The tunnel diode (also called the Esaki diode), discovered in 1957 by L. Esaki, demonstrated the first convincing evidence of electron tunneling in solids.

Negative resistance phenomena can be observed in novel tunnel structures in semiconductors. Double tunnel barriers and periodic structures with a combination of semiconductors exhibit resonant tunneling and negative resistance effects. See SEMICONDUCTOR HETEROSTRUCTURES.

Tunneling had been considered to be a possible electron transport mechanism between metal electrodes separated by either a narrow vacuum or a thin insulating film usually made of metal oxides. In 1960, I. Giaever demonstrated that, if one or both of the metals were in a superconducting state, the current-voltage curve in such metal tunnel junctions revealed many details of that state.

In 1962, B. Josephson made a penetrating theoretical analysis of tunneling between two superconductors by treating the two superconductors and the coupling process as a single system, which would be valid if the insulating oxide were sufficiently thin, say 2 nanometers. His theory predicted the existence of a supercurrent, arising from tunneling of the bound electron pairs. This led to two startling conclusions: the dc and ac Josephson effects. The dc effect implies that a supercurrent may flow even if no voltage is applied to the junction. The ac effect implies that, at finite voltage V, there is an alternating component of the supercurrent which oscillates at a frequency of 483.6 MHz per microvolt of voltage across the junction, and is typically in the microwave range. See JOSEPHSON EFFECT.

[L.E.]

Turbulent flow   A fluid motion in which velocity, pressure, and other flow quantities fluctuate irregularly in time and space. The illustration shows a slice of a water jet emerging from a circular orifice into a tank of still water. A small amount of fluorescent dye mixed in the jet makes it visible when suitably illuminated by laser light, and tags the water entering the tank. There is a small region close to the orifice where the dye concentration does not vary with position, or with time at a given position. This represents a steady laminar state. Generally in laminar motion, all variations of flow quantities, such as dye concentration, fluid velocity, and pressure, are smooth and gradual in time and space. Farther downstream, the jet undergoes a transition to a new state in which the eddy patterns are complex, and flow quantities (including vorticity)

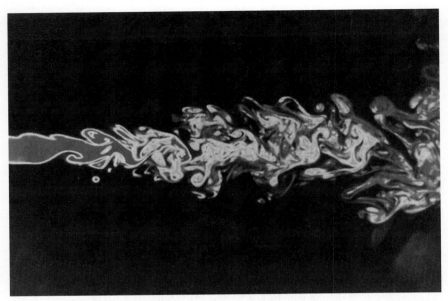

Two-dimensional image of an axisymmetric water jet, obtained by the laser-induced fluorescence technique. (*From R. R. Prasad and K. R. Sreenivasan, Measurement and interpretation of fractal dimension of the scalar interface in turbulent flows, Phys. Fluids A, 2:792–807, 1990*)

fluctuate randomly in time and three-dimensional space. This is the turbulent state. *See* JET FLOW; LAMINAR FLOW.

Turbulence occurs nearly everywhere in nature. It is characterized by the efficient dispersion and mixing of vorticity, heat, and contaminants. In flows over solid bodies such as airplane wings or turbine blades, or in confined flows through ducts and pipelines, turbulence is responsible for increased drag and heat transfer. Turbulence is therefore a subject of great engineering interest. On the other hand, as an example of collective interaction of many coupled degrees of freedom, it is also a subject at the forefront of classical physics. *See* DEGREE OF FREEDOM (MECHANICS); DIFFUSION; HEAT TRANSFER.

The illustration demonstrates the principal issues associated with turbulent flows. The first is the mechanism (or mechanisms) responsible for transition from the steady laminar state to the turbulent state. A second issue concerns the description of fully developed turbulence typified by the complex state far downstream of the orifice. Finally, it is of technological importance to be able to alter the flow behavior to suit particular needs. Less is known about eddy motions on the scale of centimeters and millimeters than about atomic structure on the subnanometer scale, reflecting the complexity of the turbulence problem. *See* NAVIER-STOKES EQUATION.

Origin of turbulence. A central role in determining the state of fluid motion is played by the Reynolds number. In general, a given flow undergoes a succession of instabilities with increasing Reynolds number and, at some point, turbulence appears more or less abruptly. It has long been thought that the origin of turbulence can be understood by sequentially examining the instabilities. This sequence depends on the particular flow and, in many circumstances, is sensitive to a number of details. A careful analysis of the perturbed equations of motion has resulted in a good understanding of the first two instabilities in a variety of circumstances. *See* REYNOLDS NUMBER.

Fully developed turbulence. Quite often in engineering, the detailed motion is not of interest, but only the long-time averages or means, such as the mean velocity in a boundary layer, the mean drag of an airplane or pressure loss in a pipeline, or the mean spread rate of a jet. It is therefore desirable to rewrite the Navier-Stokes equations for the mean motion. The basis for doing this is the Reynolds decomposition, which splits the overall motion into the time mean and fluctuations about the mean. These macroscopic fluctuations transport mass, momentum, and matter (in fact, by orders of magnitude more efficiently than molecular motion), and their overall effect is thus perceived to be in the form of additional transport or stress. This physical effect manifests itself as an additional stress (called the Reynolds stress) when the Navier-Stokes equations are rewritten for the mean motion (the Reynolds equations). The problem then is one of prescribing the Reynolds stress, which contains the unknown fluctuations in quadratic form. A property of turbulence is that the Reynolds stress terms are comparable to the other terms in the Reynolds equation, even when fluctuations are a small part of the overall motion. An equation for the Reynolds stress itself can be obtained by suitably manipulating the Navier-Stokes equations, but this contains third-order terms involving fluctuations, and an equation for third-order terms involves fourth-order quantities, and so forth. This is the closure problem in turbulence. The Navier-Stokes equations are themselves closed, but the presence of nonlinearity and the process of averaging result in nonclosure.

Given this situation, much of the progress in the field has been due to (1) exploratory experiments and numerical simulations of the Navier-Stokes equations at low Reynolds numbers; and (2) plausible hypotheses in conjunction with dimensional reasoning, scaling arguments, and their experimental verification.

Control of turbulent flows. Some typical objectives of flow control are the reduction of drag of an object such as an airplane wing, the suppression of combustion instabilities, and the suppression of vortex shedding behind bluff bodies. Interest in flow control has been stimulated by the discovery that some turbulent flows possess a certain degree of spatial coherence at large scales. Successful control has also been achieved through the reduction of the skin friction on a flat plate by making small longitudinal grooves, the so-called riblets, on the plate surface, imitating shark skin. *See* FLUID FLOW.

[K.R.S.]

Twinning (crystallography) A process in which two or more crystals, or parts of crystals, assume orientations such that one may be brought to coincidence with the other by reflection across a plane or by rotation about an axis. Crystal twins represent a particularly symmetric kind of grain boundary; however, the energy of the twin boundary is much lower than that of the general grain boundary because some of the atoms in the twin interface are in the correct positions relative to each other. *See* GRAIN BOUNDARIES.

[R.M.T.]

U

Ultrasonics The science of sound waves having frequencies above the audible range, that is, above about 20,000 Hz. Original workers in this field adopted the term supersonics. However, this name was also used in the study of airflow for velocities faster than the speed of sound. The present convention is to use the term ultrasonics as defined above. Since there is no marked distinction between the propagation and the uses of sound waves above and below 20,000 Hz, the division is artificial. *See* SOUND.

Ultrasonic generators and detectors. Ultrasonic transducers have two functions: transmission and reception. There may be separate transducers for each function or a single transducer for both functions. The usual types of generators and detectors for air, liquids, and solids are piezoelectric and magnetostrictive transducers. Quartz and lithium niobate ($LiNbO_3$) crystals are used to produce longitudinal and transverse waves; thin-film zinc oxide (ZnO) transducers can generate longitudinal waves at frequencies up to 96 GHz. Another class of materials used to generate ultrasonic signals is the piezoelectric ceramics. In contrast to the naturally occurring piezoelectric crystals, these ceramics have a polycrystalline structure. The most commonly produced piezoelectric ceramics are lead zirconate titanate (PZT), barium titanate ($BaTiO_3$), lead titanate ($PbTiO_3$), and lead metaniobate ($PbNb_2O_6$). Composite transducers are transducers in which the radiating or receiving element is a diced piezoelectric plate with filler between the elements. They are called "composite" to account for the two disparate elements, the piezoelectric diced into rods and the compliant adhesive filler. *See* MAGNETOSTRICTION; PIEZOELECTRICITY.

High-power ultrasound (typically 600 W) can be obtained with sonicators, consisting of a converter, horn, and tip. The converter transforms electrical energy to mechanical energy at a frequency of 20 kHz. Oscillation of piezoelectric transducers is transmitted and focused by a titanium horn that radiates energy into the liquid being treated. Horn and tip sizes are determined by the volume to be processed and the intensity desired. As the tip diameter increases, intensity or amplitude decreases.

Engineering applications. The engineering applications of ultrasonics can be divided into those dealing with low-amplitude sound waves and those dealing with high-amplitude (usually called macrosonics) waves.

Low-amplitude applications. Low-amplitude applications are in sonar (an underwater-detection apparatus), in the measurement of the elastic constants of gases, liquids, and solids by a determination of the velocity of propagation of sound waves, in the measurement of acoustic emission, and in a number of ultrasonic devices such as delay lines, mechanical filters, inspectoscopes, thickness gages, and surface-acoustic-wave devices. All these applications depend on the modifications that boundaries and imperfections in the materials cause in wave propagation properties. The attenuation and scattering of the sound in the media are important factors in determining the frequencies used and the sizes of the pieces that can be utilized or investigated.

High-amplitude applications. High-amplitude acoustic waves (macrosonic) have been used in a variety of applications involving gases, liquids, and solids. Some common applications are mentioned below.

A liquid subjected to high-amplitude acoustic waves can rupture, resulting in the formation of gas- and vapor-filled bubbles. When such a cavity collapses, extremely high pressures and temperatures are produced. The process, called cavitation, is the origin of a number of mechanical, chemical, and biological effects.

Cavitation plays an integral role in a wide range of processes such as ultrasonic cleaning and machining, catalysis of chemical reactions, disruption of cells, erosion of solids, degassing of liquids, emulsification of immiscible liquids, and dispersion of solids in liquids. Cavitation can also result in weak emission of light, called sonoluminescence. *See* CAVITATION.

One of the principal applications of ultrasonics to gases is particle agglomeration. This technique has been used in industry to collect fumes, dust, sulfuric acid mist, carbon black, and other substances.

Another industrial use of ultrasonics has been to produce alloys, such as lead-aluminum and lead-tin-zinc, that could not be produced by conventional metallurgical techniques. Shaking by ultrasonic means causes lead, tin, and zinc to mix.

Analytical uses. In addition to their engineering applications, high-frequency sound waves have been used to determine the specific types of motions that can occur in gaseous, liquid, and solid mediums. Both the velocity and attenuation of a sound wave are functions of the sound frequency. By studying the changes in these properties with changes of frequency, temperature, and pressure, indications of the motions taking place can be obtained. *See* SOUND ABSORPTION.

Medical applications. Application of ultrasonics in medicine can be generally classified as diagnostic and therapeutic. The more common of these at present is the diagnostic use of ultrasound, specifically ultrasonic imaging. *See* NONLINEAR ACOUSTICS.

Ultrasonic fields of sufficient amplitude can generate bioeffects in tissues. Although diagnostic ultrasound systems try to limit the potential for these effects, therapeutic levels of ultrasound have been used in medicine for a number of applications. Conventional therapeutic ultrasound is a commonly available technique used in physical therapy. High-frequency acoustic fields (typically 1 MHz) are applied through the skin to the affected area in either a continuous wave or long pulses.

Extracorporeal shock-wave lithotripsy (ESWL) disintegrates kidney stones with a high-amplitude acoustic pulse passing through the skin of the patient. The procedure eliminates the need for extensive surgery. Bioeffects are limited to the location of the stone by using highly focused fields which are targeted on the stone by imaging techniques such as ultrasound or fluoroscopy.   [H.E.Ba.; J.B.Fo.; V.M.K.]

Ultraviolet radiation   Electromagnetic radiation in the wavelength range 4–400 nanometers. The ultraviolet region begins at the short wavelength (violet) limit of visibility and extends to the wavelength of long x-rays. It is loosely divided into the near (400–300 nm), far (300–200 nm), and extreme (below 200 nm) ultraviolet regions (see illustration). In the extreme ultraviolet, strong absorption of the radiation by air requires the use of evacuated apparatus; hence this region is called the vacuum ultraviolet. Important phenomena associated with ultraviolet radiation include biological effects and applications, the generation of fluorescence, and chemical analysis through characteristic absorption or fluorescence.

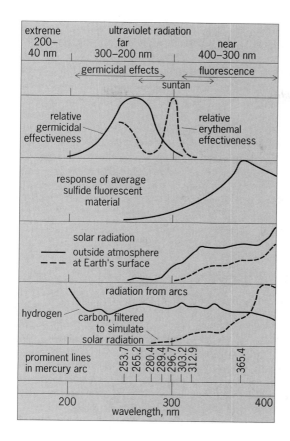

Phenomena associated with
ultraviolet radiation. (*After L. R.
Koller and General Electric*)

Sources of ultraviolet radiation include the Sun (although much solar ultraviolet radiation is absorbed in the atmosphere); arcs of elements such as carbon, hydrogen, and mercury; and incandescent bodies.

[F.W.B.]

Umklapp process A concept in the theory of transport properties of solids which has to do with the interaction of three or more waves in the solid, such as lattice waves or electron waves. In a continuum, such interactions occur only among waves described by wave vectors $\mathbf{k}_1$, $\mathbf{k}_2$, and so on, such that the interference condition, given by Eq. (1), is satisfied. The sign of $\mathbf{k}$ depends on whether the wave absorbs or

$$\mathbf{k}_1 + \mathbf{k}_2 + \mathbf{k}_3 = 0 \tag{1}$$

emits energy. Since $\hbar\mathbf{k}$ is the momentum of a quantum (or particle) described by the wave, Eq. (1) corresponds to conservation of momentum. In a crystal lattice further interactions occur, satisfying Eq. (2), where $\mathbf{b}$ is any integral combination of the three

$$\mathbf{k}_1 + \mathbf{k}_2 + \mathbf{k}_3 = \mathbf{b} \tag{2}$$

inverse lattice vectors $\mathbf{b}_i$, defined by $\mathbf{a} \cdot \mathbf{b}_j = 2\pi\delta_{ij}$, the $\mathbf{a}$'s being the periodicity vectors. The group of processes described in Eq. (2) are the Umklapp processes or flip-over processes, so called because the total momentum of the initial particles or quanta is reversed. *See* CRYSTAL.   [P.G.Kl.]

Uncertainty principle   A fundamental principle of quantum mechanics, which asserts that it is not possible to know both the position and momentum of an object with arbitrary accuracy. This contrasts with classical physics, where the position and momentum of an object can both be known exactly. In quantum mechanics, this is no longer possible, even in principle. More precisely, the indeterminacy or uncertainty principle, derived by W. Heisenberg, asserts that the product of Δx and Δp—measures of indeterminacy of a coordinate and of momentum along that coordinate—must satisfy inequality (1). The Planck constant, $h \simeq 6.63 \times 10^{-34}$ joule-second, is very small,

$$\Delta x \times \Delta p \gtrsim \hbar = \frac{h}{2\pi} \tag{1}$$

which makes inequality (1) unimportant for the measurements that are carried out in everyday life. Nevertheless, the consequences of the inequality are critically important for the interactions between the elementary constituents of matter, and are reflected in many of the properties of matter that are ordinarily taken for granted. For example, the density of solids and liquids is set to a large degree by the uncertainty principle, because the sizes of atoms are determined with decisive help of inequality (1).

In classical physics, simultaneous knowledge of position and momentum can be used to predict the future trajectory of a particle. Quantum indeterminacy and the limitations it imposes force such classical notions of causality to be abandoned.

Another well-known example of indeterminacy involves energy and time, as given by inequality (2). Physically, its origins are somewhat different from those of inequality

$$\Delta E \times \Delta t \gtrsim \hbar \tag{2}$$

(1). Inequality (2) relates, for example, lifetimes of unstable states with the widths of their lines. *See* LINEWIDTH.

In quantum physics, relations similar to inequalities (1) and (2) hold for pairs of many other quantities. They demonstrate that the acquisition of the information about a quantum object cannot be usually achieved without altering its state. Much of the strangeness of quantum physics can be traced to this impossibility of separating the information about the state from the state itself. *See* QUANTUM MECHANICS.   [W.H.Z.]

Underwater sound   The production, propagation, reflection, scattering, and reception of sound in seawater. The sea covers approximately 75% of the Earth's surface. In terms of exploration, visible observation of the sea is limited due to the high attenuation of light, and radar has very poor penetrability into salt water. Because of the extraordinary properties that sound has in the sea, and because of some of the inherent characteristics of the sea, acoustics is the principal means by which the sea has been explored.

Absorption. Sound has a remarkably low loss of energy in seawater, and it is that property above all others that allows it to be used in research and other application. Absorption is the loss of energy due to internal causes, such as viscosity. Over the frequency range from about 100 Hz (cycles per second) to 100 kHz, absorption is dominated by the reactions of two molecules, magnesium sulfate ($MgSO_4$) and boric acid [$B(OH)_3$]. These molecules are normally in equilibrium with their ionic constituents. The pressure variation caused by an acoustic wave changes the ionic balance and, during the passage of the pressure-varying acoustic field, it cannot return to the same equilibrium, and energy is given up. This is called chemical relaxation. At about 65 kHz magnesium sulfate dominates absorption, and boric acid is important near 1 kHz. *See* SOUND; SOUND ABSORPTION.

Sound speed. The speed of sound in seawater and its dependence on the parameters of the sea, such as temperature, salinity, and density, have an enormous effect on acoustics in the sea. Generally the environmental parameter that dominates acoustic processes in oceans is the temperature, because it varies both spatially and temporally. Solar heating of the upper ocean has one of the most important effects on sound propagation. As the temperature of the upper ocean increases, so does the sound speed. Winds mix the upper layer, giving rise to a layer of water of approximately constant temperature, below which is a region called the thermocline. Below that, most seawater reaches a constant temperature. All these layers depend on the season and the geographical location, and there is considerable local variation, depending on winds, cloud cover, atmospheric stability, and so on. Shallow water is even more variable due to tides, fresh-water mixing, and interactions with the sea floor. Major ocean currents, such as the Gulf Stream and Kuroshio, have major effects on acoustics. The cold and warm eddies that are spun off from these currents are present in abundance and significantly affect acoustic propagation.

Pressure waves. The science of underwater sound is the study of pressure waves in the sea over the frequency range from a few hertz to a few megahertz. The International System (SI) units are the pascal (Pa) for pressure (equal to one newton per square meter) and the watt per square meter (W/m^2) for sound intensity (the flow of energy through a unit area normal to the direction of wave propagation). In acoustics, it is more convenient to refer to pressures, which are usually much smaller than a pascal, and the consequent intensities with a different reference, the decibel. Intensity in decibels (dB) is ten times the logarithm to the base ten of the measured intensity divided by a reference intensity. *See* DECIBEL; SOUND INTENSITY; SOUND PRESSURE.

Wave propagation. The mathematical equation that sound obeys is known as the wave equation. Its derivation is based on the mathematical statements of Newton's second law for fluids (the Navier-Stokes equation), the equation of continuity (which essentially states that when a fluid is compressed, its mass is conserved), and a law of compression, relating a change of volume to a change in pressure. By the mathematical manipulation of these three equations, and the assumption that only very small physical changes in the fluid are taking place, it is possible to obtain a single differential equation that connects the acoustic pressure changes in time to those in space by a single quantity, the square of the sound speed (c), which is usually a slowly varying function of both space and time. *See* NAVIER-STOKES EQUATION; WAVE EQUATION.

Knowing the sound speed as a function of space and time allows for the investigation of the spatial and temporal properties of sound, at least in principle. The mathematics used to find solutions to the wave equation are the same as those that are used in other fields of physics, such as optics, radar, and seismics. *See* WAVE MOTION.

In addition to knowing the speed of sound, it is necessary to know the location and nature of the sources of sound, the location and features of the sea surface, the depth to the sea floor, and, in many applications, the physical structure of the sea floor. It is not possible to know the sound speed throughout the water column or know the boundaries exactly. Thus the solutions to the wave equation are never exact representations of nature, but estimates, with an accuracy that depends on both the quality of the knowledge of the environment and the degree to which the mathematical or numerical solutions to the wave equation represent the actual physical situation.

Ambient noise. A consequence of the remarkable transmission of sound is that unwanted sounds are transmitted just as efficiently. One of the ultimate limitations to the use of underwater sound is the ability to detect a signal above the noise. In the ocean, there are four distinct categories of ambient sound: biological, oceanographic physical processes, seismic, and anthropogenic. *See* ACOUSTIC NOISE; INFRASOUND.

Scattering and reverberation. The other source of unwanted sound is reverberation. Sound that is transmitted inevitably finds something to scatter from in the water column, at the sea surface, or at the sea floor. The scatter is usually in all directions, and some of it will return to the system that processes the return signals. Sources of scattering in the water column are fish, particulates, and physical inhomogeneities. The sea surface is, under normal sea conditions, agitated by winds and has the characteristic roughness associated with the prevailing atmospheric conditions. Rough surfaces scatter sound with scattering strengths that depend on the roughness, the acoustic frequency (or wavelength), and the direction of the signal. The scattering is highly time-dependent, and needs to be studied with an appropriate statistical approach. The sea floor has inherent roughness and is usually inhomogeneous, both properties causing scatter. Although scatter degrades the performance of sonars, the characteristics of the return can be determined to enable its cancellation through signal processing or array design. Scattering can also be used to study the sea surface, the sea floor, fish types and distribution, and inhomogeneities in the water column.   [R.R.G.]

Unitary symmetry   A type of symmetry law, an important example of which is flavor symmetry, one of the approximate internal symmetry laws obeyed by the strong interactions of elementary particles. According to the successful theory of strong interactions, quantum chromodynamics, flavor symmetry is the consequence of the fact that the so-called glue force (mediated by the SU_3^{color} gauge field) is the same between all the kinds (flavors) of quarks. If the quarks all had the same mass, they then would be dynamically equivalent constituents of hadrons, and hadrons would occur as degenerate multiplets of the group SU_N, where N is the number of quark flavors. The lightest quarks (u and d) have similar masses, so the lightest hadrons, made of u and d quarks, do exhibit an SU_2^{flavor} symmetry known as i-spin invariance. The mass of the next heavier quark (s) is much larger than the masses of the u and d quarks, but much smaller than the masses of the yet heavier quarks (c, b, ...); consequently the hadrons that contain no quarks heavier than the s quark clearly may be grouped into SU_3^{flavor} multiplets. *See* COLOR (QUANTUM MECHANICS); FLAVOR; HADRON; QUANTUM CHROMODYNAMICS; QUARKS.

An example of unitary symmetry is the approximate spin independence of the forces on electrons (as in an atom): There is a fundamental doublet, comprising the spin-up electron and the spin-down electron. Denoting these two states by $|u\rangle$ and $|d\rangle$, all physical properties (energy eigenvalues, charge density, and so on) are unchanged by the replacements shown in the equations below, where α and β are complex numbers.

$$|u\rangle \rightarrow \alpha|u\rangle + \beta|d\rangle$$

$$|d\rangle \rightarrow -\beta^*|u\rangle + \alpha^*|d\rangle$$

$$|\alpha|^2 + |\beta|^2 = 1$$

The group of all the transformations of two states that preserves their hermitean scalar products $[\langle u|d\rangle = 0, \langle u|u\rangle = \langle d|d\rangle = 1]$ is known as the two-dimensional unitary group, U_2; the transformations of the equations above form a subgroup SU_2 which merely lacks the uninteresting transformations of the form $|u\rangle \rightarrow e^{i\varphi}|u\rangle$ and $|d\rangle \rightarrow e^{i\varphi}|d\rangle$, that is, an equal change of phase of the two states.

The strong interactions are approximately invariant to an SU_2 group; the fundamental doublet can be taken to be the nucleon, with the up and down states proton and neutron. This SU_2 symmetry is known as charge independence, or, loosely, as *i*-spin conservation, the analog to the electron spin being known as *i*-spin **I**. *See* I-SPIN.

When a sufficient number of strange particles had been observed, it was seen that they, together with the old nonstrange particles, were grouped into multiplets of particles with the same space-time quantum numbers (except for mass; the masses of the members are only similar, not equal). This suggested the existence of a yet larger symmetry; it has turned out that this symmetry is the group of all unitary transformations of a triplet of fundamental particles, U_3, or SU_3 if the uninteresting equal phase change of all particles is omitted. This symmetry is often loosely called unitary symmetry. *See* STRANGE PARTICLES.

A striking difference in the manifestations of SU_2 and SU_3 is that whereas all possible multiplets of the former appear in nature, only those multiplets of the latter appear that can be regarded as compounds of the fundamental triplet in which the net number of component fundamental particles (number of particles minus number of antiparticles) is an integral multiple of 3. In particular, no particle that could be regarded as the fundamental triplet is found. Despite this nonappearance, it turns out that a great deal about the strongly interacting particles (hadrons) is at least qualitatively explained if they are regarded as physical compounds of a fundamental triplet of particles, to which the name quark has been given. The color theory (quantum chromodynamics) of strong interactions explains why single quarks are never observed.

According to the argument given above, hadrons have the approximate symmetry SU_N, where N is the number of kinds of quarks, or flavors. Six flavors of quark are known; in addition to the quarks with the flavors up, down, and strange described above, three more quarks, charm, bottom, and top, have been found. *See* CHARM; ELEMENTARY PARTICLE; J/PSI PARTICLE; UPSILON PARTICLES.
[C.J.G.]

Units of measurement

Values, quantities, or magnitudes in terms of which other such are expressed. Units are grouped into systems, suitable for use in the measurement of physical quantities and in the convenient statement of laws relating physical quantities. A quantity is a measurable attribute of phenomena or matter.

A given physical quantity A, such as length, time, or energy, is the product of a numerical value or measure $\{A\}$ and a unit $[A]$. Thus Eq. (1) holds.

$$A = \{A\}[A] \tag{1}$$

The unit $[A]$ can be chosen arbitrarily, but it is desirable to define units in such a way that they are derived from a few base units by equations without numerical factors other than unity, and that the equations between numerical values of quantities have exactly the same form as the equations between the quantities. For example, the kinetic energy E of a body is given in terms of its mass M and speed V by Eq. (2),

$$E = \tfrac{1}{2}MV^2 \tag{2}$$

where $E = \{E\}[E]$, $M = \{M\}[M]$, $V = \{V\}[V]$, and $\tfrac{1}{2}$ is called a definitional factor and is dimensionless. If the units of E, M, and V are defined in such a way that Eq. (3) holds, then the equation between the numerical values is Eq. (4). A system of units

$$[E] = [M][V]^2 \tag{3}$$

$$\{E\} = \tfrac{1}{2}\{M\}\{V\}^2 \tag{4}$$

defined in this way is called a coherent system. It is constructed by defining the units of a few base quantities independently; these are called base units. The units of all other quantities are defined by equations similar to Eq. (3) with no numerical factors other than unity, and are called derived units.

In 1960 the General Conference on Weights and Measures (CGPM) gave official status to a single practical system, the International System of Units, abbreviated SI in all languages. The system is a modernized version of the metric system. The SI, as subsequently extended, includes seven base units and twenty-two derived units with special names. These derived units, and others without special names, are derived from the base units in a coherent manner. A set of prefixes is used to form decimal multiples and submultiples of the SI units. Certain units which are not part of the SI but which are widely used or are useful in specialized fields have been accepted for use with the SI or for temporary use in those fields. *See* METRIC SYSTEM.

Geometrical units. Units of plane angle and solid angle are purely geometrical. The SI units of plane and solid angle are regarded as dimensionless derived units.

Plane angle units. The radian (rad), the SI unit of plane angle, is the plane angle between two radii of a circle which cut off on the circumference an arc equal in length to the radius. Since the circumference of a circle is 2π times the radius, the complete angle about a point is 2π rad.

The degree and its decimal submultiples can be used with the SI when the radian is not a convenient unit. By definition, 2π rad $= 360°$. The minute [$1' = (1/60)°$] and the second [$1'' = (1/60)'$] can also be used.

Steradian. The steradian (sr), the SI unit of solid angle, is the solid angle which, having its vertex at the center of a sphere, cuts off an area on the surface of the sphere equal to that of a square with sides of length equal to the radius of the sphere.

Mechanical units. In mechanics, it is convenient to have three base quantities, and two of these are generally chosen to be length and time. Systems of mechanical units may be classified as absolute systems, in which the third base quantity is mass, and gravitational systems, in which the third base quantity is force.

Two absolute systems of metric units are commonly employed, each named for its base units of length, mass, and time: the mks (meter-kilogram-second) absolute system, and the cgs (centimeter-gram-second) absolute system. The mks absolute system is the mechanical portion of the SI. A coherent absolute system of British units is based on the foot, the pound (1 lb $\cong$ 0.4536 kg), and the second.

Gravitational systems, in which the base quantities are length, force, and time, have been frequently employed by engineers, and are therefore sometimes called technical systems.

Length units. The meter (m) is the SI base unit of length. The use of special names for decimal submultiples of the meter should be avoided, and units formed by attaching appropriate SI prefixes to the meter should be used instead.

The angstrom (Å) is equal to 10^{-10} m. Although it has been accepted for temporary use with the SI, it is preferable to replace this unit with the nanometer, using the relation 1 Å = 0.1 nm.

The nautical mile (nmi), equal to 1852 m, has been accepted for temporary use with the SI in navigation.

The foot (ft) is, as discussed above, the unit of length in the British systems of units, and it is also in customary use in the United States. Since 1959 the foot has been defined as exactly 0.3048 m. The yard (yd) is defined as exactly 3 ft or 0.9144 m. *See* LENGTH; MEASURE.

Relative measurements of x-ray wavelengths can be made to a higher accuracy than absolute measurements. Before 1965, most x-ray wavelengths were expressed in terms of the X-unit, which is approximately 10^{-13} m. The X-unit has been superseded by the

A* unit, which is based on the tungsten $K\alpha_1$ line as a standard. The peak of this line is defined as exactly 0.2090100 A*. X-ray wavelength tables have been published in terms of this unit. At the time the A* unit was defined, it was thought to equal 10^{-10} m (the angstrom unit, Å) to within 5 parts per million, but the A* unit is now believed to be 20 ± 5 parts per million larger than 10^{-10} m.

Special units whose values are obtained experimentally are used in astronomy. For their definitions.

Area units. The square meter (m^2), the SI unit of area, is the area of a square with sides of length 1 m. Other area units are defined by forming squares of various length units in the same manner.

Cross sections, which measure the probability of interaction between an atomic nucleus, atom, or molecule and an incident particle, have the dimensions of area, and the appropriate SI unit for expressing them is therefore the square meter. The barn (b), a unit of cross section equal to 10^{-28} m^2, has been accepted for temporary use with the SI.

Units of volume. The cubic meter (m^3), the SI unit of volume, is the volume of a cube with sides of length 1 m. Other units of volume are defined by forming cubes of various length units in the same manner. The liter (symbol L in the United States) is equal to 1 cubic decimeter (1 dm^3), or equivalently to 10^{-3} m^3. It has been accepted for use with the SI for measuring volumes of liquids and gases.

Time units. The second (s) is the SI base unit of time. However, other units of time in customary use, such as the minute (1 min = 60 s), hour (1 h = 60 min), and day (1 d = 24 h), are acceptable for use with the SI. *See* TIME.

Frequency units. The hertz (Hz), the SI unit of frequency, is equal to 1 cycle per second. A periodic oscillation has a frequency of n hertz if it goes through n cycles in 1 s. *See* FREQUENCY (WAVE MOTION).

Speed and velocity units. The meter per second (m/s), the SI unit of speed or velocity, is the magnitude of the constant velocity at which a body traverses 1 m in 1 s. Other speed and velocity units are defined by dividing a unit of length by a unit of time in the same manner. *See* SPEED; VELOCITY.

The knot (kn) is equal to 1 nautical mile per hour (1 nmi/h); it has been accepted for temporary use with the SI.

Acceleration units. The meter per second squared (m/s^2), the SI unit of acceleration, is the acceleration of a body whose velocity changes by 1 m/s in 1 s. Other units of acceleration are defined by dividing a unit of velocity by a unit of time in the same manner. *See* ACCELERATION.

The gal or galileo (symbol Gal) is equal to 1 cm/s^2, or equivalently to 10^{-2} m/s^2.

Mass units. The kilogram (kg), the SI base unit of mass, is the only SI unit whose name, for historical reasons, contains a prefix. Names of decimal multiples and sub-multiples of the kilogram are formed by attaching prefixes to the word gram (g). The metric ton (t), which is equal to 10^3 kg or 1 megagram (Mg), is permitted in commercial usage of the SI.

The pound (lb), the unit of mass in British absolute system, is also in customary use in the United States. In 1959 the pound was defined to be exactly 0.45359237 kg.

The slug is the unit of mass in the British gravitational system. By definition 1 pound force (lbf) acting on a body of mass 1 slug produces an acceleration of 1 foot per second squared (1 ft/s^2). The slug is equal to approximately 32.174 lb or 14.594 kg. *See* MASS.

Force units. The newton (N), the SI unit of force, is the force which imparts an acceleration of 1 meter per second squared (1 m/s^2) to a body having a mass of 1 kg.

The dyne, the cgs absolute unit of force, is the force which imparts an acceleration of 1 centimeter per second squared (1 cm/s^2) to a body having a mass of 1 g.

The unit of force in the British absolute system is the poundal (pdl), the force which imparts an acceleration of 1 foot per second squared (1 ft/s^2) when applied to a body of mass 1 lb. One poundal is approximately 0.13825 N.

The units of force in the mks gravitational, cgs gravitational, and British gravitational systems are the forces which impart an acceleration equal to the standard acceleration of gravity, $g_n = 9.80665$ m/s$^2 \cong 32.174$ ft/s^2, when applied to bodies having masses of 1 kg, 1 g, and 1 lb, respectively. These units are named the kilogram force (kgf), gram force (gf), and pound force (lbf), respectively. Unfortunately, these units have also been called simply the kilogram, gram, and pound, giving rise to confusion with the mass units of the same name. *See* FORCE.

Pressure and stress units. The pascal (Pa), the SI unit of pressure and stress, is the pressure or stress of 1 newton per square meter (N/m^2). Other units of pressure can also be formed by dividing various units of force by various units of area, such as the pound force per square inch (lbf/in.^{2}, frequently abbreviated psi).

Pressure has been frequently expressed in terms of the bar and its decimal submultiples, where 1 bar = 10^6 dynes/cm^2 = 10^5 Pa. Pressures are also frequently expressed in terms of the height of a column of either mercury or water which the pressure will support.

Two other units which have been frequently used for measuring pressure are the standard atmosphere and the torr. The standard atmosphere (atm) is exactly 101,325 Pa, which is approximately the average value of atmospheric pressure at sea level. The torr is exactly 1/760 atmosphere, or approximately 133.322 Pa. To within 1 part per million, the torr equal to the pressure of a column of mercury of height 1 millimeter (1 mmHg) at a temperature of 0°C when the acceleration due to gravity has the standard value $g_n = 9.80665$ m/s^2. *See* PRESSURE; PRESSURE MEASUREMENT.

Energy and work units. The joule (J), the SI unit of energy or work, is the work done by a force of magnitude 1 newton when the point at which the force is applied is displaced 1 m in the direction of the force. Thus, joule is a short name for newton-meter (N-m) of energy or work. *See* ENERGY; WORK.

Units of energy or work in other systems are defined by forming the product of a unit of force and a unit of length in precisely the same manner as in the definition of the joule. Thus, the erg, the cgs absolute unit of energy or work, is the product of 1 dyne and 1 cm.

The foot-poundal (ft-pdl), the British absolute unit of energy or work, is the product of 1 poundal and 1 foot. The foot-pound, or, more properly, the foot-pound force (ft-lbf), the British gravitational unit of energy or work, is the product of 1 lbf and 1 ft.

Sometimes energy is measured in units which are products of a unit of power and a unit of time. Since 1 watt (W) of power equals 1 joule per second (1 J/s), as discussed below, the joule is equivalent to 1 watt-second (1 W · s). In electrical power applications, energy is frequently measured in kilowatthours (kWh), where 1 kWh = (10^3 W) (3600 s) = 3.6 × 10^6 J.

The calorie was originally defined as the quantity of heat required to raise the temperature of 1 g of air-free water 1°C under a constant pressure of 1 atm. However, the magnitude of the calorie, so defined, depends on the place on the Celsius temperature scale at which the measurement is made. The International (Steam) Table calorie is defined as exactly 4.1868 J; this is the type of calorie most frequently used in mechanical engineering. The thermochemical calorie, which has been used in thermochemistry in preference to the other types of calorie, is exactly 4.184 J.

The British thermal unit (Btu) was originally defined as the quantity of heat required to raise the temperature of 1 lb of air-free water 1°F under a constant pressure of 1 atm.

The International Table Btu is approximately 1055.056 J, and the thermochemical Btu is approximately 1054.350 J.

Power units. The watt (W), the SI unit of power, is the power which gives rise to the production of energy at the rate of 1 joule per second (1 J/s). Other units of power can be defined by forming the ratio of a unit of energy to a unit of time in the same manner. *See* POWER.

The horsepower (hp) is equal to exactly 550 ft · lbf/s, or approximately 745.700 W.

Torque units. The newton-meter (N · m), the SI unit of torque, is the magnitude of the torque produced by a force of 1 newton acting at a perpendicular distance of 1 m from a specified axis of rotation. The joule should never be used as a synonym for this unit.

Units of torque in other systems are defined by forming the product of a unit of force and a unit of length in precisely the same manner as in the definition of the newton-meter.

The foot-poundal (ft · pdl), the British absolute unit of torque, is the product of 1 poundal and 1 foot. The foot-pound (ft · lbf), the British gravitational unit of torque, is the product of 1 lbf and 1 ft. These units are sometimes called the poundal-foot (pdl · ft) and pound-foot (lbf · ft) to distinguish them from the units of energy or work. *See* TORQUE.

Electrical units. For a general discussion of electrical units, including the SI or mks system, three cgs systems [electrostatic system of units (esu), electromagnetic system of units (emu), and gaussian system], and definitions of the SI units ampere (A), volt (V), ohm (Ω), coulomb (C), farad (F), henry (H), weber (Wb), and tesla (T) *see* ELECTRICAL UNITS AND STANDARDS.

This section discusses some additional SI units and some units in the cgs electromagnetic system which are frequently encountered in scientific literature in spite of the fact that their use has been discouraged.

Siemens. The siemens (S), the SI unit of electrical conductance, is the electrical conductance of a conductor in which a current of 1 ampere is produced by an electric potential difference of 1 volt. *See* ELECTRICAL RESISTANCE.

The siemens was formerly called the mho (Ω) to illustrate the fact the unit is the reciprocal of the ohm.

Abampere. The abampere (abA), the cgs electromagnetic unit of current, is that current which, if maintained in two straight, parallel conductors of infinite length, of negligible circular cross section, and placed 1 cm apart in vacuum, would produce between these conductors a force equal to 2 dynes per centimeter of length. The abampere is equal to exactly 10 A.

Abvolt. The abvolt (abV), the cgs electromagnetic unit of electrical potential difference and electromotive force, is the difference of electrical potential between two points of a conductor carrying a constant current of 1 abA, when the power dissipated between these points is equal to 1 erg per second. Then 1 abV = 10^{-8} V.

Maxwell. The maxwell (Mx), the cgs electromagnetic unit of magnetic flux, is the magnetic flux which, linking a circuit of one turn, produces in it an electromotive force of 1 abV as it is reduced to zero in 1 s. Then 1 maxwell = 10^{-8} weber.

Units of magnetic flux density. The gauss (Gs), the cgs electromagnetic unit of magnetic flux density (also called magnetic induction), is a magnetic flux density of 1 maxwell per square centimeter (1 Mx/cm^2). Then 1 gauss = 10^{-4} tesla.

Units of magnetic field strength. The SI unit of magnetic field strength is 1 ampere per meter (1 A/m), which is the magnetic field strength at a distance of 1 m from a straight conductor of infinite length and negligible circular cross section which carries a current of 2π A. This definition is based on the definition, in the SI, of the magnetic field strength H_{SI}. At a distance r from a long straight conductor carrying I, H_{SI} is given by Eq. (5). The left-hand side of this equation is the line integral of H_{SI} around a circular

$$2\pi r H_{SI} = I \tag{5}$$

path, all of whose points are at distance r from the conductor. Substituting $r = 1$ m, and $I = 2\pi$ A in this equation gives $H_{SI} = 1$ A/m.

The oersted (Oe), the cgs electromagnetic unit of magnetic field strength, is the magnetic field strength at a distance of 1 cm from a straight conductor of infinite length and negligible circular cross section which carries a current of 0.5 abA.

Units of magnetic potential and mmf. The ampere serves as the SI unit of magnetic potential difference and magnetomotive force (mmf), as well as the unit of current. In the SI, the magnetomotive force around a closed path equals the current passing through a surface enclosed by the path. Thus, 1 A is the magnetomotive force around a closed path when a current of 1 A passes through an enclosed surface.

The gilbert (Gb), the cgs electromagnetic unit of magnetic potential difference and magnetomotive force, is the magnetomotive force around a closed path enclosing a surface through which flows a current of $(1/4\pi)$ abA.

Photometric units. Photometric units involve a new base quantity, luminous intensity. For the definition of the candela (cd), the SI unit of luminous intensity, *see* PHOTOMETRY; PHYSICAL MEASUREMENT.

Lumen. The lumen (lm), the SI unit of luminous flux, is the luminous flux emitted within a unit solid angle (1 steradian) by a point source having a uniform intensity of 1 candela. *See* LUMINOUS FLUX.

Luminous energy units. The lumen-second (lm·s), the SI unit of luminous energy (also called quantity of light), is the luminous energy radiated or received over a period of 1 s by a luminous flux of 1 lumen. This unit is also called the talbot. *See* LUMINOUS ENERGY.

Radiation units. Certain quantities and units are used particularly in the area of ionizing radiation. The special units curie, roentgen, rad, and rem, which were previously adopted for use in this area, are not coherent with the SI, but their temporary use with the SI has been approved while the transition to SI units takes place.

Activity units. The becquerel (Bq), the SI unit of activity (radioactive disintegration rate), is the activity of a radionuclide decaying at the rate of one spontaneous nuclear transition per second. Thus 1 Bq = 1 s^{-1}.

The curie (Ci), the special unit of activity, is equal to 3.7×10^{10} Bq. *See* RADIOACTIVITY.

Exposure units. The SI unit of exposure to ionizing radiation, 1 coulomb per kilogram (1 C/kg), is the amount of electromagnetic radiation (x-radiation or gamma radiation) which in 1 kg of pure dry air produces ion pairs carrying 1 coulomb of charge of either sign. (The ionization arising from the absorption of bremsstrahlung emitted by electrons is not to be included in measuring the charge.) *See* BREMSSTRAHLUNG.

The roentgen (R), the special unit of exposure, is equal to 2.58×10^{-4} C/kg.

Absorbed dose units. The gray (Gy), the SI unit of absorbed dose, is the absorbed dose when the energy per unit mass imparted to matter by ionizing radiation is 1 joule per kilogram (1 J/kg).

The rad (rd), the special unit of absorbed dose, is equal to 10^{-2} Gy.

Dose equivalent units. Different types of radiation cause slightly different effects in biological tissue. For this reason, a weighted absorbed dose called the dose equivalent is used in comparing the effects of radiation on living systems. The dose equivalent is the product of the absorbed dose and various dimensionless modifying factors.

The sievert (Sv), the SI unit of dose equivalent, is the dose equivalent when the absorbed dose of ionizing radiation multiplied by the stipulated dimensionless factors is 1 joule per kilogram (1 J/kg).

The rem, the special unit of dose equivalent, is equal to 10^{-2} Sv.   [J.F.We.]

Upsilon particles   A family of elementary particles whose first three members were discovered in 1977. The upsilon mesons, Υ, are the heaviest known vector mesons, with masses greater than 10 times that of the proton. They are bound states of a heavy quark and its antiquark. The quarks which bind to form the upsilons carry a new quantum number called beauty or bottomness, and they are called b-quarks or b. The mass of the b-quark is around 5 GeV. The anti-b-quark or b carries antibeauty, and therefore the upsilons carry no beauty and are often called hidden beauty states. Direct proof of the existence of the b-quark was obtained by observing the existence of B mesons which consist of a b-quark bound to a lighter quark. Thirteen bb-mesons have also been observed thus far. See ELEMENTARY PARTICLE; MESON; QUARKS.   [P.F.]

V

Vacuum measurement The determination of a gas pressure that is less in magnitude than the pressure of the atmosphere. This low pressure can be expressed in terms of the height in millimeters of a column of mercury which the given pressure (vacuum) will support, referenced to zero pressure. The height of the column of mercury which the pressure will support may also be expressed in micrometers. The unit most commonly used is the torr, equal to 1 mm (0.03937 in.) of mercury (mmHg). Less common units of measurement are fractions of an atmosphere and direct measure of force per unit area. The unit of pressure in the International System (SI) is the pascal (Pa), equal to 1 newton per square meter (1 torr = 133.322 Pa). Atmospheric pressure is sometimes used as a reference. The pressure of the standard atmosphere is 29.92 in. or 760 mm of mercury (101,325 Pa or 14.696 lbf/in.^{2}).

Pressures above 1 torr can be easily measured by familiar pressure gages, such as liquid-column gages, diaphragm-pressure gages, bellows gages, and bourdon-spring gages. At pressures below 1 torr, mechanical effects such as hysteresis, ambient errors, and vibration make these gages impractical. *See* MANOMETER; PRESSURE MEASUREMENT.

Pressures below 1 torr are best measured by gages which infer the pressure from the measurement of some other property of the gas, such as thermal conductivity or ionization. The thermocouple gage, in combination with a hot- or cold-cathode gage (ionization type), is the most widely used method of vacuum measurement today. *See* IONIZATION GAGE.

Other gages used to measure vacuum in the range of 1 torr or below are the McLeod gage, the Pirani gage, and the Knudsen gage. The McLeod gage is used as an absolute standard of vacuum measurement in the 10–10^{-4} torr (10^3–10^{-2} Pa) range. *See* MCLEOD GAGE; PIRANI GAGE.

The Knudsen gage is used to measure very low pressures. It measures pressure in terms of the net rate of transfer of momentum (force) by molecules between two surfaces maintained at different temperatures (cold and hot plates) and separated by a distance smaller than the mean free path of the gas molecules. [R.C.]

Valence band The highest electronic energy band in a semiconductor or insulator which can be filled with electrons. The electrons in the valence band correspond to the valence electrons of the constituent atoms. In a semiconductor or insulator, at sufficiently low temperatures, the valence band is completely filled and the conduction band is empty of electrons. Some of the high energy levels in the valence band may become vacant as a result of thermal excitation of electrons to higher energy bands or as a result of the presence of impurities. The net effect of the valence band is then equivalent to that of a few particles which are equal in number and similar in motion to the missing electrons but each of which carries a positive electronic charge. These "particles" are referred to as holes. *See* BAND THEORY OF SOLIDS; CONDUCTION BAND; ELECTRIC INSULATOR; HOLE STATES IN SOLIDS; SEMICONDUCTOR.   [H.Y.F.]

Van der Waals equation   An equation of state of gases and liquids proposed by J. D. van der Waals in 1873 that takes into account the nonzero size of molecules and the attractive forces between them. He expressed the pressure p as a function of the absolute temperature T and the molar volume $V_m = V/n$, where n is the number of moles of gas molecules in a volume V (see equation below). Here $R = 8.3145$ J K^{-1} mol^{-1}

$$p = \frac{RT}{V_m - b} - \frac{a}{V_m^2}$$

is the universal gas constant, and a and b are parameters that depend on the nature of the gas. Parameter a is a measure of the strength of the attractive forces between the molecules, and b is approximately equal to four times the volume of the molecules in one mole, if those molecules can be represented as elastic spheres. The equation has no rigorous theoretical basis for real molecular systems, but is important because it was the first to take reasonable account of molecular attractions and repulsions, and to emphasize the fact that the intermolecular forces acted in the same way in both gases and liquids. It is accurate enough to account for the fact that all gases have a critical temperature T_c above which they cannot be condensed to a liquid. The expression that follows from this equation is $T_c = 8a/27\ Rb$. *See* CRITICAL PHENOMENA; INTERMOLECULAR FORCES.

   In a gas mixture, the parameters a and b are taken to be quadratic functions of the mole fractions of the components since they are supposed to arise from the interaction of the molecules in pairs. The resulting equation for a binary mixture accounts in a qualitative but surprisingly complete way for the many kinds of gas-gas, gas-liquid, and liquid-liquid phase equilibria that have been observed in mixtures.

   The equation is too simple to represent quantitatively the behavior of real gases, and so the parameters a and b cannot be determined uniquely; their values depend on the ranges of density and temperature used in their determination. For this reason, the equation now has little practical value, but it remains important for its historical interest and for the concepts that led to its derivation. *See* THERMODYNAMIC PRINCIPLES.   [J.S.Ro.]

Variational methods (physics)   Methods based on the principle that, among all possible configurations or histories of a physical system, the system realizes the one that minimizes some specified quantity. Variational methods are used in physics both for theory construction and for calculational purposes.

   The earliest use of a variational principle for physics is Fermat's principle in optics, which states that when a light ray traverses a medium with nonuniform index of refraction its path is such as to minimize its travel time. An integral expresses the time that the light takes to travel from one point to another along a particular path, and an application of the calculus of variations to this integral makes it possible to determine the particular path for which the travel time is a minimum. This problem is mathematically identical to the variational principle that determines a geodesic, the path of shortest distance, in a given geometry. In that form, the same principle determines the world lines of all objects in the general theory of relativity. *See* RELATIVITY.

Similarly, in mechanics, Hamilton's principle for the action is defined for any system of point particles by an integral (called the action) that extends over an arbitrarily prescribed path Γ in configuration space. Hamilton's principle asserts that the trajectories of all the particles are determined by the requirement that Γ be such that, for given initial and final times, the action is a minimum; for this reason it is also called the principle of least action. If the calculus of variations is applied to implement this principle, the corresponding Euler-Lagrange equations are obtained. These are the lagrangian equations of motion, that is, Newton's equations of motion in lagrangian form. See ACTION; HAMILTON'S PRINCIPLE; LAGRANGIAN FUNCTION; LEAST-ACTION PRINCIPLE.

The principle of least action has been generalized to systems with infinitely many degrees of freedom, that is, fields. A Lagrange density function is then defined, which is a function of the fields and their time derivatives at any given point in space and time. For any field theory, only the Lagrange density needs to be given; the field equations are then derivable as the corresponding Euler-Lagrange equations. A similar technique makes it possible to derive the Schrödinger equation and the Dirac equation in quantum mechanics from specific Lagrange density functions. See QUANTUM MECHANICS; QUANTUM THEORY OF MATTER; RELATIVISTIC QUANTUM THEORY.

This method has great procedural advantages. For example, it facilitates a check of whether the theory satisfies certain invariance principles (such as relativistic invariance or rotational invariance) by simply ascertaining whether the Lagrange density satisfies them. The corresponding conservation laws can also be derived directly from the lagrangian. See CONSERVATION LAWS (PHYSICS); QUANTUM FIELD THEORY; SYMMETRY LAWS (PHYSICS).

The variational method also plays an important role in quantum-mechanical calculations. For the computation of needed quantities in terms of functions that result from the solution of differential equations, it is always of great advantage to use formulas that have the special form required to make them stationary with respect to small variations of the input functions in the vicinity of the unknown, exact solutions. See MINIMAL PRINCIPLES.

[R.G.Ne.]

Vector methods (physics)   Methods that make use of the behavior of physical quantities under coordinate transformations.

From the point of view of physics, the most appropriate definition of a vector in three-dimensional space is a quantity that has three components which transform under rotations of the coordinate system like the coordinates of a point in space. What characterizes rotations is that the distance from the origin—of all points **x** with cartesian

$$\sqrt{x_1^2 + x_2^2 + x_3^2}$$

coordinates x_i, $i = 1, 2, 3$—remains unchanged. Specifically, if the rotation takes the x_i to new coordinates x_i', given by Eqs. (1) [in which $\det\{a_{ij}\}$ is the determinant of the

$$x_i' = \sum_{j=1}^{3} a_{ij} x_j \qquad i = 1, 2, 3$$

$$\sum_{k=1}^{3} a_{ik} a_{jk} = \begin{cases} 1 & \text{if } i = j, \\ 0 & \text{if } i \neq j, \end{cases} \qquad \det\{a_{ij}\} = 1 \tag{1}$$

matrix $\{a_{ij}\}]$, then the three quantities V_i, $i = 1, 2, 3$, form the components of a vector **V**, if in the new coordinate system the transformed coordinates are given by Eq. (2).

$$V'_i = \sum_{j=1}^{3} a_{ij} V_j \qquad i = 1, 2, 3 \tag{2}$$

If the coordinate transformation of Eqs. (1) is such that $|a_{ij}| = 1$ for $i = j$ and $a_{ij} = 0$ for $i \neq j$, but such that $\det\{a_{ij}\} = -1$ rather than $+1$ as in Eqs. (1), then it describes a reflection, in which a right-handed coordinate system is replaced by a left-handed one. If, for such a transformation, Eq. (2) also holds, then **V** is called a polar vector, whereas if the components of **V** do not change sign, it is called an axial vector or pseudovector. The vector **V** can also be looked upon as a quantity with a direction, with the magnitude

$$\sqrt{V_1^2 + V_2^2 + V_3^2}$$

pointing from the origin of the coordinate system to the point in space with the cartesian coordinates (V_1, V_2, V_3).

A quantity that remains invariant under a rotation of the coordinate system is called a scalar. The importance of vectors and scalars in physics derives from the assumed isotropy of the universe, which implies that all general physical laws should have the same form in any two coordinate systems that differ only by a rotation. It is therefore useful to classify physical quantities according to their transformation properties under coordinate rotations. Examples of scalars include the mass of an object, its electric charge, its volume, its surface area, the energy of a system, and its temperature. Other quantities have a direction and thus are vectors, such as the force exerted on a body, its velocity, its acceleration, its angular momentum, and the electric and magnetic fields. Since the sum of two scalars is a scalar, and the sum of two vectors is a vector, it is important in the formulation of physical laws not to mix quantities that have different transformation properties under coordinate rotations. The sum of a vector and a scalar has no simple transformation properties; a law that equated a vector to a scalar would have different forms in different coordinate systems and would thus not be acceptable.   [R.G.Ne.]

Velocity   The time rate of change of position of a body in a particular direction. Linear velocity is velocity along a straight line, and its magnitude is commonly measured in such units as meters per second (m/s), feet per second (ft/s), and miles per hour (mi/h). Since both a magnitude and a direction are implied in a measurement of velocity, velocity is a directed or vector quantity, and to specify a velocity completely, the direction must always be given. The magnitude only is called the speed. *See* SPEED.

A body need not move in a straight line path to possess linear velocity. When a body is constrained to move along a curved path, it possesses at any point an instantaneous linear velocity in the direction of the tangent to the curve at that point. The average value of the linear velocity is defined as the ratio of the displacement to the elapsed time interval during which the displacement took place.

The representation of angular velocity ω as a vector is shown in the illustration. The vector is taken along the axis of spin. Its length is proportional to the angular speed and its direction is that in which a right-hand screw would move. If a body rotates

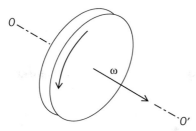

Angular velocity shown as an axial vector. Axis of rotation is OO'.

simultaneously about two or more rectangular axes, the resultant angular velocity is the vector sum of the individual angular velocities.

[R.D.Ru.]

Venturi tube   A device that consists of a gradually decreasing nozzle through which the fluid in a pipe is accelerated, followed by a gradually increasing diffuser section that allows the fluid to nearly regain its original pressure head (see illustration). It can be used to measure the flow rate in the pipe, or it can be used to pump a secondary fluid by aspirating it at the nozzle exit. The ability of the venturi tube to regain much of the original pressure head makes it especially useful in measuring the flow rate in systems which have a low pressure differential or pressure head that drives the fluid through the pipe or where the cost of pumping the fluid is an important factor. Conserving the pressure head decreases the amount of energy required to pump the fluid through the pipe.

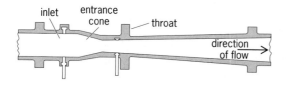

Proportions of a Herschel-type venturi tube for standard fluid-flow measurement.

A gradual expansion of flow downstream of a nozzle eliminates flow separation, allowing recovery of most of the original pressure head. In the case where the main flow separates from the wall, a large percentage of the fluid energy is lost in the eddies caused by the separation.

The flow through the device obeys Bernoulli's equation, and the formula for calculating the flow is similar to the equation for orifices. The venturi meter belongs to the class of differential pressure-sensing devices that are used to indicate flow. *See* BERNOULLI'S THEOREM.

[M.P.W.]

Vibration   The term used to describe a continuing periodic change in the magnitude of a displacement with respect to a specified central reference. The periodic motion may range from the simple to-and-fro oscillations of a pendulum, through the more complicated vibrations of a steel plate when struck with a hammer, to the extremely complicated vibrations of large structures such as an automobile on a rough road. Vibrations are also experienced by atoms, molecules, and nuclei. *See* PENDULUM.

A mechanical system must possess the properties of mass and stiffness or their equivalents in order to be capable of self-supported free vibration. Stiffness implies that an alteration in the normal configuration of the system will result in a restoring force tending to return it to this configuration. Mass or inertia implies that the velocity imparted to the system in being restored to its normal configuration will cause it to overshoot this configuration. It is in consequence of the interplay of mass and stiffness that periodic vibrations in mechanical systems are possible.

Mechanical vibration is the term used to describe the continuing periodic motion of a solid body at any frequency. When the rate of vibration of the solid body ranges between 20 and 20,000 hertz (Hz), it may also be referred to as an acoustic vibration, for if these vibrations are transmitted to a human ear they will produce the sensation of sound. The vibration of such a solid body in contact with a fluid medium such as air or water induces the molecules of the medium to vibrate in a similar fashion and thereby transmit energy in the form of an acoustic wave. Finally, when such an acoustic wave impinges on a material body, it forces the latter into a similar acoustic vibration. In the case of the human ear it produces the sensation of sound. *See* SOUND.

Systems with one degree of freedom are those for which one space coordinate alone is sufficient to specify the system's displacement from its normal configuration. An idealized example known as a simple oscillator consists of a point mass m fastened to one end of a massless spring and constrained to move back and forth in a line about its undisturbed position (Fig. 1). Although no actual acoustic vibrator is identical with this idealized example, the actual behavior of many vibrating systems when vibrating at low frequencies is similar and may be specified by giving values of a single space coordinate.

When the restoring force of the spring of a simple oscillator on its mass is directly proportional to the displacement of the latter from its normal position, the system vibrates in a sinusoidal manner called simple harmonic motion. This motion is identical with the projection of uniform circular motion on a diameter of a circle. *See* HARMONIC MOTION.

When two simple vibrating systems are interconnected by a flexible connection, the combined system has two degrees of freedom (Fig. 2). Such a system has two normal modes of vibration of two frequencies. Both of these frequencies differ from the respective natural frequencies of the individual uncoupled oscillators.

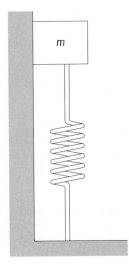

Fig. 1. Simple oscillator.

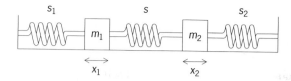

Fig. 2. Simple oscillator with two degrees of freedom. Masses m_1 and m_2, with displacements x_1 and x_2, are connected by springs s_1, s, and s_2.

A vibrating system is said to have several degrees of freedom if many space coordinates are required to describe its motion. One example is n masses $m_1, m_2, \ldots, m_n$ constrained to move in a line and interconnected by $(n-1)$ coupling springs with additional terminal springs leading from m_1 and m_n to rigid supports. This system has n normal modes of vibration, each of a distinct frequency. *See* DAMPING.   [L.E.K.]

Virial equation   An equation of state of gases that has additional terms beyond that for an ideal gas, which account for the interactions between the molecules. The pressure p can be expressed in terms of the molar volume $V_m = V/n$ (where n is the number of moles of gas molecules in a volume V), the absolute temperature T, and the universal gas constant $R = 8.3145 \, \text{J K}^{-1} \, \text{mol}^{-1}$ or, in more commonly used practical units, $0.082058 \, \text{L atm K}^{-1} \, \text{mol}^{-1}$ [Eq. (1)]. In the equation the virial coefficients $B_n(T)$

$$\frac{V_m}{RT} = 1 + \frac{B_2(T)}{V_m} + \frac{B_3(T)}{V_m^2} + \frac{B_4(T)}{V_m^3} + \cdots \tag{1}$$

are functions only of the temperature and depend on the nature of the gas. In an ideal gas, in which all interactions between the molecules can be neglected because V_m is sufficiently large, only the first term, unity, survives on the right-hand side.

The equation is important because there are rigorous relations between the coefficients B_2, B_3, and so on, as well as the interactions of the molecules in pairs, triplets, and so forth. It provides a valuable route to a knowledge of the intermolecular forces. Thus if the intermolecular energy of a pair of molecules at a separation r is $u(r)$, then the second virial coefficient can be expressed as Eq. (2). Here $N_A = 6.0221 \times 10^{23} \, \text{mol}^{-1}$

$$B_2(T) = 2\pi N_A \int_0^\infty (1 - e^{-u(r)/kT}) r^2 \, dr \tag{2}$$

is the Avogadro constant, and $k = R/N_A$. In a gas mixture, B_n is a polynomial of order n in the mole fractions of the components. *See* INTERMOLECULAR FORCES.

The virial equation is useful in practice because it represents the pressure accurately at low and moderate gas densities, for example, up to about 4 mol L^{-1} for nitrogen at room temperature, which corresponds to a pressure of about 100 atm (10 MPa). It is not useful at very high densities, where the series may diverge, and is inapplicable to liquids. It can be rearranged to give the ratio pV_m/RT as an expansion in powers of the pressure instead of the density, which is equally useful empirically, but the coefficients of the pressure expansion are not usually called virial coefficients, and lack any simple relation to the intermolecular forces or, in a mixture, to the composition of the gas. *See* VAN DER WAALS EQUATION.   [J.S.Row.]

Virial theorem   A theorem in classical mechanics which relates the kinetic energy of a system to the virial of Clausius, as defined below. The theorem can be generalized to quantum mechanics and has widespread application. It connects the average kinetic and potential energies for systems in which the potential is a power of the radius. Since the theorem involves integral quantities such as the total kinetic energy, rather than the kinetic energies of the individual particles that may be involved, it gives valuable information on the behavior of complex systems. For example, in statistical mechanics the virial theorem is intimately connected to the equipartition theorem; in astrophysics it may be used to connect the internal temperature, mass, and radius of a star and to discuss stellar stability. The virial theorem makes possible a very easy derivation of the counterintuitive result that as a star radiates energy and contracts it heats up rather than cooling down. *See* STATISTICAL MECHANICS.

The virial theorem states that the time-averaged value of the kinetic energy in a confined system (that is, a system in which the velocities and position vectors of all the particles remain finite) is equal to the virial of Clausius. The virial of Clausius is defined to equal $-1/2$ times the time-averaged value of a sum over all the particles in the system. The term in this sum associated with a particular particle is the dot product of the particle's position vector and the force acting on the particle. Alternatively, this term is the product of the distance, r, of the particle from the origin of coordinates and the radial component of the force acting on the particle.

In the common case that the forces are derivable from a power-law potential, V, proportional to r^k, where k is a constant, the virial is just $-k/2$ times the potential energy. Thus, in this case the virial theorem simply states that the kinetic energy is $k/2$ times the potential energy. For a system connected by Hooke's-law springs, $k = 2$, and the average kinetic and potential energies are equal. For $k = 1$, that is, for gravitational or Coulomb forces, the potential energy is minus twice the kinetic energy. *See* COULOMB'S LAW; GRAVITATION; HARMONIC MOTION.   [A.G.P.]

Virtual work principle   The principle stating that the total virtual work done by all the forces acting on a system in static equilibrium is zero for a set of infinitesimal virtual displacements from equilibrium. The infinitesimal displacements are called virtual because they need not be obtained by a displacement that actually occurs in the system. The virtual work is the work done by the virtual displacements, which can be arbitrary, provided they are consistent with the constraints of the system. *See* CONSTRAINT.

The principle of virtual work is equivalent to the conditions for static equilibrium of a rigid body expressed in terms of the total forces and torques. That is, the principle of virtual work can be derived from these conditions, and conversely. *See* EQUILIBRIUM OF FORCES; STATICS.

One advantage of the principle of virtual work is that it can serve as a basis for all of statics. In the solution of problems the principle of virtual work is often useful for eliminating the need for consideration of the forces of constraint, since these forces often are perpendicular to the virtual displacements and consequently do no work.   [P.W.S.]

Viscosity   The material property that measures a fluid's resistance to flowing. For example, water flows from a tilted jar more quickly and easily than honey does. Honey is more viscous than water, so although gravity creates nearly the same stresses in honey and water, the more viscous fluid flows more slowly.

The viscosity can be measured where the fluid of interest is sheared between two flat plates which are parallel to one another (see illustration). This is known as planar Couette flow. The shear stress is the ratio of the tangential force F needed to maintain the moving plate at a constant velocity V to the plate area A. The shear flow created

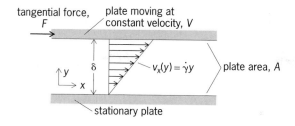

Planar Couette flow. $v_x =$ fluid velocity at distance y above the stationary plate, $\dot{\gamma} =$ velocity gradient or shear rate, $\delta =$ distance between plates.

between the plates has the velocity profile given by Eq. (1), where v_x is the velocity

$$v_x = \dot{\gamma}\, y \qquad (1)$$

parallel to the plates at a perpendicular distance y above the stationary plate. The coefficient $\dot{\gamma}$, called the velocity gradient or shear rate, is given by V/δ, where δ is the distance between the plates. It is expected that the shear stress increases with increasing shear rate but that the ratio of these two quantities depends only on the fluid between the plates. This ratio is used to define the shear viscosity, η, as in Eq. (2). The shear

$$\eta \equiv \frac{\text{shear stress}}{\text{shear rate}} = \frac{F/A}{V/\delta} \qquad (2)$$

viscosity may depend on temperature, pressure, and shear rate.

Isaac Newton is credited with first suggesting a model for the viscous property of fluids in 1687. Newton proposed that the resistance to flow caused by viscosity is proportional to the velocity at which the parts of the fluid are being separated from one another because of the flow. Although Newton's law of viscosity is an empirical idealization, many fluids, such as low-molecular-weight liquids and dilute gases, are well characterized by it over a large range of conditions. However, many other fluids, such as polymer solution and melts, blood, ink, liquid crystals, and colloidal suspensions, are not described well by Newton's law. Such fluids are referred to as non-newtonian.

For planar Couette flow, Newton's law of viscosity is given mathematically by Eq. (3),

$$\tau_{yx} = \mu \frac{dv_x}{dy} = \mu \dot{\gamma} \qquad (3)$$

where is the shear stress, and μ, a function of temperature and pressure, is the coefficient of viscosity or simply the viscosity. Therefore, by comparing Eqs. (2) and (3) the shear viscosity is equal to the coefficient of viscosity (that is, $\eta = \mu$) for a newtonian fluid. Because of this relation the shear viscosity is also often referred to as the viscosity. However, it should be clear that the two quantities are not equivalent; μ is a newtonian-model parameter, which varies only with temperature and pressure, while η is a more general material property which may vary nonlinearly with shear rate. See FLUID FLOW; NEWTONIAN FLUID.

From Eqs. (2) and (3), the units of viscosity are given by force per area per inverse time. If in planar Couette flow, for example, 1 dyne of tangential force is applied for every 1 cm^2 area of plate to create a velocity gradient of 1 s^{-1}, then the fluid between the plates has a viscosity of 1 poise ($=1$ dyne $\cdot$ s/cm^2). Several viscosity units are in common use (see table). Comparison of the viscosities of different fluids demonstrates some general trends. For example, the viscosity of gases is generally much less than that of liquids. Whereas gases tend to become more viscous as temperature is increased, the opposite is true of liquids. Other data also show that increasing pressure tends to

Viscosity conversions

Unit	poise	cp	Pa · s	$lb_m/(ft \cdot s)$	$lb_f \cdot s/ft^2$
1 poise*	1	100	0.1	6.72×10^{-2}	2.089×10^{-3}
1 centipoise	0.01	1	0.001	6.72×10^{-4}	2.089×10^{-5}
1 pascal-second†	10	1000	1	0.672	2.089×10^{-2}
1 $lb_m/(ft \cdot s)$	14.88	1488	1.488	1	3.108×10^{-2}
1 $lb_f \cdot s/ft^2$	478.8	4.788×10^4	47.88	32.17	1

*1 poise = 1 dyne · s/cm² = 1 g/(cm · s).
†1 Pa · s = 1 kg/(m · s).

increase the viscosity of dense gases, but pressure has only a small effect on the viscosity of dilute gases and liquids.

Whereas dilute gas molecules interact primarily in pairs as they collide, molecules in the liquid phase are in continuous interaction with many neighboring molecules. The concepts of velocity and mean free path have little meaning for liquids. It is clear, however, that increasing temperature increases the mobility of molecules, thus allowing neighboring molecules to more easily overcome energy barriers and slip past one another. Such arguments lead to an exponential relation for the dependence of viscosity on temperature.

Many non-newtonian fluids not only exhibit a viscosity which depends on shear rate (pseudoplastic or dilantant) but also exhibit elastic properties. These viscoelastic fluids require a large number of strain-rate-dependent material properties in addition to the shear viscosity to characterize them. The situation can become more complex when the material properties are time dependent (thixotropic or rheopectic). Fluids that are nonhomogeneous or nonisotropic require even more sophisticated analysis. The field of rheology attempts to deal with these complexities. *See* RHEOLOGY.   [L.E.W.]

Voltage measurement   Determination of the difference in electrostatic potential between two points. The unit of voltage in the International System of Units (SI) is the volt, defined as the potential difference between two points of a conducting wire carrying a constant current of 1 ampere when the power dissipated between these two points is equal to 1 watt.

Direct-current voltage measurement. The chief types of instruments for measuring direct-current (constant) voltage are potentiometers, resistive voltage dividers, pointer instruments, and electronic voltmeters.

The most fundamental dc voltage measurements from 0 to a little over 10 V can now be made by direct comparison against Josephson systems. At a slightly lower accuracy level and in the range 0 to 2 V, precision potentiometers are used in conjunction with very low-noise electronic amplifiers or photocoupled galvanometer detectors. Potentiometers are capable of self-calibration, since only linearity is important, and can give accurate measurements down to a few nanovolts. When electronic amplifiers are used, it may often be more convenient to measure small residual unbalance voltages, rather than to seek an exact balance. *See* GALVANOMETER; JOSEPHSON EFFECT.

Voltage measurements of voltages above 2 V are made by using resistive dividers. These are tapped chains of wire-wound resistors, often immersed in oil, which can be self-calibrated for linearity by using a buildup method. Instruments for use up to 1 kV, with tappings typically in a binary or binary-coded decimal series from 1 V, are known as volt ratio boxes, and normally provide uncertainties down to a few parts per million. Another configuration allows the equalization of a string of resistors, all operating at their appropriate power level, by means of an internal bridge. The use of series-parallel arrangements can provide certain easily adjusted ratios.

Higher voltages can be measured by extending such chains, but as the voltage increases above about 15 kV, increasing attention must be paid to avoid any sharp edges or corners, which could give rise to corona discharges or breakdown. High-voltage dividers for use up to 100 kV with an uncertainty of about 1 in 10^5, and to 1 MV with an uncertainty of about 1 in 10^4, have been made. *See* ELECTRICAL BREAK-DOWN.

For most of the twentieth century the principal dc indicating voltmeters have been moving-coil milliammeters, usually giving full-scale deflection with a current between 20 microamperes and 1 milliampere and provided with a suitable series resistor. Many of these will certainly continue to be used for many years, giving an uncertainty of about 1% of full-scale deflection.

The digital voltmeter has become the principal means used for voltage measurement at all levels of accuracy, even beyond one part in 10^7, and at all voltages up to 1 kV. Essentially, digital voltmeters consist of a power supply, which may be fed by either mains or batteries; a voltage reference, usually provided by a Zener diode; an analog-to-digital converter; and a digital display system. This design provides measurement over a basic range from zero to a few volts, or up to 20 V. Additional lower ranges may be provided by amplifiers, and higher ranges by resistive attenuators. The accuracy on the basic range is limited to that of the analog-to-digital converter.

Most modern digital voltmeters use an analog-to-digital converter based on a version of the charge balance principle. In such converters the charge accumulated from the input signal during a fixed time by an integrator is balanced by a reference current of opposite polarity. This current is applied for the time necessary to reach charge balance, which is proportional to the input signal. The time is measured by counting clock pulses, suitably scaled and displayed. Microprocessors are used extensively in these instruments.

Alternating-current voltage measurements. Since the working standards of voltage are of the direct-current type, all ac measurements have to be referred to dc through transfer devices or conversion systems. A variety of techniques can be used to convert an ac signal into a dc equivalent automatically. All multimeters and most ac meters make use of ac-dc conversion to provide ac ranges. These are usually based on electronic circuits. Rectifiers provide the most simple example. *See* MULTIMETER.

In a commonly used system, the signal to be measured is applied, through a relay contact, to a thermal converter. In order to improve sensitivity, a modified single-junction thermal converter may be used in which there are two or three elements in a single package, each with its own thermocouple. The output of the thermal converter is measured by a very sensitive, high-resolution analog-to-digital converter, and the digital value memorized. When a measurement is required, the relay is operated, and the thermal converter receives its input, through a different relay contact, from a dc power supply, the amplitude of which is controlled by a digital and analog feedback loop in order to bring the analog-to-digital converter output back to the memorized level. The dc signal is a converted value of the ac input and can be measured. Modern versions of this type of instrument make use of microprocessors to control the conversion process, enhance the speed of operation, and include corrections for some of the errors in the device and range-setting components.

As in the dc case, digital voltmeters are now probably the instruments in widest use for ac voltage measurement. The simplest use diode rectification of the ac to provide a dc signal, which is then amplified and displayed as in dc instruments. This provides a signal proportional to the rectified mean. For most purposes an arithmetic adjustment is made, and the root-mean-square value of a sinusoidal voltage that would give the same signal is displayed. Several application-specific analog integrated circuits have been developed for use in instruments that are required to respond to the root-mean-square value of the ac input. More refined circuits, based on the logarithmic properties of transistors or the Gilbert analog multiplier circuit, have been developed for use in precision instruments. The best design, in which changes in the gain of the conversion circuit are automatically compensated, achieves errors less than 10 ppm at low and audio frequencies.

Sampling digital voltmeters are also used, in which the applied voltage is switched for a time very short compared with the period of the signal into a sample-and-hold circuit, of which the essential element is a small capacitor. The voltage retained can then be digitized without any need for haste. At low frequencies this approach offers high accuracy and great versatility, since the voltages can be processed or analyzed as desired. At higher frequencies, for example, in the microwave region, it also makes possible the presentation and processing of fast voltage waveforms using conventional circuits. See OSCILLOSCOPE.

Voltage measurements at radio frequencies are made by the use of rectifier instruments at frequencies up to a few hundred megahertz, single-junction converters at frequencies up to 500 MHz, or matched bolometers or calorimeters. At these higher frequencies the use of a voltage at a point must be linked to information regarding the transmission system in which it is measured, and most instruments effectively measure the power in a matched transmission line, usually of 50 ohms characteristic impedance, and deduce the voltage from it. See BOLOMETER; MICROWAVE MEASUREMENTS.

Pulse voltage measurements are made most simply by transferring the pulse waveform to an oscilloscope, the deflection sensitivity of which can be calibrated by using low-frequency sine waves or dc. Digital sampling techniques may also be used. See ELECTRICAL MEASUREMENTS; VOLTMETER.   [R.B.D.K.]

Voltmeter   An instrument for the measurement of the electric potential difference between two conductors. Many different kinds of instruments are available to suit different purposes. Voltages of the order of picovolts (10^{-12} V) to megavolts (10^6 V) can be measured. Frequencies from zero (dc) to many megahertz and accuracies in the range from a fraction of part per million (ppm) to a few percent may be covered. See ELECTRICAL UNITS AND STANDARDS; VOLTAGE MEASUREMENT.

Analog voltmeters. Where no great accuracy is required, a voltage may be indicated by a mechanical displacement of a pointer against a scale. There is a wide variety of principles on which instruments of this type can be based. The d'Arsonval movement (see illustration) is one of the most popular constructions. This is basically a current-sensing instrument and is used in conjunction with a suitable resistance in series to measure voltage. A further variant, taut-band suspension, uses a pair of resilient strips under tension to carry the current to the coil, locate it, and provide the rotational restoring force. See AMMETER; MULTIMETER.

The permanent-magnet, moving-coil instrument is very sensitive, but by its nature is responsive only to the average value of the current flowing through the coil. It is therefore unsuitable for ac. A rectifier circuit can be used in order to combine the sensitivity of the movement with ac response. A transformer can be used to reduce the nonlinearity that results from the forward voltage drop of the diode rectifiers, at the expense of current drain.

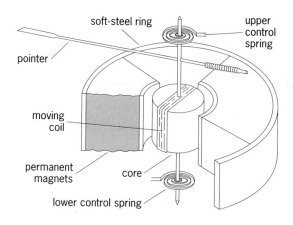

soft-steel ring
upper
control
spring
pointer
moving
coil
permanent
magnets
core
lower control spring

D'Arsonval moving-coil
instrument. (*General Electric
Co.*)

Electronic voltmeters. The movements so far described require energy from the signal being measured to cause the deflection. The resulting current is liable to modify the voltage at the measurement point. To reduce this loading effect, active circuits are often used between the input terminals and the indicating movement. Once an independent source of power is available, electronic circuits can be used to provide other features, including a variety of kinds of signal processing and digital presentation of the results.

Digital voltmeters. Digital voltmeters (DVMs) are now the preferred instruments for ac and dc measurements at all levels of accuracy and at all voltages up to 1 kV. Essentially a digital voltmeter consists of a voltage reference, usually provided by a Zener diode, an analog-to-digital converter and digital display system, and a power supply, which may be derived from either the mains or a battery. The basic range of the instrument provides measurement from zero to 10 or 20 V. Additional lower ranges may be provided by amplifiers, whose gain is stabilized by precision resistors. These electronic input amplifiers often provide a very high input impedance, perhaps exceeding 10^{10} Ω. Since this impedance is obtained by active means, a much lower impedance may be found when the instrument is switched off. Higher voltage ranges are provided by the use of resistive attenuators, usually limited to a value of 10 MΩ by economic restraints. The best accuracy is always obtained on the basic range, where it is limited to that of the analog-to-digital converter.

Sampling voltmeters. A sampling voltmeter is an instrument that uses sampling techniques and has advantages at very low frequencies, that is, below 1 Hz, and also at very high frequencies, where conventional measuring circuits become difficult or even impossible. Low-frequency sampling instruments achieve uncertainties as small as 50 ppm with 10-V signals; high-frequency instruments can achieve a few percent with frequencies as high as 12 GHz and amplitudes as small as 1 mV. Measurements are generally of rectified-mean or root-mean-square voltage. Modern digital sampling voltmeters may also be capable of calculating and displaying voltages or energy density as a function of frequency. Sampling voltmeters, like conventional voltmeters, may use scale and pointer meters, graphic recorders, cathode-ray tubes, or digital indicators for readout of measured quantities. *See* WAVEFORM DETERMINATION.   [R.B.D.K.]

Vortex In common usage, a fluid motion dominated by rotation about an isolated curved line in space, as in a tornado, a whirlpool, a hurricane, or a similar natural

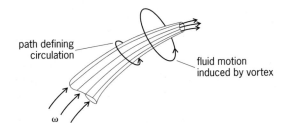

fluid motion induced by vortex

ω

Vortex tube; ω is the vorticity.

phenomenon. The importance of vortices is due to two characteristics: general fluid flows can be represented by a superposition of vortices; and vortices, once created, have a persistence that increases as the effects of viscosity are reduced. The aerodynamic lift forces and most other contributors to the forces and moments on aircraft and other bodies moving through fluids do not exist in the absence of vortices. *See* AERODYNAMIC FORCE.

The strength of rotation is measured by a vector called the vorticity, ω, defined as the curl of the velocity vector. A region of flow devoid of vorticity is known as irrotational. The spatial distribution of the vorticity vector provides a precise characterization of the rotation effects in fluids, and the nature of what subjectively and popularly would be called a vortex. *See* LAPLACE'S IRROTATIONAL MOTION.

The vorticity vector field can be constructed by measuring the instantaneous angular velocity of small masses of fluid. The vorticity vector is twice the local angular velocity vector. Starting at any arbitrary point in the fluid, a line, called a vortex line, can be drawn everywhere parallel to the vorticity vector.

A bundle of vortex lines defines a tubular region of space, called a vortex tube, with a boundary surface that no vortex line crosses.

Two simple rules follow from the definitions: (1) a vortex tube must either close on itself or end on a boundary of the fluid (including extending to "infinity" if the fluid is imagined to fill all space); and (2) at every cross section of a given vortex tube, the area integral of the normal vorticity has the same value at any given instant. The area integral is, by Stokes' theorem, equal to a line integral around the periphery of the tube, namely, the line integral of the velocity component parallel to the direction of the line integral. This quantity is also known as the circulation around the line, so at an instant of time a vortex tube has a unique value of the circulation applying to all cross sections (see illustration).

Vortex lines confined to a layer rather than a tube describe fluid motion of a different character. This is most easily visualized when the direction of the vorticity does not vary, so all of the vortex lines are straight and parallel. Assuming the vorticity has zero magnitude outside the layer, this vortex layer represents a flow with a different speed and direction on either side of the layer. Such a change in speed occurs at the edge of wakes produced by wind passing over an obstacle. Reducing the thickness of this layer of vorticity to zero leads to an idealization known as a vortex sheet, a surface in space across which there is a finite jump in velocity tangent to the surface. Vortex sheets have a tendency to roll up, because of self-induction.   [S.Lei.]

Vorticity   A vector proportional to the local angular velocity of a fluid flow. The vorticity, $\vec{\omega}$, is a derived quantity in fluid mechanics, defined, for a flow field with velocity $\vec{u}$, by Eq. (1). As the curl of the velocity vector, the vorticity is a vector with the

$$\vec{\omega} = \nabla \times \vec{u} \tag{1}$$

dimensions of both a frequency and an angular velocity $[\vec{\omega}] = [s^{-1}]$. The component of vorticity along a particular axis is related to the rate of rotation of the fluid about the axis. For this reason, flows for which $\vec{\omega} = 0$ are described as irrotational. *See* DIMENSIONAL ANALYSIS; VELOCITY.

Circulation. Closely related to vorticity is the fluid circulation, Γ, defined, for any closed contour, C, in a fluid, by Eq. (2). In this definition, $\oint_C$ indicates the conventional

$$\Gamma = \oint_C \vec{u} \cdot d\vec{l} = \iint_S \vec{\omega} \cdot \vec{n} \, dS \tag{2}$$

counterclockwise contour integral around the contour C, $\vec{l}$ is a unit vector tangent to the contour, and S is an arbitrary curved surface bounded by the contour C. [The equality of the two integrals in Eq. (2) may be deduced by the application of Stokes' theorem.] The circulation is thus a scalar quantity equal to the integrated component of vorticity normal to the surface around which Γ is taken. Circulation is important because the Kutta-Joukowski law of aerodynamics states that the lift generated by a two-dimensional airfoil is $L = \rho U \Gamma$. In this expression, ρ is the fluid density, U is the free-stream velocity, and Γ is the bound circulation of the airfoil, defined conventionally as the negative of the definition above. *See* AERODYNAMICS; AIRFOIL.

Vortex line and vortex tube. A vortex line is defined as a line that is everywhere tangent to the local vorticity vector (analogous to a streamline). A series of adjacent vortex lines is referred to as a vortex tube. The first Helmholtz vortex law states that at any instant in time the circulation about all loops taken around the exterior of a vortex tube is the same. Thus, vortex tubes must either form loops entirely within a fluid or terminate at some fluid boundary. *See* VORTEX.

Kelvin's theorem. Kelvin's theorem considers how the circulation Γ around a material loop in a fluid (a loop that moves with the fluid) varies in time. Starting with the Navier-Stokes equations, Lord Kelvin showed that if (1) the fluid is inviscid along the loop, (2) the fluid is subject only to potential body forces, and (3) the fluid pressure is a function of density alone, then the rate of change of Γ is 0. In other words, the circulation around a material loop is time-independent. Kelvin's theorem may also be stated slightly differently: subject to the above three constraints, vortex lines are material lines, convected with the local fluid velocity. *See* KELVIN'S CIRCULATION THEOREM.

Generation. Kelvin's theorem can tell what happens when vorticity is already present in a flow, but it sheds no light on how vorticity is generated. To answer this question, it is useful to consider situations for which Kelvin's theorem is inapplicable: flow with viscosity, with nonpotential body forces, and for which the pressure is not solely a function of the density.

The action of viscosity has two effects on vorticity. One effect of viscosity is to cause the diffusion of vorticity in a fluid. A second effect of viscosity is the generation of vorticity at a wall where there is a pressure gradient at the wall. *See* VISCOSITY.

A common example of a nonpotential body force is the Coriolis force, which is present in a rotating frame of reference. This force generates vorticity in a fluid, and is a major cause of the large-scale circulation in the atmosphere and oceans.

There are many flows for which the pressure may not be solely a function of the density (so-called baroclinic flows), such as the flow of gas with heat addition and the flow of water with salinity variations. Pressure gradients in such flows generate vorticity. This source of vorticity is called baroclinic torque, and is important in atmospheric flow, buoyancy-driven flow, and oceanographic flow.   [S.I.G.]

Wake flow   The flow downstream of a body immersed in a stream or the flow behind a body propagating through a fluid. Wakes are narrow elongated regions, filled with large and small eddies. The wake eddies of a bridge pier immersed in a river stream, or of a ship propelled through the water, are often visible on the surface. On windy days, similar wakes form downstream of smoke stacks or other structures, but the eddies in the air are not visible unless some smoke or dust is entrained in them.

Turbulence in the wake of bluff bodies consists of all sizes of eddies, which interact with each other in their unruly motion. Yet, out of this chaos emerges some organization, whereby large groups of eddies form a well-ordered sequence of vortices. The sense of rotation of these vortices alternates, and their spacing is quite regular. As a result, they can drive a structure that they encounter, or they can exert on the body that created them a force alternating in sign with the same frequency as that of the formation of the vortices. Such forces can impose on structures unwanted vibrations which often lead to serious damage. Flow-induced forces can be catastrophic if they are in tune with the frequency of vibration of the structure. See FLUID FLOW; TURBULENT FLOW.

Wakes are sustained for very large distances downstream of a body. Ship wakes retain their turbulent character for miles behind a vessel and can be detected by special satellites hours after their generation. Similarly, condensation in the wake of aircraft sometimes makes it look like a narrow braided cloud, traversing the sky.   [D.P.T.]

Water hammer   The propagation in a liquid of an acoustic wave that is caused by a rapid change in fluid velocity. Such relatively sudden changes in the liquid velocity are due to events such as the operation of pumps or valves in pipelines, the collapse of vapor bubbles within the liquid, underwater explosions, or the impact of water following the rapid expulsion of air from a vent or a partially open valve. Alternative terms such as pressure transients, pressure surge, hydraulic transients, and hydraulic shock are often employed. Although the physics and mathematical characterization of water hammer and underwater acoustics (employed in sonar) are identical, underwater sound is always associated with very small pressure changes compared to the potential of moderate to very large pressure differences associated with water hammer. See CAVITATION; SOUND; UNDERWATER SOUND.

A pressure change Δp is always associated with the rapid velocity change ΔV across a water hammer wave, as formulated from the basic physics of mass and momentum conservation by the Joukowsky equation, $\Delta p = -\rho a \Delta V$. Here ρ is the liquid mass density and a is the sonic velocity of the pressure wave in the fluid medium. In a pipe, this velocity depends on the ratio of the bulk modulus of the liquid to the elastic modulus of the pipe wall, and on the ratio of the inside diameter of the pipe to the wall thickness. In water in a very rigid pipe or in a tank, or even the sea, the acoustic velocity is approximately 1440 m/s (4720 ft/s), a value many times that of any liquid velocity.

Liquid-handling systems are designed so that water hammer does not result from sudden closure, but is limited to more gradual flow changes initiated by valves or other devices. The dramatic pressure rise (or drop) results can be significantly reduced by reflections of the original wave from pipe-area changes, tanks, reservoirs, and so forth. Although the Joukowsky equation applies across every wavelet, the effect of complete valve closure over a period of time greater than a minimum critical time can be quite beneficial. This critical time is the time required for an acoustic wave to propagate twice the distance along the pipe from the point of wave creation to the location of the first pipe-area change. *See* HYDRODYNAMICS.   [C.S.Ma.]

Wave (physics)   The general term applied to the description of a disturbance which propagates from one point in a medium to other points without giving the medium as a whole any permanent displacement.

Waves are generally described in terms of their amplitude, and how the amplitude varies with both space and time. The actual description of the wave amplitude involves a solution of the wave equation and the particular boundary conditions for the case being studied. *See* WAVE EQUATION; WAVE MOTION.

Acoustic waves, or sound waves, are a particular kind of the general class of elastic waves. Elastic waves are propagated in media having two properties, inertia and elasticity. Electromagnetic waves (for example, light waves and radio waves) are not elastic waves and therefore can travel through a vacuum. The velocity of the wave depends on the medium through which the wave travels. *See* ELECTROMAGNETIC WAVE.   [W.J.G.]

Wave equation   The name given to certain partial differential equations in classical and quantum physics which relate the spatial and time dependence of physical functions. In this article the classical and quantum wave equations are discussed separately, with the classical equations first for historical reasons.

In classical physics the name wave equation is given to the linear, homogeneous partial differential equations which have the form of Eq. (1). Here v is a parameter

$$\left[\nabla^2 - \frac{1}{v^2} \frac{\partial^2}{\partial t^2} \right] f(\mathbf{r}, t) = 0 \tag{1}$$

with the dimensions of velocity; $\mathbf{r}$ represents the space coordinates x, y, z; t is the time; and ∇^2 is Laplace's operator defined by Eq. (2). The function $f(\mathbf{r}, t)$ is a physical

$$\nabla^2 = \frac{\partial^2}{\partial x^2} + \frac{\partial^2}{\partial y^2} + \frac{\partial^2}{\partial z^2} \tag{2}$$

observable; that is, it can be measured and consequently must be a real function.

The simplest example of a wave equation in classical physics is that governing the transverse motion of a string under tension and constrained to move in a plane.

A second type of classical physical situation in which the wave equation (1) supplies a mathematical description of the physical reality is the propagation of pressure waves in a fluid medium. Such waves are called acoustical waves, the propagation of sound being an example. A third example of a classical physical situation in which Eq. (1) gives a description of the phenomena is afforded by electromagnetic waves. In a region of space in which the charge and current densities are zero, Maxwell's equations for the photon lead to the wave equations (3). Here **E** is the electric field strength and **B** is the

$$\left[\nabla^2 - \frac{1}{c^2} \frac{\partial^2}{\partial t^2} \right] E(\mathbf{r},t) = 0$$

$$\left[\nabla^2 - \frac{1}{c^2} \frac{\partial^2}{\partial t^2} \right] B(\mathbf{r},t) = 0$$

(3)

magnetic flux density; they are both vectors in ordinary space. The parameter c is the speed of light in vacuum. *See* ELECTROMAGNETIC RADIATION; MAXWELL'S EQUATIONS.

The nonrelativistic Schrödinger equation is an example of a quantum wave equation. Relativistic quantum-mechanical wave equations include the Schrödinger-Klein-Gordon equation and the Dirac equation. *See* QUANTUM MECHANICS; RELATIVISTIC QUANTUM THEORY.

[D.L.We.]

Wave mechanics   The modern theory of matter holding that elementary particles (such as electrons, protons, and neutrons) have wavelike properties. In 1924 L. de Broglie postulated that the wave-particle duality which had been demonstrated for electromagnetic radiation also was a property of the elementary particles making up the atoms and molecules forming ordinary matter. In particular, de Broglie postulated that a particle has an associated wavelength obeying the same relation as was found to hold for photons, namely: the wavelength equals Planck's constant divided by the particle's momentum (as customarily defined in elementary mechanics). This hypothesis was verified in 1927 in an experiment in which a beam of electrons having known momentum is diffracted by a crystal into special directions. Such diffraction seems understandable only on the hypothesis that the electrons are waves. Furthermore, the wavelength of the electrons in the incident beam, computed via the same formula as was used to derive x-ray wavelengths in x-ray diffraction experiments, agreed precisely with the de Broglie relation. *See* ELECTRON DIFFRACTION; X-RAY DIFFRACTION.

Subsequent experiments have confirmed that not merely electrons but material particles in general, such as neutrons and neutral sodium atoms, manifest the wave-particle duality and obey the de Broglie relation. The de Broglie relation and the qualitative wave-particle duality concept have been incorporated into the highly successful modern theory of quantum mechanics. *See* ATOM OPTICS; DE BROGLIE WAVELENGTH; INTERFERENCE OF WAVES; QUANTUM MECHANICS.

[E.G.]

Wave motion   The process by which a disturbance at one point in space is propagated to another point more remote from the source with no net transport of the material of the medium itself. For example, sound is a form of wave motion; wind is not. Wave motion can occur only in a medium in which energy can be stored in both kinetic and potential form. In a mechanical medium, kinetic energy results from inertia and is stored in the velocity of the molecules, while potential energy results from elasticity and is stored in the displacement of the molecules.

Media.   In a free traveling wave (as distinguished from a stationary or standing wave) one part of the medium disturbs an adjacent part, thereby imparting energy to it. This portion of the medium, in turn, disturbs another part, thereby causing a flow of energy in a given direction away from the source. More technically, wave propagation is the result of kinetic energy at one point being transferred into potential energy at an adjacent point, and vice versa. The rate of travel of the disturbance, or velocity of propagation, is determined by the constants of the medium. A stationary wave is the combination of two waves of the same frequency and strength traveling in opposite directions so that no net transfer of energy away from the source takes place. A standing wave is the same but with the returning wave (toward the source) being of lesser intensity than the outwardly traveling wave so that a net transfer of energy away from the source does take place.

Wave motion can occur in a vacuum (electromagnetic waves), in gases (sound waves), in liquids (hydrodynamic waves), and in solids (vibration waves). Electromagnetic waves can also travel in gases, liquids, and solids provided that the electrical conductivity of the medium is not perfect or that the imaginary part of the dielectric constant is not infinitely great. By current usage, elastic waves propagated in gases, liquids, and solids, regardless of whether one can hear them or not, are called acoustic waves.

Fundamental relations. A wave is commonly referred to in terms of either its wavelength or its frequency. In any type of wave motion, these two quantities are related to a third quantity, velocity of propagation, by the simple relation $f\lambda = c$, where f = frequency, λ = wavelength, and c = velocity of propagation. The period T is the reciprocal of the frequency, and the amplitude A is the maximum magnitude taken on by the variable of the wave at a given point in space. It is a basic property of wave motion that the frequency of a wave remains constant under all circumstances except for a relative motion between the source of the wave and the observer. The velocity of propagation is dependent on the properties of the medium (and, sometimes, also on the frequency) and the wavelength will vary with the velocity in accordance with the equation above.

Electromagnetic waves. The media in which electromagnetic waves travel possess no elasticity or inertia, but rather the ability to store energy in the electric and magnetic fields. J. C. Maxwell recognized in about 1863 that the basic equations governing these fields could be combined to yield an equation resembling the wave equation for mechanical wave motion. Thus he predicted the existence of electromagnetic waves which had not been suspected theretofore. Later, electromagnetic waves proved to be identical with light waves. *See* ELECTROMAGNETIC RADIATION; LIGHT; MAXWELL'S EQUATIONS; WAVE EQUATION.   [L.L.B.]

Motion in fluids. Wave motion within a fluid is generated by successive compression and expansion of adjacent volume elements. Because compression and expansion of an ordinary fluid can only proceed along the direction of propagation of the disturbance, waves within a fluid are mostly longitudinal waves.

Waves in a fluid can be classified as compression waves and expansion waves, according to whether the disturbance is a compression or an expansion. They can further be classified according to the amplitude of the disturbance and the chemical nature of the fluid. For example, waves of small amplitude are called acoustic (or sound) waves; compression waves propagating in chemically inert fluids are called shock waves; waves propagating in the Earth are seismic waves; and waves of large amplitude generated by rapid chemical reactions in explosive fluids are called detonation waves and can propagate much faster than sound waves. Waves in an electrically conducting fluid in the presence of strong magnetic fields are called magnetohydrodynamic waves. *See* MAGNETOHYDRODYNAMICS; SHOCK WAVE; SOUND.   [S.C.Li.]

Motion in liquids. Disturbances propagated at a gas-liquid interface are primarily dependent upon the gravitational fluid property (surface tension and viscosity being of secondary importance). Wave motions which occur in confined fluids (either liquid or gaseous) are primarily dependent upon the elastic property of the medium.

Oscillatory waves may be generated in a rectangular channel by a simple harmonic translation of a vertical wall forming one end of the flume. A standing wave can be considered to be composed of two equal oscillatory wave trains traveling in opposite directions.

A solitary wave consists of a single crest above the original liquid surface which is neither preceded nor followed by another elevation or depression of the surface. Such a wave is generated by the translation of a vertical wall starting from an initial position at rest and coming to rest again some distance downstream. In practice, solitary waves are generated by a motion of barges in narrow waterways or by a sudden change in the rate of inflow into a river; they are therefore related to a form of flood wave.

[D.R.F.H.]

Wave optics   The branch of optics which treats of light (or electromagnetic radiation in general) with explicit recognition of its wave nature. The counterpart to wave optics is ray optics or geometrical optics, which does not assume any wave character but treats the propagation of light as a straight-line phenomenon except for changes of direction induced by reflection or refraction. See ELECTROMAGNETIC RADIATION; GEOMETRICAL OPTICS; OPTICS.

Any optical phenomenon which is correctly describable in terms of geometrical optics can also be correctly described in terms of wave optics. However, the many phenomena of interference, diffraction, and polarization are incontrovertible evidence of the wave nature of light, and geometrical optics often gives an incomplete or incorrect description of the behavior of light in an optical system. This is especially true if changes of refractive index occur within a space which is of the order of several wavelengths of the light. See DIFFRACTION; INTERFERENCE OF WAVES; POLARIZED LIGHT.

[R.C.L.]

Waveform determination   The definition of a waveform, which describes the variation of a quantity with respect to time. The necessary measurements are normally carried out and presented in one of two ways: the amplitude may be presented as a function of time (time domain), or an analysis may be given of the relative amplitudes and phases of the frequency components (frequency domain). Although the simplest instruments measure and display the information in the same domain, it is possible to convert the data in either direction by mathematical processing.

Waveforms may be divided into two classes, depending on whether the signal is repeated at regular intervals or represents a unique event. The former signal is defined as a periodic or continuous wave, the latter as an aperiodic signal or transient.

The oscilloscope is an example of an instrument that measures and displays directly in the time domain, by deflecting an electron beam in a vertical direction in accordance with the signal while scanning at a uniform rate in the horizontal direction. The position of the beam is revealed by a fluorescent screen. See OSCILLOSCOPE.

Several methods may be used to obtain the spectral content of a waveform. In the simplest, the signal is applied to a filter that is manually tuned in turn to each frequency that is expected to be present. In order to automate the measurement, the tuning of a filter may be varied by a linear, logarithmic, or other sweep and the resulting output displayed. However, there are important restrictions that limit the technique to continuous waveforms. A fleeting appearance of a signal at a frequency away from that to which the filter happens to be tuned at the instant will be completely missed. In order to provide high resolution, the tuned circuit must have high selectivity, or high Q. Its response to changes in amplitude is therefore slow, and the rate of sweeping has to be limited. By using this technique, it is possible to construct instruments that cover extremely wide frequency ranges.

In order to overcome the limitations of the swept filter, an array of separate fixed tuned filters may be used, each adjusted to respond to a slightly different frequency. The amplitude of the signal in each filter is sampled in turn and displayed, giving a histogram of the frequency components of the waveform. The number of filters and their selectivities depend on the resolution required. Though simple in concept, such instruments are inclined to be bulky if high resolution is needed. They have the great advantage that all frequency components are taken into account throughout the measurement and their amplitudes can be examined continuously. Such instruments are useful for the analysis of music and speech.

Since the purpose of such instruments is to display the frequency components of the signal, they are often called spectrum or harmonic analyzers.

Many modern instruments use techniques in which the signal to be measured is sampled and digitized. In sampling oscilloscopes a sufficiently large number of samples is taken to define the waveform. The sampling rate is not necessarily high; where the waveform is repetitive, the Nyquist requirement that more than two samples should be available per cycle of the highest frequency of interest can be satisfied by obtaining the samples over a large number of periods of the signal. Each sample is digitized and stored. The waveform is then displayed by plotting the data on a cathode-ray tube or similar display in the correct order. As the data can be read out of digital storage at any convenient speed, this process can be relatively slow and low bandwidth display circuits are adequate.

Once the data have been collected in digital form, they can be processed in many different ways. The application of a discrete Fourier transformation (DFT) to the amplitude data enables the information to be presented in the frequency domain. In this way, harmonic distortion which was invisible on a directly displayed waveform can be made obvious.   [R.B.D.K.]

Wavelength   The distance between two points on a wave which have the same value and the same rate of change of the value of a parameter, for example, electric intensity, characterizing the wave. The wavelength, usually designated by the Greek letter λ, is equal to the speed of propagation c of the wave divided by the

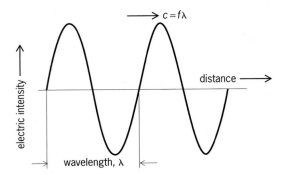

Wavelength λ and related quantities.

frequency of vibration f; that is, $\lambda = c/f$ (see illustration). *See* WAVE (PHYSICS); WAVE MOTION.   [W.J.G.]

Wavelength measurement Determination of the distance between successive wavefronts of equal phase of a wave. *See* ELECTROMAGNETIC RADIATION; WAVE MOTION; WAVELENGTH.

From the relation $\lambda = c/f$ between wavelength λ, speed c, and frequency f, the wavelength of a wave motion can be calculated if the speed is known and its frequency is measured. The ease and accuracy of electronic counting and timing make frequency measurement the most precise of all physical measurements. This method of wavelength determination is thus one of the most accurate, but only if the speed (phase velocity) is known. In free space the speed of an electromagnetic wave c_0 is, through the 1983 definition of the meter, fixed at exactly 299,792,458 m/s (186,282.397 mi/s), or roughly 300,000 km/s. Unless otherwise specified, it is general practice to quote the wavelength of an electromagnetic wave as the free-space value λ_0, given by the equation below.

$$\lambda_0 = \frac{c_0}{f}$$

See FREQUENCY MEASUREMENT; PHASE VELOCITY.

Radio and microwave regions. The presence of any dielectric material (such as air) or any magnetic matter with a permeability greater then unity will cause the wave to travel at a velocity lower than its free-space value. The speed is also altered if the waves pass through an aperture, are focused by a lens or mirror, or are constrained by a wave-guide or transmission line. In such cases it may be more appropriate to measure the wavelength directly. In the pioneering experiments on radio waves, it was found that standing waves existed in space whenever reflections occurred and that these provided a convenient means of measuring the wavelength. It thus became the convention to characterize waves by their wavelength, rather than by their frequency as is now more commonly the case. Specifying the frequency is preferred because, unlike the wavelength, it is independent of the speed of propagation and does not change as the wave moves from one medium to another.

In the microwave region the wavelengths are sufficiently short that it is convenient to measure them by using interferometer techniques directly analogous to those used with light.

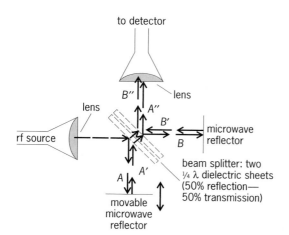

Wavelength measurement by
the Michelson interferometer
used at millimeter wavelengths.

In a typical interferometer used in the millimeter wavelength range (see illustration), a microwave beam is directed at a beam splitter, which splits the beam into two parts, A and B, by partial reflection. The A beam is reflected to a movable reflector and reflected again as A'. The beam splitter transmits part of this as A''. The transmitted part B of the original beam is reflected by a fixed microwave reflector as B'. This is partially reflected by the beam splitter as B''. The beams A'' and B'' combine to form standing waves, which are then detected. Movement of the movable reflector causes the position of the standing wave to move, which causes the detected signal to pass through successive points of maximum and minimum amplitude. The distance between points of successive maxima or minima is one-half wavelength. This distance may be determined from the motion of the movable reflector. *See* INTERFEROMETRY.

Infrared and optical regions. The free-space wavelengths of monochromatic visible and infrared radiations can be derived directly, if their corresponding frequencies are known. The wavelengths of a few stabilized laser radiations have been precisely determined in this way, by reference to the cesium-133 primary frequency standard. These wavelength standards, with uncertainties between 4.5×10^{-10} and 3×10^{-12}, are used particularly for the practical realization of the unit of length, but they also serve as reference standards for interferometric wavelength measurements and for the calibration of wavelength comparators based on dispersion methods. *See* LENGTH; LIGHT; PHYSICAL MEASUREMENT; WAVELENGTH STANDARDS.

Wavelength values to an accuracy of 1 part in 10^5 can be determined with a spectrometer, spectrograph, or monochromator, in which a prism or diffraction grating is used as a dispersive element. Each wavelength forms a line image of the entrance slit at a particular angle. An unknown wavelength can be determined by interpolation with the pattern formed by a lamp emitting the tabulated characteristic wavelengths of a particular element. *See* DIFFRACTION GRATING; OPTICAL PRISM; SPECTROSCOPY.

The most precise wavelength measurements use an interferometer to compare the unknown wavelength λ_1 with a standard wavelength λ_2. Usually either the two-beam Michelson form or the multiple-beam Fabry-Perot form of interferometer is used.

When a number of wavelengths are mixed together in the input to a moving-carriage two-beam interferometer, the output signal is the summation of the many separate sine-wave signals having different periods. A Fourier analysis of this composite signal enables the separate wavelengths to be identified. This Fourier transform method is particularly useful for the measurement of complex spectra in the infrared. *See* INFRARED SPECTROSCOPY. [W.R.C.R.]

Wavelength standards Accurately known wavelengths of spectral radiation emitted from specified sources that are used to measure the wavelengths of other spectra. In the past, the radiation from the standard source and the source under study were superimposed on the slit of a spectrometer (prism or grating) and then the unknown wavelengths could be determined from the standard wavelengths by using interpolation. This technique has evolved into the modern computer-controlled photoelectric recording spectrometer. Accuracy of many more orders of magnitude can be obtained by the use of interferometric techniques, of which Fabry-Perot and Michelson interferometers are two of the most common. *See* INTERFEROMETRY; SPECTROSCOPY.

The newest definition of the meter is in terms of the second. The wavelength of radiation from the cesium atomic clock is not used to realize length because diffraction problems at this wavelength are severe. Instead, lasers at shorter wavelengths whose frequencies have been measured are used. Frequency measurements can now be made even into the visible spectral region with great accuracy. Hence, when the 1983 Conférence Général des Poids et Mesures redefined the meter, it also gave a list of very accurate wavelengths of selected stabilized lasers which may be used as wavelength standards; these are shown in the table. Nearly ten times better accuracy can be achieved by using these wavelengths than by using the radiation from the krypton lamp which provided the previous standard. *See* FREQUENCY MEASUREMENT; HYPERFINE STRUCTURE; LASER; LASER SPECTROSCOPY; LENGTH; MOLECULAR STRUCTURE AND SPECTRA; PHYSICAL MEASUREMENT.

The progress in laser frequency measurements since 1974 has established wavelength standards throughout the infrared spectral region. This has been accomplished with the accurate frequency measurement of rotational-vibrational transitions of selected molecules. The OCS molecule is used in the 5-micrometer spectral region. At 9–10 μm, the carbon dioxide (CO_2) laser itself with over 300 accurately known lines is used. From 10 to 100 μm, rotational transitions of various molecules are used; most are optically pumped laser transitions. The increased accuracy of frequency measurements makes this technique mandatory where ultimate accuracy is needed. *See* WAVELENGTH.

[D.A.J.]

Wavelets   The elementary building blocks in a mathematical tool for analyzing functions. The functions can be very diverse; examples are solutions of a differential equation, and one- and two-dimensional signals. The tool itself, the wavelet transform, is the result of a synthesis of ideas from many different fields, ranging from pure mathematics to quantum physics and electrical engineering.

In many practical applications, it is desirable to extract frequency information from a signal—in particular, which frequencies are present and their respective importance. An example is the decomposition into spectral lines in spectroscopy. The tool that is generally used to achieve this is the Fourier transform. Many applications, however, concern nonstationary signals, in which the makeup of the different frequency components is constantly shifting. An example is music, where this shifting nature has been recognized for centuries by the standard notation, which tells a musician which note (frequency information) to play when and how long (time information). For signals of this nature, a time-frequency representation is needed.

There exist many different mathematical tools leading to a time-frequency representation of a given signal, each with its own strengths and weaknesses. The wavelet transform is such a time-frequency analysis tool. Its strength lies in its ability to deal well with transient high-frequency phenomena, such as sudden peaks or discontinuities, as well as with the smoother portions of the signal. (An example is a crack in the sound from a damaged record, or the attack at the start of a music note.) The wavelet transform is less well adapted to harmonically oscillating parts in the signal, for which Fourier-type methods are more indicated.

Applications of wavelets include various forms of data compression (such as for images and fingerprints), data analysis (nuclear magnetic resonance, radar, seismograms, and sound), and numerical analysis (fast solvers for partial differential equations).

[I.D.]

Weak nuclear interactions   Fundamental interactions of nature that play a significant role in elementary particle and nuclear physics, and are distinguished from other such interactions by special properties such as participation of all the fundamental fermions and failure to conserve parity. The weak force has very short range (less than 10^{-17} m) and is extremely feeble compared to strong and electromagnetic forces, but can be distinguished from these two by its special character. For example, according to the present view, all of matter consists of certain fundamental spin-$\frac{1}{2}$ constituents, the quarks and leptons, collectively called the fundamental fermions. While only the quarks participate in strong interactions, and only the quarks and charged leptons e, μ, and τ participate in electromagnetic interactions, all of the fundamental fermions, including neutrinos, engage in weak interactions. Also, the strong and electromagnetic interactions respect spatial inversion symmetry (they conserve parity) and are also particle-antiparticle (charge conjugation) symmetric, whereas the weak interaction violates these two symmetries. *See* FUNDAMENTAL INTERACTIONS; LEPTON; PARITY (QUANTUM MECHANICS); QUARKS; SYMMETRY LAWS (PHYSICS).

Weak interactions are classified as "charged" or "neutral," depending on whether or not a particle participating in a weak reaction suffers a change of electric charge of one electronic unit. Observed charged weak interactions include nuclear beta decay and electron capture, muon capture on nuclei, and the slow decays of unstable elementary particles such as the μ and τ leptons, π, K, and charmed mesons, and hyperons and charmed baryons. Also, there are the charged neutrino-nucleon and neutrino-lepton scattering reactions. Neutral weak interactions include neutrino-nucleon and neutrino-lepton scattering as well as the electron-nucleon reaction which can also occur by electromagnetic interaction. *See* BARYON; ELEMENTARY PARTICLE; HYPERON; MESON.

The most important development in the study of weak interactions has been the creation of a successful theory based on the principles of local gauge invariance and spontaneous symmetry breaking. This theory proposes a single basis for the weak and electromagnetic interactions, and indeed, despite striking differences in the observed characteristics of strong, electromagnetic, and weak interactions, important theoretical ideas of a similar type suggest that all these interactions possess a common origin.   [E.D.C.]

Weakly interacting massive particle (WIMP)

A hypothetical elementary particle that might make up most of the matter in the universe, and that is also predicted to exist in supersymmetry theory. Most matter is detected only through its gravitational effects; this "dark matter" has not been observed to emit, absorb, or reflect light of any wavelength. The total amount of dark matter appears to be approximately ten times as great as all the ordinary matter in the universe, and about one hundred times as great as all the visible matter. The nature of the dark matter is not yet known, although many experiments are under way to try to discover it directly or indirectly.

Almost all the currently available data in elementary particle physics can be accounted for by a theory called the standard model, in which matter is made of quarks (the building blocks of protons and neutrons) and leptons (including electrons and neutrinos), while the strong, weak, and electromagnetic forces are transmitted by particles like the photon (the carrier of electromagnetic forces). However, the standard model does not predict the existence of any particle—say, X—that could be the dark matter. Most efforts to go beyond the standard model of particle physics have been based on the idea of supersymmetry, and most versions of supersymmetry predict that there will be a stable weakly interacting massive particle (WIMP) that would be a natural candidate for the X particles. Dark matter made of WIMPs would be "cold" dark matter (CDM), and a version of CDM theory has become the standard theory of structure formation in cosmology. *See* ELEMENTARY PARTICLE; STANDARD MODEL; SUPERSYMMETRY.

There is now abundant evidence for dark matter around galaxies and clusters of galaxies, and on larger scales in the universe. Gas and satellites at large distances from galaxies have orbital velocities similar to those at smaller distances from the center, which indicates that most of the mass in the galaxy must not be near the center, where most stars are, but in a roughly spherical dark matter halo that extends to perhaps ten times the optical size of the galaxy and has a mass at least ten times that of all the stars.

Confirmation of the existence of such dark-matter halos has come from gravitational lensing observations, showing that light from more distant galaxies is bent by the gravity of nearer galaxies.

There is also much evidence for dark matter in clusters of galaxies. The astronomer Fritz Zwicky pointed out in 1933 that the galaxies in one nearby cluster were moving at such high speeds that they would not be held together gravitationally unless there was much more mass than was indicated by the light from their stars. This same was subsequently found to be true of other clusters. Later, similar conclusions were reached from x-ray observations and gravitational lensing observations of clusters.

Supersymmetry is the hypothesis that there is a relationship between the two known classes of particles, bosons and fermions. According to supersymmetry, for every kind of boson in the universe, there must also be a corresponding fermion with the same electric charge and very similar interactions with other particles. Since these hypothetical sypersymmetric partner particles have not been discovered yet, if supersymmetry is right their masses must be too large for them to have been produced at current particle accelerators. Thus far, the evidence for supersymmetry is only indirect, but if the theory is right many supersymmetric partner particles should be produced at accelerators such as the Large Hadron Collider (LHC) being built in Geneva, Switzerland. *See* PARTICLE ACCELERATOR; QUANTUM STATISTICS.

Efforts to detect WIMPs directly are based on detecting their scattering from nuclei. WIMPs can also be detected indirectly, for example by looking for particles coming from their annihilation. WIMPs are also expected to be produced at accelerators such as the LHC from rapid decays of heavier supersymmetric partner particles, and this could be where they are discovered first if they are not seen before that in direct or indirect search experiments. Failure to see supersymmetric particles at LHC energies would mean that current ideas about supersymmetry are wrong.   [J.R.Pr.]

Weight   The gravitational weight of a body is the force with which the Earth attracts the body. By extension, the term is also used for the attraction of the Sun or a planet on a nearby body. This force is proportional to the body's mass and depends on the location. Because the distance from the surface to the center of the Earth decreases at higher latitudes, and because the centrifugal force of the Earth's rotation is greatest at the Equator, the observed weight of a body is smallest at the Equator and largest at the poles. The difference is sizable, about 1 part in 300. At a given location, the weight of a body is highest at the surface of the Earth. Weight is measured by several procedures. *See* MASS.
                                                                        [H.S.B.]

Wentzel-Kramers-Brillouin method   A special technique for obtaining an approximation to the solutions of the one-dimensional time-independent Schrödinger equation, valid when the wavelength of the solution varies slowly with position. It is named after G. Wentzel, H. A. Kramers, and L. Brillouin, who independently in 1926 contributed to its understanding in the quantum-mechanical application. It is also called the WKB method, BWK method, the classical approximation, the quasi-classical approximation, and the phase integral method. *See* QUANTUM MECHANICS; SCHRÖDINGER'S WAVE EQUATION.
                                                                        [R.H.Go.]

Wheatstone bridge   A device used to measure the electrical resistance of an unknown resistor by comparing it with a known standard resistance. This method was first described by S. H. Christie in 1833. Since 1843 when Sir Charles Wheatstone

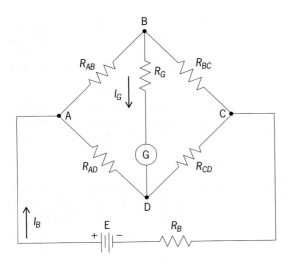

B

R_{AB}

R_G

R_{BC}

I_G

A

C

R_{AD}

G

R_{CD}

D

I_B

E

R_B

+ −

Wheatstone bridge circuit.

called attention to Christie's work, Wheatstone's name has been associated with this network.

The Wheatstone bridge network consists of four resistors R_{AB}, R_{BC}, R_{CD}, and R_{AD} interconnected as shown in the illustration to form the bridge. A detector G, having an internal resistance R_G, is connected between the B and D bridge points; and a power supply, having an open-circuit voltage E and internal resistance R_B, is connected between the A and C bridge points. *See* BRIDGE CIRCUIT.

If the network is adjusted so that Eq. (1) is satisfied, the detector current will be zero

$$R_{BC} R_{AD} - R_{AB} R_{CD} = 0 \qquad (1)$$

and this adjustment will be independent of the supply voltage, the supply resistance, and the detector resistance. Thus, when the bridge is balanced, Eq. (2) holds, and, if it

$$R_{BC} R_{AD} = R_{AB} R_{CD} \qquad (2)$$

is assumed that the unknown resistance is the one in the CD arm of the bridge, then it is given by Eq. (3).

$$R_{CD} = \left(\frac{R_{BC}}{R_{AB}} \right) \times R_{AD} \qquad (3)$$

See RESISTANCE MEASUREMENT. [C.E.A.]

Wiedemann-Franz law An empirical law of physics which states that the ratio of the thermal conductivity of a metal to its electrical conductivity is a constant times the absolute temperature, as given by the equation below. Here K_c is the thermal

$$K_c = L_0 \sigma T$$

conductivity due to the conduction electrons, σ is the electrical conductivity, T is the absolute temperature, and L_0 is known as the Lorentz number. The Wiedemann-Franz law provides an important check on theories of electrical and thermal conductivity. *See* CONDUCTION (HEAT); THERMAL CONDUCTION IN SOLIDS. [F.J.B.]

Work   In physics, the term work refers to the transference of energy that occurs when a force is applied to a body that is moving in such a way that the force has a component in the direction of the body's motion. Thus work is done on a weight that is being lifted, or on a spring that is being stretched or compressed, or on a gas that is undergoing compression in a cylinder.

When the force acting on a moving body is constant in magnitude and direction, the amount of work done is defined as the product of just two factors: the component of the force in the direction of motion, and the distance moved by the point of application of the force. Thus the defining equation for work W is given below, where f and s

$$W = f \cos\phi \cdot s$$

are the magnitudes of the force and displacement, respectively, and ϕ is the angle between these two vector quantities (see illustration). Because $f \cos\phi \cdot s = f \cdot s \cos\phi$, work may be defined alternatively as the product of the force and the component of the displacement in the direction of the force.

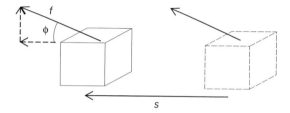

Work of constant force f is fs $\cos\phi$.

The work done is positive in sign whenever the force or any component of it is in the same direction as the displacement; one then says that work is being done by the agent exerting the force and *on* the moving body. The work is said to be negative whenever the direction of the force or force component is opposite to that of the displacement; then work is said to be done *on* the agent and *by* the moving body. From the point of view of energy, an agent doing positive work is losing energy to the body on which the work is done, and one doing negative work is gaining energy from that body.

The work principle, which is a generalization from experiments on many types of machines, asserts that, during any given time, the work of the forces applied to the machine is equal to the work of the forces resisting the motion of the machine, whether these resisting forces arise from gravity, friction, molecular interactions, or inertia.

The work done by any conservative force, such as a gravitational, elastic, or electrostatic force, during a displacement of a body from one point to another has the important property of being path-independent: Its value depends only on the initial and final positions of the body, not upon the path traversed between these two positions. On the other hand, the work done by any nonconservative force, such as friction due to air, depends on the path followed and not alone on the initial and final positions, for the direction of such a force varies with the path, being at every point of the path tangential to it. *See* ENERGY; FORCE.   [L.N.]

Work function (electronics)   A quantity with the dimensions of energy which determines the thermionic emission of a solid at a given temperature. For metals, the work function may also be determined by measuring the photoemission as a function of the frequency of the incident electromagnetic radiation: the work function is then equal to the minimum (threshold) frequency for which electon emission is observed times Planck's constant h ($= 6.63 \times 10^{-34}$ joule second). The work function of a solid is usually expressed in electronvolts.

The work function of metals varies from one crystal plane to another and also varies slightly with temperature. For a metal, the work function has a simple interpretation. At absolute zero, the energy of the most energetic electrons in a metal is referred to as the Fermi energy; the work function of a metal is then equal to the energy required to raise an electron with the Fermi energy to the energy level corresponding to an electron at rest in vacuum. The work function of a semiconductor or an insulator has the same interpretation, but in these materials the Fermi level is in general not occupied by electrons and thus has a more abstract meaning. *See* FIELD EMISSION; PHOTOEMISSION; THERMIONIC EMISSION. [A.J.D.]

Work function (thermodynamics) The thermodynamic function better known as the Helmholtz energy, $A = U - TS$, where U is the internal energy, T is the thermodynamic (absolute) temperature, and S is the entropy of the system. At constant temperature, the change in work function is equal to the maximum work that can be done by a system ($\Delta A = w_{max}$). *See* FREE ENERGY. [P.W.A.]

X, Z

X-ray crystallography   The study of crystal structure by x-ray diffraction techniques. For the experimental aspects of x-ray diffraction *See* X-RAY DIFFRACTION.

Structurally, a crystal is a three-dimensional periodic arrangement in space of atoms, groups of atoms, or molecules. If the periodicity of this pattern extends throughout a given piece of material, one speaks of a single crystal. The exact structure of any given crystal is determined if the locations of all atoms making up the three-dimensional periodic pattern called the unit cell are known. The very close and periodic arrangement of the atoms in a crystal permits it to act as a diffraction grating for x-rays. *See* CRYSTALLOGRAPHY.

[L.F.D.]

X-ray diffraction   The scattering of x-rays by matter with accompanying variation in intensity in different directions due to interference effects. X-ray diffraction is one of the most important tools of solid-state chemistry, since it constitutes a powerful and readily available method for determining atomic arrangements in matter. X-ray diffraction methods depend upon the fact that x-ray wavelengths of the order of 1 nanometer are readily available and that this is the order of magnitude of atomic dimensions. When an x-ray beam falls on matter, scattered x-radiation is produced by all the atoms. These scattered waves spread out spherically from all the atoms in the sample, and the interference effects of the scattered radiation from the different atoms cause the intensity of the scattered radiation to exhibit maxima and minima in various directions. *See* DIFFRACTION.

Uses. Some of the uses of x-ray diffraction are: (1) differentiation between crystalline and amorphous materials; (2) determination of the structure of crystalline materials (crystal axes, size and shape of the unit cell, positions of the atoms in the unit cell); (3) determination of electron distribution within the atoms, and throughout the unit cell; (4) determination of the orientation of single crystals; (5) determination of the texture of polygrained materials; (6) identification of crystalline phases and measurement of the relative proportions; (7) measurement of limits of solid solubility, and determination of phase diagrams; (8) measurement of strain and small grain size; (9) measurement of various kinds of randomness, disorder, and imperfections in crystals; and (10) determination of radial distribution functions for amorphous solids and liquids.

Techniques. The techniques employed in the study of crystalline substances, gases, and liquids are discussed below.

Laue method. The Laue pattern uses polychromatic x-rays provided by the continuous spectrum from an x-ray tube operated at 35–50 kV. The different diffracted beams have different wavelengths, and their directions are determined solely by the orientations of the set of planes with Miller indices *hkl*. Transmission Laue patterns were once used for structure determinations, but their many disadvantages have made them practically obsolete. On the other hand, the back-reflection Laue pattern is used a great deal in the study of the orientation of crystals.

Rotating crystal method. The original rotating crystal method was employed in the Bragg spectrometer. A sufficiently monochromatic beam, of wavelength of the order of 1 Å, is collimated by a system of slits and then falls on the large extended face of a single crystal as shown by Fig. 1. The Bragg spectrometer has been used extensively in obtaining quantitative measurements of the integrated intensity from planes parallel to the face of the crystal. The chamber is set at the correct 2θ angle with a slit so wide that all of the radiation reflected from the crystal can enter and be measured. The crystal is turned at constant angular speed through the Bragg law position, and the total diffracted energy received by the ionization chamber during this process is measured. Similar readings with the chamber set on either side of the peak give a background correction.

The rotation camera, which is frequently used for structure determinations, is illustrated in Fig. 2. The monochromatic primary beam $\mathbf{s}_0$ falls on a small single crystal at O. The crystal is mounted with one of its axes (say, $\mathbf{a}_3$) vertical, and it rotates with constant velocity about the vertical axis during the exposure. The various diffracted beams are registered on a cylindrical film concentric with the axis of rotation.

Powder method. The powder method involves the diffraction of a collimated monochromatic beam from a sample containing an enormous number of tiny crystals having random orientation. Powder pattern studies are made with Geiger counter, or proportional counter, diffractometers. The apparatus is shown schematically in Fig. 3. X-rays diverging from a target at T fall on the sample at O, the sample being a flat-faced briquet of powder. Diffracted radiation from the sample passes through the receiving slit at s and enters the Geiger counter. During the operation the sample turns at angular velocity ω and the counter at 2ω. The distances TO and OS are made equal to satisfy approximate focusing conditions. A filter F before the receiving slit gives the effect of a sufficiently monochromatic beam. A chart recording of the amplified output of the Geiger counter gives directly a plot of intensity versus scattering angle 2θ.   [B.E.W.]

Gases. Gases and liquids are found to give rise to x-ray diffraction patterns characterized by one or more halos or interference rings which are usually somewhat diffuse. These diffraction patterns, which are similar to those for glasses and amorphous solids,

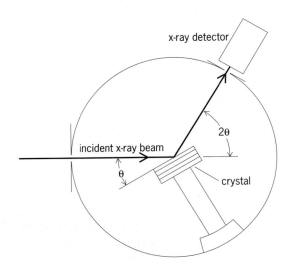

Fig. 1. Schematic of Bragg spectrometer.

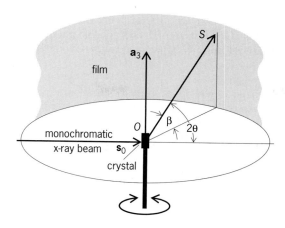

Fig. 2. Schematic of rotation camera.

are due to interference effects depending both upon the electronic distribution of each of the individual atoms or molecules and upon their relative positions in the system.

For monatomic gases the only appreciable interference effects giving rise to a distribution of scattered intensities are those produced by the electronic distribution about each nucleus. These interference effects giving rise to so-called coherent intensities are the result of the interference of the individual waves scattered by electrons in different parts of the atom. The electronic distribution of an atom is described in terms of a characteristic atomic scattering factor which is defined as the ratio of the resultant amplitude scattered by an atom to the amplitude that a free electron would scatter under the same conditions.

Liquids. One cannot, as in the cases of dilute gases and crystalline solids, derive unambiguous, detailed descriptions of liquid structures from diffraction data. Nevertheless, diffraction studies of liquids do provide most useful information. Instead of comparing the experimental intensity distributions with theoretical distributions computed for various models, the experimental results are usually provided in the form of a radial distribution function which specifies the density of atoms or electrons as a function of the radial distance from any reference atom or electron in the system without any prior assumptions about the structure. From the radial distribution function

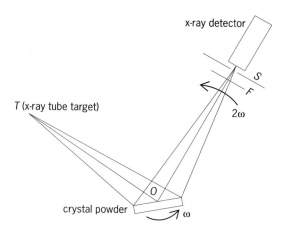

Fig. 3. Schematic representation of the Geiger counter diffractometer for powder samples.

one can obtain (1) the average interatomic distances most frequently occurring in the structure corresponding to the positions of the first, second, and possibly third nearest neighbors; (2) the distribution of distances; and (3) the average coordination number for each interatomic distance.   [L.F.D.]

X-ray optics   By analogy with the science of optics, those aspects of x-ray physics in which x-rays exhibit properties similar to those of light waves. X-ray optics may also be defined as the science of manipulating x-rays with instruments analogous to those used in visible-light optics. These instruments employ optical elements such as mirrors to focus and deflect x-rays, zone plates to form images, and diffraction gratings to analyze x-rays into their spectral components. X-ray optics is important in many fields, including x-ray astronomy, biology, medical research, thermonuclear fusion, and x-ray microlithography. It is essential to the construction of instruments that manipulate and analyze x-rays from synchrotrons and particle storage rings for synchrotron radiation research. *See* GEOMETRICAL OPTICS; OPTICS; PHYSICAL OPTICS; X-RAYS.

When W. C. Roentgen discovered x-rays in 1895, he unsuccessfully attempted to reflect, refract, and focus them with mirrors, prisms, and lenses of various materials. The reason for his lack of success became evident after it was established that x-rays are electromagnetic waves of very short wavelength for which the refractive index of all materials is smaller than unity by a only a small decrement. In addition, x-rays are absorbed by materials. The refractive index can be written as a complex quantity, as in Eq. (1), where $1 - \delta$ represents the real part, n, of the refractive index and β is

$$\bar{n} = 1 - \delta - i\beta \tag{1}$$

the absorption index. These quantities are strongly dependent on the wavelength of the x-rays and the material. X-rays of wavelength about 0.1 nanometer or less are called hard x-rays and are relatively penetrating, while x-rays of wavelength 1–10 nm are less penetrating and are called soft x-rays. Radiation in the wavelength range 10–50 nm, called the extreme-ultraviolet (EUV) region, is very strongly absorbed by most materials. Values of δ remain very small throughout the x-ray and extreme-ultraviolet regions with the consequence that radiation is very weakly refracted by any material. Thus lenses for x-rays would have to be very strongly curved and very thick to achieve an appreciable focusing effect. However, because the absorption index, β, is so high in comparison, such thick lenses would absorb most of the incident radiation, making such lenses impractical. *See* ABSORPTION OF ELECTROMAGNETIC RADIATION; REFRACTION OF WAVES; ULTRAVIOLET RADIATION.

If radiation is incident normally (that is, perpendicular) to a surface between two media of differing refractive index, the fraction of the energy that is reflected is $^1/_4(\delta^2 + \beta^2)$. This is clearly impractically small for a normal-incidence mirror for x-rays. However, useful mirrors can be constructed by using the principle of total reflection. If electromagnetic waves are incident on the boundary between one material of refractive index n_1 and another of lower refractive index n_2, there exists an angle of incidence I_c, called the critical angle, given by Eq. (2). If the angle of incidence (the angle of incident ra-

$$\sin I_c = \frac{n_2}{n_1} \tag{2}$$

diation with respect to the normal to the surface) is greater than this critical angle, all the wave energy is reflected back into the first medium. This phenomenon can be seen when looking upward into an aquarium tank; objects in the tank are reflected in the surface of the water, which acts as a perfect mirror. An analogous situation occurs for x-rays. Since the refractive index for all materials is slightly less than 1, x-rays incident from vacuum (or air) on a polished surface of, say, a metal encounter a lower refractive index and there exists a critical angle given by $\sin I_c = 1 - \delta$. Since δ is very small, I_c is very close to $90°$. In this case the angle of incidence is customarily measured from the tangent to the surface rather than from the normal, and the angle $\theta_c = 90° - I_c$ is termed the angle of glancing (or grazing) incidence. This angle is typically in the range $0.1–1.0°$. *See* REFLECTION OF ELECTROMAGNETIC RADIATION.

Although the reflectivity of surfaces at glancing angles greater than the critical angle is very small, this reflectivity can be enhanced by depositing a stack of ultrathin films having alternately high and low values of δ on the surface. The individual thicknesses of these films is adjusted so that the reflections from each interface add in phase at the top of the stack in exact analogy to the multilayer mirrors used for visible light. However, whereas visible multilayers require film thicknesses of hundreds of nanometers, in the x-ray region the thickness of each film must be between 1 and 100 nm. Such ultrathin films can be made by a variety of vacuum deposition methods, commonly sputtering and evaporation. The response of these artificial multilayers is strongly wavelength-selective. *See* X-RAY DIFFRACTION.

As a coating for glancing-incidence optics, multilayers allow a mirror to be used at a shorter wavelength (higher x-ray energy) for a given glancing angle, increasing the projected area and thus the collection efficiency of the mirror. At wavelengths longer than 3 or 4 nm, multilayer mirrors can be used to make normal-incidence mirrors of relatively high reflecting power. For example, stacks consisting of alternating layers of molybdenum and silicon can have reflectivities as high as 65% at wavelengths of 13 nm and longer. These mirrors have been used to construct optical systems that are exact analogs of mirror optics used for visible light. For example, normal-incidence x-ray tele-scopes have photographed the Sun's hot outer atmosphere at wavelengths of around 18 nm. Multilayer optics at a wavelength of 13.5 nm can be used to perform x-ray microlithography by the projection method to print features of dimensions less than 100 nm.

Crystals are natural multilayer structures and thus can reflect x-rays. Many crystals can be bent elastically (mica, quartz, silicon) or plastically (lithium fluoride) to make x-ray focusing reflectors. These are used in devices such as x-ray spectrometers, electron-beam microprobes, and diffraction cameras to focus the radiation from a small source or specimen on a film or detector. Until the advent of image-forming optics based on mirrors and zone plates, the subject of x-ray diffraction by crystals was called x-ray optics. *See* X-RAY CRYSTALLOGRAPHY.

Zone plates are diffraction devices that focus x-rays and form images. They are diffracting masks consisting of concentric circular zones of equal area, and are alter-nately transparent and opaque to x-rays. Whereas mirrors and lenses focus radiation by adjusting the phase at each point of the wavefront, zone plates act by blocking out those regions of the wavefront whose phase is more than a half-period different from that at the plate center. Thus a zone plate acts as a kind of x-ray lens. Zone-plate microscopy is the most promising candidate method for x-ray microscopy of biological specimens. *See* DIFFRACTION.   [J.H.U.]

X-ray powder methods   Physical techniques used for the identification of substances, and for other types of analyses, principally for crystalline materials in the solid state. In these techniques, a monochromatic beam of x-rays is directed onto a polycrystalline (powder) specimen, producing a diffraction pattern that is recorded on film or with a diffractometer. This x-ray pattern is a fundamental and unique property resulting from the atomic arrangement of the diffracting substance. Different substances have different atomic arrangements or crystal structures, and hence no two chemically distinct substances give identical diffraction patterns. Identification may be made by comparing the pattern of the unknown substance with patterns of known substances in a manner analogous to the identification of people by their fingerprints. The analytical information is different from that obtained by chemical or spectrographic analysis. X-ray identification of chemical compounds indicates the constituent elements and shows how they are combined.

The x-ray powder method is widely used in fundamental and applied research; for instance, it is used in the analysis of raw materials and finished products, in phase-diagram investigations, in following the course of solid-state chemical reactions, and in the study of minerals, ores, rocks, metals, chemicals, and many other types of material. The use of x-ray powder diffraction methods to determine the actual atomic arrangement, which has been important in the study of chemical bonds, crystal physics, and crystal chemistry, is described in related articles. *See* X-RAY CRYSTALLOGRAPHY; X-RAY DIFFRACTION.

There are many types of powder diffractometer available ranging from simple laboratory instruments to versatile and complex instruments using a synchrotron source. Specialized instruments allow recording of diffraction patterns under nonambient conditions, including variable temperature, pressure, and atmosphere. Completely automated equipment for x-ray analysis is available. Most laboratory instruments consist of a high-voltage generator which provides stabilized voltage for the x-ray tube, so that the x-ray source intensity varies by less than 1%. A diffractometer goniometer is mounted on a table in front of the x-ray tube window. Electronic circuits use an x-ray detector to convert the diffracted x-ray photons to measurable voltage pulses, and to record the diffraction data. [R.Je.]

X-ray tube An electronic device used for the generation of x-rays. X-rays are produced in the x-ray tube by accelerating electrons to a high velocity by an electrostatic field and then suddenly stopping them by collision with a solid body, the so-called target, interposed in their path. The x-rays radiate in all directions from the spot on the target where the collisions take place. The x-rays are due to the mutual interaction of the fast-moving electrons with the electrons and positively charged nuclei which constitute the atoms of the target. Depending upon the method used in generating the electrons, x-ray tubes may all be classified in two general groups, gas tubes and high-vacuum tubes. *See* X-RAYS.

In gas tubes electrons are freed from a cold cathode by positive ion bombardment. For the existence of the positive ions a certain gas pressure is required without which the tube will allow no current to pass. Metals, such as platinum and tungsten, are placed in the path of the electron beam to serve as the target. Concave metal cathodes are used to focus the electrons on a small area of the metal target and increase the sharpness of the resulting shadows on the fluorescent screen or the photographic film. Many designs of gas tubes have been built for useful application, particularly in the medical field.

The operational difficulties and erratic behavior of gas x-ray tubes are inherently associated with the gas itself and the positive ion bombardment that takes place during operation. The high-vacuum x-ray tube eliminates these difficulties by using other means of emitting electrons from the cathode. The original type of high-vacuum x-ray tube had a hot tungsten-filament cathode and a solid tungsten target. This tube permitted stable and reproducible operation with relatively high voltages and large masses of metals. A modern commercial hot-cathode high-vacuum x-ray tube is built with a liquid-cooled, copper-backed tungsten target. [E.E.C]

X-rays   X-rays, or roentgen rays, are electromagnetic waves in which periodically variable electric and magnetic fields are perpendicular to each other and to the direction of propagation. Thus they are identical in nature with visible light and all the other types of radiation that constitute the electromagnetic spectrum. In general, x-rays are generated as the result of energy transitions of atomic electrons caused by the bombardment of a material of high atomic weight by high-energy electrons. *See* ELECTROMAGNETIC RADIATION.

Following W. R. Röntgen's discovery of "a new kind of ray" in 1895, other scientists found the essential experimental conditions to prove that x-rays can be polarized, diffracted by crystals, refracted in prisms and in crystals, reflected by mirrors, and diffracted by ruled gratings. *See* X-RAY OPTICS.

The range of x-rays in the electromagnetic spectrum, as excited in x-ray tubes by the bombardment of anode targets by cathode electrons under a high accelerating potential, overlaps the ultraviolet range on the order of 100 nanometers on the long-wavelength side, and the shortest-wavelength limit moves downward as voltages increase. An accelerating potential of 10^9 volts, now readily generated, produces a wavelength of 10^{-15} m (10^{-6} nm). An average wavelength used in research is 0.1 nm, or about 1/6000 the wavelength of yellow light. *See* X-RAY TUBE.

In diffraction, refraction, polarization, and interference phenomena, x-rays, together with all other related radiations, appear to act as waves. In other phenomena—such as the appearance of sharp spectral lines, a definite short-wavelength limit of the continuous "white" spectrum, the shift in wavelength of x-rays scattered by electrons in atoms (Compton effect), and the photoelectric effect—the energy seems to be propagated and transferred in quanta, called photons. *See* COMPTON EFFECT; ELECTRON DIFFRACTION; NEUTRON DIFFRACTION; PHOTOEMISSION; QUANTUM MECHANICS.

Important uses have been found for x-rays in many fields of scientific endeavor, for example, roentgen spectrometry and roentgen diffractometry. Extensive tables of the wavelengths of x-ray emission lines in series (K, L, M, and so on) and so-called absorption edges, characteristic of the chemical elements, afford the necessary information for chemical analyses, exactly as in the case of optical emission spectra and for derivation of theories of atomic structure to account for the origin of spectra. *See* X-RAY CRYSTALLOGRAPHY; X-RAY DIFFRACTION; X-RAY POWDER METHODS.   [G.L.Cl.]

Zeeman effect   A splitting of spectral lines when the light source being studied is placed in a magnetic field. Discovered by P. Zeeman in 1896, the effect furnishes information of prime importance in the analysis of spectra. Each kind of spectral term has its characteristic mode of splitting, and the types of terms are most definitely identified by this property, Furthermore, the effect allows an evaluation of the ratio of charge to mass of the electron and an evaluation of its precise magnetic moment.

The normal Zeeman effect is a splitting into two or three lines, depending on the direction of observation, as shown in the illustration. The light of these components is polarized in ways indicated in the illustration. The normal effect is observed for all lines belonging to singlet systems, those for which the spin quantum number $S = 0$. The change of frequency of the shifted components can be evaluated on classical electromagnetic principles.

The anomalous Zeeman effect is a more complicated type of line splitting, so named because it did not agree with the predictions of classical theory. It occurs for any spectral line arising from a combination of terms of multiplicity greater than one. Since

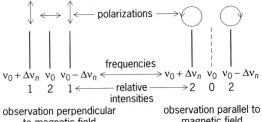

Triplet observed in normal Zeeman effect. v_0 = **unshifted frequency;** Δv_n = **frequency shift.**

multiplicity in spectral lines is caused by the presence of a resultant spin vector S of the electrons, the anomalous effect must be attributed to a nonclassical magnetic behavior of the electron spin.

The quadratic Zeeman effect, which depends on the square of the field strength, is of two kinds. The first results from second-order terms, and the second from the diamagnetic reaction of the electron when revolving in large orbits.

The inverse Zeeman effect is the Zeeman effect of absorption lines. It is closely related to the Faraday effect, the rotation of plane-polarized light by matter situated in a magnetic field. *See* FARADAY EFFECT.

The Zeeman effect in molecules is, in general, so small as to be unobservable, even for molecules which have a permanent magnetic moment. An exception occurs for some light molecules where the magnetic moment is coupled so lightly to the frame of the molecule that it can orient itself freely in the magnetic field just as for atoms.

A clear Zeeman effect also can be observed in many crystals with sharp spectrum lines in absorption or fluorescence. Such crystals are found particularly among rare-earth salts.

The magnetic moment of the nucleus causes a Zeeman splitting in atomic spectra which is of an order of magnitude a thousand times smaller than the ordinary Zeeman effect. This Zeeman effect of the hyperfine structure usually is modified by a nuclear Paschen-Back effect. *See* PASCHEN-BACK EFFECT. [F.A.J.; G.H.Di./W.W.W.]

1
Appendix

2
Contributors

3
Index

BIBLIOGRAPHIES

ACOUSTICS

Berg, R.E., and D.G. Stork, *The Physics of Sound*, 2d ed., 1994.
Blackstock, D.T., *Fundamentals of Physical Acoustics*, 2000.
Crocker, M., *Handbook of Acoustics*, 1998.
Frisk, G.V., *Ocean and Seabed Acoustics: A Theory of Wave Propagation*, 1998.
Kinsler, L.E., et al., *Fundamentals of Acoustics*, 4th ed., 1999.
Levy, M., H.E. Bass, and R.R. Stern (eds.), *Handbook of Elastic Properties of Solids, Liquids, and Gases* (4-vol. set), 2000.
Mason, W.P., R.N. Thurston, and A.D. Pierce (eds.), *Physical Acoustics: Principles and Methods*, vols. 1–25, 1964–1999.
Mechel, F.P., *Formulas of Acoustics*, 2002.
Pierce, A.D., *Acoustics: An Introduction to Its Physical Principles and Applications*, 1981, reprint 1989.
Rossing, T.D., et al., *The Science of Sound*, 3d ed., 2001.
Speaks, C.E., *Introduction to Sound Acoustics for Hearing and Speech Science*, 3d ed., 1999.
Urick, R.J., *Principles of Underwater Sound*, 3d ed., 1983, reprint 1996.

Journals:
Journal of the Acoustical Society of America, American Institute of Physics, monthly.

AERODYNAMICS

Anderson, J.D., Jr., *Fundamentals of Aerodynamics*, 3d ed., 2001.
Bertin Smith, J.J., *Aerodynamics for Engineers*, 4th ed., 2001.
Houghton, E., et al., *Aerodynamics for Engineering Students*, 5th ed., 2001.
Kuethe, A.M., and C.Y. Chow, *Foundations of Aerodynamics: Bases of Aerodynamics Design*, 5th ed., 1997.
McCormick, B.W., *Aerodynamics, Aeronautics and Flight Mechanics*, 2d ed., 1994.
Smith, H.C., *The Illustrated Guide to Aerodynamics*, 2d ed., 1982.
Wegener, P., *What Makes Airplanes Fly?: History, Science and Applications of Aerodynamics*, 2d ed., 1996.

ASTRONOMY

Beatty, J.K., C. Collins Petersen, and A. L. Chaikin (eds.), *The New Solar System*, 4th ed., 1998.
Chaisson, E., and S. McMillan, *Astronomy Today*, 4th ed., 2004.
Foukal, P., *Solar Astrophysics*, 2004.
Fraknoi, A., D. Morrison, and S. Wolff, *Voyages Through the Universe*, 3d ed., 2004.
Golub, L., and J.M. Pasachoff, *Nearest Star: The Exciting Science of Our Sun*, 2002.
Kutner, M., *Astronomy: A Physical Perspective*, 2d ed., 2003.
Pasachoff, J.M., *Astronomy: From the Earth to the Universe*, 6th ed., 2002.
Pasachoff, J.M., and A. Filippenko, *The Cosmos: Astronomy in the New Millennium*, 2d ed., 2004.
Seeds, M., *Foundations of Astronomy*, 8th ed., 2005.
Zeilik, M., *Astronomy: The Evolving Universe*, 9th ed., 2002.

ATOMIC PHYSICS

Adam, S., *Frontiers*, 2000.
Beiser, A., *Concepts of Modern Physics*, 6th ed., 2002.
Brehm, J.J., and W.J. Mullins, *Introduction to the Structure of Matter: A Course in Modern Physics*, 1989.
Cowan, R.D., *The Theory of Atomic Structure and Spectra*, 1981.
Haken, H., H.C. Wolf, and W. Brewer, *The Physics of Atoms and Quanta: Introduction to Experiments and Theory*, 6th rev. ed., 2000.
Ohanian, H.C., *Modern Physics*, 2d ed., 1996.
Serway, R.A., and J. Faughn, *Modern Physics*, 2d ed., 1996.
Thornton, T., and A. Rex, *Modern Physics for Scientists and Engineers*, 2d ed., 1999.
Yang, F., and J.H. Hamilton, *Modern Atomic and Nuclear Physics*, 1996.

Journals:
Physical Review A, American Physical Society, monthly.

CELESTIAL MECHANICS

See Astronomy.

CLASSICAL MECHANICS

Ayra, A.P., *Introduction to Classical Mechanics*, 1997.
Baierlein, R., *Newtonian Dynamics*, 1983.
Barger, V.D., and M.G. Olsson, *Classical Mechanics: A Modern Perspective*, 2d ed., 1995.
Chow, T.L., *Classical Mechanics*, 1995.
Halliday, D., and R. Resnick, *Fundamentals of Physics*, 6th ed., 2002.
José, J.V., and E.J. Saletan, *Classical Dynamics, A Contemporary Approach*, 1998.
Goldstein, H., C.P. Poole, and J.L. Safko, *Classical Mechanics*, 3d ed., 2002.
Reichert J.F., *A Modern Introduction to Mechanics*, 1991.
Shepley, L., and R. Matzner, *Classical Mechanics*, 1991.

Journals:
Archive for Rational Mechanics and Analysis, 16 times a year.
Quarterly Journal of Mechanics and Applied Mathematics, quarterly.

COSMOLOGY

Danielson, D., *The Book of the Cosmos*, 2000.
Goldsmith, D., *Connecting with the Cosmos*, 2002.
Kirshner, R. P., *The Extravagant Universe*, 2002.
Lemonick, M., *Echo of the Big Bang*, 2003.
Rees, M., *Our Cosmic Habitat*, 2001.
Rowan-Robinson, M., *Cosmology*, 4th ed., 2003.
Stern, S. A., *Our Universe: The Thrill of Extragalactic Exploration as Told by Leading Experts*, 2001.

CRYSTALLOGRAPHY

Borchardt-Ott, W., *Crystallography*, 2d ed., 1995.
Engel, P., *Geometric Crystallography: An Axiomatic Introduction to Crystallography*, 1986.
Giacovazzo, C. (ed.), *Fundamentals of Crystallography*, 1992.
Glusker, J.P., and K.N. Trueblood, *Crystal Structure Analysis: A Primer*, 2d ed., 1985.
Guinier, A., *X-Ray Diffraction*, reprint ed., 1994.

Hammond, C., *The Basics of Crystallography and Diffraction* (International Union on Crystallography Texts on Crystallography), 2d ed., 2001.
Klug, H., and L. Alexander, *X-Ray Diffraction Procedures*, 1974.
O'Keefe, M., and B.G. Hyde, *Symmetry and Structures of Crystal*, 1999.
Vainshtein, B.K., *Fundamentals of Crystals*, 2d ed., 1995.
Journals:
Acta Crystallographica, Sections A, B, and D, bimonthly; Section C, monthly.

ELECTRICITY AND MAGNETISM

Cottingham, W.N., and D.A. Greenwood, *Electricity and Magnetism*, 1991.
Duffin, W.J., *Electricity and Magnetism*, 4th ed., 1990.
Fowler, R.J., *Electricity: Principles and Applications*, 4th ed., 1993.
Jackson, J.D., *Classical Electrodynamics*, 3d ed., 1998.
Morley, A., and E. Hughes, *Principles of Electricity*, 5th ed., 1994.
Purcell, E.M., *Electricity and Magnetism*, Berkeley Physics Course, vol. 2, 2d ed., 1985.
Schwartz, M., *Principles of Electrodynamics*, reprint, 1987.
Journals:
Journal of Magnetism and Magnetic Materials, 3 issues per month.

ELECTROMAGNETIC FIELDS AND RADIATION

Guru, B., and H. Hiziroğlu, *Electromagnetic Field Theory Fundamentals*, 2d ed., 2004.
Inan, U., and A. Inan, *Engineering Electromagnetics*, 1999.
Lorrain, P., *Electromagnetic Fields and Waves*, 3d ed., 1995.
Paul, C.R., S.A. Nasar, and K. Whites, *Introduction to Electromagnetic Fields*, 3d ed., 1997.
Popović, Z., and B.D. Popović, *Introductory Electromagnetics*, 2000.
Rohan, P., *Introduction to Electromagnetic Wave Propagation*, 1991.
Sadiku, M., *Elements of Electromagnetics* (Oxford Series in Electrical and Computer Engineering), 3d ed., 2000.
Smith, G.S., *An Introduction to Classical Electromagnetic Radiation*, 1997.

ELEMENTARY PARTICLES

Alliday, J., *Quarks, Leptons and the Big Bang*, 2d ed., 2002.
Coughlan, G.D., and J.E. Dodd, *The Ideas of Particle Physics: An Introduction for Scientists*, 2d ed., 1991.
Griffiths, D., *Introduction to Elementary Particles*, 1987.
Halzen, F., and A.D. Martin, *Quarks and Leptons: An Introductory Course in Modern Particle Physics*, 1984.
Huang, K., *Quarks, Leptons and Gauge Fields*, 1992.
Hughes, I.S., *Elementary Particles*, 3d ed., 1991.
Kane, G., *Modern Elementary Particle Physics: The Fundamental Particle and Forces*, updated ed., 1993.
Perkins, D.H., *An Introduction to High Energy Physics*, 4th ed., 2000.
Riazuddin, F., *A Modern Introduction to Particle Physics*, 1994.
Rolnick, W.B., *The Fundamental Particles and Their Interactions*, 1994.
Weinberg, S., *The Discovery of Subatomic Particles*, revised, 2d ed., 2003.

Journals:
Nuclear Physics, Section B, 81 issues per year.
Physical Review D, American Physical Society, semimonthly.

FLUID MECHANICS

Currie, I.G., *Fundamental Mechanics of Fluids* (Mechanical Engineering), 3d ed., 2002.
Douglas, J.F., J.M. Gaisorek, and J.A. Swaffield, *Fluid Mechanics*, 3d ed., 1996.
Evett, J., and C. Liu, *Fundamentals of Fluid Mechanics*, 1987.
Fay, J.A., *Introduction to Fluid Mechanics*, 1994.
Fox, R.W., A.T. McDonald, and P.J. Pritchard, *Introduction to Fluid Mechanics*, 6th ed., 2003.
Franzini, J.B., and J. E. Finnemore, *Fluid Mechanics with Engineering Applications*, 10th ed., 2001.
C.T. Crowe, D.F. Elger, and J.A. Roberson, *Engineering Fluid Mechanics*, 7th ed., 2000.
Kindu, P.K., and I.M. Cohen, *Fluid Mechanics*, 3d ed., 2004.
Munsor, B.R., D.F. Young, and T.H. Dkishi, *Fundamentals of Fluid Mechanics*, 4th ed., 2002.
Shames, I.H., *Mechanics of Fluids*, 4th ed., 2002.
Streeter, V.L., E.B. Wylie, and K.W. Bedford, *Fluid Mechanics*, 9th ed., 1997.
White, F.M., *Fluid Mechanics with Student Resources* (CD-ROM), 5th ed., 2002.

Journals:
Applied Scientific Research, Central National Organization for Applied Scientific Research in the Netherlands, quarterly.
Physics of Fluids, American Institute of Physics, monthly.

LASERS

P.W. Milonni., and J.H. Eberly, *Lasers* (Wiley Series in Pure and Applied Optics), 1988.
Hecht, J.,*The Laser Guidebook*, 2d ed., 1992.
Hecht, J., *Understanding Lasers: An Entry-Level Guide* (IEEE Press Understanding Science & Technology Series), 2d ed., 1993.
Meschede, D., *Optics Light and Lasers*, 2004.
Saleh, B.E.A., and M.C. Teich, *Fundamentals of Photonics*, 1991.
Shimoda, K., *Introduction to Laser Physics*, 2d ed., 1991.
Siegman, A.E., *Lasers*, 1986.
Svelto, O., et al., *Principles of Lasers*, 4th ed., 1998.
Verdeyen, J.T., *Laser Electronics*, 3d ed., 1994.
Yariv, A., *Optical Electronics*, 5th ed., 1995.
Young, M., *Optics and Lasers: Including Fibers and Optical Waveguides* (Advanced Texts in Physics), 5th rev. ed., 2000.

Journals:
Applied Optics, Optical Society of America, American Institute of Physics, 36 issues per year.
Applied Physics B, Deutsche Physikalische Gesellschaft, 12 issues per year.
Journal of the Optical Society of America, A, B, American Institute of Physics, monthly.

LOW-TEMPERATURE PHYSICS

Betts, D.S., et al., *Introduction to Millikelvin Technology* (Cambridge Studies in Low Temperature Physics), 1989.

Brewer, D.F., et al. (eds.), *Progress in Low Temperature Physics*, vols. 1–14, 1956–1995.

Clark, A.F., et al. (eds.), *Advances in Cryogenic Engineering*, vols. 1–14, 1956–1996.

Dahl, P.F., *Superconductivity: Its Historical Roots and Development from Mercury to the Ceramic Oxides*, 1992.

De Gennes, P.G., *Superconductivity of Metals and Alloys* (Advanced Book Classics), 1999.

Flynn, T.M., *Cryogenic Engineering*, 1996.

Khalatnikov, I.M., *Introduction to the Theory of Superfluidity* (Advanced Book Classics), 1989.

Pobell, F., *Matter and Methods at Low Temperatures*, 2d ed., 1996.

Poole, C.P., et al., *Superconductivity*, 1996.

Richardson, R.C., and E.N. Smith, *Experimental Techniques in Condensed Matter Physics at Low Temperatures* (Advanced Book Classics), 1998.

Schrieffer, J.R., *Theory of Superconductivity* (Advanced Book Classics), 1999.

Tilley, D.R., and J. Tilley, *Superfluidity and Superconductivity*, 3d ed., 1990.

Tinkham, M., *Introduction to Superconductivity* (Dover Books on Physics), 2d ed., 2004.

Wilks, J., and D.S. Betts, *An Introduction to Liquid Helium*, 2d reprint ed., 1990.

Journals:
Cryogenics, monthly.

Journal of Low Temperature Physics, 24 issues per year.

MAGNETISM

See Electricity and magnetism.

MECHANICS

See Classical mechanics; Fluid mechanics; Quantum mechanics; Statistical mechanics.

MOLECULAR PHYSICS

Atkins, P.W., *Atkins' Molecules*, 2d ed., 2003.

Atkins, P.W., and R.S. Friedman, *Molecular Quantum Mechanics*, 3d ed., 1999.

Banwell, C.N., and E.M. McCash, *Fundamentals of Molecular Spectroscopy*, 4th ed., 1995.

Bernath, P.F., *Spectra of Atoms and Molecules* (Topics in Physical Chemistry), 1995.

Bransden, B.H., and C.J. Joachain, *Physics of Atoms and Molecules*, 2d ed., 2003.

Erkoc, S., and T. Uzer, *Lecture Notes on Atomic and Molecular Physics*, 1996.

R. McWeeny, *Methods of Molecular Quantum Mechanics* (Theoretical Chemistry), 2d ed., 1992.

Richards, W.G., and P.R. Scott, *Structure and Spectra of Molecules*, 1985.

Struve, W., *Fundamentals of Molecular Spectroscopy*, 1989.

Svanberg, S., *Atomic and Molecular Spectroscopy: Basic Aspects and Practical Applications*, 3d ed., 2001.

Journals:
Journal of Molecular Structure, 48 issues per year.
Molecular Physics, 18 issues per year.

NUCLEAR PHYSICS

Burcham, W.E., and M. Jobes, *Nuclear and Particle Physics*, 1995.
Cottingham, W.N., and D.A. Greenwood, *An Introduction to Nuclear Physics*, 2d ed., 2001.
Das, A., and T. Ferbel, *Introduction to Nuclear and Particle Physics*, 2d ed., 2004.
Eisenberg, J.M., and W. Greiner, *Nuclear Theory*, vol. 1, 3d ed., 1988; vol. 2, 3d ed., 1988; and vol. 3, 2d ed., 1976.
Enge, H.A., and R.P. Redwine, *Introduction to Nuclear Physics*, 2d ed., 1999.
Krane, K.S., *Introductory Nuclear Physics*, 1987.
Lilley, J.S., *Nuclear Physics*, 2001.
Walecka, J.D., *Theoretical Nuclear and Subnuclear Physics* (Oxford Studies in Nuclear Physics), 1995.
Wong, S.S.M., *Introductory Nuclear Physics*, 2d ed., 1999.

Journals:
Annual Review of Nuclear and Particle Science, annually.
Nuclear Physics, Section A, 68 issues per year.
Physical Review C, American Physical Society, monthly.

OPTICS

Born, M., and E. Wolf, *Principles of Optics*, 7th ed., 1999.
Hecht, E., and K. Guardino, *Optics*, 4th ed., 2001.
Jenkins, F.A., and H.E. White, *Fundamentals of Optics*, 4th ed., 2001.
Meyer-Arendt, J., *Introduction to Classical and Modern Optics*, 4th ed., 1994.
Optical Society of America, *Handbook of Optics*, 2d ed., 4 vols., 1995–2001.
Pedrotti, F.J., and L.S., Pedrotti, *Introduction to Optics*, 2d ed., 1992.
Smith, W.J., *Modern Optical Engineering*, 3d ed., 2000.
Yu, F.T., and X. Yang, *Introduction to Optical Engineering*, 1997.

Journals:
Applied Optics, Optical Society of America, American Institute of Physics, 36 issues per year.
Journal of the Optical Society of America, A, B, American Institute of Physics, monthly.

PHYSICAL ELECTRONICS

Enderlein, R., and N.J. Horing, *Fundamentals of Semiconductor Physics and Devices*, 1997.
Ferendeci, A.M., *Physical Foundations of Solid State and Electron Devices*, 1991.
Ghandi, S.K., *VLSI Fabrication Principles: Silicon and Gallium Arsenide*, 2d ed., 1994.
Grasserbauer, M., and H.W. Werner (eds.), *Analysis of Microelectronic Materials and Devices*, 1991.
Neaman, D., *Semiconductor Physics and Devices*, 3d ed., 2002.
Ng, K.K., *Complete Guide to Semiconductor Devices*, 2d ed., 2002.
Pierret, R.F., *Semiconductor Device Fundamentals*, 1995.
Seeger, K., *Semiconductor Physics: An Introduction*, 8th ed., 2002.

Singh, J., *Semiconductor Devices: An Introduction*, 1994.
Sze, S.M., *Semiconductor Devices: Physics and Technology*, 2d ed., 2001.
Yu, P.Y., and M. Cardona, *Fundamentals of Semiconductors*, 3d ed., 2001.

Journals:

Electrical and Electronics Abstracts (Science Abstracts, Section B), INSPEC, Institution of Electrical Engineers, monthly.
IEEE Spectrum, Institute of Electrical and Electronics Engineers, Inc., monthly.
IEEE Transactions on Electron Devices, Institute of Electrical and Electronics Engineers, Electron Devices Society, monthly.
Sold State Technology, monthly.

PHYSICS

Beiser, A., *Concepts of Modern Physics*, 6th ed., 2002.
Brown, L., A. Pais, and B. Pippard, *Twentieth Century Physics*, 1995.
Bueche, F., *Principles of Physics* (McGraw-Hill Schaum's Outline Series in Science), rev. 6th ed., 1994.
Halliday, D., R. Resnick, and J. Walker, *Fundamentals of Physics*, 6th ed., 2002.
Hartle, J.B., *Gravity: An Introduction to Einstein's General Relativity*, 2003.
Lightman, A.P., *Great Ideas in Physics*, 3d ed., 2000.
Serway, R.A., *College Physics* (with PhysicsNow), 6th ed., 2003.
Young, H.D., and R.A. Freedman, *University Physics: Solutions Manual*, 11th ed., 2003.

Journals:

American Journal of Physics, American Institute of Physics, monthly.
Annals of Physics, 18 issues per year.
Journal of Applied Physics, American Institute of Physics, monthly.
Journal of Physics, A, B, D, and G, Institute of Physics, Section G, monthly; other sections, semimonthly.
Physical Review, American Physical Society, American Institute of Physics, Section B, 4 issues per month; Section D, semimonthly; Sections A, C, E, monthly.
Physical Review Letters, American Physical Society, American Institute of Physics, weekly.
Physics Abstracts (Science Abstracts, Section A), INSPEC, Institution of Electrical Engineers, biweekly.
Physical Letters, Section A, 78 issues per year; Section B, 108 issues per year.
Physics Reports, 90 issues per year.
Physics Today, American Institute of Physics, monthly.
Proceedings of the Royal Society of London, Series A: Mathematical, Physical and Engineering Sciences, monthly.
Reviews of Modern Physics, American Physical Society, American Institute of Physics, quarterly.

PLASMA PHYSICS

Chen, F.F., *Introduction to Plasma Physics and Controlled Fusion* (Plasma Physics), 2d ed., 1984.
Dendy, R.O. (ed.), *Plasma Physics: An Introductory Course*, new edited ed., 1995.
Goldston, R., and P.H. Rutherford, *Introduction to Plasma Physics* (Plasma Physics Series), 1995.

Hutchinson, I.H., *Principles of Plasma Diagnostics*, 2d ed., 2002.
Leontovich, M.A., and B.B. Kadomstev (eds.), *Reviews of Plasma Physics*, vols. 1–19, 1965–1996.
Manos, D.M., and D.L. Flam (eds.), *Plasma Etching*, 1989.
Parks, G.K., *Physics of Space Plasmas*, 2d ed., 2004.
Stix, T.H., *Waves in Plasmas*, 1992.
Sturrock, P.A., *Plasma Physics: An Introduction to the Theory of Astrophysical, Geophysical, and Laboratory Plasmas*, new edited ed., 1994.
Wesson, J., et al., *Tokamaks* (The International Series on Monographs on Physics), 3d ed., 2004.

Journals:
IEEE Transactions on Plasma Science, Institute of Electrical and Electronics Engineers, Plasma Sciences Society, bimonthly.
Plasma Physics and Controlled Fusion, Institute of Physics, monthly.

QUANTUM MECHANICS

Bohm, A., *Quantum Mechanics: Foundation and Applications*, 3d ed., 1994.
Dirac, P.A., *Principles of Quantum Mechanics*, 4th ed., 1982.
Eisberg, R., and R. Resnick, *Quantum Physics: Of Atoms, Molecules, Solids, Nuclei and Particles*, 2d ed., 1985.
Gottfried, K., and T.M. Yan, *Quantum Mechanics: Fundamentals*, 2d ed., 2004.
Liboff, R., *Introductory Quantum Mechanics*, 4th ed., 2002.
Mandl, F., *Quantum Mechanics*, 1992, reprint, 1994.
Morrison, M., *Understanding Quantum Mechanics*, 1996.
Ohanian, H.C., *Principles of Quantum Mechanics*, 1989.
Park, D., *Introduction to the Quantum Theory* (International Series in Pure and Applied Physics), 3d ed., 1992.
Polkinghorne, J., *Quantum Mechanics: A Very Short Introduction*, 2002.
Shankar, R., *Principles of Quantum Mechanics*, 2d ed., 1994.
Zettili, N., *Quantum Mechanics: Concepts and Applications*, 2001.

RELATIVITY

Bohm, D., *The Special Theory of Relativity*, reprint, 1996.
Hartle, J.B., *Gravity: An Introduction to Einstein's General Relativity*, 2003.
Misner, C., K. Thorne, and J. Wheeler, *Gravitation*, 1973.
Rindler, W., *Essential Relativity*, rev. ed., reprint 2000.
Schutz, B., *Gravity From the Ground Up*, 2003.
Schwarz, P.M., and J.H. Schwarz, *Special Relativity: From Einstein to Strings*, 2004.
Stephani, H., *General Relativity: An Introduction to Special and General Relativity*, 4th ed., 2004.
Taylor, E.F., and J. A. Wheeler, *Spacetime Physics: Introduction to Special Relativity*, 2d ed., 1992.
Wald, R., *General Relativity*, 1984.
Will, C.M., *Was Einstein Right? Putting General Relativity to the Test*, 2d rev. ed., 1993.

Journals:
General Relativity and Gravitation, International Committee on General *Relativity and Gravitation*, monthly.

SOLID-STATE PHYSICS

Chaikin, P.M., and T.C. Lubensky, *Principles of Condensed Matter Physics*, 1995, paperback 2000.
Ehrenreich, H., et al. (eds.), *Solid State Physics*, vols. 1–56, 1955–2001.
Hook, J.R., and H.E. Hall, *Solid State Physics*, 2d ed., 1995.
Ibach, H., and H. Luth, *Solid State Physics: An Introduction to Principles of Materials Science* (Advanced Texts in Physics), 3d ed., 2003.
Kittel, C., *Introduction to Solid State Physics*, 7th ed., 1995.
Marder, M.P., *Condensed Matter Physics*, 2000.
Myers, H.P., *Introductory Solid State Physics*, 2d ed., 1997.
Omar, M.A., *Elementary Solid State Physics*, 1996.
Rudden, M.N., and J. Wilson, *Elements of Solid State Physics*, 2d ed., 1993.
Tanner, B.K., *Introduction to the Physics of Electrons in Solids*, 1995.

Journals:
Applied Physics A, Deutsche Physikalische Gesellschaft, 12 issues per year.
Journal of Physics: Condensed Matter, Institute of Physics, 50 issues per year.
Journal of Physics and Chemistry of Solids, monthly.
Physica B, 60 issues per year.
Physical Review B, American Institute of Physics, 4 issues per month.
Solid State Communications, 48 issues per year.

SPECTROSCOPY

Andrews, D.L., and A.A. Demidov (eds.), *An Introduction to Laser Spectroscopy*, 1995.
Bernath P.F., *Spectra of Atoms and Molecules* (Topics in Physical Chemistry), 1995.
Briggs, D., and M.P. Seah (eds.), *Practical Surface Analysis: Auger and X-Ray Photoelectron Spectroscopy*, 2d ed., 1990.
De Hoffman, E., and V. Stroobant, *Mass Spectrometry: Principles and Applications*, 2d ed., 2002.
Demtröder, W., *Laser Spectroscopy*, 3d ed., 1996.
Dickson, D.P., and F.J. Berry (eds.), *Mössbauer Spectroscopy*, 1987.
Graybeal, W.D., *Molecular Spectroscopy*, rev. ed., 1993.
Günther, H., *NMR Spectroscopy*, 2d ed., 1995.
Hollas, J.M., *Modern Spectroscopy*, 4th ed., 2004.
Hüfner, S., *Photoelectron Spectroscopy*, 3d ed., 2003.
Laserna, J.J. (ed.), *Modern Techniques in Raman Spectroscopy*, 1996.
Schmidt, V., et. al., *Electron Spectrometry of Atoms Using Synchrotron Radiation*, 1997.
Smith, B.C., *Fundamentals of Fourier Transform Infrared Spectroscopy*, 1995.
Tsuji, K., J. Injuk, and R. Van Grieken (eds.), *X-Ray Spectrometry: Recent Technological Advances*, 2004.

Journals:
Applied Spectroscopy, Society for Applied Spectroscopy, monthly.
Journal of Magnetic Resonance Spectroscopy, 4 copies per year.
Spectrochimica Acta, Parts A and B, monthly.

STATISTICAL MECHANICS

Baracca, A., and R. Livi, *Statistical Mechanics: Foundations, Problems, Perspectives*, 1998.

Betts, D.S., and R.E. Turner, *Introductory Statistical Mechanics*, 1992.

Bowley, R., and M. Sanchez, *Introductory Statistical Mechanics*, 2d ed., 1999.

Garrod, C., *Statistical Mechanics and Thermodynamics* (Book and Windows CE-ROM IBM ed.), 1995.

Kubo, R., *Statistical Mechanics*, 1990.

McQuarrie, D.A., *Statistical Mechanics*, 2000.

Pathria, R.K., *Statistical Mechanics*, 2d ed., 1996.

Ter Harr, D., *Elements of Statistical Mechanics*, 3d ed., 1995.

Zwanzig, R., *Nonequilibrium Statistical Mechanics*, 2001.

Journals:
Journal of Statistical Physics, 12 issues per year.
Physica A, 52 issues per year.

THEORETICAL PHYSICS

Arfken, G.B., and H.J. Weber (eds.), *Mathematical Methods for Physicists*, 5th ed., 2000.

Courant, R., and D. Hilbert, *Methods of Mathematical Physics*, vol 1., 1989.

Honerkamp, J., H. Pollack, and H. Romer, *Theoretical Physics: A Classical Approach*, 1993.

Lam, L. (ed.), *Nonlinear Physics for Beginners: Fractals, Chaos, Solitons, Pattern Formation, Cellular Automata, and Complex Systems*, 1998.

Landau, L.D., et al., *Course of Theoretical Physics*, 9 vols., 1959–1987.

Reed, M., and B. Simon, *Methods of Modern Mathematical Physics*, 4 vols., 1975–1980.

Thirring, W.E., *Course in Mathematical Physics*, 4 vols., 1997.

Journals:
Communications in Mathematical Physics, 18 issues per year.
Journal of Mathematical Physics, American Institute of Physics, monthly.

THERMODYNAMICS AND HEAT

Black, W.Z., and J.G. Hartley, *Thermodynamics* (English/SI Version), 3d ed., 1997.

Cengel, Y.A., and M.A. Boles, *Thermodynamics: An Engineering Approach*, 4th ed., 2001.

Gocken, N.A., and R.G. Reddy, *Thermodynamics*, 2d ed., 1999.

Holman, J.P., *Thermodynamics*, 4th ed., 1988.

Modell, M., and J.W. Tester, *Thermodynamics and Its Applications*, 3d ed., 1996.

Moran, M.J., and H.N. Shapiro, *Fundamentals of Engineering Thermodynamics*, 5th ed., 2003.

Soontag, R.E., C. Borgnakke, and G. J. Van Wylen, *Fundamentals of Thermodynamics*, 6th ed., 2002.

Wark, K., and D.E. Richards, *Thermodynamics*, 6th ed., 1999.

Wood, S.E., and R. Battino, *Thermodynamics of Chemical Systems*, 1990.

Zemansky, M.W., and R.H. Dittman, *Heat and Thermodynamics*, 7th ed., 1997.

Base units of the International System

Quantity	Name of unit	Unit symbol
length	meter	m
mass	kilogram	kg
time	second	s
electric current	ampere	A
temperature	kelvin	K
luminous intensity	candela	cd
amount of substance	mole	mol

Prefixes for units in the International System

Prefix	Symbol	Power	Example	Prefix	Symbol	Power	Example
yotta	Y	10^{24}		deci	d	10^{-1}	
zetta	Z	10^{21}		centi	c	10^{-2}	centimeter (cm)
exa	E	10^{18}		milli	m	10^{-3}	milligram (mg)
peta	P	10^{15}		micro	μ	10^{-6}	microgram (μg)
tera	T	10^{12}	terawatt (TW)	nano	n	10^{-9}	nanosecond (ns)
giga	G	10^{9}	gigawatt (GW)	pico	p	10^{-15}	picofarad (pF)
mega	M	10^{6}	megahertz (MHz)	femto	f	10^{-15}	femtosecond (fs)
kilo	k	10^{3}	kilometer (km)	atto	a	10^{-18}	
hecto	h	10^{2}		zepto	z	10^{-21}	
deka	da	10^{1}		yocto	y	10^{-24}	

Derived units of the International System*

Quantity	Name of unit	Unit symbol, or unit expressed in terms of other SI units	Unit expressed in terms of SI base units
plane angle	radian	rad	$m/m = 1$
solid angle	steradian	sr	$m^2/m^2 = 1$
area	square meter		m^2
volume	cubic meter		m^3
frequency	hertz	Hz	s^{-1}
density	kilogram per cubic meter		kg/m^3
velocity	meter per second		m/s
angular velocity	radian per second	rad/s	$m/(m \cdot s) = s^{-1}$
acceleration	meter per second squared		m/s^2
angular acceleration	radian per second squared	rad/s²	$m/(m \cdot s^2) = s^{-2}$
volumetric flow rate	cubic meter per second		m^3/s
force	newton	N	$kg \cdot m/s^2$
surface tension	newton per meter, joule per square meter	N/m, J/m²	kg/s^2
pressure	pascal, newton per square meter	Pa, N/m²	$kg/(m \cdot s^2)$
viscosity, dynamic	pascal-second, newton-second per square meter	Pa · s, N · s/m²	$kg/(m \cdot s)$
viscosity, kinematic	meter squared per second		m^2/s
work, torque, energy, quantity of heat	joule, newton-meter, watt-second	J, N · m, W · s	$kg \cdot m^2/s^3$
power, heat flux	watt, joule per second	W, J/s	$kg \cdot m^2/s^3$
heat flux density	watt per square meter	W/m²	kg/s^3
volumetric heat release rate	watt per cubic meter	W/m³	$kg/(m \cdot s^3)$
heat transfer coefficient	watt per square meter kelvin	W/(m² · K)	$kg/(s^3 \cdot K)$
heat capacity (specific)	joule per kilogram kelvin	J/(kg · K)	$m^2/(s^2 \cdot K)$
capacity rate	watt per kelvin	W/K	$kg \cdot m^2/(s^3 \cdot K)$
thermal conductivity	watt per meter kelvin	W/(m · K), $\dfrac{J \cdot m}{s \cdot m^2 \cdot K}$	$kg \cdot m/(s^3 \cdot K)$
quantity of electricity	coulomb	C	$A \cdot s$
electromotive force	volt	V, W/A	$kg \cdot m^2/(A \cdot s^3)$
electric field strength	volt per meter	V/m	$kg \cdot m/(A \cdot s^3)$
electric resistance	ohm	Ω, V/A	$kg \cdot m^2/(A^2 \cdot s^3)$
electric conductance	siemens	S, A/V	$A^2 \cdot s^3/(kg \cdot m^2)$
electric conductivity	ampere per volt meter	A/(V · m)	$A^2 \cdot s^3/(kg \cdot m^3)$
electric capacitance	farad	F, A · s/V	$A^2 \cdot s^4/(kg \cdot m^2)$
magnetic flux	weber	Wb, V · s	$kg \cdot m^2/A \cdot s^2)$
inductance	henry	H, V · s/A	$kg \cdot m^2/(A^2 \cdot s^2)$
magnetic permeability	henry per meter	H/m	$kg \cdot m/(A^2 \cdot s^2)$
magnetic flux density	tesla, weber per square meter	T, Wb/m²	$kg/(A \cdot s^2)$
magnetic field strength	ampere per meter		A/m
magnetomotive force	ampere		A
luminous flux	lumen	lm, cd · sr	$cd \cdot m^2/m^2 = cd$
luminance	candela per square meter		cd/m^2
illumination	lux, lumen per square meter	lx, lm/m², cd · sr/m²	$cd \cdot m^2/m^4 = cd/m^2$
activity (of radionuclides)	becquerel	Bq	s^{-1}
absorbed dose	gray	Gy, J/kg	m^2/s^2
dose equivalent	sievert	Sv, J/Kg	m^2/s^2
catalytic activity	katal	kat	mol/s

*The degree Celsius (°C) is also a derived unit of the International System; see "Some common units defined in terms of SI units."

Some common units defined in terms of SI units

Quantity	Name of unit	Unit symbol	Definition of unit
length	inch	in.	2.54×10^{-2} m
mass	pound (avoirdupois)	lb	0.45359237 kg
force	kilogram-force	kgf	9.80665 N
pressure	atmosphere	atm	101325 Pa
pressure	torr	torr	(101325/760) Pa
pressure	conventional millimeter of mercury*	mmHg	$13.5951 \times 980.665 \times 10^{-2}$ Pa
energy	kilowatt-hour	kWh	3.6×10^6 J
energy	thermochemical calorie	cal	4.184 J
energy	international steam table calorie	cal_{IT}	4.1868 J
thermodynamic temperature (T)	degree Rankine	°R	(5/9) K
customary temperature (t)	degree Celsius	°C	$t(°C) = T(K) - 273.15$
customary temperature (t)	degree Fahrenheit	°F	$t(°F) = [1.8 \times t(°C)] + 32 = T(°R) - 459.67$
radioactivity	curie	Ci	3.7×10^{10} Bq
energy†	electronvolt	eV	$eV = 1.60218 \times 10^{-19}$ J
mass†	unified atomic mass unit	u	$u = 1.66054 \times 10^{-27}$ kg

*The conventional millimeter of mercury, symbol mmHg (not mm Hg), is the pressure exerted by a column exactly 1 mm high of a fluid of density exactly 13.5951 g cm^{-3} in a place where the gravitational acceleration is exactly 980.665 cm · s^{-2}. The mmHg differs from the torr by less than 2×10^{-7} torr.

†These units defined in terms of the best available experimental values of certain physical constants may be converted to SI units. The factors for conversion of these units are subject of change in the light of new experimental measurements of the constants involved.

Conversion factors for the U.S. Customary System, metric system, and International System

A. Units of length

Units	cm	m	in.	ft	yd	mi
1 cm =	1	0.01*	0.3937008	0.03280840	0.01093613	6.213712×10^{-6}
1 m =	100.	1	39.37008	3.280840	1.093613	6.213712×10^{-4}
1 in. =	2.54*	0.0254	1	0.08333333...	0.02777777....	1.578283×10^{-5}
1 ft =	30.48	0.3048	12.*	1	0.3333333...	$1.893939... \times 10^{-4}$
1 yd =	91.44	0.9144	36.	3.*	1	$5.681818... \times 10^{-4}$
1 mi =	1.609344×10^{5}	1.609344×10^{3}	6.336×10^{4}	5280.*	1760.	1

B. Units of area

Units	cm^2	m^2	$in.^2$	ft^2	yd^2	mi^2
1 cm^2 =	1	10^{-4}*	0.1550003	1.076391×10^{-3}	1.195990×10^{-4}	3.861022×10^{-11}
1 m^2 =	10^4	1	1550.003	10.76391	1.195990	3.861022×10^{-7}
1 $in.^2$ =	6.4516*	6.4516×10^{-4}	1	$6.944444... \times 10^{-3}$	7.716049×10^{-4}	2.490977×10^{-10}
1 ft^2 =	929.0304	0.09290304	144.*	1	0.7777777....	3.587007×10^{-8}
1 yd^2 =	8361.273	0.8361273	1296.	9.*	1	3.228306×10^{-7}
1 mi^2 =	2.589988×10^{10}	2.589988×10^{6}	4.014490×10^{9}	2.78784×10^{7}*	3.0976×10^{6}	1

C. Units of volume

Units	m^3	cm^3	liter	$in.^3$	ft^3	qt	gal
1 m^3	= 1	10^6	10^3	6.102374×10^4	35.31467	1.056688×10^3	264.1721
1 cm^3	= 10^{-6}	1	10^{-3}	0.06102374	3.531467×10^{-5}	1.056688×10^{-3}	2.641721×10^{-4}
1 liter	= 10^{-3}	1000.*	1	61.02374	0.03531467	1.056688	0.2641721
1 $in.^3$	= 1.638706×10^{-5}	16.38706*	0.01638706	1	5.787037×10^{-4}	0.01731602	4.329004×10^{-3}
1 ft^3	= 2.831685×10^{-2}	28316.85	28.31685	1728.*	1	2.992208	7.480520
1 qt	= 9.463529×10^{-4}	946.3529	0.9463529	57.75	0.0342014	1	0.25
1 gal (U.S.)	= 3.785412×10^{-3}	3785.412	3.785412	231.*	0.1336806	4.*	1

D. Units of mass

Units	g	kg	oz	lb	metric ton	ton
1 g	= 1	10^{-3}	0.03527396	2.204623×10^{-3}	10^{-6}	1.102311×10^{-6}
1 kg	= 1000.	1	35.27396	2.204623	10^{-3}	1.102311×10^{-3}
1 oz (avdp)	= 28.34952	0.02834952	1	0.0625	2.834952×10^{-5}	3.125×10^{-5}
1 lb (avdp)	= 453.5924	0.4535924	16.*	1	4.535924×10^{-4}	$5. \times 10^{-4}$
1 metric ton	= 10^6	1000.*	35273.96	2204.623	1	1.102311
1 ton	= 907184.7	907.1847	32000.	2000.*	0.9071847	1

Conversion factors for the U.S. Customary System, metric system, and International System (cont.)

E. Units of density

Units	$g \cdot cm^{-3}$	$g \cdot L^{-1}, kg \cdot m^{-3}$	$oz \cdot in.^{-3}$	$lb \cdot in.^{-3}$	$lb \cdot ft^{-3}$	$lb \cdot gal^{-1}$
$1\ g \cdot cm^{-3}$ = 1	1000.	0.5780365	0.03612728	62.42795	8.345403
$1\ g \cdot L^{-1}, kg \cdot m^{-3}$ = 10^{-3}	1	5.780365×10^{-4}	3.612728×10^{-5}	0.06242795	8.345403×10^{-3}
$1\ oz \cdot in.^{-3}$ = 1.729994	1729.994	1	0.0625	108.	14.4375
$1\ lb \cdot in.^{-3}$ = 27.67991	27679.91	16.	1	1728.	231.
$1\ lb \cdot ft^{-3}$ = 0.01601847	16.01847	9.259259×10^{-3}	5.787037×10^{-4}	1	0.1336806
$1\ lb \cdot gal^{-1}$ = 0.1198264	119.8264	4.749536×10^{-3}	4.329004×10^{-3}	7.480519	1

F. Units of pressure

Units	$Pa, N \cdot m^{-2}$	$dyn \cdot cm^{-2}$	bar	atm	$kgf \cdot cm^{-2}$	mmHg (torr)	in. Hg	$lbf \cdot in.^{-2}$
$1\ Pa, 1\ N \cdot m^{-2}$ = 1	10	10^{-5}	9.869233×10^{-6}	1.019716×10^{-5}	7.500617×10^{-3}	2.952999×10^{-4}	1.450377×10^{-4}
$1\ dyn \cdot cm^{-2}$ = 0.1	1	10^{-6}	9.869233×10^{-7}	1.019716×10^{-6}	7.500617×10^{-4}	2.952999×10^{-5}	1.450377×10^{-5}
1 bar = 10^{5*}	10^{6}	1	0.9869233	1.019716	750.0617	29.52999	14.50377
1 atm = 101325*	1013250	1.01325	1	1.033227	760.	29.92126	14.69595
$1\ kgf \cdot cm^{-2}$ = 98066.5	980665	0.980665	0.9678411	1	735.5592	28.95903	14.22334
1 mmHg (torr) = 133.3224	1333.224	1.333224×10^{3}	1.315789×10^{-3}	1.3595510×10^{-3}	1	0.03937008	0.01933678
1 in. Hg = 3386.388	33863.88	0.03386388	0.03342105	0.03453155	25.4	1	0.4911541
$1\ lbf \cdot in.^{-2}$ = 6894.757	68947.57	0.06894757	0.06804596	0.07030696	51.71493	2.036021	1

G. Units of energy

Units	g mass (energy equiv)	J	eV	cal	cal$_{IT}$	Btu$_{IT}$	kWh	hp-h	ft-lbf	ft^3 · lbf · in.$^{-2}$	liter-atm
1 g mass (energy equiv) =	1	8.987552×10^{13}	5.609589×10^{32}	2.148076×10^{13}	2.146640×10^{13}	8.518555×10^{10}	2.496542×10^{7}	3.347918×10^{7}	6.628878×10^{13}	4.603388×10^{13}	8.870024×10^{11}
1 J =	1.112650×10^{-14}	1	6.241509×10^{18}	0.2390057	0.2388459	9.478172×10^{-4}	$2.777777\ldots \times 10^{-7}$	3.725062×10^{-7}	0.7375622	5.121960×10^{-3}	9.869233×10^{-3}
1 eV =	1.782662×10^{-33}	1.602177×10^{-19}	1	3.829294×10^{-20}	3.826733×10^{-20}	1.518570×10^{-22}	4.450490×10^{-26}	5.968206×10^{-26}	1.181705×10^{-19}	8.206283×10^{-22}	1.581225×10^{-21}
1 cal =	4.655328×10^{-14}	4.184*	2.611448×10^{19}	1	0.9993312	3.965667×10^{-3}	$1.1622222\ldots \times 10^{-6}$	1.558562×10^{-6}	3.085960	2.143028×10^{-2}	0.04129287
1 cal$_{IT}$ =	4.658443×10^{-14}	4.1868*	2.613195×10^{19}	1.000669	1	3.968321×10^{-3}	1.163×10^{-6}	1.559609×10^{-6}	3.088025	2.144462×10^{-2}	0.04132050
1 Btu$_{IT}$ =	1.173908×10^{-11}	1055.056	6.585141×10^{21}	252.1644	251.9958	1	2.930711×10^{-4}	3.930148×10^{-4}	778.1693	5.403953	10.41259
1 kWh =	4.005540×10^{-8}	3600000.*	2.246943×10^{25}	860420.7	859845.2	3412.142	1	1.341022	2655224.	18349.06	35529.24
1 hp-h =	2.986931×10^{-8}	2384519.	1.675545×10^{25}	641615.6	641186.5	2544.33	0.7456998	1	1980000.*	13750.	26494.15
1 ft-lbf =	1.508551×10^{-14}	1.355818	8.462351×10^{18}	0.3240483	0.3238315	1.285067×10^{-3}	3.766161×10^{-7}	$5.050505\ldots \times 10^{-7}$	1	$6.944444\ldots \times 10^{-3}$	0.01338088
1 ft^3 lbf · in.$^{-2}$ =	2.172313×10^{-12}	195.2378	1.218578×10^{21}	46.66295.	46.63174	0.1850497	5.423272×10^{-5}	$7.272727\ldots \times 10^{-5}$	144.*	1	1.926847
1 liter-atm =	1.127393×10^{-12}	101.325	6.3242109×10^{20}	24.21726	24.20106	0.09603757	2.814583×10^{-5}	3.774419×10^{-5}	74.73349	0.5189825	1

*Numbers followed by an asterisk are definitions of the relation between the two units.

Fundamental constants

Recommended values (2002) of selected fundamental constants of physics and chemistry[a]

Quantity	Symbol	Numerical value[b]	Unit[c]	Relative uncertainty (standard deviation)
UNIVERSAL CONSTANTS				
Speed of light in vacuum	c, c_0	299 792 458	$m\ s^{-1}$	(exact)
Magnetic constant	μ_0	$4\pi \times 10^{-7}$	$N\ A^{-2}$	(exact)
		$= 12.566\ 370\ 614 \ldots \times 10^{-7}$	$N\ A^{-2}$	(exact)
Electric constant, $1/\mu_0 c^2$	ε_0	$8.854\ 187\ 817 \ldots \times 10^{-12}$	$F\ m^{-1}$	(exact)
Characteristic impedance of vacuum, $\sqrt{\mu_0/\varepsilon_0} = \mu_0 c$	Z_0	$376.730\ 313\ 461 \ldots$	Ω	(exact)
Newtonian constant of gravitation	G	$6.6742(10) \times 10^{-11}$	$m^3\ kg^{-1}\ s^{-2}$	1.5×10^{-4}
	$G/\hbar c$	$6.7087(10) \times 10^{-39}$	$(GeV/c^2)^{-2}$	1.5×10^{-4}
Planck constant	h	$6.626\ 0693(11) \times 10^{-34}$	$J\ s$	1.7×10^{-7}
in eV s		$4.135\ 667\ 43(35) \times 10^{-15}$	$eV\ s$	8.5×10^{-8}
hc in eV m		$1.239\ 841\ 91(11) \times 10^{-6}$	$eV\ m$	8.5×10^{-8}
$h/2\pi$	$\hbar$	$1.054\ 571\ 68(18) \times 10^{-34}$	$J\ s$	1.7×10^{-7}
in eV s		$6.582\ 119\ 15(56) \times 10^{-16}$	$eV\ s$	8.5×10^{-8}
$\hbar c$ in eV m		$197.326\ 968(17) \times 10^{-9}$	$eV\ m$	8.5×10^{-8}
Planck mass, $(\hbar c/G)^{1/2}$	m_p	$2.176\ 45(16) \times 10^{-8}$	kg	7.5×10^{-5}
Planck length, $\hbar/m_p c = (\hbar G/c^3)^{1/2}$	l_p	$1.616\ 24(12) \times 10^{-35}$	m	7.5×10^{-5}
Planck time, $l_p/c = (\hbar G/c^5)^{1/2}$	t_p	$5.391\ 21(40) \times 10^{-44}$	s	7.5×10^{-5}
ELECTROMAGNETIC CONSTANTS				
Elementary charge	e	$1.602\ 176\ 53(14) \times 10^{-19}$	C	8.5×10^{-8}
Magnetic flux quantum, $h/2e$	Φ_0	$2.067\ 833\ 72(18) \times 10^{-15}$	Wb	8.5×10^{-8}
Josephson constant[d], $2e/h$	K_J	$483\ 597.879(41) \times 10^9$	$Hz\ V^{-1}$	8.5×10^{-8}
von Klitzing constant[e], $h/e^2 = \mu_0 c/2\alpha$	R_K	$25\ 812.807\ 449(86)$	Ω	3.3×10^{-9}
Bohr magneton, $e\hbar/2m_e$	μ_B	$927.400\ 949(80) \times 10^{-26}$	$J\ T^{-1}$	8.6×10^{-8}
in eV T^{-1}		$5.788\ 381\ 804(39) \times 10^{-5}$	$eV\ T^{-1}$	6.7×10^{-9}
	μ_B/h	$13.996\ 2458(12) \times 10^9$	$Hz\ T^{-1}$	8.6×10^{-8}
	μ_B/hc	$46.686\ 4507(40)$	$m^{-1}\ T^{-1}$	8.6×10^{-8}
	$\mu_B k$	$0.671\ 7131(12)$	$K\ T^{-1}$	1.8×10^{-6}
Nuclear magneton, $e\hbar/2m_p$	μ_N	$5.050\ 783\ 43(43) \times 10^{-27}$	$J\ T^{-1}$	8.6×10^{-8}
in eV T^{-1}		$3.152\ 451\ 259(21) \times 10^{-8}$	$eV\ T^{-1}$	6.7×10^{-9}

Quantity	Symbol	Value	Units	Relative uncertainty
	μ_N/h	7.622 593 71(65)	MHz T^{-1}	8.6×10^{-8}
	μ_N/hc	2.542 623 58(22) × 10^{-2}	m^{-1}T^{-1}	8.6×10^{-8}
	μ_N/k	3.658 2637(64) × 10^{-4}	K T^{-1}	1.8×10^{-6}

ATOMIC AND NUCLEAR CONSTANTS

General

Quantity	Symbol	Value	Units	Relative uncertainty
Fine-structure constant, $e^2/4\pi\varepsilon_0\hbar c$	α	7.297 352 568(24) × 10^{-3}		3.3×10^{-9}
Inverse fine-structure constant	α^{-1}	137.035 999 11(46)		3.3×10^{-9}
Rydberg constant, $\alpha^2 m_e c/2h$	R_∞	10 973 731.568 525(73)	m^{-1}	6.6×10^{-12}
	$R_\infty c$	3.289 841 960 360(22) × 10^{15}	Hz	6.6×10^{-12}
$R_\infty hc$ in eV		13.605 6923(12)	eV	8.5×10^{-8}
Bohr radius, $\alpha/4\pi R_\infty = 4\pi\varepsilon_0\hbar^2/m_e e^2$	a_0	0.529 177 2108(18) × 10^{-10}	m	3.3×10^{-9}
Hartree energy, $e^2/4\pi\varepsilon_0 a_0 = 2R_\infty hc = \alpha^2 m_e c^2 =$	E_h	4.359 744 17(75) × 10^{-18}	J	1.7×10^{-7}
in eV		27.211 3845(23)	eV	8.5×10^{-8}

Electroweak

Quantity	Symbol	Value	Units	Relative uncertainty
Fermi coupling constant[f]	$G_F/(\hbar c)^3$	1.166 39(1) × 10^{-5}	GeV^{-2}	8.6×10^{-6}

Electron, e$^-$

Quantity	Symbol	Value	Units	Relative uncertainty
Electron mass	m_e	9.109 3826(16) × 10^{-31}	kg	1.7×10^{-7}
in u		5.485 799 0945(24) × 10^{-4}	u	4.4×10^{-10}
Energy equivalent in MeV	$m_e c^2$	0.510 998 918(44)	MeV	8.6×10^{-8}
Electron charge to mass quotient	$-e/m_e$	$-1.758\ 820\ 12(15)$ × 10^{11}	C kg^{-1}	8.6×10^{-8}
Compton wavelength, $h/m_e c$	λ_C	2.426 310 238(16) × 10^{-12}	m	6.7×10^{-9}
$\lambda_C/2\pi = \alpha a_0 = \alpha^2/4\pi R_\infty$	$\lambdabar_C$	386.159 2678(26) × 10^{-15}	m	6.7×10^{-9}
Classical electron radius, $\alpha^2 a_0$	r_e	2.817 940 325(28) × 10^{-15}	m	1.0×10^{-8}
Thomson cross section, $(8\pi/3)r_e^2$	σ_e	0.665 245 873(13) × 10^{-28}	m^2	2.0×10^{-8}
Electron magnetic moment	μ_e	$-928.476\ 412(80)$ × 10^{-26}	J T^{-1}	8.6×10^{-8}
Electron magnetic moment anomaly, $\lvert\mu_e\rvert/\mu_B - 1$	a_e	1.159 652 1859(38) × 10^{-3}		3.2×10^{-9}
Electron gyromagnetic ratio, $2\lvert\mu_e\rvert/\hbar$	γ_e	1.760 859 74(15) × 10^{11}	s^{-1} T^{-1}	8.6×10^{-8}
	$\gamma_e/2\pi$	28 024.95 32(24)	MHz T^{-1}	8.6×10^{-8}

Muon, μ^-

Quantity	Symbol	Value	Units	Relative uncertainty
Muon mass	m_μ	1.883 531 40(83) × 10^{-28}	kg	1.7×10^{-7}
in u		0.113 428 9264(30)	u	2.6×10^{-8}
Energy equivalent in MeV	$m_\mu c^2$	105.658 3692(94)	MeV	8.9×10^{-8}

[a]Footnotes are at table end.

(cont.)

Recommended values (2002) of selected fundamental constants of physics and chemistry[a] (cont.)

Quantity	Symbol	Numerical value[b]	Unit[c]	Relative uncertainty (standard deviation)
ATOMIC AND NUCLEAR CONSTANTS (cont.)				
Muon, μ^- (cont.)				
Muon-electron mass ratio	m_μ/m_e	206.768 2838(54)		2.6×10^{-8}
Muon Compton wavelength, $h/m_\mu c$	$\lambda_{C,\mu}$	11.734 441 05(30) × 10^{-15}	m	2.5×10^{-8}
$\lambda_{C,\mu}/2\pi$	$\lambda_{C,\mu}/2\pi$	1.867 594 298(47) × 10^{-15}	m	2.5×10^{-8}
Muon magnetic moment	μ_μ	−4.490 447 99(40) × 10^{-26}	J T^{-1}	8.9×10^{-8}
Muon magnetic moment anomaly, $|\mu_\mu|/(e\hbar/2m_\mu) - 1$	a_μ	1.165 919 81(62) × 10^{-3}		5.3×10^{-7}
Tau, τ^-				
Tau mass[g]	m_τ	3.167 77(52) × 10^{-27}	kg	1.6×10^{-4}
in u		1.907 68(31)	u	1.6×10^{-4}
Energy equivalent in Mev	$m_\tau c^2$	1776.99(29)	MeV	1.6×10^{-4}
Tau-electron mass ratio	m_τ/m_e	3 477.48(57)		1.6×10^{-4}
Proton, p				
Proton mass	m_p	1.672 621 71(29) × 10^{-27}	kg	1.7×10^{-7}
in u		1.007 276 466 88(13)	u	1.3×10^{-10}
Energy equivalent in MeV	$m_p c^2$	938.272 029(80)	MeV	8.6×10^{-8}
Proton-electron mass ratio	m_p/m_e	1836.152 672 61(85)		4.6×10^{-10}
Proton charge to mass quotient	e/m_p	9.578 833 76(82) × 10^7	C kg^{-1}	8.6×10^{-8}
Proton Compton wavelength, $h/m_p c$	$\lambda_{C,p}$	1.321 409 8555(88) × 10^{-15}	m	6.7×10^{-9}
$\lambda_{C,p}/2\pi$	$\lambda_{C,p}/2\pi$	0.210 308 9104(14) × 10^{-15}	m	6.7×10^{-9}
Proton magnetic moment	μ_p	1.410 606 71(12) × 10^{-26}	J T^{-1}	8.7×10^{-8}
to nuclear magneton ratio	μ_p/μ_N	2.792 847 351(28)		1.0×10^{-8}
Shielded proton magnetic moment[h]	μ'_p	1.410 570 47(12) × 10^{-26}	J T^{-1}	8.7×10^{-8}
to nuclear magneton ratio	μ'_p/μ_N	2.792 775 604(30)		1.1×10^{-8}
Proton gyromagnetic ratio, $2\mu_p/\hbar$	γ_p	2.675 222 05(23) × 10^8	s^{-1} T^{-1}	8.6×10^{-8}
$\gamma_p/2\pi$	$\gamma_p/2\pi$	42.577 4813(37)	MHz T^{-1}	8.6×10^{-8}
Shielded proton gyromagnetic ratio[h], $2\mu'_p/\hbar$	γ'_p	2.675 153 33(23) × 10^8	s^{-1} T^{-1}	8.6×10^{-8}
$\gamma'_p/2\pi$	$\gamma'_p/2\pi$	42.576 3875(37)	MHz T^{-1}	8.6×10^{-8}

Neutron, n				
Neutron mass	m_n	1.674 927 28(29) × 10^{-27}	kg	1.7 × 10^{-7}
in u		1.008 664 915 60(55)	u	5.5 × 10^{-10}
Energy equivalent in MeV	m_nc^2	939.565 360(81)	MeV	8.6 × 10^{-8}
Neutron magnetic moment	μ_n	−0.966 236 45(24) × 10^{-26}	J T^{-1}	2.5 × 10^{-7}
to nuclear magneton ratio	μ_n/μ_N	−1.913 042 73(45)		2.4 × 10^{-7}
Deuteron, d				
Deuteron mass	m_d	3.343 583 35(57) × 10^{-27}	kg	1.7 × 10^{-7}
in u		2.013 553 212 70(35)	u	1.7 × 10^{-10}
Energy equivalent in MeV	m_dc^2	1 875.612 82(16)	MeV	8.6 × 10^{-8}
Deuteron magnetic moment	μ_d	0.433 073 482(38) × 10^{-26}	J T^{-1}	8.7 × 10^{-8}
to nuclear magneton ratio	μ_d/μ_N	0.857 438 2329(92)		1.1 × 10^{-8}

PHYSICOCHEMICAL CONSTANTS

Avogadro constant	N_A, L	6.022 1415(10) × 10^{23}	mol^{-1}	1.7 × 10^{-7}
Atomic mass constant, $m_u = \frac{1}{12}m(^{12}C) = 1$ u $= 10^{-3}$ kg mol$^{-1}/N_A$	m_u	1.660 538 86(28) × 10^{-27}	kg	1.7 × 10^{-7}
Energy equivalent in MeV	m_uc^2	931.494 043(80)	MeV	8.6 × 10^{-8}
Faraday constant, N_Ae	F	96 485.3383(83)	C mol^{-1}	8.6 × 10^{-8}
Molar Planck constant	N_Ah	3.990 312 716(27) × 10^{-10}	J s mol^{-1}	6.7 × 10^{-9}
	N_Ahc	0.119 626 565 72(80)	J m mol^{-1}	6.7 × 10^{-9}
Molar gas constant	R	8.314 472(15)	J K^{-1} mol^{-1}	1.7 × 10^{-6}
Boltzmann constant, R/N_A	k	1.380 6505(24) × 10^{-23}	J K^{-1}	1.8 × 10^{-6}
in eV K^{-1}		8.617 343(15) × 10^{-5}	eV K^{-1}	1.8 × 10^{-6}
	k/h	2.083 6644(36) × 10^{10}	Hz K^{-1}	1.7 × 10^{-6}
	k/hc	69.503 56(12)	m^{-1}K^{-1}	1.7 × 10^{-6}
k^{-1} in K eV^{-1}		11 604.505(20)	K eV^{-1}	1.8 × 10^{-6}
Molar volume of ideal gas, RT/p, for $T = 273.15$ K, $p = 101.325$ kPa	V_m	22.413 996(39) × 10^{-3}	m^3 mol^{-1}	1.7 × 10^{-6}
Loschmidt constant, N_A/V_m	n_o	2.686 7773(47) × 10^{25}	m^{-3}	1.8 × 10^{-6}
Stefan-Boltzmann constant, $(\pi^2/60)k^4/\hbar^3c^2$	σ	5.670 400(40) × 10^{-8}	W m^{-2} K^{-4}	7.0 × 10^{-6}
First radiation constant, $2\pi hc^2$	c_1	3.741 771 38(64) × 10^{-16}	W m^2	1.7 × 10^{-7}
Second radiation constant, hc/k	c_2	1.438 7752(25) × 10^{-2}	m K	1.7 × 10^{-8}
Wien displacement law constant, $b = \lambda_{max} T = c_2/4.965\ 114\ 231\ldots$	b	2.897 7685(51) × 10^{-3}	m K	1.7 × 10^{-6}

(cont.)

^{a}Footnotes are at table end.

Recommended values (2002) of selected fundamental constants of physics and chemistry[a] (cont.)

Quantity	Symbol	Numerical value[b]	Unit[c]	Relative uncertainty (standard deviation)
NON-SI UNITS ACCEPTED FOR USE WITH THE SI				
Electronvolt: (e/C) J	eV	$1.602\ 176\ 53(14) \times 10^{-19}$	J	8.5×10^{-8}
(Unified) atomic mass unit: $1\ u = m_u = \frac{1}{12} m(^{12}C)$	u	$1.660\ 538\ 86(28) \times 10^{-27}$	kg	1.7×10^{-7}

[a]This table presents a selection of the values of the fundamental constants recommended by the Committee on Data for Science and Technology (CODATA). These "2002 CODATA recommended values" form a self-consistent set based on the data available through December 31, 2002, and are generally recognized for use in all fields of science and technology. A detailed description of the data and analysis that led to these results will be published in 2005. The 2002 adjustment was carried out under the auspices of the CODATA Task Group on Fundamental Constants. The recommended values are available on the World Wide Web at http://physics.nist.gov/cuu/Constants/index.html.

[b]The digits in parentheses represent one-standard-deviation uncertainties in the final two digits of the quoted values.

[c]A = ampere, C = coulomb, F = farad, Hz = hertz, J = joule, K = kelvin, kg = kilogram, m = meter, mol = mole, N = newton, Pa = pascal, s = second, T = tesla, W = watt, Wb = weber, Ω = ohm, eV = electronvolt, u = (unified) atomic mass unit. Prefixes: k = 10^3, M = 10^6, G = 10^9.

[d]The conventional value of the Josephson constant, adopted internationally for realizing representations of the volt using the Josephson effect, is K_{J-90} = 483 597.9 GHz V^{-1}.

[e]The conventional value of the von Klitzing constant, adopted internationally for realizing representations of the ohm using the quantum Hall effect, is R_{K-90} = 25 812.807 Ω.

[f]Value recommended by the Particle Data Group in the 2002 Review of Particle Physics [K. Hagiwara et al. (Particle Data Group), *Physical Review D*, vol. 66, Paper 010001, 2002 (http://pdg.lbl.gov)].

[g]This and all other values involving m_τ are based on the value of $m_\tau c^2$ in MeV recommended by the Particle Data Group in the 2002 Review of Particle Physics (*ibid.*), but with a standard uncertainty of 0.29 MeV rather than the quoted uncertainty of -0.26 MeV, $+0.29$ MeV.

[h]Based on nuclear magnetic resonance (NMR) frequency of protons in a sphere of pure water (H$_2$O) at 25°C surrounded by vaccum.

[i]The numerical value of F to be used in coulometric chemical measurements is 96 485.336(16) [relative uncertainty = 1.7×10^{-7}] when the relevant current is measured in terms of representations of the volt and ohm based on the Josephson and quantum Hall effects and the conventional values of the Josephson and von Klitzing constants, K_{J-90} and R_{K-90}.

The fundamental particles[a]

Gauge bosons $J_C^P = 1^-_-$ Self-conjugate except $\overline{W^+} = W^-$.

Name	Symbol	Charge[b]	Mass and width, GeV	Couplings
Photon	γ	0	0	$A \Rightarrow \gamma A$
Gluon[c]	g	0	0	$A \Rightarrow gA'$
Weak bosons				
Charged[d]	$W^\pm$	± 1	80.4, 2.1	$U \Rightarrow W^+D$
Neutral[e]	Z^0	0	91.2, 2.5	$A \Rightarrow Z^0A$

Fermions[f] $J = {}^1/_2$ All have distinct antiparticles, except perhaps the neutrinos.

Name	Charge[b]	Symbol and mass, GeV		Symbol and mass, GeV		Symbol and mass, GeV	
Leptons							
Neutrinos	0	ν_e	$< 3 \times 10^{-9}$	ν_μ	$<.0002$	ν_τ	$<.02$
Charged leptons[g]	-1	e	.00051	μ	.106[h]	τ	1.78[h]
Quarks[c]							
Up type	${}^2/_3$	u	.0015–.004	c	1.55–1.35	t	170–180[i]
Down type	$-{}^1/_3$	d	.004–.008	s	.08–.13	b	4.1–4.4

[a] The graviton, with $J_C^P = 2^+_+$, has been omitted, since it plays no role in high-energy particle physics.
[b] In units of the proton charge.
[c] The gluon is a color SU_3 octet (8); each quark is a color triplet (3). These colored particles are confined constituents of hadrons; they do not appear as free particles.
[d] The branching ratios (%) of the decay modes of the W^+ are:
  $u\bar{d},\ c\bar{s}$ 34 each
  $\nu_e e^+,\ \nu_\mu \mu^+,\ \nu_\tau \tau^+$ 11 each
[e] The branching ratios (%) of the decay modes of the Z^0 are:
  $d\bar{d},\ s\bar{s},\ b\bar{b}$ 16.6 each
  $u\bar{u},\ c\bar{c}$ 10.1 each
  $\nu_e \bar{\nu}_e,\ \nu_\mu \bar{\nu}_\mu,\ \nu_\tau \bar{\nu}_\tau$ 6.7 each
  $e^+e^-,\ \mu^+\mu^-,\ \tau^+\tau^-$ 3.4 each
[f] The three known families (generations) of fermions are displayed in three columns.
[g] Any further charged leptons have mass greater than 40 GeV.
[h] The μ and τ leptons are unstable, with the following mean life and principal decay modes (branching ratios in %):
  μ $\tau_\mu = 2.2 \times 10^{-6}$ s $e\bar{\nu}_e \nu_\mu$ 100
  τ $\tau_\tau = 2.9 \times 10^{-13}$ s $\mu\bar{\nu}_\mu \nu_\tau$ 17, $e\bar{\nu}_e \nu_\tau$ 18, (hadrons)$\bar{\nu}_\tau$ 63
[i] The t quark has a width ≈ 1.4 GeV, with dominant decay to Wb.

Internal energy and generalized work

Type of energy	Intensive factor	Extensive factor	Element of work
Mechanical			
Expansion	Pressure (P)	Volume (V)	$-PdV$
Stretching	Surface tension (γ)	Area (A)	γdA
Extension	Tensile stretch (F)	Length (l)	Fdl
Thermal	Temperature (T)	Entropy (S)	TdS
Chemical	Chemical potential (μ)	Amount (n)	μdn
Electrical	Electric potential (E)	Charge (Q)	EdQ
Gravitational	Gravitational field strength (mg)	Height (h)	$mgdh$
Polarization			
Electrostatic	Electric field strength ($\mathcal{E}$)	Total electric polarization (dP)	$\mathcal{E}dP$
Magnetic	Magnetic field strength ($\mathcal{H}$)	Total magnetic polarization (M)	$\mathcal{H}dM$

The 14 Bravais lattices, derived by centering of the seven crystal classes (P and R) defined by symmetry operators

Bravais lattice cells	Axes and interaxial angles	Examples
Cubic P Cubic I Cubic F	Three axes at right angles; all equal: $a = b = c;\ \alpha = \beta = \gamma = 90°$	Copper (Cu), silver (Ag), sodium chloride (NaCl)
Tetragonal P Tetragonal I	Three axes at right angles; two equal: $a = b \neq c;\ \alpha = \beta = \gamma = 90°$	White tin (Sn), rutile (TiO_2), β-spodumene ($LiAlSi_2O_6$)
P C I F Orthorhombic	Three axes at right angles; all unequal: $a \neq b \neq c;\ \alpha = \beta = \gamma = 90°$	Gallium (Ga), perovskite ($CaTiO_3$)
Monoclinic P Monoclinic C	Three axes, one pair not at right angles, of any lengths: $a \neq b \neq c;\ \alpha = \gamma = 90° \neq \beta$	Gypsum ($CaSO_4 \cdot 2H_2O$)
Triclinic P	Three axes not at right angles, of any lengths: $a \neq b \neq c;\ \alpha \neq \beta \neq \gamma \neq 90°$	Potassium chromate (K_2CrO_7)
Trigonal R (rhombohedral)	Rhombohedral: three axes equally inclined, not at right angles; all equal: $a = b = c;\ \alpha = \beta = \gamma \neq 90°$	Calcite ($CaCO_3$), arsenic (As), bismuth (Bi)
Trigonal and hexagonal C (or P)	Hexagonal: three equal axes coplanar at 120°, fourth axis at right angles to these: $a_1 = a_2 = a_3 \neq c;$ $\alpha = \beta = 90°,\ \gamma = 120°$	Zinc (Zn), cadmium (Cd), quartz (SiO_2) [P]

Dimensional formulas of common quantities

Quantity	Symbol	Dimensional formula
Mass	Fundamental	M
Length	Fundamental	L
Time	Fundamental	T
Velocity	Distance/time	LT^{-1}
Acceleration	Velocity/time	LT^{-2}
Force	Mass × acceleration	MLT^{-2}
Momentum	Mass × velocity	MLT^{-1}
Energy	Force × distance	ML^2T^{-2}
Angle	Arc/radius	0
Angular velocity	Angle/time	T^{-1}
Angular acceleration	Angular velocity/time	T^{-2}
Torque	Force × lever arm	ML^2T^{-2}
Angular momentum	Momentum × lever arm	ML^2T^{-1}
Moment of inertia	Mass × radius squared	ML^2
Area	Length squared	L^2
Volume	Length cubed	L^3
Density	Mass/volume	ML^{-3}
Pressure	Force/area	$ML^{-1}T^{-2}$
Action	Energy × time	ML^2T^{-1}
Viscosity	Force per unit area per unit velocity gradient	$ML^{-1}T^{-1}$

Physical properties of optical materials*

Crystalline materials

Material	Symbol	Refractive index (wavelength = 500 nm)	Density, g/cm³	Hardness: Knoop, kg/mm², or Mohs (M)	Solubility, g/100 g H_2O	Specific heat, cal/(g-K)‡	Melting or softening temp., K	Thermal conductivity, W/(m·K)‡	Linear expansion coefficient, 10^{-6}/K‡	Young's modulus,‡ GPa
Germanium	Ge	4	5.33	800	Insoluble	0.074	1210	59	6.1	102.66
Lithium fluoride	LiF	1.394	3.5	100	0.27	0.37	1140	11.3	34.4	64.77
Magnesium fluoride	MgF2	1.39	3.18	415	Insoluble	0.24	1528	21	14(P), 8.9(S)	138.5
Sodium chloride	NaCl	1.53	2.17	15.2	35.7	0.2	1070	6.5	40	39.96
Zinc sulfide	ZnS	2.42	4.08	230	Insoluble	0.112	2100	17	6.6	74.5
Zinc selenide	ZnSe	2.43	5.42	137	0.001	0.0090	1790	19	7	70.97
Barium titanate	BaTiO3	–	5.9	200–580	–	0.103	1870	1.34	19	33.76
Cesium iodide	CsI	1.75	4.51	–	44	0.048	894	1.1	48.3	5.3
Diamond	C	2.4	3.51	5700–10400	Insoluble	0.124	3770	2600	0.8	1050
Lanthanum fluoride	LaF3	1.6	5.94	4.5(M)	Insoluble	0.121	1766	5.1	11(P), 17(S)	–
Magnesia	MgO	1.74	3.585	910	Insoluble	0.24	3053	40.6	8	–
Potassium chloride	KCl	1.49	1.98	7.2	34.7	0.162	1050	6.7	36.6	29.63
Sapphire	Al2O3	1.77	3.98	1370	Insoluble	0.18	2300	33.0	8(P) 5(S)	335
Crystal quartz	SiO2	1.55	2.65	741	Insoluble	0.17	1740	10.7(P), 6.2(S)	8.0(P), 13.4(S)	97.2(P), 76.5(S)
Barium fluoride	BaF2	1.47	4.89	0.82	0.17	0.096	1550	11.7	19.9	53.05
Calcium fluoride	CaF2	1.43	3.18	140	0.0017	0.204	1630	10	18.9	75.79
Cadmium telluride	CdTe	2.69	6.2	56	Insoluble	0.056	1320	6.3	5.9	36.52
Calcite	CaCO3	n_o = 1.665, n_e = 1.490	2.710	3(M)	0.0014	0.203	1612	5.526(P), 4.646(S)	25(P), –5.8(S)	72.35(P), 88.19(S)
Cuprous chloride	CuCl	2.0	4.14	2–2.5(M)	0.0061	–	695	–	10	–
Gallium phosphide	GaP	3.65	4.13	845	Insoluble	0.2	1623	54	4.7	102.6
Indium arsenide	InAs	4.5	–	330	Insoluble	0.06	1215	50	5.3	–
Lead fluoride	PbF2	1.78	8.24	200	0.064	0.085	1100	–	29	–
Lead sulfide	PbS	4.3 at 3 μm	7.5	–	–	0.050	1387	0.67	18	–
Silicon carbide	SiC	2.68	3.217	2130–2755	Insoluble	0.165	3000	490	2.8	386
Selenium	Se	2.83	4.82	2.6(M)	Insoluble	0.077	490	1.3	48.7	–
Silicon	Si	3.45	2.329	1100	Insoluble	0.18	1690	163	2.6	130.91

Infrared glasses

Arsenic trisulfide glass	As₂S₃	2.7	3.43	109	Insoluble	0.109	483	0.1674	24.62	15.85
Germanate glass (Schott IRG2)	—	1.9	5.00	481	0.012	0.108	—	0.91	8.8	95.9
Chalcogenide glass (Schott IRG 100)	—	2.73	4.67	150	—	—	624	0.3	15	21
Lead silicate glass (Schott IRG7)	—	1.57	3.06	379	0.171	0.151	—	0.73	9.6	59.7
Ultra-low-expansion (ULE) titanium silicate (Corning 7971)	—	1.484	2.205	459	Insoluble	0.183	1763	1.31	0.015	67.52
Germanate glass (Corning 9754)	—	1.67	3.581	560	—	0.13	1147	1.0	6.2	84.1
Phosphate laser glass (Schott LG750)	—	—	—	—	—	—	—	0.52	13.2	—

Ultraviolet glasses

Fused silica	SiO₂	1.43	2.203	461	Insoluble	0.22	1448	1.38	0.55	73.1
Fused silica (Corning 7940)	—	1.46	2.202	500	—	0.177	1858	1.38	0.52	73

Nonlinear/photorefractive optical materials

Ammonium dihydrogen phosphate (ADP)	$NH_4H_2PO_4$	1.51	1.803	—	22.7	—	4.63	.71162(P), 1.2558(S)	39.3	—
Lithium niobate	$LiNbO_3$	$n_o = 2.286$, $n_e = 2.203$ (0.6328 μm)	4.64	5(M)	Insoluble	—	1523	38	2.2(P), 2.0(S)	—
Potassium dihydrogen phosphate (KDP)	KH_2PO_4	1.5	2.338	—	33	0.21	525.6	1.3395	21.6	—
Yttrium vanadate	YVO_4	$n_o = 1.9929$, $n_e = 2.2154$ (0.6328 μm)	4.22	5(M)	Insoluble	—	—	5.23(P), 5.10(S)	11.37(P), 4.43(S)	—
Iron-doped lithium niobate	$Fe:LiNbO_3$	$n_o = 2.286$, $n_e = 2.203$ (0.6328 μm)	4.64	5(M)	Insoluble	—	1523	38	2.2(P), 2.0(S)	—

(cont.)

Physical properties of optical materials* (cont.)

Material	Symbol	Refractive index (wavelength = 500 nm)	Density, g/cm³	Hardness: Knoop, kg/mm² or Mohs (M)	Solubility, g/100 g H₂O	Specific heat, cal/(g-K)†	Melting or softening temp., K	Thermal conductivity, W/(m · K)‡	Linear expansion coefficient, 10⁻⁶/K‡	Young's modulus,‡ GPa
Silver gallium sulfide	$AgGaS_2$	$n_o = 2.4521$, $n_e = 2.3990$ (1.064 μm)	4.702	—	—	—	1270	0.015	12.5(P), −13.2(S)	—
Silver gallium selenide	$AgGaSe_2$	$n_o = 2.7010$, $n_e = 2.6792$ (1.064 μm)	5.700	—	—	—	1124	—	16.8(P), −7.8(S)	—
Magnesium oxide-doped lithium niobate	$MgO{:}LiNbO_3$	$n_o = 2.286$, $n_e = 2.203$ (0.6328 μm)	4.64	5(M)	Insoluble	—	1523	38	2.2(P), 2.0(S)	—
Potassium titanyl phosphate (KTP)	$KTiOPO_4$	$n_x = 1.78$, $n_y = 1.79$, $n_z = 1.89$	3.01	5(M)	Insoluble	0.1643	1445	13	—	—
Lithium triborate (LBO)	LiB_3O_5	$n_x = 1.58$, $n_y = 1.61$, $n_z = 1.62$	2.47	6(M)	—	—	1107	—	—	—
Cadmium sulfide	CdS	2.6	4.82	122	Insoluble	0.0882	1773	15.91	2.1(P), 4(S)	—
Gallium arsenide	GaAs	3.35	5.32	731	Insoluble	0.076	1511	55	5.7	82.68
Beta-barium borate (BBO)	$\beta\text{-}BaB_2O_4$	$n_o = 1.6749$, $n_e = 1.5555$	3.85	4.5(M)	—	—	1363	1.6(P), 1.2(S)	36(P), 4(S)	—

*All values are given for temperatures near room temperature.

†1 cal = 4.18 J.

‡ P denotes values measured parallel to the c axis of the crystal, and S denotes values measured perpendicular to the c axis.

Types of radioactivity

Type	Symbol	Particles emitted	Change in atomic number, ΔZ	Change in atomic mass number, ΔA	Example
Alpha	α	Helium nucleus	-2	-4	$^{226}_{86}\text{Ra} \rightarrow {}^{222}_{84}\text{Rn} + \alpha$
Beta negatron	β^-	Negative electron and antineutrino[a]	$+1$	0	$^{24}_{11}\text{Na} \rightarrow {}^{24}_{12}\text{Mg} + e^- + \bar{\nu}$
Beta positron	β^+	Positive electron and neutrino[a]	-1	0	$^{22}_{11}\text{Na} \rightarrow {}^{22}_{10}\text{Ne} + e^+ + \nu$
Electron capture	EC	Neutrino[a]	-1	0	$^{7}_{4}\text{Be} + e^- \rightarrow {}^{7}_{3}\text{Li} + \nu$
Isomeric transition[b]	IT	Gamma rays or conversion electrons or both (and positive-negative electron pair)[c]	0	0	$^{137m}_{56}\text{Ba} \rightarrow {}^{137}_{56}\text{Ba} + \gamma$ or conversion electrons
Proton	p	Proton	-1	-1	$^{151}_{71}\text{La} \rightarrow {}^{150}_{70}\text{Yb} + p$
Spontaneous fission (hot)	SF	Two intermediate-mass nuclei and 1–10 neutrons	Various	Various	$^{238}_{92}\text{U} \rightarrow {}^{133}_{50}\text{Sn} + {}^{103}_{42}\text{Mo} + 2n$
Spontaneous fission (cold)	SF	Two intermediate-mass nuclei (zero neutrons)	Various	Various	$^{252}_{98}\text{Cf} \rightarrow {}^{106}_{42}\text{Mo} + {}^{146}_{56}\text{Ba}$
Ternary spontaneous fission (hot)	TSF	Two intermediate-mass nuclei, a light particle (^{2}H, α, up to ^{10}Be), and neutrons	Various	Various	$^{252}_{98}\text{Cf} \rightarrow {}^{100}_{40}\text{Zr} + {}^{146}_{56}\text{Ba} + \alpha + 2n$
Ternary spontaneous fission (cold)	TSF	Two intermediate-mass nuclei and a light particle	Various	Various	$^{252}_{98}\text{Cf} \rightarrow {}^{96}_{38}\text{Sr} + {}^{146}_{56}\text{Ba} + {}^{10}_{4}\text{Be}$
Isomeric spontaneous fission	ISF	Heavy fragments and neutrons	Various	Various	$^{244f}_{95}\text{Am} \rightarrow {}^{134}_{53}\text{I} + {}^{107}_{42}\text{Mo} + 3n$

(cont.)

Types of radioactivity (cont.)

Type	Symbol	Particles emitted	Change in atomic number, ΔZ	Change in atomic mass number, ΔA	Example
Beta-delayed spontaneous fission	$(EC + \beta^+)SF$	Positive electron, neutrino, heavy fragments, and neutrons	Various	Various	$^{246}_{99}Es \rightarrow \beta^+ + \nu + ^{246}_{98}Cf \rightarrow$ $^{138}_{54}Xe + ^{107}_{44}Ru + n$
	β^-SF	Negative electron, antineutrino, heavy fragments, and neutrons	Various	Various	$^{236}_{91}Pa \rightarrow \beta^- + \bar{\nu} + ^{236}_{92}U \rightarrow$ $^{139}_{53}I + ^{94}_{39}Y + 3$
Beta-delayed neutron	$\beta^- n$	Negative electron, and antineutrino, neutron	$+1$	-1	$^{11}_{3}Li \rightarrow \beta^- + \bar{\nu} + ^{11}_{4}Be^* \rightarrow$ $^{10}_{4}Be + n$
Beta-delayed two-neutron (three-, four-neutron)	$\beta^- 2n(3n, 4n)$	Negative electron, antineutrino, and two (three, four) neutrons	$+1$	$-2\ (-3, -4)$	$^{11}_{3}Li \rightarrow \beta^- + \bar{\nu} + ^{11}_{4}Be^* \rightarrow$ $^{9(8)}_{4}Be + 2n(3n)$
Beta-delayed proton	$\beta^+ p$ or $(\beta^+ + EC)p$	Positive electron, neutrino, and proton	-2	-1	$^{114}_{55}Cs \rightarrow \beta^+ + \nu + ^{114}_{54}Xe^* \rightarrow$ $^{113}_{53}I + p$
Beta-delayed two-proton	$\beta^+ 2p$	Positive electron, neutrino, and two protons	-3	-2	$^{22}_{13}Al \rightarrow \beta^+ + \nu + ^{22}_{12}Mg^* \rightarrow$ $^{20}_{10}Ne + 2p$
Beta-delayed triton	$\beta^- {}^3_1H$	Negative electron, antineutrino and triton	0	-3	$^{11}_{3}Li \rightarrow \beta^- + \bar{\nu} + ^{11}_{4}B^* \rightarrow$ $^{8}_{3}Li + ^{3}_{1}H$
Beta-delayed alpha	$\beta^+ \alpha$	Positive electron, neutrino, and alpha	-3	-4	$^{114}_{55}Cs \rightarrow \beta^+ + \nu + ^{114}_{54}Xe^* \rightarrow$ $^{110}_{52}Te + \alpha$
Beta-delayed alpha	$\beta^- \alpha$	Negative electron, antineutrino, and alpha	-1	-4	$^{214}_{83}Bi \rightarrow \beta^- + \bar{\nu} + ^{214}_{84}Po^* \rightarrow$ $^{210}_{82}Pb + \alpha$
Beta-delayed alpha-neutron	$\beta^- \alpha, n$	Negative electron, antineutrino, alpha, and neutron	-1	-5	$^{11}_{3}Li \rightarrow \beta^- + \bar{\nu} + ^{11}_{4}B^* \rightarrow$ $^{6}_{2}He + \alpha + n$
Double beta decay	$\beta^- \beta^-$	Two negative electrons and two antineutrinos	$+2$	0	$^{82}_{34}Se \rightarrow ^{82}_{36}Kr + 2\beta^- + 2\bar{\nu}$

	Symbol	Description			Example
	$\beta^+\beta^+$	Two positive electrons and two neutrinos	-2	0	[d] $^{130}_{56}\text{Ba} \rightarrow {}^{130}_{54}\text{Xe} + 2\beta^+ + 2\nu$
Double electron capture[d]	EC EC	Two neutrinos	-2	0	[d] $^{130}_{56}\text{Ba} + 2e^- \rightarrow {}^{130}_{54}\text{Xe} + 2\nu$
Neutrinoless double beta decay[e]	$\beta^-\beta^-$	Two negative electrons	$+2$	0	[e] $^{82}_{34}\text{Se} \rightarrow {}^{82}_{36}\text{Kr} + 2\beta^-$
Two-proton	$2p$	Two protons	-2	-2	$^{45}_{26}\text{Fa} \rightarrow {}^{43}_{24}\text{Cr} + 2p$
Neutron[d]	n	Neutron	0	-1	
Two-neutron[d]	$2n$	Two neutrons	0	-2	
Heavy clusters[f]	$^{14}_{6}\text{C}$	$^{14}_{6}\text{C}$ nucleus	-6	-14	$^{223}_{88}\text{Ra} \rightarrow {}^{209}_{82}\text{Pb} + {}^{14}_{6}\text{C}$
	$^{20}_{8}\text{O}$	$^{20}_{8}\text{O}$ nucleus	-8	-20	$^{227}_{89}\text{Ac} \rightarrow {}^{207}_{81}\text{Tl} + {}^{20}_{8}\text{O}$
	$^{24}_{10}\text{Ne}$	$^{24}_{10}\text{Ne}$ nucleus	-10	-24	$^{232}_{92}\text{U} \rightarrow {}^{208}_{82}\text{Pb} + {}^{24}_{10}\text{Ne}$

[a] The neutrinos and antineutrinos emitted in beta decay are electron neutrinos and antineutrinos.

[b] Excited states with relatively long measured half-lives are called isomeric and are identified by placing the symbol m for metastable after the mass number, as in ^{137m}Ba. Excited states with essentially prompt decay are identified by asterisks, as in $^{11}\text{Be}^*$.

[c] Pair emission occurs as an additional competing decay mode when the decay energy exceeds 1.022 MeV.

[d] Theoretically predicted but not established experimentally.

[e] Theoretically possible and predicted in some grand unified theories.

[f] There are other possible clusters in addition to those shown. Cold SF is also a type of cluster radioactivity.

Principal spectral regions and fields of spectroscopy

Spectral region	Approximate wavelength range	Typical source	Typical detector	Energy transitions studied in matter
Gamma	1–100 pm	Radioactive nuclei	Geiger counter; scintillation counter	Nuclear transitions and disintegrations
X-rays	6 pm–100 nm	X-ray tube (electron bombardment of metals)	Geiger counter	Ionization by inner electron removal
Vacuum ultraviolet	10–200 nm	High-voltage discharge; high-vacuum spark	Photomultiplier	Ionization by outer electron removal
Ultraviolet	200–400 nm	Hydrogen-discharge lamp	Photomultiplier	Excitation of valence electrons
Visible	400–800 nm	Tungsten lamp	Phototubes	Excitation of valence electrons
Near-infrared	0.8–2.5 μm	Tungsten lamp	Photocells	Excitation of valence electrons; molecular vibrational overtones
Infrared	2.5–50 μm	Nernst glower; Globar lamp	Thermocouple; bolometer	Molecular vibrations: stretching, bending, and rocking
Far-infrared	50–1000 μm	Mercury lamp (high-pressure)	Thermocouple; bolometer	Molecular rotations
Microwave	0.1–30 cm	Klystrons; magnetrons	Silicon-tungsten crystal; bolometer	Molecular rotations; electron spin resonance
Radio-frequency	10^{-1}–10^3 m	Radio transmitter	Radio receiver	Molecular rotations; nuclear magnetic resonance

Defining fixed points of the International Temperature Scale of 1990 (ITS-90)

Equilibrium state*	Temperature	
	T_{90}, K	t_{90}, °C
Vapor pressure equation of helium	3 to 5	−270.15 to −268.15
Triple point of equilibrium hydrogen[†]	13.8033	−259.3467
Vapor pressure point of equilibrium hydrogen[†] (or constant volume gas thermometer point of helium)	≈17	≈−256.15
Vapor pressure point of equilibrium hydrogen[†] (or constant volume gas thermometer point of helium)	≈20.3	≈−252.85
Triple point of neon	24.5561	−248.5939
Triple point of oxygen	54.3584	−218.7916
Triple point of argon	83.8058	−189.3442
Triple point of mercury	234.3156	−38.8344
Triple point of water	273.16	0.01
Melting point of gallium	302.9146	29.7646
Freezing point of indium	429.7485	156.5985
Freezing point of tin	505.078	231.928
Freezing point of zinc	692.677	419.527
Freezing point of aluminum	933.473	660.323
Freezing point of silver	1234.93	961.78
Freezing point of gold	1337.33	1064.18
Freezing point of copper	1357.77	1084.62

*The triple point is the equilibrium temperature at which the solid, liquid, and vapor phases coexist. The freezing point and the melting point are the equilibrium temperatures at which the solid and liquid phases coexist under a pressure of 101,325 Pa, 1 standard atmosphere. The isotopic composition is that naturally occurring.

[†] Equilibrium hydrogen is hydrogen with the equilibrium distribution of its ortho and para states at the corresponding temperatures. Normal hydrogen at room temperature contains 25% para and 75% ortho hydrogen.

Primary thermometry methods

Method	Approximate useful range of T, K	Principal measured variables	Relation of measured variables to T	Remarks
Gas thermometry	1.3–950	Pressure P and volume V	Ideal gas law plus correction: $PV \propto k_B T$ plus corrections	Careful determination of corrections necessary, but capable of high accuracy
Acoustic interferometry	1.5–3000	Speed of sound W	$W^2 \propto k_B T$ plus corrections	
Magnetic thermometry 1. Electron paramagnetism 2. Nuclear paramagnetism	0.001–35 0.000001–1	Magnetic susceptibility	Curie's law plus corrections: $\chi \propto 1/k_B T$ plus corrections	
Gamma-ray anisotropy or nuclear orientation thermometry	0.01–1	Spatial distribution of gamma-ray emission	Spatial distribution related to Boltzmann factor for nuclear spin states	Useful standard for $T < 1$ K
Thermal electric noise thermometry 1. Josephson junction point contact 2. Conventional amplifier	0.001–1 4–1400	Mean square voltage fluctuation $\overline{V}^2$	Nyquist's law: $\overline{V}^2 \propto k_B T$	Other sources of noise serious problem for $T > 4$ K
Radiation thermometry (visual, photoelectric, or, photodiode)	500–50,000	Spectral intensity J at wavelength λ	Planck's radiation law, related to Boltzmann factor for radiation quanta	Needs blackbody conditions or well-defined emittance
Infrared spectroscopy	100–1500	Intensity I of rotational lines of light molecules	Boltzmann factor for rotational levels related to I	Also Doppler line broadening ($\propto \sqrt{k_B T}$) useful; principal applications to plasmas and astrophysical observations; proper sampling, lack of equilibrium, atmospheric absorption often problems
Ultraviolet and x-ray spectroscopy	5000–2,000,000	Emission spectra from ionized atoms—H,He, Fe, Ca, and so on	Boltzmann factor for electron states related to band structure and line density	

BIOGRAPHICAL LISTING

Abbe, Ernst (1840–1905), German physicist. Developed optical instruments, such as an apochromatic objective and a crystal refractometer.

Abell, George Ogden (1927–1983), American astronomer. Research in problems relating to organization, structure, and distribution of galaxies; observational cosmology; and planetary nebulae.

Abney, William de Wiveleslie (1843–1920), English photographic chemist and physicist. Photographed the infrared solar spectrum.

Abrikosov, Alexei A. (1928–), Russian-born American physicist. Applied the Ginzburg-Landau theory to explain the behavior of type II superconductors. Nobel Prize, 2003.

Aitken, John (1839–1919), Scottish physicist. Studied dust particles in the atmosphere, known as Aitken nuclei.

Alembert, Jean le Rond d' (1717–1783), French mathematician. Developed d'Alembert's principle and the calculus of partial differences.

Alferov, Zhores Ivanovich (1930–), Russian physicist and electronics engineer. Developed semiconductor heterostructures used in high-speed and opto-electronics, including fast transistors, laser diodes, and light-emitting diodes; Nobel Prize, 2000.

Alfvén, Hannes Olof Gösta (1908–1995), Swedish physicist. Studies in magnetohydrodynamics, planetary physics, antiferromagnetism, and ferrimagnetism; Nobel Prize, 1970.

Alhazen (965–1038), Arab mathematician and astronomer. Provided the first accounts of atmospheric refraction and reflection from concave surfaces; constructed spherical and parabolic mirrors.

Alvarez, Luis Walter (1911–1988), American physicist. Pioneer in building liquid hydrogen bubble chambers, and in developing measurement devices and computer systems to analyze data from these chambers; discovered large numbers of short-lived elementary particles; Nobel Prize, 1968.

Amagat, Émile (1841–1915), French physicist. Investigated relationship of pressure, density, and temperature in gases and liquids, particularly at high pressure.

Amici, Giovanni Battista (1786–1863), Italian astronomer, optician, and naturalist. Invented the Amici microscope; designed parabolic mirrors for reflecting telescopes.

Ampère, André Marie (1775–1836), French physicist and mathematician. Founder of electrodynamics; formulated Ampère's law; invented the astatic needle.

Anaximander (611–547 B.C.), Greek astronomer and mathematician. Reputed inventor of geographical maps; formulated the concept of the universe as infinite (apeiron).

Anderson, Carl David (1905–1991), American physicist. Discovered the meson in cosmic rays; discovered the positron; Nobel Prize, 1936.

Anderson, Philip Warren (1923–), American physicist. Demonstrated existence of electronic localization in disordered solids, and of localized magnetism in metals; Nobel Prize, 1977.

Andrade, Edward Neville da Costa (1887–1971), English physicist. Discovered Andrade's creep law and a law governing variation of viscosity of liquids with temperature.

Angström, Anders Jonas (1814–1874), Swedish physicist. Mapped the solar spectrum; discovered hydrogen in the solar atmosphere.

Appleton, Edward Victor (1892–1965), English physicist. Demonstrated the existence of the ionosphere and discovered its region known as the Appleton layer; contributed to the development of radar; Nobel Prize, 1947.

Arago, Dominique François (1786–1853), French astronomer and physicist. Discovered the magnetic properties of nonferrous materials, and the production of magnetism by electricity.

Archimedes (287–212 B.C.), Greek physicist and mathematician. Formulated Archimedes' principle; invented the compound pulley and Archimedes' screw.

Aristotle (384–322 B.C.), Greek philosopher. Exponent of the methodology and division of sciences; contributed to physics, astronomy, meteorology, psychology, and biology.

Arrhenius, Svante August (1859–1927), Swedish physicist and chemist. Developed theory of electrolytic dissociation; investigated osmosis and viscosity of solutions; Nobel Prize, 1903.

Arsonval, Jacques Arsène d' (1851–1940), French physicist and physiologist. Pioneered in electrotherapy; invented d'Arsonval galvanometer.

Aston, Francis William (1877–1945), English physicist and chemist. Discovered isotopes in nonradioactive elements by using the mass spectrograph he invented; Nobel Prize, 1922.

Atiyah, Michael Francis (1929–), British mathematician. Work centered on the interaction between geometry and analysis; developed K theory in collaboration with F. Hirzebruch; with I. M. Singer, proved the index theorem concerning elliptic differential operators on compact differentiable manifolds, which was later seen to have applications to theoretical physics; Fields Medal, 1966; Abel Prize, 2004.

Auger, Pierre Victor (1899–1993), French physicist. Discovered the Auger effect.

Avogadro, Amedeo (1776–1856), Italian physicist. Formulated Avogadro's law.

Ayrton, William Edward (1847–1908), English physicist and electrical engineer. Invented the ammeter, voltmeter, and other electrical measuring instruments.

Baade, Walter (1893–1960), German-born American astronomer. Formulated concept of stellar populations; increased distance scale of universe by factor of 2.

Babinet, Jacques (1794–1872), French physicist. Invented a polariscope and a goniometer.

Back, Ernst E. A. (1881–1959), German physicist. Developed improved spectrographs; made spectroscopic observations leading to Paschen-Back effect.

Balmer, Johann Jakob (1825–1898), Swiss physicist. Expressed the mathematical formula for frequencies of hydrogen lines in the visible spectrum.

Bardeen, John (1908–1991), American physicist. With L. N. Cooper and J. R. Schrieffer, formulated a theory of superconductivity; invented the transistor; Nobel Prize, 1956 and 1972.

Barkhausen, Heinrich Georg (1881–1956), German electronic engineer and physicist. Contributed to theory and application of electron tubes; with K. Kurz, developed Barkhausen-Kurz oscillator; discovered Barkhausen effect.

Barkla, Charles Glover (1877–1944), English physicist. Described characteristics of x-rays and other short-wave emissions of elements.

Barnett, Samuel Jackson (1873–1956), American physicist. Discovered Barnett effect and used it to measure the gyromagnetic ratio of ferromagnetic materials; gave experimental proof of existence of ionosphere.

Bartlett, James Holly (1904–2000), American physicist. Introduced concept of Bartlett force; did research on nuclear shell model, electrochemical potentiostat, and restricted three-body problem.

Basov, Nicolai Gennediyevich (1922–2001), Soviet physicist. Conducted fundamental studies in quantum electronics; with A. M. Prokhorov, developed quantum optical generators; Nobel Prize, 1964.

Beams, Jesse Wakefield (1898–1977), American physicist. Developed vacuum-type ultracentrifuges, used in purification and molecular weight determination of large-molecular-weight substances, isotope separation, and determination of the gravitational constant.

Beattle, James Alexander (1895–1981), American chemist and physicist. Studied ionic theory and thermodynamics; with P. W. Bridgman, proposed Beattie and Bridgman equation for gases.

Becquerel, Antoine César (1788–1878), French physicist. Pioneer in electrochemistry; first to extract metals from ore by electrolysis.

Becquerel, Antoine Henri (1852–1908), French physicist. A discoverer of radioactivity in uranium.

Bednorz, Johannes Georg (1950–), German physicist. With K. A. Müller, discovered high-temperature superconductivity in copper oxide ceramic materials; Nobel Prize, 1987.

Beer, August (1825–1863), German physicist. Discovered Beer's law of light absorption.

Békésy, Georg von (1889–1972), Hungarian-born American physicist. Studied hearing processes, especially inner-ear mechanics; Nobel Prize, 1961.

Bernoulli, Daniel (1700–1782), Swiss mathematician born in the Netherlands. Founder of mathematical physics; worked on hydrodynamics and differential equations; formulated the Bernoulli equation.

Bethe, Hans Albrecht (1906–), German-born American physicist. Formulated the theory of energy production in stars; research on nuclear physics; Nobel Prize, 1967.

Bhabha, Homi Jehangir (1909–1966), Indian physicist. With W. Heitler, developed theory of cascade showers of cosmic rays; observed slowing of decay rate of high-velocity mesons.

Binnig, Gerd (1947–), German physicist. With H. Rohrer, developed scanning tunneling microscope; Nobel Prize, 1986.

Biot, Jean Baptiste (1774–1862), French mathematician and physicist. Discovered circular polarization of light; invented a polariscope; with D. Brewster, discovered biaxial crystals; helped formulate Biot-Savart law.

Bjerknes, Vilhelm Fremann Doren (1862–1951), Norwegian physicist. Research on electric waves; originated the polar-front theory in meteorology.

Black, Joseph (1728–1799) Scottish physicist and chemist. Established the concepts of latent heat and specific heat, and discovered carbon dioxide.

Blackett, Patrick Maynard Stuart (1897–1974), English physicist. Built an improved cloud chamber used to photograph tracks of a nuclear disintegration and of a cosmic-ray shower; discovered the positron; Nobel Prize, 1948.

Bloch, Felix (1905–1983), Swiss-born American physicist. Discovered a technique for studying magnetism of atomic nuclei in normal matter; Nobel Prize, 1952.

Bloembergen, Nicholaas (1920–), Netherlands-born American physicist. Contributed to development of maser; made extensive contributions to theoretical and experimental development of nonlinear optics; Nobel Prize, 1981.

Bobillier, Étienne (1798–1840), French mathematician and physicist. Contributed to geometry and statics; discovered Bobillier's law.

Bohm, David (1917–1992), American-born British physicist. Research in quantum theory and new modes of description in physics.

Bohr, Aage (1922–), Danish physicist. With B. R. Mottelson, developed theory which unifies shell and liquid-drop models of the atomic nucleus, and which explains nonspherical nuclei; Nobel Prize, 1975.

Bohr, Niels (1885–1962), Danish physicist. Devised an atomic model; codeveloped the quantum theory, applying it to atomic structure in Bohr's theory; Nobel Prize, 1922.

Boltzmann, Ludwig Eduard (1844–1906), Austrian physicist. An authority on the kinetic theory of gases; demonstrated the Stefan-Boltzmann law of blackbody radiation, Boltzmann's law of energy, and the Boltzmann constant.

Borda, Jean Charles (1733–1799), French physicist and mathematician. Introduced Borda mouthpiece; developed instruments for navigation, geodesy, and determination of weights and measures.

Born, Max (1882–1970), German-born British theoretical physicist. Pioneered in the development of quantum mechanics; Nobel Prize, 1954.

Bose, Satyendra Nath (1894–1974), Indian physicist. Originated Bose-Einstein statistics to describe photons.

Bothe, Walter (1891–1957), German physicist. Devised the coincidence method for the investigation of nuclear reactions and cosmic radiation; Nobel Prize, 1954.

Bouguer, Pierre (1698–1758), French geodesist, hydrographer, and physicist. Laid foundations of photometry; discovered Bouguer-Lambert law of light intensity.

Bourdon, Eugène (1808–1884), French inventor. Invented Bourdon pressure gage.

Boussinesq, Joseph Valentin (1842–1929), French mathematical physicist. Research in hydrodynamics; introduced Boussinesq approximation.

Boyle, Robert (1627–1691), British physicist and chemist. Conducted experiments on properties of the air pump; his law concerning gases is named for him; advanced the atomistic theory of matter.

Bragg, William Henry (1862–1942), English physicist. Codeveloper, with W. L. Bragg, of the x-ray spectrometer; used x-ray diffraction to determine crystal structure; Nobel Prize, 1915.

Bragg, William Lawrence (1890–1971), British physicist. With W. H. Bragg, developed x-ray analysis of the atomic arrangement in crystalline structures; Nobel Prize, 1915.

Brattain, Walter Houser (1902–1987), American physicist. Investigated properties of semiconductors; research on surface properties of solids; Nobel Prize, 1956.

Braun, Karl Ferdinand (1850–1918), German physicist. Research on cathode rays and wireless telegraphy; Nobel Prize, 1909.

Bravais, Auguste (1811–1863), French physicist. Studied relationship between crystal form and structure; derived Bravais lattices.

Breit, Gregory (1899–1981), Russian-born American physicist. Research on quantum theory, quantum electrodynamics, hyperfine structure, and ionosphere.

Brewster, David (1781–1868), Scottish physicist. Formulated Brewster's law on polarization of light; codiscoverer, with J. B. Biot, of biaxial crystals.

Bridgman, Percy Williams (1882–1961), American physicist. Worked in high-pressure physics and thermodynamics of liquids; Nobel Prize, 1946.

Brillouin, Leon (1889–1969), French physicist. With G. Wentzel and H. A. Kramers, developed Wentzel-Kramers-Brillouin method; originated concept of Brillouin zones.

Brillouin, Louis Marcel (1854–1948), French physicist. Work on crystal structure, viscosity of liquids and gases, radiotelegraphy, and relativity.

Brockhouse, Bertram Neville (1918–2003), Canadian physicist. Developed slow neutron spectroscopy technique for studying dynamics of atoms in solids and liquids; Nobel Prize, 1994.

Broglie, Louis Victor de (1892–1987), French physicist. Worked in nuclear physics; first to link wave and corpuscular theory; Nobel Prize, 1929.

Buckingham, Edgar (1867–1940), American physicist. Worked on thermodynamics and dimensional analysis; derived Buckingham's π theorem.

Callendar, Hugh Longbourne (1863–1930), English physicist and engineer. Developed platinum resistance thermometer and continuous-flow calorimeter.

Carnot, Nicolas Léonard Sadi (1796–1832), French physicist. Formulated Carnot's theorems in thermodynamics.

Cassegrain, N. (17th century), French physician. Designed the Cassegrain reflecting telescope.

Cavendish, Henry (1731–1810), English physicist and chemist. Determined the density of the Earth and the composition of the atmosphere; studied properties of carbon dioxide and hydrogen.

Celsius, Anders (1701–1744), Swedish astronomer. Constructed the thermometer using the Celsius (centigrade) scale.

Cerenkov, Pavel Alexeyevich (1904–1990), Soviet physicist. Discovered the Cerenkov effect of radiation; devised the Cerenkov counter for particle detection; Nobel Prize, 1958.

Chadwick, James (1891–1974), English physicist. Established experimentally the existence of the neutron; Nobel Prize, 1935.

Chamberlain, Owen (1920–), American physicist. With E. G. Segrè, demonstrated the existence of the antiproton; Nobel Prize. 1959.

Chandrasekhar, Subrahmanyan (1910–1995), Indian astrophysicist. Developed a theory of white dwarf stars; Nobel Prize, 1983.

Chaplygin, Sergel Alekseevich (1869–1942), Russian physicist, engineer, and mathematician. Made contributions to fluid mechanics, particularly aerodynamics.

Chapman, Sydney (1888–1970), English mathematician and physicist. Discovered (independently of D. Enskog) gaseous thermal diffusion; studied the daily variations of the geomagnetic field and magnetic storms.

Charles, Jacques Alexandre César (1746–1823), French physicist, chemist, and inventor. Formulated Charles' law, relating gas volume to pressure.

Charpak, Georges (1924–), French physicist. Invented the multiwire proportional chamber, used as a detector in high-energy physics experiments; Nobel Prize, 1992.

Chladini, Ernst Florenz Friedrich (1756–1827), German physicist. Discovered Chladini's figures and used them to study vibrations of solid plates.

Christoffel, Elwin Bruno (1829–1900), Swiss mathematician. Worked in higher analysis, geometry, mathematical physics, and geodesy.

Chu, Ching-Wu (1941–), Chinese-born American physicist. Discovered superconductivity at temperatures over 90 K ($-298°$F) in yttrium-barium-copper-oxygen compounds.

Chu, Steven (1948–), American physicist. Developed a system of opposed laser beams to cool atoms to extremely low temperatures, and a magnetooptical trap to capture them; Nobel Prize, 1997.

Clapeyron, Benoit Paul Émile (1799–1864), French engineer. Developed N. L. S. Carnot's concept of a universal function of temperature.

Clausius, Rudolf Julius Emmanuel (1822–1888), German physicist. A founder of thermodynamics: worked out the Clausius-Clapeyron equation for the universal temperature function.

Cockcroft, John Douglas (1897–1967), English physicist. With E. T. S. Walton, split nuclei by bombarding them with accelerated protons.

Cohen-Tannoudji, Claude (1933–), French physicist. Helped develop methods to cool and trap atoms with laser light, and explained how atoms could be cooled to temperatures lower than the previously calculated theoretical limits; Nobel Prize, 1997.

Cole, Kenneth Stewart (1900–1984), American biophysicist. Research on structure and function of living cell membranes and nerve membranes in particular, concentrating on electrical approach; with brother, R. H. Cole, introduced Cole-Cole plot of dielectric behavior.

Cole, Robert Hugh (1914–1990), American chemist and physicist. Research on dielectric properties of matter and intermolecular forces; with brother, K. S. Cole, introduced Cole-Cole plot of dielectric behavior.

Collins, Samuel Cornette (1898–1984), American engineer. Invented Collins helium liquefier.

Compton, Arthur Holly (1892–1962), American physicist. Discovered the Compton effect of x-rays; studied cosmic rays; helped develop the atomic bomb; Nobel Prize, 1927.

Condon, Edward Uhler (1902–1974). American physicist. Contributed to the Franck-Condon principle, by extending and giving quantum-mechanical treatment to J. Franck's concept of nuclear motion to molecules in transition from one energy level to another.

Coolidge, William David (1873–1975), American physicist. Invented Coolidge tube; discovered method for making tungsten strong and ductile.

Cooper, Leon N. (1930–), American physicist. Showed that electrons could form Cooper pairs; with J. R. Schrieffer and J. Bardeen. formulated a theory of superconductivity; Nobel Prize, 1972.

Copernicus, Nicolaus (1473–1543), Polish (or Prussian) astronomer. Proposed the Copernican system, with the Sun as the center of planetary orbits.

Coriolis, Gaspard Gustave de (1792–1843), French physicist. Contributed to theoretical and applied mechanics; clarified and supplied concepts of work and kinetic energy; derived Coriolis acceleration.

Cormack, Allan MacLeod (1924–1998), American physicist. Contributed to the development of computerized axial tomography; Nobel Prize, 1979.

Cornell, Eric Allin (1961–), American physicist. With C. E. Wieman, succeeded for the first time in producing Bose-Einstein condensates in a dilute gas of alkali (rubidium) atoms, and carried out fundamental studies of their properties, including studies of collective excitations and vortex formation in condensates; Nobel Prize, 2001.

Cornu, Marie Alfred (1841–1902), French physicist. Used Cornu spiral for determination of intensities in interference phenomena.

Coster, Dirk (1889–1950), Dutch physicist. Work in x-ray spectroscopy; with G. von Hevesy, discovered hafnium; with R. Kronig, discovered Coster-Kronig transitions.

Coulomb, Charles Augustin de (1736–1806), French physicist. Formulated Coulomb's law of electric charges.

Cowan, Clyde L., Jr. (1919–1974), American physicist. With F. Reines, made first detection of the neutrino, a fundamental particle.

Cronin, James Watson (1931–), American physicist. Collaborated with V. L. Fitch on experiment showing that the principle of time-reversal invariance is violated in the decay of neutral K mesons; Nobel Prize, 1980.

Crookes, William (1832–1919), English physicist and chemist. Invented Crookes tube to study electrical discharges in high vacuum, and a radiometer; discovered thallium.

Curie, Marie, born Marya Sklodowska (1867–1934), Polish physical chemist in France. Explored nature of radioactivity; codiscoverer of radium, and first to separate polonium; Nobel Prize, 1903 and 1911.

Curie, Pierre (1859–1906), French chemist and physicist. Codiscoverer of radium; formulated the Curie point, relating magnetic properties and temperature; discovered the piezoelectric effect; Nobel Prize, 1903.

Dalén, Nils Gustaf (1869–1937), Swedish physicist. Invented automatic gas lighting for unsupervised lighthouses and railroad signals; Nobel Prize, 1912.

Dalitz, Richard Henry (1925–), Australian-born British theoretical physicist. Research on properties of mesons and baryons and nuclear interactions of the lambda hyperon; proposed models for elementary particles; introduced the Dalitz plot.

Dalton, John (1766–1844), English chemist and physicist. Proposed the atomic theory of chemical reactions; developed the law of partial pressures of gases; studied color-blindness.

Daniell, John Frederic (1790–1845), English physicist and chemist. Invented the Daniell cell.

Davis, Raymond, Jr. (1914–), American physicist. Developed pioneering experiments to detect solar neutrinos, and observed deficit in their number. Nobel Prize, 2002.

Davisson, Clinton Joseph (1881–1958), American physicist. Studied magnetism, radiant energy, and electricity; independent of G. P. Thomson, discovered electron diffraction by crystals; Nobel Prize, 1937.

Debye, Peter Joseph William (1884–1966), American physical chemist born in the Netherlands. Worked on dipole moments and the diffraction of x-rays in gases; formulated Debye-Hückel theory on the behavior of strong electrolytes; Nobel Prize, 1936.

de Gennes, Pierre-Gilles (1932–), French physicist. Applied physical principles to the study of complex systems, including liquid crystals and polymers; Nobel Prize, 1991.

de Haas, Wander Johannes (1878–1960), Dutch physicist. Demonstrated the Einstein-de Haas effect; worked on production of extremely low temperatures by adiabatic demagnetization; with P. Van Alphen, discovered de Haas-Van Alphen effect.

Dehmelt, Hans Georg (1922–), German-born American physicist. Developed the Penning trap, which uses magnetic and electric fields to hold ions in a small volume; used the traps to isolate a single electron and carry out extremely accurate measurements of atomic properties; Nobel Prize, 1989.

de Sitter, Willem (1872–1934), Dutch astronomer. Worked on the application of Einstein's theory to the problems of the universe; computed the size of the universe.

Dicke, Robert Henry (1916–1997), American physicist. Developed new relativistic theory of gravitation with C. Brans; investigated cosmic blackbody radiation; worked on development of radar.

Dirac, Paul Adrien Maurice (1902–1984), English physicist. Worked in quantum mechanics;

his theory of negative-energy holes predicted existence of the positron; Nobel Prize, 1933.

Doppler, Christian Johann (1803–1853), Austrian physicist and mathematician. Formulated Doppler's principle, relating the frequency of wave motion to velocity; described the Doppler effect.

Drude, Paul Karl Ludwig (1863–1906), German physicist. Attempted to correlate and account for optical, electrical, thermal, and chemical properties of substances; developed theory of properties of metals based on free electrons treated as a gas.

Duane, William (1872–1935), American physicist and radiologist. Developed treatment of cancer by radioisotopes and x-rays; with F. L. Hunt, discovered Duane-Hunt law of x-rays.

DuBridge, Lee Alvin (1901–1994), American physicist. Developed Fowler-DuBridge theory of photoelectric emission.

Dulong, Pierre Louis (1785–1838), French chemist and physicist. With A. T. Petit, formulated the law of the constancy of atomic heats; developed the Dulong formula for heat value of fuels.

Ehrenfest, Paul (1880–1933), Austrian-born Dutch theoretical physicist. Contributed to statistical mechanics and quantum mechanics; developed Ehrenfest's principle; proved Ehrenfest's theorem.

Einstein, Albert (1879–1955), German-born American physicist. Proposed the theory of relativity; extended the application of quantum theory; Nobel Prize, 1921.

Elster, Johann Philipp Ludwig Julius (1854–1920), German experimental physicist. With H. F. Geitel, studied atmospheric electricity, radioactivity, and photoelectricity, and invented photocell.

Enskog, David (1884–1947), Swedish physicist. With S. Chapman, developed the Chapman-Enskog theory for solving the Boltzmann transport equation.

Eötvös, Roland, Baron (1848–1919), Hungarian physicist. Research on gravitation and terrestrial magnetism; formulated a law which relates surface tension to temperature of liquids; designed the Eötvös torsion balance.

Esaki, Leo (1925–), Japanese physicist. Discovered a new negative-resistance characteristic in semiconductor pn junctions, leading to the discovery of the tunnel, or Esaki. diode; Nobel Prize, 1973.

Euler, Leonhard (1707–1783), Swiss mathematician. Contributed to algebraic series and differential and integral calculus; realized the significance of coefficients (Euler numbers) of certain trigonometrical expansions.

Ewald, Paul Peter (1888–1985), German-born. American physicist. Developed dynamic theory of x-ray interference in crystals.

Fabry, Charles (1867–1945), French physicist. With A. Pérot, invented Fabry-Perot interferometer; experimentally verified Doppler broadening and Doppler effect.

Fahrenheit, Gabriel Daniel (1686–1736), German physicist. Constructed thermometers; invented the Fahrenheit temperature scale.

Faraday, Michael (1791–1867), English chemist and physicist. Discovered electromagnetic induction; formulated two laws of electrolysis; invented the dynamo.

Fermat, Pierre de (1601–1665), French mathematician. Founder of the modern theory of numbers; originated Fermat's last theorem, and Fermat's principle in optics.

Fermi, Enrico (1901–1954), Italian-born American physicist. Research on producing radioactive isotopes by neutron bombardment; directed construction of the first atomic pile; Nobel Prize, 1938.

Feynman, Richard Phillips (1918–1988), American physicist. Proposed a theory to eliminate difficulties that had arisen in the study of the interaction of electrons, positrons, and radiation; Nobel Prize, 1965.

Fitch, Val Logsdon (1923–), American physicist. Collaborated with J. W. Cronin on experiment showing that the principle of time-reversal invariance is violated in the decay of neutral K mesons; Nobel Prize, 1980.

FitzGerald, George Francis (1851–1901), Irish physicist. Proposed Lorentz-FitzGerald contraction, relating to a material moving through an electromagnetic field.

Fizeau, Armand Hippolyte Louis (1819–1896), French physicist. First to accurately measure the velocity of light; conducted experiments on the velocity of electricity, use of light wavelength to measure length, and measurement of diameter of stars through the method of interference.

Fock (Fok), Vladimir Alexandrovitch (1898–1974), Soviet theoretical physicist. Contributions to quantum electrodynamics, quantum field theory, electromagnetic diffraction and propagation, and general relativity; developed Hartree-Fock approximation of wave functions.

Foucault, Jean Bernard Léon (1819–1868), French physicist. Accurately determined the velocity of light; constructed the Foucault pendulum and the Foucault prism; determined experimentally the rotation of the Earth.

Fourier, Jean Baptiste Joseph, Baron (1768–1830), French geometrician and physicist. Proposed the Fourier series on arbitrary functions; formulated the law of heat propagation.

Fowler, Ralph Howard (1889–1944), English physicist. Applied statistical mechanics to matter at high temperatures and high pressures; explained structure of white dwarf stars; with E. A. Guggenheim, R. F. Peierls, and others, developed Ising model.

Fowler, William Alfred (1911–1995), American physicist. Fundamental contributions to understanding of nuclear reactions that generate the energy of stars and synthesize the elements of the universe; Nobel Prize, 1983.

Franck, James (1882–1964), German physicist. With G. Hertz, studied energy transfer in collisions of molecules; formulated Franck-Condon principle of transition from one energy state to another; Nobel Prize, 1925.

Frank, Ilya Milkhallovich (1908–1990), Soviet physicist. With I. Y. Tamm, proposed a theoretical interpretation of Cerenkov radiation; Nobel Prize, 1958.

Franklin, Benjamin (1706–1790), American physicist, oceanographer, meteorologist, and inventor. Formulated a theory of general electrical "action"; introduced principle of conservation of charge; showed that lightning is an electrical phenomenon; invented lighting rod.

Fraunhofer, Joseph von (1787–1826), German optician and physicist. First to study the dark lines in the solar spectrum (Fraunhofer lines); invented a heliometer; improved the spectroscope.

Frenkel, Yakov Ilyich (1894–1954), Soviet physicist. Pioneered in modern atomic theory of solids; developed quantum-mechanical explanations for electron mean free path in metals, and for paramagnetism and ferromagnetism; postulated excitons, Frenkel excitons, and Frenkel defects.

Fresnel, Augustin Jean (1788–1827), French physicist. Investigated effects (Fresnel's fringes) due to the interference of light; developed a wave theory of light; originated Fresnel's reflection formula.

Friedman, Jerome Isaac (1930–), American physicist. Collaborated in experiments that demonstrated that protons, neutrons, and similar particles are made up of quarks; Nobel Prize, 1990.

Gabor, Dennis (1900–1979), Hungarian-born British physicist and engineer. Invented holography; Nobel Prize, 1971.

Galileo Galilei (1564–1642), Italian astronomer. First to use the telescope for observational purposes; made many discoveries related to the planets and the Sun; did theoretical work on classical physics.

Gamow, George (1904–1968), Russian-born American physicist. Made theoretical contributions to nuclear physics, astronomy, and biology; with E. Teller, formulated the selection rule for beta emission; proposed theoretically the genetic code.

Garvey, Gerald Thomas (1935–), American physicist. Research in experimental nuclear physics, particularly nuclear reactions, isobaric spin studies, and weak interactions in nuclear systems.

Gauss, Karl Friedrich (1777–1855), German mathematician, astronomer, and physicist. Formulated the Gauss theorem in the mathematics of electricity; made many contributions to pure and applied mathematics; determined orbits of planets and comets from observational data.

Gay-Lussac, Joseph Louis (1778–1850), French chemist and physicist. Discovered the law of expansion of gases by heat, and the law of combining volumes of gases; studied chemistry of iodine and cyanogen.

Geiger, Hans Wilhelm (1882–1945), German physicist and inventor. Invented the Geiger counter to detect alpha particles; investigated properties of alpha particles, cosmic rays, and artificial radiation.

Geissler, Johann Heinrich Wilhelm (1815–1879), German instrument maker. Developed Geissler pump and Geissler tube.

Geitel, Hans Friedrich (1855–1923), German experimental physicist. With J. Elster, studied atmospheric electricity, radioactivity, and photoelectricity, and invented photocell.

Gell-Mann, Murray (1929–), American physicist. Proposed law of conservation of strangeness; used unitary symmetry to classify and explain elementary particles; postulated concept of quarks; Nobel Prize, 1969.

Giacconi, Riccardo (1931–), Italian-born American astrophysicist. Pioneered the field of x-ray astronomy, working out the principles of an x-ray telescope and leading the development of the early x-ray telescopes. Nobel Prize, 2002.

Giaever, Ivar (1929–), Norwegian-born American physicist. Discovered that current-voltage characteristics of an electron tunneling across a thin insulating film separating two metals, one or both of which is in a superconducting state, can be used to obtain electron density of states of superconductors; Nobel Prize, 1973.

Giauque, William Francis (1895–1982), Canadian-born American chemist. Developed adiabatic demagnetization technique for production of extremely low temperatures; collaborated in discovery of isotopes of oxygen; Nobel Prize, 1949.

Gibbs, Josiah Willard (1839–1903), American mathematician and physicist. Made a mathematical treatment of chemical subjects, notably thermodynamics; worked on statistical mechanics, leading to the basis for the phase rule of heterogeneous equilibria.

Ginzburg, Vitaly Lazarevich (1916–), Soviet physicist. Developed Ginzburg-Landau and Ginzburg-London theories of superconductivity; Nobel Prize, 2003.

Giorgi, Giovanni (1871–1950), Italian electrical engineer, physicist, and mathematician. Developed the meter-kilogram-second-ampere system of units.

Glaser, Donald Arthur (1926–), American physicist. Invented the bubble chamber for detecting the paths of high-energy atomic particles; Nobel Prize, 1960.

Glashow, Sheldon Lee (1932–), American physicist. Contributed to development of theory uniting electromagnetism and weak nuclear interactions; postulated existence of charmed particles; Nobel Prize, 1979.

Goldhaber, Maurice (1911–), Austrian-born American physicist. With J. Chadwick, discovered photodisintegration and disintegration of light elements by slow neutrons; with L. Grodzins and A. W. Sunyar, discovered that the neutrino has left-handed spin.

Gordon, Walter (1893–1940), German-born Swedish physicist. Contributed to relativistic quantum theory; with O. B. Klein, originated the Klein-Gordon equation.

Goudsmit, Samuel Abraham (1902–1978), Dutch-born American physicist. With G. E. Uhlenbeck, discovered electron spin.

Green, George (1793–1841), English mathematician. Worked in analysis; derived Green's theorem and Green's identities; introduced Green's function.

Gregory, James (1638–1675), Scottish geometer. Provided first proof of the theorem of calculus; gave first description of the reflecting telescope; discovered the series from which π can be calculated.

Gross, David J. (1941–), American physicist. Discovered asymptotic freedom in the strong interactions (in collaboration with F. Wilczek, and independent of H. D. Politzer). Nobel Prize, 2004.

Gruneisen, Eduard (1877–1949), German physicist. Formulated laws relating specific heat and other properties of solids.

Guillaume, Charles Édouard (1861–1938), Swiss-born French physicist. Studied nickel-steel alloys and invented Invar; Nobel Prize, 1920.

Gunn, John Battiscombe (1928–), Egyptian-born American physicist. Discovered Gunn effect and used it to develop Gunn oscillator.

Hagen, Carl Ernst Bessel (1851–1923), German physicist. With H. Rubens, conducted experiments confirming Maxwell's electromagnetic theory of light, permitting determination of electrical conductivity of metals by optical measurements alone.

Hagen, Gotthilf Heinrich Ludwig (1797–1884), German hydraulic engineer. Discovered Hagen-Poiseuille law independently of J. L. M. Poiseuille; directed construction of dikes, harbor installations, and dune fortifications.

Hahn, Otto (1879–1968), German chemist. With L. Meitner and F. Strassman, discovered that fission of heavy nuclei was possible by irradiation with neutrons; discovered protactinium with Meitner; Nobel Prize, 1944.

Hall, Edwin Herbert (1855–1938), American physicist. Discovered Hall effect and conducted studies of this and other galvanomagnetic and thermomagnetic effects.

Hamilton, William Rowan (1805–1865), Irish mathematician and mathematical physicist. Discovered quaternions; developed mathematical theories encompassing wave and particle optics and mechanics; introduced Hamilton's principle and a form of the Hamilton-Jacobi theory.

Harker, David (1906–1991), American crystallographer. Completed development of Patterson-Harker method of x-ray diffraction analysis of crystal structure.

Hartmann, Johannes Franz (1865–1936), German astronomer. Derived Hartmann dispersion formula relating index of refraction and wavelengths; devised Hartmann test for telescope mirrors; gave first observational proof of interstellar matter.

Hartree, Douglas Rayner (1897–1958). English mathematician and mathematical physicist. Developed methods of numerical analysis which made it possible to apply Hartree method to calculation of atomic wave functions.

Hauptman, Herbert A. (1917–), American chemist. With J. Karle, developed computer-aided mathematical techniques for use in x-ray crystallography to determine three-dimensional structures of molecules: Nobel Prize, 1985.

Haüy, Réné Just, Abbé (1743–1822), French mineralogist. Formulated the geometrical law of crystallization; pioneer in the science of crystallography.

Heaviside, Oliver (1850–1925), English physicist. Proposed the Heaviside layer in the upper atmosphere.

Heeger, Alan J. (1936–), American physicist. Discovered and developed conductive polymers with Alan MacDiarmid and Hideki Shirakawa; Nobel Prize, 2000.

Hefner-Alteneck, Friedrich Franz von (1845–1904), German engineer. Invented Hefner candle as a standard of luminous intensity.

Heisenberg, Werner (1901–1976), German physicist. Founder of quantum mechanics; studied structure of the atom and the Zeeman effect; formulated the principle of indeterminancy in nuclear physics; Nobel Prize, 1932.

Heitler, Watler Heinrich (1904–1981), German-born Swiss theoretical physicist. Developed Heitler-London covalence theory of chemical bonding.

Helmholtz, Hermann Ludwig Ferdinand von (1821–1894), German physicist. anatomist, and physiologist. Physiological research on the nervous system and the human eye and ear, and theoretical work on conservation of force in physics; invented the ophthalmoscope.

Henry, Joseph (1797–1878), American physicist. Studied electromagnetic induction, solar phenomena, meteorology, and acoustics.

Hero of Alexandria (3d century or earlier), Greek mathematician. Wrote on the geometry of plane and solid figures, mechanics, and simple machines; showed that the angle of incidence equals the angle of reflection.

Herschel, John Frederick William (1792–1871), English mathematician, physicist, and astronomer. Discovered many nebulae and clusters; pioneered in celestial photography.

Herschel, William or Friedrich Wilhelm (1738–1822), German-born English astronomer. Discovered Uranus, two of its satellites, and two satellites of Saturn; discovered the Sun's intrinsic motion; proposed the concept of the form of the Milky Way.

Hertz, Gustav (1887–1975), German physicist. With J. Franck, studied effects of electron impacts on atoms; Nobel Prize, 1925.

Hertz, Heinrich Rudolph (1857–1894), German physicist. Discovered Hertzian waves in the ether; proved experimentally Maxwell's theories of electricity and magnetism.

Herzberg, Gerhard (1904–1999), German-born Canadian physicist. Determined electronic structure and geometry of diatomic and polyatomic molecules, particularly free radicals; Nobel Prize, 1971.

Hess, Victor Franz (1883–1964), Austrian physicist. Studied alpha particles from radium; discovered cosmic rays; Nobel Prize, 1936.

Hewish, Antony (1924–), British astronomer. Pioneered in discovery of pulsars, by means of radio telescopes; Nobel Prize, 1974.

Hittorf, Johann Wilhelm (1824–1914), German physicist. Described effects of Hittorf rays in vacuum tubes; studied electrolysis, and electrical discharge in rarefied gases with the Hittorf tube.

Hofstadter, Robert (1915–1990), American physicist. Investigated the properties and behavior of the proton and neutron; determined the size and shape of many nuclei; discovered the construction scheme of fundamental atomic nuclei; Nobel Prize, 1961.

Hubble, Edwin Powell (1889–1953), American astronomer. Studied nebulae; formulated Hubble's law of extragalactic nebulae.

Hugoniot, Pierre Henry (1851–1887), French physicist. Developed theory of shock waves.

Hulse, Russell A. (1950–), American astronomer and physicist. With J. H. Taylor, discovered the binary pulsar and studied it to observe phenomena predicted by general relativity; worked in plasma physics; Nobel Prize, 1993.

Hunt, Franklin Livingston (1883–1973), American physicist. Research on x-ray spectroscopy; with W. Duane, discovered and applied Duane-Hunt law.

Huygens, Christiaan (1629–1695), Dutch mathematician, physicist, and astronomer. Discovered Saturn's rings; contributed to dynamics and optics; proposed the wave theory of light.

Hylleraas, Egil Andersen (1898–1965), Norwegian physicist. Applied quantum theory to helium atom, negative hydrogen ion, and other atoms, molecules, and crystals; developed variational method and other methods for mathematical solution of quantum-mechanical problems.

Ising, Ernest (1900–1998), German-born American physicist. Introduced Ising model of ferromagnetic material; research in solid-state physics and ferromagnetism.

Jacobi, Karl Gustav Jacob (1804–1851), German mathematician. Worked on elliptic functions and differential equations; developed the theory of determinants.

Jeans, James Hopwood (1877–1946), English physicist and astronomer. Worked in stellar dynamics; proposed the tidal theory of the origin of planets.

Jensen, J. Hans D. (1906–1973), German physicist. With M. G. Mayer, formulated the nuclear shell model; Nobel Prize, 1963.

Joliot-Curie, Irène (1897–1956), French physicist. With M. Curie, discovered projection of atomic nuclei by neutrons; with J. F. Joliot-Curie, discovered artificial radiation; Nobel Prize, 1935.

Joliot-Curie, Jean Frédéric (1900–1958), French physicist. With I. Joliot-Curie, produced an artificial radioactive substance by bombarding boron with fast alpha particles; Nobel Prize, 1935.

Jordan, Pascual (1902–1980), German physicist. Contributed to formulation of quantum mechanics; introduced Jordan algebra in an attempt to generalize quantum mechanics.

Josephson, Brian David (1940–), British physicist. Predicted the Josephson effect concerning electron pairs; Nobel Prize, 1973.

Joukowski, Nikolai Jegorowitch (1847–1921), Russian applied mathematician and aerodynamicist. Helped to introduce concept of Kutta-Joukowski airfoil and to prove Kutta-Joukowski theorem.

Joule, James Prescott (1818–1889), English physicist. Formulated a mechanical theory of heat; demonstrated Joule-Thomson effect relating to the fall in temperature of a gas; first to estimate the velocity of a gas molecule.

Kaluza, Theodor Franz Eduard (1885–1954), German mathematical physicist. Developed theory which attempted to unify gravitation and electromagnetism.

Kamerlingh Onnes, Heike (1853–1926), Dutch physicist. Research on cryogenics, critical phenomena, and low temperatures; discovered the phenomenon of superconductivity; Nobel Prize, 1913.

Kapitza, Pjotr Leonidovich (1894–1984), Russian physicist. Studied magnetism and low temperature; designed hydrogen and helium liquefaction plants; Nobel Prize, 1978.

Kapteyn, Jacobus Cornelius (1851–1922), Dutch astronomer. Studied the proper motion of stars; with P. J. van Rhijn, evolved a theory of the universe.

Karle, Jerome (1918–), American crystallographer. With H. A. Hauptman, developed computer-aided mathematical techniques for use in x-ray crystallography to determine three-dimensional structures of molecules; Nobel Prize, 1985.

Kastler, Alfred (1902–1984). French physicist. Developed a double-resonance method to study energy levels of atoms in excited states; Nobel Prize, 1966.

Keesom, Willem Hendrik (1876–1956), Dutch physicist. Worked in low-temperature physics; first to solidify helium; studied molecular structure of liquids and compressed gases.

Kelvin, William Thomson, 1st Baron (1824–1907), British mathematician and physicist. Invented the Kelvin balance; formulated Kelvin's laws concerning electric cables; contributed to thermodynamics.

Kendall, Henry Way (1926–1999), American physicist. Collaborated in experiments that demonstrated that protons, neutrons, and similar particles are made up of quarks; Nobel Prize, 1990.

Kepler, Johannes (1571–1630), German astronomer. Proposed Kepler's three laws of planetary motion; worked in optics.

Kerr, John (1824–1907), Scottish physicist. Discovered the Kerr magnetooptic effect.

Ketterle, Wolfgang (1957–), German physicist. Produced Bose-Einstein condensates in a dilute gas of sodium atoms (independent of the work of E. A. Cornell and C. E. Wieman) and carried out fundamental studies of their properties, including the production of interference patterns and atom lasers; Nobel Prize, 2001.

Kilby, Jack St. Clair (1923–), American physicist, electronics engineer, and inventor. Participated in the invention of the integrated circuit; Nobel Prize, 2000.

Kirchhoff, Gustav Robert (1824–1887), German physicist. With R. W. Bunsen, discovered method of spectrum analysis; formulated Kirchhoff's law of electric currents and electromotive forces in a network.

Klein, Oskar Benjamin (1894–1977), Swedish physicist. Codeveloper of Klein-Gordon equation. Klein-Nishina formula, and Klein-Rydberg method; proposed theory of overall structure of the universe.

Klitzing, Klaus von (1943–), German physicist. Discovered quantum Hall effect; Nobel Prize, 1985.

Knudsen, Martin Hans Christian (1871–1949), Danish physicist and hydrographer. Studied flow and diffusion of gases at low pressure; developed Knudsen cell and Knudsen gage; developed methods to measure the properties of seawater.

Kohn, Walter (1923–), Austrian-born American physicist. Developed density-functional theory, which solves equations for electron density rather than positions of individual electrons; it is one of the developments that has significantly sped up computational quantum chemistry; Nobel Prize, 1998.

Kontsevich, Maxim (1964–), Russian mathematician and mathematical physicist. Worked in algebraic geometry, algebraic topology, string theory, and quantum field theory; demonstrated equivalence of two models of quantum gravitation; discovered an invariant for classifying knots; Fields Medal, 1998.

Korteweg, Diederik Johannes (1848–1941), Dutch mathematician. Work in applied mathematics, mechanics, and hydrodynamics; with G. de Vries, proposed equation of wave motion with soliton solution.

Kosiba, Masatoshi (1926–), Japanese physicist. Developed large water detectors (Kamiokande and Super Kamiokande) that observed solar and atmospheric neutrinos and neutrinos from Supernova 1987A, and found evidence for neutrino oscillations. Nobel Prize, 2002.

Kramers, Hendrik Anthony (1894–1952), Dutch physicist. Developed quantum theory of dispersion, establishing Kramers-Kronig relation; with G. Wentzel and L. Brillouin, developed Wentzel-Kramers-Brillouin method.

Kroemer, Herbert (1928–), German-American physicist and electronics engineer. Developed semiconductor heterostructures used in high-speed and opto-electronics, including fast transistors, laser diodes, and light-emitting diodes; Nobel Prize, 2000.

Kruskal, Martin David (1925–), American mathematician and physicist. Research in plasma physics, asymptotic phenomena, relativity, and minimal surfaces.

Kundt, August Adolph (1839–1894), German physicist. Used Kundt tube to determine speed of sound in gases; determined ratio of specific heats of monatomic gases; with W. K. Röntgen, demonstrated Faraday effect in gases.

Kurchatov, Igor Vasilievich (1903–1960), Soviet physicist. Discovered nuclear isomers; studied nuclear reactions; developed nuclear weapons and nuclear power.

Kusch, Polykarp (1911–1993), German-born American physicist. Precisely determined the magnetic moment of the electron; Nobel Prize, 1955.

Kutta, Wilhelm Martin (1867–1944), German applied mathematician. Helped introduce concept of Kutta-Joukowski airfoil, prove Kutta-Joukowski theorem, and develop Runge-Kutta method.

Lagrange, Joseph Louis, Count (1736–1813), French geometer and astronomer. Invented the calculus of variations; studied the mathematics of sound; wrote *Mécanique Analytique*, concerning statics and dynamics.

Lamb, Willis Eugene, Jr. (1913–), American physicist. Made precise atomic measurements leading to a new understanding of the theory of electron interactions and electromagnetic radiation; Nobel Prize, 1955.

Lambert, Johann Heinrich (1728–1777), German physicist. Formulated the Lambert theorem concerning the illumination of a surface.

Lamé, Gabriel (1795–1870), French mathematician, physicist, and engineer. Introduced curvilinear coordinates and applied them to differential equations, elasticity, thermodynamics, and number theory.

Landau, Lev Davydovich (1908–1968), Soviet physicist. Made theoretical explanation of the nature and properties of liquid helium; investigated condensed matter; Nobel Prize, 1962.

Landé, Alfred (1888–1975), German-born American physicist. Introduced Landé g factor; discovered Landé interval rule and Landé Γ-permanence rule.

Langevin, Paul (1872–1946), French physicist. Developed quantitative theories of paramagnetism and diamagnetism; helped to elucidate the theory of relativity; contributed to the development of sonar.

Langmuir, Irving (1881–1957), American chemist. With G. N. Lewis, proposed the Lewis-Langmuir atomic theory; studied surface chemistry and thermionic emission; Nobel Prize, 1932.

Laplace, Pierre Simon, Marquis de (1749–1827), French astronomer and mathematician. Contributed to celestial mechanics, especially to the study of the Moon, Saturn, and Jupiter; formulated the theory of probability; discovered the Laplace differential equation.

Larmor, Joseph (1857–1942), British physicist. Developed electron theory which fused electromagnetic and optical concepts; introduced Larmor precession and derived Larmor formula.

Laue, Max Theodor Felix von (1879–1960), German physicist. Proposed the theory of x-ray diffraction by crystals; developed the Laue method of investigating crystal structure; Nobel Prize, 1914.

Laughlin, Robert Betts (1950–), American physicist. Provided a theoretical explanation for the fractional quantum Hall effect by showing how electrons acting together in strong magnetic fields can form new types of quasiparticles with charges that are fractions of electron charges; Nobel Prize, 1998.

Laurent, Pierre Alphonse (1813–1854), French mathematician and physicist. Introduced Laurent series; research on wave theory of light.

Lawrence, Ernest Orlando (1901–1958), American physicist. Discovery, development, and use of the cyclotron; Nobel Prize, 1939.

Lederman, Leon Max (1922–), American physicist. Collaborated in experiment that demonstrated the existence of two types of neutrino; led an experiment that discovered the upsilon particle; Nobel Prize, 1988.

Lee, David Morris (1931–), American physicist. With D. D. Osheroff and R. C, Richardson, discovered superfluidity in helium-3; Nobel Prize, 1996.

Lee, Tsung-Dao (1926–), Chinese-born American physicist. With C. N. Yang disproved the parity principle; worked on statistical mechanics, astrophysics, nuclear and subnuclear physics, and field theory; Nobel Prize, 1957.

Leggett, Anthony J. (1938–), British and American physicist. Developed a theory explaining the complex behavior of superfluid helium-3. Nobel Prize, 2003.

Lenard, Phillipp Eduard Anton (1862–1947), Hungarian-born German physicist. Studied cathode rays outside the discharge tube; worked on photoelectricity; Nobel Prize, 1905.

Lennard-Jones, John Edward (1894–1954), English physicist and chemist. Proposed Lennard-Jones potential for interatomic forces; contributed to quantum theory of molecular structure and statistical mechanics of liquids, gases, and surfaces.

Lenz, Heinrich Friedrich Emil (1804–1865), German physicist. Formulated Lenz's law governing induced current.

Levi-Civita, Tullio (1873–1941), Italian mathematician and mathematical physicist. With G. Ricci-Curbastro, developed tensor analysis; introduced concept of parallelism in curved spaces.

Lippmann, Gabriel (1845–1921), French physicist. Produced the first colored photograph of the light spectrum; invented the Lippmann capillary electrometer; Nobel Prize, 1908.

Lissajous, Jules Antoine (1822–1880), French physicist. Invented the vibration microscope, involving Lissajous figures.

Littlewood, John Endensor (1885–1977), British mathematician. Work on diophantine approximation, Tauberian theorems, Fourier series and associated function theory, the zeta function, additive number theory, and inequalities.

Littrow, Joseph Johann von (1781–1840), Austrian astronomer. Studied light refraction; worked on telescope construction.

Lloyd, Humphrey (1800–1881), Irish physicist. Discovered Lloyd's mirror interference; verified W. R. Hamilton's prediction of conical refraction.

London, Fritz (1900–1954), German-born American physicist. Developed, with W. Heitler, theory of covalent bonding; with H. London, theory of superconductivity; and theory of superfluidity.

London, Heinz (1907–1970), German-born English physicist. Research on electrodynamic and ther-modynamic behavior of superconductors and properties of superfluid helium.

Lorentz, Hendrik Antoon (1853–1928), Dutch physicist. Proposed the electron theory to explain electromagnetic properties of materials; proposed the Lorentz-FitzGerald contraction and the Lorentz transformation, contributing to the theory of relativity; studied Zeeman effect; Nobel Prize, 1902.

Loschmidt, Johann Joseph (1821–1895), Austrian physicist and chemist, born in Bohemia. Worked on graphical and structural molecular formulas; attempted to estimate size of air molecules and number of air molecules per unit volume.

Lummer, Otto Richard (1860–1925), German physicist. Codeveloper of Lummer-Brodhun sight box and Lummer-Gehrcke plate; constructed an improved bolometer.

Lyapunov, Aleksandr Mikhailovich (1857–1918), Soviet mathematician and physicist. Determined in what cases linear approximations can be used to solve the problem of stability of a mechanical system with a finite number of degrees of freedom; proved existence of various figures of equilibrium for a rotating liquid.

Lyman, Theodore (1874–1954), American physicist. Observed ultraviolet spectra; clarified nature of Lyman ghosts; discovered Lyman series.

Lyot, Bernard Ferdinand (1879–1952), French astronomer. Invented the coronagraph and developed monochromatic filters that greatly extended knowledge of the solar corona.

MacDiarmid, Alan G. (1927–), New Zealand-born American chemist. Discovered and developed conductive polymers with Hideki Shirakawa and Alan Heeger; Nobel Prize, 2000.

Mach, Ernst (1838–1916), Austrian physicist. Research on supersonic flight, leading to Mach angle and Mach number; studied airflow over objects at high speeds.

Magnus, Heinrich Gustav (1802–1872), German physicist and chemist. Made first quantitative analysis of blood gases; showed that arterial blood has higher oxygen content than venous blood; discovered Magnus effect.

Majorana, Ettore (1906–1938), Italian physicist. Studied properties of elementary particles; postulated Majorana force.

Maksutov, Dmitry Dmitrievich (1896–1964), Soviet physicist and astronomer. Developed general theory of aplanatic optical systems; developed Maksutov system.

Malus, Étienne Louis (1775–1812), French engineer and physicist. Formulated Malus' cosine-squared law concerning polarized light and Malus' law of rays.

Mandelstam, Stanley (1928–), American physicist, born in South Africa. Research in theoretical physics of elementary particles; introduced Mandelstam plane and Mandelstam representation.

Mansfield, Peter (1933–), British physicist. Developed mathematical techniques for capturing, analyzing, and processing magnetic resonance signals more efficiently, making it possible to produce three-dimensional images of internal organs. Nobel Prize, 2003.

Mariotte, Edmé (?-1684), French physicist and physiologist. Discovered blind spot; studied circulation of sap in plants, collisions of bodies, properties of air, refraction and color of light, hydrostatics, hydraulics, and meteorology.

Mathieu, Emile Leonard (1835–1890), French mathematician and physicist. Worked on solution of partial differential equations; research in celestial and analytical mechanics; studied Mathieu equation and introduced Mathieu functions.

Matthias, Bernd Teo (1919–1980), German-born American physicist. Tested metals and alloys for superconductivity; developed empirical rules to predict new superconducting materials.

Maupertuis, Pierre Louis Moreau de (1698–1759), French mathematician and astronomer. Discovered the principle of least action: mathematical writings on the properties of curves.

Maxwell, James Clerk (1831–1879), Scottish physicist. Formulated the electromagnetic theory of light and the Maxwell distribution of molecular velocities of gases; invented the Maxwell disk concerning color vision.

Mayer, Julius Robert von (1814–1878), German physicist. Discovered the principle of conservation of energy.

Mayer, Maria Goeppert (1906–1972), German-born American nuclear physicist. With J. H. D.

Jensen, discovered nuclear shell structure; Nobel Prize, 1963.

McMillan, Edwin Mattison (1907–1991), American physicist. Discovered element 93 (neptunium), which led to the creation of element 94 (plutonium); conceived the theory of phase stability; Nobel Prize, 1951.

McMullen, Curtis Tracy (1958–), American mathematician. Worked in hyperbolic geometry and in complex dynamics, also known as chaos theory; Fields Medal, 1998.

Meissner, Walther (1882–1974), German physicist. Research in low-temperature physics; discovered Meissner effect.

Meitner, Lise (1878–1968), German physicist. With O. Hahn, discovered protactinium; found evidence of four other radioactive elements; with Hahn and F. Strassmann, accomplished fission of uranium.

Mersenne, Marin (1588–1648), French physicist. Showed that pitch is proportional to frequency and calculated frequencies of musical notes; discovered Mersenne's law for vibrating strings, and similar relations for wind and percussion instruments.

Michelson, Albert Abraham (1852–1931), American physicist. Experimented on the velocity of light with S. Newcomb; invented the Michelson interferometer; performed, with E. W. Morley, an experiment to determine the Earth's motion through the ether; Nobel Prize, 1907.

Mie, Gustav (1868–1957), German physicist. Carried out rigorous electrodynamic calculation of Mie scattering; attempted to formulate theory of matter.

Miller, William Hallowes (1801–1880), British crystallographer and mineralogist. Introduced Miller indices for identifying crystallographic planes.

Millikan, Robert Andrews (1868–1953), American physicist. Determined an accurate value for Planck's constant; originated the "oil drop" experiment to measure electronic charge; work on x-rays and cosmic rays; Nobel Prize, 1923.

Minkowski, Hermann (1864–1909), Russian-born German mathematician. Studied the mathematical basis of relativity, notably the concept of the space-time continuum.

Mollier, Richard (1863–1935), German physicist and engineer. Presented properties of thermodynamic media in form of charts and diagrams; introduced concept of enthalpy and Mollier diagram.

Morley, Edward Williams (1838–1923), American chemist and physicist. Associated with A. A. Michelson in an experiment on ether drift; research on variations of atmospheric oxygen content.

Moseley, Henry Gwyn Jeffries (1887–1915), English physicist. Discovered Moseley's law for frequency of x-ray spectral lines.

Mössbauer, Rudolf Ludwig (1929–), German physicist. Discovered the property of recoilless resonance absorption, the ability of some nuclei to emit and absorb gamma rays without energy loss; Nobel Prize, 1961.

Mossotti, Ottaviano Fabrizio (1791–1863), Italian physicist. Developed theory of dielectrics, from which he derived the Clausius-Mossotti equation.

Mott, Nevill Francis (1905–1996), British physicist. Applied quantum mechanics to study of charged particle scattering; with R. W. Gurney, developed Gurney-Mott theory of photographic process; introduced fundamental concepts elucidating electronic properties of disordered materials; Nobel Prize, 1977.

Mottelson, Ben Roy (1926–), American-born Danish physicist. With A. Bohr, developed theory which unifies shell and liquid-drop models of atomic nucleus, and which explains nonspherical nuclei; Nobel Prize, 1975.

Müller, Karl Alex (1927–), Swiss physicist. With J. G. Bednorz, discovered high-temperature superconductivity in copper oxide ceramic materials; Nobel Prize, 1987.

Navier, Claude Louis Marie Henri (1785–1836), French physicist and engineer. Studied analytical mechanics and its application to strength of materials, machines, and motion of solid and liquid bodies; formulated Navier-Stokes equations.

Néel, Louis Eugène Félix (1904–2000), French physicist. Proposed the theory of behavior of antiferromagnetic and other ferrimagnetic materials in which the crystal lattice is divided into one or more sublattices; Nobel Prize, 1970.

Nernst, Hermann Walther (1864–1941), German chemist. Proposed the heat theorem (third law of thermodynamics); determined the specific heat of solids at low temperatures; proposed the chain reaction theory in photochemistry; Nobel Prize, 1920.

Neumann, Carl Gottfried (1832–1925), German mathematician. Believed to be founder of logarithmic potentials; developed the potential theory.

Newton, Isaac (1642–1727), English mathematician. Proposed a dynamical theory of gravitation; discovered three basic laws of motion which are the foundation of practical mechanics; made discoveries in optics and mathematics.

Nicholson, Seth Barnes (1891–1963), American astronomer. Discovered four satellites of Jupiter; with E. Petit, invented a thermocouple to measure surface temperature of planets.

Nicol, William (1763–1851), Scottish physicist. Invented the Nicol prism for investigating the polarization of light.

Nishina, Yoshio (1890–1951), Japanese physicist. Pioneer in study of cosmic rays; with O. B. Klein, originated the Klein-Nishina formula.

Noyce, Robert Norton (1927–1990), American physicist, electronics engineer, and inventor. Participated in the invention of the integrated circuit.

Nusselt, Ernst Kraft Wilhelm (1882–1957), German mechanical engineer and physicist. Used dimensional analysis to derive functional form of solutions to equations for heat flux in a flowing fluid.

Nyquist, Harry (1889–1976), Swedish-born American physicist and engineer. Discovered conditions necessary to keep feedback control circuits stable; determined Nyquist rate for communications channels.

Oersted, Hans Christian (1777–1851), Danish physicist, chemist, and electromagnetist. Discovered a fundamental principle of electromagnetism: a magnetic needle turns at right angles to an electric current.

Ohm, Georg Simon (1787–1854), German physicist. Discovered Ohm's law relating electrical resistance to voltage and current.

Onsager, Lars (1903–1976), Norwegian-born American chemist. Laid the foundation of irreversible thermodynamics; contributed to theories of dielectrics, electrolytes, and cooperative phenomena; Nobel Prize, 1968.

Oppenheimer, J. Robert (1904–1967), American physicist. Research on nuclear disintegration, quantum theory, cosmic rays, and relativity; directed production of the atomic bomb.

Osheroff, Douglas D. (1945–), American physicist. With D. M. Lee and R. C. Richardson, discovered superfluidity in helium-3; Nobel Prize, 1996.

Pascal, Blaise (1623–1662), French mathematician and physicist. Contributed to the geometry of conies; formulated Pascal's law, relating to the pressure of a liquid at rest; applied Pascal's triangle to the calculation of probabilities.

Paschen, Louis Carl Heinrich Friedrich (1865–1947), German physicist. Established Paschen's law; with E. Back, discovered Paschen-Back effect; verified predictions of relativistic fine structure made by Bohr-Sommerfeld theory.

Patterson, Arthur Lindo (1902–1966), New Zealand-born American physicist and crystallographer. Developed Patterson-Harker method of x-ray diffraction analysis of crystal structure.

Paul, Wolfgang (1913–1993), German physicist. Invented the Paul trap, which uses radio-frequency radiation to hold ions in a small volume; Nobel Prize, 1989.

Pauli, Wolfgang (1900–1958), Austrian-born American physicist. Worked on quantum theory; formulated the Pauli exclusion principle; contributed to matrix mechanics; Nobel Prize, 1945.

Pauling, Linus Carl (1901–1994), American chemist. Applied quantum theory to chemistry; research on molecular structure and chemical bonds; contributed to electrochemical theory of valency; Nobel Prize, 1954; Nobel Peace Prize, 1963.

Peierls, Rudolf Ernst (1907–1995). German-born British physicist. Developed theory of heat conduction in nonmetallic crystals; with O. R. Frisch, calculated critical mass of uranium-235.

Peirce, Charles Santiago Sanders (1839–1914), American mathematician, logician, and physicist. Laid foundation for logical analysis of mathematics; contributed to probability theory.

Peltier, Jean Charles Athanase (1785–1845), French physicist. Discovered the Peltier effect in thermoelectricity.

Penrose, Roger (1931–), British mathematician and physicist. Developed twistor theory of space-time geometry; studied singularities in classical general relativity theory.

Penzias, Arno A. (1933–), American astrophysicist. With R. W. Wilson, discovered cosmic background radiation, confirming the big bang theory of the origin of the universe; Nobel Prize, 1978.

Perl, Martin L. (1927–), American physicist. Discovered the tau lepton, a fundamental particle; Nobel Prize, 1995.

Pérot, Jean Baptiste Gaspard Gustav Alfred (1863–1925), French physicist. With C. Fabry, developed Fabry-Pérot interferometer.

Perrin, Jean Baptiste (1870–1942), French physicist. Research on the particle nature of cathode rays; found values for Avogadro's number, thereby proving the existence of molecules; Nobel Prize, 1926.

Petit, Alexis Thérèse (1791–1820), French physicist. With P. L. Dulong, formulated the law of constancy of atomic heats; devised methods for determining thermal expansion and specific heats of solids.

Pettit, Edison (1890–1962). American astronomer. Studied the Sun and formulated laws alleged to govern the movement of prominences; constructed the interference polarizing monochromator; with S. B. Nicholson, devised a sensitive thermocouple to measure the surface temperatures of planets.

Phillips, William Daniel (1948–), American physicist. Developed a method of slowing and trapping atoms in an atomic beam by using an opposed laser beam and a magnetic trap, and cooled these atoms to temperatures lower than the previously calculated theoretical limits: Nobel Prize, 1997.

Pierce, George Washington (1872–1956). American physicist and electronic engineer. Developed theoretical basis of electrical communications: developed Pierce oscillator; with A. E. Kennelly, discovered concept of motional impedance.

Planck, Max Karl Ernst Ludwig (1858–1947), German physicist. Presented the quantum theory; introduced Planck's constant, or quantum of action.

Planté, Gaston (1834–1889), French physicist. Constructed a storage battery, the first primitive accumulator.

Plateau, Joseph Antoine Ferdinand (1801–1883), Belgian physicist. Experimented · with soapy films bounded by wires, noting that the surfaces formed were minimal surfaces; from this he formulated the Plateau problem (the problem of determining the existence of a minimal surface with a given space curve as its boundary).

Podolsky, Boris (1896–1966), Russian-born American physicist. Collaborated in formulation of Einstein-Podolsky-Rosen paradox; research on quantum electrodynamics.

Poincaré, Jules Henri (1854–1912), French mathematician. Worked on the theory of functions, on differential equations, and on the theory of orbits in astronomy.

Poinsot, Louis (1777–1859), French mathematician. Originated theory of couples.

Poiseuille, Jean Léonard Marie (1797–1869), French physiologist and physicist. Studied physiology of arterial circulation; invented improved methods for measuring blood pressure; discovered Hagen-Poiseuille law independently of G. H. L. Hagen.

Poisson, Siméon Denis (1781–1840), French mathematician. Worked on mathematical physics; contributed to the wave theory of light; formulated the Poisson ratio concerning the elasticity of materials.

Politzer, H. David (1949–), Discovered asymptotic freedom in the strong interactions (independent of D. J. Gross and F. Wilczek). Nobel Prize, 2004.

Pomeranchuk, Isaak Yakolevich (1913–1966), Soviet physicist. Showed that energy of cosmic-ray electrons reaching the atmosphere is limited by their radiation in Earth's magnetic field; proved the Pomeranchuk theorem for scattering cross sections.

Porro, Ignazio (1801–1875), Italian topographer, geodesist, and physicist. Invented optical surveying instruments, Porro prism erecting system, and modern prism binoculars.

Powell, Cecil Frank (1903–1969), British physicist. Made practical the use of photographic emulsions in nuclear research; with G. P. S. Occhialini and others, discovered and investigated production of pions from cosmic radiation in the Earth's atmosphere; Nobel Prize, 1950.

Poynting, John Henry (1852–1914), English physicist. Determined the constant of gravitation and explained why a comet's tail points away from the Sun.

Prandtl, Ludwig (1875–1953), German physicist. Contributed to fluid mechanics, particularly aerodynamics; introduced concept of boundary layer.

Prevost, Pierre (1751–1839), Swiss physicist. Developed theory of exchanges, explaining nature of heat.

Priestley, Joseph (1733–1804), English chemist and physicist. Discovered oxygen, ammonia, oxides of nitrogen, hydrochloric acid gas, nitrogen, carbon monoxide, and sulfur dioxide.

Prigogine, Ilya (1917–2003), Soviet-born Belgian chemist. Contributed to nonequilibrium thermodynamics, particularly the theory of dissipative structures; Nobel Prize, 1977.

Prokhorov, Aleksandr Mikhailovich (1916–2002), Soviet physicist. With N. G. Basov, devised a new method for amplifying electromagnetic radiation; Nobel Prize, 1964.

Pupin, Michael (1858–1935), Yugoslavian-born American physicist and electrical engineer. Developed inductance coils for telephone lines; contributed to x-ray fluoroscopy, design of radio transmitters, and network theory.

Purcell, Edward Mills (1912–1997), American physicist. Developed the method of nuclear resonance absorption; Nobel Prize, 1952.

Rabi, Isidor Isaac (1898–1988), Austrian-born American physicist. Research on neutrons, magnetism, quantum mechanics, and nuclear physics; Nobel Prize, 1944.

Rainwater, Leo James (1917–1986), American physicist. Suggested that shell-model potentials of certain atomic nuclei are not spherical but are deformed into spheroids, and proposed mechanism for this distortion; Nobel Prize, 1975.

Raman, Chandrasekhara Venkata (1888–1970), Indian physicist. Research on diffraction and oscillation; discovered the Raman effect; Nobel Prize, 1930.

Ramsden, Jesse (1735–1800), English mathematical-instrument maker. Invented an eyepiece containing cross-wires as a measuring scale; introduced equatorial mounting for telescopes.

Ramsey, Norman Foster (1915–), American physicist. Invented an accurate method of measuring differences between atomic energy levels that formed the basis for the cesium atomic clock; worked on the hydrogen maser; Nobel Prize, 1989.

Rankine, William John Macquorn (1820–1872), Scottish civil engineer. Contributed to thermodynamics and theories of elasticity and waves; wrote textbooks on the steam engine and civil engineering.

Raoult, François Marie (1830–1901), French chemist. Formulated Raoult's law concerning vapor pressure of a solution.

Rayleigh, John William Strutt, 3d Baron (1842–1919), English physicist. Worked on the theory of sound and on physical optics; with W. Ramsay, discovered argon; Nobel Prize, 1904.

Regge, Tullio (1931–), Italian physicist. Played a role in introducing the idea of complex angular momenta into elementary particle physics.

Reines, Frederick (1918–1998), American physicist. With C. L. Cowan, made first detection of the neutrino, a fundamental particle; Nobel Prize, 1995.

Reynolds, Osborne (1842–1912), British engineer and physicist. Demonstrated streamline and turbulent flow in pipes, and showed that transition between them occurs at a critical velocity determined by Reynolds' number; introduced Reynolds' analogy.

Ricci-Curbastro, Gregorio (1853–1924), Italian mathematician and mathematical physicist. Developed theory of tensor analysis, providing mathematical foundation for general relativity.

Richardson, Owen Willans (1879–1959), English physicist. Studied the emission of electricity from hot bodies and the electron theory of matter; Nobel Prize, 1928.

Richardson, Robert Coleman (1937–), American physicist. With D. M. Lee and D. D. Osheroff, discovered superfluidity in helium-3; Nobel Prize, 1996.

Richter, Burton (1931–), American physicist. Independently of S. C. C. Ting, discovered a new heavy elementary particle, which he named the psi particle; Nobel Prize, 1976.

Riemann, Georg Friedrich Bernhard (1826–1866), German mathematician. Originated Riemannian geometry, a noneuclidean system.

Righi, Augusto (1850–1920), Italian physicist. Discovered magnetic hysteresis and Righi-Leduc effect, independently of S. A. Leduc; demonstrated that microwaves have all properties characteristic of light waves.

Ritchey, George Wills (1864–1945), American astronomer. Made important astronomical observations, particularly on the Andromeda nebula; with H. Chrétien, developed Ritchey-Chrétien optics.

Ritz, Walter (1878–1909), Swiss-born German physicist. Introduced Ritz combination principle; developed Ritz method for numerical solution of boundary-value problems.

Roche, Edouard Adelbert (1820–1883), French physicist, mathematician, and meteorologist. Studied the internal structure and free-surface form of the celestial bodies; applied results to study of cosmogonic hypotheses.

Rohrer, Heinrich (1933–), Swiss physicist. With G. Binnig, developed scanning tunneling microscope; Nobel Prize, 1986.

Röntgen, Wilhelm Konrad (1845–1923), German physicist. Discovered x-rays; Nobel Prize, 1901.

Rosen, Nathan (1909–1995), American-born Israeli physicist. Collaborated in formulation of Einstein-Podolsky-Rosen paradox; research on general relativity and gravitational waves.

Routh, Edward John (1831–1907), British mathematical physicist. Made contributions to classical mechanics, including procedure for eliminating cyclic coordinates from equations of motion.

Rowland, Henry Augustus (1848–1901), American physicist. Developed the Rowland grating in spectroscopy; studied electromagnetism and heat.

Rubbia, Carlo (1934–), Italian physicist. Principal architect of experiment that first detected intermediate vector bosons, an important step in confirming theory uniting electromagnetic and weak nuclear interactions; Nobel Prize, 1984.

Rubens, Heinrich (1865–1922), German physicist. With E. B. Hagen, conducted electromagnetic experiments; built new types of galvanometer and bolometer.

Rumford, Benjamin Thompson, Count (1753–1814), British physicist. Carried out research on heat.

Runge, Carl David Tolme (1856–1927), German mathematician and physicist. Research on theoretical and experimental spectroscopy, particularly data reduction and development of series formulas; developed methods for numerical and graphical computation, including Runge-Kutta method.

Ruska, Ernst (1906–1988), German electronic engineer. Developed the electron microscope; Nobel Prize, 1986.

Russell, Henry Norris (1877–1957), American astronomer and physicist. Analyzed eclipsing binary stars: with E. Hertzsprung, introduced Hertzsprung-Russell diagram; determined abundance of chemical elements in solar atmosphere; with F. A. Saunders, devised theory of Russell-Saunders coupling.

Rutherford, Ernest, 1st Baron (1871–1937), British physicist. Discovered alpha, beta, and gamma rays; suggested the divisible nuclear atom; effected the transmutation of an atom: Nobel Prize, 1908.

Rydberg, Johannes Robert (1854–1919), Swedish physicist. Developed a formula for series of spectral lines, involving Rydberg's constant.

Sabine, Edward (1788–1883), British physicist and astronomer. Headed a magnetic survey of the world which discovered a connection between sunspots and terrestrial magnetic disturbances.

Sabine, Wallace Clement Ware (1868–1919), American physicist. Pioneered in architectural acoustics; discovered law determining reverberation time in acoustics.

Saha, Meghnad (1894–1956), Indian physicist. Developed theory for degree of ionization of hot gases, a basic component of modern astrophysics.

Salam, Abdus (1926–1996), Pakistani physicist. Independently of S. Weinberg, developed theory uniting two of the basic forces of nature, electromagnetism and the weak nuclear interactions; Nobel Prize, 1979.

Salpeter, Edwin Ernest (1924–), Austrian-born American physicist. Research in quantum theory of atoms, quantum electrodynamics, nuclear theory, energy production of stars, and theoretical astrophysics; with H. A. Bethe, introduced Bethe-Salpeter equation.

Savart, Félix (1791–1841), French physicist. Helped formulate the Biot-Savart law in electromagnetism.

Schawlow, Arthur Leonard (1921–1999), American physicist. Contributed to invention of laser; made numerous contributions to laser spectroscopy, particularly the development of Doppler-free spectroscopy; Nobel Prize, 1981.

Schmidt, Bernhard Voldemar (1879–1935), Estonian-born German astronomer. Invented Schmidt system for astronomical telescopes.

Schottky, Walter (1886–1976), Swiss-born German physicist. Discovered Schottky effect; invented screen grid and tetrode; developed Schottky theory of semiconductor-metal junctions.

Schrieffer, John Robert (1931–), American physicist. With J. Bardeen and L. N. Cooper, formulated a theory of superconductivity: Nobel Prize, 1972.

Schrödinger, Erwin (1887–1961), German physicist. Proposed concept of atomic structure based on wave mechanics; contributed to quantum theory and color theory; Nobel Prize, 1933.

Schwartz, Melvin (1932–), American physicist. Collaborated in an experiment that demonstrated the existence of two types of neutrino; Nobel Prize, 1988.

Schwarzschild, Karl (1873–1916), German astronomer. Developed photographic methods for measuring brightness of stars; discovered Schwarzschild solution of equations of general relativity.

Schwinger, Julian Seymour (1918–1994), American physicist. Made fundamental contributions to the quantum theory of radiation; worked out the mathematical formalism of interaction between charged particles and an electromagnetic field; Nobel Prize, 1965.

Seebeck, Thomas Johann (1770–1831), German physicist. Investigated thermoelectricity and invented the thermocouple.

Segrè, Emilio Gino (1905–1989), Italian-born American physicist. Codiscovered the elements technetium, astatine, and plutonium, slow neutrons, and the antiproton; Nobel Prize, 1959.

Seidel, Philipp Ludwig von (1821–1896), German astronomer and mathematician. Developed theory of aberrations; made first accurate photometric measurements of stars and planets, and evaluated them with probability theory.

Serber, Robert (1909–1997), American physicist. Laid foundations of orbit theory of high-energy particle accelerators; introduced Serber potential to describe nuclear forces.

Shirakawa, Hideki (1936–), Japanese polymer scientist. Discovered and developed conductive polymers with Alan Heeger and Alan MacDiarmid; Nobel Prize, 2000.

Shockley, William (1910–1989), English-born American physicist. Discovered the transistor effect for electronic amplification by means of solid-state semiconductors; Nobel Prize, 1956.

Shubnikov, Aleksei Vasilevich (1887–1970), Soviet crystallographer. Classified Shubnikov groups; developed techniques for growing crystals, including synthetic rubies used in lasers.

Shull, Clifford G. (1915–2001), American physicist. Developed the neutron diffraction technique for studying the atomic structure of solids and liquids; Nobel Prize, 1994.

Siegbahn, Kai Manne Börje (1918–), Swedish physicist. Pioneered the development of high-resolution electron spectroscopy; Nobel Prize, 1981.

Siegbahn, Karl Manne Georg (1886–1978), Swedish physicist. Studied x-ray spectroscopy, in which he discovered the M series; Nobel Prize, 1924.

Siemens, Ernst Werner von (1816–1892), German engineer and electrician. Developed telegraphy and self-acting dynamo.

Singer, Isadore Manual (1924–), American mathematician. Worked in global analysis, especially the theory of elliptic operators and their applications to topology and geometry, and in mathematical physics; collaborated with M. F. Atiyah in proving the index theorem; Abel Prize, 2004.

Slater, John Clarke (1900–1976), American physicist. Introduced Slater determinant describing many-electron systems; developed theory of magnetrons.

Smale, Stephen (1930–), American mathematician. Worked in differential topology, differential equations, and dynamical systems; proved that the sphere can be turned inside out and that the generalized Poincaré conjecture is valid for dimensions greater than 4; discovered strange attractors that lead to chaotic dynamical systems; Fields Medal, 1966.

Smith, Robert (1689–1768), English physicist. Developed a particulate theory of light; developed geometric propositions for computing properties of optical systems; derived a special case of the Smith-Helmholtz law.

Snell, Willebrod van Roijen (1591–1626), Dutch mathematician. Formulated Snell laws concerning angles of incidence and refraction; conceived the idea of measuring the Earth by triangulation.

Soddy, Frederick (1877–1956), English chemist. With E. Rutherford, developed theory of atomic disintegration of radioactive substances; research on isotopes; Nobel Prize, 1921.

Sommerfeld, Arnold (1868–1951), German physicist. Developed quantum theory, especially in its application to sprectral lines and the Bohr atomic model.

Stark, Johannes (1874–1957), German physicist. Studied radiation and atomic theory; discovered the Stark effect on spectrum lines and the Doppler effect in canal rays; Nobel Prize, 1919.

Stefan, Josef (1835–1893), Austrian physicist. Originated Stefan's (or Stefan-Boltzmann) law of blackbody radiation; proposed theory of diffusion of gases; studied gas conductivity.

Steinberger, Jack (1931–), German-born American physicist. Collaborated in an experiment that demonstrated the existence of two types of neutrino; Nobel Prize, 1988.

Steinmetz, Charles Proteus (1865–1923), German-born American electrical engineer. Developed complex number technique for analyzing alternating-current circuits; made numerous electrical inventions; applied mathematical methods to solution of electrical engineering problems.

Stern, Otto (1888–1969), German-born American physicist. Developed the molecular beam method and used it to prove directly the existence of the magnetic moment of atoms and nuclei and to measure their magnitudes; Nobel Prize, 1943.

Stokes, George Gabriel (1819–1903), British mathematician and physicist. Originated the idea of determining the chemical composition of the Sun and stars from their spectra; studied double refraction and electromagnetic waves.

Störmer, Horst Ludwig (1949–), German-American physicist. With D. C. Tsui, discovered the fractional quantum Hall effect, a manifestation of a new form of quantum fluid with fractionally charged excitations; Nobel Prize, 1998.

Strassman, Fritz (1902–1980), German chemist. With O. Hahn and L. Meitner, discovered nuclear fission; research on uranium and thorium isotopes.

Sturgeon, William (1783–1850), English electrician and inventor. Constructed the first useful electromagnet and the first moving-coil galvanometer.

Suhl, Harry (1922–), German-born American physicist. Discovered Suhl effect; invented Suhl amplifier; studied resonance in magnetic materials, superconductivity, and general theory of magnetism.

Tamm, Igor Yevgenevich (1895–1971), Soviet physicist. With I. M. Frank, formulated the mathematical theory explaining the physical origin and properties of Cerenkov radiation; Nobel Prize, 1958.

Taylor, Geoffrey Ingram (1886–1975), British mathematician. Work in theoretical hydrodynamics, particularly turbulence and effect of rotation on fluid flow.

Taylor, Joseph Hooten, Jr. (1941–), American astronomer. With R. A. Hulse, discovered a binary pulsar and studied it to observe phenomena predicted by general relativity; Nobel Prize, 1993.

Taylor, Richard Edward (1929–), American physicist. Collaborated in experiments that demonstrated that protons, neutrons, and similar particles are made up of quarks; Nobel Prize, 1990.

Teller, Edward (1908–2003), Hungarian-born American physicist. With associates, developed the concept which led to the construction of the first hydrogen bomb; with G. Gamow, proposed the Gamow-Teller interaction and Gamow-Teller selection rules.

Tesla, Nikola (1856–1943), American inventor born in what is now Croatia of Serbian parents. Invented the induction motor, a high-frequency electric coil; improved design of dynamos, transformers, and electric bulbs.

Thales (ca. 640-ca. 546 B.C.), Greek mathematician and astronomer. First to scientifically predict an eclipse of the Sun; discovered static electricity; credited with formulating several theorems.

Thomas, Llewellyn Hilleth (1903–1992), English-born American physicist. Discovered Thomas precession; with E. Fermi, developed Thomas-Fermi atomic model; developed basic theory for Thomas cyclotron.

Thomson, George Paget (1892–1975), English physicist. Discovered, independently of C. J.

Davisson, the diffraction of electrons by crystals; Nobel Prize, 1937.

Thomson, Joseph John (1856–1940), English physicist. Discovered that cathode rays consist of negatively charged particles, or electrons; Nobel Prize, 1906.

't Hooft, Gerardus (1946–), Dutch physicist. With M. J. G. Veltman, elucidated the quantum structure of the electroweak interactions, placing the theory of these interactions and similar particle physics theories on a firmer mathematical foundation by showing how they may be used for precise calculations of physical quantities; Nobel Prize, 1999.

Thouless, David James (1934–), British physicist. Studied many-body problem and its applications to nuclear and condensed matter physics, including phase transitions in superfluid helium films and electrons in disordered systems.

Ting, Samuel Chao Chung (1936–), American physicist. Independently of B. Richter, discovered a new heavy elementary particle, which he named the J particle; Nobel Prize, 1976.

Tomonaga, Sin-ltiro (1906–1979), Japanese physicist. Showed the modern theory of quantum electrodynamics to be quantitatively consistent with observed physical phenomena; Nobel Prize, 1965.

Torricelli, Evangelista (1608–1647), Italian physicist. Invented the mercury barometer.

Townes, Charles Hard (1915–), American physicist. Invented the maser; Nobel Prize, 1964.

Townsend, John Sealy Edward (1868–1957), British physicist. Developed collision theory of ionization of gases in an electric field.

Tsui, Daniel Chee (1939–), Chinese-born American physicist. With H. L. Störmer, discovered the fractional quantum Hall effect, a manifestation of a new form of quantum fluid with fractionally charged excitations; Nobel Prize, 1998.

Tyndall, John (1820–1893), British physicist. Studied temperature waves in metals and diathermancy of gases; discovered the effect of atmospheric density on sound transmission.

Uhlenbeck, George Eugene (1900–1988), Javanese-born American physicist. With S. Goudsmit, developed hypothesis of electron spin.

Van Allen, James Alfred (1914–), American physicist. Discovered that the Earth is circled by two high-energy radiation belts, leading to major revisions in concepts of the Earth's atmosphere and magnetic field.

Van de Graaff, Robert Jemison (1901–1967), American physicist. Contributed to the development of the direct particle accelerator and invented the electrostatic belt generator.

van der Meer, Simon (1925–), Dutch physicist. Devised method to ensure frequent and efficient collision of accelerated protons and antiprotons in the superproton synchrotron at CERN, contributing to discovery of intermediate vector bosons; Nobel Prize, 1984.

van der Waals, Johannes Diderik (1837–1923), Dutch physicist. Formulated van der Waals equation; investigated van der Waals forces, concerning intermolecular attraction; Nobel Prize, 1910.

van Rhijn, Pieter Johannes (1886–1960), Dutch astrophysicist. With J. C. Kapetyn, evolved a theory of the universe.

Van Vleck, Jan Hasbrouck (1899–1980), American mathematical physicist. Pioneer in the development of the modern quantum-mechanical theory of magnetism; Nobel Prize, 1977.

Veltman, Martinus J. G. (1931–), Dutch physicist. With G. 't Hooft, elucidated the quantum structure of the electroweak interactions, placing the theory of these interactions and similar particle physics theories on a firmer mathematical foundation by showing how they may be used for precise calculations of physical quantities; Nobel Prize, 1999.

Verdet, Marcel Emile (1824–1866), French physicist. Determined dependence of Faraday effect on magnetic field strength, wavelength of the light, and index of refraction of the material.

Voigt, Woldemar (1850–1919), German physicist. Introduced transformation equations (later known as Lorentz transformations).

Volta, Alessandro, Count (1745–1827), Italian physicist. Invented the voltaic pile; developed the theory of current electricity.

von Kármán, Theodore (1881–1963), American aerodynamicist. Theoretical contributions to aerodynamics; formulated von Kármán's theory of vortex streets, an early step in the mathematical treatment of turbulent motion.

von Neumann, John (1903–1954), Hungarian-born American mathematician. Research in logic, theory of quantum mechanics, theory of high-speed computing machines, and mathematical theory of games and strategy.

Walton, Ernest Thomas Sinton (1903–1995), British physicist. With J. D. Cockcroft, devised high-voltage apparatus capable of producing fast atomic particles with energies up to 700,000 electronvolts; showed the capability of these particles to disintegrate many light elements; Nobel Prize, 1951.

Wannier, Gregory Hugh (1911–1983), Swiss-born American physicist. Developed harmonization of localized and nonlocalized descriptions of electrons in solids.

Weber, Wilhelm Eduard (1804–1891), German physicist. Devised instruments for measurement of electrical and magnetic quantities; formulated absolute electrical and magnetic units.

Weinberg, Steven (1933–), American physicist. Independently of A. Salam, developed theory uniting two of the basic forces of nature, electromagnetism and the weak nuclear interactions; Nobel Prize, 1979.

Weiss, Pierre (1865–1940), French physicist. Developed phenomenological theory of ferromagnetism.

Weizsäcker, Carl Friedrich von (1912–), German physicist. Helped develop method for calculating bremsstrahlung in high-energy collisions; developed a theory of origin of solar system.

Wentzel, Gregor (1898–1978), German-born American physicist. Helped develop Wentzel-Kramers-Brillouin method; research on theory of atomic spectra, wave mechanics, quantum electrodynamics, meson field theories, and statistical mechanics of many-body problems, especially superconductivity.

Weyl, Hermann (1885–1955), German-born American mathematician and mathematical physicist. Basic research on group representations and Riemann surfaces.

Wheatstone, Charles (1802–1875), English physicist and inventor. Conducted experiments on sound; invented Wheatstone's bridge, an instrument for comparing electrical resistances.

Wheeler, John Archibald (1911–), American physicist. Introduced the concepts of the scattering matrix and resonating group structure into nuclear physics; with N. Bohr, elucidated the mechanism of nuclear fission and predicted the fissibility of plutonium.

Whitehead, Alfred North (1861–1947), English mathematician, physicist, and philosopher. With B. Russell, pioneered in mathematical logic and foundations of mathematics.

Whittaker, Edmund Taylor (1873–1956), British mathematician and physicist. Studied special functions of mathematical physics and equations satisfied by them, particularly Whittaker's differential equation; found general integral representation for harmonic functions; made major contributions to analytical dynamics.

Wiedemann, Gustave Heinrich (1826–1899), German physicist and physical chemist. With R. Franz, discovered Wiedemann-Franz law of thermal conductivity of metals; discovered Wiedemann effect.

Wieman, Carl Edwin (1951–), American physicist. With E. A. Cornell, succeeded in producing Bose-Einstein condensates in a dilute gas of alkali (rubidium) atoms, and carried out fundamental studies of their properties, including studies of collective excitations and vortex formation in condensates; Nobel Prize, 2001.

Wien, Wilhelm (1864–1928), German physicist. Formulated the two Wien laws pertaining to radiation from blackbodies; Nobel Prize, 1911.

Wiener, Norbert (1894–1964), American mathematician. Formulated a mathematical theory of Brownian motion; founded science of cybernetics.

Wigner, Eugene Paul (1902–1995). Hungarian-born American mathematical physicist. With G. Breit, worked out the Breit-Wigner formula for resonant nuclear reactions; proposed the Wigner theorem of conservation of the angular momentum of electron spin; Nobel Prize, 1963.

Wilczek, Frank (1951–), Discovered asymptotic freedom in the strong interactions (in collaboration with D. J. Gross, and independent of H. D. Politzer). Nobel Prize, 2004.

Wilson, Charles Thomson Rees (1869–1959), British physicist. Worked on ionization; originated the cloud chamber method of studying ionized particles; Nobel Prize, 1927.

Wilson, Kenneth Geddes (1936–), American physicist. Used renormalization group theory to analyze critical phenomena in the behavior of matter at phase transitions; Nobel Prize, 1982.

Wilson, Robert Woodrow (1936–), American astrophysicist. With A. A. Penzias, discovered cosmic background radiation, confirming the big bang theory of the origin of the universe; Nobel Prize, 1978.

Witten, Edward (1951–), American mathematical physicist. Applied advanced mathematical tools to theoretical physics, particularly quantum field theory, supersymmetry, and string theory; his physical insights were the basis for major developments in mathematics; Fields Medal, 1990.

Wollaston, William Hyde (1766–1828), English chemist and physicist. Discovered the lines in the solar spectrum; discovered palladium and rhodium; invented the Wollaston lens.

Yang, Chen Ning (1922–), Chinese-born American physicist. With T. Lee, disproved the law of conservation of parity for weak interactions; Nobel Prize, 1957.

Yau, Shing-Tung (1949–), Chinese-born American mathematician. Worked in differential geometry and partial differential equations; solved the Calabi conjecture in algebraic geometry and the positive mass conjecture of general relativity theory; Fields Medal, 1982.

Young, Thomas (1773–1829), English physicist and physician. Discovered the effect of the ciliary muscle on the shape of the eye lens (the mechanism of accommodation).

Yukawa, Hideki (1907–1981), Japanese physicist. Postulated the existence of a new fundamental particle, the meson; Nobel Prize, 1949.

Zeeman, Pieter (1865–1943), Dutch physicist. Discovered the Zeeman effect in magnetooptics; Nobel Prize, 1902.

Zener, Clarence Melvin (1905–1993), American physicist. Proposed mechanism of Zener breakdown.

Zernike, Fritz (1888–1966), Dutch physicist. Developed the phase-contrast microscope, making possible the first microscopic examination of the internal structure of living cells; Nobel Prize, 1953.

Zworykin, Vladimir Kosma (1889–1982), Russian-born American physicist. Pioneer in the development of television and the electron microscope.

Contributor Initials

Each article in the Encyclopedia is signed with the contributor's initials. This section gives all such initials. The contributor's name is provided. The contributor's affiliation can then be found in the next section.

A

A.Bi.   Arthur Bienenstock
A.B.C.   All B. Cambel
A.B.Co.   Alan B. Coppens
A.Ch.   Alan Chodos
A.C.Go.   Arthur C. Gossard
A.C.P.   Andrew C. Pike
A.D.S.   A. Douglas Stone
A.D.Sk.   A. Douglas Skinner
A.E.Ba.   A. Earle Bailey
A.E.Br.   Arthur E. Bryson, Jr.
A.E.Ca.   Anders E. Carlsson
A.E.D.   A. E. Drake
A.F.Ru.   Arthur F. Rupp
A.G.B.   A. G. Bailey
A.G.H.   Arthur G. Hansen
A.G.P.   Albert G. Petschek
A.Had.   A. Hadni
A.H.K.   Alan H. Karp
A.H.Sn.   Arthur H. Snell
A.H.So.   Alfred H. Sommer
A.J.Ca.   Angelo J. Campanella
A.J.D.   Adrianus J. Dekker
A.J.L.   Anthony J. Leggett
A.J.S.   Anthony J. Stone
A.K.   Adriah Korpel
A.K.M.   Alfred K. Mann
A.K.R.   Anant K. Ramdas
A.L.S.   Arthur L. Schawlow
A.M.Da.   Alberta M. Dawson
A.M.S.   Aephraim M. Steinberg: Coherence
A.Mot.   Allen Mottershead
A.O.   Albert Overhauser
A.O.N.   Alfred O. Nier
A.P.An.   Alan P. Anderson
A.Ri.   Arthur Rich
A.R.P.R.   A. R. P. Rau
A.S.   Abdus Salam
A.Sco.   Alwyn Scott
A.T.McD.   Alan T. McDonald
A.W.O.   Albert W. Overhauser

B

B.B.   Benjamin Bederson
B.B.Da.   Benjamin B. Dayton
B.Ca.   Blas Cabrera
B.DeF.   Brian DeFacio
B.E.W.   Bertram E. Warren
B.G.   Bernard Goodman
B.G.B.   Brian G. Bagley
B.G.D.   B. Gale Dick
B.H.Bi.   Bruce H. Billings
B.J.F.   Barry J. Feldman
B.L.R.   Burtis L. Robertson
B.Ma.   Bahram Mashhoon
B.M.McC.   Barry M. McCoy

B.N.B.   Bertram N. Brockhouse
B.Po.   Bogdan Povh
B.P.K.   Bryan P. Kibble
B.R.M.   Bruce R. Munson
B.Si.   Barry Simon
B.S.S.   Bellave S. Shivaram
B.W.M.   B. W. Mangum

C

C.B.   Charles Baltay
C.Bin.   Carrol Bingham
C.B.D.   C. B. Duke
C.C.K.   Clifford C. Klick
C.D.   Carl Dover
C.D.C.   C. Denise Caldwell
C.E.A.   Charles E. Applegate
C.E.H.   Carl E. Howe
C.E.W.   Clyde E. Wiegand
C.F.G.   Clarence F. Goodheart
C.G.   Charles Goebel
C.H.Di.   Cyril H. Dix
C.H.He.   Charles H. Henry
C.H.T.   Charles H. Townes
C.J.C.   C. J. Carpenter
C.J.G.   Charles J. Goebel
C.J.Li.   C. J. Lister
C.K.J.   C. K. Jen
C.M.H.   Cyril M. Harris
C.P.S.   Charles P. Slichter
C.Q.   Chris Quigg
C.R.   Claudio Rebbi
C.S.Ma.   C. Samuel Martin
C.W.K.   Chung W. Kim
C.Y.W.   C. Y. Wang

D

D.A.B.   D. Allan Bromley
D.A.B.M.   David A. B. Miller
D.A.Ca.   David A. Caughey
D.A.J.   Donald A. Jennings
D.Br.   Dirk Brouwer
D.B.L.   Don B. Lichtenberg
D.B.R.K.   D. B. R. Kenning
D.Cl.   Douglas Cline
D.D.McC.   Dennis D. McCarthy
D.Ev.   Doris Evans
D.E.As.   David E. Aspnes
D.E.D.   D. E. Daney
D.E.P.   David E. Pritchard
D.E.R.   Duane E. Roller
D.E.St.   David E. Stock
D.Gr.   David Gross
D.G.K.   Dennis G. Kovar
D.G.L.   Derek G. Leaist
D.H.W.   Denys H. Wilkinson
D.J.Ho.   D. J. Horen
D.J.Pe.   David J. Pegg

D.J.S.   David J. Sellmyer
D.Kl.   Daniel Kleppner
D.L.We.   David L. Weaver
D.Mi.   Dennis Mills
D.M.Bu.   Dennis M. Bushnell
D.M.Fr.   David M. Fradkin
D.M.Gre.   Daniel M. Greenberger
D.M.L.   David M. Larsen
D.M.Le.   David M. Lee
D.M.R.   David M. Rust
D.N.L.   D. N. Langenberg
D.P.   David Park
D.P.T.   Demetri P. Telionis
D.R.A.   David R. Andersen
D.R.F.H.   Donald R. F. Harleman
D.R.N.   David R. Nygren
D.S.   Dave Scott
D.Sha.   Dudley Shapere
D.Sn.   David Snoke
D.S.Ge.   Donald S. Gemmel
D.T.   David Turnbull
D.Wa.   David Ward
D.Wi.   Dudley Williams
D.Wo.   David Wollesheim
D.W.K.   Donald W. Kerst
D.W.P.   Darrell W. Pepper
D.Z.A.   Dana Z. Anderson

E

E.A.   Elihu Abrahams
E.Ad.   Elijah Adams
E.B.   E. Bauer
E.Ca.   Eleanor Campbell
E.C.St.   Edward C. Stevenson
E.D.A.   E. Dwight Adams
E.D.C.   Eugene D. Commins
E.E.C.   Ernest E. Charlton
E.F.   Eric Fawcett
E.G.   Edward Gerjuoy
E.G.Wi.   Earl G. Williams
Eva.Li.   Eva Lindroth
E.L.Hi.   Edward L. Hill
E.M.Co.   Esther M. Conwell
E.N.H.   E. Newton Harvey
E.Pa.   Eric Paterson
E.R.Co.   E. Richard Cohen
E.So.   E. So
E.S.F.   Edward S. Fry
E.Tr.   Elmar Träbert
E.W.Ba.   Er-Wei Bai
E.W.O.   Ernst W. Otten

F

F.A.Be.   Frank A. Benson
F.A.J.   Francis A. Jenkins

F.A.S.   Fay Ajzenberg-Selove
F.C.Br.   Frederick C. Brown
F.D.B.   Frank D. Brooks
F.D.L.   Frank D. Lewis
F.E.L.   F. E. Luborsky
F.E.N.   Fred E. Nicodemus
F.H.R.   Frank H. Rockett
F.I.   Franco Iachello
F.J.B.   Frank J. Blatt
F.Ke.   Frederic Keffer
F.K.H.   Forest K. Harris
F.L.H.   F. L. Hermach
F.L.P.   Fred L. Palmer
F.M.   Freydoon Mansouri
F.M.P.   Francis M. Pipkin
F.M.W.   Floyd M. Wahl
F.M.Wh.   Frank M. White
F.Ro.   F. Rohrlich
F.Ru.   Fritz Ruch
F.R.Ko.   F. Ralph Kotter
F.Ste.   Fred Stern
F.S.G.   Frederick S. Goulding
F.Wil.   Franz Wilczek
F.W.B.   Fred W. Billmeyer
F.W.K.F.   Frank W. K. Firk

G

G.A.Da.   Gilles A. Daigle
G.A.Ho.   G. A. Horton
G.A.W.   Gary A. Williams
G.D.F.   George D. Fulford
G.E.P.   George E. Pake
G.F.Be.   George F. Bertsch
G.G.   Gerson Goldhaber
G.Ho.   Gerald Holton
G.H.Di.   G. H. Dieke
G.H.M.   Glenn H. Miller
G.H.Ra.   G. H. Rayner
G.K.C.   George K. Celler
G.L.Cl.   George L. Clark
G.L.Ee.   Gary L. Eesley
G.Mu.   Glenn Murphy
G.M.C.   Gerald M. Clemence
G.O.   Gerald Oster
G.R.H.   George R. Harrison
G.She.   Gopal Shenoy
G.Swi.   Greg Swift
G.S.G.   Gary S. Grest
G.S.H.   G. S. Hurst
G.S.K.W.   George S. K. Wong
G.W.N.   Gerold W. Neudeck
G.W.S.   George W. Stroke

H

H.A.Be.   Hans A. Bethe
H.A.F.   Henry A. Fairbank
H.Bi.   Hans Bichsel
H.B.H.   H. B. Huntington
H.C.C.   H. C. Casey, Jr.
H.C.Co.   Herbert C. Corben
H.C.W.   Harold C. Weber
H.De.   Hans Dehmelt
H.E.Ba.   Henry E. Bass
H.E.D.   Henry E. Duckworth
H.E.W.   Harvey E. Wegner

H.F.   Herman Feshbach
H.G.   Harold Goodglass
H.Gr.   H. Granicher
H.G.S.   Heinz G. Sell
H.J.Ma.   Humphrey J. Maris
H.K.   Herwig Kogelnik
H.Mar.   Harvey Marshak
H.Mark.   Hershel Markovitz
H.Me.   Helmut Mehrer
H.N.H.   Herbert N. Hersh
H.P.Yu.   Horace P. Yuen
H.S.B.   Howard S. Bean
H.S.La.   Hugh S. Landes
H.W.   Hermann Wollnik
H.Win.   Herman Winick
H.W.M.   Hans W. Meissner
H.W.Ru.   Howard W. Russell
H.Y.F.   Hsu Y. Fan

I

I.A.S.   Ivan A. Sellin
I.Ba.   Itzhak Bars
I.D.   Ingrid Daubechies
I.F.K.   Isaac F. Kinnard
I.K.M.   Innes K. Mackenzie
I.K.S.   Ivan K. Schuller
I.P.K.   Ivan P. Kaminow
I.T.   Isao Tanihata

J

J.A.Da.   John A. Davies
J.A.H.   John A. Harvey
J.A.Li.   James A. Liggett
J.A.Ph.   James A. Phillips
J.A.Sc.   Joseph A. Schetz
J.A.V.   Jose A. Ventura
J.B.   Jack Bass
J.Be.   Jesse Beams
J.B.Fo.   J. Brian Fowlkes
J.B.Fr.   J. Bruce French
J.B.G.   John B. Goodenough
J.B.K.   J. B. Ketterson
J.C.   Joseph Callaway
J.Ca.   John Catchpole
J.Cl.   John Clarke
J.C.Wy.   James C. Wyant
J.D.A.   John D. Anderson, Jr.
J.D.Tr.   James D. Trolinger
J.E.B.   James E. Bayfield
J.E.Bo.   Joseph E. Borovsky
J.E.Fi.   J. E. Fischer
J.E.N.   James E. Nordman
J.Fo.   Joseph Ford
J.F.Di.   J. F. Dillon, Jr.
J.F.He.   J. F. Herbst
J.F.R.   John F. Reintjes
J.F.S.   James F. Scott
J.F.Sc.   James F. Schooley
J.F.We.   Jonathan F. Weil
J.Gei.   Jon Geist
J.Gre.   Jack Greenberg
J.Hard.   John Hardy
J.Hur.   J. Hurll
J.H.Ca.   James H. Calderwood
J.H.H.   Joseph H. Hamilton

J.H.S.   James H. Schulman
J.H.U.   James H. Underwood
J.H.Z.   John H. Zifcak
J.Mar.   John Markus
J.Men.   Joshua Menkes
J.M.A.D.   J. M. A. Danby
J.M.D.   James M. Daniels
J.M.Ke.   Joseph M. Keller
J.M.McK.   James M. McKenzie
J.M.Pa.   Jeevak M. Parpia
J.M.R.   John March-Russell
J.M.Sa.   Jose M. Sasian
J.M.Wie.   John M. Wiest
J.M.Woo.   Jerry M. Woodall
J.N.Ba.   J. Norman Bardsley
J.O.S.   J. O. Scanlan
J.Pol.   Joseph Polchinski
J.P.C.   James P. Chambers
J.P.deN.   John P. de Neufville
J.P.Go.   James P. Gordon
J.P.H.   John P. Hagen
J.P.Ha.   James Penner-Hahn:
Kapitza resistance
J.P.Ha.   John P. Harrison:
Extended x-ray
absorption fine
structure (EXAFS)
J.P.O'C.   John P. O'Connell
J.R.Hu.   John R. Huizenga
J.R.K.   Joseph R. Kinard
J.R.Pr.   Joel R. Primack
J.R.Se.   John R. Sellars
J.S.Bro.   James Steve Browder
J.S.H.   Jeffrey S. Hanor
J.S.Ro.   James S. Robertson
J.S.Row.   J. S. Rowlinson
J.Ti.   Jiri Tichy
J.V.H.   Joseph V. Hollweg
J.W.C.   J. W. Costerton
J.W.Goo.   Joseph W. Goodman
J.W.Ha.   John W. Hardy
J.W.Ol.   J. W. Olness
J.W.R.   James W. Rohlf
J.W.St.   John W. Stewart

K

K.Arn.   Karl Arnstein
K.A.Er.   K. A. Erb
K.A.J.   Kenneth A. Jackson
K.A.McC.   Kathryn A.
McCarthy
K.A.Sn.   Kurt A. Smoker
K.D.K.   Karl D. Kryter
K.D.P.   Kirk D. Peterson
K.E.L.   Kenneth E. Lassila
K.F.K.   Karl F. Koopman
K.M.E.   Kenneth M. Evenson
K.R.S.   K. R. Sreenivasan
K.V.M.   Kenneth V. Manning
K.Y.T.   K. Y. Tang

L

L.Ap.   L. Apker
L.B.H.   Louis B. Holdeman

L.C.F.   Leonard C. Feldman
L.C.P.   Leanne C. Pitchford
L.C.T.   Lee C. Teng
L.D.R.   Louis D. Roberts
L.E.   Leo Esaki
L.E.K.   Lawrence E. Kinsler
L.E.W.   Lewis E. Wedgewood
L.F.D.   Lawrence F. Dahl
L.G.C.   L. G. Christophorou
L.G.H.   Llewellyn G. Hoxton
L.J.C.   Lawrence J. Campbell
L.J.G.   Louis J. Guido
L.L.B.   Leo L. Beranek
L.N.   Leo Nedelsky
L.O.B.   Lawrence O. Brockway
L.R.W.   Laurence R. Walkep
L.S.S.   Lee S. Schroeder
L.T.   Lawrence Talbot
L.W.J.   Lawrence W. Jones

M

M.A.D.   Michel A. Duguay
M.A.S.   Michael A. Saad
M.Bu.   Martin Bucher
M.Bun.   Mario Bunge
M.B.M.   Mark B. Moffett
M.B.Ma.   M. Brian Maple
M.B.P.   Morton B. Panish
McA.H.H.   McAllister H. Hull
M.C.G.   Martin C. Gutzwiller
M.Di.   Michael Dine
M.Dr.   Max Dresden
M.E.L.   Malcolm E. Lines
M.E.Pe.   Michael E. Peskin
M.E.Z.   Michael E. Zeller
M.F.H.   Mark F. Hamilton
M.Gr.   Martin Greenspan
M.G.H.   Malcolm G. Haines
M.H.W.C.   Moses H. W. Chan
M.J.A.   Malcolm J. Abzug
M.J.Cr.   Malcolm J. Crocker
M.J.F.   Mitchell J. Feigenbaum
M.J.H.   Max J. Herzberger
M.K.   Matti Krusius
M.Ka.   Margarita Karovska
M.K.W.   Michael K. Wilkinson
M.L.Bi.   Michael L. Billet
M.L.K.   M. L. Knotek
M.L.P.   Martin L. Perl
M.L.Sw.   M. L. Swanson
M.M.Bu.   Maurice M. Bursey
M.O.   Michael Osofsky
M.P.W.   Mason P. Wilson, Jr.
M.Sol.   Mark Soldate
M.S.S.   Michael S. Smith
M.V.P.   Milind V. Purohit
M.We.   Martin Weitz

N

N.A.C.   Noel A. Clark
N.C.R.   Norman C. Rasmussen
N.F.R.   Norman Ramsey
N.He.   Noah Hershkowitz

N.Ko.   Noemie Holler
N.L.K.   N. L. Kusters
N.P.S.   Nicholas P. Samios
N.S.F.   Nelson S. Fisk
N.S.G.   Newell S. Gingrich

O

O.C.J.   Owen C. Jones
O.G.S.   Orest G. Symko
O.W.R.   Oscar W. Richards

P

P.A.He.   Paul A. Heiney
P.A.S.   Paul A. Souder
P.B.Co.   Peter B. Coates
P.B.F.   Philip B. Fraundorf
P.C.Be.   Philippe C. Becker
P.E.Bl.   Philip E. Bloomfield
P.E.R.   Peter E. Raad
P.F.   Paulo Franzini
P.G.B.   Peter G. Bergmann
P.G.Kl.   Paul G. Klemens
P.H.E.M.   Paul H. E. Meijer
P.H.Oo.   Patrick H. Oosthuizen
P.I.S.   Peter I. Somlo
P.J.B.   Paul J. Bender
P.J.M.   Philip J. Morris
P.J.W.   Peter J. Walsh
P.Ku.   Polykarp Kusch
P.M.   Peter Mohr
P.M.A.   Paul M. Anderson
P.M.K.   Peter M. Koch
P.M.S.   Philip M. Stehle
P.O.E.   Patrick O. Egan
P.Sh.   Peter Shaw
P.S.DeC.   P. S. De Carli
P.V.D.   Piet Van Duppen
P.W.A.   P. W. Atkins
P.St.   Philip Steele
P.W.Br.   Percy W. Bridgman
P.W.S.   Paul W. Schmidt

R

R.A.Bu.   Ralph A. Burton
R.A.Buc.   Richard A. Buchroeder
R.A.F.   Robert A. Fisher
R.A.Fi.   Russell A. Fisher
R.A.Hu.   Roland A. Hultsch
R.A.Ph.   Ronald A. Phaneuf
R.A.Web.   Richard A. Webb
R.Bet.   Russell Betts
R.B.D.K.   R. B. D. Knight
R.B.L.   R. Bruce Lindsay
R.C.   Richard Comeau
R.Ch.   Robert Chen
R.Cl.   Robin Cleveland
R.C.L.   Richard C. Lord
R.C.La.   Richard C. Lamb
R.C.Po.   Robert C. Powell
R.C.Pu.   Russell C. Putnam

R.D.Ru.   Rogers D. Rusk
R.D.W.   Robert D. Waldron
R.E.A.   Robert E. Apfel
R.E.P.   Raymond E. Peck
R.E.W.   R. E. Walstedt
R.F.   Ronald Fuchs
R.F.Cl.   Richard F. Clark
R.F.Dz.   Ronald F. Dziuba
R.F.P.   Richard F. Post
R.Gom.   Robert Gomer
R.G.Jon.   R. Gareth Jones
R.G.Ne.   Roger G. Newton
R.G.Wi.   Rolf G. Winter
R.Hi.   Ron Hilliard
R.Ho.   Robert Hofstadter
R.H.Bu.   Richard H. Bube
R.H.D.   Richard H. Dalitz
R.H.Go.   Roland H. Good
R.H.He.   Rolfe H. Herber
R.H.L.   Ralph H. Luebbers
R.H.Ri.   Reinhard H. Richter
R.H.Wa.   Robert H. Wald
R.J.   Roman Jackiw
R.Je.   Ron Jenkins
R.J.A.   Ronald J. Adler
R.J.K.   R. J. Keddy
R.J.S.   R. J. Stephenson
R.J.So.   Robert J. Soulen, Jr.
R.K.A.   Robert K. Adair
R.K.C.   Richard K. Cook
R.K.L.   Ronald K. Linde
R.K.MacC.   Robert K. MacCrone
R.L.Du.   Raynor L. Duncombe
R.L.Ja.   Robert L. Jaffe
R.L.R.   Robert L. Ramey
R.M.   Roy Middleton
R.M.Bo.   Robert M. Boynton
R.M.T.   Robb M. Thomson
R.M.Wal.   Robert M. Wald
R.M.Wr.   Raymond M. Wright
R.O.P.   Robert O. Pohl
R.Pan.   Richard Pantell
R.Pe.   Roger Penrose
R.Pow.   Robert Powell
R.P.Hu.   Ralph P. Hudson
R.P.Wi.   Ralph P. Winch
R.R.G.   Ralph R. Goodman
R.R.S.   Robert R. Shannon
R.S.M.   Robert S. Mulliken
R.S.McD.   Robin S. McDowell
R.S.R.   Robert S. Ross
R.S.Sy.   Robert S. Symons
R.S.Wa.   Richard S. Wagner
R.T.W.   Robert T. Weil, Jr.
R.T.Wh.   Richard T. Whitcomb
R.V.S.   Roland V. Shack
R.W.J.B.   R. W. J. Barker
R.W.Y.   Robert W. Young

S

S.A.Wi.   S. A. Williams
S.B.E.   Stuart B. Elston
S.Ch.   Steven Chu
S.C.C.T.   Samuel C. C. Ting

S.C.Li. Shao-Chi Lin
S.Fo. Simon Foner
S.F.J. Stephen F. Jacobs
S.H.L. Sam H. Liu
S.I.G. Shelton I. Green
S.I.M. Sidney I. Miller
S.I.Ma. Sameer I. Madanshetty
S.I.S. Stanley I. Sandier
S.J. S. Jafarey
S.Lei. Sidney Leibovich
S.L.S. Stephen L. Sass
S.M.G. Steven M. Girvin
S.N. Stephen Nygren
S.P.S. Soren P. Sorensen
S.R.N. Suzanne R. Nagel
S.S.C. Shoei-Sheng Chen
S.Tr. Sam Treiman
S.T.M. Steven T. Manson
S.W.MacD. Samuel W. MacDowell
S.Y.C. Shih-Yuan Chen

T

T.Ap. Thomas Appelquist
T.A.D. T. A. Deacon
T.A.K. Thomas A. Kaplan
T.A.M. Terry A. Miller
T.Bar. Ted Barnes
T.D.R. Thomas D. Rossing
T.E.C. Thomas E. Cowan
T.G.G. Thomas G. Giallorenzi
T.H.F. Thomas H. Fields
T.J.Q. T. J. Quinn

T.K. Taku Komai
T.Ki. Toichiro Kinoshita
T.L.F. Thomas L. Ferrell
T.M.F. Thomas M. Flynn
T.M.Gu. Turgut M. Gür
T.P.M. Thomas P. Murray
T.T.W. Tai Tsun Wu
T.W.Ha. Theo W. Hansch

U, V

U.F. Ugo Fano
V.D.B. Vernon D. Barger
V.E.B. Vladimir E. Bondybey
V.F. Val L. Fitch
V.K.D. V. K. Dhir
V.L.S. Victor L. Streeter
V.M.K. Veerle M. Keppens

W

W.A. William Allan
W.Ar. Walter Aron
W.A.H. Walter A. Harrison
W.A.L. William A. Lanford
W.A.S. William A. Steele
W.A.Si. William A. Sirignano
W.A.Wi. William A. Wildhack
W.A.Y. W. A. Yost
W.B.F. W. Beall Fowler
W.B.Fr. William B. Fretter
W.D.P. William D. Phillips
W.D.W. W. Dexter Whitehead
W.E.K. William E. Keller

W.F.J. William F. Jaep
W.F.M. William F. Meggers
W.Gr. Walter Greiner
W.G.M. William G. Metcalf
W.G.P. William G. Pollard
W.Hap. William Happer
W.H.G. Walter H. Gardner
W.H.G. Warren H. Giedt
W.H.Z. Wojciech Hubert Zurek
W.J.M.M. W. J. M. Moore
W.J.W. William J. Willis
W.K. Werner Kanzig
W.Ket. Wolfgang Ketterle
W.L. Willy Ley
W.L.Wo. William L. Wolfe
W.M. William Markowitz
W.M.C. William M. Carey
W.M.H. Waren M. Hagist
W.M.I. Wayne M. Itano
W.M.S. William M. Sinton
W.M.W. Walter M. Walsh
W.P.K. Wiley P. Kirk
W.P.W. W. P. Wolf
W.R.C.R. W. R. C. Rowley
W.R.Sm. William R. Smythe
W.S.J. William S. Janna
W.S.P. Wilson S. Pritchett
W.W. William West
W.W.W. William W. Watson

Y

Y.M.G. Y. M. Gupta
Y.S. Yoichiro Suzuki

Contributor Affiliations

This list comprises an alphabetical sequence of all contributors to the Encyclopedia. A brief affiliation is provided for each author. This list may be used in conjunction with the previous section to fully identify the contributor of each article.

A

Abrahams, Prof. Elihu. Department of Physics, Rutgers University.

Abzug, Dr. Malcolm J. Senior Staff Engineer, Systems Group, TRW, Inc., Redondo Beach, California.

Adair, Dr. Robert K. Department of Physics, Yale University.

Adams, Dr. E. Dwight. Department of Physics, University of Florida.

Adams, Dr. Elijah. Professor and Head, Department of Biological Chemistry, School of Medicine, University of Maryland.

Adler, Dr. Ronald J. Lockheed Palo Alto Research Laboratories, Palo Alto, California.

Ajzenberg-Selove, Prof. Fay. Physics Department, University of Pennsylvania.

Allan, William. Dean (retired), School of Engineering, City College of the City University of New York.

Andersen, Prof. David R. Department of Electrical and Computer Engineering, University of Iowa, Iowa City.

Anderson, Prof. Alan P. Department of Electronic and Electrical Engineering, University of Sheffield, England.

Anderson, Prof. Dana Z. Joint Institute for Laboratory Astrophysics, University of Colorado.

Anderson, Jr., Dr. John D. Jet Propulsion Laboratory, California Institute of Technology.

Anderson, Dr. Paul M. Department of Electrical and Computer Engineering, Arizona State University.

Apfel, Prof. Robert E. Department of Engineering and Applied Science, Yale University.

Apker, Dr. L. General Electric Research Laboratory, Schenectady, New York.

Applegate, Charles E. Consulting Engineer, Weston, Maryland.

Applequist, Dr. Thomas. Sloane Laboratory, Department of Physics, Yale University.

Arnstein, Dr. Karl. Vice President in Charge of Engineering (retired), Goodyear Aircraft Corporation, Akron, Ohio.

Aron, Dr. Walter. Consultant, Menlo Park, California.

Aspnes, Dr. David E. Department of Physics, North Carolina State University, Raleigh.

Atkins, Dr. P. W. Department of Chemistry, Oxford University, England.

B

Bagley, Dr. Brian G. Bell Laboratories, Murray Hill, New Jersey.

Bai, Dr. Er-Wei. Department of Electrical & Computer Engineering, University of Iowa, Iowa City.

Bailey, Prof. A. Earle. Department of Electrical Engineering, University of Southampton, England.

Bailey, Prof. A. G. Department of Electrical Engineering, University of Southampton, England.

Baltay, Dr. Charles. Department of Physics, Yale University.

Bardsley, Dr. James Norman. Lawrence Livermore National Laboratory, University of California.

Barger, Prof. Vernon D. Department of Physics, University of Wisconsin, Madison.

Barker, Dr. R. W. J. Department of Electrical and Electronic Engineering, Trent Polytechnic, Peterborough, Ontario, Canada.

Barnes, Dr. Ted. Physics Division, Oak Ridge National Laboratory, Oak Ridge, Tennessee.

Bars, Dr. Itzhak. Department of Physics, Yale University.

Bass, Prof. Henry E. National Center for Physical Acoustics, University of Mississippi.

Bass, Prof. Jack. Department of Physics and Astronomy, Michigan State University.

Bauer, Prof. E. Department of Physics, Arizona State University, Tempe.

Bayfield, Dr. James E. Department of Physics, University of Pittsburgh.

Beams, Dr. Jesse. Deceased; formerly, Department of Physics, University of Virginia.

Bean, Howard S. Deceased; formerly, Consultant on Fluid Metering, Liquids and Gases, Sedona, Arizona.

Becker, Dr. Philippe. AT&T Bell Laboratories, Murray Hill, New Jersey.

Bederson, Dr. Benjamin. Department of Physics, New York University.

Bender, Prof. Paul J. Professor of Physical Chemistry, University of Wisconsin.

Benson, Prof. Frank A. Department of Electronic and Electrical Engineering, University of Sheffield, England.

Beranek, Dr. Leo L. Chief Scientist (retired), Bolt Beranek and Newman, Inc., Cambridge, Massachusetts.

Bergmann, Prof. Peter G. Department of Physics, Syracuse University.

Bertsch, Dr. George F. Department of Physics, Michigan State University.

Bethe, Prof. Hans A. Floyd R. Newman Laboratory of Nuclear Studies, Cornell University.

Betts, Dr. Russell. Argonne National Laboratory, Argonne, Illinois.

Bichsel, Dr. Hans. Department of Radiology, University of Washington.

Bienenstock, Dr. Arthur. Professor of Applied Physics, Stanford University; Director, Stanford Synchrotron Radiation Laboratory.

Billet, Prof. Michael. Head, Fluid Dynamics Department, Applied Research Laboratory, Pennsylvania State University, State College.

Billings, Dr. Bruce H. Special Assistant to the Ambassador of Science and Technology, Embassy of the United States of America, Taipei.

Bingham, Prof. Carrol. Department of Physics, University of Tennessee, Knoxville.

Blatt, Prof. Frank J. Department of Physics, Michigan State University.

Bloomfield, Prof. Philip E. Department of Physics, University of Pennsylvania; City College of the City University of New York, Bloomfield.

Bondybey, Dr. Vladimir E. Bell Laboratories, Murray Hill, New Jersey.

Boynton, Dr. Robert M. Department of Psychology, University of California, San Diego.

Bridgman, Prof. Percy W. Deceased; formerly, Harvard University.

Brockhouse, Dr. Bertram N. Institute for Materials Research, McMaster University.

Brockway, Prof. Lawrence O. Deceased; formerly, Department of Chemistry, University of Michigan.

Bromley, Prof. D. Allan. Science Adviser to the President of the United States.

Brooks, Dr. Frank D. Department of Physics, University of Capetown, Rondebosch, South Africa.

Brouwer, Prof. Dirk. Deceased; formerly, Director, Observatory, Yale University.

Browder, Dr. James Steve. Department of Physics, Division of Science and Mathematics, Jacksonville University, Jacksonville, Florida.

Brown, Prof. Frederick C. Department of Physics, University of Illinois.

Bryson, Prof. Arthur E., Jr. Chairman, Department of Aeronautics and Astronautics, Stanford University.

Bube, Dr. Richard H. Professor of Electrical Engineering, Department of Material Sciences, Stanford University.

Bucher, Dr. Martin A. Department of Physics, Princeton University.

Buchroeder, Dr. Richard A. Optical Design Service, Tucson, Arizona.

Bunge, Prof. Mario. Foundations and Philosophy of Science Unit, McGill University, Montreal, Canada.

Bursey, Prof. Maurice M. Department of Chemistry, University of North Carolina.

Burton, Prof. Ralph A. Chairman, Department of Mechanical Engineering and Astronautical Sciences, Northwestern University.

Bushnell, Dennis M. Langley Research Center, National Aeronautics and Space Administration, Hampton, Virginia.

C

Cabrera, Dr. Blas. Department of Physics, Stanford University.

Calderwood, Dr. James H. Department of Electronic Engineering, University College, Galway, Ireland.

Caldwell, Prof. C. Denise. Department of Physics, Yale University.

Callaway, Prof. Joseph. Department of Physics and Astronomy, Louisiana State University.

Cambel, Dr. Ali B. Executive Vice President for Academic Affairs, Wayne State University.

Campanella, Dr. Angelo J. Campanella Associates, Hilliard, Ohio.

Campbell, Dr. Eleanor. Max-Born-Institut, Berlin, Germany.

Campbell, Dr. Laurence J. Low Temperature Physics Group, Los Alamos Scientific Laboratory, Los Alamos, New Mexico.

Carlsson, Prof. Anders E. Department of Physics, Washington University, St. Louis, Missouri.

Carpenter, Dr. C. J. Bath, Avon, England.

Casey, Dr. H. C., Jr. Department of Electrical Engineering, Duke University.

Catchpole, Dr. John. Admiralty Materials Laboratory, Dorset, England.

Caughey, Prof. David A. Sibley School of Mechanical and Aerospace Engineering, Cornell University, Ithaca, New York.

Celler, Dr. George K. AT&T Laboratories, Murray Hill, New Jersey.

Chambers, Dr. James P. Research Scientist and Research Assistant Professor of Mechanical Engineering, University of Mississippi.

Chan, Dr. Moses H. W. Department of Physics, Pennsylvania State University.

Charlton, Dr. Ernest E. Research Engineer (retired), Schenectady, New York.

Chen, Dr. Robert T. Medical Epidemiologist, Infant Immunization Section, Centers for Disease Control, Surveillance Investigations and Research Branch, Atlanta, Georgia.

Chen, Dr. Shoei-Sheng. Argonne National Laboratory, Argonne, Illinois.

Chodos, Dr. Alan. Department of Physics, Yale University.

Christophorou, Dr. L. G. Head, Atomic, Molecular and High Voltage Physics Group, Oak Ridge National Laboratory, Oak Ridge, Tennessee.

Chu, Dr. Steven. Department of Physics, Stanford University.

Clark, Prof. George L. Deceased; formerly, Emeritus Research Professor of Analytical Chemistry, University of Illinois.

Clark, Noel A. Department of Physics, University of Colorado at Boulder.

Clark, Richard F. Division of Physics, National Research Council of Canada, Ottawa, Ontario.

Clarke, Prof. John. Department of Physics, University of California, Berkeley.

Clemence, Dr. Gerald M. Deceased; formerly, Observatory, Yale University.

Cleveland, Dr. Robin. Department of Aerospace And Mechanical Engineering, Boston University, Boston, Massachusetts.

Cline, Dr. Douglas. Nuclear Structure Laboratory, University Of Rochester, New York.

Coates, Dr. Peter B. Retired; formerly, Division of Electrical Science, National Physical Laboratory, Teddington, United Kingdom.

Cohen, Dr. E. Richard. Rockwell International Science Center, Thousand Oaks, California.

Comeau, Richard. The Foxboro Company, Foxboro, Massachusetts.

Commins, Prof. Eugene D. Department of Physics, University of California, Berkeley.

Conwell, Dr. Esther M. Xerox Webster Research Center, Rochester, New York.

Cook, Dr. Richard K. Formerly, National Bureau of Standards.

Coppens, Dr. Alan B. Associate Professor of Physics, U.S. Navy Post Graduate School, Pacific Grove, California.

Corben, Dr. Herbert C. Deceased; formerly, Ramo-Wooldridge Corporation, Los Angeles, California.

Costerton, Dr. J. William F. Department of Biology, University of Calgary, Alberta, Canada.

Cowan, Dr. Thomas E. General Atomics, Photonics Division, San Diego, California.

Crocker, Prof. Malcolm J. Department of Mechanical Engineering, Purdue University.

D

Dahl, Dr. Lawrence F. Department of Chemistry, University of Wisconsin.

Daigle, Dr. Gilles A. National Research Council, Institute for Microstructural Sciences, Ottawa, Canada.

Dalitz, Prof. Richard H. Department of Theoretical Physics, Oxford University.

Danby, Dr. J. M. A. Department of Mathematics, North Carolina State University.

Daney, Dr. David E. Los Alamos National Laboratory, Los Alamos, New Mexico.

Daniels, Prof. James M. Department of Physics, University of Toronto.

Daubechies, Dr. Ingrid. AT&T Bell Laboratories, Murray Hill, New Jersey.

Dayton, Benjamin B. Consulting Physicist, East Flatrock, North Carolina.

Deacon, T. A. Deceased; formerly, Division of Electrical Science, National Physical Laboratory, Teddington, Middlesex, England.

DeCarli, Dr. P. S. Poulter Laboratory, Stanford Research Institute.

DeFacio, Prof. Brian. Department of Physics and Astronomy, University of Missouri.

Dehmelt, Dr. Hans. Department of Physics, University of Washington.

Dekker, Dr. Adrianus J. Professor of Solid State Physics, University of Groningen, Netherlands.

de Neufville, Dr. John P. President, Voltaix, Inc., North Branch, New Jersey.

Dhir, Prof. Vijay K. Department of Mechanical and Aerospace Engineering, University of California, Los Angeles.

Dick, Prof. B. Gale. Department of Physics, University of Utah.

Dieke, Prof. G. H. Deceased; formerly, Chairman, Department of Physics, Johns Hopkins University.

Dillon, J. F., Jr. Bell Telephone Laboratories, Murray Hill, New Jersey.

Dine, Dr. Michael. Physics Department, City College of New York.

Dix, Dr. Cyril H. Bishop Monkton, North Yorkshire, England.

Dover, Dr. Carl B. Department of Physics, Brookhaven National Laboratory, Upton, New York.

Drake, A. E. Division of Electrical Science, National Physical Laboratory, United Kingdom.

Dresden, Prof. Max. Institute for Theoretical Physics, State University of New York, Stony Brook.

Duckworth, Dr. Henry E. Department of Physics, University of Manitoba, Canada.

Duguay, Dr. Michel A. Laser Development, Sandia Laboratories, Albuquerque, New Mexico.

Duke, Dr. C. B. Manager, Molecular and Organic Materials Area, Xerox Corporation, Rochester, New York.

Duncombe, Dr. Raynor L. Department of Aerospace Engineering, University of Texas, Austin.

Dzuiba, Ronald. Senior Member, Electrical Measurements Laboratory, National Institute of Standards and Technology, Gaithersburg, Maryland.

E

Eesley, Dr. Gary L. GM Research and Development Center, Warren, Michigan.

Egan, Patrick. Department of Physics, Yale University.

Elston, Dr. Stuart B. Department of Physics and Astronomy, University of Tennessee, Knoxville.

Erb, Dr. Karl A. Oak Ridge National Laboratory, Oak Ridge, Tennessee.

Esaki, Leo. Thomas J. Watson Research Center, IBM, Yorktown Heights, New York.

Evans, Doris L. Deceased; formerly, Research and Development Division, Corning Glass Works, Corning, New York.

Evenson, Dr. Kenneth M. Time and Frequency Division, National Institute of Standards and Technology, Boulder, Colorado.

F

Fairbank, Prof. Henry A. Department of Physics, Duke University.

Fan, Prof. Hsu Y. Department of Physics, Purdue University.

Fano, Prof. U. Deceased; formerly, James Franck Institute, University of Chicago.

Fawcett, Prof. Eric. Department of Physics, University of Toronto, Ontario, Canada.

Feigenbaum, Mitchell J. Professor, Department of Physics, Cornell University.

Feldman, Dr. Barry J. Los Alamos Scientific Laboratory, Los Alamos, New Mexico.

Feldman, Dr. Leonard C. AT&T Bell Laboratories, Murray Hill, New Jersey.

Ferrell, Prof. Thomas L. Health and Safety Research Division, Oak Ridge National Laboratory, Oak Ridge, Tennessee.

Feshbach, Dr. Herman. Department of Physics, Massachusetts Institute of Technology.

Fields, Thomas H. Physics Division, Argonne National Laboratory, Argonne, Illinois.

Firk, Prof. Frank W. K. Electron Accelerator Laboratory, Yale University.

Fischer, Prof. J. E. Department of Materials Science and Engineering, University of Pennsylvania, Philadelphia.

Fisher, Dr. Robert A. Los Alamos National Laboratory, Los Alamos, New Mexico.

Fisher, Prof. Russell A. Department of Physics, Northwestern University.

Fisk, Prof. Nelson S. Department of Civil Engineering, Columbia University.

Fitch, Prof. Val. Department of Physics, Princeton University.

Flynn, Dr. Thomas M. Consultant, Cryogenic Engineering, Boulder, Colorado.

Fowler, Prof. W. Beall. Department of Physics, Lehigh University.

Fradkin, Dr. David M. Department of Physics, Wayne State University.

Franzini, Dr. Paolo. Dipartimento di Fisica, Universita di Roma I, Italy.

French, Dr. J. Bruce. Department of Physics, University of Rochester.

Fretter, Prof. William B. Department of Physics, University of California, Berkeley.

Fry, Prof. Edward S. Department of Physics, Texas A&M University.

Fuchs, Prof. Ronald. Department of Physics, Iowa State University.

Fulford, Prof. George D. Department of Chemical Engineering, University of Waterloo, Ontario, Canada.

G

Gardner, Dr. Walter H. Department of Agronomy and Soils, Washington State University.

Geist, Jon. Radiometric Physics Division, National Institute of Standards and Technology, Washington, D.C.

Gemmell, Dr. Donald S. Physics Division, Argonne National Laboratory, Argonne, Illinois.

Gerjuoy, Dr. Edward. Department of Physics, University of Pittsburgh.

Giallorenzi, Dr. Thomas G. Superintendent, Optical Science Division, Department of the Navy, Naval Research Laboratory, Washington, D.C.

Gingrich, Prof. Newell S. Department of Physics, University of Missouri.

Girvin, Prof. Steven M. Department of Physics, Indiana University, Bloomington.

Goebel, Prof. Charles. Department of Physics, University of Wisconsin.

Goldhaber, Dr. Gerson. Lawrence Berkeley Laboratory, University of California, Berkeley.

Gomer, Prof. Robert. James Franck Institute, University of Chicago.

Good, Prof. Roland H., Jr. Department of Physics, Pennsylvania State University.

Goodenough, Prof. John B. Center for Materials Science and Engineering, University of Texas, Austin.

Goodglass, Dr. Harold. Director, Psychology Research, Veterans Administration Hospital, Boston, Massachusetts.

Goodheart, Prof. Clarence F. Department of Electrical Engineering, Union College.

Goodman, Prof. Bernard. Department of Physics, University of Cincinnati.

Goodman, Dr. Joseph W. Stanford Electronics Laboratories, Stanford University.

Goodman, Dr. Ralph R. Naval Research Laboratory, Stennis Space Center, Mississippi.

Gossard, Prof. Arthur C. Materials Department, University of California, Santa Barbara.

Goulding, Dr. Frederick S. Lawrence Berkeley Laboratory, University of California.

Granicher, Prof. H. Laboratory of Solid State Physics, Swiss Federal Institute of Technology, Zurich.

Green, Prof. Sheldon I. Department of Mechanical Engineering, University of British Columbia, Vancouver, Canada.

Greenberger, Prof. Daniel M. Department of Physics, City College of the City University of New York.

Greenspan, Martin. National Bureau of Standards.

Greiner, Dr. Walter. Institut für Theoretische Physik, Universitat Frankfurt, Germany.

Grest, Dr. Gary S. Exxon Research and Engineering Company, Annandale, New Jersey.

Guido, Prof. Louis. Associate Professor, Electrical and Computer Engineering, Virginia Tech, Blacksburg.

Gupta, Dr. Y. M. Department of Physics, Shock Dynamics Laboratory, Washington State University.

Gutzwiller, Dr. Martin C. T. J. Watson Research Center, IBM Corporation, Yorktown Heights, New York.

H

Hadni, A. Laboratoire Infrarouge Lointain, Université de Nancy I, France.

Hagen, Dr. John P. Department of Astronomy, Pennsylvania State University.

Hagist, Prof. Warren M. Retired; formerly, Department of Mechanical Engineering, University of Rhode Island.

Haines, Prof. Malcolm G. Blackett Laboratory, Imperial College of Science and Technology, London, England.

Hamilton, Dr. Joseph H. Department of Physics-Astronomy, Vanderbilt University.

Hamilton, Dr. Mark F. Department of Mechanical Engineering, University of Texas, Austin.

Hanor, Jeffrey S. Department of Geology, Louisiana State University.

Hansch, Dr. Theo W. Department of Physics, Stanford University.

Hansen, Arthur G. President, Georgia Institute of Technology.

Happer, Dr. William. Department of Physics, Columbia University.

Hardy, Prof. John. Cyclotron Institute, Texas A & M University, College Station.

Hardy, Dr. John W. Lexington, Massachusetts.

Harris, Dr. Cyril M. Professor of Electrical Engineering and Architecture, Department of Electrical Engineering, Columbia University.

Harris, Dr. Forest K. Electrical Measurements Laboratory, National Institute of Standards and Technology, Gaithersburg, Maryland.

Harrison, George R. Deceased; formerly, Dean Emeritus, School of Science, Massachusetts Institute of Technology.

Harrison, Prof. John P. Department of Physics, Queen's University, Ontario, Canada.

Harrison, Dr. Walter A. Department of Applied Physics, Stanford University.

Harvey, Dr. E. Newton. Deceased, formerly, Henry Fairfield Osborne Professor of Biology, Princeton University.

Harvey, Dr. John A. Oak Ridge National Laboratory, Oak Ridge, Tennessee.

Heiney, Prof. Paul A. Department of Physics, University of Pennsylvania, Philadelphia.

Henry, Dr. Charles H. Bell Laboratories, Murray Hill, New Jersey.

Herber, Prof. Rolfe H. Department of Chemistry, Rutgers University.

Herbst, Dr. Jan F. Physics Department, General Motors Research Laboratories, Warren, Michigan.

Hermach, Dr. F. L. National Institute of Standards and Technology, Gaithersburg, Maryland.

Hershkowitz, Prof. Noah. Department of Nuclear Engineering, University of Wisconsin, Madison.

Herzberger, Max J. Deceased; formerly, Consulting Professor, Department of Physics, Louisiana State University.

Hill, Prof. Edward L. Deceased; formerly, Professor of Physics and Mathematics, School of Physics, University of Minnesota.

Hilliard, Dr. Ron. Optomechanics Research, Inc., Vail, Arizona.

Hofstadter, Dr. Robert. Department of Physics, Stanford University.

Holdeman, Dr. Louis B. Communications Satellite Corporation (COMSAT), Clarksburg, Maryland.

Hollweg, Prof. Joseph V. Space Science Center, University of New Hampshire.

Horen, Dr. D. J. Nuclear Division, Oak Ridge National Laboratory, Oak Ridge, Tennessee.

Howe, Prof. Carl E. Deceased; formerly, Professor Emeritus of Physics, Oberlin College.

Hoxton, Prof. Llewellyn G. Deceased; formerly, Professor Emeritus of Physics, University of Virginia.

Hudson, Dr. Ralph P. Retired; formerly, Bureau International des Poids et Mésures, Sevres, France.

Huizenga, Dr. John R. Nuclear Structure Research Laboratory, University of Rochester.

Hultsch, Dr. Roland A. Department of Physics, University of Missouri.

Huntington, Prof. H. B. Department of Physics, Rensselaer Polytechnic Institute.

Hurll, John. Measurements Standards Laboratory, Marconi Instruments Ltd., Stevenage, Hertfordshire, England.

Hurst, Dr. G. S. Department of Physics, Oak Ridge National Laboratory, Oak Ridge, Tennessee.

I

Iachello, Dr. Franco. Department of Physics, Yale University.

Itano, Dr. Wayne M. U.S. Department of Commerce, Time and Frequency Division, National Institute of Standards and Technology, Boulder, Colorado.

J

Jackiw, Prof. Roman. Department of Physics, Massachusetts Institute of Technology.

Jackson, Dr. Kenneth A. AT&T Bell Telephone Laboratories, Murray Hill, New Jersey.

Jacobs, Prof. Stephen F. Optical Sciences Center, University of Arizona.

Jaep, William F. Central Research Department, Experimental Station, E. I. du Pont de Nemours and Company, Wilmington, Delaware.

Jafarey, Prof. S. Behlen Laboratory of Physics, University of Nebraska.

Jaffe, Prof. Robert L. Department of Physics, Massachusetts Institute of Technology, Cambridge.

Janna, Prof. William S. Department of Mechanical Engineering, Memphis State University, Tennessee.

Jen, Dr. C. K. Supervisor, Microwave Physics Group, Applied Physics Laboratory, Johns Hopkins University.

Jennings, Dr. Donald A. Time and Frequency Division, National Institute of Standards and Technology, Boulder, Colorado.

Jones, Prof. Lawrence W. Harrison M. Randall Laboratory of Physics, University of Michigan.

Jones, Dr. Owen C. Superintendent, Division of Electrical Science, National Physical Laboratory, Teddington, England.

Jones, Dr. R. Gareth. Division of Electrical Science, National Physical Laboratory, Teddington, England.

K

Kaminow, Dr. Ivan P. Bell Telephone Laboratories, Holmdel, New Jersey.

Kanzig, Prof. Werner. Department of Physics, Massachusetts Institute of Technology.

Kaplan, Prof. Thomas A. Department of Physics and Astronomy, Michigan State University, East Lansing.

Karovska, Dr. Margarita. Harvard- Smithsonian Center for Astrophysics, Cambridge, Massachusetts.

Karp, Dr. Alan H. IBM Scientific Center, Palo Alto, California.

Keddy, Prof. Rex J. Schonland Research Center for Nuclear Sciences, University of Witwatersrand, Johannesburg, South Africa.

Keffer, Prof. Frederic. Department of Physics, University of Pittsburgh.

Keller, Dr. Joseph M. Department of Physics, Iowa State University.

Keller, Dr. William E. Los Alamos Scientific Laboratory, Los Alamos, New Mexico.

Kenning, Dr. David B. B. Department of Engineering Science, Oxford, United Kingdom.

Kerst, Prof. Donald W. Department of Physics, University of Wisconsin.

Ketterle, Prof. Wolfgang. Department of Physics, Massachusetts Institute of Technology, Cambridge.

Ketterson, Dr. J. B. Department of Physics, Northwestern University.

Kibble, Dr. Bryan P. Division of Electrical Science, National Physical Laboratory, Teddington, Middlesex, England.

Kim, Prof. Chung W. Department of Physics, Johns Hopkins University.

Kinard, Joseph R. Consultant, Darnestown, Maryland.

Kinnard, Dr. Isaac F. Deceased; formerly, Manager of Engineering, Instrument Department, General Electric Company, Lynn, Massachusetts.

Kinoshita, Prof. Toichiro. Laboratory of Nuclear Studies, Cornell University.

Kinsler, Prof. Lawrence E. Professor of Physics, U.S. Naval Postgraduates School, Monterey, California.

Kirk, Prof. Wiley P. Department of Physics, Texas A & M University.

Klemens, Dr. Paul G. Department of Physics, University of Connecticut.

Kleppner, Prof. Daniel. Department of Physics, Massachusetts Institute of Technology, Cambridge.

Klick, Dr. Clifford C. Superintendent, Solid Sate Division, U.S. Naval Research Laboratory.

Knight, Dr. R. B. D. Retired; formerly, Division of Electrical Science, National Physical Laboratory, Teddington, Middlesex, England.

Knoteck, Dr. M. L. Sandia National Laboratories, Albuquerque, New Mexico.

Koch, Dr. Peter M. Department of Physics, State University of New York at Stony Brook.

Koglenik, Dr. Herwig. Crawford Hill Laboratory, Bell Laboratories, Holmdel, New Jersey.

Komai, Dr. Taku. Professor Emeritus, Kyoto University, Japan.

Koopman, Dr. Karl F. Department of Mammalogy, American Museum of Natural History, New York.

Kotter, Dr. F. Ralph. National Bureau of Standards.

Kovar, Dr. Dennis. Hahn-Meitner Institut für Kernsforschung, Berlin, Germany.

Krusius, Prof. Matti. Low Temperature Laboratory, Helsinki University of Technology, Finland.

Kryter, Dr. Karl D. Director, Sensory Sciences Research Center, Stanford Research Institute, Menlo Park, California.

Kusch, Prof. Polykarp. Department of Physics, Columbia Radiation Laboratory, Columbia University.

Kusters, Dr., N. L. Division of Electrical Engineering, National Research Council, Ottawa, Ontario, Canada.

L

Lamb, Prof. Richard C. Space Radiation Laboratory, California Institute of Technology, Pasadena.

Landes, Dr. Hugh S. Department of Electrical Engineering, University of Virginia.

Lanford, Prof. William A. Department of Physics, State University of New York, Albany.

Langenberg, Prof. D. N. Department of Physics, University of Pennsylvania.

Larsen, Dr. David M. Department of Physics and Applied Physics, University of Lowell, Massachusetts.

Lassila, Prof. Kenneth E. Department of Physics, Iowa State University.

Lee, Prof. David M. Department of Physics, Cornell University; Nobelist.

Legget, Prof. Anthony. Department of Physics, University of Illinois, Urbana.

Leibovich, Prof. Sidney. Mechanical & Aerospace Engineering, Cornell University.

Lewis, Frank D. Consultant, Lexington, Massachusetts.

Ley, Dr. Willy. Deceased; formerly, Fairleigh Dickinson University.

Lichtenberg, Dr. Don B. Department of Physics, Indiana University.

Liggett, Prof. James A. School of Civil and Environmental Engineering, Cornell University, Ithaca, New York.

Lin, Wen-Hwang. Rocketdyne Propulsion & Power, The Boeing Company, Canoga Park, California.

Linde, Dr. Ronald K. President, Envirodyne, Inc., Los Angeles, California.

Lindsay, Prof. R. Bruce. Deceased; formerly, Hazard Professor of Physics, Emeritus, Brown University.

Lines, Malcom E. Bell Telephone Laboratories, Murray Hill, New Jersey.

Lister, Prof. C. J. A. W. Wright Nuclear Structure Laboratory, Physics Department, Yale University.

Liu, Dr. Sam H. Department of Physics, Iowa State University.

Lord, Prof. Richard C. Department of Chemistry, Massachusetts Institute of Technology.

Luborsky, Dr. Fred E. Research and Development Center, General Electric Company, Schenectady, New York.

Luebbers, Dr. Ralph H. Department of Chemical Engineering, University of Missouri.

M

MacCrone, Prof. Robert K. Department of Physics, Rensselaer Polytechnic Institute.

MacDowell, Dr. Samuel W. Department of Physics, Yale University.

MacKenzie, Prof. Innes K. Department of Physics, University of Guelph, Ontario, Canada.

Madanshetty, Prof. Sameer I. Department of Aerospace and Mechanical Engineering, Boston University, Boston, Massachusetts.

Mangum, B. W. Leader, Thermometry Group, Prouss Measurement Division, National Institute of Standards and Technology, Gaithersburg, Maryland.

Mann, Prof. Alfred K. Department of Physics, University of Pennsylvania.

Manning, Dr. Kenneth V. Professor Emeritus, Pennsylvania State University.

Mansouri, Freydoon. Department of Physics, Yale University.

Maple, M. Brian. Department of Physics, University of California, San Diego.

March-Russell, Dr. John. Theory Division, CERN, Geneva, Switzerland

Maris, Prof. Humphrey J. Department of Physics, Brown University.

Markovitz, Dr. Hershel. Department of Chemistry, Carnegie-Mellon University.

Markowitz, Dr. William. Oceanographic Center, Nova University; Editor, "Geophysical Surveys." Marks, Lawrence E. Pierce Foundation, New Haven, Connecticut.

Markus, John. Deceased; formerly, Consultant (retired), Sunnyvale, California.

Marshak, Dr. Harvey. National Bureau of Standards.

Martin, Prof. C. Samuel. School of Civil Engineering, Georgia Institute of Technology, Atlanta.

Mashhoon, Prof. Bahram. Department of Physics and Astronomy, College of Arts and Sciences, University of Missouri—Columbia.

McCarthy, Dr. Dennis D. Time Service Division, U. S. Naval Observatory, Washington, D.C.

McCarthy, Prof. Kathryn A. Professor of Physics, and Dean of Graduate School of Arts and Sciences, Tufts University.

McCoy, Prof. Barry M. Division of Engineering and Applied Physics, Harvard University.

McDonald, Prof. Alan T. School of Mechanical Engineering, Purdue University.

McDowell, Dr. Robin S. Senior Chief Scientist, Chemical Structure and Dynamics, Molecular Science Research Center, Battelle Pacific Northwest Laboratories, Richland, Washington.

Meggers, Dr. William F. Deceased; formerly, National Bureau of Standards.

Meijer, Prof. Paul H. E. Department of Physics, Catholic University of America.

Meissner, Dr. Hans W. Department of Physics, Stevens Institute of Technology, Hoboken, New Jersey.

Menkes, Joshua. Science and Technology Division, Institute for Defense Analyses, Arlington, Virginia.

Metcalf, William G. Woods Hole Oceanographic Institution, Woods Hole, Massachusetts.

Miller, Dr. David A. B. AT&T Bell Laboratories, Holmdel, New Jersey.

Miller, Dr. Glenn H. Weapons Effects Division, Sandia National Laboratories, Albuquerque, New Mexico.

Miller, Prof. Sidney I. Department of Chemistry, Illinois Institute of Technology.

Miller, Terry A. AT&T Bell Laboratories, Murray Hill, New Jersey.

Mills, Dr. Dennis. Advanced Photon Source, Argonne National Laboratory, Argonne, Illinois.

Moffett, Dr. Mark B. Naval Underwater Systems Center, New London Laboratory, New London, Connecticut.

Mohr, Prof. Peter. Department of Physics, Yale University.

Moore, Dr. W. J. M. Division of Electrical Engineering, National Research Council of Canada, Ottawa, Ontario.

Morris, Prof. Philip J. Department of Aerospace Engineering, Pennsylvania State University, University Park.

Mottershead, Dr. Allen. Physical Science Department, Cypress College, California.

Mulliken, Prof. Robert S. Institute of Molecular Biophysics, Florida State University.

Munson, Prof. Bruce R. Department of Aerospace Engineering and Mechanics, Iowa State University, Ames.

Murphy, Dr. Glenn. Department of Nuclear Engineering, Iowa State University.

Murray, Dr. Thomas P. Applied Research Laboratory, United States Steel Corporation, Monroeville, Pennsylvania.

N

Nagel, Suzanne R. Engineering Research Center, AT&T Bell Laboratories, Princeton, New Jersey.

Nedelsky, Prof. Leo. Department of Physical Science, University of Chicago.

Neudeck, Prof. Gerald W. Department of Electrical Engineering, Purdue University.

Newton, Prof. Roger G. Department of Physics, Indiana University, Bloomington.

Nicodemus, Dr. Fred E. Deceased; formerly, Physicist, Optical Radiation Section, Heat Division, Institute for Basic Standards, National Bureau of Standards.

Nier, Prof. Alfred O. School of Physics and Astronomy, University of Minnesota.

Nordman, Prof. James E. Department of Electrical and Computer Engineering, University of Wisconsin, Madison.

Nygren, Dr. David R. Physics Division, Lawrence Berkeley Laboratory, University of California, Berkeley.

Nygren, Stephen. Bell Telephone Laboratories, Reading, Pennsylvania.

O

O'Connell, Prof. John P. Department of Chemical Engineering, University of Florida.

Olness, Dr. John W. Brookhaven National Laboratory, Associated Universities, Inc., Upton, New York.

Oosthuizen, Dr. Patrick H. Department of Mechanical Engineering, Queen's University, Kingston, Ontario, Canada.

Osofsky, Dr. Michael. Naval Research laboratory Washington, D.C.

Oster, Prof. Gerald. Mount Sinai School of Medicine, City University of New York.

Otten, Prof. Ernst Wilhelm. Institut für Physik, Johannes Gutenberg-Universität Mainz, Germany.

Overhauser, Dr. Albert W. Department of Physics, Purdue University.

P

Pake, Dr. George E. Vice President, Xerox Corporation; General Manager, Xerox Palo Alto Research Center, Palo Alto, California.

Palmer, Prof. Fred. Lighting Research Center, Osram Sylvania, Beverly, Massachusetts.

Panish, Dr. Morton B. Head, Materials Science Research Department, AT&T Bell Laboratories, Murray Hill, New Jersey.

Pantell, Dr. Richard. Electrical Engineering Department, Stanford University.

Park, Prof. David. Department of Physics, Williams College.

Parpia, Prof. Jeevak M. Department of Physics, Cornell University.

Paterson, Eric. Iowa Institute of Hydraulic Research and Department of Mechanical Engineering, University of Iowa, Iowa City.

Peck, Prof. Raymond E. Department of Geology, University of Missouri.

Pegg, Prof. David J. Department of Physics and Astronomy, University of Tennessee, Knoxville.

Penner-Hahn, Prof. James E. Department of Chemistry, University of Michigan, Ann Arbor.

Penrose, Prof. Roger. Mathematical Institute, University of Oxford, England.

Pepper, Prof. Darrell W. Department of Mechanical Engineering, University of Nevada, Las Vegas.

Perl, Prof. Martin L. Stanford Linear Accelerator Center, Stanford, California.

Peskin, Prof. Michael E. Stanford Linear Accelerator Center, Menlo Park, California.

Peterson, Dr. Kirk D. Department of Engineering, Calvin College, Grand Rapids, Michigan.

Petschek, Prof. Albert G. Department of Physics, New Mexico Institute of Mining and Technology, Socorro.

Phaneuf, Prof. Ronald A. Chair, Department of Physics, University of Nevada, Reno.

Phillips, Dr. James A. CTR-Divison Office, Los Alamos Scientific Laboratory, Los Alamos, New Mexico.

Phillips, Dr. William D. National Institute of Standards and Technology, Gaithersburg, Maryland; Nobelist.

Pike, Dr. Andrew C. Neotronics Scientific Ltd., United Kingdom.

Pipkin, Prof. Francis M. Department of Physics, Harvard University.

Pitchford, Dr. Leanne C. Sandia National Laboratories, Albuquerque, New Mexico.

Pohl, Prof. Robert O. Laboratory of Atomic and Solid State Physics, Cornell University.

Polchinski, Prof. Joseph. Institute for Theoretical Physics, University of California, Santa Barbara.

Pollard, Dr. William G. Oak Ridge Associated Universities.

Povh, Dr. Bogdan. Max Planck Institut für Kernphysik, Saupfercheckweg, Germany.

Powell, Robert C. Consultant, Physics, Gaithersburg, Maryland.

Primack, Prof. Joel R. Physics Department, University of California, Santa Cruz, California.

Pritchard, Prof. David E. Department of Physics, Massachusetts Institute of Technology, Cambridge.

Pritchett, Prof. Wilson S. Senior Project Engineer, Noller Control Systems, Inc., Richmond, California.

Putnam, Prof. Russell C. Professor Emeritus, Case Institute of Technology.

Q

Quigg, Dr. Chris. Lawrence Berkeley Laboratory, University of California, Berkeley.

Quinn, Dr. T. J. Bureau International des Poids et Mésures, Pavillion de Breteuil, Sèvres, France.

R

Raad, Prof. Peter E. Mechanical Engineering Department, School of Engineering and Applied Science, Southern Methodist University, Dallas, Texas.

Ramdas, Prof. Anant K. Department of Physics, Purdue University, West Lafayette, Indiana.

Ramey, Dr. Robert L. Department of Electrical Engineering, University of Virginia.

Ramsey, Dr. Norman F. Department of Physics, Harvard University.

Rasmussen, Dr. Norman C. Department of Nuclear Engineering, Massachusetts Institute of Technology.

Rau, Prof. A. R. P. Department of Physics, Louisiana State University, Baton Rouge.

Rayner, Dr. G. H. Consultant, Teddington, Middlesex, England.

Rebbi, Dr. Claudio. Department of Physics, Brookhaven National Laboratory, Upton, New York.

Reintjes, John F. Naval Research Laboratory, Washington, D.C.

Rich, Prof. Arthur. Deceased; formerly, Department of Physics, University of Michigan.

Richards, Dr. Oscar W. (Retired) College of Optometry, Pacific University.

Richter, Reinhard H. Manager, Polyurethane Research, Dow Chemical U.S.A., North Haven Laboratories, North Haven, Connecticut.

Roberts, Prof. Louis D. Department of Physics, University of North Carolina.

Robertson, Dr. James S. Medical Department, Medical Physics Division, Brookhaven National Laboratory, Upton, New York.

Rockett, Frank H. Engineering Consultant, Charlottesville, Virginia.

Rohlf, Dr. James W. Department of Physics, Boston University.

Rohrlich, Prof. F. Department of Physics, Syracuse University.

Roller, Dr. Duane E. Deceased; formerly, Harvey Mudd College, Claremont, California.

Ross, Dr. Robert S. Environmental Structures, Inc., Cleveland, Ohio.

Rossing, Dr. Thomas D. Department of Physics, Northern Illinois University, DeKalb.

Rowley, Dr. W. R. C. Department of Trade and Industry, Division of Mechanical and Optical Metrology, National Physical Laboratory, Teddington, Middlesex, England.

Ruch, Prof. Fritz. Department of General Botany, Swiss Federal Institute of Technology, Zurich.

Rupp, Arthur F. Oak Ridge National Laboratory, Oak Ridge, Tennessee.

Rusk, Dr. Rogers D. Mount Holyoke College.

Russell, Dr. Howard W. Deceased; formerly, Technical Director, Battelle Memorial Institute, Columbus, Ohio.

Rust, Dr. David M. Applied Physics Laboratory, Johns Hopkins University, Laurel, Maryland.

S

Saad, Prof. Michael L. Department of Mechanical Engineering, Santa Clara University, Santa Clara, California.

Salam, Prof. Abdus. Deceased; formerly, Director, International Centre for Theoretical Physics, Trieste, Italy.

Samios, Dr. Nicholas P. Brookhaven National Laboratory, Upton, New York.

Sandier, Prof. Stanley I. Department of Chemical Engineering, University of Delaware.

Sasian, Prof. Jose M. Optical Sciences Center, University of Arizona, Tucson.

Sass, Prof. Stephen L. Department of Materials Science and Engineering, Cornell University.

Scanlan, Prof. J. O. Department of Electrical and Electronic Engineering, University College, Dublin, Ireland.

Schawlow, Prof. Arthur L. Department of Physics, Stanford University.

Schetz, Prof. Joseph A. Department of Aerospace and Ocean Engineering, Virginia Polytechnic Institute and State University, Blacksburg.

Schmidt, Dr. Paul W. Department of Physics, University of Missouri.

Schroeder, Dr. Lee S. Lawrence Berkeley Laboratory, Berkeley, California.

Schuller, Dr. Ivan K. Physics Department, University of California, La Jolla.

Schulman, Dr. James H. U.S. Naval Research Laboratory.

Scott, Prof. Alwyn. Program in Applied Mathematics, University of Arizona, Tucson.

Scott, Dave. Industrial Engineering Manager, Modular Computer Systems, Inc., Fort Lauderdale, Florida.

Scott, Prof. James F. Department of Physics, University of Colorado.

Sell, Dr. Heinz G. Metals Development Section, Westinghouse Lamp Divisions, Bloomfield, New Jersey.

Sellars, Dr. John R. Manager, Engineering Mechanics Operations, TRW, Inc., Redondo Beach, California.

Sellin, Dr. Ivan A. Department of Physics, University of Tennessee.

Sellmyer, Prof. David J. Behlen Laboratory of Physics, University of Nebraska.

Shack, Prof. Roland V. Optics Science Center, University of Arizona.

Shannon, Prof. Robert R. Optical Sciences Center, University of Arizona.

Shaw, Prof. Peter. Department of Physics, Pennsylvania State University.

Shenoy, Dr. Gopal. Senior Scientific Director, The Advanced Photon Source, Argonne National Laboratory, Illinois.

Shivaram, Dr. Bellave S. Department of Physics, John S. Beams Laboratory, University of Virginia.

Simon, Dr. Barry. Department of Mathematics, Princeton University.

Sinton, Dr. William M. Department of Astronomy, University of Hawaii.

Sirignano, Prof. William A. Dean of Engineering, University of California, Irvine.

Skinner, A. Douglas. Head of Measurement Standards Laboratory, Marconi Instruments, Ltd., Stevenage, England.

Slichter, Dr. Charles P. Department of Physics, University of Illinois, Urbana.

Smith, Dr. Michael S. Leader, Nuclear Astrophysics Research Group, Physics Division, Oak Ridge National Laboratory, Tennessee.

Smythe, Dr. William R. Department of Physics, California Institute of Technology.

Snell, Dr. Arthur H. Associate Director, Oak Ridge National Laboratory, Oak Ridge, Tennessee.

Snoke, Dr. David. Los Angeles, California.

So, Dr. E. Institute for National Measurement Standards, National Research Council of Canada, Ottawa, Ontario.

Soldate, Dr. Mark. Department of Physics, Yale University.

Somlo, Dr. Peter I. CSIRO Division of Applied Physics, Lindfield, New South Wales, Australia.

Sommer, Dr. Alfred H. Thermo Electron Corporation, Waltham, Massachusetts.

Sorenson, Prof. S. Department of Physics, University of Tennessee, Knoxville.

Souder, Dr. Paul A. Department of Physics, Yale University.

Soulen, Dr. Robert J., Jr. Naval Research Laboratory, Washington, D.C.

Sreenivasan, Prof. K. R. Engineering and Applied Sciences, Yale University.

Steele, Dr. William A. Department of Chemistry, Pennsylvania State University.

Steinberg, Dr. Aephraim M. Department of Physics, University of California, Berkeley.

Stephenson, Dr. R. J. Department of Physics, Wooster College.

Stern, Prof. Fred. Iowa Institute of Hydraulic Research, University of Iowa, Iowa City.

Stevenson, Dr. Edward C. Deceased; formerly, Department of Electrical Engineering, School of Engineering and Applied Sciences, University of Virginia.

Steward, Dr. John W. Department of Physics, University of Virginia.

Stock, Prof. David. School of Mechanical and Materials Engineering, Washington State University, Pullman, Washington.

Stone, Prof. A. Douglas. Department of Applied Physics, Yale University.

Stone, Dr. Anthony J. Department of Chemistry, University of Cambridge, University Chemical Laboratory, Cambridge, England.

Streeter, Prof. Victor L. Department of Civil Engineering, College of Engineering, University of Michigan.

Stroke, Dr. George W. Department of Electrical Sciences and Head, Electro-Optical Sciences Center, State University of New York, Stony Brook.

Suzuki, Dr. Yoichiro. Kamioka Observatory, Institute for Cosmic Ray Research, University of Tokyo, Higashi-Mozumi, Kamioka Gifu, Japan.

Swanson, Dr. M. L. Department of Physics and Astronomy, University of North Carolina.

Swift, Dr. Gregory W. Condensed Matter and Thermal Physics, Los Alamos National Laboratory, New Mexico.

Symko, Orest G. Department of Physics, University of Utah.

Symons, Dr. Robert S. Litton Electrical Devices Division, San Carlos, California.

T

Talbot, Prof. Lawrence. Department of Mechanical Engineering, University of California, Berkeley.

Tang, Prof. K. Y. Deceased; formerly, Department of Electrical Engineering, Ohio State University.

Tanihata, Dr. Isao. Accelerator Facility, Institute of Physical and Chemical Research, Saritama, Japan.

Telionis, Prof. Demetri P. Department of Engineering Science and Mechanics, Virginia Polytechnic Institute and State University, Blacksburg.

Teng, Lee C. Fermi National Accelerator Laboratory, Batavia, Illinois.

Thomson, Prof. Robb M. Chairman, Department of Materials Science, State University of New York, Stony Brook.

Tichy, Prof. Jiri. Graduate Program in Acoustics, Pennsylvania State University, State College.

Ting, Prof. Samuel C. C. Department of Physics, Massachusetts Institute of Technology.

Townes, Prof. Charles H. Department of Physics, University of California, Berkeley.

Träbert, Prof. Elmar. Experimentalphysik III, Ruhr-Universität Bochum, Germany.

Treiman, Prof. Sam B. Deceased; formerly, Department of Physics, Princeton University, Princeton, New Jersey.

Trolinger, Dr. James D. Vice President, Spectron Development Laboratories, Costa Mesa, California.

Turnbull, Prof. David. Department of Applied Physics, Harvard University.

U, V, W

Underwood, Dr. James H. Center for X-Ray Optics, Lawrence Berkeley National Laboratory, University of California, Berkeley.

Ventura, Dr. Jose A. Department of Industrial and Management Systems Engineering, Pennsylvania State University, University Park.

Wagner, Dr. Richard S. AT&T Telephone Laboratories, Murray Hill, New Jersey.

Wahl, Prof. Floyd M. Professor and Chairman, Department of Geology, University of Florida.

Wald, Prof. Robert M. Department of Physics, University of Chicago, Illinois.

Walkup, Dr. Robert D. Department of Chemistry and Biochemistry, Texas Tech University.

Walsh, Dr. Peter J. Department of Physics, Fairleigh Dickinson University.

Walstedt, Dr. R. E. Bell Laboratories, Murray Hill, New Jersey.

Wang, Prof. C. Y. Departments of Mathematics and Mechanical Engineering, Michigan State University, East Lansing.

Ward, Dr. David. Atomic Energy of Canada Limited, Chalk River Nuclear Laboratories, Ontario, Canada.

Warren, Prof. Bertram E. Department of Physics, Massachusetts Institute of Technology.

Watson, William W. Professor Emeritus of Physics, Yale University.

Weaver, Prof. David L. Department of Physics, Tufts University.

Webb, Dr. Richard A. IBM Research Division, Thomas J. Watson Research Center, IBM Corporation, Yorktown Heights, New York.

Weber, Harold C. Chemical Engineer, Boston, Massachusetts.

Wedgewood, Prof. Lewis E. Department of Chemical Engineering, University of Washington, Seattle.

Wegner, Dr. Harvey E. Physics Division, Brookhaven National Laboratory, Upton, New York.

Weil, Jonathan F. Staff Editor, "McGraw-Hill Encyclopedia of Science and Technology," McGraw-Hill, New York.

Weil, Robert T., Jr. Deceased; formerly, Dean, School of Engineering, Manhattan College.

Weitz, Dr. Martin. Sektion Physik, Ludwig-Maximilians-Universität, München, Germany.

West, Dr. William. Eastman Kodak Company, Rochester, New York.

Whitcomb, Dr. Richard T. NASA Langley Research Center, Langley Field, Virginia.

Whitehead, Prof. W. Dexter. Department of Physics, Center for Advanced Studies, University of Virginia.

Wiegand, Dr. Clyde E. Lawrence Berkeley Laboratory, University of California, Berkeley.

Wiest, Dr. John M. Department of Chemical Engineering, University of Alabama, Tuscaloosa.

Wilczek, Prof. Frank. Institute for Advanced Study, Princeton, New Jersey.

Wildhack, William A. Consultant; formerly, Associate Director, Institute for Basic Standards, National Bureau of Standards.

Wilkinson, Prof. Denys H. Department of Nuclear Physics, Oxford University, England.

Wilkinson, Dr. Michael K. Associate Director, Solid State Division, Oak Ridge National Laboratory, Oak Ridge, Tennessee.

Williams, Prof. Dudley. Department of Physics, Kansas State University.

Williams, Dr. Earl G. Structural Acoustics, Naval Research Laboratory, Washington, D.C.

Williams, Prof. Gary A. Department of Physics, University of California, Los Angeles.

Williams, S. A. Department of Physics, Iowa State University.

Willis, Dr. William J. European Organization for Nuclear Research (CERN), Geneva, Switzerland.

Wilson, Prof. Mason P., Jr. Department of Mechanical Engineering, University of Rhode Island, Kingston.

Winch, Prof. Ralph P. Department of Physics, Williams College.

Winick, Dr. Herman. Deputy Director, Stanford Synchrotron Radiation Laboratory, Stanford, California.

Winter, Prof. Rolf G. Department of Physics, College of William and Mary.

Wolf, Dr. Werner P. Department of Physics, Yale University.

Wolfe, Dr. William L. Optical Sciences Center, University of Arizona.

Wollersheim, Prof. David. Department of Mechanical and Aerospace Engineering, University of Missouri.

Wollnik, Prof. H. Physikalisches Institut, Justus Liebig-Universität, Giessen, Germany.

Wong, Dr. George S. K. Acoustical Standards, Institute for National Measurement Standards, Ottawa, Ontario, Canada.

Woodall, James E. Division Manufacturing Manager, Westinghouse Electric Corporation, Pittsburgh, Pennsylvania.

Wright, Prof. Raymond. Department of Civil and Environmental Engineering, University of Rhode Island, Kingston.

Wu, Prof. Tai Tsun. Division of Engineering and Applied Physics, Harvard University.

Wyant, Prof. James C. Optical Sciences Center, University of Arizona.

Y, Z

Yost, Dr. William A. Director, Parmly Hearing Institute, Loyola University.

Young, Dr. Robert W. Associate Editor, "Journal of the Acoustical Society of America," San Diego, California.

Yuen, Prof. Horace P. Department of Electrical Engineering and Computer Science, Northwestern University.

Zeller, Prof. Michael E. Department of Physics, Yale University.

Zifcak, John H. The Foxboro Company, Foxboro, Massachusetts.

Zurek, Dr. Wojciech H. Los Alamos National Laboratory, Los Alamos, New Mexico.

Index

The asterisk indicates page numbers of an article title.

A

A15 compounds, 635
A15 phases, 1–2*
A15 structure, 637
Abampere (abA), 701
Aberration (optics), 2–3*
  chromatic, 96–97*
  geometrical optics, 264
Absolute cgs system, 175
Absolute zero, 3*
  liquid helium, 351
  low-temperature acoustics, 352
  low-temperature physics, 354
  temperature, 661
  thermodynamic principles,
    671–672
Absorption:
  dichroism, 152*
  radiative transfer, 538
  sound, 613–614*
  underwater sound, 694–695
Absorption of electromagnetic
    radiation, 3–5*
  color filter, 106–107*
  crystal absorption spectra, 126*
  EXAFS, 220–221*
Absorption spectra, 126
Absorptive optical bistability, 446
Abvolt (abV), 701
Acceleration, 5–6*
  force, 238*
  units of measurement, 699
Acceptor atom, 6–7*
Ac-dc comparator, 664
Acoustic impedance, 7*
Acoustic interferometer, 7–8*
Acoustic levitation, 8*
Acoustic noise, 8*, 385
Acoustic pressure, 610
Acoustic radiation pressure, 8–9*
Acoustic radiometer, 9*
Acoustic reactance, 7
Acoustic resistance, 7
Acoustic resonator, 9–10*
Acoustic streaming, 421
Acoustic wave, 50
Acoustical holography, 10–11*
Acoustics, 11*
  atmospheric, 33–35*
  cavitation, 85
  fuzzy-structure, 251–252*
  low-temperature, 352–354*
  musical, 411–412*
  nonlinear, 420–422*
  quantum, 518–519*
  thermoacoustics, 666–668*
  water hammer, 721–722*
Acoustooptics, 11–12*
Action, 12*
Adaptive optics, 12–13*
Adiabatic demagnetization, 13*
Adiabatic process, 14*, 320
Admittance, 14*
  immittance, 300*
  short circuit, 604–605
  susceptance, 650*
Adsorption, 236
Aerodynamic force, 14–15*
Aerodynamic heating, 120
Aerodynamic sound, 15–16*
Aerodynamic wave drag, 16*

Aerodynamics, 16–17*
Aerostatics, 18*
Aerothermodynamics, 18*
AES (Auger electron spectroscopy),
  189
AFM (atomic force microscopy), 649
Aharonov, Y., 19
Aharonov-Bohm effect, 18–19*
Aircraft wing flutter, 232–233
Airfoil, 14–15, 19–20*
Alfvén, H., 378
Alfvén waves, 20–22*
Allotropes, 501
Alloys, 335
Alnico, 228
Alpha particles, 22–23*, 540–541
Alternating current:
  electricity, 178
  potentiometer, 505
  skin effect, 606*
  voltage measurement, 715–716
Alternating-current circuit theory:
  admittance, 30*
  conductance, 110–111*
  immittance, 300*
  reactance, 549–550*
  susceptance, 650*
Alternating-current measurement,
  663–664
Alternating-current voltage, 173–174
Aluminum:
  acceptor atom, 6
  free-electron theory of metals, 242
  light-emitting diode, 348
Ambient noise, 695–696
Ammeter, 24*
  current measurement, 140–141*
  resistance measurement, 568–570*
  resistor, 570–572*
Ammonia, 384
Ammonium cadmium sulfate, 226
Ammonium sulfate, 226
Amorphous semiconductor, 598–599
Amorphous solid, 24–25*
  glass transition, 268–269*
  low-temperature acoustics,
    352–353
Ampère, A. M.:
  Ampère's law, 25
  displacement current, 163
  Maxwell's equations, 390
  pinch effect, 490
  relativistic electrodynamics,
    554–555
Ampere (A):
  current measurement, 140
  electrical units and standards, 176
  physical measurement, 487
Ampère's law, 25–26*
  current comparator, 139
  electromagnetism, 182
Analog states, 26*
Analog voltmeter, 716–717
Anderson, C. D., 30
Anemometer, 26–27*
Angular acceleration, 6
Angular correlations, 27*
Angular momentum, 27–28*
  analog states, 26*
  electroweak interaction, 202–203
  exclusion principle, 218
  gyromagnetic effect, 276

Angular momentum—cont.
  gyromagnetic ratio, 276
  multipole radiation, 410
  spin, 621–622*
Anharmonic oscillator, 28–29*
Anode, 194
Antiferromagnetism, 29–30*
  magnetic ferroelectrics, 362
  magnon, 381–382*
  superconductivity, 637
Antimatter, 30–32*
  elementary particle, 206
  relativistic quantum theory, 557
Antimony, 6
Antiparticles, 206
Antiproton atoms, 278
Anyons, 32*
Archimedes' principle, 32–33*, 80
Architecture, 10
Area (units of measurement), 699
Arnold, V., 427
Arsenic (As):
  crystallography, 137
  light-emitting diode, 348
Artificially layered structures, 33*,
  392–393
Astronomical maser, 384–385
Atmospheric acoustics, 33–35*
  infrasound, 308
  sound absorption, 613
Atmospheric refraction, 554
Atom, 6–7, 35*
  antimatter, 30–32*
  donor, 164*
  electron capture, 185*
  hadronic, 277–278*
  isobar, 321*
  isotone, 322*
  mass defect, 386*
  mass number, 386*
  neutron, 415–416*
  proton, 509–510*
  quasiatom, 535*
  Rydberg, 582–583*
  scattering experiments, 586*
Atom cluster, 35–36*
Atom laser, 36–38*
Atom optics, 38–39*
Atomic beams, 39*, 193
Atomic clock, 39–41*, 342
Atomic force microscopy (AFM), 649
Atomic mass, 41*, 405
Atomic mass unit, 41–42*
Atomic nucleus, 42*
  exotic nuclei, 218–220*
  nuclear physics, 435–436*
  nuclear radiation, 437*
  nuclear reaction, 437–439*
  nuclear structure, 440–443*
  radioactive beams, 538–539
  radioactivity, 539–543*
  relativistic heavy-ion collisions,
    555–556
  scattering experiments, 586–587*
Atomic number, 42*
Atomic physics, 42–43*
Atomic structure and spectra, 43–48*
  Auger effect, 50–51*
  Bohr atom, 43–45
  cooling and stopping atoms and
    ions, 48
  correspondence principle, 120*